MW00582848

Structural Dynamics
for Structural Engineers

Structural Dynamics
for Structural Engineers

Gary C. Hart
Department of Civil and Environmental Engineering
School of Engineering and Applied Science
University of California, Los Angeles, USA

Kevin Wong
School of Civil and Structural Engineering
Nanyang Technological University, Singapore

John Wiley & Sons, Inc.
New York / Chichester / Weinheim / Brisbane / Singapore / Toronto

Acquisitions Editor   *Wayne Anderson*
Marketing Manager   *Katherine Hepburn*
Senior Production Editor   *Robin Factor*
Illustration Editor   *Sigmund Malinowski*
Senior Designer   *Kevin Murphy*
COVER PHOTO © Rafael Macia/Photo Researchers

This book was set in Times by Argosy and printed and bound by Donnelley/Willard. The cover was printed by Phoenix Color.

This book is printed on acid-free paper.

Copyright ©2000 John Wiley & Sons, Inc. All rights reserved.

No part of this publication may be reproduced, stored in a retrieval system, or transmitted in any form or by any means, electronic, mechanical, photocopying, recording, scanning, or otherwise, except as permitted under Sections 107 or 108 of the 1976 United States Copyright Act, without either the prior written permission of the Publisher, or authorization through payment of the appropriate per-copy fee to the Copyright Clearance Center, 222 Rosewood Drive, Danvers, MA 01923, (978) 750-8400, fax (978) 750-4470. Requests to the Publisher for permission should be addressed to the Permissions Department, John Wiley & Sons, Inc., 605 Third Avenue, New York, NY 10158-0012, (212) 850-6011, fax (212) 850-6008, E-Mail: PERMREQ@WILEY.COM. To order books please call 1 (800) 225-5945.

Library of Congress Cataloging in Publication Data:
Hart, Gary C.
    Structural dynamics for structural engineers   /   Gary C. Hart, Kevin
Wong.
        p.   cm.
    ISBN 0–471–36169–0 (cloth : alk. paper)
    1. Structural dynamics.   I. Wong, Kevin Kai Fai, 1969–
II. Title.
TA654.H37   1999
624.1'7—dc21   99–34623
                                                                    CIP

Printed in the United States of America

10 9 8 7 6 5 4 3 2 1

for

Marianne McDermott Hart

and

Estelle Yee

Contents

Part Two Earthquake Response of Structures

Preface

The first work on this book started in the early 1970s when I presented at UCLA a 9-month, once a week, 3-hour evening lecture course to practicing structural engineers. The course was offered for several years, and much of the material contained in Chapters 2 through 6 was first written and presented in that course. In the almost 30 years since that first lecture series, I have gained considerable experience in measuring the wind, harmonic steady state machine-excited, and earthquake response of buildings. In addition, I have continually tried to develop more accurate computer models to calculate the earthquake response of buildings. This experience includes buildings of most conventional materials (concrete, masonry, wood, and steel), new high-tech fiber composite materials, and also such new high-tech manufactured systems as viscous dampers and base isolators. The goal in all of these projects was to use or extend the theory of structural dynamics to develop computer models to estimate building performance.

In 1996, my Ph.D. student Kevin Wong finished his dissertation at UCLA. Kevin, who is now on the faculty at Nanyang Technological University in Singapore, is a very special and gifted structural engineer, and therefore I approached him about coauthoring a book with me. This book is our combined effort ,and it represents our attempt to present to structural engineers the key parts of the broad field of structural dynamics. It also presents our perspective of how new high-tech manufactured systems can be modeled and studied. I hope we have come close to our goal, and any suggestions for improvement would be welcome and sincerely appreciated.

I would like to express my appreciation to Dr. Sampson Huang who, for over two decades, has helped me to learn structural dynamics by performing structural dynamic analyses and designs of buildings and other structures.

Kevin and I would also like to thank Kendra Saunders for her assistance in the typing of this book and also Chukwuma Ekwueme, Anurag Jain, Matt Skokan, and Sutat Leelataviwat for their technical review of selected chapters in the book. The helpful comments of Professor Barry Goodno and Max Porter are also acknowledged. Finally, many sections of this book were researched and written over many years in the library of the Civil Engineering Department at Imperial College in London. A special thank you to Mrs. Crooks, the librarian at Imperial College.

Professor Gary C. Hart
Department of Civil and Environmental Engineering
University of California, Los Angeles

Structural Dynamics
for Structural Engineers

Part One

Foundation of Structural Dynamics

Chapter 1

Introduction

1.1 OVERVIEW

Structural dynamics is the topic of the 21st century. It took almost the entire 20th century to bring the civil structural engineering profession to the point where buildings and other structures can be analyzed with a reasonable degree of confidence to evaluate their performance to real civil engineering types of forcing functions. The first part of the 20th century was primarily devoted to developing a basic theoretical foundation that incorporated the principles of mathematics and mechanics using very simple idealizations of structures. The result was that starting in the 1960s, structural analysis moved onto a much more sophisticated and accurate phase where it incorporated the mathematics of matrix methods and numerical analysis to develop analytical models of structures. This led to modern finite element methods of structural analysis using high-speed digital computers.

The loads that act upon structures received special attention starting in the 1970s with the aid of modern data recording and acquisition systems. For example, the understanding of the characteristics of strong earthquake motions and building response began with the 1971 San Fernando earthquake. In that earthquake, for the first time, a significant number of acceleration versus time histories were accurately recorded on the ground and in buildings. During the remainder of the 20th century, more and increasingly accurate records of ground and building motions during earthquakes have been produced. Therefore, structural dynamics will enter the 21st century with a foundation of good and improving structural analysis methods and earthquake ground motion versus time records. In addition, very significant advances in the understanding of wind, wave, and other structural loads were made in the latter half of the 20th century.

Structural dynamics in the 21st century will be very exciting for structural engineers. New materials are being developed, and new elements that are manufactured in factories and installed in structures are being developed and will greatly expand in use and type in the 21st century. For example, Chapter 8 discusses base isolators of the type that are now being installed at the interface between the structure and the foundation to reduce inertia forces on the structure from earthquakes. Also, Chapter 9 discusses another type of high-tech manufactured element called a viscous damper. This element is manufactured in a factory and placed in buildings and bridges to dissipate the energy input to the structure from either wind or earthquake forces. These exciting passive structural elements that dissipate energy will be supplemented with structural elements which have electrical and computer-based parts that monitor the motion and then in almost real time apply man-made forces to the structure to reduce structural response. Chapter 10 provides a discussion of structural control, which is the first real use of computer smart-structural elements.

Modern structural dynamics requires sophisticated mathematical models if the structural engineer is to accurately estimate the response of real as-built civil engineering structures. The goal of this book is to present some of these methods in a clear and systematic manner. The book is self-contained and assumes no prior reading or knowledge in vibration theory or structural dynamics. Also, it provides visual images of real structural engineering problems. Almost all examples are presented using matrix models of frame structures for two primary reasons. First, the spring mass system, while convenient to introduce the mathematics of structural dynamics, is not a real structural engineering model. Second, frame structures (especially in two dimensions) are the structural backbone of many lateral force-resisting systems and can often help bridge the gap between mathematics and the real world. It is not the goal of this book to present structural analysis methods for formulating structural stiffness and mass matrices. The goal is to develop methods that structural engineers can use to solve real structural dynamics problems.

Many readers can correctly view this book as a first book in structural dynamics. We believe that the reader will benefit from reading other books, reports, and monographs, and these are noted at the ends of many of the chapters.

1.2 ORGANIZATION OF THE BOOK

The book is organized into three parts. The first part of the book is essential reading and should be the basis for an undergraduate course and perhaps part of a graduate course in structural dynamics. The material on proportional and nonproportional damping, state space analysis, and sensitivity analyses is typically not covered under essential reading. However, the future of structural dynamics in the 21st century is such that this material must be included in the essential category.

The second and third parts of the book can be read in many possible orders. Also, especially in Part III, these chapters start with one or more sections on the basics of the topic (e.g., base isolation systems), which allows a first reading that does not include all of the material in the chapters. For example, in an undergraduate course the instructor may assign reading, or may lecture, on only the first few sections of Chapters 8, 9, and 10. Such an approach will enable students to learn the basics of these modern systems and perhaps will excite them as graduate students or as professional structural engineers to finish reading the chapter and go even further to learn about these exciting new areas of structural dynamics.

Part I is entitled Foundation of Structural Dynamics. Our intent is to provide in Chapters 2 and 3 a basic block of material that can be learned or taught within a reasonable time frame. For example, at UCLA, the material in Chapters 2 and 3 is covered in approximately 15 to 20 lecture hours. Chapter 4 is a generalization of Chapter 3.

Part II, Earthquake Response of Structures, is devoted to structural dynamics in earthquake engineering. The organization of this part follows the organization and classification of structural dynamics as defined in earthquake engineering design criteria and codes. Chapter 5 presents the methods of structural dynamics called response spectra analysis, an analysis method in which the time variable is eliminated from the response. Chapter 6 presents the methods of structural dynamics called linear time history analysis, a type of analysis in which the stiffness, mass, and damping characteristics of the structure are not a function of time. The member-stiffness-related forces, mass-related inertial forces, and damping forces are linearly related to the time variation of the displacement, acceleration, and velocity of the structure, respectively. The nonlinear time history analysis method is presented in Chapter 7. This chapter is devoted to the methods used to calculate the nonlinear response of structures with nonlinear and inelastic material properties.

Table 1.1 Selected Topics

Topic	Sections
Sensitivity	2.9, 3.9–3.10, 4.12–4.14
Proportional damping	3.5, 4.6–4.8
State space response method	2.5, 2.8, 3.6, 3.8, 4.10–4.11
Energy methods	2.10
Problem size reduction	4.15, 6.5–6.6, 6.8, 7.8
Base isolation	All of Chapter 8
Viscous dampers	All of Chapter 9
Structural control	All of Chapter 10

Part III presents special structural topics in structural dynamics. Chapter 8 is devoted to base isolation. Chapter 9 presents the important new area of structural dynamics where manufactured elements are placed in the structure to dissipate energy through viscous damping. The damping forces in this chapter are not a linear function of velocity. Finally, Chapter 10 introduces structural control, including a discussion of what active control is and how to calculate the response of a structure with control forces.

Most of the examples and illustrations in this book are frame structures. Therefore, the primary focus is on civil engineering and building type structures. A basic undergraduate course that includes the matrix, or the direct stiffness, method of analysis of structures is desirable before reading this book. The next section provides an overview of this type of structural analysis, which is required in many sections of the book.

Structural dynamics presents a beautiful balance between applied mathematics and structural engineering. Therefore, the reader or instructor is encouraged to exercise flexibility in the order in which different topics—and their related sections—are read. Chapters 8, 9, and 10 present easily identified topics and, as previously noted, considerable understanding can be gained by reading only the first few sections of any of these three chapters. Other topics are not so easily identified, and therefore Table 1.1 provides a grouping of the sections related to these other topics.

1.3   THE STRUCTURAL MODEL

There are many different ways that a real structure can be idealized for the purposes of a book of this type. Modern structural engineering almost always formulates the structural response characteristics in terms of matrix structural analysis methods. Therefore, this is the method we use in this book. The remainder of this section briefly discusses structural analysis and structural model development. The reader is encouraged to study this topic in more detail, and the next section provides recommended additional reading.

Figure 1.1 shows a fixed-base, single-story, moment-resisting frame. There are four nodes or "joints" in this structure, each identified by a number with a circle around the number. Forces are considered to act in the plane of the frame, and thus the frame is a two-dimensional (2-D) model. The degrees of freedom (DOF) for this structure are the translation in the x- and z-directions plus the rotation at each joint, θ_y. Therefore, the structure has six degrees of freedom. The structural engineer's professional experience and the assumptions he or she makes define the final number of degrees of freedom used to model the structure for a structural dynamic analysis.

Consider first the assumption that the beam has a very large axial stiffness. This assumption means that the horizontal displacement at Node 2 is equal to the horizontal displacement

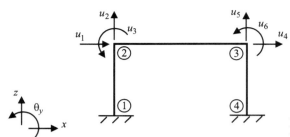

Figure 1.1 Single-Story Moment-Resisting Frame

at Node 3. If it is now assumed that the columns have a very large axial stiffness, then the vertical displacements at Nodes 2 and 3 are zero (since the vertical displacement at Nodes 1 and 4 are zero). The result of these two assumptions is a structure with three independent DOF.

Professor Moshe Rubenstein at UCLA, as part of his Ph.D. research in the early 1960s, assumed that the joint rotation at Node 2 is equal to the joint rotation at Node 3. He showed that in many situations this is a good assumption and results in very little loss in the accuracy of system response. If this assumption is made, the resultant system is a two DOF system.

If the beam is assumed to have a very large moment of inertia relative to the column, then the joint rotations at Nodes 2 and 3 can be assumed to be zero, and the frame is then modeled as a single degree of freedom (SDOF) system.

In matrix notation, the mass and stiffness matrices for the six DOF representation of the frame can be written as

$$\mathbf{m}\ddot{\mathbf{U}}(t) + \mathbf{k}\mathbf{U}(t) = 0 \tag{1.1}$$

where $\mathbf{m}$ is a 6×6 mass matrix, $\mathbf{k}$ is a 6×6 stiffness matrix, $\mathbf{U}(t)$ is a 6×1 displacement vector, and $\ddot{\mathbf{U}}(t)$ is a 6×1 acceleration vector. Note that

$$\mathbf{U}(t) = \begin{bmatrix} u_1(t) & u_2(t) & u_3(t) & u_4(t) & u_5(t) & u_6(t) \end{bmatrix}^T. \tag{1.2}$$

The mathematical representation of the beam having a very large axial stiffness is

$$u_1(t) = u_4(t). \tag{1.3}$$

A very large axial stiffness of the columns implies that

$$u_2(t) = u_5(t) = 0. \tag{1.4}$$

If the rotation at Joint 2 is assumed to be equal to the rotation at Joint 3, then

$$u_3(t) = u_6(t). \tag{1.5}$$

With these assumptions the system has two independent degrees of freedom denoted $x_1(t)$ and $x_2(t)$, and they are obtained by defining

$$x_1(t) \equiv u_1(t), \quad x_2(t) \equiv u_3(t). \tag{1.6}$$

It then follows that

$$\mathbf{U}(t) = \begin{Bmatrix} u_1(t) \\ u_2(t) \\ u_3(t) \\ u_4(t) \\ u_5(t) \\ u_6(t) \end{Bmatrix} = \begin{bmatrix} 1 & 0 \\ 0 & 0 \\ 0 & 1 \\ 1 & 0 \\ 0 & 0 \\ 0 & 1 \end{bmatrix} \begin{Bmatrix} x_1(t) \\ x_2(t) \end{Bmatrix} = \mathbf{TX}(t). \tag{1.7}$$

A very important quantity in structural dynamics is the *strain energy* that is contained in the structure when it is in a deformed state. The strain energy of the six degrees of freedom representation of the frame shown in Figure 1.1 in terms of a displacement shape defined by the **U** coordinate system is equal to

$$E_s = \frac{1}{2} \mathbf{U}^T \mathbf{k} \mathbf{U} \tag{1.8}$$

where E_s is the strain energy of the structure. In terms of the new **X** coordinate system, the strain energy can be expressed as

$$E_s = \frac{1}{2} \mathbf{X}^T \mathbf{K} \mathbf{X}. \tag{1.9}$$

Note that in Eq. 1.9 the stiffness matrix (2×2) in the **X** coordinate system (i.e., **K**) is unknown. Because the strain energy of the frame is the same in the **U** and the **X** systems, Eq. 1.8 is equal to Eq. 1.9. If Eq. 1.7 is substituted into Eq. 1.8 and then this is equated to Eq. 1.9, it follows from Eqs. 1.8 and 1.9 that

$$E_s = \frac{1}{2} \mathbf{U}^T \mathbf{k} \mathbf{U} = \frac{1}{2} \mathbf{X}^T \mathbf{T}^T \mathbf{k} \mathbf{T} \mathbf{X} = \frac{1}{2} \mathbf{X}^T \mathbf{K} \mathbf{X}. \tag{1.10}$$

For any displacement shape in the reduced **X** coordinate system, it follows from Eq. 1.10 that the stiffness matrix in the **X** coordinate system is

$$\mathbf{K} = \mathbf{T}^T \mathbf{k} \mathbf{T}. \tag{1.11}$$

Similarly, the corresponding equation for the *kinetic energy* of the structure when the structure is in a deformed state defined by $\dot{\mathbf{U}}$ is equal to

$$E_k = \frac{1}{2} \dot{\mathbf{U}}^T \mathbf{m} \dot{\mathbf{U}} = \frac{1}{2} \dot{\mathbf{X}}^T \mathbf{T}^T \mathbf{m} \mathbf{T} \dot{\mathbf{X}} = \frac{1}{2} \dot{\mathbf{X}}^T \mathbf{M} \dot{\mathbf{X}} \tag{1.12}$$

where E_k is the kinetic energy of the structure and **M** is the mass matrix in the reduced **X** coordinate system. For any velocity field in the reduced **X** coordinate system, it follows that the mass matrix is

$$\mathbf{M} = \mathbf{T}^T \mathbf{m} \mathbf{T}. \tag{1.13}$$

A similar development follows for the case where the rotations of the joints at Nodes 2 and 3 are zero. In this case, the system only has one degree of freedom and the transformation matrix **T** is equal to

$$\mathbf{U}(t) = \begin{Bmatrix} u_1(t) \\ u_2(t) \\ u_3(t) \\ u_4(t) \\ u_5(t) \\ u_6(t) \end{Bmatrix} = \begin{bmatrix} 1 \\ 0 \\ 0 \\ 1 \\ 0 \\ 0 \end{bmatrix} x(t) = \mathbf{T}x(t). \tag{1.14}$$

Now the matrices **M** and **K** are 1×1 matrices, or scalars.

Figure 1.2 shows a two-story frame. If it is assumed that

$$u_1(t) = u_4(t) = x_1(t), \quad u_7(t) = u_{10}(t) = x_2(t)$$
$$u_2(t) = u_3(t) = u_5(t) = u_6(t) = u_8(t) = u_9(t) = u_{11}(t) = u_{12}(t) = 0 \tag{1.15}$$

then it follows that

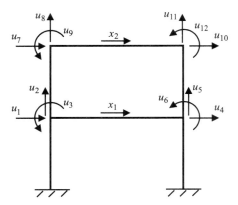

Figure 1.2 Two-Story Moment-Resisting Frame

$$
\mathbf{U}(t) =
\begin{Bmatrix}
u_1(t) \\
u_2(t) \\
u_3(t) \\
u_4(t) \\
u_5(t) \\
u_6(t) \\
u_7(t) \\
u_8(t) \\
u_9(t) \\
u_{10}(t) \\
u_{11}(t) \\
u_{12}(t)
\end{Bmatrix}
=
\begin{bmatrix}
1 & 0 \\
0 & 0 \\
0 & 0 \\
1 & 0 \\
0 & 0 \\
0 & 0 \\
0 & 1 \\
0 & 0 \\
0 & 0 \\
0 & 1 \\
0 & 0 \\
0 & 0
\end{bmatrix}
\begin{Bmatrix}
x_1(t) \\
x_2(t)
\end{Bmatrix}
= \mathbf{TX}(t).
\tag{1.16}
$$

The matrix $\mathbf{T}$ is a 12×2 matrix. It follows that the system mass and stiffness matrices (i.e., $\mathbf{M}$ and $\mathbf{K}$) in the reduced $\mathbf{X}$ coordinate system are 2×2 matrices.

EXAMPLE 1   Single Element Structural Model

Consider a single element as shown in Figure 1.3a with six degrees of freedom at the two ends of the element. The stiffness matrix is

$$
[\mathbf{k}] =
\begin{bmatrix}
\dfrac{12EI}{L^3} & 0 & \dfrac{6EI}{L^2} & -\dfrac{12EI}{L^3} & 0 & \dfrac{6EI}{L^2} \\[2mm]
0 & \dfrac{AE}{L} & 0 & 0 & -\dfrac{AE}{L} & 0 \\[2mm]
\dfrac{6EI}{L^2} & 0 & \dfrac{4EI}{L} & -\dfrac{6EI}{L^2} & 0 & \dfrac{2EI}{L} \\[2mm]
-\dfrac{12EI}{L^3} & 0 & -\dfrac{6EI}{L^2} & \dfrac{12EI}{L^3} & 0 & -\dfrac{6EI}{L^2} \\[2mm]
0 & -\dfrac{AE}{L} & 0 & 0 & \dfrac{AE}{L} & 0 \\[2mm]
\dfrac{6EI}{L^2} & 0 & \dfrac{2EI}{L} & -\dfrac{6EI}{L^2} & 0 & \dfrac{4EI}{L}
\end{bmatrix}.
\tag{1.17}
$$

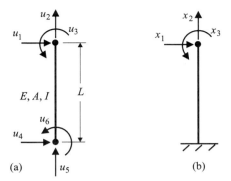

(a)

(b)   **Figure 1.3** Single-Element Structure

Now consider that the element is fixed at the base as shown in Figure 1.3b. Only three degrees of freedom remain, and these three degrees of freedom correspond directly to the u_i's in Figure 1.3a as follows:

$$x_1 = u_1, \quad x_2 = u_2, \quad x_3 = u_3. \tag{1.18}$$

Therefore, the matrix **T** becomes

$$\mathbf{T} = \begin{bmatrix} 1 & 0 & 0 \\ 0 & 1 & 0 \\ 0 & 0 & 1 \\ 0 & 0 & 0 \\ 0 & 0 & 0 \\ 0 & 0 & 0 \end{bmatrix}. \tag{1.19}$$

It follows from Eq. 1.11 that

$$\mathbf{K} = \mathbf{T}^T \mathbf{kT} = \begin{bmatrix} \dfrac{12EI}{L^3} & 0 & \dfrac{6EI}{L^2} \\ 0 & \dfrac{AE}{L} & 0 \\ \dfrac{6EI}{L^2} & 0 & \dfrac{4EI}{L} \end{bmatrix}. \tag{1.20}$$

Now assume that the single element is axially rigid, that is, $A \rightarrow \infty$. Then it follows that $x_2 = 0$. Thus Eq. 1.18 becomes

$$x_1 = u_1, \quad x_3 = u_3 \tag{1.21}$$

and the matrix **T** becomes

$$\mathbf{T} = \begin{bmatrix} 1 & 0 \\ 0 & 0 \\ 0 & 1 \\ 0 & 0 \\ 0 & 0 \\ 0 & 0 \end{bmatrix}. \tag{1.22}$$

It follows from Eq. 1.11 that

$$\mathbf{K} = \mathbf{T}^T \mathbf{kT} = \begin{bmatrix} \dfrac{12EI}{L^3} & \dfrac{6EI}{L^2} \\ \dfrac{6EI}{L^2} & \dfrac{4EI}{L} \end{bmatrix}. \tag{1.23}$$

EXAMPLE 2   *Single-Story Moment-Resisting Frame*

Consider again the single-story moment-resisting frame as shown in Figure 1.1. Assume that the beam has a length of L_b and material properties of A_b, I_b, and E, and that the two columns have the same length of L_c and the same material property of A_c, I_c, and E. Then it follows that the stiffness matrix is

$$\mathbf{k} = \begin{bmatrix} \dfrac{A_b E}{L_b} + \dfrac{12EI_c}{L_c^3} & 0 & \dfrac{6EI_c}{L_c^2} & -\dfrac{A_b E}{L_b} & 0 & 0 \\[2mm] 0 & \dfrac{12EI_b}{L_b^3} + \dfrac{A_c E}{L_c} & \dfrac{6EI_b}{L_b^2} & 0 & -\dfrac{12EI_b}{L_b^3} & \dfrac{6EI_b}{L_b^2} \\[2mm] \dfrac{6EI_c}{L_c^2} & \dfrac{6EI_b}{L_b^2} & \dfrac{4EI_b}{L_b} + \dfrac{4EI_c}{L_c} & 0 & -\dfrac{6EI_b}{L_b^2} & \dfrac{2EI_b}{L_b} \\[2mm] -\dfrac{A_b E}{L_b} & 0 & 0 & \dfrac{A_b E}{L_b} + \dfrac{12EI_c}{L_c^3} & 0 & \dfrac{6EI_c}{L_c^2} \\[2mm] 0 & -\dfrac{12EI_b}{L_b^3} & -\dfrac{6EI_b}{L_b^2} & 0 & \dfrac{12EI_b}{L_b^3} + \dfrac{A_c E}{L_c} & -\dfrac{6EI_b}{L_b^2} \\[2mm] 0 & \dfrac{6EI_b}{L_b^2} & \dfrac{2EI_b}{L_b} & \dfrac{6EI_c}{L_c^2} & -\dfrac{6EI_b}{L_b^2} & \dfrac{4EI_b}{L_b} + \dfrac{4EI_c}{L_c} \end{bmatrix} \quad (1.24)$$

First assume that both the beam and the columns are axially rigid. Then it follows that

$$u_1 = u_4, \quad u_2 = u_5 = 0. \quad (1.25)$$

The result is that only three degrees of freedom remain. Let these three degrees of freedom be defined as

$$x_1 = u_1 = u_4, \quad x_2 = u_3, \quad x_3 = u_6. \quad (1.26)$$

Then the matrix **T** becomes

$$\mathbf{T} = \begin{bmatrix} 1 & 0 & 0 \\ 0 & 0 & 0 \\ 0 & 1 & 0 \\ 1 & 0 & 0 \\ 0 & 0 & 0 \\ 0 & 0 & 1 \end{bmatrix}. \quad (1.27)$$

It follows from Eq. 1.11 that

$$\mathbf{K} = \mathbf{T}^T \mathbf{k} \mathbf{T} = \begin{bmatrix} \dfrac{24EI_c}{L_c^3} & \dfrac{6EI_c}{L_c^2} & \dfrac{6EI_c}{L_c^2} \\[2mm] \dfrac{6EI_c}{L_c^2} & \dfrac{4EI_b}{L_b} + \dfrac{4EI_c}{L_c} & \dfrac{2EI_b}{L_b} \\[2mm] \dfrac{6EI_c}{L_c^2} & \dfrac{2EI_b}{L_b} & \dfrac{4EI_b}{L_b} + \dfrac{4EI_c}{L_c} \end{bmatrix}. \quad (1.28)$$

Now assume that, in addition to both the beam and columns being axially rigid, the beam is also flexurally rigid. Thus there will be no rotation at the two ends of the beam, that is,

$$u_1 = u_4, \quad u_2 = u_3 = u_5 = u_6 = 0. \quad (1.28)$$

The result is that only one degree of freedom remains. Let this degree of freedom be defined as

$$x_1 = u_1 = u_4. \tag{1.30}$$

Then the matrix **T** becomes

$$\mathbf{T} = \begin{bmatrix} 1 \\ 0 \\ 0 \\ 1 \\ 0 \\ 0 \end{bmatrix}. \tag{1.31}$$

It follows from Eq. 1.11 that

$$\mathbf{K} = \mathbf{T}^T \mathbf{k} \mathbf{T} = \left[\frac{24EI_c}{L_c^3} \right]. \tag{1.32}$$

RECOMMENDED ADDITIONAL READING

1. L. P. Felton and R. B. Nelson, *Matrix Structural Analysis*, John Wiley and Sons, 1997.

Chapter 2

Single Degree of Freedom Linear System Response

2.1 OVERVIEW

There are many reasons why it is important to understand the response of a single degree of freedom (SDOF) system. One reason is that the interdependence of the properties of the structure and the time history of ground motion is best demonstrated using an SDOF system. Another reason is that, as will be shown later, many building code equations are based on the response of an SDOF system.

An SDOF system can be represented in many different forms, three of which are shown in Figure 2.1. Figure 2.1a shows a spring-mass system, which is the most commonly used representation of an SDOF system in textbooks used in mechanical vibration courses. In Figure 2.1b, the SDOF system is represented by a mass at the top of a column, and the rotation and vertical deflection at the end of the column are assumed to be so small that these deformations are assumed to be zero. For the structural representation shown in Figure 2.1c, the beam is assumed to be axially rigid, and the vertical displacement and the rotations of both ends of the columns are assumed to be zero.

Consider the SDOF system as shown in Figure 2.1c. Four types of forces are acting on the mass, as shown in Figure 2.2. These forces include the following:

1. *Stiffness Force*: This force acts on the floor when there is a lateral displacement of the mass. For a linear system, this force is directly proportional to the relative lateral displacement of the top and bottom of the column, x, and the constant of

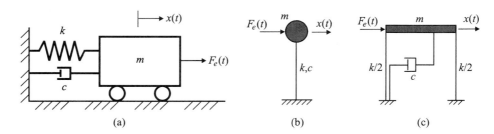

Figure 2.1 Single Degree of Freedom System Representation

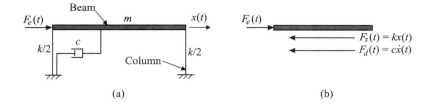

Figure 2.2 Single Degree of Freedom System

proportionality is the stiffness provided by the beam and the columns, denoted by k. When a positive lateral displacement occurs, this force always acts to "pull" the floor back to the original undeformed position. Therefore, this force always acts in the opposite direction of the displacement (i.e., $F_s = -kx$).

2.   *Damping Force*: This force acts on the floor when there is a relative lateral velocity between the mass and the ground. For a system with linear viscous damping, this force is directly proportional to the velocity, $\dot{x}$, and the constant of proportionality is the damping coefficient, c. Similar to the elastic force, when a positive relative velocity occurs, this force always acts to reduce the velocity. Therefore, this force always acts in the opposite direction of the velocity (i.e., $F_d = -c\dot{x}$).

3.   *External Force*: This force, F_e, is an external force applied to the floor. This force generally varies as a function of time.

4.   *Inertial Force*: This force represents the acceleration of the floor due to the imbalance of the preceding three types of forces.

From Newton's law, where the sum of forces is equal to the mass times the acceleration, it follows that

$$\sum F = F_s + F_d + F_e = -kx - c\dot{x} + F_e = m\ddot{x}. \tag{2.1}$$

Rearranging the terms in Eq. 2.1 gives

$$m\ddot{x}(t) + c\dot{x}(t) + kx(t) = F_e(t). \tag{2.2}$$

Equation 2.2 represents the *dynamic equilibrium equation* and is often called the *equation of motion* for the system. It is a second-order linear differential equation, and therefore it requires two initial conditions to define its response. These initial conditions generally are the initial displacement of the mass, $x(0) = x_o$, and the initial velocity of the mass, $\dot{x}(0) = \dot{x}_o$. The solution to this second-order linear differential equation involves two parts: the homogeneous solution, $x_h(t)$, and the particular solution, $x_p(t)$. The homogeneous solution, $x_h(t)$, satisfies the differential equation without the applied force, which is written

$$m\ddot{x}_h(t) + c\dot{x}_h(t) + kx_h(t) = 0. \tag{2.3}$$

The homogeneous solution response is a function of the initial conditions (i.e., initial displacement, x_o, and initial velocity, $\dot{x}_o$). The particular solution response, $x_p(t)$, is a function of the external applied force and satisfies the differential equation

$$m\ddot{x}_p(t) + c\dot{x}_p(t) + kx_p(t) = F_e(t). \tag{2.4}$$

Together, the homogeneous and the particular solution result in the response of the system, $x(t)$. Because the equation of motion (i.e., Eq. 2.2) for the response $x(t)$ is a linear differential equation, it follows that the solution is the summation of the homogeneous solution $x_h(t)$ and the particular solution $x_p(t)$, that is, $x(t) = x_h(t) + x_p(t)$, and

$$m\ddot{x}(t) + c\dot{x}(t) + kx(t) = m\left(\ddot{x}_h + \ddot{x}_p\right) + c\left(\dot{x}_h + \dot{x}_p\right) + k\left(x_h + x_p\right)$$

$$= \left(m\ddot{x}_h + c\dot{x}_h + kx_h\right) + \left(m\ddot{x}_p + c\dot{x}_p + kx_p\right) \qquad (2.5)$$

$$= 0 + F_e(t)$$

$$= F_e(t).$$

2.2 FREE VIBRATION

If the single degree of freedom system has no external applied force, it will respond in *free vibration* when its initial conditions are nonzero. Because the external force is zero, the particular solution response is zero, and the homogeneous solution, $x_h(t)$, represents the free vibration of the structure. The free vibration response is determined by solving Eq. 2.3, which is rewritten here as

$$m\ddot{x}_h(t) + c\dot{x}_h(t) + kx_h(t) = 0, \quad x_h(0) = x_o, \quad \dot{x}_h(0) = \dot{x}_o. \qquad (2.6)$$

Assume that the homogeneous solution is of the form $x_h(t) = Ce^{st}$, where C and s are constants and not a function of time. The goal here is to find a set of values for C and s that satisfies Eq. 2.6. Because Eq. 2.6 is a linear second-order differential equation with constant coefficients, two sets of values for C and s are expected. Denote these two sets as (C_1, s_1) and (C_2, s_2). Differentiating $x_h(t)$ gives $\dot{x}_h(t) = Cse^{st}$ and $\ddot{x}_h(t) = Cs^2e^{st}$. It follows from Eq. 2.6 that

$$ms^2Ce^{st} + csCe^{st} + kCe^{st} = 0. \qquad (2.7)$$

For any nonzero Ce^{st}, that is, $x_h(t) \neq 0$, it follows that

$$ms^2 + cs + k = 0. \qquad (2.8)$$

The roots of Eq. 2.8 are obtained from the quadratic equation solution and are

$$s_1 = \frac{-c + \sqrt{c^2 - 4mk}}{2m}, \quad s_2 = \frac{-c - \sqrt{c^2 - 4mk}}{2m}. \qquad (2.9)$$

It then follows that the free vibration response can be written as

$$x_h(t) = C_1 e^{\left(\frac{-c + \sqrt{c^2 - 4mk}}{2m}\right)t} + C_2 e^{\left(\frac{-c - \sqrt{c^2 - 4mk}}{2m}\right)t}. \qquad (2.10)$$

In structural dynamics, Eq. 2.6 is generally written in a normalized form. Let

$$\omega_n^2 = \frac{k}{m}, \quad 2\zeta\omega_n = \frac{c}{m}. \qquad (2.11)$$

The variable ω_n is *the undamped natural frequency of vibration*. The variable ζ is called the *critical damping ratio*, which represents the damping in the structure. It follows from Eq. 2.6 that

$$\ddot{x}_h(t) + 2\zeta\omega_n \dot{x}_h(t) + \omega_n^2 x_h(t) = 0, \quad x(0) = x_o, \quad \dot{x}(0) = \dot{x}_o. \qquad (2.12)$$

As before, let $x_h(t) = Ce^{st}$. Eq. 2.12 becomes

$$s^2Ce^{st} + 2\zeta\omega_n sCe^{st} + \omega_n^2 Ce^{st} = 0. \qquad (2.13)$$

Solving for s gives

$$s_1 = -\zeta\omega_n + \omega_n\sqrt{\zeta^2 - 1}, \quad s_2 = -\zeta\omega_n - \omega_n\sqrt{\zeta^2 - 1} \qquad (2.14)$$

and the free vibration response becomes

$$x_h(t) = C_1 e^{\left(-\zeta\omega_n + \omega_n\sqrt{\zeta^2-1}\right)t} + C_2 e^{\left(-\zeta\omega_n - \omega_n\sqrt{\zeta^2-1}\right)t}. \tag{2.15}$$

Eq. 2.15 is the free vibration solution for $x_h(t)$ when $\zeta > 1$. This solution is seldom of interest to structural engineers because structures generally have a value of critical damping ratio in the range of 2% to 10% (i.e., $0.02 < \zeta < 0.10$). Therefore, Eq. 2.15 must be reconsidered because the quantity under the square root sign is less than zero.

For $\zeta < 1$, it follows that $\sqrt{\zeta^2 - 1} = i\sqrt{1 - \zeta^2}$, and then Eq. 2.15 becomes

$$x_h(t) = C_1 e^{\left(-\zeta\omega_n + \omega_n i\sqrt{1-\zeta^2}\right)t} + C_2 e^{\left(-\zeta\omega_n - \omega_n i\sqrt{1-\zeta^2}\right)t}. \tag{2.16}$$

Let $\omega_d = \omega_n \sqrt{1 - \zeta^2}$, where ω_d is called the *damped natural frequency of vibration*. It follows from Eq. 2.16 that

$$\begin{aligned}
x_h(t) &= C_1 e^{\left(-\zeta\omega_n + \omega_n i\sqrt{1-\zeta^2}\right)t} + C_2 e^{\left(-\zeta\omega_n - \omega_n i\sqrt{1-\zeta^2}\right)t} \\
&= C_1 e^{(-\zeta\omega_n + i\omega_d)t} + C_2 e^{(-\zeta\omega_n - i\omega_d)t} \\
&= C_1 e^{-\zeta\omega_n t}\left(\cos \omega_d t + i \sin \omega_d t\right) + C_2 e^{-\zeta\omega_n t}\left(\cos \omega_d t - i \sin \omega_d t\right) \\
&= (C_1 + C_2) e^{-\zeta\omega_n t} \cos \omega_d t + i(C_1 - C_2) e^{-\zeta\omega_n t} \sin \omega_d t \\
&= C_1' e^{-\zeta\omega_n t} \cos \omega_d t + C_2' e^{-\zeta\omega_n t} \sin \omega_d t.
\end{aligned} \tag{2.17}$$

Note that for notational convenience, the two constants C_1 and C_2 have been replaced by two related constants C_1' and C_2' (i.e., $C_1' = C_1 + C_2$ and $C_2' = i(C_1 - C_2)$). The velocity response can be obtained by differentiating Eq. 2.17, that is,

$$\dot{x}_h(t) = \left(-\zeta\omega_n C_1' + \omega_d C_2'\right) e^{-\zeta\omega_n t} \cos \omega_d t + \left(-\omega_d C_1' - \zeta\omega_n C_2'\right) e^{-\zeta\omega_n t} \sin \omega_d t. \tag{2.18}$$

Now, the values for C_1' and C_2' are solved using the initial conditions given in Eq. 2.12. Substituting $x_h(0) = x_o$ into Eq. 2.17 and $\dot{x}_h(0) = \dot{x}_o$ into Eq. 2.18 gives

$$x_o = C_1', \qquad \dot{x}_o = -\zeta\omega_n C_1' + \omega_d C_2'. \tag{2.19}$$

Solving for C_1' and C_2' in Eq. 2.19 gives

$$C_1' = x_o, \qquad C_2' = \frac{\zeta\omega_n x_o + \dot{x}_o}{\omega_d}. \tag{2.20}$$

Finally, Eqs. 2.17 and 2.18 become

$$x(t) = x_h(t) = e^{-\zeta\omega_n t}\left[x_o \cos \omega_d t + \left(\frac{\zeta\omega_n x_o + \dot{x}_o}{\omega_n\sqrt{1-\zeta^2}}\right) \sin \omega_d t\right] \tag{2.21}$$

$$\dot{x}(t) = \dot{x}_h(t) = e^{-\zeta\omega_n t}\left[\dot{x}_o \cos \omega_d t - \left(\frac{\omega_n x_o + \zeta\dot{x}_o}{\sqrt{1-\zeta^2}}\right) \sin \omega_d t\right]. \tag{2.22}$$

Note in Eqs. 2.21 and 2.22 that the terms inside the brackets oscillate from a positive value to a negative value with time because of the sine and cosine functions. The exponential function that premultiplies the bracketed terms does not change sign with time, but it decreases in value with time. Therefore, the free vibration response always decreases in magnitude as time progresses whenever damping exists in the structure.

The undamped natural frequency of vibration, ω_n, has the units of radians/second. This is often not preferred. The units that structural engineers often use are cycles and seconds. Therefore, it is convenient to define the *undamped natural frequency of vibration*, f_n, in cycles/second or hertz, Hz, and the corresponding *undamped natural period of vibration*, T_n, in seconds, to describe the rates of structural oscillations. These rates are

$$f_n = \frac{\omega_n}{2\pi}, \quad T_n = \frac{1}{f_n} = \frac{2\pi}{\omega_n}. \tag{2.23}$$

The undamped natural period of vibration is a very important quantity in structural dynamics. This quantity will be discussed in detail later. The *damped natural frequency of vibration*, f_d, and the *damped natural period of vibration*, T_d, are also defined here for a similar reason, where

$$f_d = \frac{\omega_d}{2\pi} = \frac{\omega_n\sqrt{1-\zeta^2}}{2\pi} = f_n\sqrt{1-\zeta^2}, \quad T_d = \frac{1}{f_d} = \frac{2\pi}{\omega_d} = \frac{2\pi}{\omega_n\sqrt{1-\zeta^2}} = \frac{T_n}{\sqrt{1-\zeta^2}}. \tag{2.24}$$

Note that as the damping increases, the damped natural period of vibration also increases.

When a system has a nonzero initial condition, it will respond in a manner where the general amplitude of motion will decrease with time. The *logarithmic decrement* is used to quantify this rate of decay in the free vibration. Due to the presence of damping, the general amplitude of structural response always decreases with time because of the term $e^{-\zeta\omega_n t}$ in Eqs. 2.21 and 2.22.

Let t_1 be the time at which the displacement response x_1 is calculated. Now, let t_2 be the time $t_2 = t_1 + T_d$ and x_2 be the displacement at time t_2. Since the sine and cosine terms in Eq. 2.21 oscillate with a period of T_d, it follows that

$$\cos\omega_d t_2 = \cos\omega_d(t_1 + nT_d) = \cos(\omega_d t_1 + n\omega_d T_d) = \cos(\omega_d t_1 + 2\pi n) = \cos\omega_d t_1$$

$$\sin\omega_d t_2 = \sin\omega_d(t_1 + nT_d) = \sin(\omega_d t_1 + n\omega_d T_d) = \sin(\omega_d t_1 + 2\pi n) = \sin\omega_d t_1$$

and therefore the bracketed term in Eq. 2.21 is the same for both x_1 and x_2. Now, dividing these two amplitudes, x_1 and x_2, gives

$$\frac{x_1}{x_2} = \left(\frac{x_o\, e^{-\zeta\omega_n t_1}}{x_o\, e^{-\zeta\omega_n t_2}}\right) = e^{\zeta\omega_n(t_2-t_1)} = e^{\zeta\omega_n T_d}. \tag{2.25}$$

The logarithmic decrement, δ_1, is defined by taking the natural logarithm of the quotient (x_1/x_2). This gives

$$\delta_1 = \ln\left(\frac{x_1}{x_2}\right) = \zeta\omega_n T_d = \frac{2\pi\zeta\omega_n}{\omega_d} = \frac{2\pi\zeta}{\sqrt{1-\zeta^2}}. \tag{2.26}$$

EXAMPLE 1   *Free Vibration of System Released from Rest*

The free vibration displacement response of the SDOF system is given in Eq. 2.21. Now let the initial velocity be zero and the initial displacement be x_o. The equation for the response is

$$\frac{x(t)}{x_o} = e^{-\zeta\omega_n t}\left[\cos\omega_d t + \left(\frac{\zeta}{\sqrt{1-\zeta^2}}\right)\sin\omega_d t\right]. \tag{2.27}$$

Note that $\omega_n = 2\pi/T_n$, and therefore $\omega_n t = 2\pi(t/T_n)$ and $\omega_d t = \omega_n\sqrt{1-\zeta^2}\,t = 2\pi\sqrt{1-\zeta^2}\,(t/T_n)$. It follows from Eq. 2.27 that

$$\frac{x(t)}{x_o} = e^{-2\pi\zeta(t/T_n)}\left[\cos\left(2\pi\sqrt{1-\zeta^2}\,(t/T_n)\right) + \left(\frac{\zeta}{\sqrt{1-\zeta^2}}\right)\sin\left(2\pi\sqrt{1-\zeta^2}\,(t\,T_n)\right)\right]. \tag{2.28}$$

If the time variable is expressed in multiples of the damped natural period of vibration where $t = nT_d$, then $\omega_d t = \omega_d nT_d = \omega_d n(2\pi/\omega_d) = 2\pi n$. It follows that

$$\frac{x(nT_d)}{x_o} = e^{-\zeta\omega_n nT_d}\left[\cos 2\pi n + \left(\frac{\zeta\omega_n}{\omega_n\sqrt{1-\zeta^2}}\right)\sin 2\pi n\right] = e^{-\zeta\omega_n nT_d}(1+0) = e^{-\zeta\omega_n nT_d}. \quad (2.29)$$

Again, recall that $\omega_d = \omega_n\sqrt{1-\zeta^2}$; then

$$\omega_n t = \frac{\omega_d}{\sqrt{1-\zeta^2}}(nT_d) = \frac{2\pi T_d}{\sqrt{1-\zeta^2}}(nT_d) = \frac{2\pi n}{\sqrt{1-\zeta^2}}. \quad (2.30)$$

Finally,

$$\frac{x(nT_d)}{x_o} = e^{-\zeta\omega_n nT_d} = e^{-2\pi n\zeta/\sqrt{1-\zeta^2}}. \quad (2.31)$$

Figure 2.3 shows a plot of Eq. 2.31 versus n (i.e., t/T_d) for different critical damping ratios, and Figure 2.4 shows a plot of $T_d/T_n = (2\pi/\omega_d)/(2\pi/\omega_n) = \omega_n/\omega_d = 1/\sqrt{1-\zeta^2}$ and also $\left(\zeta/\sqrt{1-\zeta^2}\right)$ versus the critical damping ratio ζ.

EXAMPLE 2 *Logarithmic Decrement*

The logarithmic decrement is the natural logarithm of the ratio of the response at time t_1 and time $t_1 + T_d$. Recall from Eq. 2.31 with $n = 1$ that

$$\frac{x(t_1 + T_d)}{x(t_1)} = e^{-2\pi\zeta/\sqrt{1-\zeta^2}}. \quad (2.32)$$

Figure 2.5 shows a plot of Eq. 2.32 as a function of damping.
   Now let $t_2 = t_1 + nT_d$; then

$$\cos\omega_d t_2 = \cos\omega_d(t_1 + nT_d) = \cos(\omega_d t_1 + n\omega_d T_d) = \cos(\omega_d t_1 + 2\pi n) = \cos\omega_d t_1$$

$$\sin\omega_d t_2 = \sin\omega_d(t_1 + nT_d) = \sin(\omega_d t_1 + n\omega_d T_d) = \sin(\omega_d t_1 + 2\pi n) = \sin\omega_d t_1$$

and therefore Eq. 2.25 becomes

$$\frac{x_1}{x_2} = \frac{x_o\,e^{-\zeta\omega_n t_1}}{x_o\,e^{-\zeta\omega_n t_2}} = e^{\zeta\omega_n(t_2-t_1)} = e^{\zeta\omega_n nT_d} = e^{2\pi n\zeta/\sqrt{1-\zeta^2}}. \quad (2.33)$$

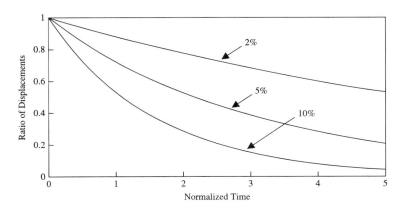

Figure 2.3 Response Decay with Damping

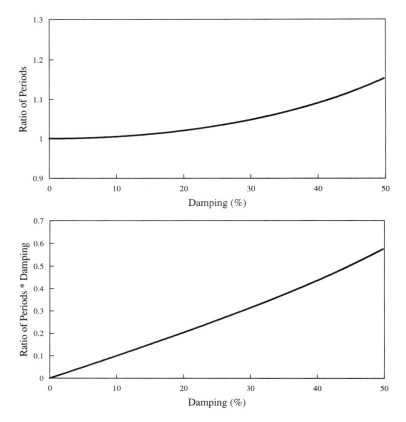

Figure 2.4 Variation of Damped Natural Period of Vibration with Damping

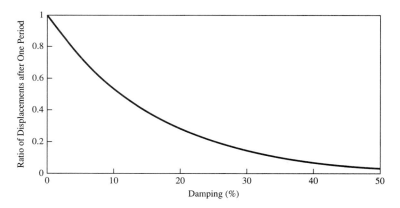

Figure 2.5 Response Decay in One Cycle as a Function of Damping

The decay in response amplitude between time t_1 and t_2 is the inverse of Eq. 2.33 and is

$$\frac{x_2}{x_1} = e^{-2\pi n\zeta/\sqrt{1-\zeta^2}}. \tag{2.34}$$

Note that Eq. 2.34 is the same as Eq. 2.31. Now define δ_n as the natural logarithm of the quotient (x_1/x_2), where the subscript n represents the number of periods separating t_1 and t_2. It follows from Eq. 2.33 that

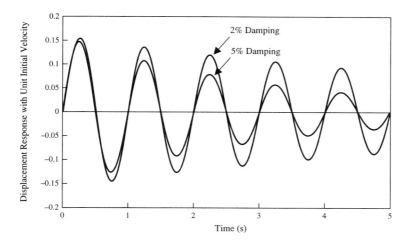

Figure 2.6 Free Vibration with Initial Velocity

$$\delta_n = \ln\left(\frac{x_1}{x_2}\right) = \zeta\omega_n n T_d = \frac{2\pi n\zeta}{\sqrt{1-\zeta^2}} = n\delta_1. \tag{2.35}$$

This shows that the logarithmic decrement after n periods is equal to n times the logarithmic decrement after one period.

EXAMPLE 3   *Free Vibration with Initial Velocity*

Consider Eqs. 2.21 and 2.22 with the initial displacement equal to zero and the initial velocity equal to $\dot{x}_o$. It follows that

$$\frac{x(t)}{\dot{x}_o} = \left(\frac{1}{\omega_n\sqrt{1-\zeta^2}}\right)e^{-\zeta\omega_n t}\sin\omega_d t. \tag{2.36}$$

Note that $\omega_n = 2\pi/T_n$, $\omega_n t = 2\pi(t/T_n)$, then $\omega_d t = \omega_n\sqrt{1-\zeta^2}\,t = 2\pi\sqrt{1-\zeta^2}\,(t/T_n)$. It then follows from Eq. 2.36 that

$$\frac{x(t)}{\dot{x}_o} = \left(\frac{T_n}{2\pi\sqrt{1-\zeta^2}}\right)e^{-2\pi\zeta(t/T_n)}\sin\left(2\pi\sqrt{1-\zeta^2}\,(t/T_n)\right). \tag{2.37}$$

The response decay is expressed in terms of a normalized time, which is (t/T_n). In Example 1, the response is a function of the sine and cosine function, except for the special case where the damping is zero. In Eq. 2.37, the response is a sine function. Figure 2.6 shows a plot of $\left[x(t)/\dot{x}_o\right]$ with a natural period of vibration of 1.0 s for different damping values.

2.3   SPECIAL CASES OF FORCED RESPONSE

Once the homogeneous (or free vibration) solution for the response of the system is known, it is then necessary to calculate the particular (or forced response) solution in order to complete the general solution of Eq. 2.2. The particular solution is known as the *forced vibration*. This particular solution must satisfy the differential equation of Eq. 2.4, which is written here and also in normalized form:

$$m\ddot{x}_p(t) + c\dot{x}_p(t) + kx_p(t) = F_e(t) \tag{2.38a}$$

$$\ddot{x}_p(t) + 2\zeta\omega_n\dot{x}_p(t) + \omega_n^2 x_p(t) = F_e(t)/m. \tag{2.38b}$$

Any one solution that satisfies Eq. 2.38 can be the particular solution. This poses a problem in satisfying the initial conditions when the system is subjected to an external applied force. If the initial conditions of the particular solution satisfy $x_p(0) = 0$ and $\dot{x}_p(0) = 0$, then Eqs. 2.21 and 2.22 represent the true homogeneous solution. However, if $x_p(0) \neq 0$ or $\dot{x}_p(0) \neq 0$, then Eqs. 2.21 and 2.22 must be modified. The method of incorporating this initial condition into the particular solution is to use Eqs. 2.17 and 2.18. This procedure is outlined as follows. Start with the general solutions for $x(t)$ and $\dot{x}(t)$, where

$$x(t) = x_h(t) + x_p(t) = C_1' e^{-\zeta\omega_n t}\cos\omega_d t + C_2' e^{-\zeta\omega_n t}\sin\omega_d t + x_p(t) \tag{2.39a}$$

$$\begin{aligned}\dot{x}(t) &= \dot{x}_h(t) + \dot{x}_p(t) \\ &= \left(-\zeta\omega_n C_1' + \omega_d C_2'\right)e^{-\zeta\omega_n t}\cos\omega_d t + \left(-\omega_d C_1' - \zeta\omega_n C_2'\right)e^{-\zeta\omega_n t}\sin\omega_d t + \dot{x}_p(t).\end{aligned} \tag{2.39b}$$

Now, substitute $x(0) = x_o$ into Eq. 2.39a and $\dot{x}(0) = \dot{x}_o$ into Eq. 2.39b, and it follows that

$$C_1' + x_p(0) = x_o \tag{2.40a}$$

$$-\zeta\omega_n C_1' + \omega_d C_2' + \dot{x}_p(0) = \dot{x}_o. \tag{2.40b}$$

Solving for C_1' and C_2' in Eq. (2.40) gives

$$C_1' = x_o - x_p(0) \tag{2.41a}$$

$$C_2' = \frac{\zeta\omega_n x_o - \zeta\omega_n x_p(0) + \dot{x}_o - \dot{x}_p(0)}{\omega_d}. \tag{2.41b}$$

Then the solution becomes

$$x(t) = \left[x_o - x_p(0)\right]e^{-\zeta\omega_n t}\cos\omega_d t + \left\{\frac{\zeta\omega_n\left[x_o - x_p(0)\right] + \left[\dot{x}_o - \dot{x}_p(0)\right]}{\omega_n\sqrt{1-\zeta^2}}\right\}e^{-\zeta\omega_n t}\sin\omega_d t + x_p(t) \tag{2.42a}$$

$$\dot{x}(t) = \left[\dot{x}_o - \dot{x}_p(0)\right]e^{-\zeta\omega_n t}\cos\omega_d t - \left\{\frac{\omega_n\left[x_o - x_p(0)\right] + \zeta\left[\dot{x}_o - \dot{x}_p(0)\right]}{\sqrt{1-\zeta^2}}\right\}e^{-\zeta\omega_n t}\sin\omega_d t + \dot{x}_p(t) \tag{2.42b}$$

Equation 2.42 represents the general solution of the SDOF system response. This equation is rather complicated because both the homogeneous and the particular solutions are combined in a condensed form. However, the homogeneous solution can be isolated from the particular solution as follows:

$$\begin{aligned}x(t) = &\left\{x_o e^{-\zeta\omega_n t}\cos\omega_d t + \left[\frac{\zeta\omega_n x_o + \dot{x}_o}{\omega_n\sqrt{1-\zeta^2}}\right]e^{-\zeta\omega_n t}\sin\omega_d t\right\} \\ &+ \left\{x_p(t) - x_p(0)e^{-\zeta\omega_n t}\cos\omega_d t - \left[\frac{\zeta\omega_n x_p(0) + \dot{x}_p(0)}{\omega_n\sqrt{1-\zeta^2}}\right]e^{-\zeta\omega_n t}\sin\omega_d t\right\}\end{aligned} \tag{2.43a}$$

$$\begin{aligned}\dot{x}(t) = &\left\{\dot{x}_o e^{-\zeta\omega_n t}\cos\omega_d t - \left[\frac{\omega_n x_o + \zeta\dot{x}_o}{\sqrt{1-\zeta^2}}\right]e^{-\zeta\omega_n t}\sin\omega_d t\right\} \\ &+ \left\{\dot{x}_p(t) - \dot{x}_p(0)e^{-\zeta\omega_n t}\cos\omega_d t + \left[\frac{\omega_n x_p(0) + \zeta\dot{x}_p(0)}{\sqrt{1-\zeta^2}}\right]e^{-\zeta\omega_n t}\sin\omega_d t\right\}.\end{aligned} \tag{2.43b}$$

The term inside the first braces of Eq. 2.43 represents the homogeneous solution that satisfies any given initial conditions, whereas the term inside the second braces of Eq. 2.43 represents the particular solution with zero displacement and velocity at time equal to zero.

Note in Eq. 2.43 that x_o and $\dot{x}_o$ are the system initial displacement and velocity, respectively.

The *steady-state response* is defined as the system response for large value of time (i.e., as $t \to \infty$). Note in Eq. 2.43 that the term $e^{-\zeta\omega_n t} \to 0$ as $t \to \infty$ for any nonzero damping. Define the *steady-state displacement* and the *steady-state velocity* as $x_{ss}(t)$ and $\dot{x}_{ss}(t)$, respectively. It then follows from Eq. 2.43 that

$$x_{ss}(t) = x_p(t) \tag{2.44a}$$

$$\dot{x}_{ss}(t) = \dot{x}_p(t). \tag{2.44b}$$

This shows that the steady-state response is the same as the particular solution. Note that any solution satisfying Eq. 2.38 is a particular solution, and the steady-state response is the one particular solution obtained by taking the limit as $t \to \infty$.

EXAMPLE 1   *Uniform Forcing Function*

Consider the special case of the forcing function, $F_e(t)$, in Eq. 2.38a to be a uniform function with constant magnitude of F_o. Because $F_e(t)$ does not vary with time, it follows that the particular solution for this forcing function also does not vary with time and is $x_p(t) = F_o/k$, and therefore $x_p(0) = F_o/k$ and $\dot{x}_p(t) = \dot{x}_p(0) = 0$. It follows from Eq. 2.42, with the initial conditions $x(0) = x_o$ and $\dot{x}(0) = \dot{x}_o$, that

$$x(t) = \left(x_o - \frac{F_o}{k}\right)e^{-\zeta\omega_n t}\cos\omega_d t + \left[\frac{\zeta\omega_n(x_o - F_o/k) + \dot{x}_o}{\omega_n\sqrt{1-\zeta^2}}\right]e^{-\zeta\omega_n t}\sin\omega_d t + \frac{F_o}{k} \tag{2.45}$$

$$\dot{x}(t) = \dot{x}_o e^{-\zeta\omega_n t}\cos\omega_d t - \left[\frac{\omega_n(x_o - F_o/k) + \zeta\dot{x}_o}{\sqrt{1-\zeta^2}}\right]e^{-\zeta\omega_n t}\sin\omega_d t. \tag{2.46}$$

For the special case with no damping, the solution is

$$x(t) = \left(x_o - \frac{F_o}{k}\right)\cos\omega_n t + \left(\frac{\dot{x}_o}{\omega_n}\right)\sin\omega_n t + \frac{F_o}{k} \tag{2.47}$$

$$\dot{x}(t) = \dot{x}_o\cos\omega_n t - \left[\omega_n\left(x_o - \frac{F_o}{k}\right)\right]\sin\omega_n t. \tag{2.48}$$

Note that for the special case where the initial displacement and initial velocity are zero, that is, $x(0) = x_o = 0$ and $\dot{x}(0) = \dot{x}_o = 0$, it follows from Eqs. 2.45 and 2.46 that

$$x(t) = -\left(\frac{F_o}{k}\right)e^{-\zeta\omega_n t}\cos\omega_d t - \left(\frac{F_o}{k}\right)\left(\frac{\zeta}{\sqrt{1-\zeta^2}}\right)e^{-\zeta\omega_n t}\sin\omega_d t + \left(\frac{F_o}{k}\right) \tag{2.49}$$

$$\dot{x}(t) = \left(\frac{F_o}{k}\right)\left(\frac{\omega_n}{\sqrt{1-\zeta^2}}\right)e^{-\zeta\omega_n t}\sin\omega_d t. \tag{2.50}$$

The steady-state response is obtained from Eqs. 2.49 and 2.50 by taking the limit as $t \to \infty$. Since $e^{-\zeta\omega_n t} \to 0$ as $t \to \infty$, it follows that

$$x_{ss}(t) = F_o/k \tag{2.51}$$

$$\dot{x}_{ss}(t) = 0. \tag{2.52}$$

Note that the steady-state displacement is a constant and therefore is known as the *static displacement*, that is, the displacement obtained by performing static analysis using the forcing function as a static applied load. If Eqs. 2.49 and 2.50 are divided by the static displacement given in Eq. 2.51, it follows that

$$\frac{x(t)}{x_{ss}(t)} = 1 - e^{-\zeta \omega_n t} \cos \omega_d t - \left(\frac{\zeta}{\sqrt{1-\zeta^2}} \right) e^{-\zeta \omega_n t} \sin \omega_d t \qquad (2.53)$$

$$\frac{\dot{x}(t)}{\omega_n x_{ss}(t)} = \left(\frac{1}{\sqrt{1-\zeta^2}} \right) e^{-\zeta \omega_n t} \sin \omega_d t. \qquad (2.54)$$

The responses given in Eqs. 2.53 and 2.54 are shown graphically in Figures 2.7 and 2.8, with a natural period of vibration $T_n = 1.0$ s and for 2%, 5%, and 10% damping.

EXAMPLE 2 *Forcing Function That Rises to Static*

This example provides insight into when a forcing function can be considered to be a static force. Consider the forcing function, $F_e(t)$, shown in Figure 2.9, defined as

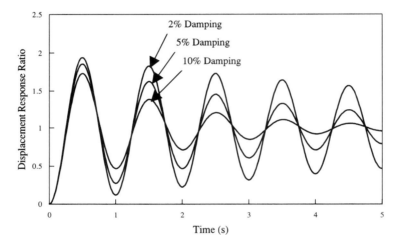

Figure 2.7 $x(t)/x_{ss}(t)$ for a Uniform Forcing Function

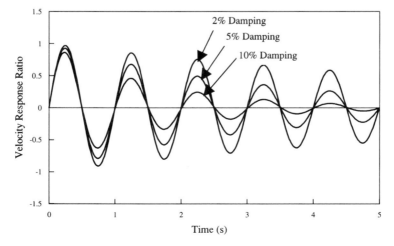

Figure 2.8 $\dot{x}(t)/\omega_n x_{ss}(t)$ for a Uniform Forcing Function

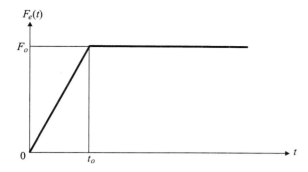

Figure 2.9 Rise to Static Forcing Function

$$F_e(t) = \begin{cases} F_o\, t/t_o & 0 \le t < t_o \\ F_o & t \ge t_o \\ 0 & \text{otherwise.} \end{cases}$$ (2.55)

The particular solution of the system before the force reaches its maximum value F_o (i.e., $t < t_o$) can be computed by letting $x_p(t) = C_1 t + C_2$. Then, for $t < t_o$, Eq. 2.38b becomes

$$0 + 2\zeta\omega_n C_1 + \omega_n^2 (C_1 t + C_2) = \frac{F_o\, t}{m\, t_o}.$$ (2.56)

This equation contains two unknowns, C_1 and C_2, which must satisfy Eq. 2.56 at all times. These unknowns can be found by setting the term on the left-hand side of Eq. 2.56 that is multiplied by t equal to the term on the right side that is multiplied by t. Similarly, all of the constant terms on the left side of the equation are equal to the constant term on the right side, which is zero. When this is done, two equations are obtained containing C_1 and C_2, and the equations are

$$\omega_n^2 C_1 = \frac{F_o}{m\, t_o}, \quad 2\zeta\omega_n C_1 + \omega_n^2 C_2 = 0.$$ (2.57)

Solving for C_1 and C_2 gives

$$C_1 = \left(\frac{F_o}{m\,\omega_n^2\, t_o}\right) = \frac{F_o}{k\, t_o}, \quad C_2 = -\left(\frac{2\zeta\omega_n C_1}{\omega_n^2}\right) = -\left(\frac{2\zeta F_o}{k\,\omega_n t_o}\right).$$ (2.58)

Therefore, the particular solution is

$$x_p(t) = \left(\frac{F_o}{k\, t_o}\right)t - \left(\frac{2\zeta F_o}{k\,\omega_n t_o}\right) = \frac{F_o}{k\, t_o}\left(t - \frac{2\zeta}{\omega_n}\right) = \frac{F_o}{k\, t_o}\left(t - \frac{\zeta T_n}{\pi}\right)$$ (2.59)

and

$$\dot{x}_p(t) = \frac{F_o}{k\, t_o}.$$ (2.60)

At $t = 0$, the initial displacement and velocity from the particular solution are

$$x_p(0) = -\left(\frac{2\zeta F_o}{k\,\omega_n t_o}\right), \quad \dot{x}_p(0) = \frac{F_o}{k\, t_o}.$$ (2.61)

Substituting these initial conditions for the particular solution into Eq. 2.42 provides the general solution for the response. Consider the special case where the system is at rest at $t = 0$, that is, $x(0) = x_o = 0$ and $\dot{x}(0) = \dot{x}_o = 0$. Then the response for $x(t)$ and $\dot{x}(t)$ for $t < t_o$ becomes

$$x(t) = \left(\frac{2\zeta F_o}{k\,\omega_n t_o}\right)e^{-\zeta\omega_n t}\cos\omega_d t + \left[\frac{F_o(2\zeta^2 - 1)}{k\,\omega_n t_o\sqrt{1-\zeta^2}}\right]e^{-\zeta\omega_n t}\sin\omega_d t + \left(\frac{F_o}{k\, t_o}\right)t - \left(\frac{2\zeta F_o}{k\,\omega_n t_o}\right)$$ (2.62)

$$\dot{x}(t) = -\left(\frac{F_o}{k t_o}\right) e^{-\zeta\omega_n t} \cos\omega_d t - \left(\frac{\zeta}{\sqrt{1-\zeta^2}} \frac{F_o}{k t_o}\right) e^{-\zeta\omega_n t} \sin\omega_d t + \left(\frac{F_o}{k t_o}\right). \tag{2.63}$$

At the time when the rise in force reaches its static value, that is, $t = t_o$, the response is

$$x(t_o) = \left(\frac{2\zeta F_o}{k\omega_n t_o}\right) e^{-\zeta\omega_n t_o} \cos\omega_d t_o + \left[\frac{F_o(2\zeta^2-1)}{k\omega_n t_o\sqrt{1-\zeta^2}}\right] e^{-\zeta\omega_n t_o} \sin\omega_d t_o + \left(\frac{F_o}{k} - \frac{2\zeta F_o}{k\omega_n t_o}\right) \tag{2.64}$$

$$\dot{x}(t_o) = -\left(\frac{F_o}{k t_o}\right) e^{-\zeta\omega_n t_o} \cos\omega_d t_o - \left(\frac{\zeta}{\sqrt{1-\zeta^2}} \frac{F_o}{k t_o}\right) e^{-\zeta\omega_n t_o} \sin\omega_d t_o + \left(\frac{F_o}{k t_o}\right). \tag{2.65}$$

Note that in each term of Eq. 2.62 or 2.64, the static response (i.e., F_o/k) is present, and therefore the equations can be viewed as the response normalized with respect to the static response, that is, $x(t)/(F_o/k)$. The resultant equations provide insight into the often-used structural engineering phrase *static load*. If the load is applied extremely slowly in terms of the undamped natural period of vibration, the ratio (t_o/T_n) becomes large. It follows that the term $\omega_n t_o$ also becomes large by recognizing that $\omega_n t_o = 2\pi(t_o/T_n)$. It is particularly apparent in Eq. 2.64 that in this limit for large (t_o/T_n), the last two terms do not contain (t_o/T_n), and therefore these two terms will dominate the response. The other important factor in defining a static load is the value of damping. As the damping value increases, the required time (t_o/T_n) to have a static load decreases.

Finally, at time $t = t_o$, the displacement and velocity of the system are $x(t_o)$ and $\dot{x}(t_o)$ and the response is as given in Eqs. 2.64 and 2.65. These values of $x(t_o)$ and $\dot{x}(t_o)$ represent the initial conditions for the response at $t \geq t_o$. Since the forcing function for time $t > t_o$ is uniform, the response takes the form as discussed in Example 1. This response for $t > t_o$ is

$$x(t) = \left[x(t_o) - \frac{F_o}{k}\right] e^{-\zeta\omega_n(t-t_o)} \cos\omega_d(t-t_o)$$

$$+ \left[\frac{\zeta\omega_n(x(t_o) - F_o/k) + \dot{x}(t_o)}{\omega_n\sqrt{1-\zeta^2}}\right] e^{-\zeta\omega_n(t-t_o)} \sin\omega_d(t-t_o) + \left(\frac{F_o}{k}\right) \tag{2.66}$$

$$\dot{x}(t) = \dot{x}(t_o) e^{-\zeta\omega_n(t-t_o)} \cos\omega_d(t-t_o)$$

$$- \left[\frac{\omega_n(x(t_o) - F_o/k) + \zeta\dot{x}(t_o)}{\sqrt{1-\zeta^2}}\right] e^{-\zeta\omega_n(t-t_o)} \sin\omega_d(t-t_o). \tag{2.67}$$

The steady-state response is obtained from Eqs. 2.66 and 2.67 by taking the limit as $t \to \infty$. Since $e^{-\zeta\omega_n(t-t_o)} \to 0$ as $t \to \infty$, it follows that

$$x_{ss}(t) = F_o/k \tag{2.68}$$

$$\dot{x}_{ss}(t) = 0. \tag{2.69}$$

If the displacement response is divided by the steady-state response, it follows from Eqs. 2.62 and 2.66 that

$$\frac{x(t)}{x_{ss}(t)} = \begin{cases} \left(\dfrac{2\zeta}{\omega_n t_o}\right) e^{-\zeta\omega_n t} \cos\omega_d t + \left(\dfrac{2\zeta^2-1}{\omega_n t_o\sqrt{1-\zeta^2}}\right) e^{-\zeta\omega_n t} \sin\omega_d t + \left(\dfrac{1}{t_o}\right) t - \left(\dfrac{2\zeta}{\omega_n t_o}\right) & t \leq t_o \\[2em] \left(\dfrac{x(t_o)}{x_{ss}(t)} - 1\right) e^{-\zeta\omega_n(t-t_o)} \cos\omega_d(t-t_o) \\[1em] \quad + \left[\dfrac{\zeta\omega_n(x(t_o)/x_{ss}(t)-1) + \dot{x}(t_o)/x_{ss}(t)}{\omega_n\sqrt{1-\zeta^2}}\right] e^{-\zeta\omega_n(t-t_o)} \sin\omega_d(t-t_o) + 1 & t > t_o \end{cases} \tag{2.70}$$

where

$$\frac{x(t_o)}{x_{ss}(t)} = \left(\frac{2\zeta}{\omega_n t_o}\right)e^{-\zeta\omega_n t_o}\cos\omega_d t_o + \left(\frac{2\zeta^2-1}{\omega_n t_o\sqrt{1-\zeta^2}}\right)e^{-\zeta\omega_n t_o}\sin\omega_d t_o + \left(1-\frac{2\zeta}{\omega_n t_o}\right) \quad (2.71)$$

$$\frac{\dot{x}(t_o)}{x_{ss}(t)} = -\left(\frac{1}{t_o}\right)e^{-\zeta\omega_n t_o}\cos\omega_d t_o - \left(\frac{\zeta}{t_o\sqrt{1-\zeta^2}}\right)e^{-\zeta\omega_n t_o}\sin\omega_d t_o + \left(\frac{1}{t_o}\right). \quad (2.72)$$

Figure 2.10 shows a plot of the response divided by the steady-state response, $x(t)/x_{ss}(t)$, for $T_n = 1.0$ s and 5% damping, and different values of t_o. Figure 2.11 shows a plot of the same ratio for $t_o = 1.0$ s and $T_n = 1.0$ s, and different values of damping.

EXAMPLE 3   *Harmonic Forcing Function*

Consider the forcing function of the form

$$F_e(t) = F_o \sin\omega_o t. \quad (2.73)$$

This type of oscillating forcing function is often called a *steady-state forcing function*. The dynamic equilibrium equation in Eq. 2.38b becomes

$$\ddot{x}_p(t) + 2\zeta\omega_n\dot{x}_p(t) + \omega_n^2 x_p(t) = F_o \sin\omega_o t/m. \quad (2.74)$$

The particular solution takes the form

$$x_p(t) = A\cos\omega_o t + B\sin\omega_o t. \quad (2.75)$$

Substituting Eq. 2.75 into Eq. 2.74 gives

$$-\omega_o^2\left(A\cos\omega_o t + B\sin\omega_o t\right) + 2\zeta\omega_n\omega_o\left(-A\sin\omega_o t + B\cos\omega_o t\right)$$
$$+ \omega_n^2\left(A\cos\omega_o t + B\sin\omega_o t\right) = F_o \sin\omega_o t/m. \quad (2.76)$$

There are two unknowns in this equation, A and B. Collecting the cosine terms into one equation and the sine terms into another equation gives the two simultaneous equations that are needed to solve for A and B. These equations are

$$-\omega_o^2 A + 2\zeta\omega_n\omega_o B + \omega_n^2 A = 0 \quad (2.77)$$

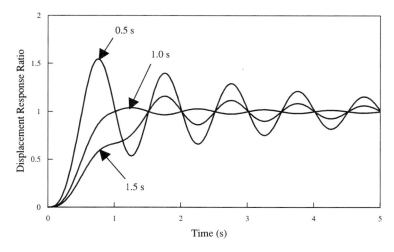

Figure 2.10 $x(t)/x_{ss}(t)$ for t_o Equal to 0.5, 1.0, and 1.5 s

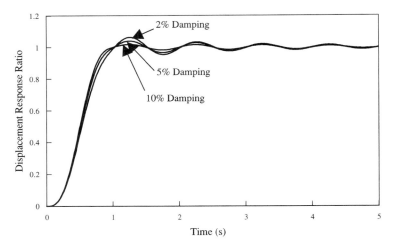

Figure 2.11 $x(t)/x_{ss}(t)$ for Damping Values Equal to 2%, 5%, and 10%

$$-\omega_o^2 B - 2\zeta\omega_n\omega_o A + \omega_n^2 B = (F_o/m). \tag{2.78}$$

For convenience, let $r = \omega_o/\omega_n$, which is the ratio of the forcing frequency to the system undamped natural frequency of vibration. It now follows that A and B can be calculated using Eqs. 2.77 and 2.78 to be

$$A = -\frac{F_o}{k}\left(\frac{2\zeta r}{\left(1-r^2\right)^2+\left(2\zeta r\right)^2}\right), \quad B = \frac{F_o}{k}\left(\frac{1-r^2}{\left(1-r^2\right)^2+\left(2\zeta r\right)^2}\right). \tag{2.79}$$

It follows from Eq. 2.75 that

$$x_p(t) = -\frac{F_o}{k}\left(\frac{2\zeta r}{\left(1-r^2\right)^2+\left(2\zeta r\right)^2}\right)\cos\omega_o t + \frac{F_o}{k}\left(\frac{1-r^2}{\left(1-r^2\right)^2+\left(2\zeta r\right)^2}\right)\sin\omega_o t. \tag{2.80}$$

The velocity of the particular solution can be calculated by differentiating Eq. 2.80. Doing so gives

$$\dot{x}_p(t) = \frac{F_o\omega_o}{k}\left[\frac{2\zeta r}{\left(1-r^2\right)^2+\left(2\zeta r\right)^2}\right]\sin\omega_o t + \frac{F_o\omega_o}{k}\left[\frac{1-r^2}{\left(1-r^2\right)^2+\left(2\zeta r\right)^2}\right]\cos\omega_o t. \tag{2.81}$$

The initial conditions of the particular solution are

$$x_p(0) = -\frac{F_o}{k}\left[\frac{2\zeta r}{\left(1-r^2\right)^2+\left(2\zeta r\right)^2}\right], \quad \dot{x}_p(0) = \frac{F_o\omega_o}{k}\left[\frac{1-r^2}{\left(1-r^2\right)^2+\left(2\zeta r\right)^2}\right]. \tag{2.82}$$

The general solution for the response is then given in Eq. 2.42. For the case where the initial conditions of the system are zero, that is, $x_o = 0$ and $\dot{x}_o = 0$, the general solution for the response becomes

$$x(t) = \frac{F_o/k}{\left(1-r^2\right)^2+\left(2\zeta r\right)^2}\left\{2\zeta r\, e^{-\zeta\omega_n t}\cos\omega_d t - \left[r\frac{\left(1-r^2\right)-2\zeta^2}{\sqrt{1-\zeta^2}}\right]e^{-\zeta\omega_n t}\sin\omega_d t\right.$$
$$\left. - 2\zeta r\cos\omega_o t + \left(1-r^2\right)\sin\omega_o t\right\} \tag{2.83}$$

$$\dot{x}(t) = \frac{F_o \omega_o / k}{\left(1-r^2\right)^2 + \left(2\zeta r\right)^2} \left\{ -\left(1-r^2\right) e^{-\zeta \omega_n t} \cos \omega_d t + \left[\frac{\zeta\left(1-r^2\right) - 2\zeta}{\sqrt{1-\zeta^2}}\right] e^{-\zeta \omega_n t} \sin \omega_d t \right.$$

$$\left. + \left(1-r^2\right) \cos \omega_o t + 2\zeta r \sin \omega_o t \right\}. \tag{2.84}$$

Note that the static response of the system is F_o/k. The dynamic response is the static response multiplied by a constant term that is only a function of r and ζ and a time-varying term. The first two time-varying terms in the bracket decrease exponentially and thus become small after sufficient time has elapsed. The third and fourth terms oscillate with constant amplitude. If the damping is zero or very small, then the displacement response is a sine function like the forcing function. As the damping increases, the cosine term becomes a more significant part of the response.

The steady-state response is obtained from Eqs. 2.83 and 2.84 by taking the limit as $t \to \infty$. Since $e^{-\zeta \omega_n t} \to 0$ as $t \to \infty$, it follows that

$$x_{ss}(t) = \frac{F_o/k}{\left(1-r^2\right)^2 + \left(2\zeta r\right)^2} \left[-2\zeta r \cos \omega_o t + \left(1-r^2\right) \sin \omega_o t\right]. \tag{2.85}$$

$$\dot{x}_{ss}(t) = \frac{F_o \omega_o/k}{\left(1-r^2\right)^2 + \left(2\zeta r\right)^2} \left[\left(1-r^2\right) \cos \omega_o t + 2\zeta r \sin \omega_o t\right]. \tag{2.86}$$

Note that the steady-state response is an oscillating function with time. This is because the applied forcing function is oscillating with time, which is the reason why this type of forcing function is called a steady-state forcing function. For any values of r and ζ, the maximum amplitude of displacement response after sufficient time has elapsed can be calculated from Eq. 2.85, which is

$$\max_t \left|x_{ss}(t)\right| = \frac{F_o/k}{\sqrt{\left(1-r^2\right)^2 + \left(2\zeta r\right)^2}}. \tag{2.87}$$

For $r \approx 1$ and a small damping ratio, the denominator of Eq. 2.87 will become small, and thus the response can be magnified significantly. The *magnification factor* (f_{mag}) can be represented as

$$f_{mag} = \frac{\max_t \left|x_{ss}(t)\right|}{F_o/k} = \frac{1}{\sqrt{\left(1-r^2\right)^2 + \left(2\zeta r\right)^2}}. \tag{2.88}$$

This magnification factor for various damping ratios is shown in Figure 2.12.

2.4 CONVOLUTION INTEGRAL SOLUTION FOR RESPONSE

The equation of motion for the system is given in Eq. 2.2. The general solution for this equation is developed by starting with the special type of forcing function discussed in Example 1 in Section 2.3. Consider the forcing function, $F_e(t)$, to be equal to

$$F_e(t) = \begin{cases} F_o & 0 \le t \le t_o \\ 0 & \text{otherwise.} \end{cases} \tag{2.89}$$

The solution for the response of the system at a time less than or equal to t_o is given in Eqs. 2.45 and 2.46. At time t_o, the response is

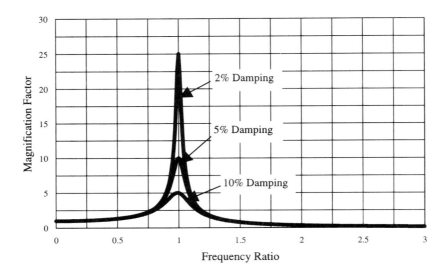

Figure 2.12 Steady-State Response Magnification Factor

$$x(t_o) = \left(x_o - \frac{F_o}{k}\right)e^{-\zeta\omega_n t_o}\cos\omega_d t_o + \left[\frac{\zeta\omega_n\left(x_o - F_o/k\right) + \dot{x}_o}{\omega_n\sqrt{1-\zeta^2}}\right]e^{-\zeta\omega_n t_o}\sin\omega_d t_o + \left(\frac{F_o}{k}\right) \quad (2.90a)$$

$$\dot{x}(t_o) = \dot{x}_o\, e^{-\zeta\omega_n t_o}\cos\omega_d t_o - \left[\frac{\omega_n\left(x_o - F_o/k\right) + \zeta\dot{x}_o}{\sqrt{1-\zeta^2}}\right]e^{-\zeta\omega_n t_o}\sin\omega_d t_o. \quad (2.90b)$$

Assume that the initial displacement and velocity are zero (i.e., $x_o = \dot{x}_o = 0$), and then it follows from Eq. 2.90 that

$$x(t_o) = \frac{F_o}{k}\left[1 - e^{-\zeta\omega_n t_o}\cos\omega_d t_o - \left(\frac{\zeta}{\sqrt{1-\zeta^2}}\right)e^{-\zeta\omega_n t_o}\sin\omega_d t_o\right] \quad (2.91a)$$

$$\dot{x}(t_o) = \frac{F_o}{k}\left[\left(\frac{\omega_n}{\sqrt{1-\zeta^2}}\right)e^{-\zeta\omega_n t_o}\sin\omega_d t_o\right]. \quad (2.91b)$$

The responses at t_o, that is, $x(t_o)$ and $\dot{x}(t_o)$, in Eq. 2.91 are the initial conditions for the free vibration response at times greater than t_o, that is,

$$x(t) = x(t_o)e^{-\zeta\omega_n(t-t_o)}\cos\omega_d(t-t_o) + \left[\frac{\zeta\omega_n x(t_o) + \dot{x}(t_o)}{\omega_n\sqrt{1-\zeta^2}}\right]e^{-\zeta\omega_n(t-t_o)}\sin\omega_d(t-t_o), \quad t > t_o \quad (2.92a)$$

$$\dot{x}(t) = \dot{x}(t_o)e^{-\zeta\omega_n(t-t_o)}\cos\omega_d(t-t_o) - \left[\frac{\omega_n x(t_o) + \zeta\dot{x}(t_o)}{\sqrt{1-\zeta^2}}\right]e^{-\zeta\omega_n(t-t_o)}\sin\omega_d(t-t_o), \quad t > t_o. \quad (2.92b)$$

Imagine now that the rectangular-shaped forcing function becomes very tall and very thin (i.e., $t_o \to \infty$ and $F_o \to \infty$) but that the area inside the rectangle is set equal to 1 (i.e., $F_o = 1/t_o$). This type of forcing function is called an *impulse*, and in mathematical terms it is called a *Delta function*. Substituting $F_o = 1/t_o$ into Eq. 2.91 and taking the limit as $t_o \to 0$, it follows from L'Hospital's Rule that

$$\lim_{t_o \to 0} x(t_o) = \lim_{t_o \to 0} \frac{F_o}{k} \left[1 - e^{-\zeta \omega_n t_o} \cos \omega_d t_o - \left(\frac{\zeta}{\sqrt{1-\zeta^2}} \right) e^{-\zeta \omega_n t_o} \sin \omega_d t_o \right]$$

$$= \frac{1}{k} \lim_{t_o \to 0} \left(\frac{1 - e^{-\zeta \omega_n t_o} \cos \omega_d t_o - \left(\zeta / \sqrt{1-\zeta^2} \right) e^{-\zeta \omega_n t_o} \sin \omega_d t_o}{t_o} \right)$$

$$\overset{*}{=} \frac{1}{k} \lim_{t_o \to 0} \left(\frac{\omega_n e^{-\zeta \omega_n t_o} \sin \omega_d t_o / \sqrt{1-\zeta^2}}{1} \right) \qquad (2.93a)$$

$$= \frac{1}{k}(0)$$

$$= 0$$

$$\lim_{t_o \to 0} \dot{x}(t_o) = \lim_{t_o \to 0} \frac{F_o}{k} \left[\left(\frac{\omega_n}{\sqrt{1-\zeta^2}} \right) e^{-\zeta \omega_n t_o} \sin \omega_d t_o \right]$$

$$= \frac{\omega_n}{k\sqrt{1-\zeta^2}} \lim_{t_o \to 0} \left(\frac{e^{-\zeta \omega_n t_o} \sin \omega_d t_o}{t_o} \right)$$

$$\overset{*}{=} \frac{\omega_n}{k\sqrt{1-\zeta^2}} \lim_{t_o \to 0} \left(\frac{-\zeta \omega_n e^{-\zeta \omega_n t_o} \sin \omega_d t_o + \omega_d e^{-\zeta \omega_n t_o} \cos \omega_d t_o}{1} \right) \qquad (2.93b)$$

$$= \frac{\omega_n}{k\sqrt{1-\zeta^2}} (0 + \omega_d)$$

$$= \frac{1}{m}$$

where the asterisk above the equal sign denotes L'Hospital operation. Now, when the results from Eq. 2.93 are substituted into Eq. 2.92 it follows that

$$x(t) = \left(\frac{1}{m\omega_d} \right) e^{-\zeta \omega_n t} \sin \omega_d t \qquad (2.94a)$$

$$\dot{x}(t) = \left(\frac{1}{m} \right) e^{-\zeta \omega_n t} \cos \omega_d t - \left(\frac{\zeta}{m\sqrt{1-\zeta^2}} \right) e^{-\zeta \omega_n t} \sin \omega_d t. \qquad (2.94b)$$

Equation 2.94 is known as the *impulse response function*, and it represents the response of the system to a Delta function applied at time equal zero.

The Delta function (sometimes called Dirac Delta function), $\delta(\cdot)$, is defined as

$$\delta(t) = \begin{cases} 0 & t \neq 0 \\ \infty & t = 0, \end{cases} \qquad \int_{-\infty}^{\infty} \delta(t)\, dt = \int_{0-}^{0+} \delta(t)\, dt = 1. \qquad (2.95)$$

Based on the definition in Eq. 2.95, a Delta function located at another instant in time, s, is represented by $\delta(t - s)$, and this Delta function has the following important property:

$$\int_{-\infty}^{\infty} f(s)\delta(s-t)\, ds = \int_{-\infty}^{\infty} f(s-t)\delta(s)\, ds = f(t). \qquad (2.96)$$

Equation 2.94 represents the special case of the system response where the Delta function is applied at $t_o = 0$. If the Delta function is applied at time t_o, the general equation of the response becomes

$$x(t) = \begin{cases} \left[\left(\dfrac{1}{m\omega_d}\right)e^{-\zeta\omega_n(t-t_o)}\sin\omega_d(t-t_o)\right] & t \ge t_o \\[4mm] 0 & t < t_o \end{cases} \tag{2.97a}$$

$$\dot{x}(t) = \begin{cases} \left[\left(\dfrac{1}{m}\right)e^{-\zeta\omega_n(t-t_o)}\cos\omega_d(t-t_o) - \left(\dfrac{\zeta}{m\sqrt{1-\zeta^2}}\right)e^{-\zeta\omega_n(t-t_o)}\sin\omega_d(t-t_o)\right] & t \ge t_o \\[4mm] 0 & t < t_o. \end{cases} \tag{2.97b}$$

Note that in Eq. 2.97, no response exists at any time before the occurrence of the impulse.

Now, consider that the forcing function is represented by n long and thin rectangles of heights $F_1, F_2, \ldots, F_n$ that occur at times $t_1, t_2, \ldots, t_n$. The response at time t, where $t > t_n > t_{n-1} > \ldots > t_1$, follows from Eq. 2.92 and is

$$x(t) = \sum_{j=1}^{n}\left[x_o(t_j)e^{-\zeta\omega_n(t-t_j)}\cos\omega_d(t-t_j) + \left(\frac{\zeta\omega_n x_o(t_j) + \dot{x}_o(t_j)}{\omega_n\sqrt{1-\zeta^2}}\right)e^{-\zeta\omega_n(t-t_j)}\sin\omega_d(t-t_j)\right] \tag{2.98a}$$

$$\dot{x}(t) = \sum_{j=1}^{n}\left[\dot{x}_o(t_j)e^{-\zeta\omega_n(t-t_j)}\cos\omega_d(t-t_j) - \left(\frac{\omega_n x_o(t_j) + \zeta\dot{x}_o(t_j)}{\sqrt{1-\zeta^2}}\right)e^{-\zeta\omega_n(t-t_j)}\sin\omega_d(t-t_j)\right] \tag{2.98b}$$

where $x_o(t_j)$ and $\dot{x}_o(t_j)$ represent the initial conditions used to calculate the response at time $t = t_o$ due to the force F_j, that is,

$$x_o(t_j) = \frac{F_j}{k}\left[1 - e^{-\zeta\omega_n t_o}\cos\omega_d t_o - \left(\frac{\zeta}{\sqrt{1-\zeta^2}}\right)e^{-\zeta\omega_n t_o}\sin\omega_d t_o\right] \tag{2.99a}$$

$$\dot{x}_o(t_j) = \frac{F_j}{k}\left[\left(\frac{\omega_n}{\sqrt{1-\zeta^2}}\right)e^{-\zeta\omega_n t_o}\sin\omega_d t_o\right]. \tag{2.99b}$$

Note that the static response is F_j/k, and the second and third terms in Eq. 2.99a are functions of t_o. In the limit as t_o becomes very small, based on the previously discussed rectangular representation of the forcing function, the collection of rectangles will result in a continuous forcing function, $F_e(t)$. The response equation for the forcing function starting at time equal zero is represented by the particular solution for the response and, therefore, in the limit Eq. 2.99 becomes

$$x_p(t) = \int_0^t \left(\frac{F_e(s)}{m\omega_d}\right)e^{-\zeta\omega_n(t-s)}\sin\omega_d(t-s)\,ds \tag{2.100a}$$

$$\dot{x}_p(t) = \int_0^t \left(\frac{F_e(s)}{m}\right)e^{-\zeta\omega_n(t-s)}\left[\cos\omega_d(t-s) - \left(\frac{\zeta}{\sqrt{1-\zeta^2}}\right)\sin\omega_d(t-s)\right]ds. \tag{2.100b}$$

Equation 2.100 is known as the *convolution integral*, where the forcing function is *convolved* with the impulse response function. It is also often referred to as a *Duhamel's integral* or *superposition integral*.

Recall from Eq. 2.90 and Eq. 2.91 that it was assumed that $x_o = \dot{x}_o = 0$. Therefore, the use of the convolution integral always gives initial conditions of the particular solution equal to zero, that is, $x_p(0) = \dot{x}_p(0) = 0$. It thus follows that the homogeneous solution for Eq. 2.38 is given by Eqs. 2.21 and 2.22 and that the general solution for the response of the system is

$$x(t) = x_o e^{-\zeta\omega_n t}\cos\omega_d t + \left(\frac{\zeta\omega_n x_o + \dot{x}_o}{\omega_n\sqrt{1-\zeta^2}}\right)e^{-\zeta\omega_n t}\sin\omega_d t + \int_0^t\left(\frac{F_e(s)}{m\omega_d}\right)e^{-\zeta\omega_n(t-s)}\sin\omega_d(t-s)\,ds \quad (2.101a)$$

$$\dot{x}(t) = \dot{x}_o e^{-\zeta\omega_n t}\cos\omega_d t - \left(\frac{\omega_n x_o + \zeta\dot{x}_o}{\sqrt{1-\zeta^2}}\right)e^{-\zeta\omega_n t}\sin\omega_d t$$

$$+\int_0^t\left(\frac{F_e(s)}{m}\right)e^{-\zeta\omega_n(t-s)}\left[\cos\omega_d(t-s)-\left(\frac{\zeta}{\sqrt{1-\zeta^2}}\right)\sin\omega_d(t-s)\right]ds. \qquad (2.101b)$$

Comparing Eq. 2.101 with Eq. 2.43 shows that both of these equations have the same homogeneous solution, but the particular solution is expressed in different forms. Once the forcing function is specified, the results from integration of Eq. 2.101 will always give the second part of Eq. 2.43. This will be demonstrated in the following examples.

EXAMPLE 1   *Delta Forcing Function*

Consider that a Delta function of magnitude F_o is applied to the structure at time $t = 0$. This forcing function is described as

$$F_e(t) = F_o\,\delta(t). \qquad (2.102)$$

The units of F_o are force-time. It follows from the convolution integral and the property of the Delta function as given in Eq. 2.96 that the response is

$$x_p(t) = \int_0^t\left(\frac{F_o\,\delta(s)}{m\omega_d}\right)e^{-\zeta\omega_n(t-s)}\sin\omega_d(t-s)\,ds = \left(\frac{F_o}{m\omega_d}\right)e^{-\zeta\omega_n t}\sin\omega_d t \qquad (2.103)$$

$$\dot{x}_p(t) = \int_0^t\left(\frac{F_o\,\delta(s)}{m}\right)e^{-\zeta\omega_n(t-s)}\left[\cos\omega_d(t-s)-\left(\frac{\zeta}{\sqrt{1-\zeta^2}}\right)\sin\omega_d(t-s)\right]ds$$

$$= \left(\frac{F_o}{m}\right)e^{-\zeta\omega_n t}\left[\cos\omega_d t - \left(\frac{\zeta}{\sqrt{1-\zeta^2}}\right)\sin\omega_d t\right]. \qquad (2.104)$$

Note that the response is the same as the impulse response function in Eq. 2.94 with a multiplication factor of F_o. This verifies that the impulse response function is the response due to a Delta function (i.e., an impulse). It follows from Eq. 2.101 that

$$x(t) = x_o e^{-\zeta\omega_n t}\cos\omega_d t + \left[\frac{\zeta\omega_n x_o + (\dot{x}_o + F_o/m)}{\omega_n\sqrt{1-\zeta^2}}\right]e^{-\zeta\omega_n t}\sin\omega_d t \qquad (2.105)$$

$$\dot{x}(t) = (\dot{x}_o + F_o/m)e^{-\zeta\omega_n t}\cos\omega_d t - \left[\frac{\omega_n x_o + \zeta(\dot{x}_o + F_o/m)}{\sqrt{1-\zeta^2}}\right]e^{-\zeta\omega_n t}\sin\omega_d t. \qquad (2.106)$$

Note that Eqs. 2.105 and 2.106 are similar to Eqs. 2.21 and 2.22 with replacement of $\dot{x}_o$ in Eqs. 2.21 and 2.22 by $(\dot{x}_o + F_o/m)$. This shows that the impulse increases the initial velocity by F_o/m. The result agrees with the impulse–momentum relationship in physics, that is,

$$m\dot{x}_p(0) = \int_0^t F_o\delta(t)\,dt = F_o \qquad (2.107)$$

and solving for the velocity, $\dot{x}_p(0)$, gives

$$\dot{x}_p(0) = F_o/m. \qquad (2.108)$$

EXAMPLE 2   *Uniform Forcing Function*

Example 1 in Section 2.3 considered a forcing function with a constant magnitude of F_o. Using the convolution integral, it follows from Eq. 2.100 that

$$x_p(t) = \left(\frac{F_o}{m\omega_d}\right)\int_0^t e^{-\zeta\omega_n(t-s)} \sin\omega_d(t-s)\,ds$$

$$= \frac{F_o}{m\omega_d}\left[\frac{\omega_d e^{-\zeta\omega_n(t-s)}\cos\omega_d(t-s)+\zeta\omega_n e^{-\zeta\omega_n(t-s)}\sin\omega_d(t-s)}{\omega_n^2}\right]_{s=0}^{s=t} \qquad (2.109)$$

$$= \frac{F_o}{k\omega_d}\left[\omega_d - \omega_d e^{-\zeta\omega_n t}\cos\omega_d t - \zeta\omega_n e^{-\zeta\omega_n t}\sin\omega_d t\right]$$

$$= -\left(\frac{F_o}{k}\right)e^{-\zeta\omega_n t}\cos\omega_d t - \left(\frac{\zeta F_o}{k\sqrt{1-\zeta^2}}\right)e^{-\zeta\omega_n t}\sin\omega_d t + \left(\frac{F_o}{k}\right)$$

$$\dot{x}_p(t) = \left(\frac{F_o}{m}\right)\int_0^t e^{-\zeta\omega_n(t-s)}\left[\cos\omega_d(t-s) - \left(\frac{\zeta}{\sqrt{1-\zeta^2}}\right)\sin\omega_d(t-s)\right]ds$$

$$= \frac{F_o}{m}\left[\frac{-\omega_d e^{-\zeta\omega_n(t-s)}\sin\omega_d(t-s)+\zeta\omega_n e^{-\zeta\omega_n(t-s)}\cos\omega_d(t-s)}{\omega_n^2}\right.$$

$$\left. -\frac{\zeta\omega_n e^{-\zeta\omega_n(t-s)}\sin\omega_d(t-s)+\zeta^2\omega_n e^{-\zeta\omega_n(t-s)}\cos\omega_d(t-s)/\sqrt{1-\zeta^2}}{\omega_n^2}\right]_{s=0}^{s=t} \qquad (2.110)$$

$$= \frac{F_o}{k}\left[-\left(\frac{\omega_n}{\sqrt{1-\zeta^2}}\right)e^{-\zeta\omega_n(t-s)}\sin\omega_d(t-s)\right]_{s=0}^{s=t}$$

$$= \left(\frac{\omega_n F_o/k}{\sqrt{1-\zeta^2}}\right)e^{-\zeta\omega_n t}\sin\omega_d t.$$

Combining Eqs. 2.109 and 2.110 with the homogeneous solution given in Eqs. 2.21 and 2.22, it follows that

$$x(t) = \left[x_o - \frac{F_o}{k}\right]e^{-\zeta\omega_n t}\cos\omega_d t + \left[\frac{\zeta\omega_n(x_o - F_o/k)+\dot{x}_o}{\omega_n\sqrt{1-\zeta^2}}\right]e^{-\zeta\omega_n t}\sin\omega_d t + \left(\frac{F_o}{k}\right) \qquad (2.111)$$

$$\dot{x}(t) = \dot{x}_o e^{-\zeta\omega_n t}\cos\omega_d t - \left[\frac{\omega_n(x_o - F_o/k)+\zeta\dot{x}_o}{\sqrt{1-\zeta^2}}\right]e^{-\zeta\omega_n t}\sin\omega_d t \qquad (2.112)$$

which is the same solution as Eqs. 2.45 and 2.46.

EXAMPLE 3   *Forcing Function that Rises to Static*

Example 2 in Section 2.3 calculated the response to this forcing function. The solution using the convolution integral is now presented. For the forcing function

$$F_e(t) = \begin{cases} F_o\, t/t_o & 0 \le t < t_o \\ F_o & t \ge t_o \\ 0 & \text{otherwise,} \end{cases} \tag{2.113}$$

let the initial conditions of the system be equal to zero. Then the response for $t \le t_o$ is

$$x_p(t) = \int_0^t \left(\frac{F_o}{m\,\omega_d t_o} \right) s e^{-\zeta\omega_n(t-s)} \sin\omega_d(t-s)\, ds \tag{2.114}$$

$$\dot{x}_p(t) = \int_0^t \left(\frac{F_o}{m\,t_o} \right) s e^{-\zeta\omega_n(t-s)} \left[\cos\omega_d(t-s) - \left(\frac{\zeta}{\sqrt{1-\zeta^2}} \right) \sin\omega_d(t-s) \right] ds. \tag{2.115}$$

First, consider the displacement response as given in Eq. 2.114. Performing the integration gives

$$x_p(t) = \int_0^t \left(\frac{F_o}{m\,\omega_d t_o} \right) s e^{-\zeta\omega_n(t-s)} \sin\omega_d(t-s)\, ds$$

$$= \frac{F_o}{m\,\omega_d t_o} \left\{ \frac{e^{-\zeta\omega_n(t-s)}}{\left(\omega_d^2 + \zeta^2\omega_n^2\right)^2} \left[\left(-2\zeta\omega_n\omega_d + (\omega_d^2 + \zeta^2\omega_n^2)\omega_d s\right)\cos\omega_d(t-s) \right. \right.$$

$$\left. \left. + \left(\omega_d^2 - \zeta^2\omega_n^2 + (\omega_d^2 + \zeta^2\omega_n^2)\zeta\omega_n s\right)\sin\omega_d(t-s) \right] \right\} \Bigg|_{s=0}^{s=t}. \tag{2.116}$$

Recall that $\omega_d^2 = \omega_n^2(1-\zeta^2)$; therefore,

$$x_p(t) = \frac{F_o}{m\,\omega_d \omega_n^4 t_o} \left\{ -2\zeta\omega_n\omega_d + (\omega_d^2 + \zeta^2\omega_n^2)\omega_d t \right.$$

$$\left. - e^{-\zeta\omega_n t} \left[-2\zeta\omega_n\omega_d \cos\omega_d t + (\omega_d^2 - \zeta^2\omega_n^2)\sin\omega_d t \right] \right\}. \tag{2.117}$$

Simplifying Eq. 2.117 and recalling that $m\omega_n^2 = k$, it follows that

$$x_p(t) = \left(\frac{2\zeta F_o}{k\,\omega_n t_o} \right) e^{-\zeta\omega_n t} \cos\omega_d t + \left[\frac{F_o(2\zeta^2 - 1)}{k\,\omega_n t_o \sqrt{1-\zeta^2}} \right] e^{-\zeta\omega_n t} \sin\omega_d t + \left(\frac{F_o}{k\,t_o} \right) t - \left(\frac{2\zeta F_o}{k\,\omega_n t_o} \right). \tag{2.118}$$

The velocity also follows from Eq. 2.115. Performing the integration gives

$$\dot{x}_p(t) = \int_0^t \left(\frac{F_o}{m\,t_o} \right) s e^{-\zeta\omega_n(t-s)} \cos\omega_d(t-s)\, ds - \frac{\zeta}{\sqrt{1-\zeta^2}} \int_0^t \left(\frac{F_o}{m\,t_o} \right) s e^{-\zeta\omega_n(t-s)} \sin\omega_d(t-s)\, ds$$

$$= \frac{F_o}{m\,t_o} \left(\frac{e^{-\zeta\omega_n(t-s)}}{\left(\omega_d^2 + \zeta^2\omega_n^2\right)^2} \left\{ \left[(\omega_d^2 - \zeta^2\omega_n^2) + (\omega_d^2 + \zeta^2\omega_n^2)\zeta\omega_n s \right]\cos\omega_d(t-s) \right. \right.$$

$$\left. \left. + \left[2\zeta\omega_n\omega_d - (\omega_d^2 + \zeta^2\omega_n^2)\omega_d s \right]\sin\omega_d(t-s) \right\} \right) \Bigg|_{s=0}^{s=t} \tag{2.119}$$

$$- \frac{\zeta}{\sqrt{1-\zeta^2}} \frac{F_o}{m\,t_o} \left(\frac{e^{-\zeta\omega_n(t-s)}}{\left(\omega_d^2 + \zeta^2\omega_n^2\right)^2} \left\{ \left[-2\zeta\omega_n\omega_d + (\omega_d^2 + \zeta^2\omega_n^2)\omega_d s \right]\cos\omega_d(t-s) \right. \right.$$

$$\left. \left. + \left[(\omega_d^2 - \zeta^2\omega_n^2) + (\omega_d^2 + \zeta^2\omega_n^2)\zeta\omega_n s \right]\sin\omega_d(t-s) \right\} \right) \Bigg|_{s=0}^{s=t}.$$

Evaluating the limits and substituting $\omega_n^2(1-\zeta^2)=\omega_d^2$ and $k=m\omega_n^2$, then it follows that

$$\dot{x}_p(t)=\frac{F_o}{k\,\omega_n^2 t_o}\Big\{(\omega_d^2-\zeta^2\omega_n^2)+(\omega_d^2+\zeta^2\omega_n^2)\zeta\omega_n t$$

$$-e^{-\zeta\omega_n t}\big[(\omega_d^2-\zeta^2\omega_n^2)\cos\omega_d t+2\zeta\omega_n\omega_d\sin\omega_d t\big]\Big\}$$

$$-\frac{\zeta}{\sqrt{1-\zeta^2}}\frac{F_o}{k\,\omega_n^2 t_o}\Big\{-2\zeta\omega_n\omega_d+(\omega_d^2+\zeta^2\omega_n^2)\omega_d t \tag{2.120}$$

$$-e^{-\zeta\omega_n t}\big[-2\zeta\omega_n\omega_d\cos\omega_d t+(\omega_d^2-\zeta^2\omega_n^2)\sin\omega_d t\big]\Big\}.$$

Finally, simplifying Eq. (2.120) gives

$$\dot{x}_p(t)=-\left(\frac{F_o}{k\,t_o}\right)e^{-\zeta\omega_n t}\cos\omega_d t-\left[\frac{\zeta}{\sqrt{1-\zeta^2}}\frac{F_o}{k\,t_o}\right]e^{-\zeta\omega_n t}\sin\omega_d t+\left(\frac{F_o}{k\,t_o}\right). \tag{2.121}$$

The solutions given by Eq. 2.62 for displacement response and Eq. 2.63 for velocity response are the same as Eqs. 2.118 and 2.121 by using the standard integration relations.

EXAMPLE 4 *Earthquake Response*

The equation of motion for the SDOF system with an applied force is given in Eq. 2.2 and is

$$m\ddot{x}(t)+c\dot{x}(t)+kx(t)=F_e(t). \tag{2.122}$$

Now consider the structure shown in Figure 2.13a, which is subjected not to an external force but to a ground motion from an earthquake. The elastic forces from the columns are now equal to

$$F_s=-k\big(y(t)-u_g(t)\big) \tag{2.123}$$

where $y(t)$ is the absolute displacement of the mass and $u_g(t)$ is the absolute displacement of the ground. If the relative displacement between the mass and the ground is denoted by $x(t)$, where $x(t)=y(t)-u_g(t)$ (see Figure 2.13b), then

$$F_s=-k\,x(t). \tag{2.124}$$

Similarly, the damping force is

$$F_d=-c\big(\dot{y}(t)-\dot{u}_g(t)\big)=-c\dot{x}(t). \tag{2.125}$$

The external force is zero, and thus $F_e(t)=0$. The inertial force is the mass times the absolute acceleration and thus $m\ddot{y}(t)=m\ddot{x}(t)+m\ddot{u}_g(t)$. It then follows from Eq. 2.1 that

$$\sum F=F_s+F_d+F_e=-kx(t)-c\dot{x}(t)=m\ddot{x}(t)+m\ddot{u}_g(t) \tag{2.126}$$

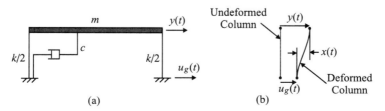

Figure 2.13 Single Degree of Freedom System Subjected to Earthquake Excitation

and rearranging terms gives

$$m\ddot{x}(t) + c\dot{x}(t) + kx(t) = -m\ddot{u}_g(t).$$ (2.127)

It is generally accepted that the ground acceleration be indicated as $a(t) = \ddot{u}_g(t)$. It then follows from Eq. 2.127 that

$$m\ddot{x}(t) + c\dot{x}(t) + kx(t) = -ma(t)$$ (2.128)

and in normalized form,

$$\ddot{x}(t) + 2\zeta\omega_n\dot{x}(t) + \omega_n^2 x(t) = -a(t).$$ (2.129)

Here, the ground acceleration is considered to be the forcing function, and therefore the solution response follows from Eq. 2.101 that

$$x(t) = x_o e^{-\zeta\omega_n t}\cos\omega_d t + \left(\frac{\zeta\omega_n x_o + \dot{x}_o}{\omega_n\sqrt{1-\zeta^2}}\right)e^{-\zeta\omega_n t}\sin\omega_d t - \int_0^t \left(\frac{a(s)}{\omega_d}\right)e^{-\zeta\omega_n(t-s)}\sin\omega_d(t-s)\,ds$$ (2.130)

$$\dot{x}(t) = \dot{x}_o e^{-\zeta\omega_n t}\cos\omega_d t - \left(\frac{\omega_n x_o + \zeta\dot{x}_o}{\sqrt{1-\zeta^2}}\right)e^{-\zeta\omega_n t}\sin\omega_d t$$

$$- \int_0^t a(s)e^{-\zeta\omega_n(t-s)}\left[\cos\omega_d(t-s) - \left(\frac{\zeta}{\sqrt{1-\zeta^2}}\right)\sin\omega_d(t-s)\right]ds.$$ (2.131)

Before the structure is subjected to ground acceleration, it is very reasonable to assume that the structure is at rest, that is, both initial displacement and velocity are equal to zero. Therefore, it follows from Eqs. 2.130 and 2.131 that

$$x(t) = -\int_0^t \left(\frac{a(s)}{\omega_d}\right)e^{-\zeta\omega_n(t-s)}\sin\omega_d(t-s)\,ds$$ (2.132)

$$\dot{x}(t) = -\int_0^t a(s)e^{-\zeta\omega_n(t-s)}\left[\cos\omega_d(t-s) - \left(\frac{\zeta}{\sqrt{1-\zeta^2}}\right)\sin\omega_d(t-s)\right]ds.$$ (2.133)

EXAMPLE 5   *Response Using Incremental Approach*

Consider the dynamic equilibrium equation

$$m\ddot{x}(t) + c\dot{x}(t) + kx(t) = F_e(t)$$ (2.134)

where the external force $F_e(t)$ is as shown in Figure 2.14. Imagine that the response of the system is known at time t_k, and the goal is to calculate the response at time t_{k+1}. Let the time increment be

$$\Delta t = t_{k+1} - t_k.$$ (2.135)

For a time increment τ that is less than Δt, the external force is

$$F_e(\tau) = F_k + \left(\frac{F_{k+1} - F_k}{t_{k+1} - t_k}\right)\tau.$$ (2.136)

The response at a time $t_k + \tau$ is equal to the summation of four parts, which are as follows:

1.   Free vibration due to initial displacement at $t = t_k$.
2.   Free vibration due to initial velocity at $t = t_k$.

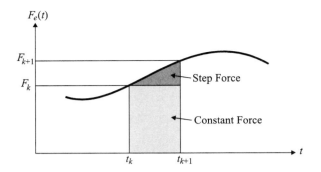

Figure 2.14 General Forcing Function

3. Forced response due to a constant force of magnitude

$$F_e(\tau) = F_k.$$ (2.137)

4. Forced response due to a step force with magnitude

$$F_e(\tau) = \left(\frac{F_{k+1} - F_k}{t_{k+1} - t_k} \right) \tau.$$ (2.138)

Therefore, the response of the system is

$$x_{k+1} = C_1 x_k + C_2 \dot{x}_k + C_3 F_k + C_4 F_{k+1}$$ (2.139)

$$\dot{x}_{k+1} = D_1 x_k + D_2 \dot{x}_k + D_3 F_k + D_4 F_{k+1}$$ (2.140)

$$\ddot{x}_{k+1} = F_{k+1}/m - 2\zeta\omega_n \dot{x}_{k+1} - \omega_n^2 x_{k+1}$$ (2.141)

where

$$C_1 = e^{-\zeta\omega_n\Delta t}\left[\cos\omega_d\Delta t + \left(\frac{\zeta}{\sqrt{1-\zeta^2}} \right) \sin\omega_d\Delta t \right]$$ (2.142)

$$C_2 = e^{-\zeta\omega_n\Delta t}\left[\left(\frac{1}{\omega_d} \right) \sin\omega_d\Delta t \right]$$ (2.143)

$$C_3 = \frac{1}{k}\left\{ \left(\frac{2\zeta}{\omega_n\Delta t} \right) + e^{-\zeta\omega_n\Delta t}\left[-\left(1 + \frac{2\zeta}{\omega_n\Delta t} \right)\cos\omega_d\Delta t + \left(\frac{1-2\zeta^2}{\omega_d\Delta t} - \frac{\zeta}{\sqrt{1-\zeta^2}} \right)\sin\omega_d\Delta t \right] \right\}$$ (2.144)

$$C_4 = \frac{1}{k}\left\{ 1 - \left(\frac{2\zeta}{\omega_n\Delta t} \right) + e^{-\zeta\omega_n\Delta t}\left[\left(\frac{2\zeta}{\omega_n\Delta t} \right)\cos\omega_d\Delta t + \left(\frac{2\zeta^2-1}{\omega_d\Delta t} \right)\sin\omega_d\Delta t \right] \right\}$$ (2.145)

$$D_1 = e^{-\zeta\omega_n\Delta t}\left[-\left(\frac{\omega_n}{\sqrt{1-\zeta^2}} \right) \sin\omega_d\Delta t \right]$$ (2.146)

$$D_2 = e^{-\zeta\omega_n\Delta t}\left[\cos\omega_d\Delta t - \left(\frac{\zeta}{\sqrt{1-\zeta^2}} \right) \sin\omega_d\Delta t \right]$$ (2.147)

$$D_3 = \frac{1}{k}\left\{ -\left(\frac{1}{\Delta t} \right) + e^{-\zeta\omega_n\Delta t}\left[\left(\frac{1}{\Delta t} \right)\cos\omega_d\Delta t + \left(\frac{\omega_n}{\sqrt{1-\zeta^2}} - \frac{\zeta}{\Delta t\sqrt{1-\zeta^2}} \right)\sin\omega_d\Delta t \right] \right\}$$ (2.148)

$$D_4 = \frac{1}{k\Delta t}\left\{ 1 - e^{-\zeta\omega_n\Delta t}\left[\cos\omega_d\Delta t + \left(\frac{\zeta}{\sqrt{1-\zeta^2}} \right)\sin\omega_d\Delta t \right] \right\}.$$ (2.149)

This solution method can be used for any digitized forcing function. For a linear system ω_n, ζ, and also Δt are all constants, and therefore the eight constants given from Eqs. 2.142 to 2.149 need to be calculated only once.

Equations 2.139 to 2.141 can be written in the matrix form. First note that the dynamic equilibrium equation of motion must be satisfied at all time; it follows that

$$F_k = m\ddot{x}_k + c\dot{x}_k + kx_k. \tag{2.150}$$

Substituting Eq. 2.150 into Eqs. 2.139 through 2.141 gives

$$\begin{aligned} x_{k+1} &= C_1 x_k + C_2 \dot{x}_k + C_3\left(m\ddot{x}_k + c\dot{x}_k + kx_k\right) + C_4 F_{k+1} \\ &= \left(C_1 + kC_3\right)x_k + \left(C_2 + cC_3\right)\dot{x}_k + mC_3\ddot{x}_k + C_4 F_{k+1} \end{aligned} \tag{2.151}$$

$$\begin{aligned} \dot{x}_{k+1} &= D_1 x_k + D_2 \dot{x}_k + D_3\left(m\ddot{x}_k + c\dot{x}_k + kx_k\right) + D_4 F_{k+1} \\ &= \left(D_1 + kD_3\right)x_k + \left(D_2 + cD_3\right)\dot{x}_k + mD_3\ddot{x}_k + D_4 F_{k+1} \end{aligned} \tag{2.152}$$

$$\ddot{x}_{k+1} = F_{k+1}/m - 2\zeta\omega_n \dot{x}_k - \omega_n^2 x_k. \tag{2.153}$$

Now representing Eqs. 2.151 to 2.153 in matrix form, it follows that

$$\begin{Bmatrix} x_{k+1} \\ \dot{x}_{k+1} \\ \ddot{x}_{k+1} \end{Bmatrix} = \begin{bmatrix} C_1 + kC_3 & C_2 + cC_3 & mC_3 \\ D_1 + kD_3 & D_2 + cD_3 & mD_3 \\ -\omega_n^2 & -2\zeta\omega_n & 0 \end{bmatrix} \begin{Bmatrix} x_k \\ \dot{x}_k \\ \ddot{x}_k \end{Bmatrix} + \begin{bmatrix} C_4 \\ D_4 \\ 1/m \end{bmatrix} F_{k+1} \tag{2.154}$$

or using matrix notation,

$$\mathbf{q}_{k+1} = \mathbf{F}_c \mathbf{q}_k + \mathbf{H}_c F_{k+1} \tag{2.155}$$

where

$$\mathbf{q}_k = \begin{Bmatrix} x_k \\ \dot{x}_k \\ \ddot{x}_k \end{Bmatrix}, \quad \mathbf{F}_c = \begin{bmatrix} C_1 + kC_3 & C_2 + cC_3 & mC_3 \\ D_1 + kD_3 & D_2 + cD_3 & mD_3 \\ -\omega_n^2 & -2\zeta\omega_n & 0 \end{bmatrix}, \quad \mathbf{H}_c = \begin{bmatrix} C_4 \\ D_4 \\ 1/m \end{bmatrix}. \tag{2.156}$$

EXAMPLE 6   *Impulse Derived from Taylor Series*

Equation 2.93 shows that in the limit,

$$\lim_{t_o \to 0} x(t_o) = \lim_{t_o \to 0} \frac{F_o}{k}\left[1 - e^{-\zeta\omega_n t_o}\cos\omega_d t_o - \left(\frac{\zeta}{\sqrt{1-\zeta^2}}\right)e^{-\zeta\omega_n t_o}\sin\omega_d t_o\right] = 0 \tag{2.157}$$

$$\lim_{t_o \to 0} \dot{x}(t_o) = \lim_{t_o \to 0} \frac{F_o}{k}\left[\left(\frac{\omega_n}{\sqrt{1-\zeta^2}}\right)e^{-\zeta\omega_n t_o}\sin\omega_d t_o\right] = \frac{1}{m}. \tag{2.158}$$

The same result can be derived using Taylor series expansion. Recall that

$$e^x = 1 + x + \tfrac{1}{2!}x^2 + \tfrac{1}{3!}x^3 + \dots$$

$$\sin x = x - \tfrac{1}{3!}x^3 + \tfrac{1}{5!}x^5 - \tfrac{1}{7!}x^7 + \dots$$

$$\cos x = 1 - \tfrac{1}{2!}x^2 + \tfrac{1}{4!}x^4 - \tfrac{1}{6!}x^6 + \dots.$$

Therefore,

$$1 - e^{-\zeta\omega_n t_o}\cos\omega_d t_o = 1 - \left(1 - \zeta\omega_n t_o + \frac{\zeta^2\omega_n^2}{2!}t_o^2 + \ldots\right)\left(1 - \frac{\omega_d^2}{2!}t_o^2 + \ldots\right)$$

$$= \zeta\omega_n t_o + \left(\frac{\omega_d^2}{2!} - \frac{\zeta^2\omega_n^2}{2!}\right)t_o^2 + \ldots. \tag{2.159}$$

Multiplying by (F_o/k), recalling that $F_o = 1/t_o$, and taking the limit as $t_o \to \infty$, it follows that

$$\lim_{t_o\to0} x(t_o) = \lim_{t_o\to0}\frac{F_o}{k}\left[1 - e^{-\zeta\omega_n t_o}\cos\omega_d t_o\right] = \frac{1}{k}\lim_{t_o\to0}\left[\zeta\omega_n + \left(\frac{\omega_d^2}{2!} - \frac{\zeta^2\omega_n^2}{2!}\right)t_o + \ldots\right] = \frac{\zeta\omega_n}{k}. \tag{2.160}$$

Similarly,

$$e^{-\zeta\omega_n t_o}\sin\omega_d t_o = \left(1 - \zeta\omega_n t_o + \frac{\zeta^2\omega_n^2}{2!}t_o^2 + \ldots\right)\left(\omega_d t_o - \frac{\omega_d^3}{3!}t_o^3 + \ldots\right)$$

$$= \omega_d t_o - \zeta\omega_n\omega_d t_o^2 + \ldots. \tag{2.161}$$

Multiplying by (F_o/k), recalling that $F_o = 1/t_o$, and taking the limit as $t_o \to \infty$, it follows that

$$\lim_{t_o\to0} x(t_o) = \lim_{t_o\to0}\frac{F_o}{k}\left[1 - e^{-\zeta\omega_n t_o}\sin\omega_d t_o\right] = \frac{1}{k}\lim_{t_o\to0}\left[\omega_d + \zeta\omega_n\omega_d t_o + \ldots\right] = \frac{\omega_d}{k}. \tag{2.162}$$

Finally, it follows from Eqs. 2.157 and 2.158 that

$$\lim_{t_o\to0} x(t_o) = \lim_{t_o\to0}\frac{F_o}{k}\left[1 - e^{-\zeta\omega_n t_o}\cos\omega_d t_o - \left(\frac{\zeta}{\sqrt{1-\zeta^2}}\right)e^{-\zeta\omega_n t_o}\sin\omega_d t_o\right]$$

$$= \lim_{t_o\to0}\left[\frac{F_o}{k}\left(1 - e^{-\zeta\omega_n t_o}\cos\omega_d t_o\right)\right] - \left(\frac{\zeta}{\sqrt{1-\zeta^2}}\right)\lim_{t_o\to0}\left[\frac{F_o}{k}\left(\frac{\zeta}{\sqrt{1-\zeta^2}}\right)e^{-\zeta\omega_n t_o}\sin\omega_d t_o\right] \tag{2.163}$$

$$= \frac{\zeta\omega_n}{k} - \left(\frac{\zeta}{\sqrt{1-\zeta^2}}\right)\frac{\omega_d}{k} = \frac{1}{k}(\zeta\omega_n - \zeta\omega_n) = 0$$

$$\lim_{t_o\to0}\dot{x}(t_o) = \lim_{t_o\to0}\frac{F_o}{k}\left[\left(\frac{\omega_n}{\sqrt{1-\zeta^2}}\right)e^{-\zeta\omega_n t_o}\sin\omega_d t_o\right] = \left(\frac{\omega_n}{\sqrt{1-\zeta^2}}\right)\lim_{t_o\to0}\left[\frac{F_o}{k}\left(e^{-\zeta\omega_n t_o}\sin\omega_d t_o\right)\right] \tag{2.164}$$

$$= \left(\frac{\omega_n}{\sqrt{1-\zeta^2}}\right)\frac{\omega_d}{k} = \frac{\omega_n^2}{k} = \frac{1}{m}.$$

This example gives insight into the contributions involved in each term of Eqs. 2.157 and 2.158. Note that in Eq. 2.163, the contribution from the cosine term exactly cancels that from the sine term, and the result of the limit is equal to zero.

EXAMPLE 7 *Bilinear Earthquake Ground Motion Pulse*

Consider the earthquake ground motion that is a triangular (bilinear) pulse as shown in Figure 2.15.

In the time domain $t = 0$ to $t = t_p$, the equation for the earthquake ground motion is

$$a(t) = \left(\frac{a_p}{t_p}\right)t, \quad 0 \le t \le t_p. \tag{2.165}$$

In the time domain $t = t_p$ to $t = 2t_p$, the earthquake ground motion is the summation of two linear functions, which are

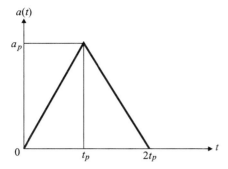

Figure 2.15 Bilinear Earthquake Ground Motion

$$a(t) = -\left(\frac{a_p}{t_p}\right)t + 2a_p = a_1(t) + a_2(t) \tag{2.166}$$

where

$$a_1(t) = \left(\frac{a_p}{t_p}\right)t, \quad t_p < t \le 2t_p \tag{2.167}$$

$$a_2(t) = -2\left(\frac{a_p}{t_p}\right)(t - t_p), \quad t_p < t \le 2t_p. \tag{2.168}$$

The function $a_1(t)$ is the same as Eq. 2.165 without considering the limits on the time domain t. Note that $a_1(t)$ increases in magnitude, and $a_2(t)$ provides a negative contribution that is sufficient to decrease the slope to a negative value.

In the time domain $t > 2t_p$, the earthquake ground motion is zero and can be represented as

$$a(t) = 0 = a_1(t) + a_2(t) + a_3(t) \tag{2.169}$$

where

$$a_1(t) = \left(\frac{a_p}{t_p}\right)t, \quad t > 2t_p \tag{2.170}$$

$$a_2(t) = -2\left(\frac{a_p}{t_p}\right)(t - t_p), \quad t > 2t_p \tag{2.171}$$

$$a_3(t) = \left(\frac{a_p}{t_p}\right)(t - 2t_p), \quad t > 2t_p. \tag{2.172}$$

Again, note that $a_1(t)$ and $a_2(t)$ are the same as Eqs. 2.167 and 2.168 without considering the limits on the time domain t.

The response of an SDOF system is obtained for each of the three linear functions $a_1(t)$, $a_2(t)$, and $a_3(t)$ and summed appropriately for each time domain. It follows for Eq. 2.118 that for the time domain $t = 0$ to $t = t_p$, the response with zero damping and zero initial conditions is

$$x(t) = \left(\frac{a_p}{\omega_n^2}\right)\left[\left(\frac{\sin \omega_n t}{\omega_n t_p}\right) - \left(\frac{t}{t_p}\right)\right], \quad 0 \le t \le t_p \tag{2.173}$$

where F_o is replaced by $-ma_p$. For the time domain $t = t_p$ to $t = 2t_p$, it follows that

$$x(t) = \left(\frac{a_p}{\omega_n^2}\right)\left[\left(\frac{\sin\omega_n t}{\omega_n t_p}\right) - \left(\frac{t}{t_p}\right)\right] - 2\left(\frac{a_p}{\omega_n^2}\right)\left[\left(\frac{\sin\omega_n(t-t_p)}{\omega_n t_p}\right) - \left(\frac{t-t_p}{t_p}\right)\right]$$

$$= \left(\frac{a_p}{\omega_n^2}\right)\left[-2 + \left(\frac{t}{t_p}\right) + \left(\frac{\sin\omega_n t}{\omega_n t_p}\right) - \left(\frac{2\sin\omega_n(t-t_p)}{\omega_n t_p}\right)\right], \quad t_p < t \le 2t_p. \tag{2.174}$$

Similarly, for the time domain $t > 2t_p$,

$$x(t) = \left(\frac{a_p}{\omega_n^2}\right)\left[-2 + \left(\frac{t}{t_p}\right) + \left(\frac{\sin\omega_n t}{\omega_n t_p}\right) - \left(\frac{2\sin\omega_n(t-t_p)}{\omega_n t_p}\right)\right]$$

$$+ \left(\frac{a_p}{\omega_n^2}\right)\left[\left(\frac{\sin\omega_n(t-2t_p)}{\omega_n t_p}\right) - \left(\frac{t-2t_p}{t_p}\right)\right]$$

$$= \left(\frac{a_p}{\omega_n^2}\right)\left[\left(\frac{\sin\omega_n t}{\omega_n t_p}\right) - \left(\frac{2\sin\omega_n(t-t_p)}{\omega_n t_p}\right) + \left(\frac{\sin\omega_n(t-2t_p)}{\omega_n t_p}\right)\right], \quad t > 2t_p. \tag{2.175}$$

In summary, the response is

$$x(t) = \begin{cases} \left(\frac{a_p}{\omega_n^2}\right)\left[\left(\frac{\sin\omega_n t}{\omega_n t_p}\right) - \left(\frac{t}{t_p}\right)\right] & 0 \le t \le t_p \\ \left(\frac{a_p}{\omega_n^2}\right)\left[-2 + \left(\frac{t}{t_p}\right) + \left(\frac{\sin\omega_n t}{\omega_n t_p}\right) - 2\left(\frac{\sin\omega_n(t-t_p)}{\omega_n t_p}\right)\right] & t_p < t \le 2t_p \\ \left(\frac{a_p}{\omega_n^2}\right)\left[\left(\frac{\sin\omega_n t}{\omega_n t_p}\right) - \left(\frac{2\sin\omega_n(t-t_p)}{\omega_n t_p}\right) + \left(\frac{\sin\omega_n(t-2t_p)}{\omega_n t_p}\right)\right] & t > 2t_p. \end{cases} \tag{2.176}$$

2.5   STATE SPACE SOLUTION FOR RESPONSE

In recent years, calculating the response of the system using an alternate method called the *state space method* has become essential for a broad range of dynamic problems. These structural engineering situations include the use of manufactured dampers in buildings, the placement of base isolators in buildings, and the use of structural control methods to reduce earthquake and wind responses. Each of these topics is discussed in Part III of this book. The state space method analyzes the response of the system using both the displacement and the velocity as independent variables, and these variables are called *states*. Recall that the dynamic equilibrium equation is given in Eq. 2.2 and is

$$\ddot{x}(t) + 2\zeta\omega_n\dot{x}(t) + \omega_n^2 x(t) = F_e(t)/m. \tag{2.177}$$

Define the vector $\mathbf{z}$ to represent both the displacement and the velocity of the system. These two independent response variables are expressed as

$$\mathbf{z}(t) = \begin{Bmatrix} x(t) \\ \dot{x}(t) \end{Bmatrix}. \tag{2.178}$$

It follows that Eq. 2.177 can be written in the following equivalent form:

$$\dot{\mathbf{z}}(t) = \begin{Bmatrix} \dot{x}(t) \\ \ddot{x}(t) \end{Bmatrix} = \begin{bmatrix} 0 & 1 \\ -\omega_n^2 & -2\zeta_n\omega_n \end{bmatrix}\begin{Bmatrix} x(t) \\ \dot{x}(t) \end{Bmatrix} + \begin{Bmatrix} 0 \\ F_e(t)/m \end{Bmatrix} = \begin{bmatrix} 0 & 1 \\ -\omega_n^2 & -2\zeta_n\omega_n \end{bmatrix}\mathbf{z}(t) + \begin{Bmatrix} 0 \\ F_e(t)/m \end{Bmatrix}. \tag{2.179}$$

Define

$$\mathbf{A} = \begin{bmatrix} 0 & 1 \\ -\omega_n^2 & -2\zeta\omega_n \end{bmatrix}, \quad \mathbf{F}(t) = \begin{Bmatrix} 0 \\ F_e(t)/m \end{Bmatrix} \quad (2.180)$$

and then it follows that

$$\dot{\mathbf{z}}(t) = \mathbf{A}\mathbf{z}(t) + \mathbf{F}(t). \quad (2.181)$$

Equation 2.181 is a first-order linear matrix differential equation of motion, and it is called the *continuous state space equation of motion*. In contrast, Eq. 2.177 is a second-order linear differential equation. Although this differential equation is in a matrix form, the method of solving this equation is the same as for solving the corresponding scalar first-order linear differential equation of motion. Thus, in general, the solution for any time $t \geq t_o$, where t_o represents the time when the initial displacement and velocity are given, can be written as

$$\mathbf{z}(t) = \mathbf{e}^{\mathbf{A}(t-t_o)}\mathbf{z}(t_o) + \mathbf{e}^{\mathbf{A}t} \int_{t_o}^{t} \mathbf{e}^{-\mathbf{A}s} \, \mathbf{F}(s) \, ds \quad (2.182)$$

where $\mathbf{e}^{\mathbf{A}t}$ is defined as

$$\mathbf{e}^{\mathbf{A}t} = \mathbf{I} + \mathbf{A}t + \frac{1}{2!}(\mathbf{A}t)^2 + \frac{1}{3!}(\mathbf{A}t)^3 + \dots . \quad (2.183)$$

If the initial conditions are given at time equal to zero (i.e., $t_o = 0$), then Eq. 2.182 can be written as

$$\mathbf{z}(t) = \mathbf{e}^{\mathbf{A}t}\mathbf{z}_o + \int_{0}^{t} \mathbf{e}^{\mathbf{A}(t-s)}\mathbf{F}(s) \, ds \quad (2.184)$$

where

$$\mathbf{z}_o = \mathbf{z}(0) = \begin{Bmatrix} x(0) \\ \dot{x}(0) \end{Bmatrix} = \begin{Bmatrix} x_o \\ \dot{x}_o \end{Bmatrix}. \quad (2.185)$$

Equation 2.184 can be viewed as a combination of the homogeneous and the particular solutions, very similar to the general solution discussed in Section 2.3. The first term in Eq. 2.184 represents the homogeneous solution with the initial condition taken into consideration. The second term in Eq. 2.184 represents the particular solution, which is expressed in terms of the time-integration of forcing function. This particular solution has zero for both the initial displacement and initial velocity, because

$$\mathbf{z}_p(0) = \int_{0}^{0} \mathbf{e}^{\mathbf{A}(t-s)}\mathbf{F}(s) \, ds = \mathbf{e}^{\mathbf{A}t}(\mathbf{0}) = \mathbf{0}. \quad (2.186)$$

The matrix $\mathbf{e}^{\mathbf{A}t}$ is called the *state transition matrix*, and it has the same dimension as the $\mathbf{A}$ matrix. The best way to compute the value of this matrix for any given value of time, t, natural frequency, ω_n, and critical damping ratio, ζ, is to use a computer to calculate Eq. 2.183. However, for an SDOF system with a 2×2 state transition matrix, it can be calculated in a closed form, which will be done later in this section.

When a diagonal matrix is operated on by an exponential function, the exponent can be applied to each term on the diagonal of the matrix, that is,

$$\exp \begin{bmatrix} \lambda_1 t & 0 & \cdots & 0 \\ 0 & \lambda_2 t & \ddots & \vdots \\ \vdots & \ddots & \ddots & 0 \\ 0 & \cdots & 0 & \lambda_n t \end{bmatrix} = \begin{bmatrix} e^{\lambda_1 t} & 0 & \cdots & 0 \\ 0 & e^{\lambda_2 t} & \ddots & \vdots \\ \vdots & \ddots & \ddots & 0 \\ 0 & \cdots & 0 & e^{\lambda_n t} \end{bmatrix}. \quad (2.187)$$

This relation can be proved by first expressing the left-hand side of Eq. 2.187 using the exponential expansion in Eq. 2.183. Since each term of the expansion involves a diagonal matrix raised to a certain power (i.e., a matrix multiplying itself several times), carrying out this operation will simply raise each diagonal term to this power. Once this is done, each term in the expansion is then added together, and the result will become the expression shown on the right-hand side of Eq. 2.187. This procedure is discussed as follows:

$$
\exp\begin{bmatrix} \lambda_1 t & 0 & \cdots & 0 \\ 0 & \lambda_2 t & \ddots & \vdots \\ \vdots & \ddots & \ddots & 0 \\ 0 & \cdots & 0 & \lambda_n t \end{bmatrix} = \mathbf{I} + \begin{bmatrix} \lambda_1 t & 0 & \cdots & 0 \\ 0 & \lambda_2 t & \ddots & \vdots \\ \vdots & \ddots & \ddots & 0 \\ 0 & \cdots & 0 & \lambda_n t \end{bmatrix} + \frac{1}{2!}\begin{bmatrix} \lambda_1 t & 0 & \cdots & 0 \\ 0 & \lambda_2 t & \ddots & \vdots \\ \vdots & \ddots & \ddots & 0 \\ 0 & \cdots & 0 & \lambda_n t \end{bmatrix}^2 + \cdots
$$

$$
= \mathbf{I} + \begin{bmatrix} \lambda_1 t & 0 & \cdots & 0 \\ 0 & \lambda_2 t & \ddots & \vdots \\ \vdots & \ddots & \ddots & 0 \\ 0 & \cdots & 0 & \lambda_n t \end{bmatrix} + \frac{1}{2!}\begin{bmatrix} \lambda_1^2 t^2 & 0 & \cdots & 0 \\ 0 & \lambda_2^2 t^2 & \ddots & \vdots \\ \vdots & \ddots & \ddots & 0 \\ 0 & \cdots & 0 & \lambda_n^2 t^2 \end{bmatrix} + \cdots
$$

$$
= \begin{bmatrix} 1 + \lambda_1 t + \frac{1}{2!}\lambda_1^2 t^2 + \ldots & 0 & \cdots & 0 \\ 0 & 1 + \lambda_2 t + \frac{1}{2!}\lambda_2^2 t^2 + \ldots & \ddots & \vdots \\ \vdots & \ddots & \ddots & 0 \\ 0 & 0 & \cdots & 0 & 1 + \lambda_n t + \frac{1}{2!}\lambda_n^2 t^2 + \ldots \end{bmatrix}
$$

$$
= \begin{bmatrix} e^{\lambda_1 t} & 0 & \cdots & 0 \\ 0 & e^{\lambda_2 t} & \ddots & \vdots \\ \vdots & \ddots & \ddots & 0 \\ 0 & \cdots & 0 & e^{\lambda_n t} \end{bmatrix}.
$$

Therefore, it is important to diagonalize the matrix $\mathbf{A}$, which can be done by using the eigenvalues and eigenvectors of the matrix $\mathbf{A}$.

Define the eigenvalues of $\mathbf{A}$ to be λ_1 and λ_2, and then define the corresponding eigenvectors to be $\boldsymbol{\phi}_1$ and $\boldsymbol{\phi}_2$. It then follows that

$$
\mathbf{A}\boldsymbol{\phi}_1 = \lambda_1 \boldsymbol{\phi}_1, \quad \mathbf{A}\boldsymbol{\phi}_2 = \lambda_2 \boldsymbol{\phi}_2 \tag{2.188}
$$

Define the diagonal matrix $\boldsymbol{\Lambda}$ to be the *eigenvalue matrix* with the eigenvalues of $\mathbf{A}$ on the diagonal, and define the matrix $\mathbf{T}$ to be the *eigenvector matrix* with columns the representing the eigenvectors of $\mathbf{A}$, that is,

$$
\boldsymbol{\Lambda} = \begin{bmatrix} \lambda_1 & 0 \\ 0 & \lambda_2 \end{bmatrix}, \quad \mathbf{T} = \begin{bmatrix} \boldsymbol{\phi}_1 & \boldsymbol{\phi}_2 \end{bmatrix} \tag{2.189}
$$

It now follows from Eq. 2.188 that

$$
\mathbf{AT} = \mathbf{A}\begin{bmatrix} \boldsymbol{\phi}_1 & \boldsymbol{\phi}_2 \end{bmatrix} = \begin{bmatrix} \boldsymbol{\phi}_1 & \boldsymbol{\phi}_2 \end{bmatrix}\begin{bmatrix} \lambda_1 & 0 \\ 0 & \lambda_2 \end{bmatrix} = \mathbf{T\Lambda} \tag{2.190}
$$

and when Eq. 2.190 is postmultiplied by $\mathbf{T}^{-1}$ that

$$
\mathbf{A} = \mathbf{T\Lambda T}^{-1}. \tag{2.191}
$$

Equation 2.183 can now be written as

$$\mathbf{e}^{\mathbf{A}t} = \mathbf{I} + \mathbf{A}t + \frac{1}{2!}(\mathbf{A}t)^2 + \frac{1}{3!}(\mathbf{A}t)^3 + \dots$$

$$= \mathbf{T}\mathbf{T}^{-1} + (\mathbf{T}\mathbf{\Lambda}\mathbf{T}^{-1})t + \frac{1}{2!}(\mathbf{T}\mathbf{\Lambda}\mathbf{T}^{-1})(\mathbf{T}\mathbf{\Lambda}\mathbf{T}^{-1})t^2 + \frac{1}{3!}(\mathbf{T}\mathbf{\Lambda}\mathbf{T}^{-1})(\mathbf{T}\mathbf{\Lambda}\mathbf{T}^{-1})(\mathbf{T}\mathbf{\Lambda}\mathbf{T}^{-1})t^3 + \dots$$

$$= \mathbf{T}\mathbf{I}\mathbf{T}^{-1} + \mathbf{T}(\mathbf{\Lambda}t)\mathbf{T}^{-1} + \frac{1}{2!}\mathbf{T}(\mathbf{\Lambda}t)^2\mathbf{T}^{-1} + \frac{1}{3!}\mathbf{T}(\mathbf{\Lambda}t)^3\mathbf{T}^{-1} + \dots \qquad (2.192)$$

$$= \mathbf{T}\left(\mathbf{I} + \mathbf{\Lambda}t + \frac{1}{2!}(\mathbf{\Lambda}t)^2 + \frac{1}{3!}(\mathbf{\Lambda}t)^3 + \dots\right)\mathbf{T}^{-1}$$

$$= \mathbf{T}\mathbf{e}^{\mathbf{\Lambda}t}\mathbf{T}^{-1}$$

The state transition matrix for an SDOF system, $\mathbf{e}^{\mathbf{A}t}$, is calculated first by finding all the eigenvalues and eigenvectors of $\mathbf{A}$. This is done by setting the determinant of the matrix $(\mathbf{A} - \lambda\mathbf{I})$ equal to zero, that is,

$$\det(\mathbf{A} - \lambda\mathbf{I}) = \begin{vmatrix} -\lambda & 1 \\ -\omega_n^2 & -2\zeta\omega_n - \lambda \end{vmatrix} = \lambda^2 + 2\zeta\omega_n\lambda + \omega_n^2 = 0. \qquad (2.193)$$

It follows from Eq. 2.193 that

$$\lambda_1 = -\zeta\omega_n + i\omega_d, \quad \lambda_2 = -\zeta\omega_n - i\omega_d. \qquad (2.194)$$

Substituting λ_1 and λ_2 into Eq. 2.188 and solving for $\boldsymbol{\phi}_1$ and $\boldsymbol{\phi}_2$ gives

$$\boldsymbol{\phi}_1 = \left\{\begin{matrix} 1 \\ -\zeta\omega_n + i\omega_d \end{matrix}\right\}, \quad \boldsymbol{\phi}_2 = \left\{\begin{matrix} 1 \\ -\zeta\omega_n - i\omega_d \end{matrix}\right\}. \qquad (2.195)$$

It follows that

$$\mathbf{\Lambda} = \begin{bmatrix} -\zeta\omega_n + i\omega_d & 0 \\ 0 & -\zeta\omega_n - i\omega_d \end{bmatrix}, \quad \mathbf{T} = \begin{bmatrix} 1 & 1 \\ -\zeta\omega_n + i\omega_d & -\zeta\omega_n - i\omega_d \end{bmatrix}. \qquad (2.196)$$

Inverting the matrix $\mathbf{T}$ gives

$$\mathbf{T}^{-1} = \frac{-1}{2i\omega_d}\begin{bmatrix} -\zeta\omega_n - i\omega_d & -1 \\ \zeta\omega_n - i\omega_d & 1 \end{bmatrix}. \qquad (2.197)$$

The state transition matrix is given by Eq. 2.192 and is

$$\mathbf{e}^{\mathbf{A}t} = \frac{-1}{2i\omega_d}\begin{bmatrix} 1 & 1 \\ -\zeta\omega_n + i\omega_d & -\zeta\omega_n - i\omega_d \end{bmatrix}\begin{bmatrix} e^{-\zeta\omega_n t}e^{i\omega_d t} & 0 \\ 0 & e^{-\zeta\omega_n t}e^{-i\omega_d t} \end{bmatrix}\begin{bmatrix} -\zeta\omega_n - i\omega_d & -1 \\ \zeta\omega_n - i\omega_d & 1 \end{bmatrix}. \qquad (2.198)$$

After the matrix triple product is performed, Eq. 2.198 becomes

$$\mathbf{e}^{\mathbf{A}t} = e^{-\zeta\omega_n t}\begin{bmatrix} \cos\omega_d t + \left(\dfrac{\zeta}{\sqrt{1-\zeta^2}}\right)\sin\omega_d t & \dfrac{\sin\omega_d t}{\omega_n\sqrt{1-\zeta^2}} \\ -\dfrac{\omega_n\sin\omega_d t}{\sqrt{1-\zeta^2}} & \cos\omega_d t - \left(\dfrac{\zeta}{\sqrt{1-\zeta^2}}\right)\sin\omega_d t \end{bmatrix}. \qquad (2.199)$$

If Eq. 2.199 is substituted into Eq. 2.184, noting that

$$\mathbf{e}^{\mathbf{A}(t-s)} = e^{-\zeta\omega_n(t-s)}\begin{bmatrix} \cos\omega_d(t-s) + \dfrac{\zeta\sin\omega_d(t-s)}{\sqrt{1-\zeta^2}} & \dfrac{\sin\omega_d(t-s)}{\omega_n\sqrt{1-\zeta^2}} \\ -\dfrac{\omega_n\sin\omega_d(t-s)}{\sqrt{1-\zeta^2}} & \cos\omega_d(t-s) - \dfrac{\zeta\sin\omega_d(t-s)}{\sqrt{1-\zeta^2}} \end{bmatrix}$$

$$\mathbf{F}(s) = \begin{Bmatrix} 0 \\ F_e(s)/m \end{Bmatrix}, \quad e^{\mathbf{A}(t-s)}\mathbf{F}(s) = \left(\frac{F_e(s)}{m}\right)e^{-\zeta\omega_n(t-s)} \begin{Bmatrix} \dfrac{\sin\omega_d(t-s)}{\omega_n\sqrt{1-\zeta^2}} \\ \cos\omega_d(t-s) - \dfrac{\zeta\sin\omega_d(t-s)}{\sqrt{1-\zeta^2}} \end{Bmatrix}$$

it follows that

$$\begin{Bmatrix} x(t) \\ \dot{x}(t) \end{Bmatrix} = e^{-\zeta\omega_n t} \begin{bmatrix} \cos\omega_d t + \left(\dfrac{\zeta}{\sqrt{1-\zeta^2}}\right)\sin\omega_d t & \dfrac{\sin\omega_d t}{\omega_n\sqrt{1-\zeta^2}} \\ -\dfrac{\omega_n\sin\omega_d t}{\sqrt{1-\zeta^2}} & \cos\omega_d t - \left(\dfrac{\zeta}{\sqrt{1-\zeta^2}}\right)\sin\omega_d t \end{bmatrix} \begin{Bmatrix} x_o \\ \dot{x}_o \end{Bmatrix}$$

$$+ \int_0^t \left(\frac{F_e(s)}{m}\right)e^{-\zeta\omega_n(t-s)} \begin{bmatrix} \dfrac{\sin\omega_d(t-s)}{\omega_n\sqrt{1-\zeta^2}} \\ \cos\omega_d(t-s) - \left(\dfrac{\zeta}{\sqrt{1-\zeta^2}}\right)\sin\omega_d(t-s) \end{bmatrix} ds. \quad (2.200)$$

Equation 2.200 represents the response of the SDOF structure in state space form. Expanding Eq. 2.200 gives

$$x(t) = x_o e^{-\zeta\omega_n t}\cos\omega_d t + \left(\frac{\zeta\omega_n x_o + \dot{x}_o}{\omega_n\sqrt{1-\zeta^2}}\right)e^{-\zeta\omega_n t}\sin\omega_d t + \int_0^t \left(\frac{F_e(s)}{m\omega_d}\right)e^{-\zeta\omega_n(t-s)}\sin\omega_d(t-s)\,ds \quad (2.201\text{a})$$

$$\dot{x}(t) = \dot{x}_o e^{-\zeta\omega_n t}\cos\omega_d t - \left(\frac{\omega_n x_o + \zeta\dot{x}_o}{\sqrt{1-\zeta^2}}\right)e^{-\zeta\omega_n t}\sin\omega_d t$$

$$\hspace{5cm} (2.202\text{b})$$

$$+ \int_0^t \left(\frac{F_e(s)}{m}\right)e^{-\zeta\omega_n(t-s)} \left[\cos\omega_d(t-s) - \left(\frac{\zeta}{\sqrt{1-\zeta^2}}\right)\sin\omega_d(t-s)\right]ds,$$

which is the same as Eq. 2.101.

EXAMPLE 1 **Free Vibration**

The free vibration of motion is given in Eq. 2.181 with the forcing function equal to zero. Therefore,

$$\dot{z}(t) = \mathbf{A}z(t). \quad (2.202)$$

The solution for the response is given by Eq. 2.182 and is

$$z(t) = e^{\mathbf{A}(t-t_o)}z(t_o). \quad (2.203)$$

The state transition matrix for an SDOF system is given in Eq. 2.199. Therefore, substituting Eq. 2.199 into Eq. 2.203, it follows for $t_o = 0$ that

$$\begin{Bmatrix} x(t) \\ \dot{x}(t) \end{Bmatrix} = e^{-\zeta\omega_n t} \begin{bmatrix} \cos\omega_d t + \left(\dfrac{\zeta}{\sqrt{1-\zeta^2}}\right)\sin\omega_d t & \dfrac{\sin\omega_d t}{\omega_n\sqrt{1-\zeta^2}} \\ -\dfrac{\omega_n\sin\omega_d t}{\sqrt{1-\zeta^2}} & \cos\omega_d t - \left(\dfrac{\zeta}{\sqrt{1-\zeta^2}}\right)\sin\omega_d t \end{bmatrix} \begin{Bmatrix} x_o \\ \dot{x}_o \end{Bmatrix}. \quad (2.204)$$

EXAMPLE 2 *Forcing Function That Rises to Static*

Example 2 in Section 2.3 and Example 3 in Section 2.4 presented the solution to this problem. The state space solution is now presented. Recall that for $t \le t_o$ and for zero initial displacement and velocity, it follows that

$$F_e(t) = F_o(t/t_o).$$ (2.205)

Equation 2.184 becomes

$$\mathbf{z}(t) = \begin{Bmatrix} x(t) \\ \dot{x}(t) \end{Bmatrix} = \int_0^t \left(\frac{F_o}{mt_o}\right) se^{-\zeta\omega_n(t-s)} \begin{Bmatrix} \dfrac{\sin\omega_d(t-s)}{\omega_n\sqrt{1-\zeta^2}} \\ \cos\omega_d(t-s) - \left(\dfrac{\zeta}{\sqrt{1-\zeta^2}}\right)\sin\omega_d(t-s) \end{Bmatrix} ds.$$ (2.206)

Finally,

$$\begin{Bmatrix} x(t) \\ \dot{x}(t) \end{Bmatrix} = \begin{Bmatrix} \int_0^t \left(\dfrac{F_o}{m\omega_d t_o}\right) se^{-\zeta\omega_n(t-s)} \sin\omega_d(t-s)\,ds \\ \int_0^t \left(\dfrac{F_o}{mt_o}\right) se^{-\zeta\omega_n(t-s)}\left[\cos\omega_d(t-s) - \left(\dfrac{\zeta}{\sqrt{1-\zeta^2}}\right)\sin\omega_d(t-s)\right] ds \end{Bmatrix}.$$ (2.207)

Note that Eq. 2.207 is the same as given in Eqs. 2.114 and 2.115 and therefore the solution is given in Eqs. 2.118 and 2.121:

$$\begin{Bmatrix} x(t) \\ \dot{x}(t) \end{Bmatrix} = \begin{Bmatrix} \left(\dfrac{2\zeta F_o}{k\omega_n t_o}\right)e^{-\zeta\omega_n t}\cos\omega_d t + \left[\dfrac{F_o(2\zeta^2-1)}{k\omega_n t_o\sqrt{1-\zeta^2}}\right]e^{-\zeta\omega_n t}\sin\omega_d t + \left(\dfrac{F_o}{kt_o}\right)t - \left(\dfrac{2\zeta F_o}{k\omega_n t_o}\right) \\ \left(\dfrac{F_o}{kt_o}\right)e^{-\zeta\omega_n t}\cos\omega_d t - \left(\dfrac{\zeta}{\sqrt{1-\zeta^2}}\dfrac{F_o}{kt_o}\right)e^{-\zeta\omega_n t}\sin\omega_d t + \left(\dfrac{F_o}{kt_o}\right) \end{Bmatrix}.$$ (2.208)

EXAMPLE 3 *Bilinear Earthquake Ground Motion Pulse*

Consider again the earthquake ground motion that is a triangular (bilinear) pulse as shown in Figure 2.15. Similar to what is done in Example 7 of Section 2.4, let the ground motion be

$$a(t) = \begin{cases} a_1(t) & 0 \le t \le t_p \\ a_1(t) + a_2(t) & t_p < t \le 2t_p \\ a_1(t) + a_2(t) + a_3(t) & t > 2t_p \end{cases}$$ (2.209)

where

$$a_1(t) = \left(\frac{a_p}{t_p}\right)t, \quad a_2(t) = -2\left(\frac{a_p}{t_p}\right)(t-t_p), \quad a_3(t) = \left(\frac{a_p}{t_p}\right)(t-2t_p).$$ (2.210)

Then it follows from Eq. 2.208 that the response for the time domain $t = 0$ to $t = t_p$ with zero damping and zero initial conditions is

$$\begin{Bmatrix} x(t) \\ \dot{x}(t) \end{Bmatrix} = \left\{ \begin{array}{l} \left[\left(\dfrac{a_p}{\omega_n^2} \right) \left[\left(\dfrac{\sin \omega_n t}{\omega_n t_p} \right) - \left(\dfrac{t}{t_p} \right) \right] \right] \\[20pt] \left[\left(\dfrac{a_p}{\omega_n^2} \right) \left[\left(\dfrac{\cos \omega_n t}{t_p} \right) - \left(\dfrac{1}{t_p} \right) \right] \right] \end{array} \right\}, \quad 0 \le t \le t_p \tag{2.211}$$

where F_o is replaced by $-ma_p$. For the time domain $t = t_p$ to $t = 2t_p$, it follows that

$$\begin{Bmatrix} x(t) \\ \dot{x}(t) \end{Bmatrix} = \left\{ \begin{array}{l} \left[\left(\dfrac{a_p}{\omega_n^2} \right) \left[-2 + \left(\dfrac{t}{t_p} \right) + \left(\dfrac{\sin \omega_n t}{\omega_n t_p} \right) - \left(\dfrac{2\sin \omega_n (t - t_p)}{\omega_n t_p} \right) \right] \right] \\[20pt] \left[\left(\dfrac{a_p}{\omega_n^2} \right) \left[\left(\dfrac{1}{t_p} \right) + \left(\dfrac{\cos \omega_n t}{t_p} \right) - \left(\dfrac{2\cos \omega_n (t - t_p)}{t_p} \right) \right] \right] \end{array} \right\}, \quad t_p < t \le 2t_p. \tag{2.212}$$

Similarly, for the time domain $t > 2t_p$,

$$\begin{Bmatrix} x(t) \\ \dot{x}(t) \end{Bmatrix} = \left\{ \begin{array}{l} \left[\left(\dfrac{a_p}{\omega_n^2} \right) \left[\left(\dfrac{\sin \omega_n t}{\omega_n t_p} \right) - \left(\dfrac{2\sin \omega_n (t - t_p)}{\omega_n t_p} \right) + \left(\dfrac{\sin \omega_n (t - 2t_p)}{\omega_n t_p} \right) \right] \right] \\[20pt] \left[\left(\dfrac{a_p}{\omega_n^2} \right) \left[\left(\dfrac{\cos \omega_n t}{t_p} \right) - \left(\dfrac{2\cos \omega_n (t - t_p)}{t_p} \right) + \left(\dfrac{\cos \omega_n (t - 2t_p)}{t_p} \right) \right] \right] \end{array} \right\}, \quad t > 2t_p. \tag{2.213}$$

2.6 NUMERICAL SOLUTION FOR RESPONSE USING NEWMARK β-METHOD

Numerical methods are used in the computer age to perform most structural response analyses. To perform a numerical analysis, it is necessary to discretize the response and the forcing function. The forcing function is usually provided to the structural engineer at discrete instants in time. Therefore, the major concern in performing a numerical analysis is the discretization of the system response.

Given all the response and force information at time t_k, the three response quantities that are desired at time t_{k+1} are the displacement, velocity, and acceleration of the mass. The dynamic equilibrium equation given in Eq. 2.2 is one of the three simultaneous equations necessary to compute the response at time t_{k+1}. This equation is written here and also in normalized form:

$$m\ddot{x}(t) + c\dot{x}(t) + kx(t) = F_e(t) \tag{2.214a}$$

$$\ddot{x}(t) + 2\zeta\omega_n\dot{x}(t) + \omega_n^2 x(t) = F_e(t)/m. \tag{2.214b}$$

The two additional equations are

$$\dot{x}(t) = \frac{dx(t)}{dt}, \quad \ddot{x}(t) = \frac{d\dot{x}(t)}{dt} = \frac{d^2 x(t)}{dt^2}. \tag{2.215}$$

When accuracy is involved, discretization is generally not done to the differential equation given by Eq. 2.215, but instead discretization is performed to an integral equation. The integral equation solution is preferred because high accuracy can be achieved in calculating the area under a curve. Therefore, Eq. 2.215 is rewritten as

$$\dot{x}(t) = \dot{x}(t_o) + \int_{t_o}^{t} \ddot{x}(s)ds, \quad x(t) = x(t_o) + \int_{t_o}^{t} \dot{x}(s)ds. \tag{2.216}$$

Let

$$t_{k+1} = t, \quad t_k = t_o, \quad \Delta t = t_{k+1} - t_k.$$

It then follows from Eq. 2.216 that

$$\dot{x}_{k+1} = \dot{x}_k + \int_{t_k}^{t_{k+1}} \ddot{x}(s)ds, \quad x_{k+1} = x_k + \int_{t_k}^{t_{k+1}} \dot{x}(s)ds. \tag{2.217}$$

To discretize the integrand in Eq. 2.217, it is necessary to approximate the continuous acceleration function, $\ddot{x}(s)$. There are many ways of discretizing a continuous function, and the three ways that are most commonly used by structural engineers are constant acceleration, constant average acceleration, and linear acceleration.

Constant acceleration assumes that the acceleration is constant within the time interval t_k to t_{k+1}. The magnitude of the acceleration in the interval is equal to the magnitude at the beginning of the time interval, that is,

$$\ddot{x}(t) = \ddot{x}(t_k) = \ddot{x}_k, \quad t_k \le t < t_{k+1}. \tag{2.218}$$

In this case, the velocity and displacement are obtained from Eq. 2.217 and are

$$\dot{x}_{k+1} = \dot{x}_k + \int_{t_k}^{t_{k+1}} \ddot{x}_k dt = \dot{x}_k + \ddot{x}_k \Delta t \tag{2.219a}$$

$$x_{k+1} = x_k + \int_{t_k}^{t_{k+1}} \left[\dot{x}_k + \ddot{x}_k (t - t_k) \right] dt = x_k + \dot{x}_k \Delta t + \tfrac{1}{2} \ddot{x}_k (\Delta t)^2. \tag{2.219b}$$

It is very important to note that the right side of Eq. 2.219 contains information only at time step k.

Constant average acceleration assumes that the acceleration is constant within each time interval, and the magnitude of the acceleration is equal to the average of the accelerations at the beginning and at the end of the time interval, that is,

$$\ddot{x}(t) = \tfrac{1}{2}\left(\ddot{x}(t_k) + \ddot{x}(t_{k+1}) \right) = \tfrac{1}{2}\left(\ddot{x}_k + \ddot{x}_{k+1} \right), \quad t_k \le t < t_{k+1}. \tag{2.220}$$

In this case, the velocity and displacement are

$$\dot{x}_{k+1} = \dot{x}_k + \int_{t_k}^{t_{k+1}} \tfrac{1}{2}\left(\ddot{x}_k + \ddot{x}_{k+1} \right) dt = \dot{x}_k + \tfrac{1}{2}\ddot{x}_k \Delta t + \tfrac{1}{2}\ddot{x}_{k+1}\Delta t \tag{2.221a}$$

$$x_{k+1} = x_k + \int_{t_k}^{t_{k+1}} \left[\dot{x}_k + \tfrac{1}{2}\left(\ddot{x}_k + \ddot{x}_{k+1} \right)(t - t_k) \right] dt = x_k + \dot{x}_k \Delta t + \tfrac{1}{4}\ddot{x}_k(\Delta t)^2 + \tfrac{1}{4}\ddot{x}_{k+1}(\Delta t)^2. \tag{2.221b}$$

With *linear acceleration*, the acceleration is assumed to vary within each time interval in a linear manner from the beginning of the time step to the end of the time step, that is,

$$\ddot{x}(t) = \ddot{x}(t_k) + \left[\frac{\ddot{x}(t_{k+1}) - \ddot{x}(t_k)}{t_{k+1} - t_k} \right](t - t_k) = \ddot{x}_k + \left[\frac{\ddot{x}_{k+1} - \ddot{x}_k}{t_{k+1} - t_k} \right](t - t_k), \quad t_k \le t < t_{k+1}. \tag{2.222}$$

In this case, the velocity and displacement are

$$\dot{x}_{k+1} = \dot{x}_k + \int_{t_k}^{t_{k+1}} \left[\ddot{x}_k + \left(\frac{\ddot{x}_{k+1} - \ddot{x}_k}{t_{k+1} - t_k} \right)(t - t_k) \right] dt = \dot{x}_k + \tfrac{1}{2}\ddot{x}_k \Delta t + \tfrac{1}{2}\ddot{x}_{k+1}\Delta t \tag{2.223a}$$

$$x_{k+1} = x_k + \int_{t_k}^{t_{k+1}} \left[\dot{x}_k + \ddot{x}_k(t - t_k) + \tfrac{1}{2}\left(\frac{\ddot{x}_{k+1} - \ddot{x}_k}{t_{k+1} - t_k} \right)(t - t_k)^2 \right] dt$$

$$= x_k + \dot{x}_k \Delta t + \tfrac{1}{3}\ddot{x}_k(\Delta t)^2 + \tfrac{1}{6}\ddot{x}_{k+1}(\Delta t)^2. \tag{2.223b}$$

The most common numerical integration method used by structural engineers is called the *Newmark β-method*. This method is a generalization of the three cases discussed previously, where the set of three simultaneous equations takes the form

$$\ddot{x}_{k+1} + 2\zeta\omega_n \dot{x}_{k+1} + \omega_n^2 x_{k+1} = F_{k+1}/m \tag{2.224a}$$

$$\dot{x}_{k+1} = \dot{x}_k + (1 - \delta)\Delta t \ddot{x}_k + \delta \Delta t \ddot{x}_{k+1} \tag{2.224b}$$

$$x_{k+1} = x_k + \dot{x}_k \Delta t + \left(\tfrac{1}{2} - \alpha \right)(\Delta t)^2 \ddot{x}_k + \alpha(\Delta t)^2 \ddot{x}_{k+1} \tag{2.224c}$$

where F_{k+1} is the digitized value of the continuous function $F_e(t)$. Note the similarities between Eqs. 2.224b and 2.224c and the three cases, that is,

1. constant acceleration: $\delta = 0$ and $\alpha = 0$.
2. constant average acceleration: $\delta = \frac{1}{2}$ and $\alpha = \frac{1}{4}$.
3. linear acceleration: $\delta = \frac{1}{2}$ and $\alpha = \frac{1}{6}$.

Equation 2.224 is solved by substituting Eqs. 2.224b and 2.224c into Eq. 2.224a. Doing so gives

$$\ddot{x}_{k+1} + 2\zeta\omega_n\left[\dot{x}_k + (1-\delta)\Delta t\ddot{x}_k + \delta\Delta t\ddot{x}_{k+1}\right]$$
$$+\omega_n^2\left[x_k + \dot{x}_k\Delta t + (\tfrac{1}{2} - \alpha)(\Delta t)^2\ddot{x}_k + \alpha(\Delta t)^2\ddot{x}_{k+1}\right] = F_{k+1}/m. \tag{2.225}$$

Rearranging the terms in Eq. 2.225 gives

$$\left(1 + 2\zeta\omega_n\delta\Delta t + \omega_n^2\alpha(\Delta t)^2\right)\ddot{x}_{k+1} = -\omega_n^2 x_k - \left[2\zeta\omega_n + \omega_n^2\Delta t\right]\dot{x}_k$$
$$-\left[2\zeta\omega_n(1-\delta)\Delta t + \omega_n^2(\tfrac{1}{2}-\alpha)(\Delta t)^2\right]\ddot{x}_k + F_{k+1}/m. \tag{2.226}$$

Solving for $\ddot{x}_{k+1}$ gives

$$\ddot{x}_{k+1} = -\left(\frac{\omega_n^2}{\beta}\right)x_k - \left(\frac{2\zeta\omega_n + \omega_n^2\Delta t}{\beta}\right)\dot{x}_k - \left(\frac{\gamma}{\beta}\right)\ddot{x}_k + \left(\frac{1}{m\beta}\right)F_{k+1} \tag{2.227}$$

where

$$\beta = 1 + 2\zeta\omega_n\delta\Delta t + \omega_n^2\alpha(\Delta t)^2$$
$$\gamma = 2\zeta\omega_n(1-\delta)\Delta t + \omega_n^2(\tfrac{1}{2}-\alpha)(\Delta t)^2. \tag{2.228}$$

Substituting Eq. 2.227 into Eqs. 2.224b and 2.224c gives

$$\dot{x}_{k+1} = -\left(\frac{\omega_n^2\delta\Delta t}{\beta}\right)x_k + \left(\frac{\beta - 2\zeta\omega_n\delta\Delta t - \omega_n^2\delta(\Delta t)^2}{\beta}\right)\dot{x}_k + \left(\frac{\beta\Delta t - \delta(\beta+\gamma)\Delta t}{\beta}\right)\ddot{x}_k + \left(\frac{\delta\Delta t}{m\beta}\right)F_{k+1} \tag{2.229a}$$

$$x_{k+1} = \left(\frac{\beta - \omega_n^2\alpha(\Delta t)^2}{\beta}\right)x_k + \left(\frac{\beta\Delta t - 2\zeta\omega_n\alpha(\Delta t)^2 - \omega_n^2\alpha(\Delta t)^3}{\beta}\right)\dot{x}_k$$
$$+ \left(\frac{\tfrac{1}{2}\beta(\Delta t)^2 - \alpha(\beta+\gamma)(\Delta t)^2}{\beta}\right)\ddot{x}_k + \left(\frac{\alpha(\Delta t)^2}{m\beta}\right)F_{k+1}. \tag{2.229b}$$

In summary, the procedure for computing the response at time step $k + 1$ given all of the information at time step k and given F_{k+1} is obtained using Eqs. 2.227, 2.229a, and 2.229b. These equations can be represented in matrix form:

$$\begin{Bmatrix} x_{k+1} \\ \dot{x}_{k+1} \\ \ddot{x}_{k+1} \end{Bmatrix} = \mathbf{F}_N \begin{Bmatrix} x_k \\ \dot{x}_k \\ \ddot{x}_k \end{Bmatrix} + \mathbf{H}_N F_{k+1} \tag{2.230}$$

where

$$\mathbf{F}_N = \frac{1}{\beta}\begin{bmatrix} \beta - \omega_n^2\alpha(\Delta t)^2 & \beta\Delta t - 2\zeta\omega_n\alpha(\Delta t)^2 - \omega_n^2\alpha(\Delta t)^3 & \tfrac{1}{2}\beta(\Delta t)^2 - \alpha(\beta+\gamma)(\Delta t)^2 \\ -\omega_n^2\delta\Delta t & \beta - 2\zeta\omega_n\delta\Delta t - \omega_n^2\delta(\Delta t)^2 & \beta\Delta t - \delta(\beta+\gamma)\Delta t \\ -\omega_n^2 & -2\zeta\omega_n - \omega_n^2\Delta t & -\gamma \end{bmatrix}$$

$$\mathbf{H}_N = \left(\frac{1}{m\beta}\right)\begin{Bmatrix} \alpha(\Delta t)^2 \\ \delta\Delta t \\ 1 \end{Bmatrix}$$

and the subscript N denotes the Newmark β-method. Simplifying Eq. 2.230, let

$$\mathbf{q}_k = \begin{Bmatrix} x_k \\ \dot{x}_k \\ \ddot{x}_k \end{Bmatrix}. \tag{2.231}$$

It follows that

$$\mathbf{q}_{k+1} = \mathbf{F}_N \mathbf{q}_k + \mathbf{H}_N F_{k+1}. \tag{2.232}$$

Note that for a linear system with a constant time step, Δt, and for selected values of α and δ, the matrices $\mathbf{F}_N$ and $\mathbf{H}_N$ need to be calculated only once.

One major concern that must be addressed when performing a numerical analysis is *numerical stability*. In the Newmark β-method, the constant average acceleration (i.e., $\delta = \frac{1}{2}$ and $\alpha = \frac{1}{4}$) is generally used in structural dynamics because it has been shown to have a high degree of numerical stability.

Another concern in performing numerical analysis is *numerical accuracy*. Reducing the integration time step will increase the accuracy of the calculation. Research has shown that a time step size of $\Delta t \le (T_n/10)$ gives very accurate result. For example, for an SDOF system with an undamped natural period of vibration of 0.5 s, a time step equal to or less than 0.05 s should be used in the numerical analysis. In earthquake analysis, the earthquake forcing function is usually discretized with a time step of 0.01 or 0.02 s, and because of the concern of numerical accuracy a time step of 0.01 s is usually selected by structural engineers when they perform numerical analyses.

EXAMPLE 1   *Constant Acceleration Solution Method for Rise to Static Forcing Function*

Consider Example 2 in Section 2.3, Example 3 in Section 2.4, and Example 2 in Section 2.5, with a forcing function that rises to static. Now select the following specific values for the structural variables:

$$m = 1.0 \text{ k-s}^2/\text{in.}, \quad \zeta = 0.05, \quad T_n = 1.0 \text{ s}$$

$$F_o = 10 \text{ kip}, \quad t_o = 1.0 \text{ s}.$$

Assume $\Delta t = 0.02$ s. The response of the system is now calculated using the constant acceleration method. The values of α and δ are zero for the constant acceleration method. Therefore, from Eq. 2.228, it follows that

$$\beta = 1 \tag{2.233}$$

$$\gamma = 2\zeta\omega_n \Delta t + \tfrac{1}{2}\omega_n^2(\Delta t)^2 = 0.020462. \tag{2.234}$$

Equation 2.230 becomes

$$\begin{Bmatrix} x_{k+1} \\ \dot{x}_{k+1} \\ \ddot{x}_{k+1} \end{Bmatrix} = \frac{1}{\beta}\begin{bmatrix} \beta & \beta\Delta t & \tfrac{1}{2}\beta(\Delta t)^2 \\ 0 & \beta & \beta\Delta t \\ -\omega_n^2 & -2\zeta\omega_n - \omega_n^2\Delta t & -\gamma \end{bmatrix}\begin{Bmatrix} x_k \\ \dot{x}_k \\ \ddot{x}_k \end{Bmatrix} + \begin{pmatrix} \dfrac{1}{m\beta} \end{pmatrix}\begin{Bmatrix} 0 \\ 0 \\ 1 \end{Bmatrix} F_{k+1}. \tag{2.235}$$

When the values for β and γ calculated in Eq. 2.234 are substituted into Eq. 2.235, it follows that

$$\begin{Bmatrix} x_{k+1} \\ \dot{x}_{k+1} \\ \ddot{x}_{k+1} \end{Bmatrix} = \begin{bmatrix} 1 & 0.02 & 0.0002 \\ 0 & 1 & 0.02 \\ -39.4784 & -1.4179 & 0.020462 \end{bmatrix}\begin{Bmatrix} x_k \\ \dot{x}_k \\ \ddot{x}_k \end{Bmatrix} + \begin{Bmatrix} 0 \\ 0 \\ 1 \end{Bmatrix} F_{k+1}. \tag{2.236}$$

Assume that the initial conditions are zero (i.e., $x_o = \dot{x}_o = 0$), and since the forcing function at time zero is equal to zero (i.e., $F_o = 0$), it follows that

$$\ddot{x}_o = \left(F_o/m\right) - 2\zeta\omega_n\dot{x}_o - \omega_n^2 x_o = 0. \tag{2.237}$$

Then the solution for the first time step (i.e., $t = \Delta t$), with the value of forcing function equal to 0.20 kip, is

$$\begin{Bmatrix} x_1 \\ \dot{x}_1 \\ \ddot{x}_1 \end{Bmatrix} = \begin{bmatrix} 1 & 0.02 & 0.0002 \\ 0 & 1 & 0.02 \\ -39.4784 & -1.4179 & 0.020462 \end{bmatrix} \begin{Bmatrix} 0 \\ 0 \\ 0 \end{Bmatrix} + \begin{Bmatrix} 0 \\ 0 \\ 1 \end{Bmatrix}(0.20) = \begin{Bmatrix} 0 \\ 0 \\ 0.20 \end{Bmatrix}. \tag{2.238}$$

Then it follows that the solution for the second time step (i.e., $t = 2\Delta t$) using Eq. 2.236 with the results from Eq. 2.238 and the value of forcing function equal to 0.40 kip is

$$\begin{Bmatrix} x_2 \\ \dot{x}_2 \\ \ddot{x}_2 \end{Bmatrix} = \begin{bmatrix} 1 & 0.02 & 0.0002 \\ 0 & 1 & 0.02 \\ -39.4784 & -1.4179 & 0.020462 \end{bmatrix} \begin{Bmatrix} 0 \\ 0 \\ 0.20 \end{Bmatrix} + \begin{Bmatrix} 0 \\ 0 \\ 1 \end{Bmatrix}(0.40) = \begin{Bmatrix} 0.00006 \\ 0.00400 \\ 0.40409 \end{Bmatrix}. \tag{2.239}$$

EXAMPLE 2 *Constant Average Acceleration Solution for Rise to Static Forcing Function*

The constant average acceleration method is a special case of the Newmark β-method with $\delta = 1/2$ and $\alpha = 1/4$. Example 1 can now be repeated with this solution method using Eq. 2.230. First, the values of β and γ are obtained from Eq. 2.228:

$$\beta = 1 + 2(0.05)(2\pi)(0.5)(0.02) + (2\pi)^2(0.25)(0.02)^2 = 1.010231 \tag{2.240}$$

$$\gamma = 2(0.05)(2\pi)(0.5)(0.02) + (2\pi)^2(0.25)(0.02)^2 = 0.010231. \tag{2.241}$$

Then it follows that

$$\begin{Bmatrix} x_{k+1} \\ \dot{x}_{k+1} \\ \ddot{x}_{k+1} \end{Bmatrix} = \begin{bmatrix} 0.99609 & 0.01986 & 0.000099 \\ -0.39079 & 0.98596 & 0.00990 \\ -39.0786 & -1.40353 & -0.01013 \end{bmatrix} \begin{Bmatrix} x_k \\ \dot{x}_k \\ \ddot{x}_k \end{Bmatrix} + \begin{Bmatrix} 0.000099 \\ 0.009899 \\ 0.98987 \end{Bmatrix} F_{k+1}. \tag{2.242}$$

Again, assume that the initial conditions are zero (i.e., $x_o = \dot{x}_o = 0$), and since the forcing function at time zero is equal to zero (i.e., $F_o = 0$), it follows that

$$\ddot{x}_o = \left(F_o/m\right) - 2\zeta\omega_n\dot{x}_o - \omega_n^2 x_o = 0. \tag{2.243}$$

The solution for the first time step (i.e., $t = \Delta t$), with the value of forcing function equal to 0.20 kip, is

$$\begin{Bmatrix} x_1 \\ \dot{x}_1 \\ \ddot{x}_1 \end{Bmatrix} = \begin{bmatrix} 0.99609 & 0.01986 & 0.000099 \\ -0.39079 & 0.98596 & 0.00990 \\ -39.0786 & -1.40353 & -0.01013 \end{bmatrix} \begin{Bmatrix} 0 \\ 0 \\ 0 \end{Bmatrix} + \begin{Bmatrix} 0.000099 \\ 0.009899 \\ 0.98987 \end{Bmatrix}(0.20) = \begin{Bmatrix} 0.00002 \\ 0.00198 \\ 0.1980 \end{Bmatrix}. \tag{2.244}$$

Then it follows that the solution for the second time step (i.e., $t = 2\Delta t$) using Eq. 2.242 with the results from Eq. 2.244 and the value of forcing function equal to 0.40 kip is

$$\begin{Bmatrix} x_2 \\ \dot{x}_2 \\ \ddot{x}_2 \end{Bmatrix} = \begin{bmatrix} 0.99609 & 0.01986 & 0.000099 \\ -0.39079 & 0.98596 & 0.00990 \\ -39.0786 & -1.40353 & -0.01013 \end{bmatrix} \begin{Bmatrix} 0.00002 \\ 0.00198 \\ 0.1980 \end{Bmatrix} + \begin{Bmatrix} 0.000099 \\ 0.009899 \\ 0.98987 \end{Bmatrix}(0.40) = \begin{Bmatrix} 0.00012 \\ 0.00786 \\ 0.39039 \end{Bmatrix}. \tag{2.245}$$

The calculation process continues, and Table 2.1 shows the results for the first six time steps.

Table 2.1 Constant Average Acceleration Calculation for First Six Time Steps

k	Time (s)	F_k (kip)	x_k (in.)	$\dot{x}_k$ (in./s)	$\ddot{x}_k$ (in./s²)
0	0.00	0.0	0.0	0.0	0.0
1	0.02	0.2	0.00002	0.00198	0.19797
2	0.04	0.4	0.00012	0.00798	0.39039
3	0.06	0.6	0.00037	0.01751	0.57431
4	0.08	0.8	0.00085	0.03072	0.74697
5	0.10	1.0	0.00163	0.04725	0.90580
6	0.12	1.2	0.00277	0.06679	1.04850

EXAMPLE 3   *Earthquake Response*

Consider the case where the earthquake ground motion excites the SDOF system. The forcing function is given by

$$F_k = -ma_k \tag{2.246}$$

where a_k is the ground motion. It follows from Example 4 in Section 2.4 that the dynamic equilibrium equation is

$$\ddot{x}_{k+1} + 2\zeta\omega_n\dot{x}_{k+1} + \omega_n^2 x_{k+1} = -a_{k+1}. \tag{2.247}$$

The result of the response for the earthquake ground motion using the Newmark β-method is

$$\mathbf{q}_{k+1} = \mathbf{F}_N\mathbf{q}_k + \mathbf{H}_N^{(EQ)}a_{k+1} \tag{2.248}$$

where

$$\mathbf{H}_N^{(EQ)} = -\left(\frac{1}{\beta}\right)\begin{Bmatrix} \alpha(\Delta t)^2 \\ \delta\Delta t \\ 1 \end{Bmatrix}. \tag{2.249}$$

EXAMPLE 4   *Time Step Size*

The value of β in Eq. 2.228 is a function of the time step size, Δt, as it relates to the period of vibration of the system. To illustrate this, consider

$$\omega_n = 2\pi/T_n$$

and therefore

$$\beta = 1 + 4\pi^2\delta\zeta\left(\frac{\Delta t}{T_n}\right) + 4\pi^2\alpha\left(\frac{\Delta t}{T_n}\right)^2. \tag{2.250}$$

Equation 2.250 shows how the ratio $(\Delta t/T_n)$ is the important variable and not just Δt; also note that the value of β is a function of the damping in the system.

EXAMPLE 5   *Constant Average Acceleration Solution for Earthquake Ground Motion*

Consider again Example 2 but with an earthquake ground motion that is a bilinear pulse as shown in Figure 2.16. Assume $\Delta t = 0.02$ s; it follows from Eq. 2.248 that

$$\mathbf{q}_{k+1} = \mathbf{F}_N\mathbf{q}_k + \mathbf{H}_N^{(EQ)}a_{k+1} \tag{2.251}$$

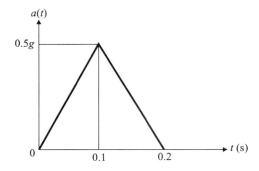

Figure 2.16 Bilinear Earthquake Ground Motion

where

$$
\mathbf{F}_N =
\begin{bmatrix}
0.99609 & 0.01986 & 0.000099 \\
-0.39079 & 0.98596 & 0.00990 \\
-39.0786 & -1.40353 & -0.01013
\end{bmatrix},
\quad
\mathbf{H}_N^{(EQ)} =
\begin{Bmatrix}
-0.000099 \\
-0.009899 \\
-0.98987
\end{Bmatrix}.
\tag{2.252}
$$

Assume that the initial conditions are zero (i.e., $x_o = \dot{x}_o = 0$), and since the forcing function at time zero is equal to zero (i.e., $F_o = 0$), it follows that

$$
\ddot{x}_o = \left(F_o/m\right) - 2\zeta\omega_n\dot{x}_o - \omega_n^2 x_o = 0.
\tag{2.253}
$$

The earthquake ground acceleration at the first time step (i.e., $t = \Delta t$) is

$$
a_1 = \left(\frac{0.5g}{0.1}\right)0.02 = 0.1g = 38.64 \text{ in./s}^2.
\tag{2.254}
$$

It follows from Eq. (2.251) that

$$
\begin{Bmatrix} x_1 \\ \dot{x}_1 \\ \ddot{x}_1 \end{Bmatrix} =
\begin{bmatrix}
0.99609 & 0.01986 & 0.000099 \\
-0.39079 & 0.98596 & 0.00990 \\
-39.0786 & -1.40353 & -0.01013
\end{bmatrix}
\begin{Bmatrix} 0 \\ 0 \\ 0 \end{Bmatrix}
+
\begin{Bmatrix} -0.000099 \\ -0.009899 \\ -0.98987 \end{Bmatrix}(38.64)
=
\begin{Bmatrix} -0.00382 \\ -0.38221 \\ -38.2211 \end{Bmatrix}.
\tag{2.255}
$$

Similarly, the earthquake ground acceleration at the second time step (i.e., $t = 2\Delta t$) is

$$
a_2 = \left(\frac{0.5g}{0.1}\right)0.04 = 0.2g = 77.28 \text{ in./s}^2.
\tag{2.256}
$$

It follows from Eq. (2.251) that

$$
\begin{Bmatrix} x_2 \\ \dot{x}_2 \\ \ddot{x}_2 \end{Bmatrix} =
\begin{bmatrix}
0.99609 & 0.01986 & 0.000099 \\
-0.39079 & 0.98596 & 0.00990 \\
-39.0786 & -1.40353 & -0.01013
\end{bmatrix}
\begin{Bmatrix} -0.00382 \\ -0.38221 \\ -38.2211 \end{Bmatrix}
+
\begin{Bmatrix} -0.000099 \\ -0.009899 \\ -0.98987 \end{Bmatrix}(77.28)
=
\begin{Bmatrix} -0.02283 \\ -1.51812 \\ -75.3693 \end{Bmatrix}.
$$

The calculation process continues, and Table 2.2 shows the results for the first 10 time steps.

EXAMPLE 6 *SDOF System Subjected to a Modified 1940 El-Centro Earthquake*

The modified NS component of the El-Centro earthquake time history is shown in Figure 2.17.
Consider an SDOF structure with a mass of 1.0 k-s^2/in. and an undamped natural period of vibration equal to 0.5 s ($\omega_n = 12.566$ rad/s). Assume that the system critical damping ratio is 5% ($\zeta = 0.05$), and let $\Delta t = 0.02$ s.
The response of the system, obtained using the Newmark β-method with $\delta = \frac{1}{2}$ and $\alpha = \frac{1}{4}$, is

Table 2.2 Constant Average Acceleration Calculation for First 10 Time Steps

k	Time (s)	a_k (in./s²)	x_k (in.)	$\dot{x}_k$ (in./s)	$\ddot{x}_k$ (in./s²)	$\ddot{y}_k$ (in./s²)
0	0.00	0.0	0.0	0.0	0.0	0.0
1	0.02	38.64	−0.00382	−0.38221	−38.2211	0.4189
2	0.04	77.28	−0.02283	−1.51812	−75.3693	1.9107
3	0.06	115.92	−0.07181	−3.38058	−110.8774	5.0426
4	0.08	154.56	−0.16493	−5.93146	−144.2105	10.3495
5	0.10	193.20	−0.31547	−9.12231	−174.8748	18.3252
6	0.12	154.56	−0.52800	−12.13088	−125.9818	28.5782
7	0.14	115.92	−0.79791	−14.14798	−75.7279	40.1921
8	0.16	77.28	−1.08382	−15.15441	−24.9152	52.3648
9	0.18	38.64	−1.38683	−15.14701	25.6548	64.2948
10	0.20	0.0	−1.67968	−14.13852	75.1948	75.1948

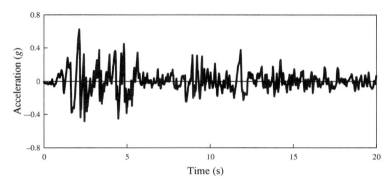

Figure 2.17 Modified 1940 N-S Component of the El-Centro Earthquake Ground Motion

$$\begin{Bmatrix} x_{k+1} \\ \dot{x}_{k+1} \\ \ddot{x}_{k+1} \end{Bmatrix} = \begin{bmatrix} 0.984644 & 0.019571 & 0.000097 \\ -1.53559 & 0.957068 & 0.009724 \\ -153.559 & -4.29317 & -0.02758 \end{bmatrix} \begin{Bmatrix} x_k \\ \dot{x}_k \\ \ddot{x}_k \end{Bmatrix} + \begin{bmatrix} -0.000097 \\ -0.00972 \\ -0.97242 \end{bmatrix} a_{k+1}. \tag{2.257}$$

Table 2.3 gives the response for the first 10 time steps. Note that $\ddot{y}_k$ in this table is the absolute acceleration of the mass, which is equal to

$$\ddot{y}_k = \ddot{x}_k + a_k. \tag{2.258}$$

2.7   NUMERICAL SOLUTION FOR RESPONSE USING WILSON θ-METHOD

Another numerical solution used by structural engineers is the *Wilson θ-method*. This method assumes that the acceleration varies linearly over the time interval t to $t + \theta\Delta t$, where the value of θ need not be an integer and is usually greater than 1.0. Based on this assumption, it follows from Eq. 2.224, after Δt is replaced by $\theta\Delta t$ and for $\delta = \frac{1}{2}$ and $\alpha = \frac{1}{6}$, that

$$\ddot{x}_{k+\theta} + 2\zeta\omega_n\dot{x}_{k+\theta} + \omega_n^2 x_{k+\theta} = F_{k+\theta}/m \tag{2.259a}$$

$$\dot{x}_{k+\theta} = \dot{x}_k + \tfrac{1}{2}\ddot{x}_k\theta\Delta t + \tfrac{1}{2}\ddot{x}_{k+\theta}\theta\Delta t \tag{2.259b}$$

Table 2.3 SDOF System Response Using the Newmark β-Method

k	Time (s)	a_k (g)	x_k (in.)	$\dot{x}_k$ (in./s)	$\ddot{x}_k$ (in./s^2)	$\ddot{y}_k$ (in./s^2)
0	0.00	-0.0026	0.0000	0.0000	0.9930	0.0000
1	0.02	-0.0198	0.0008	0.0841	7.4124	-0.2384
2	0.04	-0.0185	0.0039	0.2207	6.2569	-0.8915
3	0.06	-0.0162	0.0094	0.3270	4.3697	-1.8900
4	0.08	-0.0174	0.0167	0.4065	3.5753	-3.1481
5	0.10	-0.0220	0.0256	0.4808	3.8582	-4.6426
6	0.12	-0.0261	0.0359	0.5565	3.7093	-6.3757
7	0.14	-0.0235	0.0475	0.6018	0.8188	-8.2616
8	0.16	-0.0202	0.0594	0.5868	-2.3144	-10.1197
9	0.18	-0.0156	0.0703	0.5065	-5.7174	-11.7452
10	0.20	-0.0156	0.0792	0.3797	-6.9575	-12.9854

$$x_{k+\theta} = x_k + \dot{x}_k \theta \Delta t + \tfrac{1}{3}\ddot{x}_k \theta^2 (\Delta t)^2 + \tfrac{1}{6}\ddot{x}_{k+\theta}\theta^2 (\Delta t)^2. \tag{2.259c}$$

The forcing function, $F_{k+\theta}$, is obtained by linearly extrapolating or projecting from F_k to F_{k+1}, and is

$$F_{k+\theta} = F_k + \frac{t_{k+\theta} - t_k}{t_{k+1} - t_k}\left(F_{k+1} - F_k\right) = F_k + \frac{\theta \Delta t}{\Delta t}\left(F_{k+1} - F_k\right) = F_k + \theta\left(F_{k+1} - F_k\right). \tag{2.260}$$

Substituting Eqs. 2.259b and 2.259c into Eq. 2.259a gives

$$\ddot{x}_{k+\theta} + 2\zeta\omega_n\left(\dot{x}_k + \tfrac{1}{2}\ddot{x}_k\theta \Delta t + \tfrac{1}{2}\ddot{x}_{k+\theta}\theta \Delta t\right)$$
$$+\omega_n^2\left(x_k + \dot{x}_k\theta \Delta t + \tfrac{1}{3}\ddot{x}_k\theta^2(\Delta t)^2 + \tfrac{1}{6}\ddot{x}_{k+\theta}\theta^2(\Delta t)^2\right) = F_{k+\theta}/m. \tag{2.261}$$

Rearranging the terms in Eq. 2.261 gives

$$\left(1 + \zeta\omega_n\theta \Delta t + \tfrac{1}{6}\omega_n^2\theta^2(\Delta t)^2\right)\ddot{x}_{k+\theta} = -\omega_n^2 x_k - \left(2\zeta\omega_n + \omega_n^2\theta \Delta t\right)\dot{x}_k$$
$$-\left(\zeta\omega_n\theta \Delta t + \tfrac{1}{3}\omega_n^2\theta^2(\Delta t)^2\right)\ddot{x}_k + F_{k+\theta}/m. \tag{2.262}$$

Substituting Eq. 2.260 into Eq. 2.262 and solving for F_{k+1}, it follows that

$$\ddot{x}_{k+\theta} = -\left(\frac{\omega_n^2}{\beta}\right)x_k - \left(\frac{2\zeta\omega_n + \omega_n^2\theta \Delta t}{\beta}\right)\dot{x}_k - \left(\frac{\gamma}{\beta}\right)\ddot{x}_k + \frac{1}{m\beta}\left[F_k + \theta\left(F_{k+1} - F_k\right)\right] \tag{2.263}$$

where

$$\beta = 1 + \zeta\omega_n\theta \Delta t + \tfrac{1}{6}\omega_n^2\theta^2(\Delta t)^2$$
$$\gamma = \zeta\omega_n\theta \Delta t + \tfrac{1}{3}\omega_n^2\theta^2(\Delta t)^2. \tag{2.264}$$

Recall from the dynamic equilibrium equation that

$$F_k/m = \ddot{x}_k + 2\zeta\omega_n\dot{x}_k + \omega_n^2 x_k \tag{2.265}$$

and then when Eq. 2.265 is substituted into Eq. 2.263, it follows that

$$\ddot{x}_{k+\theta} = -\left(\frac{\omega_n^2}{\beta}\right)x_k - \left(\frac{2\zeta\omega_n + \omega_n^2\theta \Delta t}{\beta}\right)\dot{x}_k - \left(\frac{\gamma}{\beta}\right)\ddot{x}_k + \frac{1}{\beta}(1-\theta)\left(\ddot{x}_k + 2\zeta\omega_n\dot{x}_k + \omega_n^2 x_k\right) + \frac{\theta F_{k+1}}{m\beta}. \tag{2.266}$$

Rearranging terms in Eq. 2.266 gives

$$\ddot{x}_{k+\theta} = -\left(\frac{\theta\omega_n^2}{\beta}\right)x_k - \left(\frac{2\zeta\omega_n\theta + \omega_n^2\theta\Delta t}{\beta}\right)\dot{x}_k - \left(\frac{\gamma - 1 + \theta}{\beta}\right)\ddot{x}_k + \left(\frac{\theta}{m\beta}\right)F_{k+1}. \qquad (2.267)$$

The response at the next time step, t_{k+1}, is given for θ greater than one by a linear interpolation within the interval $(t_k, t_{k+\theta})$. Doing so gives

$$\ddot{x}_{k+1} = \ddot{x}_k + \frac{(t_{k+1} - t_k)}{(t_{k+\theta} - t_k)}\left(\ddot{x}_{k+\theta} - \ddot{x}_k\right) = \ddot{x}_k + \frac{\Delta t}{\theta\Delta t}\left(\ddot{x}_{k+\theta} - \ddot{x}_k\right) = \ddot{x}_k + \frac{1}{\theta}\left(\ddot{x}_{k+\theta} - \ddot{x}_k\right) \qquad (2.268a)$$

$$\dot{x}_{k+1} = \dot{x}_k + \frac{(t_{k+1} - t_k)}{(t_{k+\theta} - t_k)}\left(\dot{x}_{k+\theta} - \dot{x}_k\right) = \dot{x}_k + \frac{\Delta t}{\theta\Delta t}\left(\dot{x}_{k+\theta} - \dot{x}_k\right) = x_k + \frac{1}{\theta}\left(\dot{x}_{k+\theta} - \dot{x}_k\right) \qquad (2.268b)$$

$$x_{k+1} = x_k + \frac{(t_{k+1} - t_k)}{(t_{k+\theta} - t_k)}\left(x_{k+\theta} - x_k\right) = x_k + \frac{\Delta t}{\theta\Delta t}\left(x_{k+\theta} - x_k\right) = x_k + \frac{1}{\theta}\left(x_{k+\theta} - x_k\right). \qquad (2.268c)$$

Substituting Eqs. 2.259b and 2.259c into Eqs. 2.268b and 2.268c, it follows that

$$\ddot{x}_{k+1} = \ddot{x}_k + \frac{1}{\theta}\left(\ddot{x}_{k+\theta} - \ddot{x}_k\right) \qquad (2.269a)$$

$$\dot{x}_{k+1} = \dot{x}_k + \frac{1}{\theta}\left(\dot{x}_{k+\theta} - \dot{x}_k\right) = \dot{x}_k + \tfrac{1}{2}\ddot{x}_k\Delta t + \tfrac{1}{2}\ddot{x}_{k+\theta}\Delta t \qquad (2.269b)$$

$$x_{k+1} = x_k + \frac{1}{\theta}\left(x_{k+\theta} - x_k\right) = x_k + \dot{x}_k\Delta t + \tfrac{1}{3}\ddot{x}_k\theta(\Delta t)^2 + \tfrac{1}{6}\ddot{x}_{k+\theta}\theta(\Delta t)^2. \qquad (2.269c)$$

Substituting Eq. 2.266 into Eq. 2.269 gives the response of the system at time step $k + 1$:

$$\ddot{x}_{k+1} = -\left(\frac{\omega_n^2}{\beta}\right)x_k - \left(\frac{2\zeta\omega_n + \omega_n^2\Delta t}{\beta}\right)\dot{x}_k + \left(\frac{\beta\theta - \theta - \gamma - \beta + 1}{\beta\theta}\right)\ddot{x}_k + \left(\frac{1}{m\beta}\right)F_{k+1} \qquad (2.270a)$$

$$\dot{x}_{k+1} = -\left(\frac{\tfrac{1}{2}\omega_n^2\theta\Delta t}{\beta}\right)x_k + \left(\frac{\beta - \zeta\omega_n\theta\Delta t - \tfrac{1}{2}\omega_n^2\theta(\Delta t)^2}{\beta}\right)\dot{x}_k$$
$$+ \left(\frac{\tfrac{1}{2}(\beta - \theta - \gamma + 1)\Delta t}{\beta}\right)\ddot{x}_k + \left(\frac{\tfrac{1}{2}\theta\Delta t}{m\beta}\right)F_{k+1} \qquad (2.270b)$$

$$x_{k+1} = \left(\frac{\beta - \tfrac{1}{6}\omega_n^2\theta^2(\Delta t)^2}{\beta}\right)x_k + \left(\frac{\beta\Delta t - \tfrac{1}{3}\zeta\omega_n\theta^2(\Delta t)^2 - \tfrac{1}{6}\omega_n^2\theta^2(\Delta t)^3}{\beta}\right)\dot{x}_k$$
$$+ \left(\frac{\tfrac{1}{6}(2\beta - \theta - \gamma + 1)\theta(\Delta t)^2}{\beta}\right)\ddot{x}_k + \left(\frac{\tfrac{1}{6}\theta^2(\Delta t)^2}{m\beta}\right)F_{k+1}. \qquad (2.270c)$$

In summary, the procedure for computing the response at time step $k + 1$ given all the information at time step k and F_{k+1} is presented in Eq. 2.270. Equation 2.270, as was done for the Newmark β-method in Eq. 2.230, can be represented in the following matrix form:

$$\mathbf{q}_{k+1} = \begin{Bmatrix} x_{k+1} \\ \dot{x}_{k+1} \\ \ddot{x}_{k+1} \end{Bmatrix} = \mathbf{F}_W \begin{Bmatrix} x_k \\ \dot{x}_k \\ \ddot{x}_k \end{Bmatrix} + \mathbf{H}_W F_{k+1} = \mathbf{F}_W \mathbf{q}_k + \mathbf{H}_W F_{k+1} \qquad (2.271)$$

where

$$\mathbf{F}_W = \frac{1}{\beta}\begin{bmatrix} \beta - \frac{1}{6}\omega_n^2\theta^2(\Delta t)^2 & \beta\Delta t - \frac{1}{3}\zeta\omega_n\theta^2(\Delta t)^2 - \frac{1}{6}\omega_n^2\theta^2(\Delta t)^3 & \frac{1}{6}(2\beta - \theta - \gamma + 1)\theta(\Delta t)^2 \\ -\frac{1}{2}\omega_n^2\theta\Delta t & \beta - \zeta\omega_n\theta(\Delta t) - \frac{1}{2}\omega_n^2\theta(\Delta t)^2 & \frac{1}{2}(\beta - \theta - \gamma + 1)\Delta t \\ -\omega_n^2 & -2\zeta\omega_n - \omega_n^2\Delta t & \frac{1}{6}(\beta\theta - \theta - \gamma - \beta + 1) \end{bmatrix}$$

$$\mathbf{H}_W = \left(\frac{1}{m\beta}\right)\begin{Bmatrix} \frac{1}{6}\theta^2(\Delta t)^2 \\ \frac{1}{2}\theta\Delta t \\ 1 \end{Bmatrix}$$

and the subscript W denotes the Wilson θ-method. As previously noted for the Newmark β-method, the matrices $\mathbf{F}_W$ and $\mathbf{H}_W$ need to be calculated only once.

Note that formulation of Wilson θ-method ensures the dynamic equilibrium equation is satisfied at time $t + \theta\Delta t$, as shown in Eq. 2.259a. However, no dynamic equilibrium has been discussed or enforced at time $t + \Delta t$ in the derivation. The dynamic equilibrium equation at time $t + \Delta t$ can be computed using the following procedure:

$$\begin{aligned} \ddot{x}_{k+1} &+ 2\zeta\omega_n\dot{x}_{k+1} + \omega_n^2 x_{k+1} \\ &= \left[\ddot{x}_k + \frac{1}{\theta}(\ddot{x}_{k+\theta} - \ddot{x}_k)\right] + 2\zeta\omega_n\left[\dot{x}_k + \frac{1}{\theta}(\dot{x}_{k+\theta} - \dot{x}_k)\right] + \omega_n^2\left[x_k + \frac{1}{\theta}(x_{k+\theta} - x_k)\right] \\ &= \left(\ddot{x}_k + 2\zeta\omega_n\dot{x}_k + \omega_n^2 x_k\right) + \frac{1}{\theta}\left[\left(\ddot{x}_{k+\theta} + 2\zeta\omega_n\dot{x}_{k+\theta} + \omega_n^2 x_{k+\theta}\right) - \left(\ddot{x}_k + 2\zeta\omega_n\dot{x}_k + \omega_n^2 x_k\right)\right] \\ &= F_k/m + \frac{1}{\theta}\left[F_{k+\theta}/m - F_k/m\right] \\ &= F_{k+1}/m. \end{aligned} \tag{2.272}$$

This shows that the dynamic equilibrium equation is automatically satisfied at time $t + \Delta t$ as long as this equation is satisfied at both time t and time $t + \theta\Delta t$. Therefore, this method is applicable to solving the dynamic response of a linear system.

Research has shown that for the Wilson θ-method, a value of $\theta \geq 1.37$ gives a numerically stable result. Many engineers use a value of $\theta = 1.40$ for this reason. As for numerical accuracy, research has shown that a time step size of $\Delta t \leq (T_n/10)$ gives a very accurate result, which is the same as using the Newmark β-method.

EXAMPLE 1 Rises to Static Response

Consider Example 1 in Section 2.6, where the forcing function rises to static with the structural variables

$$m = 1.0 \text{ k-s}^2/\text{in.}, \qquad \zeta = 0.05, \qquad \omega_n = 2\pi \text{ rad/s}$$

$$F_o = 10 \text{ kip}, \qquad t_o = 1.0 \text{ s}$$

Consider a time step of $\Delta t = 0.02$ s. Eq. 2.271 gives the solution for the Wilson θ-method. Assume $\theta = 1.40$. First, using Eq. 2.264, it follows that

$$\beta = 1 + (0.05)(2\pi)(1.4)(0.02) + \frac{1}{6}(2\pi)^2(1.4)^2(0.02)^2 = 1.013955 \tag{2.273}$$

$$\gamma = (0.05)(2\pi)(1.4)(0.02) + \frac{1}{3}(2\pi)^2(1.4)^2(0.02)^2 = 0.0191135. \tag{2.274}$$

Then Eq. 2.271 becomes

$$\begin{Bmatrix} x_{k+1} \\ \dot{x}_{k+1} \\ \ddot{x}_{k+1} \end{Bmatrix} = \begin{bmatrix} 0.969475 & 0.019817 & 0.000148 \\ -0.54509 & 0.980423 & 0.005867 \\ -38.9351 & -1.39837 & -0.009532 \end{bmatrix}\begin{Bmatrix} x_k \\ \dot{x}_k \\ \ddot{x}_k \end{Bmatrix} + \begin{Bmatrix} 0.000129 \\ 0.013807 \\ 0.986237 \end{Bmatrix}F_{k+1}. \tag{2.275}$$

Assume that the initial conditions are zero (i.e., $x_o = \dot{x}_o = 0$); because the forcing function at time zero is equal to zero (i.e., $F_o = 0$), it follows that

$$\ddot{x}_o = F_o/m - 2\zeta\omega_n\dot{x}_o - \omega_n^2 x_o = 0. \tag{2.276}$$

The solution for the first time step (i.e., $t = \Delta t$), with the value of forcing function equal to 0.20 kip, is

$$\begin{Bmatrix} x_1 \\ \dot{x}_1 \\ \ddot{x}_1 \end{Bmatrix} = \begin{bmatrix} 0.969475 & 0.019817 & 0.000148 \\ -0.54509 & 0.980423 & 0.005867 \\ -38.9351 & -1.39837 & -0.009532 \end{bmatrix} \begin{Bmatrix} 0 \\ 0 \\ 0 \end{Bmatrix} + \begin{Bmatrix} 0.000129 \\ 0.013807 \\ 0.986237 \end{Bmatrix} 0.20 = \begin{Bmatrix} 0.00002 \\ 0.00276 \\ 0.19725 \end{Bmatrix}. \tag{2.277}$$

Then the solution for the second time step (i.e., $t = 2\Delta t$) follows from Eq. 2.275 with the results from Eq. 2.277 and the value of forcing function equal to 0.40 kip:

$$\begin{Bmatrix} x_2 \\ \dot{x}_2 \\ \ddot{x}_2 \end{Bmatrix} = \begin{bmatrix} 0.969475 & 0.019817 & 0.000148 \\ -0.54509 & 0.980423 & 0.005867 \\ -38.9351 & -1.39837 & -0.009532 \end{bmatrix} \begin{Bmatrix} 0.00002 \\ 0.00276 \\ 0.19725 \end{Bmatrix} + \begin{Bmatrix} 0.000129 \\ 0.013807 \\ 0.986237 \end{Bmatrix} 0.40 = \begin{Bmatrix} 0.00016 \\ 0.00937 \\ 0.38775 \end{Bmatrix}. \tag{2.278}$$

EXAMPLE 2   *Earthquake Response*

Consider Example 3 in Section 2.6, where the earthquake ground motion excites the SDOF system, but now using the Wilson θ-method. The forcing function is given by

$$F_k = -ma_k \tag{2.279}$$

where a_k is the ground motion. It follows that

$$\ddot{x}_{k+1} + 2\zeta\omega_n\dot{x}_{k+1} + \omega_n^2 x_{k+1} = -a_{k+1}. \tag{2.280}$$

The results of the response for the earthquake ground motion using the Wilson θ-method is

$$\mathbf{q}_{k+1} = \mathbf{F}_W \mathbf{q}_k + \mathbf{H}_W^{(EQ)} a_{k+1} \tag{2.281}$$

where

$$\mathbf{H}_W^{(EQ)} = -\left(\frac{1}{\beta}\right) \begin{Bmatrix} \frac{1}{6}\theta^2(\Delta t)^2 \\ \frac{1}{2}\theta\Delta t \\ 1 \end{Bmatrix}. \tag{2.282}$$

EXAMPLE 3   *SDOF System Subjected to a Modified 1940 El-Centro Earthquake*

Consider Example 6 discussed in Section 2.6 subjected to the modified El-Centro earthquake time history as shown in Figure 2.17. The response of the system can be calculated using the Wilson θ-method with θ = 1.4. To illustrate this solution, the response is

$$\begin{Bmatrix} x_{k+1} \\ \dot{x}_{k+1} \\ \ddot{x}_{k+1} \end{Bmatrix} = \begin{bmatrix} 0.980457 & 0.019454 & 0.000146 \\ -2.09391 & 0.941459 & 0.005654 \\ -149.565 & -4.1815 & -0.02472 \end{bmatrix} \begin{Bmatrix} x_k \\ \dot{x}_k \\ \ddot{x}_k \end{Bmatrix} + \begin{bmatrix} -0.00012 \\ -0.01326 \\ -0.94713 \end{bmatrix} a_{k+1}. \tag{2.283}$$

Table 2.4 shows the relative displacement, $x(t)$, relative velocity, $\dot{x}(t)$, and absolute acceleration, $\ddot{y}(t)$, for the first 10 time steps. Figures 2.18 to 2.20 show the relative displacement, relative velocity, and absolute acceleration time histories.

Table 2.4 SDOF System Response Using the Wilson θ-Method

k	Time (s)	$a_k(g)$	x_k(in.)	$\dot{x}_k$(in./s)	$\ddot{x}_k$(in./s²)	$\ddot{y}_k$(in./s²)
0	0.00	−0.0026	0.0000	0.0000	0.9930	0.0000
1	0.02	−0.0198	0.0011	0.1071	7.2217	−0.4290
2	0.04	−0.0185	0.0051	0.2341	5.9810	−1.1674
3	0.06	−0.0162	0.0112	0.3266	4.0402	−2.2195
4	0.08	−0.0174	0.0188	0.3960	3.2278	−3.4955
5	0.10	−0.0220	0.0276	0.4645	3.5109	−4.9899
6	0.12	−0.0261	0.0379	0.5331	3.3926	−6.6925
7	0.14	−0.0235	0.0491	0.5622	0.6230	−8.4574
8	0.16	−0.0202	0.0602	0.5334	−2.3204	−10.1257
9	0.18	−0.0156	0.0698	0.4430	−5.4611	−11.4889
10	0.20	−0.0156	0.0770	0.3201	−6.4426	−12.4704

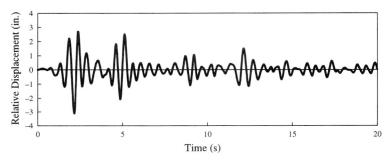

Figure 2.18 Relative Displacement Time History Response

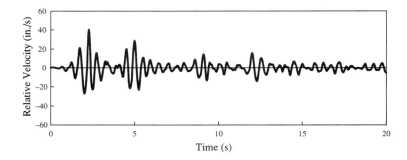

Figure 2.19 Relative Velocity Time History Response

2.8 NUMERICAL SOLUTION FOR STATE SPACE RESPONSE USING DIRECT INTEGRATION

Another method that can be used for numerical analysis is where the forcing function expressed in the integral form is integrated directly to obtain the general solution of the response. Equation 2.182 gives the state space response solution in integral form, and it is

$$\mathbf{z}(t) = \mathbf{e}^{\mathbf{A}(t-t_o)}\mathbf{z}(t_o) + \mathbf{e}^{\mathbf{A}t} \int_{t_o}^{t} \mathbf{e}^{-\mathbf{A}s}\,\mathbf{F}(s)\,ds \qquad (2.284)$$

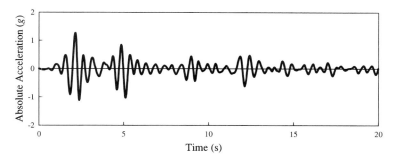

Figure 2.20 Absolute Acceleration Time History Response

where

$$\mathbf{z}(t) = \begin{Bmatrix} x(t) \\ \dot{x}(t) \end{Bmatrix}, \quad \mathbf{A} = \begin{bmatrix} 0 & 1 \\ -\omega_n^2 & -2\zeta\omega_n \end{bmatrix}, \quad \mathbf{F}(t) = \begin{Bmatrix} 0 \\ F_e(t)/m \end{Bmatrix}$$

$$\mathbf{e}^{\mathbf{A}t} = e^{-\zeta\omega_n t} \begin{bmatrix} \cos\omega_d t + \left(\dfrac{\zeta}{\sqrt{1-\zeta^2}} \right)\sin\omega_d t & \dfrac{\sin\omega_d t}{\omega_n\sqrt{1-\zeta^2}} \\ -\dfrac{\omega_n \sin\omega_d t}{\sqrt{1-\zeta^2}} & \cos\omega_d t - \left(\dfrac{\zeta}{\sqrt{1-\zeta^2}} \right)\sin\omega_d t \end{bmatrix}. \quad (2.285)$$

Let

$$t_{k+1} = t, \quad t_k = t_o, \quad \Delta t = t - t_o \quad (2.286)$$

and then it follows from Eq. 2.284 that

$$\mathbf{z}_{k+1} = \mathbf{e}^{\mathbf{A}\Delta t}\mathbf{z}_k + \mathbf{e}^{\mathbf{A}t_{k+1}} \int_{t_k}^{t_{k+1}} \mathbf{e}^{-\mathbf{A}s}\,\mathbf{F}(s)\,ds. \quad (2.287)$$

The objective of the numerical analysis using the integration method is to integrate the forcing function in Eq. 2.287. Since the forcing function is usually given in digitized form, approximation of this forcing function within the time interval is necessary. Because this method involves direct integration (i.e., calculation of the area under the curve), any reasonable approximation of the forcing function can be expected to give accurate results. Two methods used by structural engineers to represent the forcing function are the Delta forcing function method and the constant forcing function method.

With the Delta forcing function, the forcing function is digitized using a series of Delta functions. The forcing function is represented by

$$\mathbf{F}(s) = \mathbf{F}_k \delta(s - t_k)\Delta t = \begin{Bmatrix} 0 \\ F_k/m \end{Bmatrix} \delta(s - t_k)\Delta t, \quad t_k \le s < t_{k+1}. \quad (2.288)$$

Substituting Eq. 2.288 into Eq. 2.287 gives

$$\mathbf{z}_{k+1} = \mathbf{e}^{\mathbf{A}\Delta t}\mathbf{z}_k + \mathbf{e}^{\mathbf{A}t_{k+1}} \int_{t_k}^{t_{k+1}} \mathbf{e}^{-\mathbf{A}s}\mathbf{F}_k\delta(s - t_k)\Delta t\, ds$$

$$= \mathbf{e}^{\mathbf{A}\Delta t}\mathbf{z}_k + \mathbf{e}^{\mathbf{A}t_{k+1}} \left[\int_{t_k}^{t_{k+1}} \mathbf{e}^{-\mathbf{A}s}\delta(s - t_k)\,ds \right]\mathbf{F}_k\Delta t$$

$$= \mathbf{e}^{\mathbf{A}\Delta t}\mathbf{z}_k + \mathbf{e}^{\mathbf{A}t_{k+1}}\mathbf{e}^{-\mathbf{A}t_k}\mathbf{F}_k\Delta t \quad (2.289)$$

$$= \mathbf{e}^{\mathbf{A}\Delta t}\mathbf{z}_k + \Delta t \mathbf{e}^{\mathbf{A}\Delta t}\mathbf{F}_k$$

$$= \mathbf{F}_s\mathbf{z}_k + \mathbf{H}_{sd}\mathbf{F}_k$$

where

$$\mathbf{F}_s = e^{\mathbf{A}\Delta t}$$

$$\mathbf{H}_{sd} = \Delta t e^{\mathbf{A}\Delta t}.$$

With the constant forcing function, the forcing function is assumed to be constant within the time interval. The value of the force in the interval is equal to the value of the force at the beginning of the interval. Therefore, it follows that

$$\mathbf{F}(s) = \mathbf{F}_k = \begin{Bmatrix} 0 \\ F_k/m \end{Bmatrix}, \quad t_k \leq s < t_{k+1}. \tag{2.290}$$

Substituting Eq. 2.290 into Eq. 2.287 gives

$$\mathbf{z}_{k+1} = e^{\mathbf{A}\Delta t}\mathbf{z}_k + e^{\mathbf{A}t_{k+1}} \int_{t_k}^{t_{k+1}} e^{-\mathbf{A}s}\mathbf{F}_k \, ds$$

$$= e^{\mathbf{A}\Delta t}\mathbf{z}_k + e^{\mathbf{A}t_{k+1}}\mathbf{A}^{-1}\left(e^{-\mathbf{A}t_k} - e^{-\mathbf{A}t_{k+1}}\right)\mathbf{F}_k \tag{2.291}$$

$$= e^{\mathbf{A}\Delta t}\mathbf{z}_k + \mathbf{A}^{-1}\left(e^{\mathbf{A}\Delta t} - \mathbf{I}\right)\mathbf{F}_k$$

$$= \mathbf{F}_s\mathbf{z}_k + \mathbf{H}_{sc}\mathbf{F}_k$$

where

$$\mathbf{F}_s = e^{\mathbf{A}\Delta t}$$

$$\mathbf{H}_{sc} = \mathbf{A}^{-1}(e^{\mathbf{A}\Delta t} - \mathbf{I}).$$

Note that $\mathbf{F}_k$ in Eqs. 2.288 and 2.290 is the digitized value of the continuous function $F_e(t)$. Equations 2.289 and 2.291 give the displacement and velocity at time step $k + 1$. Once these values are calculated, the acceleration at time step $k + 1$ can be obtained by using the dynamic equilibrium equation, which is

$$\ddot{x}_{k+1} = \left(F_{k+1}/m\right) - 2\zeta\omega_n\dot{x}_{k+1} - \omega_n^2 x_{k+1}. \tag{2.292}$$

Note that in Eqs. 2.289 and 2.291, the value of $\mathbf{z}_{k+1}$ (i.e., displacement and velocity of the system at time step $k + 1$) is expressed only in terms of the information at time step k. This is different from the discretization method, as discussed in Section 2.6, where the forcing function at time step $k + 1$ is needed to calculate the response at the same time step.

Also, note that Eq. 2.291 is more difficult to calculate in comparison to Eq. 2.289 because the matrix $\mathbf{A}$ must be inverted. Although this method has computational difficulty, structural engineers prefer it because the Delta forcing function contains infinity at the origin (see Eq. 2.95), and structural engineers are generally not comfortable working with infinite values. However, as will be shown in Example 3, both methods give good results.

Again, the two major concerns in a numerical analysis using this method are the numerical stability and the numerical accuracy. Note that in calculating the structural response from time step t_k (i.e., $\mathbf{z}_k$) to time step t_{k+1} (i.e., $\mathbf{z}_{k+1}$), the operation is exact. In addition, integration of the forcing function introduces very small numerical errors when the time step size is small. Therefore, this method can provide very accurate results when the time step is as large as $\Delta t \leq (T_n/5)$. However, in general, it is recommended that $\Delta t \leq (T_n/10)$.

The stability of this analysis method depends on the accuracy of the calculation of the state transition matrix, $e^{\mathbf{A}\Delta t}$. Since this matrix, as given in Eq. 2.285, is exact for an SDOF system, numerical stability will not be an issue. However, if Eq. 2.183 is used to compute the state transition matrix, then the accuracy depends on the number of terms involved in the exponential expansion of $e^{\mathbf{A}\Delta t}$, which can have a major impact on the numerical stability.

Finally, note that the first terms in Eqs. 2.289 and 2.291 are the same. However, the second term in the two equations is quite different, showing the influence of the solution method on numerical analysis.

EXAMPLE 1 *Delta Forcing Function Solution*

Consider Example 1 in Section 2.6, with a forcing function that rises to static and the following specific values for the structural variables:

$$m = 1.0 \text{ k-s}^2/\text{in.}, \qquad \zeta = 0.05, \qquad T_n = 1.0 \text{ s}$$

$$F_o = 10 \text{ kip}, \qquad t_o = 1.0 \text{ s.}$$

Assume $\Delta t = 0.02$ s. The state transition matrix, $e^{A\Delta t}$, can be calculated directly using Eq. 2.285, which is

$$e^{A\Delta t} = e^{-\zeta\omega_n t}(\Delta t) \begin{bmatrix} \cos\omega_d\Delta t + \left(\dfrac{\zeta}{\sqrt{1-\zeta^2}}\right)\sin\omega_d\Delta t & \dfrac{\sin\omega_d\Delta t}{\omega_n\sqrt{1-\zeta^2}} \\[2ex] -\dfrac{\omega_n\sin\omega_d\Delta t}{\sqrt{1-\zeta^2}} & \cos\omega_d\Delta t - \left(\dfrac{\zeta}{\sqrt{1-\zeta^2}}\right)\sin\omega_d\Delta t \end{bmatrix}. \quad (2.293)$$

If we substitute the values of ζ, ω_n, and Δt into Eq. 2.293 and recall that $\omega_d = \omega_n\sqrt{1-\zeta^2}$, it follows that

$$e^{A\Delta t} = \begin{bmatrix} 0.992148 & 0.019823 \\ -0.782565 & 0.979693 \end{bmatrix}. \qquad (2.294)$$

Recall that

$$\mathbf{F}_k = \begin{Bmatrix} 0 \\ F_k/m \end{Bmatrix} = \begin{Bmatrix} 0 \\ 1/m \end{Bmatrix} F_k. \qquad (2.295)$$

Then it follows that

$$\mathbf{H}_{sd}\mathbf{F}_k = \Delta t e^{A\Delta t}\mathbf{F}_k = (0.02)\begin{bmatrix} 0.992148 & 0.019823 \\ -0.782565 & 0.979693 \end{bmatrix}\begin{Bmatrix} 0 \\ 1/1.0 \end{Bmatrix}F_k = \begin{Bmatrix} 0.000396 \\ 0.019594 \end{Bmatrix}F_k \quad (2.296)$$

and Eq. 2.289 now gives

$$\begin{Bmatrix} x_{k+1} \\ \dot{x}_{k+1} \end{Bmatrix} = \begin{bmatrix} 0.992148 & 0.019823 \\ -0.782565 & 0.979693 \end{bmatrix}\begin{Bmatrix} x_k \\ \dot{x}_k \end{Bmatrix} + \begin{Bmatrix} 0.000396 \\ 0.019594 \end{Bmatrix}F_k. \qquad (2.297)$$

Assume that the initial conditions are zero (i.e., $x_o = \dot{x}_o = 0$). Since the forcing function at time zero is equal to zero (i.e., $F_o = 0$), the acceleration at time zero can be calculated using Eq. 2.292 and is

$$\ddot{x}_o = (F_o/m) - 2\zeta\omega_n\dot{x}_o - \omega_n^2 x_o = 0. \qquad (2.298)$$

The solution for the first time step (i.e., $t = \Delta t$) is

$$\begin{Bmatrix} x_1 \\ \dot{x}_1 \end{Bmatrix} = \begin{bmatrix} 0.992148 & 0.019823 \\ -0.782565 & 0.979693 \end{bmatrix}\begin{Bmatrix} 0 \\ 0 \end{Bmatrix} + \begin{Bmatrix} 0.000396 \\ 0.019594 \end{Bmatrix}0.0 = \begin{Bmatrix} 0 \\ 0 \end{Bmatrix} \qquad (2.299)$$

and the acceleration at the first time step, with the forcing function, F_1, equal to 0.20 kip and the displacement and velocity calculated in Eq. 2.299, is

$$\ddot{x}_1 = (0.20/1.0) - 2(0.05)(2\pi)(0) - (2\pi)^2(0) = 0.20. \qquad (2.300)$$

Then the solution for the second time step (i.e., $t = 2\Delta t$) follows from Eq. 2.297 with the results from Eq. 2.299,

$$\begin{Bmatrix} x_2 \\ \dot{x}_2 \end{Bmatrix} = \begin{bmatrix} 0.992148 & 0.019823 \\ -0.782565 & 0.979693 \end{bmatrix} \begin{Bmatrix} 0 \\ 0 \end{Bmatrix} + \begin{Bmatrix} 0.000396 \\ 0.019594 \end{Bmatrix} 0.20 = \begin{Bmatrix} 0.00008 \\ 0.00392 \end{Bmatrix} \tag{2.301}$$

and therefore the acceleration at the second time step, with the forcing function, F_2, equal to 0.40 kip and the displacement and velocity calculated in Eq. 2.301, is

$$\ddot{x}_2 = (0.40/1.0) - 2(0.05)(2\pi)(0.00392) - (2\pi)^2(0.00008) = 0.39441. \tag{2.302}$$

Note that the acceleration at each time step is computed only for completeness of this example, and it is not a necessary quantity to be used later in the computation for the subsequent time steps.

EXAMPLE 2 Constant Forcing Function Solution

Consider the same SDOF system as discussed in Example 1. The state transition matrix, $\mathbf{e}^{A\Delta t}$, is defined in Example 1 and is

$$\mathbf{e}^{A\Delta t} = \begin{bmatrix} 0.992148 & 0.019823 \\ -0.782565 & 0.979693 \end{bmatrix}. \tag{2.303}$$

To calculate the inverse of $\mathbf{A}$ matrix, recall that

$$\mathbf{A} = \begin{bmatrix} 0 & 1 \\ -\omega_n^2 & -2\zeta\omega_n \end{bmatrix} = \begin{bmatrix} 0 & 1 \\ -(2\pi)^2 & -2(0.05)(2\pi) \end{bmatrix} \tag{2.304}$$

and therefore

$$\mathbf{A}^{-1} = \begin{bmatrix} -0.05/\pi & -1/(2\pi)^2 \\ 1 & 0 \end{bmatrix}. \tag{2.305}$$

It follows that

$$\mathbf{H}_{sc} = \mathbf{A}^{-1}(\mathbf{e}^{A\Delta t} - \mathbf{I}) = \begin{bmatrix} -0.05/\pi & -1/(2\pi)^2 \\ 1 & 0 \end{bmatrix} \left(\begin{bmatrix} 0.992148 & 0.019823 \\ -0.782565 & 0.979693 \end{bmatrix} - \begin{bmatrix} 1 & 0 \\ 0 & 1 \end{bmatrix} \right)$$
$$= \begin{bmatrix} 0.019948 & 0.000199 \\ -0.00785 & 0.019823 \end{bmatrix}. \tag{2.306}$$

Let

$$\mathbf{F}_k = \begin{Bmatrix} 0 \\ F_k/m \end{Bmatrix} = \begin{Bmatrix} 0 \\ 1/m \end{Bmatrix} F_k. \tag{2.307}$$

Then it follows that

$$\mathbf{H}_{sc}\mathbf{F}_k = \mathbf{A}^{-1}(\mathbf{e}^{A\Delta t} - \mathbf{I})\mathbf{F}_k = \begin{bmatrix} 0.019948 & 0.000199 \\ -0.00785 & 0.019823 \end{bmatrix} \begin{Bmatrix} 0 \\ 1/1.0 \end{Bmatrix} F_k = \begin{Bmatrix} 0.000199 \\ 0.019823 \end{Bmatrix} F_k. \tag{2.308}$$

Equation 2.291 now gives

$$\begin{Bmatrix} x_{k+1} \\ \dot{x}_{k+1} \end{Bmatrix} = \begin{bmatrix} 0.992148 & 0.019823 \\ -0.782565 & 0.979693 \end{bmatrix} \begin{Bmatrix} x_k \\ \dot{x}_k \end{Bmatrix} + \begin{Bmatrix} 0.000199 \\ 0.019823 \end{Bmatrix} F_k. \tag{2.309}$$

Assume that the initial conditions are zero (i.e., $x_o = \dot{x}_o = 0$). The solution for the first time step (i.e., $t = \Delta t$), with the forcing function at time zero is equal to zero (i.e., $F_o = 0$), is

$$\begin{Bmatrix} x_1 \\ \dot{x}_1 \end{Bmatrix} = \begin{bmatrix} 0.992148 & 0.019823 \\ -0.782565 & 0.979693 \end{bmatrix} \begin{Bmatrix} 0 \\ 0 \end{Bmatrix} + \begin{Bmatrix} 0.000199 \\ 0.019823 \end{Bmatrix} 0.0 = \begin{Bmatrix} 0 \\ 0 \end{Bmatrix}. \tag{2.310}$$

Then the solution for the second time step (i.e., $t = 2\Delta t$) follows from Eq. 2.309, with the results from Eq. 2.310 and the forcing function, F_1, equal to 0.20 kip:

$$\begin{Bmatrix} x_2 \\ \dot{x}_2 \end{Bmatrix} = \begin{bmatrix} 0.992148 & 0.019823 \\ -0.782565 & 0.979693 \end{bmatrix} \begin{Bmatrix} 0 \\ 0 \end{Bmatrix} + \begin{Bmatrix} 0.000199 \\ 0.019823 \end{Bmatrix} 0.20 = \begin{Bmatrix} 0.00004 \\ 0.00397 \end{Bmatrix}. \qquad (2.311)$$

The solution for the third time step (i.e., $t = 3\Delta t$) follows from Eq. 2.309, with the results from Eq. 2.311 and the forcing function, F_2, equal to 0.400 kip,

$$\begin{Bmatrix} x_3 \\ \dot{x}_3 \end{Bmatrix} = \begin{bmatrix} 0.992148 & 0.019823 \\ -0.782565 & 0.979693 \end{bmatrix} \begin{Bmatrix} 0.00004 \\ 0.00397 \end{Bmatrix} + \begin{Bmatrix} 0.000199 \\ 0.019823 \end{Bmatrix} 0.40 = \begin{Bmatrix} 0.00020 \\ 0.01178 \end{Bmatrix}. \qquad (2.312)$$

EXAMPLE 3   *Comparison between Delta Forcing Function and Constant Forcing Function*

Consider again Examples 1 and 2, where

$$m = 1.0 \text{ k-s}^2/\text{in.}, \qquad \zeta = 0.05, \qquad T_n = 1.0 \text{ s.}$$

The response when the Delta forcing function is used is given in Eq. 2.297, which is

$$\begin{Bmatrix} x_{k+1} \\ \dot{x}_{k+1} \end{Bmatrix} = \begin{bmatrix} 0.992148 & 0.019823 \\ -0.782565 & 0.979693 \end{bmatrix} \begin{Bmatrix} x_k \\ \dot{x}_k \end{Bmatrix} + \begin{Bmatrix} 0.000396 \\ 0.019594 \end{Bmatrix} F_k \qquad (2.313)$$

and the response when constant forcing function is used is given in Eq. 2.309, which is

$$\begin{Bmatrix} x_{k+1} \\ \dot{x}_{k+1} \end{Bmatrix} = \begin{bmatrix} 0.992148 & 0.019823 \\ -0.782565 & 0.979693 \end{bmatrix} \begin{Bmatrix} x_k \\ \dot{x}_k \end{Bmatrix} + \begin{Bmatrix} 0.000199 \\ 0.019823 \end{Bmatrix} F_k. \qquad (2.314)$$

If a uniform forcing function of $F_k = 10$ kip is applied, the response can be calculated using Eq. 2.313 and Eq. 2.314. The displacement and velocity time histories are shown in Figures 2.21 and 2.22, respectively.

    Note that the constant forcing function solution as given in Eq. 2.314 gives the exact result because the forcing function is indeed uniform and the digitization method of the constant forcing function is also uniform. Comparing the displacement response between the Delta forcing function solution and the constant forcing function solution, as shown in Figure 2.21, shows that the Delta forcing function gives very accurate results. However, comparing the velocity response between the Delta forcing function solution and the constant forcing function solution, as shown in Figure 2.22, shows that the Delta forcing function deviates slightly from the true solution (i.e., the constant forcing function solution).

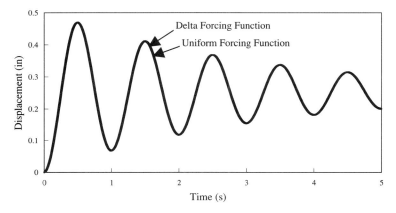

Figure 2.21 Comparison of Displacement Time History Response

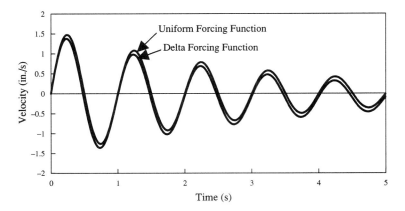

Figure 2.22 Comparison of Velocity Time History Response

One way to improve the accuracy of the Delta forcing function solution is to reduce the time step size.

Although the velocity responses between the Delta forcing function and constant forcing function deviate slightly, the forcing function is in general not a regular function (e.g., earthquake or wind loading), and therefore neither the Delta forcing function nor the constant forcing function gives exact results. The degree of accuracy, in addition to the dependence of the time step size used, also depends on the digitization scheme used. The implication of the difference in the digitization schemes is discussed in Example 7.

EXAMPLE 4   *Earthquake Response*

Consider the case where the earthquake ground motion excites the SDOF system. The forcing function is given by

$$F_k = -ma_k \tag{2.315}$$

where a_k is the ground motion. It follows that

$$\mathbf{F}_k = \left\{ \begin{matrix} 0 \\ F_k/m \end{matrix} \right\} = \left\{ \begin{matrix} 0 \\ -a_k \end{matrix} \right\} = \left\{ \begin{matrix} 0 \\ -1 \end{matrix} \right\} a_k. \tag{2.316}$$

Substituting Eq. 2.316 into Eq. 2.289, it follows from using the Delta forcing function that

$$\mathbf{z}_{k+1} = e^{\mathbf{A}\Delta t}\mathbf{z}_k + \Delta t e^{\mathbf{A}\Delta t}\left\{ \begin{matrix} 0 \\ -1 \end{matrix} \right\} a_k$$

$$= \mathbf{F}_s \mathbf{z}_k + \mathbf{H}_{sd}^{(EQ)} a_k \tag{2.317}$$

where

$$\mathbf{H}_{sd}^{(EQ)} = \Delta t e^{\mathbf{A}\Delta t}\left\{ \begin{matrix} 0 \\ -1 \end{matrix} \right\}. \tag{2.318}$$

If Eq. 2.316 is substituted into Eq. 2.291, it follows from using the constant forcing function that

$$\mathbf{z}_{k+1} = e^{\mathbf{A}\Delta t}\mathbf{z}_k + \mathbf{A}^{-1}\left(e^{\mathbf{A}\Delta t} - \mathbf{I} \right)\left\{ \begin{matrix} 0 \\ -1 \end{matrix} \right\} a_k$$

$$= \mathbf{F}_s \mathbf{z}_k + \mathbf{H}_{sc}^{(EQ)} a_k \tag{2.319}$$

where

$$\mathbf{H}_{sc}^{(EQ)} = \mathbf{A}^{-1}\left(e^{\mathbf{A}\Delta t} - \mathbf{I}\right)\begin{Bmatrix} 0 \\ -1 \end{Bmatrix}.$$

(2.320)

EXAMPLE 5   *Delta Forcing Function Solution for Earthquake Ground Motion*

Consider again the structure used in Example 1 with structural variables

$$m = 1.0 \text{ k–s}^2/\text{in.,} \qquad \zeta = 0.05, \qquad T_n = 1.0 \text{ s.}$$

Let this structure be subjected to an earthquake ground motion as shown in Figure 2.16. Assume $\Delta t = 0.02$ s. The state transition matrix, $e^{\mathbf{A}\Delta t}$, is calculated in Eq. 2.294, which is

$$e^{\mathbf{A}\Delta t} = \begin{bmatrix} 0.992148 & 0.019823 \\ -0.782565 & 0.979693 \end{bmatrix}.$$

(2.321)

In addition, $\mathbf{H}_{sd}^{(EQ)}$ can be calculated using Eq. 2.318, which is

$$\mathbf{H}_{sd}^{(EQ)} = \Delta t e^{\mathbf{A}\Delta t} \begin{Bmatrix} 0 \\ -1 \end{Bmatrix} = (0.02) \begin{bmatrix} 0.992148 & 0.019823 \\ -0.782565 & 0.979693 \end{bmatrix} \begin{Bmatrix} 0 \\ -1 \end{Bmatrix} = \begin{Bmatrix} -0.000396 \\ -0.019594 \end{Bmatrix}.$$

(2.322)

Therefore, it follows from Eq. 2.317 that

$$\begin{Bmatrix} x_{k+1} \\ \dot{x}_{k+1} \end{Bmatrix} = \begin{bmatrix} 0.992148 & 0.019823 \\ -0.782565 & 0.979693 \end{bmatrix} \begin{Bmatrix} x_k \\ \dot{x}_k \end{Bmatrix} + \begin{Bmatrix} -0.000396 \\ -0.019594 \end{Bmatrix} a_k.$$

(2.323)

Assume that the initial conditions are zero (i.e., $x_o = \dot{x}_o = 0$). The solution for the first time step (i.e., $t = \Delta t$), with the earthquake ground motion at time zero equal to zero (i.e., $a_o = 0$), is

$$\begin{Bmatrix} x_1 \\ \dot{x}_1 \end{Bmatrix} = \begin{bmatrix} 0.992148 & 0.019823 \\ -0.782565 & 0.979693 \end{bmatrix} \begin{Bmatrix} 0 \\ 0 \end{Bmatrix} + \begin{Bmatrix} -0.000396 \\ -0.019594 \end{Bmatrix} 0.0 = \begin{Bmatrix} 0 \\ 0 \end{Bmatrix}.$$

(2.324)

Then the solution for the second time step (i.e., $t = 2\Delta t$) follows from Eq. 2.323, with the results from Eq. 2.324 and the earthquake ground motion, a_1, equal to 38.64 in./s^2:

$$\begin{Bmatrix} x_2 \\ \dot{x}_2 \end{Bmatrix} = \begin{bmatrix} 0.992148 & 0.019823 \\ -0.782565 & 0.979693 \end{bmatrix} \begin{Bmatrix} 0 \\ 0 \end{Bmatrix} + \begin{Bmatrix} -0.000396 \\ -0.019594 \end{Bmatrix} (38.64) = \begin{Bmatrix} -0.01532 \\ -0.75711 \end{Bmatrix}.$$

(2.325)

The calculation process continues, and the displacement response for the first 10 seconds is shown in Figure 2.23.

EXAMPLE 6   *Constant Forcing Function Solution for Earthquake Ground Motion*

Consider again the structure used in Example 2 with structural variables

$$m = 1.0 \text{ k-s}^2/\text{in.,} \quad \zeta = 0.05, \quad T_n = 1.0 \text{ s.}$$

Let this structure be subjected to an earthquake ground motion as shown in Figure 2.16. Assume $\Delta t = 0.02$ s. The state transition matrix, $e^{\mathbf{A}\Delta t}$, is calculated in Eq. 2.294, which is

$$e^{\mathbf{A}\Delta t} = \begin{bmatrix} 0.992148 & 0.019823 \\ -0.782565 & 0.979693 \end{bmatrix}.$$

(2.326)

In addition, $\mathbf{H}_{sc}^{(EQ)}$ can be calculated using Eq. 2.320, which is

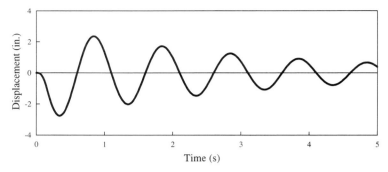

Figure 2.23 Delta Forcing Function Response Due to Earthquake Bilinear Pulse

$$\mathbf{H}_{sc}^{(EQ)} = \mathbf{A}^{-1}\left(\mathbf{e}^{\mathbf{A}\Delta t} - \mathbf{I}\right)\begin{Bmatrix} 0 \\ -1 \end{Bmatrix} = \mathbf{H}_{sc}\begin{Bmatrix} 0 \\ -1 \end{Bmatrix}. \tag{2.327}$$

Note that $\mathbf{H}_{sc}$ has been computed in Eq. 2.306. Therefore, Eq. 2.327 becomes

$$\mathbf{H}_{sc}^{(EQ)} = \begin{bmatrix} 0.019948 & 0.000199 \\ -0.00785 & 0.019823 \end{bmatrix}\begin{Bmatrix} 0 \\ -1 \end{Bmatrix} = \begin{Bmatrix} -0.000199 \\ -0.019823 \end{Bmatrix} \tag{2.328}$$

and it follows from Eq. 2.319 that

$$\begin{Bmatrix} x_{k+1} \\ \dot{x}_{k+1} \end{Bmatrix} = \begin{bmatrix} 0.992148 & 0.019823 \\ -0.782565 & 0.979693 \end{bmatrix}\begin{Bmatrix} x_k \\ \dot{x}_k \end{Bmatrix} + \begin{Bmatrix} -0.000199 \\ -0.019823 \end{Bmatrix}a_k. \tag{2.329}$$

Assume that the initial conditions are zero (i.e., $x_o = \dot{x}_o = 0$). The solution for the first time step (i.e., $t = \Delta t$), with the earthquake ground motion at time zero equal to zero (i.e., $a_o = 0$), is

$$\begin{Bmatrix} x_1 \\ \dot{x}_1 \end{Bmatrix} = \begin{bmatrix} 0.992148 & 0.019823 \\ -0.782565 & 0.979693 \end{bmatrix}\begin{Bmatrix} 0 \\ 0 \end{Bmatrix} + \begin{Bmatrix} -0.000199 \\ -0.019823 \end{Bmatrix}0.0 = \begin{Bmatrix} 0 \\ 0 \end{Bmatrix}. \tag{2.330}$$

Then the solution for the second time step (i.e., $t = 2\Delta t$) follows from Eq. 2.329, with the results from Eq. 2.330 and the earthquake ground motion, a_1, equal to 38.64 in./s²:

$$\begin{Bmatrix} x_2 \\ \dot{x}_2 \end{Bmatrix} = \begin{bmatrix} 0.992148 & 0.019823 \\ -0.782565 & 0.979693 \end{bmatrix}\begin{Bmatrix} 0 \\ 0 \end{Bmatrix} + \begin{Bmatrix} -0.000199 \\ -0.019823 \end{Bmatrix}(38.64) = \begin{Bmatrix} -0.00769 \\ -0.76595 \end{Bmatrix}. \tag{2.331}$$

The calculation process continues, and the displacement response for the first 10 seconds is shown in Figure 2.24.

EXAMPLE 7 *Interpretation of Different Digitization Method*

In Eq. 2.287, the integral

$$\int_{t_k}^{t_{k+1}} \mathbf{e}^{-\mathbf{A}s}\,\mathbf{F}(s)\,ds \tag{2.332}$$

is evaluated by numerically digitizing the forcing function $\mathbf{F}(s)$. Here, this forcing function can be interpreted as a weighting function to the integral in Eq. 2.332. To see this, let

$$\mathbf{F}(s) = \mathbf{F}_o W(s) \tag{2.333}$$

where $W(s)$ is defined in the interval from t_k to t_{k+1} with the criteria

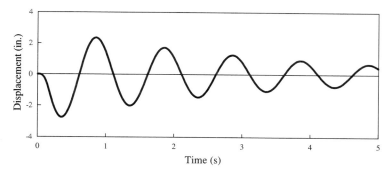

Figure 2.24 Constant Forcing Function Response Due to Earthquake Bilinear Pulse

$$\int_{t_k}^{t_{k+1}} W(s)\,ds = t_{k+1} - t_k = \Delta t. \tag{2.334}$$

Substituting Eq. 2.333 into Eq. 2.332 gives

$$\int_{t_k}^{t_{k+1}} \mathbf{e}^{-\mathbf{A}s}\mathbf{F}(s)\,ds = \int_{t_k}^{t_{k+1}} \mathbf{e}^{-\mathbf{A}s}\mathbf{F}_o W(s)\,ds = \left(\int_{t_k}^{t_{k+1}} \mathbf{e}^{-\mathbf{A}s}\,W(s)\,ds\right)\mathbf{F}_o. \tag{2.335}$$

Note that $W(s)$ weighs the exponential function before integration. Now, if the Delta forcing function is used, that is,

$$W(s) = \delta(s - t_k)\Delta t, \tag{2.336}$$

this means the exponential function is fully weighted at time t_k and zero elsewhere in the interval. If constant forcing function is used, that is,

$$W(s) = 1, \tag{2.337}$$

this means the exponential function is weighted uniformly over the interval.

2.9 SENSITIVITY ANALYSIS OF RESPONSE

Consider the equation for the undamped natural frequency of vibration given in Eq. 2.11, which is

$$\omega_n = \sqrt{\frac{k}{m}}. \tag{2.338}$$

It is essential for the structural engineer to identify the *response variables* that are important (e.g., ω_n) to the response of the system. Then, it is essential to understand how any change in the physically identifiable *design variable* of the structural system (e.g., k and m) affects the response variable. When the structural engineer studies the relationship between the response variables and design variables, it is possible to move the value of the response variable toward a desired (or target) value. The analysis to understand the relationship is called a *sensitivity analysis.*

The most direct way to perform a sensitivity analysis is to perturb one design variable at a time and study the change in the response variable, which is called a *perturbation sensitivity analysis*. It is often convenient for communication purposes to perturb the design variable by 10% and calculate the percentage change to the response variable. For example, consider the case where k = design variable of interest and ω_n = response variable of interest. It then follows that

$$k_\Delta = \text{perturbed stiffness} = k + 0.1k = 1.1k$$

$$\omega_{n\Delta} = \text{resultant effect} = \sqrt{1.1k/m} = \sqrt{1.1}\,\omega_n = 1.049\,\omega_n.$$

Therefore, a 10% change in k results in a 4.9% change in undamped natural frequency of vibration. Another way of expressing this sensitivity is

$$\Delta k = k_\Delta - k = 1.1k - k = 0.1k$$

$$\Delta\omega_n = \omega_{n\Delta} - \omega_n = 1.0492\omega_n - \omega_n = 0.049\omega_n$$

$$\left(\frac{\Delta\omega_n}{\Delta k}\right) = \left(\frac{0.049\omega_n}{0.1k}\right) = 0.49\,\sqrt{1/km}.$$

The perturbation sensitivity analysis method is often very useful when the structural engineer uses very complex structural analysis computer programs that are written by other engineers. This topic is discussed further in Chapter 4.

Now consider the dynamic equilibrium equation for an external applied force discussed in Section 2.1, which is written as

$$m\ddot{x}(t) + c\dot{x}(t) + kx(t) = F_e(t). \tag{2.339}$$

A perturbation sensitivity analysis can also be used to study the sensitivity of the displacement, velocity, or acceleration response variables (x, $\dot{x}$, or $\ddot{x}$) to the design variables (m, c, or k). The sensitivity of the response variables to the forcing function, $F_e(t)$, can also be analyzed if the underlying variables that define the forcing function are considered to be design variables and are perturbed. Consider as an example the rise to static forcing function discussed in Example 2 of Section 2.3. For the case where the damping is zero, the displacement at time $t \le t_o$ with zero initial conditions is given by

$$x(t) = \left(\frac{F_o}{k\,t_o}\right)\left[t - \left(\sqrt{\frac{m}{k}}\right)\sin\sqrt{\frac{k}{m}}\,t\right]. \tag{2.340}$$

A perturbation sensitivity analysis first identifies the design variable of interest (e.g., k) and then substitutes a perturbed stiffness value into Eq. 2.340 to evaluate the resultant effect on the response variable, $x(t)$. Therefore,

$$k_\Delta = \text{perturbed stiffness} = k + 0.1k = 1.1k$$

$$x_\Delta(t) = \left(\frac{F_o}{1.1k\,t_o}\right)\left[t - \left(\sqrt{\frac{m}{1.1k}}\right)\sin\sqrt{\frac{1.1k}{m}}\,t\right]. \tag{2.341}$$

The resultant sensitivity is $[\Delta x(t)/\Delta k] = \{[x_\Delta(t) - x(t)]/(k_\Delta - k)\}$.

A more elegant way to perform a sensitivity analysis involves viewing the solution as a limiting case where the change in the design variable decreases to an infinitesimal value. The change in the design variable becomes the slope of mathematical function relating the response variable and the design variable. For example, when the response variable is the undamped natural frequency of vibration and the design variable is the stiffness, then

$$\omega_n = \sqrt{\frac{k}{m}} \tag{2.342}$$

and

$$\left.\frac{\partial\omega_n}{\partial k}\right|_{k=\bar{k},m=\bar{m}} = \frac{1}{2}\left(\frac{\bar{k}}{\bar{m}}\right)^{-0.5}\left(\frac{1}{\bar{m}}\right) = \frac{1}{2\sqrt{\bar{k}\bar{m}}}. \tag{2.343}$$

Note that the partial derivative is calculated and then evaluated at a point in the design variable space. In this case, the point is denoted $\bar{k}$ and $\bar{m}$.

A partial derivative sensitivity analysis can be performed when the response variable is the system's displacement, velocity, or acceleration (x, $\dot{x}$, or $\ddot{x}$) and the design variable is the mass, damping, or stiffness matrices (m, c, or k). One method to perform this analysis is to calculate the equation relating the response variable and the design variable in a closed-form equation and then take the partial derivative of this equation. For example, again consider Example 2 in Section 2.3 at time $t \leq t_o$ for the case with no damping and zero initial conditions:

$$x(t) = \left(\frac{F_o}{k\,t_o}\right)\left[t - \left(\sqrt{\frac{m}{k}}\right)\sin\sqrt{\frac{k}{m}}\,t\right].$$

(2.344)

If the design variable of interest is the stiffness k, then the sensitivity of the response variable $x(t)$ with respect to k is $\partial x(t)/\partial k$. This method is not possible for many structural engineering problems because it is not possible or practical to develop the closed-form response parameter equation, and numerical integration methods must be used to calculate the response. An alternate method is more desirable in these situations.

Consider the dynamic equilibrium equation

$$m\ddot{x}(t) + c\dot{x}(t) + kx(t) = F_e(t).$$

(2.345)

The desired objective is to calculate the sensitivity of the response variable [e.g., $x(t)$] to a design variable (e.g., k). Therefore, the goal is to calculate $\partial x(t)/\partial k$. Equation 2.345 is used to start the solution, and the partial derivative of each term with respect to the design variable, k, is calculated to obtain

$$\left(\frac{\partial m}{\partial k}\right)\ddot{x}(t) + m\left(\frac{\partial \ddot{x}(t)}{\partial k}\right) + \left(\frac{\partial c}{\partial k}\right)\dot{x}(t) + c\left(\frac{\partial \dot{x}(t)}{\partial k}\right) + \left(\frac{\partial k}{\partial k}\right)x(t) + k\left(\frac{\partial x(t)}{\partial k}\right) = \frac{\partial F_e(t)}{\partial k}.$$

(2.346)

It is very critical to note the following differential calculus relationship:

$$\frac{\partial \dot{x}(t)}{\partial k} = \frac{\partial}{\partial k}\left(\frac{dx(t)}{dt}\right) = \frac{d}{dt}\left(\frac{\partial x(t)}{\partial k}\right)$$

(2.347a)

$$\frac{\partial \ddot{x}(t)}{\partial k} = \frac{\partial}{\partial k}\left(\frac{d^2x(t)}{d^2t}\right) = \frac{d^2}{d^2t}\left(\frac{\partial x(t)}{\partial k}\right).$$

(2.347b)

Substituting Eq. 2.347 into Eq. 2.346 and assuming other parameters, that is, m, c, and $F_e(t)$, are not a function of the stiffness k, it follows that

$$m\left[\frac{d^2}{dt^2}\left(\frac{\partial x(t)}{\partial k}\right)\right] + c\left[\frac{d}{dt}\left(\frac{\partial x(t)}{\partial k}\right)\right] + k\frac{\partial x(t)}{\partial k} = -x(t)$$

(2.348)

where

$$\frac{\partial m}{\partial k} = \frac{\partial c}{\partial k} = \frac{\partial F_e(t)}{\partial k} = 0.$$

(2.349)

Define

$$p(t) = \frac{\partial x(t)}{\partial k};$$

(2.350)

then Eq. 2.348 becomes

$$m\ddot{p}(t) + c\dot{p}(t) + kp(t) = -x(t).$$

(2.351)

The solution for $p(t)$ is obtained using the response solution methods discussed in Section 2.3, where the forcing function is the response $-x(t)$. Therefore, the solution for $p(t)$ involves two steps:

1. Using Eq. 2.345 to calculate the response $x(t)$ for the given m, c, k, and $F_e(t)$.
2. Using Eq. 2.351 to calculate the sensitivity function $p(t)$ subject to the forcing function $-x(t)$ obtained in Step (1).

EXAMPLE 1 *Sensitivity Analysis of Damping Ratio*

Consider the dynamic equilibrium equation for an SDOF system excited by a general forcing function,

$$m\ddot{x}(t) + c\dot{x}(t) + kx(t) = F_e(t) \tag{2.352}$$

or

$$\ddot{x}(t) + 2\zeta\omega_n\dot{x}(t) + \omega_n^2 x(t) = F_e(t)/m. \tag{2.353}$$

The sensitivity of the response to the damping coefficient, ζ, is

$$\frac{\partial \ddot{x}(t)}{\partial \zeta} + 2\left(\frac{\partial \zeta}{\partial \zeta}\right)\omega_n\dot{x}(t) + 2\zeta\left(\frac{\partial \omega_n}{\partial \zeta}\right)\dot{x}(t) + 2\zeta\omega_n\left(\frac{\partial \dot{x}(t)}{\partial \zeta}\right)$$
$$+ \left(\frac{\partial \omega_n^2}{\partial \zeta}\right)x(t) + \omega_n^2\left(\frac{\partial x(t)}{\partial \zeta}\right) = \frac{\partial\left(F_e(t)/m\right)}{\partial \zeta}. \tag{2.354}$$

Define

$$p(t) = \frac{\partial x(t)}{\partial \zeta} \tag{2.355}$$

and since

$$\frac{\partial \omega_n}{\partial \zeta} = \frac{\partial \omega_n^2}{\partial \zeta} = \frac{\partial\left(F_e(t)/m\right)}{\partial \zeta} = 0 \tag{2.356}$$

it follows from Eq. 2.354 that

$$\ddot{p}(t) + 2\zeta\omega_n\dot{p}(t) + \omega_n^2 p(t) = -2\omega_n\dot{x}(t). \tag{2.357}$$

Equation 2.357 is very similar to Eq. 2.351 except that the right side of Eq. 2.357 is now a function of the system velocity versus time.

EXAMPLE 2 *Rise to Static Sensitivity Analysis*

In Example 2 of Section 2.3, the velocity response of an SDOF system was calculated for a rise to static forcing function, which is

$$\dot{x}(t) = -\left[\frac{F_o}{kt_o}\right]e^{-\zeta\omega_n t}\cos\omega_d t - \left[\frac{\zeta}{\sqrt{1-\zeta^2}}\frac{F_o}{kt_o}\right]e^{-\zeta\omega_n t}\sin\omega_d t + \left(\frac{F_o}{kt_o}\right), \quad t \le t_o. \tag{2.358}$$

It is clear from this equation that considerable effort is needed to calculate $(\partial\dot{x}(t)/\partial\zeta)$ or $(\partial\dot{x}(t)/\partial\omega_n)$ because of the exponential sine and cosine functions. However, the reward for this effort is that this can be done in closed form and the sensitivity function can be calculated for any values of ζ, ω_n, and so forth. The perturbation method can be used in the computer age with relative ease to solve for these sensitivity functions, but before any calculation software is used, numerical values must be assigned for all of the design variables. Because, in practice, structural engineers usually focus on a particular situation, the assignment of design values presents no hardship.

EXAMPLE 3   *State Space Sensitivity Analysis*

The state space equation of motion for an SDOF system is

$$\dot{\mathbf{z}}(t) = \mathbf{A}\mathbf{z}(t) + \mathbf{F}(t). \tag{2.359}$$

The sensitivity of the response, $\mathbf{z}(t)$, to k follows by taking the derivative of Eq. 2.359 with respect to k, that is,

$$\frac{\partial \dot{\mathbf{z}}(t)}{\partial k} = \left(\frac{\partial \mathbf{A}}{\partial k}\right)\mathbf{z}(t) + \mathbf{A}\left(\frac{\partial \mathbf{z}(t)}{\partial k}\right) + \frac{\partial \mathbf{F}(t)}{\partial k}. \tag{2.360}$$

As before, define

$$\mathbf{p}(t) = \frac{\partial \mathbf{z}(t)}{\partial k} \tag{2.361}$$

and it follows that

$$\dot{\mathbf{p}}(t) = \frac{d\mathbf{p}(t)}{dt} = \frac{d}{dt}\left(\frac{\partial \mathbf{z}(t)}{\partial k}\right) = \frac{\partial}{\partial k}\left(\frac{d\mathbf{z}(t)}{dt}\right) = \frac{\partial \dot{\mathbf{z}}(t)}{\partial k}. \tag{2.362}$$

Since

$$\frac{\partial \mathbf{F}(t)}{\partial k} = 0, \tag{2.363}$$

Eq. 2.360 becomes

$$\dot{\mathbf{p}}(t) = \mathbf{A}\mathbf{p}(t) + \left(\frac{\partial \mathbf{A}}{\partial k}\right)\mathbf{z}(t) \tag{2.364}$$

where

$$\frac{\partial \mathbf{A}}{\partial k} = \frac{\partial}{\partial k}\begin{bmatrix} 0 & 1 \\ -\omega_n^2 & -2\zeta\omega_n \end{bmatrix} = \begin{bmatrix} 0 & 1 \\ -2\omega_n(\partial\omega_n/\partial k) & -2\zeta(\partial\omega_n/\partial k) \end{bmatrix} \tag{2.365}$$

and

$$\frac{\partial \omega_n}{\partial k} = \frac{\partial \sqrt{k/m}}{\partial k} = \frac{1}{2\sqrt{km}}. \tag{2.366}$$

The solution for Eq. 2.364 follows the same procedure as discussed in Section 2.5.

EXAMPLE 4   *General Solution Sensitivity Analysis*

Example 5 in Section 2.4 presented the general solution for the response of an SDOF system. The response solution for the displacement and velocity at any time $t = t_{k+1}$ is

$$x_{k+1} = C_1 x_k + C_2 \dot{x}_k + C_3 F_k + C_4 F_{k+1} \tag{2.367}$$

$$\dot{x}_{k+1} = D_1 x_k + D_2 \dot{x}_k + D_3 F_k + D_4 F_{k+1}. \tag{2.368}$$

The sensitivity of the response to the natural frequency of vibration, ω_n, is

$$\frac{\partial x_{k+1}}{\partial \omega_n} = \left(\frac{\partial C_1}{\partial \omega_n}\right)x_k + C_1\left(\frac{\partial x_k}{\partial \omega_n}\right) + \left(\frac{\partial C_2}{\partial \omega_n}\right)\dot{x}_k + C_2\left(\frac{\partial \dot{x}_k}{\partial \omega_n}\right) + \left(\frac{\partial C_3}{\partial \omega_n}\right)F_k + \left(\frac{\partial C_4}{\partial \omega_n}\right)F_{k+1} \quad (2.369)$$

and

$$\frac{\partial \dot{x}_{k+1}}{\partial \omega_n} = \left(\frac{\partial D_1}{\partial \omega_n}\right)x_k + D_1\left(\frac{\partial x_k}{\partial \omega_n}\right) + \left(\frac{\partial D_2}{\partial \omega_n}\right)\dot{x}_k + D_2\left(\frac{\partial \dot{x}_k}{\partial \omega_n}\right) + \left(\frac{\partial D_3}{\partial \omega_n}\right)F_k + \left(\frac{\partial D_4}{\partial \omega_n}\right)F_{k+1}. \quad (2.370)$$

The partial derivatives on the right side of Eq. 2.370 follow directly.

EXAMPLE 5 Sensitivity of Stiffness to a Bilinear Pulse

Consider again Example 7 in Section 2.4 and Example 3 in Section 2.5, where the response of an SDOF system with zero damping and zero initial conditions for the time domain $t = 0$ to $t = t_p$ is

$$x(t) = \left(\frac{a_p}{\omega_n^2}\right)\left[\left(\frac{\sin\omega_n t}{\omega_n t_p}\right) - \left(\frac{t}{t_p}\right)\right], \quad 0 \le t \le t_p. \quad (2.371)$$

To study the sensitivity of stiffness to this response, it follows from Eq. 2.351 with zero damping that

$$m\ddot{p}(t) + kp(t) = -\left(\frac{a_p}{\omega_n^2}\right)\left[\left(\frac{\sin\omega_n t}{\omega_n t_p}\right) - \left(\frac{t}{t_p}\right)\right] \quad (2.372)$$

or in normalized form,

$$\ddot{p}(t) + \omega_n^2 p(t) = -\left(\frac{a_p}{k}\right)\left[\left(\frac{\sin\omega_n t}{\omega_n t_p}\right) - \left(\frac{t}{t_p}\right)\right]. \quad (2.373)$$

To solve for Eq. 2.373, let

$$p(t) = p_1(t) + p_2(t) \quad (2.374)$$

where $p_1(t)$ and $p_2(t)$ satisfy the second order linear differential equations

$$\ddot{p}_1(t) + \omega_n^2 p_1(t) = \frac{a_p}{k}\left(\frac{t}{t_p}\right), \quad p_1(0) = 0, \quad \dot{p}_1(0) = 0, \quad 0 \le t \le t_p \quad (2.375)$$

$$\ddot{p}_2(t) + \omega_n^2 p_2(t) = -\frac{a_p}{k}\left(\frac{\sin\omega_n t}{\omega_n t_p}\right), \quad p_2(0) = 0, \quad \dot{p}_2(0) = 0, \quad 0 \le t \le t_p. \quad (2.376)$$

The solution for $p_1(t)$ in Eq. 2.375 is similar to the first part of a rise to static solution or a bilinear pulse solution; that is, from Eq. 2.173:

$$p_1(t) = \frac{a_p}{k\omega_n^2}\left[\left(\frac{t}{t_p}\right) - \left(\frac{\sin\omega_n t}{\omega_n t_p}\right)\right], \quad 0 \le t \le t_p. \quad (2.377)$$

The solution for Eq. 2.376 can be obtained using the solution given in Example 3 of Section 2.3, and then setting $\zeta = 0$ and taking the limit as $r \to 1$. A more direct way of computing the solution is to go back to Eq. 2.376 and assume a particular solution for $p_2(t)$. Let the particular solution of $p_2(t)$, denoted by $p_{2p}(t)$, be of the form

$$p_{2p}(t) = Ct\cos\omega_n t. \quad (2.378)$$

Taking the derivative of Eq. 2.378 gives

$$\dot{p}_{2p}(t) = C\big(\cos\omega_n t - \omega_n t \sin\omega_n t\big) \tag{2.379}$$

$$\ddot{p}_{2p}(t) = C\big(-2\omega_n \sin\omega_n t - \omega_n^2 t \cos\omega_n t\big). \tag{2.380}$$

Substituting Eqs. 2.378 and 2.380 into Eq. 2.376, it follows that

$$C\big(-2\omega_n \sin\omega_n t - \omega_n^2 t \cos\omega_n t\big) + \omega_n^2 C t \cos\omega_n t = -\frac{a_p}{k}\left(\frac{\sin\omega_n t}{\omega_n t_p}\right). \tag{2.381}$$

Simplifying Eq. 2.381 gives

$$-2C\omega_n \sin\omega_n t = -\frac{a_p}{k}\left(\frac{\sin\omega_n t}{\omega_n t_p}\right) \tag{2.382}$$

and solving for C in Eq. 2.382 gives

$$C = \frac{a_p}{2k\omega_n^2 t_p}. \tag{2.383}$$

Substituting Eq. 2.383 back into Eq. 2.378, it follows that

$$p_{2p}(t) = \frac{a_p}{2k\omega_n^2}\left(\frac{t}{t_p}\right)\cos\omega_n t. \tag{2.384}$$

Therefore, the general solution for $p_2(t)$ is given by

$$p_2(t) = A\cos\omega_n t + \left(\frac{B}{\omega_n}\right)\sin\omega_n t + \frac{a_p}{2k\omega_n^2}\left(\frac{t}{t_p}\right)\cos\omega_n t \tag{2.385}$$

and taking the derivative of Eq. 2.385 gives

$$\dot{p}_2(t) = A\omega_n \sin\omega_n t + B\cos\omega_n t + \frac{a_p}{2k\omega_n^2 t_p}\big(\cos\omega_n t - \omega_n t \sin\omega_n t\big). \tag{2.386}$$

Substituting the initial conditions given in Eq. 2.376 into Eqs. 2.385 and 2.386 gives

$$p_2(0) = A = 0, \quad \dot{p}_2(0) = B + \frac{a_p}{2k\omega_n^2 t_p} = 0. \tag{2.387}$$

If we solve for A and B in Eq. 2.387, it follows that

$$A = 0, \quad B = -\frac{a_p}{2k\omega_n^2 t_p}. \tag{2.388}$$

It then follows from Eq. 2.385 that

$$p_2(t) = \frac{a_p}{2k\omega_n^2 t_p}\left(t\cos\omega_n t - \frac{\sin\omega_n t}{\omega_n}\right), \quad 0 \le t \le t_p. \tag{2.389}$$

Finally, substituting Eqs. 2.377 and 2.389 into Eq. 2.374, it follows that

$$p(t) = \frac{a_p}{k\,\omega_n^2}\left[\left(\frac{t}{t_p}\right) - \left(\frac{\sin\omega_n t}{\omega_n t_p}\right)\right] + \frac{a_p}{2k\omega_n^2 t_p}\left(t\cos\omega_n t - \frac{\sin\omega_n t}{\omega_n}\right) \tag{2.390}$$

and simplifying Eq. 2.390 gives

$$p(t) = \frac{a_p}{k\,\omega_n^2}\left[\left(\frac{t}{t_p}\right) - \left(\frac{3\sin\omega_n t}{2\omega_n t_p}\right) + \left(\frac{t\cos\omega_n t}{2t_p}\right)\right]. \tag{2.391}$$

Equation 2.391 represents the sensitivity of $x(t)$ with respect to the stiffness.

2.10   STRUCTURAL DYNAMICS FROM AN ENERGY CONSERVATION PERSPECTIVE

The preceding sections of this chapter presented a view of structural dynamics that is based on a *force perspective*. In this view, the forces acting on the mass are first defined, then the equilibrium equations are written in the form of the equation of motion, and finally the mathematical theory of differential equations is used to solve for the response of the mass as a function of time. Typically, a first exposure to dynamics in a physics course presents a different view of dynamics, where the energy of the system is conserved and a balance is maintained between the *kinetic energy* and the *strain energy*. In recent years, with the introduction into buildings of new, high-technology structural components such as dampers and base isolators, the view of damping as an energy-dissipating source has become more common and for many structural engineers is the preferred view of structural dynamics. This section presents structural dynamics from an energy conservation perspective.

Consider, as a starting point for the discussion of energy in structural dynamics, the undamped free vibration equation of motion, which is

$$m\ddot{x}(t) + kx(t) = 0. \tag{2.392}$$

The first term represents the *inertial force*, F_i. The second term is the force related to the stiffness of the structure and is called the *stiffness force*, F_s (or *elastic force* as defined in Section 2.1). Therefore, Eq. 2.392 can be written as the force balance equation

$$F_i + F_s = 0. \tag{2.393}$$

The *work* done by a force going through a displacement is defined as

$$W = \int_{x'=x_o}^{x'=x} F(x')\,dx' \tag{2.394}$$

which is the area under a force versus displacement curve. The work associated with a differential displacement, $F(x')dx'$, and the total work up to displacement $x' = x$ is shown in a schematic form in Figure 2.25. The work performed by the stiffness force, denoted by W_s, as the mass moves from its starting position at time $t = t_o$ (i.e., x_o) to its final position at time t (i.e., $x(t)$) is

$$W_s = \int_{x'=x_o}^{x'=x} F_s\,dx' = \int_{x'=x_o}^{x'=x} kx'\,dx' = \frac{1}{2}kx(t)^2 - \frac{1}{2}kx_o^2. \tag{2.395}$$

Define the strain energy at time t to be

$$E_s(t) = \frac{1}{2}k\left[x(t)\right]^2 \tag{2.396}$$

and the *strain energy* at time $t = t_o$ to be

$$E_s(t_o) = \frac{1}{2}k\,x_o^2. \tag{2.397}$$

Therefore, the work performed by the stiffness force is equal to the change in the strain energy, that is,

$$W_s = E_s(t) - E_s(t_o). \tag{2.398}$$

Note that this work is only a function of the change in strain energy.

The work performed by the inertial force, denoted by W_i, as the mass moves from its starting position at time $t = t_o$ to its final position at time t is

$$W_i = \int_{x'=x_o}^{x'=x} F_i\,dx' = \int_{x'=x_o}^{x'=x} m\ddot{x}'\,dx'. \tag{2.399}$$

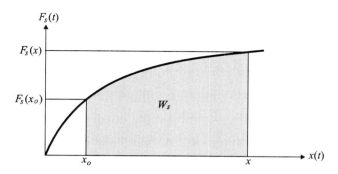

Figure 2.25 Force versus Displacement Relationship

Unlike Eq. 2.395, where the function under the integral was x', the function under the integral of Eq. 2.399 is the acceleration. Therefore, a mathematical reformulation is needed. Recall that

$$\dot{x}' = \frac{dx'}{dt'} \tag{2.400}$$

and therefore

$$dx' = \dot{x}' \, dt'. \tag{2.401}$$

Also, recall that

$$\ddot{x}' = \frac{d\dot{x}'}{dt'}. \tag{2.402}$$

If we substitute Eqs. 2.401 and 2.402 into Eq. 2.399, and note that dt' appears in both the numerator and denominator, it follows that

$$W_i = \int_{x'=x_o}^{x'=x} m\ddot{x}' dx' = \int m\left(\frac{d\dot{x}'}{dt'}\right)\dot{x}' dt' = \int_{\dot{x}'=\dot{x}_o}^{\dot{x}'=\dot{x}} m\dot{x}' d\dot{x}' = \frac{1}{2}m\dot{x}(t)^2 - \frac{1}{2}m\dot{x}_o^2. \tag{2.403}$$

Now define the *kinetic energy* at time t to be

$$E_k(t) = \frac{1}{2}m[\dot{x}(t)]^2 \tag{2.404}$$

and the kinetic energy at time $t = t_o$ to be

$$E_k(t_o) = \frac{1}{2}m\dot{x}_o^2. \tag{2.405}$$

Therefore, the work performed by the inertial force is equal to the change in the kinetic energy, that is,

$$W_i = E_k(t) - E_k(t_o). \tag{2.406}$$

The force balance equation (i.e., Eq. 2.393) can now be operated on with the integral over the displacement that the mass moves between time t_o and time t. Doing so gives

$$\int_{x'=x_o}^{x'=x} F_i \, dx' + \int_{x'=x_o}^{x'=x} F_s \, dx' = W_i + W_s = 0. \tag{2.407}$$

Equation 2.407 is called the *work balance equation*. Substituting Eqs. 2.395 and 2.403 into Eq. 2.407 gives

$$\frac{1}{2}m[\dot{x}(t)]^2 - \frac{1}{2}m\dot{x}_o^2 + \frac{1}{2}k[x(t)]^2 - \frac{1}{2}kx_o^2 = 0. \tag{2.408}$$

Rearranging terms in Eq. 2.408 leads to the equation

$$\frac{1}{2}m[\dot{x}(t)]^2 + \frac{1}{2}k[x(t)]^2 = \frac{1}{2}m\dot{x}_o^2 + \frac{1}{2}kx_o^2. \tag{2.409}$$

The left side of Eq. 2.408 represents the total energy at time t and it is equal to the total energy at time t_o. Define the total energy at time t by

$$E(t) = \frac{1}{2}m[\dot{x}(t)]^2 + \frac{1}{2}k[x(t)]^2 \tag{2.410}$$

and define the total energy at time t_o by

$$E(t_o) = \frac{1}{2}m\dot{x}_o^2 + \frac{1}{2}kx_o^2. \tag{2.411}$$

It then follows from Eq. 2.409 that

$$E(t) = E(t_o). \tag{2.412}$$

Equation 2.412 is called the *energy balance equation*.

Equation 2.409 can be verified by expressing it using the solution obtained in Section 2.2. With zero damping, the solution is given in Eqs. 2.21 and 2.22, which is

$$x(t) = x_o \cos \omega_n t + \left(\frac{\dot{x}_o}{\omega_n}\right) \sin \omega_n t \tag{2.413a}$$

$$\dot{x}(t) = \dot{x}_o \cos \omega_n t - \omega_n x_o \sin \omega_n t. \tag{2.413b}$$

It follows from Eq. 2.409, and recalling $\omega_n^2 = k/m$, that

$$\frac{1}{2}m[\dot{x}(t)]^2 + \frac{1}{2}k[x(t)]^2 = \frac{1}{2}m[\dot{x}_o \cos \omega_n t - (\omega_n x_o)\sin \omega_n t]^2 + \frac{1}{2}k\left[x_o \cos \omega_n t + \left(\frac{\dot{x}_o}{\omega_n}\right)\sin \omega_n t\right]^2$$

$$= \frac{1}{2}m\left[\dot{x}_o^2 \cos^2 \omega_n t - (2\omega_n x_o \dot{x}_o)\cos \omega_n t \sin \omega_n t + (\omega_n^2 x_o^2)\sin^2 \omega_n t\right]$$

$$+ \frac{1}{2}k\left[x_o^2 \cos^2 \omega_n t - 2\left(\frac{x_o \dot{x}_o}{\omega_n}\right)\cos \omega_n t \sin \omega_n t + \left(\frac{\dot{x}_o^2}{\omega_n^2}\right)\sin^2 \omega_n t\right] \tag{2.414}$$

$$= \frac{1}{2}m\dot{x}_o^2(\cos^2 \omega_n t + \sin^2 \omega_n t) + \frac{1}{2}kx_o^2(\cos^2 \omega_n t + \sin^2 \omega_n t)$$

$$= \frac{1}{2}m\dot{x}_o^2 + \frac{1}{2}kx_o^2.$$

Figure 2.26 shows a plot of this strain energy and kinetic energy as a function of time for the case where $m = 1.0$; $T_n = 1.0$ s, that is, $k = m(2\pi/T_n)^2 = 39.478$; $x_o = x(t_o) = x(0) = 1$; and $\dot{x}_o = \dot{x}(t_o) = \dot{x}(0) = 0$. Note that the total energy at time t_o is given by Eq. 2.411, which is

$$E(t_o) = \frac{1}{2}m\dot{x}_o^2 + \frac{1}{2}kx_o^2 = \frac{1}{2}(1.0)(0)^2 + \frac{1}{2}(39.478)(1)^2 = 19.739. \tag{2.415}$$

From Eq. 2.412, $E(t) = E(t_o)$ and therefore the total energy is

$$E(t) = E(t_o) = 19.739. \tag{2.416}$$

This total energy is shown in Figure 2.26, which is a constant function in time.

The right side of Eq. 2.409 is not a function of time; therefore, the only way that Eq. 2.409 is satisfied is if the summation of the kinetic energy and strain energy is independent of time, that is, the total energy at any time is constant. Another important point worth noting is that the strain energy and kinetic energy are only a function of the displacement and velocity at the instant in time when they are calculated. The strain energy, for example, does not depend on the history of the displacement response. Similarly, the kinetic energy does not depend on the history of the velocity response.

Now consider the addition of damping to the system. The work performed by the damping force, F_d, as the mass moves from its starting position at time $t = t_o$ to its final position at time t is

$$W_d = \int_{x'=x_o}^{x'=x} F_d \, dx'. \tag{2.417}$$

If the damping force is linearly related to the velocity, then

$$F_d = c\dot{x}(t). \tag{2.418}$$

Substituting Eqs. 2.401 and 2.418 into Eq. 2.417 gives

$$W_d = \int_{x'=x_o}^{x'=x} F_d \, dx' = \int_{x'=x_o}^{x'=x} c\dot{x}' \, dx' = \int_{t'=t_o}^{t'=t} c\big[\dot{x}(t')\big]^2 \, dt' \tag{2.419}$$

The damped free vibration equation of motion is

$$m\ddot{x}(t) + c\dot{x}(t) + kx(t) = 0. \tag{2.420}$$

This equation can now be operated on with the integral over the displacement that the mass moves between time t_o and time t. The work balance equation becomes

$$\int_{x'=x_o}^{x'=x} F_i \, dx' + \int_{x'=x_o}^{x'=x} F_d \, dx' + \int_{x'=x_o}^{x'=x} F_s \, dx' = W_i + W_d + W_s = 0. \tag{2.421}$$

Substituting Eqs. 2.395, 2.403, and 2.419 into Eq. 2.421 gives

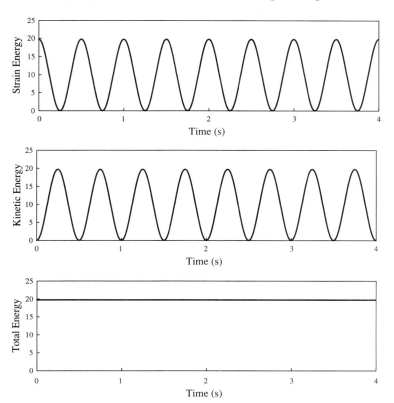

Figure 2.26 Strain Energy, Kinetic Energy, and Total Energy as a Function of Time

$$\left\{ \frac{1}{2} m[\dot{x}(t)]^2 - \frac{1}{2} m\dot{x}_o^2 \right\} + \int_{t'=t_o}^{t'=t} c\dot{x}(t')^2 \, dt' + \left\{ \frac{1}{2} k[x(t)]^2 - \frac{1}{2} kx_o^2 \right\} = 0. \tag{2.422}$$

Rearranging the terms in Eq. 2.422 leads to

$$\frac{1}{2} m[\dot{x}(t)]^2 + \frac{1}{2} k[x(t)]^2 = \frac{1}{2} m\dot{x}_o^2 + \frac{1}{2} kx_o^2 - \int_{t'=t_o}^{t'=t} c[\dot{x}(t')]^2 \, dt'. \tag{2.423}$$

The new energy balance equation (i.e., Eq. 2.423) is fundamentally different from the previous energy balance equation (i.e., Eq. 2.409). The right side of Eq. 2.409 is a constant, whereas the right side of Eq. 2.423 is a function of time. This time function involves an integral whose integrand is always a positive, since the velocity is squared in the expression. Therefore, because the velocity is not always zero, the total energy at time t is always less than the total energy at time t_o. In the limit as $t \to \infty$, the work done by the damping force will always be equal to the summation of the initial kinetic energy and the initial strain energy. This means that the total energy becomes zero and the system comes to rest.

Finally, Eq. 2.420 can be expanded to include the external applied force, and therefore it becomes

$$m\ddot{x}(t) + c\dot{x}(t) + kx(t) = F_e(t). \tag{2.424}$$

The work equation, which is obtained by operating on the external applied force with the integral over the displacement that the mass moves between time t_o and time t, is

$$\int_{x'=x_o}^{x'=x} F_i dx' + \int_{x'=x_o}^{x'=x} F_d dx' + \int_{x'=x_o}^{x'=x} F_s dx' = \int_{x'=x_o}^{x'=x} F_e dx' \tag{2.425}$$

and the work balance equation is

$$W_i + W_d + W_s = W_e. \tag{2.426}$$

The work performed by the applied force, F_e, as the mass moves from its starting position at time $t = t_o$ to its final position at time t is

$$W_e = \int_{x'=x_o}^{x'=x} F_e \, dx'. \tag{2.427}$$

Using Eqs. 2.422 and 2.423, it follows from Eq. 2.426 that

$$\frac{1}{2} m[\dot{x}(t)]^2 + \frac{1}{2} k[x(t)]^2 = \frac{1}{2} m\dot{x}_o^2 + \frac{1}{2} kx_o^2 - \int_{t'=t_o}^{t'=t} c[\dot{x}(t')]^2 \, dt' + \int_{x'=x_o}^{x'=x} F_e \, dx' \tag{2.428}$$

and in terms of total energy, Eq. 2.428 becomes

$$E(t) = E_k(t) + E_s(t) = E_k(t_o) + E_s(t_o) - W_d(t) + W_e(t). \tag{2.429}$$

Equations 2.428 and 2.429 show how the kinetic energy and strain energy of the system at any instant in time are the result of the following four factors:

1. The initial velocity of the system $\dot{x}_o$ as reflected through the initial kinetic energy $E_k(t_o)$.

2. The initial displacement of the system x_o as reflected through the initial strain energy $E_s(t_o)$.

3. The work performed by the system to overcome the damping force present in the system, $W_d(t)$.

4. The work put into the system by the external applied force, $W_e(t)$.

EXAMPLE 1   *Free Vibration with Initial Velocity*

The kinetic energy and strain energy of a system with initial conditions $x(t_o) = x(0) = 0$ and $\dot{x}(t_o) = \dot{x}(0) = \dot{x}_o$ and with viscous damping and no external applied force are

$$x(t) = \dot{x}_o e^{-\zeta\omega_n t}\left(\frac{1}{\omega_n\sqrt{1-\zeta^2}}\right)\sin\omega_d t \qquad (2.430)$$

$$\dot{x}(t) = \dot{x}_o e^{-\zeta\omega_n t}\left[\cos\omega_d t - \left(\frac{\zeta}{\sqrt{1-\zeta^2}}\right)\sin\omega_d t\right]. \qquad (2.431)$$

The stiffness force and the damping force are

$$F_s(t) = kx(t) = k\dot{x}_o e^{-\zeta\omega_n t}\left(\frac{1}{\omega_n\sqrt{1-\zeta^2}}\right)\sin\omega_d t \qquad (2.432)$$

$$F_d(t) = c\dot{x}(t) = c\dot{x}_o e^{-\zeta\omega_n t}\left[\cos\omega_d t - \left(\frac{\zeta}{\sqrt{1-\zeta^2}}\right)\sin\omega_d t\right]. \qquad (2.433)$$

The kinetic energy is

$$E_k(0) = \frac{1}{2}m\dot{x}_o^2 \qquad (2.434)$$

$$E_k(t) = \frac{1}{2}m\dot{x}_o^2\left[e^{-2\zeta\omega_n t}\left(\cos\omega_d t - \frac{\zeta}{\sqrt{1-\zeta^2}}\sin\omega_d t\right)\right]^2 \qquad (2.435)$$

and the strain energy is

$$E_s(0) = 0 \qquad (2.436)$$

$$E_s(t) = \frac{1}{2}k\dot{x}_o^2\left[e^{-\zeta\omega_n t}\left(\frac{1}{\omega_n\sqrt{1-\zeta^2}}\right)\sin\omega_d t\right]^2. \qquad (2.437)$$

The energy balance equation is

$$E_k(t) + E_s(t) = E_k(0) - W_d(t). \qquad (2.438)$$

Therefore, it follows that the work done by the damping force as a function of time is

$$W_d(t) = E_i(0) - E_k(t) - E_s(t). \qquad (2.439)$$

Figure 2.27 shows a plot of $E_k(t)$, $E_s(t)$, and $W_d(t)$ as given in Eqs. 2.435, 2.437, and 2.439, respectively, for the case where $m = 1.0$, $\zeta = 0.05$, $T_n = 1.0$ s, and $\dot{x}_o = 1.0$. Figure 2.28 shows a plot of $F_d(t)$ versus $x(t)$ for the first four free vibration cycles of motion. Let the sum of stiffness force and damping force be denoted by F_{sd}, where

$$F_{sd}(t) = F_s(t) + F_d(t). \qquad (2.440)$$

Figure 2.29 shows a plot of $F_{sd}(t)$ versus $x(t)$ for the first four free vibration cycles of motion.

EXAMPLE 2   *Harmonic Forcing Function*

Consider the special case where the system is excited by a harmonic forcing function. From Example 3 in Section 2.3, when the initial conditions are zero (i.e., $x_o = \dot{x}_o = 0$), the forcing function

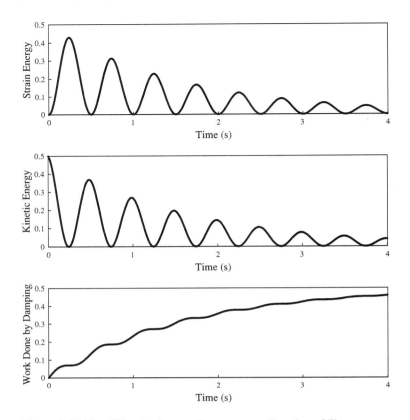

Figure 2.27 Free-Vibration Energy Response as a Function of Time

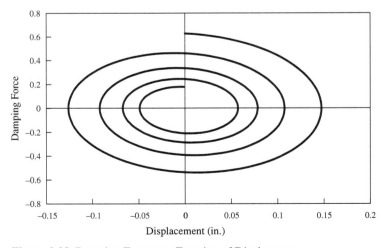

Figure 2.28 Damping Force as a Function of Displacement

$$F_e(t) = F_o \sin \omega_o t \qquad (2.441)$$

produces displacement and velocity response as defined in Eqs. 2.83 and 2.84. If sufficient time has taken place such that the terms associated with the exponential function can be assumed to be zero, then the steady-state solution is given in Eqs. 2.85 and 2.86, which are

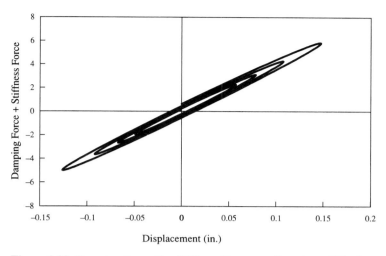

Figure 2.29 Damping Force Plus Stiffness Force as a Function of Displacement

$$x(t) = \frac{F_o/k}{\left(1 - r^2\right)^2 + \left(2\zeta r\right)^2}\left[-2\zeta r \cos \omega_o t + \left(1 - r^2\right)\sin \omega_o t\right] \qquad (2.442)$$

$$\dot{x}(t) = \frac{F_o \omega_o/k}{\left(1 - r^2\right)^2 + \left(2\zeta r\right)^2}\left[\left(1 - r^2\right)\cos \omega_o t + 2\zeta r \sin \omega_o t\right] \qquad (2.443)$$

where $r = \omega_o/\omega_n$. The stiffness force for this steady-state condition is

$$F_s(t) = kx(t) = \frac{F_o}{\left(1 - r^2\right)^2 + \left(2\zeta r\right)^2}\left[-2\zeta r \cos \omega_o t + \left(1 - r^2\right)\sin \omega_o t\right] \qquad (2.444)$$

and the damping force is

$$F_d(t) = c\,\dot{x}(t) = \frac{F_o\left(2\zeta r\right)}{\left(1 - r^2\right)^2 + \left(2\zeta r\right)^2}\left[\left(1 - r^2\right)\cos \omega_o t + 2\zeta r \sin \omega_o t\right] \qquad (2.445)$$

where $c\omega_o/k = 2\zeta r$.

Let the sum of stiffness force and damping force be denoted by F_{sd}, where

$$F_{sd}(t) = F_s(t) + F_d(t). \qquad (2.446)$$

For $r = 1.0$, it follows from Eqs. 2.442 and 2.444 to 2.446 that

$$x(t) = -\frac{F_o/k}{2\zeta}\cos \omega_o t \qquad (2.447)$$

$$F_s(t) = -\frac{F_o}{2\zeta}\cos \omega_o t \qquad (2.448)$$

$$F_d(t) = F_o \sin \omega_o t \qquad (2.449)$$

$$F_{sd}(t) = F_o\left(\sin \omega_o t - \frac{1}{2\zeta}\cos \omega_o t\right). \qquad (2.450)$$

EXAMPLE 3 *Input Energy due to Earthquake*

Consider the equation of motion of the earthquake ground motion excitation of an SDOF system. From Example 4 in Section 2.4, the equation of motion is presented in Eq. 2.128 and is written here as

$$m\ddot{x}(t) + c\dot{x}(t) + kx(t) = -ma(t). \tag{2.451}$$

The stiffness force and damping force are as previously defined in this section:

$$F_s(t) = kx(t), \quad F_d(t) = c\dot{x}(t). \tag{2.452}$$

The work performed by the stiffness force and damping forces follows from Eq. 2.452:

$$W_d = \int_{t'=t_o}^{t'=t} c\dot{x}(t')^2 \, dt', \quad W_s = \frac{1}{2}kx(t)^2 - \frac{1}{2}kx_o^2. \tag{2.453}$$

The inertia force, F_i, is equal to the mass times the *absolute acceleration* of the mass, and therefore,

$$F_i(t) = m\left[\ddot{x}(t) + a(t)\right]. \tag{2.454}$$

Equation 2.454 can be divided into its two component terms and written as

$$F_i(t) = m\ddot{x}(t) + ma(t) = F_{i1}(t) + F_a(t) \tag{2.455}$$

where $F_{i1}(t) = m\ddot{x}(t)$ and $F_a(t) = ma(t)$. The work performed by the relative acceleration inertial force, F_{i1}, as the mass moves from its starting position at time $t = t_o$ to its final position at time t is

$$W_i = \int_{x'=x_o}^{x'=x} m\ddot{x}'dx' = \int_{t'=t_o}^{t'=t} m\frac{d\dot{x}'}{dt'}\dot{x}'dt' = \int_{\dot{x}'=\dot{x}_o}^{\dot{x}'=\dot{x}} m\dot{x}'d\dot{x}' = \frac{1}{2}m\left[\dot{x}(t)\right]^2 - \frac{1}{2}m\dot{x}_o^2. \tag{2.456}$$

The work performed by the ground acceleration inertial force, $F_a(t)$, is

$$W_a = \int_{x'=x_o}^{x'=x} ma(t)dx' \tag{2.457}$$

and thus it is very similar to the work performed by the external force where the mass times the acceleration is the force.

Now, as was done in this section, the new force balance equation (i.e., Eq. 2.451) is operated on with the integral over the displacement that the mass moves between time t_o and time t:

$$\int_{x'=x_o}^{x'=x} F_{i1}dx' + \int_{x'=x_o}^{x'=x} F_d dx' + \int_{x'=x_o}^{x'=x} F_s dx' = -\int_{x'=x_o}^{x'=x} F_a dx' \tag{2.458}$$

and therefore

$$W_{i1} + W_d + W_s = -W_a. \tag{2.459}$$

Note that Eq. 2.459 is very similar to Eq. 2.426. Thus the energy balance equation now becomes

$$\frac{1}{2}m\dot{x}(t)^2 + \frac{1}{2}kx(t)^2 = \frac{1}{2}m\dot{x}_o^2 + \frac{1}{2}kx_o^2 - \int_{t'=t_o}^{t'=t} c\left[\dot{x}(t')\right]^2 dt' - \int_{x'=x_o}^{x'=x} F_a \, dx'. \tag{2.460}$$

EXAMPLE 4 *Earthquake Ground Motion that is a Bilinear Pulse*

Consider the earthquake ground motion represented by a bilinear pulse used in Example 3 of Section 2.5, where the response for the time domain $t = 0$ to $t = t_p$ with zero damping and zero initial conditions is given in Eq. 2.211:

$$x(t) = \left(\frac{a_p}{\omega_n^2}\right)\left[\left(\frac{\sin \omega_n t}{\omega_n t_p}\right) - \left(\frac{t}{t_p}\right)\right], \quad 0 \le t \le t_p \tag{2.461}$$

$$\dot{x}(t) = \left(\frac{a_p}{\omega_n^2}\right)\left[\left(\frac{\cos \omega_n t}{t_p}\right) - \left(\frac{1}{t_p}\right)\right], \quad 0 \le t \le t_p. \tag{2.462}$$

The strain energy and kinetic energy are given by Eqs. 2.396 and 2.404, where

$$E_s(t) = \frac{1}{2}k\left(\frac{a_p}{\omega_n^2}\right)^2\left[\left(\frac{\sin \omega_n t}{\omega_n t_p}\right)^2 - 2\left(\frac{t}{t_p}\right)\left(\frac{\sin \omega_n t}{\omega_n t_p}\right) + \left(\frac{t}{t_p}\right)^2\right] \tag{2.463}$$

$$E_k(t) = \frac{1}{2}m\left(\frac{a_p}{\omega_n^2}\right)^2\left[\left(\frac{\cos \omega_n t}{t_p}\right)^2 - 2\left(\frac{1}{t_p}\right)\left(\frac{\cos \omega_n t}{t_p}\right) + \left(\frac{1}{t_p}\right)^2\right]. \tag{2.464}$$

PROBLEMS

2.1 Consider a single degree of freedom system with a stiffness equal to 10 k/in. and a mass equal to 1.0 k-s^2/in.

(a) Calculate the undamped natural frequency of vibration.

(b) Calculate the undamped natural period of vibration.

(c) If the stiffness is doubled, how much will the undamped natural period of vibration decrease?

2.2 Consider the single degree of freedom system defined in Problem 2.1 but now with a value of damping coefficient equal to 1.0 k-s/in.

(a) Calculate the damped natural frequency of vibration.

(b) Calculate the damped natural period of vibration.

(c) If the damping is doubled, how much will the damped natural period of vibration increase?

2.3 If the single degree of freedom system defined in Problem 2.1 has a critical damping ratio equal to 20%, what is the value of the damping coefficient?

2.4 Consider a single degree of freedom system with an undamped natural period of vibration equal to 1.0 s and a critical damping ratio equal to 5%.

(a) Calculate and plot the free vibration response for the time 0 to 5 s. Assume that the system is released from rest with an initial displacement equal to 1.0 in.

(b) Repeat Part (a) but assume that the initial displacement is equal to 1.0 in. and initial velocity is equal to 1.0 in./s.

2.5 Repeat Problem 2.4 but assume that the undamped period of vibration is equal to 0.5 s.

2.6 Repeat Problem 2.4 but assume that the critical damping ratio is equal to 20%.

2.7 Consider the system in Problem 2.1. If the response at time t_1 is equal to 1.0 in. and the response at time $t_2 = t_1 + T_d$ is equal to 0.8 in., what is the value of the logarithmic decrement and the critical damping ratio?

2.8 Consider the system in Problem 2.1. If it is required to have a response at time $t_2 = t_1 + T_d$ that is half of the response at time t_1, then what must be the value of the critical damping ratio?

2.9 Consider the system in Problem 2.1 with zero initial conditions subjected to a uniform forcing function of magnitude 10 kip.

(a) Calculate and plot the system response for the time 0 to 5 s.

(b) Repeat Part (a) but assume a 5% critical damping ratio.

(c) Repeat Part (a) but assume a 20% critical damping ratio.

2.10 A "box car" forcing function starts at time zero with a uniform force until it reaches time 2.0 s, when it becomes zero. Consider the system in Problem 2.1 with zero initial conditions to be subjected to a box car forcing function of force magnitude equal to 10 kip. Calculate and plot the displacement response from time 0 to 5 s.

2.11 Consider the system in Problem 2.1 with zero initial conditions. Assume that the force is a rise to static force that reaches the static force level at time 2.0 s (i.e., $t_o = 2.0$ s).

(a) Calculate and plot the system response for the time 0 to 5 s.

(b) Repeat Part (a) but assume a 5% critical damping ratio.

(c) Repeat Part (a) but assume a 20% critical damping ratio.

2.12 Consider the system in Problem 2.1 with zero initial conditions. Assume that the force starts at time zero and rises in a linear manner to a magnitude equal to 10 kip at 2.0 s. Then assume that the force instantaneously goes to zero at time $t = 2.0$ s. Calculate and plot the displacement response for the time 0 to 5 s.

2.13 Consider the system in Problem 2.1 with zero initial conditions. Assume that the harmonic forcing function is of the form $F_e(t) = F_o \sin \omega_o t$, where $F_o = 10$ kip, and that the frequency of the function is 2π rad/s.

(a) Calculate and plot the system response for the time 0 to 20 s.

(b) Repeat Part (a) but assume a 5% critical damping ratio.

(c) Repeat Part (a) but assume a 20% critical damping ratio.

(d) Calculate the magnification factor for 0%, 5%, and 20% critical damping ratios.

2.14 Consider the system in Problem 2.1. Calculate and plot the impulse response for the time 0 to 5 s for 0%, 5%, 10%, and 20% critical damping ratios.

2.15 Consider the system in Problem 2.11. Let the time increment be $\Delta t = 0.02$ s and use the incremental response approach discussed in Section 2.4, Example 5 to calculate the response for the first five time increments.

2.16 Consider Example 7 in Section 2.4 and the system in Problem 2.1 with zero initial conditions. Calculate and plot the response for each of the following cases for $a_p = 1.0g$:

(a) Zero damping and $t_p = 1.0$ s.

(b) Zero damping and $t_p = 0.1$ s.

(c) Zero damping and $t_p = 10.0$ s.

(d) Repeat (a) to (c) but for critical damping ratio equal to 5%.

(e) Repeat (a) to (c) but for critical damping ratio equal to 20%.

2.17 Consider the system in Problem 2.1.

(a) Determine the matrix **A** in Eq. 2.180.

(b) Write and define each term in the power series definition of the state transition matrix as given in Eq. 2.165.

(c) Determine the eigenvalue matrix.

(d) Determine the eigenvector matrix.

(e) Write and define the state transition matrix using Eq. 2.192.

2.18 Repeat Example 1 in Section 2.6 but for a 10% critical damping ratio. Calculate the $\mathbf{F}_N$ and $\mathbf{H}_N$ matrices.

2.19 Repeat Example 2 in Section 2.6 but for a 10% critical damping ratio. Calculate the $\mathbf{F}_N$ and $\mathbf{H}_N$ matrices.

2.20 Repeat Example 2 in Section 2.6 but use the linear acceleration discretization method (i.e., $\delta = \frac{1}{2}$ and $\alpha = \frac{1}{6}$). Calculate the $\mathbf{F}_N$ and $\mathbf{H}_N$ matrices.

2.21 Repeat Example 5 in Section 2.6 but for a 10% critical damping ratio. Calculate the $\mathbf{F}_N$ and $\mathbf{H}_N^{(EQ)}$ matrices.

2.22 Repeat Example 1 in Section 2.7 but for a 10% critical damping ratio. Calculate the $\mathbf{F}_W$ and $\mathbf{H}_W$ matrices.

2.23 Repeat Example 2 in Section 2.6 but using the Wilson θ-method with θ = 1.40. Calculate the $\mathbf{F}_W$ and $\mathbf{H}_W$ matrices.

2.24 Repeat Example 5 in Section 2.6 but using the Wilson θ-method with θ = 1.40. Calculate the $\mathbf{F}_W$ and $\mathbf{H}_W^{(EQ)}$ matrices.

2.25 Repeat Example 1 in Section 2.8 but for a 10% critical damping ratio. Calculate the $\mathbf{F}_s$ and $\mathbf{H}_{sd}$ matrices.

2.26 Repeat Example 2 in Section 2.8 but for a 10% critical damping ratio. Calculate the $\mathbf{F}_s$ and $\mathbf{H}_{sc}$ matrices.

2.27 Repeat Example 5 in Section 2.8 but for a 10% critical damping ratio. Calculate the $\mathbf{F}_s$ and $\mathbf{H}_{sd}^{(EQ)}$ matrices.

2.28 Repeat Example 6 in Section 2.8 but for a 10% critical damping ratio. Calculate the $\mathbf{F}_s$ and $\mathbf{H}_{sc}^{(EQ)}$ matrices.

2.29 Consider the system in Problem 2.1. Perform a perturbation sensitivity analysis to calculate the sensitivity of the natural frequency to the system stiffness.

2.30 Repeat Problem 2.29 but calculate the sensitivity of the natural frequency to the system mass.

2.31 Consider the system in Example 2 of Section 2.6. Perform a perturbation sensitivity analysis for the times given in Table 2.1 of the example to calculate the sensitivity of the displacement response, $x(t)$, with respect to the system stiffness.

2.32 Consider the system in Example 5 of Section 2.6. Perform a perturbation sensitivity analysis, for the times given in Table 2.2 of the example to calculate the sensitivity of the displacement response, $x(t)$, with respect to the system stiffness.

2.33 Repeat Problem 2.32 but calculate the sensitivity of the absolute mass acceleration response, $\ddot{y}(t)$, with respect to the system stiffness.

2.34 Repeat Problem 2.29 but use Eq. 2.343.

2.35 Repeat Problem 2.31 but use Eq. 2.351.

2.36 Repeat Problem 2.32 but use Eq. 2.351.

2.37 Repeat Problem 2.33 but use Eq. 2.351.

2.38 Consider the system in Problem 2.1 but with a stiffness equal to 20 k/in. Assume that the system is released from rest with an initial displacement equal to 1.0 in. Use Eq. 2.414 to plot the strain energy, kinetic energy, and total energy versus time (e.g., see Figure 2.26).

2.39 Repeat Problem 2.38 but assume that the initial conditions are $x(0) = 1.0$ in. and $\dot{x}(0) = 10.0$ in./s.

2.40 Repeat Problem 2.38 but use 5% critical damping ratio and also plot the work done by the damping force as a function of displacement (e.g., see Figure 2.28). Also plot the damping force plus the stiffness as a function of displacement (e.g., see Figure 2.29).

2.41 Consider the system in Example 2 of Section 2.6. For the times given in Table 2.1, calculate the following:

    (a) Kinetic energy—$E_k(t)$.

    (b) Strain energy—$E_s(t)$.

    (c) Work done by the damping force—$W_d(t)$.

    (d) Work done by the external applied force—$W_e(t)$.

    (e) Total energy—$E(t)$.

Chapter 3

Two Degrees of Freedom Linear System Response

3.1 OVERVIEW

In many texts the concepts presented in the previous chapter are extended from an SDOF system to a multi–degree of freedom system without first presenting the solutions for a two degrees of freedom (2DOF) system. Such an approach does not allow readers to gradually learn the methodology and also does not allow for the presentation of many important structural engineering concepts. Therefore, this chapter is devoted to 2DOF systems. There are many ways that a 2DOF system can be represented, three of which are shown in Figure 3.1.

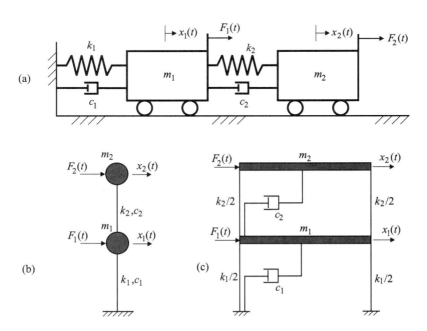

Figure 3.1 Two Degrees of Freedom System Representation

The 2DOF system shown in Figure 3.1c is used to study the dynamic forces acting on this system. The notations used in the 2DOF system are shown in Figure 3.2a. As with an SDOF system, four types of forces act on each floor mass, and these forces include stiffness, damping, external, and inertial forces. Figure 3.2b shows the free body diagrams for each of the two floors. Let the displacement $x_2(t)$ be greater than $x_1(t)$; then the reaction for the columns between the two floors is to pull $x_2(t)$ toward $x_1(t)$, and therefore the stiffness force acting on m_2 is in the negative x_2 direction. Similarly, the reaction for the columns between the two floors is to pull $x_1(t)$ toward $x_2(t)$, and therefore the stiffness force acting on m_1 is in the positive x_1 direction. The same reasoning applies to the damping force between the two floors using $\dot{x}_1(t)$ and $\dot{x}_2(t)$. Note that the stiffness and damping forces applied between the two floors are equal in magnitude but opposite in direction.

From the free body diagrams shown in Figure 3.2b, with 1 used to denote the first story above ground and 2 to denote the second story above ground, the dynamic equilibrium equations of motion are

$$-c_1\dot{x}_1(t) - c_2\big(\dot{x}_1(t) - \dot{x}_2(t)\big) - k_1 x_1(t) - k_2\big(x_1(t) - x_2(t)\big) + F_1(t) = m_1\ddot{x}_1(t) \qquad (3.1a)$$

$$-c_2\big(\dot{x}_2(t) - \dot{x}_1(t)\big) - k_2\big(x_2(t) - x_1(t)\big) + F_2(t) = m_2\ddot{x}_2(t). \qquad (3.1b)$$

Note that Eq. 3.1 represents two *coupled second-order linear differential equations*; that is, the terms $x_1(t)$ and $x_2(t)$ appear in both Eq. 3.1a and Eq. 3.1b. Rearranging terms in Eq. 3.1 and representing this equation in matrix form gives

$$\begin{bmatrix} m_1 & 0 \\ 0 & m_2 \end{bmatrix}\begin{Bmatrix} \ddot{x}_1 \\ \ddot{x}_2 \end{Bmatrix} + \begin{bmatrix} c_1 + c_2 & -c_2 \\ -c_2 & c_2 \end{bmatrix}\begin{Bmatrix} \dot{x}_1 \\ \dot{x}_2 \end{Bmatrix} + \begin{bmatrix} k_1 + k_2 & -k_2 \\ -k_2 & k_2 \end{bmatrix}\begin{Bmatrix} x_1 \\ x_2 \end{Bmatrix} = \begin{Bmatrix} F_1(t) \\ F_2(t) \end{Bmatrix}. \qquad (3.2)$$

Now define

$$\mathbf{M} = \begin{bmatrix} m_1 & 0 \\ 0 & m_2 \end{bmatrix}, \quad \mathbf{C} = \begin{bmatrix} c_1 + c_2 & -c_2 \\ -c_2 & c_2 \end{bmatrix}, \quad \mathbf{K} = \begin{bmatrix} k_1 + k_2 & -k_2 \\ -k_2 & k_2 \end{bmatrix}$$

$$\mathbf{X}(t) = \begin{Bmatrix} x_1 \\ x_2 \end{Bmatrix}, \quad \mathbf{F}(t) = \begin{Bmatrix} F_1(t) \\ F_2(t) \end{Bmatrix}$$

and then it follows from Eq. 3.2 that

$$\mathbf{M}\ddot{\mathbf{X}}(t) + \mathbf{C}\dot{\mathbf{X}}(t) + \mathbf{K}\mathbf{X}(t) = \mathbf{F}(t). \qquad (3.3)$$

Note that the mass, damping, and stiffness matrices (i.e., **M**, **C**, and **K**, respectively) are symmetrical matrices. In addition, these matrices are *positive definite*.

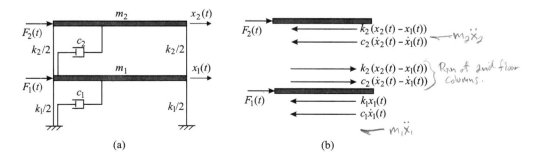

(a) (b)

Figure 3.2 Two Degrees of Freedom System

3.2 UNDAMPED FREE VIBRATION

Undamped free vibration is defined as the response solution for Eq. 3.3 with no damping (i.e., $\mathbf{C} = \mathbf{0}$) and no external force (i.e., $\mathbf{F} = \mathbf{0}$). The result is the following two simultaneous second-order linear differential equations:

$$\begin{bmatrix} m_1 & 0 \\ 0 & m_2 \end{bmatrix}\begin{Bmatrix} \ddot{x}_1 \\ \ddot{x}_2 \end{Bmatrix} + \begin{bmatrix} k_1 + k_2 & -k_2 \\ -k_2 & k_2 \end{bmatrix}\begin{Bmatrix} x_1 \\ x_2 \end{Bmatrix} = \begin{Bmatrix} 0 \\ 0 \end{Bmatrix}, \quad \begin{Bmatrix} x_1(0) \\ x_2(0) \end{Bmatrix} = \begin{Bmatrix} x_{1o} \\ x_{2o} \end{Bmatrix}, \quad \begin{Bmatrix} \dot{x}_1(0) \\ \dot{x}_2(0) \end{Bmatrix} = \begin{Bmatrix} \dot{x}_{1o} \\ \dot{x}_{2o} \end{Bmatrix}. \quad (3.4)$$

The solution to this matrix equation is

$$\begin{Bmatrix} x_1(t) \\ x_2(t) \end{Bmatrix} = \begin{Bmatrix} X_{11} \\ X_{21} \end{Bmatrix}\cos \omega t + \begin{Bmatrix} X_{12} \\ X_{22} \end{Bmatrix}\sin \omega t. \quad (3.5)$$

Substituting Eq. 3.5 into Eq. 3.4 gives

$$\left(-m_1\omega^2 + k_1 + k_2\right)\left(X_{11}\cos \omega t + X_{12}\sin \omega t\right) - k_2\left(X_{21}\cos \omega t + X_{22}\sin \omega t\right) = 0 \quad (3.6a)$$

$$-k_2\left(X_{11}\cos \omega t + X_{12}\sin \omega t\right) + \left(-m_2\omega^2 + k_2\right)\left(X_{21}\cos \omega t + X_{22}\sin \omega t\right) = 0 \quad (3.6b)$$

and in matrix form, Eq. 3.6 becomes

$$\begin{bmatrix} -m_1\omega^2 + k_1 + k_2 & -k_2 \\ -k_2 & -m_2\omega^2 + k_2 \end{bmatrix}\left(\begin{Bmatrix} X_{11} \\ X_{21} \end{Bmatrix}\cos \omega t + \begin{Bmatrix} X_{12} \\ X_{22} \end{Bmatrix}\sin \omega t\right) = 0. \quad (3.7)$$

A solution to Eq. 3.7 is $X_{11} = X_{12} = X_{21} = X_{22} = 0$. However, this solution is not of interest. Another solution to Eq. 3.7 is obtained when the determinant of the 2×2 matrix is set equal to zero, and therefore it follows that

$$\begin{vmatrix} -m_1\omega^2 + k_1 + k_2 & -k_2 \\ -k_2 & -m_2\omega^2 + k_2 \end{vmatrix} = \left|-\mathbf{M}\omega^2 + \mathbf{K}\right| = 0. \quad (3.8)$$

The following quadratic equation is then obtained for ω^2:

$$m_1 m_2 \omega^4 - \left[(k_1 + k_2)m_2 + k_2 m_1\right]\omega^2 + k_1 k_2 = 0. \quad (3.9)$$

Solving this quadratic equation for ω^2 gives

$$\omega_1^2 = \frac{\left[(k_1 + k_2)m_2 + k_2 m_1\right] - \sqrt{\left[(k_1 + k_2)m_2 + k_2 m_1\right]^2 - 4m_1 m_2 k_1 k_2}}{2m_1 m_2} \quad (3.10a)$$

$$\omega_2^2 = \frac{\left[(k_1 + k_2)m_2 + k_2 m_1\right] + \sqrt{\left[(k_1 + k_2)m_2 + k_2 m_1\right]^2 - 4m_1 m_2 k_1 k_2}}{2m_1 m_2}. \quad (3.10b)$$

Note that $\omega_1^2 \leq \omega_2^2$. Substituting Eq. 3.10 into Eq. 3.7, the value of X_{21} can be determined in terms of X_{11}, and X_{22} in terms of X_{12} for each ω^2 (i.e., ω_1^2 and ω_2^2). Let these values be denoted by $X_{11}^{(1)}$, $X_{12}^{(1)}$, $X_{21}^{(1)}$, and $X_{22}^{(1)}$ when $\omega^2 = \omega_1^2$, and $X_{11}^{(2)}$, $X_{12}^{(2)}$, $X_{21}^{(2)}$, and $X_{22}^{(2)}$ when $\omega^2 = \omega_2^2$. It follows that

$$\omega^2 = \omega_1^2: \ X_{21}^{(1)} = \left(\frac{-m_1\omega_1^2 + k_1 + k_2}{k_2}\right)X_{11}^{(1)}, \quad X_{22}^{(1)} = \left(\frac{-m_1\omega_1^2 + k_1 + k_2}{k_2}\right)X_{12}^{(1)} \quad (3.11a)$$

$$\omega^2 = \omega_2^2: \ X_{21}^{(2)} = \left(\frac{-m_1\omega_2^2 + k_1 + k_2}{k_2}\right)X_{11}^{(2)}, \quad X_{22}^{(2)} = \left(\frac{-m_1\omega_2^2 + k_1 + k_2}{k_2}\right)X_{12}^{(2)}. \quad (3.11b)$$

Now define for simplicity the two variables r_1 and r_2 as

$$r_1 = \left(\frac{-m_1\omega_1^2 + k_1 + k_2}{k_2} \right) = \left(\frac{k_2}{-m_2\omega_1^2 + k_2} \right), \quad r_2 = \left(\frac{-m_1\omega_2^2 + k_1 + k_2}{k_2} \right) = \left(\frac{k_2}{-m_2\omega_2^2 + k_2} \right). \quad (3.12)$$

Then Eq. 3.11 becomes

$$X_{21}^{(1)} = r_1 X_{11}^{(1)}, \quad X_{22}^{(1)} = r_1 X_{12}^{(1)}, \quad X_{21}^{(2)} = r_2 X_{11}^{(2)}, \quad X_{22}^{(2)} = r_2 X_{12}^{(2)}. \quad (3.13)$$

Substituting the results from Eq. 3.13 into Eq. 3.7 gives

$$\begin{bmatrix} -m_1\omega_1^2 + k_1 + k_2 & -k_2 \\ -k_2 & -m_2\omega_1^2 + k_2 \end{bmatrix} \left(\begin{Bmatrix} 1 \\ r_1 \end{Bmatrix} X_{11}^{(1)} \cos \omega t + \begin{Bmatrix} 1 \\ r_1 \end{Bmatrix} X_{12}^{(1)} \sin \omega t \right) = 0 \quad (3.14a)$$

$$\begin{bmatrix} -m_1\omega_2^2 + k_1 + k_2 & -k_2 \\ -k_2 & -m_2\omega_2^2 + k_2 \end{bmatrix} \left(\begin{Bmatrix} 1 \\ r_2 \end{Bmatrix} X_{11}^{(2)} \cos \omega t + \begin{Bmatrix} 1 \\ r_2 \end{Bmatrix} X_{12}^{(2)} \sin \omega t \right) = 0 \quad (3.14b)$$

and for any X_{11}, X_{12}, X_{21}, and X_{22}, it follows from Eq. 3.14 that

$$\begin{bmatrix} -m_1\omega_1^2 + k_1 + k_2 & -k_2 \\ -k_2 & -m_2\omega_1^2 + k_2 \end{bmatrix} \begin{Bmatrix} 1 \\ r_1 \end{Bmatrix} = 0 \quad (3.15a)$$

$$\begin{bmatrix} -m_1\omega_2^2 + k_1 + k_2 & -k_2 \\ -k_2 & -m_2\omega_2^2 + k_2 \end{bmatrix} \begin{Bmatrix} 1 \\ r_2 \end{Bmatrix} = 0. \quad (3.15b)$$

The solution of the response is now obtained using Eq. 3.5. However, since there are two values of ω^2 that satisfy the zero value of the determinant in Eq. 3.8, the response becomes the summation of the two terms of the form similar to Eq. 3.5. Therefore, Eq. 3.5 becomes

$$x_1(t) = \left[X_{11}^{(1)} \cos \omega_1 t + X_{12}^{(1)} \sin \omega_1 t \right] + \left[X_{11}^{(2)} \cos \omega_2 t + X_{12}^{(2)} \sin \omega_2 t \right] \quad (3.16a)$$

$$x_2(t) = \left[r_1 X_{11}^{(1)} \cos \omega_1 t + r_1 X_{12}^{(1)} \sin \omega_1 t \right] + \left[r_2 X_{11}^{(2)} \cos \omega_2 t + r_2 X_{12}^{(2)} \sin \omega_2 t \right]. \quad (3.16b)$$

The first bracket of Eq. 3.16 represents the portion of the response corresponding to the first natural frequency (i.e., $\omega = \omega_1$), and the second bracket of Eq. 3.16 represents the portion of the response corresponding to the second natural frequency (i.e., $\omega = \omega_2$). The velocity of the system response can be obtained by taking the derivative of Eq. 3.16 with respect to the time variable. Doing so gives

$$\dot{x}_1(t) = \left[\omega_1 X_{12}^{(1)} \cos \omega_1 t - \omega_1 X_{11}^{(1)} \sin \omega_1 t \right] + \left[\omega_2 X_{12}^{(2)} \cos \omega_2 t - \omega_2 X_{11}^{(2)} \sin \omega_2 t \right] \quad (3.17a)$$

$$\dot{x}_2(t) = \left[r_1\omega_1 X_{12}^{(1)} \cos \omega_1 t - r_1\omega_1 X_{11}^{(1)} \sin \omega_1 t \right] + \left[r_2\omega_2 X_{12}^{(2)} \cos \omega_2 t - r_2\omega_2 X_{11}^{(2)} \sin \omega_2 t \right]. \quad (3.17b)$$

Note that Eqs. 3.16 and 3.17 contain four unknown variables: $X_{11}^{(1)}$, $X_{12}^{(1)}$, $X_{11}^{(2)}$, and $X_{12}^{(2)}$. These four variables can be expressed in terms of the initial conditions of the system, that is, $x_1(0)$, $x_2(0)$, $\dot{x}_1(0)$, and $\dot{x}_2(0)$. If these initial conditions are substituted into the left-hand side of Eqs. 3.16 and 3.17 at time equal to zero, it follows that

$$x_1(0) \equiv x_{1o} = X_{11}^{(1)} + X_{11}^{(2)} \qquad x_2(0) \equiv x_{2o} = r_1 X_{11}^{(1)} + r_2 X_{11}^{(2)} \quad (3.18a)$$

$$\dot{x}_1(0) \equiv \dot{x}_{1o} = \omega_1 X_{12}^{(1)} + \omega_2 X_{12}^{(2)} \qquad \dot{x}_2(0) \equiv \dot{x}_{2o} = \omega_1 r_1 X_{12}^{(1)} + \omega_2 r_2 X_{12}^{(2)}. \quad (3.18b)$$

Solving for $X_{11}^{(1)}$, $X_{12}^{(1)}$, $X_{11}^{(2)}$, and $X_{12}^{(2)}$ in Eq. 3.18 gives

$$X_{11}^{(1)} = \frac{x_{2o} - r_2 x_{1o}}{r_1 - r_2} \qquad X_{11}^{(2)} = \frac{r_1 x_{1o} - x_{2o}}{r_1 - r_2} \quad (3.19a)$$

$$X_{12}^{(1)} = \frac{\dot{x}_{2o} - r_2 \dot{x}_{1o}}{\omega_1 (r_1 - r_2)} \qquad X_{12}^{(2)} = \frac{r_1 \dot{x}_{1o} - \dot{x}_{2o}}{\omega_2 (r_1 - r_2)}. \tag{3.19b}$$

Finally, it follows from Eqs. 3.16 and 3.17 that

$$\begin{Bmatrix} x_1(t) \\ x_2(t) \end{Bmatrix} = \begin{Bmatrix} 1 \\ r_1 \end{Bmatrix} \left[\left(\frac{x_{2o} - r_2 x_{1o}}{r_1 - r_2} \right) \cos \omega_1 t + \left(\frac{\dot{x}_{2o} - r_2 \dot{x}_{1o}}{\omega_1 (r_1 - r_2)} \right) \sin \omega_1 t \right]$$

$$+ \begin{Bmatrix} 1 \\ r_2 \end{Bmatrix} \left[\left(\frac{r_1 x_{1o} - x_{2o}}{r_1 - r_2} \right) \cos \omega_2 t + \left(\frac{r_1 \dot{x}_{1o} - \dot{x}_{2o}}{\omega_2 (r_1 - r_2)} \right) \sin \omega_2 t \right] \tag{3.20a}$$

$$\begin{Bmatrix} \dot{x}_1(t) \\ \dot{x}_2(t) \end{Bmatrix} = \begin{Bmatrix} 1 \\ r_1 \end{Bmatrix} \left[\left(\frac{\dot{x}_{2o} - r_2 \dot{x}_{1o}}{r_1 - r_2} \right) \cos \omega_1 t - \left(\frac{x_{2o} - r_2 x_{1o}}{r_1 - r_2} \right) \omega_1 \sin \omega_1 t \right]$$

$$+ \begin{Bmatrix} 1 \\ r_2 \end{Bmatrix} \left[\left(\frac{r_1 \dot{x}_{1o} - \dot{x}_{2o}}{r_1 - r_2} \right) \cos \omega_2 t - \left(\frac{r_1 x_{1o} - x_{2o}}{r_1 - r_2} \right) \omega_2 \sin \omega_2 t \right]. \tag{3.20b}$$

Equation 3.20 is the matrix equation of the free vibration response.

The variables ω_1 and ω_2 in Eq. 3.10, which are very important in structural engineering, are called the *undamped natural frequencies of vibration*. The variable ω_1 is the *fundamental undamped natural frequency of vibration*, and the variable ω_2 is the *second undamped natural frequency of vibration*. The units of ω_1 and ω_2 are radians/second (rad/s); these values are often divided by 2π and then the units are in cycles/second or, as it is commonly referred to, hertz (Hz). The corresponding *undamped natural periods of vibration* are defined to be

$$T_1 = \frac{2\pi}{\omega_1}, \qquad T_2 = \frac{2\pi}{\omega_2}. \tag{3.21}$$

The variable T_1 is the *fundamental undamped natural period of vibration*, and the variable T_2 is the *second undamped natural period of vibration*. Recall that $\omega_1 \le \omega_2$, and therefore it follows that $T_1 \ge T_2$.

In mathematics, ω_1^2 and ω_2^2 are the *eigenvalues* of the system, and the corresponding *eigenvectors* of the system, denoted by $\boldsymbol{\phi}_1$ and $\boldsymbol{\phi}_2$, are

$$\boldsymbol{\phi}_1 = \begin{Bmatrix} \varphi_{11} \\ \varphi_{21} \end{Bmatrix} = \begin{Bmatrix} 1 \\ r_1 \end{Bmatrix}, \qquad \boldsymbol{\phi}_2 = \begin{Bmatrix} \varphi_{12} \\ \varphi_{22} \end{Bmatrix} = \begin{Bmatrix} 1 \\ r_2 \end{Bmatrix}. \tag{3.22}$$

These eigenvectors relate to the eigenvalues by the following relationship discussed in Eq. 3.15, which is

$$\left[-\mathbf{M}\omega_1^2 + \mathbf{K} \right] \boldsymbol{\phi}_1 = 0, \qquad \left[-\mathbf{M}\omega_2^2 + \mathbf{K} \right] \boldsymbol{\phi}_2 = 0. \tag{3.23}$$

In structural engineering, $\boldsymbol{\phi}_1$ and $\boldsymbol{\phi}_2$ are called the *mode shapes of vibration* of the structure. The mode shape $\boldsymbol{\phi}_1$ is a vector and is called the *fundamental mode shape of vibration* (or *first mode shape of vibration*), and the mode shape $\boldsymbol{\phi}_2$ is called the *second mode shape of vibration*. Equation 3.20a shows that $x_1(t)$ and $x_2(t)$ are linear combinations of the mode shapes $\boldsymbol{\phi}_1$ and $\boldsymbol{\phi}_2$. Now define

$$q_1(t) = \left(\frac{x_{2o} - r_2 x_{1o}}{r_1 - r_2} \right) \cos \omega_1 t + \left(\frac{\dot{x}_{2o} - r_2 \dot{x}_{1o}}{\omega_1 (r_1 - r_2)} \right) \sin \omega_1 t \tag{3.24a}$$

$$q_2(t) = \left(\frac{r_1 x_{1o} - x_{2o}}{r_1 - r_2} \right) \cos \omega_2 t + \left(\frac{r_1 \dot{x}_{1o} - \dot{x}_{2o}}{\omega_2 (r_1 - r_2)} \right) \sin \omega_2 t. \tag{3.24b}$$

Equation 3.20a can now be written as

$$\mathbf{X}(t) = \begin{Bmatrix} x_1(t) \\ x_2(t) \end{Bmatrix} = \begin{Bmatrix} 1 \\ r_1 \end{Bmatrix} q_1(t) + \begin{Bmatrix} 1 \\ r_2 \end{Bmatrix} q_2(t) = \boldsymbol{\phi}_1 q_1(t) + \boldsymbol{\phi}_2 q_2(t). \tag{3.25}$$

Observe that $q_1(t)$ oscillates at an angular frequency of ω_1, and $q_2(t)$ oscillates at an angular frequency of ω_2.

EXAMPLE 1   *Natural Frequencies of Vibration*

Consider the special case where $k_1 = k_2 = k$ and $m_1 = m_2 = m$. It then follows that the first and second undamped natural frequencies of vibration are obtained from Eq. 3.10, where

$$\omega_1^2 = \frac{3km - \sqrt{9k^2m^2 - 4k^2m^2}}{2m^2} = \left(\frac{3 - \sqrt{5}}{2} \right) \frac{k}{m} \tag{3.26}$$

$$\omega_2^2 = \frac{3km + \sqrt{9k^2m^2 - 4k^2m^2}}{2m^2} = \left(\frac{3 + \sqrt{5}}{2} \right) \frac{k}{m}. \tag{3.27}$$

Then solving for ω_1 and ω_2 gives

$$\omega_1 = 0.618 \sqrt{\frac{k}{m}}, \qquad \omega_2 = 1.618 \sqrt{\frac{k}{m}}. \tag{3.28}$$

Note that for a single degree of freedom system of mass m and stiffness k, the undamped natural frequency of vibration is $\omega_n = \sqrt{k/m}$. Therefore, it follows that

$$\omega_1 = 0.618\,\omega_n, \qquad \omega_2 = 1.618\,\omega_n. \tag{3.29}$$

This example shows that if two identical SDOF systems, each with an undamped natural frequency of vibration of ω_n, are added together, then the new 2DOF system will have one natural frequency (i.e., the fundamental natural frequency) of vibration that is less than ω_n and one natural frequency that is greater than ω_n.

EXAMPLE 2   *Mode Shapes of Vibration*

For the special case where $k_1 = k_2 = k$ and $m_1 = m_2 = m$, the mode shapes of vibration are obtained from Eq. 3.13:

$$X_{21}^{(1)} = r_1 X_{11}^{(1)}, \quad X_{22}^{(1)} = r_1 X_{12}^{(1)}, \quad X_{21}^{(2)} = r_2 X_{11}^{(2)}, \quad X_{22}^{(2)} = r_2 X_{12}^{(2)} \tag{3.30}$$

where

$$r_1 = \left(\frac{-m(0.618\omega_n)^2 + 2k}{k} \right) = 1.618, \quad r_2 = \left(\frac{-m_1(1.618\omega_n)^2 + 2k}{k} \right) = -0.618. \tag{3.31}$$

Note that $X_{2i}^{(j)}$ can only be determined in terms of $X_{1i}^{(j)}$ in Eq. 3.30. However, there are several possible ways to provide the extra equation necessary to obtain unique values. Four ways are noted here:

1. $X_{1i}^{(j)} = 1$. Then $X_{2i}^{(j)} = r_j$.
2. $\text{Max}(X_{1i}^{(j)}, X_{2i}^{(j)}) = 1$. If the greater of the two values is set equal to one, it then follows that the other value will be less than one. This is often done when preparing plots of mode shapes.

3. $\boldsymbol{\phi}_i^T \mathbf{M} \boldsymbol{\phi}_i = 1$. This is referred to as normalizing the mode shape such that the generalized mass is unity. The advantage of this normalization is that $\boldsymbol{\phi}_i^T \mathbf{K} \boldsymbol{\phi}_i = \omega_i^2$.

4. $\boldsymbol{\phi}_i^T \boldsymbol{\phi}_i = 1$. This normalization method sets the length of the mode shape (or eigenvector) equal to unity.

Using these four ways to normalize the mode shapes, it follows that

1. $\boldsymbol{\phi}_1 = \{1.0 \quad 1.618\}^T$ $\boldsymbol{\phi}_2 = \{1.0 \quad -0.618\}^T$

2. $\boldsymbol{\phi}_1 = \{0.618 \quad 1.0\}^T$ $\boldsymbol{\phi}_2 = \{1.0 \quad -0.618\}^T$

3. $\boldsymbol{\phi}_1 = \{0.525/\sqrt{m} \quad 0.851/\sqrt{m}\}^T$ $\boldsymbol{\phi}_2 = \{0.851/\sqrt{m} \quad -0.526/\sqrt{m}\}^T$

4. $\boldsymbol{\phi}_1 = \{0.525 \quad 0.851\}^T$ $\boldsymbol{\phi}_2 = \{0.851 \quad -0.526\}^T$

EXAMPLE 3 *Sensitivity of Natural Frequency of Vibration to Bottom Story Stiffness*

Now consider the case where $k_2 = k$, $m_1 = m_2 = m$, and $k_1 = \alpha k_2 = \alpha k$. It follows from Eq. 3.10 that

$$\omega_1^2 = \frac{(\alpha + 2)km - \sqrt{[(\alpha + 2)km]^2 - 4\alpha m^2 k^2}}{2m^2} = \left(\frac{\alpha + 2 - \sqrt{\alpha^2 + 4}}{2}\right)\omega_n^2 \qquad (3.32)$$

$$\omega_2^2 = \frac{(\alpha + 2)km + \sqrt{[(\alpha + 2)km]^2 - 4\alpha m^2 k^2}}{2m^2} = \left(\frac{\alpha + 2 + \sqrt{\alpha^2 + 4}}{2}\right)\omega_n^2. \qquad (3.33)$$

Note that when $\alpha = 1$, the solution is the same as that obtained in Example 1. Now consider the solution for the fundamental natural frequency of vibration as a function of α. If α is equal to 2, it means that the stiffness of the lower floor is twice the stiffness of the upper floor. It then follows for this case that

$$\omega_1 = \sqrt{2 - \sqrt{2}} \sqrt{\frac{k}{m}} = 0.765 \, \omega_n, \qquad \omega_2 = \sqrt{2 + \sqrt{2}} \sqrt{\frac{k}{m}} = 1.848 \, \omega_n. \qquad (3.34)$$

Similarly, if $\alpha = 0.5$, then the stiffness of the first floor is one half the stiffness of the upper floor. In structural engineering, a structure with small values of α is often referred to as a *soft story*. It follows that

$$\omega_1 = \sqrt{1.25 - \sqrt{1.0625}} \sqrt{\frac{k}{m}} = 0.468 \, \omega_n, \qquad \omega_2 = \sqrt{1.25 + \sqrt{1.0625}} \sqrt{\frac{k}{m}} = 1.510 \, \omega_n. \quad (3.35)$$

EXAMPLE 4 *Free Vibration Response for a System Released from Rest*

Equations 3.20a and 3.25 give the response due to any set of initial conditions. For the special case of $\dot{x}_{1o} = \dot{x}_{2o} = 0$, Eq. 3.20a becomes

$$\begin{Bmatrix} x_1(t) \\ x_2(t) \end{Bmatrix} = \begin{Bmatrix} 1 \\ r_1 \end{Bmatrix} \left(\frac{x_{2o} - r_2 x_{1o}}{r_1 - r_2} \right) \cos \omega_1 t + \begin{Bmatrix} 1 \\ r_2 \end{Bmatrix} \left(\frac{r_1 x_{1o} - x_{2o}}{r_1 - r_2} \right) \cos \omega_2 t. \qquad (3.36)$$

Note that the time variation of the response is the summation of two cosine functions. One cosine function oscillates with the frequency equal to the first natural frequency of vibration, and the other cosine function oscillates with the second natural frequency of vibration.

In general, the initial displacement can be any value and thus the response is a function of both cos $\omega_1 t$ and cos $\omega_2 t$. The response will only be a function of cos $\omega_1 t$ if a special set of initial conditions is selected such that term that multiplies cos $\omega_2 t$ is equal to zero, that is,

$$\left(\frac{r_1 x_{1o} - x_{2o}}{r_1 - r_2}\right) = 0. \tag{3.37}$$

Equation 3.37 is only satisfied if

$$x_{2o} = r_1 x_{1o}. \tag{3.38}$$

Therefore, if x_{2o} and x_{1o} are of the same ratio as the corresponding elements in the fundamental (first) mode shape, then and only then will the response only vary with cos $\omega_1 t$ (i.e., the first natural frequency of vibration), and Eq. 3.36 now becomes

$$\mathbf{X}(t) = \begin{Bmatrix} x_1(t) \\ x_2(t) \end{Bmatrix} = \boldsymbol{\phi}_1 q_1(t) = \begin{Bmatrix} 1 \\ r_1 \end{Bmatrix} x_{1o} \cos \omega_1 t. \tag{3.39}$$

EXAMPLE 5 *Response Derived from Complex Variables*

Equation 3.25 expresses the response in terms of the mode shapes of vibrations. In the development of this section, the response was expressed in terms of sine and cosine functions in Eq. 3.5. An alternate way to develop the key equations in this section is to recognize that the solution to Eq. 3.4 can be obtained by assuming a solution of the form

$$\mathbf{X}(t) = \begin{Bmatrix} x_1(t) \\ x_2(t) \end{Bmatrix} = \begin{Bmatrix} c_1 \\ c_2 \end{Bmatrix} e^{i\omega t}. \tag{3.40}$$

It follows that

$$\dot{\mathbf{X}}(t) = \begin{Bmatrix} \dot{x}_1(t) \\ \dot{x}_2(t) \end{Bmatrix} = i\omega \begin{Bmatrix} c_1 \\ c_2 \end{Bmatrix} e^{i\omega t} \tag{3.41}$$

and

$$\ddot{\mathbf{X}}(t) = \begin{Bmatrix} \ddot{x}_1(t) \\ \ddot{x}_2(t) \end{Bmatrix} = -\omega^2 \begin{Bmatrix} c_1 \\ c_2 \end{Bmatrix} e^{i\omega t}. \tag{3.42}$$

Substituting Eqs. 3.40 to 3.42 into Eq. 3.4, it follows that

$$\left[-\omega^2 \mathbf{M} + \mathbf{K}\right] \begin{Bmatrix} c_1 \\ c_2 \end{Bmatrix} e^{i\omega t} = \mathbf{0}. \tag{3.43}$$

The solution to Eq. 3.43 requires that the determinant of the term in the brackets be equal to zero. This leads to the same equation as Eq. 3.8 and values for ω_1^2 and ω_2^2 in Eq. 3.10. Using ω_1 in Eq. 3.43, it follows that

$$\left[-\omega_1^2 \mathbf{M} + \mathbf{K}\right] \begin{Bmatrix} c_1^{(1)} \\ c_2^{(1)} \end{Bmatrix} e^{i\omega_1 t} = \mathbf{0}. \tag{3.44}$$

The solution of Eq. 3.44 for $c_1^{(1)}$ and $c_2^{(1)}$ follows the same procedure as discussed in this section. This solution will be proportional to the fundamental mode shape of vibration, which is

$$\left\{ \begin{matrix} c_1^{(1)} \\ c_2^{(1)} \end{matrix} \right\} = c^{(1)} \left\{ \begin{matrix} 1 \\ r_1 \end{matrix} \right\} = c^{(1)} \boldsymbol{\phi}_1 \tag{3.45}$$

where $c^{(1)}$ is the constant of proportionality that is to be determined from the given initial condition. Substituting Eq. 3.45 into Eq. 3.44, it follows for any value of $c^{(1)}$ and at any instant in time that

$$\left[-\omega_1^2 \mathbf{M} + \mathbf{K} \right] \boldsymbol{\phi}_1 = \mathbf{0} \tag{3.46}$$

which is the same result as Eq. 3.23. In addition to ω_1, $-\omega_1$ is also a root to Eq. 3.43. It follows from Eq. 3.43 that

$$\left[-\omega_1^2 \mathbf{M} + \mathbf{K} \right] \left\{ \begin{matrix} c_1^{(-1)} \\ c_2^{(-1)} \end{matrix} \right\} e^{-i\omega_1 t} = \mathbf{0}. \tag{3.47}$$

Since the term inside the bracket of Eq. 3.47 is the same as that in Eq. 3.44, it follows that $c_1^{(-1)}$ and $c_2^{(-1)}$ must also be proportional to the fundamental mode shape of vibration, that is,

$$\left\{ \begin{matrix} c_1^{(-1)} \\ c_2^{(-1)} \end{matrix} \right\} = c^{(-1)} \left\{ \begin{matrix} 1 \\ r_1 \end{matrix} \right\} = c^{(-1)} \boldsymbol{\phi}_1 \tag{3.48}$$

where $c^{(-1)}$ is the constant of proportionality that is determined from the given initial condition.

Following the discussion in this example by replacing ω_1^2 with ω_2^2 in Eqs. 3.44 and 3.46 gives

$$\left[-\omega_2^2 \mathbf{M} + \mathbf{K} \right] \left\{ \begin{matrix} c_1^{(2)} \\ c_2^{(2)} \end{matrix} \right\} e^{i\omega_2 t} = \mathbf{0}, \quad \left[-\omega_2^2 \mathbf{M} + \mathbf{K} \right] \left\{ \begin{matrix} c_1^{(-2)} \\ c_2^{(-2)} \end{matrix} \right\} e^{-i\omega_2 t} = \mathbf{0} \tag{3.49}$$

where $\left\{ c_1^{(2)} \quad c_2^{(2)} \right\}^T$ and $\left\{ c_1^{(-2)} \quad c_2^{(-2)} \right\}^T$ must be proportional to the second mode shape of vibration, that is,

$$\left\{ \begin{matrix} c_1^{(2)} \\ c_2^{(2)} \end{matrix} \right\} = c^{(2)} \left\{ \begin{matrix} 1 \\ r_2 \end{matrix} \right\} = c^{(2)} \boldsymbol{\phi}_2, \quad \left\{ \begin{matrix} c_1^{(-2)} \\ c_2^{(-2)} \end{matrix} \right\} = c^{(-2)} \left\{ \begin{matrix} 1 \\ r_2 \end{matrix} \right\} = c^{(-2)} \boldsymbol{\phi}_2. \tag{3.50}$$

Substituting Eq. 3.50 into Eq. 3.49, it follows for any value of $c^{(2)}$ or $c^{(-2)}$ and at any instant in time that

$$\left[-\omega_2^2 \mathbf{M} + \mathbf{K} \right] \boldsymbol{\phi}_2 = \mathbf{0}. \tag{3.51}$$

Equations 3.40 and 3.41 can now be expressed as

$$\left\{ \begin{matrix} x_1(t) \\ x_2(t) \end{matrix} \right\} = \left\{ \begin{matrix} 1 \\ r_1 \end{matrix} \right\} c^{(1)} e^{i\omega_1 t} + \left\{ \begin{matrix} 1 \\ r_1 \end{matrix} \right\} c^{(-1)} e^{-i\omega_1 t} + \left\{ \begin{matrix} 1 \\ r_2 \end{matrix} \right\} c^{(2)} e^{i\omega_2 t} + \left\{ \begin{matrix} 1 \\ r_2 \end{matrix} \right\} c^{(-2)} e^{-i\omega_2 t} \tag{3.52}$$

$$\left\{ \begin{matrix} \dot{x}_1(t) \\ \dot{x}_2(t) \end{matrix} \right\} = \left\{ \begin{matrix} 1 \\ r_1 \end{matrix} \right\} i\omega_1 c^{(1)} e^{i\omega_1 t} - \left\{ \begin{matrix} 1 \\ r_1 \end{matrix} \right\} i\omega_1 c^{(-1)} e^{-i\omega_1 t} + \left\{ \begin{matrix} 1 \\ r_2 \end{matrix} \right\} i\omega_2 c^{(2)} e^{i\omega_2 t} - \left\{ \begin{matrix} 1 \\ r_2 \end{matrix} \right\} i\omega_2 c^{(-2)} e^{-i\omega_2 t}. \tag{3.53}$$

Substituting the initial conditions as given in Eq. 3.18, that is, $x_1(0) = x_{1o}$, $x_2(0) = x_{2o}$, $\dot{x}_1(0) = \dot{x}_{1o}$, and $\dot{x}_2(0) = \dot{x}_{2o}$, into Eqs. 3.52 and 3.53 gives

$$x_{1o} = c^{(1)} + c^{(-1)} + c^{(2)} + c^{(-2)} \tag{3.54}$$

$$x_{2o} = r_1 c^{(1)} + r_1 c^{(-1)} + r_2 c^{(2)} + r_2 c^{(-2)} \tag{3.55}$$

$$\dot{x}_{1o} = c^{(1)} i\omega_1 - c^{(-1)} i\omega_1 + c^{(2)} i\omega_2 - c^{(-2)} i\omega_2 \tag{3.56}$$

$$\dot{x}_{2o} = r_1 c^{(1)} i\omega_1 - r_1 c^{(-1)} i\omega_1 + r_2 c^{(2)} i\omega_2 - r_2 c^{(-2)} i\omega_2. \tag{3.57}$$

Solving for $(c^{(1)} + c^{(-1)})$ and $(c^{(2)} + c^{(-2)})$ from Eqs. 3.54 and 3.55 gives

$$c^{(1)} + c^{(-1)} = \frac{x_{2o} - r_2 x_{1o}}{r_1 - r_2}, \qquad c^{(2)} + c^{(-2)} = \frac{r_1 x_{1o} - x_{2o}}{r_1 - r_2}. \tag{3.58}$$

Similarly, solving for $(c^{(1)} - c^{(-1)})i\omega_1$ and $(c^{(2)} - c^{(-2)})i\omega_2$ from Eqs. 3.56 and 3.57 gives

$$\left(c^{(1)} - c^{(-1)} \right)i\omega_1 = \frac{\dot{x}_{2o} - r_2 \dot{x}_{1o}}{r_1 - r_2}, \qquad \left(c^{(2)} - c^{(-2)} \right)i\omega_2 = \frac{r_1 \dot{x}_{1o} - \dot{x}_{2o}}{r_1 - r_2}. \tag{3.59}$$

Note that Eqs. 3.52 and 3.53 can be expressed as

$$\begin{aligned}
\begin{Bmatrix} x_1(t) \\ x_2(t) \end{Bmatrix} &= \begin{Bmatrix} 1 \\ r_1 \end{Bmatrix} c^{(1)} e^{i\omega_1 t} + \begin{Bmatrix} 1 \\ r_1 \end{Bmatrix} c^{(-1)} e^{-i\omega_1 t} + \begin{Bmatrix} 1 \\ r_2 \end{Bmatrix} c^{(2)} e^{i\omega_2 t} + \begin{Bmatrix} 1 \\ r_2 \end{Bmatrix} c^{(-2)} e^{-i\omega_2 t} \\
&= \begin{Bmatrix} 1 \\ r_1 \end{Bmatrix} \left[\left(c^{(1)} + c^{(-1)} \right) \cos \omega_1 t + \left(c^{(1)} - c^{(-1)} \right)i \sin \omega_1 t \right] \\
&\quad + \begin{Bmatrix} 1 \\ r_2 \end{Bmatrix} \left[\left(c^{(2)} + c^{(-2)} \right) \cos \omega_2 t + \left(c^{(2)} - c^{(-2)} \right)i \sin \omega_2 t \right]
\end{aligned} \tag{3.60}$$

$$\begin{aligned}
\begin{Bmatrix} \dot{x}_1(t) \\ \dot{x}_2(t) \end{Bmatrix} &= \begin{Bmatrix} 1 \\ r_1 \end{Bmatrix} i\omega_1 c^{(1)} e^{i\omega_1 t} - \begin{Bmatrix} 1 \\ r_1 \end{Bmatrix} i\omega_1 c^{(-1)} e^{-i\omega_1 t} + \begin{Bmatrix} 1 \\ r_2 \end{Bmatrix} i\omega_2 c^{(2)} e^{i\omega_2 t} - \begin{Bmatrix} 1 \\ r_2 \end{Bmatrix} i\omega_2 c^{(-2)} e^{-i\omega_2 t} \\
&= \begin{Bmatrix} 1 \\ r_1 \end{Bmatrix} \left[\left(c^{(1)} - c^{(-1)} \right) i\omega_1 \cos \omega_1 t - \left(c^{(1)} + c^{(-1)} \right) \omega_1 \sin \omega_1 t \right] \\
&\quad + \begin{Bmatrix} 1 \\ r_2 \end{Bmatrix} \left[\left(c^{(2)} - c^{(-2)} \right) i\omega_2 \cos \omega_2 t - \left(c^{(2)} + c^{(-2)} \right) \omega_2 \sin \omega_2 t \right].
\end{aligned} \tag{3.61}$$

Finally, substituting the results given in Eqs. 3.58 and 3.59 into Eqs. 3.60 and 3.61, it follows that

$$\begin{aligned}
\begin{Bmatrix} x_1(t) \\ x_2(t) \end{Bmatrix} &= \begin{Bmatrix} 1 \\ r_1 \end{Bmatrix} \left[\left(\frac{x_{2o} - r_2 x_{1o}}{r_1 - r_2} \right) \cos \omega_1 t + \left(\frac{\dot{x}_{2o} - r_2 \dot{x}_{1o}}{\omega_1(r_1 - r_2)} \right) \sin \omega_1 t \right] \\
&\quad + \begin{Bmatrix} 1 \\ r_2 \end{Bmatrix} \left[\left(\frac{r_1 x_{1o} - x_{2o}}{r_1 - r_2} \right) \cos \omega_2 t + \left(\frac{r_1 \dot{x}_{1o} - \dot{x}_{2o}}{\omega_2(r_1 - r_2)} \right) \sin \omega_2 t \right]
\end{aligned} \tag{3.62}$$

$$\begin{aligned}
\begin{Bmatrix} \dot{x}_1(t) \\ \dot{x}_2(t) \end{Bmatrix} &= \begin{Bmatrix} 1 \\ r_1 \end{Bmatrix} \left[\left(\frac{\dot{x}_{2o} - r_2 \dot{x}_{1o}}{r_1 - r_2} \right) \cos \omega_1 t - \left(\frac{x_{2o} - r_2 x_{1o}}{r_1 - r_2} \right) \omega_1 \sin \omega_1 t \right] \\
&\quad + \begin{Bmatrix} 1 \\ r_2 \end{Bmatrix} \left[\left(\frac{r_1 \dot{x}_{1o} - \dot{x}_{2o}}{r_1 - r_2} \right) \cos \omega_2 t - \left(\frac{r_1 x_{1o} - x_{2o}}{r_1 - r_2} \right) \omega_2 \sin \omega_2 t \right]
\end{aligned} \tag{3.63}$$

which are the same as Eq. 3.20.

3.3 UNDAMPED FREE VIBRATION USING THE NORMAL MODE METHOD

A very common method for calculating the response of the structure is called the *normal mode method*. In this method, the response of the structure is represented by a linear combination of the mode shapes and is expressed as

$$\begin{Bmatrix} x_1(t) \\ x_2(t) \end{Bmatrix} = \boldsymbol{\phi}_1 q_1(t) + \boldsymbol{\phi}_2 q_2(t) = \begin{Bmatrix} 1 \\ r_1 \end{Bmatrix} q_1(t) + \begin{Bmatrix} 1 \\ r_2 \end{Bmatrix} q_2(t). \tag{3.64}$$

Equation 3.64 represents a transformation of the description of the structural response from a set of physically understandable Cartesian coordinates (x_1 and x_2) into another set of coordinates that are without physical meaning and are called *normal mode coordinates*. The special transformation in Eq. 3.64 converts the coupled second-order linear differential equations of motion in Eq. 3.2 to two uncoupled second-order linear differential equation of motion. Equation 3.64 is referred to as transforming a set of system coordinates from one space (the x space) to a new space (the q space).

Define $\mathbf{\Phi}$ to be the collection of mode shapes (or eigenvectors) of the 2DOF system as follows:

$$\mathbf{\Phi} = \begin{bmatrix} \mathbf{\phi}_1 & \mathbf{\phi}_2 \end{bmatrix} = \begin{bmatrix} 1 & 1 \\ r_1 & r_2 \end{bmatrix}. \tag{3.65}$$

Then it follows that

$$\begin{Bmatrix} x_1(t) \\ x_2(t) \end{Bmatrix} = \mathbf{\Phi} \begin{Bmatrix} q_1(t) \\ q_2(t) \end{Bmatrix} = \begin{bmatrix} \mathbf{\phi}_1 & \mathbf{\phi}_2 \end{bmatrix} \begin{Bmatrix} q_1(t) \\ q_2(t) \end{Bmatrix} = \mathbf{\phi}_1 q_1(t) + \mathbf{\phi}_2 q_2(t). \tag{3.66}$$

The matrix $\mathbf{\Phi}$ is called the *modal matrix*. The initial conditions for the normal mode coordinates $q_1(t)$ and $q_2(t)$ are obtained in terms of the initial conditions of $x_1(t)$ and $x_2(t)$ using Eq. 3.64 and its time derivative. It follows that

$$\begin{Bmatrix} x_{1o} \\ x_{2o} \end{Bmatrix} = \begin{bmatrix} 1 & 1 \\ r_1 & r_2 \end{bmatrix} \begin{Bmatrix} q_{1o} \\ q_{2o} \end{Bmatrix}, \quad \begin{Bmatrix} \dot{x}_{1o} \\ \dot{x}_{2o} \end{Bmatrix} = \begin{bmatrix} 1 & 1 \\ r_1 & r_2 \end{bmatrix} \begin{Bmatrix} \dot{q}_{1o} \\ \dot{q}_{2o} \end{Bmatrix}. \tag{3.67}$$

The initial conditions of $q_1(t)$ and $q_2(t)$, obtained by solving Eq. 3.67, are

$$q_{1o} = -\left(\frac{r_2}{r_1 - r_2}\right) x_{1o} + \left(\frac{1}{r_1 - r_2}\right) x_{2o}, \quad q_{2o} = \left(\frac{r_1}{r_1 - r_2}\right) x_{1o} - \left(\frac{1}{r_1 - r_2}\right) x_{2o}$$

$$\dot{q}_{1o} = -\left(\frac{r_2}{r_1 - r_2}\right) \dot{x}_{1o} + \left(\frac{1}{r_1 - r_2}\right) \dot{x}_{2o}, \quad \dot{q}_{2o} = \left(\frac{r_1}{r_1 - r_2}\right) \dot{x}_{1o} - \left(\frac{1}{r_1 - r_2}\right) \dot{x}_{2o}. \tag{3.68}$$

The mode shapes of vibration have special features, which make the transformation in Eq. 3.64 special. To show this, consider Eq. 3.23, where

$$\begin{bmatrix} -\mathbf{M}\omega_1^2 + \mathbf{K} \end{bmatrix} \mathbf{\phi}_1 = 0, \quad \begin{bmatrix} -\mathbf{M}\omega_2^2 + \mathbf{K} \end{bmatrix} \mathbf{\phi}_2 = 0. \tag{3.69}$$

When the first equation is premultiplied by $\mathbf{\phi}_2^T$ and the second equation is premultiplied by $\mathbf{\phi}_1^T$, it follows that

$$\mathbf{\phi}_2^T \begin{bmatrix} \mathbf{M}\omega_1^2 - \mathbf{K} \end{bmatrix} \mathbf{\phi}_1 = \omega_1^2 \mathbf{\phi}_2^T \mathbf{M} \mathbf{\phi}_1 - \mathbf{\phi}_2^T \mathbf{K} \mathbf{\phi}_1 = 0 \tag{3.70a}$$

$$\mathbf{\phi}_1^T \begin{bmatrix} \mathbf{M}\omega_2^2 - \mathbf{K} \end{bmatrix} \mathbf{\phi}_2 = \omega_2^2 \mathbf{\phi}_1^T \mathbf{M} \mathbf{\phi}_2 - \mathbf{\phi}_1^T \mathbf{K} \mathbf{\phi}_2 = 0. \tag{3.70b}$$

The matrices $\mathbf{M}$ and $\mathbf{K}$ are symmetrical, and thus the important property of symmetrical matrices holds, which is

$$\mathbf{a}^T \mathbf{M} \mathbf{b} = \mathbf{b}^T \mathbf{M} \mathbf{a}, \quad \mathbf{a}^T \mathbf{K} \mathbf{b} = \mathbf{b}^T \mathbf{K} \mathbf{a} \tag{3.71}$$

for any vectors $\mathbf{a}$ and $\mathbf{b}$. Therefore, subtracting Eq. 3.70a from Eq. 3.70b gives

$$\left(\omega_2^2 - \omega_1^2\right) \mathbf{\phi}_1^T \mathbf{M} \mathbf{\phi}_2 = 0. \tag{3.72}$$

Since $\omega_i \neq \omega_j$, then

$$\mathbf{\phi}_1^T \mathbf{M} \mathbf{\phi}_2 = 0. \tag{3.73}$$

It can be shown that a similar result follows for $\mathbf{K}$, and therefore

$$\boldsymbol{\phi}_1^T \mathbf{K} \boldsymbol{\phi}_2 = 0. \tag{3.74}$$

This is a very important property in structural dynamics and is generally known as the *orthogonal property of eigenvectors* (or *orthogonal property of mode shapes*).

Substituting Eq. 3.64 into the dynamic equilibrium equation given in Eq. 3.3, it follows that

$$\mathbf{M}\big(\boldsymbol{\phi}_1 \ddot{q}_1(t) + \boldsymbol{\phi}_2 \ddot{q}_2(t)\big) + \mathbf{K}\big(\boldsymbol{\phi}_1 q_1(t) + \boldsymbol{\phi}_2 q_2(t)\big) = \mathbf{0}. \tag{3.75}$$

Premultiplying Eq. 3.75 by $\boldsymbol{\phi}_1^T$ gives

$$\big(\boldsymbol{\phi}_1^T \mathbf{M} \boldsymbol{\phi}_1\big)\ddot{q}_1(t) + \big(\boldsymbol{\phi}_1^T \mathbf{M} \boldsymbol{\phi}_2\big)\ddot{q}_2(t) + \big(\boldsymbol{\phi}_1^T \mathbf{K} \boldsymbol{\phi}_1\big)q_1(t) + \big(\boldsymbol{\phi}_1^T \mathbf{K} \boldsymbol{\phi}_2\big)q_2(t) = 0. \tag{3.76}$$

Using Eqs. 3.73 and 3.74, it follows that

$$\big(\boldsymbol{\phi}_1^T \mathbf{M} \boldsymbol{\phi}_1\big)\ddot{q}_1(t) + \big(\boldsymbol{\phi}_1^T \mathbf{K} \boldsymbol{\phi}_1\big)q_1(t) = 0. \tag{3.77}$$

Similarly, premultiplying Eq. 3.75 by $\boldsymbol{\phi}_2^T$ gives

$$\big(\boldsymbol{\phi}_2^T \mathbf{M} \boldsymbol{\phi}_2\big)\ddot{q}_2(t) + \big(\boldsymbol{\phi}_2^T \mathbf{K} \boldsymbol{\phi}_2\big)q_2(t) = 0. \tag{3.78}$$

Note that Eqs. 3.77 and 3.78 are two uncoupled equations where $q_2(t)$ does not appear in Eq. 3.77 nor $q_1(t)$ in Eq. 3.78. Now, define

$$\boldsymbol{\phi}_1^T \mathbf{M} \boldsymbol{\phi}_1 = m_1^*, \quad \boldsymbol{\phi}_2^T \mathbf{M} \boldsymbol{\phi}_2 = m_2^* \tag{3.79a}$$

$$\boldsymbol{\phi}_1^T \mathbf{K} \boldsymbol{\phi}_1 = k_1^*, \quad \boldsymbol{\phi}_2^T \mathbf{K} \boldsymbol{\phi}_2 = k_2^*. \tag{3.79b}$$

It follows from Eqs. 3.77 to 3.79 that

$$m_1^* \ddot{q}_1(t) + k_1^* q_1(t) = 0 \tag{3.80a}$$

$$m_2^* \ddot{q}_2(t) + k_2^* q_2(t) = 0. \tag{3.80b}$$

Expressing Eq. 3.80 in matrix form gives

$$\begin{bmatrix} m_1^* & 0 \\ 0 & m_2^* \end{bmatrix} \begin{Bmatrix} \ddot{q}_1(t) \\ \ddot{q}_2(t) \end{Bmatrix} + \begin{bmatrix} k_1^* & 0 \\ 0 & k_2^* \end{bmatrix} \begin{Bmatrix} q_1(t) \\ q_2(t) \end{Bmatrix} = \mathbf{0}. \tag{3.81}$$

Note that these equations are identical in form to Eq. 2.2 with $c = 0$ and $F_e(t) = 0$. Dividing Eq. 3.80a by m_1^* and Eq. 3.80b by m_2^* gives

$$\ddot{q}_1(t) + \omega_1^2 q_1(t) = 0 \tag{3.82a}$$

$$\ddot{q}_2(t) + \omega_2^2 q_2(t) = 0 \tag{3.82b}$$

where $\omega_1 = \sqrt{k_1^*/m_1^*}$ and $\omega_2 = \sqrt{k_2^*/m_2^*}$. Equation 2.12 with no damping (i.e., $\zeta = 0$) is the same as Eq. 3.82a with $\omega_n = \omega_1$ or Eq. 3.82b with $\omega_n = \omega_2$. Note that Eq. 3.82a is a differential equation containing $q_1(t)$ only, and Eq. 3.82b is another differential equation containing $q_2(t)$ only. Each of these equations can now be solved separately for $q_1(t)$ and $q_2(t)$ using the method discussed in Section 2.2. For a damping ratio of $\zeta = 0$, it follows from Eq. 2.21 that

$$q_1(t) = q_{1o} \cos \omega_1 t + \frac{1}{\omega_1} \dot{q}_{1o} \sin \omega_1 t \tag{3.83a}$$

$$q_2(t) = q_{2o} \cos \omega_2 t + \frac{1}{\omega_2} \dot{q}_{2o} \sin \omega_2 t. \tag{3.83b}$$

Now using Eq. 3.68, it follows that

$$q_1(t) = \left[-\left(\frac{r_2}{r_1 - r_2} \right) x_{1o} + \left(\frac{1}{r_1 - r_2} \right) x_{2o} \right] \cos \omega_1 t + \frac{1}{\omega_1} \left[-\left(\frac{r_2}{r_1 - r_2} \right) \dot{x}_{1o} + \left(\frac{1}{r_1 - r_2} \right) \dot{x}_{2o} \right] \sin \omega_1 t \quad (3.84a)$$

$$q_2(t) = \left[\left(\frac{r_1}{r_1 - r_2} \right) x_{1o} - \left(\frac{1}{r_1 - r_2} \right) x_{2o} \right] \cos \omega_2 t + \frac{1}{\omega_2} \left[\left(\frac{r_1}{r_1 - r_2} \right) \dot{x}_{1o} - \left(\frac{1}{r_1 - r_2} \right) \dot{x}_{2o} \right] \sin \omega_2 t. \quad (3.84b)$$

Substituting Eq. 3.84 into Eq. 3.64 gives

$$\begin{Bmatrix} x_1(t) \\ x_2(t) \end{Bmatrix} = \begin{Bmatrix} 1 \\ r_1 \end{Bmatrix} \left[\left(\frac{x_{2o} - r_2 x_{1o}}{r_1 - r_2} \right) \cos \omega_1 t + \left(\frac{\dot{x}_{2o} - r_2 \dot{x}_{1o}}{\omega_1 (r_1 - r_2)} \right) \sin \omega_1 t \right]$$
$$+ \begin{Bmatrix} 1 \\ r_2 \end{Bmatrix} \left[\left(\frac{r_1 x_{1o} - x_{2o}}{r_1 - r_2} \right) \cos \omega_2 t + \left(\frac{r_1 \dot{x}_{1o} - \dot{x}_{2o}}{\omega_2 (r_1 - r_2)} \right) \sin \omega_2 t \right]. \quad (3.85)$$

Equation 3.85 is the same as Eq. 3.20a.

EXAMPLE 1   *Free Vibration of a System Released from Rest at First Mode Displacement*

Consider Example 2 in Section 3.2 for the special case where $k_1 = k_2 = k$ and $m_1 = m_2 = m$. If the system is released from rest with initial displacement in the ratio of the first mode shape, then

$$\begin{Bmatrix} x_1(0) \\ x_2(0) \end{Bmatrix} = \begin{Bmatrix} x_{1o} \\ x_{2o} \end{Bmatrix} = \begin{Bmatrix} 1 \\ r_1 \end{Bmatrix}, \quad \begin{Bmatrix} \dot{x}_1(0) \\ \dot{x}_2(0) \end{Bmatrix} = \begin{Bmatrix} \dot{x}_{1o} \\ \dot{x}_{2o} \end{Bmatrix} = \begin{Bmatrix} 0 \\ 0 \end{Bmatrix}. \quad (3.86)$$

It then follows from Eq. 3.84 that

$$q_1(t) = \left[-\left(\frac{r_2}{r_1 - r_2} \right)(1) + \left(\frac{1}{r_1 - r_2} \right)(r_1) \right] \cos \omega_1 t + (0) \sin \omega_1 t = \cos \omega_1 t \quad (3.87)$$

$$q_2(t) = \left[\left(\frac{r_1}{r_1 - r_2} \right)(1) - \left(\frac{1}{r_1 - r_2} \right)(r_1) \right] \cos \omega_2 t + (0) \sin \omega_2 t = 0 \quad (3.88)$$

where $\omega_1 = 0.618 \sqrt{k/m}$ and r_1 and r_2 are given in Eq. 3.31. Substituting Eqs. 3.87 and 3.88 into Eq. 3.64 gives

$$x_1(t) = \cos \omega_1 t \quad (3.89)$$

$$x_2(t) = r_1 \cos \omega_1 t. \quad (3.90)$$

EXAMPLE 2   *Free Vibration of a System Released from Rest*

Consider again Example 2 in Section 3.2 for the special case where $k_1 = k_2 = k$ and $m_1 = m_2 = m$ and with initial conditions

$$\begin{Bmatrix} x_1(0) \\ x_2(0) \end{Bmatrix} = \begin{Bmatrix} x_{1o} \\ x_{2o} \end{Bmatrix} = \begin{Bmatrix} 1 \\ 1 \end{Bmatrix}, \quad \begin{Bmatrix} \dot{x}_1(0) \\ \dot{x}_2(0) \end{Bmatrix} = \begin{Bmatrix} \dot{x}_{1o} \\ \dot{x}_{2o} \end{Bmatrix} = \begin{Bmatrix} 0 \\ 0 \end{Bmatrix}. \quad (3.91)$$

It then follows from Eq. 3.84 that

$$q_1(t) = \left[-\left(\frac{r_2}{r_1 - r_2} \right)(1) + \left(\frac{1}{r_1 - r_2} \right)(1) \right] \cos \omega_1 t + (0) \sin \omega_1 t = \left(\frac{1 - r_2}{r_1 - r_2} \right) \cos \omega_1 t \qquad (3.92)$$

$$q_2(t) = \left[\left(\frac{r_1}{r_1 - r_2} \right)(1) - \left(\frac{1}{r_1 - r_2} \right)(1) \right] \cos \omega_2 t + (0) \sin \omega_2 t = \left(\frac{r_1 - 1}{r_1 - r_2} \right) \cos \omega_2 t \qquad (3.93)$$

where $\omega_1 = 0.618\sqrt{k/m}$ and $\omega_2 = 1.618\sqrt{k/m}$, and r_1 and r_2 are given in Eq. 3.31. Substituting Eqs. 3.92 and 3.93 into Eq. 3.64 gives

$$x_1(t) = \left(\frac{1 - r_2}{r_1 - r_2} \right) \cos \omega_1 t + \left(\frac{r_1 - 1}{r_1 - r_2} \right) \cos \omega_2 t \qquad (3.94)$$

$$x_2(t) = r_1 \left(\frac{1 - r_2}{r_1 - r_2} \right) \cos \omega_1 t + r_2 \left(\frac{r_1 - 1}{r_1 - r_2} \right) \cos \omega_2 t. \qquad (3.95)$$

3.4   UNDAMPED RESPONSE USING THE NORMAL MODE METHOD

Consider the dynamic equilibrium equation for a 2DOF system given in Eq. 3.2 but without damping. It follows that

$$\begin{bmatrix} m_1 & 0 \\ 0 & m_2 \end{bmatrix} \begin{Bmatrix} \ddot{x}_1 \\ \ddot{x}_2 \end{Bmatrix} + \begin{bmatrix} k_1 + k_2 & -k_2 \\ -k_2 & k_2 \end{bmatrix} \begin{Bmatrix} x_1 \\ x_2 \end{Bmatrix} = \begin{Bmatrix} F_1(t) \\ F_2(t) \end{Bmatrix}. \qquad (3.96)$$

Define

$$\mathbf{M} = \begin{bmatrix} m_1 & 0 \\ 0 & m_2 \end{bmatrix}, \quad \mathbf{K} = \begin{bmatrix} k_1 + k_2 & -k_2 \\ -k_2 & k_2 \end{bmatrix}, \quad \mathbf{X}(t) = \begin{Bmatrix} x_1(t) \\ x_2(t) \end{Bmatrix}, \quad \mathbf{F}_e(t) = \begin{Bmatrix} F_1(t) \\ F_2(t) \end{Bmatrix}.$$

It follows that

$$\mathbf{M}\ddot{\mathbf{X}}(t) + \mathbf{K}\mathbf{X}(t) = \mathbf{F}_e(t). \qquad (3.97)$$

As discussed in the previous section, the normal mode method defines a special coordinate transformation from the physically recognizable Cartesian coordinate system to the special normal mode coordinates. Define

$$\mathbf{X}(t) = \boldsymbol{\phi}_1 q_1(t) + \boldsymbol{\phi}_2 q_2(t) = \boldsymbol{\Phi} \mathbf{Q}(t) \qquad (3.98)$$

where

$$\boldsymbol{\Phi} = \begin{bmatrix} \boldsymbol{\phi}_1 & \boldsymbol{\phi}_2 \end{bmatrix} = \begin{bmatrix} 1 & 1 \\ r_1 & r_2 \end{bmatrix}, \quad \mathbf{Q}(t) = \begin{Bmatrix} q_1(t) \\ q_2(t) \end{Bmatrix}. \qquad (3.99)$$

It follows after Eq. 3.98 is substituted into Eq. 3.97 that

$$\mathbf{M}\big(\boldsymbol{\phi}_1 \ddot{q}_1(t) + \boldsymbol{\phi}_2 \ddot{q}_2(t)\big) + \mathbf{K}\big(\boldsymbol{\phi}_1 q_1(t) + \boldsymbol{\phi}_2 q_2(t)\big) = \mathbf{F}_e(t). \qquad (3.100)$$

As was done in Section 3.3 from Eqs. 3.75 to 3.78, premultiplying Eq. 3.100 by $\boldsymbol{\phi}_1^T$ gives

$$\big(\boldsymbol{\phi}_1^T \mathbf{M} \boldsymbol{\phi}_1\big) \ddot{q}_1(t) + \big(\boldsymbol{\phi}_1^T \mathbf{M} \boldsymbol{\phi}_2\big) \ddot{q}_2(t) + \big(\boldsymbol{\phi}_1^T \mathbf{K} \boldsymbol{\phi}_1\big) q_1(t) + \big(\boldsymbol{\phi}_1^T \mathbf{K} \boldsymbol{\phi}_2\big) q_2(t) = \boldsymbol{\phi}_1^T \mathbf{F}_e(t) \qquad (3.101)$$

and premultiplying Eq. 3.100 by $\boldsymbol{\phi}_2^T$ gives

$$\big(\boldsymbol{\phi}_2^T \mathbf{M} \boldsymbol{\phi}_1\big) \ddot{q}_1(t) + \big(\boldsymbol{\phi}_2^T \mathbf{M} \boldsymbol{\phi}_2\big) \ddot{q}_2(t) + \big(\boldsymbol{\phi}_2^T \mathbf{K} \boldsymbol{\phi}_1\big) q_1(t) + \big(\boldsymbol{\phi}_2^T \mathbf{K} \boldsymbol{\phi}_2\big) q_2(t) = \boldsymbol{\phi}_2^T \mathbf{F}_e(t). \qquad (3.102)$$

Using the special properties of the mode shape discussed in Eqs. 3.73, 3.74, and 3.79, which are

$$\boldsymbol{\phi}_1^T \mathbf{M} \boldsymbol{\phi}_1 = m_1^*, \quad \boldsymbol{\phi}_2^T \mathbf{M} \boldsymbol{\phi}_2 = m_2^*, \quad \boldsymbol{\phi}_1^T \mathbf{M} \boldsymbol{\phi}_2 = 0, \quad \boldsymbol{\phi}_2^T \mathbf{M} \boldsymbol{\phi}_1 = 0 \tag{3.103a}$$

$$\boldsymbol{\phi}_1^T \mathbf{K} \boldsymbol{\phi}_1 = k_1^*, \quad \boldsymbol{\phi}_2^T \mathbf{K} \boldsymbol{\phi}_2 = k_2^*, \quad \boldsymbol{\phi}_1^T \mathbf{K} \boldsymbol{\phi}_2 = 0, \quad \boldsymbol{\phi}_2^T \mathbf{K} \boldsymbol{\phi}_1 = 0 \tag{3.103b}$$

then substituting Eq. 3.103 into Eqs. 3.101 and 3.102 gives

$$m_1^* \ddot{q}_1(t) + k_1^* q_1(t) = \boldsymbol{\phi}_1^T \mathbf{F}_e(t) \tag{3.104a}$$

$$m_2^* \ddot{q}_2(t) + k_2^* q_2(t) = \boldsymbol{\phi}_2^T \mathbf{F}_e(t). \tag{3.104b}$$

Representing Eq. 3.104 in matrix form gives

$$\begin{bmatrix} m_1^* & 0 \\ 0 & m_2^* \end{bmatrix} \begin{Bmatrix} \ddot{q}_1(t) \\ \ddot{q}_2(t) \end{Bmatrix} + \begin{bmatrix} k_1^* & 0 \\ 0 & k_2^* \end{bmatrix} \begin{Bmatrix} q_1(t) \\ q_2(t) \end{Bmatrix} = \begin{bmatrix} \boldsymbol{\phi}_1 & \boldsymbol{\phi}_2 \end{bmatrix}^T \begin{Bmatrix} F_1(t) \\ F_2(t) \end{Bmatrix} = \begin{bmatrix} 1 & r_1 \\ 1 & r_2 \end{bmatrix} \begin{Bmatrix} F_1(t) \\ F_2(t) \end{Bmatrix}. \tag{3.105}$$

Expanding Eq. 3.105 gives

$$m_1^* \ddot{q}_1(t) + k_1^* q_1(t) = F_1(t) + r_1 F_2(t) \tag{3.106a}$$

$$m_2^* \ddot{q}_2(t) + k_2^* q_2(t) = F_1(t) + r_2 F_2(t). \tag{3.106b}$$

Dividing Eq. 3.106a by m_1^* and Eq. 3.106b by m_2^*, it follows that

$$\ddot{q}_1(t) + \omega_1^2 q_1(t) = \left[F_1(t) + r_1 F_2(t) \right]/m_1^* \tag{3.107a}$$

$$\ddot{q}_2(t) + \omega_2^2 q_2(t) = \left[F_1(t) + r_2 F_2(t) \right]/m_2^* \tag{3.107b}$$

where $\omega_1 = \sqrt{k_1^*/m_1^*}$ and $\omega_2 = \sqrt{k_2^*/m_2^*}$. Again, the 2DOF system has been decoupled into two uncoupled differential equations of motion. Each equation is like an SDOF system and it can be solved using the solution methods discussed in Chapter 2. The initial conditions for the normal mode coordinates $q_1(t)$ and $q_2(t)$ are obtained from the initial conditions for $x_1(t)$ and $x_2(t)$ (i.e. x_{1o}, $\dot{x}_{1o}$, x_{2o}, and $\dot{x}_{2o}$) using Eq. 3.68.

EXAMPLE 1 Uniform Forcing Function

Let

$$F_1(t) = \begin{cases} F_1 & t \geq 0 \\ 0 & \text{otherwise} \end{cases}, \quad F_2(t) = \begin{cases} F_2 & t \geq 0 \\ 0 & \text{otherwise.} \end{cases} \tag{3.108}$$

The matrix dynamic equilibrium equation in the physically understandable Cartesian coordinate system is

$$\begin{bmatrix} m_1 & 0 \\ 0 & m_2 \end{bmatrix} \begin{Bmatrix} \ddot{x}_1(t) \\ \ddot{x}_2(t) \end{Bmatrix} + \begin{bmatrix} k_1 + k_2 & -k_2 \\ -k_2 & k_2 \end{bmatrix} \begin{Bmatrix} x_1(t) \\ x_2(t) \end{Bmatrix} = \begin{Bmatrix} F_1 \\ F_2 \end{Bmatrix}. \tag{3.109}$$

Now express the response in terms of the normal mode coordinates using

$$\mathbf{X}(t) = \begin{Bmatrix} x_1(t) \\ x_2(t) \end{Bmatrix} = \boldsymbol{\phi}_1 q_1(t) + \boldsymbol{\phi}_2 q_2(t). \tag{3.110}$$

It follows that

$$m_1^* \ddot{q}_1(t) + k_1^* q_1(t) = F_1 + r_1 F_2 \tag{3.111}$$

$$m_2^* \ddot{q}_2(t) + k_2^* q_2(t) = F_1 + r_2 F_2. \tag{3.112}$$

If the initial conditions are zero, it follows from Eq. 2.47 and the properties of linear systems that the response due to F_1 can be added to the response due to F_2:

$$q_1(t) = \left(\frac{F_1 + r_1 F_2}{k_1^*}\right)(1 - \cos\omega_1 t) \tag{3.113}$$

$$q_2(t) = \left(\frac{F_1 + r_2 F_2}{k_2^*}\right)(1 - \cos\omega_2 t). \tag{3.114}$$

Now substituting Eqs. 3.113 and 3.114 into Eq. 3.110 gives

$$x_1(t) = \left(\frac{F_1 + r_1 F_2}{k_1^*}\right)(1 - \cos\omega_1 t) + \left(\frac{F_1 + r_2 F_2}{k_2^*}\right)(1 - \cos\omega_2 t) \tag{3.115}$$

$$x_2(t) = r_1\left(\frac{F_1 + r_1 F_2}{k_1^*}\right)(1 - \cos\omega_1 t) + r_2\left(\frac{F_1 + r_2 F_2}{k_2^*}\right)(1 - \cos\omega_2 t). \tag{3.116}$$

For an arbitrary set of initial conditions, Eqs. 3.115 and 3.116 must be modified with an additional term from the homogeneous solution as given in Eq. 3.20a, and thus

$$\begin{aligned}
\begin{Bmatrix} x_1(t) \\ x_2(t) \end{Bmatrix} &= \begin{Bmatrix} 1 \\ r_1 \end{Bmatrix}\left[\left(\frac{x_{2o} - r_2 x_{1o}}{r_1 - r_2}\right)\cos\omega_1 t + \left(\frac{\dot{x}_{2o} - r_2 \dot{x}_{1o}}{\omega_1(r_1 - r_2)}\right)\sin\omega_1 t + \left(\frac{F_1 + r_1 F_2}{k_1^*}\right)(1 - \cos\omega_1 t)\right] \\
&+ \begin{Bmatrix} 1 \\ r_2 \end{Bmatrix}\left[\left(\frac{r_1 x_{1o} - x_{2o}}{r_1 - r_2}\right)\cos\omega_2 t + \left(\frac{r_1 \dot{x}_{1o} - \dot{x}_{2o}}{\omega_2(r_1 - r_2)}\right)\sin\omega_2 t + \left(\frac{F_1 + r_2 F_2}{k_2^*}\right)(1 - \cos\omega_2 t)\right].
\end{aligned} \tag{3.117}$$

EXAMPLE 2   *Harmonic Forcing Function*

Let

$$F_1(t) = \begin{cases} F_1 \sin\omega_o t & t \geq 0 \\ 0 & \text{otherwise,} \end{cases} \qquad F_2(t) = \begin{cases} F_2 \sin\omega_o t & t \geq 0 \\ 0 & \text{otherwise.} \end{cases} \tag{3.118}$$

The matrix dynamic equilibrium equations in the physically understandable Cartesian coordinate system is

$$\begin{bmatrix} m_1 & 0 \\ 0 & m_2 \end{bmatrix}\begin{Bmatrix} \ddot{x}_1(t) \\ \ddot{x}_2(t) \end{Bmatrix} + \begin{bmatrix} k_1 + k_2 & -k_2 \\ -k_2 & k_2 \end{bmatrix}\begin{Bmatrix} x_1(t) \\ x_2(t) \end{Bmatrix} = \begin{Bmatrix} F_1 \sin\omega_o t \\ F_2 \sin\omega_o t \end{Bmatrix}. \tag{3.119}$$

Now express the response in terms of the normal mode coordinates using

$$\mathbf{X}(t) = \begin{Bmatrix} x_1(t) \\ x_2(t) \end{Bmatrix} = \boldsymbol{\phi}_1 q_1(t) + \boldsymbol{\phi}_2 q_2(t). \tag{3.120}$$

It follows that

$$m_1^* \ddot{q}_1(t) + k_1^* q_1(t) = (F_1 + r_1 F_2)\sin\omega_o t \tag{3.121}$$

$$m_2^* \ddot{q}_2(t) + k_2^* q_2(t) = (F_1 + r_2 F_2)\sin\omega_o t. \tag{3.122}$$

The solution for $q_1(t)$ takes the form

$$q_1(t) = A_1 \sin\omega_o t \tag{3.123}$$

and substituting Eq. 3.123 into Eq. 3.121 gives

$$\left(-m_1^* \omega_o^2 A_1 + k_1^* A_1\right)\sin\omega_o t = (F_1 + r_1 F_2)\sin\omega_o t. \tag{3.124}$$

Solving for A_1 in Eq. 3.124 gives

$$A_1 = \left(\frac{F_1 + r_1 F_2}{k_1^* - m_1^* \omega_o^2} \right) = \left(\frac{(F_1 + r_1 F_2)/k_1^*}{1 - (\omega_o/\omega_1)^2} \right). \tag{3.125}$$

Substituting Eq. 3.125 into Eq. 3.123 and incorporating the initial conditions give

$$q_1(t) = C_1 \cos \omega_1 t + D_1 \sin \omega_1 t + \left[\frac{(F_1 + r_1 F_2)/k_1^*}{1 - (\omega_o/\omega_1)^2} \right] \sin \omega_o t. \tag{3.126}$$

If the initial conditions are zero, it then follows from Eq. 3.126 that

$$C_1 = 0, \quad D_1 = -\left(\frac{\omega_o}{\omega_1} \right) \left(\frac{(F_1 + r_1 F_2)/k_1^*}{1 - (\omega_o/\omega_1)^2} \right). \tag{3.127}$$

Substituting Eq. 3.127 into Eq. 3.126 gives

$$q_1(t) = \left(\frac{(F_1 + r_1 F_2)/k_1^*}{1 - (\omega_o/\omega_1)^2} \right) \left(\sin \omega_o t - \frac{\omega_o}{\omega_1} \sin \omega_1 t \right). \tag{3.128}$$

The same solution method is used to obtain

$$q_2(t) = \left(\frac{(F_1 + r_2 F_2)/k_2^*}{1 - (\omega_o/\omega_2)^2} \right) \left(\sin \omega_o t - \frac{\omega_o}{\omega_2} \sin \omega_2 t \right). \tag{3.129}$$

Finally, substituting Eqs. 3.128 and 3.129 into Eq. 3.120 gives

$$
\begin{aligned}
x_1(t) = {} & \left[\frac{(F_1 + r_1 F_2)/k_1^*}{1 - (\omega_o/\omega_1)^2} \right] \left(\sin \omega_o t - \left(\frac{\omega_o}{\omega_1} \right) \sin \omega_1 t \right) \\
& + \left[\frac{(F_1 + r_2 F_2)/k_2^*}{1 - (\omega_o/\omega_2)^2} \right] \left(\sin \omega_o t - \left(\frac{\omega_o}{\omega_2} \right) \sin \omega_2 t \right)
\end{aligned}
\tag{3.130}
$$

$$
\begin{aligned}
x_2(t) = {} & \left[\frac{r_1(F_1 + r_1 F_2)/k_1^*}{1 - (\omega_o/\omega_1)^2} \right] \left(\sin \omega_o t - \left(\frac{\omega_o}{\omega_1} \right) \sin \omega_1 t \right) \\
& + \left[\frac{r_2(F_1 + r_2 F_2)/k_2^*}{1 - (\omega_o/\omega_2)^2} \right] \left(\sin \omega_o t - \left(\frac{\omega_o}{\omega_2} \right) \sin \omega_2 t \right).
\end{aligned}
\tag{3.131}
$$

EXAMPLE 3   *A Proportional Forcing Function*

Consider the special type of forcing function called a *proportional forcing function*. This forcing function has the special form

$$\mathbf{F}_e(t) = \begin{Bmatrix} F_1(t) \\ F_2(t) \end{Bmatrix} = \begin{Bmatrix} A_1 f(t) \\ A_2 f(t) \end{Bmatrix} = \begin{Bmatrix} A_1 \\ A_2 \end{Bmatrix} f(t) \tag{3.132}$$

where A_1 and A_2 are constants. The two forces, $F_1(t)$ and $F_2(t)$, are proportional to one function of time, which is $f(t)$. It follows that the response of the system from Eq. 3.98 is

$$\mathbf{X}(t) = \begin{Bmatrix} x_1(t) \\ x_2(t) \end{Bmatrix} = \boldsymbol{\phi}_1 q_1(t) + \boldsymbol{\phi}_2 q_2(t). \tag{3.133}$$

The normal mode responses are obtained from Eq. 3.107. When Eq. 3.132 is substituted into Eq. 3.107, it follows that

$$\ddot{q}_1(t) + \omega_1^2 q_1(t) = \left[F_1(t) + r_1 F_2(t)\right]/m_1^* = \Gamma_1 f(t) \tag{3.134}$$

$$\ddot{q}_2(t) + \omega_2^2 q_2(t) = \left[F_1(t) + r_2 F_2(t)\right]/m_2^* = \Gamma_2 f(t) \tag{3.135}$$

where

$$\Gamma_1 = \frac{A_1 + r_1 A_2}{m_1^*}, \qquad \Gamma_2 = \frac{A_1 + r_2 A_2}{m_2^*}. \tag{3.136}$$

3.5   DAMPED RESPONSE WITH RAYLEIGH DAMPING USING THE NORMAL MODE METHOD

Consider the dynamic equilibrium equation from Eq. 3.2:

$$\begin{bmatrix} m_1 & 0 \\ 0 & m_2 \end{bmatrix}\begin{Bmatrix} \ddot{x}_1 \\ \ddot{x}_2 \end{Bmatrix} + \begin{bmatrix} c_1 + c_2 & -c_2 \\ -c_2 & c_2 \end{bmatrix}\begin{Bmatrix} \dot{x}_1 \\ \dot{x}_2 \end{Bmatrix} + \begin{bmatrix} k_1 + k_2 & -k_2 \\ -k_2 & k_2 \end{bmatrix}\begin{Bmatrix} x_1 \\ x_2 \end{Bmatrix} = \begin{Bmatrix} F_1(t) \\ F_2(t) \end{Bmatrix}. \tag{3.137}$$

Let

$$\mathbf{M} = \begin{bmatrix} m_1 & 0 \\ 0 & m_2 \end{bmatrix}, \qquad \mathbf{C} = \begin{bmatrix} c_1 + c_2 & -c_2 \\ -c_2 & c_2 \end{bmatrix}, \qquad \mathbf{K} = \begin{bmatrix} k_1 + k_2 & -k_2 \\ -k_2 & k_2 \end{bmatrix}$$

$$\mathbf{X}(t) = \begin{Bmatrix} x_1(t) \\ x_2(t) \end{Bmatrix}, \qquad \mathbf{F}_e(t) = \begin{Bmatrix} F_1(t) \\ F_2(t) \end{Bmatrix}.$$

It follows from Eq. 3.137 that

$$\mathbf{M}\ddot{\mathbf{X}}(t) + \mathbf{C}\dot{\mathbf{X}}(t) + \mathbf{K}\mathbf{X}(t) = \mathbf{F}_e(t). \tag{3.138}$$

In the previous section, the normal mode method was used to transform a system of coupled equations of motion in the Cartesian coordinates to uncoupled equations of motion in the normal mode coordinate system. However, because of the presence of the damping matrix, it is generally not possible to use the normal mode method unless this damping matrix is of a special form, called *proportional damping*. One type of proportional damping is where the damping matrix is a linear combination of the mass and stiffness matrices, that is,

$$\mathbf{C} = \alpha\mathbf{M} + \beta\mathbf{K}. \tag{3.139}$$

This type of proportional damping is called *Rayleigh damping*. Consider the relationship between the Cartesian coordinates and the normal mode coordinates as before:

$$\mathbf{X}(t) = \begin{Bmatrix} x_1(t) \\ x_2(t) \end{Bmatrix} = \boldsymbol{\phi}_1 q_1(t) + \boldsymbol{\phi}_2 q_2(t). \tag{3.140}$$

Substituting Eq. 3.140 into Eq. 3.137 and premultiplying by $\boldsymbol{\phi}_1^T$ gives

$$\left(\boldsymbol{\phi}_1^T\mathbf{M}\boldsymbol{\phi}_1\right)\ddot{q}_1(t) + \left(\boldsymbol{\phi}_1^T\mathbf{C}\boldsymbol{\phi}_1\right)\dot{q}_1(t) + \left(\boldsymbol{\phi}_1^T\mathbf{C}\boldsymbol{\phi}_2\right)\dot{q}_2(t) + \left(\boldsymbol{\phi}_1^T\mathbf{K}\boldsymbol{\phi}_1\right)q_1(t) = \boldsymbol{\phi}_1^T\mathbf{F}_e(t). \tag{3.141}$$

Note that the term $(\boldsymbol{\phi}_1^T\mathbf{C}\boldsymbol{\phi}_2)$ is in general not equal to zero. However, if the damping matrix $\mathbf{C}$ is of the Rayleigh damping type, then it follows from the orthogonality of mode shapes that

$$\boldsymbol{\phi}_1^T\mathbf{C}\boldsymbol{\phi}_1 = \boldsymbol{\phi}_1^T\left(\alpha\mathbf{M} + \beta\mathbf{K}\right)\boldsymbol{\phi}_1 = \alpha\left(\boldsymbol{\phi}_1^T\mathbf{M}\boldsymbol{\phi}_1\right) + \beta\left(\boldsymbol{\phi}_1^T\mathbf{K}\boldsymbol{\phi}_1\right) = \alpha m_1^* + \beta k_1^* \tag{3.142a}$$

$$\boldsymbol{\phi}_1^T\mathbf{C}\boldsymbol{\phi}_2 = \boldsymbol{\phi}_1^T\left(\alpha\mathbf{M} + \beta\mathbf{K}\right)\boldsymbol{\phi}_2 = \alpha\left(\boldsymbol{\phi}_1^T\mathbf{M}\boldsymbol{\phi}_2\right) + \beta\left(\boldsymbol{\phi}_1^T\mathbf{K}\boldsymbol{\phi}_2\right) = 0 + 0 = 0. \tag{3.142b}$$

Therefore, Eq. 3.141 becomes

$$m_1^* \ddot{q}_1(t) + \left(\alpha m_1^* + \beta k_1^*\right) \dot{q}_1(t) + k_1^* q_1(t) = \boldsymbol{\phi}_1^T \mathbf{F}_e(t). \tag{3.143}$$

Define

$$c_1^* = \alpha m_1^* + \beta k_1^*. \tag{3.144}$$

It follows from Eq. 3.143 that

$$m_1^* \ddot{q}_1(t) + c_1^* \dot{q}_1(t) + k_1^* q_1(t) = \boldsymbol{\phi}_1^T \mathbf{F}_e(t). \tag{3.145}$$

Similarly, it follows when substituting Eq. 3.140 into Eq. 3.137 and premultiplying by $\boldsymbol{\phi}_2^T$ that

$$m_2^* \ddot{q}_2(t) + c_2^* \dot{q}_2(t) + k_2^* q_2(t) = \boldsymbol{\phi}_2^T \mathbf{F}_e(t) \tag{3.146}$$

where

$$c_2^* = \alpha m_2^* + \beta k_2^*. \tag{3.147}$$

Dividing Eq. 3.144 by m_1^* and Eq. 3.147 by m_2^* gives

$$\boxed{2\zeta_1\omega_1 = \alpha + \beta\omega_1^2, \quad 2\zeta_2\omega_2 = \alpha + \beta\omega_2^2} \tag{3.148}$$

where $\omega_1 = \sqrt{k_1^*/m_1^*}$, $\omega_2 = \sqrt{k_2^*/m_2^*}$, $2\zeta_1\omega_1 = c_1^*/m_1^*$, and $2\zeta_2\omega_2 = c_2^*/m_2^*$.

In many structural engineering applications, the damping ratios for the first and second modes are estimated by the structural engineers based primarily on their professional experience. Then the proportionality constants (i.e., α and β) can be solved using Eq. 3.148 to give

$$\alpha = \frac{2\omega_1\omega_2(\zeta_1\omega_2 - \zeta_2\omega_1)}{\omega_2^2 - \omega_1^2}, \quad \beta = \frac{2(\zeta_2\omega_2 - \zeta_1\omega_1)}{\omega_2^2 - \omega_1^2}. \tag{3.149}$$

The normal mode method can now be applied to calculate the response of the system using the normal mode coordinates and Eqs. 3.145 and 3.146. For convenience, repeating Eqs. 3.145 and 3.146 gives

$$m_1^* \ddot{q}_1(t) + c_1^* \dot{q}_1(t) + k_1^* q_1(t) = \boldsymbol{\phi}_1^T \mathbf{F}_e(t) = F_1(t) + r_1 F_2(t) \tag{3.150a}$$

$$m_2^* \ddot{q}_2(t) + c_2^* \dot{q}_2(t) + k_2^* q_2(t) = \boldsymbol{\phi}_2^T \mathbf{F}_e(t) = F_1(t) + r_2 F_2(t). \tag{3.150b}$$

Again, the transformation to the normal mode coordinates gives two independent, linear, second-order differential equations. Recall that $\omega_1 = \sqrt{k_1^*/m_1^*}$, $\omega_2 = \sqrt{k_2^*/m_2^*}$, $2\zeta_1\omega_1 = c_1^*/m_1^*$, and $2\zeta_2\omega_2 = c_2^*/m_2^*$. Therefore, Eq. 3.150 can be written in normal form as

$$\ddot{q}_1(t) + 2\zeta_1\omega_1\dot{q}_1(t) + \omega_1^2 q_1(t) = \left(\boldsymbol{\phi}_1^T \mathbf{F}_e(t)/m_1^*\right) = \left(F_1(t) + r_1 F_2(t)\right)/m_1^* \tag{3.151a}$$

$$\ddot{q}_2(t) + 2\zeta_2\omega_2\dot{q}_2(t) + \omega_2^2 q_2(t) = \left(\boldsymbol{\phi}_2^T \mathbf{F}_e(t)/m_2^*\right) = \left(F_1(t) + r_2 F_2(t)\right)/m_2^*. \tag{3.151b}$$

The solution to Eq. 3.151 was discussed in Chapter 2. Again, recall that the solutions to $F_1(t)$ and $F_2(t)$ can be calculated separately and then added because Eq. 3.151 is a linear differential equation.

EXAMPLE 1   *Uniform Forcing Function*

Let

$$F_1(t) = \begin{cases} F_1 & t \geq 0 \\ 0 & \text{otherwise,} \end{cases} \quad F_2(t) = \begin{cases} F_2 & t \geq 0 \\ 0 & \text{otherwise.} \end{cases} \tag{3.152}$$

It follows from Eq. 3.150 that

$$m_1^* \ddot{q}_1(t) + c_1^* \dot{q}_1(t) + k_1^* q_1(t) = F_1 + r_1 F_2 \tag{3.153}$$

$$m_2^* \ddot{q}_2(t) + c_2^* \dot{q}_2(t) + k_2^* q_2(t) = F_1 + r_2 F_2. \tag{3.154}$$

Assuming that the initial conditions are zero, it follows from Eq. 2.45 that

$$q_1(t) = \left(\frac{F_1 + r_1 F_2}{k_1^*} \right) \left[1 - e^{-\zeta_1 \omega_1 t} \cos \omega_{1d} t - \left(\frac{\zeta_1}{\sqrt{1 - \zeta_1^2}} \right) e^{-\zeta_1 \omega_1 t} \sin \omega_{1d} t \right] \tag{3.155}$$

$$q_2(t) = \left(\frac{F_1 + r_2 F_2}{k_2^*} \right) \left[1 - e^{-\zeta_2 \omega_2 t} \cos \omega_{2d} t - \left(\frac{\zeta_2}{\sqrt{1 - \zeta_2^2}} \right) e^{-\zeta_2 \omega_2 t} \sin \omega_{2d} t \right] \tag{3.156}$$

where $\omega_{1d} = \omega_1 \sqrt{1 - \zeta_1^2}$ and $\omega_{2d} = \omega_2 \sqrt{1 - \zeta_2^2}$. Substituting Eqs. 3.155 and 3.156 into Eq. 3.140 gives

$$\begin{Bmatrix} x_1(t) \\ x_2(t) \end{Bmatrix} = \begin{Bmatrix} 1 \\ r_1 \end{Bmatrix} \left(\frac{F_1 + r_1 F_2}{k_1^*} \right) \left[1 - e^{-\zeta_1 \omega_1 t} \cos \omega_{1d} t - \left(\frac{\zeta_1}{\sqrt{1 - \zeta_1^2}} \right) e^{-\zeta_1 \omega_1 t} \sin \omega_{1d} t \right]$$

$$+ \begin{Bmatrix} 1 \\ r_2 \end{Bmatrix} \left(\frac{F_1 + r_2 F_2}{k_2^*} \right) \left[1 - e^{-\zeta_2 \omega_2 t} \cos \omega_{2d} t - \left(\frac{\zeta_2}{\sqrt{1 - \zeta_2^2}} \right) e^{-\zeta_2 \omega_2 t} \sin \omega_{2d} t \right]. \tag{3.157}$$

EXAMPLE 2   *Earthquake Ground Motion*

Consider the two-story frame shown in Figure 3.3 is subjected to an earthquake ground motion. The summation of forces in the free body diagrams leads to

$$m_1 \ddot{y}_1 + c_2 (\dot{y}_1 - \dot{y}_2) + c_1 (\dot{y}_1 - \dot{u}_g) + k_2 (y_1 - y_2) + c_1 (y_1 - u_g) = 0 \tag{3.158}$$

$$m_2 \ddot{y}_2 + c_2 (\dot{y}_2 - \dot{y}_1) + k_2 (y_2 - y_1) = 0 \tag{3.159}$$

where u_g is the absolute ground displacement and $\dot{u}_g$ is the absolute ground velocity. Define the relative displacement between the mass and the ground as

$$x_1 = y_1 - u_g, \quad x_2 = y_2 - u_g. \tag{3.160}$$

Then

$$\dot{x}_1 = \dot{y}_1 - \dot{u}_g, \quad \ddot{x}_1 = \ddot{y}_1 - \ddot{u}_g, \quad \dot{x}_2 = \dot{y}_2 - \dot{u}_g, \quad \ddot{x}_2 = \ddot{y}_2 - \ddot{u}_g \tag{3.161}$$

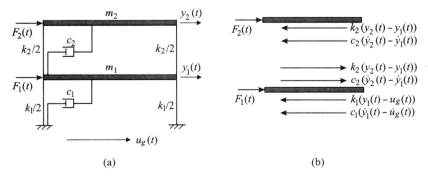

(a)   (b)

Figure 3.3 Two Degree of Freedom System with Earthquake Ground Motion

where $\ddot{u}_g$ is the absolute ground acceleration. Substituting Eqs. 3.160 and 3.161 into Eqs. 3.158 and 3.159 and then representing the result in a matrix form gives

$$\begin{bmatrix} m_1 & 0 \\ 0 & m_2 \end{bmatrix}\begin{Bmatrix} \ddot{x}_1 \\ \ddot{x}_2 \end{Bmatrix} + \begin{bmatrix} c_1 + c_2 & -c_2 \\ -c_2 & c_2 \end{bmatrix}\begin{Bmatrix} \dot{x}_1 \\ \dot{x}_2 \end{Bmatrix} + \begin{bmatrix} k_1 + k_2 & -k_2 \\ -k_2 & k_2 \end{bmatrix}\begin{Bmatrix} x_1 \\ x_2 \end{Bmatrix} = -\begin{bmatrix} m_1 & 0 \\ 0 & m_2 \end{bmatrix}\begin{Bmatrix} 1 \\ 1 \end{Bmatrix}\ddot{u}_g. \quad (3.162)$$

Assume that Rayleigh damping is used then

$$\mathbf{C} = \alpha\mathbf{M} + \beta\mathbf{K}. \quad (3.163)$$

Define the normal mode coordinates as

$$\mathbf{X}(t) = \begin{Bmatrix} x_1(t) \\ x_2(t) \end{Bmatrix} = \boldsymbol{\phi}_1 q_1(t) + \boldsymbol{\phi}_2 q_2(t). \quad (3.164)$$

As before, substituting Eq. 3.164 into Eq. 3.162 and premultiplying by $\boldsymbol{\phi}_1^T$ and then premultiplying Eq. 3.162 by $\boldsymbol{\phi}_2^T$ gives the two uncoupled equations

$$m_1^* \ddot{q}_1(t) + c_1^* \dot{q}_1(t) + k_1^* q_1(t) = -\left(m_1 + r_1 m_2\right)\ddot{u}_g \quad (3.165)$$

$$m_2^* \ddot{q}_2(t) + c_2^* \dot{q}_2(t) + k_2^* q_2(t) = -\left(m_1 + r_2 m_2\right)\ddot{u}_g. \quad (3.166)$$

The solution to Eqs. 3.165 and 3.166 was discussed in Chapter 2. Once the normal mode response, $q_1(t)$ and $q_2(t)$, is calculated, then Eq. 3.164 is used to obtain the response of the system, $x_1(t)$ and $x_2(t)$.

EXAMPLE 3   *Calculation of Rayleigh Damping Coefficients*

Consider the 2DOF system with $k_1 = k_2 = k$ and $m_1 = m_2 = m$. Recall from Section 3.2, Examples 1 and 2, that

$$\omega_1 = 0.618\,\omega_n, \qquad \omega_2 = 1.618\,\omega_n \quad (3.167)$$

where $\omega_n = \sqrt{k/m}$. Assume that the damping in the first mode of vibration is 2% and the damping in the second mode of vibration is 5%. Then the structure viscous damping matrix is

$$\mathbf{C} = \alpha\mathbf{M} + \beta\mathbf{K} \quad (3.168)$$

and from Eq. 3.149,

$$\alpha = \left(\frac{2(0.618\omega_n)(1.618\omega_n)\left[0.02(1.618\omega_n) - 0.05(0.618\omega_n)\right]}{(1.618\omega_n)^2 - (0.618\omega_n)^2}\right) = 0.001306\omega_n \quad (3.169)$$

$$\beta = \left(\frac{2\left[0.05(1.618\omega_n) - 0.02(0.618\omega_n)\right]}{(1.618\omega_n)^2 - (0.618\omega_n)^2}\right) = \frac{0.06131}{\omega_n}. \quad (3.170)$$

Therefore, Eq. 3.168 becomes

$$\mathbf{C} = 0.001306\omega_n \begin{bmatrix} m & 0 \\ 0 & m \end{bmatrix} + \left(\frac{0.06131}{\omega_n}\right)\begin{bmatrix} 2k & -k \\ -k & k \end{bmatrix} = m\omega_n \begin{bmatrix} 0.1245 & -0.0613 \\ -0.0613 & 0.0626 \end{bmatrix}. \quad (3.171)$$

EXAMPLE 4   *A Proportional Forcing Function*

Consider Example 3 in Section 3.4, where the forcing function has the special form

$$\mathbf{F}_e(t) = \begin{Bmatrix} F_1(t) \\ F_2(t) \end{Bmatrix} = \begin{Bmatrix} A_1 f(t) \\ A_2 f(t) \end{Bmatrix} = \begin{Bmatrix} A_1 \\ A_2 \end{Bmatrix} f(t) \tag{3.172}$$

where A_1 and A_2 are constants. For a 2DOF system with proportional damping, Eq. 3.151 becomes

$$\ddot{q}_1(t) + 2\zeta_1 \omega_1 \dot{q}_1(t) + \omega_1^2 q_1(t) = \Gamma_1 f(t) \tag{3.173}$$

$$\ddot{q}_2(t) + 2\zeta_2 \omega_2 \dot{q}_2(t) + \omega_2^2 q_2(t) = \Gamma_2 f(t) \tag{3.174}$$

where

$$\Gamma_1 = \frac{A_1 + r_1 A_2}{m_1^*}, \qquad \Gamma_2 = \frac{A_1 + r_2 A_2}{m_2^*}. \tag{3.175}$$

Now consider $f(t)$ to be a rise to static forcing function, where

$$f(t) = \begin{cases} f_o \, t/t_o & 0 \le t < t_o \\ f_o & t \ge t_o \\ 0 & \text{otherwise.} \end{cases} \tag{3.176}$$

Assume that $\mathbf{X}(0) = \mathbf{X}_o = \mathbf{0}$ and $\dot{\mathbf{X}}(0) = \dot{\mathbf{X}}_o = \mathbf{0}$, and therefore it follows that the initial conditions of the normal mode response are also zero, that is, $q_1(0) = \dot{q}_1(0) = 0$ and $q_2(0) = \dot{q}_2(0) = 0$. The first normal mode response, $q_1(t)$, for $0 \le t < t_o$ follows from Eq. 2.62:

$$q_1(t) = \left[\frac{2\zeta_1 \Gamma_1 f_o}{k_1^* \omega_1 t_o} \right] e^{-\zeta_1 \omega_1 t} \cos \omega_{1d} t$$

$$+ \left[\frac{\Gamma_1 f_o (2\zeta_1^2 - 1)}{k_1^* \omega_1 t_o \sqrt{1 - \zeta_1^2}} \right] e^{-\zeta_1 \omega_1 t} \sin \omega_{1d} t + \left(\frac{\Gamma_1 f_o}{k_1^* t_o} \right) t - \left(\frac{2\zeta_1 \Gamma_1 f_o}{k_1^* \omega_1 t_o} \right), \quad 0 \le t < t_o \tag{3.177}$$

where $\omega_{1d} = \omega_1 \sqrt{1 - \zeta_1^2}$, and the second normal mode response, $q_2(t)$, is

$$q_2(t) = \left[\frac{2\zeta_2 \Gamma_2 f_o}{k_2^* \omega_2 t_o} \right] e^{-\zeta_2 \omega_2 t} \cos \omega_{2d} t$$

$$+ \left[\frac{\Gamma_2 f_o (2\zeta_2^2 - 1)}{k_2^* \omega_2 t_o \sqrt{1 - \zeta_2^2}} \right] e^{-\zeta_2 \omega_2 t} \sin \omega_{2d} t + \left(\frac{\Gamma_2 f_o}{k_2^* t_o} \right) t - \left(\frac{2\zeta_2 \Gamma_2 f_o}{k_2^* \omega_2 t_o} \right), \quad 0 \le t < t_o \tag{3.178}$$

where $\omega_{2d} = \omega_2 \sqrt{1 - \zeta_2^2}$. The response for $t \ge t_o$ follows from Eq. 2.66:

$$q_1(t) = \left[q_1(t_o) - \frac{\Gamma_1 f_o}{k_1^*} \right] e^{-\zeta_1 \omega_1 (t - t_o)} \cos \omega_{1d} (t - t_o)$$

$$+ \left[\frac{\zeta_1 \omega_1 \left(q_1(t_o) - \Gamma_1 f_o / k_1^* \right) + \dot{q}_1(t_o)}{\omega_1 \sqrt{1 - \zeta_1^2}} \right] e^{-\zeta_1 \omega_1 (t - t_o)} \sin \omega_{1d} (t - t_o) + \left(\frac{\Gamma_1 f_o}{k_1^*} \right), \quad t \ge t_o \tag{3.179}$$

$$q_2(t) = \left[q_2(t_o) - \frac{\Gamma_2 f_o}{k_2^*} \right] e^{-\zeta_2 \omega_2 (t - t_o)} \cos \omega_{2d} (t - t_o)$$

$$+ \left[\frac{\zeta_2 \omega_2 \left(q_2(t_o) - \Gamma_2 f_o / k_2^* \right) + \dot{q}_2(t_o)}{\omega_2 \sqrt{1 - \zeta_2^2}} \right] e^{-\zeta_2 \omega_2 (t - t_o)} \sin \omega_{2d} (t - t_o) + \left(\frac{\Gamma_2 f_o}{k_2^*} \right), \quad t \ge t_o \tag{3.180}$$

where $q_1(t_o)$, $q_2(t_o)$, $\dot{q}_1(t_o)$, and $\dot{q}_2(t_o)$ follow from Eqs. 2.64 and 2.65, which are

$$q_1(t_o) = \left[\frac{2\zeta_1\Gamma_1 f_o}{k_1^* \omega_1 t_o}\right] e^{-\zeta_1\omega_1 t_o} \cos\omega_{1d}t_o$$

$$+\left[\frac{\Gamma_1 f_o(2\zeta_1^2 - 1)}{k_1^* \omega_1 t_o \sqrt{1-\zeta_1^2}}\right] e^{-\zeta_1\omega_1 t_o} \sin\omega_{1d}t_o + \left(\frac{\Gamma_1 f_o}{k_1^*} - \frac{2\zeta_1\Gamma_1 f_o}{k_1^* \omega_1 t_o}\right) \quad (3.181)$$

$$q_2(t_o) = \left[\frac{2\zeta_2\Gamma_2 f_o}{k_2^* \omega_2 t_o}\right] e^{-\zeta_2\omega_2 t_o} \cos\omega_{2d}t_o$$

$$+\left[\frac{\Gamma_2 f_o(2\zeta_2^2 - 1)}{k_2^* \omega_2 t_o \sqrt{1-\zeta_2^2}}\right] e^{-\zeta_2\omega_2 t_o} \sin\omega_{2d}t_o + \left(\frac{\Gamma_2 f_o}{k_2^*} - \frac{2\zeta_2\Gamma_2 f_o}{k_2^* \omega_2 t_o}\right) \quad (3.182)$$

$$\dot{q}_1(t_o) = -\left[\frac{\Gamma_1 f_o}{k_1^* t_o}\right] e^{-\zeta_1\omega_1 t_o} \cos\omega_{1d}t_o - \left[\frac{\zeta_1}{\sqrt{1-\zeta_1^2}}\frac{\Gamma_1 f_o}{k_1^* t_o}\right] e^{-\zeta_1\omega_1 t_o} \sin\omega_{1d}t_o + \left(\frac{\Gamma_1 f_o}{k_1^* t_o}\right) \quad (3.183)$$

$$\dot{q}_2(t_o) = -\left[\frac{\Gamma_2 f_o}{k_2^* t_o}\right] e^{-\zeta_2\omega_2 t_o} \cos\omega_{2d}t_o - \left[\frac{\zeta_2}{\sqrt{1-\zeta_2^2}}\frac{\Gamma_2 f_o}{k_2^* t_o}\right] e^{-\zeta_2\omega_2 t_o} \sin\omega_{2d}t_o + \left(\frac{\Gamma_2 f_o}{k_2^* t_o}\right). \quad (3.184)$$

Then the response of the system is

$$\mathbf{X}(t) = \begin{Bmatrix} x_1(t) \\ x_2(t) \end{Bmatrix} = \begin{Bmatrix} 1 \\ r_1 \end{Bmatrix} q_1(t) + \begin{Bmatrix} 1 \\ r_2 \end{Bmatrix} q_2(t) \quad (3.185)$$

where

$$q_1(t) = \begin{cases} \left[\dfrac{2\zeta_1\Gamma_1 f_o}{k_1^* \omega_1 t_o}\right] e^{-\zeta_1\omega_1 t} \cos\omega_{1d}t \\[2ex] \quad +\left[\dfrac{\Gamma_1 f_o(2\zeta_1^2 - 1)}{k_1^* \omega_1 t_o \sqrt{1-\zeta_1^2}}\right] e^{-\zeta_1\omega_1 t} \sin\omega_{1d}t + \left(\dfrac{\Gamma_1 f_o}{k_1^* t_o}\right)t - \left(\dfrac{2\zeta_1\Gamma_1 f_o}{k_1^* \omega_1 t_o}\right), \qquad 0 \le t < t_o \\[3ex] \left[q_1(t_o) - \dfrac{\Gamma_1 f_o}{k_1^*}\right] e^{-\zeta_1\omega_1(t-t_o)} \cos\omega_{1d}(t - t_o) \\[2ex] \quad +\left[\dfrac{\zeta_1\omega_1\left(q_1(t_o) - \Gamma_1 f_o/k_1^*\right) + \dot{q}_1(t_o)}{\omega_1\sqrt{1-\zeta_1^2}}\right] e^{-\zeta_1\omega_1(t-t_o)} \sin\omega_{1d}(t - t_o) + \left(\dfrac{\Gamma_1 f_o}{k_1^*}\right), \qquad t \ge t_o \end{cases}$$

$$q_2(t) = \begin{cases} \left[\dfrac{2\zeta_2\Gamma_2 f_o}{k_2^* \omega_2 t_o}\right] e^{-\zeta_2\omega_2 t} \cos\omega_{2d}t \\[2ex] \quad +\left[\dfrac{\Gamma_2 f_o(2\zeta_2^2 - 1)}{k_2^* \omega_2 t_o \sqrt{1-\zeta_2^2}}\right] e^{-\zeta_2\omega_2 t} \sin\omega_{2d}t + \left(\dfrac{\Gamma_2 f_o}{k_2^* t_o}\right)t - \left(\dfrac{2\zeta_2\Gamma_2 f_o}{k_2^* \omega_2 t_o}\right), \qquad 0 \le t < t_o \\[3ex] \left[q_2(t_o) - \dfrac{\Gamma_2 f_o}{k_2^*}\right] e^{-\zeta_2\omega_2(t-t_o)} \cos\omega_{2d}(t - t_o) \\[2ex] \quad +\left[\dfrac{\zeta_2\omega_2\left(q_2(t_o) - \Gamma_2 f_o/k_2^*\right) + \dot{q}_2(t_o)}{\omega_2\sqrt{1-\zeta_2^2}}\right] e^{-\zeta_2\omega_2(t-t_o)} \sin\omega_{2d}(t - t_o) + \left(\dfrac{\Gamma_2 f_o}{k_2^*}\right), \qquad t \ge t_o. \end{cases}$$

3.6   STATE SPACE SOLUTION FOR RESPONSE

The state space solution approach for a 2DOF system is very similar to the solution approach for an SDOF system. In an SDOF system, there are a total of two states: one displacement and one velocity. For a 2DOF system, there are four states: two displacements and two velocities.

Consider the dynamic equilibrium equation of motion, as shown in Eq. 3.138 and rewritten here as

$$\mathbf{M}\ddot{\mathbf{X}}(t) + \mathbf{C}\dot{\mathbf{X}}(t) + \mathbf{K}\mathbf{X}(t) = \mathbf{F}_e(t) \qquad (3.186)$$

where

$$\mathbf{M} = \begin{bmatrix} m_1 & 0 \\ 0 & m_2 \end{bmatrix}, \quad \mathbf{C} = \begin{bmatrix} c_1+c_2 & -c_2 \\ -c_2 & c_2 \end{bmatrix}, \quad \mathbf{K} = \begin{bmatrix} k_1+k_2 & -k_2 \\ -k_2 & k_2 \end{bmatrix}$$

$$\mathbf{X}(t) = \begin{Bmatrix} x_1(t) \\ x_2(t) \end{Bmatrix}, \quad \mathbf{F}_e(t) = \begin{Bmatrix} F_1(t) \\ F_2(t) \end{Bmatrix}.$$

Let the vector $\mathbf{z}$ represent the independent variables, where

$$\mathbf{z}(t) = \begin{Bmatrix} \mathbf{X}(t) \\ \dot{\mathbf{X}}(t) \end{Bmatrix} = \begin{Bmatrix} x_1(t) & x_2(t) & \dot{x}_1(t) & \dot{x}_2(t) \end{Bmatrix}^T. \qquad (3.187)$$

It follows from Eq. 3.186 that

$$\dot{\mathbf{z}}(t) = \begin{Bmatrix} \dot{\mathbf{X}}(t) \\ \ddot{\mathbf{X}}(t) \end{Bmatrix} = \begin{bmatrix} \mathbf{0} & \mathbf{I} \\ -\mathbf{M}^{-1}\mathbf{K} & -\mathbf{M}^{-1}\mathbf{C} \end{bmatrix} \begin{Bmatrix} \mathbf{X}(t) \\ \dot{\mathbf{X}}(t) \end{Bmatrix} + \begin{Bmatrix} \mathbf{0} \\ \mathbf{M}^{-1}\mathbf{F}_e(t) \end{Bmatrix} \qquad (3.188)$$

where $\mathbf{0}$ is a matrix with all zero entries and $\mathbf{I}$ is a 2×2 identity matrix. To simplify Eq. 3.188, let

$$\mathbf{A} = \begin{bmatrix} \mathbf{0} & \mathbf{I} \\ -\mathbf{M}^{-1}\mathbf{K} & -\mathbf{M}^{-1}\mathbf{C} \end{bmatrix}, \quad \mathbf{F}(t) = \begin{Bmatrix} \mathbf{0} \\ \mathbf{M}^{-1}\mathbf{F}_e(t) \end{Bmatrix}.$$

Then Eq. 3.188 becomes

$$\dot{\mathbf{z}}(t) = \mathbf{A}\mathbf{z}(t) + \mathbf{F}(t). \qquad (3.189)$$

The matrix $\mathbf{A}$ is a 4×4 matrix. Consider the previously described mass, stiffness, and damping matrices; it follows that

$$-\mathbf{M}^{-1}\mathbf{K} = -\begin{bmatrix} 1/m_1 & 0 \\ 0 & 1/m_2 \end{bmatrix} \begin{bmatrix} k_1+k_2 & -k_2 \\ -k_2 & k_2 \end{bmatrix} = \begin{bmatrix} -(k_1+k_2)/m_1 & k_2/m_1 \\ k_2/m_2 & -k_2/m_2 \end{bmatrix} \qquad (3.190)$$

$$-\mathbf{M}^{-1}\mathbf{C} = -\begin{bmatrix} 1/m_1 & 0 \\ 0 & 1/m_2 \end{bmatrix} \begin{bmatrix} c_1+c_2 & -c_2 \\ -c_2 & c_2 \end{bmatrix} = \begin{bmatrix} -(c_1+c_2)/m_1 & c_2/m_1 \\ c_2/m_2 & -c_2/m_2 \end{bmatrix} \qquad (3.191)$$

$$\mathbf{M}^{-1}\mathbf{F}_e(t) = \begin{bmatrix} 1/m_1 & 0 \\ 0 & 1/m_2 \end{bmatrix} \begin{Bmatrix} F_1(t) \\ F_2(t) \end{Bmatrix} = \begin{Bmatrix} F_1(t)/m_1 \\ F_2(t)/m_2 \end{Bmatrix}. \qquad (3.192)$$

It follows that

$$\mathbf{A} = \begin{bmatrix} 0 & 0 & 1 & 0 \\ 0 & 0 & 0 & 1 \\ -(k_1+k_2)/m_1 & k_2/m_1 & -(c_1+c_2)/m_1 & c_2/m_1 \\ k_2/m_2 & -k_2/m_2 & c_2/m_2 & -c_2/m_2 \end{bmatrix}, \quad \mathbf{F}(t) = \begin{Bmatrix} 0 \\ 0 \\ F_1(t)/m_1 \\ F_2(t)/m_2 \end{Bmatrix}. \qquad (3.193)$$

Therefore, it follows from Eq. 3.188 that

$$\frac{d}{dt}\begin{Bmatrix} x_1(t) \\ x_2(t) \\ \dot{x}_1(t) \\ \dot{x}_2(t) \end{Bmatrix} = \begin{bmatrix} 0 & 0 & 1 & 0 \\ 0 & 0 & 0 & 1 \\ -(k_1+k_2)/m_1 & k_2/m_1 & -(c_1+c_2)/m_1 & c_2/m_1 \\ k_2/m_2 & -k_2/m_2 & c_2/m_2 & -c_2/m_2 \end{bmatrix}\begin{Bmatrix} x_1(t) \\ x_2(t) \\ \dot{x}_1(t) \\ \dot{x}_2(t) \end{Bmatrix} + \begin{Bmatrix} 0 \\ 0 \\ F_1(t)/m_1 \\ F_2(t)/m_2 \end{Bmatrix}. \quad (3.194)$$

To solve for Eq. 3.194, note that Eq. 3.189 is the same as Eq. 2.181. Thus, the solution can be written as

$$\mathbf{z}(t) = \mathbf{e}^{\mathbf{A}(t-t_o)}\mathbf{z}(t_o) + \mathbf{e}^{\mathbf{A}t}\int_{t_o}^{t} \mathbf{e}^{-\mathbf{A}s}\mathbf{F}(s)\,ds \quad (3.195)$$

where $\mathbf{z}(t_o)$ represents the displacement and velocity at time $t = t_o$, and $\mathbf{e}^{\mathbf{A}t}$ is called the *state transition matrix* and is defined as

$$\mathbf{e}^{\mathbf{A}t} = \mathbf{I} + \mathbf{A}t + \frac{1}{2!}(\mathbf{A}t)^2 + \frac{1}{3!}(\mathbf{A}t)^3 + \dots. \quad (3.196)$$

When the initial conditions are given at time equal to zero, then setting $t_o = 0$ and Eq. 3.195 becomes

$$\mathbf{z}(t) = \mathbf{e}^{\mathbf{A}t}\mathbf{z}(0) + \int_{0}^{t} \mathbf{e}^{-\mathbf{A}(t-s)}\mathbf{F}(s)\,ds. \quad (3.197)$$

Note that for the state space method, the solution does not require any discussion of whether the damping matrix is proportional or nonproportional. However, it is usually necessary to define the magnitude of damping in a building first in terms of the damping in the normal modes of vibration. When this is the case, the proportional damping assumption is then necessary to obtain the matrix $\mathbf{C}$ from Eq. 3.139, where α and β are determined using Eq. 3.149.

As discussed in Section 2.5, calculating the state transition matrix, $\mathbf{e}^{\mathbf{A}t}$, requires calculation of all the eigenvalues and eigenvectors of the matrix $\mathbf{A}$. Once this is done, the transformation matrix $\mathbf{T}$ must be inverted. This makes the calculation of a system with more than one degree of freedom undesirable because of the large computational effort. In addition, for all the structures that have dampings less than 1.0, the eigenvalues and eigenvectors of $\mathbf{A}$ always involve complex numbers, very similar to the case for an SDOF system. Therefore, expressing $\mathbf{e}^{\mathbf{A}t}$ in a closed-form solution is very difficult, and most of the time numerical calculation is preferred. The method of numerically calculating the matrix $\mathbf{e}^{\mathbf{A}t}$ is discussed in detail in Section 3.8.

EXAMPLE 1   *Free Vibration with Zero Damping*

Consider the 2DOF system discussed in Example 1 of Section 3.5 with $k_1 = k_2 = k$, $m_1 = m_2 = m$, no damping, and no external force, then

$$\mathbf{M} = \begin{bmatrix} m & 0 \\ 0 & m \end{bmatrix}, \quad \mathbf{C} = \begin{bmatrix} 0 & 0 \\ 0 & 0 \end{bmatrix}, \quad \mathbf{K} = \begin{bmatrix} 2k & -k \\ -k & k \end{bmatrix}, \quad \mathbf{F}_e(t) = \begin{Bmatrix} 0 \\ 0 \end{Bmatrix}.$$

It follows from Eq. 3.189 that

$$\dot{\mathbf{z}}(t) = \mathbf{A}\mathbf{z}(t) \quad (3.198)$$

where

$$\mathbf{M} = m \begin{bmatrix} 1 & 0 \\ 0 & 1 \end{bmatrix}, \quad \mathbf{M}^{-1} = \left(\frac{1}{m} \right) \begin{bmatrix} 1 & 0 \\ 0 & 1 \end{bmatrix}$$

(3.199)

$$\mathbf{M}^{-1}\mathbf{K} = \left(\frac{k}{m} \right) \begin{bmatrix} 1 & 0 \\ 0 & 1 \end{bmatrix} \begin{bmatrix} 2 & -1 \\ -1 & 1 \end{bmatrix} = \omega_n^2 \begin{bmatrix} 2 & -1 \\ -1 & 1 \end{bmatrix}, \quad \mathbf{M}^{-1}\mathbf{C} = \left(\frac{1}{m} \right) \begin{bmatrix} 0 & 0 \\ 0 & 0 \end{bmatrix} = \begin{bmatrix} 0 & 0 \\ 0 & 0 \end{bmatrix}$$

and $\omega_n^2 = k/m$. Finally,

$$\mathbf{A} = \begin{bmatrix} 0 & 0 & 1 & 0 \\ 0 & 0 & 0 & 1 \\ -2\omega_n^2 & \omega_n^2 & 0 & 0 \\ \omega_n^2 & -\omega_n^2 & 0 & 0 \end{bmatrix}.$$

(3.200)

The response solution requires calculation of the state transition matrix, which can be done using Eq. 3.196 or Eq. 2.192. Consider the latter equation, which is

$$\mathbf{e}^{\mathbf{A}t} = \mathbf{T}\mathbf{e}^{\Lambda t}\mathbf{T}^{-1}.$$

(3.201)

The matrices $\mathbf{T}$ and Λ are obtained from the eigenvalue problem

$$\mathbf{A}\boldsymbol{\phi}_i = \lambda_i \boldsymbol{\phi}_i$$

(3.202)

where λ_i's are the diagonal entries of the eigenvalue matrix, that is

$$\Lambda = \begin{bmatrix} \lambda_1 & 0 & 0 & 0 \\ 0 & \lambda_2 & 0 & 0 \\ 0 & 0 & \lambda_3 & 0 \\ 0 & 0 & 0 & \lambda_4 \end{bmatrix}.$$

(3.203)

The matrix $\mathbf{T}$ is the eigenvector matrix

$$\mathbf{T} = \begin{bmatrix} \boldsymbol{\phi}_1 & \boldsymbol{\phi}_2 & \boldsymbol{\phi}_3 & \boldsymbol{\phi}_4 \end{bmatrix}.$$

(3.204)

The four eigenvalues of $\mathbf{A}$ are obtained by setting the determinant of the eigenvalue equation equal to zero, that is,

$$\det(\mathbf{A} - \lambda \mathbf{I}) = \begin{vmatrix} -\lambda & 0 & 1 & 0 \\ 0 & -\lambda & 0 & 1 \\ -2\omega_n^2 & \omega_n^2 & -\lambda & 0 \\ \omega_n^2 & -\omega_n^2 & 0 & -\lambda \end{vmatrix} = \lambda^4 + 3\omega_n^2\lambda^2 + \omega_n^4 = 0.$$

(3.205)

By treating Eq. 3.205 as a quadratic equation with the unknown λ^2, it follows that

$$\lambda^2 = \left(\frac{-3 \pm \sqrt{5}}{2} \right) \omega_n^2.$$

(3.206)

Note that the value inside the parentheses of Eq. 3.206 is always a negative number, and therefore the eigenvalues (i.e., λ_i's) are complex numbers:

$$\lambda_1 = i\omega_n \left(\frac{\sqrt{5}+1}{2} \right), \quad \lambda_2 = i\omega_n \left(\frac{\sqrt{5}-1}{2} \right), \quad \lambda_3 = -i\omega_n \left(\frac{\sqrt{5}+1}{2} \right), \quad \lambda_4 = -i\omega_n \left(\frac{\sqrt{5}-1}{2} \right).$$

(3.207)

If the damping is not zero, then Eq. 3.205 would still be a quadratic equation but the solution would be more complicated than Eq. 3.207. The eigenvectors corresponding to each of these four eigenvalues are computed using the equation

$$\begin{bmatrix} -\lambda_i & 0 & 1 & 0 \\ 0 & -\lambda_i & 0 & 1 \\ -2\omega_n^2 & \omega_n^2 & -\lambda_i & 0 \\ \omega_n^2 & -\omega_n^2 & 0 & -\lambda_i \end{bmatrix} \phi_i = 0, \quad i = 1, 2, 3, 4. \tag{3.208}$$

Solving for the eigenvectors in Eq. 3.208 for each eigenvalue gives

$$\lambda_1 = i\omega_n \left(\frac{\sqrt{5}+1}{2} \right): \quad \phi_1 = \begin{Bmatrix} 1 \\ \left(1-\sqrt{5}\right)\Big/2 \\ i\omega_n\left(1+\sqrt{5}\right)\Big/2 \\ -i\omega_n \end{Bmatrix}, \quad \lambda_2 = i\omega_n \left(\frac{\sqrt{5}-1}{2} \right): \quad \phi_2 = \begin{Bmatrix} 1 \\ \left(1+\sqrt{5}\right)\Big/2 \\ -i\omega_n\left(1-\sqrt{5}\right)\Big/2 \\ i\omega_n \end{Bmatrix}$$

$$\lambda_3 = -i\omega_n \left(\frac{\sqrt{5}+1}{2} \right): \quad \phi_3 = \begin{Bmatrix} 1 \\ \left(1-\sqrt{5}\right)\Big/2 \\ -i\omega_n\left(1+\sqrt{5}\right)\Big/2 \\ i\omega_n \end{Bmatrix}, \quad \lambda_4 = -i\omega_n \left(\frac{\sqrt{5}-1}{2} \right): \quad \phi_4 = \begin{Bmatrix} 1 \\ \left(1+\sqrt{5}\right)\Big/2 \\ i\omega_n\left(1-\sqrt{5}\right)\Big/2 \\ -i\omega_n \end{Bmatrix}$$

and therefore it follows from Eq. 3.204 that

$$\mathbf{T} = \begin{bmatrix} 1 & 1 & 1 & 1 \\ \left(1-\sqrt{5}\right)\Big/2 & \left(1+\sqrt{5}\right)\Big/2 & \left(1-\sqrt{5}\right)\Big/2 & \left(1+\sqrt{5}\right)\Big/2 \\ i\omega_n\left(1+\sqrt{5}\right)\Big/2 & -i\omega_n\left(1-\sqrt{5}\right)\Big/2 & -i\omega_n\left(1+\sqrt{5}\right)\Big/2 & i\omega_n\left(1-\sqrt{5}\right)\Big/2 \\ -i\omega_n & i\omega_n & i\omega_n & -i\omega_n \end{bmatrix}. \tag{3.209}$$

Inverting the matrix $\mathbf{T}$ as given in Eq. 3.209 gives

$$\mathbf{T}^{-1} = \frac{1}{20} \begin{bmatrix} 5+\sqrt{5} & -2\sqrt{5} & -2i\sqrt{5}\,\omega_n & i\left(5-\sqrt{5}\right)\Big/\omega_n \\ 5-\sqrt{5} & 2\sqrt{5} & -2i\sqrt{5}\Big/\omega_n & -i\left(5+\sqrt{5}\right)\Big/\omega_n \\ 5+\sqrt{5} & -2\sqrt{5} & 2i\sqrt{5}\Big/\omega_n & -i\left(5-\sqrt{5}\right)\Big/\omega_n \\ 5-\sqrt{5} & 2\sqrt{5} & 2i\sqrt{5}\Big/\omega_n & i\left(5+\sqrt{5}\right)\Big/\omega_n \end{bmatrix}. \tag{3.210}$$

Recall that

$$e^{\Lambda t} = \begin{bmatrix} \exp\left\{ \left(\frac{\sqrt{5}+1}{2}\right)i\omega_n t \right\} & 0 & 0 & 0 \\ 0 & \exp\left\{ \left(\frac{\sqrt{5}-1}{2}\right)i\omega_n t \right\} & 0 & 0 \\ 0 & 0 & \exp\left\{ -\left(\frac{\sqrt{5}+1}{2}\right)i\omega_n t \right\} & 0 \\ 0 & 0 & 0 & \exp\left\{ -\left(\frac{\sqrt{5}-1}{2}\right)i\omega_n t \right\} \end{bmatrix} \tag{3.211}$$

and therefore Eq. 3.201 can be used to calculate the state transition matrix. Finally, if the forcing functions are zero, then the free vibration solution follows from Eq. 3.198:

$$\begin{Bmatrix} x_1(t) \\ x_2(t) \\ \dot{x}_1(t) \\ \dot{x}_2(t) \end{Bmatrix} = \mathbf{T} e^{\Lambda t} \mathbf{T}^{-1} \begin{Bmatrix} x_{1o} \\ x_{2o} \\ \dot{x}_{1o} \\ \dot{x}_{2o} \end{Bmatrix}. \tag{3.212}$$

Assume that the initial conditions are

$$x_{1o} = 1, \quad x_{2o} = 0, \quad \dot{x}_{1o} = 0, \quad \dot{x}_{2o} = 0. \tag{3.213}$$

Then substituting Eqs. 3.209 to 3.211 and 3.213 into Eq. 3.212 and performing the matrix triple products step by step, it follows that

$$\mathbf{T}^{-1}\begin{Bmatrix}1\\0\\0\\0\end{Bmatrix}=\frac{1}{20}\begin{bmatrix}5+\sqrt{5} & -2\sqrt{5} & -2i\sqrt{5}\,\omega_n & i\left(5-\sqrt{5}\right)/\omega_n\\5-\sqrt{5} & 2\sqrt{5} & -2i\sqrt{5}/\omega_n & -i\left(5+\sqrt{5}\right)/\omega_n\\5+\sqrt{5} & -2\sqrt{5} & 2i\sqrt{5}/\omega_n & -i\left(5-\sqrt{5}\right)/\omega_n\\5-\sqrt{5} & 2\sqrt{5} & 2i\sqrt{5}/\omega_n & i\left(5+\sqrt{5}\right)/\omega_n\end{bmatrix}\begin{Bmatrix}1\\0\\0\\0\end{Bmatrix}=\frac{1}{20}\begin{Bmatrix}5+\sqrt{5}\\5-\sqrt{5}\\5+\sqrt{5}\\5-\sqrt{5}\end{Bmatrix}$$

$$e^{\Lambda t}\mathbf{T}^{-1}\begin{Bmatrix}1\\0\\0\\0\end{Bmatrix}=\frac{1}{20}\begin{Bmatrix}\left(5+\sqrt{5}\right)\exp\left\{\left(\frac{\sqrt{5}+1}{2}\right)i\omega_n t\right\}\\\left(5-\sqrt{5}\right)\exp\left\{\left(\frac{\sqrt{5}-1}{2}\right)i\omega_n t\right\}\\\left(5+\sqrt{5}\right)\exp\left\{-\left(\frac{\sqrt{5}+1}{2}\right)i\omega_n t\right\}\\\left(5-\sqrt{5}\right)\exp\left\{-\left(\frac{\sqrt{5}-1}{2}\right)i\omega_n t\right\}\end{Bmatrix}=\frac{1}{20}\begin{Bmatrix}\left(5+\sqrt{5}\right)\left[\cos\left(\frac{\sqrt{5}+1}{2}\right)\omega_n t+i\sin\left(\frac{\sqrt{5}+1}{2}\right)\omega_n t\right]\\\left(5-\sqrt{5}\right)\left[\cos\left(\frac{\sqrt{5}-1}{2}\right)\omega_n t+i\sin\left(\frac{\sqrt{5}-1}{2}\right)\omega_n t\right]\\\left(5+\sqrt{5}\right)\left[\cos\left(\frac{\sqrt{5}+1}{2}\right)\omega_n t-i\sin\left(\frac{\sqrt{5}+1}{2}\right)\omega_n t\right]\\\left(5-\sqrt{5}\right)\left[\cos\left(\frac{\sqrt{5}-1}{2}\right)\omega_n t-i\sin\left(\frac{\sqrt{5}-1}{2}\right)\omega_n t\right]\end{Bmatrix}$$

and therefore

$$\begin{Bmatrix}x_1(t)\\x_2(t)\\\dot{x}_1(t)\\\dot{x}_2(t)\end{Bmatrix}=\mathbf{T}e^{\Lambda t}\mathbf{T}^{-1}\begin{Bmatrix}1\\0\\0\\0\end{Bmatrix}=\frac{1}{10}\begin{Bmatrix}\left(5+\sqrt{5}\right)\cos\left(\frac{\sqrt{5}+1}{2}\right)\omega_n t+\left(5-\sqrt{5}\right)\cos\left(\frac{\sqrt{5}-1}{2}\right)\omega_n t\\-2\sqrt{5}\cos\left(\frac{\sqrt{5}+1}{2}\right)\omega_n t+2\sqrt{5}\cos\left(\frac{\sqrt{5}-1}{2}\right)\omega_n t\\-\left(5+3\sqrt{5}\right)\omega_n\sin\left(\frac{\sqrt{5}+1}{2}\right)\omega_n t+\left(5-3\sqrt{5}\right)\omega_n\sin\left(\frac{\sqrt{5}-1}{2}\right)\omega_n t\\\left(5+\sqrt{5}\right)\omega_n\sin\left(\frac{\sqrt{5}+1}{2}\right)\omega_n t-\left(5-\sqrt{5}\right)\omega_n\sin\left(\frac{\sqrt{5}-1}{2}\right)\omega_n t\end{Bmatrix}.$$

EXAMPLE 2   *Earthquake Ground Motion*

Consider the two-story frame subjected to the earthquake ground motion, which was discussed in Example 2 of Section 3.5. Following the derivation in that example, the dynamic equilibrium equation is shown in Eq. 3.162:

$$\begin{bmatrix}m_1 & 0\\0 & m_2\end{bmatrix}\begin{Bmatrix}\ddot{x}_1\\\ddot{x}_2\end{Bmatrix}+\begin{bmatrix}c_1+c_2 & -c_2\\-c_2 & c_2\end{bmatrix}\begin{Bmatrix}\dot{x}_1\\\dot{x}_2\end{Bmatrix}+\begin{bmatrix}k_1+k_2 & -k_2\\-k_2 & k_2\end{bmatrix}\begin{Bmatrix}x_1\\x_2\end{Bmatrix}=-\begin{bmatrix}m_1 & 0\\0 & m_2\end{bmatrix}\begin{Bmatrix}1\\1\end{Bmatrix}\ddot{u}_g \quad (3.214)$$

where $\ddot{u}_g$ is the ground acceleration. Let

$$\mathbf{M}=\begin{bmatrix}m_1 & 0\\0 & m_2\end{bmatrix},\quad \mathbf{C}=\begin{bmatrix}c_1+c_2 & -c_2\\-c_2 & c_2\end{bmatrix},\quad \mathbf{K}=\begin{bmatrix}k_1+k_2 & -k_2\\-k_2 & k_2\end{bmatrix}$$

$$\mathbf{X}(t)=\begin{Bmatrix}x_1(t)\\x_2(t)\end{Bmatrix},\quad \{\mathbf{I}\}=\begin{Bmatrix}1\\1\end{Bmatrix}.$$

It follows from Eq. 3.214 that

$$\mathbf{M}\ddot{\mathbf{X}}(t)+\mathbf{C}\dot{\mathbf{X}}(t)+\mathbf{K}\mathbf{X}(t)=-\mathbf{M}\{\mathbf{I}\}\ddot{u}_g. \quad (3.215)$$

Representing Eq. 3.215 in state space form, it follows that

$$\dot{z}(t)=\begin{Bmatrix}\dot{\mathbf{X}}(t)\\\ddot{\mathbf{X}}(t)\end{Bmatrix}=\begin{bmatrix}\mathbf{0} & \mathbf{I}\\-\mathbf{M}^{-1}\mathbf{K} & -\mathbf{M}^{-1}\mathbf{C}\end{bmatrix}\begin{Bmatrix}\mathbf{X}(t)\\\dot{\mathbf{X}}(t)\end{Bmatrix}+\begin{Bmatrix}\mathbf{0}\\-\mathbf{I}\end{Bmatrix}\ddot{u}_g(t). \quad (3.216)$$

Recall from Eq. 3.193 that

$$\mathbf{A}=\begin{bmatrix}\mathbf{0} & \mathbf{I}\\-\mathbf{M}^{-1}\mathbf{K} & -\mathbf{M}^{-1}\mathbf{C}\end{bmatrix}=\begin{bmatrix}0 & 0 & 1 & 0\\0 & 0 & 0 & 1\\-\left(k_1+k_2\right)/m_1 & k_2/m_1 & -\left(c_1+c_2\right)/m_1 & c_2/m_1\\k_2/m_2 & -k_2/m_2 & c_2/m_2 & -c_2/m_2\end{bmatrix}. \quad (3.217)$$

Substituting Eq. 3.217 into Eq. 3.216 gives

$$
\frac{d}{dt}\begin{Bmatrix} x_1(t) \\ x_2(t) \\ \dot{x}_1(t) \\ \dot{x}_2(t) \end{Bmatrix} = \begin{bmatrix} 0 & 0 & 1 & 0 \\ 0 & 0 & 0 & 1 \\ -(k_1+k_2)/m_1 & k_2/m_1 & -(c_1+c_2)/m_1 & c_2/m_1 \\ k_2/m_2 & -k_2/m_2 & c_2/m_2 & -c_2/m_2 \end{bmatrix}\begin{Bmatrix} x_1(t) \\ x_2(t) \\ \dot{x}_1(t) \\ \dot{x}_2(t) \end{Bmatrix} + \begin{Bmatrix} 0 \\ 0 \\ -1 \\ -1 \end{Bmatrix}\ddot{u}_g(t). \quad (3.218)
$$

The solution follows from Eq. 3.197:

$$
\begin{Bmatrix} x_1(t) \\ x_2(t) \\ \dot{x}_1(t) \\ \dot{x}_2(t) \end{Bmatrix} = e^{At}\begin{Bmatrix} x_{1o} \\ x_{2o} \\ \dot{x}_{1o} \\ \dot{x}_{2o} \end{Bmatrix} + \int_0^t e^{-A(t-s)}\begin{Bmatrix} 0 \\ 0 \\ -1 \\ -1 \end{Bmatrix}\ddot{u}_g(s)\,ds. \quad (3.219)
$$

3.7 NUMERICAL SOLUTION FOR RESPONSE USING THE DISCRETIZATION METHOD

The response of a 2DOF system with Rayleigh damping was discussed in Section 3.5. Because of the special form of this damping matrix, the normal mode method can be used to develop two uncoupled SDOF equations of motion in normal mode coordinates. If the system does not have Rayleigh damping, then either the state space or the numerical integration method in this section must be used to calculate the system response.

The response of a 2DOF system can be calculated with the discretization method of numerical analysis using a method similar to that discussed in Section 2.6. Given all the response and force information at time t_k, the three response quantities that are desired at time t_{k+1} are the displacement vector, $\mathbf{X}(t)$, velocity vector, $\dot{\mathbf{X}}(t)$, and acceleration vector, $\ddot{\mathbf{X}}(t)$. The dynamic equilibrium equation given in Eq. 3.3 presents one of the three simultaneous equations:

$$
\mathbf{M}\ddot{\mathbf{X}}(t) + \mathbf{C}\dot{\mathbf{X}}(t) + \mathbf{K}\mathbf{X}(t) = \mathbf{F}(t). \quad (3.220)
$$

The two additional equations are provided using the integral equations:

$$
\dot{\mathbf{X}}(t) = \dot{\mathbf{X}}(t_o) + \int_{t_o}^t \ddot{\mathbf{X}}(s)\,ds, \quad \mathbf{X}(t) = \mathbf{X}(t_o) + \int_{t_o}^t \dot{\mathbf{X}}(s)\,ds. \quad (3.221)
$$

As noted in Section 2.6, the integral equations are preferred because a high accuracy can be achieved in calculating the area under a curve. Let

$$
t_{k+1} = t, \quad t_k = t_o, \quad \Delta t = t_{k+1} - t_k.
$$

It follows from Eq. 3.221 that

$$
\dot{\mathbf{X}}_{k+1} = \dot{\mathbf{X}}_k + \int_{t_k}^{t_{k+1}} \ddot{\mathbf{X}}(s)\,ds, \quad \mathbf{X}_{k+1} = \mathbf{X}_k + \int_{t_k}^{t_{k+1}} \dot{\mathbf{X}}(s)\,ds \quad (3.222)
$$

where $\mathbf{X}_k$, $\dot{\mathbf{X}}_k$, and $\ddot{\mathbf{X}}_k$ represent the displacement, velocity, and acceleration vectors, respectively, at time step k, that is,

$$
\mathbf{X}_k = \begin{Bmatrix} x_1(t=k\Delta t) \\ x_2(t=k\Delta t) \end{Bmatrix} = \begin{Bmatrix} x_1 \\ x_2 \end{Bmatrix}_k, \quad \dot{\mathbf{X}}_k = \begin{Bmatrix} \dot{x}_1(t=k\Delta t) \\ \dot{x}_2(t=k\Delta t) \end{Bmatrix} = \begin{Bmatrix} \dot{x}_1 \\ \dot{x}_2 \end{Bmatrix}_k, \quad \ddot{\mathbf{X}}_k = \begin{Bmatrix} \ddot{x}_1(t=k\Delta t) \\ \ddot{x}_2(t=k\Delta t) \end{Bmatrix} = \begin{Bmatrix} \ddot{x}_1 \\ \ddot{x}_2 \end{Bmatrix}_k. \quad (3.223)
$$

To discretize the integrand in Eq. 3.222, approximation of the continuous acceleration function, $\ddot{\mathbf{X}}(s)$, is necessary. As discussed in Chapter 2, Section 2.6, there are many ways of discretizing a continuous function, and the Newmark β-method presents a general solution.

The Newmark β-method uses a set of three simultaneous equations of the form

$$\mathbf{M}\ddot{\mathbf{X}}_{k+1} + \mathbf{C}\dot{\mathbf{X}}_{k+1} + \mathbf{K}\mathbf{X}_{k+1} = \mathbf{F}_{k+1} \tag{3.224a}$$

$$\dot{\mathbf{X}}_{k+1} = \dot{\mathbf{X}}_k + (1-\delta)\ddot{\mathbf{X}}_k \Delta t + \delta\ddot{\mathbf{X}}_{k+1}\Delta t \tag{3.224b}$$

$$\mathbf{X}_{k+1} = \mathbf{X}_k + \dot{\mathbf{X}}_k \Delta t + \left(\tfrac{1}{2} - \alpha\right)\ddot{\mathbf{X}}_k (\Delta t)^2 + \alpha\ddot{\mathbf{X}}_{k+1}(\Delta t)^2. \tag{3.224c}$$

For the previously noted 2DOF system, it follows that

$$\begin{bmatrix} m_1 & 0 \\ 0 & m_2 \end{bmatrix}\begin{Bmatrix} \ddot{x}_1 \\ \ddot{x}_2 \end{Bmatrix}_{k+1} + \begin{bmatrix} c_1+c_2 & -c_2 \\ -c_2 & c_2 \end{bmatrix}\begin{Bmatrix} \dot{x}_1 \\ \dot{x}_2 \end{Bmatrix}_{k+1} + \begin{bmatrix} k_1+k_2 & -k_2 \\ -k_2 & k_2 \end{bmatrix}\begin{Bmatrix} x_1 \\ x_2 \end{Bmatrix}_{k+1} = \begin{Bmatrix} F_1 \\ F_2 \end{Bmatrix}_{k+1} \tag{3.225}$$

$$\begin{Bmatrix} \dot{x}_1 \\ \dot{x}_2 \end{Bmatrix}_{k+1} = \begin{Bmatrix} \dot{x}_1 \\ \dot{x}_2 \end{Bmatrix}_k + (1-\delta)\Delta t \begin{Bmatrix} \ddot{x}_1 \\ \ddot{x}_2 \end{Bmatrix}_k + \delta\Delta t \begin{Bmatrix} \ddot{x}_1 \\ \ddot{x}_2 \end{Bmatrix}_{k+1} \tag{3.226}$$

$$\begin{Bmatrix} x_1 \\ x_2 \end{Bmatrix}_{k+1} = \begin{Bmatrix} x_1 \\ x_2 \end{Bmatrix}_k + (\Delta t)\begin{Bmatrix} \dot{x}_1 \\ \dot{x}_2 \end{Bmatrix}_k + \left(\frac{1}{2} - \alpha\right)(\Delta t)^2\begin{Bmatrix} \ddot{x}_1 \\ \ddot{x}_2 \end{Bmatrix}_k + \alpha(\Delta t)^2\begin{Bmatrix} \ddot{x}_1 \\ \ddot{x}_2 \end{Bmatrix}_{k+1}. \tag{3.227}$$

Three special cases as discussed in Section 2.6 are

1. Constant acceleration: $\qquad\qquad \delta = 0$ and $\alpha = 0$.
2. Constant average acceleration: $\quad \delta = \tfrac{1}{2}$ and $\alpha = \tfrac{1}{4}$.
3. Linear acceleration: $\qquad\qquad\quad \delta = \tfrac{1}{2}$ and $\alpha = \tfrac{1}{6}$.

Substituting Eqs. 3.224b and 3.224c into Eq. 3.224a gives

$$\mathbf{M}\ddot{\mathbf{X}}_{k+1} + \mathbf{C}\left[\dot{\mathbf{X}}_k + (1-\delta)\ddot{\mathbf{X}}_k \Delta t + \delta\ddot{\mathbf{X}}_{k+1}\Delta t \right]$$
$$+ \mathbf{K}\left[\mathbf{X}_k + \dot{\mathbf{X}}_k \Delta t + \left(\tfrac{1}{2} - \alpha\right)\ddot{\mathbf{X}}_k (\Delta t)^2 + \alpha\ddot{\mathbf{X}}_{k+1}(\Delta t)^2 \right] = \mathbf{F}_{k+1}. \tag{3.228}$$

Rearranging the terms in Eq. 3.228 gives

$$\left(\mathbf{M} + \delta\mathbf{C}\Delta t + \alpha\mathbf{K}(\Delta t)^2\right)\ddot{\mathbf{X}}_{k+1} = -\mathbf{K}\mathbf{X}_k - [\mathbf{C} + \mathbf{K}\Delta t]\dot{\mathbf{X}}_k$$
$$-\left[(1-\delta)\mathbf{C}(\Delta t) + \left(\tfrac{1}{2} - \alpha\right)\mathbf{K}(\Delta t)^2\right]\ddot{\mathbf{X}}_k + \mathbf{F}_{k+1}. \tag{3.229}$$

Solving for $\ddot{\mathbf{X}}_{k+1}$ gives

$$\ddot{\mathbf{X}}_{k+1} = -\left(\mathbf{B}^{-1}\mathbf{K}\right)\mathbf{X}_k - \left(\mathbf{B}^{-1}\mathbf{C} + \mathbf{B}^{-1}\mathbf{K}\Delta t\right)\dot{\mathbf{X}}_k - \left(\mathbf{B}^{-1}\mathbf{G}\right)\ddot{\mathbf{X}}_k + \left(\mathbf{B}^{-1}\right)\mathbf{F}_{k+1} \tag{3.230}$$

where

$$\mathbf{B} = \mathbf{M} + \delta\mathbf{C}\Delta t + \alpha\mathbf{K}(\Delta t)^2 \tag{3.231a}$$

$$\mathbf{G} = (1-\delta)\mathbf{C}\Delta t + \left(\tfrac{1}{2} - \alpha\right)\mathbf{K}(\Delta t)^2. \tag{3.231b}$$

Substituting Eq. 3.230 into Eqs. 3.224b and 3.224c gives

$$\dot{\mathbf{X}}_{k+1} = -\left(\delta\mathbf{B}^{-1}\mathbf{K}\Delta t\right)\mathbf{X}_k + \left(\mathbf{I} - \delta\mathbf{B}^{-1}\mathbf{C}\Delta t - \delta\mathbf{B}^{-1}\mathbf{K}(\Delta t)^2\right)\dot{\mathbf{X}}_k$$
$$+ \left((1-\delta)\mathbf{I}\Delta t - \delta\mathbf{B}^{-1}\mathbf{G}\Delta t\right)\ddot{\mathbf{X}}_k + \left(\delta\mathbf{B}^{-1}\Delta t\right)\mathbf{F}_{k+1} \tag{3.232a}$$

$$\mathbf{X}_{k+1} = \left(\mathbf{I} - \alpha\mathbf{B}^{-1}\mathbf{K}(\Delta t)^2\right)\mathbf{X}_k + \left(\mathbf{I}\Delta t - \alpha\mathbf{B}^{-1}\mathbf{C}(\Delta t)^2 - \alpha\mathbf{B}^{-1}\mathbf{K}(\Delta t)^3\right)\dot{\mathbf{X}}_k$$
$$+ \left((\tfrac{1}{2} - \alpha)\mathbf{I}(\Delta t)^2 - \alpha\mathbf{B}^{-1}\mathbf{G}(\Delta t)^2\right)\ddot{\mathbf{X}}_k + \left(\alpha\mathbf{B}^{-1}(\Delta t)^2\right)\mathbf{F}_{k+1}. \tag{3.232b}$$

In summary, the procedure for computing the response at time step $k + 1$ given all of the information at time step k and given $\mathbf{F}_{k+1}$ is obtained using Eqs. 3.230, 3.232a, and 3.232b. These equations can be represented in a matrix form, that is,

$$\begin{Bmatrix} \mathbf{X}_{k+1} \\ \dot{\mathbf{X}}_{k+1} \\ \ddot{\mathbf{X}}_{k+1} \end{Bmatrix} = \mathbf{F}_N \begin{Bmatrix} \mathbf{X}_k \\ \dot{\mathbf{X}}_k \\ \ddot{\mathbf{X}}_k \end{Bmatrix} + \mathbf{H}_N \mathbf{F}_{k+1} \tag{3.233}$$

where

$$\mathbf{F}_N = \begin{bmatrix} \mathbf{I} - \alpha \mathbf{B}^{-1} \mathbf{K} (\Delta t)^2 & \mathbf{I} \Delta t - \alpha \mathbf{B}^{-1} \mathbf{C} (\Delta t)^2 - \alpha \mathbf{B}^{-1} \mathbf{K} (\Delta t)^3 & \left(\frac{1}{2} - \alpha \right) \mathbf{I} (\Delta t)^2 - \alpha \mathbf{B}^{-1} \mathbf{G} (\Delta t)^2 \\ -\delta \mathbf{B}^{-1} \mathbf{K} \Delta t & \mathbf{I} - \delta \mathbf{B}^{-1} \mathbf{C} \Delta t - \delta \mathbf{B}^{-1} \mathbf{K} (\Delta t)^2 & (1 - \delta) \mathbf{I} \Delta t - \delta \mathbf{B}^{-1} \mathbf{G} \Delta t \\ -\mathbf{B}^{-1} \mathbf{K} & -\mathbf{B}^{-1} \mathbf{C} - \mathbf{B}^{-1} \mathbf{K} \Delta t & -\mathbf{B}^{-1} \mathbf{G} \end{bmatrix}$$

$$\mathbf{H}_N = \begin{bmatrix} \alpha \mathbf{B}^{-1} (\Delta t)^2 \\ \delta \mathbf{B}^{-1} \Delta t \\ \mathbf{B}^{-1} \end{bmatrix}$$

and the subscript N denotes the Newmark β-method. For convenience, define

$$\mathbf{q}_k = \begin{Bmatrix} \mathbf{X}_k \\ \dot{\mathbf{X}}_k \\ \ddot{\mathbf{X}}_k \end{Bmatrix} \tag{3.234}$$

and then it follows from Eq. 3.233 that

$$\mathbf{q}_{k+1} = \mathbf{F}_N \mathbf{q}_k + \mathbf{H}_N \mathbf{F}_{k+1}. \tag{3.235}$$

Similar to the SDOF system, the Wilson θ-method can also be used to numerically analyze the 2DOF system. Assume that the acceleration varies linearly over the time interval t to $t = \theta \Delta t$, where the value of θ need not be an integer and is usually greater than 1.0. Consider Eq. 3.224 with Δt replaced by $\theta \Delta t$, $\delta = \frac{1}{2}$, and $\alpha = \frac{1}{6}$; then it follows that

$$\mathbf{M}\ddot{\mathbf{X}}_{k+\theta} + \mathbf{C}\dot{\mathbf{X}}_{k+\theta} + \mathbf{K}\mathbf{X}_{k+\theta} = \mathbf{F}_{k+\theta} \tag{3.236a}$$

$$\dot{\mathbf{X}}_{k+\theta} = \dot{\mathbf{X}}_k + \tfrac{1}{2}\ddot{\mathbf{X}}_k \theta \Delta t + \tfrac{1}{2}\ddot{\mathbf{X}}_{k+\theta} \theta \Delta t \tag{3.236b}$$

$$\mathbf{X}_{k+\theta} = \mathbf{X}_k + \dot{\mathbf{X}}_k \theta \Delta t + \tfrac{1}{3}\ddot{\mathbf{X}}_k \theta^2 (\Delta t)^2 + \tfrac{1}{6}\ddot{\mathbf{X}}_{k+\theta} \theta^2 (\Delta t)^2. \tag{3.236c}$$

As discussed in Section 2.7, the forcing function, $\mathbf{F}_{k+\theta}$, is obtained by linearly extrapolating from $\mathbf{F}_k$ to $\mathbf{F}_{k+1}$, which is

$$\mathbf{F}_{k+\theta} = \mathbf{F}_k + \theta \left(\mathbf{F}_{k+1} - \mathbf{F}_k \right). \tag{3.237}$$

Substituting Eqs. 3.236b and 3.236c into Eq. 3.236a gives

$$\mathbf{M}\ddot{\mathbf{X}}_{k+\theta} + \mathbf{C}\left[\dot{\mathbf{X}}_k + \tfrac{1}{2}\ddot{\mathbf{X}}_k \theta \Delta t + \tfrac{1}{2}\ddot{\mathbf{X}}_{k+\theta} \theta \Delta t \right]$$
$$+ \mathbf{K}\left[\mathbf{X}_k + \dot{\mathbf{X}}_k \theta \Delta t + \tfrac{1}{3}\ddot{\mathbf{X}}_k \theta^2 (\Delta t)^2 + \tfrac{1}{6}\ddot{\mathbf{X}}_{k+\theta} \theta^2 (\Delta t)^2 \right] = \mathbf{F}_{k+\theta}. \tag{3.238}$$

Rearranging the terms in Eq. 3.238, substituting Eq. 3.237 into Eq. 3.238, and solving for $\ddot{\mathbf{X}}_{k+\theta}$, it follows that

$$\ddot{\mathbf{X}}_{k+\theta} = -\left(\mathbf{B}^{-1} \mathbf{K} \right) \mathbf{X}_k - \left(\mathbf{B}^{-1} \mathbf{C} + \mathbf{B}^{-1} \mathbf{K} \theta \Delta t \right) \dot{\mathbf{X}}_k - \left(\mathbf{B}^{-1} \mathbf{G} \right) \ddot{\mathbf{X}}_k + \mathbf{B}^{-1} \left[\mathbf{F}_k + \theta \left(\mathbf{F}_{k+1} - \mathbf{F}_k \right) \right] \tag{3.239}$$

where

$$\mathbf{B} = \mathbf{M} + \tfrac{1}{2}\mathbf{C}\theta \Delta t + \tfrac{1}{6}\mathbf{K}\theta^2 (\Delta t)^2 \tag{3.240a}$$

$$\mathbf{G} = \tfrac{1}{2}\mathbf{C}\theta \Delta t + \tfrac{1}{3}\mathbf{K}\theta^2 (\Delta t)^2. \tag{3.240b}$$

Recall from the dynamic equilibrium equation at time step k that

$$\mathbf{F}_k = \mathbf{M}\ddot{\mathbf{X}}_k + \mathbf{C}\dot{\mathbf{X}}_k + \mathbf{K}\mathbf{X}_k. \tag{3.241}$$

Substituting Eq. 3.241 into Eq. 3.239 and rearranging terms gives

$$\ddot{\mathbf{X}}_{k+\theta} = -\left(\mathbf{B}^{-1}\mathbf{K}\theta\right)\mathbf{X}_k - \left(\mathbf{B}^{-1}\mathbf{C}\theta + \mathbf{B}^{-1}\mathbf{K}\theta\Delta t\right)\dot{\mathbf{X}}_k - \left(\mathbf{B}^{-1}\mathbf{G} + (1-\theta)\mathbf{B}^{-1}\mathbf{M}\right)\ddot{\mathbf{X}}_k + \left(\mathbf{B}^{-1}\theta\right)\mathbf{F}_{k+1}. \quad (3.242)$$

The response at the next time step, t_{k+1}, can be calculated using a linear interpolation within the interval $(t_k, t_{k+\theta})$, which is

$$\ddot{\mathbf{X}}_{k+1} = \ddot{\mathbf{X}}_k + \tfrac{1}{\theta}\left(\ddot{\mathbf{X}}_{k+\theta} - \ddot{\mathbf{X}}_k\right) \qquad\qquad (3.243a)$$

$$\dot{\mathbf{X}}_{k+1} = \dot{\mathbf{X}}_k + \tfrac{1}{\theta}\left(\dot{\mathbf{X}}_{k+\theta} - \dot{\mathbf{X}}_k\right) = \dot{\mathbf{X}}_k + \tfrac{1}{2}\ddot{\mathbf{X}}_k\Delta t + \tfrac{1}{2}\ddot{\mathbf{X}}_{k+\theta}\Delta t \qquad (3.243b)$$

$$\mathbf{X}_{k+1} = \mathbf{X}_k + \tfrac{1}{\theta}\left(\mathbf{X}_{k+\theta} - \mathbf{X}_k\right) = \mathbf{X}_k + \dot{\mathbf{X}}_k\Delta t + \tfrac{1}{3}\ddot{\mathbf{X}}_k\theta(\Delta t)^2 + \tfrac{1}{6}\ddot{\mathbf{X}}_{k+\theta}\theta(\Delta t)^2. \qquad (3.243c)$$

Finally, substituting Eq. 3.241 into Eq. 3.243 gives the response of the system at time step $k + 1$:

$$\ddot{\mathbf{X}}_{k+1} = -\left(\mathbf{B}^{-1}\mathbf{K}\right)\mathbf{X}_k - \left(\mathbf{B}^{-1}\mathbf{C} + \mathbf{B}^{-1}\mathbf{K}\Delta t\right)\dot{\mathbf{X}}_k$$
$$+\left((1-\tfrac{1}{\theta})\mathbf{I} - \mathbf{B}^{-1}\mathbf{G} + (1-\theta)\mathbf{B}^{-1}\mathbf{M}\right)\ddot{\mathbf{X}}_k + \mathbf{B}^{-1}\mathbf{F}_{k+1} \qquad (3.244a)$$

$$\dot{\mathbf{X}}_{k+1} = -\left(\tfrac{1}{2}\mathbf{B}^{-1}\mathbf{K}\theta\Delta t\right)\mathbf{X}_k + \left(\mathbf{I} - \tfrac{1}{2}\mathbf{B}^{-1}\mathbf{C}\theta\Delta t - \tfrac{1}{2}\mathbf{B}^{-1}\mathbf{K}\theta(\Delta t)^2\right)\dot{\mathbf{X}}_k$$
$$+ \tfrac{1}{2}\Delta t\left(\mathbf{I} - \mathbf{B}^{-1}\mathbf{G} - (1-\theta)\mathbf{B}^{-1}\mathbf{M}\right)\ddot{\mathbf{X}}_k + \left(\tfrac{1}{2}\mathbf{B}^{-1}\theta\Delta t\right)\mathbf{F}_{k+1} \qquad (3.244b)$$

$$\mathbf{X}_{k+1} = \left(\mathbf{I} - \tfrac{1}{6}\mathbf{B}^{-1}\mathbf{K}\theta^2(\Delta t)^2\right)\mathbf{X}_k + \left(\mathbf{I}\Delta t - \tfrac{1}{6}\mathbf{B}^{-1}\mathbf{C}\theta^2(\Delta t)^2 - \tfrac{1}{6}\mathbf{B}^{-1}\mathbf{K}\theta^2(\Delta t)^3\right)\dot{\mathbf{X}}_k$$
$$+ \tfrac{1}{6}\theta(\Delta t)^2\left(2\mathbf{I} - \mathbf{B}^{-1}\mathbf{G} - (1-\theta)\mathbf{B}^{-1}\mathbf{M}\right)\ddot{\mathbf{X}}_k + \left(\tfrac{1}{6}\mathbf{B}^{-1}\theta^2(\Delta t)^2\right)\mathbf{F}_{k+1}. \qquad (3.244c)$$

Equation 3.244 can be represented in the following matrix form:

$$\mathbf{q}_{k+1} = \begin{Bmatrix} \mathbf{X}_{k+1} \\ \dot{\mathbf{X}}_{k+1} \\ \ddot{\mathbf{X}}_{k+1} \end{Bmatrix} = \mathbf{F}_W \begin{Bmatrix} \mathbf{X}_k \\ \dot{\mathbf{X}}_k \\ \ddot{\mathbf{X}}_k \end{Bmatrix} + \mathbf{H}_W\,\mathbf{F}_{k+1} = \mathbf{F}_W\mathbf{q}_k + \mathbf{H}_W\,\mathbf{F}_{k+1} \qquad (3.245)$$

where

$$\mathbf{F}_W = \begin{bmatrix} \mathbf{I} - \tfrac{1}{6}\mathbf{B}^{-1}\mathbf{K}\theta^2(\Delta t)^2 & \mathbf{I}\Delta t - \tfrac{1}{6}\mathbf{B}^{-1}\mathbf{C}\theta^2(\Delta t)^2 - \tfrac{1}{6}\mathbf{B}^{-1}\mathbf{K}\theta^2(\Delta t)^3 & \tfrac{1}{6}\theta(\Delta t)^2\left(2\mathbf{I} - \overline{\mathbf{B}}\right) \\ -\tfrac{1}{2}\mathbf{B}^{-1}\mathbf{K}\theta\Delta t & \mathbf{I} - \mathbf{B}^{-1}\mathbf{C}\theta\Delta t - \tfrac{1}{2}\mathbf{B}^{-1}\mathbf{K}\theta(\Delta t)^2 & \tfrac{1}{2}\Delta t\left(\mathbf{I} - \overline{\mathbf{B}}\right) \\ -\mathbf{B}^{-1}\mathbf{K} & -\mathbf{B}^{-1}\mathbf{C} - \mathbf{B}^{-1}\mathbf{K}\Delta t & \left(1-\tfrac{1}{\theta}\right)\mathbf{I} - \overline{\mathbf{B}} \end{bmatrix}$$

$$\mathbf{H}_W = \begin{bmatrix} \tfrac{1}{6}\mathbf{B}^{-1}\theta^2(\Delta t)^2 \\ \tfrac{1}{2}\mathbf{B}^{-1}\theta\Delta t \\ \mathbf{B}^{-1} \end{bmatrix}$$

and

$$\overline{\mathbf{B}} = \mathbf{B}^{-1}\mathbf{G} - (1-\theta)\mathbf{B}^{-1}\mathbf{M}. \qquad (3.246)$$

The subscript W denotes the Wilson θ-method.

Again, two major concerns in performing numerical analysis are the questions of numerical stability and numerical accuracy. In the Newmark method, the constant average acceleration (i.e., $\delta = \tfrac{1}{2}$ and $\alpha = \tfrac{1}{4}$) is generally used in structural dynamics because of its high degree of numerical stability. For the Wilson θ-method, a value of $\theta = 1.40$ is generally used by structural engineers. A time step size of $\Delta t \leq T_n/10$ gives very accurate results, where T_n denotes the smallest significant period of the structure.

EXAMPLE 1   *Constant Acceleration Method*   UN web

Consider a 2DOF system as shown in Figure 3.2 with structural parameters

$$\mathbf{M} = \begin{bmatrix} 1 & 0 \\ 0 & 1 \end{bmatrix}, \quad \mathbf{C} = \begin{bmatrix} 2.0 & -1.0 \\ -1.0 & 1.0 \end{bmatrix}, \quad \mathbf{K} = \begin{bmatrix} 200 & -100 \\ -100 & 100 \end{bmatrix}, \quad \mathbf{F}(t) = \begin{Bmatrix} 0 \\ f(t) \end{Bmatrix}$$

where $f(t)$ is a rise to static forcing function applied at the upper story and takes the form

$$f(t) = \begin{cases} 10t & 0.0 \le t < 1.0 \\ 10 & t \ge 1.0 \\ 0 & t < 0.0. \end{cases}$$

Assume $\Delta t = 0.02$ s. The response of the system is now calculated using the constant acceleration method. The values of α and δ are zero for the constant acceleration method. Therefore, it follows from Eq. 3.227 that

$$\mathbf{B} = \mathbf{M} + (0)\mathbf{C}\Delta t + (0)\mathbf{K}(\Delta t)^2 = \mathbf{M} = \begin{bmatrix} 1 & 0 \\ 0 & 1 \end{bmatrix} \tag{3.247}$$

$$\mathbf{G} = \begin{bmatrix} 2.0 & -1.0 \\ -1.0 & 1.0 \end{bmatrix}(0.02) + \frac{1}{2}\begin{bmatrix} 200 & -100 \\ -100 & 100 \end{bmatrix}(0.02)^2 = \begin{bmatrix} 0.08 & -0.04 \\ -0.04 & 0.04 \end{bmatrix} \tag{3.248}$$

and the matrices in Eqs. 3.229 and 3.233 are

$$\mathbf{F}_N = \begin{bmatrix} \mathbf{I} & \mathbf{I}\Delta t & \frac{1}{2}\mathbf{I}(\Delta t)^2 \\ \mathbf{0} & \mathbf{I} & \mathbf{I}\Delta t \\ -\mathbf{B}^{-1}\mathbf{K} & -\mathbf{B}^{-1}\mathbf{C} - \mathbf{B}^{-1}\mathbf{K}\Delta t & -\mathbf{B}^{-1}\mathbf{G} \end{bmatrix}, \quad \mathbf{H}_N = \begin{bmatrix} \mathbf{0} \\ \mathbf{0} \\ \mathbf{B}^{-1} \end{bmatrix}. \tag{3.249}$$

It follows from Eq. 3.247 that

$$\mathbf{B}^{-1} = \begin{bmatrix} 1 & 0 \\ 0 & 1 \end{bmatrix}^{-1} = \begin{bmatrix} 1 & 0 \\ 0 & 1 \end{bmatrix}. \tag{3.250}$$

Therefore,

$$-\mathbf{B}^{-1}\mathbf{K} = \begin{bmatrix} -1000 & 500 \\ 500 & -500 \end{bmatrix}, \quad -\mathbf{B}^{-1}\mathbf{G} = \begin{bmatrix} -0.08 & 0.04 \\ 0.04 & -0.04 \end{bmatrix}$$

$$-\mathbf{B}^{-1}\mathbf{C} - \mathbf{B}^{-1}\mathbf{K}\Delta t = \begin{bmatrix} -2.0 & 1.0 \\ 1.0 & -1.0 \end{bmatrix} - \begin{bmatrix} 200 & -100 \\ -100 & 100 \end{bmatrix}(0.02) = \begin{bmatrix} -6.0 & 3.0 \\ 3.0 & -3.0 \end{bmatrix} \tag{3.251}$$

and Eq. 3.249 becomes

$$\mathbf{F}_N = \begin{bmatrix} 1 & 0 & 0.02 & 0 & 0.0002 & 0 \\ 0 & 1 & 0 & 0.02 & 0 & 0.0002 \\ 0 & 0 & 1 & 0 & 0.02 & 0 \\ 0 & 0 & 0 & 1 & 0 & 0.02 \\ -200 & 100 & -6.0 & 3.0 & -0.08 & 0.04 \\ 100 & -100 & 3.0 & -3.0 & 0.04 & -0.04 \end{bmatrix}, \quad \mathbf{H}_N = \begin{bmatrix} 0 & 0 \\ 0 & 0 \\ 0 & 0 \\ 0 & 0 \\ 1 & 0 \\ 0 & 1 \end{bmatrix}. \tag{3.252}$$

Finally, Eq. 3.233 becomes

$$\begin{Bmatrix} x_1 \\ x_2 \\ \dot{x}_1 \\ \dot{x}_2 \\ \ddot{x}_1 \\ \ddot{x}_2 \end{Bmatrix}_{k+1} = \begin{bmatrix} 1 & 0 & 0.02 & 0 & 0.0002 & 0 \\ 0 & 1 & 0 & 0.02 & 0 & 0.0002 \\ 0 & 0 & 1 & 0 & 0.02 & 0 \\ 0 & 0 & 0 & 1 & 0 & 0.02 \\ -200 & 100 & -6.0 & 3.0 & -0.08 & 0.04 \\ 100 & -100 & 3.0 & -3.0 & 0.04 & -0.04 \end{bmatrix} \begin{Bmatrix} x_1 \\ x_2 \\ \dot{x}_1 \\ \dot{x}_2 \\ \ddot{x}_1 \\ \ddot{x}_2 \end{Bmatrix}_{k} + \begin{bmatrix} 0 & 0 \\ 0 & 0 \\ 0 & 0 \\ 0 & 0 \\ 1 & 0 \\ 0 & 1 \end{bmatrix} \begin{Bmatrix} 0 \\ f_{k+1} \end{Bmatrix}. \quad (3.253)$$

Assume that the initial conditions are zero (i.e., $\mathbf{X}_o = \dot{\mathbf{X}}_o = \mathbf{0}$), and since the forcing function at time zero is equal to zero, that is, $f(0) = 0$, it follows that

$$\ddot{\mathbf{X}}_o = \mathbf{M}^{-1}\mathbf{F}_o - \mathbf{M}^{-1}\mathbf{C}\dot{\mathbf{X}}_o - \mathbf{M}^{-1}\mathbf{K}\mathbf{X}_o = \mathbf{M}^{-1}(0) - \mathbf{M}^{-1}\mathbf{C}(0) - \mathbf{M}^{-1}\mathbf{K}(0) = \mathbf{0}. \quad (3.254)$$

The solution for the first time step (i.e., $t = \Delta t$), with the value of forcing function equal to 0.20, is

$$\begin{Bmatrix} x_1 \\ x_2 \\ \dot{x}_1 \\ \dot{x}_2 \\ \ddot{x}_1 \\ \ddot{x}_2 \end{Bmatrix}_{1} = \begin{bmatrix} 1 & 0 & 0.02 & 0 & 0.0002 & 0 \\ 0 & 1 & 0 & 0.02 & 0 & 0.0002 \\ 0 & 0 & 1 & 0 & 0.02 & 0 \\ 0 & 0 & 0 & 1 & 0 & 0.02 \\ -200 & 100 & -6.0 & 3.0 & -0.08 & 0.04 \\ 100 & -100 & 3.0 & -3.0 & 0.04 & -0.04 \end{bmatrix} \begin{Bmatrix} 0 \\ 0 \\ 0 \\ 0 \\ 0 \\ 0 \end{Bmatrix} + \begin{bmatrix} 0 & 0 \\ 0 & 0 \\ 0 & 0 \\ 0 & 0 \\ 1 & 0 \\ 0 & 1 \end{bmatrix} \begin{Bmatrix} 0 \\ 0.20 \end{Bmatrix} = \begin{Bmatrix} 0 \\ 0 \\ 0 \\ 0 \\ 0 \\ 0.2 \end{Bmatrix}. \quad (3.255)$$

Then follows the solution for the second time step (i.e., $t = 2\Delta t$) using Eq. 3.253 with the results from Eq. 3.255 and the value of forcing function equal to 0.40:

$$\begin{Bmatrix} x_1 \\ x_2 \\ \dot{x}_1 \\ \dot{x}_2 \\ \ddot{x}_1 \\ \ddot{x}_2 \end{Bmatrix}_{2} = \begin{bmatrix} 1 & 0 & 0.02 & 0 & 0.0002 & 0 \\ 0 & 1 & 0 & 0.02 & 0 & 0.0002 \\ 0 & 0 & 1 & 0 & 0.02 & 0 \\ 0 & 0 & 0 & 1 & 0 & 0.02 \\ -200 & 100 & -6.0 & 3.0 & -0.08 & 0.04 \\ 100 & -100 & 3.0 & -3.0 & 0.04 & -0.04 \end{bmatrix} \begin{Bmatrix} 0 \\ 0 \\ 0 \\ 0 \\ 0 \\ 0.2 \end{Bmatrix} + \begin{bmatrix} 0 & 0 \\ 0 & 0 \\ 0 & 0 \\ 0 & 0 \\ 1 & 0 \\ 0 & 1 \end{bmatrix} \begin{Bmatrix} 0 \\ 0.40 \end{Bmatrix} = \begin{Bmatrix} 0 \\ 0.00004 \\ 0 \\ 0.0040 \\ 0.0080 \\ 0.392 \end{Bmatrix}.$$

It follows that the solution for the third time step (i.e., $t = 3\Delta t$) with the value of forcing function equal to 0.60 is

$$\begin{Bmatrix} x_1 \\ x_2 \\ \dot{x}_1 \\ \dot{x}_2 \\ \ddot{x}_1 \\ \ddot{x}_2 \end{Bmatrix}_{3} = \begin{bmatrix} 1 & 0 & 0.02 & 0 & 0.0002 & 0 \\ 0 & 1 & 0 & 0.02 & 0 & 0.0002 \\ 0 & 0 & 1 & 0 & 0.02 & 0 \\ 0 & 0 & 0 & 1 & 0 & 0.02 \\ -200 & 100 & -6.0 & 3.0 & -0.08 & 0.04 \\ 100 & -100 & 3.0 & -3.0 & 0.04 & -0.04 \end{bmatrix} \begin{Bmatrix} 0 \\ 0.00004 \\ 0 \\ 0.0040 \\ 0.0080 \\ 0.392 \end{Bmatrix} + \begin{Bmatrix} 0 \\ 0 \\ 0 \\ 0 \\ 0 \\ 0.60 \end{Bmatrix} = \begin{Bmatrix} 0.00000 \\ 0.00020 \\ 0.00016 \\ 0.01184 \\ 0.03104 \\ 0.56864 \end{Bmatrix}.$$

and the calculation process continues. The displacement and velocity responses are shown in Figure 3.4.

EXAMPLE 2 *Response Dependence on Time Step*

The sensitivity of the response to the time step in a 2DOF system can be seen by first defining $k_1 = k$, $k_2 = \hat{\alpha}k$, $c_1 = c$, and $c_2 = \hat{\beta}c$. Assume that $m_1 = m_2 = m$. Then the dynamic matrices become

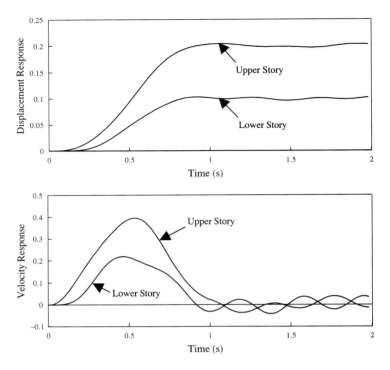

Figure 3.4 Displacement and Velocity Response of a 2DOF System

$$\mathbf{M} = m\begin{bmatrix} 1 & 0 \\ 0 & 1 \end{bmatrix}, \quad \mathbf{K} = k\begin{bmatrix} 1+\hat{\alpha} & -1 \\ -1 & 1 \end{bmatrix}, \quad \mathbf{C} = c\begin{bmatrix} 1+\hat{\beta} & -1 \\ -1 & 1 \end{bmatrix}. \tag{3.256}$$

The matrices $\mathbf{B}$ and $\mathbf{G}$ are important contributions to the response (see Eq. 3.231). These two matrices are highly dependent on the time step size. From Eq. 3.231,

$$\begin{aligned} \mathbf{B} &= \mathbf{M} + \delta\Delta t\mathbf{C} + \alpha(\Delta t)^2 \mathbf{K} \\ &= m\begin{bmatrix} 1 & 0 \\ 0 & 1 \end{bmatrix} + \delta\Delta tc\begin{bmatrix} 1+\hat{\beta} & -1 \\ -1 & 1 \end{bmatrix} + \alpha(\Delta t)^2 k\begin{bmatrix} 1+\hat{\alpha} & -1 \\ -1 & 1 \end{bmatrix} \\ &= m\left(\begin{bmatrix} 1 & 0 \\ 0 & 1 \end{bmatrix} + 2\pi\zeta\delta\left(\frac{\Delta t}{T_n}\right)\begin{bmatrix} 1+\hat{\beta} & -1 \\ -1 & 1 \end{bmatrix} + 4\pi^2\left(\frac{\Delta t}{T_n}\right)^2\begin{bmatrix} 1+\hat{\alpha} & -1 \\ -1 & 1 \end{bmatrix} \right) \end{aligned} \tag{3.257}$$

and

$$\begin{aligned} \mathbf{G} &= (1+\delta)\Delta tc\begin{bmatrix} 1+\hat{\beta} & -1 \\ -1 & 1 \end{bmatrix} + \left(\frac{1}{2}-\alpha\right)(\Delta t)^2 k\begin{bmatrix} 1+\hat{\alpha} & -1 \\ -1 & 1 \end{bmatrix} \\ &= m\left(2\pi\zeta(1+\delta)\left(\frac{\Delta t}{T_n}\right)\begin{bmatrix} 1+\hat{\beta} & -1 \\ -1 & 1 \end{bmatrix} + 4\pi^2\left(\frac{1}{2}-\alpha\right)\left(\frac{\Delta t}{T_n}\right)^2\begin{bmatrix} 1+\hat{\alpha} & -1 \\ -1 & 1 \end{bmatrix} \right). \end{aligned} \tag{3.258}$$

As in Chapter 2, the key term is $(\Delta t/T_n)$, and as this term becomes smaller the difference between the numerical integration solution methods decreases.

EXAMPLE 3   ***Earthquake Response***

Consider the case where the earthquake ground motion excites the 2DOF system as discussed in Example 2 of Section 3.5. The forcing function is given by

$$\mathbf{F}_k = -\begin{bmatrix} m_1 & 0 \\ 0 & m_2 \end{bmatrix}\begin{Bmatrix} 1 \\ 1 \end{Bmatrix} a_k \tag{3.259}$$

where a_k is the absolute ground acceleration. It follows from the dynamic equilibrium equation of motion that

$$\mathbf{M}\ddot{\mathbf{X}}_{k+1} + \mathbf{C}\dot{\mathbf{X}}_{k+1} + \mathbf{K}\mathbf{X}_{k+1} = -\mathbf{M}\{\mathbf{I}\}a_{k+1} \tag{3.260}$$

where $\{\mathbf{I}\}$ is a column vector with 1's in all entries. The results of the response for the earthquake ground motion case using the Newmark β-method or the Wilson θ-method are

$$\mathbf{q}_{k+1} = \mathbf{F}_N \mathbf{q}_k + \mathbf{H}_N^{(EQ)} a_{k+1} \tag{3.261}$$

$$\mathbf{q}_{k+1} = \mathbf{F}_W \mathbf{q}_k + \mathbf{H}_W^{(EQ)} a_{k+1} \tag{3.262}$$

where

$$\mathbf{H}_N^{(EQ)} = -\begin{bmatrix} \alpha\mathbf{B}^{-1}(\Delta t)^2 \\ \delta\mathbf{B}^{-1}\Delta t \\ \mathbf{B}^{-1} \end{bmatrix}\begin{Bmatrix} m_1 \\ m_2 \end{Bmatrix}, \quad \mathbf{H}_W^{(EQ)} = -\begin{bmatrix} \frac{1}{6}\mathbf{B}^{-1}\theta^2(\Delta t)^2 \\ \frac{1}{2}\mathbf{B}^{-1}\theta\Delta t \\ \mathbf{B}^{-1} \end{bmatrix}\begin{Bmatrix} m_1 \\ m_2 \end{Bmatrix}. \tag{3.263}$$

3.8   NUMERICAL SOLUTION FOR STATE SPACE RESPONSE USING THE DIRECT INTEGRATION METHOD

Section 2.8 presented the numerical solution using direct integration for an SDOF system and is here extended to a 2DOF system. Equation 3.195 gives the state space response solution in integral form:

$$\mathbf{z}(t) = e^{\mathbf{A}(t-t_o)}\mathbf{z}(t_o) + e^{\mathbf{A}t}\int_{t_o}^{t} e^{-\mathbf{A}s}\mathbf{F}(s)\,ds \tag{3.264}$$

where

$$\mathbf{z}(t) = \begin{Bmatrix} \mathbf{X}(t) \\ \dot{\mathbf{X}}(t) \end{Bmatrix}, \quad \mathbf{A} = \begin{bmatrix} \mathbf{0} & \mathbf{I} \\ -\mathbf{M}^{-1}\mathbf{K} & -\mathbf{M}^{-1}\mathbf{C} \end{bmatrix}, \quad \mathbf{F}(t) = \begin{Bmatrix} \mathbf{0} \\ \mathbf{M}^{-1}\mathbf{F}_e(t) \end{Bmatrix}.$$

Let

$$t_{k+1} = t, \quad t_k = t_o, \quad \Delta t = t_{k+1} - t_k. \tag{3.265}$$

It follows from Eq. 3.264 that

$$\mathbf{z}_{k+1} = e^{\mathbf{A}\Delta t}\mathbf{z}_k + e^{\mathbf{A}t_{k+1}}\int_{t_k}^{t_{k+1}} e^{-\mathbf{A}s}\mathbf{F}(s)\,ds. \tag{3.266}$$

The objective for numerical analysis using the integration method is to integrate the forcing function in Eq. 3.266. Following the discussion in Section 2.8, two methods used by structural engineers to represent the forcing function are the Delta forcing function and the constant forcing function.

With the Delta forcing function, the forcing function is digitized using a series of Delta functions. The forcing function is represented by

$$\mathbf{F}(s) = \mathbf{F}_k\delta(s - t_k)\Delta t = \begin{Bmatrix} \mathbf{0} \\ \mathbf{M}^{-1}\mathbf{F}_{ek} \end{Bmatrix}\delta(s - t_k)\Delta t, \quad t_k \le s < t_{k+1}. \tag{3.267}$$

Substituting Eq. 3.267 into Eq. 3.268 gives

$$
\begin{aligned}
\mathbf{z}_{k+1} &= \mathbf{e}^{\mathbf{A}\Delta t}\,\mathbf{z}_k + \mathbf{e}^{\mathbf{A} t_{k+1}} \int_{t_k}^{t_{k+1}} \mathbf{e}^{-\mathbf{A}s}\,\mathbf{F}_k \delta(s - t_k)\Delta t\,ds \\
&= \mathbf{e}^{\mathbf{A}\Delta t}\,\mathbf{z}_k + \mathbf{e}^{\mathbf{A} t_{k+1}} \mathbf{e}^{-\mathbf{A} t_k} \mathbf{F}_k \Delta t \\
&= \mathbf{e}^{\mathbf{A}\Delta t}\,\mathbf{z}_k + \mathbf{e}^{\mathbf{A}\Delta t} \Delta t\,\mathbf{F}_k.
\end{aligned}
\tag{3.268}
$$

With the constant forcing function, the forcing function is digitized by assuming that the forcing function is constant within the time interval. The magnitude of the forcing value is equal to the value of the force at the beginning of the interval. Therefore, it follows that

$$
\mathbf{F}(s) = \mathbf{F}_k = \left\{ \begin{array}{c} \mathbf{0} \\ \mathbf{M}^{-1}\mathbf{F}_{ek} \end{array} \right\}, \qquad t_k \le s < t_{k+1}.
\tag{3.269}
$$

Substituting Eq. 3.269 into Eq. 3.266 gives

$$
\begin{aligned}
\mathbf{z}_{k+1} &= \mathbf{e}^{\mathbf{A}\Delta t}\,\mathbf{z}_k + \mathbf{e}^{\mathbf{A} t_{k+1}} \int_{t_k}^{t_{k+1}} \mathbf{e}^{-\mathbf{A}s}\,\mathbf{F}_k\,ds \\
&= \mathbf{e}^{\mathbf{A}\Delta t}\,\mathbf{z}_k + \mathbf{e}^{\mathbf{A} t_{k+1}} \mathbf{A}^{-1}\left(\mathbf{e}^{-\mathbf{A} t_k} - \mathbf{e}^{-\mathbf{A} t_{k+1}}\right)\mathbf{F}_k \\
&= \mathbf{e}^{\mathbf{A}\Delta t}\,\mathbf{z}_k + \mathbf{A}^{-1}\left(\mathbf{e}^{\mathbf{A}\Delta t} - \mathbf{I}\right)\mathbf{F}_k.
\end{aligned}
\tag{3.270}
$$

Note that in Eqs. 3.268 and 3.270, the value of $\mathbf{z}_{k+1}$ (i.e., displacement and velocity of the system at time step $k + 1$) is expressed only in terms of the information at time step k.

Extensive research has been done on how to efficiently calculate the state transition matrix, $\mathbf{e}^{\mathbf{A}t}$. One method is to expand Eq. 3.196 directly with a specified time step, that is,

$$
\mathbf{e}^{\mathbf{A}\Delta t} = \mathbf{I} + \mathbf{A}\Delta t + \frac{1}{2!}\left(\mathbf{A}\Delta t\right)^2 + \frac{1}{3!}\left(\mathbf{A}\Delta t\right)^3 + \dots.
\tag{3.271}
$$

In numerical calculation of the matrix $\mathbf{e}^{\mathbf{A}\Delta t}$, the number of terms required in the solution is large if the value in the exponential is large. In 1994, Moler and Van Loan [1] evaluated different methods of calculating the state transition matrix, $\mathbf{e}^{\mathbf{A}\Delta t}$. The method they recommended is the *scaling and squaring method*. The key to this method is the mathematical property that

$$
\mathbf{e}^{\mathbf{A}\Delta t} = \left(\mathbf{e}^{\mathbf{A}\Delta t/m}\right)^m
\tag{3.272}
$$

where m is an integer. When the matrix $\mathbf{A}\Delta t$ is divided by m, the result is a matrix whose entries contain a maximum value of ε_m, which is usually less than the predefined limit, $\varepsilon_{\lim}$, where $\varepsilon_{\lim}$ is usually taken to be 1.0. Consider the Taylor series expansion,

$$
\mathbf{e}^{\mathbf{A}\Delta t/m} = \mathbf{I} + \frac{\mathbf{A}\Delta t}{m} + \frac{1}{2!}\left(\frac{\mathbf{A}\Delta t}{m}\right)^2 + \frac{1}{3!}\left(\frac{\mathbf{A}\Delta t}{m}\right)^3 + \dots.
\tag{3.273}
$$

The matrix $\mathbf{e}^{\mathbf{A}\Delta t/m}$ usually will converge very fast, and then the matrix $\mathbf{e}^{\mathbf{A}\Delta t/m}$ is multiplied by itself m times.

Recall from Eq. 3.266 that when calculating the structural response from time step t_k (i.e., $\mathbf{z}_k$) to time step t_{k+1} (i.e., $\mathbf{z}_{k+1}$), the operation is exact, and the integration of the forcing function (i.e., $\mathbf{F}_k$) introduces very small numerical errors when the time step size is small. Therefore, accurate results can be obtained when the time step is as large as $\Delta t \le (T_n/5)$. However, in general it is recommended that $\Delta t \le (T_n/10)$.

EXAMPLE 1 *Exponent of a Scalar*

To demonstrate the speed of using the scaling and squaring method, the exponent of a scalar is computed here. Let the exponent be -5.0; the true solution is

$$e^{-5.0} = 0.006738. \tag{3.274}$$

If Eq. 3.271 is used directly, then the terms used, denoted by n, are summarized as follows:

$$n = 0: \qquad (-5.0)^0/0! = 1.0, \qquad e^{-5.0} = 1.0 \tag{3.275a}$$

$$n = 1: \qquad (-5.0)^1/1! = -5.0, \qquad e^{-5.0} = -4.0 \tag{3.275b}$$

$$n = 2: \qquad (-5.0)^2/2! = 12.5, \qquad e^{-5.0} = 8.5 \tag{3.275c}$$

$$n = 3: \qquad (-5.0)^3/3! = -20.83, \qquad e^{-5.0} = -12.33 \tag{3.275d}$$

$$n = 4: \qquad (-5.0)^4/4! = 26.04, \qquad e^{-5.0} = 13.71 \tag{3.275e}$$

$$n = 5: \qquad (-5.0)^5/5! = -26.04, \qquad e^{-5.0} = -12.33 \tag{3.275f}$$

$$n = 6: \qquad (-5.0)^6/6! = 21.70, \qquad e^{-5.0} = 9.37 \tag{3.275g}$$

$$n = 7: \qquad (-5.0)^7/7! = -15.50, \qquad e^{-5.0} = -6.13 \tag{3.275h}$$

$$n = 8: \qquad (-5.0)^8/8! = 9.69, \qquad e^{-5.0} = 3.56 \tag{3.275i}$$

and so forth. This procedure requires a large number of terms before the value of $e^{A\Delta t}$ converges, and therefore it is very time consuming. In addition, if the computer uses only 5 or 6 significant digits, then significant error will be introduced at $n = 3$, 4, and 5 because the number of significant digits should be greater than 10. This will prevent the inaccuracy of two large numbers subtracting one another, leaving behind digits that are considered insignificant when embedded inside the large numbers.

Now consider the scaling and squaring method given in Eq. 3.273. The exponent is again -5.0. Let $\varepsilon_{lim} = 1.0$; it follows that in order to reduce the exponent of $e^{A\Delta t/m}$ less than 1.0, $m = 6$. The terms with different value of exponent, n, are summarized as follows:

$$n = 0: \qquad \left(\tfrac{-5.0}{6}\right)^0/0! = 1.0, \qquad e^{-5.0/6} = 1.0, \qquad e^{-5.0} = 1.0 \tag{3.276a}$$

$$n = 1: \qquad \left(\tfrac{-5.0}{6}\right)^1/1! = -0.83, \qquad e^{-5.0/6} = 0.17, \qquad e^{-5.0} = 0.00002 \tag{3.276b}$$

$$n = 2: \qquad \left(\tfrac{-5.0}{6}\right)^2/2! = 0.35, \qquad e^{-5.0/6} = 0.52, \qquad e^{-5.0} = 0.020 \tag{3.276c}$$

$$n = 3: \qquad \left(\tfrac{-5.0}{6}\right)^3/3! = -0.096, \qquad e^{-5.0/6} = 0.42, \qquad e^{-5.0} = 0.0058 \tag{3.276d}$$

and so forth. The solution using this method gives very close results after using only 4 terms (i.e., $n = 3$). In addition, since the exponent of the exponential function is less than 1.0, each term in the expansion will also be less than 1.0. Numerical error (i.e., raising a large number to certain power, dividing by another large number, and finally subtracting two large numbers) will not appear in this method. Therefore, this example shows that the scaling and squaring method has become very popular because of its speed and accuracy.

EXAMPLE 2 *Numerical Accuracy of an SDOF*

The numerical accuracy and speed of the scaling and squaring method appear to be very attractive for computing the exponent of a scalar. However, this method needs further study when the exponent is a matrix to establish confidence using this method. From Eq. 3.273, let the increment be defined as

$$I_k = \sum_{\text{Rows}} \sum_{\text{Columns}} \left| \frac{1}{k!} \left(\frac{\mathbf{A}\Delta t}{m} \right)^k \right|. \tag{3.277}$$

Since each entry in $(\mathbf{A}\Delta t/m)$ is less than 1.0, the sequence of I's in Eq. 3.277 will have the following relationship:

$$\ldots < I_{k-1} < I_k < I_{k+1} < \ldots. \tag{3.278}$$

Therefore, depending on the number of terms used in Eq. 3.273 to calculate the state transition matrix, if I_k is less than some predefined limit, I_{lim}, then the series has converged.

In this study, an SDOF system is selected with a fundamental period of 0.08 s. The period of this magnitude is usually a good representation of the smallest significant period (highest significant frequency) of a building. Let $\varepsilon_{\text{lim}} = 2.0$, $\zeta = 0.0$, and $\Delta t = 2.0$ s; with an initial displacement of 1.0 unit and an initial velocity of 0.0 unit, free vibration during the interval of 20 s is studied. Table 3.1 summarizes the numerical accuracy of this study. Without any damping in the system, the actual displacement should be 1.0 unit at 20 s, and the velocity should be 0.0 unit at 20 s. Table 3.1 shows that only 4 terms are needed to obtain accurate results.

EXAMPLE 3   *State Transition Matrix of a 2DOF System*

Consider a 2DOF system with Rayleigh proportional damping, where

$$\mathbf{M} = \begin{bmatrix} 1.0 & 0.0 \\ 0.0 & 1.0 \end{bmatrix}, \quad \mathbf{C} = \begin{bmatrix} 3.372 & -1.124 \\ -1.124 & 2.248 \end{bmatrix}, \quad \mathbf{K} = \begin{bmatrix} 1263.36 & -631.68 \\ -631.68 & 631.68 \end{bmatrix}. \tag{3.279}$$

Then it follows that

$$\mathbf{A} = \begin{bmatrix} 0.0 & 0.0 & 1.0 & 0.0 \\ 0.0 & 0.0 & 0.0 & 1.0 \\ -1263.36 & 631.68 & -3.372 & 1.124 \\ 631.68 & -631.68 & 1.124 & -2.248 \end{bmatrix} \tag{3.280}$$

and if $\Delta t = 0.02$ s, then

Table 3.1 Numerical Accuracy Study of the State Transition Matrix

Criteria I_{lim}	Number of terms (k)	I_k	Displacement at 20.0 s	Velocity at 20.0 s
2.0	1	1.003	Unstable	Unstable
1.0×10^{-2}	2	8.155×10^{-5}	0.99949737	−3.35361232
5.0×10^{-5}	3	2.726×10^{-5}	0.99986370	−0.00011255
1.0×10^{-8}	4	1.108×10^{-9}	0.99999999	0.00002415
1.0×10^{-12}	6	6.025×10^{-15}	1.00000000	−0.00000319
1.0×10^{-15}	8	1.755×10^{-20}	1.00000000	−0.00000319

$$\mathbf{e}^{\mathbf{A}\Delta t} = \begin{bmatrix} 0.766425 & 0.115076 & 0.017773 & 0.000992 \\ 0.115076 & 0.881501 & 0.000992 & 0.018764 \\ -21.8268 & 10.60017 & 0.70761 & 0.132823 \\ 10.60017 & -11.2266 & 0.132823 & 0.840433 \end{bmatrix}. \tag{3.281}$$

EXAMPLE 4   *Earthquake Response*

Consider the case where the earthquake ground motion excites the 2DOF system. The forcing function is given by

$$\mathbf{F}_{ek} = -\begin{bmatrix} m_1 & 0 \\ 0 & m_2 \end{bmatrix} \begin{Bmatrix} 1 \\ 1 \end{Bmatrix} a_k \tag{3.282}$$

where a_k is the absolute ground acceleration. It follows from Eqs. 3.267 and 3.269 that

$$\mathbf{F}_k = \begin{Bmatrix} \mathbf{0} \\ \mathbf{M}^{-1}\mathbf{F}_{ek} \end{Bmatrix} = \begin{Bmatrix} 0 \\ 0 \\ -1 \\ -1 \end{Bmatrix} a_k, \quad t_k \le s < t_{k+1}. \tag{3.283}$$

From Eq. 3.283 and using the Delta forcing function, Eq. 3.268 becomes

$$\mathbf{z}_{k+1} = \mathbf{e}^{\mathbf{A}\Delta t}\mathbf{z}_k + \mathbf{e}^{\mathbf{A}\Delta t}\Delta t \begin{Bmatrix} 0 \\ 0 \\ -1 \\ -1 \end{Bmatrix} a_k \tag{3.284}$$

and using the constant forcing function, Eq. 3.270 becomes

$$\mathbf{z}_{k+1} = \mathbf{e}^{\mathbf{A}\Delta t}\mathbf{z}_k + \mathbf{A}^{-1}\left(\mathbf{e}^{\mathbf{A}\Delta t} - \mathbf{I}\right) \begin{Bmatrix} 0 \\ 0 \\ -1 \\ -1 \end{Bmatrix} a_k. \tag{3.285}$$

3.9   SENSITIVITY ANALYSIS OF NATURAL FREQUENCIES AND MODE SHAPES

Section 2.9 presented methods for performing a sensitivity analysis of the response of an SDOF system. In a 2DOF system, the natural frequencies and mode shapes are often the response variables of interest. A perturbation sensitivity analysis is performed in the same way as discussed in Chapter 2.

Equation 3.10 gives the equations for the two undamped natural frequencies of vibration (i.e., the response parameters) in terms of the two masses and two stiffnesses (i.e., the four design variables). Therefore, a partial derivative sensitivity analysis for following response and design variable combination is

$$\frac{\partial \omega_1}{\partial m_1}, \quad \frac{\partial \omega_1}{\partial m_2}, \quad \frac{\partial \omega_1}{\partial k_1}, \quad \frac{\partial \omega_1}{\partial k_2}, \quad \frac{\partial \omega_2}{\partial m_1}, \quad \frac{\partial \omega_2}{\partial m_2}, \quad \frac{\partial \omega_2}{\partial k_1}, \quad \frac{\partial \omega_2}{\partial k_2}. \tag{3.286}$$

A partial derivative sensitivity analysis of the mode shapes can be performed by taking the partial derivative of the mode shapes in Eq. 3.22. Recall that the second element of a mode shape is not unique, but the ratio of the first element to the second element is unique.

EXAMPLE 1 *Sensitivity of Natural Frequencies to Stiffnesses*

Consider a two-story building frame whose computer model is a 2DOF system. The coordinates $x_1(t)$ and $x_2(t)$ are the lateral displacement of the first and second floors, respectively. It follows that the equation of motion is

$$\mathbf{M}\ddot{\mathbf{X}}(t) + \mathbf{C}\dot{\mathbf{X}}(t) + \mathbf{K}\mathbf{X}(t) = \mathbf{F}(t). \tag{3.287}$$

Consider the free-vibration case with no damping, and let the mass of the first and second floors be

$$m_1 = 1.5 \text{ k-s}^2/\text{in.}$$

$$m_2 = 1.0 \text{ k-s}^2/\text{in.}$$

It follows from Eq. 3.287 that

$$\begin{bmatrix} 1.5 & 0 \\ 0 & 1.0 \end{bmatrix} \begin{Bmatrix} \ddot{x}_1 \\ \ddot{x}_2 \end{Bmatrix} + \begin{bmatrix} k_1 + k_2 & -k_2 \\ -k_2 & k_2 \end{bmatrix} \begin{Bmatrix} x_1 \\ x_2 \end{Bmatrix} = \begin{Bmatrix} 0 \\ 0 \end{Bmatrix} \tag{3.288}$$

where k_1 and k_2 are the first- and second-story stiffnesses, respectively.

The eigenvalue problem for this 2DOF system is

$$\left(\mathbf{K} - \mathbf{M}\omega_i^2\right)\boldsymbol{\phi}_i = 0 \tag{3.289}$$

and for this example, it follows that

$$\left(\begin{bmatrix} k_1 + k_2 & -k_2 \\ -k_2 & k_2 \end{bmatrix} - \omega_i^2 \begin{bmatrix} 1.5 & 0 \\ 0 & 1.0 \end{bmatrix}\right)\boldsymbol{\phi}_i = \begin{Bmatrix} 0 \\ 0 \end{Bmatrix}. \tag{3.290}$$

The calculation of the eigenvalues of this system, ω_i^2, was discussed from Eq. 3.8 to Eq. 3.10, and the results following Eq. 3.10 are

$$\omega_1^2 = \frac{(1.0(k_1 + k_2) + 1.5k_2) - \sqrt{(1.0(k_1 + k_2) + 1.5k_2)^2 - 4(1.5)(1.0)k_1k_2}}{2(1.5)(1.0)} \tag{3.291}$$

$$\omega_2^2 = \frac{(1.0(k_1 + k_2) + 1.5k_2) + \sqrt{(1.0(k_1 + k_2) + 1.5k_2)^2 - 4(1.5)(1.0)k_1k_2}}{2(1.5)(1.0)}. \tag{3.292}$$

Simplifying Eqs. 3.291 and 3.292 gives

$$\omega_1^2 = \frac{1}{3}\left(k_1 + 2.5k_2 - \sqrt{(k_1 + 2.5k_2)^2 - 6k_1k_2}\right) \tag{3.293}$$

$$\omega_2^2 = \frac{1}{3}\left(k_1 + 2.5k_2 + \sqrt{(k_1 + 2.5k_2)^2 - 6k_1k_2}\right). \tag{3.294}$$

Now let $k_1 = \alpha k_2$. If the first-story stiffness is "large" relative to the second-story stiffness, then the 2DOF responds like an SDOF system. In this case, the second story responds as if it were on a "fixed" first-story base. Then, the SDOF natural frequency of vibration, denoted ω_{no}, for such a fixed first-story base would be

$$\omega_{no} = \sqrt{\frac{k_2}{m_2}} = \sqrt{k_2} . \tag{3.295}$$

Substituting $k_1 = \alpha k_2$ into Eqs. 3.293 and 3.294 gives

$$\omega_1^2 = \frac{k_2}{3}\left(\alpha + 2.5 - \sqrt{(\alpha + 2.5)^2 - 6\alpha} \right) \tag{3.296}$$

$$\omega_2^2 = \frac{k_2}{3}\left(\alpha + 2.5 + \sqrt{(\alpha + 2.5)^2 - 6\alpha} \right) \tag{3.297}$$

and normalizing Eqs. 3.296 and 3.297 by dividing these equations by ω_{no}^2 gives

$$\frac{\omega_1^2}{\omega_{no}^2} = \frac{1}{3}\left(\alpha + 2.5 - \sqrt{(\alpha + 2.5)^2 - 6\alpha} \right) \tag{3.298}$$

$$\frac{\omega_2^2}{\omega_{no}^2} = \frac{1}{3}\left(\alpha + 2.5 + \sqrt{(\alpha + 2.5)^2 - 6\alpha} \right). \tag{3.299}$$

Note that Eqs. 3.298 and 3.299 are in a dimensionless form. It is always very important in structural engineering to normalize the solution whenever possible, because this maximizes the opportunity to apply the problem to a similar problem in the future, and therefore the knowledge learned in one situation is transferred to a new situation. It is clear that the value of α that is required to consider the first story as "rigid" is a structural engineering decision because the difference between ω_i^2 and ω_{no}^2 is a function of α. However, many structural engineers would be satisfied with the "rigid" first story if α (i.e., k_1/k_2) were equal to or greater than 10.

The sensitivity of the fundamental system natural frequency to the stiffness at the second story can now be obtained by taking the derivative of ω_1 with respect to k_2 (i.e., $\partial\omega_1/\partial k_2$). When the sensitivity is calculated for one parameter (e.g., k_2), the values for all other parameters (e.g., k_1 or α) are set equal to a constant value. Therefore, the sensitivity is said to be calculated about a normal point where the point is the value of the set parameters (i.e., k_2 = a set value).

Consider first that α is constant (i.e., the ratio of k_1/k_2 is taken to be constant); it follows from Eqs. 3.296 and 3.297 that

$$\frac{\partial\omega_1}{\partial k_2} = \frac{\partial}{\partial k_2}\left[\sqrt{\frac{k_2}{3}\left(\alpha + 2.5 - \sqrt{(\alpha + 2.5)^2 - 6\alpha} \right)}\right] = \frac{1}{2\sqrt{3k_2}}\sqrt{\alpha + 2.5 - \sqrt{(\alpha + 2.5)^2 - 6\alpha}} \tag{3.300}$$

$$\frac{\partial\omega_2}{\partial k_2} = \frac{\partial}{\partial k_2}\left[\sqrt{\frac{k_2}{3}\left(\alpha + 2.5 + \sqrt{(\alpha + 2.5)^2 - 6\alpha} \right)}\right] = \frac{1}{2\sqrt{3k_2}}\sqrt{\alpha + 2.5 + \sqrt{(\alpha + 2.5)^2 - 6\alpha}}. \tag{3.301}$$

Now, substituting the set value of k_2 and the constant α into Eqs. 3.300 and 3.301, we can determine the sensitivity of the two natural frequencies of vibration with respect to k_2.

Now, instead of setting α to be a constant, set k_1 to be constant. It follows from Eqs. 3.293 and 3.294 that

$$\frac{\partial\omega_1}{\partial k_2} = \frac{\partial}{\partial k_2}\left[\sqrt{\frac{1}{3}\left(k_1 + 2.5k_2 - \sqrt{(k_1 + 2.5k_2)^2 - 6k_1 k_2} \right)}\right]$$

$$= \frac{\sqrt{3}}{6\sqrt{k_1 + 2.5k_2 - \sqrt{(k_1 + 2.5k_2)^2 - 6k_1 k_2}}}\left(2.5 - \frac{2.5(k_1 + 2.5k_2) - 3k_1}{\sqrt{(k_1 + 2.5k_2)^2 - 6k_1 k_2}} \right) \tag{3.302}$$

$$\frac{\partial \omega_2}{\partial k_2} = \frac{\partial}{\partial k_2}\left[\sqrt{\frac{1}{3}\left(k_1 + 2.5k_2 + \sqrt{(k_1 + 2.5k_2)^2 - 6k_1k_2}\right)}\right]$$

$$= \frac{\sqrt{3}}{6\sqrt{k_1 + 2.5k_2 + \sqrt{(k_1 + 2.5k_2)^2 - 6k_1k_2}}}\left(2.5 + \frac{2.5(k_1 + 2.5k_2) - 3k_1}{\sqrt{(k_1 + 2.5k_2)^2 - 6k_1k_2}}\right). \quad (3.303)$$

Now, substituting the set value of k_2 and the constant k_1 into Eqs. 3.302 and 3.303, we can determine the sensitivity of the two natural frequencies of vibration with respect to k_2.

The other derivatives, $(\partial\omega_1/\partial k_1)$ and $(\partial\omega_2/\partial k_1)$, follow in exactly the same way.

EXAMPLE 2 *Numerical Example of Natural Frequency Sensitivity*

In Example 1, the sensitivity of the natural frequencies was calculated with respect to k_2. In this example, a numerical example is performed to study the difference between holding α constant and holding k_1 constant. Let

$$k_1 = k_2 = 10 \text{ k/in.} \quad (3.304)$$

Then $\alpha = k_1/k_2 = 1.0$, and it follows from Eqs. 3.300 and 3.301, by holding α constant, that

$$\left.\frac{\partial \omega_1}{\partial k_2}\right|_{\alpha=\text{constant}} = \frac{1}{2\sqrt{3(10)}}\sqrt{1 + 2.5 - \sqrt{(1 + 2.5)^2 - 6(1)}} = 0.09129 \quad (3.305)$$

$$\left.\frac{\partial \omega_2}{\partial k_2}\right|_{\alpha=\text{constant}} = \frac{1}{2\sqrt{3(10)}}\sqrt{1 + 2.5 + \sqrt{(1 + 2.5)^2 - 6(1)}} = 0.22361. \quad (3.306)$$

Also, it follows from Eqs. 3.302 and 3.303, by holding k_1 constant, that

$$\left.\frac{\partial \omega_1}{\partial k_2}\right|_{k_1=\text{constant}} = \frac{\sqrt{3}\left(2.5 - \dfrac{2.5((10) + 2.5(10)) - 3(10)}{\sqrt{((10) + 2.5(10))^2 - 6(10)(10)}}\right)}{6\sqrt{10 + 2.5(10) - \sqrt{((10) + 2.5(10))^2 - 6(10)(10)}}} = 0.01826 \quad (3.307)$$

$$\left.\frac{\partial \omega_2}{\partial k_2}\right|_{k_1=\text{constant}} = \frac{\sqrt{3}\left(2.5 + \dfrac{2.5((10) + 2.5(10)) - 3(10)}{\sqrt{((10) + 2.5(10))^2 - 6(10)(10)}}\right)}{6\sqrt{10 + 2.5(10) + \sqrt{((10) + 2.5(10))^2 - 6(10)(10)}}} = 0.17889. \quad (3.308)$$

Both cases (i.e., holding α constant and holding k_1 constant) show that the effect of variation in k_2 on ω_2 is greater than that on ω_1. This shows that a higher natural frequency of vibration is more sensitive to a perturbation in the stiffness k_2.

EXAMPLE 3 *Sensitivity of Mode Shapes to Stiffnesses*

Example 1 presented a 2DOF frame system and calculated the natural frequencies of vibration and their sensitivities to a change in stiffness k_2. In this example, the sensitivity of the two mode shapes with respect to the stiffness k_2 is studied. Let the mass of the first and second floors be

$$m_1 = 1.5 \text{ k-s}^2/\text{in.}$$

$$m_2 = 1.0 \text{ k-s}^2/\text{in.}$$

Following the derivation from Eq. 3.10 to Eq. 3.12, the mode shapes for a 2DOF system are

$$\boldsymbol{\phi}_1 = \begin{Bmatrix} 1 \\ r_1 \end{Bmatrix}, \quad \boldsymbol{\phi}_2 = \begin{Bmatrix} 1 \\ r_2 \end{Bmatrix} \tag{3.309}$$

where

$$r_1 = \frac{-1.5\omega_1^2 + k_1 + k_2}{k_2}, \quad r_2 = \frac{-1.5\omega_2^2 + k_1 + k_2}{k_2} \tag{3.310}$$

and ω_1^2 and ω_2^2 are given in Eqs. 3.293 and 3.294, which are functions of k_2. Consider the case where $k_1 = \alpha k_2$; then ω_1^2 and ω_2^2 are given in Eqs. 3.296 and 3.297, which are

$$\omega_1^2 = \frac{k_2}{3}\left(\alpha + 2.5 - \sqrt{(\alpha + 2.5)^2 - 6\alpha}\right) \tag{3.311}$$

$$\omega_2^2 = \frac{k_2}{3}\left(\alpha + 2.5 + \sqrt{(\alpha + 2.5)^2 - 6\alpha}\right) \tag{3.312}$$

and therefore

$$r_1 = -0.5\left(\alpha + 2.5 - \sqrt{(\alpha + 2.5)^2 - 6\alpha}\right) + 1 + \alpha \tag{3.313}$$

$$r_2 = -0.5\left(\alpha + 2.5 + \sqrt{(\alpha + 2.5)^2 - 6\alpha}\right) + 1 + \alpha. \tag{3.314}$$

Eqs. 3.313 and 3.314 are independent of the variable k_2, and therefore the mode shapes are also independent of the variable k_2. This shows that if the ratio of the stiffnesses is kept constant (i.e., $\alpha = k_1/k_2$ constant), then the mode shapes of vibration will remain constant.

3.10   SENSITIVITY ANALYSIS OF RESPONSE

Section 2.9 starting with Eq. 2.345 presented a method that can be used to calculate the sensitivity of the response, that is, $x(t)$, as a function of time with respect to a design variable (e.g., k). The response sensitivity $p(t)$ is obtained by solving a second-order linear differential equation (i.e., Eq. 2.351) that is very similar to the original equation of motion (i.e., Eq. 2.345). The only difference is that the response sensitivity $p(t)$ replaces the response $x(t)$ and the forcing function $F_e(t)$ is replaced by a sensitivity "forcing function" $-x(t)$ in Eq. 2.351. A 2DOF system will in general have an equation of motion of the form given in Eq. 3.3:

$$\mathbf{M}\ddot{\mathbf{X}}(t) + \mathbf{C}\dot{\mathbf{X}}(t) + \mathbf{K}\mathbf{X}(t) = \mathbf{F}(t). \tag{3.315}$$

Imagine that the sensitivity of the response is desired for a design variable (e.g., k_1). Then the same procedure as used in Section 2.9 can be followed:

$$\frac{\partial \mathbf{M}}{\partial k_1}\ddot{\mathbf{X}}(t) + \mathbf{M}\frac{\partial \ddot{\mathbf{X}}(t)}{\partial k_1} + \frac{\partial \mathbf{C}}{\partial k_1}\dot{\mathbf{X}}(t) + \mathbf{C}\frac{\partial \dot{\mathbf{X}}(t)}{\partial k_1} + \frac{\partial \mathbf{K}}{\partial k_1}\mathbf{X}(t) + \mathbf{K}\frac{\partial \mathbf{X}(t)}{\partial k_1} = \frac{\partial \mathbf{F}(t)}{\partial k_1}. \tag{3.316}$$

If the damping is of the Rayleigh damping form, then

$$\mathbf{C} = \alpha\mathbf{M} + \beta\mathbf{K} = \alpha\begin{bmatrix} m_1 & 0 \\ 0 & m_2 \end{bmatrix} + \beta\begin{bmatrix} k_1 + k_2 & -k_2 \\ -k_2 & k_2 \end{bmatrix} \tag{3.317}$$

and therefore

$$\frac{\partial \mathbf{C}}{\partial k_1} = \beta \begin{bmatrix} 1 & 0 \\ 0 & 0 \end{bmatrix} = \mathbf{C}_1.$$

(3.318)

Similarly,

$$\frac{\partial \mathbf{K}}{\partial k_1} = \begin{bmatrix} 1 & 0 \\ 0 & 0 \end{bmatrix} = \mathbf{C}_2, \qquad \frac{\partial \mathbf{M}}{\partial k_1} = \mathbf{0}, \qquad \frac{\partial \mathbf{F}(t)}{\partial k_1} = \mathbf{0}.$$

(3.319)

If the response sensitivity vector is defined to be

$$\mathbf{p}(t) = \frac{\partial \mathbf{X}(t)}{\partial k_1}$$

(3.320)

then Eq. 3.316 becomes

$$\mathbf{M}\ddot{\mathbf{p}}(t) + \mathbf{C}\dot{\mathbf{p}}(t) + \mathbf{K}\mathbf{p}(t) = -\mathbf{C}_1\dot{\mathbf{X}}(t) - \mathbf{C}_2\mathbf{X}(t).$$

(3.321)

Therefore, the solution method is the same as discussed in Section 2.9, except that now the equations are in matrix form.

EXAMPLE 1 *Free Vibration Sensitivity Analysis*

Consider Example 2 in Section 3.3. The equation for the response $x_1(t)$ is

$$x_1(t) = \left(\frac{1 - r_2}{r_1 - r_2} \right) \cos \omega_1 t + \left(\frac{r_1 - 1}{r_1 - r_2} \right) \cos \omega_2 t$$

(3.322)

where, from Eq. 3.12,

$$r_1 = \frac{-m_1\omega_1^2 + k_1 + k_2}{k_2}, \qquad r_2 = \frac{-m_1\omega_2^2 + k_1 + k_2}{k_2}$$

(3.323)

with ω_1^2 and ω_2^2 given in Eq. 3.10, which are

$$\omega_1^2 = \frac{((k_1 + k_2)m_2 + k_2 m_1) - \sqrt{((k_1 + k_2)m_2 + k_2 m_1)^2 - 4m_1 m_2 k_1 k_2}}{2m_1 m_2}$$

(3.324)

$$\omega_2^2 = \frac{((k_1 + k_2)m_2 + k_2 m_1) + \sqrt{((k_1 + k_2)m_2 + k_2 m_1)^2 - 4m_1 m_2 k_1 k_2}}{2m_1 m_2}.$$

(3.325)

The sensitivity of the response $x_1(t)$ to k_1 is

$$\frac{\partial x_1(t)}{\partial k_1} = \left[\frac{\partial}{\partial k_1} \left(\frac{1 - r_2}{r_1 - r_2} \right) \right] \cos \omega_1 t + \left(\frac{1 - r_2}{r_1 - r_2} \right) \frac{\partial (\cos \omega_1 t)}{\partial k_1}$$

$$+ \left[\frac{\partial}{\partial k_1} \left(\frac{r_1 - 1}{r_1 - r_2} \right) \right] \cos \omega_2 t + \left(\frac{r_1 - 1}{r_1 - r_2} \right) \frac{\partial (\cos \omega_2 t)}{\partial k_1}.$$

(3.326)

Expanding Eq. 3.326 gives

$$\frac{\partial x_1(t)}{\partial k_1} = \left[\frac{\partial r_2}{\partial k_1} (1 - r_1) - \frac{\partial r_1}{\partial k_1} (1 - r_2) \right] \frac{\cos \omega_1 t}{(r_1 - r_2)^2} + \left(\frac{1 - r_2}{r_1 - r_2} \right) t \sin \omega_1 t \frac{\partial \omega_1}{\partial k_1}$$

$$+ \left[\frac{\partial r_1}{\partial k_1} (1 - r_2) - \frac{\partial r_2}{\partial k_1} (1 - r_1) \right] \cos \omega_2 t + \left(\frac{r_1 - 1}{r_1 - r_2} \right) t \sin \omega_2 t \frac{\partial \omega_2}{\partial k_1}$$

(3.327)

where

$$\frac{\partial r_1}{\partial k_1} = -\left[\frac{m_1}{k_2}\right]\frac{\partial}{\partial k_1}(\omega_1^2) + \left(\frac{1}{k_2}\right), \quad \frac{\partial r_2}{\partial k_1} = -\left[\frac{m_1}{k_2}\right]\frac{\partial}{\partial k_1}(\omega_2^2) + \left(\frac{1}{k_2}\right) \qquad (3.328)$$

and

$$\frac{\partial}{\partial k_1}(\omega_1^2) = \frac{1}{2m_1}\left[1 - \frac{(k_1+k_2)m_2 - m_1 k_2}{\sqrt{((k_1+k_2)m_2 + k_2 m_1)^2 - 4m_1 m_2 k_1 k_2}}\right] \qquad (3.329)$$

$$\frac{\partial}{\partial k_1}(\omega_2^2) = \frac{1}{2m_1}\left[1 + \frac{(k_1+k_2)m_2 - m_1 k_2}{\sqrt{((k_1+k_2)m_2 + k_2 m_1)^2 - 4m_1 m_2 k_1 k_2}}\right] \qquad (3.330)$$

$$\frac{\partial\omega_1}{\partial k_1} = \frac{\partial\omega_1^2/\partial k_1}{2\omega_1} = \frac{1}{4m_1\omega_1}\left[1 - \frac{(k_1+k_2)m_2 - m_1 k_2}{\sqrt{((k_1+k_2)m_2 + k_2 m_1)^2 - 4m_1 m_2 k_1 k_2}}\right] \qquad (3.331)$$

$$\frac{\partial\omega_2}{\partial k_1} = \frac{\partial\omega_2^2/\partial k_1}{2\omega_2} = \frac{1}{4m_1\omega_2}\left[1 + \frac{(k_1+k_2)m_2 - m_1 k_2}{\sqrt{((k_1+k_2)m_2 + k_2 m_1)^2 - 4m_1 m_2 k_1 k_2}}\right]. \qquad (3.332)$$

EXAMPLE 2 *Sensitivity Analysis of Normal Mode Response*

Consider a response solution obtained using the normal mode method. As discussed in Section 3.3, the response is described by Eq. 3.64:

$$\mathbf{X}(t) = \begin{Bmatrix} x_1(t) \\ x_2(t) \end{Bmatrix} = \boldsymbol{\phi}_1 q_1(t) + \boldsymbol{\phi}_2 q_2(t) = \begin{Bmatrix} 1 \\ r_1 \end{Bmatrix} q_1(t) + \begin{Bmatrix} 1 \\ r_2 \end{Bmatrix} q_2(t). \qquad (3.333)$$

The sensitivity of the response with respect to a design variable (e.g., k_1) is obtained by taking the derivative of Eq. 3.333 with respect to the design variable. Therefore, it follows that

$$\frac{\partial\mathbf{X}(t)}{\partial k_1} = \frac{\partial\boldsymbol{\phi}_1}{\partial k_1}q_1(t) + \boldsymbol{\phi}_1\frac{\partial q_1(t)}{\partial k_1} + \frac{\partial\boldsymbol{\phi}_2}{\partial k_1}q_2(t) + \boldsymbol{\phi}_2\frac{\partial q_2(t)}{\partial k_1}. \qquad (3.334)$$

Note that

$$\frac{\partial\boldsymbol{\phi}_1}{\partial k_1} = \frac{\partial}{\partial k_1}\begin{Bmatrix} 1 \\ r_1 \end{Bmatrix} = \begin{Bmatrix} 0 \\ \partial r_1/\partial k_1 \end{Bmatrix}, \quad \frac{\partial\boldsymbol{\phi}_2}{\partial k_1} = \frac{\partial}{\partial k_1}\begin{Bmatrix} 1 \\ r_2 \end{Bmatrix} = \begin{Bmatrix} 0 \\ \partial r_2/\partial k_1 \end{Bmatrix}. \qquad (3.335)$$

The sensitivity of the response $q_1(t)$ with respect to k_1 follows as described in Section 2.9, but now with m_1^*, c_1^*, and k_1^* (where Rayleigh damping is assumed; see Section 3.5). The resultant equation is

$$m_1^*\ddot{p}_1(t) + c_1^*\dot{p}_1(t) + k_1^* p_1(t) = -c_{11}\dot{q}_1(t) - c_{12}\dot{q}_2(t) \qquad (3.336)$$

where

$$m_1^* = \boldsymbol{\phi}_1^T\mathbf{M}\boldsymbol{\phi}_1, \quad c_1^* = \boldsymbol{\phi}_1^T\mathbf{C}\boldsymbol{\phi}_1, \quad k_1^* = \boldsymbol{\phi}_1^T\mathbf{K}\boldsymbol{\phi}_1 \qquad (3.337)$$

$$c_{11} = \frac{\partial(\boldsymbol{\phi}_1^T\mathbf{C}\boldsymbol{\phi}_1)}{\partial k_1}, \quad c_{12} = \frac{\partial(\boldsymbol{\phi}_1^T\mathbf{K}\boldsymbol{\phi}_1)}{\partial k_1}. \qquad (3.338)$$

REFERENCES

1. C. Moler and C. Van Loan, *Numerical Linear Algebra Techniques for Systems and Control*, IEEE Press, IEEE Control Systems Society, 1994.

PROBLEMS

3.1 Consider the 2DOF system in Example 1 in Section 3.2. Let $m = 1.0$ k-s^2/in. and $k = 10.0$ k/in.

   (a) Calculate the first and second natural frequencies of vibration in units of radians per second.

   (b) Repeat Part (a) but express the answer in cycles per second.

   (c) Calculate the first, or fundamental, undamped natural period of vibration.

   (d) Calculate the second undamped natural period of vibration.

3.2 Repeat Problem 3.1, but let $k = 100.0$ k/in.

3.3 Repeat Problem 3.1, but let $k = 0.1$ k/in.

3.4 Consider the system in Problem 3.1. Assume that the mass does not change. What must the value of k be to have an undamped fundamental period of vibration equal to 1.0 s?

3.5 Consider Example 2 in Section 3.2. Are the mode shapes a function of the mass m or the stiffness k? Explain your answer.

3.6 Consider Problem 3.1. Let $k_1 = 10.0$ k/in. and $k_2 = 10k_1 = 100.0$ k/in. Calculate the following:

   (a) The natural periods of vibration.

   (b) The mode shapes of vibration for each of the normalization methods defined in Example 2 of Section 3.2.

3.7 Repeat Problem 3.6, but now $k_1 = 10.0$ k/in. and $k_2 = 0.1k_1 = 1.0$ k/in.

3.8 Consider the system in Example 2 of Section 3.2. Calculate the response of the system if it is released from rest with the following initial conditions:

   (a) $x_1(0) = 0.618$ in. and $x_2(0) = 1.00$ in. (i.e., the first mode shape).

   (b) $x_1(0) = 1.00$ in. and $x_2(0) = -0.618$ in. (i.e., the first mode shape).

   (c) $x_1(0) = 1.00$ in. and $x_2(0) = 1.00$ in.

3.9 Consider Example 3 in Section 3.2. If the ratio of the fundamental undamped natural period of vibration to the second undamped natural period of vibration is 2.0, then what must be the value of α (i.e., k_1/k_2)?

3.10 Repeat Problem 3.9 but with a desired ratio value of 3.0.

3.11 Consider the system in Problem 3.1. Calculate the following:

   (a) The modal matrix.

   (b) $\boldsymbol{\phi}_1^T \boldsymbol{\phi}_2$.

   (c) $\boldsymbol{\phi}_1^T \mathbf{M} \boldsymbol{\phi}_2$.

   (d) $\boldsymbol{\phi}_1^T \mathbf{K} \boldsymbol{\phi}_2$.

   (e) m_1^* and m_2^* in Eq. 3.79a.

   (f) k_1^* and k_2^* in Eq. 3.79b.

   (g) (k_1^*/m_1^*) and (k_2^*/m_2^*) and compare with ω_1^2 and ω_2^2.

3.12 Consider the system in Problem 3.1. If $x_1(0) = 0.618$ in. and $x_2(0) = 1.00$ in., calculate the corresponding initial displacements in the two normal mode coordinates, that is, $q_1(0) = q_{1o}$ and $q_2(0) = q_{2o}$.

3.13 Repeat Problem 3.12, but let $x_1(0) = 1.00$ in. and $x_2(0) = -0.618$ in.

3.14 Repeat Problem 3.12, but let $x_1(0) = 1.00$ in. and $x_2(0) = 1.00$ in.

3.15 If the system in Problem 3.12 is released from rest, then calculate and plot $q_1(t)$, $q_2(t)$, $x_1(t)$, and $x_2(t)$.

3.16 If the system in Problem 3.13 is released from rest, then calculate and plot $q_1(t)$, $q_2(t)$, $x_1(t)$, and $x_2(t)$.

3.17 If the system in Problem 3.14 is released from rest, then calculate and plot $q_1(t)$, $q_2(t)$, $x_1(t)$, and $x_2(t)$.

3.18 Consider Example 1 in Section 3.4. Let the system be as defined in Problem 3.1 and assume that $F_1 = 10.0$ kip, $F_2 = 0$, and $x_1(0) = x_2(0) = \dot{x}_1(0) = \dot{x}_2(0) = 0$. Calculate and plot $q_1(t)$, $q_2(t)$, $x_1(t)$, and $x_2(t)$.

3.19 Repeat Problem 3.18 but let $F_1 = 10.0$ kip, $F_2 = 0$, $x_1(0) = x_2(0) = 1.0$ in., and $\dot{x}_1(0) = \dot{x}_2(0) = 0$.

3.20 Consider Example 2 in Section 3.4. Let the system be as defined in Problem 3.1. Assume that the initial conditions are zero. Let $F_1 = 10.0$ kip, $F_2 = 0$, and $\omega_o = \omega_1$. Calculate and plot $q_1(t)$, $q_2(t)$, $x_1(t)$, and $x_2(t)$.

3.21 Repeat Problem 3.20 but let $F_2 = 10.0$ kip and $F_1 = 0.618 F_2 = 6.18$ kip (i.e., the same ratio as the fundamental mode shape). Calculate and plot $q_1(t)$, $q_2(t)$, $x_1(t)$, and $x_2(t)$.

3.22 Consider Example 3 in Section 3.4. Let the system be as defined in Problem 3.1 and assume $A_1 = 0.618 A_2 = A$ (i.e., the same ratio as the fundamental mode shape). Calculate Γ_1 and Γ_2 in Eq. 3.136.

3.23 Consider the system in Problem 3.1.

(a) Assume that the critical damping ratios in the two modes of vibration are 5% and 5%. Calculate the proportionality constants α and β in the Rayleigh proportional damping matrix. Write the damping matrix.

(b) Assume that the critical damping ratio in the first mode of vibration is 5% and also that the damping matrix is proportional to the stiffness matrix (i.e., $\alpha = 0$). Calculate β and the implied damping in the second mode of vibration, and write the damping matrix.

(c) Assume that the critical damping ratio in the first mode of vibration is 5% and also that the damping matrix is proportional to the mass matrix (i.e., $\beta = 0$). Calculate α and the implied damping in the second mode of vibration, and write the damping matrix.

3.24 Consider the system in Example 1 in Section 3.6. Let $m = 1.0$ k-s^2/in. and $k = 10.0$ k/in.

(a) Calculate the **A** matrix in Eq. 3.189.

(b) Calculate the **T** matrix.

(c) Use Eq. 3.201 to calculate the state transition matrix.

(d) Use Eq. 3.196 with two terms to calculate the state transition matrix.

(e) Repeat Part (d) but use three terms.

(f) Repeat Part (d) but use four terms.

(g) If the system is released from rest with initial displacements $x_1(0) = 1.0$ in. and $x_2(0) = 1.0$ in., calculate the free vibration response using the state space solution procedure.

3.25 Consider the system in Example 1 in Section 3.6. Let $m = 1.0$ k-s^2/in. and $k = 10.0$ k/in. Calculate the response using the state space approach if the system is released from rest and with initial displacements $x_1(0) = 1.0$ in. and $x_2(0) = \frac{1}{2}\left(1 - \sqrt{5}\right)$ in.

3.26 Consider Problem 3.24.

(a) Calculate the damping matrix assuming Rayleigh damping and 5% modal damping ratio in the two modes of vibration.

(b) Repeat Problem 3.24 but with the damping matrix from Part (a).

(c) Repeat Problem 3.24 but with the damping matrix from Part (a) multiplied by 1.5.

3.27 Repeat Example 1 in Section 3.7 but multiply the stiffness matrix by 1.5.

3.28 Repeat Example 1 in Section 3.7 but use the constant average acceleration method (i.e., $\delta = \frac{1}{2}$ and $\alpha = \frac{1}{4}$).

3.29 Repeat Example 1 in Section 3.7 but use the linear acceleration method (i.e., $\delta = \frac{1}{2}$ and $\alpha = \frac{1}{6}$).

3.30 Repeat Example 3 in Section 3.8 but multiply the stiffness matrix by 1.5.

3.31 Calculate the state transition matrix in Example 3 in Section 3.8 using the results of Example 5 in Section 3.8.

3.32 Consider Example 1 in Section 3.9 with $k_1 = k_2 = k = 10.0$ k/in. Calculate the sensitivity of the first and second undamped natural frequencies of vibration with respect to k_1.

3.33 Calculate the sensitivity of the first and second mode shapes of vibration with respect to k_1 for Problem 3.32.

3.34 Repeat Example 1 in Section 3.10 with $m_1 = m_2 = 1.0$ k-s^2/in. and $k_1 = k_2 = 10.0$ k/in.

3.35 Repeat Example 1 in Section 3.10 with $m_1 = m_2 = 1.0$ k-s^2/in., $k_1 = 100.0$ k/in., and $k_2 = 10.0$ k/in.

3.36 Repeat Example 1 in Section 3.10 with $m_1 = m_2 = 1.0$ k-s^2/in., $k_1 = 1.0$ k/in., and $k_2 = 10.0$ k/in.

Chapter 4

Multi–Degree of Freedom Linear System Response

4.1 OVERVIEW

Chapter 3 presented the analysis methods used for two degrees of freedom systems. In this chapter, the two degrees of freedom system is extended to a system beyond two, and typically in modern structural dynamics the system may have thousands of degrees of freedom. This chapter also ties together the response package for large structural systems, which involves degrees of freedom that translate and rotate in three dimensions. Finally, for large systems with thousands of degrees of freedom, there are special methods to reduce the size of the problem and also provide structural engineering insight.

The structural properties (i.e., mass, damping, and stiffness) must be determined to represent the motion in the dynamic equilibrium equation. The determination of these properties is as follows:

1. The *stiffness matrix* is determined in the same way as in the static analysis. This matrix includes the properties of each structural member, including the length, cross-sectional area, moment of inertia, elastic constants, torsional rigidity, and so forth.

2. The *mass matrix* is generally determined by estimating the mass of the entire structure and lumping these masses to the appropriate degrees of freedom. The resulting mass matrix is known as the *lumped mass matrix* and is a diagonal matrix. However, the lumping method can be replaced by considering the variation of the deformed shapes of each structural member and therefore the effect of one mass on the other. The resulting mass matrix is known as the *consistent mass matrix*.

3. The *damping matrix* is determined by considering the damping of the structure. This is generally not done in structural engineering because no method has been developed to accurately compute the damping of each structural member. Therefore, the damping matrix is generally assumed to be proportional to the mass and stiffness matrices. This is discussed in detail in Sections 4.6 to 4.8.

Once these matrices are determined, the dynamic equilibrium equation for a multi–degree of freedom (MDOF) system is

$$\mathbf{M}\ddot{\mathbf{X}}(t) + \mathbf{C}\dot{\mathbf{X}}(t) + \mathbf{K}\mathbf{X}(t) = \mathbf{F}_e(t) \tag{4.1}$$

where

$$\mathbf{M} = \begin{bmatrix} m_{11} & m_{12} & \cdots & & m_{1n} \\ m_{21} & m_{22} & \ddots & & \vdots \\ \vdots & \ddots & \ddots & m_{n-1,n} \\ m_{n1} & \cdots & m_{n,n-1} & m_{nn} \end{bmatrix}, \quad \mathbf{C} = \begin{bmatrix} c_{11} & c_{12} & \cdots & & c_{1n} \\ c_{21} & c_{22} & \ddots & & \vdots \\ \vdots & \ddots & \ddots & c_{n-1,n} \\ c_{n1} & \cdots & c_{n,n-1} & c_{nn} \end{bmatrix}, \quad \mathbf{K} = \begin{bmatrix} k_{11} & k_{12} & \cdots & & k_{1n} \\ k_{21} & k_{22} & \ddots & & \vdots \\ \vdots & \ddots & \ddots & k_{n-1,n} \\ k_{n1} & \cdots & k_{n,n-1} & k_{nn} \end{bmatrix}$$

$$\mathbf{X}(t) = \{x_1(t) \quad x_2(t) \quad \cdots \quad x_n(t)\}^T, \quad \mathbf{F}_e(t) = \{F_1(t) \quad F_2(t) \quad \cdots \quad F_n(t)\}^T.$$

Here, n represents the total number of degrees of freedom, generally denoted as n-DOF. **M, C,** and **K** represent the mass, damping, and stiffness, respectively, and they are symmetrical positive definite matrices.

The displacement vector, $\mathbf{X}(t)$, contains terms that include translations in the x-, y-, and z-directions and rotations in the θ_x-, θ_y-, and θ_z-directions. It is usually convenient to represent this vector in a partitioned form, in which the displacements in each translational and rotational direction are grouped together. Let

$\mathbf{X}_x(t)$ = displacement vector containing all displacements in the x-direction
$\mathbf{X}_y(t)$ = displacement vector containing all displacements in the y-direction
$\mathbf{X}_z(t)$ = displacement vector containing all displacements in the z-direction
$\mathbf{X}_r(t)$ = displacement vector containing all rotations in the θ_x-direction
$\mathbf{X}_s(t)$ = displacement vector containing all rotations in the θ_y-direction
$\mathbf{X}_t(t)$ = displacement vector containing all rotations in the θ_z-direction.

Then

$$\mathbf{X}(t) = \{\mathbf{X}_x(t) \quad \mathbf{X}_y(t) \quad \mathbf{X}_z(t) \quad \mathbf{X}_r(t) \quad \mathbf{X}_s(t) \quad \mathbf{X}_t(t)\}^T. \tag{4.2}$$

Following a similar argument and using the same notation, the terms in Eq. 4.1 become

$$\dot{\mathbf{X}}(t) = \{\dot{\mathbf{X}}_x(t) \quad \dot{\mathbf{X}}_y(t) \quad \dot{\mathbf{X}}_z(t) \quad \dot{\mathbf{X}}_r(t) \quad \dot{\mathbf{X}}_s(t) \quad \dot{\mathbf{X}}_t(t)\}^T \tag{4.3a}$$

$$\ddot{\mathbf{X}}(t) = \{\ddot{\mathbf{X}}_x(t) \quad \ddot{\mathbf{X}}_y(t) \quad \ddot{\mathbf{X}}_z(t) \quad \ddot{\mathbf{X}}_r(t) \quad \ddot{\mathbf{X}}_s(t) \quad \ddot{\mathbf{X}}_t(t)\}^T \tag{4.3b}$$

$$\mathbf{F}_e(t) = \{\mathbf{F}_x(t) \quad \mathbf{F}_y(t) \quad \mathbf{F}_z(t) \quad \mathbf{F}_r(t) \quad \mathbf{F}_s(t) \quad \mathbf{F}_t(t)\}^T. \tag{4.3c}$$

Note that $\dot{\mathbf{X}}_r(t)$, $\dot{\mathbf{X}}_s(t)$, and $\dot{\mathbf{X}}_t(t)$ are the rotational velocity (or angular velocity) with units of rad/s and $\ddot{\mathbf{X}}_r(t)$, $\ddot{\mathbf{X}}_s(t)$, and $\ddot{\mathbf{X}}_t(t)$ are the rotational acceleration (or angular acceleration) with units of rad/s^2. Using the similar notations, it follows that the dynamic matrices are

$$\mathbf{M} = \begin{bmatrix} \mathbf{M}_{xx} & \mathbf{M}_{xy} & \mathbf{M}_{xz} & \mathbf{M}_{xr} & \mathbf{M}_{xs} & \mathbf{M}_{xt} \\ \mathbf{M}_{yx} & \mathbf{M}_{yy} & \mathbf{M}_{yz} & \mathbf{M}_{yr} & \mathbf{M}_{ys} & \mathbf{M}_{yt} \\ \mathbf{M}_{zx} & \mathbf{M}_{zy} & \mathbf{M}_{zz} & \mathbf{M}_{zr} & \mathbf{M}_{zs} & \mathbf{M}_{zt} \\ \mathbf{M}_{rx} & \mathbf{M}_{ry} & \mathbf{M}_{rz} & \mathbf{M}_{rr} & \mathbf{M}_{rs} & \mathbf{M}_{rt} \\ \mathbf{M}_{sx} & \mathbf{M}_{sy} & \mathbf{M}_{sz} & \mathbf{M}_{sr} & \mathbf{M}_{ss} & \mathbf{M}_{st} \\ \mathbf{M}_{tx} & \mathbf{M}_{ty} & \mathbf{M}_{tz} & \mathbf{M}_{tr} & \mathbf{M}_{ts} & \mathbf{M}_{tt} \end{bmatrix} \tag{4.4}$$

$$\mathbf{C} = \begin{bmatrix} \mathbf{C}_{xx} & \mathbf{C}_{xy} & \mathbf{C}_{xz} & \mathbf{C}_{xr} & \mathbf{C}_{xs} & \mathbf{C}_{xt} \\ \mathbf{C}_{yx} & \mathbf{C}_{yy} & \mathbf{C}_{yz} & \mathbf{C}_{yr} & \mathbf{C}_{ys} & \mathbf{C}_{yt} \\ \mathbf{C}_{zx} & \mathbf{C}_{zy} & \mathbf{C}_{zz} & \mathbf{C}_{zr} & \mathbf{C}_{zs} & \mathbf{C}_{zt} \\ \mathbf{C}_{rx} & \mathbf{C}_{ry} & \mathbf{C}_{rz} & \mathbf{C}_{rr} & \mathbf{C}_{rs} & \mathbf{C}_{rt} \\ \mathbf{C}_{sx} & \mathbf{C}_{sy} & \mathbf{C}_{sz} & \mathbf{C}_{sr} & \mathbf{C}_{ss} & \mathbf{C}_{st} \\ \mathbf{C}_{tx} & \mathbf{C}_{ty} & \mathbf{C}_{tz} & \mathbf{C}_{tr} & \mathbf{C}_{ts} & \mathbf{C}_{tt} \end{bmatrix}, \quad \mathbf{K} = \begin{bmatrix} \mathbf{K}_{xx} & \mathbf{K}_{xy} & \mathbf{K}_{xz} & \mathbf{K}_{xr} & \mathbf{K}_{xs} & \mathbf{K}_{xt} \\ \mathbf{K}_{yx} & \mathbf{K}_{yy} & \mathbf{K}_{yz} & \mathbf{K}_{yr} & \mathbf{K}_{ys} & \mathbf{K}_{yt} \\ \mathbf{K}_{zx} & \mathbf{K}_{zy} & \mathbf{K}_{zz} & \mathbf{K}_{zr} & \mathbf{K}_{zs} & \mathbf{K}_{zt} \\ \mathbf{K}_{rx} & \mathbf{K}_{ry} & \mathbf{K}_{rz} & \mathbf{K}_{rr} & \mathbf{K}_{rs} & \mathbf{K}_{rt} \\ \mathbf{K}_{sx} & \mathbf{K}_{sy} & \mathbf{K}_{sz} & \mathbf{K}_{sr} & \mathbf{K}_{ss} & \mathbf{K}_{st} \\ \mathbf{K}_{tx} & \mathbf{K}_{ty} & \mathbf{K}_{tz} & \mathbf{K}_{tr} & \mathbf{K}_{ts} & \mathbf{K}_{tt} \end{bmatrix}.$$

Note that

$$(DOF)_x + (DOF)_y + (DOF)_z + (DOF)_r + (DOF)_s + (DOF)_t = n. \tag{4.5}$$

Since $\mathbf{M}$, $\mathbf{C}$, and $\mathbf{K}$ are symmetric matrices, it follows that

$$\mathbf{M}_{ij} = \mathbf{M}_{ij}^T, \quad \mathbf{C}_{ij} = \mathbf{C}_{ij}^T, \quad \mathbf{K}_{ij} = \mathbf{K}_{ij}^T. \tag{4.6}$$

Note that when a lumped mass model is used, the mass matrix in Eq. 4.4 will contain terms such that

$$\mathbf{M}_{ij} = \begin{cases} \mathbf{M}_{ii} & i = j \\ \mathbf{0} & i \ne j \end{cases} \tag{4.7}$$

where $\mathbf{M}_{ii}$ is a diagonal matrix. Calculation of the mass matrix is discussed in the following section.

4.2   THE MASS MATRIX

To calculate the mass matrix of a structure, the most commonly used method is to balance the kinetic energy of the structure. A particle of mass m moving at a velocity of

$$\mathbf{V} = \dot{x}_x \hat{\mathbf{i}} + \dot{x}_y \hat{\mathbf{j}} + \dot{x}_z \hat{\mathbf{k}} \tag{4.8}$$

where $\dot{x}_x$, $\dot{x}_y$, and $\dot{x}_z$ are the speeds in the x-, y-, and z-directions, respectively, will have a kinetic energy equal to

$$E_k = \frac{1}{2} m \dot{x}^2 \tag{4.9}$$

where $\dot{x}$ is the speed of the particle and is equal to

$$\dot{x}^2 = \mathbf{V} \cdot \mathbf{V} = \dot{x}_x^2 + \dot{x}_y^2 + \dot{x}_z^2 \tag{4.10}$$

where the dot ($\cdot$) denotes a dot multiplication of the two $\mathbf{V}$ vectors. Substituting Eq. 4.10 into Eq. 4.9 gives

$$E_k = \frac{1}{2} m \left(\dot{x}_x^2 + \dot{x}_y^2 + \dot{x}_z^2 \right). \tag{4.11}$$

For a structure with distributed mass on each floor, it follows from Eqs. 4.9 and 4.11 that the kinetic energy is equal to

$$E_k = \int \frac{1}{2} \dot{x}^2 \, dm = \int \frac{1}{2} \left(\dot{x}_x^2 + \dot{x}_y^2 + \dot{x}_z^2 \right) dm. \tag{4.12}$$

At the same time, the structure is represented by thousands of degrees of freedom, and (using the notations discussed in Section 4.1) the kinetic energy is equal to

$$E_k = \frac{1}{2} \dot{\mathbf{X}}^T \mathbf{M} \dot{\mathbf{X}}. \tag{4.13}$$

Equating Eq. 4.12 and Eq. 4.13 gives

$$\frac{1}{2} \dot{\mathbf{X}}^T \mathbf{M} \dot{\mathbf{X}} = \int \frac{1}{2} \left(\dot{x}_x^2 + \dot{x}_y^2 + \dot{x}_z^2 \right) dm. \tag{4.14}$$

The following examples show how one can extract the mass matrix based on the result given in Eq. 4.14.

EXAMPLE 1   *Mass Matrix of an Axial Member*

A classic example of a multi–degree of freedom system is the vibration of an axial member, as shown in Figure 4.1a. As shown in Figure 4.1b, because the member is continuous,
it has an infinite number of degrees of freedom. One way of computing the mass matrix
of this member is to express the information within this member in terms of boundary
response using *interpolation functions*. Figure 4.1c shows the interpolation function with
a unit velocity at the left end and zero at the right end, whereas Figure 4.1d shows the
interpolation function with a unit velocity at the right end and zero at the left end. These
interpolation functions are expressed as

$$\dot{x}(\xi) = \left(\frac{L-\xi}{L}\right)\dot{x}_1 + \left(\frac{\xi}{L}\right)\dot{x}_2 \tag{4.15}$$

and in matrix form

$$\dot{x}(\xi) = \left[\frac{L-\xi}{L} \quad \frac{\xi}{L}\right]\begin{Bmatrix} \dot{x}_1 \\ \dot{x}_2 \end{Bmatrix}. \tag{4.16}$$

The kinetic energy of this member is

$$E_k = \int_{\xi=0}^{\xi=L} \frac{1}{2}\dot{x}(\xi)^2\, dm. \tag{4.17}$$

Let ρ_L be the mass per unit length of the member; it follows that

$$dm = \rho_L d\xi. \tag{4.18}$$

Then substituting Eqs. 4.16 and 4.18 into Eq. 4.17 gives

$$\begin{aligned}
E_k &= \frac{\rho_L}{2}\begin{Bmatrix} \dot{x}_1 \\ \dot{x}_2 \end{Bmatrix}^T \left(\int_{\xi=0}^{\xi=L}\begin{bmatrix} (L-\xi)/L \\ \xi/L \end{bmatrix}\Big[(L-\xi)/L \quad \xi/L\ \Big]d\xi\right)\begin{Bmatrix} \dot{x}_1 \\ \dot{x}_2 \end{Bmatrix} \\
&= \frac{\rho_L}{2L^2}\begin{Bmatrix} \dot{x}_1 \\ \dot{x}_2 \end{Bmatrix}^T \left(\int_{\xi=0}^{\xi=L}\begin{bmatrix} (L-\xi)^2 & \xi(L-\xi) \\ \xi(L-\xi) & \xi^2 \end{bmatrix}d\xi\right)\begin{Bmatrix} \dot{x}_1 \\ \dot{x}_2 \end{Bmatrix}.
\end{aligned} \tag{4.19}$$

Performing the integration gives

$$E_k = \frac{\rho_L}{2L^2}\begin{Bmatrix} \dot{x}_1 \\ \dot{x}_2 \end{Bmatrix}^T \begin{bmatrix} L^3/3 & L^3/6 \\ L^3/6 & L^3/3 \end{bmatrix}\begin{Bmatrix} \dot{x}_1 \\ \dot{x}_2 \end{Bmatrix}. \tag{4.20}$$

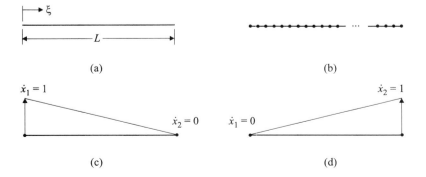

(a) (b) (c) (d)

Figure 4.1 Calculation of the Mass Matrix Using Interpolation Function

Recall that $m = \rho_L L$; simplifying Eq. 4.20, it follows that

$$E_k = \frac{1}{2}\begin{Bmatrix}\dot{x}_1\\\dot{x}_2\end{Bmatrix}^T\begin{bmatrix}1/3 & 1/6\\1/6 & 1/3\end{bmatrix}\begin{Bmatrix}\dot{x}_1\\\dot{x}_2\end{Bmatrix}. \tag{4.21}$$

The kinetic energy in the reduced coordinate system is

$$E_k = \frac{1}{2}\begin{Bmatrix}\dot{x}_1\\\dot{x}_2\end{Bmatrix}^T \mathbf{M}\begin{Bmatrix}\dot{x}_1\\\dot{x}_2\end{Bmatrix} = \frac{1}{2}\dot{\mathbf{x}}^T\mathbf{M}\dot{\mathbf{x}}. \tag{4.22}$$

Equating Eq. 4.21 and Eq. 4.22 and for any $\dot{\mathbf{x}}$, it follows that

$$\mathbf{M} = \begin{bmatrix}1/3 & 1/6\\1/6 & 1/3\end{bmatrix}. \tag{4.23}$$

This mass matrix is known as a consistent mass matrix.

   The lumped mass matrix, on the other hand, is calculated by lumping the mass on the left half of the member to the left end, and lumping the mass on the right half of the member to the right end. Therefore, the mass matrix for a lumped mass model is

$$\mathbf{M} = \begin{bmatrix}1/2 & 0\\0 & 1/2\end{bmatrix}. \tag{4.24}$$

EXAMPLE 2   *Mass Matrix of a Flexural Member*

Another classic example of a multi–degree of freedom system is the vibration of a flexural member, as shown in Figure 4.2. There are four degrees of freedom, two end displacements, and two end rotations. The interpolation functions of a flexural member are

$$\dot{x}(\xi) = \left(1 - \frac{3\xi^2}{L^2} + \frac{2\xi^3}{L^3}\right)\dot{x}_1 + \left(\xi - \frac{2\xi^2}{L} + \frac{\xi^3}{L^2}\right)\dot{x}_2 + \left(\frac{3\xi^2}{L^2} - \frac{2\xi^3}{L^3}\right)\dot{x}_3 + \left(-\frac{\xi^2}{L} + \frac{\xi^3}{L^2}\right)\dot{x}_4 \tag{4.25}$$

and in matrix form

$$\dot{x}(\xi) = \left[\left(1 - \frac{3\xi^2}{L^2} + \frac{2\xi^3}{L^3}\right) \quad \left(\xi - \frac{2\xi^2}{L} + \frac{\xi^3}{L^2}\right) \quad \left(\frac{3\xi^2}{L^2} - \frac{2\xi^3}{L^3}\right) \quad \left(-\frac{\xi^2}{L} + \frac{\xi^3}{L^2}\right)\right]\begin{Bmatrix}\dot{x}_1\\\dot{x}_2\\\dot{x}_3\\\dot{x}_4\end{Bmatrix}. \tag{4.26}$$

The terms in the parentheses of Eq. 4.25 are known as the *Hermite interpolation polynomials* and can be imagined as the "displacement" of the beam due to a succession of unit velocities that induce this velocity imagined as a "displacement."

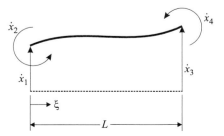

Figure 4.2 Calculation of the Mass Matrix of a Flexural Member

The kinetic energy of this member is

$$E_k = \int_{\xi=0}^{\xi=L} \frac{1}{2} \dot{x}(\xi)^2 \, dm. \tag{4.27}$$

Let ρ_L be the mass per unit length of the member; it follows that

$$dm = \rho_L d\xi. \tag{4.28}$$

Substituting Eqs. 4.26 and 4.28 into Eq. 4.27 gives

$$E_k = \frac{\rho_L}{2} \begin{Bmatrix} \dot{x}_1 \\ \dot{x}_2 \\ \dot{x}_3 \\ \dot{x}_4 \end{Bmatrix}^T \left(\int_{\xi=0}^{\xi=L} \begin{bmatrix} 1 - \frac{3\xi^2}{L^2} + \frac{2\xi^3}{L^3} \\ \xi - \frac{2\xi^2}{L} + \frac{\xi^3}{L^2} \\ \frac{3\xi^2}{L^2} - \frac{2\xi^3}{L^3} \\ -\frac{\xi^2}{L} + \frac{\xi^3}{L^2} \end{bmatrix} \begin{bmatrix} 1 - \frac{3\xi^2}{L^2} + \frac{2\xi^3}{L^3} \\ \xi - \frac{2\xi^2}{L} + \frac{\xi^3}{L^2} \\ \frac{3\xi^2}{L^2} - \frac{2\xi^3}{L^3} \\ -\frac{\xi}{L} + \frac{\xi^3}{L^2} \end{bmatrix}^T d\xi \right) \begin{Bmatrix} \dot{x}_1 \\ \dot{x}_2 \\ \dot{x}_3 \\ \dot{x}_4 \end{Bmatrix}. \tag{4.29}$$

Integrating Eq. 4.29 gives

$$E_k = \frac{\rho_L}{2} \begin{Bmatrix} \dot{x}_1 \\ \dot{x}_2 \\ \dot{x}_3 \\ \dot{x}_4 \end{Bmatrix}^T \begin{bmatrix} 13L/35 & 11L^2/210 & 9L/70 & -13L^2/420 \\ 11L^2/210 & L^3/105 & 13L^2/420 & -L^3/140 \\ 9L/70 & 13L^2/420 & 13L/35 & -11L^2/210 \\ -13L^2/420 & -L^3/140 & -11L^2/210 & L^3/105 \end{bmatrix} \begin{Bmatrix} \dot{x}_1 \\ \dot{x}_2 \\ \dot{x}_3 \\ \dot{x}_4 \end{Bmatrix}. \tag{4.30}$$

Recall that $m = \rho_L L$; simplifying Eq. 4.30, it follows that

$$E_k = \frac{1}{2} m \begin{Bmatrix} \dot{x}_1 \\ \dot{x}_2 \\ \dot{x}_3 \\ \dot{x}_4 \end{Bmatrix}^T \begin{bmatrix} 13/35 & 11L/210 & 9/70 & -13L/420 \\ 11L/210 & L^2/105 & 13L/420 & -L^2/140 \\ 9/70 & 13L/420 & 13/35 & -11L/210 \\ -13L/420 & -L^2/140 & -11L/210 & L^2/105 \end{bmatrix} \begin{Bmatrix} \dot{x}_1 \\ \dot{x}_2 \\ \dot{x}_3 \\ \dot{x}_4 \end{Bmatrix}. \tag{4.31}$$

The kinetic energy in the reduced coordinate system is

$$E_k = \frac{1}{2} \begin{Bmatrix} \dot{x}_1 \\ \dot{x}_2 \\ \dot{x}_3 \\ \dot{x}_4 \end{Bmatrix}^T \mathbf{M} \begin{Bmatrix} \dot{x}_1 \\ \dot{x}_2 \\ \dot{x}_3 \\ \dot{x}_4 \end{Bmatrix} = \frac{1}{2} \dot{\mathbf{x}}^T \mathbf{M} \dot{\mathbf{x}}. \tag{4.32}$$

Equating Eq. 4.31 and Eq. 4.32 and for any $\dot{\mathbf{x}}$, it follows that

$$\mathbf{M} = m \begin{bmatrix} 13/35 & 11L/210 & 9/70 & -13L/420 \\ 11L/210 & L^2/105 & 13L/420 & -L^2/140 \\ 9/70 & 13L/420 & 13/35 & -11L/210 \\ -13L/420 & -L^2/140 & -11L/210 & L^2/105 \end{bmatrix} \tag{4.33}$$

which is the consistent mass matrix of a flexural member.

EXAMPLE 3 *Mass Matrix of a Rigid Rectangular Floor Diaphragm*

Consider a rigid rectangular floor diaphragm with the plan view as shown in Figure 4.3. The kinetic energy at point (ξ, η) is given by

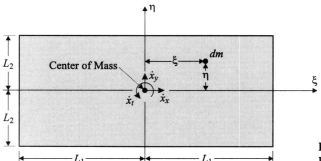

Figure 4.3 Plan View of a Rigid Rectangular Floor Diaphragm

$$dE_k = \frac{1}{2}\left[\dot{x}(\xi,\eta)\right]^2 dm \tag{4.34}$$

where $\dot{x}(\xi, \eta)$ is the speed at the point (ξ, η) and is given by

$$\left[\dot{x}(\xi,\eta)\right]^2 = \dot{\mathbf{x}} \cdot \dot{\mathbf{x}} = \left[\dot{x}_\xi(\xi,\eta)\right]^2 + \left[\dot{x}_\eta(\xi,\eta)\right]^2 \tag{4.35}$$

where $\dot{x}_\xi(\xi, \eta)$ is the speed in the ξ-direction and $\dot{x}_\eta(\xi, \eta)$ is the speed in the η-direction. Let ρ_A be defined as the mass per unit area; it follows that

$$dm = \rho_A d\xi d\eta. \tag{4.36}$$

Substituting Eqs. 4.35 and 4.36 into Eq. 4.34 and integrating over the entire floor area gives

$$E_k = \frac{\rho_A}{2}\int_{\eta=-L_2}^{\eta=L_2}\int_{\xi=-L_1}^{\xi=L_1}\left\{\left[\dot{x}_\xi(\xi,\eta)\right]^2 + \left[\dot{x}_\eta(\xi,\eta)\right]^2\right\}d\xi\,d\eta. \tag{4.37}$$

The speeds $\dot{x}_\xi(\xi, \eta)$ and $\dot{x}_\eta(\xi, \eta)$ can be expressed in terms of the velocities at the center of mass, that is,

$$\dot{x}_\xi(\xi,\eta) = \dot{x}_x - \eta\dot{x}_t = \begin{bmatrix} 1 & 0 & -\eta \end{bmatrix}\begin{Bmatrix} \dot{x}_x \\ \dot{x}_y \\ \dot{x}_t \end{Bmatrix} \tag{4.38}$$

$$\dot{x}_\eta(\xi,\eta) = \dot{x}_y + \xi\dot{x}_t = \begin{bmatrix} 0 & 1 & \xi \end{bmatrix}\begin{Bmatrix} \dot{x}_x \\ \dot{x}_y \\ \dot{x}_t \end{Bmatrix}. \tag{4.39}$$

Substituting Eqs. 4.38 and 4.39 into Eq. 4.37, it follows that

$$
\begin{aligned}
E_k &= \frac{\rho_A}{2}\begin{Bmatrix} \dot{x}_x \\ \dot{x}_y \\ \dot{x}_t \end{Bmatrix}^T \left(\int_{\eta=-L_2}^{\eta=L_2}\int_{\xi=-L_1}^{\xi=L_1}\left\{\begin{bmatrix} 1 \\ 0 \\ -\eta \end{bmatrix}\begin{bmatrix} 1 \\ 0 \\ -\eta \end{bmatrix}^T + \begin{bmatrix} 0 \\ 1 \\ \xi \end{bmatrix}\begin{bmatrix} 0 \\ 1 \\ \xi \end{bmatrix}^T\right\}d\xi\,d\eta\right)\begin{Bmatrix} \dot{x}_x \\ \dot{x}_y \\ \dot{x}_t \end{Bmatrix} \\
&= \frac{\rho_A}{2}\begin{Bmatrix} \dot{x}_x \\ \dot{x}_y \\ \dot{x}_t \end{Bmatrix}^T \left(\int_{\eta=-L_2}^{\eta=L_2}\int_{\xi=-L_1}^{\xi=L_1}\begin{bmatrix} 1 & 0 & -\eta \\ 0 & 1 & \xi \\ -\eta & \xi & \eta^2+\xi^2 \end{bmatrix}d\xi\,d\eta\right)\begin{Bmatrix} \dot{x}_x \\ \dot{x}_y \\ \dot{x}_t \end{Bmatrix}.
\end{aligned}
\tag{4.40}
$$

Performing the integration gives

$$E_k = \frac{\rho_A}{2} \begin{Bmatrix} \dot{x}_x \\ \dot{x}_y \\ \dot{x}_t \end{Bmatrix}^T \begin{bmatrix} 4L_1L_2 & 0 & 0 \\ 0 & 4L_1L_2 & 0 \\ 0 & 0 & (L_1L_2^3 + L_1^3L_2)/3 \end{bmatrix} \begin{Bmatrix} \dot{x}_x \\ \dot{x}_y \\ \dot{x}_t \end{Bmatrix}. \tag{4.41}$$

Note that the total mass of the floor is $m = 4\rho_A L_1 L_2$, and therefore

$$E_k = \frac{1}{2} \begin{Bmatrix} \dot{x}_x \\ \dot{x}_y \\ \dot{x}_t \end{Bmatrix}^T \begin{bmatrix} m & 0 & 0 \\ 0 & m & 0 \\ 0 & 0 & m(L_1^2 + L_2^2)/3 \end{bmatrix} \begin{Bmatrix} \dot{x}_x \\ \dot{x}_y \\ \dot{x}_t \end{Bmatrix}. \tag{4.42}$$

The kinetic energy in the reduced coordinate system is

$$E_k = \frac{1}{2} \begin{Bmatrix} \dot{x}_x \\ \dot{x}_y \\ \dot{x}_t \end{Bmatrix}^T \mathbf{M} \begin{Bmatrix} \dot{x}_x \\ \dot{x}_y \\ \dot{x}_t \end{Bmatrix} = \frac{1}{2} \dot{\mathbf{x}}^T \mathbf{M} \dot{\mathbf{x}}. \tag{4.43}$$

Equating Eq. 4.42 and Eq. 4.43 and for any $\dot{\mathbf{x}}$, it follows that

$$\mathbf{M} = \begin{bmatrix} m & 0 & 0 \\ 0 & m & 0 \\ 0 & 0 & m(L_1^2 + L_2^2)/3 \end{bmatrix} \tag{4.44}$$

which is a diagonal matrix. This shows that the consistent mass matrix is the same as the lumped mass matrix for a rigid floor diaphragm.

Note that the mass moment of inertia of the floor $m_{tt} = m(L_1^2 + L_2^2)/3$ is computed by performing the integration where

$$m_{tt} = \rho_A \int_{\eta=-L_2}^{\eta=L_2} \int_{\xi=-L_1}^{\xi=L_1} (\eta^2 + \xi^2) \, d\xi \, d\eta = \rho_A J = \frac{mJ}{A} \tag{4.45}$$

where J is the polar moment of inertia and A is the floor area.

EXAMPLE 4   *Mass Matrix of a Rigid L-Shaped Floor Diaphragm*

Consider a rigid L-shaped floor diaphragm with the plan view as shown in Figure 4.4. The total kinetic energy E_k is computed by

$$E_k = E_k^{(1)} + E_k^{(2)} \tag{4.46}$$

where $E_k^{(1)}$ is the kinetic energy in Leg 1 and $E_k^{(2)}$ is the kinetic energy in Leg 2.

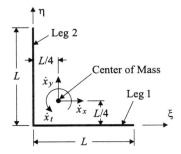

Figure 4.4 Plan View of a Rigid L-Shaped Floor Diaphragm

First consider Leg 1, where the kinetic energy is given by

$$E_k^{(1)} = \frac{1}{2}\int \left[\dot{x}(\xi,\eta)\right]^2 dm. \tag{4.47}$$

The term $\dot{x}(\xi,\eta)$ is the speed at the point (ξ,η) and is given by

$$\left[\dot{x}(\xi,\eta)\right]^2 = \dot{\mathbf{x}}\cdot\dot{\mathbf{x}} = \left[\dot{x}_\xi(\xi,\eta)\right]^2 + \left[\dot{x}_\eta(\xi,\eta)\right]^2 \tag{4.48}$$

where $\dot{x}_\xi(\xi,\eta)$ is the speed in the ξ-direction and $\dot{x}_\eta(\xi,\eta)$ is the speed in the η-direction. Let ρ_L be defined as the mass per unit length; it follows that

$$dm = \rho_L d\xi. \tag{4.49}$$

Substituting Eqs. 4.48 and 4.49 into Eq. 4.47 gives

$$E_k^{(1)} = \frac{\rho_L}{2}\int_{\xi=0}^{\xi=L}\left\{\left[\dot{x}_\xi(\xi,\eta)\right]^2 + \left[\dot{x}_\eta(\xi,\eta)\right]^2\right\}d\xi. \tag{4.50}$$

The speed $\dot{x}_\xi(\xi,\eta)$ and $\dot{x}_\eta(\xi,\eta)$ can be expressed in terms of the velocities at the center of mass, that is,

$$\dot{x}_\xi(\xi,\eta) = \dot{x}_x + \frac{L}{4}\dot{x}_t = \begin{bmatrix} 1 & 0 & L/4 \end{bmatrix}\begin{Bmatrix} \dot{x}_x \\ \dot{x}_y \\ \dot{x}_t \end{Bmatrix} \tag{4.51}$$

$$\dot{x}_\eta(\xi,\eta) = \dot{x}_y + \left(\xi - \frac{L}{4}\right)\dot{x}_t = \begin{bmatrix} 0 & 1 & \xi - L/4 \end{bmatrix}\begin{Bmatrix} \dot{x}_x \\ \dot{x}_y \\ \dot{x}_t \end{Bmatrix}. \tag{4.52}$$

Substituting Eqs. 4.51 and 4.52 into Eq. 4.50, it follows that

$$E_k^{(1)} = \frac{\rho_L}{2}\begin{Bmatrix} \dot{x}_x \\ \dot{x}_y \\ \dot{x}_t \end{Bmatrix}^T\left(\int_{\xi=0}^{\xi=L}\left[\begin{bmatrix} 1 \\ 0 \\ L/4 \end{bmatrix}\begin{bmatrix} 1 \\ 0 \\ L/4 \end{bmatrix}^T + \begin{bmatrix} 0 \\ 1 \\ \xi-L/4 \end{bmatrix}\begin{bmatrix} 0 \\ 1 \\ \xi-L/4 \end{bmatrix}^T\right]d\xi\right)\begin{Bmatrix} \dot{x}_x \\ \dot{x}_y \\ \dot{x}_t \end{Bmatrix}$$

$$= \frac{\rho_L}{2}\begin{Bmatrix} \dot{x}_x \\ \dot{x}_y \\ \dot{x}_t \end{Bmatrix}^T\left(\int_{\xi=0}^{\xi=L}\begin{bmatrix} 1 & 0 & L/4 \\ 0 & 1 & \xi-L/4 \\ L/4 & \xi-L/4 & (L/4)^2+(\xi-L/4)^2 \end{bmatrix}d\xi\right)\begin{Bmatrix} \dot{x}_x \\ \dot{x}_y \\ \dot{x}_t \end{Bmatrix}. \tag{4.53}$$

Performing the integration gives

$$E_k^{(1)} = \frac{\rho_L}{2}\begin{Bmatrix} \dot{x}_x \\ \dot{x}_y \\ \dot{x}_t \end{Bmatrix}^T\begin{bmatrix} L & 0 & L^2/4 \\ 0 & L & L^2/4 \\ L^2/4 & L^2/4 & 5L^3/24 \end{bmatrix}\begin{Bmatrix} \dot{x}_x \\ \dot{x}_y \\ \dot{x}_t \end{Bmatrix}. \tag{4.54}$$

Now consider Leg 2, where the kinetic energy is again given by

$$E_k^{(2)} = \frac{1}{2}\int \left[\dot{x}(\xi,\eta)\right]^2 dm = \frac{1}{2}\int\left\{\left[\dot{x}_\xi(\xi,\eta)\right]^2 + \left[\dot{x}_\eta(\xi,\eta)\right]^2\right\}dm. \tag{4.55}$$

Since Leg 2 is oriented along η-direction, it follows that

$$dm = \rho_L d\eta. \tag{4.56}$$

Substituting Eq. 4.56 into Eq. 4.55 gives

$$E_k^{(2)} = \frac{\rho_L}{2} \int_{\eta=0}^{\eta=L} \left\{ \left[\dot{x}_\xi(\xi,\eta) \right]^2 + \left[\dot{x}_\eta(\xi,\eta) \right]^2 \right\} d\eta. \tag{4.57}$$

The speeds $\dot{x}_\xi(\xi,\eta)$ and $\dot{x}_\eta(\xi,\eta)$ can be expressed in terms of the velocities at the center of mass, that is,

$$\dot{x}_\xi(\xi,\eta) = \dot{x}_x + \left(\frac{L}{4} - \eta \right) \dot{x}_t = \begin{bmatrix} 1 & 0 & L/4 - \eta \end{bmatrix} \begin{Bmatrix} \dot{x}_x \\ \dot{x}_y \\ \dot{x}_t \end{Bmatrix} \tag{4.58}$$

$$\dot{x}_\eta(\xi,\eta) = \dot{x}_y - \frac{L}{4} \dot{x}_t = \begin{bmatrix} 0 & 1 & -L/4 \end{bmatrix} \begin{Bmatrix} \dot{x}_x \\ \dot{x}_y \\ \dot{x}_t \end{Bmatrix}. \tag{4.59}$$

Substituting Eqs. 4.58 and 4.59 into Eq. 4.57, it follows that

$$\begin{aligned}
E_k^{(2)} &= \frac{\rho_L}{2} \begin{Bmatrix} \dot{x}_x \\ \dot{x}_y \\ \dot{x}_t \end{Bmatrix}^T \left(\int_{\eta=0}^{\eta=L} \left[\begin{bmatrix} 1 \\ 0 \\ L/4-\eta \end{bmatrix} \begin{bmatrix} 1 \\ 0 \\ L/4-\eta \end{bmatrix}^T + \begin{bmatrix} 0 \\ 1 \\ -L/4 \end{bmatrix} \begin{bmatrix} 0 \\ 1 \\ -L/4 \end{bmatrix}^T \right] d\eta \right) \begin{Bmatrix} \dot{x}_x \\ \dot{x}_y \\ \dot{x}_t \end{Bmatrix} \\
&= \frac{\rho_L}{2} \begin{Bmatrix} \dot{x}_x \\ \dot{x}_y \\ \dot{x}_t \end{Bmatrix}^T \left(\int_{\eta=0}^{\eta=L} \begin{bmatrix} 1 & 0 & L/4-\eta \\ 0 & 1 & -L/4 \\ L/4-\eta & -L/4 & (L/4-\eta)^2 + (-L/4)^2 \end{bmatrix} d\eta \right) \begin{Bmatrix} \dot{x}_x \\ \dot{x}_y \\ \dot{x}_t \end{Bmatrix}.
\end{aligned} \tag{4.60}$$

Performing the integration gives

$$E_k^{(2)} = \frac{\rho_L}{2} \begin{Bmatrix} \dot{x}_x \\ \dot{x}_y \\ \dot{x}_t \end{Bmatrix}^T \begin{bmatrix} L & 0 & -L^2/4 \\ 0 & L & -L^2/4 \\ -L^2/4 & -L^2/4 & 5L^3/24 \end{bmatrix} \begin{Bmatrix} \dot{x}_x \\ \dot{x}_y \\ \dot{x}_t \end{Bmatrix}. \tag{4.61}$$

Substituting Eqs. 4.54 and 4.61 into Eq. 4.46, it follows that the total kinetic energy is

$$\begin{aligned}
E_k &= \frac{\rho_L}{2} \begin{Bmatrix} \dot{x}_x \\ \dot{x}_y \\ \dot{x}_t \end{Bmatrix}^T \begin{bmatrix} L & 0 & L^2/4 \\ 0 & L & L^2/4 \\ L^2/4 & L^2/4 & 5L^3/24 \end{bmatrix} \begin{Bmatrix} \dot{x}_x \\ \dot{x}_y \\ \dot{x}_t \end{Bmatrix} + \frac{\rho_L}{2} \begin{Bmatrix} \dot{x}_x \\ \dot{x}_y \\ \dot{x}_t \end{Bmatrix}^T \begin{bmatrix} L & 0 & -L^2/4 \\ 0 & L & -L^2/4 \\ -L^2/4 & -L^2/4 & 5L^3/24 \end{bmatrix} \begin{Bmatrix} \dot{x}_x \\ \dot{x}_y \\ \dot{x}_t \end{Bmatrix} \\
&= \frac{\rho_L}{2} \begin{Bmatrix} \dot{x}_x \\ \dot{x}_y \\ \dot{x}_t \end{Bmatrix}^T \begin{bmatrix} 2L & 0 & 0 \\ 0 & 2L & 0 \\ 0 & 0 & 5L^3/12 \end{bmatrix} \begin{Bmatrix} \dot{x}_x \\ \dot{x}_y \\ \dot{x}_t \end{Bmatrix}.
\end{aligned} \tag{4.62}$$

Recall that the total mass of the floor is $m = 2\rho_L L$; therefore,

$$E_k = \frac{1}{2} \begin{Bmatrix} \dot{x}_x \\ \dot{x}_y \\ \dot{x}_t \end{Bmatrix}^T \begin{bmatrix} m & 0 & 0 \\ 0 & m & 0 \\ 0 & 0 & 5mL^2/24 \end{bmatrix} \begin{Bmatrix} \dot{x}_x \\ \dot{x}_y \\ \dot{x}_t \end{Bmatrix}. \tag{4.63}$$

The kinetic energy in the reduced coordinate system is

$$E_k = \frac{1}{2} \begin{Bmatrix} \dot{x}_x \\ \dot{x}_y \\ \dot{x}_t \end{Bmatrix}^T \mathbf{M} \begin{Bmatrix} \dot{x}_x \\ \dot{x}_y \\ \dot{x}_t \end{Bmatrix} = \frac{1}{2} \dot{\mathbf{x}}^T \mathbf{M} \dot{\mathbf{x}}. \tag{4.64}$$

Equating Eq. 4.63 and Eq. 4.64 and for any $\dot{\mathbf{x}}$, it follows that

$$\mathbf{M} = \begin{bmatrix} m & 0 & 0 \\ 0 & m & 0 \\ 0 & 0 & 5mL^2/24 \end{bmatrix} \tag{4.65}$$

which is a diagonal matrix. Similar to Example 3, this example shows that the consistent mass matrix is the same as the lumped mass matrix for a rigid floor diaphragm.

Note that the mass moment of inertia in Eq. 4.65 is $m_{tt} = 5mL^2/24$. The same result can be obtained using the formula $m_{tt} = \rho_L J$ as in Example 3, that is,

$$
\begin{aligned}
\rho_L J &= (\rho_L J)_{\text{Leg}1} + (\rho_L J)_{\text{Leg}2} \\
&= (\rho_L I_x)_{\text{Leg}1} + (\rho_L I_y)_{\text{Leg}1} + (\rho_L I_x)_{\text{Leg}2} + (\rho_L I_y)_{\text{Leg}2} \\
&= \left[\frac{\rho_L L^3}{12} + \rho_L L \left(\frac{L}{4} \right)^2 \right] + \left[0 + \rho_L L \left(\frac{L}{4} \right)^2 \right] + \left[0 + \rho_L L \left(\frac{L}{4} \right)^2 \right] + \left[\frac{\rho_L L^3}{12} + \rho_L L \left(\frac{L}{4} \right)^2 \right] \\
&= \frac{5}{12} \rho_L L^3 \\
&= \frac{5}{24} mL^2
\end{aligned} \tag{4.66}
$$

which is the same result as Eq. 4.65.

The assumption of a rigid diaphragm is vital to the resulting diagonal mass matrix. The legs of the L-shaped structure may actually not be as rigid as assumed. Considering Leg 1, axial rigidity along the ξ-direction is a reasonable assumption. However, assuming flexural rigidity along the η-direction may not be appropriate. It follows that Eq. 4.54 must be modified in the terms that contain $\dot{x}_\eta(\xi, \eta)$, that is,

$$\dot{x}_\eta(\xi, \eta) = \alpha_1 \dot{x}_y + \beta_1 \left(\xi - \frac{L}{4} \right) \dot{x}_t = \begin{bmatrix} 0 & \alpha_1 & \beta_1(\xi - L/4) \end{bmatrix} \begin{Bmatrix} \dot{x}_x \\ \dot{x}_y \\ \dot{x}_t \end{Bmatrix} \tag{4.67}$$

where α_1 and β_1 are the modification factors for Leg 1. Therefore,

$$E_k^{(1)} = \frac{\rho_L}{2} \begin{Bmatrix} \dot{x}_x \\ \dot{x}_y \\ \dot{x}_t \end{Bmatrix}^T \begin{bmatrix} L & 0 & L^2/4 \\ 0 & \alpha_1^2 L & \alpha_1 \beta_1 L^2/4 \\ L^2/4 & \alpha_1 \beta_1 L^2/4 & 5\beta_1^2 L^3/24 \end{bmatrix} \begin{Bmatrix} \dot{x}_x \\ \dot{x}_y \\ \dot{x}_t \end{Bmatrix}. \tag{4.68}$$

Similarly, for Leg 2, assuming axial rigidity along the η-direction is reasonable, but assuming flexural rigidity along the ξ-direction may also be appropriate. It follows that Eq. 4.61 must be modified in the terms that contain $\dot{x}_\xi(\xi, \eta)$, that is,

$$\dot{x}_\xi(\xi, \eta) = \alpha_2 \dot{x}_x + \beta_2 \left(\frac{L}{4} - \eta \right) \dot{x}_t = \begin{bmatrix} \alpha_2 & 0 & \beta_2(L/4 - \eta) \end{bmatrix} \begin{Bmatrix} \dot{x}_x \\ \dot{x}_y \\ \dot{x}_t \end{Bmatrix} \tag{4.69}$$

where α_2 and β_2 are the modification factors for Leg 2. Therefore,

$$E_k^{(2)} = \frac{\rho_L}{2} \begin{Bmatrix} \dot{x}_x \\ \dot{x}_y \\ \dot{x}_t \end{Bmatrix}^T \begin{bmatrix} \alpha_2^2 L & 0 & -\alpha_2 \beta_2 L^2/4 \\ 0 & L & -L^2/4 \\ -\alpha_2 \beta_2 L^2/4 & -L^2/4 & 5\beta_2^2 L^3/24 \end{bmatrix} \begin{Bmatrix} \dot{x}_x \\ \dot{x}_y \\ \dot{x}_t \end{Bmatrix}. \tag{4.70}$$

The total kinetic energy follows from Eq. 4.46:

$$E_k = \frac{\rho_L}{2} \left\{ \begin{matrix} \dot{x}_x \\ \dot{x}_y \\ \dot{x}_t \end{matrix} \right\}^T \begin{bmatrix} (\alpha_2^2 + 1)L/2 & 0 & (1 - \alpha_2\beta_2)L^2/8 \\ 0 & (\alpha_1^2 + 1)L/2 & (\alpha_1\beta_1 - 1)L^2/8 \\ (1 - \alpha_2\beta_2)L^2/8 & (\alpha_1\beta_1 - 1)L^2/8 & 5(\beta_1^2 + \beta_1^2)L^3/24 \end{bmatrix} \left\{ \begin{matrix} \dot{x}_x \\ \dot{x}_y \\ \dot{x}_t \end{matrix} \right\}. \tag{4.71}$$

It follows that the mass matrix becomes

$$\mathbf{M} = \begin{bmatrix} (\alpha_2^2 + 1)L/2 & 0 & (1 - \alpha_2\beta_2)L^2/8 \\ 0 & (\alpha_1^2 + 1)L/2 & (\alpha_1\beta_1 - 1)L^2/8 \\ (1 - \alpha_2\beta_2)L^2/8 & (\alpha_1\beta_1 - 1)L^2/8 & 5(\beta_1^2 + \beta_1^2)L^3/24 \end{bmatrix}. \tag{4.72}$$

Note that the mass matrix in Eq. 4.72 contains nonzero terms in the off-diagonal. This shows that couplings between $\dot{x}_x$ and $\dot{x}_t$ as well as $\dot{x}_y$ and $\dot{x}_t$ will result when the diaphragm is not rigid.

EXAMPLE 5   *Mass Matrix of a Rigid C-Shaped Floor Diaphragm*

Consider a rigid C-shaped floor diaphragm with the plan view as shown in Figure 4.5. The total kinetic energy E_k is computed by

$$E_k = E_k^{(w)} + E_k^{(lf)} + E_k^{(rf)} \tag{4.73}$$

where $E_k^{(lf)}$ is the kinetic energy in the left flange, $E_k^{(rf)}$ is the kinetic energy in the right flange, and $E_k^{(w)}$ is the kinetic energy in the web.

First consider the web, where the kinetic energy is again given by

$$E_k^{(w)} = \frac{1}{2}\int [\dot{x}(\xi, \eta)]^2\, dm = \frac{1}{2}\int\left([\dot{x}_\xi(\xi, \eta)]^2 + [\dot{x}_\eta(\xi, \eta)]^2 \right) dm. \tag{4.74}$$

Since the web is oriented along the ξ-direction, it follows that

$$dm = \rho_L d\xi. \tag{4.75}$$

Substituting Eq. 4.75 into Eq. 4.74 gives

$$E_k^{(w)} = \frac{\rho_L}{2}\int_{\xi=-L}^{\xi=L}\left\{ [\dot{x}_\xi(\xi, \eta)]^2 + [\dot{x}_\eta(\xi, \eta)]^2 \right\} d\xi. \tag{4.76}$$

The speeds $\dot{x}_\xi(\xi, \eta)$ and $\dot{x}_\eta(\xi, \eta)$ can be expressed in terms of the velocities at the center of mass:

$$\dot{x}_\xi(\xi, \eta) = \dot{x}_x + \frac{L}{4}\dot{x}_t = \begin{bmatrix} 1 & 0 & L/4 \end{bmatrix} \left\{ \begin{matrix} \dot{x}_x \\ \dot{x}_y \\ \dot{x}_t \end{matrix} \right\} \tag{4.77}$$

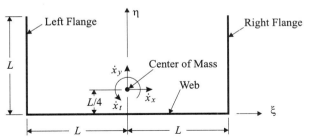

Figure 4.5 Plan View of a Rigid C-Shaped Floor Diaphragm

$$\dot{x}_\eta(\xi,\eta) = \dot{x}_y + \xi\dot{x}_t = \begin{bmatrix} 0 & 1 & \xi \end{bmatrix}\begin{Bmatrix} \dot{x}_x \\ \dot{x}_y \\ \dot{x}_t \end{Bmatrix}. \tag{4.78}$$

Substituting Eqs. 4.77 and 4.78 into Eq. 4.76, it follows that

$$
\begin{aligned}
E_k^{(w)} &= \frac{\rho_L}{2}\begin{Bmatrix} \dot{x}_x \\ \dot{x}_y \\ \dot{x}_t \end{Bmatrix}^T\left(\int_{\xi=-L}^{\xi=L}\left[\begin{bmatrix} 1 \\ 0 \\ L/4 \end{bmatrix}\begin{bmatrix} 1 \\ 0 \\ L/4 \end{bmatrix}^T + \begin{bmatrix} 0 \\ 1 \\ \xi \end{bmatrix}\begin{bmatrix} 0 \\ 1 \\ \xi \end{bmatrix}^T\right]d\xi\right)\begin{Bmatrix} \dot{x}_x \\ \dot{x}_y \\ \dot{x}_t \end{Bmatrix} \\
&= \frac{\rho_L}{2}\begin{Bmatrix} \dot{x}_x \\ \dot{x}_y \\ \dot{x}_t \end{Bmatrix}^T\left(\int_{\xi=-L}^{\xi=L}\begin{bmatrix} 1 & 0 & L/4 \\ 0 & 1 & \xi \\ L/4 & \xi & (L/4)^2+\xi^2 \end{bmatrix}d\xi\right)\begin{Bmatrix} \dot{x}_x \\ \dot{x}_y \\ \dot{x}_t \end{Bmatrix}.
\end{aligned}
\tag{4.79}
$$

Performing the integration gives

$$E_k^{(w)} = \frac{\rho_L}{2}\begin{Bmatrix} \dot{x}_x \\ \dot{x}_y \\ \dot{x}_t \end{Bmatrix}^T\begin{bmatrix} 2L & 0 & L^2/2 \\ 0 & 2L & 0 \\ L^2/2 & 0 & 19L^3/24 \end{bmatrix}\begin{Bmatrix} \dot{x}_x \\ \dot{x}_y \\ \dot{x}_t \end{Bmatrix}. \tag{4.80}$$

Similarly, the kinetic energies on the left and right sides are

$$
\begin{aligned}
E_k^{(lf)} &= \frac{\rho_L}{2}\begin{Bmatrix} \dot{x}_x \\ \dot{x}_y \\ \dot{x}_t \end{Bmatrix}^T\left(\int_{\eta=0}^{\eta=L}\left\{\begin{bmatrix} 1 \\ 0 \\ L/4-\eta \end{bmatrix}\begin{bmatrix} 1 \\ 0 \\ L/4-\eta \end{bmatrix}^T + \begin{bmatrix} 0 \\ 1 \\ L \end{bmatrix}\begin{bmatrix} 0 \\ 1 \\ L \end{bmatrix}^T\right\}d\eta\right)\begin{Bmatrix} \dot{x}_x \\ \dot{x}_y \\ \dot{x}_t \end{Bmatrix} \\
&= \frac{\rho_L}{2}\begin{Bmatrix} \dot{x}_x \\ \dot{x}_y \\ \dot{x}_t \end{Bmatrix}^T\left(\int_{\eta=0}^{\eta=L}\begin{bmatrix} 1 & 0 & L/4-\eta \\ 0 & 1 & L \\ L/4-\eta & L & (L/4-\eta)^2+L^2 \end{bmatrix}d\eta\right)\begin{Bmatrix} \dot{x}_x \\ \dot{x}_y \\ \dot{x}_t \end{Bmatrix} \\
&= \frac{\rho_L}{2}\begin{Bmatrix} \dot{x}_x \\ \dot{x}_y \\ \dot{x}_t \end{Bmatrix}^T\begin{bmatrix} L & 0 & -L^2/4 \\ 0 & L & L^2 \\ -L^2/4 & L^2 & 55L^3/48 \end{bmatrix}\begin{Bmatrix} \dot{x}_x \\ \dot{x}_y \\ \dot{x}_t \end{Bmatrix}.
\end{aligned}
\tag{4.81}
$$

$$
\begin{aligned}
E_k^{(rf)} &= \frac{\rho_L}{2}\begin{Bmatrix} \dot{x}_x \\ \dot{x}_y \\ \dot{x}_t \end{Bmatrix}^T\left(\int_{\eta=0}^{\eta=L}\left\{\begin{bmatrix} 1 \\ 0 \\ L/4-\eta \end{bmatrix}\begin{bmatrix} 1 \\ 0 \\ L/4-\eta \end{bmatrix}^T + \begin{bmatrix} 0 \\ 1 \\ -L \end{bmatrix}\begin{bmatrix} 0 \\ 1 \\ -L \end{bmatrix}^T\right\}d\eta\right)\begin{Bmatrix} \dot{x}_x \\ \dot{x}_y \\ \dot{x}_t \end{Bmatrix} \\
&= \frac{\rho_L}{2}\begin{Bmatrix} \dot{x}_x \\ \dot{x}_y \\ \dot{x}_t \end{Bmatrix}^T\left(\int_{\eta=0}^{\eta=L}\begin{bmatrix} 1 & 0 & L/4-\eta \\ 0 & 1 & -L \\ L/4-\eta & -L & (L/4-\eta)^2+L^2 \end{bmatrix}d\eta\right)\begin{Bmatrix} \dot{x}_x \\ \dot{x}_y \\ \dot{x}_t \end{Bmatrix} \\
&= \frac{\rho_L}{2}\begin{Bmatrix} \dot{x}_x \\ \dot{x}_y \\ \dot{x}_t \end{Bmatrix}^T\begin{bmatrix} L & 0 & -L^2/4 \\ 0 & L & -L^2 \\ -L^2/4 & -L^2 & 55L^3/48 \end{bmatrix}\begin{Bmatrix} \dot{x}_x \\ \dot{x}_y \\ \dot{x}_t \end{Bmatrix}.
\end{aligned}
\tag{4.82}
$$

Substituting Eqs. 4.80 to 4.82 into Eq. 4.73 gives

$$E_k = \frac{\rho_L}{2} \begin{Bmatrix} \dot{x}_x \\ \dot{x}_y \\ \dot{x}_t \end{Bmatrix}^T \begin{bmatrix} 2L & 0 & L^2/2 \\ 0 & 2L & 0 \\ L^2/2 & 0 & 19L^3/24 \end{bmatrix} \begin{Bmatrix} \dot{x}_x \\ \dot{x}_y \\ \dot{x}_t \end{Bmatrix} + \frac{\rho_L}{2} \begin{Bmatrix} \dot{x}_x \\ \dot{x}_y \\ \dot{x}_t \end{Bmatrix}^T \begin{bmatrix} L & 0 & -L^2/4 \\ 0 & L & L^2 \\ -L^2/4 & L^2 & 55L^3/48 \end{bmatrix} \begin{Bmatrix} \dot{x}_x \\ \dot{x}_y \\ \dot{x}_t \end{Bmatrix}$$

$$+ \frac{\rho_L}{2} \begin{Bmatrix} \dot{x}_x \\ \dot{x}_y \\ \dot{x}_t \end{Bmatrix}^T \begin{bmatrix} L & 0 & -L^2/4 \\ 0 & L & -L^2 \\ -L^2/4 & -L^2 & 55L^3/48 \end{bmatrix} \begin{Bmatrix} \dot{x}_x \\ \dot{x}_y \\ \dot{x}_t \end{Bmatrix}. \tag{4.83}$$

Simplifying Eq. 4.83 gives

$$E_k = \frac{\rho_L}{2} \begin{Bmatrix} \dot{x}_x \\ \dot{x}_y \\ \dot{x}_t \end{Bmatrix}^T \begin{bmatrix} 4L & 0 & 0 \\ 0 & 4L & 0 \\ 0 & 0 & 37L^3/12 \end{bmatrix} \begin{Bmatrix} \dot{x}_x \\ \dot{x}_y \\ \dot{x}_t \end{Bmatrix}. \tag{4.84}$$

Recall that the total mass of the floor is $m = 4\rho_L L$; therefore,

$$E_k = \frac{1}{2} \begin{Bmatrix} \dot{x}_x \\ \dot{x}_y \\ \dot{x}_t \end{Bmatrix}^T \begin{bmatrix} m & 0 & 0 \\ 0 & m & 0 \\ 0 & 0 & 37mL^2/48 \end{bmatrix} \begin{Bmatrix} \dot{x}_x \\ \dot{x}_y \\ \dot{x}_t \end{Bmatrix}. \tag{4.85}$$

The kinetic energy in the reduced coordinate system is

$$E_k = \frac{1}{2} \begin{Bmatrix} \dot{x}_x \\ \dot{x}_y \\ \dot{x}_t \end{Bmatrix}^T \mathbf{M} \begin{Bmatrix} \dot{x}_x \\ \dot{x}_y \\ \dot{x}_t \end{Bmatrix} = \frac{1}{2} \dot{\mathbf{x}}^T \mathbf{M} \dot{\mathbf{x}}. \tag{4.86}$$

Equating Eq. 4.85 and Eq. 4.86 and for any $\dot{\mathbf{x}}$, it follows that

$$\mathbf{M} = \begin{bmatrix} m & 0 & 0 \\ 0 & m & 0 \\ 0 & 0 & 37mL^2/48 \end{bmatrix} \tag{4.87}$$

which is a diagonal matrix for a rigid floor diaphragm.

Again, note that the mass moment of inertia $m_{tt} = 37mL^2/48$ can be computed using the polar moment of inertia:

$$\rho_L J = (\rho_L J)_{\text{Web}} + (\rho_L J)_{\text{Left Flange}} + (\rho_L J)_{\text{Right Flange}}$$

$$= \left[\frac{\rho_L (2L)^3}{12} + \rho_L (2L) \left(\frac{L}{4} \right)^2 + \rho_L (2L)(0)^2 \right]$$

$$+ \left[\frac{\rho_L L^3}{12} + \rho_L L \left(\frac{L}{4} \right)^2 + \rho_L L (L)^2 \right] + \left[\frac{\rho_L L^3}{12} + \rho_L L \left(\frac{L}{4} \right)^2 + \rho_L L (L)^2 \right] \tag{4.88}$$

$$= \frac{37}{12} \rho_L L^3$$

$$= \frac{37}{48} mL^2.$$

4.3 UNDAMPED FREE VIBRATION USING THE NORMAL MODE METHOD

The normal mode method is presented in Chapter 3 for 2DOF systems, which involves calculations of two natural frequencies and two mode shapes. In this section, this method is extended to solve structural dynamics problems with thousands of degrees of freedom. An accurate solution is obtained when all thousands of natural frequencies and mode shapes are calculated. However, this calculation will involve tremendous computing time and thus is not feasible. In addition, some modal responses do not make a difference in the overall response of the structure, especially when the structure is vibrating at high natural frequencies. Therefore, structural engineers always reduce structural dynamics problems by ignoring certain unimportant natural frequencies and mode shapes, and the number of natural frequencies and mode shapes obtained is in most cases less than the total number of degrees of freedom.

The undamped free vibration is represented by Eq. 4.1 with no damping (i.e., $\mathbf{C} = \mathbf{0}$) and no external force (i.e., $\mathbf{F} = \mathbf{0}$). The result is a matrix equation of motion of the form

$$\mathbf{M\ddot{X}}(t) + \mathbf{KX}(t) = \mathbf{0}. \tag{4.89}$$

Assume the solution to this equation takes the form

$$\mathbf{X}(t) = \begin{Bmatrix} x_1(t) \\ x_2(t) \\ \vdots \\ x_n(t) \end{Bmatrix} = \begin{Bmatrix} \alpha_1(t) \\ \alpha_2(t) \\ \vdots \\ \alpha_n(t) \end{Bmatrix} \cos \omega t + \begin{Bmatrix} \beta_1(t) \\ \beta_2(t) \\ \vdots \\ \beta_n(t) \end{Bmatrix} \sin \omega t = \mathbf{A} \cos \omega t + \mathbf{B} \sin \omega t. \tag{4.90}$$

Substituting Eq. 4.90 into Eq. 4.89 gives

$$\left(-\mathbf{M}\omega^2 + \mathbf{K}\right)\left[\mathbf{A} \cos \omega t + \mathbf{B} \sin \omega t\right] = \mathbf{0} \tag{4.91}$$

or

$$\left[-\mathbf{M}\omega^2 + \mathbf{K}\right]\mathbf{A} \cos \omega t + \left[-\mathbf{M}\omega^2 + \mathbf{K}\right]\mathbf{B} \sin \omega t = \mathbf{0}. \tag{4.92}$$

Because the sine and cosine function vary with time, they cannot be zero for all time, and therefore

$$\left[-\mathbf{M}\omega^2 + \mathbf{K}\right]\mathbf{A} = \mathbf{0} \tag{4.93a}$$

$$\left[-\mathbf{M}\omega^2 + \mathbf{K}\right]\mathbf{B} = \mathbf{0}. \tag{4.93b}$$

Equations 4.93a and 4.93b show that $\mathbf{A}$ is proportional to $\mathbf{B}$. One solution to Eq. 4.93 is to let $\mathbf{A}$ and $\mathbf{B}$ be equal to zero. However, this would result in a zero solution for $\mathbf{X}(t)$, and thus it is known as the trivial solution. Alternatively, Eq. 4.93 can be solved by setting the determinant of the matrix represented by the terms within the bracket equal to zero:

$$\left| -\mathbf{M}\omega^2 + \mathbf{K} \right| = 0. \tag{4.94}$$

Equation 4.94 is an nth-order polynomial, and therefore there are n unknown ω^2. Since both $\mathbf{M}$ and $\mathbf{K}$ are positive definite matrices, there will always be n real solutions of ω^2, denoted $\omega_1^2, \omega_2^2, \ldots, \omega_n^2$, where

$$\omega_1^2 \leq \omega_2^2 \leq \ldots \leq \omega_n^2. \tag{4.95}$$

The smallest value of ω_i (i.e., ω_1) is referred to as the *fundamental undamped natural frequency of vibration*, the second smallest value of ω_i (i.e., ω_2) is called the *second undamped natural frequency of vibration*, and so forth. The quantity ω_i^2 is called the ith eigenvalue of the matrix $[-\mathbf{M}\omega^2 + \mathbf{K}]$. Each natural frequency of the system has a corresponding eigenvector, which is denoted by $\boldsymbol{\phi}_i$, and ω_i and $\boldsymbol{\phi}_i$ are related by the eigenvalue equation

$$\left[-\mathbf{M}\omega_i^2 + \mathbf{K}\right]\boldsymbol{\phi}_i = \mathbf{0}. \tag{4.96}$$

In structural engineering, the eigenvectors are called *mode shapes of vibration* of the system. These mode shapes take the form

$$\mathbf{\phi}_i = \{\varphi_{1i} \quad \varphi_{2i} \quad \cdots \quad \varphi_{ni}\}^T. \tag{4.97}$$

The n solutions of ω^2 previously computed show that the response of the structure can be represented as the combinations of n mode shapes. To see this, consider Eq. 4.89 for each natural frequency to be written as

$$\mathbf{X}(t) = \sum_{i=1}^{n} \left[\mathbf{A}_i \cos \omega_i t + \mathbf{B}_i \sin \omega_i t\right]. \tag{4.98}$$

When Eq. 4.98 is substituted into Eq. 4.89, it follows that a summation of terms like Eq. 4.91 results:

$$\sum_{i=1}^{n} \left[-\mathbf{M}\omega_i^2 + \mathbf{K}\right]\left[\mathbf{A}_i \cos \omega_i t + \mathbf{B}_i \sin \omega_i t\right] = \mathbf{0}. \tag{4.99}$$

Therefore, to obtain a nontrivial solution, it must follow that

$$\left[-\mathbf{M}\omega_i^2 + \mathbf{K}\right]\mathbf{A}_i = \mathbf{0}, \quad \left[-\mathbf{M}\omega_i^2 + \mathbf{K}\right]\mathbf{B}_i = \mathbf{0}. \tag{4.100}$$

Comparing Eq. 4.96 and Eq. 4.100 gives the solution for $\mathbf{A}$ and $\mathbf{B}$, where

$$\mathbf{A}_i = a_i \mathbf{\phi}_i, \quad \mathbf{B}_i = b_i \mathbf{\phi}_i. \tag{4.101}$$

It follows from Eq. 4.98 that

$$\mathbf{X}(t) = \sum_{i=1}^{n} \mathbf{\phi}_i \left[a_i \cos \omega_i t + b_i \sin \omega_i t\right] = \sum_{i=1}^{n} \mathbf{\phi}_i q_i(t) \tag{4.102}$$

where

$$q_i(t) = a_i \cos \omega_i t + b_i \sin \omega_i t. \tag{4.103}$$

The values of a_i and b_i are computed based on the given initial conditions, and the general solution for these values is presented in Section 4.4, which involves the use of orthogonal property of mode shapes that will be discussed in Section 4.3.

Similar to the 2DOF system, the structural response is a linear combination of mode shapes of the structure. It is often convenient to represent the structural system response in the matrix equivalent to Eq. 4.102:

$$\mathbf{X}(t) = \mathbf{\phi}_1 q_1(t) + \mathbf{\phi}_2 q_2(t) + \ldots + \mathbf{\phi}_n q_n(t) = \mathbf{\Phi}\mathbf{Q}(t) \tag{4.104}$$

where

$$\mathbf{\Phi} = \begin{bmatrix} \mathbf{\phi}_1 & \mathbf{\phi}_2 & \cdots & \mathbf{\phi}_n \end{bmatrix} = \begin{bmatrix} \varphi_{11} & \varphi_{12} & \cdots & & \varphi_{1n} \\ \varphi_{21} & \varphi_{22} & \ddots & & \vdots \\ \vdots & \ddots & \ddots & & \varphi_{n-1,n} \\ \varphi_{n1} & & \cdots & \varphi_{n,n-1} & \varphi_{nn} \end{bmatrix}, \quad \mathbf{Q}(t) = \begin{Bmatrix} q_1(t) \\ q_2(t) \\ \vdots \\ q_n(t) \end{Bmatrix}.$$

Once again, calculation of all the natural frequencies and mode shapes may not be feasible. Consider that only k natural frequencies and mode shapes are obtained, where $k \leq n$; then Eq. 4.104 becomes

$$\mathbf{X}(t) = \mathbf{\phi}_1 q_1(t) + \mathbf{\phi}_2 q_2(t) + \ldots + \mathbf{\phi}_k q_k(t) = \mathbf{\Phi}\mathbf{Q}(t) \tag{4.105}$$

where

$$\mathbf{\Phi} = \begin{bmatrix} \mathbf{\phi}_1 & \mathbf{\phi}_2 & \cdots & \mathbf{\phi}_k \end{bmatrix} = \begin{bmatrix} \varphi_{11} & \varphi_{12} & \cdots & & \varphi_{1k} \\ \varphi_{21} & \varphi_{22} & \ddots & & \vdots \\ \vdots & \ddots & \ddots & & \varphi_{n-1,k} \\ \varphi_{n1} & & \cdots & \varphi_{n,k-1} & \varphi_{nk} \end{bmatrix}, \quad \mathbf{Q}(t) = \begin{Bmatrix} q_1(t) \\ q_2(t) \\ \vdots \\ q_k(t) \end{Bmatrix}.$$

Note that $\mathbf{\Phi}$ is an $n \times k$ matrix and is not a square matrix.

EXAMPLE 1   *Initial Conditions for a Three Degrees of Freedom (3DOF) System*

When all the mode shapes are obtained, the values of a_i and b_i in Eq. 4.103 can be determined by solving a set of n simultaneous equations with n unknowns. Consider Eq. 4.102 for a 3DOF system where

$$\mathbf{X}(t) = \sum_{i=1}^{3} \mathbf{\phi}_i \big[a_i \cos \omega_i t + b_i \sin \omega_i t \big]. \tag{4.106}$$

Taking a derivative of Eq. 4.106 with respect to time, it follows that

$$\dot{\mathbf{X}}(t) = \sum_{i=1}^{3} \mathbf{\phi}_i \big[b_i \omega_i \cos \omega_i t - a_i \omega_i \sin \omega_i t \big]. \tag{4.107}$$

At $t = 0$, it follows that

$$\mathbf{X}(0) = \mathbf{X}_o = a_1 \mathbf{\phi}_1 + a_2 \mathbf{\phi}_2 + a_3 \mathbf{\phi}_3 \tag{4.108}$$

$$\dot{\mathbf{X}}(0) = \dot{\mathbf{X}}_o = b_1 \omega_1 \mathbf{\phi}_1 + b_2 \omega_2 \mathbf{\phi}_2 + b_3 \omega_3 \mathbf{\phi}_3. \tag{4.109}$$

Recall that

$$\mathbf{\phi}_1 = \begin{Bmatrix} \varphi_{11} \\ \varphi_{21} \\ \varphi_{31} \end{Bmatrix}, \quad \mathbf{\phi}_2 = \begin{Bmatrix} \varphi_{12} \\ \varphi_{22} \\ \varphi_{32} \end{Bmatrix}, \quad \mathbf{\phi}_3 = \begin{Bmatrix} \varphi_{13} \\ \varphi_{23} \\ \varphi_{33} \end{Bmatrix}. \tag{4.110}$$

Therefore, it follows from Eq. 4.108 that

$$x_1(0) = \varphi_{11} a_1 + \varphi_{12} a_2 + \varphi_{13} a_3 \tag{4.111}$$

$$x_2(0) = \varphi_{21} a_1 + \varphi_{22} a_2 + \varphi_{23} a_3 \tag{4.112}$$

$$x_3(0) = \varphi_{31} a_1 + \varphi_{32} a_2 + \varphi_{33} a_3. \tag{4.113}$$

It is seen that a_1, a_2, and a_3 can be solved using Eqs. 4.111 to 4.113. The matrix solution is

$$\mathbf{X}(0) = \begin{Bmatrix} x_1(0) \\ x_2(0) \\ x_3(0) \end{Bmatrix} = \mathbf{\Phi} \begin{Bmatrix} a_1 \\ a_2 \\ a_3 \end{Bmatrix} \tag{4.114}$$

or

$$\begin{Bmatrix} a_1 \\ a_2 \\ a_3 \end{Bmatrix} = \mathbf{\Phi}^{-1} \begin{Bmatrix} x_1(0) \\ x_2(0) \\ x_3(0) \end{Bmatrix}. \tag{4.115}$$

Similarly, it can be shown that

$$\begin{Bmatrix} b_1 \omega_1 \\ b_2 \omega_2 \\ b_3 \omega_3 \end{Bmatrix} = \mathbf{\Phi}^{-1} \begin{Bmatrix} \dot{x}_1(0) \\ \dot{x}_2(0) \\ \dot{x}_3(0) \end{Bmatrix}. \tag{4.116}$$

This example shows that a_1 and b_1 can be calculated directly when all the mode shapes are determined. However, if not all mode shapes are determined, then the method presented here is not applicable because the matrix $\mathbf{\Phi}$ is not invertible. An alternate method will be presented in Section 4.4 using the orthogonal property of mode shapes.

EXAMPLE 2   *Response of a 3DOF System*

Consider a three-story building as shown in Figure 4.6a, with a simplified model as shown in Figure 4.6b. The dynamic matrices are

$$\mathbf{M} = \begin{bmatrix} m & 0 & 0 \\ 0 & m & 0 \\ 0 & 0 & m \end{bmatrix}, \quad \mathbf{K} = \begin{bmatrix} 3k & -k & 0 \\ -k & 2k & -k \\ 0 & -k & k \end{bmatrix} \qquad (4.117)$$

and therefore Eq. 4.94 becomes

$$\begin{vmatrix} -m\omega^2 + 3k & -k & 0 \\ -k & -m\omega^2 + 2k & -k \\ 0 & -k & -m\omega^2 + k \end{vmatrix} = 0. \qquad (4.118)$$

Let $\omega_n^2 = k/m$; if Eq. 4.118 is expressed as a cubic polynomial equation of ω^2, it follows that

$$2\omega_n^6 - 9\omega_n^4\omega^2 + 6\omega_n^2\omega^4 - \omega^6 = 0. \qquad (4.119)$$

Solving this cubic polynomial equation, the natural frequencies are

$$\omega_1^2 = \left(2 - \sqrt{3}\right)\omega_n^2, \quad \omega_2^2 = 2\omega_n^2, \quad \omega_3^2 = \left(2 + \sqrt{3}\right)\omega_n^2. \qquad (4.120)$$

The mode shape corresponding to each natural frequency is calculated using Eq. 4.96. First consider ω_1^2; it follows from Eq. 4.96 that

$$\omega_1^2: \quad \left[-\mathbf{M}\omega_1^2 + \mathbf{K}\right]\mathbf{\phi}_1 = \begin{bmatrix} \left(1 + \sqrt{3}\right)\omega_n^2 & -\omega_n^2 & 0 \\ -\omega_n^2 & \sqrt{3}\omega_n^2 & -\omega_n^2 \\ 0 & -\omega_n^2 & \left(-1 + \sqrt{3}\right)\omega_n^2 \end{bmatrix} \begin{Bmatrix} \varphi_{11} \\ \varphi_{12} \\ \varphi_{13} \end{Bmatrix} = \mathbf{0} \qquad (4.121)$$

and therefore

$$\mathbf{\phi}_1 = \begin{Bmatrix} \varphi_{11} \\ \varphi_{21} \\ \varphi_{31} \end{Bmatrix} = \begin{Bmatrix} 1 \\ 1 + \sqrt{3} \\ 2 + \sqrt{3} \end{Bmatrix}. \qquad (4.122)$$

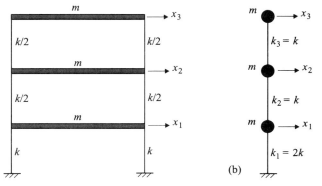

Figure 4.6 Three Degrees of Freedom System

Similarly for ω_2^2 and ω_3^2, it follows from Eq. 4.96 that

$$\omega_2^2: \quad \left[-\mathbf{M}\omega_2^2 + \mathbf{K}\right]\boldsymbol{\phi}_2 = \begin{bmatrix} \omega_n^2 & -\omega_n^2 & 0 \\ -\omega_n^2 & 0 & -\omega_n^2 \\ 0 & -\omega_n^2 & -\omega_n^2 \end{bmatrix}\begin{Bmatrix} \varphi_{21} \\ \varphi_{22} \\ \varphi_{23} \end{Bmatrix} = \mathbf{0} \tag{4.123}$$

$$\omega_3^2: \quad \left[-\mathbf{M}\omega_3^2 + \mathbf{K}\right]\boldsymbol{\phi}_3 = \begin{bmatrix} \left(1-\sqrt{3}\right)\omega_n^2 & -\omega_n^2 & 0 \\ -\omega_n^2 & -\sqrt{3}\omega_n^2 & -\omega_n^2 \\ 0 & -\omega_n^2 & \left(-1-\sqrt{3}\right)\omega_n^2 \end{bmatrix}\begin{Bmatrix} \varphi_{31} \\ \varphi_{32} \\ \varphi_{33} \end{Bmatrix} = \mathbf{0} \tag{4.124}$$

and therefore

$$\boldsymbol{\phi}_2 = \begin{Bmatrix} \varphi_{12} \\ \varphi_{22} \\ \varphi_{32} \end{Bmatrix} = \begin{Bmatrix} 1 \\ 1 \\ -1 \end{Bmatrix}, \quad \boldsymbol{\phi}_3 = \begin{Bmatrix} \varphi_{13} \\ \varphi_{23} \\ \varphi_{33} \end{Bmatrix} = \begin{Bmatrix} 1 \\ 1-\sqrt{3} \\ 2-\sqrt{3} \end{Bmatrix}. \tag{4.125}$$

These mode shapes are shown in Figure 4.7. It follows from these mode shapes that

$$\boldsymbol{\Phi} = \begin{bmatrix} \boldsymbol{\phi}_1 & \boldsymbol{\phi}_2 & \boldsymbol{\phi}_3 \end{bmatrix} = \begin{bmatrix} 1 & 1 & 1 \\ 1+\sqrt{3} & 1 & 1-\sqrt{3} \\ 2+\sqrt{3} & -1 & 2-\sqrt{3} \end{bmatrix}. \tag{4.126}$$

Note that

$$\boldsymbol{\Phi}^T\mathbf{M}\boldsymbol{\Phi} = \begin{bmatrix} \boldsymbol{\phi}_1 & \boldsymbol{\phi}_2 & \boldsymbol{\phi}_3 \end{bmatrix}^T \begin{bmatrix} m & 0 & 0 \\ 0 & m & 0 \\ 0 & 0 & m \end{bmatrix}\begin{bmatrix} \boldsymbol{\phi}_1 & \boldsymbol{\phi}_2 & \boldsymbol{\phi}_3 \end{bmatrix} = m\begin{bmatrix} 12+6\sqrt{3} & 0 & 0 \\ 0 & \sqrt{3} & 0 \\ 0 & 0 & 12-6\sqrt{3} \end{bmatrix} \tag{4.127}$$

$$\boldsymbol{\Phi}^T\mathbf{K}\boldsymbol{\Phi} = \begin{bmatrix} \boldsymbol{\phi}_1 & \boldsymbol{\phi}_2 & \boldsymbol{\phi}_3 \end{bmatrix}^T \begin{bmatrix} 3k & -k & 0 \\ -k & 2k & -k \\ 0 & -k & k \end{bmatrix}\begin{bmatrix} \boldsymbol{\phi}_1 & \boldsymbol{\phi}_2 & \boldsymbol{\phi}_3 \end{bmatrix} = \begin{bmatrix} 6k & 0 & 0 \\ 0 & 6k & 0 \\ 0 & 0 & 6k \end{bmatrix} \tag{4.128}$$

and therefore

$$\omega_1^2 = \frac{\boldsymbol{\phi}_1^T\mathbf{K}\boldsymbol{\phi}_1}{\boldsymbol{\phi}_1^T\mathbf{M}\boldsymbol{\phi}_1} = \frac{6k}{(12+6\sqrt{3})\,m} = \left(2-\sqrt{3}\right)\frac{k}{m} = \left(2-\sqrt{3}\right)\omega_n^2 \tag{4.129}$$

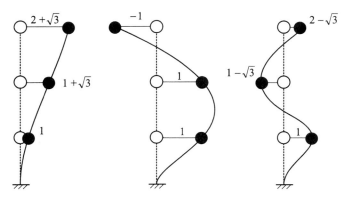

First Mode Shape   Second Mode Shape   Third Mode Shape

Figure 4.7 Mode Shapes of Vibration

$$\omega_2^2 = \frac{\boldsymbol{\phi}_2^T \mathbf{K} \boldsymbol{\phi}_2}{\boldsymbol{\phi}_2^T \mathbf{M} \boldsymbol{\phi}_2} = \frac{6k}{3m} = 2\frac{k}{m} = 2\omega_n^2 \tag{4.130}$$

$$\omega_3^2 = \frac{\boldsymbol{\phi}_3^T \mathbf{K} \boldsymbol{\phi}_3}{\boldsymbol{\phi}_3^T \mathbf{M} \boldsymbol{\phi}_3} = \frac{6k}{(12-6\sqrt{3})m} = \left(2+\sqrt{3}\right)\frac{k}{m} = \left(2+\sqrt{3}\right)\omega_n^2 \tag{4.131}$$

which are the same as obtained in Eq. 4.120. The calculation performed in Eqs. 4.129 to 4.131 is an excellent way to check the solution obtained using Eq. 4.120, and this can be done using a computer program. Finally, it follows from Eq. 4.104 that

$$\begin{Bmatrix} x_1(t) \\ x_2(t) \\ x_3(t) \end{Bmatrix} = \begin{Bmatrix} 1 \\ 1+\sqrt{3} \\ 2+\sqrt{3} \end{Bmatrix} (a_1 \cos\omega_1 t + b_1 \sin\omega_1 t)$$

$$+ \begin{Bmatrix} 1 \\ 1 \\ -1 \end{Bmatrix} (a_2 \cos\omega_2 t + b_2 \sin\omega_2 t) + \begin{Bmatrix} 1 \\ 1-\sqrt{3} \\ 2-\sqrt{3} \end{Bmatrix} (a_3 \cos\omega_3 t + b_3 \sin\omega_3 t) \tag{4.132}$$

where a_i and b_i are determined from the initial conditions using the procedure discussed in Example 1.

EXAMPLE 3   *Response of a 3DOF System Released from Rest*

Consider Example 2 with initial conditions

$$\begin{Bmatrix} x_1(0) \\ x_2(0) \\ x_3(0) \end{Bmatrix} = \begin{Bmatrix} 0 \\ 0 \\ 1 \end{Bmatrix}, \quad \begin{Bmatrix} \dot{x}_1(0) \\ \dot{x}_2(0) \\ \dot{x}_3(0) \end{Bmatrix} = \begin{Bmatrix} 0 \\ 0 \\ 0 \end{Bmatrix}. \tag{4.133}$$

Then from Example 1, Eq. 4.114 becomes

$$\boldsymbol{\Phi} \begin{Bmatrix} a_1 \\ a_2 \\ a_3 \end{Bmatrix} = \begin{bmatrix} 1 & 1 & 1 \\ 1+\sqrt{3} & 1 & 1-\sqrt{3} \\ 2+\sqrt{3} & -1 & 2-\sqrt{3} \end{bmatrix} \begin{Bmatrix} a_1 \\ a_2 \\ a_3 \end{Bmatrix} = \begin{Bmatrix} 0 \\ 0 \\ 1 \end{Bmatrix} \tag{4.134}$$

and solving for a_i gives

$$a_1 = \tfrac{1}{6}, \quad a_2 = -\tfrac{1}{3}, \quad a_3 = \tfrac{1}{6}. \tag{4.135}$$

Since the initial velocities are zero, it follows from Eq. 4.116 that

$$b_1 = b_2 = b_3 = 0. \tag{4.136}$$

Therefore, Eq. 4.132 becomes

$$\begin{Bmatrix} x_1(t) \\ x_2(t) \\ x_3(t) \end{Bmatrix} = \frac{1}{6} \begin{Bmatrix} 1 \\ 1+\sqrt{3} \\ 2+\sqrt{3} \end{Bmatrix} \cos\omega_1 t - \frac{1}{3} \begin{Bmatrix} 1 \\ 1 \\ -1 \end{Bmatrix} \cos\omega_2 t + \frac{1}{6} \begin{Bmatrix} 1 \\ 1-\sqrt{3} \\ 2-\sqrt{3} \end{Bmatrix} \cos\omega_3 t \tag{4.137}$$

where ω_1, ω_2, and ω_3 are given in Eq. 4.120.

EXAMPLE 4   *Response of a 5DOF System*

Consider the five-story building frame shown in Figure 4.8. The mass and stiffness matrices for this structure are

$$\mathbf{M} = \begin{bmatrix} m & 0 & 0 & 0 & 0 \\ 0 & m & 0 & 0 & 0 \\ 0 & 0 & m & 0 & 0 \\ 0 & 0 & 0 & m & 0 \\ 0 & 0 & 0 & 0 & m \end{bmatrix}, \quad \mathbf{K} = \begin{bmatrix} 2k & -k & 0 & 0 & 0 \\ -k & 2k & -k & 0 & 0 \\ 0 & -k & 2k & -k & 0 \\ 0 & 0 & -k & 2k & -k \\ 0 & 0 & 0 & -k & k \end{bmatrix}. \tag{4.138}$$

The five natural frequencies and mode shapes of vibration can be obtained using Eqs. 4.94 and 4.96:

$$\omega_1^2: \quad \mathbf{\Phi}_1 = \{\varphi_{11} \quad \varphi_{21} \quad \varphi_{31} \quad \varphi_{41} \quad \varphi_{51}\}^T \tag{4.139}$$

$$\omega_2^2: \quad \mathbf{\Phi}_2 = \{\varphi_{12} \quad \varphi_{22} \quad \varphi_{32} \quad \varphi_{42} \quad \varphi_{52}\}^T \tag{4.140}$$

$$\omega_3^2: \quad \mathbf{\Phi}_3 = \{\varphi_{13} \quad \varphi_{23} \quad \varphi_{33} \quad \varphi_{43} \quad \varphi_{53}\}^T \tag{4.141}$$

$$\omega_4^2: \quad \mathbf{\Phi}_4 = \{\varphi_{14} \quad \varphi_{24} \quad \varphi_{34} \quad \varphi_{44} \quad \varphi_{54}\}^T \tag{4.142}$$

$$\omega_5^2: \quad \mathbf{\Phi}_5 = \{\varphi_{15} \quad \varphi_{25} \quad \varphi_{35} \quad \varphi_{45} \quad \varphi_{55}\}^T. \tag{4.143}$$

The response at the top floor is equal to

$$x_5(t) = \varphi_{51}q_1(t) + \varphi_{52}q_2(t) + \varphi_{53}q_3(t) + \varphi_{54}q_4(t) + \varphi_{55}q_5(t). \tag{4.144}$$

The *normal mode truncation* method is a method of analysis with which the response is expressed in terms of fewer than n-degree modal coordinates. For example, in Eq. 4.144 the exact solution is obtained by using all five modes, and the approximate normal mode truncation solution can be computed using four or fewer modes. For example, if only the first three modes are used, then Eq. 4.144 becomes

$$x_5(t) = \varphi_{51}q_1(t) + \varphi_{52}q_2(t) + \varphi_{53}q_3(t). \tag{4.145}$$

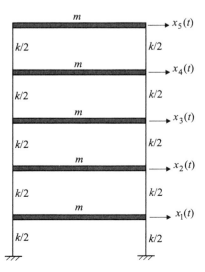

Figure 4.8 Five Story Building Frame

EXAMPLE 5   *Decoupling of a 3DOF System to a 2DOF System and an SDOF System*

Consider a one-story structure with plan view shown in Figure 4.9 and with the mass and stiffness matrices equal to

$$\mathbf{M} = \begin{bmatrix} \frac{2}{3}m & \frac{1}{3}m & 0 \\ \frac{1}{3}m & \frac{2}{3}m & 0 \\ 0 & 0 & m_t \end{bmatrix}, \quad \mathbf{K} = \begin{bmatrix} k & 0 & 0 \\ 0 & k & 0 \\ 0 & 0 & k_t \end{bmatrix}. \tag{4.146}$$

It follows from Eq. 4.89 that

$$\begin{bmatrix} \frac{2}{3}m & \frac{1}{3}m & 0 \\ \frac{1}{3}m & \frac{2}{3}m & 0 \\ 0 & 0 & m_t \end{bmatrix}\begin{Bmatrix} \ddot{x}_1(t) \\ \ddot{x}_2(t) \\ \ddot{x}_3(t) \end{Bmatrix} + \begin{bmatrix} k & 0 & 0 \\ 0 & k & 0 \\ 0 & 0 & k_t \end{bmatrix}\begin{Bmatrix} x_1(t) \\ x_2(t) \\ x_3(t) \end{Bmatrix} = \begin{Bmatrix} 0 \\ 0 \\ 0 \end{Bmatrix}. \tag{4.147}$$

Note that the third equation has already been decoupled from the first two equations, and therefore Eq. 4.147 can be written as

$$\begin{bmatrix} \frac{2}{3}m & \frac{1}{3}m \\ \frac{1}{3}m & \frac{2}{3}m \end{bmatrix}\begin{Bmatrix} \ddot{x}_1(t) \\ \ddot{x}_2(t) \end{Bmatrix} + \begin{bmatrix} k & 0 \\ 0 & k \end{bmatrix}\begin{Bmatrix} x_1(t) \\ x_2(t) \end{Bmatrix} = \begin{Bmatrix} 0 \\ 0 \end{Bmatrix} \tag{4.148}$$

$$m_t\ddot{x}_3(t) + k_t x_3(t) = 0. \tag{4.149}$$

Thus, the original 3DOF system is reduced to a 2DOF system as shown in Eq. 4.148 and a SDOF system as shown in Eq. 4.149.

The response can be calculated using the normal mode method. First consider Eq. 4.148; it follows from Eq. 4.94 that

$$\left| -\mathbf{M}\omega^2 + \mathbf{K} \right| = \begin{vmatrix} -\frac{2}{3}m\omega^2 + k & -\frac{1}{3}m\omega^2 \\ -\frac{1}{3}m\omega^2 & -\frac{2}{3}m\omega^2 + k \end{vmatrix} = 0. \tag{4.150}$$

Denote $\omega_n^2 = k/m$; expanding Eq. 4.150 gives

$$\left(-\frac{2}{3}\omega^2 + \omega_n^2 \right)^2 - \left(-\frac{1}{3}\omega^2 \right)^2 = 0. \tag{4.151}$$

Factoring Eq. 4.151 gives

$$\left(-\frac{2}{3}\omega^2 + \omega_n^2 + \frac{1}{3}\omega^2 \right)\left(-\frac{2}{3}\omega^2 + \omega_n^2 - \frac{1}{3}\omega^2 \right) = 0. \tag{4.152}$$

Solving for Eq. 4.152, it follows that

$$\omega_1^2 = \omega_n^2, \quad \omega_2^2 = 3\omega_n^2. \tag{4.153}$$

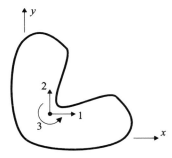

Figure 4.9 Plan View of a One-Story Structure with 3 Degrees of Freedom

The mode shapes are then computed using Eq. 4.96, which is

$$\omega_1^2: \quad \left[-\mathbf{M}\omega_1^2 + \mathbf{K}\right]\boldsymbol{\phi}_1 = m\begin{bmatrix} \frac{1}{3}\omega_n^2 & -\frac{1}{3}\omega_n^2 \\ -\frac{1}{3}\omega_n^2 & \frac{1}{3}\omega_n^2 \end{bmatrix}\begin{Bmatrix} \varphi_{11} \\ \varphi_{12} \end{Bmatrix} = \mathbf{0} \tag{4.154}$$

$$\omega_2^2: \quad \left[-\mathbf{M}\omega_2^2 + \mathbf{K}\right]\boldsymbol{\phi}_2 = m\begin{bmatrix} -\omega_n^2 & -\omega_n^2 \\ -\omega_n^2 & -\omega_n^2 \end{bmatrix}\begin{Bmatrix} \varphi_{21} \\ \varphi_{22} \end{Bmatrix} = \mathbf{0}. \tag{4.155}$$

Therefore,

$$\begin{Bmatrix} \varphi_{11} \\ \varphi_{12} \end{Bmatrix} = \begin{Bmatrix} 1 \\ 1 \end{Bmatrix}, \quad \begin{Bmatrix} \varphi_{21} \\ \varphi_{22} \end{Bmatrix} = \begin{Bmatrix} 1 \\ -1 \end{Bmatrix}. \tag{4.156}$$

Now consider Eq. 4.149, and denote $\omega_t^2 = k_t/m_t$. Because this is an SDOF system, the result can be incorporated directly into the results obtained in Eqs. 4.153 and 4.156:

$$\omega_1^2 = \omega_n^2, \quad \omega_2^2 = 3\omega_n^2, \quad \omega_3^2 = \omega_t^2 \tag{4.157}$$

and

$$\begin{Bmatrix} \varphi_{11} \\ \varphi_{21} \\ \varphi_{31} \end{Bmatrix} = \begin{Bmatrix} 1 \\ 1 \\ 0 \end{Bmatrix}, \quad \begin{Bmatrix} \varphi_{12} \\ \varphi_{22} \\ \varphi_{32} \end{Bmatrix} = \begin{Bmatrix} 1 \\ -1 \\ 0 \end{Bmatrix}, \quad \begin{Bmatrix} \varphi_{13} \\ \varphi_{23} \\ \varphi_{33} \end{Bmatrix} = \begin{Bmatrix} 0 \\ 0 \\ 1 \end{Bmatrix}. \tag{4.158}$$

These mode shapes are shown in Figure 4.10.

Let the initial conditions take a general form; it follows from Eq. 4.158 that

$$\boldsymbol{\Phi} = \begin{bmatrix} 1 & 1 & 0 \\ 1 & -1 & 0 \\ 0 & 0 & 1 \end{bmatrix}, \quad \boldsymbol{\Phi}^{-1} = \begin{bmatrix} 1/2 & 1/2 & 0 \\ 1/2 & -1/2 & 0 \\ 0 & 0 & 1 \end{bmatrix} \tag{4.159}$$

and therefore

$$\begin{Bmatrix} a_1 \\ a_2 \\ a_3 \end{Bmatrix} = \begin{bmatrix} 1/2 & 1/2 & 0 \\ 1/2 & -1/2 & 0 \\ 0 & 0 & 1 \end{bmatrix}\begin{Bmatrix} x_{1o} \\ x_{2o} \\ x_{3o} \end{Bmatrix} = \begin{Bmatrix} \frac{1}{2}x_{1o} + \frac{1}{2}x_{2o} \\ \frac{1}{2}x_{1o} - \frac{1}{2}x_{2o} \\ x_{3o} \end{Bmatrix} \tag{4.160}$$

$$\begin{Bmatrix} b_1\omega_1 \\ b_2\omega_2 \\ b_3\omega_3 \end{Bmatrix} = \begin{bmatrix} 1/2 & 1/2 & 0 \\ 1/2 & -1/2 & 0 \\ 0 & 0 & 1 \end{bmatrix}\begin{Bmatrix} \dot{x}_{1o} \\ \dot{x}_{2o} \\ \dot{x}_{3o} \end{Bmatrix} = \begin{Bmatrix} \frac{1}{2}\dot{x}_{1o} + \frac{1}{2}\dot{x}_{2o} \\ \frac{1}{2}\dot{x}_{1o} - \frac{1}{2}\dot{x}_{2o} \\ \dot{x}_{3o} \end{Bmatrix}. \tag{4.161}$$

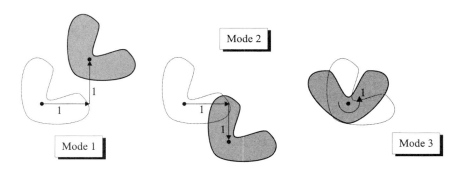

Figure 4.10 Mode Shapes

The general solution for any combination of initial conditions becomes

$$\begin{Bmatrix} x_1(t) \\ x_2(t) \\ x_3(t) \end{Bmatrix} = \begin{Bmatrix} 1 \\ 1 \\ 0 \end{Bmatrix} \left[\left(\frac{x_{1o} + x_{2o}}{2} \right) \cos \omega_n t + \left(\frac{\dot{x}_{1o} + \dot{x}_{2o}}{2\omega_n} \right) \sin \omega_n t \right]$$

$$+ \begin{Bmatrix} 1 \\ -1 \\ 0 \end{Bmatrix} \left[\left(\frac{x_{1o} - x_{2o}}{2} \right) \cos \sqrt{3}\omega_n t + \left(\frac{\dot{x}_{1o} - \dot{x}_{2o}}{2\sqrt{3}\omega_n} \right) \sin \sqrt{3}\omega_n t \right] + \begin{Bmatrix} 0 \\ 0 \\ 1 \end{Bmatrix} \left[x_{3o} \cos \omega_t t + \left(\frac{\dot{x}_{3o}}{\omega_t} \right) \sin \omega_t t \right].$$

4.4 ORTHOGONALITY OF MODE SHAPES

Natural frequencies of vibration, ω_1^2, ω_2^2, ..., ω_n^2, and the mode shapes, $\boldsymbol{\phi}_1$, $\boldsymbol{\phi}_2$, ..., $\boldsymbol{\phi}_n$, were calculated in Section 4.3, and they satisfy the eigenvalue equation

$$\left[-\mathbf{M}\omega_i^2 + \mathbf{K} \right] \boldsymbol{\phi}_i = \mathbf{0} \tag{4.162}$$

where

$$\boldsymbol{\phi}_i = \left\{ \varphi_{1i} \quad \varphi_{2i} \quad \cdots \quad \varphi_{ni} \right\}^T.$$

The *norm* (or *length*) of any mode shape is defined as

$$\left\| \boldsymbol{\phi}_i \right\| = \sqrt{\left(\varphi_{1i} \right)^2 + \left(\varphi_{2i} \right)^2 + \ldots + \left(\varphi_{ni} \right)^2}. \tag{4.163}$$

Because each mode shape has an arbitrary scaling factor, each mode shape can always be normalized such that its norm is equal to 1. This type of normalization of the mode shape is called *norm normalization* or *length normalization*. Denote the normalized mode shapes that have a unit norm as $\hat{\boldsymbol{\phi}}_1$, $\hat{\boldsymbol{\phi}}_2$, ..., $\hat{\boldsymbol{\phi}}_n$. It then follows that

$$\hat{\boldsymbol{\phi}}_i = \frac{1}{\left\| \boldsymbol{\phi}_i \right\|} \boldsymbol{\phi}_i = \left\{ \frac{\varphi_{1i}}{\left\| \boldsymbol{\phi}_i \right\|} \quad \frac{\varphi_{2i}}{\left\| \boldsymbol{\phi}_i \right\|} \quad \cdots \quad \frac{\varphi_{ni}}{\left\| \boldsymbol{\phi}_i \right\|} \right\}^T = \left\{ \hat{\varphi}_{1i} \quad \hat{\varphi}_{2i} \quad \cdots \quad \hat{\varphi}_{ni} \right\}^T. \tag{4.164}$$

One convenience with the mode shape normalized as defined by Eq. 4.164 is that the size of the elements in the mode shape do not become large and thus do not result in numerical ill-conditioning when used in other matrix equations. If Eq. 4.162 is divided by the norm of the mode shape $\boldsymbol{\phi}_i$, then it follows that

$$\frac{1}{\left\| \boldsymbol{\phi}_i \right\|} \left[\mathbf{M}\omega_i^2 - \mathbf{K} \right] \boldsymbol{\phi}_i = \left[\mathbf{M}\omega_i^2 - \mathbf{K} \right] \hat{\boldsymbol{\phi}}_i = \mathbf{0}. \tag{4.165}$$

Now, consider the natural frequencies of vibration, ω_i and ω_j, and their corresponding mode shapes, $\hat{\boldsymbol{\phi}}_i$ and $\hat{\boldsymbol{\phi}}_j$. Equation 4.162 can be written for each natural frequency, and it follows that

$$\left[\mathbf{M}\omega_i^2 - \mathbf{K} \right] \hat{\boldsymbol{\phi}}_i = \mathbf{0} \tag{4.166a}$$

$$\left[\mathbf{M}\omega_j^2 - \mathbf{K} \right] \hat{\boldsymbol{\phi}}_j = \mathbf{0}. \tag{4.166b}$$

Premultiplying Eq. 4.166a by $\hat{\boldsymbol{\phi}}_j^T$ and Eq. 4.166b by $\hat{\boldsymbol{\phi}}_i^T$ gives

$$\hat{\boldsymbol{\phi}}_j^T \left[\mathbf{M}\omega_i^2 - \mathbf{K} \right] \hat{\boldsymbol{\phi}}_i = 0 \tag{4.167a}$$

$$\hat{\boldsymbol{\phi}}_i^T \left[\mathbf{M}\omega_j^2 - \mathbf{K} \right] \hat{\boldsymbol{\phi}}_j = \mathbf{0}. \tag{4.167b}$$

Since both $\mathbf{M}$ and $\mathbf{K}$ are symmetrical matrices, it follows that

$$\hat{\boldsymbol{\phi}}_j^T \mathbf{M} \hat{\boldsymbol{\phi}}_i = \hat{\boldsymbol{\phi}}_i^T \mathbf{M} \hat{\boldsymbol{\phi}}_j, \quad \hat{\boldsymbol{\phi}}_j^T \mathbf{K} \hat{\boldsymbol{\phi}}_i = \hat{\boldsymbol{\phi}}_i^T \mathbf{K} \hat{\boldsymbol{\phi}}_j. \tag{4.168}$$

Subtracting Eq. 4.167b from Eq. 4.167a and using the property of the mass and stiffness matrices in Eq. 4.168, it follows that

$$\hat{\boldsymbol{\phi}}_j^T \mathbf{M} \hat{\boldsymbol{\phi}}_i \left(\omega_i^2 - \omega_j^2 \right) = 0. \tag{4.169}$$

For any $\omega_i \ne \omega_j$, it follows that $\hat{\boldsymbol{\phi}}_j^T \mathbf{M} \hat{\boldsymbol{\phi}}_i = 0$. This is known as the *orthogonal property* of two mode shapes with respect to the mass matrix. A similar orthogonal property of two mode shapes with respect to the stiffness matrix can be derived by first dividing Eq. 4.167a by ω_i^2 and Eq. 4.167b by ω_j^2 and then subtracting the two equations. It follows that

$$\hat{\boldsymbol{\phi}}_j^T \left[-\mathbf{M} + \frac{1}{\omega_i^2} \mathbf{K} \right] \hat{\boldsymbol{\phi}}_i - \hat{\boldsymbol{\phi}}_i^T \left[-\mathbf{M} + \frac{1}{\omega_j^2} \mathbf{K} \right] \hat{\boldsymbol{\phi}}_j = 0. \tag{4.170}$$

From Eq. 4.168, it follows that $\hat{\boldsymbol{\phi}}_j^T \mathbf{K} \hat{\boldsymbol{\phi}}_i = 0$ for any $\omega_i \ne \omega_j$. In summary, the orthogonal property for the system mass and stiffness matrices states that

$$\hat{\boldsymbol{\phi}}_i^T \mathbf{M} \hat{\boldsymbol{\phi}}_j = \hat{\boldsymbol{\phi}}_j^T \mathbf{M} \hat{\boldsymbol{\phi}}_i = \begin{cases} \hat{m} & i = j \\ 0 & i \ne j, \end{cases} \quad \hat{\boldsymbol{\phi}}_i^T \mathbf{K} \hat{\boldsymbol{\phi}}_j = \hat{\boldsymbol{\phi}}_j^T \mathbf{K} \hat{\boldsymbol{\phi}}_i = \begin{cases} \hat{k} & i = j \\ 0 & i \ne j. \end{cases} \tag{4.171}$$

Define the modal matrix of norm normalized mode shapes to be

$$\hat{\boldsymbol{\Phi}} = \begin{bmatrix} \hat{\boldsymbol{\phi}}_1 & \hat{\boldsymbol{\phi}}_2 & \cdots & \hat{\boldsymbol{\phi}}_n \end{bmatrix} = \begin{bmatrix} \hat{\varphi}_{11} & \hat{\varphi}_{12} & \cdots & & \hat{\varphi}_{1n} \\ \hat{\varphi}_{21} & \hat{\varphi}_{22} & \ddots & & \vdots \\ \vdots & \ddots & \ddots & & \hat{\varphi}_{n-1,n} \\ \hat{\varphi}_{n1} & \cdots & & \hat{\varphi}_{n,n-1} & \hat{\varphi}_{nn} \end{bmatrix} \tag{4.172}$$

and it follows that

$$\hat{\mathbf{M}} = \hat{\boldsymbol{\Phi}}^T \mathbf{M} \hat{\boldsymbol{\Phi}} = \begin{bmatrix} \hat{m}_1 & 0 & \cdots & 0 \\ 0 & \hat{m}_2 & \ddots & \vdots \\ \vdots & \ddots & \ddots & 0 \\ 0 & \cdots & 0 & \hat{m}_n \end{bmatrix}, \quad \hat{\mathbf{K}} = \hat{\boldsymbol{\Phi}}^T \mathbf{K} \hat{\boldsymbol{\Phi}} = \begin{bmatrix} \hat{k}_1 & 0 & \cdots & 0 \\ 0 & \hat{k}_2 & \ddots & \vdots \\ \vdots & \ddots & \ddots & 0 \\ 0 & \cdots & 0 & \hat{k}_n \end{bmatrix}. \tag{4.173}$$

These orthogonality relationships were derived using the normalized mode shapes as defined in Eq. 4.164. However, it is important to note that these relationships exist for any normalization of the mode shapes.

When only k natural frequencies and mode shapes are obtained, where $k \le n$, as discussed in Eq. 4.105, then Eq. 4.173 becomes

$$\hat{\mathbf{M}} = \hat{\boldsymbol{\Phi}}^T \mathbf{M} \hat{\boldsymbol{\Phi}} = \begin{bmatrix} \hat{m}_1 & 0 & \cdots & 0 \\ 0 & \hat{m}_2 & \ddots & \vdots \\ \vdots & \ddots & \ddots & 0 \\ 0 & \cdots & 0 & \hat{m}_k \end{bmatrix}, \quad \hat{\mathbf{K}} = \hat{\boldsymbol{\Phi}}^T \mathbf{K} \hat{\boldsymbol{\Phi}} = \begin{bmatrix} \hat{k}_1 & 0 & \cdots & 0 \\ 0 & \hat{k}_2 & \ddots & \vdots \\ \vdots & \ddots & \ddots & 0 \\ 0 & \cdots & 0 & \hat{k}_k \end{bmatrix} \tag{4.174}$$

where both $\hat{\mathbf{M}}$ and $\hat{\mathbf{K}}$ are $k \times k$ matrices.

Another way to express the mode shapes, called *mass normalization*, is to normalize these mode shapes with respect to the mass matrix. The mass-normalized mode shapes, denoted $\overline{\boldsymbol{\phi}}_i$, are normalized such that $\overline{\boldsymbol{\phi}}_i^T \mathbf{M} \overline{\boldsymbol{\phi}}_i = 1$, where

$$\overline{\boldsymbol{\phi}}_i = \frac{1}{C_i} \boldsymbol{\phi}_i = \left\{ \frac{\varphi_{1i}}{C_i} \quad \frac{\varphi_{2i}}{C_i} \quad \cdots \quad \frac{\varphi_{ni}}{C_i} \right\}^T = \left\{ \overline{\varphi}_{1i} \quad \overline{\varphi}_{2i} \quad \cdots \quad \overline{\varphi}_{ni} \right\}^T \tag{4.175}$$

and the constant C_i is a function of the mass matrix. It follows that

$$\overline{\boldsymbol{\phi}}_i = \frac{1}{C_i}\boldsymbol{\phi}_i = \frac{\|\boldsymbol{\phi}_i\|}{C_i}\hat{\boldsymbol{\phi}}_i. \tag{4.176}$$

Following the same steps as used in the derivation for the orthogonality of $\hat{\boldsymbol{\phi}}_i$ from Eq. 4.166 to Eq. 4.171, the orthogonality of $\overline{\boldsymbol{\phi}}_i$'s can also be shown:

$$\overline{\boldsymbol{\phi}}_j^T \mathbf{M}\,\overline{\boldsymbol{\phi}}_i\left(\omega_i^2 - \omega_j^2\right) = 0. \tag{4.177}$$

Therefore,

$$\overline{\boldsymbol{\phi}}_i^T \mathbf{M}\overline{\boldsymbol{\phi}}_j = \overline{\boldsymbol{\phi}}_j^T \mathbf{M}\overline{\boldsymbol{\phi}}_i = \begin{cases} 1 & i = j \\ 0 & i \neq j. \end{cases} \tag{4.178}$$

An important advantage of expressing the mode shape using the mass normalization method can be seen by starting with Eq. 4.162, where

$$\left[\mathbf{M}\omega_i^2 - \mathbf{K}\right]\overline{\boldsymbol{\phi}}_i = \left[\mathbf{M}\omega_i^2 - \mathbf{K}\right]\frac{\boldsymbol{\phi}_i}{C} = \mathbf{0}. \tag{4.179}$$

Premultiplying Eq. 4.179 by $\overline{\boldsymbol{\phi}}_j^T$ gives

$$\overline{\boldsymbol{\phi}}_j^T\left[\mathbf{M}\omega_i^2 - \mathbf{K}\right]\overline{\boldsymbol{\phi}}_i = \mathbf{0}. \tag{4.180}$$

Expanding Eq. 4.180, it follows that

$$\omega_i^2\left(\overline{\boldsymbol{\phi}}_j^T\mathbf{M}\overline{\boldsymbol{\phi}}_i\right) - \overline{\boldsymbol{\phi}}_j^T\mathbf{K}\,\overline{\boldsymbol{\phi}}_i = \mathbf{0}. \tag{4.181}$$

Note that the term in parentheses of Eq. 4.181 follows the results given in Eq. 4.178. Therefore, Eq. 4.181 becomes

$$\overline{\boldsymbol{\phi}}_j^T\mathbf{K}\overline{\boldsymbol{\phi}}_i = \begin{cases} \omega_i^2 & i = j \\ 0 & i \neq j. \end{cases} \tag{4.182}$$

In summary,

$$\overline{\boldsymbol{\phi}}_i^T\mathbf{M}\overline{\boldsymbol{\phi}}_j = \overline{\boldsymbol{\phi}}_j^T\mathbf{M}\overline{\boldsymbol{\phi}}_i = \begin{cases} 1 & i = j \\ 0 & i \neq j, \end{cases} \quad \overline{\boldsymbol{\phi}}_i^T\mathbf{K}\overline{\boldsymbol{\phi}}_j = \overline{\boldsymbol{\phi}}_j^T\mathbf{K}\overline{\boldsymbol{\phi}}_i = \begin{cases} \omega_i^2 & i = j \\ 0 & i \neq j. \end{cases} \tag{4.183}$$

Define the mass normalized modal matrix as

$$\overline{\boldsymbol{\Phi}} = \begin{bmatrix} \overline{\boldsymbol{\phi}}_1 & \overline{\boldsymbol{\phi}}_2 & \cdots & \overline{\boldsymbol{\phi}}_n \end{bmatrix} = \begin{bmatrix} \overline{\varphi}_{11} & \overline{\varphi}_{12} & \cdots & \overline{\varphi}_{1n} \\ \overline{\varphi}_{21} & \overline{\varphi}_{22} & \ddots & \vdots \\ \vdots & \ddots & \ddots & \overline{\varphi}_{n-1,n} \\ \overline{\varphi}_{n1} & \cdots & \overline{\varphi}_{n,n-1} & \overline{\varphi}_{nn} \end{bmatrix} \tag{4.184}$$

and then it follows that

$$\overline{\mathbf{M}} = \overline{\boldsymbol{\Phi}}^T\mathbf{M}\overline{\boldsymbol{\Phi}} = \begin{bmatrix} 1 & 0 & \cdots & 0 \\ 0 & 1 & \ddots & \vdots \\ \vdots & \ddots & \ddots & 0 \\ 0 & \cdots & 0 & 1 \end{bmatrix} = \mathbf{I}, \quad \overline{\mathbf{K}} = \overline{\boldsymbol{\Phi}}^T\mathbf{K}\overline{\boldsymbol{\Phi}} = \begin{bmatrix} \omega_1^2 & 0 & \cdots & 0 \\ 0 & \omega_2^2 & \ddots & \vdots \\ \vdots & \ddots & \ddots & 0 \\ 0 & \cdots & 0 & \omega_n^2 \end{bmatrix}. \tag{4.185}$$

Note from Eq. 4.185 that

$$\omega_i^2 = \overline{k}_{ii}/\overline{m}_{ii} \tag{4.186}$$

since $\overline{m}_{ii} = 1$ and $\overline{k}_{ii} = \omega_i^2$. This is an important advantage, because it simplifies checking the accuracy of matrix multiplications and often enables the structural engineer to directly write $\mathbf{M}$ and $\mathbf{K}$ once the natural frequency and mode shapes are calculated and mass normalized.

Finally, when only k natural frequencies and mode shapes are obtained, where $k \leq n$, as discussed in Eq. 4.105, then Eq. 4.185 becomes

$$\overline{\mathbf{M}} = \overline{\mathbf{\Phi}}^T \mathbf{M} \overline{\mathbf{\Phi}} = \begin{bmatrix} 1 & 0 & \cdots & 0 \\ 0 & 1 & \ddots & \vdots \\ \vdots & \ddots & \ddots & 0 \\ 0 & \cdots & 0 & 1 \end{bmatrix} = \mathbf{I}, \quad \overline{\mathbf{K}} = \overline{\mathbf{\Phi}}^T \mathbf{K} \overline{\mathbf{\Phi}} = \begin{bmatrix} \omega_1^2 & 0 & \cdots & 0 \\ 0 & \omega_2^2 & \ddots & \vdots \\ \vdots & \ddots & \ddots & 0 \\ 0 & \cdots & 0 & \omega_k^2 \end{bmatrix} \tag{4.187}$$

where both $\overline{\mathbf{M}}$ and $\overline{\mathbf{K}}$ are $k \times k$ matrices.

EXAMPLE 1   *Norm Normalization of Mode Shapes in a 3DOF System*

Consider Example 2 in Section 4.3, where the mode shapes are

$$\phi_1 = \begin{Bmatrix} 1 \\ 1 + \sqrt{3} \\ 2 + \sqrt{3} \end{Bmatrix}, \quad \phi_2 = \begin{Bmatrix} 1 \\ 1 \\ -1 \end{Bmatrix}, \quad \phi_3 = \begin{Bmatrix} 1 \\ 1 - \sqrt{3} \\ 2 - \sqrt{3} \end{Bmatrix}. \tag{4.188}$$

The norm of each mode shape is

$$\|\phi_1\| = \sqrt{1^2 + \left(1 + \sqrt{3}\right)^2 + \left(2 + \sqrt{3}\right)^2} = 3 + \sqrt{3} \tag{4.189}$$

$$\|\phi_2\| = \sqrt{1^2 + 1^2 + (-1)^2} = \sqrt{3} \tag{4.190}$$

$$\|\phi_3\| = \sqrt{1^2 + \left(1 - \sqrt{3}\right)^2 + \left(2 - \sqrt{3}\right)^2} = 3 - \sqrt{3}. \tag{4.191}$$

Therefore, the norm normalization follows from Eq. 4.164 and

$$\hat{\phi}_1 = \frac{1}{3 + \sqrt{3}} \begin{Bmatrix} 1 \\ 1 + \sqrt{3} \\ 2 + \sqrt{3} \end{Bmatrix}, \quad \hat{\phi}_2 = \frac{1}{\sqrt{3}} \begin{Bmatrix} 1 \\ 1 \\ -1 \end{Bmatrix}, \quad \hat{\phi}_3 = \frac{1}{3 - \sqrt{3}} \begin{Bmatrix} 1 \\ 1 - \sqrt{3} \\ 2 - \sqrt{3} \end{Bmatrix}. \tag{4.192}$$

Note that

$$\hat{\mathbf{\Phi}}^T \mathbf{M} \hat{\mathbf{\Phi}} = \begin{bmatrix} \hat{\phi}_1 & \hat{\phi}_2 & \hat{\phi}_3 \end{bmatrix}^T \begin{bmatrix} m & 0 & 0 \\ 0 & m & 0 \\ 0 & 0 & m \end{bmatrix} \begin{bmatrix} \hat{\phi}_1 & \hat{\phi}_2 & \hat{\phi}_3 \end{bmatrix} = \begin{bmatrix} m & 0 & 0 \\ 0 & m & 0 \\ 0 & 0 & m \end{bmatrix} \tag{4.193}$$

$$\hat{\mathbf{\Phi}}^T \mathbf{K} \hat{\mathbf{\Phi}} = \begin{bmatrix} \hat{\phi}_1 & \hat{\phi}_2 & \hat{\phi}_3 \end{bmatrix}^T \begin{bmatrix} 3k & -k & 0 \\ -k & 2k & -k \\ 0 & -k & k \end{bmatrix} \begin{bmatrix} \hat{\phi}_1 & \hat{\phi}_2 & \hat{\phi}_3 \end{bmatrix} = k \begin{bmatrix} 2 - \sqrt{3} & 0 & 0 \\ 0 & 2 & 0 \\ 0 & 0 & 2 + \sqrt{3} \end{bmatrix} \tag{4.194}$$

and therefore

$$\omega_1^2 = \frac{\hat{\phi}_1^T \mathbf{K} \hat{\phi}_1}{\hat{\phi}_1^T \mathbf{M} \hat{\phi}_1} = \frac{\left(2 - \sqrt{3}\right) k}{m} = \left(2 - \sqrt{3}\right) \omega_n^2 \tag{4.195}$$

$$\omega_2^2 = \frac{\hat{\phi}_2^T \mathbf{K} \hat{\phi}_2}{\hat{\phi}_2^T \mathbf{M} \hat{\phi}_2} = \frac{2k}{m} = 2\omega_n^2 \tag{4.196}$$

$$\omega_3^2 = \frac{\hat{\phi}_3^T \mathbf{K} \hat{\phi}_3}{\hat{\phi}_3^T \mathbf{M} \hat{\phi}_3} = \frac{\left(2+\sqrt{3}\right)k}{m} = \left(2+\sqrt{3}\right)\omega_n^2. \tag{4.197}$$

EXAMPLE 2 *Mass Normalization of Mode Shapes in a 3DOF System*

Consider again Example 2 in Section 4.3, where the mode shapes are

$$\phi_1 = \left\{ \begin{matrix} 1 \\ 1+\sqrt{3} \\ 2+\sqrt{3} \end{matrix} \right\}, \quad \phi_2 = \left\{ \begin{matrix} 1 \\ 1 \\ -1 \end{matrix} \right\}, \quad \phi_3 = \left\{ \begin{matrix} 1 \\ 1-\sqrt{3} \\ 2-\sqrt{3} \end{matrix} \right\}. \tag{4.198}$$

The objective is to normalize these mode shapes to have a unit modal mass. Using $\overline{\phi}_i^T \mathbf{M} \overline{\phi}_i = 1$ and Eq. 4.176, it follows that

$$\overline{\phi}_1^T \mathbf{M} \overline{\phi}_1 = \frac{1}{C_1^2} \left\{ \begin{matrix} 1 \\ 1+\sqrt{3} \\ 2+\sqrt{3} \end{matrix} \right\}^T \begin{bmatrix} m & 0 & 0 \\ 0 & m & 0 \\ 0 & 0 & m \end{bmatrix} \left\{ \begin{matrix} 1 \\ 1+\sqrt{3} \\ 2+\sqrt{3} \end{matrix} \right\} = \frac{12+6\sqrt{3}}{C_1^2} m = 1 \tag{4.199}$$

and therefore

$$C_1 = \left(3+\sqrt{3}\right)\sqrt{m}. \tag{4.200}$$

Similarly,

$$\overline{\phi}_2^T \mathbf{M} \overline{\phi}_2 = \frac{1}{C_2^2} \left\{ \begin{matrix} 1 \\ 1 \\ -1 \end{matrix} \right\}^T \begin{bmatrix} m & 0 & 0 \\ 0 & m & 0 \\ 0 & 0 & m \end{bmatrix} \left\{ \begin{matrix} 1 \\ 1 \\ -1 \end{matrix} \right\} = \frac{3}{C_2^2} m = 1 \tag{4.201}$$

and

$$\overline{\phi}_3^T \mathbf{M} \overline{\phi}_3 = \frac{1}{C_3^2} \left\{ \begin{matrix} 1 \\ 1-\sqrt{3} \\ 2-\sqrt{3} \end{matrix} \right\}^T \begin{bmatrix} m & 0 & 0 \\ 0 & m & 0 \\ 0 & 0 & m \end{bmatrix} \left\{ \begin{matrix} 1 \\ 1-\sqrt{3} \\ 2-\sqrt{3} \end{matrix} \right\} = \frac{12-6\sqrt{3}}{C_3^2} m = 1. \tag{4.202}$$

Therefore,

$$C_2 = \sqrt{3m}, \quad C_3 = \left(3-\sqrt{3}\right)\sqrt{m}. \tag{4.203}$$

Finally, the mass normalization follows from Eq. 4.176 and

$$\overline{\phi}_1 = \frac{m^{-1/2}}{3+\sqrt{3}} \left\{ \begin{matrix} 1 \\ 1+\sqrt{3} \\ 2+\sqrt{3} \end{matrix} \right\}, \quad \overline{\phi}_2 = \frac{m^{-1/2}}{\sqrt{3}} \left\{ \begin{matrix} 1 \\ 1 \\ -1 \end{matrix} \right\}, \quad \overline{\phi}_3 = \frac{m^{-1/2}}{3-\sqrt{3}} \left\{ \begin{matrix} 1 \\ 1-\sqrt{3} \\ 2-\sqrt{3} \end{matrix} \right\}. \tag{4.204}$$

Note that

$$\mathbf{\Phi}^T \mathbf{K} \mathbf{\Phi} = \begin{bmatrix} \overline{\phi}_1 & \overline{\phi}_2 & \overline{\phi}_3 \end{bmatrix}^T \begin{bmatrix} 3k & -k & 0 \\ -k & 2k & -k \\ 0 & -k & k \end{bmatrix} \begin{bmatrix} \overline{\phi}_1 & \overline{\phi}_2 & \overline{\phi}_3 \end{bmatrix} = \frac{k}{m} \begin{bmatrix} 2-\sqrt{3} & 0 & 0 \\ 0 & 2 & 0 \\ 0 & 0 & 2+\sqrt{3} \end{bmatrix} \tag{4.205}$$

and the diagonal elements of this matrix correspond to

$$\omega_1^2 = \left(2-\sqrt{3}\right)\omega_n^2, \quad \omega_2^2 = 2\omega_n^2, \quad \omega_3^2 = \left(2+\sqrt{3}\right)\omega_n^2. \tag{4.206}$$

EXAMPLE 3   *Orthogonality of Mode Shapes in a 3DOF System*

In general, the mode shapes are orthogonal with respect to each other when they are multiplied by either the mass or the stiffness matrix. The phrase "orthogonality of mode shapes" can be misleading, because only in special situations is the dot product of two mode shapes (i.e., $\boldsymbol{\phi}_i \circ \boldsymbol{\phi}_j$ or $\boldsymbol{\phi}_i^T \boldsymbol{\phi}_j$) equal to zero. For example, consider the mode shapes in Example 1, where

$$\boldsymbol{\phi}_1 = \begin{Bmatrix} 1 \\ 1+\sqrt{3} \\ 2+\sqrt{3} \end{Bmatrix}, \quad \boldsymbol{\phi}_2 = \begin{Bmatrix} 1 \\ 1 \\ -1 \end{Bmatrix}, \quad \boldsymbol{\phi}_3 = \begin{Bmatrix} 1 \\ 1-\sqrt{3} \\ 2-\sqrt{3} \end{Bmatrix}. \tag{4.207}$$

Note that the dot product (denoted by $\circ$), of any two mode shapes in Eq. 4.207 is zero:

$$\boldsymbol{\phi}_1 \circ \boldsymbol{\phi}_2 = \begin{Bmatrix} 1 \\ 1+\sqrt{3} \\ 2+\sqrt{3} \end{Bmatrix} \circ \begin{Bmatrix} 1 \\ 1 \\ -1 \end{Bmatrix} = 1+\left(1+\sqrt{3}\right)-\left(2+\sqrt{3}\right)=0 \tag{4.208}$$

$$\boldsymbol{\phi}_1 \circ \boldsymbol{\phi}_3 = \begin{Bmatrix} 1 \\ 1+\sqrt{3} \\ 2+\sqrt{3} \end{Bmatrix} \circ \begin{Bmatrix} 1 \\ 1-\sqrt{3} \\ 2-\sqrt{3} \end{Bmatrix} = 1+\left(1+\sqrt{3}\right)\left(1-\sqrt{3}\right)+\left(2+\sqrt{3}\right)\left(2-\sqrt{3}\right)=0 \tag{4.209}$$

$$\boldsymbol{\phi}_2 \circ \boldsymbol{\phi}_3 = \begin{Bmatrix} 1 \\ 1 \\ -1 \end{Bmatrix} \circ \begin{Bmatrix} 1 \\ 1-\sqrt{3} \\ 2-\sqrt{3} \end{Bmatrix} = 1+\left(1-\sqrt{3}\right)-\left(2-\sqrt{3}\right)=0. \tag{4.210}$$

This is a special case of a diagonal matrix with the mass at each floor the same, and therefore

$$\boldsymbol{\phi}_i^T \mathbf{M} \boldsymbol{\phi}_j = \boldsymbol{\phi}_i^T (m\mathbf{I}) \boldsymbol{\phi}_j = m\boldsymbol{\phi}_i^T \boldsymbol{\phi}_j = m\left(\boldsymbol{\phi}_i \circ \boldsymbol{\phi}_j\right)=0 \;\; \text{if} \;\; i \ne j. \tag{4.211}$$

In general, the dot product of two mode shapes is not equal to zero.

EXAMPLE 4   *Initial Conditions in Normal Mode Truncation Method*

The normal mode truncation method was briefly explained in Example 4 of Section 4.3. Because the number of modes determined is less than the total number of degrees of freedom (i.e., $k < n$), determining the initial conditions of each normal mode by inverting the modal matrix (see Example 1 of Section 4.3) is impossible. In this example, the orthogonality of mode shapes is used to determine the initial conditions of the normal modes.

Start with Eq. 4.105 for an undamped vibration of an n-DOF system with the response expressed in terms of k norm normalized mode shapes, that is,

$$\mathbf{X}(t) = \hat{\boldsymbol{\phi}}_1 q_1(t) + \hat{\boldsymbol{\phi}}_2 q_2(t) + \dots + \hat{\boldsymbol{\phi}}_k q_k(t) = \hat{\boldsymbol{\Phi}}\mathbf{Q}(t) \tag{4.212}$$

where

$$\hat{\boldsymbol{\Phi}} = \begin{bmatrix} \hat{\boldsymbol{\phi}}_1 & \hat{\boldsymbol{\phi}}_2 & \cdots & \hat{\boldsymbol{\phi}}_k \end{bmatrix} = \begin{bmatrix} \hat{\varphi}_{11} & \hat{\varphi}_{12} & \cdots & \hat{\varphi}_{1k} \\ \hat{\varphi}_{21} & \hat{\varphi}_{22} & & \vdots \\ \vdots & & & \hat{\varphi}_{n-1,k} \\ \hat{\varphi}_{n1} & \cdots & \hat{\varphi}_{n,k-1} & \hat{\varphi}_{nk} \end{bmatrix}, \quad \mathbf{Q}(t) = \begin{Bmatrix} q_1(t) \\ q_2(t) \\ \vdots \\ q_k(t) \end{Bmatrix}.$$

Premultiply Eq. 4.212 by the mass matrix $\mathbf{M}$, and then premultiply the resulting equation by $\hat{\boldsymbol{\phi}}_i^T$; it follows that

$$\hat{\boldsymbol{\phi}}_i^T \mathbf{M} \mathbf{X}(t) = \hat{\boldsymbol{\phi}}_i^T \mathbf{M} \hat{\boldsymbol{\phi}}_1 q_1(t) + \hat{\boldsymbol{\phi}}_i^T \mathbf{M} \hat{\boldsymbol{\phi}}_2 q_2(t) + \ldots + \hat{\boldsymbol{\phi}}_i^T \mathbf{M} \hat{\boldsymbol{\phi}}_k q_k(t). \tag{4.213}$$

Note that $\hat{\boldsymbol{\phi}}_i^T \mathbf{M} \hat{\boldsymbol{\phi}}_j = 0$ for $i \neq j$. Therefore, it follows that only the term on the right-hand side containing $q_i(t)$ is nonzero and remains in the equation:

$$\hat{\boldsymbol{\phi}}_i^T \mathbf{M} \mathbf{X}(t) = \hat{\boldsymbol{\phi}}_i^T \mathbf{M} \hat{\boldsymbol{\phi}}_i q_i(t) = \hat{m}_i q_i(t). \tag{4.214}$$

Solving for $q_i(t)$ in Eq. 4.214 gives

$$q_i(t) = \frac{\hat{\boldsymbol{\phi}}_i^T \mathbf{M} \mathbf{X}(t)}{\hat{m}_i} = \frac{\hat{\boldsymbol{\phi}}_i^T \mathbf{M} \mathbf{X}(t)}{\hat{\boldsymbol{\phi}}_i^T \mathbf{M} \hat{\boldsymbol{\phi}}_i}. \tag{4.215}$$

At $t = 0$, $\mathbf{X}(t) = \mathbf{X}(0) = \mathbf{X}_o$. It follows that the initial displacement condition for the ith mode is

$$q_i(0) = q_{io} = \frac{\hat{\boldsymbol{\phi}}_i^T \mathbf{M} \mathbf{X}_o}{\hat{\boldsymbol{\phi}}_i^T \mathbf{M} \hat{\boldsymbol{\phi}}_i}. \tag{4.216}$$

A similar result can be obtained for the initial velocity condition in the ith mode by first differentiating Eq. 4.215. Doing so gives

$$\dot{q}_i(t) = \frac{\hat{\boldsymbol{\phi}}_i^T \mathbf{M} \dot{\mathbf{X}}(t)}{\hat{m}_i} = \frac{\hat{\boldsymbol{\phi}}_i^T \mathbf{M} \dot{\mathbf{X}}(t)}{\hat{\boldsymbol{\phi}}_i^T \mathbf{M} \hat{\boldsymbol{\phi}}_i}. \tag{4.217}$$

At $t = 0$, $\dot{\mathbf{X}}(t) = \dot{\mathbf{X}}(0) = \dot{\mathbf{X}}_o$. It follows that the initial velocity condition for the ith mode is

$$\dot{q}_i(0) = \dot{q}_{io} = \frac{\hat{\boldsymbol{\phi}}_i^T \mathbf{M} \dot{\mathbf{X}}_o}{\hat{\boldsymbol{\phi}}_i^T \mathbf{M} \hat{\boldsymbol{\phi}}_i}. \tag{4.218}$$

For illustration, consider the 5DOF system discussed in Example 4 of Section 4.3. Imagine that the structure is placed in free vibration from rest (i.e., no initial velocity) and with an initial displacement vector equal to

$$\mathbf{X}_o = \left\{ x_{1o} \quad x_{2o} \quad x_{3o} \quad x_{4o} \quad x_{5o} \right\}^T. \tag{4.219}$$

Consider the case where first three modes are used. The corresponding initial displacements in the first three normal modes are obtained from Eq. 4.216:

$$q_{1o} = \frac{\hat{\boldsymbol{\phi}}_1^T \mathbf{M} \mathbf{X}_o}{\hat{\boldsymbol{\phi}}_1^T \mathbf{M} \hat{\boldsymbol{\phi}}_1}, \qquad q_{2o} = \frac{\hat{\boldsymbol{\phi}}_2^T \mathbf{M} \mathbf{X}_o}{\hat{\boldsymbol{\phi}}_2^T \mathbf{M} \hat{\boldsymbol{\phi}}_2}, \qquad q_{3o} = \frac{\hat{\boldsymbol{\phi}}_3^T \mathbf{M} \mathbf{X}_o}{\hat{\boldsymbol{\phi}}_3^T \mathbf{M} \hat{\boldsymbol{\phi}}_3}. \tag{4.220}$$

Similarly, from Eq. 4.218, the initial velocities in the first three normal modes are

$$\dot{q}_{1o} = \frac{\hat{\boldsymbol{\phi}}_1^T \mathbf{M} \dot{\mathbf{X}}_o}{\hat{\boldsymbol{\phi}}_1^T \mathbf{M} \hat{\boldsymbol{\phi}}_1} = 0, \qquad \dot{q}_{2o} = \frac{\hat{\boldsymbol{\phi}}_2^T \mathbf{M} \dot{\mathbf{X}}_o}{\hat{\boldsymbol{\phi}}_2^T \mathbf{M} \hat{\boldsymbol{\phi}}_2} = 0, \qquad \dot{q}_{3o} = \frac{\hat{\boldsymbol{\phi}}_3^T \mathbf{M} \dot{\mathbf{X}}_o}{\hat{\boldsymbol{\phi}}_3^T \mathbf{M} \hat{\boldsymbol{\phi}}_3} = 0. \tag{4.221}$$

This gives the initial conditions for the normal mode equation of motion:

$$\ddot{q}_i(t) + \omega_i^2 q_i(t) = 0, \qquad q_{io} = \frac{\hat{\boldsymbol{\phi}}_i^T \mathbf{M} \dot{\mathbf{X}}_o}{\hat{\boldsymbol{\phi}}_i^T \mathbf{M} \hat{\boldsymbol{\phi}}_i}. \tag{4.222}$$

Then from Eq. 2.21 with zero damping, it follows that

$$q_i(t) = q_{io} \cos \omega_i(t) = \left(\frac{\hat{\boldsymbol{\phi}}_i^T \mathbf{M} \dot{\mathbf{X}}_o}{\hat{\boldsymbol{\phi}}_i^T \mathbf{M} \hat{\boldsymbol{\phi}}_i} \right) \cos \omega_i(t). \tag{4.223}$$

Finally, substituting Eq. 4.223 into Eq. 4.212 gives the normal mode truncation response:

$$X(t) = \hat{\phi}_1 \left(\frac{\hat{\phi}_1^T M \dot{X}_o}{\hat{\phi}_1^T M \hat{\phi}_1} \right) \cos \omega_1(t) + \hat{\phi}_2 \left(\frac{\hat{\phi}_2^T M \dot{X}_o}{\hat{\phi}_2^T M \hat{\phi}_2} \right) \cos \omega_2(t) + \hat{\phi}_3 \left(\frac{\hat{\phi}_3^T M \dot{X}_o}{\hat{\phi}_3^T M \hat{\phi}_3} \right) \cos \omega_3(t). \quad (4.224)$$

Any floor displacement can be obtained using Eq. 4.224. For example, the displacement at the top floor (i.e., roof displacement) is

$$x_5(t) = \hat{\phi}_{51} \left(\frac{\hat{\phi}_1^T M \dot{X}_o}{\hat{\phi}_1^T M \hat{\phi}_1} \right) \cos \omega_1(t) + \hat{\phi}_{52} \left(\frac{\hat{\phi}_2^T M \dot{X}_o}{\hat{\phi}_2^T M \hat{\phi}_2} \right) \cos \omega_2(t) + \hat{\phi}_{53} \left(\frac{\hat{\phi}_3^T M \dot{X}_o}{\hat{\phi}_3^T M \hat{\phi}_3} \right) \cos \omega_3(t). \quad (4.225)$$

4.5 UNDAMPED RESPONSE USING THE NORMAL MODE METHOD

Consider the dynamic equilibrium equation of motion for the undamped n-DOF system in Eq. 4.1:

$$M \ddot{X}(t) + K X(t) = F_e(t). \quad (4.226)$$

The response of the system can be obtained using the normal mode method as discussed in Section 3.4. Assume that k natural frequencies and mode shapes are obtained, where $k \le n$. The system response in the X coordinate system can be expressed as a summation of the system response in the normal coordinate system. This is a coordinate transformation from the X coordinates to the Q coordinates, where

$$X(t) = \hat{\phi}_1 q_1(t) + \hat{\phi}_2 q_2(t) + \ \dots \ + \hat{\phi}_k q_k(t) = \hat{\Phi} Q(t). \quad (4.227)$$

In this equation,

$$\hat{\Phi} = \begin{bmatrix} \hat{\phi}_1 & \hat{\phi}_2 & \cdots & \hat{\phi}_k \end{bmatrix} = \begin{bmatrix} \hat{\phi}_{11} & \hat{\phi}_{12} & \cdots & \hat{\phi}_{1k} \\ \hat{\phi}_{21} & \hat{\phi}_{22} & \ddots & \vdots \\ \vdots & \ddots & \ddots & \hat{\phi}_{n-1,k} \\ \hat{\phi}_{n1} & \cdots & \hat{\phi}_{n,k-1} & \hat{\phi}_{nk} \end{bmatrix}, \quad Q(t) = \begin{Bmatrix} q_1(t) \\ q_2(t) \\ \vdots \\ q_k(t) \end{Bmatrix}. \quad (4.228)$$

The modal matrix $\hat{\Phi}$ is an $n \times k$ matrix. It follows from Eq. 4.226 that

$$M \hat{\Phi} \ddot{Q}(t) + K \hat{\Phi} Q(t) = F_e(t). \quad (4.229)$$

Premultiplying Eq. 4.229 by $\hat{\Phi}^T$, which is an $k \times n$ matrix, gives

$$\hat{\Phi}^T M \hat{\Phi} \ddot{Q}(t) + \hat{\Phi}^T K \hat{\Phi} Q(t) = \hat{\Phi}^T F_e(t). \quad (4.230)$$

Using the orthogonal property of M and K matrices, that is,

$$\hat{\Phi}^T M \hat{\Phi} = \hat{M} = \begin{bmatrix} \hat{m}_1 & 0 & \cdots & 0 \\ 0 & \hat{m}_2 & \ddots & \vdots \\ \vdots & \ddots & \ddots & 0 \\ 0 & \cdots & 0 & \hat{m}_k \end{bmatrix}, \quad \hat{\Phi}^T K \hat{\Phi} = \hat{K} = \begin{bmatrix} \hat{k}_1 & 0 & \cdots & 0 \\ 0 & \hat{k}_2 & \ddots & \vdots \\ \vdots & \ddots & \ddots & 0 \\ 0 & \cdots & 0 & \hat{k}_k \end{bmatrix} \quad (4.231)$$

it follows from Eq. 4.230 that

$$\begin{bmatrix} \hat{m}_1 & 0 & \cdots & 0 \\ 0 & \hat{m}_2 & \ddots & \vdots \\ \vdots & \ddots & \ddots & 0 \\ 0 & \cdots & 0 & \hat{m}_k \end{bmatrix} \begin{Bmatrix} \ddot{q}_1(t) \\ \ddot{q}_2(t) \\ \vdots \\ \ddot{q}_k(t) \end{Bmatrix} + \begin{bmatrix} \hat{k}_1 & 0 & \cdots & 0 \\ 0 & \hat{k}_2 & \ddots & \vdots \\ \vdots & \ddots & \ddots & 0 \\ 0 & \cdots & 0 & \hat{k}_k \end{bmatrix} \begin{Bmatrix} q_1(t) \\ q_2(t) \\ \vdots \\ q_k(t) \end{Bmatrix} = \begin{bmatrix} \hat{\phi}_1 & \hat{\phi}_2 & \cdots & \hat{\phi}_k \end{bmatrix}^T \begin{Bmatrix} F_1(t) \\ F_2(t) \\ \vdots \\ F_n(t) \end{Bmatrix}. \quad (4.232)$$

Expanding Eq. 4.232 gives a set of k independent equations of the form

$$\hat{m}_i \ddot{q}_i(t) + \hat{k}_i q_i(t) = \sum_{j=1}^{n} \hat{\phi}_{ji} F_j(t) \tag{4.233}$$

or

$$\ddot{q}_i(t) + \omega_i^2 q_i(t) = \sum_{j=1}^{n} \hat{\phi}_{ji} F_j(t) / m_i^*. \tag{4.234}$$

In the preceding discussion, $\hat{\boldsymbol{\phi}}_i$ was used to denote a norm normalization of the mode shape. Then $\hat{m}_i$ and $\hat{k}_i$ were obtained, and $(\hat{k}_i / \hat{m}_i) = \omega_i^2$. However, Eq. 4.231 can be used for any normalization of the mode shape, and even though values for $\hat{m}_i$ and $\hat{k}_i$ change to m_i^* and k_i^*, it always follows that $(k_i^* / m_i^*) = \omega_i^2$.

Equation 4.234 represents a set of k uncoupled, second-order, linear differential equations. In effect, each equation is like an SDOF system discussed in Chapter 2 with a unit mass and a natural frequency ω_i^2. Therefore, the solution for each $q_i(t)$ can be obtained by treating Eq. 4.234 as k SDOF systems. When this is done, the solution takes the form

$$q_i(t) = a_i \cos \omega_i t + b_i \sin \omega_i t + q_{pi}(t) \tag{4.235}$$

where a_i and b_i are arbitrary constants used to satisfy the initial conditions in the **Q** coordinates and $q_{pi}(t)$ is the particular solution to the forcing function given in Eq. 4.234 in the **Q** coordinates. Substituting this solution for each q_i into Eq. 4.227 gives

$$\mathbf{X}(t) = \sum_{i=1}^{k} \hat{\boldsymbol{\phi}}_i q_i(t) = \sum_{i=1}^{k} \hat{\boldsymbol{\phi}}_i \left[a_i \cos \omega_i t + b_i \sin \omega_i t \right] + \sum_{i=1}^{k} \hat{\boldsymbol{\phi}}_i q_{pi}(t). \tag{4.236}$$

Note the similarities between Eq. 4.102 and the first term of Eq. 4.236. This is because the first term of Eq. 4.236 is the homogeneous solution, and the second term is the particular solution.

The values of a_i and b_i depend on the initial conditions in the **X** coordinate system. When k is equal to n, the values of a_i and b_i can be determined using the method discussed in Example 1 of Section 4.3. These values follow from Eqs. 4.115 and 4.116:

$$\mathbf{a} = \boldsymbol{\Phi}^{-1} \mathbf{X}(0) \tag{4.237a}$$

$$\mathbf{b} = \boldsymbol{\Phi}^{-1} \dot{\mathbf{X}}(0). \tag{4.237b}$$

However, if the number of modes used in the normal mode solution is less than the total number of the system (i.e., $k < n$), then the eigenvector or modal matrix, $\boldsymbol{\Phi}$, is not a square matrix. Therefore, $\boldsymbol{\Phi}$ is an $n \times k$ matrix and is not invertible. However, the values of a_i and b_i can be determined using the orthogonality of mode shapes presented in Example 4 of Section 4.4 for the homogeneous solution. First write

$$\mathbf{X}(t) = \hat{\boldsymbol{\phi}}_1 q_1(t) + \hat{\boldsymbol{\phi}}_2 q_2(t) + \ \dots \ + \hat{\boldsymbol{\phi}}_k q_k(t). \tag{4.238}$$

Next, premultiply all of the terms in Eq. 4.238 by the mass matrix **M** and then by $\hat{\boldsymbol{\phi}}_i^T$. It then follows that

$$\hat{\boldsymbol{\phi}}_i^T \mathbf{M} \mathbf{X}(t) = \left(\hat{\boldsymbol{\phi}}_i^T \mathbf{M} \hat{\boldsymbol{\phi}}_1 \right) q_1(t) + \left(\hat{\boldsymbol{\phi}}_i^T \mathbf{M} \hat{\boldsymbol{\phi}}_2 \right) q_2(t) + \ \dots \ + \left(\hat{\boldsymbol{\phi}}_i^T \mathbf{M} \hat{\boldsymbol{\phi}}_k \right) q_k(t). \tag{4.239}$$

Observe that if $i = 1$, then $(\hat{\boldsymbol{\phi}}_i^T \mathbf{M} \hat{\boldsymbol{\phi}}_i) = m_1^*$ and $(\hat{\boldsymbol{\phi}}_i^T \mathbf{M} \hat{\boldsymbol{\phi}}_j) = 0$ for $j \neq 1$. Similarly, if $i = 3$, then $(\hat{\boldsymbol{\phi}}_i^T \mathbf{M} \hat{\boldsymbol{\phi}}_3) = \hat{m}_3$ and $(\hat{\boldsymbol{\phi}}_i^T \mathbf{M} \hat{\boldsymbol{\phi}}_j) = 0$ for $j \neq 3$. Therefore, Eq. 4.171 can be used, and it follows from Eq. 4.239 that

$$\hat{\boldsymbol{\phi}}_i^T \mathbf{M} \mathbf{X}(t) = \left(\hat{\boldsymbol{\phi}}_i^T \mathbf{M} \hat{\boldsymbol{\phi}}_i \right) q_i(t) = \hat{m}_i q_i(t). \tag{4.240}$$

Dividing both sides of Eq. 4.240 by $\hat{m}_i$ gives

$$q_i(t) = \frac{\hat{\boldsymbol{\phi}}_i^T \mathbf{M} \mathbf{X}(t)}{\hat{m}_i}. \tag{4.241}$$

Equating Eq. 4.235 and Eq. 4.241, it follows that

$$q_i(t) = a_i \cos \omega_i t + b_i \sin \omega_i t + q_{pi}(t) = \frac{\hat{\boldsymbol{\phi}}_i^T \mathbf{M} \mathbf{X}(t)}{\hat{m}_i}. \tag{4.242}$$

When Eq. 4.242 is differentiated with respect to time, it follows that

$$\dot{q}_i(t) = b_i \omega_i \cos \omega_i t - a_i \omega_i \sin \omega_i t + \dot{q}_{pi}(t) = \frac{\hat{\boldsymbol{\phi}}_i^T \mathbf{M} \dot{\mathbf{X}}(t)}{\hat{m}_i}. \tag{4.243}$$

In most structural dynamics problems, the particular solution $q_{pi}(t)$ is such that $q_{pi}(0) = \dot{q}_{pi}(0) = 0$. Then at time $t = 0$, it follows from Eqs. 4.242 and 4.243 that

$$a_i = \frac{\hat{\boldsymbol{\phi}}_i^T \mathbf{M} \mathbf{X}(0)}{\hat{m}_i}, \qquad b_i = \frac{\hat{\boldsymbol{\phi}}_i^T \mathbf{M} \dot{\mathbf{X}}(0)}{\hat{m}_i \omega_i}. \tag{4.244}$$

Equation 4.244 enables calculation of the values of a_i and b_i for $k < n$ because no inverse of the modal matrix is necessary.

EXAMPLE 1   *Normal Mode Equation Using a General Mode Shape*

Equation 4.236 presents the response of the system in the $\mathbf{X}$ coordinate system in terms of normal modes that have been norm normalized. In general, any form of the mode shapes with arbitrary multipliers can be used in a normal mode solution; to demonstrate this, consider the mode shapes $\tilde{\boldsymbol{\phi}}_1, \tilde{\boldsymbol{\phi}}_2, \ldots, \tilde{\boldsymbol{\phi}}_n$, where

$$\mathbf{X}(t) = \tilde{\boldsymbol{\phi}}_1 \tilde{q}_1(t) + \tilde{\boldsymbol{\phi}}_2 \tilde{q}_2(t) + \ldots + \tilde{\boldsymbol{\phi}}_n \tilde{q}_n(t) = \tilde{\boldsymbol{\Phi}} \tilde{\mathbf{Q}}(t) \tag{4.245}$$

and

$$\tilde{\boldsymbol{\Phi}} = \begin{bmatrix} \tilde{\boldsymbol{\phi}}_1 & \tilde{\boldsymbol{\phi}}_2 & \cdots & \tilde{\boldsymbol{\phi}}_n \end{bmatrix}, \quad \tilde{\mathbf{Q}}(t) = \begin{Bmatrix} \tilde{q}_1(t) & \tilde{q}_2(t) & \cdots & \tilde{q}_n(t) \end{Bmatrix}^T. \tag{4.246}$$

Because, as previously noted, a mode shape can be multiplied by any constant, it follows that

$$\tilde{\boldsymbol{\phi}}_i = D_i \hat{\boldsymbol{\phi}}_i \tag{4.247}$$

and therefore, it follows from Eq. 4.247 that

$$\tilde{\boldsymbol{\Phi}} = \hat{\boldsymbol{\Phi}} \begin{bmatrix} D_1 & 0 & \cdots & 0 \\ 0 & D_2 & \ddots & \vdots \\ \vdots & \ddots & \ddots & 0 \\ 0 & \cdots & 0 & D_n \end{bmatrix} = \hat{\boldsymbol{\Phi}} \mathbf{D}. \tag{4.248}$$

Substituting Eq. 4.245 into Eq. 4.226 gives

$$\mathbf{M} \tilde{\boldsymbol{\Phi}} \ddot{\mathbf{Q}}(t) + \mathbf{K} \tilde{\boldsymbol{\Phi}} \mathbf{Q}(t) = \mathbf{F}_e(t). \tag{4.249}$$

Premultiplying Eq. 4.249 by $\tilde{\boldsymbol{\Phi}}^T$, it follows that

$$\tilde{\boldsymbol{\Phi}}^T \mathbf{M} \tilde{\boldsymbol{\Phi}} \ddot{\mathbf{Q}}(t) + \tilde{\boldsymbol{\Phi}}^T \mathbf{K} \tilde{\boldsymbol{\Phi}} \mathbf{Q}(t) = \tilde{\boldsymbol{\Phi}}^T \mathbf{F}_e(t). \tag{4.250}$$

Using the orthogonal property of $\mathbf{M}$ and $\mathbf{K}$ matrices, and since $\mathbf{D}^T = \mathbf{D}$, it follows that

$$\boldsymbol{\tilde{\Phi}}^T \mathbf{M} \boldsymbol{\tilde{\Phi}} = \mathbf{D}\left(\boldsymbol{\hat{\Phi}}^T \mathbf{M} \boldsymbol{\hat{\Phi}}\right)\mathbf{D} = \mathbf{D}\begin{bmatrix} \hat{m}_1 & 0 & \cdots & 0 \\ 0 & \hat{m}_2 & \ddots & \vdots \\ \vdots & \ddots & \ddots & 0 \\ 0 & \cdots & 0 & \hat{m}_n \end{bmatrix}\mathbf{D} = \begin{bmatrix} \hat{m}_1 D_1^2 & 0 & \cdots & 0 \\ 0 & \hat{m}_2 D_2^2 & \ddots & \vdots \\ \vdots & \ddots & \ddots & 0 \\ 0 & \cdots & 0 & \hat{m}_n D_n^2 \end{bmatrix} \quad (4.251)$$

$$\boldsymbol{\tilde{\Phi}}^T \mathbf{K} \boldsymbol{\tilde{\Phi}} = \mathbf{D}\left(\boldsymbol{\hat{\Phi}}^T \mathbf{K} \boldsymbol{\hat{\Phi}}\right)\mathbf{D} = \mathbf{D}\begin{bmatrix} \hat{k}_1 & 0 & \cdots & 0 \\ 0 & \hat{k}_2 & \ddots & \vdots \\ \vdots & \ddots & \ddots & 0 \\ 0 & \cdots & 0 & \hat{k}_n \end{bmatrix}\mathbf{D} = \begin{bmatrix} \hat{k}_1 D_1^2 & 0 & \cdots & 0 \\ 0 & \hat{k}_2 D_2^2 & \ddots & \vdots \\ \vdots & \ddots & \ddots & 0 \\ 0 & \cdots & 0 & \hat{k}_n D_n^2 \end{bmatrix}. \quad (4.252)$$

It follows from Eq. 4.250 that

$$\begin{bmatrix} \hat{m}_1 D_1^2 & 0 & \cdots & 0 \\ 0 & \hat{m}_2 D_2^2 & \ddots & \vdots \\ \vdots & \ddots & \ddots & 0 \\ 0 & \cdots & 0 & \hat{m}_n D_n^2 \end{bmatrix}\begin{Bmatrix} \ddot{\tilde{q}}_1(t) \\ \ddot{\tilde{q}}_2(t) \\ \vdots \\ \ddot{\tilde{q}}_n(t) \end{Bmatrix} + \begin{bmatrix} \hat{k}_1 D_1^2 & 0 & \cdots & 0 \\ 0 & \hat{k}_2 D_2^2 & \ddots & \vdots \\ \vdots & \ddots & \ddots & 0 \\ 0 & \cdots & 0 & \hat{k}_n D_n^2 \end{bmatrix}\begin{Bmatrix} \tilde{q}_1(t) \\ \tilde{q}_2(t) \\ \vdots \\ \tilde{q}_n(t) \end{Bmatrix} = \begin{bmatrix} D_1 \boldsymbol{\hat{\phi}}_1^T \\ D_2 \boldsymbol{\hat{\phi}}_2^T \\ \vdots \\ D_n \boldsymbol{\hat{\phi}}_n^T \end{bmatrix}\begin{Bmatrix} F_1(t) \\ F_2(t) \\ \vdots \\ F_n(t) \end{Bmatrix}.$$

Expanding this equation gives a set of n independent equations of the form

$$m_i^* D_i^2 \ddot{\tilde{q}}_i(t) + k_i^* D_i^2 \tilde{q}_i(t) = \sum_{j=1}^{n} D_i \hat{\phi}_{ji} F_j(t). \quad (4.253)$$

Dividing Eq. 4.253 by $m_i^* D_i^2$ gives

$$\ddot{\tilde{q}}_i(t) + \omega_i^2 \tilde{q}_i(t) = \frac{\sum_{j=1}^{n} D_i \hat{\phi}_{ji} F_j(t)}{m_i^* D_i^2} = \frac{\sum_{j=1}^{n} \hat{\phi}_{ji} F_j(t)}{m_i^* D_i} \quad (4.254)$$

where $\omega_i^2 = (m_i^* D_i^2)/(k_i^* D_i^2) = m_i^*/k_i^*$. Finally, note that the solution for Eq. 4.234 is Eq. 4.235, and therefore the solution for Eq. 4.254 is

$$\tilde{q}_i(t) = \frac{a_i \cos \omega_i t + b_i \sin \omega_i t + q_{pi}(t)}{D_i} = \frac{q_i(t)}{D_i}. \quad (4.255)$$

Substituting this result in Eq. 4.245 gives

$$\begin{aligned} \mathbf{X}(t) &= \boldsymbol{\tilde{\phi}}_1 \tilde{q}_1(t) + \boldsymbol{\tilde{\phi}}_2 \tilde{q}_2(t) + \ \cdots \ + \boldsymbol{\tilde{\phi}}_n \tilde{q}_n(t) \\ &= D_1 \boldsymbol{\hat{\phi}}_1 \frac{q_1(t)}{D_1} + D_2 \boldsymbol{\hat{\phi}}_2 \frac{q_2(t)}{D_2} + \ \cdots \ + D_n \boldsymbol{\hat{\phi}}_n \frac{q_n(t)}{D_n} \\ &= \boldsymbol{\hat{\phi}}_1 q_1(t) + \boldsymbol{\hat{\phi}}_2 q_2(t) + \ \cdots \ + \boldsymbol{\hat{\phi}}_n q_n(t) \end{aligned} \quad (4.256)$$

which is the same as Eq. 4.227. This example shows that any mode shape can be used in the normal mode method of analysis.

EXAMPLE 2   *Initial Conditions Using Normal Mode Method*

Consider the case of free vibration in Example 2 of Section 4.3 with initial conditions given in Example 3 of Section 4.3:

$$\begin{Bmatrix} x_1(0) \\ x_2(0) \\ x_3(0) \end{Bmatrix} = \begin{Bmatrix} 0 \\ 0 \\ 1 \end{Bmatrix}, \quad \begin{Bmatrix} \dot{x}_1(0) \\ \dot{x}_2(0) \\ \dot{x}_3(0) \end{Bmatrix} = \begin{Bmatrix} 0 \\ 0 \\ 0 \end{Bmatrix}. \tag{4.257}$$

It then follows from Eq. 4.244 that

$$a_1 = \frac{\hat{\boldsymbol{\phi}}_1^T \mathbf{M} \mathbf{X}(0)}{\hat{m}_1} = \frac{1}{m} \left(\frac{1}{3+\sqrt{3}} \right) \begin{Bmatrix} 1 \\ 1+\sqrt{3} \\ 2+\sqrt{3} \end{Bmatrix}^T \begin{bmatrix} m & 0 & 0 \\ 0 & m & 0 \\ 0 & 0 & m \end{bmatrix} \begin{Bmatrix} 0 \\ 0 \\ 1 \end{Bmatrix} = \frac{3+\sqrt{3}}{6} \tag{4.258}$$

$$a_2 = \frac{\hat{\boldsymbol{\phi}}_2^T \mathbf{M} \mathbf{X}(0)}{\hat{m}_2} = \frac{1}{m} \left(\frac{1}{\sqrt{3}} \right) \begin{Bmatrix} 1 \\ 1 \\ -1 \end{Bmatrix}^T \begin{bmatrix} m & 0 & 0 \\ 0 & m & 0 \\ 0 & 0 & m \end{bmatrix} \begin{Bmatrix} 0 \\ 0 \\ 1 \end{Bmatrix} = -\frac{\sqrt{3}}{3} \tag{4.259}$$

$$a_3 = \frac{\hat{\boldsymbol{\phi}}_3^T \mathbf{M} \mathbf{X}(0)}{\hat{m}_3} = \frac{1}{m} \left(\frac{1}{3-\sqrt{3}} \right) \begin{Bmatrix} 1 \\ 1-\sqrt{3} \\ 2-\sqrt{3} \end{Bmatrix}^T \begin{bmatrix} m & 0 & 0 \\ 0 & m & 0 \\ 0 & 0 & m \end{bmatrix} \begin{Bmatrix} 0 \\ 0 \\ 1 \end{Bmatrix} = \frac{3-\sqrt{3}}{6}. \tag{4.260}$$

Because the initial velocities are zero, it follows that

$$b_1 = b_2 = b_3 = 0. \tag{4.261}$$

Therefore, for free vibration, that is, $\mathbf{F}_e(t) = \mathbf{0}$, it follows from Eq. 4.236 that

$$\begin{Bmatrix} x_1(t) \\ x_2(t) \\ x_3(t) \end{Bmatrix} = \hat{\boldsymbol{\phi}}_1 \left(\frac{3+\sqrt{3}}{6} \right) \cos \omega_1 t - \hat{\boldsymbol{\phi}}_2 \left(\frac{\sqrt{3}}{6} \right) \cos \omega_2 t + \hat{\boldsymbol{\phi}}_3 \left(\frac{3-\sqrt{3}}{6} \right) \cos \omega_3 t$$

$$= \frac{1}{6} \begin{Bmatrix} 1 \\ 1+\sqrt{3} \\ 2+\sqrt{3} \end{Bmatrix} \cos \omega_1 t - \frac{1}{3} \begin{Bmatrix} 1 \\ 1 \\ -1 \end{Bmatrix} \cos \omega_2 t + \frac{1}{6} \begin{Bmatrix} 1 \\ 1-\sqrt{3} \\ 2-\sqrt{3} \end{Bmatrix} \cos \omega_3 t. \tag{4.262}$$

This response solution is the same as Eq. 4.137. Note that if instead of using all three normal mode coordinates $q_1(t)$, $q_2(t)$, and $q_3(t)$, only the first two are used; then Eq. 4.262 becomes

$$\begin{Bmatrix} x_1(t) \\ x_2(t) \\ x_3(t) \end{Bmatrix} = \hat{\boldsymbol{\phi}}_1 q_1(t) + \hat{\boldsymbol{\phi}}_2 q_2(t) = \frac{1}{6} \begin{Bmatrix} 1 \\ 1+\sqrt{3} \\ 2+\sqrt{3} \end{Bmatrix} \cos \omega_1 t - \frac{1}{3} \begin{Bmatrix} 1 \\ 1 \\ -1 \end{Bmatrix} \cos \omega_2 t. \tag{4.263}$$

This solution method for $k < n$ is called normal mode truncation method, which was discussed in Example 4 of Section 4.3.

EXAMPLE 3   *Response of a 3DOF System Subjected to a Uniform Forcing Function*

Consider Example 2 of Section 4.3 with the initial conditions equal to zero, that is,

$$\begin{Bmatrix} x_1(0) \\ x_2(0) \\ x_3(0) \end{Bmatrix} = \begin{Bmatrix} 0 \\ 0 \\ 0 \end{Bmatrix}, \quad \begin{Bmatrix} \dot{x}_1(0) \\ \dot{x}_2(0) \\ \dot{x}_3(0) \end{Bmatrix} = \begin{Bmatrix} 0 \\ 0 \\ 0 \end{Bmatrix} \tag{4.264}$$

but with a uniform forcing function of magnitude F_o applied at the top floor. It follows from Eq. 4.234 that

$$\ddot{q}_1(t) + \omega_1^2 q_1(t) = \frac{1}{3+\sqrt{3}} \begin{Bmatrix} 1 \\ 1+\sqrt{3} \\ 2+\sqrt{3} \end{Bmatrix}^T \begin{Bmatrix} 0 \\ 0 \\ 1 \end{Bmatrix} \frac{F_o}{m} = \frac{3+\sqrt{3}}{6} \left(\frac{F_o}{m} \right) \tag{4.265}$$

$$\ddot{q}_2(t) + \omega_2^2 q_2(t) = \frac{1}{\sqrt{3}} \begin{Bmatrix} 1 \\ 1 \\ -1 \end{Bmatrix}^T \begin{Bmatrix} 0 \\ 0 \\ 1 \end{Bmatrix} \frac{F_o}{m} = -\frac{\sqrt{3}}{3} \left(\frac{F_o}{m} \right) \tag{4.266}$$

$$\ddot{q}_3(t) + \omega_3^2 q_3(t) = \frac{1}{3-\sqrt{3}} \begin{Bmatrix} 1 \\ 1-\sqrt{3} \\ 2-\sqrt{3} \end{Bmatrix}^T \begin{Bmatrix} 0 \\ 0 \\ 1 \end{Bmatrix} \frac{F_o}{m} = \frac{3-\sqrt{3}}{6} \left(\frac{F_o}{m} \right). \tag{4.267}$$

Since the initial conditions are zero, it follows that

$$a_1 = a_2 = a_3 = b_1 = b_2 = b_3 = 0. \tag{4.268}$$

The solutions for Eqs. 4.265 to 4.267 follow from Eq. 2.47:

$$q_1(t) = \frac{3+\sqrt{3}}{6} \left(\frac{F_o}{k} \right) (1 - \cos \omega_1 t) \tag{4.269}$$

$$q_2(t) = -\frac{\sqrt{3}}{3} \left(\frac{F_o}{k} \right) (1 - \cos \omega_2 t) \tag{4.270}$$

$$q_3(t) = \frac{3-\sqrt{3}}{6} \left(\frac{F_o}{k} \right) (1 - \cos \omega_3 t). \tag{4.271}$$

The normal mode solution follows from Eq. 4.236:

$$\begin{Bmatrix} x_1(t) \\ x_2(t) \\ x_3(t) \end{Bmatrix} = \hat{\boldsymbol{\phi}}_1 q_1(t) + \hat{\boldsymbol{\phi}}_2 q_2(t) + \hat{\boldsymbol{\phi}}_3 q_3(t)$$

$$= \frac{1}{6} \begin{Bmatrix} 1 \\ 1+\sqrt{3} \\ 2+\sqrt{3} \end{Bmatrix} \frac{F_o}{k} (1 - \cos \omega_1 t) - \frac{1}{3} \begin{Bmatrix} 1 \\ 1 \\ -1 \end{Bmatrix} \frac{F_o}{k} (1 - \cos \omega_2 t) + \frac{1}{6} \begin{Bmatrix} 1 \\ 1-\sqrt{3} \\ 2-\sqrt{3} \end{Bmatrix} \frac{F_o}{k} (1 - \cos \omega_3 t). \tag{4.272}$$

If instead of using all three normal coordinates, only two are used, then Eq. 4.272 becomes

$$\begin{Bmatrix} x_1(t) \\ x_2(t) \\ x_3(t) \end{Bmatrix} = \hat{\boldsymbol{\phi}}_1 q_1(t) + \hat{\boldsymbol{\phi}}_2 q_2(t)$$

$$= \frac{1}{6} \begin{Bmatrix} 1 \\ 1+\sqrt{3} \\ 2+\sqrt{3} \end{Bmatrix} \frac{F_o}{k} (1 - \cos \omega_1 t) - \frac{1}{3} \begin{Bmatrix} 1 \\ 1 \\ -1 \end{Bmatrix} \frac{F_o}{k} (1 - \cos \omega_2 t). \tag{4.273}$$

In this example, $n = 3$, and thus the difference between $k = 2$ and $k = 3$ is not significant as it relates to computational effort. However, in many structural dynamics problems, n can be as much as thousands and k can be as few as three times the number of stories of the structure. This significantly impacts the computation.

EXAMPLE 4 *Response of an MDOF System Subjected to Earthquake Ground Acceleration*

Consider a multistory two-dimensional structure similar to the one described in Example 3 but with more than three stories. The structure is now assumed to be excited by an earthquake ground acceleration, denoted $a(t)$. The equation of motion corresponding to Eq. 4.226 is

$$\mathbf{M\ddot{X}}(t) + \mathbf{KX}(t) = -\mathbf{M}\{\mathbf{I}\}a(t). \tag{4.274}$$

To obtain the solution using the normal mode method, first substitute Eq. 4.227 into Eq. 4.226 and then premultiply the resulting equation by $\hat{\mathbf{\Phi}}^T$ (a $k \times n$ matrix). It then follows from Eq. 4.232 that

$$
\begin{bmatrix} \hat{m}_1 & 0 & \cdots & 0 \\ 0 & \hat{m}_2 & \ddots & \vdots \\ \vdots & \ddots & \ddots & 0 \\ 0 & \cdots & 0 & \hat{m}_k \end{bmatrix}
\begin{Bmatrix} \ddot{q}_1(t) \\ \ddot{q}_2(t) \\ \vdots \\ \ddot{q}_k(t) \end{Bmatrix}
+
\begin{bmatrix} \hat{k}_1 & 0 & \cdots & 0 \\ 0 & \hat{k}_2 & \ddots & \vdots \\ \vdots & \ddots & \ddots & 0 \\ 0 & \cdots & 0 & \hat{k}_k \end{bmatrix}
\begin{Bmatrix} q_1(t) \\ q_2(t) \\ \vdots \\ q_k(t) \end{Bmatrix}
$$

$$
= -\begin{bmatrix} \hat{\mathbf{\Phi}}_1 & \hat{\mathbf{\Phi}}_2 & \cdots & \hat{\mathbf{\Phi}}_k \end{bmatrix}^T
\begin{bmatrix} m_{11} & m_{12} & \cdots & m_{1n} \\ m_{21} & m_{22} & \ddots & \vdots \\ \vdots & \ddots & \ddots & m_{n-1,n} \\ m_{n1} & \cdots & m_{n,n-1} & m_{nn} \end{bmatrix}
\begin{Bmatrix} 1 \\ 1 \\ \vdots \\ 1 \end{Bmatrix} a(t)
\tag{4.275}
$$

and

$$
\begin{Bmatrix} \ddot{q}_1(t) \\ \ddot{q}_2(t) \\ \vdots \\ \ddot{q}_k(t) \end{Bmatrix}
+
\begin{bmatrix} \omega_1^2 & 0 & \cdots & 0 \\ 0 & \omega_2^2 & \ddots & \vdots \\ \vdots & \ddots & \ddots & 0 \\ 0 & \cdots & 0 & \omega_k^2 \end{bmatrix}
\begin{Bmatrix} q_1(t) \\ q_2(t) \\ \vdots \\ q_k(t) \end{Bmatrix}
$$

$$
= -\begin{bmatrix} \hat{m}_1 & 0 & \cdots & 0 \\ 0 & \hat{m}_2 & \ddots & \vdots \\ \vdots & \ddots & \ddots & 0 \\ 0 & \cdots & 0 & \hat{m}_k \end{bmatrix}^{-1}
\begin{bmatrix} \hat{\mathbf{\Phi}}_1 & \hat{\mathbf{\Phi}}_2 & \cdots & \hat{\mathbf{\Phi}}_k \end{bmatrix}^T
\begin{bmatrix} m_{11} & m_{12} & \cdots & m_{1n} \\ m_{21} & m_{22} & \ddots & \vdots \\ \vdots & \ddots & \ddots & m_{n-1,n} \\ m_{n1} & \cdots & m_{n,n-1} & m_{nn} \end{bmatrix}
\begin{Bmatrix} a(t) \\ a(t) \\ \vdots \\ a(t) \end{Bmatrix}.
\tag{4.276}
$$

This equation can be written as

$$\mathbf{\ddot{Q}}(t) + \mathbf{\Omega Q}(t) = -\mathbf{\hat{M}}^{-1}\hat{\mathbf{\Phi}}^T\mathbf{M}\{\mathbf{I}\}a(t) \tag{4.277}$$

where

$$
\mathbf{\Omega} = \begin{bmatrix} \omega_1^2 & 0 & \cdots & 0 \\ 0 & \omega_2^2 & \ddots & \vdots \\ \vdots & \ddots & \ddots & 0 \\ 0 & \cdots & 0 & \omega_k^2 \end{bmatrix}.
$$

Equation 4.277 is as a set of k uncoupled linear simultaneous equations, and they can be written as

$$\ddot{q}_i(t) + \omega_i^2 q_i(t) = -\hat{\Gamma}_i\, a(t) \tag{4.278}$$

where

$$\hat{\Gamma}_i = \frac{\hat{\boldsymbol{\phi}}_i^T \mathbf{M}\{\mathbf{I}\}}{\hat{m}_i}.$$
(4.279)

The term $\hat{\Gamma}_i$ is called the ith mode participation factor.

4.6 DAMPED RESPONSE WITH RAYLEIGH DAMPING USING THE NORMAL MODE METHOD

Starting with the dynamic equilibrium equation,

$$\mathbf{M}\ddot{\mathbf{X}}(t) + \mathbf{C}\dot{\mathbf{X}}(t) + \mathbf{K}\mathbf{X}(t) = \mathbf{F}_e(t),$$
(4.280)

let

$$\mathbf{X}(t) = \hat{\boldsymbol{\Phi}}\mathbf{Q}(t)$$
(4.281)

where

$$\hat{\boldsymbol{\Phi}} = \begin{bmatrix} \hat{\boldsymbol{\phi}}_1 & \hat{\boldsymbol{\phi}}_2 & \cdots & \hat{\boldsymbol{\phi}}_k \end{bmatrix}, \quad \mathbf{Q}(t) = \{q_1(t) \quad q_2(t) \quad \cdots \quad q_k(t)\}^T.$$
(4.282)

It follows from Eq. 4.280 that

$$\mathbf{M}\hat{\boldsymbol{\Phi}}\ddot{\mathbf{Q}}(t) + \mathbf{C}\hat{\boldsymbol{\Phi}}\dot{\mathbf{Q}}(t) + \mathbf{K}\hat{\boldsymbol{\Phi}}\mathbf{Q}(t) = \mathbf{F}_e(t).$$
(4.283)

Premultiplying Eq. 4.283 by $\hat{\boldsymbol{\Phi}}^T$ gives

$$\hat{\boldsymbol{\Phi}}^T\mathbf{M}\hat{\boldsymbol{\Phi}}\ddot{\mathbf{Q}}(t) + \hat{\boldsymbol{\Phi}}^T\mathbf{C}\hat{\boldsymbol{\Phi}}\dot{\mathbf{Q}}(t) + \hat{\boldsymbol{\Phi}}^T\mathbf{K}\hat{\boldsymbol{\Phi}}\mathbf{Q}(t) = \hat{\boldsymbol{\Phi}}^T\mathbf{F}_e(t).$$
(4.284)

Section 3.5 discussed the damped response of a 2DOF system with Rayleigh proportional damping, where the damping matrix is a linear combination of the mass and the stiffness matrices:

$$\mathbf{C} = \alpha\mathbf{M} + \beta\mathbf{K}.$$
(4.285)

In this section, Rayleigh damping described in Eq. 4.285 is extended to an MDOF system. Because the eigenvectors are orthogonal to the mass and stiffness matrices, these eigenvectors will be orthogonal to the damping matrix. Therefore, it follows that

$$\hat{\boldsymbol{\Phi}}^T\mathbf{M}\hat{\boldsymbol{\Phi}} = \begin{bmatrix} \hat{\boldsymbol{\phi}}_1^T \\ \hat{\boldsymbol{\phi}}_2^T \\ \vdots \\ \hat{\boldsymbol{\phi}}_k^T \end{bmatrix} \begin{bmatrix} m_1 & 0 & \cdots & 0 \\ 0 & m_2 & \ddots & \vdots \\ \vdots & \ddots & \ddots & 0 \\ 0 & \cdots & 0 & m_n \end{bmatrix} \begin{bmatrix} \hat{\boldsymbol{\phi}}_1 & \hat{\boldsymbol{\phi}}_2 & \cdots & \hat{\boldsymbol{\phi}}_k \end{bmatrix} = \begin{bmatrix} \hat{m}_1 & 0 & \cdots & 0 \\ 0 & \hat{m}_2 & \ddots & \vdots \\ \vdots & \ddots & \ddots & 0 \\ 0 & \cdots & 0 & \hat{m}_k \end{bmatrix} = \hat{\mathbf{M}}$$
(4.286a)

$$\hat{\boldsymbol{\Phi}}^T\mathbf{C}\hat{\boldsymbol{\Phi}} = \begin{bmatrix} \hat{\boldsymbol{\phi}}_1^T \\ \hat{\boldsymbol{\phi}}_2^T \\ \vdots \\ \hat{\boldsymbol{\phi}}_k^T \end{bmatrix} \begin{bmatrix} c_{11} & c_{12} & \cdots & c_{1n} \\ c_{21} & c_{22} & \ddots & \vdots \\ \vdots & \ddots & \ddots & \vdots \\ c_{n1} & \cdots & \cdots & c_{nn} \end{bmatrix} \begin{bmatrix} \hat{\boldsymbol{\phi}}_1 & \hat{\boldsymbol{\phi}}_2 & \cdots & \hat{\boldsymbol{\phi}}_k \end{bmatrix} = \begin{bmatrix} \hat{c}_1 & 0 & \cdots & 0 \\ 0 & \hat{c}_2 & \ddots & \vdots \\ \vdots & \ddots & \ddots & 0 \\ 0 & \cdots & 0 & \hat{c}_k \end{bmatrix} = \hat{\mathbf{C}}$$
(4.286b)

$$\hat{\boldsymbol{\Phi}}^T\mathbf{K}\hat{\boldsymbol{\Phi}} = \begin{bmatrix} \hat{\boldsymbol{\phi}}_1^T \\ \hat{\boldsymbol{\phi}}_2^T \\ \vdots \\ \hat{\boldsymbol{\phi}}_k^T \end{bmatrix} \begin{bmatrix} k_{11} & k_{12} & \cdots & k_{1n} \\ k_{21} & k_{22} & \ddots & \vdots \\ \vdots & \ddots & \ddots & \vdots \\ k_{n1} & \cdots & \cdots & k_{nn} \end{bmatrix} \begin{bmatrix} \hat{\boldsymbol{\phi}}_1 & \hat{\boldsymbol{\phi}}_2 & \cdots & \hat{\boldsymbol{\phi}}_k \end{bmatrix} = \begin{bmatrix} \hat{k}_1 & 0 & \cdots & 0 \\ 0 & \hat{k}_2 & \ddots & \vdots \\ \vdots & \ddots & \ddots & 0 \\ 0 & \cdots & 0 & \hat{k}_k \end{bmatrix} = \hat{\mathbf{K}}.$$
(4.286c)

Substituting Eq. 4.286 into Eq. 4.284 gives

$$\hat{\mathbf{M}}\ddot{\mathbf{Q}}(t) + \hat{\mathbf{C}}\dot{\mathbf{Q}}(t) + \hat{\mathbf{K}}\mathbf{Q}(t) = \hat{\boldsymbol{\Phi}}^T\mathbf{F}_e(t).$$
(4.287)

Equation 4.287 represents k uncoupled equations, each of which takes the form

$$\hat{m}_i \ddot{q}_i(t) + \hat{c}_i \dot{q}_i(t) + \hat{k}_i q_i(t) = \boldsymbol{\phi}_i^T \mathbf{F}_e(t) \tag{4.288}$$

or alternatively

$$\ddot{q}_i(t) + 2\zeta_i \omega_i \dot{q}_i(t) + \omega_i^2 q_i(t) = \boldsymbol{\phi}_i^T \mathbf{F}_e(t)/\hat{m}_i \tag{4.289}$$

where $2\zeta_i \omega_i = \hat{c}_i/\hat{m}_i$. The solution for Eqs. 4.288 and 4.289 was discussed in Chapter 2.

When Rayleigh damping is used, the values of α and β are unknowns and must be determined through two equations. Structural engineers usually assume two damping ratios for specific modes, giving the two equations. Recall that

$$\hat{c}_i = \boldsymbol{\phi}_i^T \mathbf{C} \boldsymbol{\phi}_i = \boldsymbol{\phi}_i^T (\alpha \mathbf{M} + \beta \mathbf{K}) \boldsymbol{\phi}_i = \alpha \boldsymbol{\phi}_i^T \mathbf{M} \boldsymbol{\phi}_i + \beta \boldsymbol{\phi}_i^T \mathbf{K} \boldsymbol{\phi}_i = \alpha \hat{m}_i + \beta \hat{k}_i. \tag{4.290}$$

Dividing both sides of Eq. 4.290 by $\hat{m}_i$ gives

$$\hat{c}_i/\hat{m}_i = 2\zeta_i \omega_i = \left(\alpha \hat{m}_i + \beta \hat{k}_i\right)/\hat{m}_i = \alpha + \beta \omega_i^2. \tag{4.291}$$

Assume the damping in the first two modes of vibration:

$$2\zeta_1 \omega_1 = \alpha + \beta \omega_1^2 \tag{4.292a}$$

$$2\zeta_2 \omega_2 = \alpha + \beta \omega_2^2. \tag{4.292b}$$

Then solving for α and β gives

$$\alpha = \frac{2\omega_1 \omega_2 (\zeta_1 \omega_2 - \zeta_2 \omega_1)}{\omega_2^2 - \omega_1^2}, \quad \beta = \frac{2(\zeta_2 \omega_2 - \zeta_1 \omega_1)}{\omega_2^2 - \omega_1^2}. \tag{4.293}$$

It follows that the damping in the other normal modes is now uniquely determined because α and β are known. For example, consider the third mode, where the equations for this mode are

$$\boldsymbol{\phi}_3^T \mathbf{C} \boldsymbol{\phi}_3 = \alpha \boldsymbol{\phi}_3^T \mathbf{M} \boldsymbol{\phi}_3 + \beta \boldsymbol{\phi}_3^T \mathbf{K} \boldsymbol{\phi}_3 \tag{4.294}$$

$$\hat{c}_3 = \alpha \hat{m}_3 + \beta \hat{k}_3 \tag{4.295}$$

$$2\zeta_3 \omega_3 = \alpha + \beta \omega_3^2. \tag{4.296}$$

Solving for ζ_3 in Eq. 4.296 gives

$$\zeta_3 = \frac{\alpha + \beta \omega_3^2}{2\omega_3}. \tag{4.297}$$

In general, the damping in the jth mode is defined, following Eq. 4.297, as

$$\zeta_j = \frac{\alpha + \beta \omega_j^2}{2\omega_j}. \tag{4.298}$$

Besides using the first and second modes of vibration, structural engineers can compute the values of α and β using damping ratios from any two modes (e.g., i and j), and the procedure is as follows:

$$2\zeta_i \omega_i = \alpha + \beta \omega_i^2 \tag{4.299a}$$

$$2\zeta_j \omega_j = \alpha + \beta \omega_j^2. \tag{4.299b}$$

Then solving for α and β gives

$$\alpha = \frac{2\omega_i \omega_j (\zeta_i \omega_j - \zeta_j \omega_i)}{\omega_j^2 - \omega_i^2}, \quad \beta = \frac{2(\zeta_j \omega_j - \zeta_i \omega_i)}{\omega_j^2 - \omega_i^2}. \tag{4.300}$$

Once α and β are determined, then, as previously noted, the damping in all other modes is fixed.

EXAMPLE 1   *Mass Proportional Damping*

Assume that the damping matrix is proportional to the mass matrix only, that is,

$$\mathbf{C} = \alpha\mathbf{M}. \tag{4.301}$$

The equation of motion becomes

$$\mathbf{M}\ddot{\mathbf{X}}(t) + \alpha\mathbf{M}\dot{\mathbf{X}}(t) + \mathbf{K}\mathbf{X}(t) = \mathbf{F}_e(t). \tag{4.302}$$

Let $\mathbf{X}(t) = \hat{\mathbf{\Phi}}\mathbf{Q}(t)$; premultiplying Eq. 4.302 by $\hat{\mathbf{\Phi}}^T$ gives

$$\left(\hat{\mathbf{\Phi}}^T\mathbf{M}\hat{\mathbf{\Phi}}\right)\ddot{\mathbf{Q}}(t) + \alpha\left(\hat{\mathbf{\Phi}}^T\mathbf{M}\hat{\mathbf{\Phi}}\right)\dot{\mathbf{Q}}(t) + \left(\hat{\mathbf{\Phi}}^T\mathbf{K}\hat{\mathbf{\Phi}}\right)\mathbf{Q}(t) = \hat{\mathbf{\Phi}}^T\mathbf{F}_e(t) \tag{4.303}$$

and therefore

$$\hat{\mathbf{M}}\ddot{\mathbf{Q}}(t) + \alpha\hat{\mathbf{M}}\dot{\mathbf{Q}}(t) + \hat{\mathbf{K}}\mathbf{Q}(t) = \hat{\mathbf{\Phi}}^T\mathbf{F}_e(t) \tag{4.304}$$

and in uncoupled form,

$$\hat{m}_i\ddot{q}_i(t) + \alpha\hat{m}_i\dot{q}_i(t) + \hat{k}_iq_i(t) = \hat{\mathbf{\phi}}_i^T\mathbf{F}_e(t). \tag{4.305}$$

Equation 4.305 is the uncoupled equation of motion, and it is solved using the methods discussed in Chapter 2. Here, it is interesting to see the effect of α on the damping ratio, ζ_i, and the natural frequency of vibration, ω_i. Recall from Eq. 4.298 that

$$\zeta_j = \frac{\alpha + \beta\omega_j^2}{2\omega_j} \tag{4.306}$$

and for $\beta = 0$, it follows that

$$\zeta_j = \frac{\alpha}{2\omega_j}. \tag{4.307}$$

Figure 4.11 shows a plot of ζ_j versus ω_j for $\alpha = 0.5$.

In practice, a value of damping is selected for the first normal mode of vibration (i.e., ζ_1), and this enables the determination of α using Eq. 4.307, where

$$\alpha = 2\omega_1\zeta_1. \tag{4.308}$$

Substituting Eq. 4.308 into Eq. 4.307 gives the damping ratio associated with all other modes of vibration:

$$\zeta_j = \left(\frac{\omega_1}{\omega_j}\right)\zeta_1. \tag{4.309}$$

Consider as an example a building with 5% damping in the first mode and a fundamental undamped period of vibration equal to 1.0 s. Therefore,

$$\zeta_1 = 0.05, \quad T_1 = 1.0\,\text{s}, \quad \omega_1 = 2\pi. \tag{4.310}$$

It follows from Eq. 4.308 that

$$\alpha = 2(2\pi)(0.05) = 0.2\pi. \tag{4.311}$$

If the second undamped period of vibration is equal to 0.4 s, then

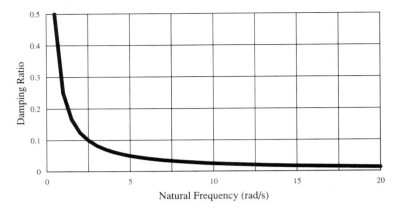

Figure 4.11 Mass Proportional Damping

$$T_2 = 0.4\,\text{s}, \quad \omega_2 = 5\pi, \quad \zeta_2 = \left(\frac{2\pi}{5\pi}\right)0.05 = 0.02 = 2.0\%. \tag{4.312}$$

EXAMPLE 2   Stiffness Proportional Damping

Assume that the damping matrix is proportional to the stiffness matrix only:

$$\mathbf{C} = \beta\mathbf{K}. \tag{4.313}$$

The equation of motion becomes

$$\mathbf{M}\ddot{\mathbf{X}}(t) + \beta\mathbf{K}\dot{\mathbf{X}}(t) + \mathbf{K}\mathbf{X}(t) = \mathbf{F}_e(t). \tag{4.314}$$

Let $\mathbf{X}(t) = \hat{\boldsymbol{\Phi}}\mathbf{Q}(t)$; premultiplying Eq. 4.314 by $\hat{\boldsymbol{\Phi}}^T$ gives

$$\left(\hat{\boldsymbol{\Phi}}^T\mathbf{M}\hat{\boldsymbol{\Phi}}\right)\ddot{\mathbf{Q}}(t) + \beta\left(\hat{\boldsymbol{\Phi}}^T\mathbf{K}\hat{\boldsymbol{\Phi}}\right)\dot{\mathbf{Q}}(t) + \left(\hat{\boldsymbol{\Phi}}^T\mathbf{K}\hat{\boldsymbol{\Phi}}\right)\mathbf{Q}(t) = \hat{\boldsymbol{\Phi}}^T\mathbf{F}_e(t) \tag{4.315}$$

and therefore

$$\hat{\mathbf{M}}\ddot{\mathbf{Q}}(t) + \beta\hat{\mathbf{K}}\dot{\mathbf{Q}}(t) + \hat{\mathbf{K}}\mathbf{Q}(t) = \hat{\boldsymbol{\Phi}}^T\mathbf{F}_e(t) \tag{4.316}$$

and in uncoupled form

$$\hat{m}_i\ddot{q}_i(t) + \beta\hat{k}_i\dot{q}_i(t) + \hat{k}_i q_i(t) = \hat{\boldsymbol{\phi}}_i^T\mathbf{F}_e(t). \tag{4.317}$$

Equation 4.317 is the uncoupled equation of motion, and it is solved using the methods discussed in Chapter 2. Recall from Eq. 4.298 that

$$\zeta_j = \frac{\alpha + \beta\omega_j^2}{2\omega_j} \tag{4.318}$$

and for $\alpha = 0$, it follows that

$$\zeta_j = \frac{\beta\omega_j}{2}. \tag{4.319}$$

Figure 4.12 shows a plot of ζ_j versus ω_j for $\beta = 0.02$.

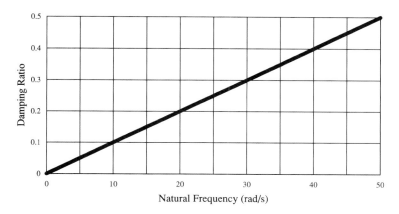

Figure 4.12 Stiffness Proportional Damping

The structural engineer can assume that the structural damping is proportional to the stiffness matrix and select a value of damping for the first normal mode of vibration (i.e., ζ_1). This enables the determination of β using Eq. 4.319, where

$$\beta = \frac{2\zeta_1}{\omega_1}. \tag{4.320}$$

Substituting Eq. 4.320 into Eq. 4.319 gives all of the damping ratios associated with the other modes of vibration:

$$\zeta_j = \left(\frac{\omega_j}{\omega_1}\right)\zeta_1. \tag{4.321}$$

Consider as an example a building with 5% damping in the first mode and a fundamental undamped period of vibration equal to 1.0 s. Therefore,

$$\zeta_1 = 0.05, \quad T_1 = 1.0\,\text{s}, \quad \omega_1 = 2\pi. \tag{4.322}$$

It follows from Eq. 4.320 that

$$\beta = \frac{2(0.05)}{2\pi} = \frac{0.05}{\pi}. \tag{4.323}$$

If the second undamped period of vibration is equal to 0.4 s, then

$$T_2 = 0.4\,\text{s}, \quad \omega_2 = 5\pi, \quad \zeta_2 = \left(\frac{5\pi}{2\pi}\right)0.05 = 0.125 = 12.5\%. \tag{4.324}$$

EXAMPLE 3 *Rayleigh Proportional Damping*

Assume that the damping matrix is proportional to the mass and stiffness matrices:

$$\mathbf{C} = \alpha\mathbf{M} + \beta\mathbf{K}. \tag{4.325}$$

The equation of motion becomes

$$\mathbf{M}\ddot{\mathbf{X}}(t) + (\alpha\mathbf{M} + \beta\mathbf{K})\dot{\mathbf{X}}(t) + \mathbf{K}\mathbf{X}(t) = \mathbf{F}_e(t). \tag{4.326}$$

Let $\mathbf{X}(t) = \hat{\mathbf{\Phi}}\mathbf{Q}(t)$; premultiplying Eq. 4.314 by $\hat{\mathbf{\Phi}}^T$ gives

$$\left(\hat{\boldsymbol{\Phi}}^T\mathbf{M}\hat{\boldsymbol{\Phi}}\right)\ddot{\mathbf{Q}}(t) + \left(\alpha\hat{\boldsymbol{\Phi}}^T\mathbf{M}\hat{\boldsymbol{\Phi}} + \beta\hat{\boldsymbol{\Phi}}^T\mathbf{K}\hat{\boldsymbol{\Phi}}\right)\dot{\mathbf{Q}}(t) + \left(\hat{\boldsymbol{\Phi}}^T\mathbf{K}\hat{\boldsymbol{\Phi}}\right)\mathbf{Q}(t) = \hat{\boldsymbol{\Phi}}^T\mathbf{F}_e(t) \tag{4.327}$$

and therefore

$$\hat{\mathbf{M}}\ddot{\mathbf{Q}}(t) + \left(\alpha\hat{\mathbf{M}} + \beta\hat{\mathbf{K}}\right)\dot{\mathbf{Q}}(t) + \hat{\mathbf{K}}\mathbf{Q}(t) = \hat{\boldsymbol{\Phi}}^T\mathbf{F}_e(t) \tag{4.328}$$

and in uncoupled form

$$\hat{m}_i\ddot{q}_i(t) + \left(\alpha\hat{m}_i + \beta\hat{k}_i\right)\dot{q}_i(t) + \hat{k}_i q_i(t) = \hat{\boldsymbol{\phi}}_i^T\mathbf{F}_e(t). \tag{4.329}$$

Equation 4.329 is the uncoupled equation of motion, and it is solved using the methods discussed in Chapter 2. Here, let's study the effect of combining both α and β on the damping ratio, ζ_i, and the natural frequency of vibration, ω_i. Recall from Eq. 4.298 that

$$\zeta_j = \frac{\alpha + \beta\omega_j^2}{2\omega_j} \tag{4.330}$$

Figure 4.13 shows a plot of ζ_j versus ω_j for $\alpha = 0.5$ and $\beta = 0.005$.

By assuming Rayleigh proportional damping, the structural engineer selects two values of damping for the first and second normal modes of vibration (i.e., ζ_1 and ζ_2). This determines the values of β and α by using Eq. 4.300, where

$$\alpha = \frac{2\omega_i\omega_j\left(\zeta_i\omega_j - \zeta_j\omega_i\right)}{\omega_j^2 - \omega_i^2}, \quad \beta = \frac{2\left(\zeta_j\omega_j - \zeta_i\omega_i\right)}{\omega_j^2 - \omega_i^2}. \tag{4.331}$$

It then follows from Eq. 4.298 for the other modes that

$$\zeta_j = \frac{\alpha + \beta\omega_j^2}{2\omega_j}. \tag{4.332}$$

Consider as an example a building with 5% damping in both the first and third modes and a fundamental undamped period of vibration equal to 1.0 s and a third undamped period of vibration of 0.2 s. Therefore,

$$\zeta_1 = 0.05, \quad T_1 = 1.0\,\text{s}, \quad \omega_1 = 2\pi \tag{4.333}$$

$$\zeta_3 = 0.05, \quad T_3 = 0.2\,\text{s}, \quad \omega_3 = 10\pi. \tag{4.334}$$

It follows from Eq. 4.331 that

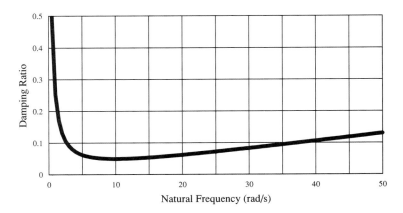

Figure 4.13 Rayleigh Damping

$$\alpha = \frac{2(2\pi)(10\pi)((0.05)(10\pi)-(0.05)(2\pi))}{(10\pi)^2-(2\pi)^2} = \frac{\pi}{6} \qquad (4.335)$$

$$\beta = \frac{2((0.05)(10\pi)-(0.05)(2\pi))}{(10\pi)^2-(2\pi)^2} = \frac{1}{120\pi}. \qquad (4.336)$$

If the second undamped period of vibration is equal to 0.4 s, then

$$T_2 = 0.4\,\text{s}, \quad \omega_2 = 5\pi, \quad \zeta_2 = \frac{1}{2(5\pi)}\left(\frac{\pi}{6}+\frac{(5\pi)^2}{120\pi}\right) = 0.0375 = 3.75\%. \qquad (4.337)$$

EXAMPLE 4   *Damping Matrix for a Three Degrees of Freedom System*

Consider again the three-story building as discussed in Example 2 of Section 4.3. Recall that

$$\mathbf{M} = \begin{bmatrix} m & 0 & 0 \\ 0 & m & 0 \\ 0 & 0 & m \end{bmatrix}, \quad \mathbf{K} = \begin{bmatrix} 3k & -k & 0 \\ -k & 2k & -k \\ 0 & -k & k \end{bmatrix} \qquad (4.338)$$

with the natural frequencies

$$\omega_1^2 = \left(2-\sqrt{3}\right)\omega_n^2, \quad \omega_2^2 = 2\omega_n^2, \quad \omega_3^2 = \left(2+\sqrt{3}\right)\omega_n^2 \qquad (4.339)$$

with the norm normalized mode shapes, given in Example 1 of Section 4.4, equal to

$$\hat{\boldsymbol{\phi}}_1 = \frac{1}{3+\sqrt{3}}\begin{Bmatrix} 1 \\ 1+\sqrt{3} \\ 2+\sqrt{3} \end{Bmatrix}, \quad \hat{\boldsymbol{\phi}}_2 = \frac{1}{\sqrt{3}}\begin{Bmatrix} 1 \\ 1 \\ -1 \end{Bmatrix}, \quad \hat{\boldsymbol{\phi}}_3 = \frac{1}{3-\sqrt{3}}\begin{Bmatrix} 1 \\ 1-\sqrt{3} \\ 2-\sqrt{3} \end{Bmatrix} \qquad (4.340)$$

and finally

$$\hat{\mathbf{M}} = \hat{\boldsymbol{\Phi}}^T\mathbf{M}\hat{\boldsymbol{\Phi}} = \begin{bmatrix} m & 0 & 0 \\ 0 & m & 0 \\ 0 & 0 & m \end{bmatrix}, \quad \hat{\mathbf{K}} = \hat{\boldsymbol{\Phi}}^T\mathbf{K}\hat{\boldsymbol{\Phi}} = k\begin{bmatrix} 2-\sqrt{3} & 0 & 0 \\ 0 & 2 & 0 \\ 0 & 0 & 2+\sqrt{3} \end{bmatrix}. \qquad (4.341)$$

Let the damping of the structure be Rayleigh damping with damping ratios $\zeta_1 = \zeta_2 = 5\%$. It follows from evaluating Eq. 4.293 that

$$\alpha = 0.03789\omega_n, \quad \beta = \frac{0.05176}{\omega_n}. \qquad (4.342)$$

Then Eq. 4.285 becomes

$$\mathbf{C} = 0.03789\,\omega_n\begin{bmatrix} m & 0 & 0 \\ 0 & m & 0 \\ 0 & 0 & m \end{bmatrix} + \frac{0.05176}{\omega_n}\begin{bmatrix} 3k & -k & 0 \\ -k & 2k & -k \\ 0 & -k & k \end{bmatrix}. \qquad (4.343)$$

Recall that $k = m\omega_n^2$. Simplifying Eq. 4.343 gives

$$\mathbf{C} = \begin{bmatrix} 0.19319 & -0.05176 & 0 \\ -0.05176 & 0.14142 & -0.05176 \\ 0 & -0.05176 & 0.08966 \end{bmatrix} m\omega_n. \qquad (4.344)$$

This **C** matrix represents the damping matrix for the structure with 5% damping in both the first and second modes, and it has the same orthogonal property as the mass and stiffness matrices from Eq. 4.286b:

$$\hat{\mathbf{C}} = \boldsymbol{\Phi}^T \mathbf{C} \boldsymbol{\Phi} = m\omega_n \begin{bmatrix} \hat{\boldsymbol{\phi}}_1^T \\ \hat{\boldsymbol{\phi}}_2^T \\ \hat{\boldsymbol{\phi}}_3^T \end{bmatrix} \begin{bmatrix} 0.19319 & -0.05176 & 0 \\ -0.05176 & 0.14142 & -0.05176 \\ 0 & -0.05176 & 0.08966 \end{bmatrix} \begin{bmatrix} \hat{\boldsymbol{\phi}}_1 & \hat{\boldsymbol{\phi}}_2 & \hat{\boldsymbol{\phi}}_3 \end{bmatrix}$$

$$= \begin{bmatrix} 0.05176m\omega_n & 0 & 0 \\ 0 & 0.14142m\omega_n & 0 \\ 0 & 0 & 0.23108m\omega_n \end{bmatrix} = \begin{bmatrix} 0.10m\omega_1 & 0 & 0 \\ 0 & 0.10m\omega_2 & 0 \\ 0 & 0 & 0.1196m\omega_3 \end{bmatrix} \quad (4.345)$$

where the relationship between ω_n and ω_i is given in Eq. 4.339. Note that the value $0.10m\omega_1$ on the right-hand side of Eq. 4.345 is $2\zeta_1 m\omega_1$, $0.10m\omega_2$ is equal to $2\zeta_2 m\omega_2$, and $0.1196m\omega_3$ is equal to $2\zeta_3 m\omega_3$ where ζ_3 can also be computed using Eq. 4.297:

$$\zeta_3 = \frac{\alpha + \beta\omega_3^2}{2\omega_3} = \frac{0.03789\omega_n + \left(0.05176/\omega_n\right)\left(2 + \sqrt{3}\right)\omega_n^2}{2\sqrt{2 + \sqrt{3}}\,\omega_n} = 0.0598 = 5.989\%. \quad (4.346)$$

4.7   DAMPED RESPONSE WITH CAUGHY DAMPING USING THE NORMAL MODE METHOD

Recall that the definition of a proportional damping matrix is one that is diagonalized by premultiplying by the transpose of the modal matrix and postmultiplying by the modal matrix, that is,

$$\hat{\mathbf{C}} = \boldsymbol{\Phi}^T \mathbf{C} \boldsymbol{\Phi} \quad (4.347)$$

where $\hat{\mathbf{C}}$ is a diagonal damping matrix. In the preceding section, it was shown that when the damping matrix **C** is of the form called Rayleigh damping, where

$$\mathbf{C} = \alpha\mathbf{M} + \beta\mathbf{K}, \quad (4.348)$$

then the matrix $\hat{\mathbf{C}}$ is diagonalized. A general extension of the Rayleigh damping concept was proposed by Caughy and is here referred to as *Caughy damping*. It is defined as

$$\mathbf{C} = \sum_{i=0}^{m-1} \alpha_i \mathbf{M}\left(\mathbf{M}^{-1}\mathbf{K}\right)^i = \sum_{i=0}^{m-1} \alpha_i \left(\mathbf{K}\mathbf{M}^{-1}\right)^{i-1}\mathbf{K}. \quad (4.349)$$

For example, when $m = 2$, then

$$\mathbf{C} = \alpha_0 \mathbf{M}\left(\mathbf{M}^{-1}\mathbf{K}\right)^0 + \alpha_1 \mathbf{M}\left(\mathbf{M}^{-1}\mathbf{K}\right)^1 = \alpha_0 \mathbf{M} + \alpha_1 \mathbf{K} \quad (4.350)$$

which is of the same form as Rayleigh damping and is very appropriate for a 2DOF system. Now, for $m = 3$, the Caughy damping matrix takes the form

$$\mathbf{C} = \alpha_0 \mathbf{M}\left(\mathbf{M}^{-1}\mathbf{K}\right)^0 + \alpha_1 \mathbf{M}\left(\mathbf{M}^{-1}\mathbf{K}\right)^1 + \alpha_2 \mathbf{M}\left(\mathbf{M}^{-1}\mathbf{K}\right)^2 = \alpha_0 \mathbf{M} + \alpha_1 \mathbf{K} + \alpha_2 \mathbf{K}\mathbf{M}^{-1}\mathbf{K}. \quad (4.351)$$

It follows that

$$\hat{\boldsymbol{\phi}}_i^T \mathbf{C} \hat{\boldsymbol{\phi}}_j = \alpha_0 \hat{\boldsymbol{\phi}}_i^T \mathbf{M} \hat{\boldsymbol{\phi}}_j + \alpha_1 \hat{\boldsymbol{\phi}}_i^T \mathbf{K} \hat{\boldsymbol{\phi}}_j + \alpha_2 \hat{\boldsymbol{\phi}}_i^T \mathbf{K}\mathbf{M}^{-1}\mathbf{K} \hat{\boldsymbol{\phi}}_j. \quad (4.352)$$

Recall that

$$\omega_j^2 \mathbf{M} \hat{\boldsymbol{\phi}}_j = \mathbf{K} \hat{\boldsymbol{\phi}}_j. \quad (4.353)$$

Premultiplying Eq. 4.353 by $\mathbf{K}\mathbf{M}^{-1}$ gives

$$\omega_j^2 \mathbf{K}\hat{\boldsymbol{\phi}}_j = \mathbf{KM}^{-1}\mathbf{K}\hat{\boldsymbol{\phi}}_j. \tag{4.354}$$

It follows from the third term on the right side of Eq. 4.352 that

$$\alpha_2 \hat{\boldsymbol{\phi}}_i^T \mathbf{KM}^{-1}\mathbf{K}\hat{\boldsymbol{\phi}}_j = \alpha_2 \omega_j^2 \hat{\boldsymbol{\phi}}_i^T \mathbf{K}\hat{\boldsymbol{\phi}}_j = \begin{cases} \alpha_2 \omega_j^2 \hat{k}_j & i = j \\ 0 & i \neq j \end{cases} \tag{4.355}$$

and for $m = 3$, Eq. 4.352 becomes

$$\hat{c}_i = \hat{\boldsymbol{\phi}}_i^T \mathbf{C}\hat{\boldsymbol{\phi}}_i = \alpha_0 \hat{m}_i + \alpha_1 \hat{k}_i + \alpha_2 \omega_i^2 \hat{k}_i. \tag{4.356}$$

For any values of m, a similar result can be obtained by first premultiplying Eq. 4.353 by $\mathbf{KM}^{-1}$ ($m - 2$) times. This gives

$$\left(\mathbf{KM}^{-1}\right)^{m-2}\mathbf{K}\hat{\boldsymbol{\phi}}_j = \omega_j^2\left(\mathbf{KM}^{-1}\right)^{m-3}\mathbf{K}\hat{\boldsymbol{\phi}}_j = \dots = \left(\omega_j^2\right)^{m-2}\mathbf{K}\hat{\boldsymbol{\phi}}_j \tag{4.357}$$

and then

$$\alpha_{m-1}\hat{\boldsymbol{\phi}}_i^T\left(\mathbf{KM}^{-1}\right)^{m-2}\mathbf{K}\hat{\boldsymbol{\phi}}_j = \alpha_{m-1}\left(\omega_j^2\right)^{m-2}\hat{\boldsymbol{\phi}}_i^T\mathbf{K}\hat{\boldsymbol{\phi}}_j = \begin{cases} \alpha_{m-1}\left(\omega_j^2\right)^{m-2}\hat{k}_j & i = j \\ 0 & i \neq j. \end{cases} \tag{4.358}$$

Finally,

$$\hat{c}_i = \hat{\boldsymbol{\phi}}_i^T \mathbf{C}\hat{\boldsymbol{\phi}}_i = \alpha_0 \hat{m}_i + \alpha_1 \hat{k}_i + \alpha_2 \omega_i^2 \hat{k}_i + \dots + \alpha_{m-1}\left(\omega_i^2\right)^{m-2}\hat{k}_i = \sum_{j=0}^{m-1}\alpha_j\left(\omega_i^2\right)^{j-1}\hat{k}_i. \tag{4.359}$$

Now consider the dynamic equilibrium equation,

$$\mathbf{M}\ddot{\mathbf{X}}(t) + \mathbf{C}\dot{\mathbf{X}}(t) + \mathbf{K}\mathbf{X}(t) = \mathbf{F}_e(t) \tag{4.360}$$

where the matrix $\mathbf{C}$ is determined by using Caughy damping. Let

$$\mathbf{X}(t) = \hat{\boldsymbol{\Phi}}\mathbf{Q}(t) \tag{4.361}$$

where

$$\hat{\boldsymbol{\Phi}} = \begin{bmatrix} \hat{\boldsymbol{\phi}}_1 & \hat{\boldsymbol{\phi}}_2 & \cdots & \hat{\boldsymbol{\phi}}_k \end{bmatrix}, \quad \mathbf{Q}(t) = \begin{Bmatrix} q_1(t) & q_2(t) & \cdots & q_k(t) \end{Bmatrix}^T \tag{4.362}$$

and then it follows from Eq. 4.360 that

$$\mathbf{M}\hat{\boldsymbol{\Phi}}\ddot{\mathbf{Q}}(t) + \mathbf{C}\hat{\boldsymbol{\Phi}}\dot{\mathbf{Q}}(t) + \mathbf{K}\hat{\boldsymbol{\Phi}}\mathbf{Q}(t) = \mathbf{F}_e(t). \tag{4.363}$$

Premultiplying Eq. 4.363 by $\hat{\boldsymbol{\Phi}}^T$ gives

$$\hat{\boldsymbol{\Phi}}^T\mathbf{M}\hat{\boldsymbol{\Phi}}\ddot{\mathbf{Q}}(t) + \hat{\boldsymbol{\Phi}}^T\mathbf{C}\hat{\boldsymbol{\Phi}}\dot{\mathbf{Q}}(t) + \hat{\boldsymbol{\Phi}}^T\mathbf{K}\hat{\boldsymbol{\Phi}}\mathbf{Q}(t) = \hat{\boldsymbol{\Phi}}^T\mathbf{F}_e(t). \tag{4.364}$$

Note that

$$\hat{\boldsymbol{\Phi}}^T\mathbf{M}\hat{\boldsymbol{\Phi}} = \begin{bmatrix} \hat{\boldsymbol{\phi}}_1^T \\ \hat{\boldsymbol{\phi}}_2^T \\ \vdots \\ \hat{\boldsymbol{\phi}}_k^T \end{bmatrix}\begin{bmatrix} m_1 & 0 & \cdots & 0 \\ 0 & m_2 & \ddots & \vdots \\ \vdots & \ddots & \ddots & 0 \\ 0 & \cdots & 0 & m_n \end{bmatrix}\begin{bmatrix} \hat{\boldsymbol{\phi}}_1 & \hat{\boldsymbol{\phi}}_2 & \cdots & \hat{\boldsymbol{\phi}}_k \end{bmatrix} = \begin{bmatrix} \hat{m}_1 & 0 & \cdots & 0 \\ 0 & \hat{m}_2 & \ddots & \vdots \\ \vdots & \ddots & \ddots & 0 \\ 0 & \cdots & 0 & \hat{m}_k \end{bmatrix} = \hat{\mathbf{M}} \tag{4.365a}$$

$$\hat{\boldsymbol{\Phi}}^T\mathbf{C}\hat{\boldsymbol{\Phi}} = \begin{bmatrix} \hat{\boldsymbol{\phi}}_1^T \\ \hat{\boldsymbol{\phi}}_2^T \\ \vdots \\ \hat{\boldsymbol{\phi}}_k^T \end{bmatrix}\begin{bmatrix} c_{11} & c_{12} & \cdots & c_{1n} \\ c_{21} & c_{22} & \ddots & \vdots \\ \vdots & \vdots & \ddots & \ddots \\ c_{n1} & \cdots & \cdots & c_{nn} \end{bmatrix}\begin{bmatrix} \hat{\boldsymbol{\phi}}_1 & \hat{\boldsymbol{\phi}}_2 & \cdots & \hat{\boldsymbol{\phi}}_k \end{bmatrix} = \begin{bmatrix} \hat{c}_1 & 0 & \cdots & 0 \\ 0 & \hat{c}_2 & \ddots & \vdots \\ \vdots & \ddots & \ddots & 0 \\ 0 & \cdots & 0 & \hat{c}_k \end{bmatrix} = \hat{\mathbf{C}} \tag{4.365b}$$

$$\hat{\Phi}^T K \hat{\Phi} = \begin{bmatrix} \hat{\Phi}_1^T \\ \hat{\Phi}_2^T \\ \vdots \\ \hat{\Phi}_k^T \end{bmatrix} \begin{bmatrix} k_{11} & k_{12} & \cdots & k_{1n} \\ k_{21} & k_{22} & & \vdots \\ \vdots & & \ddots & \vdots \\ k_{n1} & \cdots & \cdots & k_{nn} \end{bmatrix} \begin{bmatrix} \hat{\Phi}_1 & \hat{\Phi}_2 & \cdots & \hat{\Phi}_k \end{bmatrix} = \begin{bmatrix} \hat{k}_1 & 0 & \cdots & 0 \\ 0 & \hat{k}_2 & \ddots & \vdots \\ \vdots & \ddots & \ddots & 0 \\ 0 & \cdots & 0 & \hat{k}_k \end{bmatrix} = \hat{K}. \quad (4.365c)$$

It follows from Eq. 4.365 that

$$\hat{M}\ddot{Q}(t) + \hat{C}\dot{Q}(t) + \hat{K}Q(t) = \hat{\Phi}^T F_e(t). \quad (4.366)$$

Equation 4.366 represents n uncoupled equations, each of which takes the form

$$\hat{m}_i \ddot{q}_i(t) + \hat{c}_i \dot{q}_i(t) + \hat{k}_i q_i(t) = \hat{\Phi}_i^T F_e(t) \quad (4.367)$$

or

$$\ddot{q}_i(t) + 2\zeta_i \omega_i \dot{q}_i(t) + \omega_i^2 q_i(t) = \hat{\Phi}_i^T F_e(t)/\hat{m}_i \quad (4.368)$$

where $2\zeta_i \omega_i = \hat{c}_i/\hat{m}_i$. With the damping matrix C of the Caughy form, the damping coefficient in the ith mode is calculated using Eq. 4.359. Dividing this equation by $\hat{m}_i$, it follows that

$$\hat{c}_i/\hat{m}_i = 2\zeta_i \omega_i = \left[\alpha_0 \hat{m}_i + \alpha_1 \hat{k}_i + \alpha_2 \omega_i^2 \hat{k}_i + \ldots + \alpha_{m-1}\left(\omega_i^2\right)^{m-2} \hat{k}_i\right]/\hat{m}_i$$
$$= \alpha_0 + \alpha_1 \omega_i^2 + \alpha_2 \omega_i^4 + \ldots + \alpha_{m-1}\left(\omega_i^2\right)^{m-1}. \quad (4.369)$$

The most common method of forming the Caughy damping matrix is to select the value of damping in several modes of vibration. For example, the damping in the first three modes are selected. It then follows from Eq. 4.369 that

$$\alpha_0 + \alpha_1 \omega_1^2 + \alpha_2 \omega_1^4 = 2\zeta_1 \omega_1 \quad (4.370a)$$

$$\alpha_0 + \alpha_1 \omega_2^2 + \alpha_2 \omega_2^4 = 2\zeta_2 \omega_2 \quad (4.370b)$$

$$\alpha_0 + \alpha_1 \omega_3^2 + \alpha_2 \omega_3^4 = 2\zeta_3 \omega_3. \quad (4.370c)$$

In Eq. 4.370, all quantities are known except α_0, α_1, and α_2. Therefore, with three equations these three unknowns can be solved.

Note that after α_0, α_1, and α_2 are calculated, the modal damping is defined for any other mode using Eq. 4.369:

$$\zeta_i = \frac{1}{2\omega_i}\left[\alpha_0 + \alpha_1 \omega_i^2 + \alpha_2 \omega_i^4 + \ldots + \alpha_{m-1}\left(\omega_i^2\right)^{m-1}\right]. \quad (4.371)$$

A comment on the solution of equations like Eq. 4.370 is important. As the number of α_i terms increases, the natural frequency is raised to ever-increasing powers. For example, in Eq. 4.370, corresponding to α_2 is ω_i^4. Therefore, numerical ill-conditioning may result.

EXAMPLE 1   *Three Degrees of Freedom System*

Consider again the three-story building as discussed in Example 2 of Section 4.3. Recall that

$$M = \begin{bmatrix} m & 0 & 0 \\ 0 & m & 0 \\ 0 & 0 & m \end{bmatrix}, \quad K = \begin{bmatrix} 3k & -k & 0 \\ -k & 2k & -k \\ 0 & -k & k \end{bmatrix} \quad (4.372)$$

with the natural frequencies

$$\omega_1^2 = \left(2 - \sqrt{3}\right)\omega_n^2, \quad \omega_2^2 = 2\omega_n^2, \quad \omega_3^2 = \left(2 + \sqrt{3}\right)\omega_n^2 \quad (4.373)$$

with the norm normalized mode shapes, given in Example 1 of Section 4.4, equal to

$$\hat{\boldsymbol{\phi}}_1 = \frac{1}{3 + \sqrt{3}}\begin{Bmatrix} 1 \\ 1 + \sqrt{3} \\ 2 + \sqrt{3} \end{Bmatrix}, \quad \hat{\boldsymbol{\phi}}_2 = \frac{1}{\sqrt{3}}\begin{Bmatrix} 1 \\ 1 \\ -1 \end{Bmatrix}, \quad \hat{\boldsymbol{\phi}}_3 = \frac{1}{3 - \sqrt{3}}\begin{Bmatrix} 1 \\ 1 - \sqrt{3} \\ 2 - \sqrt{3} \end{Bmatrix} \quad (4.374)$$

and finally

$$\hat{\mathbf{M}} = \hat{\boldsymbol{\Phi}}^T \mathbf{M} \hat{\boldsymbol{\Phi}} = \begin{bmatrix} m & 0 & 0 \\ 0 & m & 0 \\ 0 & 0 & m \end{bmatrix}, \quad \hat{\mathbf{K}} = \hat{\boldsymbol{\Phi}}^T \mathbf{K} \hat{\boldsymbol{\Phi}} = k \begin{bmatrix} 2 - \sqrt{3} & 0 & 0 \\ 0 & 2 & 0 \\ 0 & 0 & 2 + \sqrt{3} \end{bmatrix}. \quad (4.375)$$

Let the damping of the structure be $\zeta_1 = \zeta_2 = \zeta_3 = 5\%$. It follows from Eq. 4.370 that

$$\alpha_0 + \alpha_1\left(2 - \sqrt{3}\right)\omega_n^2 + \alpha_2\left(2 - \sqrt{3}\right)^2\omega_n^4 = 2(0.05)\sqrt{2 - \sqrt{3}}\ \omega_n \quad (4.376)$$

$$\alpha_0 + \alpha_1\left(2\omega_n^2\right) + \alpha_2(2)^2\omega_n^4 = 2(0.05)\sqrt{2}\ \omega_n \quad (4.377)$$

$$\alpha_0 + \alpha_1\left(2 + \sqrt{3}\right)\omega_n^2 + \alpha_2\left(2 + \sqrt{3}\right)^2\omega_n^4 = 2(0.05)\sqrt{2 + \sqrt{3}}\ \omega_n. \quad (4.378)$$

Solving for this set of simultaneous equation gives

$$\alpha_0 = 0.03451\omega_n, \quad \alpha_1 = \frac{0.06609}{\omega_n}, \quad \alpha_2 = -\frac{0.00632}{\omega_n^3}. \quad (4.379)$$

It follows from Eq. 4.351 that

$$\mathbf{C} = 0.03451\omega_n\begin{bmatrix} m & 0 & 0 \\ 0 & m & 0 \\ 0 & 0 & m \end{bmatrix} + \frac{0.06609}{\omega_n}\begin{bmatrix} 3k & -k & 0 \\ -k & 2k & -k \\ 0 & -k & k \end{bmatrix}$$

$$- \frac{0.00632}{\omega_n^3}\begin{bmatrix} 3k & -k & 0 \\ -k & 2k & -k \\ 0 & -k & k \end{bmatrix}\begin{bmatrix} 1/m & 0 & 0 \\ 0 & 1/m & 0 \\ 0 & 0 & 1/m \end{bmatrix}\begin{bmatrix} 3k & -k & 0 \\ -k & 2k & -k \\ 0 & -k & k \end{bmatrix}. \quad (4.380)$$

Recall that $k = m\omega_n^2$. Simplifying Eq. 4.380 gives

$$\mathbf{C} = \begin{bmatrix} 0.16962 & -0.0345 & -0.0063 \\ -0.0345 & 0.12879 & -0.0471 \\ -0.0063 & -0.0471 & 0.08797 \end{bmatrix} m\omega_n. \quad (4.381)$$

From Eq. 4.356, it follows that $\hat{c}_i = \hat{\boldsymbol{\phi}}_i^T \mathbf{C} \hat{\boldsymbol{\phi}}_i = \alpha_0\hat{m}_i + \alpha_1\hat{k}_i + \alpha_2\omega_i^2\hat{k}_i$

$$\hat{c}_1 = 0.03451\omega_n m + \frac{0.06609}{\omega_n}\left(2 - \sqrt{3}\right)k - \frac{0.00632}{\omega_n^3}\left(2 - \sqrt{3}\right)\omega_n^2\left(2 - \sqrt{3}\right)k = 0.05176m\omega_n \quad (4.382)$$

$$\hat{c}_2 = 0.03451\omega_n m + \frac{0.06609}{\omega_n}(2k) - \frac{0.00632}{\omega_n^3}2\omega_n^2(2k) = 0.14142m\omega_n \quad (4.383)$$

$$\hat{c}_3 = 0.03451\omega_n m + \frac{0.06609}{\omega_n}\left(2 + \sqrt{3}\right)k - \frac{0.00632}{\omega_n^3}\left(2 + \sqrt{3}\right)\omega_n^2\left(2 + \sqrt{3}\right)k = 0.19319m\omega_n \quad (4.384)$$

and therefore the damping matrix in the normal mode coordinate space is

$$\hat{C} = \begin{bmatrix} 0.05176 & 0 & 0 \\ 0 & 0.14142 & 0 \\ 0 & 0 & 0.19319 \end{bmatrix} m\omega_n = \begin{bmatrix} 0.1m\omega_1 & 0 & 0 \\ 0 & 0.1m\omega_2 & 0 \\ 0 & 0 & 0.1m\omega_3 \end{bmatrix} \qquad (4.385)$$

where the relationship between ω_n and ω_i is given in Eq. 4.373. Note that the value 0.1 on the right-hand side of Eq. 4.385 is 2ζ.

4.8 DAMPED RESPONSE WITH PENZIEN–WILSON DAMPING USING THE NORMAL MODE METHOD

Another form of proportional damping matrix very commonly used is *Penzien–Wilson damping*. In this method, the damping for the normal modes is specified and then the damping matrix is calculated.

Consider a damping matrix that is diagonalized in the normal modes:

$$\overline{C} = \begin{bmatrix} 2\zeta_1\omega_1 & 0 & \cdots & 0 \\ 0 & 2\zeta_2\omega_2 & \ddots & \vdots \\ \vdots & \ddots & \ddots & 0 \\ 0 & \cdots & 0 & 2\zeta_k\omega_k \end{bmatrix} = \overline{\Phi}^T C \overline{\Phi}. \qquad (4.386)$$

For Eq. 4.386 to hold, the mode shapes must be normalized with respect to the mass matrix. This modal matrix is

$$\overline{\Phi} = \begin{bmatrix} \overline{\phi}_1 & \overline{\phi}_2 & \cdots & \overline{\phi}_k \end{bmatrix} = \begin{bmatrix} \overline{\phi}_{11} & \overline{\phi}_{12} & \cdots & \overline{\phi}_{1k} \\ \overline{\phi}_{21} & \overline{\phi}_{22} & \ddots & \vdots \\ \vdots & \ddots & \ddots & \overline{\phi}_{n-1,k} \\ \overline{\phi}_{n1} & \cdots & \overline{\phi}_{n,k-1} & \overline{\phi}_{nk} \end{bmatrix}. \qquad (4.387)$$

The objective is to find C for this given $\overline{C}$. Since $\overline{\Phi}$ is an $n \times k$ matrix, it is not invertible. Therefore, Eq. 4.386 cannot be solved directly. Consider first the case where all mode shapes are obtained (i.e., $k = n$); then $\overline{C}$ can be solved using Eq. 4.386 by premutliplying by $(\overline{\Phi}^T)^{-1}$ and postmultiplying by $(\overline{\Phi})^{-1}$. Doing so gives

$$C = (\overline{\Phi}^T)^{-1} \begin{bmatrix} 2\zeta_1\omega_1 & 0 & \cdots & 0 \\ 0 & 2\zeta_2\omega_2 & \ddots & \vdots \\ \vdots & \ddots & \ddots & 0 \\ 0 & \cdots & 0 & 2\zeta_n\omega_n \end{bmatrix} \overline{\Phi}^{-1}. \qquad (4.388)$$

Recall that

$$\overline{\Phi}^T M \overline{\Phi} = I \qquad (4.389)$$

and therefore premultiplying both sides of Eq. 4.389 by $(\overline{\Phi}^T)^{-1}$ gives

$$M\overline{\Phi} = (\overline{\Phi}^T)^{-1}. \qquad (4.390)$$

Similarly, postmultiplying both sides of Eq. 4.389 by $\overline{\Phi}^{-1}$ gives

$$\overline{\Phi}^T M = \overline{\Phi}^{-1}. \qquad (4.391)$$

Substituting Eqs. 4.390 and 4.391 into Eq. 4.388, it follows that

$$\mathbf{C} = \mathbf{M}\overline{\mathbf{\Phi}} \begin{bmatrix} 2\zeta_1\omega_1 & 0 & \cdots & 0 \\ 0 & 2\zeta_2\omega_2 & \ddots & \vdots \\ \vdots & \ddots & \ddots & 0 \\ 0 & \cdots & 0 & 2\zeta_n\omega_n \end{bmatrix} \overline{\mathbf{\Phi}}^T \mathbf{M} = \mathbf{M}\left(\sum_{i=1}^{n} 2\zeta_i\omega_i\overline{\mathbf{\phi}}_i\overline{\mathbf{\phi}}_i^T\right)\mathbf{M}. \qquad (4.392)$$

Equation 4.392 defines the matrix $\mathbf{C}$. Equation 4.392 assumes that all the damping values are specified. However, the damping values are usually not known in all modes because a structural model may have thousands of degrees of freedom. Therefore, now consider the original problem where only k modes are considered:

$$\overline{\mathbf{C}} = \begin{bmatrix} 2\zeta_1\omega_1 & 0 & \cdots & 0 \\ 0 & 2\zeta_2\omega_2 & \ddots & \vdots \\ \vdots & \ddots & \ddots & 0 \\ 0 & \cdots & 0 & 2\zeta_k\omega_k \end{bmatrix} = \overline{\mathbf{\Phi}}^T \mathbf{C}\overline{\mathbf{\Phi}}. \qquad (4.393)$$

Now $\overline{\mathbf{\Phi}}$ is an $n \times k$ modal matrix. Let $\mathbf{C}$ be similar to Eq. 4.392 but where

$$\mathbf{C} = \mathbf{M}\left(\sum_{i=1}^{k} 2\zeta_i\omega_i\overline{\mathbf{\phi}}_i\overline{\mathbf{\phi}}_i^T\right)\mathbf{M}. \qquad (4.394)$$

Substituting Eq. 4.394 into Eq. 4.393 gives

$$\overline{\mathbf{C}} = \overline{\mathbf{\Phi}}^T \mathbf{M}\left(\sum_{i=1}^{k} 2\zeta_i\omega_i\overline{\mathbf{\phi}}_i\overline{\mathbf{\phi}}_i^T\right)\mathbf{M}\overline{\mathbf{\Phi}} = \sum_{i=1}^{k} 2\zeta_i\omega_i\left(\overline{\mathbf{\Phi}}^T\mathbf{M}\overline{\mathbf{\phi}}_i\right)\left(\overline{\mathbf{\phi}}_i^T\mathbf{M}\overline{\mathbf{\Phi}}\right). \qquad (4.395)$$

Note that

$$\overline{\mathbf{\Phi}}^T\mathbf{M}\overline{\mathbf{\phi}}_i = \begin{bmatrix} \overline{\mathbf{\phi}}_1 & \overline{\mathbf{\phi}}_2 & \cdots & \overline{\mathbf{\phi}}_k \end{bmatrix}^T \mathbf{M}\overline{\mathbf{\phi}}_i = \{\mathbf{I}_i\} \qquad (4.396)$$

where $\{\mathbf{I}_i\}$ denotes a vector with the ith entry equal to one and all other entries equal to zero. Similarly, it follows that

$$\overline{\mathbf{\phi}}_i^T\mathbf{M}\overline{\mathbf{\Phi}} = \overline{\mathbf{\phi}}_i^T\mathbf{M}\begin{bmatrix} \overline{\mathbf{\phi}}_1 & \overline{\mathbf{\phi}}_2 & \cdots & \overline{\mathbf{\phi}}_k \end{bmatrix} = \{\mathbf{I}_i\}^T. \qquad (4.397)$$

Therefore, Eq. 4.395 becomes

$$\overline{\mathbf{C}} = \sum_{i=1}^{k} 2\zeta_i\omega_i\{\mathbf{I}_i\}\{\mathbf{I}_i\}^T = \begin{bmatrix} 2\zeta_1\omega_1 & 0 & \cdots & 0 \\ 0 & 2\zeta_2\omega_2 & \ddots & \vdots \\ \vdots & \ddots & \ddots & 0 \\ 0 & \cdots & 0 & 2\zeta_k\omega_k \end{bmatrix} \qquad (4.398)$$

which is the desired result. Structural engineers often like to choose the damping values for each mode. The freedom of this choice and the simplicity of the computation make this type of damping very attractive.

After the damping matrix is formulated, the normal mode method can be used to calculate the response of the system by transforming the response in Cartesian coordinates to the normal mode coordinates. Start with the dynamic equilibrium equation

$$\mathbf{M}\ddot{\mathbf{X}}(t) + \mathbf{C}\dot{\mathbf{X}}(t) + \mathbf{K}\mathbf{X}(t) = \mathbf{F}_e(t) \qquad (4.399)$$

where the matrix $\mathbf{C}$ is determined using Penzien–Wilson damping. Let

$$\mathbf{X}(t) = \overline{\mathbf{\Phi}}\mathbf{Q}(t) \qquad (4.400)$$

where

$$\overline{\Phi} = \begin{bmatrix} \overline{\phi}_1 & \overline{\phi}_2 & \cdots & \overline{\phi}_k \end{bmatrix}, \quad Q(t) = \begin{Bmatrix} q_1(t) & q_2(t) & \cdots & q_k(t) \end{Bmatrix}^T$$

and then it follows from Eq. 4.399 that

$$\mathbf{M}\overline{\Phi}\ddot{Q}(t) + \mathbf{C}\overline{\Phi}\dot{Q}(t) + \mathbf{K}\overline{\Phi}Q(t) = F_e(t). \tag{4.401}$$

Premultiplying Eq. 4.401 by $\overline{\Phi}^T$ gives

$$\overline{\Phi}^T\mathbf{M}\overline{\Phi}\ddot{Q}(t) + \overline{\Phi}^T\mathbf{C}\overline{\Phi}\dot{Q}(t) + \overline{\Phi}^T\mathbf{K}\overline{\Phi}Q(t) = \overline{\Phi}^T F_e(t). \tag{4.402}$$

Note that

$$\overline{\Phi}^T\mathbf{M}\overline{\Phi} = \begin{bmatrix} \overline{\phi}_1^T \\ \overline{\phi}_2^T \\ \vdots \\ \overline{\phi}_k^T \end{bmatrix} \begin{bmatrix} m_1 & 0 & \cdots & 0 \\ 0 & m_2 & \ddots & \vdots \\ \vdots & \ddots & \ddots & 0 \\ 0 & \cdots & 0 & m_n \end{bmatrix} \begin{bmatrix} \overline{\phi}_1 & \overline{\phi}_2 & \cdots & \overline{\phi}_k \end{bmatrix} = \begin{bmatrix} 1 & 0 & \cdots & 0 \\ 0 & 1 & \ddots & \vdots \\ \vdots & \ddots & \ddots & 0 \\ 0 & \cdots & 0 & 1 \end{bmatrix} = \mathbf{I} \tag{4.403a}$$

$$\overline{\Phi}^T\mathbf{C}\overline{\Phi} = \begin{bmatrix} \overline{\phi}_1^T \\ \overline{\phi}_2^T \\ \vdots \\ \overline{\phi}_k^T \end{bmatrix} \begin{bmatrix} c_{11} & c_{12} & \cdots & c_{1n} \\ c_{21} & c_{22} & \ddots & \vdots \\ \vdots & \ddots & \ddots & \vdots \\ c_{n1} & \cdots & \cdots & c_{nn} \end{bmatrix} \begin{bmatrix} \overline{\phi}_1 & \overline{\phi}_2 & \cdots & \overline{\phi}_k \end{bmatrix} = \begin{bmatrix} 2\zeta_1\omega_1 & 0 & \cdots & 0 \\ 0 & 2\zeta_2\omega_2 & \ddots & \vdots \\ \vdots & \ddots & \ddots & 0 \\ 0 & \cdots & 0 & 2\zeta_k\omega_k \end{bmatrix} = \overline{\mathbf{C}} \tag{4.403b}$$

$$\overline{\Phi}^T\mathbf{K}\overline{\Phi} = \begin{bmatrix} \overline{\phi}_1^T \\ \overline{\phi}_2^T \\ \vdots \\ \overline{\phi}_k^T \end{bmatrix} \begin{bmatrix} k_{11} & k_{12} & \cdots & k_{1n} \\ k_{21} & k_{22} & \ddots & \vdots \\ \vdots & \ddots & \ddots & \vdots \\ k_{n1} & \cdots & \cdots & k_{nn} \end{bmatrix} \begin{bmatrix} \overline{\phi}_1 & \overline{\phi}_2 & \cdots & \overline{\phi}_k \end{bmatrix} = \begin{bmatrix} \omega_1^2 & 0 & \cdots & 0 \\ 0 & \omega_2^2 & \ddots & \vdots \\ \vdots & \ddots & \ddots & 0 \\ 0 & \cdots & 0 & \omega_k^2 \end{bmatrix} = \overline{\mathbf{K}}. \tag{4.403c}$$

It follows from Eq. 4.402 that

$$\ddot{Q}(t) + \overline{\mathbf{C}}\dot{Q}(t) + \overline{\mathbf{K}}Q(t) = \overline{\Phi}^T F_e(t). \tag{4.404}$$

Equation 4.404 represents k uncoupled second-order linear differential equations, each of which takes the form

$$\ddot{q}_i(t) + 2\zeta_i\omega_i\dot{q}_i(t) + \omega_i^2 q_i(t) = \overline{\phi}_i^T F_e(t). \tag{4.405}$$

The solution to Eq. 4.405 was discussed in Chapter 2.

EXAMPLE 1   *Three Degrees of Freedom System*

Consider again the three-story building discussed in Example 2 of Section 4.3. Recall that

$$\mathbf{M} = \begin{bmatrix} m & 0 & 0 \\ 0 & m & 0 \\ 0 & 0 & m \end{bmatrix}, \quad \mathbf{K} = \begin{bmatrix} 3k & -k & 0 \\ -k & 2k & -k \\ 0 & -k & k \end{bmatrix} \tag{4.406}$$

with the natural frequencies

$$\omega_1^2 = \left(2 - \sqrt{3}\right)\omega_n^2, \quad \omega_2^2 = 2\omega_n^2, \quad \omega_3^2 = \left(2 + \sqrt{3}\right)\omega_n^2 \tag{4.407}$$

and with the mass normalized mode shapes, given in Example 2 of Section 4.4 in Eq. 4.204, equal to

$$\overline{\boldsymbol{\phi}}_1 = \frac{m^{-1/2}}{3+\sqrt{3}} \begin{Bmatrix} 1 \\ 1+\sqrt{3} \\ 2+\sqrt{3} \end{Bmatrix}, \quad \overline{\boldsymbol{\phi}}_2 = \frac{m^{-1/2}}{\sqrt{3}} \begin{Bmatrix} 1 \\ 1 \\ -1 \end{Bmatrix}, \quad \overline{\boldsymbol{\phi}}_3 = \frac{m^{-1/2}}{3-\sqrt{3}} \begin{Bmatrix} 1 \\ 1-\sqrt{3} \\ 2-\sqrt{3} \end{Bmatrix}. \tag{4.408}$$

Note that

$$\overline{\boldsymbol{\phi}}_1 \overline{\boldsymbol{\phi}}_1^T = \left[\frac{m^{-1/2}}{3+\sqrt{3}}\right]^2 \begin{Bmatrix} 1 \\ 1+\sqrt{3} \\ 2+\sqrt{3} \end{Bmatrix} \begin{Bmatrix} 1 \\ 1+\sqrt{3} \\ 2+\sqrt{3} \end{Bmatrix}^T = \frac{m^{-1}}{12+6\sqrt{3}} \begin{bmatrix} 1 & 1+\sqrt{3} & 2+\sqrt{3} \\ 1+\sqrt{3} & 4+2\sqrt{3} & 5+3\sqrt{3} \\ 2+\sqrt{3} & 5+3\sqrt{3} & 7+4\sqrt{3} \end{bmatrix} \tag{4.409}$$

$$\overline{\boldsymbol{\phi}}_2 \overline{\boldsymbol{\phi}}_2^T = \left[\frac{m^{-1/2}}{\sqrt{3}}\right]^2 \begin{Bmatrix} 1 \\ 1 \\ -1 \end{Bmatrix} \begin{Bmatrix} 1 \\ 1 \\ -1 \end{Bmatrix}^T = \frac{m^{-1}}{3} \begin{bmatrix} 1 & 1 & -1 \\ 1 & 1 & -1 \\ -1 & -1 & 1 \end{bmatrix} \tag{4.410}$$

$$\overline{\boldsymbol{\phi}}_3 \overline{\boldsymbol{\phi}}_3^T = \left[\frac{m^{-1/2}}{3-\sqrt{3}}\right]^2 \begin{Bmatrix} 1 \\ 1-\sqrt{3} \\ 2-\sqrt{3} \end{Bmatrix} \begin{Bmatrix} 1 \\ 1-\sqrt{3} \\ 2-\sqrt{3} \end{Bmatrix}^T = \frac{m^{-1}}{12-6\sqrt{3}} \begin{bmatrix} 1 & 1-\sqrt{3} & 2-\sqrt{3} \\ 1-\sqrt{3} & 4-2\sqrt{3} & 5-3\sqrt{3} \\ 2-\sqrt{3} & 5-3\sqrt{3} & 7-4\sqrt{3} \end{bmatrix}. \tag{4.411}$$

Therefore,

$$\sum_{i=1}^{3} 2\zeta_i \omega_i \overline{\boldsymbol{\phi}}_i \overline{\boldsymbol{\phi}}_i^T = \frac{2\zeta_1\omega_1 m^{-1}}{12+6\sqrt{3}} \begin{bmatrix} 1 & 1+\sqrt{3} & 2+\sqrt{3} \\ 1+\sqrt{3} & 4+2\sqrt{3} & 5+3\sqrt{3} \\ 2+\sqrt{3} & 5+3\sqrt{3} & 7+4\sqrt{3} \end{bmatrix}$$
$$+ \frac{2\zeta_2\omega_2 m^{-1}}{3} \begin{bmatrix} 1 & 1 & -1 \\ 1 & 1 & -1 \\ -1 & -1 & 1 \end{bmatrix} + \frac{2\zeta_3\omega_3 m^{-1}}{12-6\sqrt{3}} \begin{bmatrix} 1 & 1-\sqrt{3} & 2-\sqrt{3} \\ 1-\sqrt{3} & 4-2\sqrt{3} & 5-3\sqrt{3} \\ 2-\sqrt{3} & 5-3\sqrt{3} & 7-4\sqrt{3} \end{bmatrix}. \tag{4.412}$$

Finally,

$$\mathbf{C} = \mathbf{M}\left(\sum_{i=1}^{3} 2\zeta_i \omega_i \overline{\boldsymbol{\phi}}_i \overline{\boldsymbol{\phi}}_i^T\right)\mathbf{M} = m^2\mathbf{I}\left(\sum_{i=1}^{3} 2\zeta_i \omega_i \overline{\boldsymbol{\phi}}_i \overline{\boldsymbol{\phi}}_i^T\right)\mathbf{I} = m^2\left(\sum_{i=1}^{3} 2\zeta_i \omega_i \overline{\boldsymbol{\phi}}_i \overline{\boldsymbol{\phi}}_i^T\right)$$
$$= \frac{2\zeta_1\omega_1 m}{12+6\sqrt{3}} \begin{bmatrix} 1 & 1+\sqrt{3} & 2+\sqrt{3} \\ 1+\sqrt{3} & 4+2\sqrt{3} & 5+3\sqrt{3} \\ 2+\sqrt{3} & 5+3\sqrt{3} & 7+4\sqrt{3} \end{bmatrix}$$
$$+ \frac{2\zeta_2\omega_2 m}{3} \begin{bmatrix} 1 & 1 & -1 \\ 1 & 1 & -1 \\ -1 & -1 & 1 \end{bmatrix} + \frac{2\zeta_3\omega_3 m}{12-6\sqrt{3}} \begin{bmatrix} 1 & 1-\sqrt{3} & 2-\sqrt{3} \\ 1-\sqrt{3} & 4-2\sqrt{3} & 5-3\sqrt{3} \\ 2-\sqrt{3} & 5-3\sqrt{3} & 7-4\sqrt{3} \end{bmatrix}. \tag{4.413}$$

It then follows from Eq. 4.403b that

$$\overline{\mathbf{C}} = \overline{\boldsymbol{\Phi}}^T \mathbf{C} \overline{\boldsymbol{\Phi}} = \begin{bmatrix} 2\zeta_1\omega_1 & 0 & 0 \\ 0 & 2\zeta_2\omega_2 & 0 \\ 0 & 0 & 2\zeta_3\omega_3 \end{bmatrix}. \tag{4.414}$$

Now, assume that only the first two normal modes are used to calculate the damping matrix; instead of a summation from 1 to 3 in Eq. 4.413, the sum goes from only 1 to 2. It follows that

$$\mathbf{C} = \mathbf{M} \left(\sum_{i=1}^{2} 2\zeta_i \omega_i \overline{\phi}_i \overline{\phi}_i^{T} \right) \mathbf{M} = m^2 \mathbf{I} \left(\sum_{i=1}^{2} 2\zeta_i \omega_i \overline{\phi}_i \overline{\phi}_i^{T} \right) \mathbf{I} = m^2 \left(\sum_{i=1}^{2} 2\zeta_i \omega_i \overline{\phi}_i \overline{\phi}_i^{T} \right)$$

$$= \frac{2\zeta_1 \omega_1 m}{12 + 6\sqrt{3}} \begin{bmatrix} 1 & 1+\sqrt{3} & 2+\sqrt{3} \\ 1+\sqrt{3} & 4+2\sqrt{3} & 5+3\sqrt{3} \\ 2+\sqrt{3} & 5+3\sqrt{3} & 7+4\sqrt{3} \end{bmatrix} + \frac{2\zeta_2 \omega_2 m}{3} \begin{bmatrix} 1 & 1 & -1 \\ 1 & 1 & -1 \\ -1 & -1 & 1 \end{bmatrix}. \tag{4.415}$$

EXAMPLE 2 Formulation of Dynamic Matrices

Consider a three-story building with a unit mass on each floor and the same stiffness between the floors. The mass and stiffness matrices for this structure are

$$\mathbf{M} = \begin{bmatrix} 1.0 & 0.0 & 0.0 \\ 0.0 & 1.0 & 0.0 \\ 0.0 & 0.0 & 1.0 \end{bmatrix}, \quad \mathbf{K} = \begin{bmatrix} 2k & -k & 0 \\ -k & 2k & -k \\ 0 & -k & k \end{bmatrix}. \tag{4.416}$$

The value of k can be determined for any given fundamental period of vibration. Assume that $T_1 = 0.5$ s; then k can be determined from Eq. 4.94, that is,

$$\left| -\omega_i^2 \mathbf{M} + \mathbf{K} \right| = \begin{vmatrix} -\omega_i^2 + 2k & -k & 0 \\ -k & -\omega_i^2 + 2k & -k \\ 0 & -k & -\omega_i^2 + k \end{vmatrix} = 0 \tag{4.417}$$

and the eigenvalue equation is

$$k^3 - 6k^2 \omega_i^2 + 5k \omega_i^4 - \omega_i^6 = 0. \tag{4.418}$$

For $T_1 = 0.5$ s, it follows that

$$\omega_1 = 2\pi/T_1 = 2\pi/0.5 = 12.57 \text{ rad/s} \tag{4.419}$$

and therefore

$$k^3 - 6k^2 (12.57)^2 + 5k (12.57)^4 - (12.57)^6 = 0. \tag{4.420}$$

Solving Eq. 4.420 for k gives three solutions:

$$k_1 = 797.25 \text{ k/in.}, \quad k_2 = 101.55 \text{ k/in.}, \quad k_3 = 48.63 \text{ k/in.} \tag{4.421}$$

Although there are three solutions to Eq. 4.420, only $k_1 = 797.25$ k/in. gives $\omega_1 = 12.57$ rad/s. Therefore, the stiffness matrix becomes

$$\mathbf{K} = \begin{bmatrix} 1594.5 & -797.25 & 0 \\ -797.25 & 1594.5 & -797.25 \\ 0 & -797.25 & 797.25 \end{bmatrix}. \tag{4.422}$$

Substituting k into Eq. 4.418, it follows that

$$(797.25)^3 - 6(797.25)^2 \omega_i^2 + 5(797.25) \omega_i^4 - \omega_i^6 = 0 \tag{4.423}$$

and solving for ω_i^2 gives

$$\omega_1 = 12.57 \text{ rad/s}, \quad \omega_2 = 35.21 \text{ rad/s}, \quad \omega_3 = 50.88 \text{ rad/s}. \tag{4.424}$$

The corresponding undamped natural periods of vibration are

$$T_1 = 0.5 \text{ s}, \quad T_2 = 0.18 \text{ s}, \quad T_3 = 0.12 \text{ s}. \tag{4.425}$$

The mode shapes corresponding to each natural frequency can be calculated using the equation

$$\begin{bmatrix} -\omega_i^2 + 2k & -k & 0 \\ -k & -\omega_i^2 + 2k & -k \\ 0 & -k & -\omega_i^2 + k \end{bmatrix} \begin{Bmatrix} \varphi_{1i} \\ \varphi_{2i} \\ \varphi_{3i} \end{Bmatrix} = \begin{Bmatrix} 0 \\ 0 \\ 0 \end{Bmatrix} \qquad (4.426)$$

where $k_1 = 797.25$ k/in. Select for each mode of vibration $\varphi_{3i} = 1.0$, and then it follows from Eq. 4.426 that

$$\omega_i = \omega_1, \quad \begin{bmatrix} -\omega_1^2 + 2k & -k & 0 \\ -k & -\omega_1^2 + 2k & -k \\ 0 & -k & -\omega_1^2 + k \end{bmatrix} \begin{Bmatrix} \varphi_{11} \\ \varphi_{21} \\ 1.0 \end{Bmatrix} = \begin{Bmatrix} 0 \\ 0 \\ 0 \end{Bmatrix}, \quad \begin{Bmatrix} \varphi_{11} \\ \varphi_{21} \\ \varphi_{31} \end{Bmatrix} = \begin{Bmatrix} 0.445 \\ 0.802 \\ 1.0 \end{Bmatrix} \qquad (4.427)$$

$$\omega_i = \omega_2, \quad \begin{bmatrix} -\omega_2^2 + 2k & -k & 0 \\ -k & -\omega_2^2 + 2k & -k \\ 0 & -k & -\omega_2^2 + k \end{bmatrix} \begin{Bmatrix} \varphi_{12} \\ \varphi_{22} \\ 1.0 \end{Bmatrix} = \begin{Bmatrix} 0 \\ 0 \\ 0 \end{Bmatrix}, \quad \begin{Bmatrix} \varphi_{12} \\ \varphi_{22} \\ \varphi_{32} \end{Bmatrix} = \begin{Bmatrix} -1.247 \\ -0.555 \\ 1.0 \end{Bmatrix} \qquad (4.428)$$

$$\omega_i = \omega_3, \quad \begin{bmatrix} -\omega_3^2 + 2k & -k & 0 \\ -k & -\omega_3^2 + 2k & -k \\ 0 & -k & -\omega_3^2 + k \end{bmatrix} \begin{Bmatrix} \varphi_{13} \\ \varphi_{23} \\ 1.0 \end{Bmatrix} = \begin{Bmatrix} 0 \\ 0 \\ 0 \end{Bmatrix}, \quad \begin{Bmatrix} \varphi_{13} \\ \varphi_{23} \\ \varphi_{33} \end{Bmatrix} = \begin{Bmatrix} 1.802 \\ -2.247 \\ 1.0 \end{Bmatrix}. \qquad (4.429)$$

The mode shapes are shown in Figure 4.14.

Using the results from Eqs. 4.427 to 4.429, the following matrices can now be calculated:

$$\mathbf{\Phi} = \begin{bmatrix} 0.445 & -1.247 & 1.802 \\ 0.802 & -0.555 & -2.247 \\ 1.0 & 1.0 & 1.0 \end{bmatrix}, \quad \hat{\mathbf{\Phi}} = \overline{\mathbf{\Phi}} = \begin{bmatrix} 0.328 & -0.737 & 0.591 \\ 0.591 & -0.328 & -0.737 \\ 0.737 & 0.591 & 0.328 \end{bmatrix} \qquad (4.430)$$

$$\mathbf{M}^* = \begin{bmatrix} m_1^* & 0 & 0 \\ 0 & m_2^* & 0 \\ 0 & 0 & m_3^* \end{bmatrix} = \mathbf{\Phi}^T \begin{bmatrix} 1.0 & 0.0 & 0.0 \\ 0.0 & 1.0 & 0.0 \\ 0.0 & 0.0 & 1.0 \end{bmatrix} \mathbf{\Phi} = \begin{bmatrix} 1.841 & 0 & 0 \\ 0 & 2.863 & 0 \\ 0 & 0 & 9.296 \end{bmatrix} \qquad (4.431)$$

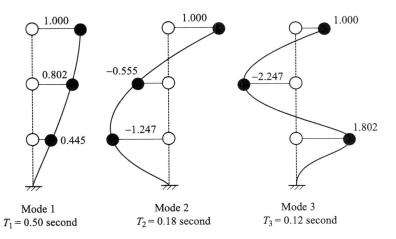

Mode 1
$T_1 = 0.50$ second

Mode 2
$T_2 = 0.18$ second

Mode 3
$T_3 = 0.12$ second

Figure 4.14 Mode Shapes of Vibration

$$\hat{\mathbf{M}} = \begin{bmatrix} \hat{m}_1 & 0 & 0 \\ 0 & \hat{m}_2 & 0 \\ 0 & 0 & \hat{m}_3 \end{bmatrix} = \overline{\mathbf{M}} = \begin{bmatrix} \overline{m}_1 & 0 & 0 \\ 0 & \overline{m}_2 & 0 \\ 0 & 0 & \overline{m}_3 \end{bmatrix} = \begin{bmatrix} 1.0 & 0.0 & 0.0 \\ 0.0 & 1.0 & 0.0 \\ 0.0 & 0.0 & 1.0 \end{bmatrix}. \tag{4.432}$$

The system damping matrix can now be calculated because the natural frequencies and mode shapes are known. First, assume that the system damping is of the Rayleigh damping type and that the objective is to find the damping matrix $\mathbf{C}$ for $\zeta_1 = \zeta_2 = 5\%$. It follows from Eq. 4.293 that

$$\alpha = \frac{2\omega_1\omega_2(\zeta_1\omega_2 - \zeta_2\omega_1)}{\omega_2^2 - \omega_1^2} = 0.9263, \quad \beta = \frac{2(\zeta_2\omega_2 - \zeta_1\omega_1)}{\omega_2^2 - \omega_1^2} = 0.002093 \tag{4.433}$$

and therefore

$$\mathbf{C} = \alpha\mathbf{M} + \beta\mathbf{K} = \begin{bmatrix} 4.264 & -1.669 & 0 \\ -1.669 & 4.264 & -1.669 \\ 0 & -1.669 & 2.595 \end{bmatrix}. \tag{4.434}$$

It follows that ζ_3 is now fixed and is as given in Eq. 4.297, that is,

$$\zeta_3 = \frac{\alpha + \beta\omega_3^2}{2\omega_3} = 0.0623 = 6.23\% \tag{4.435}$$

and then

$$\mathbf{C}^* = \begin{bmatrix} 2\zeta_1\omega_1/m_1^* & 0 & 0 \\ 0 & 2\zeta_2\omega_2/m_2^* & 0 \\ 0 & 0 & 2\zeta_3\omega_3/m_3^* \end{bmatrix} = \begin{bmatrix} 0.6828 & 0 & 0 \\ 0 & 1.2298 & 0 \\ 0 & 0 & 0.6820 \end{bmatrix} \tag{4.436}$$

$$\hat{\mathbf{C}} = \overline{\mathbf{C}} = \begin{bmatrix} 2\zeta_1\omega_1/\hat{m}_1 & 0 & 0 \\ 0 & 2\zeta_2\omega_2/\hat{m}_2 & 0 \\ 0 & 0 & 2\zeta_3\omega_3/\hat{m}_3 \end{bmatrix} = \begin{bmatrix} 1.257 & 0 & 0 \\ 0 & 3.521 & 0 \\ 0 & 0 & 6.340 \end{bmatrix}. \tag{4.437}$$

Now, instead of assuming Rayleigh damping, assume that the Caughy damping method is used with three damping ratios given: $\zeta_1 = \zeta_2 = 5\%$ and $\zeta_3 = 6.23\%$. It then follows from Eq. 4.370 that

$$\alpha_0 + \alpha_1(12.57)^2 + \alpha_2(12.57)^4 = 2(0.05)(12.57) \tag{4.438}$$

$$\alpha_0 + \alpha_1(35.21)^2 + \alpha_2(35.21)^4 = 2(0.05)(35.21) \tag{4.439}$$

$$\alpha_0 + \alpha_1(50.88)^2 + \alpha_2(50.88)^4 = 2(0.0623)(50.88). \tag{4.440}$$

Solving this set of simultaneous equations gives

$$\alpha_0 = 0.9263, \quad \alpha_1 = 0.002093, \quad \alpha_2 = 0.0 \tag{4.441}$$

and it follows from Eq. 4.351 that

$$\mathbf{C} = \alpha_0\mathbf{M} + \alpha_1\mathbf{K} + \alpha_2\mathbf{K}\mathbf{M}^{-1}\mathbf{K} = \begin{bmatrix} 4.264 & -1.669 & 0 \\ -1.669 & 4.264 & -1.669 \\ 0 & -1.669 & 2.595 \end{bmatrix}. \tag{4.442}$$

Finally, assume that the damping matrix is formed using the Penzien–Wilson damping method with damping ratios $\zeta_1 = \zeta_2 = 5\%$ and $\zeta_3 = 6.23\%$. It follows from Eq. 4.392 that

$$\mathbf{C} = \begin{bmatrix} 1.0 & 0.0 & 0.0 \\ 0.0 & 1.0 & 0.0 \\ 0.0 & 0.0 & 1.0 \end{bmatrix} \left(\sum_{i=1}^{3} 2\zeta_i \omega_i \overline{\boldsymbol{\phi}}_i \overline{\boldsymbol{\phi}}_i^T \right) \begin{bmatrix} 1.0 & 0.0 & 0.0 \\ 0.0 & 1.0 & 0.0 \\ 0.0 & 0.0 & 1.0 \end{bmatrix} = \sum_{i=1}^{3} 2\zeta_i \omega_i \overline{\boldsymbol{\phi}}_i \overline{\boldsymbol{\phi}}_i^T. \tag{4.443}$$

Therefore,

$$\mathbf{C} = 2(0.05)(12.57) \begin{Bmatrix} 0.328 \\ 0.591 \\ 0.737 \end{Bmatrix} \begin{Bmatrix} 0.328 \\ 0.591 \\ 0.737 \end{Bmatrix}^T$$

$$+ 2(0.05)(35.21) \begin{Bmatrix} -0.737 \\ -0.328 \\ 0.591 \end{Bmatrix} \begin{Bmatrix} -0.737 \\ -0.328 \\ 0.591 \end{Bmatrix}^T + 2(0.0623)(50.88) \begin{Bmatrix} 0.591 \\ -0.737 \\ 0.328 \end{Bmatrix} \begin{Bmatrix} 0.591 \\ -0.737 \\ 0.328 \end{Bmatrix}^T \tag{4.444}$$

and simplifying Eq. 4.444 gives

$$\mathbf{C} = \begin{bmatrix} 4.264 & -1.669 & 0 \\ -1.669 & 4.264 & -1.669 \\ 0 & -1.669 & 2.595 \end{bmatrix}. \tag{4.445}$$

The proportional damping matrix $\mathbf{C}$ is the same for all three methods because the damping ratios in each mode of vibration for the three methods are equal.

4.9   DAMPED RESPONSE WITH NONPROPORTIONAL DAMPING

Sections 4.6 to 4.8 presented solution methods that require the damping matrix to be diagonalized when it is premultiplied by the transpose of the modal matrix and postmultiplied by the modal matrix. Structural engineering systems with this type of damping matrix are called systems with proportional damping.

   If a structural system does not have proportional damping, two solution options exist. One option is to perform a numerical solution for the response using methods such as the Newmark β-method or the Wilson θ-method. The second option is to transform the equation of motion into first-order equations and then use the state space solution method. The first option is discussed in this section and the second option is discussed in the following section.

   Calculating the response of an MDOF system using the discretization method of numerical analysis extends directly from the method for the 2DOF system as discussed in Section 3.7. The only difference between 2DOF and MDOF systems is that the MDOF matrix presentation is much bigger than the 2DOF matrix presentation, and all equations discussed in Section 3.7 can be used with all the matrices and vectors containing as many as n terms.

   The solution for the response of a nonproportional damped system using the Newmark β-method of discretization is

$$\mathbf{q}_{k+1} = \begin{Bmatrix} \mathbf{X}_{k+1} \\ \dot{\mathbf{X}}_{k+1} \\ \ddot{\mathbf{X}}_{k+1} \end{Bmatrix} = \mathbf{F}_N \begin{Bmatrix} \mathbf{X}_k \\ \dot{\mathbf{X}}_k \\ \ddot{\mathbf{X}}_k \end{Bmatrix} + \mathbf{H}_N \mathbf{F}_{k+1} = \mathbf{F}_N \mathbf{q}_k + \mathbf{H}_N \mathbf{F}_{k+1} \tag{4.446}$$

where

$$\mathbf{F}_N = \begin{bmatrix} \mathbf{I} - \alpha\mathbf{B}^{-1}\mathbf{K}(\Delta t)^2 & \mathbf{I}\Delta t - \alpha\mathbf{B}^{-1}\mathbf{C}(\Delta t)^2 - \alpha\mathbf{B}^{-1}\mathbf{K}(\Delta t)^3 & \left(\tfrac{1}{2}-\alpha\right)\mathbf{I}(\Delta t)^2 - \alpha\mathbf{B}^{-1}\mathbf{G}(\Delta t)^2 \\ -\delta\mathbf{B}^{-1}\mathbf{K}\Delta t & \mathbf{I} - \delta\mathbf{B}^{-1}\mathbf{C}\Delta t - \delta\mathbf{B}^{-1}\mathbf{K}(\Delta t)^2 & (1-\delta)\mathbf{I}\Delta t - \delta\mathbf{B}^{-1}\mathbf{G}\Delta t \\ -\mathbf{B}^{-1}\mathbf{K} & -\mathbf{B}^{-1}\mathbf{C} - \mathbf{B}^{-1}\mathbf{K}\Delta t & -\mathbf{B}^{-1}\mathbf{G} \end{bmatrix}$$

$$\mathbf{H}_N = \begin{bmatrix} \alpha\mathbf{B}^{-1}(\Delta t)^2 \\ \delta\mathbf{B}^{-1}\Delta t \\ \mathbf{B}^{-1} \end{bmatrix}$$

$$\mathbf{B} = \mathbf{M} + \delta\mathbf{C}\Delta t + \alpha\mathbf{K}(\Delta t)^2$$

$$\mathbf{G} = (1-\delta)\mathbf{C}\Delta t + \left(\tfrac{1}{2}-\alpha\right)\mathbf{K}(\Delta t)^2.$$

Similarly, the solution using the Wilson θ-method of discretization is

$$\mathbf{q}_{k+1} = \begin{Bmatrix} \mathbf{X}_{k+1} \\ \dot{\mathbf{X}}_{k+1} \\ \ddot{\mathbf{X}}_{k+1} \end{Bmatrix} = \mathbf{F}_W \begin{Bmatrix} \mathbf{X}_k \\ \dot{\mathbf{X}}_k \\ \ddot{\mathbf{X}}_k \end{Bmatrix} + \mathbf{H}_W\,\mathbf{F}_{k+1} = \mathbf{F}_W\mathbf{q}_k + \mathbf{H}_W\,\mathbf{F}_{k+1} \qquad (4.447)$$

where

$$\mathbf{F}_W = \begin{bmatrix} \mathbf{I} - \tfrac{1}{6}\mathbf{B}^{-1}\mathbf{K}\theta^2(\Delta t)^2 & \mathbf{I}\Delta t - \tfrac{1}{6}\mathbf{B}^{-1}\mathbf{C}\theta^2(\Delta t)^2 - \tfrac{1}{6}\mathbf{B}^{-1}\mathbf{K}\theta^2(\Delta t)^3 & \tfrac{1}{6}\theta(\Delta t)^2\left(2\mathbf{I} - \overline{\mathbf{B}}\right) \\ -\tfrac{1}{2}\mathbf{B}^{-1}\mathbf{K}\theta\Delta t & \mathbf{I} - \mathbf{B}^{-1}\mathbf{C}\theta\Delta t - \tfrac{1}{2}\mathbf{B}^{-1}\mathbf{K}\theta(\Delta t)^2 & \tfrac{1}{2}\Delta t\left(\mathbf{I} - \overline{\mathbf{B}}\right) \\ -\mathbf{B}^{-1}\mathbf{K} & -\mathbf{B}^{-1}\mathbf{C} - \mathbf{B}^{-1}\mathbf{K}\Delta t & \left(1-\tfrac{1}{\theta}\right)\mathbf{I} - \overline{\mathbf{B}} \end{bmatrix}$$

$$\mathbf{H}_W = \begin{bmatrix} \tfrac{1}{6}\mathbf{B}^{-1}\theta^2(\Delta t)^2 \\ \tfrac{1}{2}\mathbf{B}^{-1}\theta\Delta t \\ \mathbf{B}^{-1} \end{bmatrix}$$

$$\mathbf{B} = \mathbf{M} + \tfrac{1}{2}\mathbf{C}\theta\Delta t + \tfrac{1}{6}\mathbf{K}\theta^2(\Delta t)^2$$

$$\overline{\mathbf{B}} = \mathbf{B}^{-1}\mathbf{G} - (1-\theta)\mathbf{B}^{-1}\mathbf{M}$$

$$\mathbf{G} = \tfrac{1}{2}\mathbf{C}\theta\Delta t + \tfrac{1}{3}\mathbf{K}\theta^2(\Delta t)^2.$$

Note that both $\mathbf{F}_N$ and $\mathbf{F}_W$ are $3n \times 3n$ matrices, whereas $\mathbf{H}_N$ and $\mathbf{H}_W$ are $3n \times n$ matrices.

EXAMPLE 1 *Earthquake Response Using Discretization Methods*

Consider the case where the earthquake ground motion excites the 3DOF system as shown in Figure 4.15. This 3DOF system is a one-story three-dimensional structure, and it is assumed that the masses in the vertical direction (i.e., m_{zz}) and the out-of-plane mass moment of inertia of the floor (i.e., m_{rr} and m_{ss}) are zero.

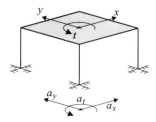

Figure 4.15 One-Story 3DOF System

The forcing function is given by

$$\mathbf{F}_k = -\begin{bmatrix} m_{xx} & 0 & 0 \\ 0 & m_{yy} & 0 \\ 0 & 0 & m_{tt} \end{bmatrix}\begin{Bmatrix} a_{xk} \\ a_{yk} \\ a_{tk} \end{Bmatrix} = -\begin{bmatrix} m_{xx} & 0 & 0 \\ 0 & m_{yy} & 0 \\ 0 & 0 & m_{tt} \end{bmatrix}\begin{Bmatrix} a_x \\ a_y \\ a_t \end{Bmatrix}_k \tag{4.448}$$

where a_{xk}, a_{yk}, and a_{tk} are the ground accelerations in the x-, y-, and θ_z-directions, respectively. It follows from the dynamic equilibrium equation that

$$\mathbf{M}\ddot{\mathbf{X}}_{k+1} + \mathbf{C}\dot{\mathbf{X}}_{k+1} + \mathbf{K}\mathbf{X}_{k+1} = -\mathbf{M}\mathbf{a}_{k+1} \tag{4.449}$$

where

$$\mathbf{M} = \begin{bmatrix} m_{xx} & 0 & 0 \\ 0 & m_{yy} & 0 \\ 0 & 0 & m_{tt} \end{bmatrix}, \quad \mathbf{C} = \begin{bmatrix} c_{xx} & c_{xy} & c_{xt} \\ c_{yx} & c_{yy} & c_{yt} \\ c_{tx} & c_{ty} & c_{tt} \end{bmatrix}, \quad \mathbf{K} = \begin{bmatrix} k_{xx} & k_{xy} & k_{xt} \\ k_{yx} & k_{yy} & k_{yt} \\ k_{tx} & k_{ty} & k_{tt} \end{bmatrix}$$

$$\mathbf{X}_k = \begin{Bmatrix} x_{xk} & x_{yk} & x_{tk} \end{Bmatrix}^T, \quad \mathbf{a}_k = \begin{Bmatrix} a_{xk} & a_{yk} & a_{tk} \end{Bmatrix}^T.$$

The results of the response for the earthquake ground motion case using the Newmark β-method or the Wilson θ-method are

$$\mathbf{q}_{k+1} = \mathbf{F}_N \mathbf{q}_k + \mathbf{H}_N^{(EQ)} \mathbf{a}_{k+1} \tag{4.450}$$

$$\mathbf{q}_{k+1} = \mathbf{F}_W \mathbf{q}_k + \mathbf{H}_W^{(EQ)} \mathbf{a}_{k+1} \tag{4.451}$$

where

$$\mathbf{H}_N^{(EQ)} = -\begin{bmatrix} \alpha\mathbf{B}^{-1}(\Delta t)^2 \\ \delta\mathbf{B}^{-1}\Delta t \\ \mathbf{B}^{-1} \end{bmatrix}\begin{bmatrix} m_{xx} & 0 & 0 \\ 0 & m_{yy} & 0 \\ 0 & 0 & m_{tt} \end{bmatrix} = -\begin{bmatrix} \alpha\mathbf{B}^{-1}\mathbf{M}(\Delta t)^2 \\ \delta\mathbf{B}^{-1}\mathbf{M}\Delta t \\ \mathbf{B}^{-1}\mathbf{M} \end{bmatrix}$$

$$\mathbf{H}_W^{(EQ)} = -\begin{bmatrix} \frac{1}{6}\mathbf{B}^{-1}\theta^2(\Delta t)^2 \\ \frac{1}{2}\mathbf{B}^{-1}\theta\Delta t \\ \mathbf{B}^{-1} \end{bmatrix}\begin{bmatrix} m_{xx} & 0 & 0 \\ 0 & m_{yy} & 0 \\ 0 & 0 & m_{tt} \end{bmatrix} = -\begin{bmatrix} \frac{1}{6}\mathbf{B}^{-1}\mathbf{M}\theta^2(\Delta t)^2 \\ \frac{1}{2}\mathbf{B}^{-1}\mathbf{M}\theta\Delta t \\ \mathbf{B}^{-1}\mathbf{M} \end{bmatrix}.$$

EXAMPLE 2   *Response of a 3DOF System*

Consider the three-story building discussed in Example 2 of Section 4.3 with zero damping:

$$\mathbf{M} = \begin{bmatrix} m & 0 & 0 \\ 0 & m & 0 \\ 0 & 0 & m \end{bmatrix}, \quad \mathbf{C} = \begin{bmatrix} 0 & 0 & 0 \\ 0 & 0 & 0 \\ 0 & 0 & 0 \end{bmatrix}, \quad \mathbf{K} = \begin{bmatrix} 3k & -k & 0 \\ -k & 2k & -k \\ 0 & -k & k \end{bmatrix}. \tag{4.452}$$

Assume that $m = 1.0$ and $k = 10.0$. Then Eq. 4.452 becomes

$$\mathbf{M} = \begin{bmatrix} 1.0 & 0.0 & 0.0 \\ 0.0 & 1.0 & 0.0 \\ 0.0 & 0.0 & 1.0 \end{bmatrix}, \quad \mathbf{C} = \begin{bmatrix} 0 & 0 & 0 \\ 0 & 0 & 0 \\ 0 & 0 & 0 \end{bmatrix}, \quad \mathbf{K} = \begin{bmatrix} 30.0 & -10.0 & 0.0 \\ -10.0 & 20.0 & -10.0 \\ 0.0 & -10.0 & 10.0 \end{bmatrix}. \tag{4.453}$$

Additionally, assume that $\Delta t = 0.01$ s with the Newmark β-method of constant acceleration (i.e., $\delta = 0$ and $\alpha = 0$). Then it follows from the matrices defined in Eq. 4.446 that

$$\mathbf{B} = \begin{bmatrix} 1.0 & 0.0 & 0.0 \\ 0.0 & 1.0 & 0.0 \\ 0.0 & 0.0 & 1.0 \end{bmatrix} + (0) \begin{bmatrix} 0 & 0 & 0 \\ 0 & 0 & 0 \\ 0 & 0 & 0 \end{bmatrix} (0.01) + (0) \begin{bmatrix} 30.0 & -10.0 & 0.0 \\ -10.0 & 20.0 & -10.0 \\ 0.0 & -10.0 & 10.0 \end{bmatrix} (0.01)^2$$

$$= \begin{bmatrix} 1.0 & 0.0 & 0.0 \\ 0.0 & 1.0 & 0.0 \\ 0.0 & 0.0 & 1.0 \end{bmatrix}$$

(4.454)

and therefore

$$\mathbf{B}^{-1} = \begin{bmatrix} 1.0 & 0.0 & 0.0 \\ 0.0 & 1.0 & 0.0 \\ 0.0 & 0.0 & 1.0 \end{bmatrix} = \mathbf{I}.$$

(4.455)

Similarly,

$$\mathbf{G} = (1-0) \begin{bmatrix} 0 & 0 & 0 \\ 0 & 0 & 0 \\ 0 & 0 & 0 \end{bmatrix} (0.01) + \left(\tfrac{1}{2} - 0\right) \begin{bmatrix} 30.0 & -10.0 & 0.0 \\ -10.0 & 20.0 & -10.0 \\ 0.0 & -10.0 & 10.0 \end{bmatrix} (0.01)^2$$

$$= \begin{bmatrix} 0.0015 & -0.0005 & 0.0 \\ -0.0005 & 0.0010 & -0.0005 \\ 0.0 & -0.0005 & 0.0005 \end{bmatrix}.$$

(4.456)

Therefore,

$$\mathbf{F}_N = \left[\begin{array}{ccc|ccc|ccc} 1.0 & 0.0 & 0.0 & 0.01 & 0.0 & 0.0 & 0.00005 & 0.0 & 0.0 \\ 0.0 & 1.0 & 0.0 & 0.0 & 0.01 & 0.0 & 0.0 & 0.00005 & 0.0 \\ 0.0 & 0.0 & 1.0 & 0.0 & 0.0 & 0.01 & 0.0 & 0.0 & 0.00005 \\ \hline 0.0 & 0.0 & 0.0 & 1.0 & 0.0 & 0.0 & 0.01 & 0.0 & 0.0 \\ 0.0 & 0.0 & 0.0 & 0.0 & 1.0 & 0.0 & 0.0 & 0.01 & 0.0 \\ 0.0 & 0.0 & 0.0 & 0.0 & 0.0 & 1.0 & 0.0 & 0.0 & 0.01 \\ \hline -30.0 & 10.0 & 0.0 & -0.3 & 0.1 & 0.0 & -0.0015 & 0.0005 & 0.0 \\ 10.0 & -20.0 & 10.0 & 0.1 & -0.2 & 0.1 & 0.0005 & -0.0010 & 0.0005 \\ 0.0 & 10.0 & -10.0 & 0.0 & 0.1 & -0.1 & 0.0 & 0.0005 & -0.0005 \end{array} \right]$$

(4.457)

$$\mathbf{H}_N = \left[\begin{array}{ccc} 0.0 & 0.0 & 0.0 \\ 0.0 & 0.0 & 0.0 \\ 0.0 & 0.0 & 0.0 \\ \hline 0.0 & 0.0 & 0.0 \\ 0.0 & 0.0 & 0.0 \\ 0.0 & 0.0 & 0.0 \\ \hline 1.0 & 0.0 & 0.0 \\ 0.0 & 1.0 & 0.0 \\ 0.0 & 0.0 & 1.0 \end{array} \right].$$

(4.458)

Finally, Eq. 4.446 becomes

$$
\begin{Bmatrix} x_1 \\ x_2 \\ x_3 \\ \dot{x}_1 \\ \dot{x}_2 \\ \dot{x}_3 \\ \ddot{x}_1 \\ \ddot{x}_2 \\ \ddot{x}_3 \end{Bmatrix}_{k+1} = \mathbf{F}_N \begin{Bmatrix} x_1 \\ x_2 \\ x_3 \\ \dot{x}_1 \\ \dot{x}_2 \\ \dot{x}_3 \\ \ddot{x}_1 \\ \ddot{x}_2 \\ \ddot{x}_3 \end{Bmatrix}_{k} + \begin{bmatrix} 0.0 & 0.0 & 0.0 \\ 0.0 & 0.0 & 0.0 \\ 0.0 & 0.0 & 0.0 \\ 0.0 & 0.0 & 0.0 \\ 0.0 & 0.0 & 0.0 \\ 0.0 & 0.0 & 0.0 \\ 1.0 & 0.0 & 0.0 \\ 0.0 & 1.0 & 0.0 \\ 0.0 & 0.0 & 1.0 \end{bmatrix} \begin{Bmatrix} f_1 \\ f_2 \\ f_3 \end{Bmatrix}_{k+1}
\tag{4.459}
$$

where $\mathbf{F}_N$ is given in Eq. 4.457.

4.10   DAMPED RESPONSE USING THE STATE SPACE METHOD

The state space solution method for an MDOF system extends directly from the solution method for a 2DOF system. For an n-DOF system, there are $2n$ states, which contain n displacements and n velocities.

Consider the dynamic equilibrium equation, as shown in Eq. 4.1, which is

$$
\mathbf{M}\ddot{\mathbf{X}}(t) + \mathbf{C}\dot{\mathbf{X}}(t) + \mathbf{K}\mathbf{X}(t) = \mathbf{F}_e(t)
\tag{4.460}
$$

where

$$
\mathbf{X}(t) = \begin{Bmatrix} x_1(t) \\ x_2(t) \\ \vdots \\ x_n(t) \end{Bmatrix}, \quad \mathbf{F}_e(t) = \begin{Bmatrix} F_1(t) \\ F_2(t) \\ \vdots \\ F_n(t) \end{Bmatrix}, \quad \mathbf{M} = \begin{bmatrix} m_{11} & m_{12} & \cdots & m_{1n} \\ m_{21} & m_{22} & \ddots & \vdots \\ \vdots & \ddots & \ddots & m_{n-1,n} \\ m_{n1} & \cdots & m_{n,n-1} & m_{nn} \end{bmatrix}
$$

$$
\mathbf{C} = \begin{bmatrix} c_{11} & c_{12} & \cdots & c_{1n} \\ c_{21} & c_{22} & \ddots & \vdots \\ \vdots & \ddots & \ddots & c_{n-1,n} \\ c_{n1} & \cdots & c_{n,n-1} & c_{nn} \end{bmatrix}, \quad \mathbf{K} = \begin{bmatrix} k_{11} & k_{12} & \cdots & k_{1n} \\ k_{21} & k_{22} & \ddots & \vdots \\ \vdots & \ddots & \ddots & k_{n-1,n} \\ k_{n1} & \cdots & k_{n,n-1} & k_{nn} \end{bmatrix}.
$$

The vector $\mathbf{z}$ represents the independent displacement and velocity vectors and is

$$
\mathbf{z}(t) = \begin{Bmatrix} \mathbf{X}(t) \\ \dot{\mathbf{X}}(t) \end{Bmatrix}.
\tag{4.461}
$$

It follows from Eq. 4.460 that

$$
\dot{\mathbf{z}}(t) = \begin{Bmatrix} \dot{\mathbf{X}}(t) \\ \ddot{\mathbf{X}}(t) \end{Bmatrix} = \begin{bmatrix} \mathbf{0} & \mathbf{I} \\ -\mathbf{M}^{-1}\mathbf{K} & -\mathbf{M}^{-1}\mathbf{C} \end{bmatrix} \begin{Bmatrix} \mathbf{X}(t) \\ \dot{\mathbf{X}}(t) \end{Bmatrix} + \begin{Bmatrix} \mathbf{0} \\ \mathbf{M}^{-1}\mathbf{F}_e(t) \end{Bmatrix}
\tag{4.462}
$$

where $\mathbf{0}$ is a matrix with all zero entries and $\mathbf{I}$ is the identity matrix. To simplify Eq. 4.462, let

$$
\mathbf{A} = \begin{bmatrix} \mathbf{0} & \mathbf{I} \\ -\mathbf{M}^{-1}\mathbf{K} & -\mathbf{M}^{-1}\mathbf{C} \end{bmatrix}, \quad \mathbf{F}(t) = \begin{Bmatrix} \mathbf{0} \\ \mathbf{M}^{-1}\mathbf{F}_e(t) \end{Bmatrix}
\tag{4.463}
$$

and then it follows that

$$
\dot{\mathbf{z}}(t) = \mathbf{A}\mathbf{z}(t) + \mathbf{F}(t).
\tag{4.464}
$$

Equation 4.464 is the same as Eq. 2.181. Thus, the solution to the equation can be written as

$$\mathbf{z}(t) = \mathbf{e}^{\mathbf{A}(t-t_o)} \mathbf{z}(t_o) + \mathbf{e}^{\mathbf{A}t} \int_{t_o}^{t} \mathbf{e}^{-\mathbf{A}s} \mathbf{F}(s)\, ds \tag{4.465}$$

where $\mathbf{z}(t_o)$ represents the displacement and velocity at time $t = t_o$.

As discussed in Section 2.5, calculating the exact values of the state transition matrix, $\mathbf{e}^{\mathbf{A}t}$, requires calculation of all the eigenvalues and eigenvectors of the matrix $\mathbf{A}$. Once this is done, the transformation matrix $\mathbf{T}$ must be inverted. Therefore, expressing $\mathbf{e}^{\mathbf{A}t}$ in a closed-form solution is very difficult, and most of the time it is preferable to do a numerical calculation. The method of numerically calculating the matrix $\mathbf{e}^{\mathbf{A}t}$ was discussed in Section 3.8.

The matrix equation for a numerical integration solution using the state space method is

$$\mathbf{z}_{k+1} = \mathbf{e}^{\mathbf{A}\Delta t} \mathbf{z}_k + \mathbf{e}^{\mathbf{A}\Delta t} \Delta t\, \mathbf{F}_k \tag{4.466}$$

when the Delta forcing function assumption is used and

$$\mathbf{z}_{k+1} = \mathbf{e}^{\mathbf{A}\Delta t} \mathbf{z}_k + \mathbf{A}^{-1}\left(\mathbf{e}^{\mathbf{A}\Delta t} - \mathbf{I}\right)\mathbf{F}_k \tag{4.467}$$

when the constant forcing function assumption is used, where

$$\mathbf{z}(t) = \begin{Bmatrix} \mathbf{X}(t) \\ \dot{\mathbf{X}}(t) \end{Bmatrix}, \quad \mathbf{A} = \begin{bmatrix} \mathbf{0} & \mathbf{I} \\ -\mathbf{M}^{-1}\mathbf{K} & -\mathbf{M}^{-1}\mathbf{C} \end{bmatrix}, \quad \mathbf{F}(t) = \begin{Bmatrix} \mathbf{0} \\ \mathbf{M}^{-1}\mathbf{F}_e(t) \end{Bmatrix}. \tag{4.468}$$

Note that $\mathbf{e}^{\mathbf{A}t}$ is a $2n \times 2n$ matrix, and thus the computation may require considerable computational effort.

EXAMPLE 1 *Earthquake Response Using the State Space Method*

Consider the 3DOF system discussed in Example 1 of Section 4.9 with an earthquake ground motion, that is,

$$\mathbf{F}_{ek} = -\begin{bmatrix} m_{xx} & 0 & 0 \\ 0 & m_{yy} & 0 \\ 0 & 0 & m_{tt} \end{bmatrix} \begin{Bmatrix} a_{xk} \\ a_{yk} \\ a_{tk} \end{Bmatrix} = -\begin{bmatrix} m_{xx} & 0 & 0 \\ 0 & m_{yy} & 0 \\ 0 & 0 & m_{tt} \end{bmatrix} \begin{Bmatrix} a_x \\ a_y \\ a_t \end{Bmatrix}_k \tag{4.469}$$

where a_{xk}, a_{yk}, and a_{tk} are the ground accelerations at time step k in the x-, y-, and θ_z-directions, respectively. It follows that

$$\mathbf{F}_k = \begin{Bmatrix} 0 & 0 & 0 & -a_{xk} & -a_{yk} & -a_{tk} \end{Bmatrix}^T. \tag{4.470}$$

The result of the response for the earthquake ground motion using the Delta forcing function assumption is

$$\begin{Bmatrix} x_x \\ x_y \\ x_t \\ \dot{x}_x \\ \dot{x}_y \\ \dot{x}_t \end{Bmatrix}_{k+1} = \mathbf{e}^{\mathbf{A}\Delta t} \begin{Bmatrix} x_x \\ x_y \\ x_t \\ \dot{x}_x \\ \dot{x}_y \\ \dot{x}_t \end{Bmatrix}_k + \mathbf{e}^{\mathbf{A}\Delta t}\Delta t \begin{Bmatrix} 0 \\ 0 \\ 0 \\ -a_x \\ -a_y \\ -a_t \end{Bmatrix}_k \tag{4.471}$$

and the result of the response using the constant forcing function assumption is

$$\begin{Bmatrix} x_x \\ x_y \\ x_t \\ \dot{x}_x \\ \dot{x}_y \\ \dot{x}_t \end{Bmatrix}_{k+1} = \mathbf{e}^{\mathbf{A}\Delta t} \begin{Bmatrix} x_x \\ x_y \\ x_t \\ \dot{x}_x \\ \dot{x}_y \\ \dot{x}_t \end{Bmatrix}_k + \mathbf{A}^{-1}\left(\mathbf{e}^{\mathbf{A}\Delta t} - \mathbf{I}\right) \begin{Bmatrix} 0 \\ 0 \\ 0 \\ -a_x \\ -a_y \\ -a_t \end{Bmatrix}_k . \tag{4.472}$$

EXAMPLE 2   *Response of a 3DOF System*

Consider the three-story building discussed in Example 2 of Section 4.3 with zero damping:

$$\mathbf{M} = \begin{bmatrix} m & 0 & 0 \\ 0 & m & 0 \\ 0 & 0 & m \end{bmatrix}, \quad \mathbf{C} = \begin{bmatrix} 0 & 0 & 0 \\ 0 & 0 & 0 \\ 0 & 0 & 0 \end{bmatrix}, \quad \mathbf{K} = \begin{bmatrix} 3k & -k & 0 \\ -k & 2k & -k \\ 0 & -k & k \end{bmatrix}. \tag{4.473}$$

Assume that $m = 1.0$ and $k = 10.0$. Then Eq. 4.473 becomes

$$\mathbf{M} = \begin{bmatrix} 1.0 & 0.0 & 0.0 \\ 0.0 & 1.0 & 0.0 \\ 0.0 & 0.0 & 1.0 \end{bmatrix}, \quad \mathbf{C} = \begin{bmatrix} 0 & 0 & 0 \\ 0 & 0 & 0 \\ 0 & 0 & 0 \end{bmatrix}, \quad \mathbf{K} = \begin{bmatrix} 30.0 & -10.0 & 0.0 \\ -10.0 & 20.0 & -10.0 \\ 0.0 & -10.0 & 10.0 \end{bmatrix}. \tag{4.474}$$

It follows from Eq. 4.468 that

$$\mathbf{z}(t) = \begin{Bmatrix} x_1(t) \\ x_2(t) \\ x_3(t) \\ \dot{x}_1(t) \\ \dot{x}_2(t) \\ \dot{x}_3(t) \end{Bmatrix}, \quad \mathbf{A} = \begin{bmatrix} 0.0 & 0.0 & 0.0 & 1.0 & 0.0 & 0.0 \\ 0.0 & 0.0 & 0.0 & 0.0 & 1.0 & 0.0 \\ 0.0 & 0.0 & 0.0 & 0.0 & 0.0 & 1.0 \\ -30.0 & 10.0 & 0.0 & 0.0 & 0.0 & 0.0 \\ 10.0 & -20.0 & 10.0 & 0.0 & 0.0 & 0.0 \\ 0.0 & 10.0 & -10.0 & 0.0 & 0.0 & 0.0 \end{bmatrix}, \quad \mathbf{F}(t) = \begin{Bmatrix} 0.0 \\ 0.0 \\ 0.0 \\ f_1(t) \\ f_2(t) \\ f_3(t) \end{Bmatrix}. \tag{4.475}$$

Assume that a time step of $\Delta = 0.001$ s is used; then it follows that the state transition matrix $\mathbf{e}^{\mathbf{A}\Delta t}$ can be computed using the method discussed in Section 3.8:

$$\mathbf{e}^{\mathbf{A}\Delta t} = \begin{bmatrix} 0.998500 & 0.000500 & 0.000000 & 0.009995 & 0.000002 & 0.000000 \\ 0.000500 & 0.999000 & 0.000500 & 0.000002 & 0.009997 & 0.000002 \\ 0.000000 & 0.00500 & 0.999500 & 0.000000 & 0.000002 & 0.009998 \\ -0.299833 & 0.099917 & 0.000017 & 0.998500 & 0.000500 & 0.000000 \\ 0.099917 & -0.199900 & 0.099950 & 0.000500 & 0.999000 & 0.000500 \\ 0.000016 & 0.099950 & -0.09997 & 0.000000 & 0.000500 & 0.999500 \end{bmatrix}. \tag{4.476}$$

When the Delta forcing function is used, then it follows from Eq. 4.466 that

$$\begin{Bmatrix} x_1 \\ x_2 \\ x_3 \\ \dot{x}_1 \\ \dot{x}_2 \\ \dot{x}_3 \end{Bmatrix}_{k+1} = \mathbf{e}^{\mathbf{A}\Delta t} \begin{Bmatrix} x_1 \\ x_2 \\ x_3 \\ \dot{x}_1 \\ \dot{x}_2 \\ \dot{x}_3 \end{Bmatrix}_k + \mathbf{e}^{\mathbf{A}\Delta t} \Delta t \begin{Bmatrix} 0.0 \\ 0.0 \\ 0.0 \\ f_1 \\ f_2 \\ f_3 \end{Bmatrix}_k \tag{4.477}$$

where $\mathbf{e}^{\mathbf{A}\Delta t}$ is given in Eq. 4.476.

4.11   DAMPED RESPONSE USING THE NORMAL MODE METHOD AND THE STATE SPACE FORMULATION

Another method of calculating the response of a system with nonproportional damping is to represent the equation of motion in the state space form and then to use the normal mode method. The advantage of this method is that the solution can be expressed as a linear combination of $2n$ uncoupled first-order differential equations. To see this, consider Eq. 4.464:

$$\dot{\mathbf{z}}(t) = \mathbf{A}\mathbf{z}(t) + \mathbf{F}(t) \tag{4.478}$$

where

$$\mathbf{z}(t) = \begin{Bmatrix} \mathbf{X}(t) \\ \dot{\mathbf{X}}(t) \end{Bmatrix}, \quad \mathbf{A} = \begin{bmatrix} \mathbf{0} & \mathbf{I} \\ -\mathbf{M}^{-1}\mathbf{K} & -\mathbf{M}^{-1}\mathbf{C} \end{bmatrix}, \quad \mathbf{F}(t) = \begin{Bmatrix} \mathbf{0} \\ \mathbf{M}^{-1}\mathbf{F}_e(t) \end{Bmatrix}. \tag{4.479}$$

The normal mode method represents the response as a summation of normal mode coordinates, and in the state space form it can be written as

$$\mathbf{z}(t) = \sum_{i=1}^{2n} \boldsymbol{\phi}_i q_{zi}(t) = \boldsymbol{\Phi}_z \mathbf{Q}_z(t) \tag{4.480}$$

where

$$\boldsymbol{\Phi}_z = \begin{bmatrix} \boldsymbol{\phi}_1 & \boldsymbol{\phi}_2 & \cdots & \boldsymbol{\phi}_{2n} \end{bmatrix}$$

with $\boldsymbol{\phi}_i$ representing the eigenvectors of the matrix $\mathbf{A}$. Substituting Eq. 4.480 into Eq. 4.478 gives

$$\boldsymbol{\Phi}_z \dot{\mathbf{Q}}_z(t) = \mathbf{A}\boldsymbol{\Phi}_z \mathbf{Q}_z(t) + \mathbf{F}(t). \tag{4.481}$$

After premultiplying Eq. 4.481 by $\boldsymbol{\Phi}_z^{-1}$, it follows that

$$\dot{\mathbf{Q}}_z(t) = \boldsymbol{\Phi}_z^{-1}\mathbf{A}\boldsymbol{\Phi}_z \mathbf{Q}_z(t) + \boldsymbol{\Phi}_z^{-1}\mathbf{F}(t). \tag{4.482}$$

Since each column of $\boldsymbol{\Phi}_z$ is an eigenvector of $\mathbf{A}$, it follows that the result of the matrix triple product $\boldsymbol{\Phi}_z^{-1}\mathbf{A}\boldsymbol{\Phi}_z$ is a diagonal matrix, that is,

$$\boldsymbol{\Phi}_z^{-1}\mathbf{A}\boldsymbol{\Phi}_z = \begin{bmatrix} \lambda_1 & 0 & \cdots & 0 \\ 0 & \lambda_2 & \ddots & \vdots \\ \vdots & \ddots & \ddots & 0 \\ 0 & \cdots & 0 & \lambda_{2n} \end{bmatrix} = \boldsymbol{\Lambda} \tag{4.483}$$

where λ_i is the ith eigenvalue of $\mathbf{A}$. It follows from Eq. 4.482 that

$$\dot{\mathbf{Q}}_z(t) = \boldsymbol{\Lambda}\mathbf{Q}_z(t) + \boldsymbol{\Phi}_z^{-1}\mathbf{F}(t). \tag{4.484}$$

Equation 4.484 is a set of $2n$ uncoupled first-order linear differential equations, that is,

$$\dot{q}_{zi}(t) = \lambda_i q_{zi}(t) + \sum_{j=1}^{n} \varphi_{ij}^{-1} f_j(t) \tag{4.485}$$

where φ_{ij}^{-1} is the entry in the ith row and jth column of $\boldsymbol{\Phi}_z^{-1}$, and $f_j(t)$ is the jth entry in the vector $\mathbf{F}(t)$. The solution to Eq. 4.485 is

$$q_{zi}(t) = e^{\lambda_1(t-t_o)}q_{zi}(t_o) + \int_{s=t_o}^{s=t} e^{\lambda_i(t-s)} \sum_{j=1}^{n} \varphi_{ij}^{-1} f_j(s)\,ds. \tag{4.486}$$

Substituting Eq. 4.486 into Eq. 4.480 gives the general solution:

$$\mathbf{z}(t) = \sum_{i=1}^{2n} \boldsymbol{\phi}_i q_{zi}(t) = \sum_{i=1}^{2n} \boldsymbol{\phi}_i \left[e^{\lambda_i(t-t_o)} q_{zi}(t_o) + \int_{s=t_o}^{s=t} e^{\lambda_i(t-s)} \left(\sum_{j=1}^{n} \varphi_{ij}^{-1} f_j(s) \right) ds \right]$$

$$= \boldsymbol{\Phi}_z e^{\Lambda(t-t_o)} \mathbf{Q}_z(t_o) + \sum_{i=1}^{2n} \left[\int_{s=t_o}^{s=t} e^{\lambda_i(t-s)} \left(\sum_{j=1}^{n} \varphi_{ij}^{-1} f_j(s) \right) ds \right].$$

(4.487)

Finally, the initial conditions $\mathbf{Q}_z(t_o)$ must be expressed in terms of $\mathbf{z}(t_o)$. Inverting Eq. 4.480 and evaluating the equation at $t = t_o$ gives

$$\mathbf{Q}_z(t_o) = \boldsymbol{\Phi}_z^{-1} \mathbf{z}(t_o).$$

(4.488)

Substituting Eq. 4.488 into Eq. 4.487, it follows that

$$\mathbf{z}(t) = \boldsymbol{\Phi}_z e^{\Lambda(t-t_o)} \boldsymbol{\Phi}_z^{-1} \mathbf{z}(t_o) + \sum_{i=1}^{2n} \left[\int_{s=t_o}^{s=t} e^{\lambda_i(t-s)} \left(\sum_{j=1}^{n} \varphi_{ij}^{-1} f_j(s) \right) ds \right].$$

(4.489)

Note that $\boldsymbol{\Phi}_z e^{\Lambda(t-t_o)} \boldsymbol{\Phi}_z^{-1} = e^{\mathbf{A}(t-t_o)}$; therefore, Eq. 4.489 becomes

$$\mathbf{z}(t) = e^{\mathbf{A}(t-t_o)} \mathbf{z}(t_o) + \sum_{i=1}^{2n} \left[\int_{s=t_o}^{s=t} e^{\lambda_i(t-s)} \left(\sum_{j=1}^{n} \varphi_{ij}^{-1} f_j(s) \right) ds \right].$$

(4.490)

This solution method requires computation of the eigenvalues and a modal matrix, whose entries are complex values.

EXAMPLE 1   *Undamped Two Degrees of Freedom System Response*

Consider the two degrees of freedom undamped system as shown in Figure 4.16, with dynamic matrices and forcing vector equal to

$$\mathbf{M} = \begin{bmatrix} 2m & 0 \\ 0 & m \end{bmatrix}, \quad \mathbf{C} = \begin{bmatrix} 0 & 0 \\ 0 & 0 \end{bmatrix}, \quad \mathbf{K} = \begin{bmatrix} 4k & -2k \\ -2k & 3k \end{bmatrix}, \quad \mathbf{F}(t) = \begin{Bmatrix} f_1(t) \\ f_2(t) \end{Bmatrix}.$$

(4.491)

Then

$$\mathbf{M}^{-1}\mathbf{K} = \begin{bmatrix} 2m & 0 \\ 0 & m \end{bmatrix}^{-1} \begin{bmatrix} 4k & -2k \\ -2k & 3k \end{bmatrix} = \begin{bmatrix} 2\omega_n^2 & -\omega_n^2 \\ -2\omega_n^2 & 3\omega_n^2 \end{bmatrix}, \quad \mathbf{A} = \begin{bmatrix} 0 & 0 & 1 & 0 \\ 0 & 0 & 0 & 1 \\ -2\omega_n^2 & \omega_n^2 & 0 & 0 \\ 2\omega_n^2 & -3\omega_n^2 & 0 & 0 \end{bmatrix}$$

(4.492)

where $\omega_n^2 = k/m$.

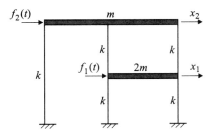

Figure 4.16 Two-Story Frame

The general expression for this 2DOF system is given in Eq. 4.478:

$$\frac{d}{dt}\begin{Bmatrix} x_1(t) \\ x_2(t) \\ \dot{x}_1(t) \\ \dot{x}_2(t) \end{Bmatrix} = \begin{bmatrix} 0 & 0 & 1 & 0 \\ 0 & 0 & 0 & 1 \\ -2\omega_n^2 & \omega_n^2 & 0 & 0 \\ 2\omega_n^2 & -3\omega_n^2 & 0 & 0 \end{bmatrix}\begin{Bmatrix} x_1(t) \\ x_2(t) \\ \dot{x}_1(t) \\ \dot{x}_2(t) \end{Bmatrix} + \begin{Bmatrix} 0 \\ 0 \\ f_1(t) \\ f_2(t) \end{Bmatrix}. \tag{4.493}$$

To compute the eigenvalues of the matrix $\mathbf{A}$, consider the determinant

$$\begin{vmatrix} -\lambda & 0 & 1 & 0 \\ 0 & -\lambda & 0 & 1 \\ -2\omega_n^2 & \omega_n^2 & -\lambda & 0 \\ 2\omega_n^2 & -3\omega_n^2 & 0 & -\lambda \end{vmatrix} = 0. \tag{4.494}$$

This gives the characteristic equation

$$\lambda^4 + 5\omega_n^2\lambda^2 + 4\omega_n^4 = 0 \tag{4.495}$$

and the solutions are

$$\lambda_1 = i\omega_n, \quad \lambda_2 = -i\omega_n, \quad \lambda_3 = 2i\omega_n, \quad \lambda_4 = -2i\omega_n. \tag{4.496}$$

The eigenvectors corresponding to these eigenvalues are

$$\lambda_1 = i\omega_n: \quad \begin{bmatrix} -i\omega_n & 0 & 1 & 0 \\ 0 & -i\omega_n & 0 & 1 \\ -2\omega_n^2 & \omega_n^2 & -i\omega_n & 0 \\ 2\omega_n^2 & -3\omega_n^2 & 0 & -i\omega_n \end{bmatrix}\begin{Bmatrix} \varphi_{11} \\ \varphi_{21} \\ \varphi_{31} \\ \varphi_{41} \end{Bmatrix} = \begin{Bmatrix} 0 \\ 0 \\ 0 \\ 0 \end{Bmatrix}, \quad \boldsymbol{\phi}_1 = \begin{Bmatrix} \varphi_{11} \\ \varphi_{21} \\ \varphi_{31} \\ \varphi_{41} \end{Bmatrix} = \begin{Bmatrix} 1 \\ 1 \\ i\omega_n \\ i\omega_n \end{Bmatrix} \tag{4.497}$$

$$\lambda_2 = -i\omega_n: \quad \begin{bmatrix} i\omega_n & 0 & 1 & 0 \\ 0 & i\omega_n & 0 & 1 \\ -2\omega_n^2 & \omega_n^2 & i\omega_n & 0 \\ 2\omega_n^2 & -3\omega_n^2 & 0 & i\omega_n \end{bmatrix}\begin{Bmatrix} \varphi_{12} \\ \varphi_{22} \\ \varphi_{32} \\ \varphi_{42} \end{Bmatrix} = \begin{Bmatrix} 0 \\ 0 \\ 0 \\ 0 \end{Bmatrix}, \quad \boldsymbol{\phi}_2 = \begin{Bmatrix} \varphi_{12} \\ \varphi_{22} \\ \varphi_{32} \\ \varphi_{42} \end{Bmatrix} = \begin{Bmatrix} 1 \\ 1 \\ -i\omega_n \\ -i\omega_n \end{Bmatrix} \tag{4.498}$$

$$\lambda_3 = 2i\omega_n: \quad \begin{bmatrix} -2i\omega_n & 0 & 1 & 0 \\ 0 & -2i\omega_n & 0 & 1 \\ -2\omega_n^2 & \omega_n^2 & -2i\omega_n & 0 \\ 2\omega_n^2 & -3\omega_n^2 & 0 & -2i\omega_n \end{bmatrix}\begin{Bmatrix} \varphi_{13} \\ \varphi_{23} \\ \varphi_{33} \\ \varphi_{43} \end{Bmatrix} = \begin{Bmatrix} 0 \\ 0 \\ 0 \\ 0 \end{Bmatrix}, \quad \boldsymbol{\phi}_3 = \begin{Bmatrix} \varphi_{13} \\ \varphi_{23} \\ \varphi_{33} \\ \varphi_{43} \end{Bmatrix} = \begin{Bmatrix} 1 \\ -2 \\ 2i\omega_n \\ -4i\omega_n \end{Bmatrix} \tag{4.499}$$

$$\lambda_4 = -2i\omega_n: \quad \begin{bmatrix} 2i\omega_n & 0 & 1 & 0 \\ 0 & 2i\omega_n & 0 & 1 \\ -2\omega_n^2 & \omega_n^2 & 2i\omega_n & 0 \\ 2\omega_n^2 & -3\omega_n^2 & 0 & 2i\omega_n \end{bmatrix}\begin{Bmatrix} \varphi_{14} \\ \varphi_{24} \\ \varphi_{34} \\ \varphi_{44} \end{Bmatrix} = \begin{Bmatrix} 0 \\ 0 \\ 0 \\ 0 \end{Bmatrix}, \quad \boldsymbol{\phi}_4 = \begin{Bmatrix} \varphi_{14} \\ \varphi_{24} \\ \varphi_{34} \\ \varphi_{44} \end{Bmatrix} = \begin{Bmatrix} 1 \\ -2 \\ -2i\omega_n \\ 4i\omega_n \end{Bmatrix}. \tag{4.500}$$

This gives the modal matrix and its inverse

$$\boldsymbol{\Phi}_z = \begin{bmatrix} 1 & 1 & 1 & 1 \\ 1 & 1 & -2 & -2 \\ i\omega_n & -i\omega_n & 2i\omega_n & -2i\omega_n \\ i\omega_n & -i\omega_n & -4i\omega_n & 4i\omega_n \end{bmatrix}, \quad \boldsymbol{\Phi}_z^{-1} = \begin{bmatrix} \frac{1}{3} & \frac{1}{6} & -\frac{i}{3\omega_n} & -\frac{i}{6\omega_n} \\ \frac{1}{3} & \frac{1}{6} & \frac{i}{3\omega_n} & \frac{i}{6\omega_n} \\ \frac{1}{6} & -\frac{1}{6} & -\frac{i}{12\omega_n} & \frac{i}{12\omega_n} \\ \frac{1}{6} & -\frac{1}{6} & \frac{i}{12\omega_n} & -\frac{i}{12\omega_n} \end{bmatrix}. \tag{4.501}$$

It follows from Eq. 4.483 that

$$\Lambda = \Phi_z^{-1} A \Phi_z = \begin{bmatrix} \frac{1}{3} & \frac{1}{6} & -\frac{i}{3\omega_n} & -\frac{i}{6\omega_n} \\ \frac{1}{3} & \frac{1}{6} & \frac{i}{3\omega_n} & \frac{i}{6\omega_n} \\ \frac{1}{6} & -\frac{1}{6} & -\frac{i}{12\omega_n} & \frac{i}{12\omega_n} \\ \frac{1}{6} & -\frac{1}{6} & \frac{i}{12\omega_n} & -\frac{i}{12\omega_n} \end{bmatrix} \begin{bmatrix} 0 & 0 & 1 & 0 \\ 0 & 0 & 0 & 1 \\ -2\omega_n^2 & \omega_n^2 & 0 & 0 \\ 2\omega_n^2 & -3\omega_n^2 & 0 & 0 \end{bmatrix} \begin{bmatrix} 1 & 1 & 1 & 1 \\ 1 & 1 & -2 & -2 \\ i\omega_n & -i\omega_n & 2i\omega_n & -2i\omega_n \\ i\omega_n & -i\omega_n & -4i\omega_n & 4i\omega_n \end{bmatrix}.$$

Performing the matrix triple product gives

$$\Lambda = \begin{bmatrix} i\omega_n & 0 & 0 & 0 \\ 0 & -i\omega_n & 0 & 0 \\ 0 & 0 & 2i\omega_n & 0 \\ 0 & 0 & 0 & -2i\omega_n \end{bmatrix}. \tag{4.502}$$

Therefore, it follows from Eq. 4.484 that

$$\frac{d}{dt} \begin{Bmatrix} q_1(t) \\ q_2(t) \\ q_3(t) \\ q_4(t) \end{Bmatrix} = \begin{bmatrix} i\omega_n & 0 & 0 & 0 \\ 0 & -i\omega_n & 0 & 0 \\ 0 & 0 & 2i\omega_n & 0 \\ 0 & 0 & 0 & -2i\omega_n \end{bmatrix} \begin{Bmatrix} q_1(t) \\ q_2(t) \\ q_3(t) \\ q_4(t) \end{Bmatrix} + \begin{bmatrix} \frac{1}{3} & \frac{1}{6} & -\frac{i}{3\omega_n} & -\frac{i}{6\omega_n} \\ \frac{1}{3} & \frac{1}{6} & \frac{i}{3\omega_n} & \frac{i}{6\omega_n} \\ \frac{1}{6} & -\frac{1}{6} & -\frac{i}{12\omega_n} & \frac{i}{12\omega_n} \\ \frac{1}{6} & -\frac{1}{6} & \frac{i}{12\omega_n} & -\frac{i}{12\omega_n} \end{bmatrix} \begin{Bmatrix} 0 \\ 0 \\ f_1(t) \\ f_2(t) \end{Bmatrix} \tag{4.503}$$

where

$$\begin{Bmatrix} x_1(t) \\ x_2(t) \\ \dot{x}_1(t) \\ \dot{x}_2(t) \end{Bmatrix} = \begin{bmatrix} 1 & 1 & 1 & 1 \\ 1 & 1 & -2 & -2 \\ i\omega_n & -i\omega_n & 2i\omega_n & -2i\omega_n \\ i\omega_n & -i\omega_n & -4i\omega_n & 4i\omega_n \end{bmatrix} \begin{Bmatrix} q_1(t) \\ q_2(t) \\ q_3(t) \\ q_4(t) \end{Bmatrix}, \quad \begin{Bmatrix} q_1(t) \\ q_2(t) \\ q_3(t) \\ q_4(t) \end{Bmatrix} = \begin{bmatrix} \frac{1}{3} & \frac{1}{6} & -\frac{i}{3\omega_n} & -\frac{i}{6\omega_n} \\ \frac{1}{3} & \frac{1}{6} & \frac{i}{3\omega_n} & \frac{i}{6\omega_n} \\ \frac{1}{6} & -\frac{1}{6} & -\frac{i}{12\omega_n} & \frac{i}{12\omega_n} \\ \frac{1}{6} & -\frac{1}{6} & \frac{i}{12\omega_n} & -\frac{i}{12\omega_n} \end{bmatrix} \begin{Bmatrix} x_1(t) \\ x_2(t) \\ \dot{x}_1(t) \\ \dot{x}_2(t) \end{Bmatrix}.$$

Equation 4.503 represents four independent linear differential equations:

$$\dot{q}_1(t) = i\omega_n q_1(t) - \frac{i}{3\omega_n} f_1(t) - \frac{i}{6\omega_n} f_2(t) \tag{4.504}$$

$$\dot{q}_2(t) = -i\omega_n q_2(t) + \frac{i}{3\omega_n} f_1(t) + \frac{i}{6\omega_n} f_2(t) \tag{4.505}$$

$$\dot{q}_3(t) = 2i\omega_n q_3(t) - \frac{i}{12\omega_n} f_1(t) + \frac{i}{12\omega_n} f_2(t) \tag{4.506}$$

$$\dot{q}_4(t) = -2i\omega_n q_4(t) + \frac{i}{12\omega_n} f_1(t) - \frac{i}{12\omega_n} f_2(t). \tag{4.507}$$

Solving for these first-order linear differential equations, assuming that initial conditions are specified at time $t = t_o = 0$, gives

$$q_1(t) = e^{i\omega_n t} q_{1o} + \int_0^t \left[-\frac{i}{3\omega_n} f_1(s) - \frac{i}{6\omega_n} f_2(s) \right] e^{i\omega_n(t-s)} \, ds \tag{4.508}$$

$$q_2(t) = e^{-i\omega_n t} q_{2o} + \int_0^t \left[\frac{i}{3\omega_n} f_1(s) + \frac{i}{6\omega_n} f_2(s) \right] e^{-i\omega_n(t-s)} \, ds \tag{4.509}$$

$$q_3(t) = e^{2i\omega_n t} q_{3o} + \int_0^t \left[-\frac{i}{12\omega_n} f_1(s) + \frac{i}{12\omega_n} f_2(s) \right] e^{2i\omega_n(t-s)} \, ds \tag{4.510}$$

$$q_4(t) = e^{-2i\omega_n t} q_{4o} + \int_0^t \left[\frac{i}{12\omega_n} f_1(s) - \frac{i}{12\omega_n} f_2(s) \right] e^{-2i\omega_n(t-s)} \, ds. \tag{4.511}$$

Finally, the response solution is

$$\begin{Bmatrix} x_1(t) \\ x_2(t) \\ \dot{x}_1(t) \\ \dot{x}_2(t) \end{Bmatrix} = \mathbf{\Phi} e^{\Lambda t} \begin{Bmatrix} q_{1o} \\ q_{2o} \\ q_{3o} \\ q_{4o} \end{Bmatrix} + \int_0^t \mathbf{\Phi} e^{\Lambda(t-s)} \begin{bmatrix} -\frac{i}{3\omega_n} & -\frac{i}{6\omega_n} \\ \frac{i}{3\omega_n} & \frac{i}{6\omega_n} \\ -\frac{i}{12\omega_n} & \frac{i}{12\omega_n} \\ \frac{i}{12\omega_n} & -\frac{i}{12\omega_n} \end{bmatrix} \begin{Bmatrix} f_1(s) \\ f_2(s) \end{Bmatrix} ds \tag{4.512}$$

where

$$\mathbf{\Phi}_z e^{\Lambda t} = \begin{bmatrix} e^{i\omega_n t} & e^{-i\omega_n t} & e^{2i\omega_n t} & e^{-2i\omega_n t} \\ e^{i\omega_n t} & e^{-i\omega_n t} & -2e^{2i\omega_n t} & -2e^{-2i\omega_n t} \\ i\omega_n e^{i\omega_n t} & -i\omega_n e^{-i\omega_n t} & 2i\omega_n e^{2i\omega_n t} & -2i\omega_n e^{-2i\omega_n t} \\ i\omega_n e^{i\omega_n t} & -i\omega_n e^{-i\omega_n t} & -4i\omega_n e^{2i\omega_n t} & 4i\omega_n e^{-2i\omega_n t} \end{bmatrix}.$$

Recall from Eq. 4.488 that

$$\begin{Bmatrix} q_{1o} \\ q_{2o} \\ q_{3o} \\ q_{4o} \end{Bmatrix} = \begin{bmatrix} \frac{1}{3} & \frac{1}{6} & -\frac{i}{3\omega_n} & -\frac{i}{6\omega_n} \\ \frac{1}{3} & \frac{1}{6} & \frac{i}{3\omega_n} & \frac{i}{6\omega_n} \\ \frac{1}{6} & -\frac{1}{6} & -\frac{i}{12\omega_n} & \frac{i}{12\omega_n} \\ \frac{1}{6} & -\frac{1}{6} & \frac{i}{12\omega_n} & -\frac{i}{12\omega_n} \end{bmatrix} \begin{Bmatrix} x_{1o} \\ x_{2o} \\ \dot{x}_{1o} \\ \dot{x}_{2o} \end{Bmatrix}. \tag{4.513}$$

It follows that

$$e^{\Lambda t} = \mathbf{\Phi}_z e^{\Lambda t} \mathbf{\Phi}_z^{-1} = \begin{bmatrix} \frac{2}{3}(c1) + \frac{1}{3}(c2) & \frac{1}{3}(c1) - \frac{1}{3}(c2) & \frac{2}{3\omega_n}(s1) + \frac{1}{6\omega_n}(s2) & \frac{1}{3\omega_n}(s1) - \frac{1}{6\omega_n}(s2) \\ \frac{2}{3}(c1) - \frac{2}{3}(c2) & \frac{1}{3}(c1) + \frac{2}{3}(c2) & \frac{2}{3\omega_n}(s1) - \frac{1}{3\omega_n}(s2) & \frac{1}{3\omega_n}(s1) + \frac{1}{3\omega_n}(s2) \\ -\frac{2\omega_n}{3}(s1) - \frac{2\omega_n}{3}(s2) & -\frac{\omega_n}{3}(s1) + \frac{2\omega_n}{3}(s2) & \frac{2}{3}(c1) + \frac{1}{3}(c2) & \frac{1}{3}(c1) - \frac{1}{3}(c2) \\ -\frac{2\omega_n}{3}(s1) + \frac{4\omega_n}{3}(s2) & -\frac{\omega_n}{3}(s1) - \frac{4\omega_n}{3}(s2) & \frac{2}{3}(c1) - \frac{2}{3}(c2) & \frac{1}{3}(c1) + \frac{2}{3}(c2) \end{bmatrix}$$

where $c1 = \cos \omega_n t$, $c2 = \cos 2\omega_n t$, $s1 = \sin \omega_n t$, and $s2 = \sin 2\omega_n t$. Substituting Eq. 4.513 into Eq. 4.512 gives

$$\begin{Bmatrix} x_1(t) \\ x_2(t) \\ \dot{x}_1(t) \\ \dot{x}_2(t) \end{Bmatrix} = \begin{bmatrix} \frac{2}{3}(c1) + \frac{1}{3}(c2) & \frac{1}{3}(c1) - \frac{1}{3}(c2) & \frac{2}{3\omega_n}(s1) + \frac{1}{6\omega_n}(s2) & \frac{1}{3\omega_n}(s1) - \frac{1}{6\omega_n}(s2) \\ \frac{2}{3}(c1) - \frac{2}{3}(c2) & \frac{1}{3}(c1) + \frac{2}{3}(c2) & \frac{2}{3\omega_n}(s1) - \frac{1}{3\omega_n}(s2) & \frac{1}{3\omega_n}(s1) + \frac{1}{3\omega_n}(s2) \\ -\frac{2\omega_n}{3}(s1) - \frac{2\omega_n}{3}(s2) & -\frac{\omega_n}{3}(s1) + \frac{2\omega_n}{3}(s2) & \frac{2}{3}(c1) + \frac{1}{3}(c2) & \frac{1}{3}(c1) - \frac{1}{3}(c2) \\ -\frac{2\omega_n}{3}(s1) + \frac{4\omega_n}{3}(s2) & -\frac{\omega_n}{3}(s1) - \frac{4\omega_n}{3}(s2) & \frac{2}{3}(c1) - \frac{2}{3}(c2) & \frac{1}{3}(c1) + \frac{2}{3}(c2) \end{bmatrix} \begin{Bmatrix} x_{1o} \\ x_{2o} \\ \dot{x}_{1o} \\ \dot{x}_{2o} \end{Bmatrix}$$

$$+ \int_0^t \begin{bmatrix} \frac{2}{3\omega_n} \sin \omega_n(t-s) + \frac{1}{6\omega_n} \sin 2\omega_n(t-s) & \frac{1}{3\omega_n} \sin \omega_n(t-s) - \frac{1}{6\omega_n} \sin 2\omega_n(t-s) \\ \frac{2}{3\omega_n} \sin \omega_n(t-s) - \frac{1}{3\omega_n} \sin 2\omega_n(t-s) & \frac{1}{3\omega_n} \sin \omega_n(t-s) + \frac{1}{3\omega_n} \sin 2\omega_n(t-s) \\ \frac{2}{3} \cos \omega_n(t-s) + \frac{1}{3} \cos 2\omega_n(t-s) & \frac{1}{3} \cos \omega_n(t-s) - \frac{1}{3} \cos 2\omega_n(t-s) \\ \frac{2}{3} \cos \omega_n(t-s) - \frac{2}{3} \cos 2\omega_n(t-s) & \frac{1}{3} \cos \omega_n(t-s) + \frac{2}{3} \cos 2\omega_n(t-s) \end{bmatrix} \begin{Bmatrix} f_1(s) \\ f_2(s) \end{Bmatrix} ds$$

which is the general solution of the response.

EXAMPLE 2 *A Relationship between the Eigenvalue and the Corresponding Eigenvector*

One interesting property for the normal mode method in the state space formulation is the relationship between the eigenvalue and the corresponding eigenvector. Consider the matrix $\mathbf{A}$ with the ith eigenvalue λ_i and the ith eigenvector $\boldsymbol{\phi}_i$. It follows from the property that

$$\mathbf{A}\boldsymbol{\phi}_i = \begin{bmatrix} \mathbf{0} & \mathbf{I} \\ -\mathbf{M}^{-1}\mathbf{K} & -\mathbf{M}^{-1}\mathbf{C} \end{bmatrix}\boldsymbol{\phi}_i = \lambda_i\boldsymbol{\phi}_i. \tag{4.514}$$

Let the eigenvector $\boldsymbol{\phi}_i$ be of the form

$$\boldsymbol{\phi}_i = \left\{ \frac{\boldsymbol{\phi}_i^{(1)}}{\boldsymbol{\phi}_i^{(2)}} \right\}, \tag{4.515}$$

where $\boldsymbol{\phi}_i^{(1)}$ and $\boldsymbol{\phi}_i^{(2)}$ are n-dimensional column vectors. Substituting Eq. 4.515 into Eq. 4.514, it follows that

$$\begin{bmatrix} \mathbf{0} & \mathbf{I} \\ -\mathbf{M}^{-1}\mathbf{K} & -\mathbf{M}^{-1}\mathbf{C} \end{bmatrix}\left\{ \frac{\boldsymbol{\phi}_i^{(1)}}{\boldsymbol{\phi}_i^{(2)}} \right\} = \lambda_i \left\{ \frac{\boldsymbol{\phi}_i^{(1)}}{\boldsymbol{\phi}_i^{(2)}} \right\}. \tag{4.516}$$

Expanding the top equation in Eq. 4.516 gives

$$\mathbf{0}\boldsymbol{\phi}_i^{(1)} + \mathbf{I}\boldsymbol{\phi}_i^{(2)} = \lambda_i\boldsymbol{\phi}_i^{(1)} \tag{4.517}$$

and simplifying Eq. 4.517 gives

$$\boldsymbol{\phi}_i^{(2)} = \lambda_i\boldsymbol{\phi}_i^{(1)}. \tag{4.518}$$

This shows that the upper n entries in each eigenvector are proportional to the lower n entries, with the proportional constant equal to the eigenvalue.

4.12 SENSITIVITY ANALYSIS OF NATURAL FREQUENCIES

The eigenvalue equation for an MDOF system is

$$\left[-\mathbf{M}\omega_i^2 + \mathbf{K}\right]\overline{\boldsymbol{\phi}}_i = \mathbf{0}. \tag{4.519}$$

Assume that the design variable under consideration is defined to be p. Then the derivative of Eq. 4.519 with respect to p is

$$\left[-\frac{\partial\mathbf{M}}{\partial p}\omega_i^2 - \mathbf{M}\frac{\partial\omega_i^2}{\partial p} + \frac{\partial\mathbf{K}}{\partial p}\right]\overline{\boldsymbol{\phi}}_i + \left[-\mathbf{M}\omega_i^2 + \mathbf{K}\right]\frac{\partial\overline{\boldsymbol{\phi}}_i}{\partial p} = \mathbf{0}. \tag{4.520}$$

Premultiply Eq. 4.520 by $\overline{\boldsymbol{\phi}}_i^T$, and it follows that

$$\overline{\boldsymbol{\phi}}_i^T\left[-\frac{\partial\mathbf{M}}{\partial p}\omega_i^2 - \mathbf{M}\frac{\partial\omega_i^2}{\partial p} + \frac{\partial\mathbf{K}}{\partial p}\right]\overline{\boldsymbol{\phi}}_i + \overline{\boldsymbol{\phi}}_i^T\left[-\mathbf{M}\omega_i^2 + \mathbf{K}\right]\frac{\partial\overline{\boldsymbol{\phi}}_i}{\partial p} = 0. \tag{4.521}$$

Because $\mathbf{K}$ and $\mathbf{M}$ are symmetrical matrices, it follows from Eq. 4.519 that

$$\left[-\mathbf{M}\omega_i^2 + \mathbf{K}\right]\overline{\boldsymbol{\phi}}_i = \left\{\overline{\boldsymbol{\phi}}_i^T\left[-\mathbf{M}\omega_i^2 + \mathbf{K}\right]\right\}^T = \mathbf{0}. \tag{4.522}$$

This implies that the second term in Eq. 4.521 is zero. Therefore, Eq. 4.521 becomes

$$-\left(\overline{\boldsymbol{\phi}}_i^T\mathbf{M}\overline{\boldsymbol{\phi}}_i\right)\frac{\partial\omega_i^2}{\partial p} + \overline{\boldsymbol{\phi}}_i^T\left[-\frac{\partial\mathbf{M}}{\partial p}\omega_i^2 + \frac{\partial\mathbf{K}}{\partial p}\right]\overline{\boldsymbol{\phi}}_i = 0. \tag{4.523}$$

Finally,

$$\frac{\partial\omega_i^2}{\partial p} = \frac{\overline{\boldsymbol{\phi}}_i^T\left[-\dfrac{\partial\mathbf{M}}{\partial p}\omega_i^2 + \dfrac{\partial\mathbf{K}}{\partial p}\right]\overline{\boldsymbol{\phi}}_i}{\overline{\boldsymbol{\phi}}_i^T\mathbf{M}\overline{\boldsymbol{\phi}}_i}. \tag{4.524}$$

The sensitivity of ω_i^2 is only a function of ω_i^2 and $\overline{\Phi}_i$ and therefore does not require the calculation of eigenvalues and eigenvectors in other modes of vibrations. Note that

$$\frac{\partial \omega_i^2}{\partial p} = 2\omega_i \frac{\partial \omega_i}{\partial p} \tag{4.525}$$

or

$$\frac{\partial \omega_i}{\partial p} = \frac{1}{2\omega_i}\left(\frac{\partial \omega_i^2}{\partial p}\right). \tag{4.526}$$

Substituting Eq. 4.524 into Eq. 4.526, it follows for a mass-normalized eigenvector that

$$\frac{\partial \omega_i}{\partial p} = \frac{\overline{\Phi}_i^T\left[-\dfrac{\partial \mathbf{M}}{\partial p}\omega_i^2 + \dfrac{\partial \mathbf{K}}{\partial p}\right]\overline{\Phi}_i}{2\omega_i\left(\overline{\Phi}_i^T \mathbf{M}\,\overline{\Phi}_i\right)} = \frac{1}{2\omega_i}\overline{\Phi}_i^T\left[-\dfrac{\partial \mathbf{M}}{\partial p}\omega_i^2 + \dfrac{\partial \mathbf{K}}{\partial p}\right]\overline{\Phi}_i. \tag{4.527}$$

EXAMPLE 1 *Sensitivity to Stiffness for a 2DOF System*

Consider the 2DOF system given in Example 1 of Section 3.9 with the following properties:

$$m_1 = 1.5 \text{ k-s}^2/\text{in.}$$

$$m_2 = 1.0 \text{ k-s}^2/\text{in.}$$

It then follows that the mass and stiffness matrices are

$$\mathbf{M} = \begin{bmatrix} 1.5 & 0 \\ 0 & 1.0 \end{bmatrix}, \quad \mathbf{K} = \begin{bmatrix} k_1 + k_2 & -k_2 \\ -k_2 & k_2 \end{bmatrix}.$$

The eigenvalues of the system are given in Eqs. 3.293 and 3.294, which are

$$\omega_1^2 = \frac{1}{3}\left(k_1 + 2.5k_2 - \sqrt{(k_1 + 2.5k_2)^2 - 6k_1k_2}\right) \tag{4.528}$$

$$\omega_2^2 = \frac{1}{3}\left(k_1 + 2.5k_2 + \sqrt{(k_1 + 2.5k_2)^2 - 6k_1k_2}\right). \tag{4.529}$$

The mode shapes are given in Eq. 3.309, and when they are mass normalized, it follows that

$$\overline{\Phi}_1 = \frac{1}{1.5 + r_1^2}\begin{Bmatrix} 1 \\ r_1 \end{Bmatrix}, \quad \overline{\Phi}_2 = \frac{1}{1.5 + r_2^2}\begin{Bmatrix} 1 \\ r_2 \end{Bmatrix} \tag{4.530}$$

where

$$r_1 = 0.5\frac{k_1}{k_2} - 0.25 + 0.5\sqrt{\left(\frac{k_1}{k_2} + 2.5\right)^2 - 6\frac{k_1}{k_2}} \tag{4.531}$$

$$r_2 = 0.5\frac{k_1}{k_2} - 0.25 - 0.5\sqrt{\left(\frac{k_1}{k_2} + 2.5\right)^2 - 6\frac{k_1}{k_2}}. \tag{4.532}$$

Let the design variable be k_1; then

$$\frac{\partial \mathbf{M}}{\partial k_1} = \mathbf{0}, \quad \frac{\partial \mathbf{K}}{\partial k_1} = \begin{bmatrix} 1 & 0 \\ 0 & 0 \end{bmatrix}. \tag{4.533}$$

Equation 4.527 becomes

$$\frac{\partial \omega_1}{\partial k_1} = \frac{1}{2\omega_1} \left(\frac{1}{1.5 + r_1^2} \right)^2 \begin{Bmatrix} 1 \\ r_1 \end{Bmatrix}^T \begin{bmatrix} 1 & 0 \\ 0 & 0 \end{bmatrix} \begin{Bmatrix} 1 \\ r_1 \end{Bmatrix} = \frac{1}{2\omega_1} \left(\frac{1}{1.5 + r_1^2} \right)^2 \tag{4.534}$$

$$\frac{\partial \omega_2}{\partial k_1} = \frac{1}{2\omega_2} \left(\frac{1}{1.5 + r_2^2} \right)^2 \begin{Bmatrix} 1 \\ r_2 \end{Bmatrix}^T \begin{bmatrix} 1 & 0 \\ 0 & 0 \end{bmatrix} \begin{Bmatrix} 1 \\ r_2 \end{Bmatrix} = \frac{1}{2\omega_2} \left(\frac{1}{1.5 + r_2^2} \right)^2. \tag{4.535}$$

Let $k_1 = k_2 = 10$; then

$$\omega_1^2 = \frac{10}{3}, \quad \omega_2^2 = 20, \quad r_1 = 1.5, \quad r_2 = -1.0. \tag{4.536}$$

Therefore, the sensitivities of ω_1 and ω_2 with respect to k_1 are

$$\frac{\partial \omega_1}{\partial k_1} = \frac{1}{2\sqrt{10/3}} \left(\frac{1}{1.5 + (1.5)^2} \right)^2 = 0.01947 \tag{4.537}$$

$$\frac{\partial \omega_2}{\partial k_1} = \frac{1}{2\sqrt{20}} \left(\frac{1}{1.5 + (-1.0)^2} \right)^2 = 0.01789. \tag{4.538}$$

EXAMPLE 2   *Sensitivity to Stiffness for a 3DOF System*

Consider the 3DOF system discussed in Example 2 of Section 4.3, where

$$\mathbf{M} = \begin{bmatrix} m & 0 & 0 \\ 0 & m & 0 \\ 0 & 0 & m \end{bmatrix}, \quad \mathbf{K} = \begin{bmatrix} 3k & -k & 0 \\ -k & 2k & -k \\ 0 & -k & k \end{bmatrix} \tag{4.539}$$

$$\omega_1^2 = \left(2 - \sqrt{3}\right)\frac{k}{m}, \quad \omega_2^2 = 2\frac{k}{m}, \quad \omega_3^2 = \left(2 + \sqrt{3}\right)\frac{k}{m} \tag{4.540}$$

and the mode shapes are given in Eq. 4.204:

$$\overline{\Phi}_1 = \frac{m^{-1/2}}{3 + \sqrt{3}} \begin{Bmatrix} 1 \\ 1 + \sqrt{3} \\ 2 + \sqrt{3} \end{Bmatrix}, \quad \overline{\Phi}_2 = \frac{m^{-1/2}}{\sqrt{3}} \begin{Bmatrix} 1 \\ 1 \\ -1 \end{Bmatrix}, \quad \overline{\Phi}_3 = \frac{m^{-1/2}}{3 - \sqrt{3}} \begin{Bmatrix} 1 \\ 1 - \sqrt{3} \\ 2 - \sqrt{3} \end{Bmatrix}. \tag{4.541}$$

Let k be the parameter of interest, and therefore $p = k$. It follows that

$$\frac{\partial \omega_1}{\partial p} = \frac{\partial \omega_1}{\partial k} = \sqrt{2 - \sqrt{3}} \left(\frac{1}{2} \right) \sqrt{\frac{1}{km}} = \frac{0.2588}{\sqrt{km}} \tag{4.542}$$

$$\frac{\partial \omega_2}{\partial p} = \frac{\partial \omega_2}{\partial k} = \sqrt{2} \left(\frac{1}{2} \right) \sqrt{\frac{1}{km}} = \frac{0.7071}{\sqrt{km}} \tag{4.543}$$

$$\frac{\partial \omega_3}{\partial p} = \frac{\partial \omega_3}{\partial k} = \sqrt{2 + \sqrt{3}} \left(\frac{1}{2} \right) \sqrt{\frac{1}{km}} = \frac{0.9659}{\sqrt{km}}. \tag{4.544}$$

Similar results can be obtained using Eq. 4.527:

$$\frac{\partial \omega_i}{\partial k} = \frac{1}{2\omega_i} \overline{\boldsymbol{\phi}}_i^T \left[-\frac{\partial \mathbf{M}}{\partial k} \omega_i^2 + \frac{\partial \mathbf{K}}{\partial k} \right] \overline{\boldsymbol{\phi}}_i. \tag{4.545}$$

Taking partial derivatives of the mass and stiffness matrices with respect to k, it follows that

$$\frac{\partial \mathbf{M}}{\partial k} = \begin{bmatrix} 0 & 0 & 0 \\ 0 & 0 & 0 \\ 0 & 0 & 0 \end{bmatrix}, \quad \frac{\partial \mathbf{K}}{\partial k} = \begin{bmatrix} 3 & -1 & 0 \\ -1 & 2 & -1 \\ 0 & -1 & 1 \end{bmatrix}. \tag{4.546}$$

Substituting Eqs. 4.541 and 4.546 into Eq. 4.545 gives

$$\frac{\partial \omega_1}{\partial k} = \frac{1}{2\omega_1} \left(\frac{m^{-1/2}}{3+\sqrt{3}} \right) \begin{Bmatrix} 1 \\ 1+\sqrt{3} \\ 2+\sqrt{3} \end{Bmatrix}^T \begin{bmatrix} 3 & -1 & 0 \\ -1 & 2 & -1 \\ 0 & -1 & 1 \end{bmatrix} \begin{Bmatrix} 1 \\ 1+\sqrt{3} \\ 2+\sqrt{3} \end{Bmatrix} \left(\frac{m^{-1/2}}{3+\sqrt{3}} \right) = \frac{1}{2+\sqrt{3}} \left(\frac{1}{2m\omega_1} \right) \tag{4.547}$$

$$\frac{\partial \omega_2}{\partial k} = \frac{1}{2\omega_2} \left(\frac{m^{-1/2}}{\sqrt{3}} \right) \begin{Bmatrix} 1 \\ 1 \\ -1 \end{Bmatrix}^T \begin{bmatrix} 3 & -1 & 0 \\ -1 & 2 & -1 \\ 0 & -1 & 1 \end{bmatrix} \begin{Bmatrix} 1 \\ 1 \\ -1 \end{Bmatrix} \left(\frac{m^{-1/2}}{\sqrt{3}} \right) = \frac{1}{m\omega_2} \tag{4.548}$$

$$\frac{\partial \omega_3}{\partial k} = \frac{1}{2\omega_3} \left(\frac{m^{-1/2}}{3-\sqrt{3}} \right) \begin{Bmatrix} 1 \\ 1-\sqrt{3} \\ 2-\sqrt{3} \end{Bmatrix}^T \begin{bmatrix} 3 & -1 & 0 \\ -1 & 2 & -1 \\ 0 & -1 & 1 \end{bmatrix} \begin{Bmatrix} 1 \\ 1-\sqrt{3} \\ 2-\sqrt{3} \end{Bmatrix} \left(\frac{m^{-1/2}}{3-\sqrt{3}} \right) = \frac{1}{2-\sqrt{3}} \left(\frac{1}{2m\omega_3} \right). \tag{4.549}$$

Substituting Eq. 4.540 into Eqs. 4.547 to 4.549, it follows that

$$\omega_1^2 = \left(2-\sqrt{3}\right)\frac{k}{m}, \quad \omega_2^2 = 2\frac{k}{m}, \quad \omega_3^2 = \left(2+\sqrt{3}\right)\frac{k}{m}$$

$$\frac{\partial \omega_1}{\partial k} = \frac{1}{2+\sqrt{3}} \left(\frac{1}{2m\omega_1} \right) = \frac{1}{2} \left(\frac{1}{2+\sqrt{3}} \right) \left(\frac{1}{\sqrt{2-\sqrt{3}}} \right) \sqrt{\frac{1}{km}} = \frac{0.2588}{\sqrt{km}} \tag{4.550}$$

$$\frac{\partial \omega_2}{\partial k} = \frac{1}{m\omega_2} = \left(\frac{1}{\sqrt{2}} \right) \sqrt{\frac{1}{km}} = \frac{0.7071}{\sqrt{km}} \tag{4.551}$$

$$\frac{\partial \omega_3}{\partial k} = \frac{1}{2-\sqrt{3}} \left(\frac{1}{m\omega_3} \right) = \frac{1}{2} \left(\frac{1}{2-\sqrt{3}} \right) \left(\frac{1}{\sqrt{2+\sqrt{3}}} \right) \sqrt{\frac{1}{km}} = \frac{0.9659}{\sqrt{km}}. \tag{4.552}$$

which is the same as given in Eqs. 4.542 to 4.544.

4.13 SENSITIVITY ANALYSIS OF MODE SHAPES

The sensitivity of a mode shape with respect to a design variable is calculated by starting with the eigenvalue equation:

$$\left[-\mathbf{M}\omega_i^2 + \mathbf{K} \right] \overline{\boldsymbol{\phi}}_i = \mathbf{0}. \tag{4.553}$$

The derivative of Eq. 4.553 with respect to the design variable, denoted p, gives

$$\left[-\frac{\partial \mathbf{M}}{\partial p} \omega_i^2 - \mathbf{M}\frac{\partial \omega_i^2}{\partial p} + \frac{\partial \mathbf{K}}{\partial p} \right] \overline{\boldsymbol{\phi}}_i + \left[-\mathbf{M}\omega_i^2 + \mathbf{K} \right] \frac{\partial \overline{\boldsymbol{\phi}}_i}{\partial p} = \mathbf{0}. \tag{4.554}$$

The partial derivative of the mode shape can be represented as a power series expansion of the mode shapes:

$$\frac{\partial \overline{\boldsymbol{\phi}}_i}{\partial p} = \alpha_1 \overline{\boldsymbol{\phi}}_1 + \alpha_2 \overline{\boldsymbol{\phi}}_2 + \alpha_3 \overline{\boldsymbol{\phi}}_3 + \ \dots \tag{4.555}$$

Consider first the premultiplication of Eq. 4.554 by $\overline{\boldsymbol{\phi}}_j^T$, where $j \neq i$. It follows that

$$\overline{\boldsymbol{\phi}}_j^T \left[-\frac{\partial \mathbf{M}}{\partial p} \omega_i^2 - \mathbf{M} \frac{\partial \omega_i^2}{\partial p} + \frac{\partial \mathbf{K}}{\partial p} \right] \overline{\boldsymbol{\phi}}_i + \overline{\boldsymbol{\phi}}_j^T \left[-\mathbf{M}\omega_i^2 + \mathbf{K} \right]\left(\alpha_1 \overline{\boldsymbol{\phi}}_1 + \alpha_2 \overline{\boldsymbol{\phi}}_2 + \ \dots \right) = 0. \tag{4.556}$$

Recall that

$$\overline{\boldsymbol{\phi}}_j^T \mathbf{M} \overline{\boldsymbol{\phi}}_i = \begin{cases} 1 & i = j \\ 0 & i \neq j \end{cases} \tag{4.557}$$

and

$$\overline{\boldsymbol{\phi}}_j^T \left[-\mathbf{M}\omega_i^2 + \mathbf{K} \right]\left(\alpha_1 \overline{\boldsymbol{\phi}}_1 + \alpha_2 \overline{\boldsymbol{\phi}}_2 + \dots \right) = \alpha_j \left(\omega_j^2 - \omega_i^2 \right). \tag{4.558}$$

Therefore, Eq. 4.556 becomes

$$\overline{\boldsymbol{\phi}}_j^T \left[-\frac{\partial \mathbf{M}}{\partial p} \omega_i^2 + \frac{\partial \mathbf{K}}{\partial p} \right] \overline{\boldsymbol{\phi}}_i + \alpha_j \left(\omega_j^2 - \omega_i^2 \right) = 0 \tag{4.559}$$

and finally,

$$\alpha_j = \frac{\overline{\boldsymbol{\phi}}_j^T \left[\dfrac{\partial \mathbf{M}}{\partial p} \omega_i^2 - \dfrac{\partial \mathbf{K}}{\partial p} \right] \overline{\boldsymbol{\phi}}_i}{\omega_j^2 - \omega_i^2}, \quad j \neq i. \tag{4.560}$$

Now consider the case where $j = i$. Taking the partial derivative of Eq. 4.557 with respect to the design variable, p, gives

$$\frac{\partial \overline{\boldsymbol{\phi}}_i^T}{\partial p} \mathbf{M} \overline{\boldsymbol{\phi}}_i + \overline{\boldsymbol{\phi}}_i^T \frac{\partial \mathbf{M}}{\partial p} \overline{\boldsymbol{\phi}}_i + \overline{\boldsymbol{\phi}}_i^T \mathbf{M} \frac{\partial \overline{\boldsymbol{\phi}}_i}{\partial p} = 0. \tag{4.561}$$

Now substituting Eq. 4.555 into Eq. 4.561, it follows that

$$\left(\alpha_1 \overline{\boldsymbol{\phi}}_1^T + \alpha_2 \overline{\boldsymbol{\phi}}_2^T + \dots \right) \mathbf{M} \overline{\boldsymbol{\phi}}_i + \overline{\boldsymbol{\phi}}_i^T \frac{\partial \mathbf{M}}{\partial p} \overline{\boldsymbol{\phi}}_i + \overline{\boldsymbol{\phi}}_i^T \mathbf{M} \left(\alpha_1 \overline{\boldsymbol{\phi}}_1 + \alpha_2 \overline{\boldsymbol{\phi}}_2 + \dots \right) = 0 \tag{4.562}$$

and therefore

$$\alpha_i(1) + \overline{\boldsymbol{\phi}}_i^T \frac{\partial \mathbf{M}}{\partial p} \overline{\boldsymbol{\phi}}_i + \alpha_i(1) = 0. \tag{4.563}$$

Finally,

$$\alpha_i = -\frac{1}{2} \left(\overline{\boldsymbol{\phi}}_i^T \frac{\partial \mathbf{M}}{\partial p} \overline{\boldsymbol{\phi}}_i \right). \tag{4.564}$$

Equations 4.560 and 4.564 define the coefficients in the power series expansion representation of the mode shape sensitivity in Eq. 4.555.

EXAMPLE 1 *Sensitivity to Stiffness for a 2DOF System*

Consider the 2DOF system given in Example 1 in Section 4.12. The mode shape sensitivity with respect to the design variable, k_1, is

$$\frac{\partial \overline{\Phi}_i}{\partial k_1} = \alpha_1 \overline{\Phi}_1 + \alpha_2 \overline{\Phi}_2. \tag{4.565}$$

Since

$$\frac{\partial \mathbf{M}}{\partial k_1} = \mathbf{0}, \quad \frac{\partial \mathbf{K}}{\partial k_1} = \begin{bmatrix} 1 & 0 \\ 0 & 0 \end{bmatrix}. \tag{4.566}$$

Let $i = 1$; it follows that

$$\alpha_1 = -\frac{1}{2}\left(\overline{\Phi}_1^T \frac{\partial \mathbf{M}}{\partial p} \overline{\Phi}_1 \right) = 0 \tag{4.567}$$

and

$$\alpha_2 = \frac{1}{\omega_2^2 - \omega_1^2}\left(\frac{1}{1+r_1^2} \right)\left(\frac{1}{1+r_2^2} \right)\begin{Bmatrix} 1 \\ r_2 \end{Bmatrix}^T \begin{bmatrix} 1 & 0 \\ 0 & 0 \end{bmatrix}\begin{Bmatrix} 1 \\ r_1 \end{Bmatrix} = \frac{1}{\omega_2^2 - \omega_1^2}\left(\frac{1}{1+r_1^2} \right)\left(\frac{1}{1+r_2^2} \right). \tag{4.568}$$

Let $k_1 = k_2 = 10$. Then from Eq. 4.536, where

$$\omega_1^2 = \frac{10}{3}, \quad \omega_1^2 = 20, \quad r_1 = 1.5, \quad r_2 = -1.0, \tag{4.569}$$

it follows that

$$\alpha_2 = \frac{1}{(20)^2 - (10/3)^2}\left(\frac{1}{1+(1.5)^2} \right)\left(\frac{1}{1+(-1.0)^2} \right) = 3.956 \times 10^{-4}. \tag{4.570}$$

EXAMPLE 2 *Sensitivity to Stiffness Ratios for a 3DOF System*

Consider again the 3DOF system in Example 2 of Section 4.3, where the three mode shapes of vibration are given in Eq. 4.204:

$$\overline{\Phi}_1 = \frac{m^{-1/2}}{3+\sqrt{3}}\begin{Bmatrix} 1 \\ 1+\sqrt{3} \\ 2+\sqrt{3} \end{Bmatrix}, \quad \overline{\Phi}_2 = \frac{m^{-1/2}}{\sqrt{3}}\begin{Bmatrix} 1 \\ 1 \\ -1 \end{Bmatrix}, \quad \overline{\Phi}_3 = \frac{m^{-1/2}}{3-\sqrt{3}}\begin{Bmatrix} 1 \\ 1-\sqrt{3} \\ 2-\sqrt{3} \end{Bmatrix}. \tag{4.571}$$

Let k be the parameter of interest, and therefore $p = k$. Note that the mode shapes are not a function of k. Therefore, it follows that

$$\frac{\partial \overline{\Phi}_1}{\partial p} = \frac{\partial \overline{\Phi}_1}{\partial k} = \begin{Bmatrix} \partial \overline{\varphi}_{11}/\partial k \\ \partial \overline{\varphi}_{21}/\partial k \\ \partial \overline{\varphi}_{31}/\partial k \end{Bmatrix} = \begin{Bmatrix} 0 \\ 0 \\ 0 \end{Bmatrix} \tag{4.572}$$

$$\frac{\partial \overline{\Phi}_2}{\partial p} = \frac{\partial \overline{\Phi}_2}{\partial k} = \begin{Bmatrix} \partial \overline{\varphi}_{12}/\partial k \\ \partial \overline{\varphi}_{22}/\partial k \\ \partial \overline{\varphi}_{32}/\partial k \end{Bmatrix} = \begin{Bmatrix} 0 \\ 0 \\ 0 \end{Bmatrix} \tag{4.573}$$

$$\frac{\partial \overline{\Phi}_3}{\partial p} = \frac{\partial \overline{\Phi}_3}{\partial k} = \left\{ \begin{array}{c} \partial \overline{\varphi}_{13}/\partial k \\ \partial \overline{\varphi}_{23}/\partial k \\ \partial \overline{\varphi}_{33}/\partial k \end{array} \right\} = \left\{ \begin{array}{c} 0 \\ 0 \\ 0 \end{array} \right\}. \tag{4.574}$$

This means that the mode shapes are insensitive to any change in the stiffness. The same result can be obtained using Eqs. 4.560 and 4.564:

$$\alpha_j = \frac{\overline{\Phi}_j^T \left[\frac{\partial \mathbf{M}}{\partial p} \omega_i^2 - \frac{\partial \mathbf{K}}{\partial p} \right] \overline{\Phi}_i}{\omega_j^2 - \omega_i^2}, \quad j \neq i \tag{4.575}$$

$$\alpha_i = -\frac{1}{2} \left(\overline{\Phi}_i^T \frac{\partial \mathbf{M}}{\partial p} \overline{\Phi}_i \right). \tag{4.576}$$

Taking the partial derivatives of the mass and stiffness matrices gives

$$\frac{\partial \mathbf{M}}{\partial k} = \frac{\partial}{\partial k} \begin{bmatrix} m & 0 & 0 \\ 0 & m & 0 \\ 0 & 0 & m \end{bmatrix} = \begin{bmatrix} 0 & 0 & 0 \\ 0 & 0 & 0 \\ 0 & 0 & 0 \end{bmatrix}, \quad \frac{\partial \mathbf{K}}{\partial k} = \frac{\partial}{\partial k} \begin{bmatrix} 3k & -k & 0 \\ -k & 2k & -k \\ 0 & -k & k \end{bmatrix} = \begin{bmatrix} 3 & -1 & 0 \\ -1 & 2 & -1 \\ 0 & -1 & 1 \end{bmatrix}. \tag{4.577}$$

First consider $i = 1$. It follows from Eqs. 4.575 and 4.576 that

$$\alpha_1 = -\frac{1}{2} \left(\overline{\Phi}_1^T \frac{\partial \mathbf{M}}{\partial k} \overline{\Phi}_1 \right) = 0 \tag{4.578}$$

$$\alpha_2 = \frac{\overline{\Phi}_2^T \left[\frac{\partial \mathbf{M}}{\partial k} \omega_1^2 - \frac{\partial \mathbf{K}}{\partial k} \right] \overline{\Phi}_1}{\omega_2^2 - \omega_1^2} = \frac{1}{\omega_2^2 - \omega_1^2} (0 - 0) = 0 \tag{4.579}$$

$$\alpha_3 = \frac{\overline{\Phi}_3^T \left[\frac{\partial \mathbf{M}}{\partial k} \omega_1^2 - \frac{\partial \mathbf{K}}{\partial k} \right] \overline{\Phi}_1}{\omega_3^2 - \omega_1^2} = \frac{1}{\omega_3^2 - \omega_1^2} (0 - 0) = 0. \tag{4.580}$$

Therefore, it follows from Eq. 4.555 that

$$\frac{\partial \overline{\Phi}_1}{\partial k} = (0)\overline{\Phi}_1 + (0)\overline{\Phi}_2 + (0)\overline{\Phi}_3 = \left\{ \begin{array}{c} 0 \\ 0 \\ 0 \end{array} \right\} \tag{4.581}$$

which is the same as Eq. 4.572. Similar procedures can be performed for $i = 2$ and $i = 3$ to obtain Eqs. 4.573 and 4.574, respectively.

EXAMPLE 3 *Sensitivity to Individual Stiffness for a 3DOF System*

Consider again the 3DOF system in Example 2 of Section 4.3, where the three mode shapes of vibration are

$$\omega_1^2 = \left(2 - \sqrt{3} \right) \omega_n^2, \quad \omega_2^2 = 2\omega_n^2, \quad \omega_3^2 = \left(2 + \sqrt{3} \right) \omega_n^2 \tag{4.582}$$

$$\overline{\Phi}_1 = \frac{m^{-1/2}}{3 + \sqrt{3}} \left\{ \begin{array}{c} 1 \\ 1 + \sqrt{3} \\ 2 + \sqrt{3} \end{array} \right\}, \quad \overline{\Phi}_2 = \frac{m^{-1/2}}{\sqrt{3}} \left\{ \begin{array}{c} 1 \\ 1 \\ -1 \end{array} \right\}, \quad \overline{\Phi}_3 = \frac{m^{-1/2}}{3 - \sqrt{3}} \left\{ \begin{array}{c} 1 \\ 1 - \sqrt{3} \\ 2 - \sqrt{3} \end{array} \right\}. \tag{4.583}$$

Let the stiffness of the first story be the parameter of interest; therefore, $p = \beta$ where β represents the ratio of the first-story stiffness to the stiffness of other stories (i.e., $k_1 = \beta k$). Consider that the sensitivity of k_1 is studied at $k_1 = k$ (i.e., $\beta = 1$); it follows from taking partial derivatives of the mass and stiffness matrices that

$$\frac{\partial \mathbf{M}}{\partial \beta} = \frac{\partial}{\partial \beta} \begin{bmatrix} m & 0 & 0 \\ 0 & m & 0 \\ 0 & 0 & m \end{bmatrix} = \begin{bmatrix} 0 & 0 & 0 \\ 0 & 0 & 0 \\ 0 & 0 & 0 \end{bmatrix} \tag{4.584}$$

$$\frac{\partial \mathbf{K}}{\partial \beta} = \frac{\partial}{\partial \beta} \begin{bmatrix} (2\beta+1)k & -k & 0 \\ -k & 2k & -k \\ 0 & -k & k \end{bmatrix} = \begin{bmatrix} 2k & 0 & 0 \\ 0 & 0 & 0 \\ 0 & 0 & 0 \end{bmatrix}. \tag{4.585}$$

First consider $i = 1$. It follows from Eqs. 4.560 and 4.564 that

$$\alpha_1 = -\frac{1}{2}\left(\overline{\boldsymbol{\phi}}_1^T \frac{\partial \mathbf{M}}{\partial \beta} \overline{\boldsymbol{\phi}}_1 \right) = 0 \tag{4.586}$$

$$\alpha_2 = \frac{\overline{\boldsymbol{\phi}}_2^T \left[\dfrac{\partial \mathbf{M}}{\partial \beta} \omega_1^2 - \dfrac{\partial \mathbf{K}}{\partial \beta} \right] \overline{\boldsymbol{\phi}}_1}{\omega_2^2 - \omega_1^2}$$

$$= \frac{1}{\sqrt{3}\omega_n^2}\left(0 - \left(\frac{m^{-1/2}}{\sqrt{3}}\right) \begin{Bmatrix} 1 \\ 1 \\ -1 \end{Bmatrix}^T \begin{bmatrix} 2k & 0 & 0 \\ 0 & 0 & 0 \\ 0 & 0 & 0 \end{bmatrix} \begin{Bmatrix} 1 \\ 1+\sqrt{3} \\ 2+\sqrt{3} \end{Bmatrix} \left(\frac{m^{-1/2}}{3+\sqrt{3}}\right) \right) = -\frac{2}{9+3\sqrt{3}} \tag{4.587}$$

$$\alpha_3 = \frac{\overline{\boldsymbol{\phi}}_3^T \left[\dfrac{\partial \mathbf{M}}{\partial \beta} \omega_1^2 - \dfrac{\partial \mathbf{K}}{\partial \beta} \right] \overline{\boldsymbol{\phi}}_1}{\omega_3^2 - \omega_1^2}$$

$$= \frac{1}{2\sqrt{3}\omega_n^2}\left(0 - \left(\frac{m^{-1/2}}{3-\sqrt{3}}\right) \begin{Bmatrix} 1 \\ 1-\sqrt{3} \\ 2-\sqrt{3} \end{Bmatrix}^T \begin{bmatrix} 2k & 0 & 0 \\ 0 & 0 & 0 \\ 0 & 0 & 0 \end{bmatrix} \begin{Bmatrix} 1 \\ 1+\sqrt{3} \\ 2+\sqrt{3} \end{Bmatrix} \left(\frac{m^{-1/2}}{3+\sqrt{3}}\right) \right) = -\frac{1}{6\sqrt{3}}. \tag{4.588}$$

Therefore, it follows from Eq. 4.555 that

$$\frac{\partial \overline{\boldsymbol{\phi}}_1}{\partial \beta} = (0)\left(\frac{m^{-1/2}}{3+\sqrt{3}}\right) \begin{Bmatrix} 1 \\ 1+\sqrt{3} \\ 2+\sqrt{3} \end{Bmatrix}$$

$$+ \left(-\frac{2}{9+3\sqrt{3}}\right)\left(\frac{m^{-1/2}}{\sqrt{3}}\right) \begin{Bmatrix} 1 \\ 1 \\ -1 \end{Bmatrix} + \left(-\frac{1}{6\sqrt{3}}\right)\left(\frac{m^{-1/2}}{3-\sqrt{3}}\right) \begin{Bmatrix} 1 \\ 1-\sqrt{3} \\ 2-\sqrt{3} \end{Bmatrix} = \frac{m^{-1/2}}{36} \begin{Bmatrix} 3-5\sqrt{3} \\ 6-4\sqrt{3} \\ 3\sqrt{3}-3 \end{Bmatrix}. \tag{4.589}$$

For $i = 2$, it follows from Eqs. 4.560 and 4.564 that

$$\alpha_1 = \frac{\overline{\boldsymbol{\phi}}_1^T \left[\dfrac{\partial \mathbf{M}}{\partial \beta} \omega_2^2 - \dfrac{\partial \mathbf{K}}{\partial \beta} \right] \overline{\boldsymbol{\phi}}_2}{\omega_1^2 - \omega_2^2}$$

$$= -\frac{1}{\sqrt{3}\omega_n^2}\left(0 - \left(\frac{m^{-1/2}}{3+\sqrt{3}}\right) \begin{Bmatrix} 1 \\ 1+\sqrt{3} \\ 2+\sqrt{3} \end{Bmatrix}^T \begin{bmatrix} 2k & 0 & 0 \\ 0 & 0 & 0 \\ 0 & 0 & 0 \end{bmatrix} \begin{Bmatrix} 1 \\ 1 \\ -1 \end{Bmatrix} \left(\frac{m^{-1/2}}{\sqrt{3}}\right) \right) = \frac{2}{9+3\sqrt{3}} \tag{4.590}$$

$$\alpha_2 = -\frac{1}{2}\left(\overline{\Phi}_2^T \frac{\partial \mathbf{M}}{\partial \beta} \overline{\Phi}_2\right) = 0 \tag{4.591}$$

$$\alpha_3 = \frac{\overline{\Phi}_3^T\left[\frac{\partial \mathbf{M}}{\partial \beta}\omega_2^2 - \frac{\partial \mathbf{K}}{\partial \beta}\right]\overline{\Phi}_2}{\omega_3^2 - \omega_2^2}$$

$$= \frac{1}{\sqrt{3}\omega_n^2}\left(0 - \left(\frac{m^{-1/2}}{3-\sqrt{3}}\right)\begin{Bmatrix}1\\1-\sqrt{3}\\2-\sqrt{3}\end{Bmatrix}^T\begin{bmatrix}2k&0&0\\0&0&0\\0&0&0\end{bmatrix}\begin{Bmatrix}1\\1\\-1\end{Bmatrix}\left(\frac{m^{-1/2}}{\sqrt{3}}\right)\right) = -\frac{2}{9-3\sqrt{3}}. \tag{4.592}$$

Therefore, it follows from Eq. 4.555 that

$$\frac{\partial \overline{\Phi}_2}{\partial \beta} = \left(\frac{2}{9+3\sqrt{3}}\right)\left(\frac{m^{-1/2}}{3+\sqrt{3}}\right)\begin{Bmatrix}1\\1+\sqrt{3}\\2+\sqrt{3}\end{Bmatrix}$$

$$+(0)\left(\frac{m^{-1/2}}{\sqrt{3}}\right)\begin{Bmatrix}1\\1\\-1\end{Bmatrix}+\left(-\frac{2}{9-3\sqrt{3}}\right)\left(\frac{m^{-1/2}}{3-\sqrt{3}}\right)\begin{Bmatrix}1\\1-\sqrt{3}\\2-\sqrt{3}\end{Bmatrix}=\frac{m^{-1/2}}{3}\begin{Bmatrix}-2\sqrt{3}\\2\sqrt{3}\\0\end{Bmatrix}. \tag{4.593}$$

Finally, for $i = 3$, it follows from Eqs. 4.560 and 4.564 that

$$\alpha_1 = \frac{\overline{\Phi}_1^T\left[\frac{\partial \mathbf{M}}{\partial \beta}\omega_3^2 - \frac{\partial \mathbf{K}}{\partial \beta}\right]\overline{\Phi}_3}{\omega_1^2 - \omega_3^2}$$

$$= -\frac{1}{2\sqrt{3}\omega_n^2}\left(0 - \left(\frac{m^{-1/2}}{3+\sqrt{3}}\right)\begin{Bmatrix}1\\1+\sqrt{3}\\2+\sqrt{3}\end{Bmatrix}^T\begin{bmatrix}2k&0&0\\0&0&0\\0&0&0\end{bmatrix}\begin{Bmatrix}1\\1-\sqrt{3}\\2-\sqrt{3}\end{Bmatrix}\left(\frac{m^{-1/2}}{3-\sqrt{3}}\right)\right) = \frac{1}{6\sqrt{3}} \tag{4.594}$$

$$\alpha_2 = \frac{\overline{\Phi}_2^T\left[\frac{\partial \mathbf{M}}{\partial \beta}\omega_3^2 - \frac{\partial \mathbf{K}}{\partial \beta}\right]\overline{\Phi}_3}{\omega_2^2 - \omega_3^2}$$

$$= -\frac{1}{\sqrt{3}\omega_n^2}\left(0 - \left(\frac{m^{-1/2}}{\sqrt{3}}\right)\begin{Bmatrix}1\\1\\-1\end{Bmatrix}^T\begin{bmatrix}2k&0&0\\0&0&0\\0&0&0\end{bmatrix}\begin{Bmatrix}1\\1-\sqrt{3}\\2-\sqrt{3}\end{Bmatrix}\left(\frac{m^{-1/2}}{3-\sqrt{3}}\right)\right) = \frac{2}{9-3\sqrt{3}}. \tag{4.595}$$

$$\alpha_3 = -\frac{1}{2}\left(\overline{\Phi}_3^T \frac{\partial \mathbf{M}}{\partial \beta} \overline{\Phi}_3\right) = 0. \tag{4.596}$$

It follows from Eq. 4.555 that

$$\frac{\partial \overline{\Phi}_3}{\partial \beta} = \left(\frac{1}{6\sqrt{3}}\right)\left(\frac{m^{-1/2}}{3+\sqrt{3}}\right)\begin{Bmatrix}1\\1+\sqrt{3}\\2+\sqrt{3}\end{Bmatrix}$$

$$+\left(\frac{2}{9-3\sqrt{3}}\right)\left(\frac{m^{-1/2}}{\sqrt{3}}\right)\begin{Bmatrix}1\\1\\-1\end{Bmatrix}+(0)\left(\frac{m^{-1/2}}{3-\sqrt{3}}\right)\begin{Bmatrix}1\\1-\sqrt{3}\\2-\sqrt{3}\end{Bmatrix}=\frac{m^{-1/2}}{36}\begin{Bmatrix}3+5\sqrt{3}\\6+4\sqrt{3}\\-3-3\sqrt{3}\end{Bmatrix}. \tag{4.597}$$

In summary,

$$\frac{\partial \overline{\Phi}_1}{\partial \beta} = \frac{m^{-1/2}}{36} \begin{Bmatrix} 3 - 5\sqrt{3} \\ 6 - 4\sqrt{3} \\ 3\sqrt{3} - 3 \end{Bmatrix}, \quad \frac{\partial \overline{\Phi}_2}{\partial \beta} = \frac{m^{-1/2}}{3} \begin{Bmatrix} -2\sqrt{3} \\ 2\sqrt{3} \\ 0 \end{Bmatrix}, \quad \frac{\partial \overline{\Phi}_3}{\partial \beta} = \frac{m^{-1/2}}{36} \begin{Bmatrix} 3 + 5\sqrt{3} \\ 6 + 4\sqrt{3} \\ -3 - 3\sqrt{3} \end{Bmatrix}. \quad (4.598)$$

4.14 SENSITIVITY ANALYSIS OF RESPONSE

Sections 2.9 and 3.10 discuss methods for calculating the sensitivity of the response of a system to a design variable. When the response is calculated using the normal mode method, the sensitivity analysis requires calculation of the derivative of a mode shape with respect to the design variable. Usually the mathematical relationship between a mode shape and a design variable cannot be expressed in a closed form because of the size of the system or the complexity of the computer program. Therefore, the method presented in Section 4.13 is used to calculate this quantity.

4.15 PROBLEM SIZE REDUCTION METHODS

The equation of motion for an MDOF system is given in Eq. 4.1:

$$\mathbf{M}\ddot{\mathbf{X}}(t) + \mathbf{C}\dot{\mathbf{X}}(t) + \mathbf{K}\mathbf{X}(t) = \mathbf{F}_e(t). \quad (4.599)$$

The reduction in coordinates from n coordinates of $\mathbf{X}(t)$ to k coordinates of $\tilde{\mathbf{X}}(t)$ can be expressed as

$$\mathbf{X}(t) = \mathbf{T}\tilde{\mathbf{X}}(t). \quad (4.600)$$

After this coordinate transformation is performed, the dynamic equilibrium equation of motion in Eq. 4.599 can be expressed using the reduced coordinate system, which is

$$\tilde{\mathbf{M}}\ddot{\tilde{\mathbf{X}}}(t) + \tilde{\mathbf{C}}\dot{\tilde{\mathbf{X}}}(t) + \tilde{\mathbf{K}}\tilde{\mathbf{X}}(t) = \tilde{\mathbf{F}}_e(t) \quad (4.601)$$

where $\tilde{\mathbf{M}}$, $\tilde{\mathbf{C}}$, $\tilde{\mathbf{K}}$, and $\tilde{\mathbf{F}}_e(t)$ are determined using the conservation of energy method. This method was discussed in Section 2.10 for an SDOF system and is here extended to an MDOF system.

The strain energy in the Cartesian coordinate system, denoted by $\mathbf{E}_s(\mathbf{X}(t))$, can be written as

$$E_s(\mathbf{X}(t)) = \frac{1}{2}\mathbf{X}^T(t)\mathbf{K}\mathbf{X}(t). \quad (4.602)$$

Therefore, substituting Eq. 4.600 into Eq. 4.602 gives

$$E_s(\mathbf{X}(t)) = \frac{1}{2}\tilde{\mathbf{X}}^T(t)\mathbf{T}^T\mathbf{K}\mathbf{T}\tilde{\mathbf{X}}(t). \quad (4.603)$$

The strain energy in the reduced coordinate system, denoted by $\mathbf{E}_s(\mathbf{X}(t))$, is

$$E_s(\tilde{\mathbf{X}}(t)) = \frac{1}{2}\tilde{\mathbf{X}}^T(t)\tilde{\mathbf{K}}\tilde{\mathbf{X}}(t). \quad (4.604)$$

The strain energy at any instant in time should be the same in both coordinate systems. Therefore,

$$E_s(\mathbf{X}(t)) = \frac{1}{2}\tilde{\mathbf{X}}^T(t)\mathbf{T}^T\mathbf{K}\mathbf{T}\tilde{\mathbf{X}}(t) = E_s(\tilde{\mathbf{X}}(t)) = \frac{1}{2}\tilde{\mathbf{X}}^T(t)\tilde{\mathbf{K}}\tilde{\mathbf{X}}(t). \quad (4.605)$$

It follows that

$$\tilde{\mathbf{X}}^T(t)\mathbf{T}^T\mathbf{K}\mathbf{T}\tilde{\mathbf{X}}(t) = \tilde{\mathbf{X}}^T(t)\tilde{\mathbf{K}}\tilde{\mathbf{X}}(t). \tag{4.606}$$

Equation 4.606 must be satisfied at all times; thus

$$\tilde{\mathbf{K}} = \mathbf{T}^T\mathbf{K}\mathbf{T}. \tag{4.607}$$

The kinetic energy can be viewed in a similar way by first defining the kinetic energy of the system in Cartesian coordinates, denoted by $\mathbf{E}_k(\mathbf{X}(t))$, to be

$$E_k(\mathbf{X}(t)) = \frac{1}{2}\dot{\mathbf{X}}^T(t)\mathbf{M}\dot{\mathbf{X}}(t). \tag{4.608}$$

Substituting Eq. 4.600 into Eq. 4.608 gives

$$E_k(\mathbf{X}(t)) = \frac{1}{2}\dot{\tilde{\mathbf{X}}}^T(t)\mathbf{T}^T\mathbf{M}\mathbf{T}\dot{\tilde{\mathbf{X}}}(t). \tag{4.609}$$

In addition, the kinetic energy in the reduced coordinate system, denoted by $\mathbf{E}_k(\tilde{\mathbf{X}}(t))$, is

$$E_k(\tilde{\mathbf{X}}(t)) = \frac{1}{2}\dot{\tilde{\mathbf{X}}}^T(t)\tilde{\mathbf{M}}\dot{\tilde{\mathbf{X}}}(t). \tag{4.610}$$

The kinetic energy at any instant in time should be independent of the coordinate system, and therefore,

$$E_k(\mathbf{X}(t)) = \frac{1}{2}\dot{\tilde{\mathbf{X}}}^T(t)\mathbf{T}^T\mathbf{M}\mathbf{T}\dot{\tilde{\mathbf{X}}}(t) = E_k(\tilde{\mathbf{X}}(t)) = \frac{1}{2}\dot{\tilde{\mathbf{X}}}^T(t)\tilde{\mathbf{M}}\dot{\tilde{\mathbf{X}}}(t). \tag{4.611}$$

It then follows that

$$\dot{\tilde{\mathbf{X}}}^T(t)\mathbf{T}^T\mathbf{M}\mathbf{T}\dot{\tilde{\mathbf{X}}}(t) = \dot{\tilde{\mathbf{X}}}^T(t)\tilde{\mathbf{M}}\dot{\tilde{\mathbf{X}}}(t). \tag{4.612}$$

Equation 4.612 must be satisfied at all values of time; thus

$$\tilde{\mathbf{M}} = \mathbf{T}^T\mathbf{M}\mathbf{T}. \tag{4.613}$$

The work performed by the damping force in the Cartesian coordinate system as the mass moves from its starting position at time $t = t_o$ to its final position at time t for an SDOF system is given in Eq. 2.417. This equation can be extended to an MDOF system, which is

$$W_d(\mathbf{X}(t)) = \int \mathbf{F}_d \circ d\mathbf{X} = \int \mathbf{F}_d^T\, d\mathbf{X}. \tag{4.614}$$

If the damping force is linearly related to the velocity, then

$$\mathbf{F}_d = \mathbf{C}\dot{\mathbf{X}}(t). \tag{4.615}$$

Substituting Eq. 4.615 and then Eq. 4.600 into Eq. 4.614 gives

$$W_d(\mathbf{X}(t)) = \int \mathbf{F}_d^T\, d\mathbf{X} = \int \dot{\mathbf{X}}^T\mathbf{C}\, d\mathbf{X} = \int \dot{\tilde{\mathbf{X}}}^T\mathbf{T}^T\mathbf{C}\mathbf{T}\, d\tilde{\mathbf{X}}. \tag{4.616}$$

Meanwhile, the work performed by the damping force in the reduced coordinate system follows directly from Eq. 4.614, which is

$$W_d(\tilde{\mathbf{X}}(t)) = \int \tilde{\mathbf{F}}_d^T\, d\tilde{\mathbf{X}} = \int \dot{\tilde{\mathbf{X}}}^T\tilde{\mathbf{C}}\, d\tilde{\mathbf{X}}. \tag{4.617}$$

The work performed by the damping force at any instant in time must be independent of the coordinate system, and therefore

$$W_d(\mathbf{X}(t)) = \int \dot{\tilde{\mathbf{X}}}^T\mathbf{T}^T\mathbf{C}\mathbf{T}\, d\tilde{\mathbf{X}} = W_d(\tilde{\mathbf{X}}(t)) = \int \dot{\tilde{\mathbf{X}}}^T\tilde{\mathbf{C}}\, d\tilde{\mathbf{X}}. \tag{4.618}$$

Since Eq. 4.618 must be satisfied at all values of time, the integrands on both sides of the equation must be equal. Therefore,

$$\tilde{\mathbf{C}} = \mathbf{T}^T \mathbf{C} \mathbf{T}. \tag{4.619}$$

Finally, the work performed by the applied force in the Cartesian coordinate system as the mass moves from its starting position at time $t = t_o$ to its final position at time t is given in Eq. 2.427 for an SDOF system. For an MDOF system, it is

$$W_e(\mathbf{X}(t)) = \int \mathbf{F}_e^T \, d\mathbf{X}. \tag{4.620}$$

Substituting Eq. 4.600 into Eq. 4.620 gives

$$W_e(\mathbf{X}(t)) = \int \mathbf{F}_e^T \mathbf{T} \, d\tilde{\mathbf{X}} = \int \left(\mathbf{T}^T \mathbf{F}_e\right)^T d\tilde{\mathbf{X}}. \tag{4.621}$$

The work performed by the applied force in the reduced coordinate system follows directly from Eq. 4.621, which is

$$W_e(\tilde{\mathbf{X}}(t)) = \int \tilde{\mathbf{F}}_e^T \, d\tilde{\mathbf{X}}. \tag{4.622}$$

The work performed by the applied force at any instant in time must be constant and is independent of the coordinate system; therefore,

$$W_e(\mathbf{X}(t)) = \int \left(\mathbf{T}^T \mathbf{F}_e\right)^T d\tilde{\mathbf{X}} = W_e(\tilde{\mathbf{X}}(t)) = \int \tilde{\mathbf{F}}_e^T d\tilde{\mathbf{X}}. \tag{4.623}$$

Equation 4.623 must be satisfied at all values of time, and thus the integrands on both sides of the equation must be equal. Therefore,

$$\tilde{\mathbf{F}}_e = \mathbf{T}^T \mathbf{F}_e. \tag{4.624}$$

In summary, the equation of motion in Eq. 4.599 can be reduced to Eq. 4.601 using the coordinate transformation in Eq. 4.600, where the matrices in Eq. 4.601 are

$$\tilde{\mathbf{M}} = \mathbf{T}^T \mathbf{M} \mathbf{T}, \quad \tilde{\mathbf{C}} = \mathbf{T}^T \mathbf{C} \mathbf{T}, \quad \tilde{\mathbf{K}} = \mathbf{T}^T \mathbf{K} \mathbf{T}, \quad \tilde{\mathbf{F}}_e(t) = \mathbf{T}^T \mathbf{F}_e(t). \tag{4.625}$$

EXAMPLE 1   *Reduction to Normal Mode Coordinates*

Recall from Section 4.5, where the normal mode method was introduced, that

$$\mathbf{X}(t) = \hat{\mathbf{\Phi}} \mathbf{Q}(t) \tag{4.626}$$

where

$$\hat{\mathbf{\Phi}} = \begin{bmatrix} \hat{\boldsymbol{\phi}}_1 & \hat{\boldsymbol{\phi}}_2 & \cdots & \hat{\boldsymbol{\phi}}_k \end{bmatrix}$$

Therefore, for the normal mode method, the matrix $\mathbf{T}$ is equal to the eigenvector matrix, $\hat{\mathbf{\Phi}}$. It then follows from Eq. 4.625 that

$$\begin{aligned} \tilde{\mathbf{M}} &= \hat{\mathbf{\Phi}}^T \mathbf{M} \hat{\mathbf{\Phi}} = \hat{\mathbf{M}}, \quad \tilde{\mathbf{C}} = \hat{\mathbf{\Phi}}^T \mathbf{C} \hat{\mathbf{\Phi}} = \hat{\mathbf{C}} \\ \tilde{\mathbf{K}} &= \hat{\mathbf{\Phi}}^T \mathbf{K} \hat{\mathbf{\Phi}} = \hat{\mathbf{K}}, \quad \tilde{\mathbf{F}}_e(t) = \hat{\mathbf{\Phi}}^T \mathbf{F}_e(t) = \hat{\mathbf{F}}_e(t). \end{aligned} \tag{4.627}$$

EXAMPLE 2   *Reduction of Mass Matrix*

Consider the special case where the mass matrix is not fully populated and has a zero sub-matrix:

$$\mathbf{M} = \begin{bmatrix} \mathbf{M}_{11} & \mathbf{M}_{12} \\ \mathbf{M}_{21} & \mathbf{M}_{22} \end{bmatrix} = \begin{bmatrix} \mathbf{M}_{11} & \mathbf{0} \\ \mathbf{0} & \mathbf{0} \end{bmatrix}. \tag{4.628}$$

When the mass matrix is in this form, structural engineers very often assume that the damping matrix follows a similar form:

$$\mathbf{C} = \begin{bmatrix} \mathbf{C}_{11} & \mathbf{C}_{12} \\ \mathbf{C}_{21} & \mathbf{C}_{22} \end{bmatrix} = \begin{bmatrix} \mathbf{C}_{11} & \mathbf{0} \\ \mathbf{0} & \mathbf{0} \end{bmatrix}. \tag{4.629}$$

It then follows from Eq. 4.599 that

$$\begin{bmatrix} \mathbf{M}_{11} & \mathbf{0} \\ \mathbf{0} & \mathbf{0} \end{bmatrix} \begin{Bmatrix} \ddot{\mathbf{X}}_1(t) \\ \ddot{\mathbf{X}}_2(t) \end{Bmatrix} + \begin{bmatrix} \mathbf{C}_{11} & \mathbf{0} \\ \mathbf{0} & \mathbf{0} \end{bmatrix} \begin{Bmatrix} \dot{\mathbf{X}}_1(t) \\ \dot{\mathbf{X}}_2(t) \end{Bmatrix} + \begin{bmatrix} \mathbf{K}_{11} & \mathbf{K}_{12} \\ \mathbf{K}_{21} & \mathbf{K}_{22} \end{bmatrix} \begin{Bmatrix} \mathbf{X}_1(t) \\ \mathbf{X}_2(t) \end{Bmatrix} = \begin{Bmatrix} \mathbf{F}_1(t) \\ \mathbf{F}_2(t) \end{Bmatrix}. \tag{4.630}$$

Expanding Eq. 4.630 gives

$$\mathbf{M}_{11}\ddot{\mathbf{X}}_1(t) + \mathbf{C}_{11}\dot{\mathbf{X}}_1(t) + \mathbf{K}_{11}\mathbf{X}_1(t) + \mathbf{K}_{12}\mathbf{X}_2(t) = \mathbf{F}_1(t) \tag{4.631}$$

$$\mathbf{K}_{21}\mathbf{X}_1(t) + \mathbf{K}_{22}\mathbf{X}_2(t) = \mathbf{F}_2(t). \tag{4.632}$$

Solving for $\mathbf{X}_2(t)$ in Eq. 4.632 gives

$$\mathbf{X}_2(t) = \mathbf{K}_{22}^{-1}\mathbf{F}_2(t) - \mathbf{K}_{22}^{-1}\mathbf{K}_{21}\mathbf{X}_1(t). \tag{4.633}$$

Substituting Eq. 4.633 into Eq. 4.631, it follows that

$$\mathbf{M}_{11}\ddot{\mathbf{X}}_1(t) + \mathbf{C}_{11}\dot{\mathbf{X}}_1(t) + \left(\mathbf{K}_{11} - \mathbf{K}_{12}\mathbf{K}_{22}^{-1}\mathbf{K}_{21}\right)\mathbf{X}_1(t) = \mathbf{F}_1(t) - \mathbf{K}_{12}\mathbf{K}_{22}^{-1}\mathbf{F}_2(t) \tag{4.634}$$

and, therefore, the reduced coordinate system is

$$\tilde{\mathbf{X}}(t) = \mathbf{X}_1(t), \quad \tilde{\mathbf{F}}_e(t) = \mathbf{F}_1(t) - \mathbf{K}_{12}\mathbf{K}_{22}^{-1}\mathbf{F}_2(t)$$
$$\tilde{\mathbf{M}} = \mathbf{M}_{11}, \quad \tilde{\mathbf{C}} = \mathbf{C}_{11}, \quad \tilde{\mathbf{K}} = \mathbf{K}_{11} - \mathbf{K}_{12}\mathbf{K}_{22}^{-1}\mathbf{K}_{21}. \tag{4.635}$$

A similar result can be obtained by using the method discussed in this section. Let

$$\mathbf{X}(t) = \begin{Bmatrix} \mathbf{X}_1(t) \\ \mathbf{X}_2(t) \end{Bmatrix} = \begin{bmatrix} \mathbf{I} \\ -\mathbf{K}_{22}^{-1}\mathbf{K}_{21} \end{bmatrix} \tilde{\mathbf{X}}(t) = \mathbf{T}\tilde{\mathbf{X}}(t). \tag{4.636}$$

Then

$$\tilde{\mathbf{M}} = \mathbf{T}^T \mathbf{M} \mathbf{T} = \begin{bmatrix} \mathbf{I} & -\mathbf{K}_{22}^{-1}\mathbf{K}_{21} \end{bmatrix} \begin{bmatrix} \mathbf{M}_{11} & \mathbf{0} \\ \mathbf{0} & \mathbf{0} \end{bmatrix} \begin{bmatrix} \mathbf{I} \\ -\mathbf{K}_{22}^{-1}\mathbf{K}_{21} \end{bmatrix} = \mathbf{M}_{11} \tag{4.637}$$

$$\tilde{\mathbf{C}} = \mathbf{T}^T \mathbf{C} \mathbf{T} = \begin{bmatrix} \mathbf{I} & -\mathbf{K}_{22}^{-1}\mathbf{K}_{21} \end{bmatrix} \begin{bmatrix} \mathbf{C}_{11} & \mathbf{0} \\ \mathbf{0} & \mathbf{0} \end{bmatrix} \begin{bmatrix} \mathbf{I} \\ -\mathbf{K}_{22}^{-1}\mathbf{K}_{21} \end{bmatrix} = \mathbf{C}_{11} \tag{4.638}$$

$$\tilde{\mathbf{K}} = \mathbf{T}^T \mathbf{K} \mathbf{T} = \begin{bmatrix} \mathbf{I} & -\mathbf{K}_{22}^{-1}\mathbf{K}_{21} \end{bmatrix} \begin{bmatrix} \mathbf{K}_{11} & \mathbf{K}_{12} \\ \mathbf{K}_{21} & \mathbf{K}_{22} \end{bmatrix} \begin{bmatrix} \mathbf{I} \\ -\mathbf{K}_{22}^{-1}\mathbf{K}_{21} \end{bmatrix} = \mathbf{K}_{11} - \mathbf{K}_{12}\mathbf{K}_{22}^{-1}\mathbf{K}_{21} \tag{4.639}$$

$$\tilde{\mathbf{F}}_e(t) = \mathbf{T}^T \mathbf{F}_e(t) = \begin{bmatrix} \mathbf{I} & -\mathbf{K}_{22}^{-1}\mathbf{K}_{21} \end{bmatrix} \begin{Bmatrix} \mathbf{F}_1(t) \\ \mathbf{F}_2(t) \end{Bmatrix} = \mathbf{F}_1(t) - \mathbf{K}_{22}^{-1}\mathbf{K}_{21}\mathbf{F}_2(t). \tag{4.640}$$

The matrices in Eqs. 4.637 through 4.640 are the same as those in Eq. 4.635.

PROBLEMS

4.1 Repeat Example 2 in Section 4.3 but assume that the bottom-story stiffness is $0.5k$ instead of $2k$.

4.2 Repeat Example 3 in Section 4.3 but assume that the system is released from rest with the initial displacement $x_1(0) = 1.0$, $x_2(0) = 1 + \sqrt{3}$, and $x_3(0) = 2 + \sqrt{3}$.

4.3 Repeat Problem 4.2 but with $x_1(0) = 1.0$, $x_2(0) = 1.0$, and $x_3(0) = -1.0$.

4.4 Repeat Problem 4.2 but with $x_1(0) = 1.0 + 1.0 = 2.0$, $x_2(0) = 1 + \sqrt{3} + 1.0 = 2 + \sqrt{3}$, and $x_3(0) = 2 + \sqrt{3} - 1.0 = 1 + \sqrt{3}$.

4.5 Repeat Example 3 in Section 4.3 but with $\dot{x}_1(0) = 1.0$, $\dot{x}_2(0) = 1.0$, and $\dot{x}_3(0) = -1.0$.

4.6 Repeat Example 1 in Section 4.4 but for the system defined in Problem 4.1.

4.7 Repeat Example 2 in Section 4.4 but for the system defined in Problem 4.1.

4.8 Repeat Example 3 in Section 4.4 but for the system defined in Problem 4.1.

4.9 Using the method in Example 4 in Section 4.4, calculate the initial displacement in only the first mode for the system defined in Example 3 in Section 4.3.

4.10 Calculate the initial displacement and velocity in only the first mode of Problem 4.5 using the method discussed in Example 4 of Section 4.4.

4.11 Repeat Example 2 in Section 4.5 but with the initial conditions from Problem 4.2.

4.12 Repeat Example 2 in Section 4.5 but with the initial conditions from Problem 4.3.

4.13 Repeat Example 2 in Section 4.5 but for the system defined in Problem 4.1.

4.14 Repeat Example 3 in Section 4.5 but for the system defined in Problem 4.1.

4.15 Repeat Example 3 in Section 4.5 but with the initial conditions from Problem 4.2.

4.16 Repeat Example 3 in Section 4.5 but with the initial conditions from Problem 4.4.

4.17 Consider the system defined in Example 2 of Section 4.3 with $m = 1.0$ k-s^2/in. and a fundamental period of vibration of 1.0 s. Calculate k and the system natural frequencies of vibration.

4.18 Consider the system in Problem 4.17 with mass proportional damping (see Example 1 in Section 4.6) and the damping ratio in the first mode equal to 5%. Calculate the values of α and β; also calculate the corresponding damping ratios in the second and third modes.

4.19 Repeat Problem 4.18 but define the damping ratio in the third mode to be 5%. Calculate the values of α and β; also calculate the corresponding damping ratios in the first and second modes.

4.20 Repeat Problem 4.18 but for stiffness proportional damping (see Example 2 in Section 4.6).

4.21 Repeat Problem 4.19 but for stiffness proportional damping.

4.22 Consider the system in Problem 4.17 with Rayleigh proportional damping. Calculate the values of α and β if the damping ratio in each of the first two modes is 5%. What is the corresponding damping ratio in the third mode of vibration?

4.23 Repeat Problem 4.22 but define the damping ratios in the first and third modes to be 5% and calculate the corresponding damping ratio in the second mode of vibration.

4.24 Repeat Problem 4.23 but define the damping ratios in the first and third modes to be 2% and 5%, respectively.

4.25 Consider the system in Problem 4.17. Let $\omega_a = 0.8\omega_1$ and $\omega_b = 0.8\omega_3$. Assume Rayleigh proportional damping and let the damping at ω_a and ω_b equal 2% and 5%, respectively. Calculate the values of α and β and the damping ratio in the three modes of vibration.

4.26 Repeat Example 1 in Section 4.7 but assume that the damping ratios in the first three modes of vibration are 2%, 2%, and 2%, respectively.

4.27 Consider the system in Problem 4.17. Assume that the damping ratio in all three modes of vibration is 5%. Calculate α_1, α_2, and α_3 and the Caughy damping matrix. Compare your solution with the values of α and β and the Rayleigh damping matrix from Problems 4.22 and 4.23.

4.28 Repeat Example 1 in Section 4.8 but for the system in Problem 4.17.

4.29 Repeat Problem 4.28 but use the damping ratio in all modes equal to 5%. Compare your results with the results from Problem 4.27.

4.30 Repeat Problem 4.26 but assume Penzien–Wilson damping.

4.31   Repeat Example 2 in Section 4.8 but assume that the damping ratios in the three modes of vibration are $\zeta_1 = \zeta_2 = 5\%$ and $\zeta_2 = 0$.

4.32   Repeat Problem 4.31 but assume that $\zeta_1 = 5\%$ and $\zeta_2 = \zeta_3 = 0$.

4.33   Repeat Problem 4.31 but assume that $\zeta_1 = 5\%$ and $\zeta_2 = \zeta_3 = 20\%$.

4.34   Repeat Example 2 in Section 4.9 but using the Newmark β-method of constant average acceleration (i.e., $\delta = \frac{1}{2}$ and $\alpha = \frac{1}{4}$).

4.35   Repeat Example 2 in Section 4.9 but using the Newmark β-method of linear acceleration (i.e., $\delta = \frac{1}{2}$ and $\alpha = \frac{1}{6}$).

4.36   Repeat Example 2 in Section 4.9 but using the Wilson θ-method with $\theta = 1.40$.

4.37   Consider the 3DOF system in Example 2 of Section 4.8. Define all the matrices in Eq. 4.446 in Section 4.9 using a time step of $\Delta t = 0.01$ s and Newmark β-method of constant acceleration (i.e., $\delta = 0$ and $\alpha = 0$).

4.38   Consider the 3DOF system in Example 2 of Section 4.8. Define all the matrices in Eq. 4.446 in Section 4.9 using a time step of $\Delta t = 0.01$ s and the Newmark β-method of constant average acceleration (i.e., $\delta = \frac{1}{2}$ and $\alpha = \frac{1}{4}$).

4.39   Consider the 3DOF system in Example 2 of Section 4.8. Define all the matrices in Eq. 4.446 in Section 4.9 using a time step of $\Delta t = 0.01$ s and the Newmark β-method of linear acceleration (i.e., $\delta = \frac{1}{2}$ and $\alpha = \frac{1}{6}$).

4.40   Consider the 3DOF system in Example 2 of Section 4.8. Define all the matrices in Eq. 4.447 in Section 4.9 using a time step of $\Delta t = 0.01$ s and the Wilson θ-method with $\theta = 1.40$.

4.41   Consider the system in Problem 4.17. Create a damping matrix $\mathbf{C}^*$ by adding the damping matrices from Problems 4.18 and 4.20. Is $\mathbf{C}^*$ a proportional damping matrix?

4.42   Repeat Problem 4.41 but define $\mathbf{C}^*$ by adding the damping matrices from Problems 4.20 and 4.29.

4.43   Consider Example 2 in Section 4.8. Define all the matrices in Eqs. 4.462, 4.466, and 4.467 in Section 4.10 by following the similar procedure discussed in Example 2 of Section 4.10.

4.44   Consider Problem 4.41. Define all the matrices in Eqs. 4.462, 4.466, and 4.467 in Section 4.10 by following the procedure discussed in Example 2 of Section 4.10.

4.45   Repeat Example 1 in Section 4.11 but for the system in Problem 3.1.

4.46   Repeat Example 1 in Section 4.11 but for the system in Problem 3.4.

4.47   Consider the system in Problem 4.20. Define all terms in Eq. 4.514.

4.48   Consider the system in Problem 4.41. Define all terms in Eq. 4.514.

4.49   Consider Example 1 in Section 4.12. Calculate the sensitivity of ω_1 and ω_2 with respect to k_2 using Eq. 4.527.

4.50   Repeat Example 2 in Section 4.12 but calculate the sensitivity with respect to the mass.

4.51   Consider the system as shown in Figure 3.1. Assume that $m_1 = 1.0$ k-s^2/in., $m_2 = 2.0$ k-s^2/in., $k_1 = 10.0$ k-s^2/in., and $k_2 = 15.0$ k/in. Use Eq. 4.527 to calculate the sensitivity of ω_1 and ω_2 with respect to k_1.

4.52   Repeat Problem 4.51 but calculate the sensitivity with respect to k_2.

4.53   Repeat Problem 4.51 but calculate the sensitivity with respect to m_1.

4.54   Consider a system similar to the one discussed in Problem 4.51 Add one story to the top such that it is now a 3DOF system, and let $m_3 = m_2$ and $k_3 = k_2$. Use Eq. 4.527 to calculate the sensitivity of ω_1, ω_2, and ω_3 with respect to k_1.

4.55   Consider Example 1 in Section 4.12. Let $k_1 = 10.0$ k/in. and $k_2 = 15.0$ k/in. Calculate the sensitivity of the first and second mode shapes with respect to k_1 using Eq. 4.555.

4.56   Repeat Problem 4.55 but now calculate the sensitivity with respect to m_1.

4.57   Consider the system in Problem 4.51. Calculate the sensitivity of the first and second mode shapes with respect to k_1.

4.58   Consider the system in Problem 4.51. Calculate the sensitivity of the first and second mode shapes with respect to m_1.

4.59   Consider the system in Problem 4.555. Use Eq. 4.555 to calculate the sensitivity of the three mode shapes with respect to k_1.

Part Two

Earthquake
Response
of Structures

Chapter 5

Response Spectra Analysis

5.1 OVERVIEW

Modern building codes, such as the *Uniform Building Code* (UBC),[1] present three options for the dynamic analysis of structures:

1. *Linear time history analysis*—In this analysis method, it is assumed that for the entire response time history, the material properties of the structure can be modeled in the linear elastic response domain (i.e., no yielding of any fiber of the material).

2. *Response spectra analysis*—In this analysis method, the structural response is still assumed to be in the linear elastic domain as it is with Option 1, but now the time variable is eliminated from the solution.

3. *Nonlinear (inelastic) time history analysis*—The response is now allowed to induce strains in the materials beyond the yield strain. This analysis method directly incorporates into the solution the magnitude and time history of all past yield behavior.

This chapter presents the response spectra analysis approach, and the next two chapters present the other two analysis methods.

The *response spectra analysis method* is popular because it provides an approximate, but relatively accurate, method of analysis that is basically a static analysis. This method focuses on estimating the maximum structural response and does not retain the time variable. Eliminating the time variable greatly reduces the quantity of response data that must be stored on the computer and/or postprocessed. If the time step for the numerical integration of response is 100 samples per second and an earthquake ground motion lasts 50 s, then 5,000 response values are obtained for each response variable that the structural engineer wishes to study. A response spectra value only looks at one response value, which is the maximum response.

Structural dynamics is not an easy technical area to learn. The beauty of structural dynamics is the analytical rigor required to accurately formulate and calculate structural response. The basic problem in applying structural dynamics to solve real-world earthquake design problems is that this analytical rigor often is overwhelming. In structural dynamics it is very difficult to develop an "engineering feel" that the structural engineer can carry from project to project. The response spectra analysis method provides an excellent vehicle for gaining this "engineering feel" and also for obtaining insight that can be applied to future projects.

The response spectra analysis method combines several components of structural dynamics and design:

1. *The normal mode method*—The natural frequencies and mode shapes of vibration are critical dynamic response variables in this analysis method. Therefore, the structure must be modeled as a linear system.

2. *Damping*—Representing damping for a particular structure is very difficult because of the lack of full-scale response measurements and also the variability of the total building design. Therefore, there is a welcome simplicity in the response spectra method, which only requires one value of damping for each normal response model retained in the solution.

3. *Earthquake ground motion*—This method separates development of the structural model that results in the natural frequencies and mode shapes from the estimation of the design value of earthquake ground motion at the site. This has an advantage when design time is short, because work can be performed simultaneously on developing the structural model and estimating the ground motion. Also, because the structural system is assumed to be linear, the extrapolation of previously calculated values of structural response is straightforward for new proposed earthquake ground motions and thus will be very time efficient.

4. *Solution speed*—Calculating natural frequencies and mode shapes can require significant computing time, but very often it only needs to be done once. Once these natural frequencies and mode shapes are calculated they can be reused as many times as necessary for different earthquake ground motions and different damping values.

How accurate are the results of a response spectra analysis compared to linear time history analysis and nonlinear time history analysis? A complete and well-developed nonlinear time history analysis is more accurate than a complete and well-developed linear time history analysis, which is more accurate than a complete and well-developed response spectra analysis. Unfortunately, for numerous reasons, funding may only exist for a complete and well-developed response spectra analysis. In this situation, many structural engineers prefer to conduct this complete analysis rather than a less than complete linear or nonlinear time history analysis.

This chapter begins by defining different response spectra derived from the response of a single degree of freedom system. It then discusses the concept of a design spectrum. Next the response of a multi–degree of freedom system is calculated using only the fundamental mode of vibration and the response spectra of the earthquake ground motion. This solution method is then extended to include two or more normal modes of vibration. The chapter concludes by presenting the response spectra analysis of a general three-dimensional structure.

5.2 SINGLE DEGREE OF FREEDOM RESPONSE

The equation of motion for a single degree of freedom system subjected to an earthquake ground motion was derived in Example 4 in Section 2.4. However, it will be derived here again for completeness. Consider the structure shown in Figure 5.1, which is subjected to an acceleration versus time ground motion from an earthquake. The total elastic force from the columns is equal to

$$F_s = -k\left(y(t) - u_g(t)\right) \tag{5.1}$$

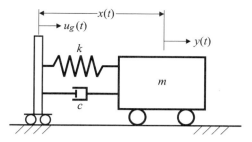

Figure 5.1 Single Degree of Freedom Building with Ground Motion

where $y(t)$ is the *absolute displacement* of the mass and $u_g(t)$ is the absolute displacement of the ground. If the *relative displacement* between the mass and the ground is denoted by $x(t)$, where $x(t) = y(t) - u_g(t)$, then

$$F_s = -kx(t). \tag{5.2}$$

The damping force for a system with linear viscous damping is

$$F_d = -c\left(\dot{y}(t) - \dot{u}_g(t)\right) = -c\dot{x}(t). \tag{5.3}$$

The external force applied to the mass is zero, and thus $F_e(t) = 0$. The inertial force is the mass times the absolute acceleration, and thus $m\ddot{y}(t) = m\ddot{x}(t) + m\ddot{u}_g(t)$. It then follows from Newton's law and from Eq. 2.1 that

$$\sum F = F_s + F_d + F_e = -kx(t) - c\dot{x}(t) = m\ddot{x}(t) + m\ddot{u}_g(t) \tag{5.4}$$

and after rearranging terms it follows that

$$m\ddot{x}(t) + c\dot{x}(t) + kx(t) = -m\ddot{u}_g(t). \tag{5.5}$$

Note that Eq. 5.5 is similar to the equation of motion for an SDOF system, as shown in Eq. 2.2, with the replacement of $F_e(t) = -m\ddot{u}_g(t)$. This means that by replacing the external applied force with the mass times the ground acceleration, the structural engineer mathematically changes the problem from a dynamic analysis due to external force to a dynamic analysis due to an earthquake ground motion. Therefore, any of the response analysis methods presented in Chapters 2 to 4 can be used to analyze the response of the structure due to an earthquake ground motion.

For convenience, the *earthquake ground acceleration* can be denoted as $a(t) = \ddot{u}_g(t)$. It then follows from Eq. 5.5 that

$$m\ddot{x}(t) + c\dot{x}(t) + kx(t) = -ma(t). \tag{5.6}$$

After dividing both sides of this equation by the mass, it follows that

$$\ddot{x}(t) + 2\zeta\omega_n\dot{x}(t) + \omega_n^2 x(t) = -a(t) \tag{5.7}$$

where $\omega_n^2 = k/m$ and $2\zeta\omega_n = c/m$. Note that the inertial force is computed as the absolute acceleration times the mass:

$$m\ddot{y}(t) = m\ddot{x}(t) + m\ddot{u}_g(t) = -c\dot{x}(t) - kx(t). \tag{5.8}$$

It follows from the response solution methods presented in Chapter 2 that the relative displacement versus time of the response, that is, $x(t)$, can be calculated for a given earthquake ground motion time history, $a(t)$, a given natural frequency, ω_n, and a given damping ratio, ζ. When this response time history, $x(t)$, is scanned for all values of time, then the maximum absolute value of the response can be determined, and this response is called the *spectral displacement* of the single degree of freedom system. The spectral displacement is denoted as $S_d(\omega_n, \zeta)$, or alternatively by $S_d(T_n, \zeta)$, or simply by S_d, and it can be written as

$$S_d(\omega_n,\zeta) = |x(t)|_{\max} = \max|x(t)| = \max[x(t)].\qquad(5.9)$$

The presence of the absolute value means that all values of S_d are positive even if the maximum value is a maximum negative value. However, structural engineers usually ignore this mathematical definition, and therefore the reader should keep in mind that $\max[x(t)]$ represents the absolute maximum value.

In a similar manner, the relative velocity, $\dot{x}(t)$, and absolute acceleration of the mass, $\ddot{y}(t) = \ddot{x}(t) + a(t)$ can be calculated and scanned, and then maximum values for each can be determined. These response quantities are called the *spectral velocity*, S_v, and *spectral acceleration*, S_a, respectively, and are denoted by

$$S_v(\omega_n,\zeta) = |\dot{x}(t)|_{\max}\qquad(5.10)$$

$$S_a(\omega_n,\zeta) = |\ddot{y}(t)|_{\max}.\qquad(5.11)$$

It is possible to very rapidly calculate by computer the spectral displacement, spectral velocity, and spectral acceleration for a given earthquake ground motion time history for different values of ω_n and ζ. If this is done, the response can be summarized in a tabular or graphic form. In either form, most structural engineers prefer to express S_d, S_v, and S_a in terms of the undamped natural period of vibration, T_n, instead of the undamped natural frequency of vibration, ω_n.

Perhaps the single most referenced earthquake record is the acceleration versus time response in the north–south (NS) direction recorded in the 1940 El-Centro earthquake. Figure 5.2 shows this acceleration versus time ground motion. Figure 5.3 shows plots of the corresponding spectral displacement, spectral velocity, and spectral acceleration versus T_n.

Figure 5.4 shows a plot of the spectral acceleration versus spectral displacement for the NS component of the 1940 El-Centro earthquake. This plot is called a *bi-spectra plot*. This figure is obtained by first setting the value of the undamped period of vibration (e.g., T_n) and then calculating S_d and S_a. Then this point is plotted as shown for $T = 1.0$ s with the undamped period of vibration denoted on the plot.

Two other functions commonly used by earthquake and structural engineers are *pseudo spectral velocity*, S_{vp}, and *pseudo spectral acceleration*, S_{ap}. These functions are derived from the spectra displacement, and they are defined as

$$S_{vp}(\omega_n,\zeta) = \omega_n S_d(\omega_n,\zeta)\qquad(5.12)$$

$$S_{ap}(\omega_n,\zeta) = \omega_n S_{vp}(\omega_n,\zeta) = \omega_n^2 S_d(\omega_n,\zeta).\qquad(5.13)$$

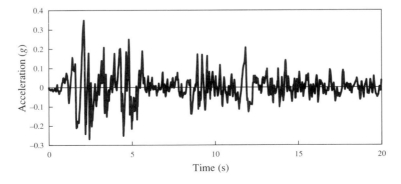

Figure 5.2 North–South Component of 1940 El-Centro Earthquake

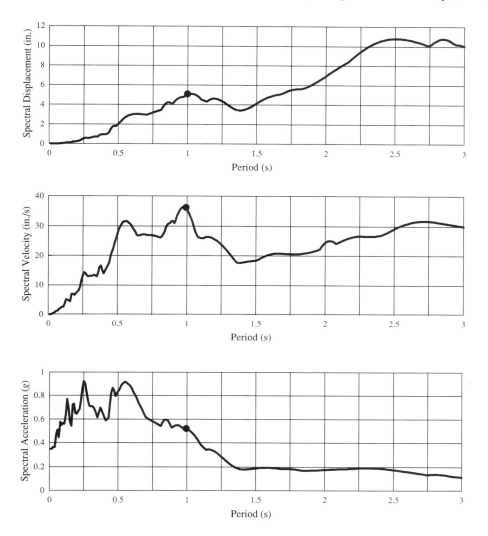

Figure 5.3 Acceleration, Velocity, and Displacement Spectra for NS Component of 1940 El-Centro Earthquake (5% Damping)

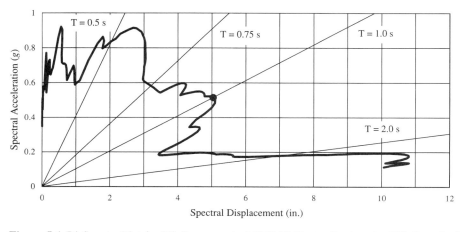

Figure 5.4 Bi-Spectra Plot for NS Component of 1940 El-Centro Earthquake (5% Damping)

A special relationship exists between the spectral acceleration and the pseudo spectral acceleration when the damping in the system is zero. To show this relationship, consider Eq. 5.8; after dividing both sides of the equation by the mass, it follows that

$$\ddot{y}(t) = -2\zeta\omega_n\dot{x}(t) - \omega_n^2 x(t). \tag{5.14}$$

For the special case where the damping is zero, Eq. 5.14 becomes

$$\ddot{y}(t) = -\omega_n^2 x(t). \tag{5.15}$$

Note from Eq. 5.15 that the maximum absolute acceleration occurs at the same instant in time as the maximum relative displacement. Therefore, it follows from Eqs. 5.9, 5.11, 5.13, and 5.15 that

$$S_a(\omega_n,\zeta) = |\ddot{y}(t)|_{max} = |-\omega_n^2 x(t)|_{max} = \omega_n^2 S_d(\omega_n,\zeta) = S_{ap}(\omega_n,\zeta). \tag{5.16}$$

This shows that the spectral acceleration and the pseudo spectral acceleration are equal when the damping is zero.

If an earthquake ground acceleration, $a(t)$, is analyzed for a given value of damping, then the response spectra can be plotted and used in the future without recalculating. If the structural engineer has a single degree of freedom system with a known period, T_n, then the maximum value of the relative displacement between the mass and the ground follows from the response spectra:

$$\max[x(t)] = S_d(T_n,\zeta). \tag{5.17}$$

This relative displacement can then be used to calculate the maximum stiffness force during the earthquake:

$$\max[F_s(t)] = \max[-kx(t)] = k\,S_d(T_n,\zeta). \tag{5.18}$$

The maximum absolute acceleration of the mass follows from

$$\max[\ddot{y}(t)] = S_a(T_n,\zeta) \tag{5.19}$$

and the maximum inertial force on the mass is

$$\max[F_i(t)] = \max[-m\ddot{y}(t)] = mS_a(T_n,\zeta). \tag{5.20}$$

The maximum inertial force is only equal to the maximum stiffness force when the damping is zero. The total force acting on the support (i.e., base of the structure) is the summation of the stiffness force and the damping force, and this is equal to the inertia force.

Consider the SDOF system shown in Figure 5.1. The inertia force is the force acting on the mass, and from equilibrium it is also transmitted by the system to the ground. This force is called the *base shear force*, V, and it is written as

$$V = \max[F_i(t)] = mS_a(T_n,\zeta). \tag{5.21}$$

Recall that

$$m = W/g \tag{5.22}$$

where W is the weight of the structure and g is the acceleration of gravity. Substituting Eq. 5.22 into Eq. 5.21 gives

$$V = \left(\frac{S_a(T_n,\zeta)}{g}\right)W. \tag{5.23}$$

Recall that the unit of $S_a(T_n, \zeta)$ is the same as acceleration (i.e., length/time^2). However, as indicated by Eq. 5.23, if $S_a(T_n, \zeta)$ is divided by the acceleration of gravity (e.g., 386.4 in./s^2 or 9.806 m/s^2), then it is in the units of g's. Structural engineers prefer this form, and therefore the spectral acceleration is almost always expressed in g's and seldom in the unit (length/s²). With $\hat{S}_a(T_n, \zeta)$ in g's, Eq. 5.23 becomes

$$V = \hat{S}_a(T_n, \zeta) W. \tag{5.24}$$

Equation 5.24 shows that the base shear force is equal to the value of the spectral acceleration times the weight of the structure. Also, the vertical axis of the bi-spectra plot, $\hat{S}_a$, is the normalized force (V/W), and thus the bi-spectra plot is a force versus deflection plot.

EXAMPLE 1   *Harmonic Ground Motion*

The response of an SDOF system to a harmonic forcing function was considered in Example 3 in Section 2.3. The equation of motion from Eq. 2.74 for this case was

$$\ddot{x}_p(t) + 2\zeta\omega_n \dot{x}_p(t) + \omega_n^2 x_p(t) = F_o \sin\omega_o t/m. \tag{5.25}$$

Assume that the ground motion of an earthquake is simplified to be

$$a(t) = (F_o/m)\sin\omega_o t = -a_o \sin\omega_o t. \tag{5.26}$$

It then follows from the response obtained in Eq. 2.83 with initial conditions equal to zero (i.e., $x_o = \dot{x}_o = 0$) that

$$x(t) = -\frac{ma_o/k}{(1-r^2)^2 + (2\zeta r)^2}\left[2\zeta r\, e^{-\zeta\omega_n t}\cos\omega_d t + \left(r\frac{(1-r^2)-2\zeta^2}{\sqrt{1-\zeta^2}}\right)e^{-\zeta\omega_n t}\sin\omega_d t\right.$$
$$\left. - 2\zeta r\cos\omega_o t + (1-r^2)\sin\omega_o t\right]. \tag{5.27}$$

The terms containing $e^{-\zeta\omega_n t}$ in Eq. 5.27 decrease and become very small as time increases, and therefore when these terms are sufficiently small so they can be neglected, the response becomes

$$x(t) = \frac{ma_o/k}{(1-r^2)^2 + (2\zeta r)^2}\left[2\zeta r\cos\omega_o t - (1-r^2)\sin\omega_o t\right]. \tag{5.28}$$

This response is called the *steady state response*. The maximum absolute value of Eq. 5.28 gives the spectral displacement, which is

$$S_d(T_n, \zeta) = \max[x(t)] = \frac{ma_o/k}{\sqrt{(1-r^2)^2 + (2\zeta r)^2}}. \tag{5.29}$$

EXAMPLE 2   *Bilinear Pulse*

Consider the earthquake ground motion to be a bilinear pulse as discussed in Example 7 in Section 2.4. The displacement response of an undamped SDOF system is calculated in that example to be

$$x(t) = \begin{cases} \left(\dfrac{a_p}{\omega_n^2}\right)\left[\left(\dfrac{\sin\omega_n t}{\omega_n t_p}\right) - \left(\dfrac{t}{t_p}\right)\right] & 0 \le t \le t_p \\ \left(\dfrac{a_p}{\omega_n^2}\right)\left[-2 + \left(\dfrac{t}{t_p}\right) + \left(\dfrac{\sin\omega_n t}{\omega_n t_p}\right) - 2\left(\dfrac{\sin\omega_n(t-t_p)}{\omega_n t_p}\right)\right] & t_p < t \le 2t_p \\ \left(\dfrac{a_p}{\omega_n^2}\right)\left[\left(\dfrac{\sin\omega_n t}{\omega_n t_p}\right) - \left(\dfrac{2\sin\omega_n(t-t_p)}{\omega_n t_p}\right) + \left(\dfrac{\sin\omega_n(t-2t_p)}{\omega_n t_p}\right)\right] & t > 2t_p. \end{cases} \tag{5.30}$$

The displacement response spectra for this earthquake ground motion can be obtained by calculating the maximum response of Eq. 5.30 as a function of the natural frequency of the system.

The first step in solving the response spectra is to calculate the time when the maximum displacement occurs, denoted t_{max}. This time is obtained by taking the time derivative of Eq. 5.30 and setting it equal to zero. Consider that the duration of this pulse is very short, and therefore the maximum displacement will occur at time $t_{max} > 2t_p$. Then it follows from Eq. 5.30 that

$$\frac{dx(t)}{dt} = \left(\frac{a_p}{\omega_n^2}\right)\left[\left(\frac{\cos\omega_n t_{max}}{t_p}\right) - \left(\frac{2\cos\omega_n(t_{max}-t_p)}{t_p}\right) + \left(\frac{\cos\omega_n(t_{max}-2t_p)}{t_p}\right)\right] = 0. \quad (5.31)$$

Equation 5.31 can be rewritten as

$$\cos\left[\omega_n(t_{max}-t_p)+\omega_n t_p\right] - 2\cos\omega_n(t_{max}-t_p) + \cos\left[\omega_n(t_{max}-t_p)-\omega_n t_p\right] = 0. \quad (5.32)$$

Expanding Eq. 5.32 gives

$$\left[\cos\omega_n(t_{max}-t_p)\cos\omega_n t_p - \sin\omega_n(t_{max}-t_p)\sin\omega_n t_p\right] - 2\cos\omega_n(t_{max}-t_p)$$
$$+\left[\cos\omega_n(t_{max}-t_p)\cos\omega_n t_p + \sin\omega_n(t_{max}-t_p)\sin\omega_n t_p\right] = 0. \quad (5.33)$$

Simplifying Eq. 5.33, it follows that

$$\cos\omega_n\left(t_{max}-t_p\right)\left(\cos\omega_n t_p - 2\right) = 0. \quad (5.34)$$

Note that $\cos\omega_n t_p$ can never equal 2. Therefore, Eq. 5.34 can only be satisfied when

$$\cos\omega_n(t_{max}-t_p) = 0. \quad (5.35)$$

Solving for Eq. 5.35 by taking the first value that cosine is equal to zero (i.e., $\pi/2$), it follows that

$$t_{max} = t_p + \frac{\pi}{2\omega_n}. \quad (5.36)$$

Recall that $T_n = 2\pi/\omega_n$; therefore, Eq. 5.36 becomes

$$t_{max} = t_p + \frac{T_n}{4}. \quad (5.37)$$

Substituting t_{max} in Eq. 5.37 into Eq. 5.30 gives

$$\max[x(t)] = \left(\frac{a_p}{\omega_n^2}\right)\left[\left(\frac{\sin(\omega_n t_p+\pi/2)}{\omega_n t_p}\right) - \left(\frac{2\sin\pi/2}{\omega_n t_p}\right) + \left(\frac{\sin(-\omega_n t_p+\pi/2)}{\omega_n t_p}\right)\right]$$
$$= \left(\frac{2a_p}{\omega_n^3 t_p}\right)\left[1-\cos\omega_n t_p\right]. \quad (5.38)$$

It can be shown that for $t_p > T_n/4$, the maximum displacement response is in the time interval $t_p < t \le 2t_p$. Figure 5.5 shows a plot of the response spectrum for $a_p = 1.0g$ and $t_p = 0.1$ s. Note that in Figure 5.5, the natural period of vibration starts at $T_n = 0.4$ s because this is when $t_p = T_n/4$.

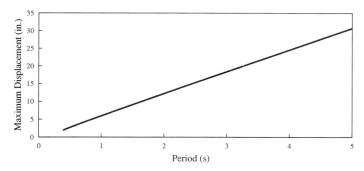

Figure 5.5 Displacement Response Spectrum (0% Damping)

EXAMPLE 3   *Response Spectra for Different Periods of Vibrations*

Figure 5.6 shows the earthquake ground motion recorded at Sylmar in the NS direction during the 1994 Northridge earthquake. The relative displacement of a single degree of freedom system can be obtained using any of the numerical solution methods presented in Chapter 2. Assume that an SDOF has 5% damping. Figures 5.7 to 5.9 show the displacement versus time response, $x(t)$, for three different periods of vibration at $T_n = 0.3$ s, $T_n = 0.6$ s, and $T_n = 1.0$ s, respectively. Scanning the response time histories from Figures 5.7 to 5.9 gives a maximum displacement response of 2.36 in., 4.79 in., and 8.49 in. Performing the same calculation and scanning for other natural periods of vibration gives the values of the displacement response spectrum. The 5% damped displacement response spectrum is shown in Figure 5.10 with the maximum response from Figures 5.7 to 5.9 noted in the figure. Similarly, the other response spectra at different natural periods of vibration can be computed the same way, and Figure 5.11 shows the corresponding spectral acceleration curve.

EXAMPLE 4   *Response Spectra for Different Values of Damping Ratios*

Figure 5.12 shows the earthquake ground motion time history in the NS direction recorded at Newhall Fire Station during the 1994 Northridge earthquake. The displacement response versus time for an SDOF system with an undamped period of vibration (T_n) of 1.0 s and 5% structural damping is shown in Figure 5.13. The response can be calculated using the procedures discussed in Chapter 2, and the maximum displacement is 11.40 in. Figure 5.14

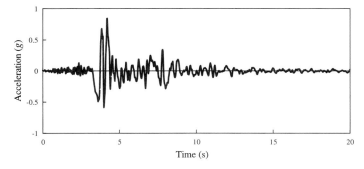

Figure 5.6 NS Component at Sylmar During 1994 Northridge Earthquake

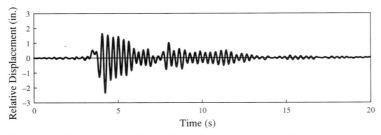

Figure 5.7 Displacement Time History Response Due to NS Component at Sylmar During 1994 Northridge Earthquake ($T_n = 0.3$ s, $\zeta = 5\%$)

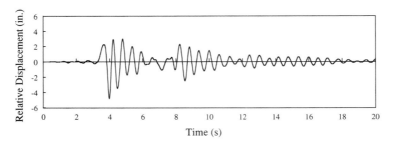

Figure 5.8 Displacement Time History Response Due to NS Component at Sylmar During 1994 Northridge Earthquake ($T_n = 0.6$ s, $\zeta = 5\%$)

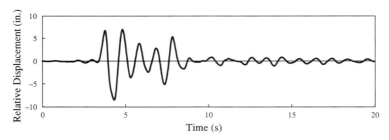

Figure 5.9 Displacement Time History Response Due to NS Component at Sylmar During 1994 Northridge Earthquake ($T_n = 1.0$ s, $\zeta = 5\%$)

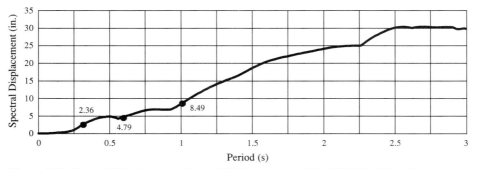

Figure 5.10 Spectral Displacement Due to NS Component of the 1994 Northridge Earthquake at Sylmar

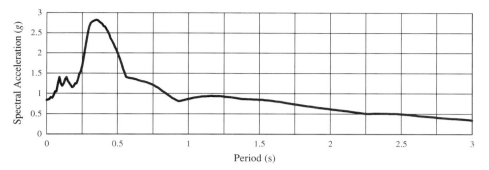

Figure 5.11 Spectral Acceleration Due to NS Component of the 1994 Northridge Earthquake at Sylmar

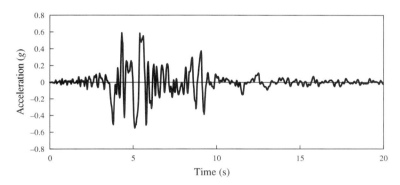

Figure 5.12 NS Component Measured at Newhall Fire Station During 1994 Northridge Earthquake

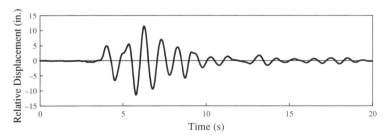

Figure 5.13 Displacement Time History Response Due to NS Component Measured at Newhall Fire Station During 1994 Northridge Earthquake ($T_n = 1.0$ s, $\zeta = 5\%$)

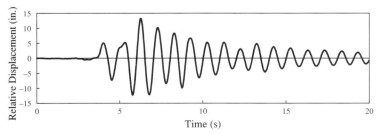

Figure 5.14 Displacement Time History Response Due to NS Component Measured at Newhall Fire Station During 1994 Northridge Earthquake ($T_n = 1.0$ s, $\zeta = 2\%$)

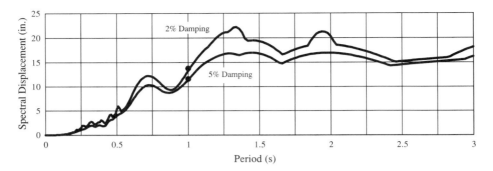

Figure 5.15 Spectral Displacement Due to NS Component Measured at Newhall Fire Station During 1994 Northridge Earthquake

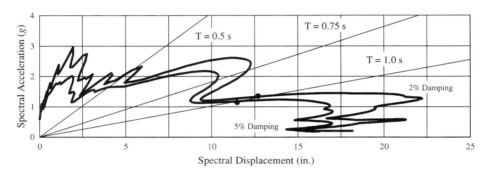

Figure 5.16 Bi-Spectra Plot of the NS Component Measured at Newhall Fire Station During 1994 Northridge Earthquake

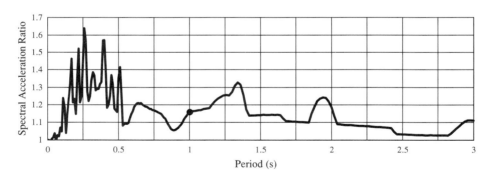

Figure 5.17 $[S_a(2\%)/S_a(5\%)]$ for NS Component Measured at Newhall Fire Station During 1994 Northridge Earthquake

shows the response for 2% damping. Figure 5.15 shows the spectral displacement response spectra (S_d) for both 2% damping and 5% damping. Figure 5.16 shows the bi-spectra plot for 2% and 5% damping. Figure 5.17 shows a plot of the ratio of the spectral accelerations between 2% damping and 5% damping.

5.3   DESIGN EARTHQUAKE RESPONSE SPECTRA

The previous section defined how response spectra are calculated for a given acceleration versus time ground motion. As shown in Figure 5.3, the shape of the response spectrum is very irregular, and structural engineers prefer for design purposes to use a smooth response spectrum curve that has a rational relationship to recorded earthquake ground motions but is not calculated from a specific recorded ground motion.

Figure 5.18 shows nine earthquake ground acceleration time histories of the Northridge earthquake developed by Dr. Paul Somerville, an engineering seismologist, using analytical methods. Based on state-of-the-art engineering seismology methods, the plots represent a best estimate of the earthquake ground acceleration expected to occur at a building site in Woodland Hills, California, if the 1994 Northridge earthquake were to repeat itself. Each of the nine earthquake ground motion time histories is equally likely to have occurred at the building site. This building site is 20 km from the epicenter of the earthquake. Figure 5.19 shows a plot of the 5% damped response spectra for these records. Figure 5.20 shows plots of the upper and lower envelopes and the mean of the 5% damped response spectra for the nine earthquake ground motion time histories shown in Figure 5.18. Note in Figure 5.20 how the plot of the mean rises as the undamped natural period of vibration increases, then reaches a peak, and finally decreases as the undamped natural period of vibration increases.

The development of a *design earthquake response spectrum*, or simply the *design spectrum*, starts with the understanding that it is not a response spectrum of any single earthquake ground motion time history that is desired, but instead a smooth representation of many response spectra. The estimation of a smooth response spectrum, usually a 5% damped acceleration spectrum, is usually performed by an engineering seismologist. Factors considered in the development of the smooth spectrum include the magnitude of possible earthquakes near the building site, the path of the earthquake waves through the

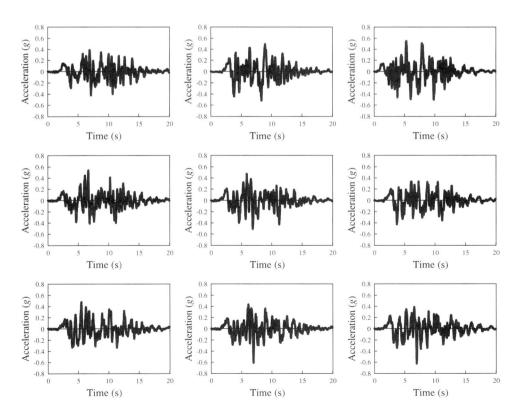

Figure 5.18 Ensemble of Nine Equally Probable Acceleration Time Histories for a Building Site (1994 Northridge Earthquake)

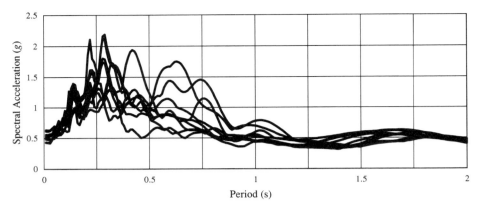

Figure 5.19 Response Spectra of Nine Acceleration Time Histories (5% Damping)

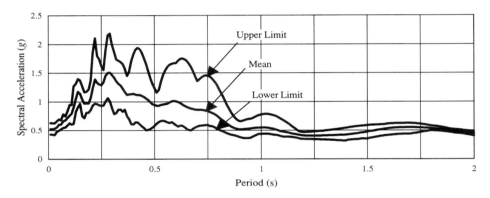

Figure 5.20 Mean and Envelopes of Nine Response Spectra (5% Damping)

earth, the properties of the earth along this path, and other geotechnical engineering factors. Because the maximum spectrum cannot be estimated with certainty, its value at any
natural period of vibration is a random variable. The design spectrum corresponds to a
level of ground motion for a specific building site that has a defined probability of being
exceeded (e.g., 10%) in a specified time (e.g., the next 50 years). This type of response
spectra is called a design spectrum.

Different levels of earthquake risk are considered by structural engineers when selecting the one or more design earthquake spectra for a building design. Table 5.1 shows four
levels of earthquake risk that have been defined by the Structural Engineers Association of
California (SEAOC). The most common risk level is a 10% probability of being exceeded
in 50 years. This is called a *rare earthquake* by SEAOC and a *design basis earthquake* in
the UBC. Figure 5.21 shows for a specific building site in Southern California the 5%
damped design response spectrum for a rare earthquake. Table 5.2 and Figure 5.22 show
for this same building site the design spectra for four earthquake risk levels.

The design earthquake has three variables in its definition. One is the *design life* or
exposure time for the building. This is the best estimate made at the time of the design of
how long the building will stand before its demolition or a change in the design of its lateral earthquake-force-resisting system. For example, most building codes assume 50 years
for a default value. The second variable is the *probability of nonoccurrence* in the design
life. The third variable is the *average return period*, usually just called the *return period*

Table 5.1 SEAOC Design Earthquake Risk Levels[2]

Earthquake	Probability of exceedance	Average return period (years)
Frequent	50% in 30 years	43
Occasional	50% in 50 years	72
Rare	10% in 50 years	475
Very rare	2% in 50 years	2,475

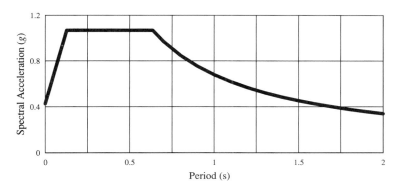

Figure 5.21 Rare Design Basis Earthquake Spectrum (5% Damping)

Table 5.2 Design Earthquake Response Spectra in g's (Los Angeles Site, 5% Damping)

Earthquake	Period of vibration (T_n)		
	0.3 s	1.0 s	2.0 s
Frequent (50% in 30 years)	0.43	0.22	0.11
Occasional (50% in 50 years)	0.51	0.29	0.15
Rare (10% in 50 years)	1.07	0.68	0.33
Very rare (2% in 50 years)	1.61	1.19	0.5

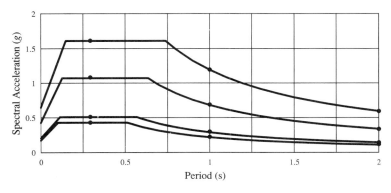

Figure 5.22 Design Spectra, Los Angeles Building Site (5% Damping)

or *recurrence interval*, which is the expected number of years between earthquakes with this response spectrum or larger. These three variables are related through the equation

$$p_o = 1 - e^{-(t/T)} \qquad (5.39)$$

where t is the design life or exposure time, p_o is the probability of occurrence, and T is the return period. For example, the UBC defines a design basis earthquake and SEAOC defines a rare earthquake as one that has a 10% probability of being exceeded in the next 50 years. Therefore, $p_o = 0.10$ (i.e., 10%) and $t = 50$ years (i.e., the design life). Rearranging the terms in Eq. 5.39 gives

$$T = -\frac{t}{\ln(1 - p_o)}. \tag{5.40}$$

Therefore, a rare earthquake has a return period of

$$T = -\frac{50}{\ln(1 - 0.10)} = 475 \text{ years}. \tag{5.41}$$

Two probabilities of occurrence are very commonly used: 50% and 10%. The 50% option makes some sense because it corresponds to a 50/50 chance of occurring or nonoccurring, which is like the outcome of flipping a fair coin. The origin of the 10% option is unclear but it is probably based on the subjective opinion of earthquake engineers in the mid-1970s. At that time structural and earthquake engineers were developing the document referred to as ATC-3, which was the source document for the first National Earthquake Hazard Reduction Program (NEHRP) seismic design provisions.[3] Ten percent reflects a small chance of something occurring and is 1 chance in 10. Recently, as the public has become more educated in probability through discussions with doctors related to medical predictions, or stockbrokers through stock purchases, a 2% probability of occurrence has come into common use to reflect an event with a very small chance of occurring. Therefore, the three values of p_o that structural engineers commonly use are $p_o = 0.50, 0.10$, and 0.02. It follows that Eq. 5.40 can be simplified as follows:

1.   50% probability of occurrence:

$$T = -\frac{t}{\ln(1 - 0.50)} = 1.44\,t \tag{5.42}$$

2.   10% probability of occurrence:

$$T = -\frac{t}{\ln(1 - 0.10)} = 9.49\,t \tag{5.43}$$

3.   2% probability of occurrence:

$$T = -\frac{t}{\ln(1 - 0.02)} = 49.50\,t. \tag{5.44}$$

Equations 5.42 to 5.44 can be used to calculate, for a given probability of occurrence and a given exposure time, the return period of the earthquake that has this probability of occurrence. Note from Eq. 5.44 that $(1/0.02)$ is 50.0, which is close to 49.5. Therefore, the return period can be approximately estimated using $(1/p_o)$. From Eq. 5.43 it similarly follows that $(1/p_o)$ is equal to 10.0, and this is close to 9.49. However, from Eq. 5.42 it follows that $(1/p_o)$ is 2.0, a substantial difference. Therefore, only for a small p_o is the estimate of T from $(1/p_o)$ reasonably accurate.

For example, if the owner wants to control the damage to the contents on a floor, then the primary response variable of interest is maximum floor acceleration. Imagine that a study has been conducted and for the building under consideration a maximum floor acceleration of $0.5g$ is the design limit for a damage control design. Now the owner must decide on a risk level for an exposure time where this damage cannot be tolerated. For example, say the selection is to only have a 10% probability of occurrence in the next 5 years. Therefore, using Eq. 5.43, this means that the design earthquake for the *damage control design* is

$$T = 9.49(5) = 47.5 \text{ years.} \tag{5.45}$$

The structural engineer then instructs the engineering seismologist to provide a design response spectrum with a 47.5-year return period (or a 10% probability of occurrence in 5 years). Then, the structural engineer will calculate the maximum floor acceleration. If this maximum floor acceleration is less than $0.5g$, the design is acceptable. If it is greater than $0.5g$, then the structure must be redesigned because it does not satisfy the owner's design objective.

A reasonable question is, if a design basis earthquake response spectrum is used in the analysis, what is the probability that the structure will experience these amplitudes of motion in the future? The design basis or rare earthquake has a 10% chance of being exceeded in the next 50 years. Therefore, the calculated responses have a 1 in 10 (i.e., 10%) chance of being exceeded in the next 50 years. Using Eq. 5.41 this also means that, on average, these responses will be exceeded once every 475 years because this is the return period for this response spectrum or one of greater magnitude. The answer to the question can be expanded upon further and in terms perhaps more easily understood. For example, using Eq. 5.39 and the earthquake with a return period of 475 years, there follows a 2% chance that this response will be exceeded in the next 10 years (i.e., a 10-year exposure time). Table 5.3 gives the values of the probability that the maximum structural responses calculated using a design spectrum for a rare earthquake with a 475-year return period will be exceeded in the next 10, 20, 30, 50, and 100 years.

Table 5.4 includes other types of design earthquakes used in structural engineering. Ten years in the business community is often the span of command when the decision maker is in charge and therefore directly or singularly accountable for what happens to the business operation in the event of the earthquake. If it is important for the structural engineer to quantify the structural response (and associated nonstructural or structural damage) during this span of command, then the response of the structure can be calculated for the different earthquake response spectra noted in Column 1 with the probability of this exceedance corresponding to the exposure time of 10 years in Column 3.

The best way to develop a design spectrum for a specific building site is to contract with a professional engineering seismologist to develop the spectrum. However, it is beneficial to have a default value of a design spectrum in the UBC or other similar design standards. For example, Figure 5.23 shows the design spectrum for a rare earthquake that is in the 1997 UBC.

Figure 5.24 shows how the magnitudes of the response spectra decrease with increased damping for a particular earthquake ground motion time history estimated to have a 10% probability of being exceeded in 50 years for the 1940 El-Centro earthquake record shown in Figure 5.2. For example, from Figure 5.24, an SDOF system with an

Table 5.3 Probability of Exceeding Calculated Response for Design Basis Earthquake

Exposure time (years)	Probability of exceedance (%)
10	2
20	4
30	6
50	10
100	19

Table 5.4 Earthquake Response Spectra

Earthquake	Return period (years)	Exposure time (years)	Probability of exceedance (%)
Business 1	14	10	50
		20	76
		30	88
Business 2	29	10	29
		20	50
		30	64
		50	82
Occasional	72	10	13
		20	24
		30	34
		50	50
Rare	475	10	2
		20	4
		30	6
		50	10
		100	19
Hospital	949	10	1
		20	2
		30	3
		50	5
		100	10
Very Rare	1,898	10	0.5
		20	1
		30	1.5
		50	3
		100	5
		200	10

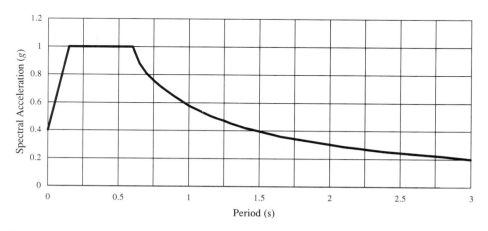

Figure 5.23 1997 Uniform Building Code Design Response Spectrum for S2 Soil (5% Damping)

undamped natural period of vibration of 1.0 s will have spectral acceleration values of 0.68g, 0.52g, and 0.36g for damping values of 2%, 5%, and 10%, respectively. The variation of response with damping can be expressed as a ratio of spectral acceleration divided by the spectral acceleration for 5% damping. For example, from Figure 5.24, it follows that

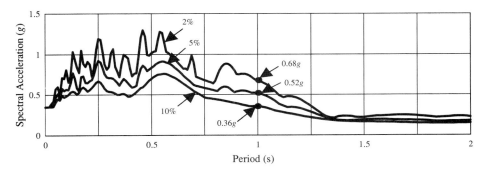

Figure 5.24 Response Spectra with 2%, 5%, and 10% Dampings

$$\left[\frac{S_a(T_n=1.0\,\mathrm{s},\zeta=2\%)}{S_a(T_n=1.0\,\mathrm{s},\zeta=5\%)}\right]=\frac{0.68g}{0.52g}=1.31 \tag{5.46}$$

and

$$\left[\frac{S_a(T_n=1.0\,\mathrm{s},\zeta=10\%)}{S_a(T_n=1.0\,\mathrm{s},\zeta=5\%)}\right]=\frac{0.36g}{0.52g}=0.69. \tag{5.47}$$

Different researchers have studied different earthquake records and have proposed formulas for calculating this response ratio. Newmark and Hall[4] proposed values of a *damping response reduction coefficient*, denoted R_d, that is used to multiply a value of the 5% damped elastic response spectrum to obtain a new spectrum value. The response reduction is

$$S_a(T_n,\zeta)=R_d S_a(T_n,\zeta=5\%). \tag{5.48}$$

Table 5.5 shows the values of R_d for different natural periods of vibration.

Table 5.6 shows the ratio of the response spectra values for a damping value ζ to the response spectra value for 5% damping developed for a given collection, or ensemble, of earthquake ground motion acceleration records.[5]

Table 5.5 Newmark–Hall Damping Response Reduction Coefficient (R_d)

Damping (%)	Natural period of vibration		
	0.3 s	1.0 s	3.0 s
5	1.00	1.00	1.00
10	0.77	0.83	0.86
20	0.55	0.65	0.73

Table 5.6 Damping Influence on Response Spectra[5]

Critical damping ratio (%)	$S_a(T_n,\zeta)/S_a(T_n,5\%)$	
	$T_n\le0.6$ s	$T_n>0.6$ s
2	1.25	1.25
5	1.00	1.00
10	0.77	0.83
20	0.56	0.67
30	0.43	0.59

Another ensemble of earthquake records was studied by Wu and Hanson,[6] who presented the following equations for R_d:

$$R_d = -0.349 \ln(0.0959\zeta), \quad T_n = 0.1\,\text{s} \tag{5.49a}$$

$$R_d = -0.547 \ln(0.417\zeta), \quad T_n = 0.5\,\text{s} \tag{5.49b}$$

$$R_d = -0.471 \ln(0.524\zeta), \quad T_n = 1.0\,\text{s} \tag{5.49c}$$

$$R_d = -0.478 \ln(0.475\zeta), \quad T_n = 3.0\,\text{s}. \tag{5.49d}$$

Table 5.7 gives the values of R_d for several damping values.

Therefore, selecting which damping ratio is best depends on the building site and the appropriate ensemble of earthquake ground motion records for that site. Again, the best damping ratio is the one developed from a building-specific analysis of an ensemble of earthquake ground acceleration time histories provided by an engineering seismologist. It is very important to consider the interaction between the soil and the structure when selecting a value of damping.

EXAMPLE 1 Maximum Interstory Displacement

Consider the design spectra shown in Figure 5.23, which is for Type S2 soil conditions in the 1997 UBC. The spectral acceleration for this design basis or rare earthquake (i.e., 10% probability of being exceeded in 50 years) at the natural period of vibration equal to 0.5 s and for 5% damping is

$$S_a(T_n = 0.5 \text{ s}, \zeta = 5\%) = 1.0g. \tag{5.50}$$

Therefore, for an SDOF system the base shear is equal to

$$V = 1.0W. \tag{5.51}$$

Assume that $x(t)$ is the relative displacement between the first floor and the ground, and it is desired to calculate the maximum relative displacement (i.e., the maximum lateral displacement of the columns). First, assume that the spectral acceleration is equal to the pseudo spectral acceleration (i.e., $S_{ap} = S_a$). Then the relative displacement between top and bottom of the column is

$$S_d(T_n = 0.5\,\text{s}, \zeta = 5\%) = \frac{S_{ap}}{\omega_n^2} = \frac{S_a}{\omega_n^2} = \frac{1.0 \times 386}{(4\pi)^2} = 2.4 \text{ in.} \tag{5.52}$$

If the period of vibration were reduced to one-half its original value due to stiffening of the columns, then the period would be 0.25 s and the displacement would be one-fourth the original value (i.e., 0.6 in. compared to 2.4 in.).

Table 5.7 Damping Response Reduction Coefficient (Wu and Hanson)

Critical damping ratio (%)	Period of vibration			
	0.1 s	0.5 s	1.0 s	3.0 s
5	1.00	1.00	1.00	1.00
10	0.87	0.82	0.81	0.81
20	0.74	0.64	0.62	0.63
30	0.67	0.53	0.51	0.52

5.4 FUNDAMENTAL MODE RESPONSE

The Uniform Building Code and other design standards offer a design procedure for structural engineers to use, called the *equivalent static lateral force procedure*, which is based on the principles of structural dynamics. Insight into this procedure can be obtained using the normal mode method with only the fundamental mode shape retained in the solution. This section presents the normal mode response procedure incorporating the response spectra representation of earthquake ground motion.

The two-dimensional, multistory structure shown in Figure 5.25 with a general forcing function has, as discussed in Chapter 4, the dynamic equation of motion

$$\mathbf{M}\ddot{\mathbf{X}}(t) + \mathbf{C}\dot{\mathbf{X}}(t) + \mathbf{K}\mathbf{X}(t) = \mathbf{F}_e(t) \tag{5.53}$$

where

$$\mathbf{M} = \begin{bmatrix} m_{11} & m_{12} & \cdots & m_{1n} \\ m_{21} & m_{22} & \ddots & \vdots \\ \vdots & \ddots & \ddots & m_{n-1,n} \\ m_{n1} & \cdots & m_{n,n-1} & m_{nn} \end{bmatrix}, \quad \mathbf{C} = \begin{bmatrix} c_{11} & c_{12} & \cdots & c_{1n} \\ c_{21} & c_{22} & \ddots & \vdots \\ \vdots & \ddots & \ddots & c_{n-1,n} \\ c_{n1} & \cdots & c_{n,n-1} & c_{nn} \end{bmatrix}, \quad \mathbf{K} = \begin{bmatrix} k_{11} & k_{12} & \cdots & k_{1n} \\ k_{21} & k_{22} & \ddots & \vdots \\ \vdots & \ddots & \ddots & k_{n-1,n} \\ k_{n1} & \cdots & k_{n,n-1} & k_{nn} \end{bmatrix}$$

$$\mathbf{X}(t) = \left\{ x_1(t) \quad x_2(t) \quad \cdots \quad x_n(t) \right\}^T, \quad \mathbf{F}_e(t) = \left\{ F_1(t) \quad F_2(t) \quad \cdots \quad F_n(t) \right\}^T.$$

When the structure has no external forcing function but is excited by an earthquake ground motion (see Figure 5.25), it is like the forcing function being replaced by the mass times the acceleration:

$$\begin{Bmatrix} F_1(t) \\ F_2(t) \\ \vdots \\ F_n(t) \end{Bmatrix} = - \begin{bmatrix} m_{11} & m_{12} & \cdots & m_{1n} \\ m_{21} & m_{22} & \ddots & \vdots \\ \vdots & \ddots & \ddots & m_{n-1,n} \\ m_{n1} & \cdots & m_{n,n-1} & m_{nn} \end{bmatrix} \begin{Bmatrix} a(t) \\ a(t) \\ \vdots \\ a(t) \end{Bmatrix} = -\mathbf{M}\{\mathbf{I}\}a(t). \tag{5.54}$$

Therefore, the dynamic equation of motion for this multi–degree of freedom system is

$$\mathbf{M}\ddot{\mathbf{X}}(t) + \mathbf{C}\dot{\mathbf{X}}(t) + \mathbf{K}\mathbf{X}(t) = -\mathbf{M}\{\mathbf{I}\}a(t). \tag{5.55}$$

Recall from the normal mode method in Chapter 4 that the response in the physical coordinates, $x_i(t)$, can be expressed in terms of normal mode coordinates, $q_i(t)$, using the equation

$$\mathbf{X}(t) = \mathbf{\Phi}\mathbf{Q}(t) = \boldsymbol{\phi}_1 q_1(t) + \boldsymbol{\phi}_2 q_2(t) + \ldots + \boldsymbol{\phi}_n q_n(t). \tag{5.56}$$

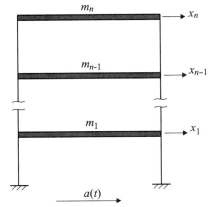

Figure 5.25 Two-Dimensional Multistory Structure

Consider the special case where only the first normal mode is retained in the normal mode solution; then Eq. 5.56 becomes

$$\mathbf{X}(t) = \boldsymbol{\phi}_1 q_1(t). \tag{5.57}$$

If Eq. 5.57 is substituted into Eq. 5.55 and both sides of the equation are premultiplied by $\boldsymbol{\phi}_1^T$, it follows that

$$\boldsymbol{\phi}_1^T \mathbf{M} \boldsymbol{\phi}_1 \ddot{q}_1(t) + \boldsymbol{\phi}_1^T \mathbf{C} \boldsymbol{\phi}_1 \dot{q}_1(t) + \boldsymbol{\phi}_1^T \mathbf{K} \boldsymbol{\phi}_1 q_1(t) = -\boldsymbol{\phi}_1^T \mathbf{M} \{\mathbf{I}\} a(t). \tag{5.58}$$

Using the definitions in Eq. 4.171 or 4.183 for a general $\boldsymbol{\phi}_1$, and assuming that the damping matrix $\mathbf{C}$ satisfies the conditions for proportional damping, then it follows from Eq. 5.58 that

$$m_1^* \ddot{q}_1(t) + c_1^* \dot{q}_1(t) + k_1^* q_1(t) = -\boldsymbol{\phi}_1^T \mathbf{M} \{\mathbf{I}\} a(t). \tag{5.59}$$

After Eq. 5.59 is divided by m_1^*, the equation of motion in terms of the first mode normal coordinate is

$$\ddot{q}_1(t) + 2\zeta_1 \omega_1 \dot{q}_1(t) + \omega_1^2 q_1(t) = -\Gamma_1 a(t). \tag{5.60}$$

The parameter Γ_1, the *fundamental mode participation factor*, is defined as

$$\Gamma_1 = \frac{\boldsymbol{\phi}_1^T \mathbf{M} \{\mathbf{I}\}}{m_1^*} = \frac{\boldsymbol{\phi}_1^T \mathbf{M} \{\mathbf{I}\}}{\boldsymbol{\phi}_1^T \mathbf{M} \boldsymbol{\phi}_1}. \tag{5.61}$$

In mathematical terms, Eq. 5.60 is the same as Eq. 5.7 with the ground acceleration on the right-hand side except for the presence of the fundamental mode participation factor, which is a scalar, on the right-hand side of Eq. 5.60. Both of these equations are second-order linear differential equations with $\omega_n = \omega_1$ and $\zeta = \zeta_1$.

In Sections 5.2 and 5.3, the response of an SDOF system was discussed in terms of response and design spectra. For example, for a given value of ω_n and ζ, and for a given $a(t)$, the spectral displacement, $S_d(\omega_n, \zeta)$, is the maximum value of $x(t)$ in an SDOF system. Because the fundamental mode participation factor, Γ_1, is a scalar, it follows that the maximum response of $q_1(t)$ can be directly obtained using

$$\max[q_1(t)] = \Gamma_1 S_d(T_n = T_1, \zeta = \zeta_1) \tag{5.62}$$

where S_d is the spectral displacement. The value obtained in Eq. 5.62 must always be positive because the response spectrum uses the maximum absolute value. The *maximum floor relative displacements* can now be obtained by substituting Eq. 5.62 into Eq. 5.57 and

$$\max[\mathbf{X}(t)] = \max[\boldsymbol{\phi}_1 q_1(t)] = \boldsymbol{\phi}_1 \Gamma_1 S_d(T_1, \zeta_1). \tag{5.63}$$

Equation 5.63 defines the displacement response at any floor as the product of three terms. One term is the spectral displacement, S_d, where the period of vibration and damping of the SDOF system is set equal to the fundamental period of vibration and modal damping of the structure. Also, the value of S_d depends on the earthquake ground motion characteristics at the site of the structure. The second term is the fundamental mode participation factor, Γ_1, which is a function of the distribution of the mass of the structure and the fundamental mode shape of vibration. Notice that the mass matrix appears both in the numeration and denomination of Eq. 5.61 for Γ_1. Therefore, the distribution of mass, rather than the total mass, impacts the value of Γ_1. Also note that there is only one mode shape term in the numerator but two mode shape terms in the denominator. Therefore, the value of Γ_1 depends on how the mode shape is normalized. For example, if $\boldsymbol{\phi}_1$ is normalized to have a unit norm, that is, $\boldsymbol{\phi}_1^T \boldsymbol{\phi}_1 = 1$ (which is denoted by $\hat{\boldsymbol{\phi}}_1$ is Section 4.4), it will result in a different value of Γ_1 than if it is normalized to have a unit value of the modal mass, that is, $\boldsymbol{\phi}_1^T \mathbf{M} \boldsymbol{\phi}_1 = 1$ (which is denoted by $\overline{\boldsymbol{\phi}}_1$ in Section 4.4).

The product of the first two terms is a scalar. The third term is the mode shape and thus it distributes the response into a vector form consistent with x. Whereas the value of Γ_1 depends on how the mode shape is normalized, the product of ϕ_1 and Γ_1 removes this dependence because $\phi_1\Gamma_1$ has two mode shapes in the numerator and two in the denominator.

Let $x_1(t)$ be the displacement of the ith degree of freedom. It can be computed by the equation

$$x_i(t) = \mathbf{T}_i^T \mathbf{X}(t) \qquad (5.64)$$

where $\mathbf{T}_i^T$ is the displacement selection vector where all elements are equal to zero except the ith element, which is equal to one.

The *roof displacement* can be expressed as

$$x_{\text{roof}}(t) = \mathbf{T}_{\text{roof}}^T \mathbf{X}(t). \qquad (5.65)$$

Substituting Eq. 5.63 into Eq. 5.65, the *maximum roof displacement* is

$$\max\left[x_{\text{roof}}(t)\right] = \max\left[\mathbf{T}_{\text{roof}}^T \mathbf{X}(t)\right] = \left(\mathbf{T}_{\text{roof}}^T \phi_1 \Gamma_1\right) S_d(T_1, \zeta_1). \qquad (5.66)$$

Equation 5.66 can be simplified by recognizing that

$$\max\left[x_{\text{roof}}(t)\right] = \left(\mathbf{T}_{\text{roof}}^T \phi_1 \Gamma_1\right) S_d(T_1, \zeta_1) = \left(\frac{\mathbf{T}_{\text{roof}}^T \phi_1 \phi_1^T \mathbf{M}\{\mathbf{I}\}}{\phi_1^T \mathbf{M}\phi_1}\right) S_d(T_1, \zeta_1). \qquad (5.67)$$

Notice that the left-hand side of Eq. 5.63 is a vector but the left-hand side of Eq. 5.66 is a scalar.

Recall that a mode shape has an arbitrary constant, and therefore it is always possible to express the element in the mode shape in terms of one that corresponds to the roof displacement equal to unity. Therefore, the mode shape ϕ_1 can be expressed as a mode shape with one corresponding to the roof displacement times at constant c_r. This equation is

$$\phi_1 = c_r \phi_{\text{roof}(1)} \qquad (5.68)$$

where $\phi_{\text{roof}(1)}$ is the mode shape corresponding to the roof displacement equal to one and c_r is the scaling factor. It follows from Eq. 5.67 that

$$\max\left[x_{\text{roof}}(t)\right] = \left(\frac{c_r^2 \mathbf{T}_{\text{roof}}^T \phi_{\text{roof}(1)} \phi_{\text{roof}(1)}^T \mathbf{M}\{\mathbf{I}\}}{c_r^2 \phi_{\text{roof}(1)}^T \mathbf{M}\phi_{\text{roof}(1)}}\right) S_d(T_1, \zeta_1) = \left(\frac{\mathbf{T}_{\text{roof}}^T \phi_{\text{roof}(1)} \phi_{\text{roof}(1)}^T \mathbf{M}\{\mathbf{I}\}}{\phi_{\text{roof}(1)}^T \mathbf{M}\phi_{\text{roof}(1)}}\right) S_d(T_1, \zeta_1). \qquad (5.69)$$

Recall that $\mathbf{T}_{\text{roof}}^T$ is the displacement selection vector where the element corresponding to the roof displacement is equal to one and all other elements are equal to zero. It follows that

$$\mathbf{T}_{\text{roof}}^T \phi_{\text{roof}(1)} = 1 \qquad (5.70)$$

and therefore recalling Eq. 5.61

$$\max\left[x_{\text{roof}}(t)\right] = \left(\frac{\phi_{\text{roof}(1)}^T \mathbf{M}\{\mathbf{I}\}}{\phi_{\text{roof}(1)}^T \mathbf{M}\phi_{\text{roof}(1)}}\right) S_d(T_1, \zeta_1) = \Gamma_{\text{roof}(1)} S_d(T_1, \zeta_1). \qquad (5.71)$$

Equation 5.71 shows that the maximum roof displacement is easy to calculate because for the special normalization where the element in the fundamental mode corresponding to the roof is equal to one, the roof displacement is equal to the fundamental mode participation factor times the spectral displacement. With experience the structural engineer can remember approximate values for this participation factor and thus can mentally scale the value of a displacement response spectrum to estimate the maximum response of the roof in the earthquake under consideration. If a displacement other than the roof is the focus of attention, the same procedure can be used to calibrate other variables in the equation.

The *relative displacement* between any two displacements, $x_i(t)$ and $x_j(t)$, denoted by $\Delta x_{ij}(t)$, can be written as

$$
\begin{aligned}
\Delta x_{ij}(t) &= x_i(t) - x_j(t) \\
&= \mathbf{T}_i^T \mathbf{X}(t) - \mathbf{T}_j^T \mathbf{X}(t) \\
&= \left(\mathbf{T}_i^T - \mathbf{T}_j^T \right) \mathbf{X}(t) \\
&= \mathbf{T}_{i-j}^T \mathbf{X}(t)
\end{aligned}
\tag{5.72}
$$

where $\mathbf{T}_{i-j}^T$ is the displacement selection vector with all zeros except for the ith entry, which is 1, and the jth entry, which is -1. Substituting Eq. 5.63 into Eq. 5.72 gives the *maximum relative displacement*:

$$
\max\left[\Delta x_{ij}(t) \right] = \max\left[\mathbf{T}_{i-j}^T \mathbf{X}(t) \right] = \left(\mathbf{T}_{i-j}^T \boldsymbol{\phi}_1 \Gamma_1 \right) S_d(T_1, \zeta_1).
\tag{5.73}
$$

The *maximum floor absolute acceleration* can be calculated starting with Eq. 5.55, where

$$
\mathbf{M}\ddot{\mathbf{Y}}(t) + \mathbf{C}\dot{\mathbf{X}}(t) + \mathbf{K}\mathbf{X}(t) = \mathbf{0}.
\tag{5.74}
$$

Multiplying Eq. 5.74 by the inverse of the mass matrix and solving for $\ddot{\mathbf{Y}}(t)$ gives

$$
\ddot{\mathbf{Y}}(t) = -\mathbf{M}^{-1}\mathbf{C}\dot{\mathbf{X}}(t) - \mathbf{M}^{-1}\mathbf{K}\mathbf{X}(t).
\tag{5.75}
$$

Substituting Eq. 5.57 into Eq. 5.75 gives

$$
\ddot{\mathbf{Y}}(t) = -\mathbf{M}^{-1}\mathbf{C}\boldsymbol{\phi}_1 \dot{q}_1(t) - \mathbf{M}^{-1}\mathbf{K}\boldsymbol{\phi}_1 q_1(t).
\tag{5.76}
$$

Recall that

$$
\left[-\omega_1^2 \mathbf{M} + \mathbf{K} \right] \boldsymbol{\phi}_1 = \mathbf{0}
\tag{5.77}
$$

and therefore

$$
\omega_1^2 \mathbf{M}\,\boldsymbol{\phi}_1 = \mathbf{K}\boldsymbol{\phi}_1.
\tag{5.78}
$$

Premultiplying Eq. 5.78 by the inverse of the mass matrix gives

$$
\omega_1^2\,\boldsymbol{\phi}_1 = \mathbf{M}^{-1}\mathbf{K}\boldsymbol{\phi}_1.
\tag{5.79}
$$

Substituting Eq. 5.79 into Eq. 5.76, it follows that

$$
\ddot{\mathbf{Y}}(t) = -\mathbf{M}^{-1}\mathbf{C}\boldsymbol{\phi}_1 \dot{q}_1(t) - \omega_1^2 \boldsymbol{\phi}_1 q_1(t).
\tag{5.80}
$$

Now assume that the damping matrix $\mathbf{C}$ satisfies the conditions of Rayleigh proportional damping. It follows that

$$
\mathbf{M}^{-1}\mathbf{C}\boldsymbol{\phi}_1 = \mathbf{M}^{-1}(\alpha\mathbf{M} + \beta\mathbf{K})\boldsymbol{\phi}_1 = \alpha\boldsymbol{\phi}_1 + \beta\mathbf{M}^{-1}\mathbf{K}\boldsymbol{\phi}_1 = \left(\alpha + \beta\omega_1^2 \right)\boldsymbol{\phi}_1 = 2\zeta_1\omega_1\boldsymbol{\phi}_1
\tag{5.81}
$$

and therefore Eq. 5.80 becomes

$$
\ddot{\mathbf{Y}}(t) = -2\zeta_1\omega_1\boldsymbol{\phi}_1 \dot{q}_1(t) - \omega_1^2 \boldsymbol{\phi}_1 q_1(t).
\tag{5.82}
$$

Equation 5.82 shows that the maximum absolute acceleration at any floor does not occur at the same time as the maximum relative displacement. The reason for this difference in time when the maximum response occur is the existence of the velocity term in Eq. 5.82. Note that it is the absolute acceleration of the floor that is related to the relative displacement between the floor and the ground.

If the damping term is small and can for structural engineering purposes be assumed to be zero, then Eq. 5.82 becomes

$$
\ddot{\mathbf{Y}}(t) = -\omega_1^2 \boldsymbol{\phi}_1 q_1(t)
\tag{5.83}
$$

and the maximum floor absolute accelerations is

$$\max\left[\ddot{\mathbf{Y}}(t)\right] = \boldsymbol{\phi}_1\Gamma_1\omega_1^2 S_d(T_1,\zeta_1) = \boldsymbol{\phi}_1\Gamma_1 S_a(T_1,\zeta_1). \tag{5.84}$$

Note that Eq. 5.84 is very similar to Eq. 5.63, and the maximum floor relative displacement and acceleration are linearly related to the spectral displacement and spectral acceleration, respectively. If the first mode shape is normalized to have a unit value at the floor of interest (i.e., roof), then Eq. 5.84 gives easily remembered terms for the maximum floor acceleration.

A comparison of Eq. 5.63 and Eq. 5.84 shows that the maximum floor displacement relative to the ground or the absolute floor acceleration is obtained by scaling the spectral displacement or spectral acceleration. It was shown in Eq. 5.71 that the maximum roof displacement is equal to the spectral displacement times a scalar that is the fundamental mode participation factor for the special case where the fundamental mode shape was normalized to have a unit value at the roof. It follows in exactly the same manner that the maximum roof absolute acceleration is

$$\max\left[\ddot{y}_{\mathrm{roof}}(t)\right] = \Gamma_{\mathrm{roof}(1)} S_a(T_1,\zeta_1). \tag{5.85}$$

The *seismic floor forces*, $\mathbf{S}(t)$, are equal to the floor mass times the floor absolute acceleration for zero damping, and thus from Eq. 5.82,

$$\mathbf{S}(t) = \mathbf{M}\ddot{\mathbf{Y}}(t) = -\omega_1^2 \mathbf{M}\boldsymbol{\phi}_1 q_1(t). \tag{5.86}$$

Therefore, the *maximum seismic floor force* can be computed by substituting Eq. 5.84 into Eq. 5.86, which gives

$$\max[\mathbf{S}(t)] = \mathbf{M}\boldsymbol{\phi}_1\Gamma_1\omega_1^2 S_d(T_1,\zeta_1) = \mathbf{M}\boldsymbol{\phi}_1\Gamma_1 S_a(T_1,\zeta_1). \tag{5.87}$$

The seismic floor forces can be used with a free body diagram to give the total seismic shear, or lateral force, at any story level. The most important story shear force is the one at the bottom, or base, of the structure because it is the total earthquake force acting on the structure. The *base shear force*, V, is the summation of the seismic floor forces, and this is calculated using the equation

$$V(t) = \mathbf{S}(t)^T\{\mathbf{I}\} = \ddot{\mathbf{Y}}(t)^T \mathbf{M}\{\mathbf{I}\} = -\omega_1^2 \boldsymbol{\phi}_1^T \mathbf{M}\{\mathbf{I}\} q_1(t). \tag{5.88}$$

Now, recall from Eq. 5.61 that

$$\boldsymbol{\phi}_1^T \mathbf{M}\{\mathbf{I}\} = m_1^* \Gamma_1 \tag{5.89}$$

and therefore the *maximum base shear force* can be computed using Eq. 5.89 and by substituting Eq. 5.84 into Eq. 5.88, which gives

$$V_{\max} = \max[V(t)] = m_1^* \Gamma_1^2 \omega_1^2 S_d(T_1,\zeta_1) = m_1^* \Gamma_1^2 S_a(T_1,\zeta_1). \tag{5.90}$$

Equations 5.58 and 5.59 show that, individually, m_1^* and Γ_1^2 are functions of how the mode shape is normalized (see Eq. 5.61). However, the product $m_1^*\Gamma_1^2$ is not a function of how the mode shape is normalized, as can be seen from Eq. 5.61. Also note that Γ_1 is not a function of the total mass of the structure but only its distribution, whereas m_1^* is a function of the total mass of the structure. Therefore, recall that

$$m_1^* = \boldsymbol{\phi}_1^T \mathbf{M}\boldsymbol{\phi}_1. \tag{5.91}$$

If each term in the mass matrix is divided by the total weight of the structure, W, where

$$W = \left(\sum_{i=1}^n m_{ii}\right)g \tag{5.92}$$

it follows that

$$\mathbf{M} = W \begin{bmatrix} m_{11}/W & m_{12}/W & \cdots & m_{1n}/W \\ m_{21}/W & m_{22}/W & \ddots & \vdots \\ \vdots & \ddots & \ddots & m_{n-1,n}/W \\ m_{n1}/W & \cdots & m_{n,n-1}/W & m_{nn}/W \end{bmatrix}.$$

If the mass is expressed as the weight divided by gravity (i.e., $W_{11}/g = m_{11}$), then it follows that

$$\mathbf{M} = \left(\frac{W}{g}\right) \begin{bmatrix} W_{11}/W & W_{12}/W & \cdots & W_{1n}/W \\ W_{21}/W & W_{22}/W & \ddots & \vdots \\ \vdots & \ddots & \ddots & W_{n-1,n}/W \\ W_{n1}/W & \cdots & W_{n,n-1}/W & W_{nn}/W \end{bmatrix} = \left(\frac{W}{g}\right)\mathbf{W} \qquad (5.93)$$

where

$$\mathbf{W} = \begin{bmatrix} W_{11}/W & W_{12}/W & \cdots & W_{1n}/W \\ W_{21}/W & W_{22}/W & \ddots & \vdots \\ \vdots & \ddots & \ddots & W_{n-1,n}/W \\ W_{n1}/W & \cdots & W_{n,n-1}/W & W_{nn}/W \end{bmatrix}.$$

Substituting Eq. 5.93 into Eq. 5.92 gives

$$m_1^* = \left(\frac{W}{g}\right)\boldsymbol{\phi}_1^T \mathbf{W}\boldsymbol{\phi}_1 = \left(\frac{W}{g}\right)W_1^* \qquad (5.94)$$

where

$$W_1^* = \boldsymbol{\phi}_1^T \mathbf{W}\boldsymbol{\phi}_1. \qquad (5.95)$$

Note that the matrix $\mathbf{W}$ is an $n \times n$ dimensionless matrix, and therefore W_1^* is a dimensionless scalar. Substituting Eq. 5.94 into Eq. 5.90 gives

$$\max[V(t)] = \left(W_1^* \Gamma_1^2\right)\left(\frac{S_a(T_1,\zeta_1)}{g}\right)W. \qquad (5.96)$$

Note that S_a is expressed in units of distance/time^2. Structural engineers often normalize S_a with respect to gravity, that is,

$$\hat{S}_a(T_1,\zeta_1) = \frac{S_a(T_1,\zeta_1)}{g} \qquad (5.97)$$

where $\hat{S}_a(T_1, \zeta_1)$ represents the normalized spectral acceleration. Therefore, the maximum base shear in Eq. 5.96 can be written as

$$V_{\max} = \max[V(t)] = C_{s1}\hat{S}_a W \qquad (5.98)$$

where

$$C_{s1} = W_1^* \Gamma_1^2. \qquad (5.99)$$

Equation 5.98 forms the basis of most earthquake design codes and standards. The term C_{s1} represents the mass distribution and vibrational characteristics of the structure and is called the *structural system fundamental mode base shear coefficient*. Recall that W_1^* is a dimensionless scalar and therefore C_{s1} is also a dimensionless quantity. The *fundamental mode earthquake base shear coefficient*, which is a function of the structural characteristics (i.e., C_{s1}) and the ground motion $\hat{S}_a$, is defined by dividing the maximum base shear by the total weight of the structure:

$$C_{e1} = \frac{V_{\max}}{W} = C_{s1}\hat{S}_a.$$

(5.100)

EXAMPLE 1   *Fundamental Mode Participation Factor*

Equation 5.66 or Eq. 5.71 can be used to calculate the maximum displacement at the top story of the structure (i.e., the roof). Equation 5.71 requires that the first mode shape be normalized such that the value corresponding to the roof is unity. When this is done, the maximum roof displacement is a function of two terms: (1) the fundamental mode participation factor, which is a function of mass distribution in the structure and the fundamental mode shape; and (2) the magnitude of earthquake ground motion as manifested through the displacement response spectra (i.e., spectral displacement). This example considers the first term and uses Eq. 5.71.

Recall from Eq. 5.71 that

$$\Gamma_{\text{roof}(1)} = \left(\frac{\boldsymbol{\phi}_{\text{roof}(1)}^T \mathbf{M}\{\mathbf{I}\}}{\boldsymbol{\phi}_{\text{roof}(1)}^T \mathbf{M} \boldsymbol{\phi}_{\text{roof}(1)}} \right).$$

(5.101)

As previously discussed, the mass matrix is in the numerator and the denominator and thus it can be normalized in any way without affecting the value of the fundamental mode participation factor. For example, each term in the mass matrix can be divided by the total mass of the structure, and the resultant matrix would be a distribution of mass matrix. Consider, for discussion, some examples discussed in earlier sections.

1. *Examples 1 and 2 in Section 3.2 (2DOF System)*

    The displacement vector and mass and stiffness matrices in this example are

    $$\mathbf{X}(t) = \begin{Bmatrix} x_1(t) \\ x_2(t) \end{Bmatrix}, \quad \mathbf{M} = \begin{bmatrix} m & 0 \\ 0 & m \end{bmatrix}, \quad \mathbf{K} = \begin{bmatrix} 2k & -k \\ -k & k \end{bmatrix}.$$

    (5.102)

    The fundamental natural frequency of vibration is

    $$\omega_1 = 0.618\omega_n$$

    where $\omega_n = \sqrt{k/m}$, and the fundamental mode shapes normalized to have a unit magnitude at the top story (i.e., roof) are

    $$\boldsymbol{\phi}_{\text{roof}(1)} = \{0.618 \quad 1.00\}^T.$$

    Therefore, the fundamental mode participation factor is

    $$\Gamma_{\text{roof}(1)} = \frac{\begin{Bmatrix} 0.618 \\ 1.000 \end{Bmatrix}^T \begin{bmatrix} m & 0 \\ 0 & m \end{bmatrix} \begin{Bmatrix} 1 \\ 1 \end{Bmatrix}}{\begin{Bmatrix} 0.618 \\ 1.000 \end{Bmatrix}^T \begin{bmatrix} m & 0 \\ 0 & m \end{bmatrix} \begin{Bmatrix} 0.618 \\ 1.000 \end{Bmatrix}} = 1.171$$

    (5.103)

    and the fundamental mode earthquake base shear coefficient is

    $$C_{s1} = W_1^* \Gamma_{\text{roof}(1)}^2 = \left(\boldsymbol{\phi}_{\text{roof}(1)}^T \mathbf{W} \boldsymbol{\phi}_{\text{roof}(1)} \right) \Gamma_{\text{roof}(1)}^2$$

    $$= \begin{Bmatrix} 0.618 \\ 1.000 \end{Bmatrix}^T \begin{bmatrix} 0.5 & 0 \\ 0 & 0.5 \end{bmatrix} \begin{Bmatrix} 0.618 \\ 1.000 \end{Bmatrix} (1.171)^2 = 0.947.$$

    (5.104)

Note in this example that the mode shape, participation factor, and earthquake base shear coefficients are not a function of the value of k.

2. *Example 3, Section 3.2 (2DOF System)*

In this example the stiffness of the bottom story was considered to be a variable:

$$m_1 = m_2 = m, \quad k_2 = k, \quad k_1 = \alpha k. \tag{5.105}$$

The fundamental natural frequency of vibration for this system is

$$\omega_1^2 = \left[\frac{\alpha + 2 - \sqrt{\alpha^2 + 4}}{2} \right] \omega_n^2. \tag{5.106}$$

It follows that the fundamental mode shape can be written as

$$\Phi_{\text{roof}(1)} = \left\{ \begin{array}{c} \left(-\alpha + \sqrt{\alpha^2 + 4}\right)/2 \\ 1.0 \end{array} \right\}. \tag{5.107}$$

Therefore, the fundamental mode participation factor is

$$\Gamma_{\text{roof}(1)} = \frac{\left\{ \begin{array}{c} \left(-\alpha + \sqrt{\alpha^2 + 4}\right)/2 \\ 1.0 \end{array} \right\}^T \left[\begin{array}{cc} m & 0 \\ 0 & m \end{array} \right] \left\{ \begin{array}{c} 1 \\ 1 \end{array} \right\}}{\left\{ \begin{array}{c} \left(-\alpha + \sqrt{\alpha^2 + 4}\right)/2 \\ 1.0 \end{array} \right\}^T \left[\begin{array}{cc} m & 0 \\ 0 & m \end{array} \right] \left\{ \begin{array}{c} \left(-\alpha + \sqrt{\alpha^2 + 4}\right)/2 \\ 1.0 \end{array} \right\}} = \frac{-\alpha + 2 + \sqrt{\alpha^2 + 4}}{\alpha^2 + 4 - \alpha\sqrt{\alpha^2 + 4}}. \tag{5.108}$$

From Eq. 5.95, W_1^* is

$$W_1^* = \left\{ \begin{array}{c} \left(-\alpha + \sqrt{\alpha^2 + 4}\right)/2 \\ 1.0 \end{array} \right\}^T \left[\begin{array}{cc} 0.5 & 0 \\ 0 & 0.5 \end{array} \right] \left\{ \begin{array}{c} \left(-\alpha + \sqrt{\alpha^2 + 4}\right)/2 \\ 1.0 \end{array} \right\} = \frac{\alpha^2 + 4 - \alpha\sqrt{\alpha^2 + 4}}{4} \tag{5.109}$$

and therefore, from Eq. 5.99, the fundamental mode base shear coefficient, C_{s1}, is

$$C_{s1} = \frac{\alpha^2 + 4 - \alpha\sqrt{\alpha^2 + 4}}{4} \left(\frac{-\alpha + 2 + \sqrt{\alpha^2 + 4}}{\alpha^2 + 4 - \alpha\sqrt{\alpha^2 + 4}} \right)^2 = \frac{\left(-\alpha + 2 + \sqrt{\alpha^2 + 4}\right)^2}{4\left(\alpha^2 + 4 - \alpha\sqrt{\alpha^2 + 4}\right)}. \tag{5.110}$$

Figure 5.26 shows a plot of $\Gamma_{\text{roof}(1)}$ versus α and Figure 5.27 shows a plot of C_{s1} versus α. Note that when $\alpha = 1$, Eq. 5.110 gives

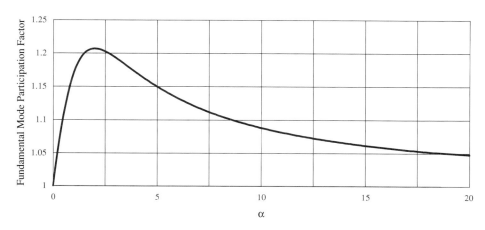

Figure 5.26 Fundamental Mode Participation Factor $\Gamma_{\text{roof}(1)}$ versus α

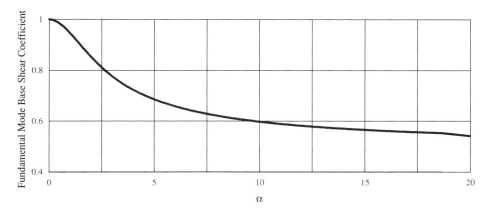

Figure 5.27 Fundamental Mode Base Shear Coefficient C_{s1} versus α

$$C_{s1} = \frac{\left(-1 + 2 + \sqrt{(1)^2 + 4}\right)^2}{4\left((1)^2 + 4 - (1)\sqrt{(1)^2 + 4}\right)} = 0.947 \tag{5.111}$$

which is the same as Eq. 5.104.

3. *Example 2, Section 4.3 (3DOF System)*

   The displacement vector and mass and stiffness matrices for this example are

$$\mathbf{X}(t) = \begin{Bmatrix} x_1(t) \\ x_2(t) \\ x_3(t) \end{Bmatrix}, \quad \mathbf{M} = \begin{bmatrix} m & 0 & 0 \\ 0 & m & 0 \\ 0 & 0 & m \end{bmatrix}, \quad \mathbf{K} = \begin{bmatrix} 3k & -k & 0 \\ -k & 2k & -k \\ 0 & -k & k \end{bmatrix}. \tag{5.112}$$

   The fundamental natural frequency is

$$\omega_1^2 = \left(2 - \sqrt{3}\right)\omega_n^2 = 0.268\omega_n^2 \tag{5.113}$$

   where $\omega_n = \sqrt{k/m}$. The fundamental mode shape is

$$\boldsymbol{\phi}_1 = \begin{Bmatrix} 1 \\ 1 + \sqrt{3} \\ 2 + \sqrt{3} \end{Bmatrix} = \begin{Bmatrix} 1.000 \\ 2.732 \\ 3.732 \end{Bmatrix}. \tag{5.114}$$

   The fundamental mode shape can be normalized to have a unit value at the roof:

$$\boldsymbol{\phi}_{\text{roof}(1)} = \begin{Bmatrix} 2 - \sqrt{3} \\ -1 + \sqrt{3} \\ 1 \end{Bmatrix} = \begin{Bmatrix} 0.268 \\ 0.732 \\ 1.000 \end{Bmatrix}. \tag{5.115}$$

   Therefore, the fundamental mode participation factor is

$$\Gamma_{\text{roof}(1)} = \frac{\begin{Bmatrix} 2 - \sqrt{3} \\ -1 + \sqrt{3} \\ 1 \end{Bmatrix}^T \begin{bmatrix} m & 0 & 0 \\ 0 & m & 0 \\ 0 & 0 & m \end{bmatrix} \begin{Bmatrix} 1 \\ 1 \\ 1 \end{Bmatrix}}{\begin{Bmatrix} 2 - \sqrt{3} \\ -1 + \sqrt{3} \\ 1 \end{Bmatrix}^T \begin{bmatrix} m & 0 & 0 \\ 0 & m & 0 \\ 0 & 0 & m \end{bmatrix} \begin{Bmatrix} 2 - \sqrt{3} \\ -1 + \sqrt{3} \\ 1 \end{Bmatrix}} = \frac{2 + \sqrt{3}}{3} = 1.244 \tag{5.116}$$

and the fundamental mode earthquake base shear coefficient is

$$C_{s1} = W_1^* \Gamma_{\text{roof}(1)}^2 = \begin{Bmatrix} 2-\sqrt{3} \\ -1+\sqrt{3} \\ 1 \end{Bmatrix}^T \begin{bmatrix} 1/3 & 0 & 0 \\ 0 & 1/3 & 0 \\ 0 & 0 & 1/3 \end{bmatrix} \begin{Bmatrix} 2-\sqrt{3} \\ -1+\sqrt{3} \\ 1 \end{Bmatrix} \left(\frac{2+\sqrt{3}}{3} \right)^2 \tag{5.117}$$

$$= \frac{4+2\sqrt{3}}{9} = 0.829.$$

EXAMPLE 2 *Two Degree of Freedom System*

Consider the two-story building discussed in Part 1 of Example 1. If the fundamental period of vibration is equal to $T_1 = 0.5$ s, then

$$\omega_1 = (0.618)\sqrt{\frac{k}{m}} = \frac{2\pi}{0.5} = 4\pi \tag{5.118}$$

Assume that $m = 1.0$ k-s^2/in. It follows from Eq. 5.118 that

$$k = m(0.618)^2 \omega_1^2 = (1)(0.618)^2 (4\pi)^2 = 60.318 \text{ k/in.} \tag{5.119}$$

Assume a design earthquake that produces a spectral acceleration of

$$S_a = S_a(T_n = T_1 = 0.5\,\text{s}, \zeta = 5\%) = 2.0\,g \tag{5.120}$$

and the pseudo acceleration is equal to the acceleration response spectra. Since $S_a = 2.0g$, it follows that $\hat{S}_a = 2.0$, and

$$S_d = \frac{S_a(T_n = T_1, \zeta = 5\%)}{\omega_1^2} = \frac{2.0(386.4)}{(4\pi)^2} = 4.89 \text{ in.} \tag{5.121}$$

Most of the time the structural engineer is provided the acceleration response spectrum by the engineering seismologist but seldom receives the pseudo acceleration response spectrum or the displacement response spectrum. The structural engineer should obtain this information but often has to assume, as was done here, that the acceleration and pseudo acceleration spectra are equal for structural engineering purposes.

The maximum floor displacements are obtained using Eq. 5.63 and the results of Example 2 in Section 3.2. Therefore,

$$\max \begin{Bmatrix} x_1 \\ x_2 \end{Bmatrix} = \begin{Bmatrix} 0.618 \\ 1.0 \end{Bmatrix} \Gamma_1 S_d. \tag{5.122}$$

The participation factor Γ_1 from Eq. 5.103 is equal to

$$\Gamma_1 = \frac{\begin{Bmatrix} 0.618 \\ 1.0 \end{Bmatrix}^T \begin{bmatrix} 1.0 & 0.0 \\ 0.0 & 1.0 \end{bmatrix} \begin{Bmatrix} 1.0 \\ 1.0 \end{Bmatrix}}{\begin{Bmatrix} 0.618 \\ 1.0 \end{Bmatrix}^T \begin{bmatrix} 1.0 & 0.0 \\ 0.0 & 1.0 \end{bmatrix} \begin{Bmatrix} 0.618 \\ 1.0 \end{Bmatrix}} = 1.171. \tag{5.123}$$

Finally,

$$\max \begin{Bmatrix} x_1 \\ x_2 \end{Bmatrix} = \begin{Bmatrix} 0.618 \\ 1.0 \end{Bmatrix} (1.171)(4.89) = \begin{Bmatrix} 3.54 \\ 5.73 \end{Bmatrix} \text{ in.} \tag{5.124}$$

The maximum absolute floor acceleration is obtained using Eq. 5.84:

$$\max\left[\ddot{\mathbf{Y}}(t)\right] = \mathbf{\phi}_1 \Gamma_1 S_a(T_1, \zeta_1) = \begin{Bmatrix} 0.618 \\ 1.00 \end{Bmatrix}(1.171)(2.00g) = \begin{Bmatrix} 1.45 \\ 2.34 \end{Bmatrix} g. \tag{5.125}$$

Note that the dimension of the maximum absolute floor acceleration in Eq. 5.125 is in g's. The maximum seismic floor force is equal to the maximum story absolute acceleration times the story mass, and using Eq. 5.87:

$$\max[\mathbf{S}(t)] = \mathbf{M}\left(\max\left[\ddot{\mathbf{Y}}(t)\right]\right) = \begin{bmatrix} m & 0 \\ 0 & m \end{bmatrix}\mathbf{\phi}_1\Gamma_1 S_a(T_1, \zeta_1). \tag{5.126}$$

A common mistake here is failure to recognize that if the maximum floor accelerations are in units of gravity (i.e., g), then the mass term that is the floor weight divided by gravity becomes in effect the floor weight because the g term cancels. In this case with $m = 1.0$ k-s^2/in., it is necessary to convert the g term back to its units of 386.4 in./s^2. From Eq. 5.126, the maximum seismic floor force is

$$\max[\mathbf{S}(t)] = \begin{bmatrix} 1 & 0 \\ 0 & 1 \end{bmatrix}\begin{Bmatrix} 0.618 \\ 1.00 \end{Bmatrix}(1.171)(2.00)(386.4) = \begin{Bmatrix} 559 \\ 905 \end{Bmatrix} \text{kip}. \tag{5.127}$$

The base shear force is obtained for this mode by summing all of the story floor forces, and it follows from Eq. 5.90, Eq. 5.96, or Eq. 5.98 that

$$\max[V(t)] = m_1^* \Gamma_1^2 S_a(T_1, \zeta_1) = C_{s1} S_a(T_1, \zeta_1) \tag{5.128}$$

where

$$m_1^* = \mathbf{\phi}_1^T \mathbf{M}\mathbf{\phi}_1 = \begin{Bmatrix} 0.618 \\ 1.00 \end{Bmatrix}^T \begin{bmatrix} 1 & 0 \\ 0 & 1 \end{bmatrix}\begin{Bmatrix} 0.618 \\ 1.00 \end{Bmatrix} = 1.382 \text{ k-s}^2/\text{in}.$$

It follows from Eq. 5.128 that

$$\max[V(t)] = (1.382)(1.171)^2(2.0)(386.4) = 1,464 \text{ kip}. \tag{5.129}$$

The maximum base shear force computed in Eq. 5.129 can also be obtained using Eq. 5.127, because the maximum base shear force is the sum of the maximum seismic floor force when only one normal mode is used:

$$\max[V(t)] = 559 + 905 = 1,464 \text{ kip}. \tag{5.130}$$

The total weight of the structure is

$$W = W_{11} + W_{22} = mg + mg = 2(386.4) = 773 \text{ kip}.$$

Therefore, an alternate expression is of the form

$$\max[V(t)] = \left[\frac{1,464}{773}\right]W = 1.89W. \tag{5.131}$$

This same result can be obtained when Eq. 5.98 is used, and it is

$$\max[V(t)] = \left(W_1^*\Gamma_1^2\right)\hat{S}_a W = C_{s1}\hat{S}_a W \tag{5.132}$$

where

$$W_1^* = \begin{Bmatrix} 0.618 \\ 1.00 \end{Bmatrix}^T \begin{bmatrix} mg/2mg & 0 \\ 0 & mg/2mg \end{bmatrix}\begin{Bmatrix} 0.618 \\ 1.00 \end{Bmatrix} = 0.691 \tag{5.133}$$

and

$$C_{s1} = W_1^*\Gamma_1^2 = (0.6910)(1.171)^2 = 0.947. \tag{5.134}$$

Finally, it follows from Eq. 5.134 that

$$\max[V(t)] = C_{s1}\hat{S}_a W = (0.947)(2.0)W = 1.89W. \tag{5.135}$$

Therefore,

$$C_{e1} = C_{s1}\hat{S}_a = \frac{\max[V(t)]}{W} = 1.89. \tag{5.136}$$

EXAMPLE 3   Three Degree of Freedom System

Consider the 3DOF system discussed in Part 3 of Example 1. Recall that

$$\mathbf{M} = \begin{bmatrix} m & 0 & 0 \\ 0 & m & 0 \\ 0 & 0 & m \end{bmatrix}, \quad \mathbf{K} = \begin{bmatrix} 3k & -k & 0 \\ -k & 2k & -k \\ 0 & -k & k \end{bmatrix}. \tag{5.137}$$

Solving the 3DOF system follows the same procedure as for the 2DOF system because in both cases only the first normal mode displacement is used to calculate the system response. Therefore, using the results from Example 1, it follows that

$$\max[\mathbf{X}(t)] = \boldsymbol{\phi}_1 \Gamma_1 S_d(T_1,\zeta_1) = \begin{Bmatrix} 2-\sqrt{3} \\ -1+\sqrt{3} \\ 1 \end{Bmatrix}\left(\frac{2+\sqrt{3}}{3}\right)S_d(T_1,\zeta_1) = \begin{Bmatrix} 1/3 \\ (1+\sqrt{3})/3 \\ (2+\sqrt{3})/3 \end{Bmatrix}S_d(T_1,\zeta_1) \tag{5.138}$$

$$\max[\ddot{\mathbf{Y}}(t)] = \boldsymbol{\phi}_1 \Gamma_1 S_a(T_1,\zeta_1) = \begin{Bmatrix} 2-\sqrt{3} \\ -1+\sqrt{3} \\ 1 \end{Bmatrix}\left(\frac{2+\sqrt{3}}{3}\right)S_a(T_1,\zeta_1) = \begin{Bmatrix} 1/3 \\ (1+\sqrt{3})/3 \\ (2+\sqrt{3})/3 \end{Bmatrix}S_a(T_1,\zeta_1) \tag{5.139}$$

$$\max[\mathbf{S}(t)] = \mathbf{M}\boldsymbol{\phi}_1 \Gamma_1 S_a(T_1,\zeta_1)$$

$$= \begin{bmatrix} m & 0 & 0 \\ 0 & m & 0 \\ 0 & 0 & m \end{bmatrix}\begin{Bmatrix} 2-\sqrt{3} \\ -1+\sqrt{3} \\ 1 \end{Bmatrix}\left(\frac{2+\sqrt{3}}{3}\right)S_a(T_1,\zeta_1) = \begin{Bmatrix} m/3 \\ (1+\sqrt{3})m/3 \\ (2+\sqrt{3})m/3 \end{Bmatrix}S_a(T_1,\zeta_1). \tag{5.140}$$

Note that the units of $\hat{S}_a(T_1, \zeta_1)$ in Eqs. 5.139 and 5.140 are in length/time^2. The maximum base shear force is computed using Eq. 5.90, that is,

$$\max[V(t)] = m_1^*\Gamma_1^2\hat{S}_a(T_1,\zeta_1) = W_1^*\Gamma_1^2\hat{S}_a(T_1,\zeta_1)W = C_{s1}\hat{S}_a(T_1,\zeta_1)W \tag{5.141}$$

where W is the total weight of the structure:

$$W = W_{11} + W_{22} + W_{33} = mg + mg + mg = 3mg.$$

It then follows that

$$W_1^* = \begin{Bmatrix} 2-\sqrt{3} \\ -1+\sqrt{3} \\ 1 \end{Bmatrix}^T\begin{bmatrix} mg/3mg & 0 & 0 \\ 0 & mg/3mg & 0 \\ 0 & 0 & mg/3mg \end{bmatrix}\begin{Bmatrix} 2-\sqrt{3} \\ -1+\sqrt{3} \\ 1 \end{Bmatrix} = 4-2\sqrt{3} = 0.536 \tag{5.142}$$

and the fundamental mode earthquake base shear coefficient is

$$C_{s1} = W_1^*\Gamma_1^2 = (4-2\sqrt{3})\left(\frac{2+\sqrt{3}}{3}\right)^2 = \frac{4+2\sqrt{3}}{9} = 0.829. \tag{5.143}$$

Finally, it follows from Eq. 5.138 that

$$\max[V(t)] = C_{s1}\hat{S}_a(T_1,\zeta_1)W = \left(\frac{4+2\sqrt{3}}{9}\right)\hat{S}_a(T_1,\zeta_1)W = (0.829)\hat{S}_a(T_1,\zeta_1)W. \quad (5.144)$$

Here, the units of $\hat{S}_a(T_1, \zeta_1)$ in Eq. 5.142 are in g's. It follows that

$$C_{el} = \frac{\max[V(t)]}{W} = C_{s1}\hat{S}_a(T_1,\zeta_1) = \left(\frac{4+2\sqrt{3}}{9}\right)\hat{S}_a(T_1,\zeta_1) = (0.829)\hat{S}_a(T_1,\zeta_1). \quad (5.145)$$

EXAMPLE 4 **Base Shear**

Consider a two-story building with the mass matrix

$$\mathbf{M} = \frac{1}{g}\begin{bmatrix} W_{11} & 0 \\ 0 & W_{22} \end{bmatrix}. \quad (5.146)$$

The earthquake forces on the floors are

$$\mathbf{S}(t) = \begin{Bmatrix} S_1(t) \\ S_2(t) \end{Bmatrix} = \mathbf{M}\ddot{\mathbf{Y}}(t) = \frac{1}{g}\begin{bmatrix} W_{11} & 0 \\ 0 & W_{22} \end{bmatrix}\begin{Bmatrix} \ddot{y}_1(t) \\ \ddot{y}_2(t) \end{Bmatrix} = \begin{Bmatrix} (W_{11}/g)\ddot{y}_1(t) \\ (W_{22}/g)\ddot{y}_2(t) \end{Bmatrix} \quad (5.147)$$

or alternately,

$$\mathbf{S}(t) = \mathbf{M}\ddot{\mathbf{Y}}(t) = \mathbf{M}\big(\ddot{\mathbf{X}}(t) + \{\mathbf{I}\}a(t)\big) = \mathbf{M}\big(\boldsymbol{\phi}_1 q_1(t) + \boldsymbol{\phi}_2 q_2(t) + \{\mathbf{I}\}a(t)\big). \quad (5.148)$$

Using only the first mode,

$$\mathbf{S}(t) = \mathbf{M}\big(\boldsymbol{\phi}_1\ddot{q}_1(t) + \{\mathbf{I}\}a(t)\big) = \begin{Bmatrix} (W_{11}/g)(\varphi_{11}\ddot{q}_1(t) + a(t)) \\ (W_{22}/g)(\varphi_{21}\ddot{q}_1(t) + a(t)) \end{Bmatrix}. \quad (5.149)$$

Using Eq. 5.87, it follows that

$$\max[\mathbf{S}(t)] = \begin{bmatrix} W_{11}/g & 0 \\ 0 & W_{22}/g \end{bmatrix}\boldsymbol{\phi}_1\Gamma_1 S_a(T_1,\zeta_1). \quad (5.150)$$

If the spectral acceleration is expressed in g's and if the total building weight is $W = W_{11} + W_{22}$, then the floor forces are

$$\max[\mathbf{S}(t)] = \left(\begin{bmatrix} W_{11}/W & 0 \\ 0 & W_{22}/W \end{bmatrix}\boldsymbol{\phi}_1\Gamma_1\hat{S}_a(T_1,\zeta_1)\right)W \quad (5.151)$$

and the base shear force is

$$\max[V(t)] = \left(\begin{Bmatrix} 1 \\ 1 \end{Bmatrix}^T\begin{bmatrix} W_{11}/W & 0 \\ 0 & W_{22}/W \end{bmatrix}\boldsymbol{\phi}_1\Gamma_1\hat{S}_a(T_1,\zeta_1)\right)W$$

$$= \left[\left(\frac{W_{11}}{W}\right)\varphi_{11} + \left(\frac{W_{22}}{W}\right)\varphi_{21}\right]\Gamma_1\hat{S}_a(T_1,\zeta_1)W \quad (5.152)$$

$$= C_{s1}\hat{S}_a(T_1,\zeta_1)W$$

$$= C_{el}W$$

where

$$C_{s1} = \left[\left(\frac{W_{11}}{W}\right)\varphi_{11} + \left(\frac{W_{22}}{W}\right)\varphi_{21}\right]\Gamma_1. \tag{5.153}$$

The coefficient C_{s1} is a function of the normalized weight of any floor (i.e., W_{11}/W) times the mode shape value at that floor. Using the same values as in Example 2, where $W_{11} = W_{22} = mg$ and $\hat{S}_a(T_1, \zeta_1) = 2.0g$, that is, $\hat{S}_a(T_1, \zeta_1) = 2.0$, the terms C_{s1} and C_{e1} become

$$C_{s1} = \left[\left(\frac{mg}{2mg}\right)(0.618) + \left(\frac{mg}{2mg}\right)(1.000)\right](1.171) = 0.947 \tag{5.154}$$

$$C_{e1} = 0.947(2.0) = 1.89 \tag{5.155}$$

and therefore, it follows from Eq. 5.152 that

$$\max[V(t)] = 1.89W \tag{5.156}$$

which is the same as Eqs. 5.131 and 5.135.

5.5 RESPONSE USING TWO NORMAL MODES OF VIBRATION

In Section 5.4, the response of the structure was calculated using only the fundamental normal mode of vibration. This solution is now extended to include the first two normal modes of vibrations.

Consider a two-dimensional multistory building like the one shown in Figure 5.25, represented by the dynamic equilibrium equation of motion in Eq. 5.55 and the normal mode response expression in Eq. 5.56. If only the first two normal modes of vibration are retained in the summation, then Eq. 5.56 becomes

$$\mathbf{X}(t) = \boldsymbol{\phi}_1 q_1(t) + \boldsymbol{\phi}_2 q_2(t). \tag{5.157}$$

Substituting Eq. 5.157 into Eq. 5.55 gives

$$\mathbf{M}\big(\boldsymbol{\phi}_1\ddot{q}_1(t) + \boldsymbol{\phi}_2\ddot{q}_2(t)\big) + \mathbf{C}\big(\boldsymbol{\phi}_1\dot{q}_1(t) + \boldsymbol{\phi}_2\dot{q}_2(t)\big) + \mathbf{K}\big(\boldsymbol{\phi}_1 q_1(t) + \boldsymbol{\phi}_2 q_2(t)\big) = -\mathbf{M}\,a(t) = -\mathbf{M}\{\mathbf{I}\}a(t). \tag{5.158}$$

Premultiply Eq. 5.158 by $\boldsymbol{\phi}_1^T$ and recall that

$$\begin{aligned}
\boldsymbol{\phi}_1^T \mathbf{M} \boldsymbol{\phi}_1 &= m_1^*, & \boldsymbol{\phi}_1^T \mathbf{C} \boldsymbol{\phi}_1 &= c_1^*, & \boldsymbol{\phi}_1^T \mathbf{K} \boldsymbol{\phi}_1 &= k_1^* \\
\boldsymbol{\phi}_1^T \mathbf{M} \boldsymbol{\phi}_2 &= 0, & \boldsymbol{\phi}_1^T \mathbf{C} \boldsymbol{\phi}_2 &= 0, & \boldsymbol{\phi}_1^T \mathbf{K} \boldsymbol{\phi}_2 &= 0.
\end{aligned} \tag{5.159}$$

It follows from Eq. 5.158 that

$$m_1^* \ddot{q}_1(t) + c_1^* \dot{q}_1(t) + k_1^* q_1(t) = -\boldsymbol{\phi}_1^T \mathbf{M}\{\mathbf{I}\}a(t). \tag{5.160}$$

Similarly, premultiply Eq. 5.158 by $\boldsymbol{\phi}_2^T$ and recall that

$$\begin{aligned}
\boldsymbol{\phi}_2^T \mathbf{M} \boldsymbol{\phi}_2 &= m_2^*, & \boldsymbol{\phi}_2^T \mathbf{C} \boldsymbol{\phi}_2 &= c_2^*, & \boldsymbol{\phi}_2^T \mathbf{K} \boldsymbol{\phi}_2 &= k_2^* \\
\boldsymbol{\phi}_2^T \mathbf{M} \boldsymbol{\phi}_1 &= 0, & \boldsymbol{\phi}_2^T \mathbf{C} \boldsymbol{\phi}_1 &= 0, & \boldsymbol{\phi}_2^T \mathbf{K} \boldsymbol{\phi}_1 &= 0.
\end{aligned} \tag{5.161}$$

It follows from Eq. 5.158 that

$$m_2^* \ddot{q}_2(t) + c_2^* \dot{q}_2(t) + k_2^* q_2(t) = -\boldsymbol{\phi}_2^T \mathbf{M}\{\mathbf{I}\}a(t). \tag{5.162}$$

Representing Eqs. 5.160 and 5.162 in matrix form, it follows that

$$\begin{bmatrix} \hat{m}_1 & 0 \\ 0 & \hat{m}_2 \end{bmatrix}\begin{Bmatrix} \ddot{q}_1(t) \\ \ddot{q}_2(t) \end{Bmatrix} + \begin{bmatrix} \hat{c}_1 & 0 \\ 0 & \hat{c}_2 \end{bmatrix}\begin{Bmatrix} \dot{q}_1(t) \\ \dot{q}_2(t) \end{Bmatrix} + \begin{bmatrix} \hat{k}_1 & 0 \\ 0 & \hat{k}_2 \end{bmatrix}\begin{Bmatrix} q_1(t) \\ q_2(t) \end{Bmatrix} = -\begin{bmatrix} \boldsymbol{\phi}_1^T \\ \boldsymbol{\phi}_2^T \end{bmatrix}\mathbf{M}\{\mathbf{I}\}a(t). \tag{5.163}$$

Dividing Eq. 5.160 by $\hat{m}_1$ and Eq. 5.162 by $\hat{m}_2$ gives

$$\ddot{q}_1(t) + 2\zeta_1\omega_1\dot{q}_1(t) + \omega_1^2 q_1(t) = -\Gamma_1\, a(t) \tag{5.164a}$$

$$\ddot{q}_2(t) + 2\zeta_2\omega_2\dot{q}_2(t) + \omega_2^2 q_2(t) = -\Gamma_2\, a(t) \qquad (5.164b)$$

where Γ_i is the ith mode participation factor, which is

$$\Gamma_i = \frac{\boldsymbol{\phi}_i^T \mathbf{M}\{\mathbf{I}\}}{m_i^*} = \frac{\boldsymbol{\phi}_i^T \mathbf{M}\{\mathbf{I}\}}{\boldsymbol{\phi}_i^T \mathbf{M}\boldsymbol{\phi}_i}. \qquad (5.165)$$

Note that Eq. 5.164a is the same as Eq. 5.60. Equation 5.164a can be used to calculate the first, or fundamental, normal mode response versus time, that is, $q_1(t)$. Similarly, Eq. 5.164b can be used to calculate the response in the second normal mode, that is, $q_2(t)$. Note that the ground acceleration time history, $a(t)$, is present on the right-hand side of both Eqs. 5.164a and 5.164b. In general, $q_1(t)$ and $q_2(t)$ will look very different and the maximum value of each will occur at different times. The maximum value of $q_1(t)$ depends on the fundamental undamped natural period of vibration, T_1, and the damping in the fundamental mode, ζ_1. Similarly, the maximum value of $q_2(t)$ depends on the second undamped natural period of vibration, T_2, and the damping in the second mode, ζ_2. It follows in the same manner as discussed in the previous section that

$$\max[q_1(t)] = \Gamma_1\, S_d(T_1, \zeta_1) \qquad (5.166)$$

$$\max[q_2(t)] = \Gamma_2\, S_d(T_2, \zeta_2) \qquad (5.167)$$

where S_d is the spectral displacement vector of the ground motion $a(t)$ discussed in Sections 5.2 and 5.3.

Similar to the generalization of the participation factor from one mode Γ_1 to the ith mode Γ_i in Eq. 5.165, the structural system fundamental mode base shear coefficient, C_{s1} in Eq. 5.99, can be generalized to the structural system base shear coefficient for the ith mode and

$$C_{si} = W_i^* \Gamma_i^2 \qquad (5.168)$$

where, similar to Eq. 5.95,

$$W_i^* = \boldsymbol{\phi}_i^T \mathbf{W}\boldsymbol{\phi}_i. \qquad (5.169)$$

In a similar way the fundamental mode earthquake base shear coefficient in Eq. 5.100 can be generalized to be

$$C_{ei} = C_{si}\hat{S}_a(T_i, \zeta_i). \qquad (5.170)$$

Returning to the basic response of the structure, recall from Eq. 5.157 that both $q_1(t)$ and $q_2(t)$ vary with time, and to obtain the system response, that is, $\mathbf{X}(t)$, the response at each instant in time, t, must be summed as indicated in Eq. 5.157. The maximum response in Eq. 5.166 will occur, in general, at a different instant in time than the maximum response in Eq. 5.167. However, the largest value of the system response that can ever occur is if both $q_1(t)$ and $q_2(t)$ reach their maximum at exactly the same instant in time. Therefore, the *maximum absolute response* (ABS) for any system response quantity, for example, $\mathbf{X}(t)$, is obtained by assuming that the maximum response in each mode occurs at the same instant in time. Thus the maximum value of the response quantity is the sum of the maximum absolute value of the response associated with each mode. Therefore, using ABS, the maximum displacement $\mathbf{X}(t)$ is

$$\max[\mathbf{X}(t)]_{\mathrm{ABS}} = \boldsymbol{\phi}_1 \max[q_1(t)] + \boldsymbol{\phi}_2 \max[q_2(t)] = |\boldsymbol{\phi}_1\Gamma_1\, S_d(T_1,\zeta_1)| + |\boldsymbol{\phi}_2\Gamma_2\, S_d(T_2,\zeta_2)|. \quad (5.171)$$

Special care must be exercised in using Eq. 5.171 because the sign of the maximum of $q_1(t)$ and $q_2(t)$ is lost in calculating the response spectra. In addition, some entries in the mode shapes vector can be negative. Therefore, the absolute value of the two terms is needed, as described in Eq. 5.171.

The ABS method can be used to calculate the response for the same response variables discussed in Section 5.4. Therefore it follows that

(a) *Maximum Displacement $x_i(t)$:*

$$x_i(t) = \mathbf{T}_i^T \mathbf{X}(t) \qquad (5.172)$$

$$\max\left[x_i(t)\right]_{\text{ABS}} = \mathbf{T}_i^T \max\left[\mathbf{X}(t)\right]_{\text{ABS}} = \mathbf{T}_i^T \left\{ \left|\boldsymbol{\phi}_1 \Gamma_1 S_d(T_1,\zeta_1)\right| + \left|\boldsymbol{\phi}_2 \Gamma_2 S_d(T_2,\zeta_2)\right| \right\} \qquad (5.173)$$

(b) *Maximum Roof Displacement $x_{\text{roof}}(t)$:*

$$x_{\text{roof}}(t) = \mathbf{T}_{\text{roof}(1)}^T \mathbf{X}(t) \qquad (5.174)$$

$$\max\left[x_{\text{roof}}(t)\right]_{\text{ABS}} = \mathbf{T}_{\text{roof}(1)}^T \max\left[\mathbf{X}(t)\right]_{\text{ABS}} = \left|\Gamma_{\text{roof}(1)} S_d(T_1,\zeta_1)\right| + \left|\Gamma_{\text{roof}(2)} S_d(T_2,\zeta_2)\right| \qquad (5.175)$$

(c) *Maximum Relative Displacement $x_{\text{roof}}(t)$:*

$$\Delta x_{ij}(t) = \mathbf{T}_{i-j}^T \mathbf{X}(t) \qquad (5.176)$$

$$\max\left[x_{ij}(t)\right]_{\text{ABS}} = \mathbf{T}_{i-j}^T \max\left[\mathbf{X}(t)\right]_{\text{ABS}} = \mathbf{T}_{i-j}^T \left\{ \left|\boldsymbol{\phi}_1 \Gamma_1 S_d(T_1,\zeta_1)\right| + \left|\boldsymbol{\phi}_1 \Gamma_2 S_d(T_2,\zeta_2)\right| \right\} \qquad (5.177)$$

(d) *Maximum Floor Acceleration $\ddot{\mathbf{Y}}(t)$:*

$$\max\left[\ddot{\mathbf{Y}}(t)\right]_{\text{ABS}} = \left|\boldsymbol{\phi}_1 \Gamma_1 S_a(T_1,\zeta_1)\right| + \left|\boldsymbol{\phi}_2 \Gamma_2 S_a(T_2,\zeta_2)\right| \qquad (5.178)$$

(e) *Maximum Seismic Floor Force $\mathbf{S}(t)$:*

$$\max\left[\mathbf{S}(t)\right]_{\text{ABS}} = \mathbf{M}\left\{ \left|\boldsymbol{\phi}_1 \Gamma_1 S_a(T_1,\zeta_1)\right| + \left|\boldsymbol{\phi}_2 \Gamma_2 S_a(T_2,\zeta_2)\right| \right\} \qquad (5.179)$$

(f) *Maximum Base Shear Force $V(t)$:*

$$\max\left[V(t)\right]_{\text{ABS}} = \left|m_1^* \Gamma_1^2 S_a(T_1,\zeta_1)\right| + \left|m_2^* \Gamma_2^2 S_a(T_2,\zeta_2)\right| \qquad (5.180)$$

or

$$\max\left[V(t)\right]_{\text{ABS}} = \left\{ \left(W_1^* \Gamma_1^2\right)\left|\frac{S_a(T_1,\zeta_1)}{g}\right| + \left(W_2^* \Gamma_2^2\right)\left|\frac{S_a(T_2,\zeta_2)}{g}\right| \right\} W \qquad (5.181)$$

or

$$\max\left[V(t)\right]_{\text{ABS}} = \left\{ C_{s1} \hat{S}_a(T_1,\zeta_1) + C_{s2} \hat{S}_a(T_2,\zeta_2) \right\} W \qquad (5.182)$$

or

$$\max\left[V(t)\right]_{\text{ABS}} = \left\{ C_{e1} + C_{e2} \right\} W. \qquad (5.183)$$

The first term in Eq. 5.175 is the same as Eq. 5.71. Recall that in the derivation of this equation, the first mode shape was normalized to give a unit value at the roof. The second term in Eq. 5.175 represents the additional roof displacement that comes from the second normal mode of vibration. In the derivation of the second term, the second mode shape was normalized to give a unit value at the roof.

Equations 5.180 through 5.183 show that the total lateral earthquake force on the structure, the base shear force, is equal to the first mode contribution as discussed in Section 5.4 and a second mode contribution. In the derivation of Eqs. 5.178 to 5.183, the damping is assumed to be small and, as was done in Section 5.4, the spectral acceleration is equal to the pseudo spectral acceleration and the extra response term due to damping was assumed to be zero.

A very common mistake is to add the response in the first normal mode, without taking the absolute value of the response, to the response in the second mode. When this is done, a minus sign reduces the value of the response. This is not correct, because the response spectrum is the absolute value of the response, and therefore all signs are positive.

Although the ABS method predicts the upper limit of the system response, it generally overestimates the actual response by such a significant amount that it is seldom used by structural engineers. Another method of calculating the system response is the *square root of the sum of the square response* (SRSS). Denote SRSS of two vectors **A** and **B** as

$$\sqrt{\mathbf{A}^2 + \mathbf{B}^2} = \sqrt{\begin{bmatrix} a_1^2 \\ a_2^2 \\ \vdots \\ a_n^2 \end{bmatrix} + \begin{bmatrix} b_1^2 \\ b_2^2 \\ \vdots \\ b_n^2 \end{bmatrix}} = \begin{Bmatrix} \sqrt{a_1^2 + b_1^2} \\ \sqrt{a_2^2 + b_2^2} \\ \vdots \\ \sqrt{a_n^2 + b_n^2} \end{Bmatrix}. \tag{5.184}$$

Then, in SRSS, the system response, for example, $\mathbf{X}(t)$, is calculated by the equation

$$\max[\mathbf{X}(t)]_{\text{SRSS}} = \sqrt{\left(\max[\boldsymbol{\phi}_1 q_1(t)]\right)^2 + \left(\max[\boldsymbol{\phi}_2 q_2(t)]\right)^2}$$
$$= \sqrt{\left[\boldsymbol{\phi}_1 \Gamma_1 S_d(T_1,\zeta_1)\right]^2 + \left[\boldsymbol{\phi}_2 \Gamma_2 S_d(T_2,\zeta_2)\right]^2}. \tag{5.185}$$

The SRSS can be used to calculate the previously discussed response quantities, and it follows that

(a) *Maximum Displacement $x_i(t)$*:

$$x_i(t) = \mathbf{T}_i^T \mathbf{X}(t) \tag{5.186}$$

$$\max[x_i(t)]_{\text{SRSS}} = \mathbf{T}_i^T \max[\mathbf{X}(t)]_{\text{SRSS}} = \mathbf{T}_i^T \sqrt{\left[\boldsymbol{\phi}_1 \Gamma_1 S_d(T_1,\zeta_1)\right]^2 + \left[\boldsymbol{\phi}_2 \Gamma_2 S_d(T_2,\zeta_2)\right]^2} \tag{5.187}$$

(b) *Maximum Roof Displacement $x_{\text{roof}}(t)$*:

$$x_{\text{roof}}(t) = \mathbf{T}_{\text{roof}(1)}^T \mathbf{X}(t) \tag{5.188}$$

$$\max[x_{\text{roof}}(t)]_{\text{SRSS}} = \mathbf{T}_{\text{roof}(1)}^T \max[\mathbf{X}(t)]_{\text{SRSS}} = \sqrt{\left[\Gamma_{\text{roof}(1)} S_d(T_1,\zeta_1)\right]^2 + \left[\Gamma_{\text{roof}(2)} S_d(T_2,\zeta_2)\right]^2} \tag{5.189}$$

(c) *Maximum Relative Displacement $\Delta x_{ij}(t)$*:

$$\Delta x_{ij}(t) = \mathbf{T}_{i-j}^T \mathbf{X}(t) \tag{5.190}$$

$$\max[\Delta x_{ij}(t)]_{\text{SRSS}} = \mathbf{T}_{i-j}^T \max[\mathbf{X}(t)]_{\text{SRSS}} = \mathbf{T}_{i-j}^T \sqrt{\left[\boldsymbol{\phi}_1 \Gamma_1 S_d(T_1,\zeta_1)\right]^2 + \left[\boldsymbol{\phi}_2 \Gamma_2 S_d(T_2,\zeta_2)\right]^2} \tag{5.191}$$

(d) *Maximum Floor Acceleration $\ddot{\mathbf{Y}}(t)$*:

$$\max[\ddot{\mathbf{Y}}(t)]_{\text{SRSS}} = \sqrt{\left[\boldsymbol{\phi}_1 \Gamma_1 S_a(T_1,\zeta_1)\right]^2 + \left[\boldsymbol{\phi}_2 \Gamma_2 S_a(T_2,\zeta_2)\right]^2} \tag{5.192}$$

(e) *Maximum Seismic Floor Force $\mathbf{S}(t)$*:

$$\max[\mathbf{S}(t)]_{\text{SRSS}} = \mathbf{M}\sqrt{\left[\boldsymbol{\phi}_1 \Gamma_1 S_a(T_1,\zeta_1)\right]^2 + \left[\boldsymbol{\phi}_2 \Gamma_2 S_a(T_2,\zeta_2)\right]^2} \tag{5.193}$$

(f) *Maximum Base Shear Force $V(t)$*:

$$\max[V(t)]_{\text{SRSS}} = \sqrt{\left[m_1^* \Gamma_1^2 S_a(T_1,\zeta_1)\right]^2 + \left[m_2^* \Gamma_2^2 S_a(T_2,\zeta_2)\right]^2} \tag{5.194}$$

or

$$\max[V(t)]_{\text{SRSS}} = W\sqrt{\left[(W_1^* \Gamma_1^2)\frac{S_a(T_1,\zeta_1)}{g}\right]^2 + \left[(W_2^* \Gamma_2^2)\frac{S_a(T_2,\zeta_2)}{g}\right]^2} \tag{5.195}$$

or

$$\max[V(t)]_{\text{SRSS}} = W\sqrt{\left[C_{s1}\hat{S}_a(T_1,\zeta_1)\right]^2 + \left[C_{s2}\hat{S}_a(T_2,\zeta_2)\right]^2} \tag{5.196}$$

or

$$\max[V(t)]_{\text{SRSS}} = W\sqrt{C_{e1}^2 + C_{e2}^2}. \tag{5.197}$$

A third method for combining the response of the individual modes is intended to extend the SRSS method to be more accurate when the values of two natural frequencies of vibration are very close to each other. This method is called the *complete quadratic combination* (CQC) response method.

The CQC method is now introduced by considering a single response variable (i.e., roof response) and in the scope of this section two modes of vibration. The roof displacements in the two normal modes are

$$x_{\text{roof}(1)} = \Gamma_{\text{roof}(1)} S_d (T_1, \zeta_1) \tag{5.198}$$

$$x_{\text{roof}(2)} = \Gamma_{\text{roof}(2)} S_d (T_2, \zeta_2). \tag{5.199}$$

The CQC method defines the combination of two mode responses to be

$$\max[x_{\text{roof}}(t)]_{\text{CQC}} = \sqrt{\rho_{11} x_{\text{roof}(1)}^2 + \rho_{12} x_{\text{roof}(1)} x_{\text{roof}(2)} + \rho_{21} x_{\text{roof}(2)} x_{\text{roof}(1)} + \rho_{22} x_{\text{roof}(2)}^2} \tag{5.200}$$

where

$$\rho_{ij} = \frac{8\sqrt{\zeta_i \zeta_j} \left(\zeta_i + \beta_{ij} \zeta_j \right) \beta_{ij}^{3/2}}{\left(1 - \beta_{ij}^2 \right)^2 + 4\zeta_i \zeta_j \beta_{ij} \left(1 + \beta_{ij}^2 \right) + 4\left(\zeta_i^2 + \zeta_j^2 \right) \beta_{ij}^2} \tag{5.201}$$

and

$$\beta_{ij} = \omega_i / \omega_j. \tag{5.202}$$

It can be shown that $\rho_{ij} = \rho_{ji}$. To show this, first substituting Eq. 5.202 into Eq. 5.201, it follows that

$$\rho_{ij} = \frac{8\sqrt{\zeta_i \zeta_j} \left[\zeta_i + \left(\dfrac{\omega_i}{\omega_j} \right) \zeta_j \right] \left(\dfrac{\omega_i}{\omega_j} \right)^{3/2}}{\left(1 - \dfrac{\omega_i^2}{\omega_j^2} \right)^2 + 4\zeta_i \zeta_j \left(\dfrac{\omega_i}{\omega_j} \right) \left(1 + \dfrac{\omega_i^2}{\omega_j^2} \right) + 4\left(\zeta_i^2 + \zeta_j^2 \right) \left(\dfrac{\omega_i^2}{\omega_j^2} \right)}. \tag{5.203}$$

Multiplying the top and bottom of Eq. 5.203 by ω_j^4 / ω_i^4 gives

$$\rho_{ij} = \frac{8\sqrt{\zeta_i \zeta_j} \left[\left(\dfrac{\omega_j}{\omega_i} \right) \zeta_i + \zeta_j \right] \left(\dfrac{\omega_j}{\omega_i} \right)^{3/2}}{\left(\dfrac{\omega_j^2}{\omega_i^2} - 1 \right)^2 + 4\zeta_i \zeta_j \left(\dfrac{\omega_j}{\omega_i} \right) \left(\dfrac{\omega_j^2}{\omega_i^2} + 1 \right) + 4\left(\zeta_i^2 + \zeta_j^2 \right) \left(\dfrac{\omega_j^2}{\omega_i^2} \right)}. \tag{5.204}$$

Rearranging the terms in Eq. 5.204 gives

$$\rho_{ij} = \frac{8\sqrt{\zeta_j \zeta_i} \left[\zeta_j + \left(\dfrac{\omega_j}{\omega_i} \right) \zeta_i \right] \left(\dfrac{\omega_j}{\omega_i} \right)^{3/2}}{\left(1 - \dfrac{\omega_j^2}{\omega_i^2} \right)^2 + 4\zeta_j \zeta_i \left(\dfrac{\omega_j}{\omega_i} \right) \left(1 + \dfrac{\omega_j^2}{\omega_i^2} \right) + 4\left(\zeta_j^2 + \zeta_i^2 \right) \left(\dfrac{\omega_j^2}{\omega_i^2} \right)} = \rho_{ji} \tag{5.205}$$

which is the desired result.

Consider the case where $i = j$. It follows from Eq. 5.202 that

$$\beta_{ii} = \omega_i / \omega_i = 1 \tag{5.206}$$

and therefore from Eq. 5.201,

$$\rho_{ii} = \frac{8\sqrt{\zeta_i \zeta_i} \left(\zeta_i + (1)\zeta_i \right)(1)}{\left(1 - (1) \right)^2 + 4\zeta_i \zeta_i (1)(1 + (1)) + 4\left(\zeta_i^2 + \zeta_i^2 \right)(1)} = 1. \tag{5.207}$$

Therefore, it follows from Eq. 5.207 that

$$\rho_{11} = \rho_{22} = 1. \tag{5.208}$$

Now consider the case where $i = 1$ and $j = 2$ (or $i = 2$ and $j = 1$); it follows from Eq. 5.203 that

$$\rho_{12} = \rho_{21} = \frac{8\sqrt{\zeta_1\zeta_2}\left[\zeta_1 + \left(\dfrac{\omega_1}{\omega_2}\right)\zeta_2\right]\left(\dfrac{\omega_1}{\omega_2}\right)^{3/2}}{\left(1 - \dfrac{\omega_1^2}{\omega_2^2}\right)^2 + 4\zeta_1\zeta_2\left(\dfrac{\omega_1}{\omega_2}\right)\left(1 + \dfrac{\omega_1^2}{\omega_2^2}\right) + 4\left(\zeta_1^2 + \zeta_2^2\right)\left(\dfrac{\omega_1^2}{\omega_2^2}\right)}. \tag{5.209}$$

Therefore, Eq. 5.200 reduces to

$$\max\left[x_{\text{roof}}(t)\right]_{\text{CQC}} = \sqrt{x_{\text{roof}(1)}^2 + x_{\text{roof}(2)}^2 + 2\rho_{12}x_{\text{roof}(1)}x_{\text{roof}(2)}} \tag{5.210}$$

Note from Eq. 5.209 that when either $\zeta_1 = 0$ or $\zeta_2 = 0$, this gives $\rho_{12} = 0$. In this case, Eq. 5.210 reduces to

$$\max\left[x_{\text{roof}}(t)\right]_{\text{CQC}} = \sqrt{x_{\text{roof}(1)}^2 + x_{\text{roof}(2)}^2} \tag{5.211}$$

which is the same as the SRSS method. Also, if ρ_{12} becomes small, then the CQC and the SRSS results are close. Figure 5.28 shows a plot of ρ_{12} versus β_{12} for the special case where $\zeta_1 = \zeta_2$. Note that when ω_1 is close to ω_2 (i.e., $\beta_{12} \approx 1$), the value of ρ_{12} is significant and the difference between CQC and SRSS solutions is also significant.

The same solution procedure can be followed for any response variable and, therefore, for close natural frequencies, the CQC method is an improvement over the SRSS solution. A more in-depth discussion of the basis for the CQC method is provided by De Kiureghian.[7] In that paper, random vibration theory is used to provide the basis for the formula presented in Eq. 5.201.

EXAMPLE 1   *Second Mode Participation Factor*

Example 1 of Section 5.4 calculated the fundamental mode participation factor for two 2DOF systems and one 3DOF system. The corresponding second mode participation factors are now calculated for the second normal mode normalized to have a unit value at the roof.

1.   *Examples 1 and 2 in Section 3.2 (2DOF System)*
     The displacement vector and mass and stiffness matrices in this example are

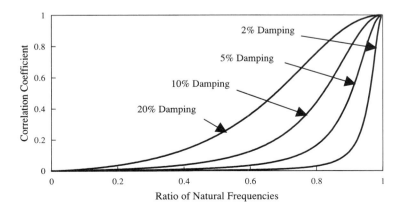

Figure 5.28 CQC Correlation Coefficient

$$\mathbf{X}(t) = \begin{Bmatrix} x_1(t) \\ x_2(t) \end{Bmatrix}, \quad \mathbf{M} = \begin{bmatrix} m & 0 \\ 0 & m \end{bmatrix}, \quad \mathbf{K} = \begin{bmatrix} 2k & -k \\ -k & k \end{bmatrix}. \tag{5.212}$$

The two natural frequencies of vibration are

$$\omega_1 = 0.618\,\omega_n, \quad \omega_2 = 1.618\,\omega_n$$

where $\omega_n = \sqrt{k/m}$, and the two mode shapes normalized to have a unit magnitude at the top story (i.e., roof) are

$$\boldsymbol{\phi}_{\text{roof}(1)} = \{0.618 \quad 1.00\}^T, \quad \boldsymbol{\phi}_{\text{roof}(2)} = \{-1.618 \quad 1.00\}^T.$$

Therefore, the second mode participation factor is

$$\Gamma_{\text{roof}(2)} = \frac{\begin{Bmatrix} -1.618 \\ 1.000 \end{Bmatrix}^T \begin{bmatrix} m & 0 \\ 0 & m \end{bmatrix} \begin{Bmatrix} 1 \\ 1 \end{Bmatrix}}{\begin{Bmatrix} -1.618 \\ 1.000 \end{Bmatrix}^T \begin{bmatrix} m & 0 \\ 0 & m \end{bmatrix} \begin{Bmatrix} -1.618 \\ 1.000 \end{Bmatrix}} = -0.171 \tag{5.213}$$

and the second mode earthquake base shear coefficient is

$$C_{s2} = W_2^* \Gamma_{\text{roof}(2)}^2 = \left(\boldsymbol{\phi}_{\text{roof}(2)}^T \mathbf{W} \boldsymbol{\phi}_{\text{roof}(2)}\right)\Gamma_{\text{roof}(2)}^2$$

$$= \begin{Bmatrix} -1.618 \\ 1.000 \end{Bmatrix}^T \begin{bmatrix} 0.5 & 0 \\ 0 & 0.5 \end{bmatrix} \begin{Bmatrix} -1.618 \\ 1.000 \end{Bmatrix}(-0.171)^2 = 0.053. \tag{5.214}$$

2.   *Example 3, Section 3.2 (2DOF System)*

In this example, the stiffness of the bottom story was considered to be a variable:

$$m_1 = m_2 = m, \quad k_2 = k, \quad k_1 = \alpha k.$$

The natural frequencies of vibration for this system were

$$\omega_1^2 = \left[\frac{\alpha + 2 - \sqrt{\alpha^2+4}}{2}\right]\omega_n^2, \quad \omega_2^2 = \left[\frac{\alpha + 2 + \sqrt{\alpha^2+4}}{2}\right]\omega_n^2.$$

It follows that the two mode shapes can be written

$$\boldsymbol{\phi}_{\text{roof}(1)} = \begin{Bmatrix} \left(-\alpha + \sqrt{\alpha^2+4}\right)/2 \\ 1.0 \end{Bmatrix}, \quad \boldsymbol{\phi}_{\text{roof}(2)} = \begin{Bmatrix} \left(-\alpha - \sqrt{\alpha^2+4}\right)/2 \\ 1.0 \end{Bmatrix}.$$

Therefore, the second mode participation factor is

$$\Gamma_{\text{roof}(2)} = \frac{\begin{Bmatrix} \left(-\alpha - \sqrt{\alpha^2+4}\right)/2 \\ 1.0 \end{Bmatrix}^T \begin{bmatrix} m & 0 \\ 0 & m \end{bmatrix} \begin{Bmatrix} 1 \\ 1 \end{Bmatrix}}{\begin{Bmatrix} \left(-\alpha - \sqrt{\alpha^2+4}\right)/2 \\ 1.0 \end{Bmatrix}^T \begin{bmatrix} m & 0 \\ 0 & m \end{bmatrix} \begin{Bmatrix} \left(-\alpha - \sqrt{\alpha^2+4}\right)/2 \\ 1.0 \end{Bmatrix}} = \frac{-\alpha + 2 - \sqrt{\alpha^2+4}}{\alpha^2+4+\alpha\sqrt{\alpha^2+4}}. \tag{5.215}$$

From Eq. 5.169, W_2^* is

$$W_2^* = \begin{Bmatrix} \left(-\alpha - \sqrt{\alpha^2+4}\right)/2 \\ 1.0 \end{Bmatrix}^T \begin{bmatrix} 0.5 & 0 \\ 0 & 0.5 \end{bmatrix} \begin{Bmatrix} \left(-\alpha - \sqrt{\alpha^2+4}\right)/2 \\ 1.0 \end{Bmatrix} = \frac{\alpha^2+4+\alpha\sqrt{\alpha^2+4}}{4} \tag{5.216}$$

and therefore, from Eq. 5.168, the second mode base shear coefficient, C_{s2}, is

$$C_{s2} = \frac{\alpha^2 + 4 + \alpha\sqrt{\alpha^2 + 4}}{4}\left(\frac{-\alpha + 2 - \sqrt{\alpha^2 + 4}}{\alpha^2 + 4 + \alpha\sqrt{\alpha^2 + 4}}\right)^2 = \frac{\left(-\alpha + 2 - \sqrt{\alpha^2 + 4}\right)^2}{4\left(\alpha^2 + 4 + \alpha\sqrt{\alpha^2 + 4}\right)}. \quad (5.217)$$

Figure 5.29 shows a plot of $\Gamma_{\text{roof}(2)}$ versus α and Figure 5.30 shows a plot of C_{s2} versus α. Note that when $\alpha = 1$, Eq. 5.217 gives

$$C_{s2} = \frac{\left(-1 + 2 - \sqrt{(1)^2 + 4}\right)^2}{4\left((1)^2 + 4 + (1)\sqrt{(1)^2 + 4}\right)} = 0.053 \quad (5.218)$$

which is the same as Eq. 5.214.

3.   *Example 2, Section 4.3 (3DOF System)*

The displacement vector and the mass and stiffness matrices for this example are

$$\mathbf{X}(t) = \begin{Bmatrix} x_1(t) \\ x_2(t) \\ x_3(t) \end{Bmatrix}, \quad \mathbf{M} = \begin{bmatrix} m & 0 & 0 \\ 0 & m & 0 \\ 0 & 0 & m \end{bmatrix}, \quad \mathbf{K} = \begin{bmatrix} 3k & -k & 0 \\ -k & 2k & -k \\ 0 & -k & k \end{bmatrix}.$$

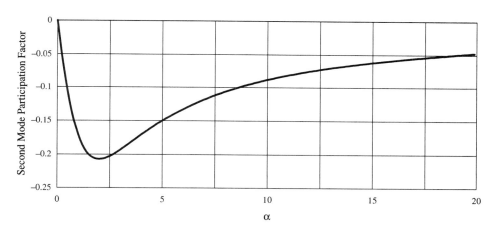

Figure 5.29   Second Mode Participation Factor $\Gamma_{\text{roof}(2)}$ versus α

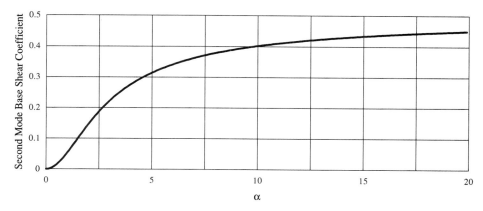

Figure 5.30   Second Mode Base Shear Coefficient C_{s2} versus α

The three natural frequencies are

$$\omega_1^2 = \left(2 - \sqrt{3}\right)\omega_n^2 = 0.268\omega_n^2, \quad \omega_2^2 = 2\omega_n^2, \quad \omega_3^2 = \left(2 + \sqrt{3}\right)\omega_n^2 = 3.732\,\omega_n^2$$

where $\omega_n = \sqrt{k/m}$ and the mode shapes normalized to have a unit value at the roof are

$$\boldsymbol{\phi}_{\text{roof}(1)} = \left\{\begin{matrix} 2-\sqrt{3} \\ -1+\sqrt{3} \\ 1 \end{matrix}\right\} = \left\{\begin{matrix} 0.268 \\ 0.732 \\ 1.000 \end{matrix}\right\}, \quad \boldsymbol{\phi}_{\text{roof}(2)} = \left\{\begin{matrix} -1 \\ -1 \\ 1 \end{matrix}\right\}, \quad \boldsymbol{\phi}_{\text{roof}(3)} = \left\{\begin{matrix} 2+\sqrt{3} \\ -1-\sqrt{3} \\ 1 \end{matrix}\right\} = \left\{\begin{matrix} 3.732 \\ -2.732 \\ 1.000 \end{matrix}\right\}.$$

Therefore, the second mode participation factor is

$$\Gamma_{\text{roof}(2)} = \frac{\left\{\begin{matrix} -1 \\ -1 \\ 1 \end{matrix}\right\}^T \begin{bmatrix} m & 0 & 0 \\ 0 & m & 0 \\ 0 & 0 & m \end{bmatrix} \left\{\begin{matrix} 1 \\ 1 \\ 1 \end{matrix}\right\}}{\left\{\begin{matrix} -1 \\ -1 \\ 1 \end{matrix}\right\}^T \begin{bmatrix} m & 0 & 0 \\ 0 & m & 0 \\ 0 & 0 & m \end{bmatrix} \left\{\begin{matrix} -1 \\ -1 \\ 1 \end{matrix}\right\}} = -0.333. \tag{5.219}$$

EXAMPLE 2   *Maximum Roof Displacement*

Consider the 2DOF system discussed Examples 1 and 2 of Section 3.2, where

$$\mathbf{X}(t) = \left\{\begin{matrix} x_1(t) \\ x_2(t) \end{matrix}\right\}, \quad \mathbf{M} = \begin{bmatrix} m & 0 \\ 0 & m \end{bmatrix}, \quad \mathbf{K} = \begin{bmatrix} 2k & -k \\ -k & k \end{bmatrix} \tag{5.220}$$

with the two natural frequencies of vibration,

$$\omega_1 = 0.618\,\omega_n, \quad \omega_2 = 1.618\,\omega_n$$

where $\omega_n = \sqrt{k/m}$, and the two mode shapes normalized to have a unit magnitude at the top story,

$$\boldsymbol{\phi}_{\text{roof}(1)} = \{0.618 \quad 1.00\}^T, \quad \boldsymbol{\phi}_{\text{roof}(2)} = \{-1.618 \quad 1.00\}^T.$$

The fundamental mode participation factor is calculated in Example 1 of Section 5.4, and the second mode participation factor is calculated in Part 1 of the previous example:

$$\Gamma_{\text{roof}(1)} = 1.171, \quad \Gamma_{\text{roof}(2)} = -0.171. \tag{5.221}$$

The normal mode method can be used to calculate the floor displacement as a function of time using the equation

$$\mathbf{X}(t) = \left\{\begin{matrix} x_1(t) \\ x_2(t) \end{matrix}\right\} = \boldsymbol{\phi}_1 q_1(t) + \boldsymbol{\phi}_2 q_2(t) = \left\{\begin{matrix} 0.618 \\ 1.00 \end{matrix}\right\} q_1(t) + \left\{\begin{matrix} -1.618 \\ 1.00 \end{matrix}\right\} q_2(t). \tag{5.222}$$

Because this 2DOF system is a set of two linear differential equations, $q_1(t)$ and $q_2(t)$ can be separately calculated and then added using Eq. 5.222 to give the displacement response. The response spectra solution method calculates the maximum value of $q_1(t)$ and $q_2(t)$, but unfortunately it does not retain the time variable. That is, when $q_1(t)$ is at its maximum value, the value of $q_2(t)$ is not known. All that is known is that it is less than or equal to the maximum of $q_2(t)$. It follows from Eqs. 5.166 and 5.167 that

$$\max[q_1(t)] = \Gamma_1 S_d(T_1, \zeta_1) = 1.171 S_d(T_1, \zeta_1) = 1.171 S_d^{(1)} \tag{5.223}$$

$$\max[q_2(t)] = \Gamma_2 S_d(T_2, \zeta_2) = 0.171 S_d(T_2, \zeta_2) = 0.171 S_d^{(2)}. \tag{5.224}$$

The participation factor of the second normal mode is smaller than the fundamental mode participation factor. This is a result of the change in sign of the terms in the second mode shape, which reduces the value of the numerator of Eq. 5.213 as compared to Eq. 5.103. The relative values of the spectral displacement for the first and second mode response are a function of the earthquake, the building site soil conditions, and the first and second periods of vibration as they relate to the dominant periods of the earthquake ground motion.

The maximum roof displacement using the different modal contribution methods is as follows:

1. *Absolute Sum*—Eq. 5.175:

$$\max\left[x_2(t)\right]_{ABS} = 1.171 S_d^{(1)} + 0.171 S_d^{(2)} \tag{5.225}$$

2. *Square Root of Sum of Squares*—Eq. 5.189:

$$\max\left[x_2(t)\right]_{SRSS} = \sqrt{\left[1.171 S_d^{(1)}\right]^2 + \left[0.171 S_d^{(2)}\right]^2} \tag{5.226}$$

3. *CQC*—Eq. 5.210:

$$\max\left[x_2(t)\right]_{CQC} = \sqrt{\left[1.171 S_d^{(1)}\right]^2 + \left[0.171 S_d^{(2)}\right]^2 + 2\rho_{12}\left[1.171 S_d^{(1)}\right]\left[0.171 S_d^{(2)}\right]} \tag{5.227}$$

where

$$\rho_{12} = \frac{8\sqrt{\zeta_1\zeta_2}\left[\zeta_1 + \left(\dfrac{0.618}{1.618}\right)\zeta_2\right]\left(\dfrac{0.618}{1.618}\right)^{3/2}}{\left(1 - \dfrac{(0.618)^2}{(1.618)^2}\right)^2 + 4\zeta_1\zeta_2\left(\dfrac{0.618}{1.618}\right)\left(1 + \dfrac{(0.618)^2}{(1.618)^2}\right) + 4(\zeta_1^2 + \zeta_2^2)\dfrac{(0.618)^2}{(1.618)^2}}$$

$$= \frac{1.889\sqrt{\zeta_1\zeta_2}\left[\zeta_1 + (0.382)\zeta_2\right]}{0.729 + 1.751\zeta_1\zeta_2 + 0.584}. \tag{5.228}$$

In general, the first mode contributes the most to the response at the roof. This can be illustrated by using the ABS method and factoring out the first mode response in Eq. 5.172:

$$\max\left[x_2(t)\right]_{ABS} = \left|\Gamma_1 S_d(T_1,\zeta_1)\right| + \left|\Gamma_2 S_d(T_2,\zeta_2)\right|$$

$$= \left(1 + \left|\frac{\Gamma_2}{\Gamma_1}\right|\frac{S_d(T_2,\zeta_2)}{S_d(T_1,\zeta_1)}\right)\Gamma_1 S_d(T_1,\zeta_1). \tag{5.229}$$

For the first-floor displacement, a similar expression can be obtained. First note that

$$\mathbf{T}_1^T = \{1 \quad 0\} \tag{5.230}$$

and therefore

$$\mathbf{T}_1^T\boldsymbol{\phi}_1 = \varphi_{11}, \quad \mathbf{T}_1^T\boldsymbol{\phi}_2 = \varphi_{21}. \tag{5.231}$$

Then it follows from Eq. 5.171 that

$$\max\left[x_1(t)\right]_{ABS} = \left|\varphi_{11}\Gamma_1 S_d(T_1,\zeta_1)\right| + \left|\varphi_{21}\Gamma_2 S_d(T_2,\zeta_2)\right|$$

$$= \left(1 + \left|\frac{\varphi_{21}\Gamma_2}{\varphi_{11}\Gamma_1}\right|\frac{S_d(T_2,\zeta_2)}{S_d(T_1,\zeta_1)}\right)\varphi_{11}\Gamma_1 S_d(T_1,\zeta_1). \tag{5.232}$$

The value of $S_d(T_2, \zeta_2)$ is usually significantly less than $S_d(T_1, \zeta_1)$ due to the typical shape of the earthquake response spectra.

EXAMPLE 3 *Maximum Roof Acceleration*

For the 2DOF system in Example 2, the maximum roof acceleration in each mode can be written as

$$\max[\ddot{q}_1(t)] = \Gamma_1 S_a(T_1,\zeta_1) \approx \Gamma_1 S_{ap}(T_1,\zeta_1) \tag{5.233}$$

$$\max[\ddot{q}_2(t)] = \Gamma_2 S_a(T_2,\zeta_2) \approx \Gamma_2 S_{ap}(T_2,\zeta_2). \tag{5.234}$$

The roof acceleration using the ABS method can be computed as

$$\max[\ddot{y}_2(t)]_{\text{ABS}} \approx \left| \Gamma_1 S_{ap}(T_1,\zeta_1) \right| + \left| \Gamma_2 S_{ap}(T_2,\zeta_2) \right|. \tag{5.235}$$

Recall that the pseudo spectral acceleration can be written in terms of the spectral displacement using the equation

$$S_{ap}(T_i,\zeta_i) = \omega_i^2 S_d(T_i,\zeta_i) = \left(\frac{2\pi}{T_i} \right)^2 S_d(T_i,\zeta_i). \tag{5.236}$$

Because the second natural period of vibration is always smaller than the fundamental period of vibration (i.e., $T_2 < T_1$), the importance of the second mode is more significant in acceleration calculations for the same relative first and second mode magnitude of spectral displacement. Substituting Eq. 5.236 into Eq. 5.235, it follows that

$$\max[\ddot{y}_2(t)]_{\text{ABS}} \approx \left| \omega_1^2 \Gamma_1 S_d(T_1,\zeta_1) \right| + \left| \omega_2^2 \Gamma_2 S_d(T_2,\zeta_2) \right|$$
$$= \left(1 + \frac{\omega_2^2}{\omega_1^2} \left| \frac{\Gamma_2}{\Gamma_1} \right| \left| \frac{S_d(T_2,\zeta_2)}{S_d(T_1,\zeta_1)} \right| \right) \Gamma_1 S_d(T_1,\zeta_1). \tag{5.237}$$

Since ω_2^2/ω_1^2 can be a large quantity, this explains why the second mode of vibration can be significant in acceleration calculations.

EXAMPLE 4 *Maximum Displacement Response of a Two-Story Structure*

Consider the structure discussed in Part 1 of Example 1. Assume that $m = 1.0$ k-s^2/in. and $k = 103.4$ k/in. It then follows that the natural periods of vibration are

$$T_1 = \frac{2\pi}{\omega_1} = \frac{2\pi}{0.618} \sqrt{\frac{1.0}{103.4}} = 1.00 \text{ s} \tag{5.238}$$

$$T_2 = \frac{2\pi}{\omega_2} = \frac{2\pi}{1.618} \sqrt{\frac{1.0}{103.4}} = 0.38 \text{ s}. \tag{5.239}$$

The mode shapes and the modal participation factors are

$$\boldsymbol{\Phi}_{\text{roof}(1)} = \left\{ \begin{matrix} \varphi_{\text{roof}(11)} \\ \varphi_{\text{roof}(21)} \end{matrix} \right\} = \left\{ \begin{matrix} 0.618 \\ 1.00 \end{matrix} \right\}, \quad \boldsymbol{\Phi}_{\text{roof}(2)} = \left\{ \begin{matrix} \varphi_{\text{roof}(12)} \\ \varphi_{\text{roof}(22)} \end{matrix} \right\} = \left\{ \begin{matrix} -0.618 \\ 1.00 \end{matrix} \right\} \tag{5.240}$$

$$\Gamma_{\text{roof}(1)} = 1.171, \quad \Gamma_{\text{roof}(2)} = -0.171. \tag{5.241}$$

Consider the design earthquake response spectrum shown in Figure 5.31. It follows that

$$S_a(T_n = T_1 = 1.00 \text{ s}, \zeta = 5\%) = 0.85g \tag{5.242}$$

$$S_a(T_n = T_2 = 0.38 \text{ s}, \zeta = 5\%) = 1.0g. \tag{5.243}$$

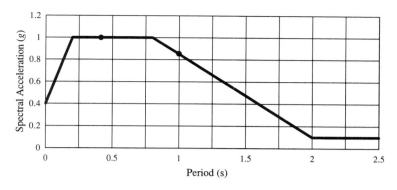

Figure 5.31 Design Earthquake Response Spectrum

Assume that the design earthquake response spectrum is equal to the pseudo acceleration response spectrum. Then

$$S_d^{(1)} = S_d(T_n = 1.00 \text{ s}, \zeta = 5\%) = \frac{S_a(T_n = 1.00 \text{ s}, \zeta = 5\%)}{\omega_1^2} = \frac{0.85(386.4)}{(2\pi/1.0)^2} = 8.32 \text{ in.} \quad (5.244)$$

$$S_d^{(2)} = S_d(T_n = 0.38 \text{ s}, \zeta = 5\%) = \frac{S_a(T_n = 0.38 \text{ s}, \zeta = 5\%)}{\omega_2^2} = \frac{1.0(386.4)}{(2\pi/0.38)^2} = 1.41 \text{ in.} \quad (5.245)$$

It follows that the roof responses in the two modes of vibration are

$$\max\left[x_2^{(1)}\right] = \varphi_{\text{roof}(21)}\Gamma_{\text{roof}(1)}S_d^{(1)} = (1.0)(1.171)(8.32) = 9.74 \text{ in.} \quad (5.246)$$

$$\max\left[x_2^{(2)}\right] = \varphi_{\text{roof}(22)}\Gamma_{\text{roof}(2)}S_d^{(2)} = (1.0)(-0.171)(1.41) = 0.24 \text{ in.} \quad (5.247)$$

Note that the minus sign in the participation factor is lost because the sign of the response spectrum is not known. Also note that because the mode shapes are normalized to have a 1.0 at the roof, the roof displacement is simply equal to the participation factor times the spectral displacement.

Similarly, the response of the first floor in each of the two modes of vibration is

$$\max\left[x_1^{(1)}\right] = \varphi_{\text{roof}(11)}\Gamma_{\text{roof}(1)}S_d^{(1)} = (0.618)(1.171)(8.32) = 6.02 \text{ in.} \quad (5.248)$$

$$\max\left[x_1^{(2)}\right] = \varphi_{\text{roof}(12)}\Gamma_{\text{roof}(2)}S_d^{(2)} = (-1.618)(-0.171)(1.41) = 0.39 \text{ in.} \quad (5.249)$$

It is important to calculate the magnitude of the interstory displacement, and this response is

$$\begin{aligned}\max\left[\Delta_{2-1}^{(1)}\right] &= \max\left[x_2^{(1)} - x_1^{(1)}\right] \\ &= \left[\varphi_{\text{roof}(21)} - \varphi_{\text{roof}(11)}\right]\Gamma_{\text{roof}(1)}S_d^{(1)} \\ &= [1.0 - 0.618](1.171)(8.32) = 3.72 \text{ in.}\end{aligned} \quad (5.250)$$

$$\begin{aligned}\max\left[\Delta_{2-1}^{(2)}\right] &= \max\left[x_2^{(2)} - x_1^{(2)}\right] \\ &= \left[\varphi_{\text{roof}(22)} - \varphi_{\text{roof}(12)}\right]\Gamma_{\text{roof}(2)}S_d^{(2)} \\ &= [1.0 - (-1.618)](-0.171)(1.41) = 0.63 \text{ in.}\end{aligned} \quad (5.251)$$

The difference in the mode shape is used to calculate interstory drift, and for the second mode the sign is important, that is, $[1.0 - (-1.618)] = 2.618$. It is not correct to subtract the resultant floor displacements. For example, the correct interstory drift in the second mode is calculated to be 0.63 in. If the second-floor displacement in the second mode (i.e., 0.24 in.) is subtracted from the first-floor displacement in the second mode (i.e., 0.39 in.), then the result is 0.15 in. This result is incorrect because the sign of the spectral displacement is lost.

The second- (i.e., roof), first-, and interstory displacements using the ABS method are

$$\max\left[x_2\right]_{ABS} = 9.72 + 0.24 = 9.98 \text{ in.} \tag{5.252}$$

$$\max\left[x_1\right]_{ABS} = 6.02 + 0.39 = 6.41 \text{ in.} \tag{5.253}$$

$$\max\left[\Delta_{2-1}\right]_{ABS} = 3.72 + 0.63 = 4.35 \text{ in.} \tag{5.254}$$

The corresponding displacements for the SRSS method are

$$\max\left[x_2\right]_{SRSS} = \sqrt{(9.72)^2 + (0.24)^2} = 9.72 \text{ in.} \tag{5.255}$$

$$\max\left[x_1\right]_{SRSS} = \sqrt{(6.02)^2 + (0.39)^2} = 6.03 \text{ in.} \tag{5.256}$$

$$\max\left[\Delta_{2-1}\right]_{SRSS} = \sqrt{(3.72)^2 + (0.63)^2} = 3.77 \text{ in.} \tag{5.257}$$

For the CQC method, the coefficients for ρ_{ij} must first be computed. Doing so gives

$$\frac{\omega_1}{\omega_2} = \frac{0.618}{1.618} = 0.382$$

$$\rho_{12} = \frac{8\sqrt{(0.05)(0.05)}\left[0.05 + (0.382)(0.05)\right](0.382)^{3/2}}{\left(1 - (0.382)^2\right)^2 + 4(0.05)^2(0.382)\left(1 + (0.382)^2\right) + 4\left((0.05)^2 + (0.05)^2\right)(0.382)^2} = 0.00885.$$

The corresponding displacements for the CQC method are

$$\max\left[x_2\right]_{CQC} = \sqrt{(9.72)^2 + (0.24)^2 + 2(0.00885)(9.72)(0.24)} = 9.73 \text{ in.} \tag{5.258}$$

$$\max\left[x_1\right]_{CQC} = \sqrt{(6.02)^2 + (0.39)^2 + 2(0.00885)(6.02)(0.39)} = 6.04 \text{ in.} \tag{5.259}$$

$$\max\left[\Delta_{2-1}\right]_{CQC} = \sqrt{(3.72)^2 + (0.63)^2 + 2(0.00885)(3.72)(0.63)} = 3.78 \text{ in.} \tag{5.260}$$

EXAMPLE 5   *Maximum Acceleration Response of a Two-Story Structure*

Consider again the structure discussed in Example 4, where $m = 1.0$ k-s^2/in. and $k = 103.4$ k/in. The mode shapes and the modal participation factors are

$$\boldsymbol{\phi}_{\text{roof}(1)} = \begin{Bmatrix} \varphi_{\text{roof}(11)} \\ \varphi_{\text{roof}(21)} \end{Bmatrix} = \begin{Bmatrix} 0.618 \\ 1.00 \end{Bmatrix}, \quad \boldsymbol{\phi}_{\text{roof}(2)} = \begin{Bmatrix} \varphi_{\text{roof}(12)} \\ \varphi_{\text{roof}(22)} \end{Bmatrix} = \begin{Bmatrix} -0.618 \\ 1.00 \end{Bmatrix} \tag{5.261}$$

$$\Gamma_{\text{roof}(1)} = 1.171, \quad \Gamma_{\text{roof}(2)} = -0.171 \tag{5.262}$$

and

$$S_a^{(1)} = S_a(T_n = T_1 = 1.00 \text{ s}, \zeta = 5\%) = 0.85g \tag{5.263}$$

$$S_a^{(2)} = S_a(T_n = T_2 = 0.38 \text{ s}, \zeta = 5\%) = 1.0g. \tag{5.264}$$

The maximum floor accelerations for each mode of vibration are

$$\max\left[\ddot{\mathbf{Y}}^{(1)}\right] = \boldsymbol{\phi}_{\text{roof}(1)} \Gamma_{\text{roof}(1)} S_a^{(1)} = \begin{Bmatrix} 0.618 \\ 1.00 \end{Bmatrix}(1.171)(0.85g) = \begin{Bmatrix} 0.62g \\ 0.99g \end{Bmatrix} \tag{5.265}$$

$$\max\left[\ddot{\mathbf{Y}}^{(2)}\right] = \boldsymbol{\phi}_{\text{roof}(2)}\Gamma_{\text{roof}(2)}S_a^{(2)} = \begin{Bmatrix} -1.618 \\ 1.00 \end{Bmatrix}(-0.171)(1.0g) = \begin{Bmatrix} 0.28g \\ -0.17g \end{Bmatrix}. \quad (5.266)$$

Note that the minus sign in the participation factor is retained in the result of Eq. 5.266 because the maximum floor accelerations for each mode of vibration are expressed in vector form.

The maximum seismic floor forces for each mode of vibration are computed by multiplying the floor acceleration by the mass of that floor:

$$\max\left[\mathbf{S}^{(1)}\right] = \mathbf{M}\boldsymbol{\phi}_{\text{roof}(1)}\Gamma_{\text{roof}(1)}S_a^{(1)} = \mathbf{M}\left\{\max\left[\ddot{\mathbf{Y}}^{(1)}\right]\right\} = \begin{bmatrix} 1.0 & 0.0 \\ 0.0 & 1.0 \end{bmatrix}\begin{bmatrix} 0.62g \\ 0.99g \end{bmatrix} = \begin{Bmatrix} 240 \\ 383 \end{Bmatrix} \text{ kip} \quad (5.267)$$

$$\max\left[\mathbf{S}^{(2)}\right] = \mathbf{M}\boldsymbol{\phi}_{\text{roof}(2)}\Gamma_{\text{roof}(2)}S_a^{(2)} = \mathbf{M}\left\{\max\left[\ddot{\mathbf{Y}}^{(2)}\right]\right\} = \begin{bmatrix} 1.0 & 0.0 \\ 0.0 & 1.0 \end{bmatrix}\begin{bmatrix} 0.28g \\ -0.17g \end{bmatrix} = \begin{Bmatrix} 108 \\ -66 \end{Bmatrix} \text{ kip.} \quad (5.268)$$

The maximum base shear force is the sum of the maximum seismic floor forces:

$$\max\left[V^{(1)}\right] = 240 + 383 = 623 \text{ kip} \quad (5.269)$$

$$\max\left[V^{(2)}\right] = 108 + (-66) = 42 \text{ kip.} \quad (5.270)$$

Because the sign is retained while computing the maximum seismic floor forces, the base shear is simply the sum of all entries in the maximum seismic floor forces vector.

The acceleration, seismic floor force, and base shear using the ABS method are

$$\max\left[\ddot{\mathbf{Y}}\right]_{\text{ABS}} = \begin{Bmatrix} 0.62g \\ 0.99g \end{Bmatrix} + \begin{Bmatrix} 0.28g \\ 0.17g \end{Bmatrix} = \begin{Bmatrix} 0.90g \\ 1.16g \end{Bmatrix} \quad (5.271)$$

$$\max[\mathbf{S}]_{\text{ABS}} = \begin{Bmatrix} 240 \\ 383 \end{Bmatrix} + \begin{Bmatrix} 108 \\ 66 \end{Bmatrix} = \begin{Bmatrix} 348 \\ 449 \end{Bmatrix} \text{ kip} \quad (5.272)$$

$$\max[V]_{\text{ABS}} = 623 + 42 = 667 \text{ kip.} \quad (5.273)$$

The corresponding values for the SRSS method are

$$\max\left[\ddot{\mathbf{Y}}\right]_{\text{SRSS}} = \sqrt{\begin{Bmatrix} 0.62g \\ 0.99g \end{Bmatrix}^2 + \begin{Bmatrix} 0.28g \\ 0.17g \end{Bmatrix}^2} = \begin{Bmatrix} 0.68g \\ 1.00g \end{Bmatrix} \quad (5.274)$$

$$\max[\mathbf{S}]_{\text{SRSS}} = \sqrt{\begin{Bmatrix} 240 \\ 383 \end{Bmatrix}^2 + \begin{Bmatrix} 108 \\ 66 \end{Bmatrix}^2} = \begin{Bmatrix} 263 \\ 389 \end{Bmatrix} \text{ kip} \quad (5.275)$$

$$\max[V]_{\text{ABS}} = \sqrt{(623)^2 + (42)^2} = 624 \text{ kip.} \quad (5.276)$$

Finally, the corresponding values for the CQC method based on $\rho_{12} = 0.00885$ computed in Example 4 are

$$\max\left[\ddot{\mathbf{Y}}\right]_{\text{CQC}} = \sqrt{\begin{Bmatrix} 0.62g \\ 0.99g \end{Bmatrix}^2 + \begin{Bmatrix} 0.28g \\ 0.17g \end{Bmatrix}^2 + 2(0.00885)\begin{Bmatrix} 0.62g \\ 0.99g \end{Bmatrix}\begin{Bmatrix} 0.28g \\ 0.17g \end{Bmatrix}} = \begin{Bmatrix} 0.68g \\ 1.01g \end{Bmatrix} \quad (5.277)$$

$$\max[\mathbf{S}]_{\text{CQC}} = \sqrt{\begin{Bmatrix} 240 \\ 383 \end{Bmatrix}^2 + \begin{Bmatrix} 108 \\ 66 \end{Bmatrix}^2 + 2(0.00885)\begin{Bmatrix} 240 \\ 383 \end{Bmatrix}\begin{Bmatrix} 108 \\ 66 \end{Bmatrix}} = \begin{Bmatrix} 264 \\ 389 \end{Bmatrix} \text{ kip} \quad (5.278)$$

$$\max[V]_{\text{CQC}} = \sqrt{(623)^2 + (42)^2 + 2(0.00885)(623)(42)} = 625 \text{ kip.} \quad (5.279)$$

Note that when ABS, SRSS, or CQC is used to combine modal response, all the signs are dropped because only the maximum response is desired.

5.6   RESPONSE OF A MULTI–DEGREE OF FREEDOM SYSTEM

The previous section discussed a structural system with two degrees of freedom. If the system is an n–degree of freedom system, then the response spectra solution procedure is a direct extension of the presentation in the previous section.

The dynamic equilibrium equation of motion is as presented in Eq. 5.55, and the response in the normal mode coordinates is as discussed in Section 4.5, that is,

$$\mathbf{X}(t) = \hat{\boldsymbol{\phi}}_1 q_1(t) + \hat{\boldsymbol{\phi}}_2 q_2(t) + \quad \dots \quad + \hat{\boldsymbol{\phi}}_k q_k(t) \tag{5.280}$$

where $k \le n$. If $k = n$, then there is no normal mode truncation. The response of the variable of interest, for example, $x_{\text{roof}}(t)$, is first calculated for each mode up to and including the kth mode. Then the ABS, SRSS, or CQC method is used to combine the individual mode response to estimate the maximum response that includes the contribution from all modes.

The previously defined equations for Γ_i, C_{si}, C_{ei}, and ρ_{ij} for any $i \le n$ are

$$\Gamma_i = \frac{\boldsymbol{\phi}_i^T \mathbf{M}\{\mathbf{I}\}}{m_i^*} = \frac{\boldsymbol{\phi}_i^T \mathbf{M}\{\mathbf{I}\}}{\boldsymbol{\phi}_i^T \mathbf{M}\boldsymbol{\phi}_i} \tag{5.281}$$

$$C_{si} = W_i^* \Gamma_i^2 = \left(\boldsymbol{\phi}_i^T \mathbf{W}\boldsymbol{\phi}_i\right)\Gamma_i^2 \tag{5.282}$$

$$C_{ei} = C_{si}\hat{S}_a(T_i, \zeta_i) \tag{5.283}$$

$$\rho_{ij} = \frac{8\sqrt{\zeta_i\zeta_j}\left[\zeta_i + \left(\dfrac{\omega_i}{\omega_j}\right)\zeta_j\right]\left(\dfrac{\omega_i}{\omega_j}\right)^{3/2}}{\left(1 - \dfrac{\omega_i^2}{\omega_j^2}\right)^2 + 4\zeta_i\zeta_j\left(\dfrac{\omega_i}{\omega_j}\right)\left(1 + \dfrac{\omega_i^2}{\omega_j^2}\right) + 4\left(\zeta_i^2 + \zeta_j^2\right)\left(\dfrac{\omega_i^2}{\omega_j^2}\right)}. \tag{5.284}$$

EXAMPLE 1   *Three Degrees of Freedom System*

Consider the 3DOF system discussed in Example 2 of Section 4.3, where the dynamic matrices, natural frequencies of vibration, and mode shapes are

$$\mathbf{M} = \begin{bmatrix} m & 0 & 0 \\ 0 & m & 0 \\ 0 & 0 & m \end{bmatrix}, \quad \mathbf{K} = \begin{bmatrix} 3k & -k & 0 \\ -k & 2k & -k \\ 0 & -k & k \end{bmatrix} \tag{5.285}$$

$$\omega_1^2 = \left(2 - \sqrt{3}\right)\omega_n^2 = 0.268\omega_n^2, \quad \omega_2^2 = 2\omega_n^2, \quad \omega_3^2 = \left(2 + \sqrt{3}\right)\omega_n^2 = 3.732\,\omega_n^2 \tag{5.286}$$

$$\boldsymbol{\phi}_1 = \begin{Bmatrix} 1 \\ 1+\sqrt{3} \\ 2+\sqrt{3} \end{Bmatrix} = \begin{Bmatrix} 1.000 \\ 2.732 \\ 3.732 \end{Bmatrix}, \quad \boldsymbol{\phi}_2 = \begin{Bmatrix} 1 \\ 1 \\ -1 \end{Bmatrix}, \quad \boldsymbol{\phi}_3 = \begin{Bmatrix} 1 \\ 1-\sqrt{3} \\ 2-\sqrt{3} \end{Bmatrix} = \begin{Bmatrix} 1.000 \\ -0.732 \\ 0.268 \end{Bmatrix}. \tag{5.287}$$

When the mode shapes are normalized with respect to the roof displacement, then it follows from Eq. 5.287 that

$$\boldsymbol{\phi}_{\text{roof}(1)} = \begin{Bmatrix} 2-\sqrt{3} \\ -1+\sqrt{3} \\ 1 \end{Bmatrix} = \begin{Bmatrix} 0.268 \\ 0.732 \\ 1.000 \end{Bmatrix}, \quad \boldsymbol{\phi}_{\text{roof}(2)} = \begin{Bmatrix} -1 \\ -1 \\ 1 \end{Bmatrix}, \quad \boldsymbol{\phi}_{\text{roof}(3)} = \begin{Bmatrix} 2+\sqrt{3} \\ -1-\sqrt{3} \\ 1 \end{Bmatrix} = \begin{Bmatrix} 3.732 \\ -2.732 \\ 1.000 \end{Bmatrix}. \tag{5.288}$$

Let $m = 1.0$ k-s²/in. and $k = 147.3$ k/in. It follows that

$$T_1 = \frac{2\pi}{\omega_1} = \frac{2\pi}{\sqrt{2-\sqrt{3}}}\sqrt{\frac{1.0}{147.3}} = 1.00 \text{ s} \tag{5.289}$$

$$T_2 = \frac{2\pi}{\omega_2} = \frac{2\pi}{\sqrt{2}}\sqrt{\frac{1.0}{147.3}} = 0.37 \text{ s} \tag{5.290}$$

$$T_3 = \frac{2\pi}{\omega_3} = \frac{2\pi}{\sqrt{2+\sqrt{3}}}\sqrt{\frac{1.0}{147.3}} = 0.27 \text{ s.} \tag{5.291}$$

Assume that this 3DOF system has 5% damping in all three modes of vibration and is subjected to an earthquake ground motion with a response spectrum shown in Figure 5.31. It follows from this figure that

$$S_a^{(1)} = S_a(T_n = T_1 = 1.00\,\text{s}, \zeta = 5\%) = 0.85g \tag{5.292}$$

$$S_a^{(2)} = S_a(T_n = T_2 = 0.37\,\text{s}, \zeta = 5\%) = 1.00g \tag{5.293}$$

$$S_a^{(3)} = S_a(T_n = T_3 = 0.27\,\text{s}, \zeta = 5\%) = 1.00g. \tag{5.294}$$

The corresponding spectral displacements, assuming that the earthquake response spectrum is equal to the pseudo acceleration response spectrum, are

$$S_d^{(1)} = \frac{S_a^{(1)}}{\omega_1^2} = \frac{0.85(386.4)}{(2\pi/1.00)^2} = 8.32 \text{ in.} \tag{5.295}$$

$$S_d^{(2)} = \frac{S_a^{(2)}}{\omega_2^2} = \frac{1.00(386.4)}{(2\pi/0.37)^2} = 1.34 \text{ in.} \tag{5.296}$$

$$S_d^{(3)} = \frac{S_a^{(3)}}{\omega_3^2} = \frac{1.00(386.4)}{(2\pi/0.27)^2} = 0.71 \text{ in.} \tag{5.297}$$

Example 3 in Section 5.4 calculated the fundamental modal participation factor, and Part 3 of Example 1 in Section 5.5 calculated second mode participation factor:

$$\Gamma_{\text{roof}(1)} = \frac{\begin{Bmatrix} 2-\sqrt{3} \\ -1+\sqrt{3} \\ 1 \end{Bmatrix}^T \begin{bmatrix} m & 0 & 0 \\ 0 & m & 0 \\ 0 & 0 & m \end{bmatrix} \begin{Bmatrix} 1 \\ 1 \\ 1 \end{Bmatrix}}{\begin{Bmatrix} 2-\sqrt{3} \\ -1+\sqrt{3} \\ 1 \end{Bmatrix}^T \begin{bmatrix} m & 0 & 0 \\ 0 & m & 0 \\ 0 & 0 & m \end{bmatrix} \begin{Bmatrix} 2-\sqrt{3} \\ -1+\sqrt{3} \\ 1 \end{Bmatrix}} = \frac{2+\sqrt{3}}{3} = 1.244 \tag{5.298}$$

$$\Gamma_{\text{roof}(2)} = \frac{\begin{Bmatrix} -1 \\ -1 \\ 1 \end{Bmatrix}^T \begin{bmatrix} m & 0 & 0 \\ 0 & m & 0 \\ 0 & 0 & m \end{bmatrix} \begin{Bmatrix} 1 \\ 1 \\ 1 \end{Bmatrix}}{\begin{Bmatrix} -1 \\ -1 \\ 1 \end{Bmatrix}^T \begin{bmatrix} m & 0 & 0 \\ 0 & m & 0 \\ 0 & 0 & m \end{bmatrix} \begin{Bmatrix} -1 \\ -1 \\ 1 \end{Bmatrix}} = -\frac{1}{3} = -0.333. \tag{5.299}$$

The third mode participation factor is

$$\Gamma_{\text{roof}(3)} = \frac{\left\{\begin{matrix} 2+\sqrt{3} \\ -1-\sqrt{3} \\ 1 \end{matrix}\right\}^T \begin{bmatrix} m & 0 & 0 \\ 0 & m & 0 \\ 0 & 0 & m \end{bmatrix} \left\{\begin{matrix} 1 \\ 1 \\ 1 \end{matrix}\right\}}{\left\{\begin{matrix} 2+\sqrt{3} \\ -1-\sqrt{3} \\ 1 \end{matrix}\right\}^T \begin{bmatrix} m & 0 & 0 \\ 0 & m & 0 \\ 0 & 0 & m \end{bmatrix} \left\{\begin{matrix} 2+\sqrt{3} \\ -1-\sqrt{3} \\ 1 \end{matrix}\right\}} = \frac{2-\sqrt{3}}{3} = 0.089. \tag{5.300}$$

The displacement responses in each floor for all three modes of vibration are

$$\max\left[\mathbf{X}^{(1)}\right] = \boldsymbol{\phi}_{\text{roof}(1)}\Gamma_{\text{roof}(1)}S_d^{(1)} = \left\{\begin{matrix} 2-\sqrt{3} \\ -1+\sqrt{3} \\ 1 \end{matrix}\right\}(1.244)(8.32) = \left\{\begin{matrix} 2.77 \\ 7.58 \\ 10.35 \end{matrix}\right\} \text{ in.} \tag{5.301}$$

$$\max\left[\mathbf{X}^{(2)}\right] = \boldsymbol{\phi}_{\text{roof}(2)}\Gamma_{\text{roof}(2)}S_d^{(2)} = \left\{\begin{matrix} -1 \\ -1 \\ 1 \end{matrix}\right\}(-0.333)(1.34) = \left\{\begin{matrix} 0.45 \\ 0.45 \\ -0.45 \end{matrix}\right\} \text{ in.} \tag{5.302}$$

$$\max\left[\mathbf{X}^{(3)}\right] = \boldsymbol{\phi}_{\text{roof}(3)}\Gamma_{\text{roof}(3)}S_d^{(3)} = \left\{\begin{matrix} 2+\sqrt{3} \\ -1-\sqrt{3} \\ 1 \end{matrix}\right\}(0.089)(0.71) = \left\{\begin{matrix} 0.24 \\ -0.17 \\ 0.06 \end{matrix}\right\} \text{ in.} \tag{5.303}$$

and therefore the relative displacements between floors are

$$\max\left[\Delta_{2-1}^{(1)}\right] = 7.58 - 2.77 = 4.81 \text{ in.}, \quad \max\left[\Delta_{3-2}^{(1)}\right] = 10.35 - 7.58 = 2.77 \text{ in.} \tag{5.304}$$

$$\max\left[\Delta_{2-1}^{(2)}\right] = 0.45 - 0.45 = 0.0 \text{ in.}, \quad \max\left[\Delta_{3-2}^{(2)}\right] = -0.45 - 0.45 = -0.90 \text{ in.} \tag{5.305}$$

$$\max\left[\Delta_{2-1}^{(3)}\right] = -0.17 - 0.24 = -0.41 \text{ in.}, \quad \max\left[\Delta_{3-2}^{(3)}\right] = 0.06 - (-0.17) = 0.23 \text{ in.} \tag{5.306}$$

The acceleration responses in each floor for all three modes of vibration are

$$\max\left[\ddot{\mathbf{Y}}^{(1)}\right] = \boldsymbol{\phi}_{\text{roof}(1)}\Gamma_{\text{roof}(1)}S_a^{(1)} = \left\{\begin{matrix} 2-\sqrt{3} \\ -1+\sqrt{3} \\ 1 \end{matrix}\right\}(1.244)(0.85g) = \left\{\begin{matrix} 0.28g \\ 0.77g \\ 1.06g \end{matrix}\right\} \tag{5.307}$$

$$\max\left[\ddot{\mathbf{Y}}^{(2)}\right] = \boldsymbol{\phi}_{\text{roof}(2)}\Gamma_{\text{roof}(2)}S_a^{(2)} = \left\{\begin{matrix} -1 \\ -1 \\ 1 \end{matrix}\right\}(-0.333)(1.00g) = \left\{\begin{matrix} 0.33g \\ 0.33g \\ -0.33g \end{matrix}\right\} \tag{5.308}$$

$$\max\left[\ddot{\mathbf{Y}}^{(3)}\right] = \boldsymbol{\phi}_{\text{roof}(3)}\Gamma_{\text{roof}(3)}S_a^{(3)} = \left\{\begin{matrix} 2+\sqrt{3} \\ -1-\sqrt{3} \\ 1 \end{matrix}\right\}(0.089)(1.00g) = \left\{\begin{matrix} 0.33g \\ -0.24g \\ 0.09g \end{matrix}\right\}. \tag{5.309}$$

The maximum seismic floor forces for each mode of vibration are computed by multiplying the floor acceleration by the mass of that floor:

$$\max\left[\mathbf{S}^{(1)}\right] = \mathbf{M}\left\{\max\left[\ddot{\mathbf{Y}}^{(1)}\right]\right\} = \begin{bmatrix} 1.0 & 0.0 & 0.0 \\ 0.0 & 1.0 & 0.0 \\ 0.0 & 0.0 & 1.0 \end{bmatrix}\left\{\begin{matrix} 0.28g \\ 0.77g \\ 1.06g \end{matrix}\right\} = \left\{\begin{matrix} 108 \\ 298 \\ 410 \end{matrix}\right\} \text{ kip} \tag{5.310}$$

$$\max\left[\mathbf{S}^{(2)}\right] = \mathbf{M}\left\{\max\left[\ddot{\mathbf{Y}}^{(2)}\right]\right\} = \begin{bmatrix} 1.0 & 0.0 & 0.0 \\ 0.0 & 1.0 & 0.0 \\ 0.0 & 0.0 & 1.0 \end{bmatrix} \begin{Bmatrix} 0.33g \\ 0.33g \\ -0.33g \end{Bmatrix} = \begin{Bmatrix} 128 \\ 128 \\ -128 \end{Bmatrix} \text{ kip} \tag{5.311}$$

$$\max\left[\mathbf{S}^{(3)}\right] = \mathbf{M}\left\{\max\left[\ddot{\mathbf{Y}}^{(3)}\right]\right\} = \begin{bmatrix} 1.0 & 0.0 & 0.0 \\ 0.0 & 1.0 & 0.0 \\ 0.0 & 0.0 & 1.0 \end{bmatrix} \begin{Bmatrix} 0.33g \\ -0.24g \\ 0.09g \end{Bmatrix} = \begin{Bmatrix} 128 \\ -93 \\ 35 \end{Bmatrix} \text{ kip.} \tag{5.312}$$

The maximum base shear force is the sum of the maximum seismic floor forces:

$$\max\left[V^{(1)}\right] = 108 + 298 + 410 = 816 \text{ kip} \tag{5.313}$$

$$\max\left[V^{(2)}\right] = 128 + 128 + (-128) = 128 \text{ kip} \tag{5.314}$$

$$\max\left[V^{(3)}\right] = 128 + (-93) + 35 = 70 \text{ kip.} \tag{5.315}$$

After the responses in each of the three modes are obtained, the total response can be calculated using the ABS method, SRSS method, or CQC method. All the signs will be dropped when any one of these three methods is used. The responses using the ABS method are

$$\max[\mathbf{X}]_{ABS} = \begin{Bmatrix} 2.77 \\ 7.58 \\ 10.35 \end{Bmatrix} + \begin{Bmatrix} 0.45 \\ 0.45 \\ 0.45 \end{Bmatrix} + \begin{Bmatrix} 0.24 \\ 0.17 \\ 0.06 \end{Bmatrix} = \begin{Bmatrix} 3.46 \\ 8.20 \\ 10.86 \end{Bmatrix} \text{ in.} \tag{5.316}$$

$$\max\left[\Delta_{2-1}\right]_{ABS} = 4.81 + 0.0 + 0.41 = 5.22 \text{ in.} \tag{5.317}$$

$$\max\left[\Delta_{3-2}\right]_{ABS} = 2.77 + 0.90 + 0.23 = 3.90 \text{ in.} \tag{5.318}$$

$$\max\left[\ddot{\mathbf{Y}}\right]_{ABS} = \begin{Bmatrix} 0.28g \\ 0.77g \\ 1.06g \end{Bmatrix} + \begin{Bmatrix} 0.33g \\ 0.33g \\ 0.33g \end{Bmatrix} + \begin{Bmatrix} 0.33g \\ 0.24g \\ 0.09g \end{Bmatrix} = \begin{Bmatrix} 0.94g \\ 1.34g \\ 1.48g \end{Bmatrix} \tag{5.319}$$

$$\max[\mathbf{S}]_{ABS} = \begin{Bmatrix} 108 \\ 298 \\ 410 \end{Bmatrix} + \begin{Bmatrix} 128 \\ 128 \\ 128 \end{Bmatrix} + \begin{Bmatrix} 128 \\ 93 \\ 35 \end{Bmatrix} = \begin{Bmatrix} 364 \\ 519 \\ 573 \end{Bmatrix} \text{ kip} \tag{5.320}$$

$$\max[V]_{ABS} = 816 + 128 + 70 = 1014 \text{ kip.} \tag{5.321}$$

The responses using the SRSS method are

$$\max[\mathbf{X}]_{SRSS} = \sqrt{\begin{Bmatrix} 2.77 \\ 7.58 \\ 10.35 \end{Bmatrix}^2 + \begin{Bmatrix} 0.45 \\ 0.45 \\ 0.45 \end{Bmatrix}^2 + \begin{Bmatrix} 0.24 \\ 0.17 \\ 0.06 \end{Bmatrix}^2} = \begin{Bmatrix} 2.82 \\ 7.60 \\ 10.36 \end{Bmatrix} \text{ in.} \tag{5.322}$$

$$\max\left[\Delta_{2-1}\right]_{SRSS} = \sqrt{(4.81)^2 + (0.0)^2 + (0.41)^2} = 4.83 \text{ in.} \tag{5.323}$$

$$\max\left[\Delta_{3-2}\right]_{SRSS} = \sqrt{(2.77)^2 + (0.90)^2 + (0.23)^2} = 2.92 \text{ in.} \tag{5.324}$$

$$\max\left[\ddot{\mathbf{Y}}\right]_{SRSS} = \sqrt{\begin{Bmatrix} 0.28g \\ 0.77g \\ 1.06g \end{Bmatrix}^2 + \begin{Bmatrix} 0.33g \\ 0.33g \\ 0.33g \end{Bmatrix}^2 + \begin{Bmatrix} 0.33g \\ 0.24g \\ 0.09g \end{Bmatrix}^2} = \begin{Bmatrix} 0.54g \\ 0.87g \\ 1.11g \end{Bmatrix} \tag{5.325}$$

$$\max[\mathbf{S}]_{SRSS} = \sqrt{\begin{Bmatrix} 108 \\ 298 \\ 410 \end{Bmatrix}^2 + \begin{Bmatrix} 128 \\ 128 \\ 128 \end{Bmatrix}^2 + \begin{Bmatrix} 128 \\ 93 \\ 35 \end{Bmatrix}^2} = \begin{Bmatrix} 211 \\ 337 \\ 431 \end{Bmatrix} \text{ kip} \tag{5.326}$$

$$\max[V]_{\text{SRSS}} = \sqrt{(816)^2 + (128)^2 + (70)^2} = 829 \text{ kip}. \tag{5.327}$$

Finally, for the CQC method, the coefficients for all ρ_{ij}'s must first be computed:

$$\frac{\omega_1}{\omega_2} = \frac{2-\sqrt{3}}{2} = 0.134, \quad \frac{\omega_1}{\omega_3} = \frac{2-\sqrt{3}}{2+\sqrt{3}} = 0.0718, \quad \frac{\omega_2}{\omega_3} = \frac{2}{2+\sqrt{3}} = 0.536$$

$\rho_{11} = \rho_{22} = \rho_{33} = 1.00.$

$$\rho_{12} = \rho_{21} = \frac{8\sqrt{(0.05)(0.05)}\left[0.05 + (0.134)(0.05)\right](0.134)^{3/2}}{\left[1-(0.134)^2\right]^2 + 4(0.05)^2(0.134)\left[1+(0.134)^2\right] + 4\left[2(0.05)^2\right](0.134)^2} = 0.00115$$

$$\rho_{13} = \rho_{31} = \frac{8\sqrt{(0.05)(0.05)}\left[0.05 + (0.0718)(0.05)\right](0.0718)^{3/2}}{\left[1-(0.0718)^2\right]^2 + 4(0.05)^2(0.0718)\left[1+(0.0718)^2\right] + 4\left[2(0.05)^2\right](0.0718)^2} = 0.00042$$

$$\rho_{23} = \rho_{32} = \frac{8\sqrt{(0.05)(0.05)}\left[0.05 + (0.536)(0.05)\right](0.536)^{3/2}}{\left[1-(0.536)^2\right]^2 + 4(0.05)^2(0.536)\left[1+(0.536)^2\right] + 4\left[2(0.05)^2\right](0.536)^2} = 0.02316.$$

The corresponding displacements for the CQC method are

$$\max[\mathbf{X}]_{\text{CQC}} = \sqrt{\sum_{j=1}^{3}\sum_{i=1}^{3}\rho_{ij}\mathbf{X}^{(i)}\mathbf{X}^{(j)}} = \begin{Bmatrix} 2.82 \\ 7.60 \\ 10.36 \end{Bmatrix} \text{ in.} \tag{5.328}$$

$$\max[\Delta_{2-1}]_{\text{CQC}} = \sqrt{\sum_{j=1}^{3}\sum_{i=1}^{3}\rho_{ij}\Delta_{2-1}^{(i)}\Delta_{2-1}^{(j)}} = 4.83 \text{ in.} \tag{5.329}$$

$$\max[\Delta_{3-2}]_{\text{CQC}} = \sqrt{\sum_{j=1}^{3}\sum_{i=1}^{3}\rho_{ij}\Delta_{3-2}^{(i)}\Delta_{3-2}^{(j)}} = 2.92 \text{ in.} \tag{5.330}$$

$$\max[\ddot{\mathbf{Y}}]_{\text{CQC}} = \sqrt{\sum_{j=1}^{3}\sum_{i=1}^{3}\rho_{ij}\mathbf{Y}^{(i)}\mathbf{Y}^{(j)}} = \begin{Bmatrix} 0.55g \\ 0.87g \\ 1.11g \end{Bmatrix} \tag{5.331}$$

$$\max[\mathbf{S}]_{\text{CQC}} = \sqrt{\sum_{j=1}^{3}\sum_{i=1}^{3}\rho_{ij}\mathbf{S}^{(i)}\mathbf{S}^{(j)}} = \begin{Bmatrix} 213 \\ 337 \\ 431 \end{Bmatrix} \text{ kip} \tag{5.332}$$

$$\max[V]_{\text{CQC}} = \sqrt{\sum_{j=1}^{3}\sum_{i=1}^{3}\rho_{ij}V^{(i)}V^{(j)}} = 829 \text{ kip}. \tag{5.333}$$

5.7   THREE-DIMENSIONAL MULTI–DEGREE OF FREEDOM RESPONSE

Consider a general three-dimensional structure with dynamic equilibrium equation of motion

$$\mathbf{M}\ddot{\mathbf{X}}(t) + \mathbf{C}\dot{\mathbf{X}}(t) + \mathbf{K}\mathbf{X}(t) = -\mathbf{M}a(t), \quad \mathbf{X}(0) = \mathbf{X}_o = \mathbf{0}, \quad \dot{\mathbf{X}}(0) = \dot{\mathbf{X}}_o = \mathbf{0} \tag{5.334}$$

where

$$\mathbf{a}(t) = \begin{Bmatrix} a_1(t) \\ a_2(t) \\ \vdots \\ a_n(t) \end{Bmatrix}. \tag{5.335}$$

A detailed discussion of the ground acceleration vector is presented in Section 6.1. This acceleration vector can be represented by each component of the ground acceleration vector, that is,

$$\begin{Bmatrix} a_1(t) \\ a_2(t) \\ \vdots \\ a_n(t) \end{Bmatrix} = \begin{bmatrix} \mathbf{I}_v \end{bmatrix} \begin{Bmatrix} a_x(t) \\ a_y(t) \\ a_z(t) \\ a_r(t) \\ a_s(t) \\ a_t(t) \end{Bmatrix} = \begin{bmatrix} \mathbf{I}_v \end{bmatrix} \mathbf{a}(t) = \begin{bmatrix} I_{11} & I_{12} & \cdots & I_{16} \\ I_{21} & I_{22} & \ddots & \vdots \\ \vdots & \ddots & \ddots & I_{n-1,6} \\ I_{n1} & \cdots & I_{n5} & I_{n6} \end{bmatrix} \begin{Bmatrix} a_x(t) \\ a_y(t) \\ a_z(t) \\ a_r(t) \\ a_s(t) \\ a_t(t) \end{Bmatrix} \tag{5.336}$$

where $[\mathbf{I}_v]$ is an $n \times 6$ matrix with entries of 0's and 1's, and each row of this matrix, denoted by $\mathbf{I}_j$, can be written as

$$\mathbf{I}_j = \begin{bmatrix} I_{j1} & I_{j2} & I_{j3} & I_{j4} & I_{j5} & I_{j6} \end{bmatrix} = \begin{cases} \begin{bmatrix} 1 & 0 & 0 & 0 & 0 & 0 \end{bmatrix} & \text{if} \quad a_j(t) = a_x(t) \\ \begin{bmatrix} 0 & 1 & 0 & 0 & 0 & 0 \end{bmatrix} & \text{if} \quad a_j(t) = a_y(t) \\ \begin{bmatrix} 0 & 0 & 1 & 0 & 0 & 0 \end{bmatrix} & \text{if} \quad a_j(t) = a_z(t) \\ \begin{bmatrix} 0 & 0 & 0 & 1 & 0 & 0 \end{bmatrix} & \text{if} \quad a_j(t) = a_r(t) \\ \begin{bmatrix} 0 & 0 & 0 & 0 & 1 & 0 \end{bmatrix} & \text{if} \quad a_j(t) = a_s(t) \\ \begin{bmatrix} 0 & 0 & 0 & 0 & 0 & 1 \end{bmatrix} & \text{if} \quad a_j(t) = a_t(t). \end{cases} \tag{5.337}$$

A more general discussion of the $[\mathbf{I}_v]$ matrix is presented in Section 6.2. It follows from Eq. 5.334 that

$$\mathbf{M}\ddot{\mathbf{X}}(t) + \mathbf{C}\dot{\mathbf{X}}(t) + \mathbf{K}\mathbf{X}(t) = -\mathbf{M}\begin{bmatrix} \mathbf{I}_v \end{bmatrix} \mathbf{a}_g(t). \tag{5.338}$$

Define

$$\mathbf{X}(t) = \mathbf{\Phi}\mathbf{Q}(t) \tag{5.339}$$

where

$$\mathbf{\Phi} = \begin{bmatrix} \mathbf{\phi}_1 & \mathbf{\phi}_2 & \cdots & \mathbf{\phi}_n \end{bmatrix} = \begin{bmatrix} \varphi_{11} & \varphi_{12} & \cdots & \varphi_{1n} \\ \varphi_{21} & \varphi_{22} & \ddots & \vdots \\ \vdots & \ddots & \ddots & \varphi_{n-1,n} \\ \varphi_{n1} & \cdots & \varphi_{n,n-1} & \varphi_{nn} \end{bmatrix}, \quad \mathbf{Q}(t) = \begin{Bmatrix} q_1(t) \\ q_2(t) \\ \vdots \\ q_n(t) \end{Bmatrix}. \tag{5.340}$$

Substituting Eq. 5.339 into Eq. 5.338, it follows that

$$\mathbf{M}\mathbf{\Phi}\ddot{\mathbf{Q}}(t) + \mathbf{C}\mathbf{\Phi}\dot{\mathbf{Q}}(t) + \mathbf{K}\mathbf{\Phi}\mathbf{Q}(t) = -\mathbf{M}\begin{bmatrix} \mathbf{I}_v \end{bmatrix} \mathbf{a}_g(t). \tag{5.341}$$

Premultiplying Eq. 5.341 by $\mathbf{\Phi}^T$ gives

$$\mathbf{\Phi}^T\mathbf{M}\mathbf{\Phi}\ddot{\mathbf{Q}}(t) + \mathbf{\Phi}^T\mathbf{C}\mathbf{\Phi}\dot{\mathbf{Q}}(t) + \mathbf{\Phi}^T\mathbf{K}\mathbf{\Phi}\mathbf{Q}(t) = -\mathbf{\Phi}^T\mathbf{M}\begin{bmatrix} \mathbf{I}_v \end{bmatrix} \mathbf{a}_g(t). \tag{5.342}$$

Using the orthogonal property of $\mathbf{M}$ and $\mathbf{K}$ matrices and the assumption of proportional damping, it follows that

$$\mathbf{\Phi}^T\mathbf{M}\mathbf{\Phi} = \begin{bmatrix} m_1^* & 0 & \cdots & 0 \\ 0 & m_2^* & \ddots & \vdots \\ \vdots & \ddots & \ddots & 0 \\ 0 & \cdots & 0 & m_n^* \end{bmatrix}, \quad \mathbf{\Phi}^T\mathbf{C}\mathbf{\Phi} = \begin{bmatrix} c_1^* & 0 & \cdots & 0 \\ 0 & c_2^* & \ddots & \vdots \\ \vdots & \ddots & \ddots & 0 \\ 0 & \cdots & 0 & c_n^* \end{bmatrix}, \quad \mathbf{\Phi}^T\mathbf{K}\mathbf{\Phi} = \begin{bmatrix} k_1^* & 0 & \cdots & 0 \\ 0 & k_2^* & \ddots & \vdots \\ \vdots & \ddots & \ddots & 0 \\ 0 & \cdots & 0 & k_n^* \end{bmatrix}. \quad (5.343)$$

It then follows from Eqs. 5.342 and 5.343 that

$$\begin{bmatrix} m_1^* & 0 & \cdots & 0 \\ 0 & m_2^* & \ddots & \vdots \\ \vdots & \ddots & \ddots & 0 \\ 0 & \cdots & 0 & m_n^* \end{bmatrix}\begin{Bmatrix} \ddot{q}_1(t) \\ \ddot{q}_2(t) \\ \vdots \\ \ddot{q}_n(t) \end{Bmatrix} + \begin{bmatrix} c_1^* & 0 & \cdots & 0 \\ 0 & c_2^* & \ddots & \vdots \\ \vdots & \ddots & \ddots & 0 \\ 0 & \cdots & 0 & c_n^* \end{bmatrix}\begin{Bmatrix} \dot{q}_1(t) \\ \dot{q}_2(t) \\ \vdots \\ \dot{q}_n(t) \end{Bmatrix} + \begin{bmatrix} k_1^* & 0 & \cdots & 0 \\ 0 & k_2^* & \ddots & \vdots \\ \vdots & \ddots & \ddots & 0 \\ 0 & \cdots & 0 & k_n^* \end{bmatrix}\begin{Bmatrix} q_1(t) \\ q_2(t) \\ \vdots \\ q_n(t) \end{Bmatrix}$$

$$= -\begin{bmatrix} \mathbf{\phi}_1 & \mathbf{\phi}_2 & \cdots & \mathbf{\phi}_n \end{bmatrix}^T \begin{bmatrix} m_{11} & m_{12} & \cdots & m_{1n} \\ m_{21} & m_{22} & \ddots & \vdots \\ \vdots & \ddots & \ddots & m_{n-1,n} \\ m_{n1} & \cdots & m_{n,n-1} & m_{nn} \end{bmatrix}\begin{bmatrix} \mathbf{I}_1 \\ \mathbf{I}_2 \\ \vdots \\ \mathbf{I}_n \end{bmatrix}\mathbf{a}_g(t). \quad (5.344)$$

Note that the transformed mass, damping, and stiffness matrices in Eq. 5.344 are all diagonal matrices. Therefore, Eq. 5.344 gives a set of n independent second-order linear differential equations of motion of the form

$$m_i^* \ddot{q}_i(t) + c_i^* \dot{q}_i(t) + k_i^* q_i(t) = -\mathbf{\phi}_i^T\mathbf{M}[\mathbf{I}_v]\mathbf{a}_g(t), \quad i = 1, 2, \ldots, n. \quad (5.345)$$

Let $\mathbf{\Gamma}_i$ be a 1×6 matrix, defined as

$$\mathbf{\Gamma}_i = \begin{bmatrix} \Gamma_{i1} & \Gamma_{i2} & \Gamma_{i3} & \Gamma_{i4} & \Gamma_{i5} & \Gamma_{i6} \end{bmatrix} = \frac{\mathbf{\phi}_i^T\mathbf{M}[\mathbf{I}_v]}{m_i^*} = \frac{\mathbf{\phi}_i^T\mathbf{M}[\mathbf{I}_v]}{\mathbf{\phi}_i^T\mathbf{M}\mathbf{\phi}_i}. \quad (5.346)$$

Then Eq. 5.345 becomes

$$m_i^* \ddot{q}_i(t) + c_i^* \dot{q}_i(t) + k_i^* q_i(t) = -\mathbf{\Gamma}_i \mathbf{a}_g(t)$$
$$= -\Gamma_{i1}a_x(t) - \Gamma_{i2}a_y(t) - \Gamma_{i3}a_z(t) - \Gamma_{i4}a_r(t) - \Gamma_{i5}a_s(t) - \Gamma_{i6}a_t(t). \quad (5.347)$$

The maximum value of $q_i(t)$ depends on the ith undamped natural period of vibration, T_i, and the damping in the ith mode, ζ_i. It follows in the same manner as discussed in the previous sections that

$$\max[q_i(t)] = \mathbf{\Gamma}_i \mathbf{S}_d(T_i, \zeta_i) \quad (5.348)$$

where

$$\mathbf{S}_d(T_i, \zeta_i) = \begin{Bmatrix} S_{dx}(T_i, \zeta_i) \\ S_{dy}(T_i, \zeta_i) \\ S_{dz}(T_i, \zeta_i) \\ S_{dr}(T_i, \zeta_i) \\ S_{ds}(T_i, \zeta_i) \\ S_{dt}(T_i, \zeta_i) \end{Bmatrix}. \quad (5.349)$$

Each of the six components of the earthquake ground motion time histories has its own response spectrum, which gives the six values in Eq. 5.349. In general, the maximum response due to one earthquake component may not be at the same time as the maximum response due to another earthquake component. Therefore, Eq. 5.348 represents the absolute maximum response due to the combination of all six earthquake components. It follows from Eq. 5.348 that the maximum floor displacement is

$$\max\left[\mathbf{X}^{(i)}(t)\right] = \boldsymbol{\phi}_i \max\left[q_i(t)\right] = \boldsymbol{\phi}_i\boldsymbol{\Gamma}_i\,\mathbf{S}_d(T_i,\zeta_i). \tag{5.350}$$

Similarly, the maximum floor acceleration of the structure can be computed using the spectral acceleration, that is,

$$\max\left[\ddot{\mathbf{Y}}_i(t)\right] = \boldsymbol{\phi}_i\boldsymbol{\Gamma}_i\,\mathbf{S}_a(T_i,\zeta_i) \tag{5.351}$$

where

$$\mathbf{S}_a(T_i,\zeta_i) = \begin{Bmatrix} S_{ax}(T_i,\zeta_i) \\ S_{ay}(T_i,\zeta_i) \\ S_{az}(T_i,\zeta_i) \\ S_{ar}(T_i,\zeta_i) \\ S_{as}(T_i,\zeta_i) \\ S_{at}(T_i,\zeta_i) \end{Bmatrix}. \tag{5.352}$$

Again, note that in general the maximum response due to one earthquake component may not be at the same time as the maximum response due to another earthquake component. Therefore, Eq. 5.351 represents the absolute maximum acceleration due to the combination of all six earthquake components.

Other responses can be computed using the method discussed in Section 5.4. The maximum relative displacement between floors p and q due to the ith mode response is

$$\max\left[\Delta x_{pq}^{(i)}(t)\right] = \max\left[\mathbf{T}_{p-q}^T\mathbf{X}^{(i)}(t)\right] = \left(\mathbf{T}_{p-q}^T\boldsymbol{\phi}_i\boldsymbol{\Gamma}_i\right)\mathbf{S}_d(T_i,\zeta_i). \tag{5.353}$$

The maximum seismic floor force due to the ith mode response is

$$\max\left[\mathbf{S}^{(i)}(t)\right] = \mathbf{M}\boldsymbol{\phi}_i\boldsymbol{\Gamma}_i\mathbf{S}_a(T_i,\zeta_i). \tag{5.354}$$

Finally, the maximum base shear in the six components is computed by summing all the story forces in the same directions. It follows from Eq. 5.354 that

$$\max\left[\mathbf{V}^{(i)}(t)\right] = \left[\mathbf{I}_v\right]^T\max\left[\mathbf{S}^{(i)}(t)\right] = \left[\mathbf{I}_v\right]^T\mathbf{M}\boldsymbol{\phi}_i\boldsymbol{\Gamma}_i\mathbf{S}_a(T_i,\zeta_i) = m_i^*\boldsymbol{\Gamma}_i^T\boldsymbol{\Gamma}_i\mathbf{S}_a(T_i,\zeta_i). \tag{5.355}$$

The maximum response for the combination of two or more modes can then be calculated using either the ABS, SRSS, or CQC method discussed in Section 5.6.

ACKNOWLEDGMENT

We would like to acknowledge the insight into the CQC method in Section 5.5 provided by Professor Joel Conte and the helpful review comments of Professor Jonathan Stewart..

REFERENCES

1. Uniform Building Code. International Congress of Building Officials, Whittier, CA, 1997.

2. *SEAOC Blue Book.* Structural Engineers Association of California, Whittier, CA, 1999.

3. ATC-3. Applied Technology Council, Redwood City, CA, 1976

4. N. Newmark and W. Hall. EERI Monograph: "Earthquake Spectra and Design." Earthquake Engineering Research Institute, 1982.

5. FEMA 273, Federal Emergency Management Agency, Washington, DC, 1997.

6. J. Wu. and R. D. Hanson "Study of Inelastic Spectra with High Damping." *Journal of the ASCE Structural Division,* Vol. 115, No. 6, pp. 1412–1431, 1989.

7. A. De Kiureghian, *CQC, SEAOC Blue Book,* Structural Engineers Association of California, Whittier, CA, 1999.

PROBLEMS

5.1 Consider a single degree of freedom system with an undamped period of vibration equal to 1.0 s. Assume that the system mass is $m = 1.0$ k-s^2/in. The spectral displacement for the system for an earthquake ground motion is 1.0 in. Calculate the corresponding pseudo spectral velocity and acceleration.

5.2 Calculate the base shear force for the system in Problem 5.1.

5.3 Consider an earthquake ground motion that produces a spectral acceleration (S_a) versus natural period of vibration (T_n) relationship given by

$$S_a = \begin{cases} S_o & 0.0 \leq T_n < T_o \text{ s} \\ S_o(T_o/T_n) & T_o \leq T_n \leq 3.0 \text{ s} \end{cases}$$

Let $S_o = 1.0g$ and $T_o = 1.0$ s. Plot S_a versus T_n. Calculate and plot the spectral displacement as a function of T_n. Assume that the spectral acceleration is equal to the pseudo spectral acceleration.

5.4 Repeat Problem 5.3 but for a spectral acceleration versus natural period of vibration relationship given by

$$S_a = \begin{cases} S_o & 0.0 \leq T_n < T_o \text{ s} \\ S_o(T_o/T_n)^2 & T_o \leq T_n \leq 3.0 \text{ s}. \end{cases}$$

with $S_o = 1.0g$ and $T_o = 1.0$ s.

5.5 Plot the bi-spectra for the earthquake ground motion given in Problem 5.3.

5.6 Plot the bi-spectra for the earthquake ground motion given in Problem 5.4.

5.7 Consider the system in Problem 5.1 with $m = 1.0$ k-s^2/in. For the earthquake in Problem 5.3, calculate the equation for and plot the base shear force (V) versus period of vibration (T_n).

5.8 Convert the bi-spectra plot in Problem 5.5 to a normalized earthquake force (V/W) versus T_n plot. (Hint: see Eq. 5.24.)

5.9 Consider the earthquake ground motion in Example 3 in Section 5.2. Consider the representational form for the spectral acceleration given in Problem 5.3. Use the fact that the spectral displacement at $T_n = 0.3$ s is equal to 2.36 in. to calculate the value for S_o. Assume $T_o > 0.6$ s.

5.10 Repeat Problem 5.9 but calculate the value of S_o using the value of spectral displacement equal to 4.79 in. at $T_n = 0.6$ s.

5.11 Assume that S_o in Problem 5.3 is equal to the average of the S_o values obtained in Problems 5.9 and 5.10. Assume that $T_o < 1.0$ s. Also, assume from Example 3 in Section 5.2 that the spectral displacement is 8.49 in. at $T_n = 1.0$ s. Calculate the value of T_o for the earthquake spectral acceleration representation in Problem 5.3.

5.12 Assume that a structure has a design life of 50 years. If the design earthquake has a return period of 100 years and the corresponding spectral acceleration is $1.5g$, then calculate the probability of the occurrence, p_o, of the design earthquake in the design life of the structure.

5.13 Consider Problem 5.12. Calculate the probability of occurrence of the design earthquake in a 20-year exposure time.

5.14 Assume that the spectral acceleration relationship in Problem 5.3 is for 5% damping. Use the Newmark–Hall damping response correction coefficient to calculate the corresponding spectral acceleration relationship for 10% damping.

5.15 Repeat Problem 5.14 but for 20% damping.

5.16 Repeat Problem 5.14 but use the response correction coefficient in Table 5.6.

5.17 Repeat Problem 5.14 but use Eq. 5.49.

5.18 Calculate the fundamental mode participation factor and the fundamental mode base shear coefficient for the system in Problem 3.1.

5.19 Repeat Problem 5.18 but for the system in Problem 3.3.

5.20 Repeat Example 2 in Section 5.4 but use the spectral acceleration relationship in Problem 5.3.

5.21 Repeat Problem 5.20 but use the system in Problem 3.1 and the results of Problem 5.18.

5.22 Repeat Problem 5.20 but use the system in Problem 3.3 and the results of Problem 5.19.

5.23 Calculate the second mode participation factor and the second mode earthquake base shear force for the system in Problem 3.1.

5.24 Repeat Problem 5.23 but for the system in Problem 3.3.

5.25 Repeat Example 4 in Section 5.5 but use the spectral acceleration relationship in Problem 5.3.

5.26 Repeat Problem 5.25 but use the system in Problem 3.1.

5.27 Repeat Problem 5.25 but use the system in Problem 3.3.

5.28 Repeat Example 5 in Section 5.5 but use the spectral acceleration relationship in Problem 5.3.

5.29 Repeat Problem 5.28 but use the system in Problem 3.1.

5.30 Repeat Problem 5.28 but use the system in Problem 3.3.

5.31 Repeat Example 4 in Section 5.5 but for a system with 10% damping in all modes. Modify Figure 5.31 using the Wu and Hanson damping response reduction coefficients in Eq. 5.49 and Table 5.7.

5.32 Repeat Problem 5.31 but for 5% damping in the fundamental mode and 10% damping in the second mode.

5.33 Repeat Problem 5.31 but for 20% damping in all modes.

5.34 Repeat Problem 5.31 but for Example 5 in Section 5.5.

5.35 Repeat Problem 5.32 but for Example 5 in Section 5.5.

5.36 Repeat Problem 5.33 but for Example 5 in Section 5.5.

5.37 Repeat Problem 5.18 but for the system in Problem 4.1 with $m = 1.0$ k-s^2/in. and $k = 10.0$ k/in.

5.38 Repeat Problem 5.20 but use the system in Problems 4.1 and 5.37.

5.39 Repeat Problem 5.23 but for the system in Problems 4.1 and 5.37 with $m = 1.0$ k-s^2/in. and $k = 10.0$ k/in.

5.40 Repeat Problem 5.25 but use the system in Problems 4.1, 5.37, and 5.39.

5.41 Repeat Problem 5.28 but use the system in Problems 4.1, 5.37, and 5.39.

5.42 Repeat Example 1 in Section 5.6 but use the spectral acceleration relationship in Problem 5.3.

5.43 Repeat Example 1 in Section 5.6 but with the bottom-story stiffness decreased from $k_1 = 2k$ to $k_1 = 1.5k$. Let $k = 147.3$ k/in.

Chapter 6

Linear Time History Analysis

6.1 OVERVIEW

Chapters 2 through 4 presented a complete development of the methods used to calculate the response of linear systems. This chapter presents a consolidation of many of the methods presented in Chapters 2 through 4 for the specific case of the structure being excited by an earthquake ground motion. It also presents in selected cases an extension of those methods where appropriate for earthquake applications.

Chapter 5 showed that the dynamic response of the structure to an earthquake ground motion can be calculated by replacing the forcing function described in Chapters 2 to 4 with the mass time the ground acceleration:

$$F(t) = -ma(t). \tag{6.1}$$

It therefore follows that

$$m\ddot{x}(t) + c\dot{x}(t) + kx(t) = -ma(t) \tag{6.2}$$

and in normalized form,

$$\ddot{x}(t) + 2\zeta\omega_n\dot{x}(t) + \omega_n^2 x(t) = -a(t). \tag{6.3}$$

For an n-DOF system, Eq. 6.2 becomes

$$\mathbf{M}\ddot{\mathbf{X}}(t) + \mathbf{C}\dot{\mathbf{X}}(t) + \mathbf{K}\mathbf{X}(t) = -\mathbf{M}\mathbf{a}(t) \tag{6.4}$$

where $\mathbf{a}(t)$ is a column vector containing the acceleration values, depending on the direction of the degree of freedom associated with the mass. In most cases, only horizontal translation of the earthquake ground motion is considered by structural engineers. Therefore, for a two-dimensional or planar representation of the structure, this corresponds to a single component of ground motion. However, for a three-dimensional structure, six components of acceleration are possible, which include $a_x(t)$, $a_y(t)$, $a_z(t)$, $a_r(t)$, $a_s(t)$, and $a_t(t)$, where x, y, and z denote translations in the x-, y-, and z-directions, respectively, and r, s, and t denote rotations in the x-, y-, and z-directions, respectively.

At the instant the earthquake occurs, the structure is generally assumed to be at rest. Therefore, the initial conditions of the structure for this case are

$$\mathbf{X}(0) = \mathbf{X}_o = \mathbf{0}, \quad \dot{\mathbf{X}}(0) = \dot{\mathbf{X}}_o = \mathbf{0}. \tag{6.5}$$

This chapter, unlike Chapter 5, considers the solution to Eq. 6.4 that retains the time variable.

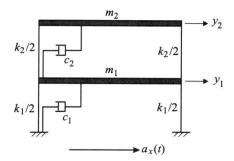

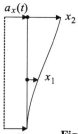

Figure 6.1 Two-Story Frame with Two Degrees of Freedom

EXAMPLE 1   *Two-Story Frame with Horizontal Ground Motion*

Consider a two-story frame as shown in Figure 6.1 with two degrees of freedom in the x-direction. The dynamic equilibrium equation of motion is

$$\begin{bmatrix} m_1 & 0 \\ 0 & m_2 \end{bmatrix} \begin{Bmatrix} \ddot{y}_1(t) \\ \ddot{y}_2(t) \end{Bmatrix} + \begin{bmatrix} c_1 + c_2 & -c_2 \\ -c_2 & c_2 \end{bmatrix} \begin{Bmatrix} \dot{x}_1(t) \\ \dot{x}_2(t) \end{Bmatrix} + \begin{bmatrix} k_1 + k_2 & -k_2 \\ -k_2 & k_2 \end{bmatrix} \begin{Bmatrix} x_1(t) \\ x_2(t) \end{Bmatrix} = \begin{Bmatrix} 0 \\ 0 \end{Bmatrix}. \tag{6.6}$$

Recall that

$$\begin{Bmatrix} \ddot{y}_1(t) \\ \ddot{y}_2(t) \end{Bmatrix} = \begin{Bmatrix} \ddot{x}_1(t) + a_x(t) \\ \ddot{x}_2(t) + a_x(t) \end{Bmatrix} = \begin{Bmatrix} \ddot{x}_1(t) \\ \ddot{x}_2(t) \end{Bmatrix} + \begin{Bmatrix} 1 \\ 1 \end{Bmatrix} a_x(t). \tag{6.7}$$

Therefore, Eq. 6.6 becomes

$$\begin{bmatrix} m_1 & 0 \\ 0 & m_2 \end{bmatrix} \begin{Bmatrix} \ddot{x}_1(t) \\ \ddot{x}_2(t) \end{Bmatrix} + \begin{bmatrix} c_{11} & c_{12} \\ c_{21} & c_{22} \end{bmatrix} \begin{Bmatrix} \dot{x}_1(t) \\ \dot{x}_2(t) \end{Bmatrix} + \begin{bmatrix} k_{11} & k_{12} \\ k_{21} & k_{22} \end{bmatrix} \begin{Bmatrix} x_1(t) \\ x_2(t) \end{Bmatrix} = -\begin{bmatrix} m_1 & 0 \\ 0 & m_2 \end{bmatrix} \begin{Bmatrix} 1 \\ 1 \end{Bmatrix} a_x(t) \tag{6.8}$$

or

$$\mathbf{M}\ddot{\mathbf{X}}(t) + \mathbf{C}\dot{\mathbf{X}}(t) + \mathbf{K}\mathbf{X}(t) = -\mathbf{M}\{\mathbf{I}\}a_x(t) \tag{6.9}$$

where

$$\{\mathbf{I}\} = \begin{Bmatrix} 1 \\ 1 \end{Bmatrix}. \tag{6.10}$$

6.2   EARTHQUAKE GROUND MOTION

The motion of a point in space can be described in terms of six components, which include three translations and three rotations. Similarly, the earthquake ground motion at the location of the structure can be expressed in terms of six components. Using the notations as in Section 6.1, these six components are $a_x(t)$, $a_y(t)$, $a_z(t)$, $a_r(t)$, $a_s(t)$, and $a_t(t)$, where x, y, and z denote translations in the x-, y-, and z-directions, respectively, and r, s, and t denote rotations in the x-, y-, and z-directions, respectively. For presentation in this book, the earthquake ground accelerations are as follows:

$a_x(t)$ and $a_y(t)$:   Translational accelerations on the plane of the earth (horizontal ground motion).   (6.11a)

$a_z(t)$:   Translational acceleration out of the plane of the earth (vertical ground motion).   (6.11b)

$a_r(t)$ and $a_s(t)$: Rotational accelerations that relate to the earth rocking.   (6.11c)

$\qquad a_t(t)$: Rotational acceleration that relates to the twisting on the earth's surface.   (6.11d)

Example 1 of Section 6.1 presented a two-dimensional structure with earthquake ground motion only in the x-direction. The result is that the equation of motion consists of the identity vector $\{\mathbf{I}\}$, which is a column vector with 1's in all entries. In this section, the identity vector will be extended to an $n \times 6$ vector whose entries contain only 0's and 1's.

Consider the MDOF system with the dynamic equilibrium equation of motion

$$\mathbf{M\ddot{X}}(t) + \mathbf{C\dot{X}}(t) + \mathbf{KX}(t) = -\mathbf{Ma}(t), \quad \mathbf{X}(0) = \mathbf{X}_o = \mathbf{0}, \quad \mathbf{\dot{X}}(0) = \mathbf{\dot{X}}_o = \mathbf{0} \qquad (6.12)$$

where

$$\mathbf{a}(t) = \begin{Bmatrix} a_1(t) \\ a_2(t) \\ \vdots \\ a_n(t) \end{Bmatrix}. \qquad (6.13)$$

Each $a_i(t)$ is an acceleration component as described by Eq. 6.11 and depends on the direction of the ith degree of freedom. The acceleration vector in Eq. 6.13 can be expressed in terms of the six components of the ground acceleration in Eq. 6.11 as

$$\begin{Bmatrix} a_1(t) \\ a_2(t) \\ \vdots \\ a_n(t) \end{Bmatrix} = \begin{bmatrix} \mathbf{I}_v \end{bmatrix} \begin{Bmatrix} a_x(t) \\ a_y(t) \\ a_z(t) \\ a_r(t) \\ a_s(t) \\ a_t(t) \end{Bmatrix} = \begin{bmatrix} \mathbf{I}_v \end{bmatrix} \mathbf{a}_g(t) = \begin{bmatrix} I_{11} & I_{12} & \cdots & I_{16} \\ I_{21} & I_{22} & \ddots & \vdots \\ \vdots & \ddots & \ddots & I_{n-1,6} \\ I_{n1} & \cdots & I_{n5} & I_{n6} \end{bmatrix} \begin{Bmatrix} a_x(t) \\ a_y(t) \\ a_z(t) \\ a_r(t) \\ a_s(t) \\ a_t(t) \end{Bmatrix} \qquad (6.14)$$

where $[\mathbf{I}_v]$ is an $n \times 6$ matrix with entries of 0's and 1's, and each row of this matrix, denoted by $\mathbf{I}_j$, can be written as

$$\mathbf{I}_j = \begin{bmatrix} I_{j1} & I_{j2} & I_{j3} & I_{j4} & I_{j5} & I_{j6} \end{bmatrix} = \begin{cases} \begin{bmatrix} 1 & 0 & 0 & 0 & 0 & 0 \end{bmatrix} & \text{if } a_j(t) = a_x(t) \\ \begin{bmatrix} 0 & 1 & 0 & 0 & 0 & 0 \end{bmatrix} & \text{if } a_j(t) = a_y(t) \\ \begin{bmatrix} 0 & 0 & 1 & 0 & 0 & 0 \end{bmatrix} & \text{if } a_j(t) = a_z(t) \\ \begin{bmatrix} 0 & 0 & 0 & 1 & 0 & 0 \end{bmatrix} & \text{if } a_j(t) = a_r(t) \\ \begin{bmatrix} 0 & 0 & 0 & 0 & 1 & 0 \end{bmatrix} & \text{if } a_j(t) = a_s(t) \\ \begin{bmatrix} 0 & 0 & 0 & 0 & 0 & 1 \end{bmatrix} & \text{if } a_j(t) = a_t(t). \end{cases} \qquad (6.15)$$

When the six components of earthquake accelerations are considered, $[\mathbf{I}_v]$ is an $n \times 6$ matrix. If only three earthquake translations are used, then $[\mathbf{I}_v]$ will be an $n \times 3$ matrix. The following examples discuss the formulation of this $[\mathbf{I}_v]$ matrix.

EXAMPLE 1 *Two-Story Frame with Both Horizontal and Vertical Degrees of Freedom*

Consider the same two-dimensional two-story frame in Example 1 of Section 6.1, but this time with the additional four degrees of freedom shown in Figure 6.2. With the degrees of freedom labeled from 1 to 6 as shown in Figure 6.2a, where the horizontal degrees of freedom are labeled 1 and 2 and the vertical degrees of freedom are labeled 3 to 6, the ground acceleration vector becomes

$$\mathbf{a}(t) = \begin{Bmatrix} a_x(t) \\ a_x(t) \\ a_z(t) \\ a_z(t) \\ a_z(t) \\ a_z(t) \end{Bmatrix} = \begin{Bmatrix} \mathbf{a}_x(t) \\ \mathbf{a}_z(t) \end{Bmatrix}. \tag{6.16}$$

The alternative form of Eq. 6.16 is

$$\mathbf{a}(t) = \begin{Bmatrix} a_x(t) \\ a_x(t) \\ a_z(t) \\ a_z(t) \\ a_z(t) \\ a_z(t) \end{Bmatrix} = \begin{bmatrix} 1 & 0 \\ 1 & 0 \\ 0 & 1 \\ 0 & 1 \\ 0 & 1 \\ 0 & 1 \end{bmatrix} \begin{Bmatrix} a_x(t) \\ a_z(t) \end{Bmatrix} = \begin{bmatrix} \mathbf{I}_v \end{bmatrix} \begin{Bmatrix} a_x(t) \\ a_z(t) \end{Bmatrix}. \tag{6.17}$$

Note that in Eq. 6.16, the accelerations in the x- and z-directions are vectors, where each component of acceleration can represent many degrees of freedom in the same direction. In Eq. 6.17, the acceleration vectors on the right-hand side are the accelerations in the x- and z-directions, which are scalar quantities.

In addition to the labeling shown in Figure 6.2a, these degrees of freedom can be labeled according to Figure 6.2b. When the horizontal degrees of freedom are labeled 1 and 4 and the vertical degrees of freedom are labeled 2, 3, 5, and 6, then the ground acceleration vector becomes

$$\mathbf{a}(t) = \begin{Bmatrix} a_x(t) \\ a_z(t) \\ a_z(t) \\ a_x(t) \\ a_z(t) \\ a_z(t) \end{Bmatrix} = \begin{bmatrix} 1 & 0 \\ 0 & 1 \\ 0 & 1 \\ 1 & 0 \\ 0 & 1 \\ 0 & 1 \end{bmatrix} \begin{Bmatrix} a_x(t) \\ a_z(t) \end{Bmatrix} = \begin{bmatrix} \mathbf{I}_v \end{bmatrix} \begin{Bmatrix} a_x(t) \\ a_z(t) \end{Bmatrix}. \tag{6.18}$$

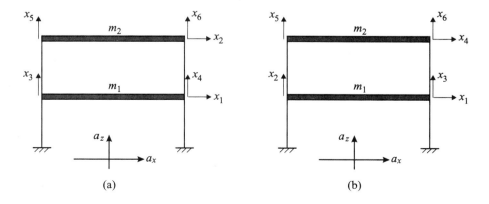

Figure 6.2 Two-Story Frame with Six Degrees of Freedom

EXAMPLE 2 *General Two-Story Frame*

Consider the same two-dimensional two-story frame in Example 1, but this time with additional four rotational degrees of freedom, as shown in Figure 6.3. With the degrees of freedom labeled from 1 to 10 as shown in Figure 6.3a, the ground acceleration vector becomes

$$
\mathbf{a}(t) = \begin{Bmatrix} a_x(t) \\ a_x(t) \\ a_z(t) \\ a_z(t) \\ a_z(t) \\ a_z(t) \\ a_s(t) \\ a_s(t) \\ a_s(t) \\ a_s(t) \end{Bmatrix} = \begin{Bmatrix} \mathbf{a}_x(t) \\ \mathbf{a}_z(t) \\ \mathbf{a}_s(t) \end{Bmatrix} = \begin{bmatrix} 1 & 0 & 0 \\ 1 & 0 & 0 \\ 0 & 1 & 0 \\ 0 & 1 & 0 \\ 0 & 1 & 0 \\ 0 & 1 & 0 \\ 0 & 0 & 1 \\ 0 & 0 & 1 \\ 0 & 0 & 1 \\ 0 & 0 & 1 \end{bmatrix} \begin{Bmatrix} a_x(t) \\ a_z(t) \\ a_s(t) \end{Bmatrix} = \begin{bmatrix} \mathbf{I}_v \end{bmatrix} \begin{Bmatrix} a_x(t) \\ a_z(t) \\ a_s(t) \end{Bmatrix}.
$$ (6.19)

For the labeling scheme shown in Figure 6.3b, the ground acceleration vector is

$$
\mathbf{a}(t) = \begin{Bmatrix} a_x(t) \\ a_x(t) \\ a_z(t) \\ a_z(t) \\ a_z(t) \\ a_z(t) \\ a_s(t) \\ a_s(t) \\ a_s(t) \\ a_s(t) \end{Bmatrix} = \begin{Bmatrix} \mathbf{a}_x(t) \\ \mathbf{a}_z(t) \\ \mathbf{a}_s(t) \end{Bmatrix} = \begin{bmatrix} 1 & 0 & 0 \\ 0 & 1 & 0 \\ 0 & 0 & 1 \\ 0 & 1 & 0 \\ 0 & 0 & 1 \\ 1 & 0 & 0 \\ 0 & 1 & 0 \\ 0 & 0 & 1 \\ 0 & 1 & 0 \\ 0 & 0 & 1 \end{bmatrix} \begin{Bmatrix} a_x(t) \\ a_z(t) \\ a_s(t) \end{Bmatrix} = \begin{bmatrix} \mathbf{I}_v \end{bmatrix} \begin{Bmatrix} a_x(t) \\ a_z(t) \\ a_s(t) \end{Bmatrix}.
$$ (6.20)

EXAMPLE 3 *Three-Dimensional Two-Story Frame with Rigid Diaphragm*

Consider the three-dimensional two-story frame as shown in Figure 6.4, where the diaphragm is assumed to be rigid and therefore there will be three degrees of freedom per

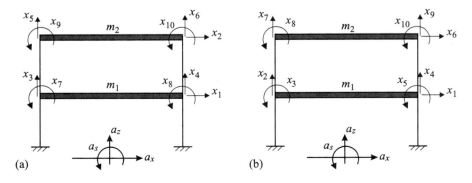

Figure 6.3 Two-Story Frame with 10 Degrees of Freedom

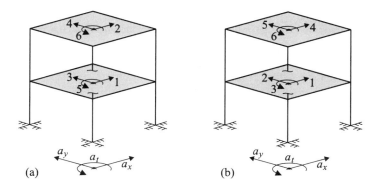

Figure 6.4 Three-Dimensional Two-Story Frame with Six Degrees of Freedom

floor. With the degrees of freedom labeled from 1 to 6 as shown in Figure 6.4a, the ground acceleration vector becomes

$$\mathbf{a}(t) = \begin{Bmatrix} a_x(t) \\ a_x(t) \\ a_y(t) \\ a_y(t) \\ a_t(t) \\ a_t(t) \end{Bmatrix} = \begin{Bmatrix} \mathbf{a}_x(t) \\ \mathbf{a}_y(t) \\ \mathbf{a}_t(t) \end{Bmatrix} = \begin{bmatrix} 1 & 0 & 0 \\ 1 & 0 & 0 \\ 0 & 1 & 0 \\ 0 & 1 & 0 \\ 0 & 0 & 1 \\ 0 & 0 & 1 \end{bmatrix} \begin{Bmatrix} a_x(t) \\ a_y(t) \\ a_t(t) \end{Bmatrix} = \begin{bmatrix} \mathbf{I}_v \end{bmatrix} \begin{Bmatrix} a_x(t) \\ a_y(t) \\ a_t(t) \end{Bmatrix}. \tag{6.21}$$

On the other hand, if the degrees of freedom are labeled in the directions as shown in Figure 6.4b, then the ground acceleration vector becomes

$$\mathbf{a}(t) = \begin{Bmatrix} a_x(t) \\ a_y(t) \\ a_t(t) \\ a_x(t) \\ a_y(t) \\ a_t(t) \end{Bmatrix} = \begin{bmatrix} 1 & 0 & 0 \\ 0 & 1 & 0 \\ 0 & 0 & 1 \\ 1 & 0 & 0 \\ 0 & 1 & 0 \\ 0 & 0 & 1 \end{bmatrix} \begin{Bmatrix} a_x(t) \\ a_y(t) \\ a_t(t) \end{Bmatrix} = \begin{bmatrix} \mathbf{I}_v \end{bmatrix} \begin{Bmatrix} a_x(t) \\ a_y(t) \\ a_t(t) \end{Bmatrix}. \tag{6.22}$$

6.3 NORMAL MODE METHOD WITH PROPORTIONAL DAMPING

Consider the MDOF system with the dynamic equilibrium equation of motion given in Eq. 6.12:

$$\mathbf{M}\ddot{\mathbf{X}}(t) + \mathbf{C}\dot{\mathbf{X}}(t) + \mathbf{K}\mathbf{X}(t) = -\mathbf{M}\mathbf{a}(t), \quad \mathbf{X}(0) = \mathbf{X}_o = \mathbf{0}, \quad \dot{\mathbf{X}}(0) = \dot{\mathbf{X}}_o = \mathbf{0}. \tag{6.23}$$

Structural engineers very often assume that the damping in the structure is such that the damping matrix is proportional, of the Rayleigh, Caughy, or Penzien–Wilson type. Very often this assumption is made based more on numerical computational convenience than reality, because if the damping is proportional, then the normal mode solution method can be used to calculate the response of the structure. This section presents such a normal mode solution procedure, and the next section presents the solution method for nonproportional damping.

The response of structures with proportional damping can be calculated using the normal mode method discussed in Sections 4.6 to 4.8. As previously noted, a very important benefit of using the normal mode method is that modes that contribute negligibly to the structural response of interest can be assumed to be zero with very little loss in accuracy. This is referred to as normal mode truncation. This assumption can significantly reduce the computer calculation time and also often simplifies the response data analysis phase of a structural engineering project. Normal mode truncation also often enables the structural engineer to derive more engineering insight from the response solution than is possible using the numerical method that is discussed in Section 6.4.

Recall that the natural frequencies are numbered in ascending magnitude as $\omega_1, \omega_2, \ldots, \omega_n$, and mode shapes, or normalized eigenvectors, are $\hat{\boldsymbol{\phi}}_1, \hat{\boldsymbol{\phi}}_2, \ldots, \hat{\boldsymbol{\phi}}_n$. The modal matrix (or eigenvector matrix) is

$$\hat{\boldsymbol{\Phi}} = \begin{bmatrix} \hat{\boldsymbol{\phi}}_1 & \hat{\boldsymbol{\phi}}_2 & \cdots & \hat{\boldsymbol{\phi}}_n \end{bmatrix} = \begin{bmatrix} \hat{\varphi}_{11} & \hat{\varphi}_{12} & \cdots & & \hat{\varphi}_{1n} \\ \hat{\varphi}_{21} & \hat{\varphi}_{22} & & \ddots & \vdots \\ \vdots & & \ddots & & \\ & & & & \hat{\varphi}_{n-1,n} \\ \hat{\varphi}_{n1} & & \cdots & \hat{\varphi}_{n,n-1} & \hat{\varphi}_{nn} \end{bmatrix}. \tag{6.24}$$

Note from Section 4.4 that the mode shapes, denoted $\hat{\boldsymbol{\phi}}_i$, are normalized such that their norms are equal to 1. The coordinate transformation from the physical $\mathbf{X}$ coordinates to the normal mode coordinate, $\mathbf{Q}$, is defined as

$$\mathbf{X}(t) = \hat{\boldsymbol{\Phi}}\mathbf{Q}(t) \tag{6.25}$$

where

$$\mathbf{Q}(t) = \begin{Bmatrix} q_1(t) & q_2(t) & \cdots & q_n(t) \end{Bmatrix}^T. \tag{6.26}$$

It follows from Eq. 6.23 that

$$\mathbf{M}\hat{\boldsymbol{\Phi}}\ddot{\mathbf{Q}}(t) + \mathbf{C}\hat{\boldsymbol{\Phi}}\dot{\mathbf{Q}}(t) + \mathbf{K}\hat{\boldsymbol{\Phi}}\mathbf{Q}(t) = -\mathbf{M}\mathbf{a}(t). \tag{6.27}$$

Premultiplying Eq. 6.27 by $\hat{\boldsymbol{\Phi}}^T$ gives

$$\hat{\boldsymbol{\Phi}}^T\mathbf{M}\hat{\boldsymbol{\Phi}}\ddot{\mathbf{Q}}(t) + \hat{\boldsymbol{\Phi}}^T\mathbf{C}\hat{\boldsymbol{\Phi}}\dot{\mathbf{Q}}(t) + \hat{\boldsymbol{\Phi}}^T\mathbf{K}\hat{\boldsymbol{\Phi}}\mathbf{Q}(t) = -\hat{\boldsymbol{\Phi}}^T\mathbf{M}\mathbf{a}(t). \tag{6.28}$$

Using the orthogonal property of $\mathbf{M}$ and $\mathbf{K}$ matrices and the assumption of proportional damping, it follows that

$$\hat{\boldsymbol{\Phi}}^T\mathbf{M}\hat{\boldsymbol{\Phi}} = \begin{bmatrix} \hat{m}_1 & 0 & \cdots & 0 \\ 0 & \hat{m}_2 & \ddots & \vdots \\ \vdots & \ddots & \ddots & 0 \\ 0 & \cdots & 0 & \hat{m}_n \end{bmatrix}, \quad \hat{\boldsymbol{\Phi}}^T\mathbf{C}\hat{\boldsymbol{\Phi}} = \begin{bmatrix} \hat{c}_1 & 0 & \cdots & 0 \\ 0 & \hat{c}_2 & \ddots & \vdots \\ \vdots & \ddots & \ddots & 0 \\ 0 & \cdots & 0 & \hat{c}_n \end{bmatrix}, \quad \hat{\boldsymbol{\Phi}}^T\mathbf{K}\hat{\boldsymbol{\Phi}} = \begin{bmatrix} \hat{k}_1 & 0 & \cdots & 0 \\ 0 & \hat{k}_2 & \ddots & \vdots \\ \vdots & \ddots & \ddots & 0 \\ 0 & \cdots & 0 & \hat{k}_n \end{bmatrix}. \tag{6.29}$$

It then follows from Eqs. 6.28 and 6.29 that

$$\begin{bmatrix} \hat{m}_1 & 0 & \cdots & 0 \\ 0 & \hat{m}_2 & \ddots & \vdots \\ \vdots & \ddots & \ddots & 0 \\ 0 & \cdots & 0 & \hat{m}_n \end{bmatrix} \begin{Bmatrix} \ddot{q}_1(t) \\ \ddot{q}_2(t) \\ \vdots \\ \ddot{q}_n(t) \end{Bmatrix} + \begin{bmatrix} \hat{c}_1 & 0 & \cdots & 0 \\ 0 & \hat{c}_2 & \ddots & \vdots \\ \vdots & \ddots & \ddots & 0 \\ 0 & \cdots & 0 & \hat{c}_n \end{bmatrix} \begin{Bmatrix} \dot{q}_1(t) \\ \dot{q}_2(t) \\ \vdots \\ \dot{q}_n(t) \end{Bmatrix} + \begin{bmatrix} \hat{k}_1 & 0 & \cdots & 0 \\ 0 & \hat{k}_2 & \ddots & \vdots \\ \vdots & \ddots & \ddots & 0 \\ 0 & \cdots & 0 & \hat{k}_n \end{bmatrix} \begin{Bmatrix} q_1(t) \\ q_2(t) \\ \vdots \\ q_n(t) \end{Bmatrix}$$

$$= -\begin{bmatrix} \hat{\boldsymbol{\phi}}_1 & \hat{\boldsymbol{\phi}}_2 & \cdots & \hat{\boldsymbol{\phi}}_n \end{bmatrix}^T \begin{bmatrix} m_{11} & m_{12} & \cdots & m_{1n} \\ m_{21} & m_{22} & & \vdots \\ \vdots & & \ddots & m_{n-1,n} \\ m_{n1} & \cdots & m_{n,n-1} & m_{nn} \end{bmatrix} \begin{Bmatrix} a_1(t) \\ a_2(t) \\ \vdots \\ a_n(t) \end{Bmatrix}. \tag{6.30}$$

Using the notations discussed in Eq. 6.14, the acceleration vector in Eq. 6.30 can be written as

$$
\begin{Bmatrix} a_1(t) \\ a_2(t) \\ \vdots \\ a_n(t) \end{Bmatrix} = [\mathbf{I}_v] \begin{Bmatrix} a_x(t) \\ a_y(t) \\ a_z(t) \\ a_r(t) \\ a_s(t) \\ a_t(t) \end{Bmatrix} = [\mathbf{I}_v] \mathbf{a}_g(t) = \begin{bmatrix} I_{11} & I_{12} & \cdots & I_{16} \\ I_{21} & I_{22} & \ddots & \vdots \\ \vdots & \ddots & \ddots & I_{n-1,6} \\ I_{n1} & \cdots & I_{n5} & I_{n6} \end{bmatrix} \begin{Bmatrix} a_x(t) \\ a_y(t) \\ a_z(t) \\ a_r(t) \\ a_s(t) \\ a_t(t) \end{Bmatrix}
$$

(6.31)

where $[\mathbf{I}_v]$ is an $n \times 6$ matrix with entries of 0's and 1's, and each row of this matrix $\mathbf{I}_j$ is given in Eq. 6.15:

$$
\mathbf{I}_j = \begin{bmatrix} I_{j1} & I_{j2} & I_{j3} & I_{j4} & I_{j5} & I_{j6} \end{bmatrix} = \begin{cases} [1 & 0 & 0 & 0 & 0 & 0] & a_j = a_x \\ [0 & 1 & 0 & 0 & 0 & 0] & a_j = a_y \\ [0 & 0 & 1 & 0 & 0 & 0] & a_j = a_z \\ [0 & 0 & 0 & 1 & 0 & 0] & a_j = a_r \\ [0 & 0 & 0 & 0 & 1 & 0] & a_j = a_s \\ [0 & 0 & 0 & 0 & 0 & 1] & a_j = a_t. \end{cases}
$$

(6.32)

Substituting Eq. 6.31 into Eq. 6.30 gives

$$
\begin{bmatrix} \hat{m}_1 & 0 & \cdots & 0 \\ 0 & \hat{m}_2 & \ddots & \vdots \\ \vdots & \ddots & \ddots & 0 \\ 0 & \cdots & 0 & \hat{m}_n \end{bmatrix} \begin{Bmatrix} \ddot{q}_1(t) \\ \ddot{q}_2(t) \\ \vdots \\ \ddot{q}_n(t) \end{Bmatrix} + \begin{bmatrix} \hat{c}_1 & 0 & \cdots & 0 \\ 0 & \hat{c}_2 & \ddots & \vdots \\ \vdots & \ddots & \ddots & 0 \\ 0 & \cdots & 0 & \hat{c}_n \end{bmatrix} \begin{Bmatrix} \dot{q}_1(t) \\ \dot{q}_2(t) \\ \vdots \\ \dot{q}_n(t) \end{Bmatrix} + \begin{bmatrix} \hat{k}_1 & 0 & \cdots & 0 \\ 0 & \hat{k}_2 & \ddots & \vdots \\ \vdots & \ddots & \ddots & 0 \\ 0 & \cdots & 0 & \hat{k}_n \end{bmatrix} \begin{Bmatrix} q_1(t) \\ q_2(t) \\ \vdots \\ q_n(t) \end{Bmatrix}
$$

$$
= -\begin{bmatrix} \hat{\boldsymbol{\phi}}_1 & \hat{\boldsymbol{\phi}}_2 & \cdots & \hat{\boldsymbol{\phi}}_n \end{bmatrix}^T \begin{bmatrix} m_{11} & m_{12} & \cdots & m_{1n} \\ m_{21} & m_{22} & \ddots & \vdots \\ \vdots & \ddots & \ddots & m_{n-1,n} \\ m_{n1} & \cdots & m_{n,n-1} & m_{nn} \end{bmatrix} \begin{bmatrix} \mathbf{I}_1 \\ \mathbf{I}_2 \\ \vdots \\ \mathbf{I}_n \end{bmatrix} \mathbf{a}_g(t).
$$

(6.33)

Note that the transformed mass, damping, and stiffness matrices in Eq. 6.30 or Eq. 6.33 are all diagonal matrices. Therefore, Eq. 6.33 gives a set of n independent second-order linear differential equations of motion of the form

$$
\hat{m}_i \ddot{q}_i(t) + \hat{c}_i \dot{q}_i(t) + \hat{k}_i q_i(t) = -\sum_{j=1}^{n} \sum_{k=1}^{n} \hat{\phi}_{ji} m_{jk} \mathbf{I}_k \mathbf{a}_g(t), \quad i = 1, 2, \ldots, n.
$$

(6.34)

Equation 6.34 can be written in the following alternate form by dividing the equation by $\hat{m}_i$, that is,

$$
\ddot{q}_i(t) + 2\zeta_i \omega_i \dot{q}_i(t) + \omega_i^2 q_i(t) = -\boldsymbol{\Gamma}_i \mathbf{a}_g(t), \quad i = 1, 2, \ldots, n
$$

(6.35)

where $(2\zeta_i \omega_i) = \hat{c}_i / \hat{m}_i$, $\omega_i^2 = \hat{k}_i / \hat{m}_i$, and $\boldsymbol{\Gamma}_i$ is a 1×6 matrix where

$$
\boldsymbol{\Gamma}_i = \begin{bmatrix} \Gamma_{i1} & \Gamma_{i2} & \Gamma_{i3} & \Gamma_{i4} & \Gamma_{i5} & \Gamma_{i6} \end{bmatrix} = \frac{1}{\hat{m}_i} \sum_{i=1}^{n} \sum_{k=1}^{n} \hat{\phi}_{ji} m_{jk} \mathbf{I}_k = \frac{\hat{\boldsymbol{\phi}}_i^T \mathbf{M} [\mathbf{I}_v]}{\hat{m}_i}.
$$

(6.36)

A general response solution is not available because the earthquake acceleration is very irregular in nature and cannot be described by mathematical function. However, since Eqs. 6.34 and 6.35 represent n independent SDOF equations, the response can be calculated directly using any numerical method for an SDOF system presented in

Sections 2.6 to 2.8. Let the solution of the ith equation be $q_i^s(t)$. Then the solution of the MDOF system becomes

$$\mathbf{X}(t) = \mathbf{\hat{\Phi}}\mathbf{Q}(t) = \begin{bmatrix} \hat{\phi}_1 & \hat{\phi}_2 & \cdots & \hat{\phi}_n \end{bmatrix} \begin{Bmatrix} q_1^s(t) \\ q_2^s(t) \\ \vdots \\ q_n^s(t) \end{Bmatrix} = \sum_{i=1}^{n} \hat{\phi}_i q_i^s(t). \tag{6.37}$$

EXAMPLE 1   *Two Degrees of Freedom Response*

Consider the 2DOF system shown in Figure 6.5. The dynamic equilibrium equation of motion is

$$\begin{bmatrix} 2m & 0 \\ 0 & m \end{bmatrix} \begin{Bmatrix} \ddot{x}_1(t) \\ \ddot{x}_2(t) \end{Bmatrix} + \begin{bmatrix} 3c & -c \\ -c & 2c \end{bmatrix} \begin{Bmatrix} \dot{x}_1(t) \\ \dot{x}_2(t) \end{Bmatrix} + \begin{bmatrix} 4k & -2k \\ -2k & 3k \end{bmatrix} \begin{Bmatrix} x_1(t) \\ x_2(t) \end{Bmatrix} = -\begin{bmatrix} 2m & 0 \\ 0 & m \end{bmatrix} \begin{Bmatrix} 1 \\ 1 \end{Bmatrix} a_x(t). \tag{6.38}$$

The eigenvalue characteristic equation is

$$\left| -\mathbf{M}\omega^2 + \mathbf{K} \right| = \begin{vmatrix} -2m\omega^2 + 4k & -2k \\ -2k & -m\omega^2 + 3k \end{vmatrix} = 2\omega^4 - 10\omega_n^2\omega^2 + 8\omega_n^4 = 0 \tag{6.39}$$

where $\omega_n = \sqrt{k/m}$, and it follows from Eq. 6.39 that

$$\omega_1^2 = \omega_n^2, \quad \omega_2^2 = 4\omega_n^2 \tag{6.40}$$

and therefore

$$\omega_1 = \omega_n, \quad \omega_2 = 2\omega_n. \tag{6.41}$$

The corresponding first mode shape is

$$\left[-\mathbf{M}\omega_1^2 + \mathbf{K} \right]\phi_1 = \frac{1}{m} \begin{bmatrix} 2\omega_n^2 & -2\omega_n^2 \\ -2\omega_n^2 & 2\omega_n^2 \end{bmatrix} \phi_1 = \begin{Bmatrix} 0 \\ 0 \end{Bmatrix}, \quad \phi_1 = \begin{Bmatrix} 1 \\ 1 \end{Bmatrix} \tag{6.42}$$

and the second mode shape is

$$\left[-\mathbf{M}\omega_2^2 + \mathbf{K} \right]\phi_2 = \frac{1}{m} \begin{bmatrix} -4\omega_n^2 & -2\omega_n^2 \\ -2\omega_n^2 & -\omega_n^2 \end{bmatrix} \phi_2 = \begin{Bmatrix} 0 \\ 0 \end{Bmatrix}, \quad \phi_2 = \begin{Bmatrix} 1 \\ -2 \end{Bmatrix}. \tag{6.43}$$

If each mode shape is normalized to have a unit length or norm, it then follows that

$$\omega_1^2 = \omega_n^2: \ \hat{\phi}_1 = \frac{1}{\sqrt{2}} \begin{Bmatrix} 1 \\ 1 \end{Bmatrix}, \quad \omega_2^2 = 4\omega_n^2: \ \hat{\phi}_2 = \frac{1}{\sqrt{5}} \begin{Bmatrix} 1 \\ -2 \end{Bmatrix}. \tag{6.44}$$

The system modal matrix is

$$\mathbf{\hat{\Phi}} = \begin{bmatrix} \hat{\phi}_1 & \hat{\phi}_2 \end{bmatrix} = \begin{bmatrix} 1/\sqrt{2} & 1/\sqrt{5} \\ 1/\sqrt{2} & -2/\sqrt{5} \end{bmatrix} \tag{6.45}$$

and the transformations of the mass, damping, and stiffness matrices into the normal mode coordinate system give

$$\mathbf{\hat{\Phi}}^T\mathbf{M}\mathbf{\hat{\Phi}} = \begin{bmatrix} 1/\sqrt{2} & 1/\sqrt{2} \\ 1/\sqrt{5} & -2/\sqrt{5} \end{bmatrix} \begin{bmatrix} 2m & 0 \\ 0 & m \end{bmatrix} \begin{bmatrix} 1/\sqrt{2} & 1/\sqrt{5} \\ 1/\sqrt{2} & -2/\sqrt{5} \end{bmatrix} = \begin{bmatrix} 3/2 & 0 \\ 0 & 6/5 \end{bmatrix} m \tag{6.46}$$

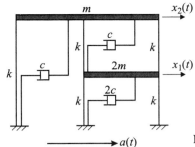

Figure 6.5 Two-Story Frame

$$\hat{\boldsymbol{\Phi}}^T \mathbf{C} \hat{\boldsymbol{\Phi}} = \begin{bmatrix} 1/\sqrt{2} & 1/\sqrt{2} \\ 1/\sqrt{5} & -2/\sqrt{5} \end{bmatrix} \begin{bmatrix} 3c & -c \\ -c & 2c \end{bmatrix} \begin{bmatrix} 1/\sqrt{2} & 1/\sqrt{5} \\ 1/\sqrt{2} & -2/\sqrt{5} \end{bmatrix} = \begin{bmatrix} 3/2 & 0 \\ 0 & 3 \end{bmatrix} c \tag{6.47}$$

$$\hat{\boldsymbol{\Phi}}^T \mathbf{K} \hat{\boldsymbol{\Phi}} = \begin{bmatrix} 1/\sqrt{2} & 1/\sqrt{2} \\ 1/\sqrt{5} & -2/\sqrt{5} \end{bmatrix} \begin{bmatrix} 4k & -2k \\ -2k & 3k \end{bmatrix} \begin{bmatrix} 1/\sqrt{2} & 1/\sqrt{5} \\ 1/\sqrt{2} & -2/\sqrt{5} \end{bmatrix} = \begin{bmatrix} 3/2 & 0 \\ 0 & 24/5 \end{bmatrix} k \tag{6.48}$$

$$\hat{\boldsymbol{\Phi}}^T \mathbf{M} \mathbf{a}(t) = \begin{bmatrix} 1/\sqrt{2} & 1/\sqrt{2} \\ 1/\sqrt{5} & -2/\sqrt{5} \end{bmatrix} \begin{bmatrix} 2m & 0 \\ 0 & m \end{bmatrix} \begin{Bmatrix} 1 \\ 1 \end{Bmatrix} a_x(t) = m \begin{Bmatrix} 3\sqrt{2}/2 \\ 0 \end{Bmatrix} a_x(t). \tag{6.49}$$

It then follows that

$$\frac{3}{2} m \ddot{q}_1(t) + \frac{3}{2} c \dot{q}_1(t) + \frac{3}{2} k q_1(t) = -\frac{3\sqrt{2}}{2} m a_x(t) \tag{6.50}$$

$$\frac{6}{5} m \ddot{q}_2(t) + 3 c \dot{q}_2(t) + \frac{24}{5} k q_2(t) = 0. \tag{6.51}$$

In Eq. 6.51, for this very specific example, the earthquake ground motion does not excite the second mode. Therefore, the response of the structure can be represented by the first mode only without any loss in accuracy. Let the solution of the first mode be $q_1^s(t)$, and then the response of the 2DOF system is

$$\mathbf{X}(t) = \hat{\boldsymbol{\phi}}_1 q_1^s(t) = \frac{1}{\sqrt{2}} \begin{Bmatrix} 1 \\ 1 \end{Bmatrix} q_1^s(t). \tag{6.52}$$

EXAMPLE 2   *2DOF System Response Due to 1994 Northridge Earthquake*

Consider again Example 1, where the response is given in Eq. 6.50, that is,

$$m \ddot{q}_1(t) + c \dot{q}_1(t) + k q_1(t) = -\sqrt{2} m a_x(t) \tag{6.53}$$

or in normalized form,

$$\ddot{q}_1(t) + 2 \zeta_1 \omega_1 \dot{q}_1(t) + \omega_1^2 q_1(t) = -\sqrt{2} a_x(t). \tag{6.54}$$

Let

$$\zeta_1 = 5\%, \quad T_1 = 1.0 \text{ s.} \tag{6.55}$$

It follows that the fundamental natural frequency of vibration is

$$\omega_1 = \frac{2\pi}{1.0} = 6.28 \text{ rad/s.} \tag{6.56}$$

Assume that the structure is subjected to the NS component at Newhall Fire Station during the 1994 Northquake earthquake, with the earthquake ground acceleration time history shown in Figure 5.12. The response is computed using the Newmark β-method. Since Eq. 6.54 represents the response of an SDOF system, Eq. 2.248 can be used, that is,

$$
\begin{Bmatrix} q_1 \\ \dot{q}_1 \\ \ddot{q}_1 \end{Bmatrix}_{k+1} = \mathbf{F}_N \begin{Bmatrix} q_1 \\ \dot{q}_1 \\ \ddot{q}_1 \end{Bmatrix}_k + \left(\frac{\sqrt{2}}{\beta} \right) \begin{Bmatrix} -\alpha(\Delta t)^2 \\ -\delta\Delta t \\ -1 \end{Bmatrix} a_{k+1}
\tag{6.57}
$$

where

$$
\mathbf{F}_N = \frac{1}{\beta} \begin{bmatrix} \beta - \omega_n^2\alpha(\Delta t)^2 & \beta\Delta t - 2\zeta\omega_n\alpha(\Delta t)^2 - \omega_n^2\alpha(\Delta t)^3 & \tfrac{1}{2}\beta(\Delta t)^2 - \alpha(\beta+\gamma)(\Delta t)^2 \\ -\omega_n^2\delta\Delta t & \beta - 2\zeta\omega_n\delta\Delta t - \omega_n^2\delta(\Delta t)^2 & \beta\Delta t - \delta(\beta+\gamma)\Delta t \\ -\omega_n^2 & -2\zeta\omega_n - \omega_n^2\Delta t & -\gamma \end{bmatrix}
$$

$$
\beta = 1 + 2\zeta\omega_n\delta\Delta t + \omega_n^2\alpha(\Delta t)^2
$$

$$
\gamma = 2\zeta\omega_n(1-\delta)\Delta t + \omega_n^2(\tfrac{1}{2}-\alpha)(\Delta t)^2.
$$

Using the constant average acceleration method where $\delta = \tfrac{1}{2}$ and $\alpha = \tfrac{1}{4}$ and a time step size of $\Delta t = 0.02$ s, it follows that Eq. 6.57 becomes

$$
\begin{Bmatrix} q_1 \\ \dot{q}_1 \\ \ddot{q}_1 \end{Bmatrix}_{k+1} = \begin{bmatrix} 0.996092 & 0.019860 & 0.000099 \\ -0.39079 & 0.985965 & 0.009899 \\ -39.0786 & -1.40353 & -0.01013 \end{bmatrix} \begin{Bmatrix} q_1 \\ \dot{q}_1 \\ \ddot{q}_1 \end{Bmatrix}_k + \sqrt{2} \begin{Bmatrix} -0.000099 \\ -0.00990 \\ -0.98987 \end{Bmatrix} a_{k+1}.
\tag{6.58}
$$

Figure 6.6 shows the first mode response for $q_1(t)$. Note that the only difference between Figure 6.6 and Figure 5.13 is that the magnitude of $q_1(t)$ in Figure 6.6 is $\sqrt{2}$ times the magnitude of the response given in Figure 5.13.

The response of the 2DOF system is given in Eq. 6.52:

$$
\mathbf{X}(t) = \hat{\boldsymbol{\phi}}_1 q_1^s(t) = \frac{1}{\sqrt{2}} \begin{Bmatrix} 1 \\ 1 \end{Bmatrix} q_1(t).
\tag{6.59}
$$

Figure 6.7 shows the displacement response for the 2DOF system. Note that the displacement response for the first story, $x_1(t)$, and the second story, $x_2(t)$, are the same because the mode shape $\hat{\boldsymbol{\phi}}_1$ is

$$
\hat{\boldsymbol{\phi}}_1 = \begin{Bmatrix} 1 \\ 1 \end{Bmatrix}.
\tag{6.60}
$$

Therefore, the relative displacement between $x_1(t)$ and $x_2(t)$ will always be equal to zero.

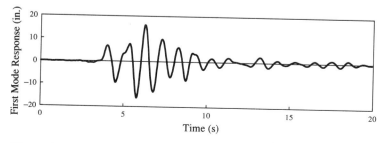

Figure 6.6 First Mode Response Due to NS Component Measured at Newhall Fire Station During 1994 Northridge Earthquake ($T_1 = 1.0$ s, $\zeta = 5\%$)

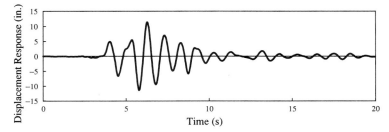

Figure 6.7 Displacement Response of 2DOF System Due to NS Component Measured at Newhall Fire Station During 1994 Northridge Earthquake ($T_1 = 1.0$ s, $\zeta = 5\%$)

6.4   NONPROPORTIONAL DAMPING

Most structural engineering problems do not have proportional damping. However, historically, most of these problems have been solved by assuming that they do have proportional damping. One reason for assuming proportional damping is the mathematical beauty of the normal mode method solution procedure. Natural frequencies and mode shapes can be described in terms of the undamped free vibration of the system. The orthogonality of the mode shapes produces a series of uncoupled single degree of freedom systems that are solved and the response combined to give the total response.

A second reason for the proportional damping assumption is the significant reduction in computer time required to solve for the response. As shown later in this section, the non-proportional damping solution can be obtained using either the Newmark β-method or the Wilson θ-method. However, structural dynamics problems typically have thousands of degrees of freedom, and for these large problems direct solution methods are very time consuming. If proportional damping is assumed, then the number of normal modes (i.e., q_i's) retained in the solution can usually be limited to modes with natural frequencies less than 50 cycles per second. The number of retained normal modes for large problems is much less than the total number of degrees of freedom. Therefore, the computer time required for the normal mode solution method is significantly less than for the direct solution method.

A very common way to convert a nonproportional damping problem is to assume that the normal mode damping coupling terms are zero. Recall that for proportional damping to exist, it must follow that

$$\hat{\mathbf{C}} = \hat{\mathbf{\Phi}}^T \mathbf{C} \hat{\mathbf{\Phi}} = \begin{bmatrix} \hat{c}_1 & 0 & \cdots & 0 \\ 0 & \hat{c}_2 & \ddots & \vdots \\ \vdots & \ddots & \ddots & 0 \\ 0 & \cdots & 0 & \hat{c}_n \end{bmatrix} \tag{6.61}$$

and

$$\left(\hat{c}_i / \hat{m}_i \right) = 2 \zeta_i \omega_i \tag{6.62}$$

where ω_i and ζ_i are the undamped natural frequency and modal damping ratio in the ith mode. The damping in the normal mode coordinates, $\hat{\mathbf{C}}$, is by definition a diagonal matrix for proportional damping. If the damping is nonproportional, then

$$\hat{\mathbf{C}} = \hat{\mathbf{\Phi}}^T \mathbf{C} \hat{\mathbf{\Phi}} = \begin{bmatrix} \hat{c}_{11} & \hat{c}_{12} & \cdots & \hat{c}_{1n} \\ \hat{c}_{21} & \hat{c}_{22} & \ddots & \vdots \\ \vdots & \ddots & \ddots & \hat{c}_{n-1,n} \\ \hat{c}_{n1} & \cdots & \hat{c}_{n,n-1} & \hat{c}_{nn} \end{bmatrix}. \tag{6.63}$$

Structural engineers desiring to solve for the response using the normal mode method often assume that all off-diagonal terms in $\hat{\mathbf{C}}$ are zero and that $\hat{c}_i = \hat{c}_{ii}$ is used in normal mode solution. The only way to determine the error made by assuming that the off-diagonal terms are zero is to solve the problem using the direct solution. Obviously, this is not typically done because it reduces the benefits derived from the assumption. Therefore, other reasons are sought for the loss of accuracy due to neglecting the off-diagonal terms.

Imagine the case where the off-diagonal terms in Eq. 6.63 are retained; then it follows that

$$\ddot{q}_i(t) + 2\zeta_i \omega_i \dot{q}_i(t) + \omega_i^2 q_i(t) = -\Gamma_i a_j(t) - c_{err}/\hat{m}_i \tag{6.64}$$

where

$$c_{err} = \hat{c}_{i1}\dot{q}_1(t) + \ldots + \hat{c}_{i,i-1}\dot{q}_{i-1}(t) + \hat{c}_{i,i+1}\dot{q}_{i+1}(t) + \ldots + \hat{c}_{in}\dot{q}_n(t). \tag{6.65}$$

The assumption that the off-diagonal terms are zero results in the lack of "artificial" forcing functions that excite the ith normal mode. These artificial forcing functions are added to $a(t)$ and excite the response in the $q_i(t)$ mode. If the amplitude or phasing of these artificially induced responses is small or not coincident, then the contribution from these response terms may be small.

If the solution to the nonproportional damping case is desired, then consider the dynamic equilibrium equation of motion

$$\mathbf{M}\ddot{\mathbf{X}}(t) + \mathbf{C}\dot{\mathbf{X}}(t) + \mathbf{K}\mathbf{X}(t) = -\mathbf{M}a(t), \quad \mathbf{X}(0) = \mathbf{X}_o = \mathbf{0}, \quad \dot{\mathbf{X}}(0) = \dot{\mathbf{X}}_o = \mathbf{0}. \tag{6.66}$$

The numerical solution using discretization was presented in Section 3.7 using the Newmark β-method and the Wilson θ-method. The matrix equation for a general forcing function solution using the Newmark β-method is given in Eq. 3.233, and for a specific earthquake ground acceleration response it is

$$\begin{Bmatrix} \mathbf{X}_{k+1} \\ \dot{\mathbf{X}}_{k+1} \\ \ddot{\mathbf{X}}_{k+1} \end{Bmatrix} = \mathbf{F}_N \begin{Bmatrix} \mathbf{X}_k \\ \dot{\mathbf{X}}_k \\ \ddot{\mathbf{X}}_k \end{Bmatrix} + \mathbf{H}_N^{(EQ)} a_{k+1} \tag{6.67}$$

where

$$\mathbf{F}_N = \begin{bmatrix} \mathbf{I} - \alpha\mathbf{B}^{-1}\mathbf{K}(\Delta t)^2 & \mathbf{I}\Delta t - \alpha\mathbf{B}^{-1}\mathbf{C}(\Delta t)^2 - \alpha\mathbf{B}^{-1}\mathbf{K}(\Delta t)^3 & \left(\frac{1}{2} - \alpha\right)\mathbf{I}(\Delta t)^2 - \alpha\mathbf{B}^{-1}\mathbf{G}(\Delta t)^2 \\ -\delta\mathbf{B}^{-1}\mathbf{K}\Delta t & \mathbf{I} - \delta\mathbf{B}^{-1}\mathbf{C}\Delta t - \delta\mathbf{B}^{-1}\mathbf{K}(\Delta t)^2 & (1 - \delta)\mathbf{I}\Delta t - \delta\mathbf{B}^{-1}\mathbf{G}\Delta t \\ -\mathbf{B}^{-1}\mathbf{K} & -\mathbf{B}^{-1}\mathbf{C} - \mathbf{B}^{-1}\mathbf{K}\Delta t & -\mathbf{B}^{-1}\mathbf{G} \end{bmatrix}$$

$$\mathbf{H}_N^{(EQ)} = \begin{bmatrix} \alpha\mathbf{B}^{-1}\mathbf{M}(\Delta t)^2 \\ \delta\mathbf{B}^{-1}\mathbf{M}\Delta t \\ \mathbf{B}^{-1}\mathbf{M} \end{bmatrix}$$

$$\mathbf{B} = \mathbf{M} + \delta\mathbf{C}\Delta t + \alpha\mathbf{K}(\Delta t)^2$$

$$\mathbf{G} = (1 - \delta)\mathbf{C}\Delta t + \left(\frac{1}{2} - \alpha\right)\mathbf{K}(\Delta t)^2.$$

Similarly, the matrix equation for the general forcing function solution using the Wilson θ-method is given in Eq. 3.245, and for a specific earthquake ground acceleration response it is

$$\begin{Bmatrix} \mathbf{X}_{k+1} \\ \dot{\mathbf{X}}_{k+1} \\ \ddot{\mathbf{X}}_{k+1} \end{Bmatrix} = \mathbf{F}_W \begin{Bmatrix} \mathbf{X}_k \\ \dot{\mathbf{X}}_k \\ \ddot{\mathbf{X}}_k \end{Bmatrix} + \mathbf{H}_W^{(EQ)} a_{k+1} \tag{6.68}$$

where

$$\mathbf{F}_W = \begin{bmatrix} \mathbf{I} - \frac{1}{6}\mathbf{B}^{-1}\mathbf{K}\theta^2(\Delta t)^2 & \mathbf{I}\Delta t - \frac{1}{6}\mathbf{B}^{-1}\mathbf{C}\theta^2(\Delta t)^2 - \frac{1}{6}\mathbf{B}^{-1}\mathbf{K}\theta^2(\Delta t)^3 & \frac{1}{6}\theta(\Delta t)^2\left(2\mathbf{I} - \overline{\mathbf{B}}\right) \\ -\frac{1}{2}\mathbf{B}^{-1}\mathbf{K}\theta\Delta t & \mathbf{I} - \mathbf{B}^{-1}\mathbf{C}\theta\Delta t - \frac{1}{2}\mathbf{B}^{-1}\mathbf{K}\theta(\Delta t)^2 & \frac{1}{2}\Delta t\left(\mathbf{I} - \overline{\mathbf{B}}\right) \\ -\mathbf{B}^{-1}\mathbf{K} & -\mathbf{B}^{-1}\mathbf{C} - \mathbf{B}^{-1}\mathbf{K}\Delta t & \left(1 - \frac{1}{\theta}\right)\mathbf{I} - \overline{\mathbf{B}} \end{bmatrix}$$

$$\mathbf{H}_W^{(EQ)} = \begin{bmatrix} \frac{1}{6}\mathbf{B}^{-1}\mathbf{M}\theta^2(\Delta t)^2 \\ \frac{1}{2}\mathbf{B}^{-1}\mathbf{M}\theta\Delta t \\ \mathbf{B}^{-1}\mathbf{M} \end{bmatrix}$$

$$\mathbf{B} = \mathbf{M} + \frac{1}{2}\mathbf{C}\theta\Delta t + \frac{1}{6}\mathbf{K}\theta^2(\Delta t)^2$$

$$\overline{\mathbf{B}} = \mathbf{B}^{-1}\mathbf{G} - (1-\theta)\mathbf{B}^{-1}\mathbf{M}$$

$$\mathbf{G} = \frac{1}{2}\mathbf{C}\theta\Delta t + \frac{1}{3}\mathbf{K}\theta^2(\Delta t)^2.$$

Once the numerical calculation is complete, by using either the Newmark β-method or the Wilson θ-method, the response gives the displacement, velocity, and acceleration vectors at discrete time steps. Some important quantities that structural engineers pay close attention to are the maximum responses. The maximum relative displacement, relative velocity, and absolute acceleration vectors are defined as

$$\max[\mathbf{X}(t)] = \max_t \begin{bmatrix} \begin{Bmatrix} x_1(t) \\ x_2(t) \\ \vdots \\ x_n(t) \end{Bmatrix} \end{bmatrix} = \begin{Bmatrix} \max_k[x_{1k}] \\ \max_k[x_{2k}] \\ \vdots \\ \max_k[x_{nk}] \end{Bmatrix} \tag{6.69a}$$

$$\max[\dot{\mathbf{X}}(t)] = \max_t \begin{bmatrix} \begin{Bmatrix} \dot{x}_1(t) \\ \dot{x}_2(t) \\ \vdots \\ \dot{x}_n(t) \end{Bmatrix} \end{bmatrix} = \begin{Bmatrix} \max_k[\dot{x}_{1k}] \\ \max_k[\dot{x}_{2k}] \\ \vdots \\ \max_k[\dot{x}_{nk}] \end{Bmatrix} \tag{6.69b}$$

$$\max[\ddot{\mathbf{Y}}(t)] = \max_t \begin{bmatrix} \begin{Bmatrix} \ddot{y}_1(t) \\ \ddot{y}_2(t) \\ \vdots \\ \ddot{y}_n(t) \end{Bmatrix} \end{bmatrix} = \begin{Bmatrix} \max_k[\ddot{y}_{1k}] \\ \max_k[\ddot{y}_{2k}] \\ \vdots \\ \max_k[\ddot{y}_{nk}] \end{Bmatrix}. \tag{6.69c}$$

In general, the maximum response at different degrees of freedom occurs at different time steps. A direct consequence is that the maximum relative displacement between two floors cannot be accurately calculated by subtracting the two maximum floor displacements. Using the same notation as defined in Eq. 5.72, it follows that

$$\Delta x_{ij}(t) = x_i(t) - x_j(t) = \mathbf{T}_i^T \mathbf{X}(t) - \mathbf{T}_j^T \mathbf{X}(t) \tag{6.70}$$

where $\Delta x_{ij}(t)$ is the relative displacement between the two displacements, $x_i(t)$ and $x_j(t)$, and $\mathbf{T}_i^T$ is the displacement selection vector where all elements are equal to zero except the ith element, which is equal to one. The maximum relative displacement is given by

$$\max[\Delta x_{ij}(t)] = \max_k[x_{ik} - x_{jk}] = \max_k[\mathbf{T}_i^T \mathbf{X}_k - \mathbf{T}_j^T \mathbf{X}_k] \tag{6.71}$$

where this maximum relative displacement is calculated for x_{ik} and x_{jk} at the same time step.

The *seismic floor forces* is calculated by multiplying the mass by the absolute acceleration:

$$\mathbf{S}(t) = \mathbf{M}\ddot{\mathbf{Y}}(t) = \mathbf{M}\left[\ddot{\mathbf{X}}(t) + \mathbf{a}(t)\right] = -\mathbf{KX}(t) - \mathbf{C}\dot{\mathbf{X}}(t). \tag{6.72}$$

The maximum seismic floor forces is

$$\max[\mathbf{S}(t)] = \max_t\left[\mathbf{M}\ddot{\mathbf{Y}}(t)\right] = \max_k\left[\mathbf{M}\ddot{\mathbf{Y}}_k\right]. \tag{6.73}$$

Similar to the relative displacement, the maximum story shear for different floors can occur at different time steps.

The base shear is the total earthquake inertia force on the building at each instant in time, and it is defined as

$$V(t) = -\{\mathbf{I}\}^T \mathbf{S}(t) = -\{\mathbf{I}\}^T \mathbf{M}\ddot{\mathbf{Y}}(t). \tag{6.74}$$

The maximum base shear is

$$\max[V(t)] = \max_k\left[\{\mathbf{I}\}^T \mathbf{M}\ddot{\mathbf{Y}}_k\right]. \tag{6.75}$$

6.5   PROBLEM SIZE REDUCTION USING STATIC CONDENSATION

Section 4.1 presented a formulation of the dynamic equilibrium equation of motion of a MDOF system where all of the degrees of freedom along the same direction were grouped together and the equation of motion for earthquake response was described by

$$\mathbf{M}\ddot{\mathbf{X}}(t) + \mathbf{C}\dot{\mathbf{X}}(t) + \mathbf{KX}(t) = -\mathbf{Ma}(t) \tag{6.76}$$

where

$$\mathbf{M} = \begin{bmatrix} \mathbf{M}_{xx} & \mathbf{M}_{xy} & \mathbf{M}_{xz} & \mathbf{M}_{xr} & \mathbf{M}_{xs} & \mathbf{M}_{xt} \\ \mathbf{M}_{yx} & \mathbf{M}_{yy} & \mathbf{M}_{yz} & \mathbf{M}_{yr} & \mathbf{M}_{ys} & \mathbf{M}_{yt} \\ \mathbf{M}_{zx} & \mathbf{M}_{zy} & \mathbf{M}_{zz} & \mathbf{M}_{zr} & \mathbf{M}_{zs} & \mathbf{M}_{zt} \\ \mathbf{M}_{rx} & \mathbf{M}_{ry} & \mathbf{M}_{rz} & \mathbf{M}_{rr} & \mathbf{M}_{rs} & \mathbf{M}_{rt} \\ \mathbf{M}_{sx} & \mathbf{M}_{sy} & \mathbf{M}_{sz} & \mathbf{M}_{sr} & \mathbf{M}_{ss} & \mathbf{M}_{st} \\ \mathbf{M}_{tx} & \mathbf{M}_{ty} & \mathbf{M}_{tz} & \mathbf{M}_{tr} & \mathbf{M}_{ts} & \mathbf{M}_{tt} \end{bmatrix}$$

$$\mathbf{C} = \begin{bmatrix} \mathbf{C}_{xx} & \mathbf{C}_{xy} & \mathbf{C}_{xz} & \mathbf{C}_{xr} & \mathbf{C}_{xs} & \mathbf{C}_{xt} \\ \mathbf{C}_{yx} & \mathbf{C}_{yy} & \mathbf{C}_{yz} & \mathbf{C}_{yr} & \mathbf{C}_{ys} & \mathbf{C}_{yt} \\ \mathbf{C}_{zx} & \mathbf{C}_{zy} & \mathbf{C}_{zz} & \mathbf{C}_{zr} & \mathbf{C}_{zs} & \mathbf{C}_{zt} \\ \mathbf{C}_{rx} & \mathbf{C}_{ry} & \mathbf{C}_{rz} & \mathbf{C}_{rr} & \mathbf{C}_{rs} & \mathbf{C}_{rt} \\ \mathbf{C}_{sx} & \mathbf{C}_{sy} & \mathbf{C}_{sz} & \mathbf{C}_{sr} & \mathbf{C}_{ss} & \mathbf{C}_{st} \\ \mathbf{C}_{tx} & \mathbf{C}_{ty} & \mathbf{C}_{tz} & \mathbf{C}_{tr} & \mathbf{C}_{ts} & \mathbf{C}_{tt} \end{bmatrix}, \quad \mathbf{K} = \begin{bmatrix} \mathbf{K}_{xx} & \mathbf{K}_{xy} & \mathbf{K}_{xz} & \mathbf{K}_{xr} & \mathbf{K}_{xs} & \mathbf{K}_{xt} \\ \mathbf{K}_{yx} & \mathbf{K}_{yy} & \mathbf{K}_{yz} & \mathbf{K}_{yr} & \mathbf{K}_{ys} & \mathbf{K}_{yt} \\ \mathbf{K}_{zx} & \mathbf{K}_{zy} & \mathbf{K}_{zz} & \mathbf{K}_{zr} & \mathbf{K}_{zs} & \mathbf{K}_{zt} \\ \mathbf{K}_{rx} & \mathbf{K}_{ry} & \mathbf{K}_{rz} & \mathbf{K}_{rr} & \mathbf{K}_{rs} & \mathbf{K}_{rt} \\ \mathbf{K}_{sx} & \mathbf{K}_{sy} & \mathbf{K}_{sz} & \mathbf{K}_{sr} & \mathbf{K}_{ss} & \mathbf{K}_{st} \\ \mathbf{K}_{tx} & \mathbf{K}_{ty} & \mathbf{K}_{tz} & \mathbf{K}_{tr} & \mathbf{K}_{ts} & \mathbf{K}_{tt} \end{bmatrix}$$

$$\mathbf{X}(t) = \left\{\mathbf{X}_x(t) \quad \mathbf{X}_y(t) \quad \mathbf{X}_z(t) \quad \mathbf{X}_r(t) \quad \mathbf{X}_s(t) \quad \mathbf{X}_t(t)\right\}^T$$

$$\dot{\mathbf{X}}(t) = \left\{\dot{\mathbf{X}}_x(t) \quad \dot{\mathbf{X}}_y(t) \quad \dot{\mathbf{X}}_z(t) \quad \dot{\mathbf{X}}_r(t) \quad \dot{\mathbf{X}}_s(t) \quad \dot{\mathbf{X}}_t(t)\right\}^T$$

$$\ddot{\mathbf{X}}(t) = \left\{\ddot{\mathbf{X}}_x(t) \quad \ddot{\mathbf{X}}_y(t) \quad \ddot{\mathbf{X}}_z(t) \quad \ddot{\mathbf{X}}_r(t) \quad \ddot{\mathbf{X}}_s(t) \quad \ddot{\mathbf{X}}_t(t)\right\}^T$$

$$\mathbf{a}(t) = \left\{\mathbf{a}_x(t) \quad \mathbf{a}_y(t) \quad \mathbf{a}_z(t) \quad \mathbf{a}_r(t) \quad \mathbf{a}_s(t) \quad \mathbf{a}_t(t)\right\}^T.$$

A building structure usually has over 1,000 degrees of freedom, some of which are insignificant in contribution to the total earthquake response of the structure. Therefore, it is often desirable to reduce the importance of these degrees of freedom in order to reduce the size of the problem. In this section, a method called *static condensation* is used to reduce the size of the problem. In Section 6.6, a method called *dynamic condensation* is used to reduce the size of the problem.

In static condensation, the damping matrix is assumed to be the same order as the mass matrix. The first step in this method is to zero out as many mass terms as possible. This is usually done by retaining only the mass terms corresponding to the translation and rotation of the floors. Figure 6.8 shows a two-dimensional frame where only the floor translations are retained while the joint rotations are ignored. In the following discussion, different cases are presented where the masses are assumed to be zero.

1. *Lumped Mass*: When a lumped mass approach is used to model the distribution of mass in the structure, it is assumed that there is no interaction between the masses at different degree of freedom. Structural engineers have used this lumped-mass approach to model the dynamics of structural response for several decades, and many research papers have been published on its accuracy for different applications. Using the lumped mass approach, the mass matrix becomes diagonal, and thus

$$\mathbf{M} = \begin{bmatrix} \mathbf{M}_{xx} & 0 & 0 & 0 & 0 & 0 \\ 0 & \mathbf{M}_{yy} & 0 & 0 & 0 & 0 \\ 0 & 0 & \mathbf{M}_{zz} & 0 & 0 & 0 \\ 0 & 0 & 0 & \mathbf{M}_{rr} & 0 & 0 \\ 0 & 0 & 0 & 0 & \mathbf{M}_{ss} & 0 \\ 0 & 0 & 0 & 0 & 0 & \mathbf{M}_{tt} \end{bmatrix} \tag{6.77}$$

where $\mathbf{M}_{xx}$, $\mathbf{M}_{yy}$, $\mathbf{M}_{zz}$, $\mathbf{M}_{rr}$, $\mathbf{M}_{ss}$, and $\mathbf{M}_{tt}$ are all diagonal matrices.

2. *Joint Rotational Mass Moment of Inertia*: The rotational mass moment of inertia of a joint can be very small compared to the translational mass inertia terms and thus can have little significance in the overall response of the structure. Thus, in this case the mass matrix in Eq. 6.77 becomes

$$\mathbf{M} = \begin{bmatrix} \mathbf{M}_{xx} & 0 & 0 & 0 & 0 & 0 \\ 0 & \mathbf{M}_{yy} & 0 & 0 & 0 & 0 \\ 0 & 0 & \mathbf{M}_{zz} & 0 & 0 & 0 \\ 0 & 0 & 0 & 0 & 0 & 0 \\ 0 & 0 & 0 & 0 & 0 & 0 \\ 0 & 0 & 0 & 0 & 0 & 0 \end{bmatrix} \tag{6.78}$$

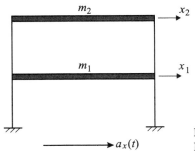

Figure 6.8 Example of a Two-Dimensional Frame with Floor Translations Only

where $\mathbf{M}_{xx}$, $\mathbf{M}_{yy}$, and $\mathbf{M}_{zz}$ are diagonal matrices. The transformation matrix for this case (see Section 4.15) is

$$\mathbf{T} = \begin{bmatrix} \mathbf{I} & \mathbf{0} & \mathbf{0} \\ \mathbf{0} & \mathbf{I} & \mathbf{0} \\ \mathbf{0} & \mathbf{0} & \mathbf{I} \\ \mathbf{0} & \mathbf{0} & \mathbf{0} \\ \mathbf{0} & \mathbf{0} & \mathbf{0} \\ \mathbf{0} & \mathbf{0} & \mathbf{0} \end{bmatrix}, \quad \begin{Bmatrix} \mathbf{X}_x(t) \\ \mathbf{X}_y(t) \\ \mathbf{X}_z(t) \\ \mathbf{X}_r(t) \\ \mathbf{X}_s(t) \\ \mathbf{X}_t(t) \end{Bmatrix} = \begin{bmatrix} \mathbf{I} & \mathbf{0} & \mathbf{0} \\ \mathbf{0} & \mathbf{I} & \mathbf{0} \\ \mathbf{0} & \mathbf{0} & \mathbf{I} \\ \mathbf{0} & \mathbf{0} & \mathbf{0} \\ \mathbf{0} & \mathbf{0} & \mathbf{0} \\ \mathbf{0} & \mathbf{0} & \mathbf{0} \end{bmatrix} \begin{Bmatrix} \tilde{\mathbf{X}}_x(t) \\ \tilde{\mathbf{X}}_y(t) \\ \tilde{\mathbf{X}}_z(t) \end{Bmatrix}. \tag{6.79}$$

Note from Eq. 6.79 that $\tilde{\mathbf{X}}_x(t) = \mathbf{X}_x(t)$, $\tilde{\mathbf{X}}_y(t) = \mathbf{X}_y(t)$, and $\tilde{\mathbf{X}}_z(t) = \mathbf{X}_z(t)$. It then follows from Eq. 4.625 that

$$\tilde{\mathbf{M}} = \mathbf{T}^T \mathbf{M} \mathbf{T} = \begin{bmatrix} \mathbf{M}_{xx} & \mathbf{0} & \mathbf{0} \\ \mathbf{0} & \mathbf{M}_{yy} & \mathbf{0} \\ \mathbf{0} & \mathbf{0} & \mathbf{M}_{zz} \end{bmatrix}, \quad \tilde{\mathbf{C}} = \mathbf{T}^T \mathbf{C} \mathbf{T} = \begin{bmatrix} \tilde{\mathbf{C}}_{xx} & \tilde{\mathbf{C}}_{xy} & \tilde{\mathbf{C}}_{xz} \\ \tilde{\mathbf{C}}_{yx} & \tilde{\mathbf{C}}_{yy} & \tilde{\mathbf{C}}_{yz} \\ \tilde{\mathbf{C}}_{zx} & \tilde{\mathbf{C}}_{zy} & \tilde{\mathbf{C}}_{zz} \end{bmatrix} \tag{6.80a}$$

$$\tilde{\mathbf{K}} = \mathbf{T}^T \mathbf{K} \mathbf{T} = \begin{bmatrix} \tilde{\mathbf{K}}_{xx} & \tilde{\mathbf{K}}_{xy} & \tilde{\mathbf{K}}_{xz} \\ \tilde{\mathbf{K}}_{yx} & \tilde{\mathbf{K}}_{yy} & \tilde{\mathbf{K}}_{yz} \\ \tilde{\mathbf{K}}_{zx} & \tilde{\mathbf{K}}_{zy} & \tilde{\mathbf{K}}_{zz} \end{bmatrix}, \quad \mathbf{T}^T \mathbf{M} = \begin{bmatrix} \mathbf{M}_{xx} & \mathbf{0} & \mathbf{0} & \mathbf{0} & \mathbf{0} & \mathbf{0} \\ \mathbf{0} & \mathbf{M}_{yy} & \mathbf{0} & \mathbf{0} & \mathbf{0} & \mathbf{0} \\ \mathbf{0} & \mathbf{0} & \mathbf{M}_{zz} & \mathbf{0} & \mathbf{0} & \mathbf{0} \end{bmatrix}. \tag{6.80b}$$

Since the last three columns of the matrix $\mathbf{T}^T\mathbf{M}$ in Eq. 6.80b are all zero, it follows from Eq. 6.76 that

$$\begin{bmatrix} \mathbf{M}_{xx} & \mathbf{0} & \mathbf{0} \\ \mathbf{0} & \mathbf{M}_{yy} & \mathbf{0} \\ \mathbf{0} & \mathbf{0} & \mathbf{M}_{zz} \end{bmatrix} \begin{Bmatrix} \ddot{\mathbf{X}}_x(t) \\ \ddot{\mathbf{X}}_y(t) \\ \ddot{\mathbf{X}}_z(t) \end{Bmatrix} + \begin{bmatrix} \tilde{\mathbf{C}}_{xx} & \tilde{\mathbf{C}}_{xy} & \tilde{\mathbf{C}}_{xz} \\ \tilde{\mathbf{C}}_{yx} & \tilde{\mathbf{C}}_{yy} & \tilde{\mathbf{C}}_{yz} \\ \tilde{\mathbf{C}}_{zx} & \tilde{\mathbf{C}}_{zy} & \tilde{\mathbf{C}}_{zz} \end{bmatrix} \begin{Bmatrix} \dot{\mathbf{X}}_x(t) \\ \dot{\mathbf{X}}_y(t) \\ \dot{\mathbf{X}}_z(t) \end{Bmatrix} + \begin{bmatrix} \tilde{\mathbf{K}}_{xx} & \tilde{\mathbf{K}}_{xy} & \tilde{\mathbf{K}}_{xz} \\ \tilde{\mathbf{K}}_{yx} & \tilde{\mathbf{K}}_{yy} & \tilde{\mathbf{K}}_{yz} \\ \tilde{\mathbf{K}}_{zx} & \tilde{\mathbf{K}}_{zy} & \tilde{\mathbf{K}}_{zz} \end{bmatrix} \begin{Bmatrix} \mathbf{X}_x(t) \\ \mathbf{X}_y(t) \\ \mathbf{X}_z(t) \end{Bmatrix}$$

$$= - \begin{bmatrix} \mathbf{M}_{xx} & \mathbf{0} & \mathbf{0} \\ \mathbf{0} & \mathbf{M}_{yy} & \mathbf{0} \\ \mathbf{0} & \mathbf{0} & \mathbf{M}_{zz} \end{bmatrix} \begin{Bmatrix} \mathbf{a}_x(t) \\ \mathbf{a}_y(t) \\ \mathbf{a}_z(t) \end{Bmatrix}. \tag{6.81}$$

Thus the problem size has been reduced by half.

3. *Vertical Mass*: The vertical mass of a joint can be small and often is assumed to be zero. If this is the case, a significant number of degrees of freedom can be reduced. The mass matrix in Eq. 6.78 becomes

$$\mathbf{M} = \begin{bmatrix} \mathbf{M}_{xx} & \mathbf{0} & \mathbf{0} & \mathbf{0} & \mathbf{0} & \mathbf{0} \\ \mathbf{0} & \mathbf{M}_{yy} & \mathbf{0} & \mathbf{0} & \mathbf{0} & \mathbf{0} \\ \mathbf{0} & \mathbf{0} & \mathbf{0} & \mathbf{0} & \mathbf{0} & \mathbf{0} \\ \mathbf{0} & \mathbf{0} & \mathbf{0} & \mathbf{0} & \mathbf{0} & \mathbf{0} \\ \mathbf{0} & \mathbf{0} & \mathbf{0} & \mathbf{0} & \mathbf{0} & \mathbf{0} \\ \mathbf{0} & \mathbf{0} & \mathbf{0} & \mathbf{0} & \mathbf{0} & \mathbf{0} \end{bmatrix} \tag{6.82}$$

where $\mathbf{M}_{xx}$ and $\mathbf{M}_{yy}$ are diagonal matrices. The transformation matrix for this case is

$$\mathbf{T} = \begin{bmatrix} \mathbf{I} & \mathbf{0} \\ \mathbf{0} & \mathbf{I} \\ \mathbf{0} & \mathbf{0} \\ \mathbf{0} & \mathbf{0} \\ \mathbf{0} & \mathbf{0} \\ \mathbf{0} & \mathbf{0} \end{bmatrix}, \quad \begin{Bmatrix} \mathbf{X}_x(t) \\ \mathbf{X}_y(t) \\ \mathbf{X}_z(t) \\ \mathbf{X}_r(t) \\ \mathbf{X}_s(t) \\ \mathbf{X}_t(t) \end{Bmatrix} = \begin{bmatrix} \mathbf{I} & \mathbf{0} \\ \mathbf{0} & \mathbf{I} \\ \mathbf{0} & \mathbf{0} \\ \mathbf{0} & \mathbf{0} \\ \mathbf{0} & \mathbf{0} \\ \mathbf{0} & \mathbf{0} \end{bmatrix} \begin{Bmatrix} \tilde{\mathbf{X}}_x(t) \\ \tilde{\mathbf{X}}_y(t) \end{Bmatrix}. \tag{6.83}$$

Note from Eq. 6.83 that $\tilde{\mathbf{X}}_x(t) = \mathbf{X}_x(t)$ and $\tilde{\mathbf{X}}_y(t) = \mathbf{X}_y(t)$. It then follows from Eq. 4.625 that

$$\tilde{\mathbf{M}} = \mathbf{T}^T\mathbf{M}\mathbf{T} = \begin{bmatrix} \mathbf{M}_{xx} & \mathbf{0} \\ \mathbf{0} & \mathbf{M}_{yy} \end{bmatrix}, \quad \tilde{\mathbf{C}} = \mathbf{T}^T\mathbf{C}\mathbf{T} = \begin{bmatrix} \tilde{\mathbf{C}}_{xx} & \tilde{\mathbf{C}}_{xy} \\ \tilde{\mathbf{C}}_{yx} & \tilde{\mathbf{C}}_{yy} \end{bmatrix} \tag{6.84a}$$

$$\tilde{\mathbf{K}} = \mathbf{T}^T\mathbf{K}\mathbf{T} = \begin{bmatrix} \tilde{\mathbf{K}}_{xx} & \tilde{\mathbf{K}}_{xy} \\ \tilde{\mathbf{K}}_{yx} & \tilde{\mathbf{K}}_{yy} \end{bmatrix}, \quad \mathbf{T}^T\mathbf{M} = \begin{bmatrix} \mathbf{M}_{xx} & \mathbf{0} & \mathbf{0} & \mathbf{0} & \mathbf{0} & \mathbf{0} \\ \mathbf{0} & \mathbf{M}_{yy} & \mathbf{0} & \mathbf{0} & \mathbf{0} & \mathbf{0} \end{bmatrix}. \tag{6.84b}$$

Since the last four columns of the matrix $\mathbf{T}^T\mathbf{M}$ in Eq. 6.84b are all zero, it follows from Eq. 6.76 that

$$\begin{bmatrix} \mathbf{M}_{xx} & \mathbf{0} \\ \mathbf{0} & \mathbf{M}_{yy} \end{bmatrix} \begin{Bmatrix} \ddot{\mathbf{X}}_x(t) \\ \ddot{\mathbf{X}}_y(t) \end{Bmatrix} + \begin{bmatrix} \tilde{\mathbf{C}}_{xx} & \tilde{\mathbf{C}}_{xy} \\ \tilde{\mathbf{C}}_{yx} & \tilde{\mathbf{C}}_{yy} \end{bmatrix} \begin{Bmatrix} \dot{\mathbf{X}}_x(t) \\ \dot{\mathbf{X}}_y(t) \end{Bmatrix} + \begin{bmatrix} \tilde{\mathbf{K}}_{xx} & \tilde{\mathbf{K}}_{xy} \\ \tilde{\mathbf{K}}_{yx} & \tilde{\mathbf{K}}_{yy} \end{bmatrix} \begin{Bmatrix} \mathbf{X}_x(t) \\ \mathbf{X}_y(t) \end{Bmatrix}$$
$$= -\begin{bmatrix} \mathbf{M}_{xx} & \mathbf{0} \\ \mathbf{0} & \mathbf{M}_{yy} \end{bmatrix} \begin{Bmatrix} \mathbf{a}_x(t) \\ \mathbf{a}_y(t) \end{Bmatrix}. \tag{6.85}$$

Thus the problem size has been reduced by two-thirds.

4. *In-Plane Joint Movement*: When floors of a structure have a very large axial stiffness, they are referred to as having a *rigid diaphragm*. In this case, the floor is assumed to rotate as a unit about a vertical axis through the center of mass of the floor. Consider a building with m floors, and assume that both the vertical mass and the joint rotational mass moment of inertia are zero; then from a problem size reduction point of view, the transformation is

$$\begin{Bmatrix} \mathbf{X}_x(t) \\ \mathbf{X}_y(t) \\ \mathbf{X}_z(t) \\ \mathbf{X}_r(t) \\ \mathbf{X}_s(t) \\ \mathbf{X}_t(t) \end{Bmatrix} = \begin{bmatrix} \mathbf{T}_{xx} & \mathbf{0} & \mathbf{T}_{xt} \\ \mathbf{0} & \mathbf{T}_{yy} & \mathbf{T}_{yt} \\ \mathbf{0} & \mathbf{0} & \mathbf{0} \\ \mathbf{0} & \mathbf{0} & \mathbf{0} \\ \mathbf{0} & \mathbf{0} & \mathbf{0} \\ \mathbf{0} & \mathbf{0} & \mathbf{T}_{tt} \end{bmatrix} \begin{Bmatrix} \tilde{\mathbf{X}}_x(t) \\ \tilde{\mathbf{X}}_y(t) \\ \tilde{\mathbf{X}}_t(t) \end{Bmatrix} \tag{6.86}$$

where $\mathbf{T}_{ij}$ represents the transformation of the distance from the center of mass to the location of the degree of freedom, and

$$\tilde{\mathbf{X}}_x(t) = \left\{ x_x^{[1]}(t) \quad x_x^{[2]}(t) \quad \cdots \quad x_x^{[m]}(t) \right\}^T \tag{6.87a}$$

$$\tilde{\mathbf{X}}_y(t) = \left\{ x_y^{[1]}(t) \quad x_y^{[2]}(t) \quad \cdots \quad x_y^{[m]}(t) \right\}^T \tag{6.87b}$$

$$\tilde{\mathbf{X}}_t(t) = \left\{ x_t^{[1]}(t) \quad x_t^{[2]}(t) \quad \cdots \quad x_t^{[m]}(t) \right\}^T \tag{6.87c}$$

where the superscript inside the brackets denotes the floor number. It then follows from Eq. 4.625 that

$$\tilde{\mathbf{M}} = \mathbf{T}^T\mathbf{M}\mathbf{T} = \begin{bmatrix} \tilde{\mathbf{M}}_{xx} & \mathbf{0} & \mathbf{0} \\ \mathbf{0} & \tilde{\mathbf{M}}_{yy} & \mathbf{0} \\ \mathbf{0} & \mathbf{0} & \tilde{\mathbf{M}}_{tt} \end{bmatrix}, \quad \tilde{\mathbf{C}} = \mathbf{T}^T\mathbf{C}\mathbf{T} = \begin{bmatrix} \tilde{\mathbf{C}}_{xx} & \tilde{\mathbf{C}}_{xy} & \tilde{\mathbf{C}}_{xt} \\ \tilde{\mathbf{C}}_{yx} & \tilde{\mathbf{C}}_{yy} & \tilde{\mathbf{C}}_{yt} \\ \tilde{\mathbf{C}}_{tx} & \tilde{\mathbf{C}}_{ty} & \tilde{\mathbf{C}}_{tt} \end{bmatrix} \tag{6.88a}$$

$$\tilde{\mathbf{K}} = \mathbf{T}^T\mathbf{K}\mathbf{T} = \begin{bmatrix} \tilde{\mathbf{K}}_{xx} & \tilde{\mathbf{K}}_{xy} & \tilde{\mathbf{K}}_{xt} \\ \tilde{\mathbf{K}}_{yx} & \tilde{\mathbf{K}}_{yy} & \tilde{\mathbf{K}}_{yt} \\ \tilde{\mathbf{K}}_{tx} & \tilde{\mathbf{K}}_{ty} & \tilde{\mathbf{K}}_{tt} \end{bmatrix}, \quad \mathbf{T}^T\mathbf{M} = \begin{bmatrix} \tilde{\mathbf{M}}_{xx} & \mathbf{0} & \mathbf{0} & \mathbf{0} & \mathbf{0} & \mathbf{0} \\ \mathbf{0} & \mathbf{M}_{yy} & \mathbf{0} & \mathbf{0} & \mathbf{0} & \mathbf{0} \\ \mathbf{0} & \mathbf{0} & \mathbf{0} & \mathbf{0} & \mathbf{0} & \tilde{\mathbf{M}}_{tt} \end{bmatrix}. \tag{6.88b}$$

Recall from Section 4.2 that when a rigid diaphragm is assumed, the mass matrix becomes diagonal. Note that the third, fourth, and fifth columns of the matrix $\mathbf{T}^T\mathbf{M}$ in Eq. 6.88b are all zero and therefore it follows from Eq. 6.76 that

$$
\begin{bmatrix} \tilde{\mathbf{M}}_{xx} & 0 & 0 \\ 0 & \tilde{\mathbf{M}}_{yy} & 0 \\ 0 & 0 & \tilde{\mathbf{M}}_{tt} \end{bmatrix}
\begin{Bmatrix} \ddot{\tilde{\mathbf{X}}}_x(t) \\ \ddot{\tilde{\mathbf{X}}}_y(t) \\ \ddot{\tilde{\mathbf{X}}}_t(t) \end{Bmatrix}
+
\begin{bmatrix} \tilde{\mathbf{C}}_{xx} & \tilde{\mathbf{C}}_{xy} & \tilde{\mathbf{C}}_{xt} \\ \tilde{\mathbf{C}}_{yx} & \tilde{\mathbf{C}}_{yy} & \tilde{\mathbf{C}}_{yt} \\ \tilde{\mathbf{C}}_{tx} & \tilde{\mathbf{C}}_{ty} & \tilde{\mathbf{C}}_{tt} \end{bmatrix}
\begin{Bmatrix} \dot{\tilde{\mathbf{X}}}_x(t) \\ \dot{\tilde{\mathbf{X}}}_y(t) \\ \dot{\tilde{\mathbf{X}}}_t(t) \end{Bmatrix}
+
\begin{bmatrix} \tilde{\mathbf{K}}_{xx} & \tilde{\mathbf{K}}_{xy} & \tilde{\mathbf{K}}_{xt} \\ \tilde{\mathbf{K}}_{yx} & \tilde{\mathbf{K}}_{yy} & \tilde{\mathbf{K}}_{yt} \\ \tilde{\mathbf{K}}_{tx} & \tilde{\mathbf{K}}_{ty} & \tilde{\mathbf{K}}_{tt} \end{bmatrix}
\begin{Bmatrix} \tilde{\mathbf{X}}_x(t) \\ \tilde{\mathbf{X}}_y(t) \\ \tilde{\mathbf{X}}_t(t) \end{Bmatrix}
$$

$$
= - \begin{bmatrix} \tilde{\mathbf{M}}_{xx} & 0 & 0 \\ 0 & \tilde{\mathbf{M}}_{yy} & 0 \\ 0 & 0 & \tilde{\mathbf{M}}_{tt} \end{bmatrix}
\begin{Bmatrix} \mathbf{a}_x(t) \\ \mathbf{a}_y(t) \\ \mathbf{a}_t(t) \end{Bmatrix}. \tag{6.89}
$$

Thus the problem size has been reduced to only $3m$ degrees of freedom.

Once the structural engineer decides on the type and number of mass terms that are retained in the solution using Eq. 6.81, 6.85, or 6.89, then the degrees of freedom corresponding to each retained mass term are called *dynamic degrees of freedom*. The degrees of freedom corresponding to zero mass terms can be calculated in terms of the dynamic degrees of freedom, and thus they are not an independent variable. Therefore, the degrees of freedom corresponding to the zero mass terms are condensed out using a static force-displacement relationship, which is why this method is called static condensation. If the degrees of freedom with mass are labeled as 1 and those without mass are labeled as 3, the equation of motion described in Eq. 6.76 can be written as

$$
\begin{bmatrix} \mathbf{M}_{11} & 0 \\ 0 & 0 \end{bmatrix}
\begin{Bmatrix} \ddot{\mathbf{X}}_1(t) \\ \ddot{\mathbf{X}}_3(t) \end{Bmatrix}
+
\begin{bmatrix} \mathbf{C}_{11} & 0 \\ 0 & 0 \end{bmatrix}
\begin{Bmatrix} \dot{\mathbf{X}}_1(t) \\ \dot{\mathbf{X}}_3(t) \end{Bmatrix}
+
\begin{bmatrix} \mathbf{K}_{11} & \mathbf{K}_{13} \\ \mathbf{K}_{31} & \mathbf{K}_{33} \end{bmatrix}
\begin{Bmatrix} \mathbf{X}_1(t) \\ \mathbf{X}_3(t) \end{Bmatrix}
= - \begin{bmatrix} \mathbf{M}_{11} & 0 \\ 0 & 0 \end{bmatrix}
\begin{Bmatrix} \mathbf{a}_1(t) \\ \mathbf{a}_3(t) \end{Bmatrix}. \tag{6.90}
$$

Expanding Eq. 6.90 into two simultaneous equations gives

$$
\mathbf{M}_{11}\ddot{\mathbf{X}}_1(t) + \mathbf{C}_{11}\dot{\mathbf{X}}_1(t) + \mathbf{K}_{11}\mathbf{X}_1(t) + \mathbf{K}_{13}\mathbf{X}_3(t) = -\mathbf{M}_{11}\mathbf{a}_1(t) \tag{6.91a}
$$

$$
\mathbf{K}_{31}\mathbf{X}_1(t) + \mathbf{K}_{33}\mathbf{X}_3(t) = \mathbf{0}. \tag{6.91b}
$$

Note that Eq. 6.91b does not contain any velocity or acceleration terms, and thus it is a static equilibrium equation. Solving for $\mathbf{X}_3(t)$ in Eq. 6.91b gives

$$
\mathbf{X}_3(t) = -\mathbf{K}_{33}^{-1}\mathbf{K}_{31}\mathbf{X}_1(t). \tag{6.92}
$$

Equation 6.92 gives the response of the degrees of freedom without mass, $\mathbf{X}_3(t)$, in terms of the response of the degrees of freedom with mass, $\mathbf{X}_1(t)$. Substituting Eq. 6.92 into Eq. 6.91a, it follows that

$$
\mathbf{M}_{11}\ddot{\mathbf{X}}_1(t) + \mathbf{C}_{11}\dot{\mathbf{X}}_1(t) + \left(\mathbf{K}_{11} - \mathbf{K}_{13}\mathbf{K}_{33}^{-1}\mathbf{K}_{31} \right)\mathbf{X}_1(t) = -\mathbf{M}_{11}\mathbf{a}_1(t). \tag{6.93}
$$

Equation 6.93 represents the reduce equation as described by Eq. 6.81, 6.85, or 6.89, where

$$
\mathbf{M}_{11} = \tilde{\mathbf{M}}, \quad \mathbf{C}_{11} = \tilde{\mathbf{C}}, \quad \mathbf{K}_{11} - \mathbf{K}_{13}\mathbf{K}_{33}^{-1}\mathbf{K}_{31} = \tilde{\mathbf{K}}. \tag{6.94}
$$

In summary, the solution procedure is as follows:

1.   Identify the degrees of freedom where the mass can be ignored. The degrees of freedom that are retained are labeled 1, and those that have zero mass are labeled 3.
2.   Formulate the dynamic equilibrium of motion in the form of Eq. 6.90, which is

$$
\begin{bmatrix} \mathbf{M}_{11} & 0 \\ 0 & 0 \end{bmatrix}
\begin{Bmatrix} \ddot{\mathbf{X}}_1(t) \\ \ddot{\mathbf{X}}_3(t) \end{Bmatrix}
+
\begin{bmatrix} \mathbf{C}_{11} & 0 \\ 0 & 0 \end{bmatrix}
\begin{Bmatrix} \dot{\mathbf{X}}_1(t) \\ \dot{\mathbf{X}}_3(t) \end{Bmatrix}
+
\begin{bmatrix} \mathbf{K}_{11} & \mathbf{K}_{13} \\ \mathbf{K}_{31} & \mathbf{K}_{33} \end{bmatrix}
\begin{Bmatrix} \mathbf{X}_1(t) \\ \mathbf{X}_3(t) \end{Bmatrix}
= - \begin{bmatrix} \mathbf{M}_{11} & 0 \\ 0 & 0 \end{bmatrix}
\begin{Bmatrix} \mathbf{a}_1(t) \\ \mathbf{a}_3(t) \end{Bmatrix}. \tag{6.95}
$$

3. Use static condensation to reduce Eq. 6.95 to the form shown in Eq. 6.93, which is

$$\mathbf{M}_{11}\ddot{\mathbf{X}}_1(t) + \mathbf{C}_{11}\dot{\mathbf{X}}_1(t) + \left(\mathbf{K}_{11} - \mathbf{K}_{13}\mathbf{K}_{33}^{-1}\mathbf{K}_{31}\right)\mathbf{X}_1(t) = -\mathbf{M}_{11}\mathbf{a}_1(t). \tag{6.96}$$

4. Perform the numerical calculation by using the method discussed in Section 6.3 or Section 6.4 to obtain the responses, that is, $\mathbf{X}_1(t)$, $\dot{\mathbf{X}}_1(t)$, and $\ddot{\mathbf{X}}_1(t)$, at discrete time steps.

5. Use Eq. 6.92 to calculate the response for those degrees of freedom without mass:

$$\mathbf{X}_3(t) = -\mathbf{K}_{33}^{-1}\mathbf{K}_{31}\mathbf{X}_1(t). \tag{6.97}$$

EXAMPLE 1   *Two Degrees of Freedom Structure with Zero Mass in Upper Story*

Consider the two degrees of freedom system shown in Figure 3.2 with zero damping. The dynamic equilibrium equation of motion is

$$\begin{bmatrix} m_1 & 0 \\ 0 & m_2 \end{bmatrix}\begin{Bmatrix} \ddot{x}_1 \\ \ddot{x}_2 \end{Bmatrix} + \begin{bmatrix} k_1 + k_2 & -k_2 \\ -k_2 & k_2 \end{bmatrix}\begin{Bmatrix} x_1 \\ x_2 \end{Bmatrix} = \begin{Bmatrix} F_1(t) \\ F_2(t) \end{Bmatrix}. \tag{6.98}$$

Assume that $m_2 = 0$ and $F_2(t) = 0$; it follows from Eq. 6.98 that

$$\begin{bmatrix} m_1 & 0 \\ 0 & 0 \end{bmatrix}\begin{Bmatrix} \ddot{x}_1 \\ \ddot{x}_2 \end{Bmatrix} + \begin{bmatrix} k_1 + k_2 & -k_2 \\ -k_2 & k_2 \end{bmatrix}\begin{Bmatrix} x_1 \\ x_2 \end{Bmatrix} = \begin{Bmatrix} F_1(t) \\ 0 \end{Bmatrix}. \tag{6.99}$$

Expanding Eq. 6.99 gives

$$m_1\ddot{x}_1(t) + \left(k_1 + k_2\right)x_1(t) - k_2 x_2(t) = F_1(t) \tag{6.100}$$

$$-k_2 x_1(t) + k_2 x_2(t) = 0. \tag{6.101}$$

Solving for $x_2(t)$ in Eq. 6.101 gives

$$x_2(t) = x_1(t). \tag{6.102}$$

Substituting the result in Eq. 6.102 into Eq. 6.100 gives

$$m_1\ddot{x}_1(t) + k_1 x_1(t) = F_1(t) \tag{6.103}$$

which is the equation of motion in the reduced coordinate system.

The same result can be obtained using the procedure discussed above. Let

$$\mathbf{M}_{11} = m_1, \quad \mathbf{C}_{11} = 0, \quad \mathbf{X}_1 = x_1, \quad \mathbf{X}_3 = x_2$$
$$\mathbf{K}_{11} = k_1 + k_2, \quad \mathbf{K}_{13} = \mathbf{K}_{31} = -k_2, \quad \mathbf{K}_{33} = k_2 \tag{6.104}$$
$$-\mathbf{M}_{11}\mathbf{a}_1(t) = F_1(t).$$

Then it follows from Eq. 6.96 that

$$m_1\ddot{x}_1(t) + \left[k_1 + k_2 - (-k_2)(k_2^{-1})(-k_2)\right]x_1(t) = F_1(t) \tag{6.105}$$

and simplifying Eq. 6.105 gives

$$m_1\ddot{x}_1(t) + k_1 x_1(t) = F_1(t) \tag{6.106}$$

which is the same as Eq. 6.103. In addition, the degree of freedom without mass can be calculated using Eq. 6.94, that is,

$$x_2(t) = -k_2^{-1}(-k_2)x_1(t) = x_1(t) \tag{6.107}$$

which is the same as Eq. 6.102.

EXAMPLE 2 *Two Degrees of Freedom Structure with Zero Mass in Lower Story*

Consider Example 1 with mass proportional damping and earthquake excitation. The dynamic equilibrium equation of motion is

$$\begin{bmatrix} m_1 & 0 \\ 0 & m_2 \end{bmatrix} \begin{Bmatrix} \ddot{x}_1 \\ \ddot{x}_2 \end{Bmatrix} + \begin{bmatrix} c_1 & 0 \\ 0 & c_2 \end{bmatrix} \begin{Bmatrix} \dot{x}_1 \\ \dot{x}_2 \end{Bmatrix} + \begin{bmatrix} k_1 + k_2 & -k_2 \\ -k_2 & k_2 \end{bmatrix} \begin{Bmatrix} x_1 \\ x_2 \end{Bmatrix} = -\begin{bmatrix} m_1 & 0 \\ 0 & m_2 \end{bmatrix} \begin{Bmatrix} a_1(t) \\ a_2(t) \end{Bmatrix}. \tag{6.108}$$

Assume now that $m_1 = 0$; it follows from Eq. 6.108 and interchanging the first and second equations that

$$\begin{bmatrix} m_2 & 0 \\ 0 & 0 \end{bmatrix} \begin{Bmatrix} \ddot{x}_2 \\ \ddot{x}_1 \end{Bmatrix} + \begin{bmatrix} c_2 & 0 \\ 0 & 0 \end{bmatrix} \begin{Bmatrix} \dot{x}_2 \\ \dot{x}_1 \end{Bmatrix} + \begin{bmatrix} k_2 & -k_2 \\ -k_2 & k_1 + k_2 \end{bmatrix} \begin{Bmatrix} x_2 \\ x_1 \end{Bmatrix} = -\begin{bmatrix} m_2 & 0 \\ 0 & 0 \end{bmatrix} \begin{Bmatrix} a_2(t) \\ a_1(t) \end{Bmatrix}. \tag{6.109}$$

Expanding Eq. 6.109 gives

$$m_2 \ddot{x}_2(t) + c_2 \dot{x}_2(t) + k_2 x_2(t) - k_2 x_1(t) = -m_2 a_2(t) \tag{6.110}$$

$$-k_2 x_2(t) + (k_1 + k_2) x_1(t) = 0. \tag{6.111}$$

Solving for $x_1(t)$ in Eq. 6.111 gives

$$x_1(t) = \left(\frac{k_2}{k_1 + k_2} \right) x_2(t). \tag{6.112}$$

Substituting the result in Eq. 6.112 into Eq. 6.110 gives

$$m_2 \ddot{x}_2(t) + c_2 \dot{x}_2(t) + \left(\frac{k_1 k_2}{k_1 + k_2} \right) x_2(t) = -m_2 a_2(t) \tag{6.113}$$

which is the equation of motion in the reduced coordinate system.

The same result can be obtained using the procedure discussed previously. Let

$$\mathbf{M}_{11} = m_2, \quad \mathbf{C}_{11} = c_2, \quad \mathbf{X}_1 = x_2, \quad \mathbf{X}_3 = x_1$$
$$\mathbf{K}_{11} = k_2, \quad \mathbf{K}_{13} = \mathbf{K}_{31} = -k_2, \quad \mathbf{K}_{33} = k_1 + k_2 \tag{6.114}$$
$$\mathbf{a}_1(t) = a_2(t), \quad \mathbf{a}_3(t) = a_1(t).$$

Then it follows from Eq. 6.96 that

$$m_2 \ddot{x}_2(t) + c_2 \dot{x}_2(t) + \left[k_2 - (-k_2)(k_1 + k_2)^{-1}(-k_2) \right] x_2(t) = -m_2 a_2(t). \tag{6.115}$$

Simplifying Eq. 6.115 gives

$$m_2 \ddot{x}_2(t) + c_2 \dot{x}_2(t) + \left(\frac{k_1 k_2}{k_1 + k_2} \right) x_2(t) = -m_2 a_2(t) \tag{6.116}$$

which is the same as Eq. 6.113. In addition, the degree of freedom without mass can be calculated using Eq. 6.97, that is,

$$x_1(t) = -(k_1 + k_2)^{-1}(k_2) x_2(t) = \left(\frac{k_2}{k_1 + k_2} \right) x_2(t) \tag{6.117}$$

which is the same as Eq. 6.112.

EXAMPLE 3 *Two Degrees of Freedom Response*

Consider the 2DOF system in Example 1 of Section 6.3 but assume that the damping is zero. The dynamic equilibrium equation of motion is

$$\begin{bmatrix} 2m & 0 \\ 0 & m \end{bmatrix}\begin{Bmatrix} \ddot{x}_1(t) \\ \ddot{x}_2(t) \end{Bmatrix} + \begin{bmatrix} 4k & -2k \\ -2k & 3k \end{bmatrix}\begin{Bmatrix} x_1(t) \\ x_2(t) \end{Bmatrix} = -\begin{bmatrix} 2m & 0 \\ 0 & m \end{bmatrix}\begin{Bmatrix} 1 \\ 1 \end{Bmatrix} a_x(t) \qquad (6.118)$$

with the natural frequencies and mode shapes of the first two modes of vibration equal to

$$\omega_1 = \sqrt{\frac{k}{m}}, \quad \omega_2 = 2\sqrt{\frac{k}{m}} \qquad (6.119)$$

$$\hat{\boldsymbol{\phi}}_1 = \frac{1}{\sqrt{2}}\begin{Bmatrix} 1 \\ 1 \end{Bmatrix}, \quad \hat{\boldsymbol{\phi}}_2 = \frac{1}{\sqrt{5}}\begin{Bmatrix} 1 \\ -2 \end{Bmatrix}. \qquad (6.120)$$

Now assume that the mass on the bottom floor is zero:

$$\begin{bmatrix} 0 & 0 \\ 0 & m \end{bmatrix}\begin{Bmatrix} \ddot{x}_1(t) \\ \ddot{x}_2(t) \end{Bmatrix} + \begin{bmatrix} 4k & -2k \\ -2k & 3k \end{bmatrix}\begin{Bmatrix} x_1(t) \\ x_2(t) \end{Bmatrix} = -\begin{bmatrix} 0 & 0 \\ 0 & m \end{bmatrix}\begin{Bmatrix} a_x(t) \\ a_x(t) \end{Bmatrix}. \qquad (6.121)$$

Let

$$\begin{aligned} \mathbf{M}_{11} &= m, \quad \mathbf{C}_{11} = 0, \quad \mathbf{X}_1 = x_2, \quad \mathbf{X}_3 = x_1 \\ \mathbf{K}_{11} &= 3k, \quad \mathbf{K}_{13} = \mathbf{K}_{31} = -2k, \quad \mathbf{K}_{33} = 4k \\ \mathbf{a}_1(t) &= a_x(t), \quad \mathbf{a}_3(t) = a_x(t). \end{aligned} \qquad (6.122)$$

Then it follows from Eq. 6.96 that

$$m\ddot{x}_2(t) + \left[3k - (-2k)(4k)^{-1}(-2k) \right] x_2(t) = -ma_x(t). \qquad (6.123)$$

Simplifying Eq. 6.123 gives

$$m\ddot{x}_2(t) + 2kx_2(t) = -ma_x(t). \qquad (6.124)$$

In addition, the degree of freedom without mass can be calculated using Eq. 6.97:

$$x_1(t) = -(4k)^{-1}(-2k)x_2(t) = \frac{1}{2}x_2(t). \qquad (6.125)$$

Note that by removing one mass in the equation of motion for the 2DOF system, the structure vibrates at one natural frequency (ω_s) with only one mode shape $(\hat{\boldsymbol{\phi}}_s)$:

$$\omega_s = \sqrt{\frac{2k}{m}} = 1.414\sqrt{\frac{k}{m}} \qquad (6.126)$$

$$\hat{\boldsymbol{\phi}}_s = \frac{1}{\sqrt{5}}\begin{Bmatrix} 1 \\ 2 \end{Bmatrix}. \qquad (6.127)$$

EXAMPLE 4   *Two-Dimensional One-Story Frame*

Consider the one-story frame shown in Figure 6.9. The dynamic equilibrium equation of motion is given by

$$\mathbf{M}\begin{Bmatrix} \ddot{x}_1(t) \\ \ddot{x}_2(t) \\ \ddot{x}_3(t) \\ \ddot{x}_4(t) \\ \ddot{x}_5(t) \\ \ddot{x}_6(t) \end{Bmatrix} + \mathbf{C}\begin{Bmatrix} \dot{x}_1(t) \\ \dot{x}_2(t) \\ \dot{x}_3(t) \\ \dot{x}_4(t) \\ \dot{x}_5(t) \\ \dot{x}_6(t) \end{Bmatrix} + \mathbf{K}\begin{Bmatrix} x_1(t) \\ x_2(t) \\ x_3(t) \\ x_4(t) \\ x_5(t) \\ x_6(t) \end{Bmatrix} = -\mathbf{M}\begin{Bmatrix} a_x(t) \\ a_z(t) \\ a_s(t) \\ a_x(t) \\ a_z(t) \\ a_s(t) \end{Bmatrix}. \qquad (6.128)$$

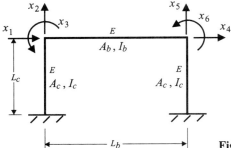

Figure 6.9 Two-Dimensional One-Story Frame

Assume that lumped mass is used then the dynamic matrices given in Eq. 6.76 are

$$
\mathbf{M} = \begin{bmatrix} m_x/2 & 0 & 0 & 0 & 0 & 0 \\ 0 & m_z/2 & 0 & 0 & 0 & 0 \\ 0 & 0 & m_s/2 & 0 & 0 & 0 \\ 0 & 0 & 0 & m_x/2 & 0 & 0 \\ 0 & 0 & 0 & 0 & m_z/2 & 0 \\ 0 & 0 & 0 & 0 & 0 & m_s/2 \end{bmatrix}, \quad
\mathbf{C} = \begin{bmatrix} c_{11} & c_{12} & c_{13} & c_{14} & c_{15} & c_{16} \\ c_{21} & c_{22} & c_{23} & c_{24} & c_{25} & c_{26} \\ c_{31} & c_{32} & c_{33} & c_{34} & c_{35} & c_{36} \\ c_{41} & c_{42} & c_{43} & c_{44} & c_{45} & c_{46} \\ c_{51} & c_{52} & c_{53} & c_{54} & c_{55} & c_{56} \\ c_{61} & c_{62} & c_{63} & c_{64} & c_{65} & c_{66} \end{bmatrix}
$$

$$
\mathbf{K} = E \begin{bmatrix}
\dfrac{A_b}{L_b} + \dfrac{12I_c}{L_c^3} & 0 & \dfrac{6I_c}{L_c^2} & -\dfrac{A_b}{L_b} & 0 & 0 \\[2mm]
0 & \dfrac{A_c}{L_c} + \dfrac{12I_b}{L_b^3} & \dfrac{6I_b}{L_b^2} & 0 & -\dfrac{12I_b}{L_b^3} & \dfrac{6I_b}{L_b^2} \\[2mm]
\dfrac{6I_c}{L_c^2} & 0 & \dfrac{4I_b}{L_b} + \dfrac{4I_c}{L_c} & 0 & \dfrac{6I_b}{L_b^2} & \dfrac{2I_b}{L_b} \\[2mm]
-\dfrac{A_b}{L_b} & 0 & 0 & \dfrac{A_b}{L_b} + \dfrac{12I_c}{L_c^3} & 0 & \dfrac{6I_c}{L_c^2} \\[2mm]
0 & -\dfrac{12I_b}{L_b^3} & \dfrac{6I_b}{L_b^2} & 0 & \dfrac{A_c}{L_c} + \dfrac{12I_b}{L_b^3} & -\dfrac{6I_b}{L_b^2} \\[2mm]
0 & \dfrac{6I_b}{L_b^2} & \dfrac{2I_b}{L_b} & \dfrac{6I_c}{L_c^2} & -\dfrac{6I_b}{L_b^2} & \dfrac{4I_b}{L_b} + \dfrac{4I_c}{L_c}
\end{bmatrix}.
$$

Now assume that the beam is rigid and that only earthquake ground motion in the horizontal direction is considered; then the transformation matrix can be expressed as

$$
\mathbf{T} = \begin{bmatrix} 1 \\ 0 \\ 0 \\ 1 \\ 0 \\ 0 \end{bmatrix}. \tag{6.129}
$$

It follows from Eq. 6.88 that

$$
\tilde{\mathbf{M}} = \mathbf{T}^T \mathbf{M} \mathbf{T} = \frac{m_{xx}}{2} + \frac{m_{xx}}{2} = m_{xx}, \quad \tilde{\mathbf{C}} = \mathbf{T}^T \mathbf{C} \mathbf{T} = c_{11} + c_{14} + c_{41} + c_{44} \tag{6.130}
$$

$$
\tilde{\mathbf{K}} = \mathbf{T}^T \mathbf{K} \mathbf{T} = \frac{EA_b}{L_b} + \frac{12EI_c}{L_c^3} - \frac{EA_b}{L_b} - \frac{EA_b}{L_b} + \frac{EA_b}{L_b} + \frac{12EI_c}{L_c^3} = \frac{24EI_c}{L_c^3}. \tag{6.131}
$$

It then follows from Eq. 6.89 that

$$m_{xx}\ddot{x}_1(t) + \left(c_{11} + c_{14} + c_{41} + c_{44}\right)\dot{x}_1(t) + \left(\frac{24EI_c}{L_c^3}\right)x_1(t) = -m_{xx}a_x(t) \tag{6.132}$$

which is the equation of motion of an SDOF system.

EXAMPLE 5   *Three-Dimensional One-Story Frame with Rigid Diaphragm*

Consider the one-story frame shown in Figure 6.10a with the plan view shown in Figure 6.10b. Both the vertical mass and the joint rotation mass moment of inertia are assumed to be zero. It follows that there are a total of four joints, each with three degrees of freedom, giving a total of 12 degrees of freedom. Thus

$$\mathbf{X}(t) = \left\{x_x^{(1)}(t) \quad \cdots \quad x_x^{(4)}(t) \quad x_y^{(1)}(t) \quad \cdots \quad x_y^{(4)}(t) \quad x_t^{(1)}(t) \quad \cdots \quad x_t^{(4)}(t)\right\}^T \tag{6.133}$$

where the superscript in parentheses denotes the joint number.

When a rigid diaphragm is assumed, the degrees of freedom represented in Eq. 6.133 can be transformed into three degrees of freedom at the center of mass, that is,

$$\tilde{\mathbf{X}}(t) = \left\{\tilde{x}_x(t) \quad \tilde{x}_y(t) \quad \tilde{x}_t(t)\right\}^T \tag{6.134}$$

with the transformation

$$
\begin{Bmatrix}
x_x^{(1)}(t) \\
x_x^{(2)}(t) \\
x_x^{(3)}(t) \\
x_x^{(4)}(t) \\
x_y^{(1)}(t) \\
x_y^{(2)}(t) \\
x_y^{(3)}(t) \\
x_y^{(4)}(t) \\
x_t^{(1)}(t) \\
x_t^{(2)}(t) \\
x_t^{(3)}(t) \\
x_t^{(4)}(t)
\end{Bmatrix}
=
\begin{bmatrix}
1 & 0 & -L/2 \\
1 & 0 & -L/2 \\
1 & 0 & L/2 \\
1 & 0 & L/2 \\
0 & 1 & L/2 \\
0 & 1 & -L/2 \\
0 & 1 & -L/2 \\
0 & 1 & L/2 \\
0 & 0 & 1 \\
0 & 0 & 1 \\
0 & 0 & 1 \\
0 & 0 & 1
\end{bmatrix}
\begin{Bmatrix}
\tilde{x}_x(t) \\
\tilde{x}_y(t) \\
\tilde{x}_t(t)
\end{Bmatrix},
\quad
\mathbf{T} =
\begin{bmatrix}
1 & 0 & -L/2 \\
1 & 0 & -L/2 \\
1 & 0 & L/2 \\
1 & 0 & L/2 \\
0 & 1 & L/2 \\
0 & 1 & -L/2 \\
0 & 1 & -L/2 \\
0 & 1 & L/2 \\
0 & 0 & 1 \\
0 & 0 & 1 \\
0 & 0 & 1 \\
0 & 0 & 1
\end{bmatrix}.
\tag{6.135}
$$

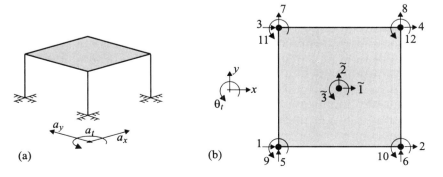

(a)   (b)

Figure 6.10 Three-Dimensional One-Story Frame

It then follows from Eq. 6.88 that

$$\tilde{\mathbf{M}} = \mathbf{T}^T \mathbf{M} \mathbf{T} = \begin{bmatrix} \tilde{m}_{xx} & 0 & 0 \\ 0 & \tilde{m}_{yy} & 0 \\ \tilde{m}_{tx} & 0 & \tilde{m}_{tt} \end{bmatrix}, \quad \tilde{\mathbf{C}} = \mathbf{T}^T \mathbf{C} \mathbf{T} = \begin{bmatrix} \tilde{c}_{xx} & \tilde{c}_{xy} & \tilde{c}_{xt} \\ \tilde{c}_{yx} & \tilde{c}_{yy} & \tilde{c}_{yt} \\ \tilde{c}_{tx} & \tilde{c}_{ty} & \tilde{c}_{tt} \end{bmatrix} \quad (6.136)$$

$$\tilde{\mathbf{K}} = \mathbf{T}^T \mathbf{K} \mathbf{T} = \begin{bmatrix} \tilde{k}_{xx} & \tilde{k}_{xy} & \tilde{k}_{xt} \\ \tilde{k}_{yx} & \tilde{k}_{yy} & \tilde{k}_{yt} \\ \tilde{k}_{tx} & \tilde{k}_{ty} & \tilde{k}_{tt} \end{bmatrix}, \quad \mathbf{T}^T \mathbf{M} = \begin{bmatrix} \tilde{m}_{xx} & 0 & 0 & 0 & 0 \\ 0 & \tilde{m}_{yy} & 0 & 0 & 0 \\ 0 & \tilde{m}_{ty} & 0 & 0 & \tilde{m}_{tt} \end{bmatrix}. \quad (6.137)$$

If we ignore the third, fourth, and fifth columns of the matrix $\mathbf{T}^T \mathbf{M}$ by removing $a_z(t)$, $a_r(t)$, and $a_s(t)$, Eq. 6.89 becomes

$$\begin{bmatrix} \tilde{m}_{xx} & 0 & 0 \\ 0 & \tilde{m}_{yy} & 0 \\ 0 & 0 & \tilde{m}_{tt} \end{bmatrix} \begin{Bmatrix} \ddot{\tilde{x}}_x(t) \\ \ddot{\tilde{x}}_y(t) \\ \ddot{\tilde{x}}_t(t) \end{Bmatrix} + \begin{bmatrix} \tilde{c}_{xx} & \tilde{c}_{xy} & \tilde{c}_{xt} \\ \tilde{c}_{yx} & \tilde{c}_{yy} & \tilde{c}_{yt} \\ \tilde{c}_{tx} & \tilde{c}_{ty} & \tilde{c}_{tt} \end{bmatrix} \begin{Bmatrix} \dot{\tilde{x}}_x(t) \\ \dot{\tilde{x}}_y(t) \\ \dot{\tilde{x}}_t(t) \end{Bmatrix} + \begin{bmatrix} \tilde{k}_{xx} & \tilde{k}_{xy} & \tilde{k}_{xt} \\ \tilde{k}_{yx} & \tilde{k}_{yy} & \tilde{k}_{yt} \\ \tilde{k}_{tx} & \tilde{k}_{ty} & \tilde{k}_{tt} \end{bmatrix} \begin{Bmatrix} \tilde{x}_x(t) \\ \tilde{x}_y(t) \\ \tilde{x}_t(t) \end{Bmatrix}$$

$$= -\begin{bmatrix} \tilde{m}_{xx} & 0 & 0 \\ 0 & \tilde{m}_{yy} & 0 \\ 0 & 0 & \tilde{m}_{tt} \end{bmatrix} \begin{Bmatrix} a_x(t) \\ a_y(t) \\ a_t(t) \end{Bmatrix}. \quad (6.138)$$

6.6 PROBLEM SIZE REDUCTION USING DYNAMIC CONDENSATION

In Section 6.5, static condensation is used to reduce the number of degrees of freedom in a structure with certain mass terms assumed to be zero. In that case, certain degrees of freedom can be expressed directly in terms of other degrees of freedom, as presented in Eq. 6.92. In this section, the same condensation is performed but on a structure with mass in all the degrees of freedom. This is known as problem size reduction using *dynamic condensation*.

Consider Eq. 6.90 with mass in all the degrees of freedom:

$$\begin{bmatrix} \mathbf{M}_{11} & \mathbf{M}_{13} \\ \mathbf{M}_{31} & \mathbf{M}_{33} \end{bmatrix} \begin{Bmatrix} \ddot{\mathbf{X}}_1(t) \\ \ddot{\mathbf{X}}_3(t) \end{Bmatrix} + \begin{bmatrix} \mathbf{C}_{11} & \mathbf{C}_{13} \\ \mathbf{C}_{31} & \mathbf{C}_{33} \end{bmatrix} \begin{Bmatrix} \dot{\mathbf{X}}_1(t) \\ \dot{\mathbf{X}}_3(t) \end{Bmatrix} + \begin{bmatrix} \mathbf{K}_{11} & \mathbf{K}_{13} \\ \mathbf{K}_{31} & \mathbf{K}_{33} \end{bmatrix} \begin{Bmatrix} \mathbf{X}_1(t) \\ \mathbf{X}_3(t) \end{Bmatrix}$$

$$= -\begin{bmatrix} \mathbf{M}_{11} & \mathbf{M}_{13} \\ \mathbf{M}_{31} & \mathbf{M}_{33} \end{bmatrix} \begin{Bmatrix} \mathbf{a}_1(t) \\ \mathbf{a}_3(t) \end{Bmatrix}. \quad (6.139)$$

In dynamic condensation, the objective is to impose the condition as given in Eq. 6.92:

$$\mathbf{X}_3(t) = -\mathbf{K}_{33}^{-1} \mathbf{K}_{31} \mathbf{X}_1(t). \quad (6.140)$$

From Eq. 6.140, the transformation matrix is given by

$$\begin{Bmatrix} \mathbf{X}_1(t) \\ \mathbf{X}_3(t) \end{Bmatrix} = \begin{bmatrix} \mathbf{I} \\ -\mathbf{K}_{33}^{-1} \mathbf{K}_{31} \end{bmatrix} \mathbf{X}_1(t) = \mathbf{T} \mathbf{X}_1(t). \quad (6.141)$$

Substituting Eq. 6.141 into Eq. 6.139, it follows that

$$\begin{bmatrix} \mathbf{M}_{11} & \mathbf{M}_{13} \\ \mathbf{M}_{31} & \mathbf{M}_{33} \end{bmatrix} \begin{bmatrix} \mathbf{I} \\ -\mathbf{K}_{33}^{-1} \mathbf{K}_{31} \end{bmatrix} \ddot{\mathbf{X}}_1(t) + \begin{bmatrix} \mathbf{C}_{11} & \mathbf{C}_{13} \\ \mathbf{C}_{31} & \mathbf{C}_{33} \end{bmatrix} \begin{bmatrix} \mathbf{I} \\ -\mathbf{K}_{33}^{-1} \mathbf{K}_{31} \end{bmatrix} \dot{\mathbf{X}}_1(t)$$

$$+ \begin{bmatrix} \mathbf{K}_{11} & \mathbf{K}_{13} \\ \mathbf{K}_{31} & \mathbf{K}_{33} \end{bmatrix} \begin{bmatrix} \mathbf{I} \\ -\mathbf{K}_{33}^{-1} \mathbf{K}_{31} \end{bmatrix} \mathbf{X}_1(t) = -\begin{bmatrix} \mathbf{M}_{11} & \mathbf{M}_{13} \\ \mathbf{M}_{31} & \mathbf{M}_{33} \end{bmatrix} \begin{Bmatrix} \mathbf{a}_1(t) \\ \mathbf{a}_3(t) \end{Bmatrix}. \quad (6.142)$$

Premultiplying Eq. 6.142 by $\mathbf{T}^T$ gives

$$\begin{bmatrix} \mathbf{I} \\ -\mathbf{K}_{33}^{-1}\mathbf{K}_{31} \end{bmatrix}^T \begin{bmatrix} \mathbf{M}_{11} & \mathbf{M}_{13} \\ \mathbf{M}_{31} & \mathbf{M}_{33} \end{bmatrix} \begin{bmatrix} \mathbf{I} \\ -\mathbf{K}_{33}^{-1}\mathbf{K}_{31} \end{bmatrix} \ddot{\mathbf{X}}_1(t) + \begin{bmatrix} \mathbf{I} \\ -\mathbf{K}_{33}^{-1}\mathbf{K}_{31} \end{bmatrix}^T \begin{bmatrix} \mathbf{C}_{11} & \mathbf{C}_{13} \\ \mathbf{C}_{31} & \mathbf{C}_{33} \end{bmatrix} \begin{bmatrix} \mathbf{I} \\ -\mathbf{K}_{33}^{-1}\mathbf{K}_{31} \end{bmatrix} \dot{\mathbf{X}}_1(t)$$

$$+ \begin{bmatrix} \mathbf{I} \\ -\mathbf{K}_{33}^{-1}\mathbf{K}_{31} \end{bmatrix}^T \begin{bmatrix} \mathbf{K}_{11} & \mathbf{K}_{13} \\ \mathbf{K}_{31} & \mathbf{K}_{33} \end{bmatrix} \begin{bmatrix} \mathbf{I} \\ -\mathbf{K}_{33}^{-1}\mathbf{K}_{31} \end{bmatrix} \mathbf{X}_1(t) = -\begin{bmatrix} \mathbf{I} \\ -\mathbf{K}_{33}^{-1}\mathbf{K}_{31} \end{bmatrix}^T \begin{bmatrix} \mathbf{M}_{11} & \mathbf{M}_{13} \\ \mathbf{M}_{31} & \mathbf{M}_{33} \end{bmatrix} \begin{Bmatrix} \mathbf{a}_1(t) \\ \mathbf{a}_3(t) \end{Bmatrix}. \tag{6.143}$$

Expanding Eq. 6.143 gives

$$\left[\mathbf{M}_{11} - \mathbf{M}_{13}\mathbf{K}_{33}^{-1}\mathbf{K}_{31} - \mathbf{K}_{13}\mathbf{K}_{33}^{-1}\mathbf{M}_{31} + \mathbf{K}_{13}\mathbf{K}_{33}^{-1}\mathbf{M}_{33}\mathbf{K}_{33}^{-1}\mathbf{K}_{31} \right] \ddot{\mathbf{X}}_1(t)$$

$$+ \left[\mathbf{C}_{11} - \mathbf{C}_{13}\mathbf{K}_{33}^{-1}\mathbf{K}_{31} - \mathbf{K}_{13}\mathbf{K}_{33}^{-1}\mathbf{C}_{31} + \mathbf{K}_{13}\mathbf{K}_{33}^{-1}\mathbf{C}_{33}\mathbf{K}_{33}^{-1}\mathbf{K}_{31} \right] \dot{\mathbf{X}}_1(t)$$

$$+ \left[\mathbf{K}_{11} - \mathbf{K}_{13}\mathbf{K}_{33}^{-1}\mathbf{K}_{31} \right] \mathbf{X}_1(t) = -\begin{bmatrix} \mathbf{M}_{11} - \mathbf{K}_{13}\mathbf{K}_{33}^{-1}\mathbf{M}_{31} & \mathbf{M}_{13} - \mathbf{K}_{13}\mathbf{K}_{33}^{-1}\mathbf{M}_{33} \end{bmatrix} \begin{Bmatrix} \mathbf{a}_1(t) \\ \mathbf{a}_3(t) \end{Bmatrix}. \tag{6.144}$$

Let

$$\tilde{\mathbf{M}} = \mathbf{M}_{11} - \mathbf{M}_{13}\mathbf{K}_{33}^{-1}\mathbf{K}_{31} - \mathbf{K}_{13}\mathbf{K}_{33}^{-1}\mathbf{M}_{31} + \mathbf{K}_{13}\mathbf{K}_{33}^{-1}\mathbf{M}_{33}\mathbf{K}_{33}^{-1}\mathbf{K}_{31} \tag{6.145a}$$

$$\tilde{\mathbf{C}} = \mathbf{C}_{11} - \mathbf{C}_{13}\mathbf{K}_{33}^{-1}\mathbf{K}_{31} - \mathbf{K}_{13}\mathbf{K}_{33}^{-1}\mathbf{C}_{31} + \mathbf{K}_{13}\mathbf{K}_{33}^{-1}\mathbf{C}_{33}\mathbf{K}_{33}^{-1}\mathbf{K}_{31} \tag{6.145b}$$

$$\tilde{\mathbf{K}} = \mathbf{K}_{11} - \mathbf{K}_{13}\mathbf{K}_{33}^{-1}\mathbf{K}_{31}. \tag{6.145c}$$

Equation 6.144 becomes

$$\tilde{\mathbf{M}}\ddot{\mathbf{X}}_1(t) + \tilde{\mathbf{C}}\dot{\mathbf{X}}_1(t) + \tilde{\mathbf{K}}\mathbf{X}_1(t) = -\begin{bmatrix} \mathbf{M}_{11} - \mathbf{K}_{13}\mathbf{K}_{33}^{-1}\mathbf{M}_{31} & \mathbf{M}_{13} - \mathbf{K}_{13}\mathbf{K}_{33}^{-1}\mathbf{M}_{33} \end{bmatrix} \begin{Bmatrix} \mathbf{a}_1(t) \\ \mathbf{a}_3(t) \end{Bmatrix}. \tag{6.146}$$

Note that when $\mathbf{M}_{13} = \mathbf{M}_{31} = \mathbf{M}_{33} = 0$ and $\mathbf{C}_{13} = \mathbf{C}_{31} = \mathbf{C}_{33} = 0$, Eq. 6.145 becomes

$$\tilde{\mathbf{M}} = \mathbf{M}_{11}, \quad \tilde{\mathbf{C}} = \mathbf{C}_{11} \tag{6.147}$$

and therefore Eq. 6.46 becomes

$$\mathbf{M}_{11}\ddot{\mathbf{X}}_1(t) + \mathbf{C}_{11}\dot{\mathbf{X}}_1(t) + \left(\mathbf{K}_{11} - \mathbf{K}_{13}\mathbf{K}_{33}^{-1}\mathbf{K}_{31} \right)\mathbf{X}_1(t) = -\mathbf{M}_{11}\mathbf{a}_1(t) \tag{6.148}$$

which is the same as Eq. 6.93.

Equation 6.141 represents a transformation matrix that defines the displacements $\mathbf{X}_1(t)$ and $\mathbf{X}_3(t)$ in terms of the displacement vector $\mathbf{X}_1(t)$ only. Therefore, it represents a constraint on the response. By the nature of the transformation in Eqs. 6.141 through 6.146, it follows that the kinetic, damping, and strain energy of Eq. 6.139 is the same as Eq. 6.146. Also, the work done by the earthquake force on the right-hand side of Eq. 6.139 is the same as the work done by the earthquake force on the right-hand side of Eq. 6.146.

EXAMPLE 1 *Two Degrees of Freedom Structure with Zero Mass in Upper Story*

Consider the two degrees of freedom system discussed in Example 1 of Section 6.5. The dynamic equilibrium equation of motion is

$$\begin{bmatrix} m_1 & 0 \\ 0 & m_2 \end{bmatrix} \begin{Bmatrix} \ddot{x}_1 \\ \ddot{x}_2 \end{Bmatrix} + \begin{bmatrix} k_1 + k_2 & -k_2 \\ -k_2 & k_2 \end{bmatrix} \begin{Bmatrix} x_1 \\ x_2 \end{Bmatrix} = \begin{Bmatrix} F_1(t) \\ F_2(t) \end{Bmatrix}. \tag{6.149}$$

Assume that Eq. 6.102 is correct, that is,

$$x_2(t) = x_1(t). \tag{6.150}$$

It follows from expanding the first equation in Eq. 6.149 that

$$m_1\ddot{x}_1(t) + (k_1 + k_2)x_1(t) - k_2x_1(t) = F_1(t). \tag{6.151}$$

Simplifying Eq. 6.151 gives

$$m_1\ddot{x}_1(t) + k_1x_1(t) = F_1(t) \tag{6.152}$$

which is the equation of motion when static condensation is used.

Consider that dynamic condensation is used, which can be obtained using the procedure discussed previously. Let

$$\mathbf{M}_{11} = m_1, \quad \mathbf{M}_{13} = \mathbf{M}_{31} = 0, \quad \mathbf{M}_{33} = m_2, \quad \mathbf{C}_{11} = \mathbf{C}_{13} = \mathbf{C}_{31} = \mathbf{C}_{33} = 0$$
$$\mathbf{K}_{11} = k_1 + k_2, \quad \mathbf{K}_{13} = \mathbf{K}_{31} = -k_2, \quad \mathbf{K}_{33} = k_2, \quad \mathbf{X}_1 = x_1, \quad \mathbf{X}_3 = x_2 \tag{6.153}$$
$$-\mathbf{M}_{11}\mathbf{a}_1(t) - \mathbf{M}_{13}\mathbf{a}_3(t) = F_1(t), \quad -\mathbf{M}_{31}\mathbf{a}_1(t) - \mathbf{M}_{33}\mathbf{a}_3(t) = F_2(t).$$

Then it follows from Eqs. 6.143 and 6.144 that

$$\left[m_1 + (-k_2)(k_2^{-1})m_2(k_2^{-1})(-k_2)\right]\ddot{x}_1(t)$$

$$+\left[k_1 + k_2 - (-k_2)(k_2^{-1})(-k_2)\right]x_1(t) = \begin{bmatrix} 1 & -(-k_2)(k_2^{-1}) \end{bmatrix}\begin{Bmatrix} F_1(t) \\ F_2(t) \end{Bmatrix}. \tag{6.154}$$

Simplifying Eq. 6.154 gives

$$(m_1 + m_2)\ddot{x}_1(t) + k_1x_1(t) = F_1(t) + F_2(t) \tag{6.155}$$

which is different from Eq. 6.152. This is because the second equation in Eq. 6.149 is not considered in Eq. 6.152, and this is the main difference between static and dynamic condensations.

For dynamic condensation, first note that since $x_2(t) = x_1(t)$ is assumed, it follows that

$$\dot{x}_2(t) = \dot{x}_1(t), \quad \ddot{x}_2(t) = \ddot{x}_1(t). \tag{6.156}$$

Substituting Eq. 6.156 into the second equation of Eq. 6.149 gives

$$m_2\ddot{x}_1 - k_2x_1 + k_2x_1 = F_2(t). \tag{6.157}$$

Simplifying Eq. 6.157, it follows that

$$m_2\ddot{x}_1 = F_2(t). \tag{6.158}$$

Since $F_1(t)$ and $F_2(t)$ are general forcing functions, no solution can satisfy both Eqs. 6.152 and 6.158 at the same time. An approximate solution is obtained by using the conservation of energy, which is composed of the energy from Eq. 6.152 and Eq. 6.158. For this example, as shown in Eq. 6.150 where $x_2(t) = x_1(t)$, the amounts of energy stored in Eqs. 6.152 and 6.158 are equal. Therefore, adding Eq. 6.152 to Eq. 6.158 gives

$$m_1\ddot{x}_1(t) + k_1x_1(t) + m_2\ddot{x}_1 = F_1(t) + F_2(t). \tag{6.159}$$

Simplifying Eq. 6.159, it follows that

$$(m_1 + m_2)\ddot{x}_1(t) + k_1x_1(t) = F_1(t) + F_2(t) \tag{6.160}$$

which is the same as Eq. 6.155.

EXAMPLE 2   *Two Degrees of Freedom Structure with Zero Mass in Lower Story*

Consider the two degrees of freedom system discussed in Example 2 of Section 6.5. The dynamic equilibrium equation of motion is

$$\begin{bmatrix} m_1 & 0 \\ 0 & m_2 \end{bmatrix}\begin{Bmatrix} \ddot{x}_1 \\ \ddot{x}_2 \end{Bmatrix} + \begin{bmatrix} c_1 & 0 \\ 0 & c_2 \end{bmatrix}\begin{Bmatrix} \dot{x}_1 \\ \dot{x}_2 \end{Bmatrix} + \begin{bmatrix} k_1+k_2 & -k_2 \\ -k_2 & k_2 \end{bmatrix}\begin{Bmatrix} x_1 \\ x_2 \end{Bmatrix} = -\begin{bmatrix} m_1 & 0 \\ 0 & m_2 \end{bmatrix}\begin{Bmatrix} a_1(t) \\ a_2(t) \end{Bmatrix}. \tag{6.161}$$

Based on dynamic condensation, assume that Eq. 6.112 is correct, that is,

$$x_1(t) = \left(\frac{k_2}{k_1+k_2}\right)x_2(t). \tag{6.162}$$

It follows that

$$\dot{x}_1(t) = \left(\frac{k_2}{k_1+k_2}\right)\dot{x}_2(t), \quad \ddot{x}_1(t) = \left(\frac{k_2}{k_1+k_2}\right)\ddot{x}_2(t). \tag{6.163}$$

Substituting Eqs. 6.162 and 6.163 into Eq. 6.161 gives two equations:

$$m_1\left(\frac{k_2}{k_1+k_2}\right)\ddot{x}_2 + c_1\left(\frac{k_2}{k_1+k_2}\right)\dot{x}_2 = -m_1 a_1(t) \tag{6.164}$$

$$m_2\ddot{x}_2 + c_2\dot{x}_2 + \left(\frac{k_1 k_2}{k_1+k_2}\right)x_2 = -m_2 a_2(t). \tag{6.165}$$

For any earthquake ground accelerations $a_1(t)$ and $a_2(t)$, no solution can satisfy both Eqs. 6.164 and 6.165 at the same time. Using the coefficient for $x_2(t)$ in Eq. 6.162 as the contribution of Eqs. 6.164 and 6.165 to the conservation of energy, multiplying Eq. 6.164 by $k_2/(k_1+k_2)$, and adding the result to Eq. 6.165 gives

$$\left[m_2 + \left(\frac{k_2}{k_1+k_2}\right)^2 m_1\right]\ddot{x}_2(t) + \left[c_2 + \left(\frac{k_2}{k_1+k_2}\right)^2 c_1\right]\dot{x}_2(t)$$
$$+\left(\frac{k_1 k_2}{k_1+k_2}\right)x_2(t) = -m_2 a_2(t) - \left(\frac{k_2}{k_1+k_2}\right)m_1 a_1(t) \tag{6.166}$$

which is the equation of motion after dynamic condensation.

The same result can be obtained using the procedure discussed previously. Let

$$\mathbf{M}_{11} = m_2, \quad \mathbf{M}_{13} = \mathbf{M}_{31} = 0, \quad \mathbf{M}_{33} = m_1$$
$$\mathbf{C}_{11} = c_2, \quad \mathbf{C}_{13} = \mathbf{C}_{31} = 0, \quad \mathbf{C}_{33} = c_1$$
$$\mathbf{K}_{11} = k_2, \quad \mathbf{K}_{13} = \mathbf{K}_{31} = -k_2, \quad \mathbf{K}_{33} = k_1+k_2$$
$$\mathbf{a}_1(t) = a_2(t), \quad \mathbf{a}_3(t) = a_1(t), \quad \mathbf{X}_1 = x_2, \quad \mathbf{X}_3 = x_1. \tag{6.167}$$

It follows from Eq. 6.144 that

$$\left[m_2 + (-k_2)(k_1+k_2)^{-1}m_1(k_1+k_2)^{-1}(-k_2)\right]\ddot{x}_2(t)$$
$$+\left[c_2 + (-k_2)(k_1+k_2)^{-1}c_1(k_1+k_2)^{-1}(-k_2)\right]\dot{x}_2(t) \tag{6.168}$$
$$+\left[k_2 - (-k_2)(k_1+k_2)^{-1}(-k_2)\right]x_2(t) = -m_2 a_2(t) - \left[(-k_2)(k_1+k_2)^{-1}m_1\right]a_1(t).$$

Simplifying Eq. 6.168 gives

$$\left[m_2 + \left(\frac{k_2}{k_1+k_2}\right)^2 m_1\right]\ddot{x}_2(t) + \left[c_2 + \left(\frac{k_2}{k_1+k_2}\right)^2 c_1\right]\dot{x}_2(t)$$
$$+\left(\frac{k_1 k_2}{k_1+k_2}\right)x_2(t) = -m_2 a_2(t) - \left(\frac{k_2}{k_1+k_2}\right)m_1 a_1(t) \tag{6.169}$$

which is the same as Eq. 6.166.

EXAMPLE 3   *Two Degrees of Freedom Response*

Consider the two degrees of freedom system discussed in Example 3 of Section 6.5. The dynamic equilibrium equation of motion is

$$\begin{bmatrix} 2m & 0 \\ 0 & m \end{bmatrix}\begin{Bmatrix} \ddot{x}_1(t) \\ \ddot{x}_2(t) \end{Bmatrix} + \begin{bmatrix} 4k & -2k \\ -2k & 3k \end{bmatrix}\begin{Bmatrix} x_1(t) \\ x_2(t) \end{Bmatrix} = -\begin{bmatrix} 2m & 0 \\ 0 & m \end{bmatrix}\begin{Bmatrix} 1 \\ 1 \end{Bmatrix} a_x(t) \tag{6.170}$$

with the natural frequencies and mode shapes of the first two modes of vibration equal to

$$\omega_1 = \sqrt{\frac{k}{m}}, \quad \omega_2 = 2\sqrt{\frac{k}{m}} \tag{6.171}$$

$$\hat{\phi}_1 = \frac{1}{\sqrt{2}}\begin{Bmatrix} 1 \\ 1 \end{Bmatrix}, \quad \hat{\phi}_2 = \frac{1}{\sqrt{5}}\begin{Bmatrix} 1 \\ -2 \end{Bmatrix}. \tag{6.172}$$

Now consider the case where the degree of freedom of the bottom floor is to be reduced using dynamic condensation. Let

$$\mathbf{M}_{11} = m, \quad \mathbf{M}_{13} = \mathbf{M}_{31} = 0, \quad \mathbf{M}_{33} = 2m, \quad \mathbf{C}_{11} = \mathbf{C}_{13} = \mathbf{C}_{31} = \mathbf{C}_{33} = 0$$
$$\mathbf{K}_{11} = 3k, \quad \mathbf{K}_{13} = \mathbf{K}_{31} = -2k, \quad \mathbf{K}_{33} = 4k \tag{6.173}$$
$$\mathbf{a}_1(t) = a_x(t), \quad \mathbf{a}_3(t) = a_x(t), \quad \mathbf{X}_1 = x_2, \quad \mathbf{X}_3 = x_1.$$

It follows from Eq. 6.144 that

$$\left[m + (-2k)(4k)^{-1}(2m)(4k)^{-1}(-2k)\right]\ddot{x}_2(t) + \left[3k - (-2k)(4k)^{-1}(-2k)\right]x_2(t)$$
$$= -ma_x(t) - (-2k)(4k)^{-1}(2m)a_x(t). \tag{6.174}$$

Simplifying Eq. 6.174 gives

$$1.5m\ddot{x}_2(t) + 2kx_2(t) = -2ma_x(t). \tag{6.175}$$

In addition, the degree of freedom that has been condensed out of the equation of motion can be calculated using Eq. 6.140, that is,

$$x_1(t) = -(4k)^{-1}(-2k)x_2(t) = \frac{1}{2}x_2(t). \tag{6.176}$$

Note that by using dynamic condensation on the equation of motion for the 2DOF system, the structure vibrates at one natural frequency (ω_d) with only one mode shape ($\hat{\phi}_d$):

$$\omega_d = \sqrt{\frac{2k}{1.5m}} = 1.155\sqrt{\frac{k}{m}} \tag{6.177}$$

$$\hat{\phi}_d = \frac{1}{\sqrt{5}}\begin{Bmatrix} 1 \\ 2 \end{Bmatrix}. \tag{6.178}$$

Recall from Eq. 6.126 that

$$\omega_s = 1.414\sqrt{\frac{k}{m}} \tag{6.179}$$

which means that using dynamic condensation gives a natural frequency of vibration that is closer to the actual fundamental frequency of vibration of a 2DOF system than using static condensation, that is,

$$\omega_1 \leq \omega_d \leq \omega_s. \tag{6.180}$$

The mode shapes of vibration for both static and dynamic condensations are the same (see Eq. 6.127), that is,

$$\hat{\boldsymbol{\phi}}_s = \hat{\boldsymbol{\phi}}_d = \frac{1}{\sqrt{5}} \begin{Bmatrix} 1 \\ 2 \end{Bmatrix} \tag{6.181}$$

because both condensation methods use the same constraint equation (i.e., Eq. 6.176). However, these mode shapes can be quite different from the first mode shape of vibration for the original system, which is given in Eq. 6.172.

6.7   STATE SPACE RESPONSE

The state space response for an SDOF system was discussed in Section 2.5, and the numerical method using the state space method was presented in Section 2.8. In addition, for an MDOF system, the state space method was presented in Sections 3.6, 3.8, and 4.10. The objective of this section is to use the state space method for earthquake time history analysis.

The general formulation presented for a structural dynamics analysis can be used for the earthquake analysis of a structure by replacing the external applied force with the mass times the ground acceleration:

$$\mathbf{F}_e(t) = -\mathbf{M}\mathbf{a}(t). \tag{6.182}$$

For an MDOF system, it therefore follows from Eq. 4.465 that

$$\mathbf{z}(t) = e^{\mathbf{A}(t-t_o)} \mathbf{z}(t) = e^{\mathbf{A}(t-t_o)} + e^{\mathbf{A}t} \int_{t_o}^{t} e^{-\mathbf{A}s} \mathbf{F}(s)ds \tag{6.183}$$

where

$$\mathbf{z}(t) = \begin{Bmatrix} \mathbf{X}(t) \\ \dot{\mathbf{X}}(t) \end{Bmatrix}, \quad \mathbf{A} = \begin{bmatrix} \mathbf{0} & \mathbf{I} \\ -\mathbf{M}^{-1}\mathbf{K} & -\mathbf{M}^{-1}\mathbf{C} \end{bmatrix}, \quad \mathbf{F}(t) = \begin{Bmatrix} \mathbf{0} \\ -\mathbf{a}(t) \end{Bmatrix}.$$

Following the procedure discussed in Section 3.8 and using numerical integration, it follows when the Delta forcing function approach is used that

$$\mathbf{z}_{k+1} = e^{\mathbf{A}\Delta t} \mathbf{z}_k + e^{\mathbf{A}\Delta t} \Delta t \, \mathbf{F}_k \tag{6.184}$$

where

$$\mathbf{F}_k = \begin{Bmatrix} \mathbf{0} \\ -\mathbf{a}_k \end{Bmatrix}, \quad t_k \leq s < t_{k+1}. \tag{6.185}$$

When the constant forcing function approach is used, then

$$\mathbf{z}_{k+1} = e^{\mathbf{A}\Delta t} \mathbf{z}_k + \mathbf{A}^{-1}\left(e^{\mathbf{A}\Delta t} - \mathbf{I}\right)\mathbf{F}_k. \tag{6.186}$$

Let

$$\mathbf{F}_s = e^{\mathbf{A}\Delta t} \tag{6.187}$$

and

$$
\mathbf{F}_k = \left\{ \begin{matrix} \mathbf{0} \\ -\mathbf{a}_k \end{matrix} \right\} = \begin{bmatrix} \mathbf{0}_v & \mathbf{0}_v & \mathbf{0}_v & \mathbf{0}_v & \mathbf{0}_v & \mathbf{0}_v \\ \mathbf{0}_v & \mathbf{0}_v & \mathbf{0}_v & \mathbf{0}_v & \mathbf{0}_v & \mathbf{0}_v \\ \mathbf{0}_v & \mathbf{0}_v & \mathbf{0}_v & \mathbf{0}_v & \mathbf{0}_v & \mathbf{0}_v \\ \mathbf{0}_v & \mathbf{0}_v & \mathbf{0}_v & \mathbf{0}_v & \mathbf{0}_v & \mathbf{0}_v \\ \mathbf{0}_v & \mathbf{0}_v & \mathbf{0}_v & \mathbf{0}_v & \mathbf{0}_v & \mathbf{0}_v \\ \mathbf{0}_v & \mathbf{0}_v & \mathbf{0}_v & \mathbf{0}_v & \mathbf{0}_v & \mathbf{0}_v \\ \hdashline -\mathbf{1}_v & \mathbf{0}_v & \mathbf{0}_v & \mathbf{0}_v & \mathbf{0}_v & \mathbf{0}_v \\ \mathbf{0}_v & -\mathbf{1}_v & \mathbf{0}_v & \mathbf{0}_v & \mathbf{0}_v & \mathbf{0}_v \\ \mathbf{0}_v & \mathbf{0}_v & -\mathbf{1}_v & \mathbf{0}_v & \mathbf{0}_v & \mathbf{0}_v \\ \mathbf{0}_v & \mathbf{0}_v & \mathbf{0}_v & -\mathbf{1}_v & \mathbf{0}_v & \mathbf{0}_v \\ \mathbf{0}_v & \mathbf{0}_v & \mathbf{0}_v & \mathbf{0}_v & -\mathbf{1}_v & \mathbf{0}_v \\ \mathbf{0}_v & \mathbf{0}_v & \mathbf{0}_v & \mathbf{0}_v & \mathbf{0}_v & -\mathbf{1}_v \end{bmatrix} \left\{ \begin{matrix} a_x \\ a_y \\ a_z \\ a_r \\ a_s \\ a_t \end{matrix} \right\}_k = \mathbf{H}\mathbf{a}_k \tag{6.188}
$$

where $\mathbf{0}_v$ is the vector containing zeroes in all entries and $-\mathbf{1}_v$ is the vector containing -1's in all entries. Here, the first six rows of matrix $\mathbf{H}$ relate to the displacements in the order $\mathbf{X}_{xk}$, $\mathbf{X}_{yk}$, $\mathbf{X}_{zk}$, $\mathbf{X}_{rk}$, $\mathbf{X}_{sk}$, and $\mathbf{X}_{tk}$, and the last six rows relate to the velocity in the order $\dot{\mathbf{X}}_{xk}$, $\dot{\mathbf{X}}_{yk}$, $\dot{\mathbf{X}}_{zk}$, $\dot{\mathbf{X}}_{rk}$, $\dot{\mathbf{X}}_{sk}$, and $\dot{\mathbf{X}}_{tk}$. Substituting Eq. 6.188 into Eq. 6.184 gives

$$
\mathbf{z}_{k+1} = e^{\mathbf{A}\Delta t}\,\mathbf{z}_k + e^{\mathbf{A}\Delta t}\mathbf{H}\Delta t\,\mathbf{a}_k. \tag{6.189}
$$

Similarly, substituting Eq. 6.188 into Eq. 6.186 gives

$$
\mathbf{z}_{k+1} = e^{\mathbf{A}\Delta t}\,\mathbf{z}_k + \mathbf{A}^{-1}\left(e^{\mathbf{A}\Delta t} - \mathbf{I}\right)\mathbf{H}\mathbf{a}_k. \tag{6.190}
$$

When the Delta forcing function approach is used, define

$$
\mathbf{H}_d^{(EQ)} = e^{\mathbf{A}\Delta t}\mathbf{H}\Delta t \tag{6.191}
$$

or when constant forcing function is used, define

$$
\mathbf{H}_d^{(EQ)} = \mathbf{A}^{-1}\left(e^{\mathbf{A}\Delta t} - \mathbf{I}\right)\mathbf{H}. \tag{6.192}
$$

It then follows from Eq. 6.189 or Eq. 6.190 that

$$
\mathbf{z}_{k+1} = \mathbf{F}_s\mathbf{z}_k + \mathbf{H}_d^{(EQ)}\mathbf{a}_k. \tag{6.193}
$$

EXAMPLE 1 *SDOF System Subjected to El-Centro Earthquake*

The NS component of the El-Centro earthquake time history is used as the input ground motion in this example. The actual El-Centro earthquake time history is shown in Figure 5.2. The El-Centro earthquake has been widely used because it was one of the first accurately measured earthquake time histories. However, this earthquake is generally too small by modern standards to represent a design basis earthquake. Therefore, it is here modified, as shown in Figure 6.11, where each acceleration value is scaled by the same scalar of 1.8, to simulate a design basis earthquake.

Consider an SDOF structure as shown in Figure 6.12 with a mass of 1.0 k-s²/in., damping ratio of 5%, and period of 0.5 s. Then

$$
\omega_n = 2\pi/T_n = 2\pi \text{ rad/s} \tag{6.194}
$$

$$
c = 2\zeta\omega_n m = (2)(0.05)(2\pi)(1.0) = 1.2567 \text{ k-s/in.} \tag{6.195}
$$

$$
k = m\omega_n^2 = (1.0)(2\pi)^2 = 157.92 \text{ k/in.} \tag{6.196}
$$

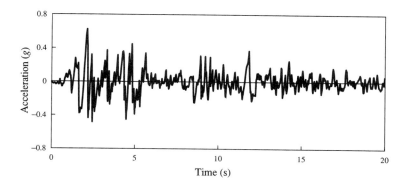

Figure 6.11 Modified 1940 NS Component of the El-Centro Earthquake Ground Motion

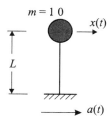

Figure 6.12 Single Degree of Freedom System

This SDOF system is subjected to the modified El-Centro earthquake in the horizontal direction. Then

$$\mathbf{A} = \begin{bmatrix} 0 & 1 \\ -157.92 & -1.2567 \end{bmatrix}, \quad \mathbf{H} = \begin{bmatrix} 0 \\ -1 \end{bmatrix}. \tag{6.197}$$

First assume that $\Delta t = 0.02$ s. It then follows when the Delta forcing function approach is used that

$$\mathbf{A}\Delta t = \begin{bmatrix} 0 & 0.02 \\ -3.1584 & -0.02513 \end{bmatrix}, \quad \mathbf{F}_s = e^{\mathbf{A}\Delta t} = \begin{bmatrix} 0.968843 & 0.019543 \\ -3.086307 & 0.944284 \end{bmatrix}$$

$$\mathbf{H}_d^{(EQ)} = e^{\mathbf{A}\Delta t}\mathbf{H}\Delta t = 0.02 \begin{Bmatrix} -0.019543 \\ -0.944284 \end{Bmatrix}, \quad a_k = a_k.$$

It follows that

$$\begin{Bmatrix} x_{k+1} \\ \dot{x}_{k+1} \end{Bmatrix} = \begin{bmatrix} 0.968843 & 0.019543 \\ -3.086307 & 0.944284 \end{bmatrix} \begin{Bmatrix} x_k \\ \dot{x}_k \end{Bmatrix} + 0.02 \begin{Bmatrix} -0.019543 \\ -0.944284 \end{Bmatrix} a_k. \tag{6.198}$$

Table 6.1 gives the solution for the first five time steps and illustrates the recursive nature of the solution. The response time history is shown in Figure 6.13.

Now assume that $\Delta t = 0.01$ s; using the Delta forcing function approach it follows from Eq. 6.189 that

$$\mathbf{A}\Delta t = \begin{bmatrix} 0 & 0.01 \\ -1.5792 & -0.01257 \end{bmatrix}, \quad \mathbf{F}_s = e^{\mathbf{A}\Delta t} = \begin{bmatrix} 0.992147 & 0.009911 \\ -1.565192 & 0.979692 \end{bmatrix}$$

Table 6.1 Sample Calculation of SDOF with $\Delta t = 0.02$ s

k	Time (s)	a_k (in./s^2)	x_k (in.)	$\dot{x}_k$ (in./s)
0	0.0	−0.9931	0.0	0.0
1	0.02	−7.6507	0.00039	0.0188
2	0.04	−7.1484	0.00373	0.1610
3	0.06	−6.2597	0.00956	0.2755
4	0.08	−6.7234	0.01709	0.3489
5	0.10	−8.5008	0.02601	0.4037

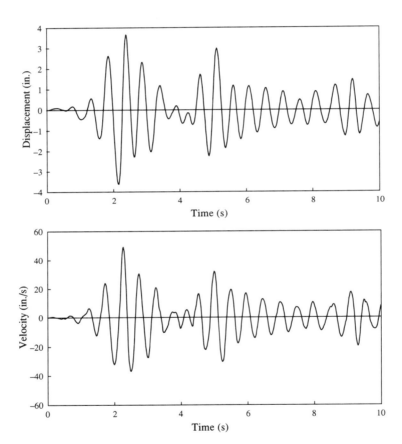

Figure 6.13 Response Time History for Time Step of 0.02 s

$$\mathbf{H}_d^{(EQ)} = e^{\mathbf{A}\Delta t}\mathbf{H}\Delta t = 0.01\begin{Bmatrix} -0.009911 \\ -0.979692 \end{Bmatrix}$$

and

$$\begin{Bmatrix} x_{k+1} \\ \dot{x}_{k+1} \end{Bmatrix} = \begin{bmatrix} 0.992147 & 0.009911 \\ -1.565192 & 0.979692 \end{bmatrix}\begin{Bmatrix} x_k \\ \dot{x}_k \end{Bmatrix} + 0.01\begin{Bmatrix} -0.009911 \\ -0.979692 \end{Bmatrix} a_k. \tag{6.199}$$

Table 6.2 gives the solution for the first 10 time steps.

Table 6.2 Sample Calculation of SDOF with $\Delta t = 0.01$ s

k	Time (s)	a_k (in./s^2)	x_k (in.)	$\dot{x}$ (in./s)
0	0.0	−0.9931	0.0	0.0
1	0.01	−4.3219	0.00010	0.0097
2	0.02	−7.6507	0.00062	0.0517
3	0.03	−7.3996	0.00189	0.1246
4	0.04	−7.1484	0.00384	0.1917
5	0.05	−6.7041	0.00642	0.2518
6	0.06	−6.2597	0.00953	0.3023
7	0.07	−6.4916	0.01307	0.3426
8	0.08	−6.7234	0.01701	0.3787
9	0.09	−7.6121	0.02129	0.4103
10	0.10	−8.5008	0.02595	0.4432

EXAMPLE 2 *2DOF System Subjected to El-Centro Earthquake*

To illustrate the numerical solution for a 2DOF system, consider the two-story structure where

$$\mathbf{M} = \begin{bmatrix} 1.0 & 0.0 \\ 0.0 & 1.0 \end{bmatrix}, \quad \mathbf{C} = \begin{bmatrix} 3.372 & -1.124 \\ -1.124 & 2.248 \end{bmatrix}, \quad \mathbf{K} = \begin{bmatrix} 1263.36 & -631.68 \\ -631.68 & 631.68 \end{bmatrix}. \quad (6.200)$$

Using the modified El-Centro earthquake given in Figure 6.11 as the input earthquake time history along the x-direction, it then follows that

$$\mathbf{A} = \begin{bmatrix} 0.0 & 0.0 & 1.0 & 0.0 \\ 0.0 & 0.0 & 0.0 & 1.0 \\ -1263.36 & 631.68 & -3.372 & 1.124 \\ 631.68 & -631.68 & 1.124 & -2.248 \end{bmatrix}, \quad \mathbf{H} = \begin{bmatrix} 0 \\ 0 \\ -1 \\ -1 \end{bmatrix}. \quad (6.201)$$

First assume that $\Delta t = 0.02$ s. It then follows when the Delta forcing function approach is used that

$$\mathbf{F}_s = e^{\mathbf{A}\Delta t} = \begin{bmatrix} 0.766425 & 0.115076 & 0.017773 & 0.000992 \\ 0.115076 & 0.881501 & 0.000992 & 0.018764 \\ -21.8268 & 10.60017 & 0.70761 & 0.132823 \\ 10.60017 & -11.2266 & 0.132823 & 0.840433 \end{bmatrix}, \quad \mathbf{H}_d^{(EQ)} = 0.02 \begin{Bmatrix} -0.018765 \\ -0.019756 \\ -0.840433 \\ -0.973256 \end{Bmatrix}$$

and

$$\begin{Bmatrix} x_1 \\ x_2 \\ \dot{x}_1 \\ \dot{x}_2 \end{Bmatrix}_{k+1} = e^{\mathbf{A}\Delta t} \begin{Bmatrix} x_1 \\ x_2 \\ \dot{x}_1 \\ \dot{x}_2 \end{Bmatrix}_k + 0.02 \begin{Bmatrix} -0.018765 \\ -0.019756 \\ -0.840433 \\ -0.973256 \end{Bmatrix} a_k. \quad (6.202)$$

Table 6.3 gives the solution for the first five time steps of this 2DOF system. The response time history for time step of 0.02 s is shown in Figure 6.14.

Table 6.3 Sample Calculation of 2DOF with $\Delta t = 0.02$ s

k	time (s)	a_k (in./s^2)	x_{1k} (in.)	x_{2k} (in.)	$\dot{x}_{1k}$ (in./s)	$\dot{x}_{2k}$ (in./s)
0	0.0	−0.9931	0.0	0.0	0.0	0.0
1	0.02	−7.6507	0.00037	0.00039	0.0167	0.0193
2	0.04	−7.1484	0.00352	0.00379	0.1390	0.1669
3	0.06	−6.2597	0.00845	0.00984	0.2041	0.2926
4	0.08	−6.7234	0.01388	0.01781	0.2084	0.3740
5	0.10	−8.5008	0.01928	0.02718	0.1961	0.4200

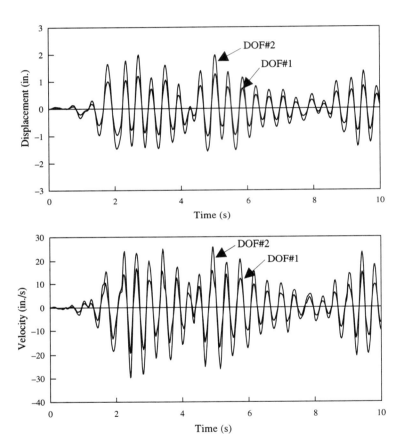

Figure 6.14 Response Time History for Time Step of 0.02 s

Now assume that $\Delta t = 0.01$ s; using the Delta forcing function approach it follows from Eq. 6.201 that

$$
\mathbf{F}_s = \mathbf{e}^{\mathbf{A}\Delta t} = \begin{bmatrix} 0.938466 & 0.030628 & 0.009629 & 0.000157 \\ 0.030628 & 0.969093 & 0.000157 & 0.009786 \\ -12.0652 & 5.982965 & 0.906174 & 0.041097 \\ 5.982965 & -6.08222 & 0.041097 & 0.947272 \end{bmatrix}, \quad \mathbf{H}_d^{(EQ)} = 0.01 \begin{Bmatrix} -0.009786 \\ -0.009943 \\ -0.947271 \\ -0.988369 \end{Bmatrix}
$$

Table 6.4 Sample Calculation of 2DOF with $\Delta t = 0.01$ s

k	Time (s)	a_k (in./s^2)	x_{1k} (in.)	x_{2k} (in.)	$\dot{x}_{1k}$ (in./s)	$\dot{x}_{2k}$ (in./s)
0	0.0	−0.9931	0.0	0.0	0.0	0.0
1	0.01	−4.3219	0.00010	0.00010	0.0094	0.0098
2	0.02	−7.6507	0.00061	0.00063	0.0493	0.0524
3	0.03	−7.3996	0.00182	0.00191	0.1157	0.1271
4	0.04	−7.1484	0.00363	0.00390	0.1696	0.1976
5	0.05	−6.7041	0.00589	0.00656	0.2091	0.2628
6	0.06	−6.2597	0.00844	0.00981	0.2320	0.3191
7	0.07	−6.4916	0.01111	0.01355	0.2396	0.3644
8	0.08	−6.7234	0.01384	0.01772	0.2405	0.4033
9	0.09	−7.6121	0.01658	0.02225	0.2373	0.4335
10	0.10	−8.5008	0.01933	0.02710	0.2380	0.4594

and

$$\left\{ \begin{array}{c} x_1 \\ x_2 \\ \dot{x}_1 \\ \dot{x}_2 \end{array} \right\}_{k+1} = e^{A\Delta t} \left\{ \begin{array}{c} x_1 \\ x_2 \\ \dot{x}_1 \\ \dot{x}_2 \end{array} \right\}_k + 0.01 \left\{ \begin{array}{c} -0.009786 \\ -0.009943 \\ -0.947271 \\ -0.988369 \end{array} \right\} a_k. \tag{6.203}$$

Table 6.4 provides the solution for the first 10 time steps.

6.8   STATE SPACE REDUCTION

Similar to the condensation methods discussed in Sections 6.5 and 6.6, the size of a structural dynamics problem can be reduced in the state space formulation. Consider, for a general case, all of the degrees of freedom with mass to be labeled as 1, those with damping but without mass to be labeled as 2, and those without damping to be labeled as 3. The dynamic equilibrium equation of motion becomes

$$\begin{bmatrix} \mathbf{M}_{11} & \mathbf{0} & \mathbf{0} \\ \mathbf{0} & \mathbf{0} & \mathbf{0} \\ \mathbf{0} & \mathbf{0} & \mathbf{0} \end{bmatrix} \begin{Bmatrix} \ddot{\mathbf{X}}_1(t) \\ \ddot{\mathbf{X}}_2(t) \\ \ddot{\mathbf{X}}_3(t) \end{Bmatrix} + \begin{bmatrix} \mathbf{C}_{11} & \mathbf{C}_{12} & \mathbf{0} \\ \mathbf{C}_{21} & \mathbf{C}_{22} & \mathbf{0} \\ \mathbf{0} & \mathbf{0} & \mathbf{0} \end{bmatrix} \begin{Bmatrix} \dot{\mathbf{X}}_1(t) \\ \dot{\mathbf{X}}_2(t) \\ \dot{\mathbf{X}}_3(t) \end{Bmatrix} + \begin{bmatrix} \mathbf{K}_{11} & \mathbf{K}_{12} & \mathbf{K}_{13} \\ \mathbf{K}_{21} & \mathbf{K}_{22} & \mathbf{K}_{23} \\ \mathbf{K}_{31} & \mathbf{K}_{32} & \mathbf{K}_{33} \end{bmatrix} \begin{Bmatrix} \mathbf{X}_1(t) \\ \mathbf{X}_2(t) \\ \mathbf{X}_3(t) \end{Bmatrix}$$

$$= - \begin{bmatrix} \mathbf{M}_{11} & \mathbf{0} & \mathbf{0} \\ \mathbf{0} & \mathbf{0} & \mathbf{0} \\ \mathbf{0} & \mathbf{0} & \mathbf{0} \end{bmatrix} \begin{Bmatrix} \mathbf{a}_1(t) \\ \mathbf{a}_2(t) \\ \mathbf{a}_3(t) \end{Bmatrix}. \tag{6.204}$$

Figure 6.15 shows a special case where the dynamic matrices in Eq. 6.204 can exist.

Expanding Eq. 6.204 gives three simultaneous equations, which are

$$\mathbf{M}_{11}\ddot{\mathbf{X}}_1(t) + \mathbf{C}_{11}\dot{\mathbf{X}}_1(t) + \mathbf{C}_{12}\dot{\mathbf{X}}_2(t) + \mathbf{K}_{11}\mathbf{X}_1(t) + \mathbf{K}_{12}\mathbf{X}_2(t) + \mathbf{K}_{13}\mathbf{X}_3(t) = -\mathbf{M}_{11}\mathbf{a}_1(t) \tag{6.205a}$$

$$\mathbf{C}_{21}\dot{\mathbf{X}}_1(t) + \mathbf{C}_{22}\dot{\mathbf{X}}_2(t) + \mathbf{K}_{21}\mathbf{X}_1(t) + \mathbf{K}_{22}\mathbf{X}_2(t) + \mathbf{K}_{23}\mathbf{X}_3(t) = \mathbf{0} \tag{6.205b}$$

$$\mathbf{K}_{31}\mathbf{X}_1(t) + \mathbf{K}_{32}\mathbf{X}_2(t) + \mathbf{K}_{33}\mathbf{X}_3(t) = \mathbf{0}. \tag{6.205c}$$

Solving for Eq. 6.205c gives

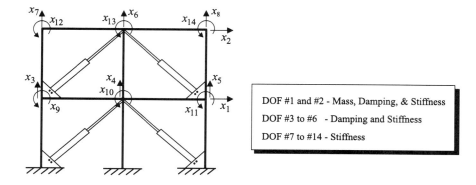

Figure 6.15 Example of a Structure with Different Types of Dynamic Matrices

$$\mathbf{X}_3(t) = -\mathbf{K}_{33}^{-1}\mathbf{K}_{31}\mathbf{X}_1(t) - \mathbf{K}_{33}^{-1}\mathbf{K}_{32}\mathbf{X}_2(t). \tag{6.206}$$

Substituting Eq. 6.206 into Eq. 6.205b, it follows that

$$\mathbf{C}_{21}\dot{\mathbf{X}}_1(t) + \mathbf{C}_{22}\dot{\mathbf{X}}_2(t) + \left(\mathbf{K}_{21} - \mathbf{K}_{23}\mathbf{K}_{33}^{-1}\mathbf{K}_{31}\right)\mathbf{X}_1(t) + \left(\mathbf{K}_{22} - \mathbf{K}_{23}\mathbf{K}_{33}^{-1}\mathbf{K}_{32}\right)\mathbf{X}_2(t) = \mathbf{0}. \tag{6.207}$$

Define for convenience

$$\mathbf{K}_{21}' = \mathbf{K}_{21} - \mathbf{K}_{23}\mathbf{K}_{33}^{-1}\mathbf{K}_{31}, \quad \mathbf{K}_{22}' = \mathbf{K}_{22} - \mathbf{K}_{23}\mathbf{K}_{33}^{-1}\mathbf{K}_{32}.$$

Substituting these two terms into Eq. 6.207 and solving for $\dot{\mathbf{X}}_2(t)$, it follows that

$$\dot{\mathbf{X}}_2(t) = -\mathbf{C}_{22}^{-1}\mathbf{C}_{21}\dot{\mathbf{X}}_1(t) - \mathbf{C}_{22}^{-1}\mathbf{K}_{21}'\mathbf{X}_1(t) - \mathbf{C}_{22}^{-1}\mathbf{K}_{22}'\mathbf{X}_2(t). \tag{6.208}$$

Substituting Eqs. 6.206 and 6.208 into Eq. 6.205a gives

$$\mathbf{M}_{11}\ddot{\mathbf{X}}_1(t) + \mathbf{C}_{11}'\dot{\mathbf{X}}_1(t) + \mathbf{K}_{11}''\mathbf{X}_1(t) + \mathbf{K}_{12}''\mathbf{X}_2(t) = -\mathbf{M}_{11}\mathbf{a}_1(t) \tag{6.209}$$

where

$$\mathbf{C}_{11}' = \mathbf{C}_{11} - \mathbf{C}_{12}\mathbf{C}_{22}^{-1}\mathbf{C}_{21}, \quad \mathbf{K}_{11}'' = \mathbf{K}_{11}' - \mathbf{C}_{12}\mathbf{C}_{22}^{-1}\mathbf{K}_{21}', \quad \mathbf{K}_{12}'' = \mathbf{K}_{12}' - \mathbf{C}_{12}\mathbf{C}_{22}^{-1}\mathbf{K}_{22}'$$

$$\mathbf{K}_{11}' = \mathbf{K}_{11} - \mathbf{K}_{13}\mathbf{K}_{33}^{-1}\mathbf{K}_{31}, \quad \mathbf{K}_{12}' = \mathbf{K}_{12} - \mathbf{K}_{13}\mathbf{K}_{33}^{-1}\mathbf{K}_{32}.$$

Note that $\mathbf{K}_{11}'$ is the same as $\tilde{\mathbf{K}}$ in Eqs. 6.94 and 6.145c. Solving for $\ddot{\mathbf{X}}_1(t)$ in Eq. 6.209 gives

$$\ddot{\mathbf{X}}_1(t) = -\mathbf{M}_{11}^{-1}\mathbf{C}_{11}'\dot{\mathbf{X}}_1(t) - \mathbf{M}_{11}^{-1}\mathbf{K}_{11}''\mathbf{X}_1(t) - \mathbf{M}_{11}^{-1}\mathbf{K}_{12}''\mathbf{X}_2(t) - \mathbf{a}_1(t). \tag{6.210}$$

Define

$$\mathbf{z}(t) = \begin{Bmatrix} \mathbf{X}_1(t) \\ \mathbf{X}_2(t) \\ \dot{\mathbf{X}}_1(t) \end{Bmatrix} \tag{6.211}$$

and then the state space equation becomes

$$\dot{\mathbf{z}}(t) = \mathbf{A}\mathbf{z}(t) + \mathbf{H}\mathbf{a}(t) \tag{6.212}$$

where

$$\mathbf{A} = \begin{bmatrix} \mathbf{0} & \mathbf{0} & \mathbf{I} \\ -\mathbf{C}_{22}^{-1}\mathbf{K}_{21}' & -\mathbf{C}_{22}^{-1}\mathbf{K}_{22}' & -\mathbf{C}_{22}^{-1}\mathbf{C}_{21} \\ -\mathbf{M}_{11}^{-1}\mathbf{K}_{11}'' & -\mathbf{M}_{11}^{-1}\mathbf{K}_{12}'' & -\mathbf{M}_{11}^{-1}\mathbf{C}_{11}' \end{bmatrix} \tag{6.213}$$

and

$$
\mathbf{H} = \begin{bmatrix}
\mathbf{0}_v & \mathbf{0}_v & \mathbf{0}_v & \mathbf{0}_v & \mathbf{0}_v & \mathbf{0}_v \\
\mathbf{0}_v & \mathbf{0}_v & \mathbf{0}_v & \mathbf{0}_v & \mathbf{0}_v & \mathbf{0}_v \\
\mathbf{0}_v & \mathbf{0}_v & \mathbf{0}_v & \mathbf{0}_v & \mathbf{0}_v & \mathbf{0}_v \\
\mathbf{0}_v & \mathbf{0}_v & \mathbf{0}_v & \mathbf{0}_v & \mathbf{0}_v & \mathbf{0}_v \\
\mathbf{0}_v & \mathbf{0}_v & \mathbf{0}_v & \mathbf{0}_v & \mathbf{0}_v & \mathbf{0}_v \\
\mathbf{0}_v & \mathbf{0}_v & \mathbf{0}_v & \mathbf{0}_v & \mathbf{0}_v & \mathbf{0}_v \\
\mathbf{0}_v & \mathbf{0}_v & \mathbf{0}_v & \mathbf{0}_v & \mathbf{0}_v & \mathbf{0}_v \\
\mathbf{0}_v & \mathbf{0}_v & \mathbf{0}_v & \mathbf{0}_v & \mathbf{0}_v & \mathbf{0}_v \\
\mathbf{0}_v & \mathbf{0}_v & \mathbf{0}_v & \mathbf{0}_v & \mathbf{0}_v & \mathbf{0}_v \\
\mathbf{0}_v & \mathbf{0}_v & \mathbf{0}_v & \mathbf{0}_v & \mathbf{0}_v & \mathbf{0}_v \\
\mathbf{0}_v & \mathbf{0}_v & \mathbf{0}_v & \mathbf{0}_v & \mathbf{0}_v & \mathbf{0}_v \\
\mathbf{0}_v & \mathbf{0}_v & \mathbf{0}_v & \mathbf{0}_v & \mathbf{0}_v & \mathbf{0}_v \\
-\mathbf{1}_v & \mathbf{0}_v & \mathbf{0}_v & \mathbf{0}_v & \mathbf{0}_v & \mathbf{0}_v \\
\mathbf{0}_v & -\mathbf{1}_v & \mathbf{0}_v & \mathbf{0}_v & \mathbf{0}_v & \mathbf{0}_v \\
\mathbf{0}_v & \mathbf{0}_v & -\mathbf{1}_v & \mathbf{0}_v & \mathbf{0}_v & \mathbf{0}_v \\
\mathbf{0}_v & \mathbf{0}_v & \mathbf{0}_v & -\mathbf{1}_v & \mathbf{0}_v & \mathbf{0}_v \\
\mathbf{0}_v & \mathbf{0}_v & \mathbf{0}_v & \mathbf{0}_v & -\mathbf{1}_v & \mathbf{0}_v \\
\mathbf{0}_v & \mathbf{0}_v & \mathbf{0}_v & \mathbf{0}_v & \mathbf{0}_v & -\mathbf{1}_v
\end{bmatrix}, \quad
\mathbf{a}_k = \begin{Bmatrix} a_x \\ a_y \\ a_z \\ a_r \\ a_s \\ a_t \end{Bmatrix}_k
\tag{6.214}
$$

where $\mathbf{0}_v$ is the vector containing zeroes in all entries and $-\mathbf{1}_v$ is the vector containing -1's in all entries. The first six rows of matrix $\mathbf{H}$ relate to the displacements in the order $\mathbf{X}_{1xk}$, $\mathbf{X}_{1yk}$, $\mathbf{X}_{1zk}$, $\mathbf{X}_{1rk}$, $\mathbf{X}_{1sk}$, and $\mathbf{X}_{1tk}$; the second six rows relate to the displacements in the order $\mathbf{X}_{2xk}$, $\mathbf{X}_{2yk}$, $\mathbf{X}_{2zk}$, $\mathbf{X}_{2rk}$, $\mathbf{X}_{2sk}$, and $\mathbf{X}_{2tk}$; and the last six rows relate to the velocity in the order $\dot{\mathbf{X}}_{1xk}$, $\dot{\mathbf{X}}_{1yk}$, $\dot{\mathbf{X}}_{1zk}$, $\dot{\mathbf{X}}_{1rk}$, $\dot{\mathbf{X}}_{1sk}$, and $\dot{\mathbf{X}}_{1tk}$. It follows from Eq. 6.212 that if

$$
t_{k+1} = t, \quad t_k = t_o, \quad \Delta t = t - t_o
$$

then

$$
\mathbf{z}_{k+1} = e^{\mathbf{A}\Delta t}\mathbf{z}_k + e^{\mathbf{A}t_{k+1}} \int_{t_k}^{t_{k+1}} e^{-\mathbf{A}s}\mathbf{H}\mathbf{a}(s)\,ds.
\tag{6.215}
$$

Therefore, when the Delta forcing function approach is used, Eq. 6.215 becomes

$$
\mathbf{z}_{k+1} = \mathbf{F}_s\mathbf{z}_k + \mathbf{H}_d^{(EQ)}\mathbf{a}_k, \quad \mathbf{F}_s = e^{\mathbf{A}\Delta t}, \quad \mathbf{H}_d^{(EQ)} = e^{\mathbf{A}\Delta t}\mathbf{H}\Delta t
\tag{6.216}
$$

and when the constant forcing function approach is used, Eq. 6.215 becomes

$$
\mathbf{z}_{k+1} = \mathbf{F}_s\mathbf{z}_k + \mathbf{H}_d^{(EQ)}\mathbf{a}_k, \quad \mathbf{F}_s = e^{\mathbf{A}\Delta t}, \quad \mathbf{H}_d^{(EQ)} = \mathbf{A}^{-1}\left(e^{\mathbf{A}\Delta t} - \mathbf{I}\right)\mathbf{H}.
\tag{6.217}
$$

Equations 6.216 and 6.217 give the recursive equations that are used in an earthquake time history analysis. Once the response of the structure is calculated using Eq. 6.216 or 6.217, $\mathbf{X}_{1k}$, $\mathbf{X}_{2k}$, and $\dot{\mathbf{X}}_{1k}$ are known, and therefore $\dot{\mathbf{X}}_{2k}$ can be computed using Eq. 6.208 and $\mathbf{X}_{3k}$ using Eq. 6.206:

$$
\dot{\mathbf{X}}_{2k} = -\mathbf{C}_{22}^{-1}\mathbf{C}_{21}\dot{\mathbf{X}}_{1k} - \mathbf{C}_{22}^{-1}\mathbf{K}_{21}'\,\mathbf{X}_{1k} - \mathbf{C}_{22}^{-1}\mathbf{K}_{22}'\,\mathbf{X}_{2k}
\tag{6.218a}
$$

$$
\mathbf{X}_{3k} = -\mathbf{K}_{33}^{-1}\mathbf{K}_{31}\mathbf{X}_{1k} - \mathbf{K}_{33}^{-1}\mathbf{K}_{32}\mathbf{X}_{2k}.
\tag{6.218b}
$$

EXAMPLE 1   *Two-Story Structure with One Mass*

Consider the two-story frame shown in Figure 6.16. The structure consists of two floors, with the mass on the bottom floor assumed to be zero. All joint rotations are assumed to

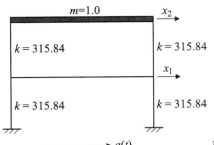

Figure 6.16 Example of a Two-Story Frame

be zero, and the only degrees of freedom are the lateral translation of each floor. Assume that the frame has nonproportional damping, where

$$\mathbf{M} = \begin{bmatrix} 1.0 & 0.0 \\ 0.0 & 0.0 \end{bmatrix}, \quad \mathbf{C} = \begin{bmatrix} 2.248 & -1.124 \\ -1.124 & 3.372 \end{bmatrix}, \quad \mathbf{K} = \begin{bmatrix} 631.68 & -631.68 \\ -631.68 & 1263.36 \end{bmatrix}. \quad (6.219)$$

Since the dimension of the damping matrix is the same as that of the stiffness matrix, there is no degree of freedom in association with the label 3. It follows that

$$\mathbf{K}'_{11} = \mathbf{K}_{11} = 631.68, \quad \mathbf{K}'_{12} = \mathbf{K}_{12} = -631.68$$
$$\mathbf{K}'_{21} = \mathbf{K}_{21} = 631.68, \quad \mathbf{K}'_{22} = \mathbf{K}_{22} = 1263.36 \quad (6.220)$$

and therefore

$$\mathbf{C}'_{11} = \mathbf{C}_{11} - \mathbf{C}_{12}\mathbf{C}_{22}^{-1}\mathbf{C}_{21} = 2.248 - \frac{(-1.124)(-1.124)}{3.372} = 1.87333 \quad (6.221)$$

$$\mathbf{K}''_{11} = \mathbf{K}'_{11} - \mathbf{C}_{12}\mathbf{C}_{22}^{-1}\mathbf{K}'_{21} = 631.68 - \frac{(-1.124)(-631.68)}{3.372} = 421.120 \quad (6.222)$$

$$\mathbf{K}''_{12} = \mathbf{K}'_{12} - \mathbf{C}_{12}\mathbf{C}_{22}^{-1}\mathbf{K}'_{22} = -631.68 - \frac{(-1.124)(1263.36)}{3.372} = -210.560. \quad (6.223)$$

Then it follows from Eq. 6.213 that

$$\mathbf{A} = \begin{bmatrix} 0.0 & 0.0 & 1.0 \\ 187.3310 & -374.662 & 0.333333 \\ -421.120 & 210.560 & -1.87333 \end{bmatrix}. \quad (6.224)$$

Assume that this structure along the x-direction is subjected to the modified El-Centro earthquake shown in Figure 6.11. It then follows that

$$\mathbf{z}(t) = \{x_1(t) \quad x_2(t) \quad \dot{x}_1(t)\}^T, \quad \mathbf{H} = \begin{bmatrix} 0 \\ 0 \\ -1 \end{bmatrix}, \quad \mathbf{a}(t) = a(t). \quad (6.225)$$

Consider the time step to be 0.02 s. It follows that

$$\mathbf{A}\Delta t = \begin{bmatrix} 0.0 & 0.0 & 0.02 \\ 3.746619 & -7.49324 & 0.006667 \\ -8.4224 & 4.2112 & -0.03747 \end{bmatrix}, \quad e^{\mathbf{A}\Delta t} = \begin{bmatrix} 0.933585 & 0.009415 & 0.019205 \\ 0.468984 & 0.005031 & 0.009194 \\ -6.32412 & 0.516356 & 0.900745 \end{bmatrix}$$

and therefore

$$\begin{Bmatrix} x_1 \\ x_2 \\ \dot{x}_1 \end{Bmatrix}_{k+1} = \begin{bmatrix} 0.933585 & 0.009415 & 0.019205 \\ 0.468984 & 0.005031 & 0.009194 \\ -6.32412 & 0.516356 & 0.900745 \end{bmatrix} \begin{Bmatrix} x_1 \\ x_2 \\ \dot{x}_1 \end{Bmatrix}_k + 0.02 \begin{Bmatrix} -0.019205 \\ -0.009194 \\ -0.900745 \end{Bmatrix} a_k \quad (6.226)$$

$$\{\dot{x}_2\}_k = \begin{bmatrix} 187.33 & -374.66 & 0.3333 \end{bmatrix} \begin{Bmatrix} x_1 \\ x_2 \\ \dot{x}_1 \end{Bmatrix}_k . \quad (6.227)$$

Table 6.5 provides the solution for the first 5 time steps. The response time history for a time step of 0.02 s is shown in Figure 6.17.

Table 6.5 Sample Calculation of 2DOF with One Zero Mass, $\Delta t = 0.02$ s

k	Time (s)	a_k (in./s²)	x_{1k} (in.)	x_{2k} (in.)	$\dot{x}_{1k}$ (in./s)	$\dot{x}_{2k}$ (in./s)
0	0.0	−0.9931	0.0	0.0	0.0	0.0
1	0.02	−7.6507	0.00038	0.00018	0.0179	0.0090
2	0.04	−7.1484	0.00364	0.00175	0.1516	0.0764
3	0.06	−6.2597	0.00907	0.00442	0.2432	0.1230
4	0.08	−6.7234	0.01559	0.00776	0.2768	0.1407
5	0.10	−8.5008	0.02252	0.01113	0.2758	0.1412

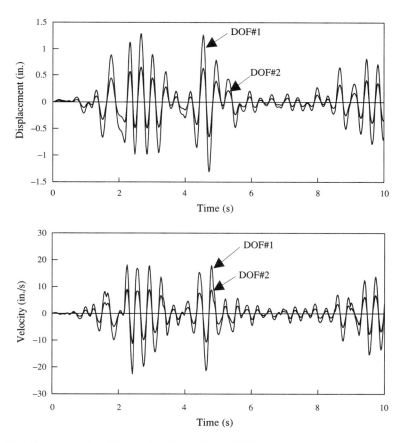

Figure 6.17 Response Time History for a Time Step of 0.02 s

EXAMPLE 2 *One-Story Frame with Stiffness Proportional Damping*

Consider the one-story frame as shown in Figure 6.18, where the mass and stiffness matrices are

$$\mathbf{M} = \begin{bmatrix} m & 0 & 0 \\ 0 & 0 & 0 \\ 0 & 0 & 0 \end{bmatrix}, \quad \mathbf{K} = \begin{bmatrix} \dfrac{24EI_1}{L_1^3} & \dfrac{6EI_1}{L_1^2} & \dfrac{6EI_1}{L_1^2} \\ \dfrac{6EI_1}{L_1^2} & \dfrac{4EI_1}{L_1} + \dfrac{4EI_2}{L_2} & \dfrac{2EI_2}{L_2} \\ \dfrac{6EI_1}{L_1^2} & \dfrac{2EI_2}{L_2} & \dfrac{4EI_1}{L_1} + \dfrac{4EI_2}{L_2} \end{bmatrix}. \tag{6.228}$$

Assume that

$$m = 1.0 \text{ k-s}^2/\text{in.}, \qquad EI_1 = EI_2 = 1,000,000 \text{ k-in.}^2, \qquad L_1 = L_2 = 100 \text{ in.}$$

and then it follows from Eq. 6.228 that

$$\mathbf{M} = \begin{bmatrix} 1.0 & 0.0 & 0.0 \\ 0.0 & 0.0 & 0.0 \\ 0.0 & 0.0 & 0.0 \end{bmatrix}, \quad \mathbf{K} = \begin{bmatrix} 24 & 600 & 600 \\ 600 & 80,000 & 20,000 \\ 600 & 20,000 & 80,000 \end{bmatrix}. \tag{6.229}$$

Let the damping of the structure be represented by Rayleigh proportional damping with $\alpha = 0$ and $\beta = 0.0005$. It then follows from Eq. 6.229 that

$$\mathbf{C} = (0)\mathbf{M} + (0.0005) \begin{bmatrix} 24 & 600 & 600 \\ 600 & 80,000 & 20,000 \\ 600 & 20,000 & 80,000 \end{bmatrix} = \begin{bmatrix} 0.012 & 0.30 & 0.30 \\ 0.30 & 40.0 & 10.0 \\ 0.30 & 10.0 & 40.0 \end{bmatrix}. \tag{6.230}$$

Since the dimension of the damping matrix is the same as that of the stiffness matrix, there is no degree of freedom in association with the label 3. It follows that

$$\mathbf{K}'_{11} = \mathbf{K}_{11} = 24, \quad \mathbf{K}'_{12} = \mathbf{K}_{12} = \begin{bmatrix} 600 & 600 \end{bmatrix}$$

$$\mathbf{K}'_{21} = \mathbf{K}_{21} = \begin{bmatrix} 600 \\ 600 \end{bmatrix}, \quad \mathbf{K}'_{22} = \mathbf{K}_{22} = \begin{bmatrix} 80,000 & 20,000 \\ 20,000 & 80,000 \end{bmatrix} \tag{6.231}$$

and therefore

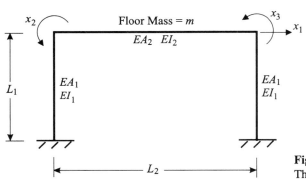

Figure 6.18 One-Story Frame with Three Degrees of Freedom

$$\mathbf{C}_{22}^{-1} = \begin{bmatrix} 40.0 & 10.0 \\ 10.0 & 40.0 \end{bmatrix}^{-1} = \begin{bmatrix} 0.026667 & -0.006667 \\ -0.006667 & 0.026667 \end{bmatrix} \tag{6.232}$$

$$\mathbf{K}_{11}'' = 24 - \begin{bmatrix} 0.30 & 0.30 \end{bmatrix} \begin{bmatrix} 0.026667 & -0.006667 \\ -0.006667 & 0.026667 \end{bmatrix} \begin{bmatrix} 600 \\ 600 \end{bmatrix} = 16.8 \tag{6.233}$$

$$\mathbf{K}_{12}'' = \begin{bmatrix} 600 & 600 \end{bmatrix} - \begin{bmatrix} 0.30 & 0.30 \end{bmatrix} \begin{bmatrix} 0.026667 & -0.006667 \\ -0.006667 & 0.026667 \end{bmatrix} \begin{bmatrix} 80,000 & 20,000 \\ 20,000 & 80,000 \end{bmatrix} = \begin{bmatrix} 0 & 0 \end{bmatrix} \tag{6.234}$$

$$\mathbf{C}_{11}' = 0.012 - \begin{bmatrix} 0.30 & 0.30 \end{bmatrix} \begin{bmatrix} 0.026667 & -0.006667 \\ -0.006667 & 0.026667 \end{bmatrix} \begin{bmatrix} 0.30 \\ 0.30 \end{bmatrix} = 0.0084 \tag{6.235}$$

$$\mathbf{C}_{22}^{-1}\mathbf{K}_{21}' = \begin{bmatrix} 0.026667 & -0.006667 \\ -0.006667 & 0.026667 \end{bmatrix} \begin{bmatrix} 600 \\ 600 \end{bmatrix} = \begin{bmatrix} 12.0 \\ 12.0 \end{bmatrix} \tag{6.236}$$

$$\mathbf{C}_{22}^{-1}\mathbf{K}_{22}' = \begin{bmatrix} 0.026667 & -0.006667 \\ -0.006667 & 0.026667 \end{bmatrix} \begin{bmatrix} 80,000 & 20,000 \\ 20,000 & 80,000 \end{bmatrix} = \begin{bmatrix} 2,000 & 0 \\ 0 & 2,000 \end{bmatrix} \tag{6.237}$$

$$\mathbf{C}_{22}^{-1}\mathbf{C}_{21} = \begin{bmatrix} 0.026667 & -0.006667 \\ -0.006667 & 0.026667 \end{bmatrix} \begin{bmatrix} 0.3 \\ 0.3 \end{bmatrix} = \begin{bmatrix} 0.006 \\ 0.006 \end{bmatrix}. \tag{6.238}$$

It follows from Eq. 6.212 that

$$\mathbf{z}(t) = \begin{Bmatrix} x_1 \\ x_2 \\ x_3 \\ \dot{x}_1 \end{Bmatrix}, \quad \mathbf{A} = \begin{bmatrix} 0.0 & 0.0 & 0.0 & 1.0 \\ -12.0 & -2,000 & 0.0 & -0.0060 \\ -12.0 & 0.0 & -2,000 & -0.0060 \\ -16.8 & 0.0 & 0.0 & -0.0084 \end{bmatrix}, \quad \mathbf{H} = \begin{bmatrix} 0 \\ 0 \\ 0 \\ -1 \end{bmatrix}, \quad a(t) = a(t). \tag{6.239}$$

Assume that a time step of 0.02 s and a Delta forcing function are used; then

$$\mathbf{e}^{\mathbf{A}\Delta t} = \begin{bmatrix} 0.996642 & 0.0 & 0.0 & 0.019976 \\ -0.00598 & 0.0 & 0.0 & -0.00012 \\ -0.00598 & 0.0 & 0.0 & -0.00012 \\ -0.33560 & 0.0 & 0.0 & 0.996474 \end{bmatrix}, \quad \mathbf{H}_d^{(EQ)} = \mathbf{e}^{\mathbf{A}\Delta t}\mathbf{H}\Delta t = \begin{bmatrix} -0.000400 \\ 0.0000024 \\ 0.0000024 \\ 0.0199295 \end{bmatrix}. \tag{6.240}$$

It follows from Eq. 6.240 that

$$\begin{Bmatrix} x_1 \\ x_2 \\ x_3 \\ \dot{x}_1 \end{Bmatrix}_{k+1} = \begin{bmatrix} 0.996642 & 0.0 & 0.0 & 0.019976 \\ -0.00598 & 0.0 & 0.0 & -0.00012 \\ -0.00598 & 0.0 & 0.0 & -0.00012 \\ -0.33560 & 0.0 & 0.0 & 0.996474 \end{bmatrix} \begin{Bmatrix} x_1 \\ x_2 \\ x_3 \\ \dot{x}_1 \end{Bmatrix}_k + \begin{bmatrix} -0.000400 \\ 0.0000024 \\ 0.0000024 \\ 0.0199295 \end{bmatrix} a_k. \tag{6.241}$$

When this structure is subjected to the El-Centro earthquake given in Figure 6.11, the response is as shown in Figure 6.19. Because the mass matrix is not invertible, the total number of states is reduced by the degree of singularity of the mass matrix.

EXAMPLE 3   *One-Story Frame with Mass Proportional Damping*

Consider Example 2 and Figure 6.18. Assume that

$$m = 1.0 \text{ k-s}^2/\text{in.}, \qquad EI_1 = EI_2 = 1,000,000 \text{ k-in.}^2, \qquad L_1 = L_2 = 100 \text{ in.}$$

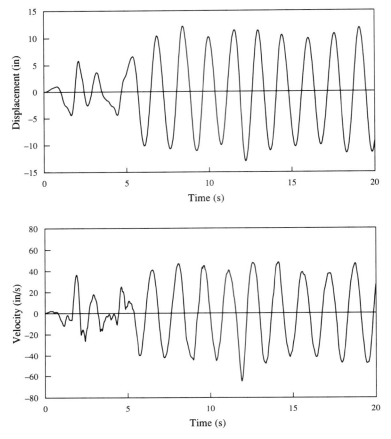

Figure 6.19 Response of an SDOF System with Stiffness Proportional Damping

and then it follows from Eq. 6.228 that

$$\mathbf{M} = \begin{bmatrix} 1.0 & 0.0 & 0.0 \\ 0.0 & 0.0 & 0.0 \\ 0.0 & 0.0 & 0.0 \end{bmatrix}, \quad \mathbf{K} = \begin{bmatrix} 24 & 600 & 600 \\ 600 & 80,000 & 20,000 \\ 600 & 20,000 & 80,000 \end{bmatrix}. \tag{6.242}$$

Now let the damping of the structure be represented by Rayleigh proportional damping with $\alpha = 0.012$ and $\beta = 0.0$. It then follows from Eq. 6.242 that

$$\mathbf{C} = (0.012) \begin{bmatrix} 1.0 & 0.0 & 0.0 \\ 0.0 & 0.0 & 0.0 \\ 0.0 & 0.0 & 0.0 \end{bmatrix} + (0)\mathbf{K} = \begin{bmatrix} 0.012 & 0.0 & 0.0 \\ 0.0 & 0.0 & 0.0 \\ 0.0 & 0.0 & 0.0 \end{bmatrix}. \tag{6.243}$$

The dimension of the damping matrix is the same as that of the mass matrix, and therefore there is no degree of freedom in association with the label 2. It follows that

$$\mathbf{K}_{33}^{-1} = \begin{bmatrix} 80,000 & 20,000 \\ 20,000 & 80,000 \end{bmatrix}^{-1} = \frac{1}{300,000} \begin{bmatrix} 4 & -1 \\ -1 & 4 \end{bmatrix} \tag{6.244}$$

$$\mathbf{K}'_{11} = 24 - \frac{1}{300,000} \begin{bmatrix} 600 & 600 \end{bmatrix} \begin{bmatrix} 4 & -1 \\ -1 & 4 \end{bmatrix} \begin{bmatrix} 600 \\ 600 \end{bmatrix} = 16.8 \qquad (6.245)$$

$$\mathbf{C}'_{11} = \mathbf{C}_{11} = 0.012, \quad \mathbf{K}''_{11} = \mathbf{K}'_{11} = 16.8 \qquad (6.246)$$

and from Eq. 6.212 that

$$\mathbf{z}(t) = \begin{Bmatrix} x_1 \\ \dot{x}_1 \end{Bmatrix}, \quad \mathbf{A} = \begin{bmatrix} 0.0 & 1.0 \\ -16.8 & -0.012 \end{bmatrix}, \quad \mathbf{H} = \begin{bmatrix} 0 \\ -1 \end{bmatrix}, \quad \mathbf{a}(t) = a(t). \qquad (6.247)$$

Assume a time step of 0.02 s; using the Delta forcing function approach, then

$$\mathbf{e}^{\mathbf{A}\Delta t} = \begin{bmatrix} 0.996642 & 0.019975 \\ -0.335584 & 0.996402 \end{bmatrix}, \quad \mathbf{H}_d^{(EQ)} = \mathbf{e}^{\mathbf{A}\Delta t} \mathbf{H}\Delta t = \begin{bmatrix} -0.00040 \\ 0.019928 \end{bmatrix}. \qquad (6.248)$$

It follows from Eq. 6.248 that

$$\begin{Bmatrix} x_1 \\ \dot{x}_1 \end{Bmatrix}_{k+1} = \begin{bmatrix} 0.996642 & 0.019975 \\ -0.335584 & 0.996402 \end{bmatrix} \begin{Bmatrix} x_1 \\ \dot{x}_1 \end{Bmatrix}_k + \begin{bmatrix} -0.00040 \\ 0.019928 \end{bmatrix} a_k. \qquad (6.249)$$

When this structure is subjected to the modified El-Centro earthquake Figure 6.11, the response is as shown in Figure 6.20. Because damping and mass matrices are not invertible, the total number of states is reduced by the degree of singularity of the damping matrix plus the mass matrix.

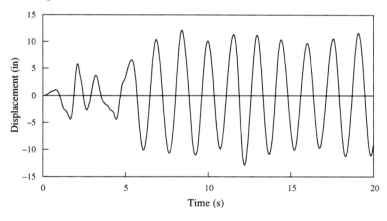

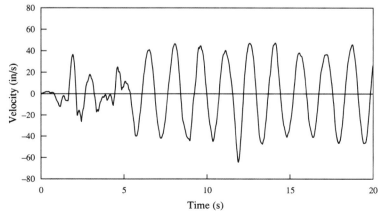

Figure 6.20 Response of an SDOF System with Mass Proportional Damping

RECOMMENDED ADDITIONAL READING

1. A. K. Chopra. *Dynamics of Structure—Theory and Applications to Earthquake Engineering*, Prentice Hall, 1995.

2. M. Paz. *Structural Dynamics—Theory and Computation* (4th edition). Chapman and Hall, 1997.

PROBLEMS

6.1 Repeat Example 1 in Section 6.3 but let the mass at the bottom level be reduced from $2m$ to m.

6.2 Repeat Example 1 in Section 6.3 but let the damping coefficient at the bottom level be reduced from $2c$ to c.

6.3 Repeat Example 1 in Section 6.5 but with $m_1 = 0$ and $m_2 \neq 0$. Condense the problem to have the equation of motion in terms of x_2.

6.4 Repeat Example 2 in Section 6.5 but with $m_1 \neq 0$ and $m_2 = 0$. Condense the problem to have the equation of motion in terms of x_1.

6.5 Repeat Example 3 in Section 6.5 but with the top mass equal to zero and the bottom mass equal to $2m$. Condense the problem to have the equation of motion in terms of x_1.

6.6 Repeat Example 1 in Section 6.6 but use dynamic condensation to have the equation of motion in terms of x_2.

6.7 Repeat Example 2 in Section 6.6 but use dynamic condensation to have the equation of motion in terms of x_1.

6.8 Repeat Example 3 in Section 6.6 but use dynamic condensation to have the equation motion in terms of x_1.

6.9 Repeat Example 1 in Section 6.7 but with a damping ratio equal to 20%.

6.10 Repeat Example 1 in Section 6.7 but with a period of vibration equal to 1.0 s.

6.11 Repeat Example 2 in Section 6.7 but with a damping matrix equal to two times the damping matrix in the example.

6.12 Repeat Example 2 in Section 6.7 but with a stiffness matrix equal to one half the stiffness matrix in the example.

6.13 Repeat Example 1 in Section 6.8 but use $m = 2.0$.

6.14 Repeat Example 1 in Section 6.8 but use $k = 2 \times 315.84 = 631.68$.

6.15 Repeat Example 1 in Section 6.8 but with a damping matrix equal to two times the damping matrix in the example.

6.16 Repeat Example 2 in Section 6.8 with Rayleigh damping and a 5% damping ratio in the first two normal modes.

Chapter 7

Nonlinear Time History Analysis

7.1 OVERVIEW

The earthquake response of most buildings and civil engineering structures will induce deformations in one or more structural elements that are beyond the yield limit. Therefore, the structure will respond with a nonlinear relationship between force and deformation. The mathematical complexity of the solution increases as the response goes from the linear to the nonlinear domain. However, the computational effort is rewarded with an accurate description of response.

Consider the steel cantilever beam shown in Figure 7.1. Assume that the stress versus strain curve for the steel is as shown in Figure 7.2. As the concentrated force at the end of the beam increases in magnitude, the moment at the support of the cantilever increases in magnitude. The theory of strength of beam materials assumes that the variation of the strain on any material for any cross-section of the beam will be zero at the neutral axis and will increase with a linear relationship with the distance from the neutral axis. This is often called "plane cross-sections remain plane." Therefore, as the force at the end of the cantilever increases, the strain in the steel at the support also increases. Because the steel stress versus strain curve is a linear function until the strain reaches its yield strain, ε_y, the relationship between the force and the deflection at the end of the cantilever is a linear function. Figure 7.3 shows a plot of the force versus vertical deflection at the end of the

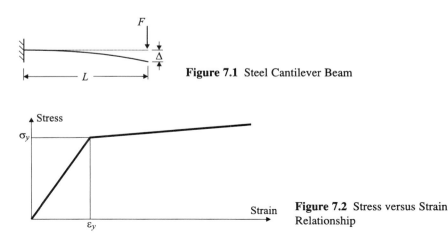

Figure 7.1 Steel Cantilever Beam

Figure 7.2 Stress versus Strain Relationship

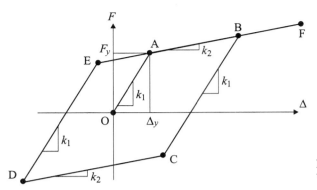

Figure 7.3 Applied Force versus Vertical Deflection

cantilever. The portion of the curve between O and A is the linear portion of the curve. If the force is less than F_y, where F_y is called the yield force, then the strain in the steel will be less than the yield strain, ε_y, and the moment will be less than the yield moment of the cross-section, M_y. The yield moment, M_y, is the moment that needs to be applied to the beam cross-section to produce a strain in a steel fiber farthest from the neutral axis equal to the yield strain. Therefore, if the force is less than the yield force or, alternately, the deflection of the end of the cantilever is less than the yield deflection, Δ_y, then the response of the structure is in the linear response domain.

Now consider the situation where the force exceeds F_y and therefore the strain in the steel exceeds ε_y. The moment that the force induces on the cross-section will increase and the deflection will increase along a path from A to B. The force versus deflection relationship will not follow the stiffness k_1 but will now have a stiffness equal to k_2. If the force is reduced before the steel is fractured, then the force versus deflection path will be from B to C. The stiffness can now be imagined to be equal to k_1. The path will then proceed along the trail C to D, D to E, and E to F.

The response of this system is called nonlinear because the stiffness does not remain equal to k_1 for all amplitudes of response. There are two key elements of a nonlinear system:

1. The *envelope curve* or *backbone curve*: This is the O–A–F curve.
2. The *hysteresis relationship*: This is the law describing how the system unloads and reloads below the backbone curve. This is the B to C and D to E path.

The description of these two key elements is a function of the structural material (e.g., steel, concrete, etc.) and the nature of the redundancy of the indeterminate structure.

EXAMPLE 1 *Bilinear Elastic Structure*

Consider the SDOF nonlinear structural system shown in Figure 7.4. Assume that the stiffness and mass for the structure are $k_e = 631.65$ k/in. and $m = 1.0$ k-s^2/in. If the structure displaces to the right 1.0 in., that is, $x(t) = 1.0$ in., it contacts a spring of stiffness k_1. The stiffness of the spring is $k_1 = 355.31$ k/in. The natural frequency and period of vibration for displacements $x(t) \le 1.0$ in. are

$$\omega_n = \sqrt{k_e/m} = \sqrt{631.65/1.0} = 25.13 \text{ rad/s} \qquad (7.1)$$

$$T_n = 2\pi/\omega_n = 2\pi/25.13 = 0.25 \text{ s.} \qquad (7.2)$$

For displacements $x(t) > 1.0$ in., the natural frequency and period of vibration are

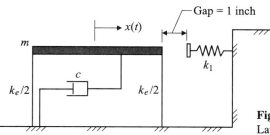

Figure 7.4 SDOF Frame with Gap to Lateral Support

$$\omega_n = \sqrt{(k_e + k_1)/m} = \sqrt{(631.65 + 355.31)/1.0} = 31.42 \text{ rad/s} \qquad (7.3)$$

$$T_n = 2\pi/\omega_n = 2\pi/31.42 = 0.20 \text{ s.} \qquad (7.4)$$

The critical damping ratio in the system is assumed to be 5% for all response amplitudes. Then the damped natural frequency of vibration is

$$\omega_d = \omega_n\sqrt{1-\zeta^2} = \begin{cases} (25.13)\sqrt{1-(0.05)^2} \\ (31.42)\sqrt{1-(0.05)^2} \end{cases} = \begin{cases} 25.10 \text{ rad/s} & x(t) \le 1.0 \text{ in.} \\ 31.38 \text{ rad/s} & x(t) > 1.0 \text{ in.} \end{cases} \qquad (7.5)$$

If the system is released from rest at a displacement of 5 in. to the left, then the initial conditions are

$$x(t=0) = -5.0 \text{ in}, \quad \dot{x}(t=0) = 0.0 \text{ in./s.} \qquad (7.6)$$

The free vibration response of an SDOF system was discussed in Section 2.2, where the equations for the displacement and velocity of the system are given in Eq. 2.21 and 2.22:

$$x(t) = e^{-\zeta\omega_n t}\left[x_o \cos\omega_d t + \left(\frac{\zeta\omega_n x_o + \dot{x}_o}{\omega_n\sqrt{1-\zeta^2}}\right)\sin\omega_d t\right] = e^{-1.257t}\left[-5.0\cos 25.10t - 0.25\sin 25.10t\right] \quad (7.7)$$

$$\dot{x}(t) = e^{-\zeta\omega_n t}\left[\dot{x}_o \cos\omega_d t - \left(\frac{\omega_n x_o + \zeta\dot{x}_o}{\sqrt{1-\zeta^2}}\right)\sin\omega_d t\right] = 125.8 e^{-1.257t}\sin 25.10t. \qquad (7.8)$$

Free vibration as defined by Eqs. 7.7 and 7.8 will occur until the gap is closed. The displacement at the gap closure is $x(t) = 1.0$ in. Therefore, using Eq. 7.7, this instant in time, denoted by t_1, is

$$e^{-1.257t_1}\left[-5.0\cos 25.10t_1 - 0.25\sin 25.10t_1\right] = 1.0 \text{ in.} \qquad (7.9)$$

Solving for t_1 in Eq. 7.9 gives

$$t_1 = 0.07337 \text{ s.} \qquad (7.10)$$

The velocity of the system at this time is

$$\dot{x}(t=t_1) = 125.8 e^{-1.257(0.07337)}\sin 25.10(0.07337) = 110.5 \text{ in./s.} \qquad (7.11)$$

The response of the system for time $t > t_1$ will be governed by a free vibration equation, and the SDOF system will now have a stiffness equal to $k_e + k_1$. The initial conditions that produce the response from $t = t_1$ until the system responds at displacements greater than $x(t) = 1.0$ in., comes to a stop, and then returns to $x(t) = 1.0$ in. are

$$x(t=t_1 = 0.07337) = 1.0 \text{ in.}, \quad \dot{x}(t=t_1 = 0.07337) = 110.5 \text{ in./s.} \qquad (7.12)$$

The free vibration equations for the displacement and velocity of the SDOF system in this time domain are

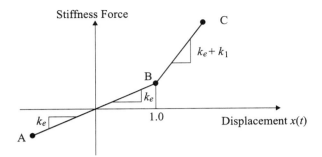

Figure 7.5 Bilinear Stiffness Force

$$x(t) = e^{-1.571(t-0.07337)}\left[\cos 31.38(t - 0.07337) + 3.571\sin 31.38(t - 0.07337)\right] \quad (7.13)$$

$$\dot{x}(t) = e^{-1.571(t-0.07337)}\left[110.5\cos 31.38(t - 0.07337) - 36.99\sin 31.38(t - 0.07337)\right]. \quad (7.14)$$

The SDOF system will go out of contact with the spring k_1 when $x(t)$ returns to and becomes less than $x(t) = 1.0$ in. This time, denoted by t_2, is obtained using Eq. 7.13, that is,

$$e^{-1.571(t_2-0.07337)}\left[\cos 31.38(t_2 - 0.07337) + 3.571\sin 31.38(t_2 - 0.07337)\right] = 1.0. \quad (7.15)$$

Solving Eq. 7.15 gives

$$t_2 = 0.15487 \text{ s}. \quad (7.16)$$

The velocity of the system at this time is computed using Eq. 7.14, and it is

$$\dot{x}(t_2) = -101.8 \text{ in./s}. \quad (7.17)$$

The SDOF system response after $t = t_2$ returns to a system with a stiffness equal to k_e. Again, the SDOF free vibration equations given in Eqs. 7.7 and 7.8 are used to calculate the response with a time lag t_2 like the time lag t_1 in Eqs. 7.13 and 7.14. The initial conditions for this phase of the response, which starts at $t = t_2$, are

$$x(t = t_2 = 0.15487) = 1.0 \text{ in.}, \quad \dot{x}(t = t_2 = 0.15487) = -101.8 \text{ in./s}. \quad (7.18)$$

Figure 7.5 shows the stiffness force versus displacement relationship for this nonlinear SDOF system.

7.2 SINGLE DEGREE OF FREEDOM RESPONSE WITH NONLINEAR STIFFNESS

Consider the equation of motion for a linear SDOF system, which is

$$m\ddot{x}(t) + c\dot{x}(t) + k_e x(t) = F_e(t) \quad (7.19)$$

where k_e denotes the elastic stiffness for a linear system. At time $t = t_k$, Eq. 7.19 can be written as

$$m\ddot{x}_k + c\dot{x}_k + k_e x_k = F_k \quad (7.20)$$

where, for simplicity, $F_k = F_{ek} = F_e(t = t_k)$. At time $t = t_{k+1}$, it similarly follows that

$$m\ddot{x}_{k+1} + c\dot{x}_{k+1} + k_e x_{k+1} = F_{k+1}. \quad (7.21)$$

When discretization methods of analysis are used (e.g., constant acceleration method, Newmark β-method, etc.), the time variation in the acceleration between time t_k and time t_{k+1} is assumed to be of a specific form, and then the variations of the velocity and displacement between t_k and t_{k+1} follow directly from integration. For example, when constant acceleration is assumed, the acceleration is assumed to be

$$\ddot{x}(t) = \ddot{x}(t_k) = \ddot{x}_k, \quad t_k \leq t < t_{k+1} \tag{7.22}$$

and it follows from integration that

$$\dot{x}_{k+1} = \dot{x}_k + \ddot{x}_k \Delta t \tag{7.23a}$$

$$x_{k+1} = x_k + \dot{x}_k \Delta t + \tfrac{1}{2} \ddot{x}_k (\Delta t)^2. \tag{7.23b}$$

In the Newmark β-method, the corresponding equations for $\dot{x}_{k+1}$ and x_{k+1} are

$$\dot{x}_{k+1} = \dot{x}_k + (1-\delta)\ddot{x}_k \Delta t + \delta \ddot{x}_{k+1} \Delta t \tag{7.24a}$$

$$x_{k+1} = x_k + \dot{x}_k \Delta t + \left(\tfrac{1}{2} - \alpha\right)\ddot{x}_k (\Delta t)^2 + \alpha \ddot{x}_{k+1}(\Delta t)^2. \tag{7.24b}$$

Equation 7.19 represents the force equilibrium for the system, and it can be alternately expressed as

$$F_i(t) + F_d(t) + F_s(t) = F_e(t) \tag{7.25}$$

where F_i, F_d, and F_s are the forces associated with the inertia, damping, and stiffness-related deformations of the system. It is clear that

$$F_i(t) = m\ddot{x}(t) \tag{7.26}$$

and therefore the inertia force is a linear function of the acceleration because the mass, m, does not vary with time. Similarly, the viscous damping force is a linear function of the velocity, that is, $F_d(t) = c\dot{x}(t)$. Now consider the force related to the system stiffness. If the system is a linear system, then

$$F_s(t) = k_e x(t) \tag{7.27}$$

and k_e is not a function of $x(t)$. Figure 7.6 shows a linear system with a straight line in the force versus displacement plot. The value of the stiffness for a linear system is

$$k_s = \left[\frac{F_s(t_{k+1}) - F_s(t_k)}{x_{k+1} - x_k} \right]. \tag{7.28}$$

Now imagine that when Eq. 7.28 is solved, k_e is not a constant because the force from the system stiffness is not a linear function of the displacement. A special solution procedure is required to solve this nonlinear stiffness problem.

Consider Eq. 7.20 but now for a system with nonlinear stiffness. Equation 7.20 can be written as

$$m\ddot{x}_k + c\dot{x}_k + F_s(x_k) = F_k. \tag{7.29}$$

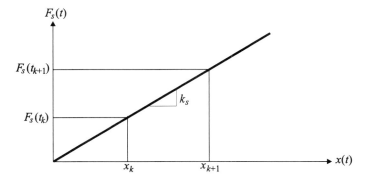

Figure 7.6 Force-Displacement Plot with Secant Stiffness

Similarly, Eq. 7.21 can be written as

$$m\ddot{x}_{k+1} + c\dot{x}_{k+1} + F_s(x_{k+1}) = F_{k+1}. \tag{7.30}$$

Subtracting Eq. 7.29 from Eq. 7.30, it follows that

$$m[\ddot{x}_{k+1} - \ddot{x}_k] + c[\dot{x}_{k+1} - \dot{x}_k] + [F_s(x_{k+1}) - F_s(x_k)] = [F_{k+1} - F_k]. \tag{7.31}$$

Define the following terms:

$$\Delta\ddot{x} = \ddot{x}_{k+1} - \ddot{x}_k \tag{7.32a}$$

$$\Delta\dot{x} = \dot{x}_{k+1} - \dot{x}_k \tag{7.32b}$$

$$\Delta F = F_{k+1} - F_k \tag{7.32c}$$

$$\Delta F_s = F_s(x_{k+1}) - F_s(x_k) \tag{7.32d}$$

where $\Delta(\cdot)$ is the change in the quantity in parenthesis between time t_k and time t_{k+1}. It then follows that

$$m\Delta\ddot{x} + c\Delta\dot{x} + \Delta F_s = \Delta F. \tag{7.33}$$

If the variation in the stiffness-related force between x_k and x_{k+1} is assumed to be a linear function, then

$$F_s(x_{k+1}) = F_s(x_k) + \left[\frac{F_s(x_{k+1}) - F_s(x_k)}{x_{k+1} - x_k}\right](x_{k+1} - x_k). \tag{7.34}$$

The *secant stiffness* is defined to be

$$k_s = \left[\frac{F_s(x_{k+1}) - F_s(x_k)}{x_{k+1} - x_k}\right] = \frac{\Delta F_s}{\Delta x}. \tag{7.35}$$

Therefore, substituting Eq. 7.35 into Eq. 7.33, it follows that

$$m\Delta\ddot{x} + c\Delta\dot{x} + k_s\Delta x = \Delta F. \tag{7.36}$$

Equation 7.36 can be solved exactly if the secant stiffness is known. Unfortunately, this is not the case because the displacement at time $t = t_{k+1}$ is not known.

The response solution procedure must make an assumption for k_s, and one possibility is that the secant stiffness is equal to the secant stiffness used in the previous time step, denoted k_k. Figure 7.7 shows these two stiffness terms. Therefore,

$$k_s = k_k = \left[\frac{F_s(x_k) - F_s(x_{k-1})}{x_k - x_{k-1}}\right]. \tag{7.37}$$

Because the secant stiffness between x_{k-1} and x_k is in general not equal to the secant stiffness between x_k and x_{k+1}, there is a force imbalance, or error, as shown in Figure 7.7. Note

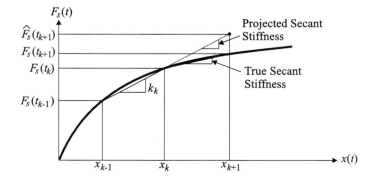

Figure 7.7 Force-Displacement Plot with Secant Stiffness

that $F_s(t_{k+1})$ is the correct force and the estimated force is $\hat{F}_s(t_{k+1})$. Substituting Eq. 7.37 into Eq. 7.36, it follows that

$$m\Delta\ddot{x} + c\Delta\dot{x} + k_s\Delta x = \Delta F. \tag{7.38}$$

The constant acceleration method, Newmark β-method, or other numerical solution methods can be used to solve Eq. 7.36 or 7.38. Consider, for example, the constant acceleration method. Recall from Eq. 7.23 that

$$\Delta\dot{x} = \dot{x}_{k+1} - \dot{x}_k = \ddot{x}_k\Delta t \tag{7.39a}$$

$$\Delta x = x_{k+1} - x_k = \dot{x}_k\Delta t + \tfrac{1}{2}\ddot{x}_k(\Delta t)^2. \tag{7.39b}$$

Substituting Eq. 7.39 into Eq. 7.36 gives

$$m\,\Delta\ddot{x} + c\left[\ddot{x}_k\Delta t\right] + k_s\left[\dot{x}_k\Delta t + \tfrac{1}{2}\ddot{x}_k(\Delta t)^2\right] = \Delta F \tag{7.40}$$

and, finally,

$$\Delta\ddot{x} = \left(\frac{\Delta F}{m}\right) - \left(\frac{c}{m}\right)\left[\ddot{x}_k\Delta t\right] - \left(\frac{k_s}{m}\right)\left[\dot{x}_k\Delta t + \tfrac{1}{2}\ddot{x}_k(\Delta t)^2\right]. \tag{7.41}$$

Using Eqs. 7.39a, 7.39b, and 7.41, it then follows that

$$x_{k+1} = x_k + \Delta x, \qquad \dot{x}_{k+1} = \dot{x}_k + \Delta\dot{x}, \qquad \ddot{x}_{k+1} = \ddot{x}_k + \Delta\ddot{x}. \tag{7.42}$$

As the time increment Δt decreases, the variation in the displacement decreases and the difference between the true stiffness and the projected stiffness decreases.

Consider now the case where the solution for the response is obtained using the Newmark β-method. Recall from Eq. 7.24 that

$$\Delta\dot{x} = \dot{x}_{k+1} - \dot{x}_k = (1-\delta)\ddot{x}_k\Delta t + \delta\ddot{x}_{k+1}\Delta t \tag{7.43a}$$

$$\Delta x = x_{k+1} - x_k = \dot{x}_k\Delta t + \left(\tfrac{1}{2}-\alpha\right)\ddot{x}_k(\Delta t)^2 + \alpha\ddot{x}_{k+1}(\Delta t)^2. \tag{7.43b}$$

Substituting Eq. 7.43 into Eq. 7.36, it follows that

$$m\Delta\ddot{x} + c\left[(1-\delta)\ddot{x}_k\Delta t + \delta\ddot{x}_{k+1}\Delta t\right] + k_s\left[\dot{x}_k\Delta t + \left(\tfrac{1}{2}-\alpha\right)\ddot{x}_k(\Delta t)^2 + \alpha\ddot{x}_{k+1}(\Delta t)^2\right] = \Delta F. \tag{7.44}$$

A comparison of Eq. 7.40 and Eq. 7.44 shows the very significant difference that Eq. 7.40 only contains the mass term with an unknown (i.e., $\Delta\ddot{x}$). Therefore, Eq. 7.44 must be developed further before a solution can be obtained. Recall from Eq. 7.32 that

$$\ddot{x}_{k+1} = \ddot{x}_k + \Delta\ddot{x}. \tag{7.45}$$

Substituting Eq. 7.45 into Eq. 7.44 gives

$$\begin{aligned} m\Delta\ddot{x} + c(1-\delta)\ddot{x}_k\Delta t + c\delta\ddot{x}_k\Delta t + c\delta\Delta\ddot{x}\Delta t + k_s\dot{x}_k\Delta t \\ + k_s\left(\tfrac{1}{2}-\alpha\right)\ddot{x}_k(\Delta t)^2 + k_s\alpha\ddot{x}_k(\Delta t)^2 + k_s\alpha\Delta\ddot{x}(\Delta t)^2 = \Delta F. \end{aligned} \tag{7.46}$$

Collecting similar terms, it follows that

$$\begin{aligned} \left[m + c\delta\Delta t + k_s\alpha(\Delta t)^2\right]\Delta\ddot{x} + c\left[(1-\delta)\ddot{x}_k\Delta t + \delta\ddot{x}_k\Delta t\right] \\ + k_s\left[\dot{x}_k\Delta t + \left(\tfrac{1}{2}-\alpha\right)\ddot{x}_k(\Delta t)^2 + \alpha\ddot{x}_k(\Delta t)^2\right] = \Delta F. \end{aligned} \tag{7.47}$$

Simplifying Eq. 7.47 gives

$$\left[m + c\delta\Delta t + k_s\alpha(\Delta t)^2\right]\Delta\ddot{x} + c\ddot{x}_k\Delta t + k_s\left[\dot{x}_k\Delta t + \tfrac{1}{2}\ddot{x}_k(\Delta t)^2\right] = \Delta F. \tag{7.48}$$

Solving for $\Delta\ddot{x}$ gives

$$\Delta\ddot{x} = \frac{\Delta F - c\ddot{x}_k\Delta t - k_s\left[\dot{x}_k\Delta t + \tfrac{1}{2}\ddot{x}_k(\Delta t)^2\right]}{m + c\delta\Delta t + k_s\alpha(\Delta t)^2}. \tag{7.49}$$

Equation 7.49 provides the solution for $\Delta\ddot{x}$, and using Eq. 7.43 it is possible to calculate Δx and $\Delta\dot{x}$. Then using Eq. 7.42, the values for x_{k+1} and $\dot{x}_{k+1}$ can be determined.

Because the value of x_{k+1} has been calculated, it is possible to calculate an improved estimate of the true secant stiffness. Denote this calculated value of x_{k+1} from the initial solution as $x_{k+1}^{(1)}$. It then follows that the secant stiffness obtained using $x_{k+1}^{(1)}$, denoted $k_s^{(1)}$, is

$$k_s^{(1)} = \left[\frac{F_s(x_{k+1}^{(1)}) - F_s(x_k)}{x_{k+1}^{(1)} - x_k} \right]. \tag{7.50}$$

The initial solution for x_{k+1} (i.e., $x_{k+1}^{(1)}$) was obtained using for the estimate of the secant stiffness the value of the secant stiffness from the previous time step (see Eq. 7.37). Therefore, $x_{k+1}^{(1)}$ is not exact. If Eq. 7.50 is used in Eqs. 7.44 through 7.49, an improved estimate will be obtained for x_{k+1}. This represents one iteration of the initial solution. If the values for x_{k+1}, $\dot{x}_{k+1}$, and $\ddot{x}_{k+1}$ obtained from this iteration solution are sufficiently close to the values obtained in the initial solution, then it is appropriate to move on and calculate the response at time step t_{k+2} given the calculated response at time t_{k+1}. However, if the response from the iteration significantly changes x_{k+1}, $\dot{x}_{k+1}$, or $\ddot{x}_{k+1}$, then either a smaller time increment should be used or a second iteration should be performed using $x_{k+1}^{(2)}$ and

$$k_s^{(2)} = \left[\frac{F_s(x_{k+1}^{(2)}) - F_s(x_k)}{x_{k+1}^{(2)} - x_k} \right]. \tag{7.51}$$

Note that any error in the calculated values of x_{k+1}, $\dot{x}_{k+1}$, and $\ddot{x}_{k+1}$ will carry forward to the next time step.

Recall from Section 2.6 that the Newmark β-method solution can be represented in a matrix form (see Eqs. 2.224 through 2.232). A comparison of Eq. 7.49 and Eq. 2.227 shows that the latter is expressed in terms of the SDOF system natural frequency of vibration ω_n and critical damping ratio ζ. In a nonlinear system, k is not constant and therefore $\omega_n = \sqrt{k/m}$ is not a constant. However, if k_s and c are used instead of ω_n and ζ, then the solution can again be represented in a matrix form. First, let

$$\beta = m + c\delta\Delta t + k_s\alpha(\Delta t)^2. \tag{7.52}$$

Then Eq. 7.49 becomes

$$\Delta\ddot{x} = \left(\frac{1}{\beta}\right)\Delta F - \left(\frac{k_s\Delta t}{\beta}\right)\dot{x}_k - \left(\frac{c\Delta t + \frac{1}{2}k_s(\Delta t)^2}{\beta}\right)\ddot{x}_k. \tag{7.53}$$

Note that Eq. 7.53 is the same as Eq. 7.41 when $\delta = 0$ and $\alpha = 0$. The acceleration at time step $k+1$ is

$$\ddot{x}_{k+1} = \ddot{x}_k + \Delta\ddot{x} = \ddot{x}_k + \left(\frac{1}{\beta}\right)\Delta F - \left(\frac{k_s\Delta t}{\beta}\right)\dot{x}_k - \left(\frac{c\Delta t + \frac{1}{2}k_s(\Delta t)^2}{\beta}\right)\ddot{x}_k$$

$$= \left(\frac{1}{\beta}\right)\Delta F - \left(\frac{k_s\Delta t}{\beta}\right)\dot{x}_k + \left(\frac{\beta - c\Delta t - \frac{1}{2}k_s(\Delta t)^2}{\beta}\right)\ddot{x}_k. \tag{7.54}$$

Substituting Eq. 7.54 into Eq. 7.24 gives

$$\dot{x}_{k+1} = \dot{x}_k + (1-\delta)\ddot{x}_k\Delta t + \delta\Delta t\left[\left(\frac{1}{\beta}\right)\Delta F - \left(\frac{k_s\Delta t}{\beta}\right)\dot{x}_k + \left(\frac{\beta - c\Delta t - \frac{1}{2}k_s(\Delta t)^2}{\beta}\right)\ddot{x}_k\right]$$

$$= \left(\frac{\delta\Delta t}{\beta}\right)\Delta F + \left(\frac{\beta - \delta k_s(\Delta t)^2}{\beta}\right)\dot{x}_k + \left(\frac{\beta\Delta t - c\delta(\Delta t)^2 - \frac{1}{2}k_s\delta(\Delta t)^3}{\beta}\right)\ddot{x}_k \tag{7.55a}$$

$$x_{k+1} = x_k + \dot{x}_k \Delta t + \left(\tfrac{1}{2} - \alpha\right)\ddot{x}_k (\Delta t)^2 + \alpha(\Delta t)^2 \left[\left(\frac{1}{\beta}\right)\Delta F - \left(\frac{k_s \Delta t}{\beta}\right)\dot{x}_k + \left(\frac{\beta - c\Delta t - \tfrac{1}{2}k_s(\Delta t)^2}{\beta}\right)\ddot{x}_k \right]$$

$$= \left(\frac{\alpha(\Delta t)^2}{\beta}\right)\Delta F + x_k + \left(\frac{\beta \Delta t - k_s \alpha(\Delta t)^3}{\beta}\right)\dot{x}_k + \left(\frac{\tfrac{1}{2}\beta(\Delta t)^2 - c\alpha(\Delta t)^3 - \tfrac{1}{2}k_s\alpha(\Delta t)^4}{\beta}\right)\ddot{x}_k. \tag{7.55b}$$

Therefore, representing Eqs. 7.54 and 7.55 in matrix form, it follows that

$$\begin{Bmatrix} x_{k+1} \\ \dot{x}_{k+1} \\ \ddot{x}_{k+1} \end{Bmatrix} = \mathbf{F}_N^{(n)} \begin{Bmatrix} x_k \\ \dot{x}_k \\ \ddot{x}_k \end{Bmatrix} + \mathbf{H}_N^{(n)} \Delta F \tag{7.56}$$

where

$$\mathbf{F}_N^{(n)} = \frac{1}{\beta}\begin{bmatrix} \beta & \beta\Delta t - k_s\alpha(\Delta t)^3 & \tfrac{1}{2}\beta(\Delta t)^2 - c\alpha(\Delta t)^3 - \tfrac{1}{2}k_s\alpha(\Delta t)^4 \\ 0 & \beta - \delta k_s(\Delta t)^2 & \beta\Delta t - c\delta(\Delta t)^2 - \tfrac{1}{2}k_s\delta(\Delta t)^3 \\ 0 & -k_s\Delta t & \beta - c\Delta t - \tfrac{1}{2}k_s(\Delta t)^2 \end{bmatrix}$$

$$\mathbf{H}_N^{(n)} = \frac{1}{\beta}\begin{Bmatrix} \alpha(\Delta t)^2 \\ \delta\Delta t \\ 1 \end{Bmatrix} \tag{7.57}$$

and the superscript (n) denotes nonlinear time history analysis. Let

$$\mathbf{q}_k = \begin{Bmatrix} x_k \\ \dot{x}_k \\ \ddot{x}_k \end{Bmatrix}.$$

Then Eq. 7.56 becomes

$$\mathbf{q}_{k+1} = \mathbf{F}_N^{(n)}\mathbf{q}_k + \mathbf{H}_N^{(n)}\Delta F = \mathbf{F}_N^{(n)}\mathbf{q}_k + \mathbf{H}_N^{(n)}F_{k+1} - \mathbf{H}_N^{(n)}F_k. \tag{7.58}$$

Example 3 in Section 2.6 showed that if the SDOF was excited by an earthquake ground motion, then Eq. 7.58 was still valid with the $\mathbf{H}_N^{(n)}$ matrix and ΔF changed to $\mathbf{H}_N^{(nEQ)}$ and Δa, where

$$\mathbf{H}_N^{(nEQ)} = -\left(\frac{m}{\beta}\right)\begin{Bmatrix} \alpha(\Delta t)^2 \\ \delta\Delta t \\ 1 \end{Bmatrix}. \tag{7.59}$$

Thus Eq. 7.58 becomes

$$\mathbf{q}_{k+1} = \mathbf{F}_N^{(n)}\mathbf{q}_k + \mathbf{H}_N^{(nEQ)}a_{k+1} - \mathbf{H}_N^{(nEQ)}a_k. \tag{7.60}$$

One important note to Eqs. 7.58 and 7.60 is that the forcing function at time step k (i.e., F_k or a_k) appears in the equation, which is different from Eq. 2.232. This is because the dynamic equilibrium equation of motion was used in Section 2.6 at time step k to replace this forcing function by the displacement, velocity, and acceleration at time step k (i.e., x_k, $\dot{x}_k$, and $\ddot{x}_k$). Therefore, the effect of F_k or a_k in the response is embedded in the $\mathbf{F}_N$ matrix in Section 2.6. In this section, the dynamic equilibrium equation of motion cannot be used because k_s represents the incremental stiffness, which is not the stiffness that goes through x_k from the origin. The result is that the $\mathbf{F}_N^{(n)}$ matrix changes in Eqs. 7.58 and 7.60 and that F_k or a_k appears in the equations.

Equation 7.49 can have a denominator that becomes very small or even zero because k_s can be negative. The denominator will be zero when

$$m + c\delta\Delta t + k_s\alpha(\Delta t)^2 = 0 \tag{7.61}$$

or dividing Eq. 7.61 by m, it follows that

$$1 + \left(\frac{c}{m}\right)\delta\Delta t + \left(\frac{k_s}{m}\right)\alpha(\Delta t)^2 = 0. \tag{7.62}$$

Solving for k_s/m in Eq. 7.62 gives

$$\left(\frac{k_s}{m}\right) = -\left(\frac{1}{\alpha(\Delta t)^2}\right) - \left(\frac{c}{m}\right)\frac{\delta}{\alpha\Delta t}. \tag{7.63}$$

As in Chapter 2 for an SDOF system, define $(c/m) = 2\zeta_e\omega_{ne}$ where $\omega_{ne} = \sqrt{k_e/m}$ and k_e is the pre-yield elastic stiffness; then

$$\left(\frac{k_s}{m}\right) = -\left(\frac{1}{\alpha(\Delta t)^2}\right) - \left(2\zeta_e\omega_{ne}\right)\frac{\delta}{\alpha\Delta t}. \tag{7.64}$$

As Δt becomes smaller, it takes a much larger negative k_s to produce a denominator that is zero. If the damping term is assumed to be zero, then a zero denominator exists when

$$\left(\frac{k_s}{m}\right) = -\left(\frac{1}{\alpha(\Delta t)^2}\right). \tag{7.65}$$

EXAMPLE 1 *Free Vibration of an SDOF System with a Stiffening Bilinear Stiffness*

Example 1 in Section 7.1 presented the exact solution for the free vibration of an SDOF system with a bilinear stiffness-related force versus deflection relationship as shown in Figure 7.5. This type of relationship is called a nonlinear stiffness force versus deflection relationship because the path on the force versus deflection curve that is followed is "up and down" the same path. For example, for all deflections less than 1.0 in., that is, $x(t) \leq$ 1.0 in., the path is along line AB. For all deflections greater than 1.0 in., that is, $x(t) > 1.0$ in., the path is along line BC.

A Newmark β-method solution for this problem requires that the stiffness and associated natural frequency of vibration be calculated for each time and then either Eq. 7.49 or Eq. 7.58 be used to solve for the response. This example considers a solution using Eq. 7.58.

The constant average acceleration approach to the Newmark β-method is used with $\delta = \frac{1}{2}$ and $\alpha = \frac{1}{4}$. The initial displacement and velocity are $x(0) = -5.0$ in. and $\dot{x}(0) = 0.0$ in/s. The mass and critical damping ratio are assumed to be 1.0 k-s^2/in. and 5%, respectively. The initial stiffness is 631.65 k/in., and therefore

$$\omega_n = \sqrt{k_s/m} = \sqrt{631.65/1.0} = 25.13 \text{ rad/s} \tag{7.66}$$

$$T_n = \frac{2\pi}{\omega_n} = \frac{2\pi}{25.13} = 0.25 \text{ s} \tag{7.67}$$

$$c = 2\zeta\omega_n m = 2(0.05)(25.13)(1.0) = 2.513 \text{ k-s/in.} \tag{7.68}$$

To aid in the selection of Δt, recall from Section 2.6 that $\Delta t \leq T_n/10 = 0.025$ s, and therefore $\Delta t = 0.02$ s can be used. However, a smaller time step size of $\Delta t = 0.005$ s. is used in this example because error may be introduced when the response is in the nonlinear domain and, as in this case, there is an abrupt change in stiffness. It follows from the derivation of the Newmark β-method and Eq. 7.52 that

$$\beta = 1 + (2.513)(\tfrac{1}{2})(0.005) + (631.65)(\tfrac{1}{4})(0.005)^2 = 1.01023. \tag{7.69}$$

It then directly follows from Eq. 7.57 that

$$\mathbf{F}_N^{(n)} = \begin{bmatrix} 1.0 & 0.004980 & 0.000012 \\ 0.0 & 0.992184 & 0.004949 \\ 0.0 & -3.12629 & 0.979745 \end{bmatrix}. \tag{7.70}$$

At time $t = 0$, the acceleration response is

$$\ddot{x}(0) = -\left(\frac{c}{m}\right)\dot{x}(0) - \left(\frac{k_s}{m}\right)x(0) = -\left(\frac{2.513}{1.0}\right)(0) - \left(\frac{631.65}{1.0}\right)(-5.0) = 3158.273 \text{ in./s}^2. \tag{7.71}$$

At $t = \Delta t = 0.005$ s, it follows from Eq. 7.58 that

$$\begin{Bmatrix} x_1 \\ \dot{x}_1 \\ \ddot{x}_1 \end{Bmatrix} = \begin{bmatrix} 1.0 & 0.004980 & 0.000012 \\ 0.0 & 0.992184 & 0.004949 \\ 0.0 & -3.12629 & 0.979745 \end{bmatrix} \begin{Bmatrix} -5.0 \\ 0 \\ 3,158.273 \end{Bmatrix} = \begin{Bmatrix} -4.96092 \\ 15.63144 \\ 3,094.303 \end{Bmatrix}. \tag{7.72}$$

At $t = 2\Delta t = 0.010$ s, it follows from Eq. 7.58 that

$$\begin{Bmatrix} x_2 \\ \dot{x}_2 \\ \ddot{x}_2 \end{Bmatrix} = \begin{bmatrix} 1.0 & 0.004980 & 0.000012 \\ 0.0 & 0.992184 & 0.004949 \\ 0.0 & -3.12629 & 0.979745 \end{bmatrix} \begin{Bmatrix} -4.96092 \\ 15.63144 \\ 3,094.303 \end{Bmatrix} = \begin{Bmatrix} -4.84478 \\ 30.82410 \\ 2,982.760 \end{Bmatrix}. \tag{7.73}$$

Calculation continues for several time steps until the displacement response reaches $x(t) = 1.0$ in. This happens at time step $t = 15\Delta t = 0.075$ s, where

$$\begin{Bmatrix} x_{15} \\ \dot{x}_{15} \\ \ddot{x}_{15} \end{Bmatrix} = \begin{bmatrix} 0.996092 & 0.004965 & 0.000006 \\ -1.56314 & 0.985965 & 0.002475 \\ -625.258 & -5.61411 & -0.01013 \end{bmatrix} \begin{Bmatrix} 0.61268 \\ 113.320 \\ -671.81 \end{Bmatrix} = \begin{Bmatrix} 1.16875 \\ 109.109 \\ -1,012.47 \end{Bmatrix}. \tag{7.74}$$

When $x(t) = 1.0$ in., the exact time t is 0.073 s (see Eq. 7.10). A more appropriate method of handling the change in stiffness within a time step is to calculate the response from 0.070 s to 0.073 s using the initial stiffness k_e (see Figure 7.5) and then the response from 0.073 s to 0.075 s using the stiffness $k_e + k_1$, where $k_1 = 355.31$ k/in. Although this is the case, approximation using the initial stiffness k_e through this time step is taken for simplicity and for illustration purposes. However, this approximation violates the dynamic equilibrium equation of motion because the stiffness force is not properly adjusted for the change in stiffness. One way to overcome this problem is to impose the condition of satisfying the dynamic equilibrium equation by updating the acceleration response, that is,

$$\begin{aligned} \ddot{x}_{15} &= -\left(\frac{c}{m}\right)\dot{x}_{15} - \left[\frac{F_s(x_{15})}{m}\right] \\ &= -\left(\frac{2.513}{1.0}\right)(109.109) - \left[\left(\frac{631.65}{1.0}\right)(1.0) + \left(\frac{631.65 + 355.31}{1.0}\right)(1.16875 - 1.0)\right] \\ &= -1,140.98. \end{aligned} \tag{7.75}$$

This new value of $\ddot{x}_{15} = -1,140.98$ in./s^2 is different from what was originally calculated, which was $-1,012.47$ in./s^2.

The calculation continues, and at time step $t = 15\Delta t = 0.075$ s, the natural frequency of vibration and damping becomes

$$\omega_n = \sqrt{(631.65 + 355.31)/1.0} = 31.42 \text{ rad/s} \tag{7.76}$$

$$c = 2(0.05)(31.42)(1.0) = 3.142 \text{ k-s/in.} \tag{7.77}$$

Note that in this example, the damping ratio is assumed to be 5%. Therefore, the damping value c must be adjusted when the value of the natural frequency of vibration changes. It follows from the derivation of the Newmark β-method and Eq. 7.52 that

$$\beta = 1 + (3.142)(\tfrac{1}{2})(0.005) + (631.65 + 355.31)(\tfrac{1}{4})(0.005)^2 = 1.01402. \qquad (7.78)$$

It then directly follows from Eq. 7.57 that

$$\mathbf{F}_N = \begin{bmatrix} 1.0 & 0.004970 & 0.000012 \\ 0.0 & 0.987834 & 0.004931 \\ 0.0 & -4.86656 & 0.972343 \end{bmatrix}. \qquad (7.79)$$

At time step $t = 16\Delta t = 0.080$ s, the response is

$$\begin{Bmatrix} x_{16} \\ \dot{x}_{16} \\ \ddot{x}_{16} \end{Bmatrix} = \begin{bmatrix} 1.0 & 0.004970 & 0.000012 \\ 0.0 & 0.987834 & 0.004931 \\ 0.0 & -4.86656 & 0.972343 \end{bmatrix} \begin{Bmatrix} 1.16875 \\ 109.109 \\ -1140.98 \end{Bmatrix} = \begin{Bmatrix} 1.69692 \\ 102.156 \\ -1640.41 \end{Bmatrix}. \qquad (7.80)$$

This calculation process continues, and the displacement and velocity response are shown in Figures 7.8 and 7.9, respectively.

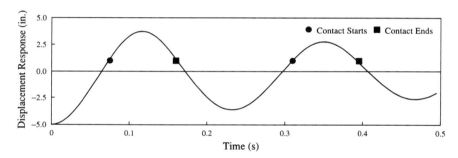

Figure 7.8 Displacement Response

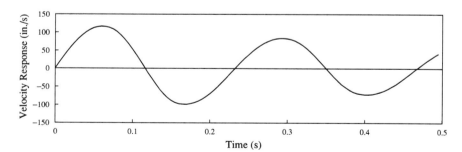

Figure 7.9 Velocity Response

EXAMPLE 2   *Free Vibration of an SDOF System with Bilinear Stiffness*

Assume that the SDOF system has following mass and damping properties:

$$m = 1.0 \text{ k-s}^2/\text{in.}, \quad \zeta = 5\%.$$

Also assume that when the absolute value of the response is less than 1.0 in., then the stiffness is 631.65 k/in. When the absolute value of the response exceeds 1.0 in., the stiffness of the system decreases to 20% of the initial stiffness. This type of nonlinear relationship is like the cantilever beam discussed in Section 7.1 and shown in Figures 7.1 to 7.3. Let this reduced stiffness be denoted as k_2. Therefore,

$$k_1 = 631.65 \text{ k/in.}, \quad k_2 = 0.2k_1 = 126.33 \text{ k/in.}, \quad x_y = 1.0 \text{ in.}$$

and

$$\omega_n = \begin{cases} \sqrt{k_1/m} & |x_k| \leq 1.0 \\ \sqrt{k_2/m} & |x_k| > 1.0 \end{cases} = \begin{cases} 25.13 \text{ rad/s} & |x_k| \leq 1.0 \\ 11.24 \text{ rad/s} & |x_k| > 1.0 \end{cases} \tag{7.81}$$

$$T_n = \frac{2\pi}{\omega_n} = \begin{cases} 0.25 \text{ s} & |x_k| \leq 1.0 \\ 0.56 \text{ s} & |x_k| > 1.0 \end{cases} \tag{7.82}$$

$$c_1 = 2(0.05)(25.13)(1.0) = 2.513 \text{ k-s/in.} \tag{7.83}$$

$$c_2 = 2(0.05)(11.24)(1.0) = 1.124 \text{ k-s/in.} \tag{7.84}$$

Figure 7.10 shows the force versus displacement curve. An example of a structure with stiffness force versus displacement characteristics shown in this figure is the base isolated system, which is discussed in Chapter 8.

The time step size should be less than $(T_n/10)$, and for this case it follows that $\Delta t \leq (T_n/10) = 0.025$ s. In this example, a time step of 0.005 s is selected for the Newmark β-method of analysis.

Consider the case where the system is initially at an undeformed position (i.e., $x_o = 0.0$) but with an initial velocity of $\dot{x}_o = 40.0$ in./s. Therefore, the initial acceleration is given by the dynamic equilibrium equation, where

$$\ddot{x}_o = -\left(\frac{c}{m}\right)\dot{x}_o - \left(\frac{k_1}{m}\right)x_o = -\left(\frac{2.513}{1.0}\right)(40) - \left(\frac{631.65}{1.0}\right)(0) = -100.531 \text{ in./s}^2 \tag{7.85}$$

The response is given in Eq. 7.58, and for the constant average acceleration method (i.e., $\delta = \frac{1}{2}$ and $\alpha = \frac{1}{4}$) it follows that

$$\beta = 1.0 + \frac{1}{2}(2.513)(0.005) + \frac{1}{4}(631.65)(0.005)^2 = 1.01023. \tag{7.86}$$

It then follows from Eq. 7.58 that

$$\begin{Bmatrix} x_{k+1} \\ \dot{x}_{k+1} \\ \ddot{x}_{k+1} \end{Bmatrix} = \begin{bmatrix} 1.0 & 0.004980 & 0.000012 \\ 0.0 & 0.992184 & 0.004949 \\ 0.0 & -3.12629 & 0.979745 \end{bmatrix} \begin{Bmatrix} x_k \\ \dot{x}_k \\ \ddot{x}_k \end{Bmatrix}. \tag{7.87}$$

Equation 7.87 represents the response of the structure if the displacement is within the range DA in Figure 7.10. At $t = \Delta t = 0.005$ s, it follows from Eq. 7.87 that

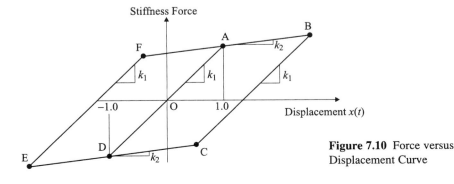

Figure 7.10 Force versus Displacement Curve

$$\begin{Bmatrix} x_1 \\ \dot{x}_1 \\ \ddot{x}_1 \end{Bmatrix} = \begin{bmatrix} 1.0 & 0.004980 & 0.000012 \\ 0.0 & 0.992184 & 0.004949 \\ 0.0 & -3.12629 & 0.979745 \end{bmatrix} \begin{Bmatrix} 0.0 \\ 40.0 \\ -100.531 \end{Bmatrix} = \begin{Bmatrix} 0.197975 \\ 39.18981 \\ -223.546 \end{Bmatrix}. \tag{7.88}$$

At this level of displacement, the stiffness force, denoted as $F_{s,k}$, is

$$F_{s,1} = 631.65(0.197975) = 125.05 \text{ kip.} \tag{7.89}$$

At $t = 2\Delta t = 0.010$ s, it follows from Eq. 7.87 that

$$\begin{Bmatrix} x_2 \\ \dot{x}_2 \\ \ddot{x}_2 \end{Bmatrix} = \begin{bmatrix} 1.0 & 0.004980 & 0.000012 \\ 0.0 & 0.992184 & 0.004949 \\ 0.0 & -3.12629 & 0.979745 \end{bmatrix} \begin{Bmatrix} 0.197975 \\ 39.18981 \\ -223.546 \end{Bmatrix} = \begin{Bmatrix} 0.390392 \\ 37.77710 \\ -341.537 \end{Bmatrix} \tag{7.90}$$

and at this level of displacement, the stiffness force is

$$F_{s,2} = 631.65(0.390392) = 246.59 \text{ kip.} \tag{7.91}$$

This calculation process continues for several time steps until the displacement response reaches $x(t) = 1.0$ in. This happens at time step $t = 6\Delta t = 0.03$ s, where

$$\begin{Bmatrix} x_6 \\ \dot{x}_6 \\ \ddot{x}_6 \end{Bmatrix} = \begin{bmatrix} 1.0 & 0.004980 & 0.000012 \\ 0.0 & 0.992184 & 0.004949 \\ 0.0 & -3.12629 & 0.979745 \end{bmatrix} \begin{Bmatrix} 0.905802 \\ 30.26189 \\ -648.210 \end{Bmatrix} = \begin{Bmatrix} 1.048499 \\ 26.81715 \\ -729.688 \end{Bmatrix} \tag{7.92}$$

and at this level of displacement, the stiffness force is

$$F_{s,6} = 631.65(1.0) + 126.33(1.048499 - 1.0) = 637.78 \text{ kip.} \tag{7.93}$$

Note that the displacement of 1.0 in. occurs at about 0.029 s, and therefore the stiffness changes from k_1 to k_2 within the time period of 0.025 s to 0.03 s. The more accurate way of solving this problem is to use $k_1 = 631.65$ k/in. for the time period 0.025 s to 0.029 s, and then use $k_2 = 126.33$ k/in. for the time period 0.029 s to 0.030 s. However, this is not done because Eq. 7.87 must be reformulated using a time step of $\Delta t = 0.004$ s with stiffness k_1, and then using a time step of $\Delta t = 0.001$ s with stiffness k_2. The resulting response obtained in Eq. 7.92 will therefore violate the dynamic equilibrium equation of motion because the stiffness force is not properly adjusted for the change in stiffness. One way to overcome this problem is to impose the condition of satisfying the dynamic equilibrium equation by updating the acceleration response, that is,

$$\begin{aligned} \ddot{x}_6 &= -\left(\frac{c}{m}\right)\dot{x}_6 - \left[\frac{F_s(x_6)}{m}\right] \\ &= -\left(\frac{1.124}{1.0}\right)(26.81715) - \left[\left(\frac{631.65}{1.0}\right)(1.0) + \left(\frac{126.33}{1.0}\right)(1.048499 - 1.0)\right] \\ &= -667.923. \end{aligned} \tag{7.94}$$

This new value of $\ddot{x}_6 = -667.923$ in./s^2 is different from what was originally calculated, which was -729.688 in./s^2.

The calculation continues, and the displacement response of the structure follows the path AB in Figure 7.10. At time step $t = 6\Delta t = 0.03$ s, the natural frequency of vibration of $\omega_n = 11.22$ rad/s and damping of $c = 1.124$ k-s/in. It follows from the derivation of the Newmark β-method and Eq. 7.58 that

$$\beta = 1 + (1.124)(\tfrac{1}{2})(0.005) + (126.33)(\tfrac{1}{4})(0.005)^2 = 1.00360. \tag{7.95}$$

It then follows from Eq. 7.57 that

$$\mathbf{F}_N^{(n)} = \begin{bmatrix} 1.0 & 0.004996 & 0.000012 \\ 0.0 & 0.998427 & 0.004982 \\ 0.0 & -0.62939 & 0.992827 \end{bmatrix}. \tag{7.96}$$

At time step $t = 7\Delta t = 0.035$ s, the response is

$$\begin{Bmatrix} x_7 \\ \dot{x}_7 \\ \ddot{x}_7 \end{Bmatrix} = \begin{bmatrix} 1.0 & 0.004996 & 0.000012 \\ 0.0 & 0.998427 & 0.004982 \\ 0.0 & -0.62939 & 0.992827 \end{bmatrix} \begin{Bmatrix} 1.048499 \\ 26.81715 \\ -667.923 \end{Bmatrix} = \begin{Bmatrix} 1.174160 \\ 23.44731 \\ -680.011 \end{Bmatrix} \tag{7.97}$$

and at this level of displacement, the stiffness force is

$$F_{s,7} = 631.65(1.0) + 126.33(1.174160 - 1.0) = 653.66 \text{ kip.} \tag{7.98}$$

This calculation process continues until $t = 14\Delta t = 0.07$ s, when the response is

$$\begin{Bmatrix} x_{14} \\ \dot{x}_{14} \\ \ddot{x}_{14} \end{Bmatrix} = \begin{bmatrix} 1.0 & 0.004996 & 0.000012 \\ 0.0 & 0.998427 & 0.004982 \\ 0.0 & -0.62939 & 0.992827 \end{bmatrix} \begin{Bmatrix} 1.564722 \\ 2.464924 \\ -705.767 \end{Bmatrix} = \begin{Bmatrix} 1.568247 \\ -1.05513 \\ -702.256 \end{Bmatrix} \tag{7.99}$$

and at this level of displacement, the stiffness force is

$$F_{s,14} = 631.65(1.0) + 126.33(1.568247 - 1.0) = 703.43 \text{ kip.} \tag{7.100}$$

At this time, the velocity response becomes a negative number, which means that a maximum displacement has been reached and the displacement at the next time step will be less than 1.568247 in. The system will now be in an unloading process, and the displacement response of the structure will follow the path BC in Figure 7.10. The stiffness for this path is $k_1 = 631.65$ k/in. It follows that at $t = 15\Delta t = 0.075$ s, the response is

$$\begin{Bmatrix} x_{15} \\ \dot{x}_{15} \\ \ddot{x}_{15} \end{Bmatrix} = \begin{bmatrix} 1.0 & 0.004980 & 0.000012 \\ 0.0 & 0.992184 & 0.004949 \\ 0.0 & -3.12629 & 0.979745 \end{bmatrix} \begin{Bmatrix} 1.568247 \\ -1.05513 \\ -702.256 \end{Bmatrix} = \begin{Bmatrix} 1.554341 \\ -4.50722 \\ -681.687 \end{Bmatrix}. \tag{7.101}$$

At this level of displacement, the unloading process occurs elastically. Therefore, a reduction of displacement from 1.568247 in. to 1.554341 in. changes the stiffness force elastically:

$$F_{s,15} = 631.65(1.0 - 1.568247 + 1.554541) + 126.33(1.568247 - 1.0) = 696.44 \text{ kip.} \tag{7.102}$$

Calculation continues using stiffness k_1. Note that the maximum displacement is equal to 1.568247 in. Therefore, the structure will begin to yield under reversed loading at a displacement equal to

$$x_y = 1.568247 - 2.0 = -0.431753 \text{ in.} \tag{7.103}$$

where the value 2 in Eq. 7.103 represents the elastic displacement envelope, which is the range between the upper and lower limits of the path corresponding to k_1 in Figure 7.10 (i.e., the displacement or horizontal distance between A and D, C and B, or E and F). This displacement level of $x_y = -0.431753$ in. is reached at time step $t = 36\Delta t = 0.18$ s. The response at this time step is

$$\begin{Bmatrix} x_{36} \\ \dot{x}_{36} \\ \ddot{x}_{36} \end{Bmatrix} = \begin{bmatrix} 1.0 & 0.004980 & 0.000012 \\ 0.0 & 0.992184 & 0.004949 \\ 0.0 & -3.12629 & 0.979745 \end{bmatrix} \begin{Bmatrix} -0.38752 \\ -11.1242 \\ 561.531 \end{Bmatrix} = \begin{Bmatrix} -0.43598 \\ -8.25808 \\ 584.935 \end{Bmatrix}. \tag{7.104}$$

At this level of displacement, the stiffness changes from k_1 to k_2, and the path is between C and E. Therefore, the stiffness force can be computed by

$$F_{s,36} = 631.65(-1.0) + 126.33(-0.43598 + 1.0) = -560.40 \text{ kip.} \qquad (7.105)$$

Similar to Eq. 7.94, imposing the condition to satisfy the dynamic equilibrium equation by updating the acceleration response gives

$$
\begin{aligned}
\ddot{x}_{36} &= -\left(\frac{c}{m}\right)\dot{x}_{36} - \left[\frac{F_s(x_{36})}{m}\right] \\
&= -\left(\frac{1.124}{1.0}\right)(-8.25808) - \left[\left(\frac{631.65}{1.0}\right)(-1.0) + \left(\frac{126.33}{1.0}\right)(-0.431753 + 1.0)\right] \qquad (7.106) \\
&= -569.683.
\end{aligned}
$$

Again, this new value of $\ddot{x}_{36} = -569.683$ in./s^2 is different from what was originally calculated, which was -584.935 in./s^2.

This calculation process continues, and Table 7.1 summarizes the calculations at different time steps. In addition, the displacement and velocity responses are shown in Figures 7.11 and 7.12, respectively, and the force versus displacement curve is shown in Figure 7.13.

Table 7.1 Summary of Calculations of Response at Selected Time Steps

k	Time (s)	x_k (in.)	$\dot{x}_k$ (in./s)	$\ddot{x}_k$ (in./s^2)	$F_{s,k}$ (kip)
0	0.000	0.00000	40.0000	−100.531	0.00
1	0.005	0.19797	39.1898	−223.546	125.05
2	0.010	0.39039	37.7771	−341.537	246.59
3	0.015	0.57431	35.7915	−452.721	362.77
4	0.020	0.74697	33.2710	−555.446	471.83
5	0.025	0.90580	30.2619	−648.210	572.15
6	0.030	1.04850	26.8171	−667.923	637.78
7	0.035	1.17416	23.4473	−680.011	653.66
⋮	⋮	⋮	⋮	⋮	⋮
13	0.065	1.56472	2.4649	−705.767	703.47
14	0.070	1.56825	−1.0551	−702.256	703.43
⋮	⋮	⋮	⋮	⋮	⋮
35	0.175	−0.38752	−11.1243	561.531	−532.04
36	0.180	−0.43598	−8.2581	569.683	−560.40

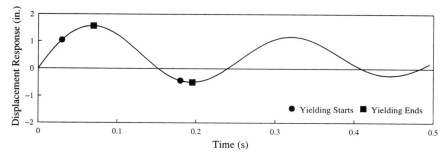

Figure 7.11 Displacement Response of a Bilinear Structure

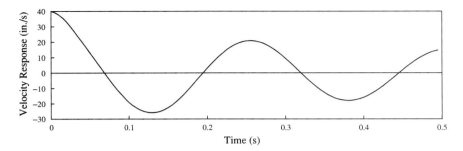

Figure 7.12 Velocity Response of a Bilinear Structure

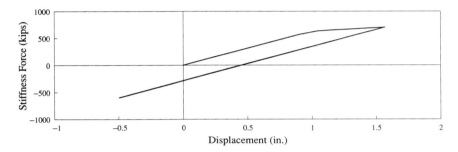

Figure 7.13 Stiffness Force versus Displacement Response of a Bilinear Structure

EXAMPLE 3 *Forced Response of an SDOF System with Bilinear Stiffness*

Consider Example 2 with bilinear stiffness of $k_1 = 631.65$ k/in. and $k_2 = 126.33$ k/in., and a yield displacement of $x_y = 0.6$ in. The structure has a mass equal to $m = 1.0$ k-s²/in, and it is assumed to have zero initial displacement and initial velocity but is subjected to a magnified bilinear pulse ground acceleration as shown in Figure 7.14. The constant acceleration method with a time step of $\Delta t = 0.005$ s is used. The equations, given in Eqs. 7.30 and 7.39 are

$$x_{k+1} = x_k + \dot{x}_k \Delta t + \tfrac{1}{2} \ddot{x}_k (\Delta t)^2 \tag{7.107}$$

$$\dot{x}_{k+1} = \dot{x}_k + \ddot{x}_k \Delta t \tag{7.108}$$

$$\ddot{x}_{k+1} = a_{k+1} - \left(\frac{c}{m} \right) \dot{x}_{k+1} - \frac{F_s(x_{k+1})}{m}. \tag{7.109}$$

For this example, the damping value is fixed and is equal to 2.513 k-s/in. This means that the damping ratio ζ changes for different values of ω_n.

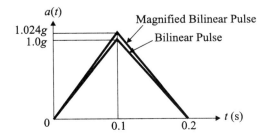

Figure 7.14 Bilinear Pulse

The ground acceleration at time zero is equal to 0. The initial acceleration can be computed using Eq. 7.109:

$$\ddot{x}_o = 0 - \left(\frac{c}{m}\right)(0) - \frac{(0)}{m} = 0 \text{ in./s}^2 \tag{7.110}$$

At $t = 2\Delta t = 0.010$ s, the ground acceleration is equal to $0.0512g$. The response can be computed using Eqs. 7.107 to 7.109:

$$x_1 = (0) + (0)(0.005) + \tfrac{1}{2}(0)(0.005)^2 = 0 \text{ in.} \tag{7.111}$$

$$\dot{x}_1 = (0) + (0)(0.005) = 0 \text{ in./s} \tag{7.112}$$

$$\ddot{x}_1 = -(0.0512 \times 386.4) - \left(\frac{2.513}{1.0}\right)(0) - \frac{631.65(0)}{1.0} = -19.7837 \text{ in./s}^2 \tag{7.113}$$

$$F_{s,1} = 631.65(0) = 0.0 \text{ kip} \tag{7.114}$$

At $t = 2\Delta t = 0.010$ s, it follows from Eqs. 7.107 to 7.109 that

$$x_2 = (0) + (0)(0.005) + \tfrac{1}{2}(-19.7837)(0.005)^2 = -0.00025 \text{ in.} \tag{7.115}$$

$$\dot{x}_2 = (0) + (-19.7837)(0.005) = -0.09892 \text{ in./s} \tag{7.116}$$

$$\ddot{x}_2 = -(0.1024 \times 386.4) - \left(\frac{2.513}{1.0}\right)(-0.09892) - \frac{631.65(-0.0005)}{1.0} = -39.1625 \text{ in./s}^2 \tag{7.117}$$

$$F_{s,2} = 631.65(-0.00025) = -0.16 \text{ kip.} \tag{7.118}$$

Calculation continues until a displacement of 0.6 in. is reached. This corresponds to a time step of $t = 23\Delta t = 0.115$ s and a ground acceleration of $0.8704g$. The responses at time step $t = 22\Delta t = 0.11$ s are $x_{22} = -0.55586$ in., $\dot{x}_{22} = -11.4312$ in./s, and $\ddot{x}_{22} = 23.7360$ in./s^2. Therefore, the displacement and velocity responses at time step 23 follow from Eqs. 7.107 to 7.109:

$$x_{23} = (-0.55586) + (-11.4312)(0.005) + \tfrac{1}{2}(23.7360)(0.005)^2 = -0.61272 \text{ in.} \tag{7.119}$$

$$\dot{x}_{23} = (-11.4312) + (23.7360)(0.005) = -11.3126 \text{ in./s} \tag{7.120}$$

$$\ddot{x}_{23} = -0.8704g - \left(\frac{2.513}{1.0}\right)(-11.3126) - \frac{631.65(-0.6) + 126.33(-0.01272)}{1.0} = 72.7089 \text{ in./s}^2 \tag{7.121}$$

$$F_{s,23} = 631.65(-0.6) + 126.33(-0.61272 + 0.6) = -380.58 \text{ kip} \tag{7.122}$$

where for the total displacement of 0.61272 in., a 0.6-in. displacement corresponds to the stiffness of $k_1 = 631.65$ k/in. and a 0.01272-in. displacement corresponds to the stiffness of $k_2 = 126.33$ k/in.

The calculation process continues. Note that the acceleration response is computed after the displacement and velocity responses are calculated. Therefore, the dynamic equilibrium equation will be satisfied and no acceleration value must be updated due to the stiffness nonlinearity, which was not the case in Examples 1 and 2. This is because the constant acceleration method is used in this example.

Table 7.2 summarizes the calculations at different time steps. In addition, the displacement and velocity responses are shown in Figures 7.15 and 7.16, respectively. The force versus displacement curve is shown in Figure 7.17.

Figure 7.18 shows a comparison of displacement response between the nonlinear response shown in Figure 7.15 and the linear response based on the calculation using stiffness $k_1 = 631.65$ k/in. only.

Table 7.2 Summary of Calculations of Response at Selected Time Steps

k	Time (s)	a_k (g)	x_k (in.)	$\dot{x}_k$ (in./s)	$\ddot{x}_k$ (in./s^2)	$F_{s,k}$ (kip)
0	0.000	0.00	0.00000	0.0000	0.000	0.00
1	0.005	0.0512	0.00000	0.0000	−19.783	0.00
2	0.010	0.1024	−0.00025	−0.0989	−39.163	−0.16
3	0.015	0.1536	−0.00123	−0.2947	−57.832	−0.78
⋮	⋮	⋮	⋮	⋮	⋮	⋮
22	0.110	0.9216	−0.55586	−11.4312	23.736	−351.11
23	0.115	0.8704	−0.61272	−11.3126	72.709	−380.58

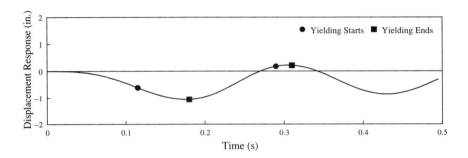

Figure 7.15 Displacement Response of a Bilinear Structure

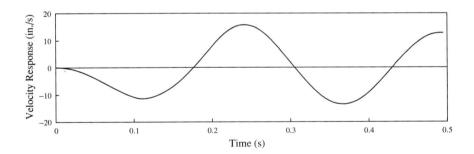

Figure 7.16 Velocity Response of a Bilinear Structure

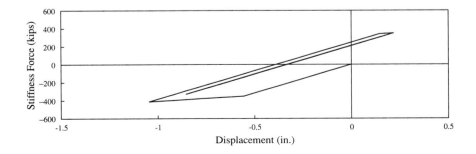

Figure 7.17 Stiffness Force versus Displacement Response of a Bilinear Structure

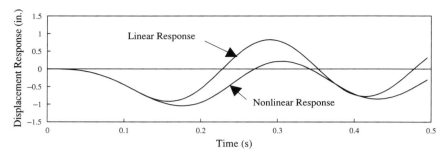

Figure 7.18 Comparison of Displacements between Linear and Nonlinear Responses

EXAMPLE 4 *Fictitious Stiffness-Related Force*

Recall that the stiffness used in the solution for time step t_{k+1} is the stiffness from time step t_{k-1} to t_k. If an iteration of this initial solution is performed, then this initial stiffness is replaced by $k_s^{(1)}$, $k_s^{(2)}$, or more.

An alternate approach to this iterative solution is to apply a "fictitious force" to the system at time t_{k+1} to compensate for the error in the initial estimate of the stiffness. Assume that Eq. 7.35 is used to estimate the stiffness of the SDOF system at time step t_{k+1}. Therefore, the equation of motion from Eq. 7.30 is

$$m\ddot{x}_{k+1} + c\dot{x}_{k+1} + F_s(x_{k+1}) = F_{k+1} \tag{7.123}$$

where from Figure 7.7

$$F_s(x_{k+1}) = F_s(x_k) + k_k(x_{k+1} - x_k) = \hat{F}_s(t_{k+1}) = \hat{F}_{s,k+1}. \tag{7.124}$$

Now imagine that a solution is obtained using the Newmark β-method, and a value for x_{k+1} is calculated to be $x_{k+1}^{(1)}$. Then from Figure 7.19,

$$\hat{F}_s(x_{k+1}^{(1)}) = \hat{F}_s^{(1)}(t_{k+1}) = \hat{F}_{s,k+1}^{(1)}. \tag{7.125}$$

The exact force is $F_s(t_{k+1}) = F_{s,k+1}$, but it is unknown. The force used in the initial solution is $\hat{F}_{s,k+1}$. The force that would be used in an iteration is $\hat{F}_{s,k+1}^{(1)}$. The difference between the initial solution force and the iterative force is

$$\Delta\hat{F}_s = \hat{F}_{s,k+1}^{(1)} - \hat{F}_{s,k+1}. \tag{7.126}$$

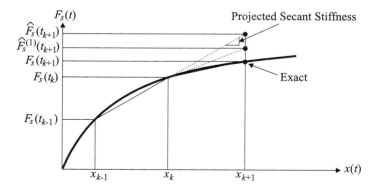

Figure 7.19 Nonlinear Force versus Deflection Curve

The equation of motion for the next time step from Eq. 7.30 is

$$m\ddot{x}_{k+2} + c\dot{x}_{k+2} + F_s(x_{k+2}) = F_{k+2} \tag{7.127}$$

where

$$F_s(x_{k+2}) = F_s(x_{k+1}) + k_{k+1}(x_{k+2} - x_{k+1}) \tag{7.128}$$

and

$$k_{k+1} = \left[\frac{F_s(x_{k+1}) - F_s(x_k)}{x_{k+1} - x_k}\right]. \tag{7.129}$$

However, in the initial solution the stiffness used and the related force were incorrect by an amount $\Delta\hat{F}_s = \hat{F}^{(1)}_{s,k+1} - \hat{F}_{s,k+1}$ because the iteration was not performed. Therefore, imagine that a fictitious force is now applied to the right-hand side of Eq. 7.127, and it follows that

$$m\ddot{x}_{k+2} + c\dot{x}_{k+2} + F_s(x_{k+2}) = F_{k+2} + \Delta\hat{F}_s. \tag{7.130}$$

The solution approach and extensions of it can be used as an alternative to a smaller time step or repeated iterations for the response at time t_{k+1}.

Consider Example 2, where the force versus deflection curve is a bilinear relationship with $k_1 = 631.65$ k/in., $k_2 = 126.33$ k/in., and a yield displacement of $x_y = 1.0$ in. Since this force versus deflection curve is linear up to the elastic limit of $x_y = 1.0$ in., there will be no imbalance of force in the response. However, at the time step when k_1 changes to k_2 (i.e., at the time when yielding occurs), there is an imbalance of force, which can be seen in Eqs. 7.92 an 7.93. From Eq. 7.92, the response is

$$\begin{Bmatrix} x_6 \\ \dot{x}_6 \\ \ddot{x}_6 \end{Bmatrix} = \begin{bmatrix} 1.0 & 0.004980 & 0.000012 \\ 0.0 & 0.992184 & 0.004949 \\ 0.0 & -3.12629 & 0.979745 \end{bmatrix} \begin{Bmatrix} 0.905802 \\ 30.26189 \\ -648.210 \end{Bmatrix} = \begin{Bmatrix} 1.048499 \\ 26.81715 \\ -729.688 \end{Bmatrix} \tag{7.131}$$

and at this level of displacement, the stiffness force is

$$\hat{F}_{s,6} = 631.65(1.0) + 126.33(1.048499 - 1.0) = 637.78 \text{ kip.} \tag{7.132}$$

However, from the dynamic equilibrium equation of motion as given in Eq. 7.123, where $m = 1.0$ k-s^2/in., $c = 2.513$ k-s/in., and $F_k = 0$ (i.e., no external force), the stiffness force is

$$\hat{F}_s(x_6^{(1)}) = -m\ddot{x}_6 - c\dot{x}_6 = -(1.0)(-729.688) - (1.124)(26.81715) = 699.55 \text{ kip} \tag{7.133}$$

and therefore Eq. 7.126 becomes

$$\Delta\hat{F}_s = \hat{F}^{(1)}_{s,6} - \hat{F}_{s,6} = 699.55 - 637.78 = 61.76 \text{ kip.} \tag{7.134}$$

Consider this $\Delta\hat{F}_s$ in Eq. 7.134 as a fictitious force; it follows from Eq. 7.130 that the equation of motion for the next time step becomes

$$m\ddot{x}_7 + c\dot{x}_7 + F_s(x_7) = \Delta\hat{F}_s. \tag{7.135}$$

Now, instead of calculating the response of the SDOF system with no external force, this fictitious force introduces a term on the right-hand side of Eq. 7.135. This $\Delta\hat{F}_s$ is considered as an external force in the next time step, and therefore Eq. 7.56 becomes

$$\begin{Bmatrix} x_7 \\ \dot{x}_7 \\ \ddot{x}_7 \end{Bmatrix} = \mathbf{F}^{(n)}_N \begin{Bmatrix} x_6 \\ \dot{x}_6 \\ \ddot{x}_6 \end{Bmatrix} + \mathbf{H}^{(n)}_N \Delta\hat{F}_s \tag{7.136}$$

where $\mathbf{F}^{(n)}_N$ was calculated in Eq. 7.96 for $k_s = k_2 = 126.33$ k/in., which is

$$\mathbf{F}_N^{(n)} = \begin{bmatrix} 1.0 & 0.004996 & 0.000012 \\ 0.0 & 0.998427 & 0.004982 \\ 0.0 & -0.62939 & 0.992827 \end{bmatrix} \tag{7.137}$$

and $\mathbf{H}_N^{(n)}$ is

$$\mathbf{H}_N^{(n)} = \frac{1}{\beta} \left\{ \begin{matrix} \alpha(\Delta t)^2 \\ \delta \Delta t \\ 1 \end{matrix} \right\} = \frac{1}{1.00360} \left\{ \begin{matrix} \tfrac{1}{4}(0.005)^2 \\ \tfrac{1}{2}(0.005) \\ 1 \end{matrix} \right\} = \left\{ \begin{matrix} 0.000006 \\ 0.002491 \\ 0.996413 \end{matrix} \right\}. \tag{7.138}$$

Therefore, the response at the next time step follows from Eqs. 7.135 to 7.138:

$$\left\{ \begin{matrix} x_7 \\ \dot{x}_7 \\ \ddot{x}_7 \end{matrix} \right\} = \begin{bmatrix} 1.0 & 0.004996 & 0.000012 \\ 0.0 & 0.998427 & 0.004982 \\ 0.0 & -0.62939 & 0.992827 \end{bmatrix} \left\{ \begin{matrix} 1.048499 \\ 26.81715 \\ -729.688 \end{matrix} \right\} + \left\{ \begin{matrix} 0.000006 \\ 0.002491 \\ 0.996413 \end{matrix} \right\} (61.76) = \left\{ \begin{matrix} 1.174106 \\ 23.29351 \\ -679.792 \end{matrix} \right\} \tag{7.139}$$

and at this level of displacement, the stiffness force is

$$\hat{F}_{s,7} = 631.65(1.0) + 126.33(1.174106 - 1.0) = 653.62 \text{ kip.} \tag{7.140}$$

However, from Eq. 7.123, the stiffness force is

$$\hat{F}_s(x_7^{(1)}) = -m\ddot{x}_7 - c\dot{x}_7 = -(1.0)(-679.792) - (1.124)(23.29351) = 653.61 \text{ kip} \tag{7.141}$$

and therefore the fictitious force in Eq. 7.126 becomes

$$\Delta \hat{F}_s = \hat{F}_{s,7}^{(1)} - \hat{F}_{s,7} = 653.61 - 653.62 = -0.01 \text{ kip.} \tag{7.142}$$

This calculation process continues as time progresses to adjust for the imbalance of force. Note that the fictitious force in Eq. 7.142 is much smaller than that in Eq. 7.134, which shows that the imbalance of force has been considered in the calculation process.

EXAMPLE 5   *Nonlinear Response Spectrum*

Section 5.2 discussed the development of a response spectrum for a linear elastic SDOF system. The spectral displacement was the maximum displacement of the system for a given earthquake ground motion acceleration versus time and a given value of damping and natural period of vibration. Wu and Hanson,[1] as discussed in Section 5.3, studied the response of an SDOF system for an ensemble of earthquake ground motions and developed an equation for a damping response reduction coefficient, R_d. This coefficient enables the structural engineer to modify the 5% damped elastic response spectrum to obtain a spectrum for any selected value of damping, ζ. Equation 5.49 presented the equation for R_d. Wu and Hanson also studied the response of an elastic–plastic SDOF system for the same ensemble of earthquake ground motions.

Define x_y to be the displacement of an SDOF system for a given earthquake ground motion and with a pre-yield natural period of vibration T_n and damping value of ζ. Now assume that the SDOF has an elastic–plastic stiffness of the type shown in Figure 7.20 and a yield force level, F_y, equal to $F_y = k_1 x_y$ where k_1 is the pre-yield stiffness of the SDOF system. Denote the maximum response of this nonlinear SDOF system as x_{max}. Define the *ductility demand* for the given earthquake ground motion acceleration versus time to be μ, where

$$\mu = \frac{x_{\text{max}}}{x_y}. \tag{7.143}$$

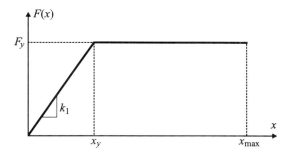

Figure 7.20 Elastic–Plastic Stiffness

Wu and Hanson studied many earthquake ground motions for a given pre-yield natural period of vibration T_n. For a given earthquake ground motion, the maximum displacement for 5% damping, that is, $S_d(T_n, 5\%)$, and the maximum displacement for damping ζ and the maximum ductility demand μ, that is $S_d(T_n, \zeta, \mu)$, are obtained. The ratio of these two maximum response quantities, denoted by $R_{d\mu}$,[1] is

$$R_{d\mu} = \frac{S_d(T_n, \zeta, \mu)}{S_d(T_n, 5\%)}. \tag{7.144}$$

Wu and Hanson then derived the following equations:

$$R_{d\mu} = -0.349\ln(0.0959\zeta)(2.89\mu - 1.89)^{-0.244}, \quad T_n = 0.1\,\text{s} \tag{7.145}$$

$$R_{d\mu} = -0.547\ln(0.417\zeta)(1.82\mu - 0.82)^{-0.562}, \quad T_n = 0.5\,\text{s} \tag{7.146}$$

$$R_{d\mu} = -0.471\ln(0.524\zeta)(1.53\mu - 0.53)^{-0.706}, \quad T_n = 1.0\,\text{s} \tag{7.147}$$

$$R_{d\mu} = -0.478\ln(0.475\zeta)\mu^{-1.06}, \quad T_n = 3.0\,\text{s}. \tag{7.148}$$

Figure 7.21 shows a plot of $R_{d\mu}$ as a function of natural period of vibration and ductility with damping ratio equal to 5%. Figure 7.22 shows a plot of $R_{d\mu}$ as a function of a natural period of vibration and damping ratio with ductility equal to 3.

Newmark and Hall did a similar study but only for an SDOF system with 5% damping[2]. They calculated a response reduction factor of

$$R_{5\%\mu} = \frac{S_d(T_n, 5\%, \mu)}{S_d(T_n, 5\%)}. \tag{7.149}$$

which is similar to Eq. 7.144. The equation they proposed was

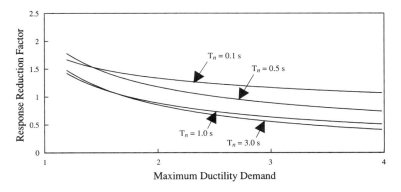

Figure 7.21 Response Reduction Factor ($R_{d\mu}$) as a Function of Natural Period of Vibration and Ductility with Damping $\zeta = 5\%$ (Wu and Hanson)

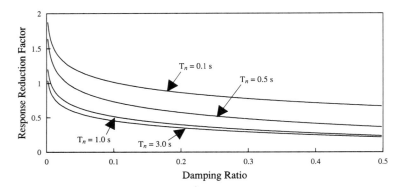

Figure 7.22 Response Reduction Factor (R_{du}) as a Function of Natural Period of Vibration and Damping Ratio with Ductility $\mu = 3$ (Wu and Hanson)

$$R_{5\%\mu} = 1.0, \quad T_n \le 0.3\,\text{s} \tag{7.150}$$

$$R_{5\%\mu} = \sqrt{2\mu - 1}, \quad 0.3 < T_n \le 1.0\,\text{s} \tag{7.151}$$

$$R_{5\%\mu} = \mu, \quad T_n > 1.0\,\text{s}. \tag{7.152}$$

Krawinkler and Nassar developed for 15 Western U.S. earthquake records the following equation for $R_{d\mu}$[3]:

$$R_{d\mu} = \left[c(\mu - 1) + 1 \right]^{1/c} \tag{7.153}$$

where

$$c = \left(\frac{T_n}{1 + T_n} \right) + \left(\frac{b}{T_n} \right). \tag{7.154}$$

The pre-yield natural period of vibration is denoted T_n. The slope of the post-yield force versus deflection curve is called the *strain hardening coefficient* and is denoted α (e.g., α = 0% is elastic–plastic as shown in Figure 7.20). The value of the coefficient b in Eq. 7.154 is a function of α, and for a 5% damped SDOF system, the value of b is

$$\alpha = 0\%, \quad b = 0.42 \tag{7.155}$$

$$\alpha = 2\%, \quad b = 0.37 \tag{7.156}$$

$$\alpha = 10\%, \quad b = 0.29. \tag{7.157}$$

Figure 7.23 shows a plot of c versus T_n. Figure 7.24 shows a plot of $R_{d\mu}$ versus T_n.

This example demonstrates how, for a given SDOF system and ensemble of earthquake ground motion records, it is possible to develop generalized design equations for use in structural engineering.

7.3 MULTI–DEGREE OF FREEDOM RESPONSE WITH NONLINEAR STIFFNESS

Consider now a response solution for a multi–degree of freedom system where any of the elements in the stiffness matrix have a nonlinear force versus deflection relationship. The matrix equation of motion for this multi-degree of freedom system is

$$\mathbf{M}\ddot{\mathbf{X}}(t) + \mathbf{C}\dot{\mathbf{X}}(t) + [\mathbf{K}(\mathbf{X})]\mathbf{X}(t) = -\mathbf{M}\mathbf{a}(t). \tag{7.158}$$

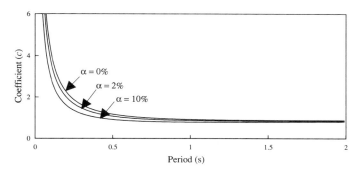

Figure 7.23 Coefficient (c) as a Function of the Natural Period of Vibration and Strain Hardening Coefficient (Krawinkler and Nassar)

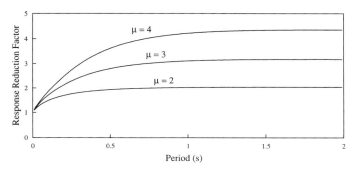

Figure 7.24 Response Reduction Factor ($R_{d\mu}$) as a Function of Natural Period of Vibration with $\alpha = 0\%$ (Krawinkler and Nassar)

In the previous chapters, the stiffness matrix $\mathbf{K}(\mathbf{X})$ in Eq. 7.158 was a constant, that is, $\mathbf{K}(\mathbf{X}) = \mathbf{K}$, which was not a function of $\mathbf{X}$ and therefore not a function of time. Now consider the stiffness matrix to be a function of the deformed position of the structure, that is, $\mathbf{X}(t)$.

Consider the case where the solution for the response is obtained by using the Newmark β-method. Recall from Eq. 3.224 that

$$\mathbf{M}\ddot{\mathbf{X}}_{k+1} + \mathbf{C}\dot{\mathbf{X}}_{k+1} + \mathbf{K}\mathbf{X}_{k+1} = \mathbf{F}_{k+1} \tag{7.159a}$$

$$\dot{\mathbf{X}}_{k+1} = \dot{\mathbf{X}}_k + (1 - \delta)\ddot{\mathbf{X}}_k\Delta t + \delta\ddot{\mathbf{X}}_{k+1}\Delta t \tag{7.159b}$$

$$\mathbf{X}_{k+1} = \mathbf{X}_k + \dot{\mathbf{X}}_k\Delta t + \left(\tfrac{1}{2} - \alpha\right)\ddot{\mathbf{X}}_k(\Delta t)^2 + \alpha\ddot{\mathbf{X}}_{k+1}(\Delta t)^2. \tag{7.159c}$$

Equation 7.159a represents the dynamic equilibrium equation of motion for the case where the stiffness matrix $\mathbf{K}$ is not a function of $\mathbf{X}$. For the case where the stiffness depends on the displacement $\mathbf{X}$, Eq. 7.159 must be written in an incremental form, that is,

$$\mathbf{M}\Delta\ddot{\mathbf{X}} + \mathbf{C}\Delta\dot{\mathbf{X}} + \mathbf{K}_s\Delta\mathbf{X} = \Delta\mathbf{F} \tag{7.160a}$$

$$\Delta\dot{\mathbf{X}} = (1 - \delta)\ddot{\mathbf{X}}_k\Delta t + \delta\ddot{\mathbf{X}}_{k+1}\Delta t \tag{7.160b}$$

$$\Delta\mathbf{X} = \dot{\mathbf{X}}_k\Delta t + \left(\tfrac{1}{2} - \alpha\right)\ddot{\mathbf{X}}_k(\Delta t)^2 + \alpha\ddot{\mathbf{X}}_{k+1}(\Delta t)^2 \tag{7.160c}$$

where $\mathbf{K}_s$ represents the stiffness corresponding to the displacement from $\mathbf{X}_k$ to $\mathbf{X}_{k+1}$, and

$$\Delta\mathbf{X} = \mathbf{X}_{k+1} - \mathbf{X}_k, \quad \Delta\dot{\mathbf{X}} = \dot{\mathbf{X}}_{k+1} - \dot{\mathbf{X}}_k, \quad \Delta\ddot{\mathbf{X}} = \ddot{\mathbf{X}}_{k+1} - \ddot{\mathbf{X}}_k, \quad \Delta\mathbf{F} = \mathbf{F}_{k+1} - \mathbf{F}_k. \tag{7.161}$$

Substituting Eqs. 7.160b and 7.160c into Eq. 7.160a, it follows that

$$\mathbf{M}\Delta\ddot{\mathbf{X}} + \mathbf{C}\left[(1-\delta)\ddot{\mathbf{X}}_k\Delta t + \delta\ddot{\mathbf{X}}_{k+1}\Delta t\right] + \mathbf{K}_s\left[\dot{\mathbf{X}}_k\Delta t + \left(\tfrac{1}{2}-\alpha\right)\ddot{\mathbf{X}}_k(\Delta t)^2 + \alpha\ddot{\mathbf{X}}_{k+1}(\Delta t)^2\right] = \Delta\mathbf{F}. \quad (7.162)$$

From Eq. 7.161,

$$\ddot{\mathbf{X}}_{k+1} = \ddot{\mathbf{X}}_k + \Delta\ddot{\mathbf{X}}. \quad (7.163)$$

Substituting Eq. 7.163 into Eq. 7.162 gives

$$\mathbf{M}\Delta\ddot{\mathbf{X}} + (1-\delta)\Delta t\mathbf{C}\ddot{\mathbf{X}}_k + \delta\Delta t\mathbf{C}\ddot{\mathbf{X}}_k + \delta\Delta t\mathbf{C}\Delta\ddot{\mathbf{X}}$$
$$+ \Delta t\mathbf{K}_s\dot{\mathbf{X}}_k + \left(\tfrac{1}{2}-\alpha\right)(\Delta t)^2\mathbf{K}_s\ddot{\mathbf{X}}_k + \alpha(\Delta t)^2\mathbf{K}_s\ddot{\mathbf{X}}_k + \alpha(\Delta t)^2\mathbf{K}_s\Delta\ddot{\mathbf{X}} = \Delta\mathbf{F}. \quad (7.164)$$

Collecting similar terms, it follows from Eq. 7.164 that

$$\left[\mathbf{M} + \delta\Delta t\mathbf{C} + \alpha(\Delta t)^2\mathbf{K}_s\right]\Delta\ddot{\mathbf{X}} + \left[(1-\delta)\Delta t + \delta\Delta t\right]\mathbf{C}\ddot{\mathbf{X}}_k$$
$$+ \Delta t\mathbf{K}_s\dot{\mathbf{X}}_k + \left[\left(\tfrac{1}{2}-\alpha\right)(\Delta t)^2 + \alpha(\Delta t)^2\right]\mathbf{K}_s\ddot{\mathbf{X}}_k = \Delta\mathbf{F}. \quad (7.165)$$

Simplifying Eq. 7.165 gives

$$\left[\mathbf{M} + \delta\Delta t\mathbf{C} + \alpha(\Delta t)^2\mathbf{K}_s\right]\Delta\ddot{\mathbf{X}} + \Delta t\mathbf{C}\ddot{\mathbf{X}}_k + \Delta t\mathbf{K}_s\dot{\mathbf{X}}_k + \tfrac{1}{2}(\Delta t)^2\mathbf{K}_s\ddot{\mathbf{X}}_k = \Delta\mathbf{F}. \quad (7.166)$$

Solving for $\Delta\ddot{\mathbf{X}}$ gives

$$\Delta\ddot{\mathbf{X}} = \left[\mathbf{M} + \delta\Delta t\mathbf{C} + \alpha(\Delta t)^2\mathbf{K}_s\right]^{-1}\left[\Delta\mathbf{F} - \Delta t\mathbf{C}\ddot{\mathbf{X}}_k - \Delta t\mathbf{K}_s\dot{\mathbf{X}}_k - \tfrac{1}{2}(\Delta t)^2\mathbf{K}_s\ddot{\mathbf{X}}_k\right]. \quad (7.167)$$

Equation 7.167 provides the solution for $\Delta\ddot{\mathbf{X}}$, and using Eq. 7.160 it is possible to calculate $\Delta\mathbf{X}$ and $\Delta\dot{\mathbf{X}}$. Then using Eq. 7.161, the values for $\mathbf{X}_{k+1}$ and $\dot{\mathbf{X}}_{k+1}$ can be determined.

To represent the solution in a matrix form, first let

$$\mathbf{B} = \mathbf{M} + \delta\Delta t\mathbf{C} + \alpha(\Delta t)^2\mathbf{K}_s. \quad (7.168)$$

Then Eq. 7.167 becomes

$$\Delta\ddot{\mathbf{X}} = \left[\mathbf{B}^{-1}\right]\Delta\mathbf{F} - \left[\Delta t\mathbf{B}^{-1}\mathbf{K}_s\right]\dot{\mathbf{X}}_k - \left[\Delta t\mathbf{B}^{-1}\mathbf{C} + \tfrac{1}{2}(\Delta t)^2\mathbf{B}^{-1}\mathbf{K}_s\right]\ddot{\mathbf{X}}_k. \quad (7.169)$$

Then the acceleration at time step $k+1$ is

$$\ddot{\mathbf{X}}_{k+1} = \ddot{\mathbf{X}}_k + \Delta\ddot{\mathbf{X}} = \ddot{\mathbf{X}}_k + \left[\mathbf{B}^{-1}\right]\Delta\mathbf{F} - \left[\Delta t\mathbf{B}^{-1}\mathbf{K}_s\right]\dot{\mathbf{X}}_k - \left[\Delta t\mathbf{B}^{-1}\mathbf{C} + \tfrac{1}{2}(\Delta t)^2\mathbf{B}^{-1}\mathbf{K}_s\right]\ddot{\mathbf{X}}_k$$
$$= \left[\mathbf{B}^{-1}\right]\Delta\mathbf{F} - \left[\Delta t\mathbf{B}^{-1}\mathbf{K}_s\right]\dot{\mathbf{X}}_k + \left[\mathbf{I} - \Delta t\mathbf{B}^{-1}\mathbf{C} - \tfrac{1}{2}(\Delta t)^2\mathbf{B}^{-1}\mathbf{K}_s\right]\ddot{\mathbf{X}}_k. \quad (7.170)$$

Substituting Eq. 7.170 into Eqs. 7.160b, 7.160c and 7.161 gives

$$\dot{\mathbf{X}}_{k+1} = \dot{\mathbf{X}}_k + (1-\delta)\ddot{\mathbf{X}}_k\Delta t + \delta\Delta t\left[\mathbf{B}^{-1}\Delta\mathbf{F} - \left(\Delta t\mathbf{B}^{-1}\mathbf{K}_s\right)\dot{\mathbf{X}}_k + \left(\mathbf{I} - \Delta t\mathbf{B}^{-1}\mathbf{C} - \tfrac{1}{2}(\Delta t)^2\mathbf{B}^{-1}\mathbf{K}_s\right)\ddot{\mathbf{X}}_k\right]$$
$$= \left[\delta\Delta t\mathbf{B}^{-1}\right]\Delta\mathbf{F} - \left[\mathbf{I} - \delta(\Delta t)^2\mathbf{B}^{-1}\mathbf{K}_s\right]\dot{\mathbf{X}}_k + \left[(\Delta t)\mathbf{I} - \delta(\Delta t)^2\mathbf{B}^{-1}\mathbf{C} - \tfrac{1}{2}\delta(\Delta t)^3\mathbf{B}^{-1}\mathbf{K}_s\right]\ddot{\mathbf{X}}_k. \quad (7.171a)$$

$$\mathbf{X}_{k+1} = \mathbf{X}_k + \Delta t\dot{\mathbf{X}}_k + \left(\tfrac{1}{2}-\alpha\right)(\Delta t)^2\ddot{\mathbf{X}}_k$$
$$+ \alpha(\Delta t)^2\left[\mathbf{B}^{-1}\Delta\mathbf{F} - \left(\Delta t\mathbf{B}^{-1}\mathbf{K}_s\right)\dot{\mathbf{X}}_k + \left(\mathbf{I} - \Delta t\mathbf{B}^{-1}\mathbf{C} - \tfrac{1}{2}(\Delta t)^2\mathbf{B}^{-1}\mathbf{K}_s\right)\ddot{\mathbf{X}}_k\right]$$
$$= \left[\alpha(\Delta t)^2\mathbf{B}^{-1}\right]\Delta\mathbf{F} + \mathbf{X}_k + \left[(\Delta t)\mathbf{I} - \alpha(\Delta t)^3\mathbf{B}^{-1}\mathbf{K}_s\right]\dot{\mathbf{X}}_k$$
$$+ \left[\tfrac{1}{2}(\Delta t)^2\mathbf{I} - \alpha(\Delta t)^3\mathbf{B}^{-1}\mathbf{C} - \tfrac{1}{2}\alpha(\Delta t)^4\mathbf{B}^{-1}\mathbf{K}_s\right]\ddot{\mathbf{X}}_k. \quad (7.171b)$$

Therefore, representing Eqs. 7.170 and 7.171 in matrix form, it follows that

$$\begin{Bmatrix}\mathbf{X}_{k+1}\\\dot{\mathbf{X}}_{k+1}\\\ddot{\mathbf{X}}_{k+1}\end{Bmatrix} = \mathbf{F}_N^{(n)}\begin{Bmatrix}\mathbf{X}_k\\\dot{\mathbf{X}}_k\\\ddot{\mathbf{X}}_k\end{Bmatrix} + \mathbf{H}_N^{(n)}\Delta\mathbf{F} \quad (7.172)$$

where

$$
\mathbf{F}_N^{(n)} = \begin{bmatrix} \mathbf{I} & (\Delta t)\mathbf{I} - \alpha(\Delta t)^3 \mathbf{B}^{-1}\mathbf{K}_s & \frac{1}{2}(\Delta t)^2 \mathbf{I} - \alpha(\Delta t)^3 \mathbf{B}^{-1}\mathbf{C} - \frac{1}{2}\alpha(\Delta t)^4 \mathbf{B}^{-1}\mathbf{K}_s \\ 0 & \mathbf{I} - \delta(\Delta t)^2 \mathbf{B}^{-1}\mathbf{K}_s & (\Delta t)\mathbf{I} - \delta(\Delta t)^2 \mathbf{B}^{-1}\mathbf{C} - \frac{1}{2}\delta(\Delta t)^3 \mathbf{B}^{-1}\mathbf{K}_s \\ 0 & -\Delta t\mathbf{B}^{-1}\mathbf{K}_s & \mathbf{I} - \Delta t\mathbf{B}^{-1}\mathbf{C} - \frac{1}{2}(\Delta t)^2 \mathbf{B}^{-1}\mathbf{K}_s \end{bmatrix}
\tag{7.173a}
$$

$$
\mathbf{H}_N^{(n)} = \begin{bmatrix} \alpha(\Delta t)^2 \mathbf{B}^{-1} \\ \delta\Delta t\mathbf{B}^{-1} \\ \mathbf{B}^{-1} \end{bmatrix}
\tag{7.173b}
$$

and the superscript (n) denotes nonlinear time history analysis. Let

$$
\mathbf{q}_k = \begin{Bmatrix} \mathbf{X}_k \\ \dot{\mathbf{X}}_k \\ \ddot{\mathbf{X}}_k \end{Bmatrix}.
$$

Then Eq. 7.172 becomes

$$
\mathbf{q}_{k+1} = \mathbf{F}_N^{(n)}\mathbf{q}_k + \mathbf{H}_N^{(n)}\Delta\mathbf{F} = \mathbf{F}_N^{(n)}\mathbf{q}_k + \mathbf{H}_N^{(n)}\mathbf{F}_{k+1} - \mathbf{H}_N^{(n)}\mathbf{F}_k.
\tag{7.174}
$$

When the structure is subjected to an earthquake ground motion, then Eq. 7.174 is still valid with the $\mathbf{H}_N^{(n)}$ matrix and $\Delta\mathbf{F}$ changed to $\mathbf{H}_N^{(n)}$ and $\Delta\mathbf{a}$, respectively, where

$$
\mathbf{H}_N^{(n\mathrm{EQ})} = -\begin{bmatrix} \alpha(\Delta t)^2 \mathbf{B}^{-1}\mathbf{M} \\ \delta\Delta t\mathbf{B}^{-1}\mathbf{M} \\ \mathbf{B}^{-1}\mathbf{M} \end{bmatrix}.
\tag{7.175}
$$

Thus Eq. 7.175 becomes

$$
\mathbf{q}_{k+1} = \mathbf{F}_N^{(n)}\mathbf{q}_k + \mathbf{H}_N^{(n\mathrm{EQ})}\mathbf{a}_{k+1} - \mathbf{H}_N^{(n\mathrm{EQ})}\mathbf{a}_k.
\tag{7.176}
$$

Note that the $\mathbf{F}_N^{(n)}$ and $\mathbf{H}_N^{(n\mathrm{EQ})}$ matrices in Eq. 7.176 are functions of time, and therefore these matrices must be computed at each time step. Also, note that computing the $\mathbf{F}_N^{(n)}$ and $\mathbf{H}_N^{(n\mathrm{EQ})}$ matrices requires the inversion of the $\mathbf{B}$ matrix at each time step.

Finally, when the constant acceleration method is used (i.e., $\delta = 0$ and $\alpha = 0$), then Eqs. 7.168, 7.173, and 7.175 become

$$
\mathbf{B} = \mathbf{M} + (0)\Delta t\mathbf{C} + (0)(\Delta t)^2 \mathbf{K}_s = \mathbf{M}
\tag{7.177a}
$$

$$
\mathbf{F}_N^{(n)} = \begin{bmatrix} \mathbf{I} & (\Delta t)\mathbf{I} & \frac{1}{2}(\Delta t)^2 \mathbf{I} \\ 0 & \mathbf{I} & (\Delta t)\mathbf{I} \\ 0 & -\Delta t\mathbf{M}^{-1}\mathbf{K}_s & \mathbf{I} - \Delta t\mathbf{M}^{-1}\mathbf{C} - \frac{1}{2}(\Delta t)^2 \mathbf{M}^{-1}\mathbf{K}_s \end{bmatrix}
\tag{7.177b}
$$

$$
\mathbf{H}_N^{(n)} = \begin{bmatrix} 0 \\ 0 \\ \mathbf{M}^{-1} \end{bmatrix}, \quad \mathbf{H}_N^{(n\mathrm{EQ})} = \begin{bmatrix} 0 \\ 0 \\ -\mathbf{I} \end{bmatrix}.
\tag{7.178}
$$

EXAMPLE 1   *Free Vibration of a 2DOF System*

Consider the two-story braced frame as shown in Figure 7.25. Assume that the material property of the braces is bilinear, of the form shown in Figure 7.26. The mass and stiffness matrices for the structure are

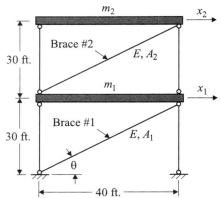

Figure 7.25 Two-Story Frame

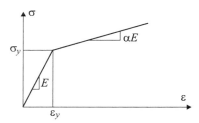

Figure 7.26 Stress versus Strain Behavior of Braces 1 and 2

$$\mathbf{M} = \begin{bmatrix} m_1 & 0.0 \\ 0.0 & m_2 \end{bmatrix}, \quad \mathbf{K} = \begin{bmatrix} \dfrac{A_1 E}{L_1}\cos^2\theta + \dfrac{A_2 E}{L_2}\cos^2\theta & -\dfrac{A_2 E}{L_2}\cos^2\theta \\ -\dfrac{A_2 E}{L_2}\cos^2\theta & \dfrac{A_2 E}{L_2}\cos^2\theta \end{bmatrix}. \tag{7.179}$$

Let $m_1 = m_2 = 1.0$ k-s^2/in., $A_1 = A_2 = A = 2.5$ in.^{2}, $E = 37,500$ ksi, and $\alpha = 20\%$ (i.e., $\alpha E = 0.2 \times 37,500 = 7,500$ ksi). The mass and initial stiffness matrices are

$$\mathbf{M} = \begin{bmatrix} 1.0 & 0.0 \\ 0.0 & 1.0 \end{bmatrix} \tag{7.180}$$

$$\mathbf{K} = \begin{bmatrix} \dfrac{2AE}{L}\cos^2\theta & -\dfrac{AE}{L}\cos^2\theta \\ -\dfrac{AE}{L}\cos^2\theta & \dfrac{AE}{L}\cos^2\theta \end{bmatrix} = \frac{(2.5)(37,500)}{(50)(12)}\left(\frac{4}{5}\right)^2\begin{bmatrix} 2 & -1 \\ -1 & 1 \end{bmatrix} = \begin{bmatrix} 200 & -100 \\ -100 & 100 \end{bmatrix}. \tag{7.181}$$

When Brace 1 yields, then the stiffness matrix becomes

$$\mathbf{K}_1 = \begin{bmatrix} \dfrac{(1+\alpha)AE}{L}\cos^2\theta & -\dfrac{AE}{L}\cos^2\theta \\ -\dfrac{AE}{L}\cos^2\theta & \dfrac{AE}{L}\cos^2\theta \end{bmatrix} = \frac{(2.5)(37,500)}{(50)(12)}\left(\frac{4}{5}\right)^2\begin{bmatrix} 1.2 & -1 \\ -1 & 1 \end{bmatrix} = \begin{bmatrix} 120 & -100 \\ -100 & 100 \end{bmatrix}. \tag{7.182}$$

If Brace 2 yields, then the stiffness matrix becomes

$$\mathbf{K}_2 = \begin{bmatrix} \dfrac{(1+\alpha)AE}{L}\cos^2\theta & -\dfrac{\alpha AE}{L}\cos^2\theta \\ -\dfrac{\alpha AE}{L}\cos^2\theta & \dfrac{\alpha AE}{L}\cos^2\theta \end{bmatrix} = \begin{bmatrix} 120 & -20 \\ -20 & 20 \end{bmatrix}. \tag{7.183}$$

Finally, if both Braces 1 and 2 yield, then the stiffness matrix becomes

$$\mathbf{K}_{12} = \begin{bmatrix} \dfrac{2\alpha AE}{L}\cos^2\theta & -\dfrac{\alpha AE}{L}\cos^2\theta \\ -\dfrac{\alpha AE}{L}\cos^2\theta & \dfrac{\alpha AE}{L}\cos^2\theta \end{bmatrix} = \begin{bmatrix} 40 & -20 \\ -20 & 20 \end{bmatrix} = \alpha\mathbf{K}. \tag{7.184}$$

Assume that the damping is proportional initial stiffness matrix with a Rayleigh proportionality constant of $\beta = 0.01$. Then the damping matrix is

$$\mathbf{C} = (0.01)\mathbf{K} = (0.01)\begin{bmatrix} 200 & -100 \\ -100 & 100 \end{bmatrix} = \begin{bmatrix} 2 & -1 \\ -1 & 1 \end{bmatrix}. \tag{7.185}$$

In addition, assume that this damping matrix will not change even when the stiffness matrix changes due to yielding of the braces. Let the yielding of the braces be equal to $\varepsilon_y = 0.002$ for both tension and compression; then these braces will yield at a relative displacement of

$$x_y = \frac{L\varepsilon_y}{\cos\theta} = \frac{(50)(12)(0.002)}{4/5} = 1.5 \text{ in}. \tag{7.186}$$

This means that Brace 1 yields at $|x_1| = 1.5$ in. and Brace 2 yields at $|x_2 - x_1| = 1.5$ in.

Recall from Eq. 3.10 that the natural frequencies of vibrations are

$$\omega_1^2 = \frac{((k_1+k_2)m_2 + k_2 m_1) - \sqrt{((k_1+k_2)m_2 + k_2 m_1)^2 - 4m_1 m_2 k_1 k_2}}{2m_1 m_2} \tag{7.187}$$

$$\omega_2^2 = \frac{((k_1+k_2)m_2 + k_2 m_1) + \sqrt{((k_1+k_2)m_2 + k_2 m_1)^2 - 4m_1 m_2 k_1 k_2}}{2m_1 m_2} \tag{7.188}$$

where $m_1 = m_2 = 1.0$ k-s^2/in. and $k_1 = k_2 = 100$ k/in. Since $k_1 = k_2$ and $m_1 = m_2$, it follows from Eq. 3.28 that

$$\omega_1 = 0.618\sqrt{100/1.0} = 6.18 \text{ rad/s} \tag{7.189}$$

$$\omega_2 = 1.618\sqrt{100/1.0} = 16.18 \text{ rad/s}. \tag{7.190}$$

The natural periods of vibration are

$$T_1 = \frac{2\pi}{\omega_1} = \frac{2\pi}{6.18} = 1.02 \text{ s} \tag{7.191}$$

$$T_2 = \frac{2\pi}{\omega_2} = \frac{2\pi}{16.18} = 0.39 \text{ s}. \tag{7.192}$$

Assume the frame has no initial displacement (i.e., $\mathbf{X}_0 = 0$) but has an initial velocity of $\dot{x}_{1o} = 30$ in./s and $\dot{x}_{2o} = 50$ in./s and no external force. The time step size should be less than $(T_n/10)$. Since $T_2 < T_1$ for this case it follows that $\Delta t \le (T_2/10) = 0.039$ s. Assume that the constant acceleration method is used with a time step of $\Delta t = 0.01$ s. Then the solution follows from Eq. 7.174 or 7.176 that

$$\mathbf{q}_{k+1} = \mathbf{F}_N^{(n)}\mathbf{q}_k \tag{7.193}$$

where, for the elastic structure, the matrices in Eq. 7.193 are

$$\mathbf{F}_N^{(n)} = \begin{bmatrix} 1.0 & 0 & | & 0.01 & 0 & | & 0.00005 & 0 \\ 0 & 1.0 & | & 0 & 0.01 & | & 0 & 0.00005 \\ \hline 0 & 0 & | & 1.0 & 0 & | & 0.01 & 0 \\ 0 & 0 & | & 0 & 1.0 & | & 0 & 0.01 \\ \hline 0 & 0 & | & -2.0 & 1.0 & | & 0.97 & 0.015 \\ 0 & 0 & | & 1.0 & -1.0 & | & 0.015 & 0.985 \end{bmatrix}, \quad \mathbf{q}_k = \begin{Bmatrix} x_1 \\ x_2 \\ \dot{x}_1 \\ \dot{x}_2 \\ \ddot{x}_1 \\ \ddot{x}_2 \end{Bmatrix}_k. \tag{7.194}$$

At time step 0, the initial acceleration is

$$\begin{Bmatrix} \ddot{x}_1 \\ \ddot{x}_2 \end{Bmatrix}_o = -\begin{bmatrix} 1 & 0 \\ 0 & 1 \end{bmatrix}^{-1}\begin{bmatrix} 2 & -1 \\ -1 & 1 \end{bmatrix}\begin{Bmatrix} 30 \\ 50 \end{Bmatrix} - \begin{bmatrix} 1 & 0 \\ 0 & 1 \end{bmatrix}^{-1}\begin{bmatrix} 200 & -100 \\ -100 & 100 \end{bmatrix}\begin{Bmatrix} 0 \\ 0 \end{Bmatrix} = \begin{Bmatrix} -10 \\ -20 \end{Bmatrix}. \tag{7.195}$$

Then using Eq. 7.195, the response at time step $k = 1$ (i.e., $t = \Delta t = 0.01$ s) is

$$\begin{Bmatrix} x_1 \\ x_2 \\ \dot{x}_1 \\ \dot{x}_2 \\ \ddot{x}_1 \\ \ddot{x}_2 \end{Bmatrix}_1 = \begin{bmatrix} 1.0 & 0 & | & 0.01 & 0 & | & 0.00005 & 0 \\ 0 & 1.0 & | & 0 & 0.01 & | & 0 & 0.00005 \\ \hline 0 & 0 & | & 1.0 & 0 & | & 0.01 & 0 \\ 0 & 0 & | & 0 & 1.0 & | & 0 & 0.01 \\ \hline 0 & 0 & | & -2.0 & 1.0 & | & 0.97 & 0.015 \\ 0 & 0 & | & 1.0 & -1.0 & | & 0.015 & 0.985 \end{bmatrix}\begin{Bmatrix} 0 \\ 0 \\ \hline 30 \\ 50 \\ \hline -10 \\ -20 \end{Bmatrix} = \begin{Bmatrix} 0.2995 \\ 0.4990 \\ 29.900 \\ 49.800 \\ -20.00 \\ -39.85 \end{Bmatrix}. \tag{7.196}$$

The relative displacement between the two floors at $t = \Delta t = 0.01$ s is

$$(\Delta_{2-1})_1 = x_2 - x_1 = 0.4990 - 0.2995 = 0.1995 \text{ in.} \tag{7.197}$$

This calculation process continues, and at $t = 6\Delta t = 0.06$ s, the response is

$$\begin{Bmatrix} x_1 \\ x_2 \\ \dot{x}_1 \\ \dot{x}_2 \\ \ddot{x}_1 \\ \ddot{x}_2 \end{Bmatrix}_6 = \begin{bmatrix} 1.0 & 0 & | & 0.01 & 0 & | & 0.00005 & 0 \\ 0 & 1.0 & | & 0 & 0.01 & | & 0 & 0.00005 \\ \hline 0 & 0 & | & 1.0 & 0 & | & 0.01 & 0 \\ 0 & 0 & | & 0 & 1.0 & | & 0 & 0.01 \\ \hline 0 & 0 & | & -2.0 & 1.0 & | & 0.97 & 0.015 \\ 0 & 0 & | & 1.0 & -1.0 & | & 0.015 & 0.985 \end{bmatrix}\begin{Bmatrix} 1.47250 \\ 2.44551 \\ 28.5005 \\ 47.0445 \\ -59.907 \\ -115.844 \end{Bmatrix} = \begin{Bmatrix} 1.75451 \\ 2.91016 \\ 27.9015 \\ 45.8860 \\ -69.804 \\ -133.549 \end{Bmatrix}. \tag{7.198}$$

At this displacement of $x_{1,6} = 1.75451$ in., Brace 1 yields. The resulting $\mathbf{F}_N^{(n)}$ matrix becomes

$$\mathbf{F}_N^{(n)} = \begin{bmatrix} 1.0 & 0 & | & 0.01 & 0 & | & 0.00005 & 0 \\ 0 & 1.0 & | & 0 & 0.01 & | & 0 & 0.00005 \\ \hline 0 & 0 & | & 1.0 & 0 & | & 0.01 & 0 \\ 0 & 0 & | & 0 & 1.0 & | & 0 & 0.01 \\ \hline 0 & 0 & | & -1.2 & 1.0 & | & 0.974 & 0.015 \\ 0 & 0 & | & 1.0 & -1.0 & | & 0.015 & 0.985 \end{bmatrix}. \tag{7.199}$$

In addition, since Brace 1 yields, the acceleration response must be calculated again using the new stiffness matrix (i.e., $\mathbf{K}_1$). Using Eq. 7.160a, it follows that

$$\ddot{\mathbf{X}}_{k+1} = \ddot{\mathbf{X}}_k - \mathbf{M}^{-1}\mathbf{C}\Delta\dot{\mathbf{X}} - \mathbf{M}^{-1}\mathbf{K}_1\Delta\mathbf{X}. \tag{7.200}$$

Substituting the numerical values into Eq. 7.200 gives

$$\begin{Bmatrix} \ddot{x}_1 \\ \ddot{x}_2 \end{Bmatrix}_6 = \begin{Bmatrix} -59.907 \\ -115.844 \end{Bmatrix} - \begin{bmatrix} 1 & 0 \\ 0 & 1 \end{bmatrix}^{-1} \begin{bmatrix} 2 & -1 \\ -1 & 1 \end{bmatrix} \begin{Bmatrix} 27.9015 - 28.5005 \\ 45.8860 - 47.0445 \end{Bmatrix}$$

$$- \begin{bmatrix} 1 & 0 \\ 0 & 1 \end{bmatrix}^{-1} \begin{bmatrix} 120 & -100 \\ -100 & 100 \end{bmatrix} \begin{Bmatrix} 1.75451 - 1.47250 \\ 2.91016 - 2.44551 \end{Bmatrix} = \begin{Bmatrix} -47.243 \\ -133.549 \end{Bmatrix}. \tag{7.201}$$

Then the calculation process continues using the $\mathbf{F}_N^{(n)}$ matrix in Eq. 7.199, and at time step $t = 9\Delta t = 0.09$ s, the response is

$$\begin{Bmatrix} x_1 \\ x_2 \\ \dot{x}_1 \\ \dot{x}_2 \\ \ddot{x}_1 \\ \ddot{x}_2 \end{Bmatrix}_9 = \begin{bmatrix} 1.0 & 0 & 0.01 & 0 & 0.00005 & 0 \\ 0 & 1.0 & 0 & 0.01 & 0 & 0.00005 \\ 0 & 0 & 1.0 & 0 & 0.01 & 0 \\ 0 & 0 & 0 & 1.0 & 0 & 0.01 \\ 0 & 0 & -1.2 & 1.0 & 0.974 & 0.015 \\ 0 & 0 & 1.0 & -1.0 & 0.015 & 0.985 \end{bmatrix} \begin{Bmatrix} 2.30368 \\ 3.80034 \\ 27.0729 \\ 43.0482 \\ -47.106 \\ -165.641 \end{Bmatrix} = \begin{Bmatrix} 2.57314 \\ 4.22253 \\ 26.8198 \\ 41.3918 \\ -38.129 \\ -179.511 \end{Bmatrix} \tag{7.202}$$

and the relative displacement between the two floors at $t = 9\Delta t = 0.09$ s is

$$(\Delta_{2-1})_9 = x_{2,9} - x_{1,9} = 4.22253 - 2.57314 = 1.64940 \text{ in.} \tag{7.203}$$

Since the relative displacement given in Eq. 7.203 is greater than the yield displacement of 1.5 in., Brace 2 yields. The resulting $\mathbf{F}_N^{(n)}$ matrix becomes

$$\mathbf{F}_N^{(n)} = \begin{bmatrix} 1.0 & 0 & 0.01 & 0 & 0.00005 & 0 \\ 0 & 1.0 & 0 & 0.01 & 0 & 0.00005 \\ 0 & 0 & 1.0 & 0 & 0.01 & 0 \\ 0 & 0 & 0 & 1.0 & 0 & 0.01 \\ 0 & 0 & -0.4 & 0.2 & 0.978 & 0.011 \\ 0 & 0 & 0.2 & -0.2 & 0.011 & 0.989 \end{bmatrix}. \tag{7.204}$$

In addition, since Brace 2 yields, the acceleration response must be calculated again using the new stiffness matrix (i.e., $\mathbf{K}_{12}$). Using Eq. 7.160a, it follows that

$$\ddot{\mathbf{X}}_{k+1} = \ddot{\mathbf{X}}_k - \mathbf{M}^{-1}\mathbf{C}\Delta\dot{\mathbf{X}} - \mathbf{M}^{-1}\mathbf{K}_{12}\Delta\mathbf{X}. \tag{7.205}$$

Substituting the numerical values into Eq. 7.205 gives

$$\begin{Bmatrix} \ddot{x}_1 \\ \ddot{x}_2 \end{Bmatrix}_9 = \begin{Bmatrix} -47.106 \\ -165.641 \end{Bmatrix} - \begin{bmatrix} 1 & 0 \\ 0 & 1 \end{bmatrix}^{-1} \begin{bmatrix} 2 & -1 \\ -1 & 1 \end{bmatrix} \begin{Bmatrix} 26.8198 - 27.0729 \\ 41.3918 - 43.0482 \end{Bmatrix}$$

$$- \begin{bmatrix} 1 & 0 \\ 0 & 1 \end{bmatrix}^{-1} \begin{bmatrix} 40 & -20 \\ -20 & 20 \end{bmatrix} \begin{Bmatrix} 2.57314 - 2.30368 \\ 4.22253 - 3.80034 \end{Bmatrix} = \begin{Bmatrix} -28.790 \\ -167.293 \end{Bmatrix}. \tag{7.206}$$

Then the calculation process continues using the $\mathbf{F}_N^{(n)}$ matrix in Eq. 7.204, and at time step $t = 22\Delta t = 0.22$ s, the response is

$$\begin{Bmatrix} x_1 \\ x_2 \\ \dot{x}_1 \\ \dot{x}_2 \\ \ddot{x}_1 \\ \ddot{x}_2 \end{Bmatrix}_{22} = \begin{bmatrix} 1.0 & 0 & 0.01 & 0 & 0.00005 & 0 \\ 0 & 1.0 & 0 & 0.01 & 0 & 0.00005 \\ 0 & 0 & 1.0 & 0 & 0.01 & 0 \\ 0 & 0 & 0 & 1.0 & 0 & 0.01 \\ 0 & 0 & -0.4 & 0.2 & 0.978 & 0.011 \\ 0 & 0 & 0.2 & -0.2 & 0.011 & 0.989 \end{bmatrix} \begin{Bmatrix} 5.48433 \\ 7.96410 \\ 20.7054 \\ 20.9052 \\ -78.664 \\ -169.528 \end{Bmatrix} = \begin{Bmatrix} 5.68745 \\ 8.16468 \\ 19.9188 \\ 19.2100 \\ -82.900 \\ -168.569 \end{Bmatrix}. \tag{7.207}$$

The relative displacements between the two floors at both time steps $t = 21\Delta t = 0.21$ s and $t = 22\Delta t = 0.22$ s are

$$(\Delta_{2-1})_{21} = x_{2,21} - x_{1,21} = 7.96410 - 5.48433 = 2.47978 \text{ in.} \qquad (7.208)$$

$$(\Delta_{2-1})_{22} = x_{2,22} - x_{1,22} = 8.16468 - 5.68745 = 2.47723 \text{ in.} \qquad (7.209)$$

Since the relative displacement between the two floors reduces from $(\Delta_{2-1})_{21} = 2.47978$ in. to $(\Delta_{2-1})_{22} = 2.47723$ in., unloading of Brace 2 occurs. This means that the stiffness matrix returns to $\mathbf{K}_1$, and the $\mathbf{F}_N^{(n)}$ matrix is given in Eq. 7.199. The acceleration response must again be calculated using Eq. 7.201:

$$
\begin{aligned}
\left\{ \begin{array}{c} \ddot{x}_1 \\ \ddot{x}_2 \end{array} \right\}_{22} &= \left\{ \begin{array}{c} -94.982 \\ -170.072 \end{array} \right\} - \begin{bmatrix} 1 & 0 \\ 0 & 1 \end{bmatrix}^{-1} \begin{bmatrix} 2 & -1 \\ -1 & 1 \end{bmatrix} \left\{ \begin{array}{c} 19.9188 - 20.7054 \\ 19.2100 - 20.9052 \end{array} \right\} \\
&\quad - \begin{bmatrix} 1 & 0 \\ 0 & 1 \end{bmatrix}^{-1} \begin{bmatrix} 120 & -100 \\ -100 & 100 \end{bmatrix} \left\{ \begin{array}{c} 5.68745 - 5.48433 \\ 8.16468 - 7.96410 \end{array} \right\} = \left\{ \begin{array}{c} -83.103 \\ -168.365 \end{array} \right\}.
\end{aligned}
\qquad (7.210)
$$

Calculation continues using the $\mathbf{F}_N^{(n)}$ matrix in Eq. 7.199, and at time step $t = 39\Delta t = 0.39$ s, the response is

$$
\left\{ \begin{array}{c} x_1 \\ x_2 \\ \dot{x}_1 \\ \dot{x}_2 \\ \ddot{x}_1 \\ \ddot{x}_2 \end{array} \right\}_{39}
=
\begin{bmatrix}
1.0 & 0 & 0.01 & 0 & 0.00005 & 0 \\
0 & 1.0 & 0 & 0.01 & 0 & 0.00005 \\
0 & 0 & 1.0 & 0 & 0.01 & 0 \\
0 & 0 & 0 & 1.0 & 0 & 0.01 \\
0 & 0 & -1.2 & 1.0 & 0.974 & 0.015 \\
0 & 0 & 1.0 & -1.0 & 0.015 & 0.985
\end{bmatrix}
\left\{ \begin{array}{c} 7.44836 \\ 9.30966 \\ -0.2559 \\ -3.0307 \\ -161.806 \\ -104.706 \end{array} \right\}
=
\left\{ \begin{array}{c} 7.43771 \\ 9.27412 \\ -1.8740 \\ -4.0778 \\ -161.893 \\ -102.788 \end{array} \right\}.
\qquad (7.211)
$$

At this time step, since the displacement of the lower floor reduces from $x_{1,38} = 7.44836$ in. to $x_{1,39} = 7.43771$ in., unloading of Brace 1 occurs. This means that the stiffness matrix returns to the original stiffness $\mathbf{K}$, and the $\mathbf{F}_N^{(n)}$ matrix is given in Eq. 7.194. The acceleration response must again be calculated using Eq. 7.160a:

$$
\begin{aligned}
\left\{ \begin{array}{c} \ddot{x}_1 \\ \ddot{x}_2 \end{array} \right\}_{39} &= \left\{ \begin{array}{c} -162.244 \\ -104.706 \end{array} \right\} - \begin{bmatrix} 1 & 0 \\ 0 & 1 \end{bmatrix}^{-1} \begin{bmatrix} 2 & -1 \\ -1 & 1 \end{bmatrix} \left\{ \begin{array}{c} -1.8740 + 0.2559 \\ -4.0778 + 3.0307 \end{array} \right\} \\
&\quad - \begin{bmatrix} 1 & 0 \\ 0 & 1 \end{bmatrix}^{-1} \begin{bmatrix} 120 & -100 \\ -100 & 100 \end{bmatrix} \left\{ \begin{array}{c} 7.43771 - 7.44836 \\ 9.27412 - 9.30966 \end{array} \right\} = \left\{ \begin{array}{c} -161.041 \\ -102.788 \end{array} \right\}.
\end{aligned}
\qquad (7.212)
$$

The calculation process continues. Note that the maximum displacement of the lower floor is $x_{1,\max} = x_{1,38} = 7.44836$ in., which causes an inelastic extension of Brace 1. Therefore, Brace 1 will begin to yield in compression at a displacement level of

$$x_{1y} = x_{1,\max} - 2x_y = 7.44836 - 2(1.5) = 4.44836 \text{ in.} \qquad (7.213)$$

Similarly, the maximum relative displacement between the upper floor and the lower floor is $\Delta_{2-1,\max} = \Delta_{2-1,21} = 2.47978$ in. Therefore, Brace 2 will begin to yield in the opposite direction at a displacement level of

$$\Delta_{2-1,y} = \Delta_{2-1,\max} - 2x_y = 2.47978 - 2(1.5) = -0.52022 \text{ in.} \qquad (7.214)$$

The displacement and velocity responses are shown in Figures 7.27 and 7.28, respectively.

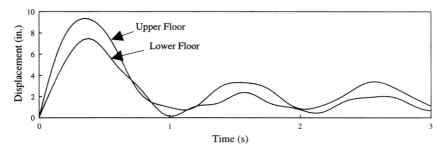

Figure 7.27 Floor Displacement Response

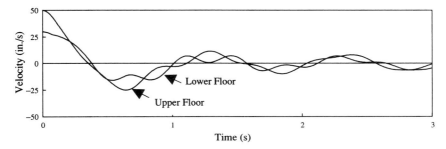

Figure 7.28 Floor Velocity Response

EXAMPLE 2   *Forced Vibration of a 2DOF System*

Consider the 2DOF system discussed in Example 1 with no initial displacement and velocity, but subjected to a bilinear pulse ground acceleration shown in Figure 7.14. Use the constant acceleration method with a time step size of $\Delta t = 0.01$ s (recall from Example 1 that $\Delta t = T_2/10 = 0.039$ s). The solution follows from Eq. 7.176 that

$$\mathbf{q}_{k+1} = \mathbf{F}_N^{(n)}\mathbf{q}_k + \mathbf{H}_N^{(nEQ)}\mathbf{a}_{k+1} - \mathbf{H}_N^{(nEQ)}\mathbf{a}_k \qquad (7.215)$$

where, for the elastic structure, the matrices in Eq. 7.215 are

$$\mathbf{F}_N^{(n)} = \begin{bmatrix} 1.0 & 0 & 0.01 & 0 & 0.00005 & 0 \\ 0 & 1.0 & 0 & 0.01 & 0 & 0.00005 \\ 0 & 0 & 1.0 & 0 & 0.01 & 0 \\ 0 & 0 & 0 & 1.0 & 0 & 0.01 \\ 0 & 0 & -2.0 & 1.0 & 0.97 & 0.015 \\ 0 & 0 & 1.0 & -1.0 & 0.015 & 0.985 \end{bmatrix}, \quad \mathbf{H}_N^{(nEQ)} = \begin{bmatrix} 0 & 0 \\ 0 & 0 \\ 0 & 0 \\ 0 & 0 \\ -1 & 0 \\ 0 & -1 \end{bmatrix} \qquad (7.216)$$

$$\mathbf{q}_k = \begin{Bmatrix} x_1 \\ x_2 \\ \dot{x}_1 \\ \dot{x}_2 \\ \ddot{x}_1 \\ \ddot{x}_2 \end{Bmatrix}_k, \quad \mathbf{a}_k = \begin{Bmatrix} a_x \\ a_x \end{Bmatrix}_k = \begin{Bmatrix} 1 \\ 1 \end{Bmatrix} a_{xk}, \quad \mathbf{H}_N^{(nEQ)}\begin{Bmatrix} 1 \\ 1 \end{Bmatrix} = \begin{Bmatrix} 0 \\ 0 \\ 0 \\ 0 \\ -1 \\ -1 \end{Bmatrix}. \qquad (7.217)$$

Therefore, Eq. 7.215 becomes

$$
\begin{Bmatrix} x_1 \\ x_2 \\ \dot{x}_1 \\ \dot{x}_2 \\ \ddot{x}_1 \\ \ddot{x}_2 \end{Bmatrix}_{k+1}
=
\begin{bmatrix}
1.0 & 0 & 0.01 & 0 & 0.00005 & 0 \\
0 & 1.0 & 0 & 0.01 & 0 & 0.00005 \\
0 & 0 & 1.0 & 0 & 0.01 & 0 \\
0 & 0 & 0 & 1.0 & 0 & 0.01 \\
0 & 0 & -2.0 & 1.0 & 0.97 & 0.015 \\
0 & 0 & 1.0 & -1.0 & 0.015 & 0.985
\end{bmatrix}
\begin{Bmatrix} x_1 \\ x_2 \\ \dot{x}_1 \\ \dot{x}_2 \\ \ddot{x}_1 \\ \ddot{x}_2 \end{Bmatrix}_{k}
+
\begin{Bmatrix} 0 \\ 0 \\ 0 \\ 0 \\ -1 \\ -1 \end{Bmatrix} a_{x,k+1}
-
\begin{Bmatrix} 0 \\ 0 \\ 0 \\ 0 \\ -1 \\ -1 \end{Bmatrix} a_{xk}. \quad (7.218)
$$

At time step 0, the initial acceleration is

$$
\ddot{\mathbf{X}}_o = \begin{Bmatrix} \ddot{x}_1 \\ \ddot{x}_2 \end{Bmatrix}_o = -\begin{Bmatrix} 1 \\ 1 \end{Bmatrix} a_{xo} - \mathbf{M}^{-1}\mathbf{C}\dot{\mathbf{X}}_o - \mathbf{M}^{-1}\mathbf{K}\mathbf{X}_o
$$

$$
= -\begin{Bmatrix} 0 \\ 0 \end{Bmatrix} - \begin{bmatrix} 1 & 0 \\ 0 & 1 \end{bmatrix}^{-1} \begin{bmatrix} 2 & -1 \\ -1 & 1 \end{bmatrix} \begin{Bmatrix} 0 \\ 0 \end{Bmatrix} - \begin{bmatrix} 1 & 0 \\ 0 & 1 \end{bmatrix}^{-1} \begin{bmatrix} 200 & -100 \\ -100 & 100 \end{bmatrix} \begin{Bmatrix} 0 \\ 0 \end{Bmatrix} = \begin{Bmatrix} 0 \\ 0 \end{Bmatrix}.
$$

$$(7.219)$$

At time step $t = \Delta t = 0.01$ s, the ground acceleration is $0.1g$. It follows from Eq. 7.218 that the response is

$$
\begin{Bmatrix} x_1 \\ x_2 \\ \dot{x}_1 \\ \dot{x}_2 \\ \ddot{x}_1 \\ \ddot{x}_2 \end{Bmatrix}_1
= \mathbf{F}_N^{(n)}
\begin{Bmatrix} 0 \\ 0 \\ 0 \\ 0 \\ 0 \\ 0 \end{Bmatrix}
+ \begin{Bmatrix} 0 \\ 0 \\ 0 \\ 0 \\ -1 \\ -1 \end{Bmatrix} (0.1 \times 386.4)
- \begin{Bmatrix} 0 \\ 0 \\ 0 \\ 0 \\ -1 \\ -1 \end{Bmatrix} (0.0)
= \begin{Bmatrix} 0 \\ 0 \\ 0 \\ 0 \\ -38.640 \\ -38.640 \end{Bmatrix}.
$$

$$(7.220)$$

Then at time step $t = 2\Delta t = 0.02$ s, the ground acceleration is $0.2g$. It follows from Eq. 7.218 that the response is

$$
\begin{Bmatrix} x_1 \\ x_2 \\ \dot{x}_1 \\ \dot{x}_2 \\ \ddot{x}_1 \\ \ddot{x}_2 \end{Bmatrix}_2
= \mathbf{F}_N^{(n)}
\begin{Bmatrix} 0 \\ 0 \\ 0 \\ 0 \\ -38.640 \\ -38.640 \end{Bmatrix}
+ \begin{Bmatrix} 0 \\ 0 \\ 0 \\ 0 \\ -1 \\ -1 \end{Bmatrix} (0.2 \times 386.4)
- \begin{Bmatrix} 0 \\ 0 \\ 0 \\ 0 \\ -1 \\ -1 \end{Bmatrix} (0.1 \times 386.4)
= \begin{Bmatrix} -0.00193 \\ -0.00193 \\ -0.3864 \\ -0.3864 \\ -76.700 \\ -77.280 \end{Bmatrix}.
$$

$$(7.221)$$

The calculation process continues, and at time step $t = 15\Delta t = 0.15$ s, the ground acceleration is $0.5g$. In addition, at time step $t = 14\Delta t = 0.14$ s, the ground acceleration is $0.6g$. It follows from Eq. 7.218 that the response is

$$
\begin{Bmatrix} x_1 \\ x_2 \\ \dot{x}_1 \\ \dot{x}_2 \\ \ddot{x}_1 \\ \ddot{x}_2 \end{Bmatrix}_{15}
= \mathbf{F}_N^{(n)}
\begin{Bmatrix} -1.37947 \\ -1.52027 \\ -25.3897 \\ -30.1399 \\ -87.334 \\ -213.009 \end{Bmatrix}
+ \begin{Bmatrix} 0 \\ 0 \\ 0 \\ 0 \\ -1 \\ -1 \end{Bmatrix} (0.5 \times 386.4)
- \begin{Bmatrix} 0 \\ 0 \\ 0 \\ 0 \\ -1 \\ -1 \end{Bmatrix} (0.6 \times 386.4)
= \begin{Bmatrix} -1.63773 \\ -1.83232 \\ -26.2630 \\ -32.2700 \\ -28.630 \\ -167.734 \end{Bmatrix}.
$$

$$(7.222)$$

At this displacement of $x_{1,15} = 1.6377$ in., Brace 1 yields. The resulting $\mathbf{F}_N^{(n)}$ matrix is

$$\mathbf{F}_N^{(n)} = \begin{bmatrix} 1.0 & 0 & 0.01 & 0 & 0.00005 & 0 \\ 0 & 1.0 & 0 & 0.01 & 0 & 0.00005 \\ 0 & 0 & 1.0 & 0 & 0.01 & 0 \\ 0 & 0 & 0 & 1.0 & 0 & 0.01 \\ 0 & 0 & -1.2 & 1.0 & 0.974 & 0.015 \\ 0 & 0 & 1.0 & -1.0 & 0.015 & 0.985 \end{bmatrix} \tag{7.223}$$

which is the same as the $\mathbf{F}_N^{(n)}$ matrix in Eq. 7.199. In addition, since Brace 1 yields, the acceleration response must be calculated again using the new stiffness matrix (i.e., $\mathbf{K}_1$ in Eq. 7.182). Using Eq. 7.160a, it follows that

$$\ddot{\mathbf{X}}_{k+1} = \ddot{\mathbf{X}}_k - \Delta\mathbf{a} - \mathbf{M}^{-1}\mathbf{C}\Delta\dot{\mathbf{X}} - \mathbf{M}^{-1}\mathbf{K}_1\Delta\mathbf{X}. \tag{7.224}$$

Substituting the numerical values into Eq. 7.224 gives

$$\begin{Bmatrix} \ddot{x}_1 \\ \ddot{x}_2 \end{Bmatrix}_{15} = \begin{Bmatrix} -87.334 \\ -213.009 \end{Bmatrix} - \begin{Bmatrix} 1 \\ 1 \end{Bmatrix}(0.5 \times 386.4 - 0.6 \times 386.4)$$

$$- \begin{bmatrix} 1 & 0 \\ 0 & 1 \end{bmatrix}^{-1} \begin{bmatrix} 2 & -1 \\ -1 & 1 \end{bmatrix} \begin{Bmatrix} -26.2630 + 25.3897 \\ -32.2700 + 30.1399 \end{Bmatrix} \tag{7.225}$$

$$- \begin{bmatrix} 1 & 0 \\ 0 & 1 \end{bmatrix}^{-1} \begin{bmatrix} 120 & -100 \\ -100 & 100 \end{bmatrix} \begin{Bmatrix} -1.63773 + 1.37947 \\ -1.83232 + 1.52027 \end{Bmatrix} = \begin{Bmatrix} -49.291 \\ -167.734 \end{Bmatrix}.$$

Calculation continues using the $\mathbf{F}_N^{(n)}$ matrix in Eq. 7.223, and at time step $t = 27\Delta t = 0.27$ s, the ground acceleration is equal to zero. Therefore, Eq. 7.218 can be used without considering the ground acceleration terms. In this case, the response is

$$\begin{Bmatrix} x_1 \\ x_2 \\ \dot{x}_1 \\ \dot{x}_2 \\ \ddot{x}_1 \\ \ddot{x}_2 \end{Bmatrix} = \begin{bmatrix} 1.0 & 0 & 0.01 & 0 & 0.00005 & 0 \\ 0 & 1.0 & 0 & 0.01 & 0 & 0.00005 \\ 0 & 0 & 1.0 & 0 & 0.01 & 0 \\ 0 & 0 & 0 & 1.0 & 0 & 0.01 \\ 0 & 0 & -1.2 & 1.0 & 0.974 & 0.015 \\ 0 & 0 & 1.0 & -1.0 & 0.015 & 0.985 \end{bmatrix} \begin{Bmatrix} -4.15977 \\ -5.57064 \\ -15.5882 \\ -29.4963 \\ 129.155 \\ 154.995 \end{Bmatrix} = \begin{Bmatrix} -4.30854 \\ -5.85785 \\ -14.1663 \\ -27.9464 \\ 130.028 \\ 168.711 \end{Bmatrix} \tag{7.226}$$

and the relative displacement between the two floors at $t = 27\Delta t = 0.27$ s is

$$(\Delta_{2-1})_{27} = x_{2,27} - x_{1,27} = -5.85785 + 4.30854 = 1.54931 \text{ in.} \tag{7.227}$$

Since the relative displacement given in Eq. 7.227 is greater than the yield displacement of 1.5 in., Brace 2 yields. The resulting $\mathbf{F}_N^{(n)}$ matrix becomes

$$\mathbf{F}_N^{(n)} = \begin{bmatrix} 1.0 & 0 & 0.01 & 0 & 0.00005 & 0 \\ 0 & 1.0 & 0 & 0.01 & 0 & 0.00005 \\ 0 & 0 & 1.0 & 0 & 0.01 & 0 \\ 0 & 0 & 0 & 1.0 & 0 & 0.01 \\ 0 & 0 & -0.4 & 0.2 & 0.978 & 0.011 \\ 0 & 0 & 0.2 & -0.2 & 0.011 & 0.989 \end{bmatrix}. \tag{7.228}$$

In addition, since Brace 2 yields, the acceleration response must be calculated again using the new stiffness matrix (i.e., $\mathbf{K}_{12}$ in Eq. 7.184). Using Eq. 7.160a, it follows that

$$\ddot{\mathbf{X}}_{k+1} = \ddot{\mathbf{X}}_k - \Delta\mathbf{a} - \mathbf{M}^{-1}\mathbf{C}\Delta\dot{\mathbf{X}} - \mathbf{M}^{-1}\mathbf{K}_{12}\Delta\mathbf{X}. \tag{7.229}$$

Substituting the numerical values into Eq. 7.229 gives

$$\begin{Bmatrix}\ddot{x}_1\\\ddot{x}_2\end{Bmatrix}_{27}=\begin{Bmatrix}129.155\\154.995\end{Bmatrix}-\begin{Bmatrix}0\\0\end{Bmatrix}-\begin{bmatrix}1&0\\0&1\end{bmatrix}^{-1}\begin{bmatrix}2&-1\\-1&1\end{bmatrix}\begin{Bmatrix}-14.1663+15.5882\\-27.9464+29.4963\end{Bmatrix}$$

$$-\begin{bmatrix}1&0\\0&1\end{bmatrix}^{-1}\begin{bmatrix}40&-20\\-20&20\end{bmatrix}\begin{Bmatrix}-4.30854+4.15977\\-5.85785+5.57064\end{Bmatrix}=\begin{Bmatrix}141.103\\157.636\end{Bmatrix}.$$

(7.230)

Calculation continues using the $\mathbf{F}_N^{(n)}$ matrix in Eq. 7.228, and at time step $t = 39\Delta t = 0.39$ s, the response is

$$\begin{Bmatrix}x_1\\x_2\\\dot{x}_1\\\dot{x}_2\\\ddot{x}_1\\\ddot{x}_2\end{Bmatrix}=\begin{bmatrix}1.0&0&0.01&0&0.00005&0\\0&1.0&0&0.01&0&0.00005\\0&0&1.0&0&0.01&0\\0&0&0&1.0&0&0.01\\0&0&-0.4&0.2&0.978&0.011\\0&0&0.2&-0.2&0.011&0.989\end{bmatrix}\begin{Bmatrix}-5.04512\\-7.93238\\0.3724\\-9.3448\\118.600\\180.332\end{Bmatrix}=\begin{Bmatrix}-5.03546\\-8.01681\\1.5584\\-7.5415\\115.956\\181.597\end{Bmatrix}.$$

(7.231)

At this time step, since the displacement of the lower floor reduces from $x_{1,38} = -5.04512$ in. to $x_{1,39} = -5.03546$ in., unloading of Brace 1 occurs. This means that the stiffness matrix is equal to $\mathbf{K}_2$ as given in Eq. 7.183, and the $\mathbf{F}_N^{(n)}$ matrix is equal to

$$\mathbf{F}_N^{(n)}=\begin{bmatrix}1.0&0&0.01&0&0.00005&0\\0&1.0&0&0.01&0&0.00005\\0&0&1.0&0&0.01&0\\0&0&0&1.0&0&0.01\\0&0&-1.2&0.2&0.974&0.011\\0&0&0.2&-0.2&0.011&0.989\end{bmatrix}.$$

(7.232)

The acceleration response must be calculated again using the new stiffness matrix (i.e., $\mathbf{K}_2$). Using Eq. 7.160a, it follows that

$$\ddot{\mathbf{X}}_{k+1}=\ddot{\mathbf{X}}_k-\Delta\mathbf{a}-\mathbf{M}^{-1}\mathbf{C}\Delta\dot{\mathbf{X}}-\mathbf{M}^{-1}\mathbf{K}_2\Delta\mathbf{X}.$$

(7.233)

Substituting the numerical values into Eq. 7.233 gives

$$\begin{Bmatrix}\ddot{x}_1\\\ddot{x}_2\end{Bmatrix}_{39}=\begin{Bmatrix}118.600\\180.332\end{Bmatrix}-\begin{Bmatrix}0\\0\end{Bmatrix}-\begin{bmatrix}1&0\\0&1\end{bmatrix}^{-1}\begin{bmatrix}2&-1\\-1&1\end{bmatrix}\begin{Bmatrix}1.5584-0.3724\\-7.5415+9.3448\end{Bmatrix}$$

$$-\begin{bmatrix}1&0\\0&1\end{bmatrix}^{-1}\begin{bmatrix}120&-20\\-20&20\end{bmatrix}\begin{Bmatrix}-5.03546+5.04512\\-8.01681+7.93238\end{Bmatrix}=\begin{Bmatrix}115.184\\181.597\end{Bmatrix}.$$

(7.234)

Then the calculation process continues using the $\mathbf{F}_N^{(n)}$ matrix in Eq. 7.232, and at time step $t = 50\Delta t = 0.50$ s, the response is

$$\begin{Bmatrix}x_1\\x_2\\\dot{x}_1\\\dot{x}_2\\\ddot{x}_1\\\ddot{x}_2\end{Bmatrix}_{50}=\begin{bmatrix}1.0&0&0.01&0&0.00005&0\\0&1.0&0&0.01&0&0.00005\\0&0&1.0&0&0.01&0\\0&0&0&1.0&0&0.01\\0&0&-1.2&0.2&0.974&0.015\\0&0&0.2&-0.2&0.015&0.985\end{bmatrix}\begin{Bmatrix}-4.38723\\-7.85334\\10.2130\\10.8374\\41.734\\181.568\end{Bmatrix}=\begin{Bmatrix}-4.28301\\-7.73589\\10.6304\\12.6531\\32.558\\179.905\end{Bmatrix}.$$

(7.235)

The relative displacements between the two floors at both time steps $t = 49\Delta t = 0.49$ s and $t = 50\Delta t = 0.50$ s are

$$(\Delta_{2-1})_{49} = x_{2,49} - x_{1,49} = -7.85334 + 4.38723 = -3.46611 \text{ in.} \tag{7.236}$$

$$(\Delta_{2-1})_{50} = x_{2,50} - x_{1,50} = -7.73589 + 4.28301 = -3.45288 \text{ in.} \tag{7.237}$$

Since the relative displacement between the two floors reduces from $(\Delta_{2-1})_{49} = -3.46611$ in. to $(\Delta_{2-1})_{50} = -3.45288$ in., unloading of Brace 2 occurs. This means that the stiffness matrix returns to $\mathbf{K}$, and the $\mathbf{F}_N^{(n)}$ matrix is given in Eq. 7.216. The acceleration response must again be calculated using Eq. 7.160a:

$$\ddot{\mathbf{X}}_{k+1} = \ddot{\mathbf{X}}_k - \Delta\mathbf{a} - \mathbf{M}^{-1}\mathbf{C}\Delta\dot{\mathbf{X}} - \mathbf{M}^{-1}\mathbf{K}\Delta\mathbf{X}. \tag{7.238}$$

Substituting the numerical values into Eq. 7.238 gives

$$
\begin{Bmatrix} \ddot{x}_1 \\ \ddot{x}_2 \end{Bmatrix}_{50} = \begin{Bmatrix} 41.734 \\ 181.568 \end{Bmatrix} - \begin{Bmatrix} 0 \\ 0 \end{Bmatrix} - \begin{bmatrix} 1 & 0 \\ 0 & 1 \end{bmatrix}^{-1} \begin{bmatrix} 2 & -1 \\ -1 & 1 \end{bmatrix} \begin{Bmatrix} 12.6531 - 10.8371 \\ 10.6304 - 10.2130 \end{Bmatrix}
$$
$$
- \begin{bmatrix} 1 & 0 \\ 0 & 1 \end{bmatrix}^{-1} \begin{bmatrix} 120 & -100 \\ -100 & 100 \end{bmatrix} \begin{Bmatrix} -4.28301 + 4.38723 \\ -7.73589 + 7.85334 \end{Bmatrix} = \begin{Bmatrix} 32.558 \\ 179.905 \end{Bmatrix}. \tag{7.239}
$$

The calculation process continues. Note that the maximum displacement of the lower floor is $x_{1,\max} = x_{1,38} = -5.04512$ in., which causes an inelastic compression of Brace 1. Therefore, Brace 1 will begin to yield in tension at a displacement level of

$$x_{1y} = x_{1,\max} + 2x_y = -5.04512 + 2(1.5) = -2.04512 \text{ in.} \tag{7.240}$$

Similarly, the maximum relative displacement between the upper floor and the lower floor is $\Delta_{2-1,\max} = \Delta_{2-1,49} = -3.46611$ in. Therefore, Brace 2 will begin to yield in the opposite direction at a displacement level of

$$\Delta_{2-1,y} = \Delta_{2-1,\max} + 2x_y = -3.46611 + 2(1.5) = -0.46611 \text{ in.} \tag{7.241}$$

The displacement and velocity responses are shown in Figures 7.29 and 7.30, respectively.

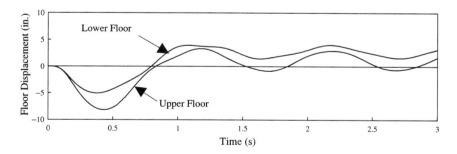

Figure 7.29 Floor Displacement Response

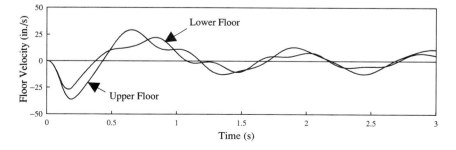

Figure 7.30 Floor Velocity Response

Consider the 2DOF system discussed in Examples 1 and 2, that is, $A_1 = A_2 = A = 2.5$ in.^{2}, $E = 37,500$ ksi, $\alpha = 20\%$, and

$$\mathbf{M} = \begin{bmatrix} 1.0 & 0.0 \\ 0.0 & 1.0 \end{bmatrix}, \quad \mathbf{C} = \begin{bmatrix} 2 & -1 \\ -1 & 1 \end{bmatrix}, \quad \mathbf{K} = \begin{bmatrix} 200 & -100 \\ -100 & 100 \end{bmatrix} \tag{7.242}$$

$$\mathbf{K}_1 = \begin{bmatrix} 120 & -100 \\ -100 & 100 \end{bmatrix}, \quad \mathbf{K}_2 = \begin{bmatrix} 120 & -20 \\ -20 & 20 \end{bmatrix}, \quad \mathbf{K}_{12} = \begin{bmatrix} 40 & -20 \\ -20 & 20 \end{bmatrix} = \alpha\mathbf{K}. \tag{7.243}$$

Let the yielding of the braces be equal to $\varepsilon_y = 0.006$ for both tension and compression; then these braces will yield at a relative displacement of

$$x_y = \frac{L\varepsilon_y}{\cos\theta} = \frac{(50)(12)(0.006)}{4/5} = 4.5 \text{ in.} \tag{7.244}$$

and therefore Brace 1 yields at $|x_1| = 4.5$ in. and Brace 2 yields at $|x_2 - x_1| = 4.5$ in. Assume that the system has no initial displacement and velocity and that it is subjected to the modified El-Centro earthquake ground motion shown in Figure 6.11.

Use the constant average acceleration method (i.e., $\delta = \frac{1}{2}$ and $\alpha = \frac{1}{4}$) with a time step size of $\Delta t = 0.02$ s (recall from Example 1 that $\Delta t = T_2/10 = 0.039$ s). Then the solution follows from Eq. 7.176 that

$$\mathbf{q}_{k+1} = \mathbf{F}_N^{(n)}\mathbf{q}_k + \mathbf{H}_N^{(nEQ)}\mathbf{a}_{k+1} - \mathbf{H}_N^{(nEQ)}\mathbf{a}_k = \mathbf{F}_N^{(n)}\mathbf{q}_k + \mathbf{H}_N^{(nEQ)}(\mathbf{a}_{k+1} - \mathbf{a}_k) \tag{7.245}$$

where

$$\mathbf{q}_k = \begin{Bmatrix} x_1 \\ x_2 \\ \dot{x}_1 \\ \dot{x}_2 \\ \ddot{x}_1 \\ \ddot{x}_2 \end{Bmatrix}_k, \quad \mathbf{a}_k = \begin{Bmatrix} a_x \\ a_x \end{Bmatrix}_k = \begin{Bmatrix} 1 \\ 1 \end{Bmatrix}a_{xk}. \tag{7.246}$$

The matrices in Eq. 7.245 can be calculated using Eqs. 7.168, 7.173 and 7.175:

$$\mathbf{B} = \begin{bmatrix} 1.0 & 0.0 \\ 0.0 & 1.0 \end{bmatrix} + \frac{1}{2}(0.02)\begin{bmatrix} 2 & -1 \\ -1 & 1 \end{bmatrix} + \frac{1}{4}(0.02)^2\begin{bmatrix} 200 & -100 \\ -100 & 100 \end{bmatrix} = \begin{bmatrix} 1.04 & -0.02 \\ -0.02 & 1.02 \end{bmatrix} \tag{7.247}$$

$$\mathbf{B}_1 = \begin{bmatrix} 1.0 & 0.0 \\ 0.0 & 1.0 \end{bmatrix} + \frac{1}{2}(0.02)\begin{bmatrix} 2 & -1 \\ -1 & 1 \end{bmatrix} + \frac{1}{4}(0.02)^2\begin{bmatrix} 120 & -100 \\ -100 & 100 \end{bmatrix} = \begin{bmatrix} 1.032 & -0.02 \\ -0.02 & 1.02 \end{bmatrix} \tag{7.248}$$

$$\mathbf{B}_2 = \begin{bmatrix} 1.0 & 0.0 \\ 0.0 & 1.0 \end{bmatrix} + \frac{1}{2}(0.02)\begin{bmatrix} 2 & -1 \\ -1 & 1 \end{bmatrix} + \frac{1}{4}(0.02)^2\begin{bmatrix} 120 & -20 \\ -20 & 20 \end{bmatrix} = \begin{bmatrix} 1.032 & -0.012 \\ -0.012 & 1.012 \end{bmatrix} \tag{7.249}$$

$$\mathbf{B}_{12} = \begin{bmatrix} 1.0 & 0.0 \\ 0.0 & 1.0 \end{bmatrix} + \frac{1}{2}(0.02)\begin{bmatrix} 2 & -1 \\ -1 & 1 \end{bmatrix} + \frac{1}{4}(0.02)^2\begin{bmatrix} 40 & -20 \\ -20 & 20 \end{bmatrix} = \begin{bmatrix} 1.024 & -0.012 \\ -0.012 & 1.012 \end{bmatrix}. \tag{7.250}$$

Inverting the matrices given in Eqs. 7.247 to 7.250 results in

$$\mathbf{B}^{-1} = \begin{bmatrix} 0.961901 & 0.018861 \\ 0.018861 & 0.980762 \end{bmatrix}, \quad \mathbf{B}_1^{-1} = \begin{bmatrix} 0.969361 & 0.019007 \\ 0.019007 & 0.980768 \end{bmatrix} \tag{7.251}$$

$$\mathbf{B}_2^{-1} = \begin{bmatrix} 0.969126 & 0.011492 \\ 0.011492 & 0.988279 \end{bmatrix}, \quad \mathbf{B}_{12}^{-1} = \begin{bmatrix} 0.976698 & 0.011581 \\ 0.011581 & 0.988280 \end{bmatrix}. \tag{7.252}$$

Therefore, it follows from Eq. 7.173a that

$$\mathbf{F}_N^{(n)} = \begin{bmatrix} 1.0 & 0.0 & 0.019619 & 0.000189 & 0.000192 & 0.000004 \\ 0.0 & 1.0 & 0.000189 & 0.019808 & 0.000004 & 0.000196 \\ 0.0 & 0.0 & 0.961901 & 0.018861 & 0.019238 & 0.000377 \\ 0.0 & 0.0 & 0.018861 & 0.980762 & 0.000377 & 0.019615 \\ 0.0 & 0.0 & -3.809883 & 1.886081 & 0.923802 & 0.037722 \\ 0.0 & 0.0 & 1.886081 & -1.923802 & 0.037722 & 0.961524 \end{bmatrix} \tag{7.253}$$

$$\mathbf{F}_1^{(n)} = \begin{bmatrix} 1.0 & 0.0 & 0.019771 & 0.000190 & 0.000194 & 0.000004 \\ 0.0 & 1.0 & 0.000192 & 0.019808 & 0.000004 & 0.000196 \\ 0.0 & 0.0 & 0.977116 & 0.019007 & 0.019387 & 0.000380 \\ 0.0 & 0.0 & 0.019159 & 0.980765 & 0.000380 & 0.019615 \\ 0.0 & 0.0 & -2.288451 & 1.900707 & 0.938721 & 0.038014 \\ 0.0 & 0.0 & 1.915913 & -1.923516 & 0.038014 & 0.961530 \end{bmatrix} \tag{7.254}$$

$$\mathbf{F}_2^{(n)} = \begin{bmatrix} 1.0 & 0.0 & 0.019768 & 0.000038 & 0.000194 & 0.000002 \\ 0.0 & 1.0 & 0.000037 & 0.019961 & 0.000002 & 0.000198 \\ 0.0 & 0.0 & 0.976787 & 0.003831 & 0.019383 & 0.000230 \\ 0.0 & 0.0 & 0.003677 & 0.996093 & 0.000230 & 0.019766 \\ 0.0 & 0.0 & -2.32131 & 0.383054 & 0.938252 & 0.022983 \\ 0.0 & 0.0 & 0.367732 & -0.390715 & 0.022983 & 0.976557 \end{bmatrix} \tag{7.255}$$

$$\mathbf{F}_{12}^{(n)} = \begin{bmatrix} 1.0 & 0.0 & 0.019922 & 0.000039 & 0.000195 & 0.000002 \\ 0.0 & 1.0 & 0.000039 & 0.019961 & 0.000002 & 0.000198 \\ 0.0 & 0.0 & 0.992233 & 0.003861 & 0.019534 & 0.000232 \\ 0.0 & 0.0 & 0.003861 & 0.996093 & 0.000232 & 0.019766 \\ 0.0 & 0.0 & -0.776726 & 0.386047 & 0.953396 & 0.023163 \\ 0.0 & 0.0 & 0.386047 & -0.390679 & 0.023163 & 0.976559 \end{bmatrix} \tag{7.256}$$

and from Eq. 7.175 that

$$\mathbf{H}_N^{(n\mathrm{EQ})} = \begin{bmatrix} -0.000096 & -0.000002 \\ -0.000002 & -0.000098 \\ -0.009619 & -0.000189 \\ -0.000189 & -0.009808 \\ -0.961901 & -0.018861 \\ -0.018861 & -0.980762 \end{bmatrix}, \quad \mathbf{H}_1^{(n\mathrm{EQ})} = \begin{bmatrix} -0.000097 & -0.000002 \\ -0.000002 & -0.000098 \\ -0.009694 & -0.000190 \\ -0.000190 & -0.009808 \\ -0.969361 & -0.019007 \\ -0.019007 & -0.980765 \end{bmatrix} \tag{7.257}$$

$$\mathbf{H}_2^{(nEQ)} = \begin{bmatrix} -0.000097 & -0.000001 \\ -0.000001 & -0.000099 \\ \hline -0.009691 & -0.000115 \\ -0.000115 & -0.009883 \\ \hline -0.969126 & -0.011492 \\ -0.011492 & -0.988279 \end{bmatrix}, \quad \mathbf{H}_{12}^{(nEQ)} = \begin{bmatrix} -0.000098 & -0.000001 \\ -0.000001 & -0.000099 \\ \hline -0.009767 & -0.000116 \\ -0.000116 & -0.009883 \\ \hline -0.976698 & -0.011581 \\ -0.011581 & -0.988280 \end{bmatrix}. \tag{7.258}$$

At time step $k = 0$, the initial acceleration is

$$\ddot{\mathbf{X}}_o = \begin{Bmatrix} \ddot{x}_1 \\ \ddot{x}_2 \end{Bmatrix}_o = -\begin{Bmatrix} 1 \\ 1 \end{Bmatrix} a_{xo} - \mathbf{M}^{-1}\mathbf{C}\dot{\mathbf{X}}_o - \mathbf{M}^{-1}\mathbf{K}\mathbf{X}_o$$

$$= -\begin{Bmatrix} 1 \\ 1 \end{Bmatrix}(-0.993) - \begin{bmatrix} 1 & 0 \\ 0 & 1 \end{bmatrix}^{-1}\begin{bmatrix} 2 & -1 \\ -1 & 1 \end{bmatrix}\begin{Bmatrix} 0 \\ 0 \end{Bmatrix} - \begin{bmatrix} 1 & 0 \\ 0 & 1 \end{bmatrix}^{-1}\begin{bmatrix} 200 & -100 \\ -100 & 100 \end{bmatrix}\begin{Bmatrix} 0 \\ 0 \end{Bmatrix} = \begin{Bmatrix} 0.993 \\ 0.993 \end{Bmatrix}. \tag{7.259}$$

At time step $t = \Delta t = 0.02$ s, the structure will respond in the elastic range. Therefore, using $\mathbf{F}_N^{(n)}$ and $\mathbf{H}_N^{(nEQ)}$ from Eqs. 7.253 and 7.257, respectively, it follows from Eq. 7.245 that the response is

$$\begin{Bmatrix} x_1 \\ x_2 \\ \hline \dot{x}_1 \\ \dot{x}_2 \\ \hline \ddot{x}_1 \\ \ddot{x}_2 \end{Bmatrix}_1 = \mathbf{F}_N^{(n)} \begin{Bmatrix} 0 \\ 0 \\ \hline 0 \\ 0 \\ \hline -0.993 \\ -0.993 \end{Bmatrix} + \mathbf{H}_N^{(nEQ)} \begin{Bmatrix} 1 \\ 1 \end{Bmatrix} (-7.651 + 0.993) = \begin{Bmatrix} 0.00085 \\ 0.00086 \\ \hline 0.0848 \\ 0.0864 \\ \hline 7.484 \\ 7.647 \end{Bmatrix} \tag{7.260}$$

$$(\Delta_{2-1})_1 = x_{2,1} - x_{1,1} = 0.00086 - 0.00085 = 0.00001 \text{ in.} \tag{7.261}$$

Then, at time step $t = 2\Delta t = 0.04$ s, the ground acceleration is -7.148 in./s^2. It follows from Eq. 7.245 that the response is

$$\begin{Bmatrix} x_1 \\ x_2 \\ \hline \dot{x}_1 \\ \dot{x}_2 \\ \hline \ddot{x}_1 \\ \ddot{x}_2 \end{Bmatrix}_2 = \mathbf{F}_N^{(n)} \begin{Bmatrix} 0.00085 \\ 0.00086 \\ \hline 0.0848 \\ 0.0864 \\ \hline 7.484 \\ 7.647 \end{Bmatrix} + \mathbf{H}_N^{(nEQ)} \begin{Bmatrix} 1 \\ 1 \end{Bmatrix} (-7.148 + 7.651) = \begin{Bmatrix} 0.00395 \\ 0.00407 \\ \hline 0.2251 \\ 0.2342 \\ \hline 6.550 \\ 7.127 \end{Bmatrix} \tag{7.262}$$

$$(\Delta_{2-1})_2 = x_{2,2} - x_{1,2} = 0.00407 - 0.00395 = 0.00012 \text{ in.} \tag{7.263}$$

The calculation process continues, and at time step $t = 139\Delta t = 2.78$ s, the response is

$$\begin{Bmatrix} x_1 \\ x_2 \\ \hline \dot{x}_1 \\ \dot{x}_2 \\ \hline \ddot{x}_1 \\ \ddot{x}_2 \end{Bmatrix}_{139} = \begin{Bmatrix} 4.31573 \\ 6.62956 \\ \hline 18.5755 \\ 40.6829 \\ \hline -209.989 \\ -266.822 \end{Bmatrix} \tag{7.264}$$

$$(\Delta_{2-1})_{139} = x_{2,139} - x_{1,139} = 6.62956 - 4.31573 = 2.31383 \text{ in.} \tag{7.265}$$

The ground acceleration at this time step is 0.0345g, and the ground acceleration at the next time step (i.e., $t = 140\Delta t = 2.80$ s) is $-0.017g$. It follows from Eq. 7.245 that the response at the next time step is

$$\begin{Bmatrix} x \\ x_2 \\ \hline \dot{x}_1 \\ \dot{x}_2 \\ \hline \ddot{x}_1 \\ \ddot{x}_2 \end{Bmatrix}_{140} = \mathbf{F}_N^{(n)} \begin{Bmatrix} 4.31573 \\ 6.62956 \\ \hline 18.5755 \\ 40.6829 \\ \hline -209.989 \\ -266.822 \end{Bmatrix} + \mathbf{H}_N^{(nEQ)} \begin{Bmatrix} 1 \\ 1 \end{Bmatrix} (-6.723 + 13.331) = \begin{Bmatrix} 4.64840 \\ 7.38777 \\ \hline 14.6914 \\ 35.1381 \\ \hline -178.424 \\ -287.661 \end{Bmatrix} \tag{7.266}$$

$$(\Delta_{2-1})_{140} = x_{2,140} - x_{1,140} = 7.38777 - 4.64840 = 2.73937 \text{ in.} \tag{7.267}$$

At this displacement of $x_{1,140} = 4.64840$ in., Brace 1 yields, and therefore the resulting $\mathbf{F}_N^{(n)}$ and $\mathbf{H}_N^{(nEQ)}$ matrices become $\mathbf{F}_1^{(n)}$ and $\mathbf{H}_1^{(nEQ)}$, as given in Eqs. 7.254 and 7.257, respectively.

Now consider the method of using fictitious forces to adjust for the imbalance of forces as discussed in Example 4 of Section 7.2. The horizontal components of forces in Braces 1 and 2 at $x_{1,140} = 4.64840$ in. and $(\Delta_{2-1})_{140} = 2.73937$ in. are

$$F_{s1,140} = (100)(4.5) + (20)(4.64840 - 4.5) = 453.30 \text{ kip} \tag{7.268}$$

$$F_{s2,140} = \left(\frac{AE}{L}\cos^2\theta\right)(\Delta_{2-1})_{140} = (100)(2.73937) = 273.94 \text{ kip.} \tag{7.269}$$

Since $F_{s1,140} = 453.30$ kip acts only on the lower floor mass whereas $F_{s2,140} = 273.94$ kip acts on both the lower and upper floor masses (i.e., +273.94 kip acting on the upper floor mass and −273.94 kip acting on the lower floor mass), the brace forces acting on each degree of freedom are

$$\begin{Bmatrix} \hat{F}_{s1} \\ \hat{F}_{s2} \end{Bmatrix}_{140} = \begin{Bmatrix} 1 \\ 0 \end{Bmatrix} F_{s1,140} + \begin{Bmatrix} -1 \\ 1 \end{Bmatrix} F_{s2,140} = \begin{Bmatrix} 453.30 - 273.94 \\ 273.94 \end{Bmatrix} = \begin{Bmatrix} 179.36 \\ 273.94 \end{Bmatrix} \text{ kip.} \tag{7.270}$$

However, from the dynamic equilibrium equation of motion, the horizontal components of forces in Braces 1 and 2 are

$$\begin{Bmatrix} \hat{F}_{s1}^{(1)} \\ \hat{F}_{s2}^{(1)} \end{Bmatrix}_{140} = -\mathbf{M}\begin{Bmatrix} \ddot{x}_1 \\ \ddot{x}_2 \end{Bmatrix}_{140} - \mathbf{C}\begin{Bmatrix} \dot{x}_1 \\ \dot{x}_2 \end{Bmatrix}_{140} - \mathbf{M}\begin{Bmatrix} 1 \\ 1 \end{Bmatrix} a_{140}$$

$$= -\begin{bmatrix} 1.0 & 0.0 \\ 0.0 & 1.0 \end{bmatrix}\begin{Bmatrix} -178.424 \\ -287.661 \end{Bmatrix} - \begin{bmatrix} 2 & -1 \\ -1 & 1 \end{bmatrix}\begin{Bmatrix} 14.6914 \\ 35.1381 \end{Bmatrix} - \begin{bmatrix} 1.0 & 0.0 \\ 0.0 & 1.0 \end{bmatrix}\begin{Bmatrix} 1 \\ 1 \end{Bmatrix}(-6.723) \tag{7.271}$$

$$= \begin{Bmatrix} 190.90 \\ 273.94 \end{Bmatrix}$$

and therefore there is an imbalance of forces when Brace 1 yields, and it is

$$\begin{Bmatrix} \Delta\hat{F}_{s1} \\ \Delta\hat{F}_{s2} \end{Bmatrix} = \begin{Bmatrix} \hat{F}_{s1}^{(1)} \\ \hat{F}_{s2}^{(1)} \end{Bmatrix}_{140} - \begin{Bmatrix} \hat{F}_{s1} \\ \hat{F}_{s2} \end{Bmatrix}_{140} = \begin{Bmatrix} 190.90 \\ 273.94 \end{Bmatrix} - \begin{Bmatrix} 179.36 \\ 273.94 \end{Bmatrix} = \begin{Bmatrix} 11.54 \\ 0.00 \end{Bmatrix} \text{ kip.} \tag{7.272}$$

It follows from Eq. 7.130 that Eq. 7.245 must be modified, and it becomes

$$\mathbf{q}_{k+1} = \mathbf{F}_N^{(n)}\mathbf{q}_k + \mathbf{H}_N^{(nEQ)}(\mathbf{a}_{k+1} - \mathbf{a}_k) + \mathbf{H}_N^{(n)}\Delta\hat{\mathbf{F}}_s \tag{7.243}$$

where $\mathbf{H}_N^{(n)}$ is given in Eq. 7.173b:

$$\mathbf{H}_N^{(n)} = \begin{bmatrix} 0.000096 & 0.000002 \\ 0.000002 & 0.000098 \\ \hline 0.009619 & 0.000189 \\ 0.000189 & 0.009808 \\ \hline 0.961901 & 0.018861 \\ 0.018861 & 0.980762 \end{bmatrix}, \quad \mathbf{H}_1^{(n)} = \begin{bmatrix} 0.000097 & 0.000002 \\ 0.000002 & 0.000098 \\ \hline 0.009694 & 0.000190 \\ 0.000190 & 0.009808 \\ \hline 0.969361 & 0.019007 \\ 0.019007 & 0.980765 \end{bmatrix} \quad (7.274)$$

$$\mathbf{H}_2^{(n)} = \begin{bmatrix} 0.000097 & 0.000001 \\ 0.000001 & 0.000099 \\ \hline 0.009691 & 0.000115 \\ 0.000115 & 0.009883 \\ \hline 0.969126 & 0.011492 \\ 0.011492 & 0.988279 \end{bmatrix}, \quad \mathbf{H}_{12}^{(n)} = \begin{bmatrix} 0.000098 & 0.000001 \\ 0.000001 & 0.000099 \\ \hline 0.009767 & 0.000116 \\ 0.000116 & 0.009883 \\ \hline 0.976698 & 0.011581 \\ 0.011581 & 0.988280 \end{bmatrix}. \quad (7.275)$$

Since Brace 1 yields at time step $t = 140\Delta t = 2.80$ s, it follows that $\mathbf{H}_1^{(n)}$ in Eq. 7.274 will be used in Eq. 7.273 to calculate the next time step:

$$\begin{Bmatrix} x_1 \\ x_2 \\ \dot{x}_1 \\ \dot{x}_2 \\ \ddot{x}_1 \\ \ddot{x}_2 \end{Bmatrix}_{141} = \mathbf{F}_1^{(n)} \begin{Bmatrix} 4.64840 \\ 7.38777 \\ 14.6914 \\ 35.1381 \\ -178.424 \\ -287.661 \end{Bmatrix} + \mathbf{H}_1^{(nEQ)} \begin{Bmatrix} 1 \\ 1 \end{Bmatrix} (-30.719 + 6.724) + \mathbf{H}_1^{(n)} \begin{Bmatrix} 11.54 \\ 0.00 \end{Bmatrix} = \begin{Bmatrix} 4.91338 \\ 8.03191 \\ 11.8068 \\ 29.2755 \\ -110.034 \\ -298.603 \end{Bmatrix} \quad (7.276)$$

$$(\Delta_{2-1})_{141} = x_{2,141} - x_{1,141} = 8.03191 - 4.91338 = 3.11853 \text{ in.} \quad (7.277)$$

$$F_{s1,141} = (100)(4.5) + (20)(4.91338 - 4.5) = 458.27 \text{ kip} \quad (7.278)$$

$$F_{s2,141} = (100)(3.11853) = 311.85 \text{ kip.} \quad (7.279)$$

The brace forces acting on each degree of freedom are

$$\begin{Bmatrix} \hat{F}_{s1} \\ \hat{F}_{s2} \end{Bmatrix}_{141} = \begin{Bmatrix} 1 \\ 0 \end{Bmatrix} F_{s1,141} + \begin{Bmatrix} -1 \\ 1 \end{Bmatrix} F_{s2,141} = \begin{Bmatrix} 458.27 - 311.85 \\ 311.85 \end{Bmatrix} = \begin{Bmatrix} 146.42 \\ 311.85 \end{Bmatrix} \text{ kip} \quad (7.280)$$

$$\begin{Bmatrix} \hat{F}_{s1}^{(1)} \\ \hat{F}_{s2}^{(1)} \end{Bmatrix}_{141} = -\begin{bmatrix} 1.0 & 0.0 \\ 0.0 & 1.0 \end{bmatrix} \begin{Bmatrix} -110.034 \\ -298.603 \end{Bmatrix} - \begin{bmatrix} 2 & -1 \\ -1 & 1 \end{bmatrix} \begin{Bmatrix} 11.8068 \\ 29.2755 \end{Bmatrix} - \begin{bmatrix} 1.0 & 0.0 \\ 0.0 & 1.0 \end{bmatrix} \begin{Bmatrix} 1 \\ 1 \end{Bmatrix} (-30.719)$$
$$= \begin{Bmatrix} 146.42 \\ 311.85 \end{Bmatrix} \quad (7.281)$$

and therefore the imbalance of forces is

$$\begin{Bmatrix} \Delta \hat{F}_{s1} \\ \Delta \hat{F}_{s2} \end{Bmatrix} = \begin{Bmatrix} \hat{F}_{s1}^{(1)} \\ \hat{F}_{s2}^{(1)} \end{Bmatrix}_{141} - \begin{Bmatrix} \hat{F}_{s1} \\ \hat{F}_{s2} \end{Bmatrix}_{141} = \begin{Bmatrix} 146.42 \\ 311.85 \end{Bmatrix} - \begin{Bmatrix} 146.42 \\ 311.85 \end{Bmatrix} = \begin{Bmatrix} 0.00 \\ 0.00 \end{Bmatrix} \text{ kip} \quad (7.282)$$

which means that the imbalance of forces has been corrected.

The calculation process continues, and at time step $t = 148\Delta t = 2.96$ s, the response is

$$\begin{Bmatrix} x_1 \\ x_2 \\ \hline \dot{x}_1 \\ \dot{x}_2 \\ \hline \ddot{x}_1 \\ \ddot{x}_2 \end{Bmatrix}_{148} = \begin{Bmatrix} 5.89769 \\ 9.07667 \\ \hline -0.1897 \\ -15.1666 \\ \hline -187.981 \\ -316.059 \end{Bmatrix} \tag{7.283}$$

$$(\Delta_{2-1})_{148} = x_{2,148} - x_{1,148} = 9.07667 - 5.89769 = 3.17898 \text{ in.} \tag{7.284}$$

At this level of response, the velocity response for both the lower and upper floor masses becomes negative, which means that the displacement has reached maximum value and thus Brace 1 begins to unload. Therefore, $\mathbf{F}_N^{(n)}$ and $\mathbf{H}_N^{(nEQ)}$ in Eqs. 7.253 and 7.257 will be used for the next time step calculation. The ground acceleration at this time step is $0.0340g$, and the ground acceleration at the next time step (i.e., $t = 149\Delta t = 2.98$ s) is $0.0771g$. It follows from Eq. 7.245 that the response at the next time step is

$$\begin{Bmatrix} x_1 \\ x_2 \\ \hline \dot{x}_1 \\ \dot{x}_2 \\ \hline \ddot{x}_1 \\ \ddot{x}_2 \end{Bmatrix}_{149} = \mathbf{F}_N^{(n)} \begin{Bmatrix} 5.89769 \\ 9.07667 \\ \hline -0.1897 \\ -15.1666 \\ \hline -187.981 \\ -316.059 \end{Bmatrix} + \mathbf{H}_N^{(nEQ)} \begin{Bmatrix} 1 \\ 1 \end{Bmatrix} (29.791 - 13.138) = \begin{Bmatrix} 5.85212 \\ 8.71185 \\ \hline -4.3674 \\ -21.3154 \\ \hline -229.796 \\ -298.816 \end{Bmatrix} \tag{7.285}$$

$$(\Delta_{2-1})_{149} = x_{2,149} - x_{1,149} = 8.71185 - 5.85212 = 2.85973 \text{ in.} \tag{7.286}$$

$$F_{s1,149} = (100)(4.5 - 5.89769 + 5.85212) + (20)(5.89769 - 4.5) = 473.39 \text{ kip} \tag{7.287}$$

$$F_{s2,149} = (100)(2.85973) = 285.97 \text{ kip.} \tag{7.288}$$

The brace forces acting on each degree of freedom are

$$\begin{Bmatrix} \hat{F}_{s1} \\ \hat{F}_{s2} \end{Bmatrix}_{149} = \begin{Bmatrix} 1 \\ 0 \end{Bmatrix} F_{s1,149} + \begin{Bmatrix} -1 \\ 1 \end{Bmatrix} F_{s2,149} = \begin{Bmatrix} 470.39 - 285.97 \\ 285.97 \end{Bmatrix} = \begin{Bmatrix} 187.42 \\ 285.97 \end{Bmatrix} \text{ kip} \tag{7.289}$$

$$\begin{Bmatrix} \hat{F}_{s1}^{(1)} \\ \hat{F}_{s2}^{(1)} \end{Bmatrix}_{149} = -\begin{bmatrix} 1.0 & 0.0 \\ 0.0 & 1.0 \end{bmatrix}\begin{Bmatrix} -229.796 \\ -298.816 \end{Bmatrix} - \begin{bmatrix} 2 & -1 \\ -1 & 1 \end{bmatrix}\begin{Bmatrix} -4.3674 \\ -21.3154 \end{Bmatrix} - \begin{bmatrix} 1.0 & 0.0 \\ 0.0 & 1.0 \end{bmatrix}\begin{Bmatrix} 1 \\ 1 \end{Bmatrix}(-29.791) \tag{7.290}$$

$$= \begin{Bmatrix} 187.42 \\ 285.97 \end{Bmatrix}$$

and therefore the imbalance of forces is

$$\begin{Bmatrix} \Delta\hat{F}_{s1} \\ \Delta\hat{F}_{s2} \end{Bmatrix} = \begin{Bmatrix} \hat{F}_{s1}^{(1)} \\ \hat{F}_{s2}^{(1)} \end{Bmatrix}_{149} - \begin{Bmatrix} \hat{F}_{s1} \\ \hat{F}_{s2} \end{Bmatrix}_{149} = \begin{Bmatrix} 187.42 \\ 285.97 \end{Bmatrix} - \begin{Bmatrix} 187.42 \\ 285.97 \end{Bmatrix} = \begin{Bmatrix} 0.00 \\ 0.00 \end{Bmatrix} \text{ kip.} \tag{7.291}$$

Note from Eq. 7.291 that there is no imbalance of forces during unloading because the change in stiffness from $\mathbf{F}_1^{(n)}$ to $\mathbf{F}_N^{(n)}$ to occurs in different time steps, whereas the change in stiffness from $\mathbf{F}_N^{(n)}$ to $\mathbf{F}_1^{(n)}$ during loading (see Eqs. 7.266 to 7.272) is within one time step from $k = 139$ to $k = 140$. Also, note that the maximum displacement of the lower floor mass has reached is 5.89769 in., which means that it will yield negatively when this lower floor mass reaches

$$x_{y1} = 5.89769 - 2(4.5) = -3.10231 \text{ in.} \tag{7.292}$$

Calculation continues using $\mathbf{F}_N^{(n)}$ and $\mathbf{H}_N^{(nEQ)}$ in Eq. 7.245 until time step $t = 165\Delta t = 3.30$ s, when the response is

$$
\begin{Bmatrix} x_1 \\ x_2 \\ \hline \dot{x}_1 \\ \dot{x}_2 \\ \hline \ddot{x}_1 \\ \ddot{x}_2 \end{Bmatrix}_{165} = \begin{Bmatrix} -3.05789 \\ -4.90826 \\ \hline -24.7757 \\ -44.7627 \\ \hline 141.529 \\ 109.197 \end{Bmatrix}
\tag{7.293}
$$

$$(\Delta_{2-1})_{165} = x_{2,165} - x_{1,165} = -4.90826 - (-3.05789) = -1.85037 \text{ in.} \tag{7.294}$$

The ground acceleration at this time step is $0.248g$, and the ground acceleration at the next time step (i.e., $t = 166\Delta t = 3.32$ s) is $0.375g$. It follows from Eq. 7.245 that the response at the next time step is

$$
\begin{Bmatrix} x_1 \\ x_2 \\ \hline \dot{x}_1 \\ \dot{x}_2 \\ \hline \ddot{x}_1 \\ \ddot{x}_2 \end{Bmatrix}_{166} = \mathbf{F}_N^{(n)} \begin{Bmatrix} -3.05789 \\ -4.90826 \\ \hline -24.7757 \\ -44.7627 \\ \hline 141.529 \\ 109.197 \end{Bmatrix} + \mathbf{H}_N^{(nEQ)} \begin{Bmatrix} 1 \\ 1 \end{Bmatrix}(144.900 - 95.827) = \begin{Bmatrix} -3.52959 \\ -5.78253 \\ \hline -22.3934 \\ -42.6641 \\ \hline 96.702 \\ 100.665 \end{Bmatrix}
\tag{7.295}
$$

$$(\Delta_{2-1})_{166} = x_{2,166} - x_{1,166} = -5.78253 - (-3.52959) = -2.25295 \text{ in.} \tag{7.296}$$

Since the displacement is $x_{1,166} = -3.52959$ in., which exceeds the yield displacement computed in Eq. 7.292, Brace 1 yields negatively. The resulting $\mathbf{F}_N^{(n)}$ and $\mathbf{H}_N^{(nEQ)}$ matrices become $\mathbf{F}_1^{(n)}$ and $\mathbf{H}_1^{(nEQ)}$ as given in Eqs. 7.254 and 7.257, respectively. The horizontal components of forces in Braces 1 and 2 at this time are

$$F_{s1,166} = (100)(-4.5) + (20)(-3.52959 + 4.5) = -430.60 \text{ kip} \tag{7.297}$$

$$F_{s2,166} = \left(\frac{AE}{L} \cos^2 \theta \right)(\Delta_{2-1})_{166} = (100)(-2.25295) = -225.30 \text{ kip.} \tag{7.298}$$

Therefore, the brace forces acting on each degree of freedom are

$$
\begin{Bmatrix} \hat{F}_{s1} \\ \hat{F}_{s2} \end{Bmatrix}_{166} = \begin{Bmatrix} 1 \\ 0 \end{Bmatrix} F_{s1,166} + \begin{Bmatrix} -1 \\ 1 \end{Bmatrix} F_{s2,166} = \begin{Bmatrix} -430.60 - (-225.30) \\ -225.30 \end{Bmatrix} = \begin{Bmatrix} -205.30 \\ -225.30 \end{Bmatrix} \text{ kip} \tag{7.299}
$$

$$
\begin{Bmatrix} \hat{F}_{s1}^{(1)} \\ \hat{F}_{s2}^{(1)} \end{Bmatrix}_{166} = -\begin{bmatrix} 1.0 & 0.0 \\ 0.0 & 1.0 \end{bmatrix} \begin{Bmatrix} 96.702 \\ 100.665 \end{Bmatrix} - \begin{bmatrix} 2 & -1 \\ -1 & 1 \end{bmatrix} \begin{Bmatrix} -22.3934 \\ -42.6641 \end{Bmatrix} - \begin{bmatrix} 1.0 & 0.0 \\ 0.0 & 1.0 \end{bmatrix} \begin{Bmatrix} 1 \\ 1 \end{Bmatrix}(144.900)
$$
$$
= \begin{Bmatrix} -239.48 \\ -225.30 \end{Bmatrix}
\tag{7.300}
$$

and therefore the imbalance of forces is

$$
\begin{Bmatrix} \Delta\hat{F}_{s1} \\ \Delta\hat{F}_{s2} \end{Bmatrix} = \begin{Bmatrix} \hat{F}_{s1}^{(1)} \\ \hat{F}_{s2}^{(1)} \end{Bmatrix}_{166} - \begin{Bmatrix} \hat{F}_{s1} \\ \hat{F}_{s2} \end{Bmatrix}_{166} = \begin{Bmatrix} -239.48 \\ -225.30 \end{Bmatrix} - \begin{Bmatrix} -205.30 \\ -225.30 \end{Bmatrix} = \begin{Bmatrix} -34.18 \\ 0.00 \end{Bmatrix} \text{ kip.} \tag{7.301}
$$

This imbalance of forces will be adjusted in the next time step.

Calculation continues, and the displacement and velocity responses within a 20-s period are shown in Figures 7.31 and 7.32.

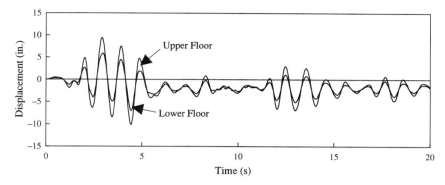

Figure 7.31 Floor Displacement Response

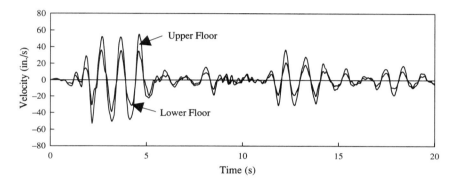

Figure 7.32 Floor Velocity Response

7.4   INELASTIC DISPLACEMENT AND FORCE ANALOGY METHOD

Previous sections of this chapter presented an inelastic structural response method where the stiffness of the structural members is the structural property that changes as a function of time. In this section and the rest of this chapter, a different method is presented, called the *force analogy method*. This method was first proposed by Professor T. H. Lin. Before the details of the force analogy method are presented, it is necessary to present the concept of an *inelastic displacement*.

The objective of changing the stiffness in the previous sections of this chapter was to capture the force in the structure after yielding. Consider the SDOF spring-mass system as shown in Figure 7.33a with the spring force versus deflection curve shown in Figure 7.33b. The force in the spring, F_s, for a displacement of the spring greater than the yield displacement can be expressed as

$$F_s = F_y + F_{is} \tag{7.302}$$

where F_y represents the yield force due to the elastic portion of the displacement and F_{is} represents the force due to the displacement after yielding. Let the initial stiffness be denoted by k_e and the post-yield stiffness be denoted by k_t. It then follows from Eq. 7.302 that

$$F_s = k_e x_y + k_t x_i \tag{7.303}$$

where x_y is the displacement of the spring at yield and x_i is the part of the total spring displacement that occurs after yielding. The total displacement of the spring, x, is

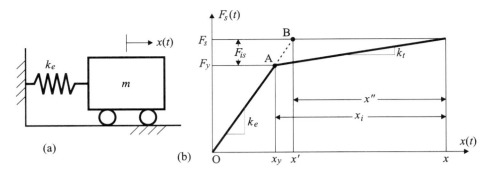

Figure 7.33 SDOF System with Bilinear Force versus Deflection Curve

$$x = x_y + x_i \tag{7.304}$$

and therefore Eq. 7.303 becomes

$$F_s = k_e x_y + k_t (x - x_y). \tag{7.305}$$

Equation 7.305 shows the relationship between the pre- and post-yield stiffness and the force in the spring.

In the force analogy method, the inelastic structural response focuses on a change in displacement, not stiffness, to give the same level of force. As shown in Figure 7.33b, the *elastic diplacement, x′,* is calculated by extending the initial stiffness line (i.e., OA) until it equals the force in the spring F_s at the total displacement x. This point is denoted as B. At this force level, the total displacement is x, and the difference between the total displacement and the elastic displacement is called the *inelastic displacement,* denoted by $x″$:

$$x = x′ + x″. \tag{7.306}$$

The force in the spring at the displacement x is given by the equation

$$F_s = k_e x′. \tag{7.307}$$

Solving for $x′$ in Eq. 7.306 and substituting the result into Eq. 7.307 gives

$$F_s = k_e (x - x″). \tag{7.308}$$

Equation 7.308 shows the effect that a change in the displacement of the system (i.e., x) has on the force level. The elastic displacement is not a constant and it changes as the force changes, and the elastic displacement is not equal to the yield displacement, that is,

$$x′ \neq x_y. \tag{7.309}$$

The basic concept of the force analogy method is to replace the inelastic displacement, $x″$, by a fictitious force (i.e., a force is an analogy to the inelastic displacement). In the SDOF spring-mass system, assume that an unknown force is applied and the resulting total displacement is x. The objective is to calculate the spring force, F_s, and the inelastic displacement, $x″$. Equation 7.308 gives one equation for solving for the two unknowns. Another equation is provided by the force deflection relationship described in Figure 7.33b, and based on this relationship both the yield force F_y and the yield displacement x_e are known. Equation 7.308 can be rewritten as

$$F_s + k_e x″ = k_e x \tag{7.310}$$

with the unknowns F_s and $x″$ on the left-hand side. To solve Eq. 7.310, the total displacement level is compared with the yield displacement. If the total displacement is less than the yield displacement, then the response will be elastic (i.e., no inelastic displacement and $x'' = 0$). Then it follows that the spring force is obtained from Eq. 7.310, where

$$F_s = k_e x, \quad x'' = 0. \tag{7.311}$$

If the total displacement is greater than the yield displacement, then F_s in Eq. 7.310 becomes

$$F_s = F_y + k_t\left(x - x_y\right). \tag{7.312}$$

Substituting Eq. 7.312 into Eq. 7.310 gives

$$F_y + k_t\left(x - x_y\right) + k_e x'' = k_e x. \tag{7.313}$$

Rearranging terms in Eq. 7.313 and solving for x'' gives

$$x'' = x - \left(\frac{F_y}{k_e}\right) - \left(\frac{k_t}{k_e}\right)\left(x - x_y\right) = \left(\frac{k_e - k_t}{k_e}\right)x + \left(\frac{k_t}{k_e}\right)x_y - \left(\frac{F_y}{k_e}\right). \tag{7.314}$$

Recall from Figure 7.33b that $F_y/k_e = x_y$; it follows from Eq. 7.314 that

$$x'' = \left(\frac{k_e - k_t}{k_e}\right)x + \left(\frac{k_t - k_e}{k_e}\right)x_y = \left(\frac{k_e - k_t}{k_e}\right)\left(x - x_y\right) = \left(1 - \frac{k_t}{k_e}\right)\left(x - x_y\right). \tag{7.315}$$

In summary, the results are

$$x'' = \begin{cases} 0 & x \le x_y \\ \left(1 - k_t/k_e\right)\left(x - x_y\right) & x > x_y, \end{cases} \quad F_s = \begin{cases} k_e x & x \le x_y \\ F_y + k_t\left(x - x_y\right) & x > x_y. \end{cases} \tag{7.316}$$

Note that the derivation of the force analogy method made no assumption as to whether the bilinear model is of strain-hardening type, elastic–plastic type, or strain-softening type. Thus, this method can be applied to any type of spring-hardening model.

In addition, note that in Eq. 7.135, $(x - x_y) = x_i$, which is the displacement after yielding. Thus it follows from Eq. 7.315 that

$$x'' = \left(1 - \frac{k_t}{k_e}\right)x_i. \tag{7.317}$$

Equation 7.317 and Figure 7.34 show that the inelastic displacement is always less than the displacement after yielding for a spring with strain hardening. When the spring is of the strain-softening type (i.e., $k_t < 0$), then Eq. 7.317 and Figure 7.34 show that the inelastic displacement is always greater than the displacement after yielding. Finally, for an elastic–plastic spring, $k_t = 0$, and therefore the inelastic displacement is equal to the displacement after yielding (i.e., $x'' = x_i$) (see Figure 7.34).

The inelastic displacement represents the inelasticity of the structure. This is because the inelastic displacement remains in the spring when the force F_s is removed from the spring-mass system. As shown in Figure 7.35, when the force F_s is unloaded from the displacement x (i.e., at point C), the path of the unloading follows line CD and the slope is equal to the initial stiffness k_e. When the force F_s is completely removed (i.e., at point D), the residual displacement is equal to inelastic displacement:

$$x'' = x - x'. \tag{7.318}$$

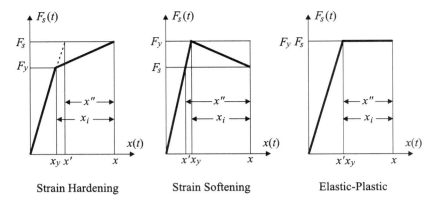

Figure 7.34 Comparison between Displacement after Yielding and Inelastic Displacement

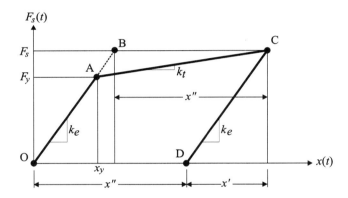

Figure 7.35 Path of Force Deflection Curve

EXAMPLE 1 *Bilinear Spring-Mass System*

Consider a bilinear spring-mass system shown in Figure 7.35, which has the stiffness properties

$$k_e = 500 \text{ k/in.,} \quad k_t = 100 \text{ k/in.,} \quad x_y = 1.0 \text{ in.} \tag{7.319}$$

At a total displacement of $x = 0.5$ in., it follows from Eq. 7.316 that

$$x'' = 0 \text{ in.} \tag{7.320}$$

$$F_s = k_e x = (500)(0.5) = 250 \text{ kip.} \tag{7.321}$$

Similarly, at a total displacement of $x = 3.0$ in., it follows from Eq. 7.316 that

$$x'' = \left(1 - k_t/k_e\right)\left(x - x_y\right) = \left(1 - \frac{100}{500}\right)(3.0 - 1.0) = 2.4 \text{ in.} \tag{7.322}$$

$$F_s = F_y + k_t\left(x - x_y\right) = (500)(1.0) + (100)(3.0 - 1.0) = 700 \text{ kip.} \tag{7.323}$$

EXAMPLE 2 *Trilinear Force-Deflection Relationship*

A powerful approach that structural engineers often use in modeling a multilinear force versus deflection curve is to combine many bilinear force versus deflection relationships. In this example, two bilinear force versus deflection relationships are combined to form a trilinear force versus deflection curve. Consider the spring-mass system shown in Figure 7.36, where two springs with different stiffnesses, denoted k_1 and k_2, are attached to the mass.

The force versus deflection relationships for Spring 1 and Spring 2 are shown in Figures 7.37a and 7.37b, respectively. The yield displacement for Spring 1 is assumed to be greater than Spring 2. Because the two springs are in parallel, a displacement of x

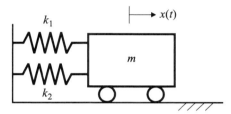

Figure 7.36 SDOF System with Two Springs

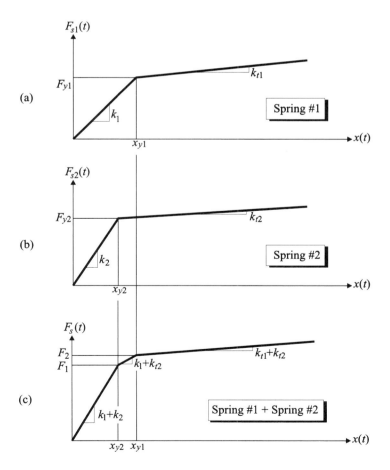

Figure 7.37 Trilinear Force-Deflection Relationship

induces a force in Spring 1, F_{s1}, and a force in Spring 2, F_{s2}. The total force exerted on the mass is

$$F_s = F_{s1} + F_{s2} = k_1 x + k_2 x = (k_1 + k_2)x. \tag{7.324}$$

Equation 7.324 shows that the total force on the mass is the sum of stiffnesses of the two springs multiplied by the displacement. This sum of stiffnesses is shown in Figure 7.37c. The forces in the two springs are

$$F_1 = F_{s1} + F_{y2} = k_1 x_{y2} + k_2 x_{y2} = (k_1 + k_2)x_{y2} \tag{7.325}$$

$$F_2 = F_{y1} + F_{s2} = k_1 x_{y1} + k_2 x_{y2} + k_{t2}(x_{y1} - x_{y2}) \tag{7.326}$$

where F_1 and F_2 are known quantities.

The force analogy method is now used to compute the forces in the system and the inelastic displacement for any given total displacement. First consider the case where $x \le x_{y2}$. Since the displacement x is less than the yield displacements for both Spring 1 and Spring 2, the system will respond elastically. Thus it follows that

$$F_s = (k_1 + k_2)x, \quad F_{s1} = k_1 x, \quad F_{s2} = k_2 x, \quad x'' = 0. \tag{7.327}$$

Now consider the case where $x_{y2} < x \le x_{y1}$. The displacement x is greater than the yield displacement for Spring 2, and therefore the stiffness of Spring 2 becomes k_{t2}. It then follows from Eq. 7.315 that

$$x'' = \left(1 - \frac{k_1 + k_{t2}}{k_1 + k_2}\right)(x - x_{y2}) = \left(\frac{k_2 - k_{t2}}{k_1 + k_2}\right)(x - x_{y2}) \tag{7.328}$$

and from Eq. 7.312 that

$$\begin{aligned} F_s &= F_1 + (k_1 + k_{t2})(x - x_{y2}) \\ &= (k_1 + k_2)x_{y2} + (k_1 + k_{t2})(x - x_{y2}) \\ &= k_1 x + k_2 x_{y2} + k_{t2}(x - x_{y2}). \end{aligned} \tag{7.329}$$

Note that Eq. 7.329 can be obtained by considering the displacement of the springs separately:

$$F_{s1} = k_1 x \tag{7.330}$$

$$F_{s2} = k_2 x_{y2} + k_{t2}(x - x_{y2}). \tag{7.331}$$

Thus the total force on the system is

$$F_s = F_{s1} + F_{s2} = k_1 x + k_2 x_{y2} + k_{t2}(x - x_{y2}) \tag{7.332}$$

which is the same as Eq. 7.329.

Now consider the case where $x > x_{y1}$. At $x = x_{y1}$, the inelastic displacement is given in Eq. 7.328, which is

$$x''_{y1} = \left(1 - \frac{k_1 + k_{t2}}{k_1 + k_2}\right)(x_{y1} - x_{y2}) = \left(\frac{k_2 - k_{t2}}{k_1 + k_2}\right)(x_{y1} - x_{y2}). \tag{7.333}$$

The additional inelastic displacement induced by a displacement greater than x_{y1} is again given in Eq. 7.315, which is

$$\Delta x'' = \left(1 - \frac{k_{t1} + k_{t2}}{k_1 + k_2}\right)(x - x_{y1}) = \left(\frac{k_1 + k_2 - k_{t1} - k_{t2}}{k_1 + k_2}\right)(x - x_{y1}). \tag{7.334}$$

Note that the denominator on the right-hand side of Eq. 7.334 uses the initial stiffness $k_1 + k_2$. This is because the force analogy method is developed based on the initial stiffness. It follows that the total inelastic displacement is

$$x'' = x''_{y1} + \Delta x''$$

$$= \left(1 - \frac{k_1 + k_{t2}}{k_1 + k_2}\right)(x_{y1} - x_{y2}) + \left(1 - \frac{k_{t1} + k_{t2}}{k_1 + k_2}\right)(x - x_{y1})$$

$$= \left(\frac{k_2 - k_{t2}}{k_1 + k_2}\right)(x_{y1} - x_{y2}) + \left(\frac{k_1 + k_2 - k_{t1} - k_{t2}}{k_1 + k_2}\right)(x - x_{y1})$$

$$= \left(\frac{k_1 + k_2 - k_{t1} - k_{t2}}{k_1 + k_2}\right)x + \left(\frac{k_{t1} - k_1}{k_1 + k_2}\right)x_{y1} + \left(\frac{k_{t2} - k_2}{k_1 + k_2}\right)x_{y2}.$$

(7.335)

The total force on the system is again given by Eq. 7.312, which is

$$F_s = F_2 + \left(k_{t1} + k_{t2}\right)\left(x - x_{y1}\right)$$

$$= k_1 x_{y1} + k_2 x_{y2} + k_{t2}\left(x_{y1} - x_{y2}\right) + \left(k_{t1} + k_{t2}\right)\left(x - x_{y1}\right)$$

$$= k_1 x_{y1} + k_{t1}\left(x - x_{y1}\right) + k_2 x_{y2} + k_{t2}\left(x - x_{y2}\right).$$

(7.336)

Again, note that Eq. 7.336 can be obtained by considering separately the displacement of the springs:

$$F_{s1} = k_1 x_{y1} + k_{t1}\left(x - x_{y1}\right)$$

(7.337)

$$F_{s2} = k_2 x_{y2} + k_{t2}\left(x - x_{y2}\right).$$

(7.338)

Thus the total force on the system is

$$F_s = F_{s1} + F_{s2} = k_1 x_{y1} + k_{t1}\left(x - x_{y1}\right) + k_2 x_{y2} + k_{t2}\left(x - x_{y2}\right)$$

(7.339)

which is the same as Eq. 7.336. Note that Eq. 7.335 can be verified by using Eq. 7.310, which is

$$F_s + \left(k_1 + k_2\right)x'' = \left(k_1 + k_2\right)x.$$

(7.340)

Solving for x'' in Eq. 7.340 gives

$$x'' = x - \frac{F_s}{k_1 + k_2}.$$

(7.341)

Substituting Eq. 7.339 into Eq. 7.341, it follows that

$$x'' = x - \frac{k_1 x_{y1} + k_{t1}\left(x - x_{y1}\right) + k_2 x_{y2} + k_{t2}\left(x - x_{y2}\right)}{k_1 + k_2}$$

$$= \left(\frac{k_1 + k_2 - k_{t1} - k_{t2}}{k_1 + k_2}\right)x + \left(\frac{k_{t1} - k_1}{k_1 + k_2}\right)x_{y1} + \left(\frac{k_{t2} - k_2}{k_1 + k_2}\right)x_{y2}$$

(7.342)

which is the same as Eq. 7.335. At total displacement x, the inelastic displacement of each spring can also be calculated using Eq. 7.315. From Figures 7.37a and 7.37b, the inelastic displacements in Spring 1 (i.e., x_1'') and Spring 2 (i.e., x_2'') are given by

$$x_1'' = \left(1 - \frac{k_{t1}}{k_1}\right)\left(x - x_{y1}\right), \quad x_2'' = \left(1 - \frac{k_{t2}}{k_2}\right)\left(x - x_{y2}\right).$$

(7.343)

In summary, the inelastic displacement and the total force on the system are

$$x'' = \begin{cases} 0 & x \le x_{y2} \\ \left(\dfrac{k_2 - k_{t2}}{k_1 + k_2}\right)x + \left(\dfrac{k_{t2} - k_2}{k_1 + k_2}\right)x_{y2} & x_{y2} < x \le x_{y1} \\ \left[\left(\dfrac{k_1 + k_2 - k_{t1} - k_{t2}}{k_1 + k_2}\right)x + \left(\dfrac{k_{t1} - k_1}{k_1 + k_2}\right)x_{y1} + \left(\dfrac{k_{t2} - k_2}{k_1 + k_2}\right)x_{y2}\right] & x_{y1} < x \end{cases} \qquad (7.344)$$

$$F_s = \begin{cases} k_1 x + k_2 x & x \le x_{y2} \\ k_1 x + k_2 x_{y2} + k_{t2}\left(x - x_{y2}\right) & x_{y2} < x \le x_{y1} \\ k_1 x_{y1} + k_{t1}\left(x - x_{y1}\right) + k_2 x_{y2} + k_{t2}\left(x - x_{y2}\right) & x_{y1} < x. \end{cases} \qquad (7.345)$$

EXAMPLE 3   *Bilinear Analogy to Trilinear Force-Deflection Relationship*

In Example 2, the inelastic displacement of a trilinear spring-mass system was calculated for the displacement $x > x_{y1}$ and was given in Eq. 7.335, which is

$$x'' = \left(\frac{k_1 + k_2 - k_{t1} - k_{t2}}{k_1 + k_2}\right)x + \left(\frac{k_{t1} - k_1}{k_1 + k_2}\right)x_{y1} + \left(\frac{k_{t2} - k_2}{k_1 + k_2}\right)x_{y2}. \qquad (7.346)$$

The inelastic displacements in each spring, given by Eq. 7.343, are

$$x_1'' = \left(1 - \frac{k_{t1}}{k_1}\right)\left(x - x_{y1}\right), \quad x_2'' = \left(1 - \frac{k_{t2}}{k_2}\right)\left(x - x_{y2}\right). \qquad (7.347)$$

In this example, the inelastic displacement will be calculated by representing the trilinear force-deflection relationship by a bilinear force-deflection curve as shown in Figure 7.38. By applying the force analogy method, this bilinear representation is called the *bilinear analogy*.

When a displacement value is given, the stiffness corresponding to this displacement is known from the force-deflection relationship. This stiffness, together with the initial stiffness, can be imagined as a bilinear force-deflection relationship with yield displacement denoted by x_y' and yield force denoted by F_y', as shown in Figure 7.38. To calculate x_y', note that

$$F_2 - F_1 = \left(k_1 + k_{t2}\right)\left(x_{y1} - x_{y2}\right). \qquad (7.348)$$

In addition,

$$F_2 - F_1 = \left(F_2 - F_y'\right) + \left(F_y' - F_1\right) = \left(k_1 + k_2\right)\left(x_y' - x_{y2}\right) + \left(k_{t1} + k_{t2}\right)\left(x_{y1} - x_y'\right). \qquad (7.349)$$

Equating Eq. 7.348 with Eq. 7.349, it follows that

$$\left(k_1 + k_{t2}\right)\left(x_{y1} - x_{y2}\right) = \left(k_1 + k_2\right)\left(x_y' - x_{y2}\right) + \left(k_{t1} + k_{t2}\right)\left(x_{y1} - x_y'\right) \qquad (7.350)$$

and solving for x_y' in Eq. 7.350 gives

$$x_y' = \left(\frac{k_1 - k_{t1}}{k_1 + k_2 - k_{t1} - k_{t2}}\right)x_{y1} + \left(\frac{k_2 + k_{t2}}{k_1 + k_2 - k_{t1} - k_{t2}}\right)x_{y2}. \qquad (7.351)$$

Then it follows from Eq. 7.315 that

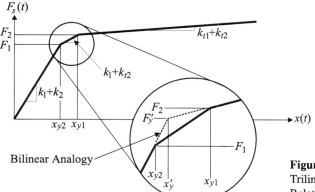

Figure 7.38 Bilinear Analogy to Trilinear Force-Deflection Relationship

$$x'' = \left(1 - \frac{k_{t1} + k_{t2}}{k_1 + k_2}\right)\left(x - \frac{k_1 - k_{t1}}{k_1 + k_2 - k_{t1} - k_{t2}}x_{y1} - \frac{k_2 - k_{t2}}{k_1 + k_2 - k_{t1} - k_{t2}}x_{y2}\right)$$
$$= \left(\frac{k_1 + k_2 - k_{t1} - k_{t2}}{k_1 + k_2}\right)x + \left(\frac{k_{t1} - k_1}{k_1 + k_2}\right)x_{y1} + \left(\frac{k_{t2} - k_2}{k_1 + k_2}\right)x_{y2}$$

(7.352)

which is the same as Eq. 7.346.

The value of x_y' is computed using the information on the force versus deflection curve and is independent of the system displacement. Therefore, the value of x_y' can be determined before calculating the system response. If the force versus deflection relationship is multilinear, then different values of x_y' can be calculated before the analysis, and the inelastic response during the analysis can be calculated using the bilinear analogy.

EXAMPLE 4   *Incremental Approach to Force Analogy Method*

Consider the trilinear force-deflection relationship used in Example 2. Assume that the total displacement at time step k is x_k and is as shown in Figure 7.39. From Eq. 7.335, the inelastic displacement due to x_k, denoted by x_k'', is given by

$$x_k'' = \left(\frac{k_1 + k_2 - k_{t1} - k_{t2}}{k_1 + k_2}\right)x_k + \left(\frac{k_{t1} - k_1}{k_1 + k_2}\right)x_{y1} + \left(\frac{k_{t2} - k_2}{k_1 + k_2}\right)x_{y2}.$$

(7.353)

Now assume that the displacement is increased to x_{k+1}, as shown in Figure 7.39, with an incremental displacement denoted by Δx. It follows that this incremental displacement causes an additional inelastic displacement, $\Delta x''$. According to Eq. 7.315, $\Delta x''$ is given by

$$\Delta x'' = \left(1 - \frac{k_{t1} + k_{t2}}{k_1 + k_2}\right)\Delta x = \left(\frac{k_1 + k_2 - k_{t1} - k_{t2}}{k_1 + k_2}\right)\Delta x.$$

(7.354)

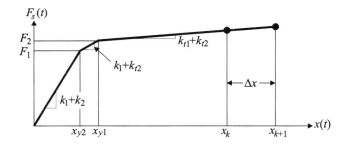

Figure 7.39 Incremental Approach of Trilinear Force-Deflection Relationship

Thus, the total inelastic displacement at $x_{k+1} = x_k + \Delta x$ is given by

$$x''_{k+1} = x''_k + \Delta x''$$

$$= \left(\frac{k_1 + k_2 - k_{t1} - k_{t2}}{k_1 + k_2}\right)x_k + \left(\frac{k_{t1} - k_1}{k_1 + k_2}\right)x_{y1} + \left(\frac{k_{t2} - k_2}{k_1 + k_2}\right)x_{y2} + \left(\frac{k_1 + k_2 - k_{t1} - k_{t2}}{k_1 + k_2}\right)\Delta x$$

$$= \left(\frac{k_1 + k_2 - k_{t1} - k_{t2}}{k_1 + k_2}\right)(x_k + \Delta x) + \left(\frac{k_{t1} - k_1}{k_1 + k_2}\right)x_{y1} + \left(\frac{k_{t2} - k_2}{k_1 + k_2}\right)x_{y2}$$

$$= \left(\frac{k_1 + k_2 - k_{t1} - k_{t2}}{k_1 + k_2}\right)x_{k+1} + \left(\frac{k_{t1} - k_1}{k_1 + k_2}\right)x_{y1} + \left(\frac{k_{t2} - k_2}{k_1 + k_2}\right)x_{y2} \qquad (7.355)$$

which is again in the same form as Eq. 7.335.

This example shows that the force analogy method can be used with an incremental analysis method of calculating structural response. To illustrate this, assume that x_k and x''_k, which are the displacement and inelastic displacement at time step k, respectively, are known. Then from an incremental analysis $x_{k+1} = x_k + \Delta x$ is calculated, and then x''_{k+1} can also be computed using the incremental approach to the force analogy method.

EXAMPLE 5 *Application of Force Analogy Method to Unloading*

Consider the trilinear force versus deflection relationship used in Example 2. Assume that the total displacement at time k is x_k, as shown in Figure 7.40. The inelastic displacement, total force in the system, and the forces in each spring are given by Eqs. 7.335 through 7.338, and they are

$$x''_k = \left(\frac{k_1 + k_2 - k_{t1} - k_{t2}}{k_1 + k_2}\right)x_k + \left(\frac{k_{t1} - k_1}{k_1 + k_2}\right)x_{y1} + \left(\frac{k_{t2} - k_2}{k_1 + k_2}\right)x_{y2} \qquad (7.356)$$

$$F_{s,k} = k_1 x_{y1} + k_{t1}\left(x_k - x_{y1}\right) + k_2 x_{y2} + k_{t2}\left(x_k - x_{y2}\right) \qquad (7.357)$$

$$F_{s1,k} = k_1 x_{y1} + k_{t1}\left(x_k - x_{y1}\right) \qquad (7.358)$$

$$F_{s2,k} = k_2 x_{y2} + k_{t2}\left(x_k - x_{y2}\right). \qquad (7.359)$$

The inelastic displacements in each spring, given in Eq. 7.343, are

$$x''_{1,k} = \left(1 - \frac{k_{t1}}{k_1}\right)\left(x_k - x_{y1}\right), \quad x''_{2,k} = \left(1 - \frac{k_{t2}}{k_2}\right)\left(x_k - x_{y2}\right). \qquad (7.360)$$

Now assume that the displacement is decreased to x_{k+1}, as shown in Figure 7.40, with an incremental displacement denoted by Δx. Since the total displacement is decreasing, the path follows a slope equal to the initial stiffness (see Figure 7.40). The resulting inelastic displacement will remain the same, that is,

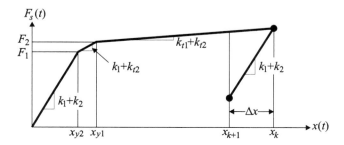

Figure 7.40 Unloading from the Trilinear Force-Deflection Relationship

$$x''_{k+1} = x''_k = \left(\frac{k_1 + k_2 - k_{t1} - k_{t2}}{k_1 + k_2}\right)x_k + \left(\frac{k_{t1} - k_1}{k_1 + k_2}\right)x_{y1} + \left(\frac{k_{t2} - k_2}{k_1 + k_2}\right)x_{y2} \qquad (7.361)$$

and the inelastic displacement of each spring will also remain the same, that is,

$$x''_{1,k+1} = x''_{1,k} = \left(1 - \frac{k_{t1}}{k_1}\right)(x_k - x_{y1}), \quad x''_{2,k+1} = x''_{2,k} = \left(1 - \frac{k_{t2}}{k_2}\right)(x_k - x_{y2}). \qquad (7.362)$$

The total incremental force on the system, ΔF_s, will be

$$\Delta F_s = -(k_1 + k_2)\Delta x. \qquad (7.363)$$

Therefore, the total force on the system is

$$\begin{aligned}
F_{s,k+1} &= F_{s,k} + \Delta F_s \\
&= k_1 x_{y1} + k_{t1}(x_k - x_{y1}) + k_2 x_{y2} + k_{t2}(x_k - x_{y2}) - (k_1 + k_2)\Delta x \\
&= k_1(x_{y1} - \Delta x) + k_{t1}(x_k - x_{y1}) + k_2(x_{y2} - \Delta x) + k_{t2}(x_k - x_{y2}).
\end{aligned} \qquad (7.364)$$

The incremental forces on each spring are

$$\Delta F_{s1} = -k_1 \Delta x, \quad \Delta F_{s2} = -k_2 \Delta x \qquad (7.365)$$

and the total forces on each spring are

$$F_{s1,k+1} = F_{s1,k} + \Delta F_{s1} = k_1(x_{y1} - \Delta x) + k_{t1}(x_k - x_{y1}) \qquad (7.366)$$

$$F_{s2,k+1} = F_{s2,k} + \Delta F_{s2} = k_2(x_{y2} - \Delta x) + k_{t2}(x_k - x_{y2}). \qquad (7.367)$$

Again, the total force on the system is the sum of forces in each spring:

$$F_{s,k+1} = F_{s1,k+1} + F_{s2,k+1}. \qquad (7.368)$$

Note that the quantity x_k remains in Eqs. 7.364, 7.366, and 7.367, which shows the effect of the hysteresis loop on the forces in the spring and the system.

EXAMPLE 6 *Equilibrium After Unloading*

Consider the spring-mass system shown in Figure 7.36 with the force versus deflection relationship shown in Figure 7.37. Assume that the total displacement is at x_k, as shown in Figure 7.41. The inelastic displacement, the total force in the system, and the forces in each spring follow from Eqs. 7.356 through 7.360:

$$x''_k = \left(\frac{k_1 + k_2 - k_{t1} - k_{t2}}{k_1 + k_2}\right)x_k + \left(\frac{k_{t1} - k_1}{k_1 + k_2}\right)x_{y1} + \left(\frac{k_{t2} - k_2}{k_1 + k_2}\right)x_{y2} \qquad (7.369)$$

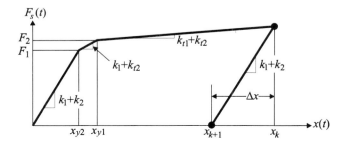

Figure 7.41 Complete Unloading

$$F_{s,k} = k_1 x_{y1} + k_{t1}(x_k - x_{y1}) + k_2 x_{y2} + k_{t2}(x_k - x_{y2}) \tag{7.370}$$

$$F_{s1,k} = k_1 x_{y1} + k_{t1}(x_k - x_{y1}) \tag{7.371}$$

$$F_{s2,k} = k_2 x_{y2} + k_{t2}(x_k - x_{y2}) \tag{7.372}$$

$$x_{1,k}'' = \left(1 - \frac{k_{t1}}{k_1}\right)(x_k - x_{y1}), \quad x_{2,k}'' = \left(1 - \frac{k_{t2}}{k_2}\right)(x_k - x_{y2}). \tag{7.373}$$

Now assume that the unloading occurs such that the total force on the system is zero, that is,

$$F_{s,k+1} = F_{s,k} + \Delta F_s = 0. \tag{7.374}$$

Then substituting Eq. 7.370 into Eq. 7.374 and solving the total incremental force ΔF_s gives

$$\Delta F_s = -F_{s,k} = -k_1 x_{y1} - k_{t1}(x_k - x_{y1}) - k_2 x_{y2} - k_{t2}(x_k - x_{y2}). \tag{7.375}$$

The incremental displacement is given by

$$\Delta x = \frac{\Delta F_s}{k_1 + k_2}. \tag{7.376}$$

Substituting Eq. 7.375 into Eq. 7.376, it follows that

$$\Delta x = -\frac{k_1}{k_1 + k_2} x_{y1} - \frac{k_{t1}}{k_1 + k_2}(x_k - x_{y1}) - \frac{k_2}{k_1 + k_2} x_{y2} - \frac{k_{t2}}{k_1 + k_2}(x_k - x_{y2}). \tag{7.377}$$

The incremental forces in the springs are given by

$$\Delta F_{s1} = k_1 \Delta x$$
$$= -\frac{k_1^2}{k_1 + k_2} x_{y1} - \frac{k_1 k_{t1}}{k_1 + k_2}(x_k - x_{y1}) - \frac{k_1 k_2}{k_1 + k_2} x_{y2} - \frac{k_1 k_{t2}}{k_1 + k_2}(x_k - x_{y2}) \tag{7.378}$$

$$\Delta F_{s2} = k_2 \Delta x$$
$$= -\frac{k_1 k_2}{k_1 + k_2} x_{y1} - \frac{k_2 k_{t1}}{k_1 + k_2}(x_k - x_{y1}) - \frac{k_2^2}{k_1 + k_2} x_{y2} - \frac{k_2 k_{t2}}{k_1 + k_2}(x_k - x_{y2}). \tag{7.379}$$

Therefore, the total forces in each spring are

$$F_{s1,k+1} = F_{s1,k} + \Delta F_{s1}$$
$$= k_1 x_{y1} + k_{t1}(x_k - x_{y1}) - \frac{k_1^2}{k_1 + k_2} x_{y1} - \frac{k_1 k_{t1}}{k_1 + k_2}(x_k - x_{y1})$$
$$- \frac{k_1 k_2}{k_1 + k_2} x_{y2} - \frac{k_1 k_{t2}}{k_1 + k_2}(x_k - x_{y2})$$
$$= \frac{k_1 k_2}{k_1 + k_2} x_{y1} + \frac{k_2 k_{t1}}{k_1 + k_2}(x_k - x_{y1}) - \frac{k_1 k_2}{k_1 + k_2} x_{y2} - \frac{k_1 k_{t2}}{k_1 + k_2}(x_k - x_{y2}) \tag{7.380}$$

$$F_{s2,k+1} = F_{s2,k} + \Delta F_{s2}$$
$$= k_2 x_{y2} + k_{t2}(x_k - x_{y2}) - \frac{k_1 k_2}{k_1 + k_2} x_{y1} - \frac{k_2 k_{t1}}{k_1 + k_2}(x_k - x_{y1})$$
$$- \frac{k_2^2}{k_1 + k_2} x_{y2} - \frac{k_2 k_{t2}}{k_1 + k_2}(x_k - x_{y2}) \tag{7.381}$$
$$= -\frac{k_1 k_2}{k_1 + k_2} x_{y1} - \frac{k_2 k_{t1}}{k_1 + k_2}(x_k - x_{y1}) + \frac{k_1 k_2}{k_1 + k_2} x_{y2} + \frac{k_2 k_{t1}}{k_1 + k_2}(x_k - x_{y2}).$$

The sum of the spring forces in Eqs. 7.380 and 7.381 represents the total force of the system, that is,

$$F_{s,k+1} = F_{s1,k+1} + F_{s2,k+1}. \tag{7.382}$$

Substituting Eqs. 7.380 and 7.381 into Eq. 7.382 gives

$$
\begin{aligned}
F_{s,k+1} &= F_{s1,k+1} + F_{s2,k+1} \\
&= \left[\frac{k_1 k_2}{k_1 + k_2} x_{y1} + \frac{k_2 k_{t1}}{k_1 + k_2} \left(x_k - x_{y1} \right) - \frac{k_1 k_2}{k_1 + k_2} x_{y2} - \frac{k_1 k_{t2}}{k_1 + k_2} \left(x_k - x_{y2} \right) \right] \\
&\quad + \left[-\frac{k_1 k_2}{k_1 + k_2} x_{y1} - \frac{k_2 k_{t1}}{k_1 + k_2} \left(x_k - x_{y1} \right) + \frac{k_1 k_2}{k_1 + k_2} x_{y2} + \frac{k_2 k_{t1}}{k_1 + k_2} \left(x_k - x_{y2} \right) \right] \\
&= 0
\end{aligned}
\tag{7.383}
$$

which is the desired result as given in Eq. 7.374. Finally, since the system is in the unloading stage, the inelastic displacement remains the same:

$$x''_{k+1} = x''_k = \left(\frac{k_1 + k_2 - k_{t1} - k_{t2}}{k_1 + k_2} \right) x_k + \left(\frac{k_{t1} - k_1}{k_1 + k_2} \right) x_{y1} + \left(\frac{k_2 - k_{t2}}{k_1 + k_2} \right) x_{y2} \tag{7.384}$$

$$x''_{1,k+1} = x''_{1,k} = \left(1 - \frac{k_{t1}}{k_1} \right) \left(x_k - x_{y1} \right), \quad x''_{2,k+1} = x''_{2,k} = \left(1 - \frac{k_{t2}}{k_2} \right) \left(x_k - x_{y2} \right). \tag{7.385}$$

Note in Figure 7.42 that because $x_{y2} < x_{y1}$, the inelastic displacement in Spring 2 will be greater than that in Spring 1 (i.e., $x''_2 > x''_1$). This shows that Spring 2 will elongate more than Spring 1.

   This example shows that when an inelastic displacement exists in the system, it induces internal force within the system. This is seen in Eqs. 7.380 and 7.381, where the two springs experience equal and opposite forces. This is because the inelastic displacement of Spring 1 is unequal to that in Spring 2 (see Figure 7.43a). To retain compatibility (i.e., same total displacement, x_k and x_{k+1}), Spring 1 must be stretched whereas Spring 2 must be compressed, as shown in Figure 7.43b. The result is a spring acting on the other spring, inducing equal and opposite forces in each. This occurs even when there is no external force applied on the system, as given in Eq. 7.383.

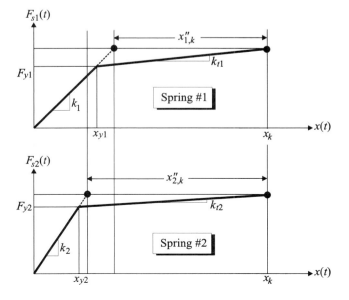

Figure 7.42 Comparison of Inelastic Displacements between Two Springs

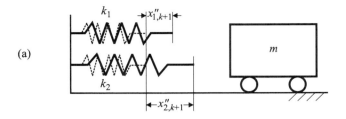

(a)

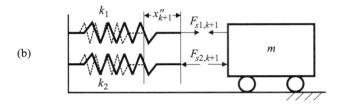

(b)

Figure 7.43 Equilibrium and Compatibility Conditions After Unloading

7.5 SINGLE DEGREE OF FREEDOM RESPONSE USING THE FORCE ANALOGY METHOD

The basic concepts of the force analogy method discussed in Section 7.4 for a spring-mass system are now extended to calculate the response of an SDOF structure. As a starting point, consider that the total displacement of the structure can be written as

$$x(t) = x'(t) + x''(t) \tag{7.386}$$

where $x(t)$ represents the total displacement, $x'(t)$ is the elastic displacement, and $x''(t)$ is the inelastic displacement. Recall that, as discussed in Section 7.4, the inelastic displacement is the portion of the total displacement that is beyond the initially linear portion of the curve. The inelastic displacement of a structure is the result of the inelastic deformations of the structural members. This section uses an example of a cantilever beam member in which *plastic rotation* of the structural members is caused by a moment demand greater than the yield moment. For this example, consider the SDOF system shown in Figure 7.44a, where the total moment at the base of the structure is expressed as

$$M(t) = M'(t) + M''(t) \tag{7.387}$$

where $M(t)$ is the total moment, $M'(t)$ is the elastic moment due to elastic displacement, and $M''(t)$ is the inelastic moment due to inelastic displacement. The relationship between the applied moment and the elastic displacement in the elastic response domain can be expressed as

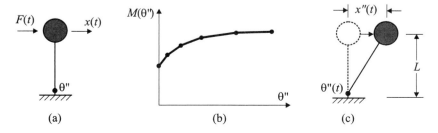

Figure 7.44 Single Degree of Freedom Inelastic Structure

$$M'(t) = L F(t) = L\left[\left(\frac{3EI}{L^3}\right)x'(t)\right] = \left(\frac{3EI}{L^2}\right)x'(t) \tag{7.388}$$

where E and I are the modulus of elasticity and the moment of inertia of the structural member.

To determine the inelastic moment, a relationship between the inelastic moment and the inelastic displacement must be developed. Consider the SDOF system (see Figure 7.44a) where the moment versus rotation relationship of the column is assumed to be multilinear, as shown in Figure 7.44b. A plastic rotation, θ'', at the base of the structure (see Figure 7.44c) causes a permanent deformation at the top of the structure, noted as

$$x''(t) = L\,\theta''(t). \tag{7.389}$$

To represent this permanent deformation as a force, imagine that a force is applied to restore the structural displacement to its original position. The restoring force that is applied (see Figure 7.45b) is

$$F_{RF}(t) = -\left(\frac{3EI}{L^3}\right)x''(t) = -\left(\frac{3EI}{L^2}\right)\theta''(t) \tag{7.390a}$$

$$M_{RF}(t) = L\,F_{RF}(t) = -\left(\frac{3EI}{L^2}\right)x''(t) = -\left(\frac{3EI}{L}\right)\theta''(t). \tag{7.390b}$$

The forces in Eq. 7.390 are not externally applied to the structure, and therefore equal and opposite forces must be applied to the structure. The force $M_{RF}(t)$ is applied at the base of the structure and thus can be ignored. The force $F_{RF}(t)$ is applied corresponding to the structure's degree of freedom, and therefore it must be compensated for by applying an equal and opposite force, denoted $F_a(t)$:

$$F_a(t) = -F_{RF}(t) = \left(\frac{3EI}{L^2}\right)\theta''(t). \tag{7.391}$$

When the force $F_a(t)$ is applied, the displacement at the top of the structure, $x''(t)$, is equal to

$$x''(t) = \left(\frac{L^3}{3EI}\right)F_a(t). \tag{7.392}$$

Substituting Eq. 7.391 into Eq. 7.392 gives Eq. 7.389. This relationship between the plastic rotations and the inelastic displacement is very important in the force analogy method, and the derivation from Eqs. 7.390 to 7.392 is essential in Section 7.6 for MDOF systems. In addition, applying the force $F_a(t)$ causes an additional moment at the base of

$$M_p(t) = L\,F_a(t) = \left(\frac{3EI}{L}\right)\theta''(t) \tag{7.393}$$

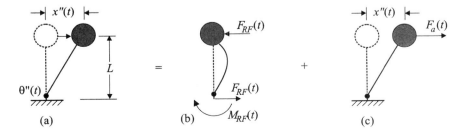

Figure 7.45 Force Analogy Method

where $M_p(t)$ is the moment at the base due to $F_a(t)$. Therefore, the inelastic moment due to the plastic rotation is the sum of Eqs. 7.390b and 7.393, which is

$$M''(t) = M_{RF}(t) + M_p(t) = -\left(\frac{3EI}{L}\right)\theta''(t) + \left(\frac{3EI}{L}\right)\theta''(t) = 0. \qquad (7.394)$$

Equation 7.394 shows that for the special case of an SDOF system with only one *plastic hinge location* (PHL), the inelastic moment due to plastic rotation is zero.

Finally, substituting Eqs. 7.386, 7.388, and 7.394 into Eq. 7.387, it follows that

$$M(t) = M'(t) + M''(t) = \left(\frac{3EI}{L^2}\right)x'(t) + 0 = \left(\frac{3EI}{L^2}\right)[x(t) - x''(t)]. \qquad (7.395)$$

Substituting Eq. 7.389 into Eq. 7.395 and rearranging terms gives

$$M(t) + \left(\frac{3EI}{L}\right)\theta''(t) = \left(\frac{3EI}{L^2}\right)x(t) \qquad (7.396)$$

which is the governing equation for inelastic analysis using the force analogy method.

EXAMPLE 1 *Elastic–Plastic Moment-Rotation Relationship*

Consider the SDOF system shown in Figure 7.44 with an elastic–plastic moment versus plastic rotation relationship, as shown in Figure 7.46. Denote the moment capacity of the member M_c. Then for any value of $x(t)$, Eq. 7.396 can be solved by first assuming that no plastic rotation exists:

$$M(t) = \left(\frac{3EI}{L^2}\right)x(t). \qquad (7.397)$$

Once the value of $M(t)$ is obtained, it is then compared with M_c. If $M(t)$ is less than M_c, then the assumption is correct and no plastic rotation exists. However, if $M(t)$ is greater than M_c, then $M(t) = M_c$. Because the material is elastic–plastic, Eq. 7.396 is again used to calculate the plastic rotation, that is,

$$M_c + \left(\frac{3EI}{L}\right)\theta''(t) = \left(\frac{3EI}{L^2}\right)x(t) \qquad (7.398)$$

and therefore

$$\theta''(t) = \left(\frac{1}{L}\right)x(t) - \left(\frac{L}{3EI}\right)M_c. \qquad (7.399)$$

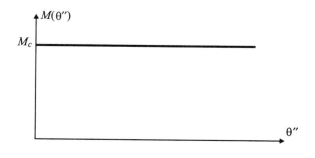

Figure 7.46 Elastic–Plastic Moment versus Plastic Rotation Relationship

Denote

$$x_c = \left(\frac{L^2}{3EI}\right)M_c \qquad (7.400)$$

and then in summary,

$$M(t) = \begin{cases} \left(3EI/L^2\right)x(t) & x(t) \le x_c \\ M_c & x(t) > x_c, \end{cases} \qquad \theta''(t) = \begin{cases} 0 & x(t) \le x_c \\ (1/L)x(t) - (L/3EI)M_c & x(t) > x_c. \end{cases} \qquad (7.401)$$

The force versus deflection curve of the structure is

$$F(t) = \frac{1}{L}M(t) = \begin{cases} \left(3EI/L^2\right)x(t) & x(t) \le x_c \\ M_c/L & x(t) > x_c. \end{cases} \qquad (7.402)$$

A plot of the force versus deflection curve is shown in Figure 7.47.
Finally, the inelastic displacement of the structure is

$$x''(t) = L\theta''(t) = \begin{cases} 0 & x(t) \le x_c \\ x(t) - \left[L^2/(3EI)\right]M_c & x(t) > x_c. \end{cases} \qquad (7.403)$$

Note that $[L^2/(3EI)]M_c$ is the elastic displacement $x'(t)$, and therefore the inelastic displacement is total displacement minus elastic displacement.

EXAMPLE 2   *Strain Hardening Moment versus Plastic Rotation Relationship*

Consider the SDOF system with a linear strain hardening moment versus plastic rotation relationship of slope α, as shown in Figure 7.48. Again, first assume that no plastic rotation exists, and therefore it follows from Eq. 7.396 that

$$M(t) = \left(\frac{3EI}{L^2}\right)x(t). \qquad (7.404)$$

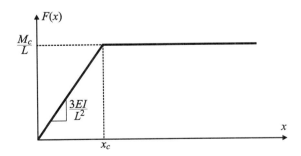

Figure 7.47 Force versus Displacement Curve

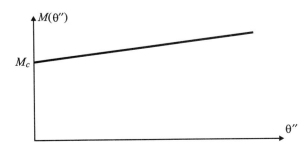

Figure 7.48 Linear Strain Hardening Moment versus Plastic Rotation Relationship

Once the value of $M(t)$ is obtained, it is then compared with M_c. If $M(t)$ is less than M_c, then the assumption that no plastic rotation exists is valid. However, if $M(t)$ is greater than M_c, then $M(t) = M_c + \alpha\theta''(t)$, and Eq. 7.396 is again used to calculate the plastic rotation:

$$M_c + \alpha\theta''(t) + \left(\frac{3EI}{L}\right)\theta''(t) = \left(\frac{3EI}{L^2}\right)x(t). \tag{7.405}$$

Solving for $\theta''(t)$ gives

$$\theta''(t) = \left(\frac{3EI}{L(3EI + \alpha L)}\right)x(t) - \left(\frac{L}{3EI + \alpha L}\right)M_c \tag{7.406}$$

and therefore the moment is

$$M(t) = M_c + \alpha\theta''(t) = \frac{3EI\alpha}{L(3EI + \alpha L)}x(t) + \left(\frac{3EI}{3EI + \alpha L}\right)M_c. \tag{7.407}$$

Denote

$$x_c = \left(\frac{L^2}{3EI}\right)M_c. \tag{7.408}$$

Then in summary,

$$M(t) = \begin{cases} \frac{3EI}{L^2}x(t) & x \le x_c \\ \frac{3EI\alpha}{L(3EI+\alpha L)}x(t) + \frac{3EI}{3EI+\alpha L}M_c & x > x_c \end{cases} \tag{7.409}$$

$$\theta''(t) = \begin{cases} 0 & x \le x_c \\ \frac{3EI}{L(3EI+\alpha L)}x(t) - \frac{L}{3EI+\alpha L}M_c & x > x_c. \end{cases} \tag{7.410}$$

The force versus deflection curve of the structure is

$$F(t) = \frac{1}{L}M(t) = \begin{cases} \frac{3EI}{L^3}x(t) & x \le x_c \\ \frac{3EI\alpha}{L^2(3EI+\alpha L)}x(t) + \frac{3EI}{L(3EI+\alpha L)}M_c & x > x_c. \end{cases} \tag{7.411}$$

A plot of the force versus deflection curve is shown in Figure 7.49. In the derivation the value of α is not specified, and therefore it can be a negative value, which represents strain softening. The limiting case is when $\alpha = -3EI/L$, in which case the denominator of Eq. 7.406 becomes zero; therefore, once the elastic limit is reached, the structure will collapse. Finally, the inelastic displacement of the structure is

$$x''(t) = L\theta''(t) = \begin{cases} 0 & x \le x_c \\ \frac{3EI}{3EI+\alpha L}x(t) - \frac{L^2}{3EI+\alpha L}M_c & x > x_c. \end{cases} \tag{7.412}$$

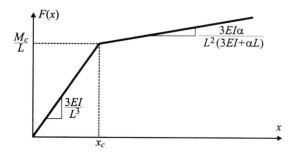

Figure 7.49 Force versus Displacement Curve

EXAMPLE 3   *SDOF System with Braced Frame*

Consider a one-story braced frame as shown in Figure 7.50a. Using simple geometry (see Figure 7.50b), the relationship between the plastic elongation of the brace and the inelastic displacement is

$$x''(t) = \left(\frac{1}{\cos\theta}\right)\delta''(t). \tag{7.413}$$

The plastic elongation requires a compressive force, $P_{RF}(t)$, to restore the brace to its original position (see Figure 7.50c). The relationship is

$$P_{RF}(t) = -\left(\frac{AE}{L}\right)\delta''(t). \tag{7.414}$$

This restoring force, $P_{RF}(t)$, is physically not present and therefore must be removed by applying equal and opposite force. The columns are assumed to be axially rigid, and therefore the vertical component of $P_{RF}(t)$ will transmit directly to the base of the structure. The horizontal component, however, must be removed by applying $F_a(t)$, an equal and opposite force, as shown in Figure 7.50d:

$$F_a(t) = -(\cos\theta)P_{RF}(t) = \left(\frac{AE\cos\theta}{L}\right)\delta''(t). \tag{7.415}$$

Applying this force on the structure gives

$$x''(t) = \left(\frac{L}{AE\cos^2\theta}\right)F_a(t) \tag{7.416}$$

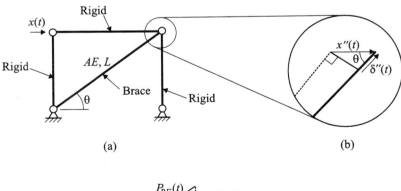

(a)   (b)

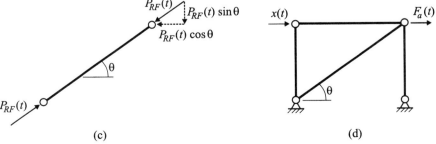

(c)   (d)

Figure 7.50 One-Story Braced Frame and Force Analogy Method

$$P_p(t) = \left(\frac{1}{\cos\theta}\right)F_a(t).$$ (7.417)

where $P_p(t)$ is the force in the brace due to $F_a(t)$. Substituting Eq. 7.415 into Eq. 7.416 gives

$$x''(t) = \left(\frac{L}{AE\cos^2\theta}\right)F_a(t) = \left(\frac{1}{\cos\theta}\right)\delta''(t)$$ (7.418)

which is the same as Eq. 7.413. Also, substituting Eq. 7.415 into Eq. 7.417 gives

$$P_p(t) = \left(\frac{1}{\cos\theta}\right)F_a(t) = \left(\frac{AE}{L}\right)\delta''(t)$$ (7.419)

Therefore, it follows from Eqs. 7.414 and 7.419 that

$$P''(t) = P_{RF}(t) + P_p(t) = -\left(\frac{1}{\cos\theta}\right)F_a(t) + \left(\frac{1}{\cos\theta}\right)F_a(t) = 0.$$ (7.420)

Finally, the total force in the truss member is

$$P(t) = P'(t) + P''(t) = \left(\frac{AE\cos\theta}{L}\right)x'(t) + 0 = \left(\frac{AE\cos\theta}{L}\right)[x(t) - x''(t)].$$ (7.421)

Substituting Eq. 7.413 into Eq. 7.421 and rearranging terms gives

$$P(t) + \left(\frac{AE}{L}\right)\delta''(t) = \left(\frac{AE\cos\theta}{L}\right)x(t)$$ (7.422)

which is the governing equation for inelastic analysis using the force analogy method.

EXAMPLE 4   *Force Analogy Method on Axial Members*

Consider the axial member as shown in Figure 7.51a, with the stress versus strain relationship as shown in Figure 7.51b. Note that α in this case is different from Example 2, in
that α here contains both elastic and inelastic strains, that is,

$$\Delta\sigma(t) = \alpha E\Delta\varepsilon(t) = E\Delta\varepsilon'(t)$$ (7.423)

where $\sigma(t)$ is the stress of the member, $\varepsilon(t)$ is the total strain, $\varepsilon'(t)$ is the elastic strain, $\varepsilon''(t)$
is the inelastic strain, and $\Delta(\cdot)$ denotes the increment change of the variable. Recall that
$\varepsilon'(t) = \varepsilon(t) - \varepsilon''(t)$. Substituting this into Eq. 7.423 gives

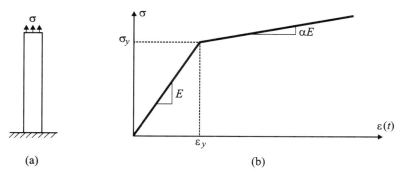

(a)   (b)

Figure 7.51 Stress versus Strain Relationship of an Axial Member

$$\alpha E \Delta\varepsilon(t) = E\Delta\varepsilon(t) - E\Delta\varepsilon''(t). \tag{7.424}$$

Solving for $\Delta\varepsilon''(t)$, it follows that

$$\Delta\varepsilon''(t) = (1-\alpha)\Delta\varepsilon(t). \tag{7.425}$$

Similarly, it follows from Eq. 7.423 that

$$\Delta\varepsilon'(t) = \alpha\Delta\varepsilon(t). \tag{7.426}$$

Equations 7.425 and 7.426 show that after the member yields, the incremental total strain contains α portion that is elastic and $(1-\alpha)$ portion that is inelastic.

The governing equation for this case is

$$\sigma(t) = E\varepsilon'(t) = E\big[\varepsilon(t) - \varepsilon''(t)\big]. \tag{7.427}$$

Rearranging terms gives

$$\sigma(t) + E\varepsilon''(t) = E\varepsilon(t). \tag{7.428}$$

Denote the yield strain by ε_y. For any $\varepsilon(t) \le \varepsilon_y$, Eq. 7.428 can be used by ignoring the term containing $\varepsilon''(t)$ because the member has not yielded. In this case, it follows that

$$\sigma(t) = E\varepsilon(t) \tag{7.429}$$

which is the classic elastic stress versus strain relationship. For $\varepsilon(t) > \varepsilon_y$, Eq. 7.424 can be used by setting

$$\sigma(t) = E\varepsilon_y + \Delta\sigma(t). \tag{7.430}$$

Substituting Eq. 7.423 into Eq. 7.430 gives

$$\sigma(t) = E\varepsilon_y + \alpha E\big[\varepsilon(t) - \varepsilon_y\big]. \tag{7.431}$$

Substituting Eq. 7.431 into Eq. 7.428, it follows that

$$E\varepsilon_y + \alpha E\big[\varepsilon(t) - \varepsilon_y\big] + E\varepsilon''(t) = E\varepsilon(t). \tag{7.432}$$

Solving for $\varepsilon''(t)$ gives

$$\varepsilon''(t) = (1-\alpha)\big[\varepsilon(t) - \varepsilon_y\big]. \tag{7.433}$$

Note that Eq. 7.433 is the same as Eq. 7.425, since $\varepsilon''(t) = \Delta\varepsilon''(t)$ and $\Delta\varepsilon(t) = \varepsilon(t) - \varepsilon_y$.

EXAMPLE 5 *Incremental Analysis*

Consider Example 2 with a strain hardening moment versus plastic rotation relationship of slope α. Assume that Eq. 7.396 is satisfied at time step k, that is,

$$M_k + \left(\frac{3EI}{L}\right)\theta_k'' = \left(\frac{3EI}{L^2}\right)x_k \tag{7.434}$$

with M_k, θ_k'', and x_k all known and the structure in the inelastic range (i.e., $\theta_k'' > 0$). Now from a time history analysis the displacement at time step $k + 1$ is calculated, where $x_{k+1} > x_k$, and the response increases and remains in the inelastic response domain. The objective is to compute M_{k+1} and θ_{k+1}'' using the equation

$$M_{k+1} + \left(\frac{3EI}{L}\right)\theta_{k+1}'' = \left(\frac{3EI}{L^2}\right)x_{k+1}. \tag{7.435}$$

To do so, first set

$$M_{k+1} = M_k + \Delta M = M_k + \alpha \Delta \theta'' \tag{7.436}$$

$$\theta''_{k+1} = \theta''_k + \Delta \theta''. \tag{7.437}$$

Then substituting Eqs. 7.436 and 7.437 into Eq. 7.435 gives

$$\left[M_k + \alpha \Delta \theta'' \right] + \frac{3EI}{L} \left[\theta''_k + \Delta \theta'' \right] = \frac{3EI}{L^2} x_{k+1}. \tag{7.438}$$

Solving for $\Delta \theta''$ gives

$$\Delta \theta'' = \left(\frac{3EI}{L(3EI + \alpha L)} \right) x_{k+1} - \left(\frac{L}{3EI + \alpha L} \right) M_k - \left(\frac{3EI}{3EI + \alpha L} \right) \theta''_k. \tag{7.439}$$

Finally, substituting Eq. 7.439 into Eqs. 7.436 and 7.437 gives

$$M_{k+1} = \left(\frac{3EI\alpha}{L(3EI + \alpha L)} \right) x_{k+1} + \left(\frac{3EI}{3EI + \alpha L} \right) M_k - \left(\frac{3EI\alpha}{3EI + \alpha L} \right) \theta''_k \tag{7.440}$$

$$\theta''_{k+1} = \left(\frac{3EI}{L(3EI + \alpha L)} \right) x_{k+1} - \left(\frac{L}{3EI + \alpha L} \right) M_k + \left(\frac{\alpha L}{3EI + \alpha L} \right) \theta''_k. \tag{7.441}$$

Now, assume that the displacement at time step $k + 1$ is obtained and $x_{k+1} < x_k$; therefore, the structure responds in the descending branch of the force versus deflection curve, which is in the elastic range. Thus $\Delta \theta'' = 0$, and it follows from Eq. 7.437 that

$$\theta''_{k+1} = \theta''_k. \tag{7.442}$$

Substituting Eq. 7.442 into Eq. 7.435 gives

$$M_{k+1} + \left(\frac{3EI}{L} \right) \theta''_k = \left(\frac{3EI}{L^2} \right) x_{k+1}. \tag{7.443}$$

Solving for M_{k+1}, it follows from Eq. 7.443 that

$$M_{k+1} = \left(\frac{3EI}{L^2} \right) x_{k+1} - \left(\frac{3EI}{L} \right) \theta''_k. \tag{7.444}$$

7.6   MULTI–DEGREE OF FREEDOM RESPONSE USING THE FORCE ANALOGY METHOD

For an n-DOF system, the inelastic displacements can be written in vector form as

$$\mathbf{X}(t) = \mathbf{X}'(t) + \mathbf{X}''(t) = \begin{Bmatrix} x'_1(t) \\ x'_2(t) \\ \vdots \\ x'_n(t) \end{Bmatrix} + \begin{Bmatrix} x''_1(t) \\ x''_2(t) \\ \vdots \\ x''_n(t) \end{Bmatrix} \tag{7.445}$$

where $\mathbf{X}(t)$ represents the total displacement vector, $\mathbf{X}'(t)$ is the elastic displacement vector, and $\mathbf{X}''(t)$ is the inelastic displacement vector.

The force in each member of the structure must be calculated to determine whether the response of the structure is in the elastic or inelastic response domain. The moments at the two ends of the members are critical because these moments due to an earthquake often exceed the moment capacity. At the locations where the moment exceeds the yield moment, yielding occurs and these locations are called *plastic hinge locations* (PHLs).

The total moments $\mathbf{M}(t)$ at the location where plastic hinges may form can be expressed as

$$\mathbf{M}(t) = \mathbf{M}'(t) + \mathbf{M}''(t) = \begin{Bmatrix} M_1'(t) \\ M_2'(t) \\ \vdots \\ M_m'(t) \end{Bmatrix} + \begin{Bmatrix} M_1''(t) \\ M_2''(t) \\ \vdots \\ M_m''(t) \end{Bmatrix} \qquad (7.446)$$

where $\mathbf{M}'(t)$ is the elastic moment vector due to elastic displacement, $\mathbf{M}''(t)$ is the residual moment vector due to inelastic displacement, and m is the total number of potential PHLs.

Consider the residual moment vector, $\mathbf{M}''(t)$. As noted in Section 7.5, the method of calculating residual moments is known as the force analogy method. When plastic rotations occur in the structure, the plastic rotations at these locations are replaced with a fictitious force. By applying this fictitious force to the structure it is possible to calculate the deformed state of the structure. To illustrate the force analogy method, let the plastic rotation vector, $\mathbf{\Theta}''(t)$, be defined as

$$\mathbf{\Theta}''(t) = \begin{Bmatrix} \theta_1''(t) \\ \theta_2''(t) \\ \vdots \\ \theta_m''(t) \end{Bmatrix}. \qquad (7.447)$$

The sign convention used in the force analogy method is shown in Figure 7.52. Figure 7.52a shows a joint with two beams and two columns. When a positive moment is applied to this joint, the beams and columns react in the opposite direction as shown Figure 7.52b. When these reactions become larger than the positive yield moments, "positive" plastic rotations form, as shown in Figure 7.52c.

An example of the PHLs is shown in Figure 7.53a. This form of the structure can never exist because it violates the compatibility condition. Thus, to make the structure deform in a compatible way, these members are first isolated and restoring forces are applied to bring these members back to the original undeformed shapes, as shown in Figure 7.53b. This induces internal restoring forces within these members. At the location of the structure's degrees of freedom, the restoring forces are

$$\mathbf{F}_{RF}(t) = \begin{Bmatrix} F_{RF1}(t) \\ F_{RF2}(t) \\ \vdots \\ F_{RFn}(t) \end{Bmatrix} = -\mathbf{K}_p \mathbf{\Theta}''(t) \qquad (7.448)$$

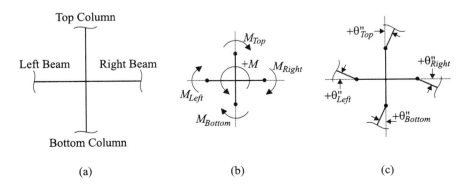

Figure 7.52 Sign Convention of Plastic Rotation

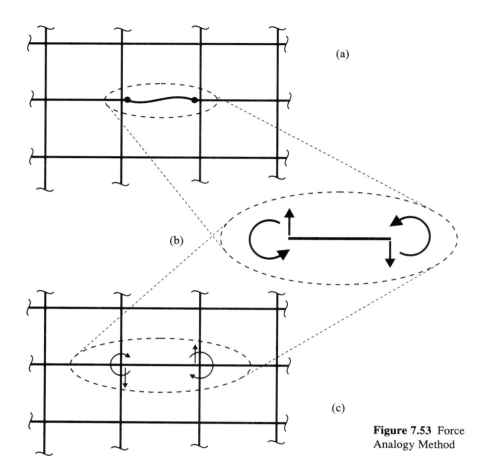

Figure 7.53 Force Analogy Method

where $\mathbf{K}_p$ is the *member force recovery matrix,* which defines the force that must be applied at each structural degree of freedom to produce a structure with compatible deformation. Similarly, at the location of the potential plastic hinges, the restoring moments due to the plastic rotation are

$$\mathbf{M}_{RF}(t) = \begin{Bmatrix} M_{RF1}(t) \\ M_{RF2}(t) \\ \vdots \\ M_{RFm}(t) \end{Bmatrix} = -\mathbf{K}_R \mathbf{\Theta}''(t) \qquad (7.449)$$

where $\mathbf{K}_R$ is the *member restoring force matrix.* In order to rotate a positive plastic rotation to zero, negative restoring forces must be applied, and this is why minus signs appear in Eqs. 7.448 and 7.449. Now the members are assembled and the displacements are compatible. However, the restoring forces as shown in Figure 7.53c are actually not present, and as a result equal and opposite forces must be applied to the structure. These forces are described in the equation

$$\mathbf{F}_a(t) = -\mathbf{F}_{RF}(t) = \mathbf{K}_p \mathbf{\Theta}''(t) = \mathbf{K}\mathbf{X}''(t) \qquad (7.450)$$

where $\mathbf{K}$ is the structure stiffness matrix. Through this process, the structure remains in equilibrium. Solving for the inelastic displacement vector in Eq. 7.450, it follows that

$$\mathbf{X}''(t) = \mathbf{K}^{-1}\mathbf{K}_p \mathbf{\Theta}''(t). \qquad (7.451)$$

The moments at the potential plastic hinge locations, $\mathbf{M}_p(t)$, as shown in Figure 7.53c, which are due to $\mathbf{F}_a(t)$, are related to the inelastic displacement field by the equation

$$\mathbf{M}_p(t) = \mathbf{K}_p^T \mathbf{X}''(t) = \mathbf{K}_p^T \mathbf{K}^{-1} \mathbf{K}_p \mathbf{\Theta}''(t). \tag{7.452}$$

The reason for the transpose of $\mathbf{K}_p$ in Eq. 7.452 can be seen by using the Maxwell reciprocal theorem, in which

$$\begin{Bmatrix} M_1 \\ M_2 \\ \vdots \\ M_m \end{Bmatrix} = \begin{bmatrix} k_{11} & k_{12} & \cdots & k_{1n} \\ k_{21} & k_{22} & \cdots & k_{2n} \\ \vdots & \vdots & \ddots & \vdots \\ k_{m1} & k_{m2} & \cdots & k_{mn} \end{bmatrix} \begin{Bmatrix} x_1 \\ x_2 \\ \vdots \\ x_n \end{Bmatrix} \quad \text{and} \quad \begin{Bmatrix} F_1 \\ F_2 \\ \vdots \\ F_n \end{Bmatrix} = \begin{bmatrix} k_{11} & k_{21} & \cdots & k_{m1} \\ k_{12} & k_{22} & \cdots & k_{m2} \\ \vdots & \vdots & \ddots & \vdots \\ k_{1n} & k_{2n} & \cdots & k_{mn} \end{bmatrix} \begin{Bmatrix} \theta_1 \\ \theta_2 \\ \vdots \\ \theta_m \end{Bmatrix}$$

where the left-hand side of the equation comes from Eq. 7.452 and the right-hand side of the equation comes from Eq. 7.450. Note that after performing the stiffness method of structural analysis, the displacements are obtained at all degrees of freedom. All the forces in the members are then recovered from the displacement field using the matrix $\mathbf{K}_p$.

Finally, because the restoring forces initially applied to the member remain on that member, these forces must be considered in the residual moment equation

$$\mathbf{M}''(t) = \mathbf{M}_{RF}(t) + \mathbf{M}_p(t) = -\left(\mathbf{K}_R - \mathbf{K}_p^T \mathbf{K}^{-1} \mathbf{K}_p\right) \mathbf{\Theta}''(t). \tag{7.453}$$

Equation 7.453 is the equation for the residual moment vector, and the moments are due to the plastic rotations within the structure with no external applied force. For example, if the earthquake causes these plastic rotations within the structure, then the residual moments are the forces remaining in the members after the earthquake.

The elastic moments are calculated using the member force recovery matrix to recover all the elastic moments, and this is done using the equation

$$\mathbf{M}'(t) = \mathbf{K}_p^T \mathbf{X}'(t). \tag{7.454}$$

If the total displacement vector is given instead of the elastic displacement, then the following extra step is necessary:

$$\mathbf{M}'(t) = \mathbf{K}_p^T \mathbf{X}'(t) = \mathbf{K}_p^T \left[\mathbf{X}(t) - \mathbf{X}''(t)\right]. \tag{7.455}$$

The elastic moments that are due to the inelastic displacement vector can be computed in a manner similar to the computation of residual moments. From Eq. 7.451, it follows that the moments at the locations of potential plastic hinges are

$$\mathbf{M}'(t) = \mathbf{K}_p^T \left[\mathbf{X}(t) - \mathbf{X}''(t)\right] = \mathbf{K}_p^T \left[\mathbf{X}(t) - \mathbf{K}^{-1} \mathbf{K}_p \mathbf{\Theta}''(t)\right]. \tag{7.456}$$

After the elastic and residual moments are obtained, the total moments at all the potential plastic hinge locations can be calculated by substituting Eqs. 7.453 and 7.455 into Eq. 7.446. Therefore,

$$\begin{aligned} \mathbf{M}(t) &= \mathbf{M}'(t) + \mathbf{M}''(t) \\ &= \mathbf{K}_p^T \left[\mathbf{X}(t) - \mathbf{K}^{-1} \mathbf{K}_p \mathbf{\Theta}''(t)\right] - \left[\mathbf{K}_R - \mathbf{K}_p^T \mathbf{K}^{-1} \mathbf{K}_p\right] \mathbf{\Theta}''(t) \\ &= \mathbf{K}_p^T \mathbf{X}(t) - \mathbf{K}_R \mathbf{\Theta}''(t). \end{aligned} \tag{7.457}$$

Rearranging the terms in Eq. 7.457, it follows that

$$\mathbf{M}(t) + \mathbf{K}_R \mathbf{\Theta}''(t) = \mathbf{K}_p^T \mathbf{X}(t). \tag{7.458}$$

Figure 7.54, which is a graphic representation of Eq. 7.458, shows a typical moment at a location for different level of displacements in the structure. A similar graphic representation can be obtained for the case of strain softening, where an increase of rotation decreases the moment.

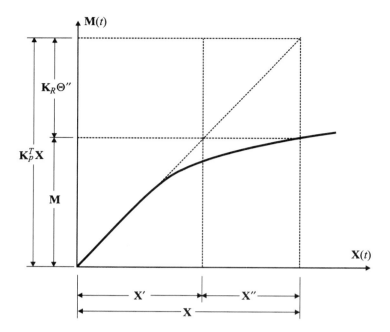

Figure 7.54 Moment at Different Levels of Structural Displacements

In Eq. 7.458, given the total displacement vector, there are m unknowns in the total moment vector as well as m unknowns in the plastic rotation vector, giving a total of $2m$ unknowns. This matrix equation gives m independent equations, together with m moment versus rotation relationships (e.g., Figure 7.44b), giving a total of $2m$ equations. This procedure is very useful in computing all the moments and plastic rotations in the members when the total displacement vector is given.

The inelastic displacement vector can be calculated using Eq. 7.451 and the plastic rotations computed in Eq. 7.458:

$$\mathbf{X}''(t) = \mathbf{K}^{-1}\mathbf{K}_p\mathbf{\Theta}''(t). \tag{7.459}$$

This inelastic displacement vector is the portion of the total displacement that exists after unloading.

EXAMPLE 1   *Two-Member Frame*

Consider the two-element frame shown in Figure 7.55a. There are four potential plastic hinge locations and three degrees of freedom. The global stiffness matrix of the structure (i.e., $\mathbf{K}$), which relates the applied forces at all three degrees of freedom with the displacements, is

$$\mathbf{K}: \quad \begin{Bmatrix} F_1(t) \\ F_2(t) \\ F_3(t) \end{Bmatrix} = \begin{bmatrix} \dfrac{12EI_1}{L_1^3} & 0 & \dfrac{6EI_1}{L_1^2} \\ 0 & \dfrac{AE}{L_1} + \dfrac{12EI_2}{L_2^3} & \dfrac{6EI_2}{L_2^2} \\ \dfrac{6EI_1}{L_1^2} & \dfrac{6EI_2}{L_2^2} & \dfrac{4EI_1}{L_1} + \dfrac{4EI_2}{L_2} \end{bmatrix} \begin{Bmatrix} x_1(t) \\ x_2(t) \\ x_3(t) \end{Bmatrix}. \tag{7.460}$$

The $\mathbf{K}_p$ matrix relates the applied forces at the three degrees of freedom with the plastic rotations at the four potential plastic hinge locations:

$$\mathbf{K}_p : \begin{Bmatrix} F_{RF1}(t) \\ F_{RF2}(t) \\ F_{RF3}(t) \end{Bmatrix} = - \begin{bmatrix} \dfrac{6EI_1}{L_1^2} & \dfrac{6EI_1}{L_1^2} & 0 & 0 \\[2mm] 0 & 0 & \dfrac{6EI_2}{L_2^2} & \dfrac{6EI_2}{L_2^2} \\[2mm] \dfrac{2EI_1}{L_1} & \dfrac{4EI_1}{L_1} & \dfrac{4EI_2}{L_2} & \dfrac{2EI_2}{L_2} \end{bmatrix} \begin{Bmatrix} \theta_1''(t) \\ \theta_2''(t) \\ \theta_3''(t) \\ \theta_4''(t) \end{Bmatrix}. \tag{7.461}$$

These plastic rotations are treated very similarly to the other rotations. For example, a unit plastic rotation of θ_1 induces a force F_1 and a moment F_3, as shown in Figure 7.55b. Similarly, unit plastic rotations in θ_2, θ_3, and θ_4 induce forces as shown in Figures 7.55c, d, and e, respectively. These forces are then entered into the $\mathbf{K}_p$ matrix.

The $\mathbf{K}_R$ matrix relates the member moments and the plastic rotations:

$$\mathbf{K}_R : \begin{Bmatrix} M_{RF1}(t) \\ M_{RF2}(t) \\ M_{RF3}(t) \\ M_{RF4}(t) \end{Bmatrix} = - \begin{bmatrix} \dfrac{4EI_1}{L_1} & \dfrac{2EI_1}{L_1} & 0 & 0 \\[2mm] \dfrac{2EI_1}{L_1} & \dfrac{4EI_1}{L_1} & 0 & 0 \\[2mm] 0 & 0 & \dfrac{4EI_2}{L_2} & \dfrac{2EI_2}{L_2} \\[2mm] 0 & 0 & \dfrac{2EI_2}{L_2} & \dfrac{4EI_2}{L_2} \end{bmatrix} \begin{Bmatrix} \theta_1''(t) \\ \theta_2''(t) \\ \theta_3''(t) \\ \theta_4''(t) \end{Bmatrix}. \tag{7.462}$$

The matrix in Eq. 7.462 can be calculated in a similar way as the $\mathbf{K}_p$ matrix. For example, a unit plastic rotation at θ_1 induces moments of $M_1 = 4EI_1/L_1$ and $M_2 = 2EI_2/L_2$. The $\mathbf{K}_R$

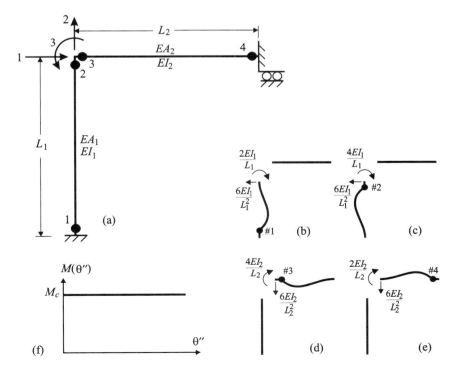

Figure 7.55 Force Analogy Method on a Two-Member Frame

matrix assumes that the plastic hinge has no length and the plastic rotation is concentrated at a point very near the joint.

Using these three matrices as described in Eqs. 7.460 to 7.462, it follows from Eq. 7.458 that

$$
\begin{Bmatrix} M_1(t) \\ M_2(t) \\ M_3(t) \\ M_4(t) \end{Bmatrix} + \begin{bmatrix} \dfrac{4EI_1}{L_1} & \dfrac{2EI_1}{L_1} & 0 & 0 \\ \dfrac{2EI_1}{L_1} & \dfrac{4EI_1}{L_1} & 0 & 0 \\ 0 & 0 & \dfrac{4EI_2}{L_2} & \dfrac{2EI_2}{L_2} \\ 0 & 0 & \dfrac{2EI_2}{L_2} & \dfrac{4EI_2}{L_2} \end{bmatrix} \begin{Bmatrix} \theta_1''(t) \\ \theta_2''(t) \\ \theta_3''(t) \\ \theta_4''(t) \end{Bmatrix} = \begin{bmatrix} \dfrac{6EI_1}{L_1^2} & 0 & \dfrac{2EI_1}{L_1} \\ \dfrac{6EI_1}{L_1^2} & 0 & \dfrac{4EI_1}{L_1} \\ 0 & \dfrac{6EI_2}{L_2^2} & \dfrac{4EI_2}{L_2} \\ 0 & \dfrac{6EI_2}{L_2^2} & \dfrac{2EI_2}{L_2} \end{bmatrix} \begin{Bmatrix} x_1(t) \\ x_2(t) \\ x_3(t) \end{Bmatrix}. \tag{7.463}
$$

As a numerical example, assume that

$$EA_1 = EA_2 = 10,000 \text{ kip}$$

$$EI_1 = EI_2 = 1,000,000 \text{ k-in.}^2$$

$$L_1 = L_2 = 100 \text{ in.}$$

Also, let

$$
\begin{Bmatrix} x_1(t) \\ x_2(t) \\ x_3(t) \end{Bmatrix} = \begin{Bmatrix} 0.393 \text{ in.} \\ -0.011 \text{ in.} \\ -0.008 \text{ rad.} \end{Bmatrix}, \qquad \begin{Bmatrix} M_{c1} \\ M_{c2} \\ M_{c3} \\ M_{c4} \end{Bmatrix} = \begin{Bmatrix} 300.0 \\ 300.0 \\ 300.0 \\ 300.0 \end{Bmatrix} \text{ k-in.}
$$

As shown in Figure 7.55f, an elastic–plastic model is assumed for the moment-rotation relationship of all potential plastic hinges. The solution to this problem is

$$
\begin{Bmatrix} \theta_1''(t) \\ \theta_2''(t) \\ \theta_3''(t) \\ \theta_4''(t) \end{Bmatrix} = \begin{Bmatrix} 0.00 \\ 0.00 \\ 0.000665 \\ 0.00 \end{Bmatrix} \text{rad}, \qquad \begin{Bmatrix} x_1''(t) \\ x_2''(t) \\ x_3''(t) \end{Bmatrix} = \begin{Bmatrix} 0.0261431 \text{ in.} \\ -0.0007615 \text{ in.} \\ -0.0005229 \text{ rad} \end{Bmatrix}
$$

$$
\begin{Bmatrix} M_1(t) \\ M_2(t) \\ M_3(t) \\ M_4(t) \end{Bmatrix} = \begin{Bmatrix} 75.80 \\ -84.20 \\ -300.0 \\ -153.3 \end{Bmatrix} \text{ k-in.} \tag{7.464}
$$

Note that for degree of freedom 1, the total displacement is 0.393 in. The inelastic displacement is 0.026 in. of this total displacement. Thus, only 0.367 in. is elastic.

To check this result, first compute the elastic displacement vector $\mathbf{X}'(t)$. The elastic moments can be calculated from this elastic displacement, and they are

$$
\begin{Bmatrix} x_1'(t) \\ x_2'(t) \\ x_3'(t) \end{Bmatrix} = \begin{Bmatrix} x_1(t) \\ x_2(t) \\ x_3(t) \end{Bmatrix} - \begin{Bmatrix} x_1''(t) \\ x_2''(t) \\ x_3''(t) \end{Bmatrix} = \begin{Bmatrix} 0.36686 \text{ in.} \\ -0.01024 \text{ in.} \\ -0.007477 \text{ rad.} \end{Bmatrix}, \qquad \mathbf{M}'(t) = \mathbf{K}_p\mathbf{X}'(t) = \begin{Bmatrix} 70.576 \\ -78.964 \\ -305.224 \\ -155.684 \end{Bmatrix} \text{ k-in.}
$$

For a unit plastic rotation at potential hinge 3, the force analogy method gives residual moments

$$\mathbf{M}''(\theta_3(t)=1.0)=\begin{Bmatrix} 7,862.6 \\ -7,862.6 \\ 7,862.6 \\ 3,587.8 \end{Bmatrix} \text{k-in.}$$

The residual moments at potential hinges 2 and 3 are equal and opposite in signs, and therefore equilibrium is satisfied at that joint. The plastic rotation at potential hinge 3 is calculated to be 0.00665 rad. Therefore,

$$\mathbf{M}''(t)=\sum_{k=1}^{4}\mathbf{M}''(\theta_k(t)=1.0)\times\theta_k(t)=\begin{Bmatrix} 5.23 \\ -5.23 \\ 5.23 \\ 2.39 \end{Bmatrix} \text{k-in.}$$

and the total moments at the potential hinges are

$$\mathbf{M}(t)=\mathbf{M}'(t)+\mathbf{M}''(t)=\begin{Bmatrix} 70.576 \\ -78.964 \\ -305.224 \\ -155.684 \end{Bmatrix}+\begin{Bmatrix} 5.23 \\ -5.23 \\ 5.23 \\ 2.39 \end{Bmatrix}=\begin{Bmatrix} 75.8 \\ -84.2 \\ -300.0 \\ -153.3 \end{Bmatrix} \text{k-in.}$$

These results are the same as the results in Eq. 7.464.

7.7   INELASTIC DYNAMIC STATE SPACE RESPONSE USING THE FORCE ANALOGY METHOD

Consider the dynamic equilibrium equation of motion

$$\mathbf{M}\ddot{\mathbf{X}}(t)+\mathbf{C}\dot{\mathbf{X}}(t)+\mathbf{K}\mathbf{X}'(t)=-\mathbf{M}\mathbf{a}(t). \qquad (7.465)$$

The elastic displacement is defined as the total force divided by the initial stiffness. Therefore, multiplying the elastic displacement by the stiffness matrix gives the total force. This total force contributes to the balancing of forces in Eq. 7.465. From Eq. 7.445, the elastic displacement can be expressed as

$$\mathbf{X}'(t)=\begin{Bmatrix} x_1'(t) \\ x_2'(t) \\ \vdots \\ x_n'(t) \end{Bmatrix}=\mathbf{X}(t)-\mathbf{X}''(t)=\begin{Bmatrix} x_1(t) \\ x_2(t) \\ \vdots \\ x_n(t) \end{Bmatrix}-\begin{Bmatrix} x_1''(t) \\ x_2''(t) \\ \vdots \\ x_n''(t) \end{Bmatrix}. \qquad (7.466)$$

It follows from Eq. 7.465 that

$$\mathbf{M}\ddot{\mathbf{X}}(t)+\mathbf{C}\dot{\mathbf{X}}(t)+\mathbf{K}\mathbf{X}(t)=-\mathbf{M}\mathbf{a}(t)+\mathbf{K}\mathbf{X}''(t). \qquad (7.467)$$

To represent Eq. 7.467 in state space form, define

$$\mathbf{z}(t)=\begin{Bmatrix} \mathbf{X}(t) \\ \dot{\mathbf{X}}(t) \end{Bmatrix}$$

and then Eq. 7.467 becomes

$$\dot{\mathbf{z}}(t)=\begin{Bmatrix} \dot{\mathbf{X}}(t) \\ \ddot{\mathbf{X}}(t) \end{Bmatrix}=\begin{bmatrix} \mathbf{0} & \mathbf{I} \\ -\mathbf{M}^{-1}\mathbf{K} & -\mathbf{M}^{-1}\mathbf{C} \end{bmatrix}\begin{Bmatrix} \mathbf{X}(t) \\ \dot{\mathbf{X}}(t) \end{Bmatrix}+\begin{Bmatrix} \mathbf{0} \\ -\mathbf{a}(t) \end{Bmatrix}+\begin{Bmatrix} \mathbf{0} \\ \mathbf{M}^{-1}\mathbf{K}\mathbf{X}''(t) \end{Bmatrix}. \qquad (7.468)$$

To simplify Eq. 7.468, let

$$\mathbf{A} = \begin{bmatrix} \mathbf{0} & \mathbf{I} \\ -\mathbf{M}^{-1}\mathbf{K} & -\mathbf{M}^{-1}\mathbf{C} \end{bmatrix}, \quad \mathbf{H} = \begin{bmatrix} \mathbf{0} \\ -\mathbf{1} \end{bmatrix}, \quad \mathbf{F}_p^c = \begin{bmatrix} \mathbf{0} \\ \mathbf{M}^{-1}\mathbf{K} \end{bmatrix}.$$

It then follows that

$$\dot{\mathbf{z}}(t) = \mathbf{A}\mathbf{z}(t) + \mathbf{H}\mathbf{a}(t) + \mathbf{F}_p^c\mathbf{X}''(t). \tag{7.469}$$

The matrix $\mathbf{H}$ contains six columns, each of which was defined in Eq. 6.188 and corresponds to the ground acceleration in six different directions. The solution to Eq. 7.469 is

$$\mathbf{z}(t) = e^{\mathbf{A}(t-t_o)}\mathbf{z}(t_o) + e^{\mathbf{A}t}\int_{t_o}^{t} e^{-\mathbf{A}s}\left[\mathbf{H}\mathbf{a}(s) + \mathbf{F}_p^c\mathbf{X}''(s)\right]ds. \tag{7.470}$$

Let

$$t_{k+1} = t, \quad t_k = t_o, \quad \Delta t = t - t_o. \tag{7.471}$$

It follows from Eq. 7.470 that

$$\mathbf{z}_{k+1} = e^{\mathbf{A}\Delta t}\mathbf{z}_k + e^{\mathbf{A}t_{k+1}}\int_{t_k}^{t_{k+1}} e^{-\mathbf{A}s}\left[\mathbf{H}\mathbf{a}(s) + \mathbf{F}_p^c\mathbf{X}''(s)\right]ds. \tag{7.472}$$

When the Delta forcing function is used to represent both the ground motion and the inelastic displacement, that is,

$$\mathbf{a}(s) = \mathbf{a}_k\delta(s - t_k)\Delta t, \quad \mathbf{X}''(s) = \mathbf{X}_k''\delta(s - t_k)\Delta t, \quad t_k \le s < t_{k+1}, \tag{7.473}$$

then Eq. 7.472 becomes

$$\mathbf{z}_{k+1} = e^{\mathbf{A}\Delta t}\mathbf{z}_k + e^{\mathbf{A}\Delta t}\mathbf{H}\Delta t\,\mathbf{a}_k + e^{\mathbf{A}\Delta t}\mathbf{F}_p^c\Delta t\,\mathbf{X}_k''. \tag{7.474}$$

When a constant forcing function is used to represent both the ground motion and the inelastic displacement, that is,

$$\mathbf{a}(s) = \mathbf{a}_k, \quad \mathbf{X}''(s) = \mathbf{X}_k'', \quad t_k \le s < t_{k+1}, \tag{7.475}$$

then Eq. 7.472 becomes

$$\mathbf{z}_{k+1} = e^{\mathbf{A}\Delta t}\mathbf{z}_k + \mathbf{A}^{-1}\left(e^{\mathbf{A}\Delta t} - \mathbf{I}\right)\mathbf{H}\,\mathbf{a}_k + \mathbf{A}^{-1}\left(e^{\mathbf{A}\Delta t} - \mathbf{I}\right)\mathbf{F}_p^c\mathbf{X}_k''. \tag{7.476}$$

To simplify Eq. 7.474 or Eq. 7.476, let

$$\mathbf{F}_s = e^{\mathbf{A}\Delta t}, \quad \mathbf{H}_d^{(EQ)} = e^{\mathbf{A}\Delta t}\mathbf{H}\Delta t, \quad \mathbf{F}_p = e^{\mathbf{A}\Delta t}\mathbf{F}_p^c\Delta t \tag{7.477}$$

when Eq. 7.474 is used, or let

$$\mathbf{F}_s = e^{\mathbf{A}\Delta t}, \quad \mathbf{H}_d^{(EQ)} = \mathbf{A}^{-1}\left(e^{\mathbf{A}\Delta t} - \mathbf{I}\right)\mathbf{H}, \quad \mathbf{F}_p = \mathbf{A}^{-1}\left(e^{\mathbf{A}\Delta t} - \mathbf{I}\right)\mathbf{F}_p^c \tag{7.478}$$

when Eq. 7.476 is used, it then follows that

$$\mathbf{z}_{k+1} = \mathbf{F}_s\mathbf{z}_k + \mathbf{H}_d^{(EQ)}\mathbf{a}_k + \mathbf{F}_p\mathbf{X}_k''. \tag{7.479}$$

Given all the information at time step k, both the displacement and velocity vectors at time step $k + 1$ can be calculated using Eq. 7.479. The displacement vector, $\mathbf{X}_{k+1}$, defines the total displacements at all degrees of freedom and is embedded in $\mathbf{z}_{k+1}$. The inelastic displacements at time step $k +1$ (i.e., $\mathbf{X}_{k+1}''$) can be determined using $\mathbf{X}_{k+1}$ and the force analogy method discussed in Section 7.6. From Eq. 7.458,

$$\mathbf{M}_{k+1} + \mathbf{K}_R\mathbf{\Theta}_{k+1}'' = \mathbf{K}_p^T\mathbf{X}_{k+1}. \tag{7.480}$$

The term $\mathbf{\Theta}_{k+1}''$, which represents the plastic rotation at each potential plastic hinge location, depends on the past history of $\mathbf{\Theta}_{k+1}''$. Let

$$\mathbf{\Theta}_{k+1}'' = \mathbf{\Theta}_k'' + \Delta\mathbf{\Theta}''. \tag{7.481}$$

It then follows from Eq. 7.480 that

$$\mathbf{M}_{k+1} + \mathbf{K}_R \Delta \mathbf{\Theta}'' = \mathbf{K}_p^T \mathbf{X}_{k+1} - \mathbf{K}_R \mathbf{\Theta}_k''. \tag{7.482}$$

With the procedure discussed in Section 7.6, $\mathbf{M}$ and $\Delta \mathbf{\Theta}''$ can be solved simultaneously. From this incremental plastic rotation, the inelastic displacement vector can be determined from Eq. 7.451:

$$\mathbf{X}_{k+1}'' = \mathbf{K}^{-1} \mathbf{K}_p \mathbf{\Theta}_{k+1}'' = \mathbf{K}^{-1} \mathbf{K}_p \left(\mathbf{\Theta}_k'' + \Delta \mathbf{\Theta}'' \right). \tag{7.483}$$

EXAMPLE 1   *One-Story Frame*

Consider the one-story frame shown in Figure 7.56a. There are three degrees of freedom (one translation and two rotations) and four potential plastic hinge locations. An elastic–plastic model is used for the moment versus rotation relationship at each potential hinge location (see Figure 7.56b), where the columns are assumed to have moment capacities of M_c and the beam has a moment capacity of M_b.

Assuming that all three members of the frame are axially rigid (i.e., no relative movement in the member direction is allowed between the two ends of each member), the three matrices, $\mathbf{K}$, $\mathbf{K}_p$, and $\mathbf{K}_R$ are as follows:

$$\mathbf{K} = \begin{bmatrix} \dfrac{24EI_1}{L_1^3} & \dfrac{6EI_1}{L_1^2} & \dfrac{6EI_1}{L_1^2} \\[3mm] \dfrac{6EI_1}{L_1^2} & \dfrac{4EI_1}{L_1} + \dfrac{4EI_2}{L_2} & \dfrac{2EI_2}{L_2} \\[3mm] \dfrac{6EI_1}{L_1^2} & \dfrac{2EI_2}{L_2} & \dfrac{4EI_1}{L_1} + \dfrac{4EI_2}{L_2} \end{bmatrix}. \tag{7.484}$$

$$\mathbf{K}_p = \begin{bmatrix} \dfrac{6EI_1}{L_1^2} & 0 & 0 & \dfrac{6EI_1}{L_1^2} \\[3mm] \dfrac{2EI_1}{L_1} & \dfrac{4EI_2}{L_2} & \dfrac{2EI_2}{L_2} & 0 \\[3mm] 0 & \dfrac{2EI_2}{L_2} & \dfrac{4EI_2}{L_2} & \dfrac{2EI_1}{L_1} \end{bmatrix} \tag{7.485}$$

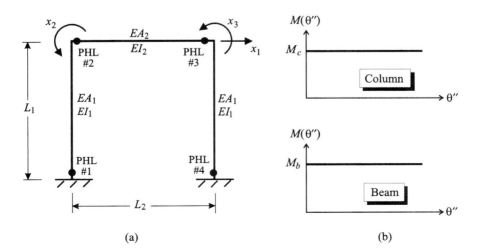

(a)   (b)

Figure 7.56 One-Story Frame

$$\mathbf{K}_R = \begin{bmatrix} \dfrac{4EI_1}{L_1} & 0 & 0 & 0 \\[2ex] 0 & \dfrac{4EI_2}{L_2} & \dfrac{2EI_2}{L_2} & 0 \\[2ex] 0 & \dfrac{2EI_2}{L_2} & \dfrac{4EI_2}{L_2} & 0 \\[2ex] 0 & 0 & 0 & \dfrac{4EI_1}{L_1} \end{bmatrix}. \tag{7.486}$$

Consider that the mass matrix, $\mathbf{M}$, and damping matrices, $\mathbf{C}$, are known. It follows that $\mathbf{F}$ and $\mathbf{F}_p$ are known with a predefined Δt. The structure is in the x–z plane, and therefore it follows that

$$\mathbf{z}_k = \begin{Bmatrix} x_1 \\ x_2 \\ x_3 \\ \dot{x}_1 \\ \dot{x}_2 \\ \dot{x}_3 \end{Bmatrix}, \quad \mathbf{H} = \begin{bmatrix} 0 & 0 \\ 0 & 0 \\ 0 & 0 \\ -1 & 0 \\ 0 & -1 \\ 0 & -1 \end{bmatrix}, \quad \mathbf{a}_k = \begin{Bmatrix} a_x \\ a_s \end{Bmatrix}_k \tag{7.487}$$

$$\mathbf{H}_d^{(EQ)} = e^{\mathbf{A}\Delta t}\mathbf{H}\Delta t, \quad \mathbf{F}_p = e^{\mathbf{A}\Delta t}\mathbf{F}_p^c\Delta t. \tag{7.488}$$

Note that the term a_z is not present in the earthquake acceleration vector because the columns are assumed to be axially rigid, and therefore no vertical degree of freedom exists. In addition, assume that all information at time step k is known and then it follows that Eqs. 7.458 and 7.459 are satisfied at time step k:

$$\begin{Bmatrix} M_1 \\ M_2 \\ M_3 \\ M_4 \end{Bmatrix}_k + \begin{bmatrix} \frac{4EI_1}{L_1} & 0 & 0 & 0 \\ 0 & \frac{4EI_2}{L_2} & \frac{2EI_2}{L_2} & 0 \\ 0 & \frac{2EI_2}{L_2} & \frac{4EI_2}{L_2} & 0 \\ 0 & 0 & 0 & \frac{4EI_1}{L_1} \end{bmatrix} \begin{Bmatrix} \theta_1'' \\ \theta_2'' \\ \theta_3'' \\ \theta_4'' \end{Bmatrix}_k = \begin{bmatrix} \frac{6EI_1}{L_1^2} & \frac{2EI_1}{L_1} & 0 \\ 0 & \frac{4EI_2}{L_2} & \frac{2EI_2}{L_2} \\ 0 & \frac{2EI_2}{L_2} & \frac{4EI_2}{L_2} \\ \frac{6EI_1}{L_1^2} & 0 & \frac{2EI_1}{L_1} \end{bmatrix} \begin{Bmatrix} x_1 \\ x_2 \\ x_3 \end{Bmatrix}_k \tag{7.489}$$

$$\begin{Bmatrix} x_1'' \\ x_2'' \\ x_3'' \end{Bmatrix}_k = \begin{bmatrix} \frac{24EI_1}{L_1^3} & \frac{6EI_1}{L_1^2} & \frac{6EI_1}{L_1^2} \\ \frac{6EI_1}{L_1^2} & \frac{4EI_1}{L_1}+\frac{4EI_2}{L_2} & \frac{2EI_2}{L_2} \\ \frac{6EI_1}{L_1^2} & \frac{2EI_2}{L_2} & \frac{4EI_1}{L_1}+\frac{4EI_2}{L_2} \end{bmatrix}^{-1} \begin{bmatrix} \frac{6EI_1}{L_1^2} & 0 & 0 & \frac{6EI_1}{L_1^2} \\ \frac{2EI_1}{L_1} & \frac{4EI_2}{L_2} & \frac{2EI_2}{L_2} & 0 \\ 0 & \frac{2EI_2}{L_2} & \frac{4EI_2}{L_2} & \frac{2EI_1}{L_1} \end{bmatrix} \begin{Bmatrix} \theta_1'' \\ \theta_2'' \\ \theta_3'' \\ \theta_4'' \end{Bmatrix}_k. \tag{7.490}$$

This information includes $\mathbf{M}_k$, $\boldsymbol{\Theta}_k''$, $\mathbf{z}_k$, and $\mathbf{X}_k''$. Measuring the earthquake ground motion at time step k gives $\mathbf{a}_k$, and therefore the goal here is to determine the information at time step $k + 1$. Using Eq. 7.479 gives the displacement and velocity vectors at time step $k + 1$:

$$\begin{Bmatrix} x_1 \\ x_2 \\ x_3 \\ \dot{x}_1 \\ \dot{x}_2 \\ \dot{x}_3 \end{Bmatrix}_{k+1} = \mathbf{F}_s \begin{Bmatrix} x_1 \\ x_2 \\ x_3 \\ \dot{x}_1 \\ \dot{x}_2 \\ \dot{x}_3 \end{Bmatrix}_k + \mathbf{H}_d^{(EQ)} \begin{Bmatrix} a_x \\ a_s \end{Bmatrix}_k + \mathbf{F}_p \begin{Bmatrix} x_1'' \\ x_2'' \\ x_3'' \end{Bmatrix}_k. \tag{7.491}$$

The response $\mathbf{X}_{k+1}$ is embedded in $\mathbf{z}_{k+1}$, and then using Eq. 7.482 at time step $k + 1$ gives the moment and the incremental plastic rotation vectors:

$$
\left\{\begin{matrix} M_1 \\ M_2 \\ M_3 \\ M_4 \end{matrix}\right\}_{k+1} + \begin{bmatrix} \frac{4EI_1}{L_1} & 0 & 0 & 0 \\ 0 & \frac{4EI_2}{L_2} & \frac{2EI_2}{L_2} & 0 \\ 0 & \frac{2EI_2}{L_2} & \frac{4EI_2}{L_2} & 0 \\ 0 & 0 & 0 & \frac{4EI_1}{L_1} \end{bmatrix} \left\{\begin{matrix} \Delta\theta_1'' \\ \Delta\theta_2'' \\ \Delta\theta_3'' \\ \Delta\theta_4'' \end{matrix}\right\}
$$

$$
= \begin{bmatrix} \frac{6EI_1}{L_1^2} & \frac{2EI_1}{L_1} & 0 \\ 0 & \frac{4EI_2}{L_2} & \frac{2EI_2}{L_2} \\ 0 & \frac{2EI_2}{L_2} & \frac{4EI_2}{L_2} \\ \frac{6EI_1}{L_1^2} & 0 & \frac{2EI_1}{L_1} \end{bmatrix} \left\{\begin{matrix} x_1 \\ x_2 \\ x_3 \end{matrix}\right\}_{k+1} - \begin{bmatrix} \frac{4EI_1}{L_1} & 0 & 0 & 0 \\ 0 & \frac{4EI_2}{L_2} & \frac{2EI_2}{L_2} & 0 \\ 0 & \frac{2EI_2}{L_2} & \frac{4EI_2}{L_2} & 0 \\ 0 & 0 & 0 & \frac{4EI_1}{L_1} \end{bmatrix} \left\{\begin{matrix} \theta_1'' \\ \theta_2'' \\ \theta_3'' \\ \theta_4'' \end{matrix}\right\}_k \qquad (7.492)
$$

and therefore the plastic rotation vector is given in Eq. 7.481, which is

$$
\left\{\begin{matrix} \theta_1'' \\ \theta_2'' \\ \theta_3'' \\ \theta_4'' \end{matrix}\right\}_{k+1} = \left\{\begin{matrix} \theta_1'' \\ \theta_2'' \\ \theta_3'' \\ \theta_4'' \end{matrix}\right\}_k + \left\{\begin{matrix} \Delta\theta_1'' \\ \Delta\theta_2'' \\ \Delta\theta_3'' \\ \Delta\theta_4'' \end{matrix}\right\}. \qquad (7.493)
$$

Finally, the inelastic displacement vector at time step $k + 1$, $\mathbf{X}_{k+1}''$, can be computed using Eq. 7.483:

$$
\left\{\begin{matrix} x_1'' \\ x_2'' \\ x_3'' \end{matrix}\right\}_{k+1} = \begin{bmatrix} \frac{24EI_1}{L_1^3} & \frac{6EI_1}{L_1^2} & \frac{6EI_1}{L_1^2} \\ \frac{6EI_1}{L_1^2} & \frac{4EI_1}{L_1} + \frac{4EI_2}{L_2} & \frac{2EI_2}{L_2} \\ \frac{6EI_1}{L_1^2} & \frac{2EI_2}{L_2} & \frac{4EI_1}{L_1} + \frac{4EI_2}{L_2} \end{bmatrix}^{-1} \begin{bmatrix} \frac{6EI_1}{L_1^2} & 0 & 0 & \frac{6EI_1}{L_1^2} \\ \frac{2EI_1}{L_1} & \frac{4EI_2}{L_2} & \frac{2EI_2}{L_2} & 0 \\ 0 & \frac{2EI_2}{L_2} & \frac{4EI_2}{L_2} & \frac{2EI_1}{L_1} \end{bmatrix} \left\{\begin{matrix} \theta_1'' \\ \theta_2'' \\ \theta_3'' \\ \theta_4'' \end{matrix}\right\}_{k+1}. \qquad (7.494)
$$

This completes the calculation for all the information at time step $k + 1$.

EXAMPLE 2   *Response of an SDOF System*

Consider the single degree of freedom system in Example 1 of Section 6.7 subjected to the modified El-Centro earthquake ground acceleration. Now assume that a potential hinge is located at the base of the column as shown in Figure 7.57. Example 1 of Section 6.7 gives

$$
m = 1.0 \text{ k-s}^2/\text{in.} \quad c = 1.2567 \text{ k-s/in.,} \quad k = 157.92 \text{ k/in.} \qquad (7.495)
$$

$$
\mathbf{A} = \begin{bmatrix} 0 & 1 \\ -157.92 & -1.2567 \end{bmatrix}, \quad \mathbf{H} = \begin{bmatrix} 0 \\ -1 \end{bmatrix}, \quad \mathbf{a}_k = a_k
$$

$$
\mathbf{F}_s = e^{\mathbf{A}\Delta t} = \begin{bmatrix} 0.968843 & 0.019543 \\ -3.086307 & 0.944284 \end{bmatrix}, \quad \mathbf{H}_d^{(EQ)} = e^{\mathbf{A}\Delta t}\mathbf{H}\Delta t = 0.02\left\{\begin{matrix} -0.019543 \\ -0.944284 \end{matrix}\right\} \qquad (7.496)
$$

$$
\mathbf{F}_p^c = \begin{bmatrix} 0 \\ 157.92 \end{bmatrix}, \quad \mathbf{F}_p = e^{\mathbf{A}\Delta t}\mathbf{F}_p^c\Delta t = 0.02\left\{\begin{matrix} 3.0862 \\ 149.12 \end{matrix}\right\}. \qquad (7.497)
$$

It follows that

$$
\left\{\begin{matrix} x_{k+1} \\ \dot{x}_{k+1} \end{matrix}\right\} = \begin{bmatrix} 0.968843 & 0.019543 \\ 3.086307 & 0.944284 \end{bmatrix}\left\{\begin{matrix} x_k \\ \dot{x}_k \end{matrix}\right\} + 0.02\left\{\begin{matrix} -0.019543 \\ -0.944284 \end{matrix}\right\}a_k + 0.02\left\{\begin{matrix} -3.0862 \\ 149.12 \end{matrix}\right\}x_k''. \qquad (7.498)
$$

The matrices relating the forces on the system and the plastic rotation are

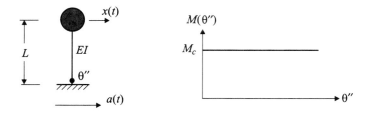

Figure 7.57 SDOF System with PHL at the Base of the Column

$$\mathbf{K} = \frac{3EI}{L^3}, \quad \mathbf{K}_R = \frac{3EI}{L}, \quad \mathbf{K}_p = \frac{3EI}{L^2} \tag{7.499}$$

and therefore

$$M_k + \frac{3EI}{L}\theta''_k = \frac{3EI}{L^2}x_k \tag{7.500}$$

$$x''_k = \left(\frac{3EI}{L^3}\right)\left(\frac{L^2}{3EI}\right)\theta''_k = L\theta''_k \tag{7.501}$$

which are the same as the equations given in Eq. 7.389 and Eq. 7.396.
    As a numerical example, let

$$L = 10 \text{ in.} \tag{7.502}$$

and then it follows from Eqs. 7.495 and 7.499 that

$$\mathbf{K}_R = L^2\mathbf{K} = (10)^2(157.92) = 15,792.0 \tag{7.503}$$

$$\mathbf{K}_p = L\mathbf{K} = (10)(157.92) = 1,579.2. \tag{7.504}$$

In addition, assume that the column can take a maximum moment corresponding to a force on the mass of $1.0g$, that is,

$$M_c = maL = (1.0)(386.4)(10) = 3,864.0 \text{ k-in.} \tag{7.505}$$

Table 7.3 presents the results of the response obtained using Eqs. 7.498, 7.500, and 7.501.
    The structure is at rest at a time step equal to zero. Using Eq. 7.498, the response at the first time step is

$$\begin{Bmatrix} x_1 \\ \dot{x}_1 \end{Bmatrix} = \begin{bmatrix} 0.968843 & 0.019543 \\ 3.086307 & 0.944284 \end{bmatrix}\begin{Bmatrix} 0 \\ 0 \end{Bmatrix}$$

$$+0.02\begin{Bmatrix} -0.019543 \\ -0.944284 \end{Bmatrix}(-0.0026)(386.4) + 0.02\begin{Bmatrix} -3.0862 \\ 149.12 \end{Bmatrix}(0) = \begin{Bmatrix} 0.0004 \\ 0.0188 \end{Bmatrix}. \tag{7.506}$$

From Eq. 7.500, it follows that

$$M_1 + (15,792.0)\theta''_1 = (1,579.2)(0.0004). \tag{7.507}$$

The solution for Eq. 7.507 is obtained by first letting the plastic rotation θ''_1 be equal to zero and then the moment is equal to

$$M_1 = (1,579.2)(0.0004) = 0.61 \text{ k-in.} \tag{7.508}$$

The moment M_1 is less than the yield moment, and therefore the plastic rotation is zero. It then follows that using Eq. 7.501 the response is

Table 7.3 Numerical Calculation of SDOF System Inelastic Response

k	$a_k(g)$	x_k (in.)	$\dot{x}_k$ (in./s)	M_k (k-in.)	θ_k'' (rad)	x_k'' (in.)
0	−0.0026	0.0	0.0	0.0	0.0	0.0
1	−0.0198	0.0004	0.0188	0.61	0.0	0.0
2	−0.0185	0.0037	0.1610	5.90	0.0	0.0
$\vdots$	$\vdots$	$\vdots$	$\vdots$	$\vdots$	$\vdots$	$\vdots$
91	−0.2990	2.4317	9.7346	3840	0.0	0.0
92	−0.2470	2.5913	3.8692	3864	0.0145	0.1445
93	−0.2000	2.6325	−2.1106	3864	0.0186	0.1856

$$x_1'' = (10)(0) = 0 \text{ in.} \tag{7.509}$$

This procedure continues, and at time step 92 there is a plastic rotation. Consider that the calculation at time step 91 has been performed and therefore all response quantities are known. It then follows that using Eq. 7.498, the response at time step 92 is

$$\begin{Bmatrix} x_{92} \\ \dot{x}_{92} \end{Bmatrix} = \begin{bmatrix} 0.968843 & 0.019543 \\ 3.086307 & 0.944284 \end{bmatrix} \begin{Bmatrix} 2.4317 \\ 9.7346 \end{Bmatrix}$$
$$+ 0.02 \begin{Bmatrix} -0.019543 \\ -0.944284 \end{Bmatrix} (-0.2470)(386.4) + 0.02 \begin{Bmatrix} -3.0862 \\ 149.12 \end{Bmatrix} (0) = \begin{Bmatrix} 2.5913 \\ 3.8692 \end{Bmatrix}. \tag{7.510}$$

Using the displacement from Eq. 7.510, Eq. 7.500 becomes

$$M_{92} + (15,792.0)\theta_{92}'' = (1,579.2)(2.5913). \tag{7.511}$$

To solve Eq. 7.511, let the plastic rotation θ_{92}'' be equal to zero, and then the moment is equal to

$$M_{92} = (1,579.2)(2.5913) = 4,092 \text{ k-in.} \tag{7.512}$$

This moment is greater than the yield moment and therefore a plastic rotation exists. The moment versus rotation relationship is elastic–plastic, and it follows that

$$M_{92} = M_c = 3,864 \text{ k-in.} \tag{7.513}$$

and using Eq. 7.500 that

$$3,864 + (15,792.0)\theta_{92}'' = (1,579.2)(2.5913). \tag{7.514}$$

The plastic rotation at time step 92 is

$$\theta_{92}'' = 0.01445 \text{ rad} \tag{7.515}$$

and the inelastic displacement at time step 92 follows from Eq. 7.501 and is equal to

$$x_{92}'' = (10)(0.01445) = 0.1445 \text{ in.} \tag{7.516}$$

This completes the calculation of the response at time step 92. Again, using Eq. 7.498, the response at time step 93 is

$$\begin{Bmatrix} x_{93} \\ \dot{x}_{93} \end{Bmatrix} = \begin{bmatrix} 0.968843 & 0.019543 \\ 3.086307 & 0.944284 \end{bmatrix} \begin{Bmatrix} 2.5913 \\ 3.8692 \end{Bmatrix}$$
$$+ 0.02 \begin{Bmatrix} -0.019543 \\ -0.944284 \end{Bmatrix} (-0.2000)(386.4) + 0.02 \begin{Bmatrix} -3.0862 \\ 149.12 \end{Bmatrix} (0.1445) = \begin{Bmatrix} 2.6325 \\ -2.1106 \end{Bmatrix}. \tag{7.517}$$

For this displacement, Eq. 7.500 becomes

$$M_{93} + (15,792.0)\Delta\theta'' = (1,579.2)(2.6325) - (15,792.0)(0.01445). \tag{7.518}$$

Let the plastic rotation $\Delta\theta''$ be equal to zero in Eq. 7.518, and then the moment is equal to

$$M_{93} = (1,579.2)(2.6325) - (15,792.0)(0.01445) = 3,929 \text{ k-in.} \tag{7.519}$$

This moment is greater than the yield moment, and therefore the plastic rotation increases. Using the moment versus rotation relationship that is elastic–plastic, it follows that

$$M_{92} = M_c = 3,864 \text{ k-in.} \tag{7.520}$$

Therefore, it follows from Eq. 7.500 that

$$3,864 + (15,792.0)\Delta\theta'' = (1,579.2)(2.6325) - (15,792.0)(0.01445) \tag{7.521}$$

and thus the plastic rotation at time step 93 becomes

$$\theta''_{93} = 0.01856 \text{ rad.} \tag{7.522}$$

Finally, the inelastic displacement at time step 93 follows from Eq. 7.501, and it is equal to

$$x''_{93} = (10)(0.01856) = 0.1856 \text{ in.} \tag{7.523}$$

This numerical solution procedure can be used for all values in time that are of interest, and the response time history is shown in Figure 7.58. The displacement time history after 5 s oscillates at an average that is below zero, which shows that a permanent deformation has developed in the structure.

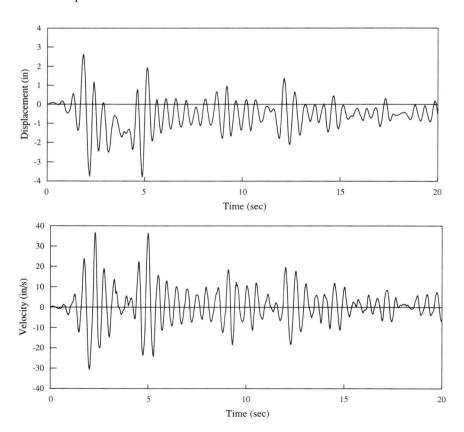

Figure 7.58 Inelastic Response of the SDOF System

7.8 INELASTIC DYNAMIC RESPONSE WITH STATE SPACE REDUCTION

The mass matrix in Section 7.7 was fully populated and therefore invertible. However, structures often contain many thousands of DOF, and thus structural engineers very often attempt to reduce the size of the problem by assuming zero for mass terms with a negligible mass as discussed in Section 6.5. Once this is done, the mass matrix becomes singular, and Eq. 7.468 is undefined. Therefore, state space reduction is often necessary, as discussed in Section 6.8 for a linear time history analysis. The objective of this section is to extend the state space reduction method to the inelastic response domain.

Using the notation used in Section 6.8, let all of the degrees of freedom with mass be labeled as 1, those with damping but without mass be labeled as 2, and those without mass or damping be labeled as 3. Then Eq. 7.465 becomes

$$
\begin{bmatrix} \mathbf{M}_{11} & 0 & 0 \\ 0 & 0 & 0 \\ 0 & 0 & 0 \end{bmatrix}
\begin{Bmatrix} \ddot{\mathbf{X}}_1(t) \\ \ddot{\mathbf{X}}_2(t) \\ \ddot{\mathbf{X}}_3(t) \end{Bmatrix}
+
\begin{bmatrix} \mathbf{C}_{11} & \mathbf{C}_{12} & 0 \\ \mathbf{C}_{21} & \mathbf{C}_{22} & 0 \\ 0 & 0 & 0 \end{bmatrix}
\begin{Bmatrix} \dot{\mathbf{X}}_1(t) \\ \dot{\mathbf{X}}_2(t) \\ \dot{\mathbf{X}}_3(t) \end{Bmatrix}
+
\begin{bmatrix} \mathbf{K}_{11} & \mathbf{K}_{12} & \mathbf{K}_{13} \\ \mathbf{K}_{21} & \mathbf{K}_{22} & \mathbf{K}_{23} \\ \mathbf{K}_{31} & \mathbf{K}_{32} & \mathbf{K}_{33} \end{bmatrix}
\begin{Bmatrix} \mathbf{X}_1'(t) \\ \mathbf{X}_2'(t) \\ \mathbf{X}_3'(t) \end{Bmatrix}
$$
$$
= -\begin{bmatrix} \mathbf{M}_{11} & 0 & 0 \\ 0 & 0 & 0 \\ 0 & 0 & 0 \end{bmatrix}
\begin{Bmatrix} \mathbf{a}_1(t) \\ \mathbf{a}_2(t) \\ \mathbf{a}_3(t) \end{Bmatrix}. \tag{7.524}
$$

Recall that

$$
\begin{Bmatrix} \mathbf{X}_1'(t) \\ \mathbf{X}_2'(t) \\ \mathbf{X}_3'(t) \end{Bmatrix}
=
\begin{Bmatrix} \mathbf{X}_1(t) \\ \mathbf{X}_2(t) \\ \mathbf{X}_3(t) \end{Bmatrix}
-
\begin{Bmatrix} \mathbf{X}_1''(t) \\ \mathbf{X}_2''(t) \\ \mathbf{X}_3''(t) \end{Bmatrix}. \tag{7.525}
$$

Substituting Eq. 7.525 into Eq. 7.524 gives

$$
\begin{bmatrix} \mathbf{M}_{11} & 0 & 0 \\ 0 & 0 & 0 \\ 0 & 0 & 0 \end{bmatrix}
\begin{Bmatrix} \ddot{\mathbf{X}}_1(t) \\ \ddot{\mathbf{X}}_2(t) \\ \ddot{\mathbf{X}}_3(t) \end{Bmatrix}
+
\begin{bmatrix} \mathbf{C}_{11} & \mathbf{C}_{12} & 0 \\ \mathbf{C}_{21} & \mathbf{C}_{22} & 0 \\ 0 & 0 & 0 \end{bmatrix}
\begin{Bmatrix} \dot{\mathbf{X}}_1(t) \\ \dot{\mathbf{X}}_2(t) \\ \dot{\mathbf{X}}_3(t) \end{Bmatrix}
+
\begin{bmatrix} \mathbf{K}_{11} & \mathbf{K}_{12} & \mathbf{K}_{13} \\ \mathbf{K}_{21} & \mathbf{K}_{22} & \mathbf{K}_{23} \\ \mathbf{K}_{31} & \mathbf{K}_{32} & \mathbf{K}_{33} \end{bmatrix}
\begin{Bmatrix} \mathbf{X}_1(t) \\ \mathbf{X}_2(t) \\ \mathbf{X}_3(t) \end{Bmatrix}
$$
$$
= -\begin{bmatrix} \mathbf{M}_{11} & 0 & 0 \\ 0 & 0 & 0 \\ 0 & 0 & 0 \end{bmatrix}
\begin{Bmatrix} \mathbf{a}_1(t) \\ \mathbf{a}_2(t) \\ \mathbf{a}_3(t) \end{Bmatrix}
+
\begin{bmatrix} \mathbf{K}_{11} & \mathbf{K}_{12} & \mathbf{K}_{13} \\ \mathbf{K}_{21} & \mathbf{K}_{22} & \mathbf{K}_{23} \\ \mathbf{K}_{31} & \mathbf{K}_{32} & \mathbf{K}_{33} \end{bmatrix}
\begin{Bmatrix} \mathbf{X}_1''(t) \\ \mathbf{X}_2''(t) \\ \mathbf{X}_3''(t) \end{Bmatrix}. \tag{7.526}
$$

Expanding Eq. 7.526 gives the following three simultaneous equations:

$$
\mathbf{M}_{11}\ddot{\mathbf{X}}_1(t) + \mathbf{C}_{11}\dot{\mathbf{X}}_1(t) + \mathbf{C}_{12}\dot{\mathbf{X}}_2(t) + \mathbf{K}_{11}\mathbf{X}_1(t) + \mathbf{K}_{12}\mathbf{X}_2(t) + \mathbf{K}_{13}\mathbf{X}_3(t)
$$
$$
= -\mathbf{M}_{11}\mathbf{a}_1(t) + \mathbf{K}_{11}\mathbf{X}_1''(t) + \mathbf{K}_{12}\mathbf{X}_2''(t) + \mathbf{K}_{13}\mathbf{X}_3''(t) \tag{7.527a}
$$

$$
\mathbf{C}_{21}\dot{\mathbf{X}}_1(t) + \mathbf{C}_{22}\dot{\mathbf{X}}_2(t) + \mathbf{K}_{21}\mathbf{X}_1(t) + \mathbf{K}_{22}\mathbf{X}_2(t) + \mathbf{K}_{23}\mathbf{X}_3(t)
$$
$$
= \mathbf{K}_{21}\mathbf{X}_1''(t) + \mathbf{K}_{22}\mathbf{X}_2''(t) + \mathbf{K}_{23}\mathbf{X}_3''(t) \tag{7.527b}
$$

$$
\mathbf{K}_{31}\mathbf{X}_1(t) + \mathbf{K}_{32}\mathbf{X}_2(t) + \mathbf{K}_{33}\mathbf{X}_3(t) = \mathbf{K}_{31}\mathbf{X}_1''(t) + \mathbf{K}_{32}\mathbf{X}_2''(t) + \mathbf{K}_{33}\mathbf{X}_3''(t). \tag{7.527c}
$$

Solving for $\mathbf{X}_3(t)$ in Eq. 7.527c gives

$$
\mathbf{X}_3(t) = -\mathbf{K}_{33}^{-1}\mathbf{K}_{31}\mathbf{X}_1(t) - \mathbf{K}_{33}^{-1}\mathbf{K}_{32}\mathbf{X}_2(t) + \mathbf{K}_{33}^{-1}\mathbf{K}_{31}\mathbf{X}_1''(t) + \mathbf{K}_{33}^{-1}\mathbf{K}_{32}\mathbf{X}_2''(t) + \mathbf{X}_3''(t). \tag{7.528}
$$

Substituting Eq. 7.528 into Eq. 7.527b, it follows that

$$
\mathbf{C}_{21}\dot{\mathbf{X}}_1(t) + \mathbf{C}_{22}\dot{\mathbf{X}}_2(t) + \mathbf{K}_{21}'\mathbf{X}_1(t) + \mathbf{K}_{22}'\mathbf{X}_2(t) = \mathbf{K}_{21}'\mathbf{X}_1''(t) + \mathbf{K}_{22}'\mathbf{X}_2''(t) \tag{7.529}
$$

where

$$\mathbf{K}'_{21} = \mathbf{K}_{21} - \mathbf{K}_{23}\mathbf{K}_{33}^{-1}\mathbf{K}_{31}, \quad \mathbf{K}'_{22} = \mathbf{K}_{22} - \mathbf{K}_{23}\mathbf{K}_{33}^{-1}\mathbf{K}_{32}.$$

Solving for $\dot{\mathbf{X}}_2(t)$ in Eq. 7.529 gives

$$\dot{\mathbf{X}}_2(t) = -\mathbf{C}_{22}^{-1}\mathbf{C}_{21}\dot{\mathbf{X}}_1(t) - \mathbf{C}_{22}^{-1}\mathbf{K}'_{21}\mathbf{X}_1(t) - \mathbf{C}_{22}^{-1}\mathbf{K}'_{22}\mathbf{X}_2(t) + \mathbf{C}_{22}^{-1}\mathbf{K}'_{21}\mathbf{X}''_1(t) + \mathbf{C}_{22}^{-1}\mathbf{K}'_{22}\mathbf{X}''_2(t). \quad (7.530)$$

Now substituting Eqs. 7.528 and 7.530 into Eq. 7.527a, it follows that

$$\mathbf{M}_{11}\ddot{\mathbf{X}}_1(t) + \mathbf{C}''_{11}\dot{\mathbf{X}}_1(t) + \mathbf{K}''_{11}\mathbf{X}_1(t) + \mathbf{K}''_{12}\mathbf{X}_2(t) = -\mathbf{M}_{11}\mathbf{a}_1(t) + \mathbf{K}''_{11}\mathbf{X}''_1(t) + \mathbf{K}''_{12}\mathbf{X}''_2(t) \quad (7.531)$$

where

$$\mathbf{C}''_{11} = \mathbf{C}_{11} - \mathbf{C}_{12}\mathbf{C}_{22}^{-1}\mathbf{C}_{21}, \quad \mathbf{K}''_{11} = \mathbf{K}'_{11} - \mathbf{C}_{12}\mathbf{C}_{22}^{-1}\mathbf{K}'_{21}, \quad \mathbf{K}''_{12} = \mathbf{K}'_{12} - \mathbf{C}_{12}\mathbf{C}_{22}^{-1}\mathbf{K}'_{22}$$

$$\mathbf{K}'_{11} = \mathbf{K}_{11} - \mathbf{K}_{13}\mathbf{K}_{33}^{-1}\mathbf{K}_{31}, \quad \mathbf{K}'_{12} = \mathbf{K}_{12} - \mathbf{K}_{13}\mathbf{K}_{33}^{-1}\mathbf{K}_{32}.$$

Solving for $\ddot{\mathbf{X}}_1(t)$ in Eq. 7.531 gives

$$\begin{aligned}\ddot{\mathbf{X}}_1(t) = &-\mathbf{a}_1(t) - \mathbf{M}_{11}^{-1}\mathbf{C}''_{11}\dot{\mathbf{X}}_1(t) - \mathbf{M}_{11}^{-1}\mathbf{K}''_{11}\mathbf{X}_1(t) - \mathbf{M}_{11}^{-1}\mathbf{K}''_{12}\mathbf{X}_2(t) \\ &+ \mathbf{M}_{11}^{-1}\mathbf{K}''_{11}\mathbf{X}''_1(t) + \mathbf{M}_{11}^{-1}\mathbf{K}''_{12}\mathbf{X}''_2(t).\end{aligned} \quad (7.532)$$

In state space form, Eqs. 7.530 and 7.532 are

$$\begin{Bmatrix} \dot{\mathbf{X}}_1(t) \\ \dot{\mathbf{X}}_2(t) \\ \ddot{\mathbf{X}}_1(t) \end{Bmatrix} = \begin{bmatrix} \mathbf{0} & \mathbf{0} & \mathbf{I} \\ -\mathbf{C}_{22}^{-1}\mathbf{K}'_{21} & -\mathbf{C}_{22}^{-1}\mathbf{K}'_{22} & -\mathbf{C}_{22}^{-1}\mathbf{C}_{21} \\ -\mathbf{M}_{11}^{-1}\mathbf{K}''_{11} & -\mathbf{M}_{11}^{-1}\mathbf{K}''_{12} & -\mathbf{M}_{11}^{-1}\mathbf{C}''_{11} \end{bmatrix} \begin{bmatrix} \mathbf{X}_1(t) \\ \mathbf{X}_2(t) \\ \dot{\mathbf{X}}_1(t) \end{bmatrix} + \begin{Bmatrix} \mathbf{0} \\ \mathbf{0} \\ -\mathbf{a}(t) \end{Bmatrix}$$

$$+ \begin{bmatrix} \mathbf{0} & \mathbf{0} \\ \mathbf{C}_{22}^{-1}\mathbf{K}'_{21} & \mathbf{C}_{22}^{-1}\mathbf{K}'_{22} \\ \mathbf{M}_{11}^{-1}\mathbf{K}''_{11} & \mathbf{M}_{11}^{-1}\mathbf{K}''_{12} \end{bmatrix} \begin{Bmatrix} \mathbf{X}''_1(t) \\ \mathbf{X}''_2(t) \end{Bmatrix}. \quad (7.533)$$

Denote

$$\mathbf{z}(t) = \begin{Bmatrix} \mathbf{X}_1(t) \\ \mathbf{X}_2(t) \\ \dot{\mathbf{X}}_1(t) \end{Bmatrix}, \quad \mathbf{X}(t) = \begin{Bmatrix} \mathbf{X}''_1(t) \\ \mathbf{X}''_2(t) \end{Bmatrix}, \quad \mathbf{A} = \begin{bmatrix} \mathbf{0} & \mathbf{0} & \mathbf{I} \\ -\mathbf{C}_{22}^{-1}\mathbf{K}'_{21} & -\mathbf{C}_{22}^{-1}\mathbf{K}'_{22} & -\mathbf{C}_{22}^{-1}\mathbf{C}_{21} \\ -\mathbf{M}_{11}^{-1}\mathbf{K}''_{11} & -\mathbf{M}_{11}^{-1}\mathbf{K}''_{12} & -\mathbf{M}_{11}^{-1}\mathbf{C}''_{11} \end{bmatrix}, \quad \mathbf{H} = \begin{bmatrix} \mathbf{0} \\ \mathbf{0} \\ -\mathbf{1} \end{bmatrix}$$

$$\mathbf{a}(t) = \begin{Bmatrix} a_x(t) & a_y(t) & a_z(t) & a_r(t) & a_s(t) & a_t(t) \end{Bmatrix}^T, \quad \mathbf{F}_p^c = \begin{bmatrix} \mathbf{0} & \mathbf{0} \\ \mathbf{C}_{22}^{-1}\mathbf{K}'_{21} & \mathbf{C}_{22}^{-1}\mathbf{K}'_{22} \\ \mathbf{M}_{11}^{-1}\mathbf{K}''_{11} & \mathbf{M}_{11}^{-1}\mathbf{K}''_{12} \end{bmatrix}$$

and then it follows that

$$\dot{\mathbf{z}}(t) = \mathbf{A}\mathbf{z}(t) + \mathbf{H}\mathbf{a}(t) + \mathbf{F}_p^c\mathbf{X}''(t) \quad (7.534)$$

which has the same form as Eq. 7.469. The matrix $\mathbf{H}$ contains six columns, where each column corresponds to the ground acceleration in six different directions as defined in Eq. 6.214. Following the same procedure discussed from Eq. 7.470 to Eq. 7.479, it follows that

$$\mathbf{z}_{k+1} = \mathbf{F}_s\,\mathbf{z}_k + \mathbf{H}_d^{(EQ)}\mathbf{a}_k + \mathbf{F}_p\mathbf{X}''_k \quad (7.535)$$

where if Delta forcing function assumption is used, then

$$\mathbf{F}_s = \mathbf{e}^{\mathbf{A}\Delta t}, \quad \mathbf{H}_d^{(EQ)} = \mathbf{e}^{\mathbf{A}\Delta t}\mathbf{H}\Delta t, \quad \mathbf{F}_p = \mathbf{e}^{\mathbf{A}\Delta t}\mathbf{F}_p^c\Delta t \quad (7.536)$$

or if the constant forcing function assumption is used, then

$$\mathbf{F}_s = \mathbf{e}^{\mathbf{A}\Delta t}, \quad \mathbf{H}_d^{(EQ)} = \mathbf{A}^{-1}\left(\mathbf{e}^{\mathbf{A}\Delta t} - \mathbf{I}\right)\mathbf{H}, \quad \mathbf{F}_p = \mathbf{A}^{-1}\left(\mathbf{e}^{\mathbf{A}\Delta t} - \mathbf{I}\right)\mathbf{F}_p^c. \quad (7.537)$$

Assuming that all information at time step k is known, then Eq. 7.535 can be used to obtain $\mathbf{z}_{k+1}$, which includes $\mathbf{X}_{1,k+1}$, $\mathbf{X}_{2,k+1}$, and $\dot{\mathbf{X}}_{1,k+1}$. The response objective is to use Eq. 7.480 to give both the moments and plastic rotations at the potential hinge locations:

$$\mathbf{M}_{k+1} + \mathbf{K}_R \mathbf{\Theta}''_{k+1} = \mathbf{K}_p^T \mathbf{X}_{k+1}. \tag{7.538}$$

However, in Eq. 7.538, $\mathbf{X}_{k+1}$ is unknown because it contains $\mathbf{X}_{3,k+1}$. Recall from Eq. 7.528 that

$$\mathbf{X}_{3,k+1} = -\mathbf{K}_{33}^{-1}\mathbf{K}_{31}\mathbf{X}_{1,k+1} - \mathbf{K}_{33}^{-1}\mathbf{K}_{32}\mathbf{X}_{2,k+1} + \mathbf{K}_{33}^{-1}\mathbf{K}_{31}\mathbf{X}''_{1,k+1} + \mathbf{K}_{33}^{-1}\mathbf{K}_{32}\mathbf{X}''_{2,k+1} + \mathbf{X}''_{3,k+1}. \tag{7.539}$$

Therefore, $\mathbf{X}_{k+1}$ in Eq. 7.538 becomes

$$\mathbf{X}_{k+1} = \begin{Bmatrix} \mathbf{X}_1 \\ \mathbf{X}_2 \\ \mathbf{X}_3 \end{Bmatrix}_{k+1} = \begin{bmatrix} \mathbf{I} & \mathbf{0} \\ \mathbf{0} & \mathbf{I} \\ -\mathbf{K}_{33}^{-1}\mathbf{K}_{31} & -\mathbf{K}_{33}^{-1}\mathbf{K}_{32} \end{bmatrix} \begin{Bmatrix} \mathbf{X}_1 \\ \mathbf{X}_2 \end{Bmatrix}_{k+1} + \begin{bmatrix} \mathbf{0} & \mathbf{0} & \mathbf{0} \\ \mathbf{0} & \mathbf{0} & \mathbf{0} \\ \mathbf{K}_{33}^{-1}\mathbf{K}_{31} & \mathbf{K}_{33}^{-1}\mathbf{K}_{32} & \mathbf{I} \end{bmatrix} \begin{Bmatrix} \mathbf{X}''_1 \\ \mathbf{X}''_2 \\ \mathbf{X}''_3 \end{Bmatrix}_{k+1}. \tag{7.540}$$

Now, recall from Eq. 7.483 that

$$\mathbf{X}''_{k+1} = \mathbf{K}^{-1}\mathbf{K}_p \mathbf{\Theta}''_{k+1}. \tag{7.541}$$

The matrix $\mathbf{K}$ in Eq. 7.541 is the global stiffness matrix. Substituting Eq. 7.541 into Eq. 7.540 gives

$$\mathbf{X}_{k+1} = \begin{Bmatrix} \mathbf{X}_1 \\ \mathbf{X}_2 \\ \mathbf{X}_3 \end{Bmatrix}_{k+1} = \begin{bmatrix} \mathbf{I} & \mathbf{0} \\ \mathbf{0} & \mathbf{I} \\ -\mathbf{K}_{33}^{-1}\mathbf{K}_{31} & -\mathbf{K}_{33}^{-1}\mathbf{K}_{32} \end{bmatrix} \begin{Bmatrix} \mathbf{X}_1 \\ \mathbf{X}_2 \end{Bmatrix}_{k+1} + \begin{bmatrix} \mathbf{0} & \mathbf{0} & \mathbf{0} \\ \mathbf{0} & \mathbf{0} & \mathbf{0} \\ \mathbf{K}_{33}^{-1}\mathbf{K}_{31} & \mathbf{K}_{33}^{-1}\mathbf{K}_{32} & \mathbf{I} \end{bmatrix} \mathbf{K}^{-1}\mathbf{K}_p \mathbf{\Theta}''_{k+1} \tag{7.542}$$

and then substituting Eq. 7.542 into Eq. 7.538, it follows that

$$\mathbf{M}_{k+1} + \mathbf{K}_R \mathbf{\Theta}''_{k+1} =$$
$$\mathbf{K}_p^T \begin{bmatrix} \mathbf{I} & \mathbf{0} \\ \mathbf{0} & \mathbf{I} \\ -\mathbf{K}_{33}^{-1}\mathbf{K}_{31} & -\mathbf{K}_{33}^{-1}\mathbf{K}_{32} \end{bmatrix} \begin{Bmatrix} \mathbf{X}_1 \\ \mathbf{X}_2 \end{Bmatrix}_{k+1} + \mathbf{K}_p^T \begin{bmatrix} \mathbf{0} & \mathbf{0} & \mathbf{0} \\ \mathbf{0} & \mathbf{0} & \mathbf{0} \\ \mathbf{K}_{33}^{-1}\mathbf{K}_{31} & \mathbf{K}_{33}^{-1}\mathbf{K}_{32} & \mathbf{I} \end{bmatrix} \mathbf{K}^{-1}\mathbf{K}_p \mathbf{\Theta}''_{k+1}. \tag{7.543}$$

Now define

$$\overline{\mathbf{K}}_R = \mathbf{K}_R - \mathbf{K}_p^T \begin{bmatrix} \mathbf{0} & \mathbf{0} & \mathbf{0} \\ \mathbf{0} & \mathbf{0} & \mathbf{0} \\ \mathbf{K}_{33}^{-1}\mathbf{K}_{31} & \mathbf{K}_{33}^{-1}\mathbf{K}_{32} & \mathbf{I} \end{bmatrix} \mathbf{K}^{-1}\mathbf{K}_p \tag{7.544a}$$

$$\overline{\mathbf{K}}_p^T = \mathbf{K}_p^T \begin{bmatrix} \mathbf{I} & \mathbf{0} \\ \mathbf{0} & \mathbf{I} \\ -\mathbf{K}_{33}^{-1}\mathbf{K}_{31} & -\mathbf{K}_{33}^{-1}\mathbf{K}_{32} \end{bmatrix}, \quad \overline{\mathbf{X}}_{k+1} = \begin{Bmatrix} \mathbf{X}_1 \\ \mathbf{X}_2 \end{Bmatrix}_{k+1} \tag{7.544b}$$

and then it follows from Eq. 7.543 that

$$\mathbf{M}_{k+1} + \overline{\mathbf{K}}_R \mathbf{\Theta}''_{k+1} = \overline{\mathbf{K}}_p^T \overline{\mathbf{X}}_{k+1}. \tag{7.545}$$

Equation 7.545 gives the moment and plastic rotations at time step $k+1$ and, as discussed in Eq. 7.482, it can be written in incremental form as

$$\mathbf{M}_{k+1} + \overline{\mathbf{K}}_R \Delta\mathbf{\Theta}'' = \overline{\mathbf{K}}_p^T \overline{\mathbf{X}}_{k+1} - \overline{\mathbf{K}}_R \mathbf{\Theta}''_k. \tag{7.546}$$

Finally, once $\mathbf{M}_{k+1}$ and $\Delta\mathbf{\Theta}''$ are obtained, the inelastic displacement vector $\mathbf{X}''_{k+1}$ can be determined using Eq. 7.541 in incremental form:

$$\mathbf{X}''_{k+1} = \mathbf{K}^{-1}\mathbf{K}_p \left(\mathbf{\Theta}''_k + \Delta\mathbf{\Theta}'' \right). \tag{7.547}$$

This completes the calculation for time step $k+1$.

EXAMPLE 1 *One-Story Frame, Three Degrees of Freedom and One Mass*

Consider the one-story frame used in Example 1 of Section 7.7. Assume that the mass moment of inertia is ignored at degrees of freedom 2 and 3 (i.e., the rotational degrees of freedom), and therefore a state space reduction is necessary. Consider the columns to have moment capacities of M_c and the beams to have a moment capacity of M_b. The three matrices, $\mathbf{K}$, $\mathbf{K}_p$, and $\mathbf{K}_R$, are presented in Eqs. 7.484 to 7.486 and are rewritten here for completeness.

$$\mathbf{K} = \begin{bmatrix} \dfrac{24EI_1}{L_1^3} & \dfrac{6EI_1}{L_1^2} & \dfrac{6EI_1}{L_1^2} \\[2ex] \dfrac{6EI_1}{L_1^2} & \dfrac{4EI_1}{L_1} + \dfrac{4EI_2}{L_2} & \dfrac{2EI_2}{L_2} \\[2ex] \dfrac{6EI_1}{L_1^2} & \dfrac{2EI_2}{L_2} & \dfrac{4EI_1}{L_1} + \dfrac{4EI_2}{L_2} \end{bmatrix} \tag{7.548}$$

$$\mathbf{K}_p = \begin{bmatrix} \dfrac{6EI_1}{L_1^2} & 0 & 0 & \dfrac{6EI_1}{L_1^2} \\[2ex] \dfrac{2EI_1}{L_1} & \dfrac{4EI_2}{L_2} & \dfrac{2EI_2}{L_2} & 0 \\[2ex] 0 & \dfrac{2EI_2}{L_2} & \dfrac{4EI_2}{L_2} & \dfrac{2EI_1}{L_1} \end{bmatrix} \tag{7.549}$$

$$\mathbf{K}_R = \begin{bmatrix} \dfrac{4EI_1}{L_1} & 0 & 0 & 0 \\[2ex] 0 & \dfrac{4EI_2}{L_2} & \dfrac{2EI_2}{L_2} & 0 \\[2ex] 0 & \dfrac{2EI_2}{L_2} & \dfrac{4EI_2}{L_2} & 0 \\[2ex] 0 & 0 & 0 & \dfrac{4EI_1}{L_1} \end{bmatrix}. \tag{7.550}$$

As a numerical example, let

$$\mathbf{M} = \begin{bmatrix} 1.0 & 0.0 & 0.0 \\ 0.0 & 0.0 & 0.0 \\ 0.0 & 0.0 & 0.0 \end{bmatrix} \tag{7.551}$$

$$EI_1 = EI_2 = 1,000,000 \text{ k-in.}^2, \qquad L_1 = L_2 = 100 \text{ in.}$$

It then follows from Eqs. 7.548 to 7.550 that

$$\mathbf{K} = \begin{bmatrix} 24 & 600 & 600 \\ 600 & 80,000 & 20,000 \\ 600 & 20,000 & 80,000 \end{bmatrix} \tag{7.552}$$

$$\mathbf{K}_p = \begin{bmatrix} 600 & 0 & 0 & 600 \\ 20,000 & 40,000 & 20,000 & 0 \\ 0 & 20,000 & 40,000 & 20,000 \end{bmatrix} \tag{7.553}$$

$$\mathbf{K}_R = \begin{bmatrix} 40,000 & 0 & 0 & 0 \\ 0 & 40,000 & 20,000 & 0 \\ 0 & 20,000 & 40,000 & 0 \\ 0 & 0 & 0 & 40,000 \end{bmatrix}. \tag{7.554}$$

Now assume that the damping in the structure is represented by Rayleigh proportional damping with $\alpha = 0$ and $\beta = 0.0005$. It then follows from Eqs. 7.551 and 7.552 that

$$\mathbf{C} = (0)\mathbf{M} + (0.0005)\begin{bmatrix} 24 & 600 & 600 \\ 600 & 80,000 & 20,000 \\ 600 & 20,000 & 80,000 \end{bmatrix} = \begin{bmatrix} 0.012 & 0.30 & 0.30 \\ 0.30 & 40.0 & 10.0 \\ 0.30 & 10.0 & 40.0 \end{bmatrix}. \tag{7.555}$$

The state transition matrix was calculated in Example 2 of Section 6.8. Based on the previously noted values in this example, an inelastic dynamic analysis can be performed for the modified El-Centro earthquake (see Figure 6.11). Three different cases of yield moments are studied in this example:

1. $M_c = 7,000$ k-in. and $M_b = 6,000$ k-in.: Both the beam and the columns remain elastic.

2. $M_c = 5,800$ k-in. and $M_b = 4,200$ k-in.: Both the beam and the columns have yielded. After yielding, the structure experiences permanent deformation.

3. $M_c = 4,500$ k-in. and $M_b = 3,500$ k-in.: Both the beam and the columns have yielded, and a permanent deformation exists.

Table 7.4 gives the calculated response at different time steps for Case 3. The structure is symmetrical and

$$x_{2k} = x_{3k}, \quad x_{2k}'' = x_{3k}'' \tag{7.556}$$

$$M_{1k} = M_{4k}, \quad \theta_{1k}'' = \theta_{4k}'' \tag{7.557}$$

$$M_{2k} = M_{3k}, \quad \theta_{2k}'' = \theta_{3k}''. \tag{7.558}$$

The solution for the first few steps in Table 7.4 shows that the moments in the beams and columns are less than the yield moment. Therefore, the structure remains elastic and the numerical solution procedure is the same as the procedures discussed in Chapter 6. Consider now that the solution has evolved with time and the solution is at time step $k = 300$. The displacements and velocity are computed and then it follows from Eq. 7.538 that

$$\begin{Bmatrix} M_1 \\ M_2 \\ M_3 \\ M_4 \end{Bmatrix}_{300} + \begin{bmatrix} 40,000 & 0 & 0 & 0 \\ 0 & 40,000 & 20,000 & 0 \\ 0 & 20,000 & 40,000 & 0 \\ 0 & 0 & 0 & 40,000 \end{bmatrix} \begin{Bmatrix} \theta_1'' \\ \theta_2'' \\ \theta_3'' \\ \theta_4'' \end{Bmatrix}_{300}$$

$$= \begin{bmatrix} 600 & 20,000 & 0 \\ 0 & 40,000 & 20,000 \\ 0 & 20,000 & 40,000 \\ 600 & 0 & 20,000 \end{bmatrix} \begin{Bmatrix} -9.4807 \\ 0.0569 \\ 0.0569 \end{Bmatrix}. \tag{7.559}$$

First assume that no plastic rotations exist, and it follows from Eq. 7.559 that

$$\begin{Bmatrix} M_1 \\ M_2 \\ M_3 \\ M_4 \end{Bmatrix}_{300} = \begin{bmatrix} 600 & 20,000 & 0 \\ 0 & 40,000 & 20,000 \\ 0 & 20,000 & 40,000 \\ 600 & 0 & 20,000 \end{bmatrix} \begin{Bmatrix} -9.4807 \\ 0.0569 \\ 0.0569 \end{Bmatrix} = \begin{Bmatrix} -4,550 \\ 3,413 \\ 3,413 \\ -4,550 \end{Bmatrix}. \tag{7.560}$$

The moments M_1 and M_4 are greater than M_c, and because the moment versus rotation relationship for the column is elastic–plastic, Eq. 7.559 becomes

Table 7.4 Numerical Calculation Procedure

k	a_k	x_{1k}	x_{2k}	$\dot{x}_{1k}$	M_{1k}	θ''_{1k}	M_{2k}	θ''_{2k}	x''_{1k}	x''_{2k}
0	−0.0003	0.0	0.0	0.0	0.0	0.0	0.0	0.0	0.0	0.0
1	−0.0020	0.0004	−0.0000	0.0198	0.19	0.0	−0.14	0.0	0.0	0.0
2	−0.0019	0.0039	−0.0000	0.1721	1.85	0.0	−1.39	0.0	0.0	0.0
⋮	⋮	⋮	⋮	⋮	⋮	⋮	⋮	⋮	⋮	⋮
299	−0.181	−9.1561	0.0549	−16.976	−4,395	0.0	3,296	0.0	0.0	0.0
300	−0.229	−9.4807	0.0569	−14.660	−4,500	−0.0013	3,413	0.0	−0.0726	0.0002
301	−0.216	−9.7496	0.0585	−11.811	−4,500	−0.0045	3,500	0.0002	−0.2639	0.0008

$$\left\{\begin{array}{c} -4,500 \\ M_2 \\ M_3 \\ -4,500 \end{array}\right\}_{300} + \begin{bmatrix} 40,000 & 0 & 0 & 0 \\ 0 & 40,000 & 20,000 & 0 \\ 0 & 20,000 & 40,000 & 0 \\ 0 & 0 & 0 & 40,000 \end{bmatrix}\left\{\begin{array}{c} \theta''_1 \\ 0 \\ 0 \\ \theta''_4 \end{array}\right\}_{300} = \left\{\begin{array}{c} -4,550 \\ 3,413 \\ 3,413 \\ -4,550 \end{array}\right\} \tag{7.561}$$

and solving for Eq. 7.561 gives

$$\left\{\begin{array}{c} M_1 \\ M_2 \\ M_3 \\ M_4 \end{array}\right\}_{300} = \left\{\begin{array}{c} -4,500 \\ 3,413 \\ 3,413 \\ -4,500 \end{array}\right\}, \quad \left\{\begin{array}{c} \theta''_1 \\ \theta''_2 \\ \theta''_3 \\ \theta''_4 \end{array}\right\}_{300} = \left\{\begin{array}{c} -0.0013 \\ 0 \\ 0 \\ -0.0013 \end{array}\right\}. \tag{7.562}$$

Finally, the inelastic displacement vector is obtained using Eq. 7.541, and

$$\left\{\begin{array}{c} x''_1 \\ x''_2 \\ x''_3 \end{array}\right\}_{300} = \begin{bmatrix} 24 & 600 & 600 \\ 600 & 80,000 & 20,000 \\ 600 & 20,000 & 80,000 \end{bmatrix}^{-1}\begin{bmatrix} 600 & 20,000 & 0 \\ 0 & 40,000 & 20,000 \\ 0 & 20,000 & 40,000 \\ 600 & 0 & 20,000 \end{bmatrix}\left\{\begin{array}{c} \theta''_1 \\ \theta''_2 \\ \theta''_3 \\ \theta''_4 \end{array}\right\}_{300} = \left\{\begin{array}{c} -0.0726 \\ 0.0002 \\ 0.0002 \end{array}\right\}. \tag{7.563}$$

This completes the calculation of the response for time step 300. The displacement time history for Cases 1, 2, and 3 can be calculated using the previously noted procedure and the responses are shown in Figures 7.59, 7.60, and 7.61, respectively.

EXAMPLE 2 *One-Story Frame with Mass Proportional Damping*

Consider the one-story frame discussed in Example 1 with

$$m = 1.0 \text{ k-s}^2/\text{in.}, \quad EI_1 = EI_2 = 1,000,000 \text{ k-in.}^2, \quad L_1 = L_2 = 100 \text{ in.}$$

It follows from Eq. 7.551 to Eq. 7.554 that

$$\mathbf{M} = \begin{bmatrix} 1.0 & 0.0 & 0.0 \\ 0.0 & 0.0 & 0.0 \\ 0.0 & 0.0 & 0.0 \end{bmatrix}, \quad \mathbf{K} = \begin{bmatrix} 24 & 600 & 600 \\ 600 & 80,000 & 20,000 \\ 600 & 20,000 & 80,000 \end{bmatrix} \tag{7.564}$$

$$\mathbf{K}_p = \begin{bmatrix} 600 & 0 & 0 & 600 \\ 20,000 & 40,000 & 20,000 & 0 \\ 0 & 20,000 & 40,000 & 20,000 \end{bmatrix} \tag{7.565}$$

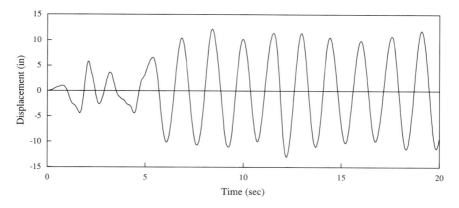

Figure 7.59 Structural Response with $M_c = 7,000$ k-in. and $M_b = 6,000$ k-in.

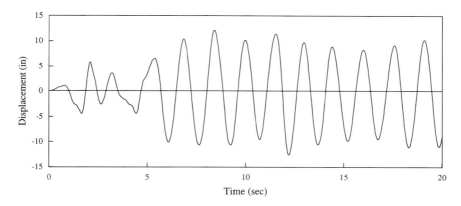

Figure 7.60 Structural Response with $M_c = 5,800$ k-in. and $M_b = 4,200$ k-in.

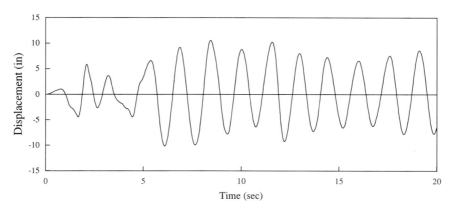

Figure 7.61 Structural Response with $M_c = 4,500$ k-in. and $M_b = 3,500$ k-in.

$$\mathbf{K}_R = \begin{bmatrix} 40,000 & 0 & 0 & 0 \\ 0 & 40,000 & 20,000 & 0 \\ 0 & 20,000 & 40,000 & 0 \\ 0 & 0 & 0 & 40,000 \end{bmatrix}. \tag{7.566}$$

Assume that the damping in the structure is represented by Rayleigh proportional damping with $\alpha = 0.012$ and $\beta = 0.0$. It then follows from Eq. 7.564 that

$$\mathbf{C} = (0.012)\begin{bmatrix} 1.0 & 0.0 & 0.0 \\ 0.0 & 0.0 & 0.0 \\ 0.0 & 0.0 & 0.0 \end{bmatrix} + (0)\mathbf{K} = \begin{bmatrix} 0.012 & 0.0 & 0.0 \\ 0.0 & 0.0 & 0.0 \\ 0.0 & 0.0 & 0.0 \end{bmatrix}. \tag{7.567}$$

The state transition matrix for this example was calculated in Example 3 of Section 6.8. Note that the dimension of the damping matrix is the same as the dimension of the mass matrix, and both of these matrices have a dimension less than the stiffness matrix. Therefore, Eq. 7.545 must be used to calculate the moments and plastic rotations and

$$\mathbf{K}_{33}^{-1} = \begin{bmatrix} 80,000 & 20,000 \\ 20,000 & 80,000 \end{bmatrix}^{-1} = \frac{1}{300,000}\begin{bmatrix} 4 & -1 \\ -1 & 4 \end{bmatrix} \tag{7.568}$$

$$\mathbf{K}_{33}^{-1}\mathbf{K}_{31} = \frac{1}{300,000}\begin{bmatrix} 4 & -1 \\ -1 & 4 \end{bmatrix}\begin{bmatrix} 600 \\ 600 \end{bmatrix} = \begin{bmatrix} 0.006 \\ 0.006 \end{bmatrix} \tag{7.569}$$

$$\mathbf{K}^{-1} = \begin{bmatrix} 24 & 600 & 600 \\ 600 & 80,000 & 20,000 \\ 600 & 20,000 & 80,000 \end{bmatrix}^{-1} = \begin{bmatrix} 0.059524 & -0.00036 & -0.00036 \\ -0.00036 & 0.000015 & -0.0000012 \\ -0.00036 & -0.0000012 & 0.000015 \end{bmatrix} \tag{7.570}$$

$$\begin{bmatrix} \mathbf{0} & \mathbf{0} & \mathbf{0} \\ \mathbf{0} & \mathbf{0} & \mathbf{0} \\ \mathbf{K}_{33}^{-1}\mathbf{K}_{31} & \mathbf{K}_{33}^{-1}\mathbf{K}_{32} & \mathbf{I} \end{bmatrix} = \begin{bmatrix} 0 & 0 & 0 \\ 0.006 & 1 & 0 \\ 0.006 & 0 & 1 \end{bmatrix}, \quad \begin{bmatrix} \mathbf{I} & \mathbf{0} \\ \mathbf{0} & \mathbf{I} \\ -\mathbf{K}_{33}^{-1}\mathbf{K}_{31} & -\mathbf{K}_{33}^{-1}\mathbf{K}_{32} \end{bmatrix} = \begin{bmatrix} 1 \\ -0.006 \\ -0.006 \end{bmatrix}. \tag{7.571}$$

It follows that

$$\mathbf{K}_p^T \begin{bmatrix} \mathbf{0} & \mathbf{0} & \mathbf{0} \\ \mathbf{0} & \mathbf{0} & \mathbf{0} \\ \mathbf{K}_{33}^{-1}\mathbf{K}_{31} & \mathbf{K}_{33}^{-1}\mathbf{K}_{32} & \mathbf{I} \end{bmatrix} \mathbf{K}^{-1}\mathbf{K}_p = \begin{bmatrix} 5,333 & 9,333 & 2,667 & -1,333 \\ 9,333 & 21,333 & 14,667 & 2,667 \\ 2,667 & 14,667 & 21,333 & 9,333 \\ -1,333 & 2,667 & 9,333 & 5,333 \end{bmatrix} \tag{7.572}$$

$$\overline{\mathbf{K}}_p^T = \begin{bmatrix} 600 & 20,000 & 0 \\ 0 & 40,000 & 20,000 \\ 0 & 20,000 & 40,000 \\ 600 & 0 & 20,000 \end{bmatrix}\begin{bmatrix} 1 \\ -0.006 \\ -0.006 \end{bmatrix} = \begin{bmatrix} 480 \\ -360 \\ -360 \\ 480 \end{bmatrix} \tag{7.573}$$

$$\overline{\mathbf{K}}_R = \begin{bmatrix} 34,667 & -9,333 & -2,667 & 1,333 \\ -9,333 & 18,667 & 5,333 & -2,667 \\ -2,667 & 5,333 & 18,667 & -9,333 \\ 1,333 & -2,667 & -9,333 & 34,667 \end{bmatrix}. \tag{7.574}$$

Therefore, Eq. 7.545 becomes

$$\begin{Bmatrix} M_1 \\ M_2 \\ M_3 \\ M_4 \end{Bmatrix}_{k+1} + \begin{bmatrix} 34,667 & -9,333 & -2,667 & 1,333 \\ -9,333 & 18,667 & 5,333 & -2,667 \\ -2,667 & 5,333 & 18,667 & -9,333 \\ 1,333 & -2,667 & -9,333 & 34,667 \end{bmatrix}\begin{Bmatrix} \theta_1'' \\ \theta_2'' \\ \theta_3'' \\ \theta_4'' \end{Bmatrix}_{k+1} = \begin{bmatrix} 480 \\ -360 \\ -360 \\ 480 \end{bmatrix}x_{1,k+1}. \tag{7.575}$$

Table 7.5 shows the numerical calculations of the response using the modified El-Centro earthquake (see Figure 6.11) with

$$M_c = 4,500 \text{ k-in.}, \quad M_b = 3,500 \text{ k-in.} \tag{7.576}$$

Table 7.5 Numerical Calculation Procedure

k	a_k	x_{1k}	$\dot{x}_{1k}$	M_{1k}	θ''_{1k}	M_{2k}	θ''_{2k}	x''_{1k}
	0–0.0003	0.0	0.0	0.0	0.0	0.0	0.0	0.0
1	−0.0020	0.0004	0.0198	0.19	0.0	−0.14	0.0	0.0
2	−0.0019	0.0039	0.1721	1.85	0.0	−1.38	0.0	0.0
⋮	⋮	⋮	⋮	⋮	⋮	⋮	⋮	⋮
299	−0.181	−9.1275	−16.968	−4381	0.0	3286	0.0	0.0
300	−0.229	−9.4521	−14.660	−4500	−0.0010	3413	0.0	−0.0587
301	−0.216	−9.7208	−11.815	−4500	−0.0046	3500	0.0	−0.2635

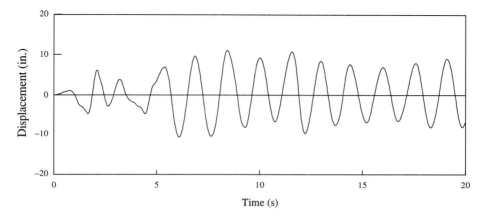

Figure 7.62 Structural Response with $M_c = 4{,}500$ k-in. and $M_b = 3{,}500$ k-in.

The displacement time history is shown in Figure 7.62. Note that the structure is symmetrical, where

$$x_{2k} = x_{3k}, \quad x''_{2k} = x''_{3k} \tag{7.577}$$

$$M_{1k} = M_{4k}, \quad \theta''_{1k} = \theta''_{4k} \tag{7.578}$$

$$M_{2k} = M_{3k}, \quad \theta''_{2k} = \theta''_{3k}. \tag{7.579}$$

RECOMMENDED READING

1. J. Wu. and R. D. Hanson "Study of Inelastic Spectra with High Damping." *Journal of the ASCE Structural Division,* Vol. 115, No. 6, pp. 1412–1431, 1989.

2. N. Newmark and W. Hall. EERI Monograph: "Earthquake Spectra and Design." Earthquake Engineering Research Institute, 1982.

3. H. Krawinkler and A. Nassar. ATC-40. Applied Technology Council, Redwood City, CA.

PROBLEMS

7.1 Repeat Example 1 in Section 7.1 but with the stiffness of the structure equal to one-half k_e.

7.2 Repeat Example 1 in Section 7.1 but with the stiffness of the spring equal to one-half k_1.

7.3 Repeat Example 2 in Section 7.2 but with 5% damping.

7.4 Repeat Example 2 in Section 7.2 but with the stiffness of the system decreasing to 10% of the initial stiffness when the response exceeds 1.0 in.

7.5 Repeat Example 2 in Section 7.2 but with the initial stiffness, k_1, equal to one-half the k_1 in the example and the yield displacement, x_y, remaining at 1.0 in.

7.6 Repeat Example 3 in Section 7.2 but use a yield displacement equal to one-half the yield displacement in the example (i.e., 0.3 in.).

7.7 Repeat Example 3 in Section 7.2 but use a damping value equal to two times the value in the example (i.e., $c = 2 \times 2.513 = 5.026$ k-s/in.).

7.8 Repeat Example 1 in Section 7.3 but with one-half the brace area (i.e., $A_1 = A_2 = A = 1.25$ in.^{2}).

7.9 Repeat Example 1 in Section 7.3 but with $A_1 = 2.50$ in.^{2} and $A_2 = 1.25$ in.^{2}

7.10 Repeat Example 1 in Section 7.3 but with $A_1 = 1.25$ in.^{2} and $A_2 = 2.50$ in.^{2}

7.11 Repeat Example 2 in Section 7.3 but use a damping value equal to two times the value in the example (i.e., $c = 2 \times 2.513 = 5.026$ k-s/in.).

7.12 Repeat Example 2 in Section 7.3 but with one-half the brace area (i.e., $A_1 = A_2 = A = 1.25$ in.^{2}).

7.13 Repeat Example 2 in Section 7.3 but with $A_1 = 2.50$ in.^{2} and $A_2 = 1.25$ in.^{2}

7.14 Repeat Example 2 in Section 7.3 but with $A_1 = 1.25$ in.^{2} and $A_2 = 2.50$ in.^{2}

7.15 Repeat Example 1 in Section 7.4 but with $k_e = 250$ k/in.

7.16 Repeat Example 1 in Section 7.4 but with $x_y = 0.5$ in.

7.17 Repeat Example 1 in Section 7.6 but with $EA_1 = 5,000$ kip, $EA_2 = 10,000$ kip, $EI_1 = 500,000$ k-in.^{2}, and $EI_2 = 1,000,000$ k-in.^{2}

7.18 Repeat Example 1 in Section 7.6 but with $L_1 = L_2 = 200$ in.

7.19 Repeat Example 2 in Section 7.7 but with $k = 0.5 \times 157.92 = 78.96$ k/in.

7.20 Repeat Example 2 in Section 7.7 but with $c = 2.0 \times 1.2567 = 2.5134$ k-s/in.

7.21 Repeat Example 1 in Section 7.8 but with $EI_1 = EI_2 = 500,000$ k-in.^{2}

7.22 Repeat Example 1 in Section 7.8 but with $L_1 = L_2 = 200$ in.

7.23 Repeat Example 1 in Section 7.8 but with $L_1 = 200$ in. and $L_2 = 100$ in.

7.24 Repeat Example 1 in Section 7.8 but with $\beta = 0.0010$.

7.25 Repeat Example 1 in Section 7.8 with Rayleigh damping, with a 5% damping ratio in the first two modes of vibration for the pre-yield state (i.e., the damping matrix remains constant before and after yielding).

Part Three

Special Topics in Structural Dynamics

Chapter 8

Base Isolation

8.1 OVERVIEW

For thousands of years, buildings and many other structures are designed to protect property and human life. The material in the first seven chapters of this book provides a theoretical foundation for analyzing a lateral force-resisting system in the elastic or inelastic response domain for earthquake ground motions. Prior to the late 1980s, all lateral force-resisting systems used in California and other geographical areas with a very severe seismic hazard were composed entirely of conventional materials (e.g., steel, concrete, and masonry). In the late 1980s, a California structural engineer, Dr. Ron Mayes, started a professional effort within the structural engineering community that resulted in the first design provisions for base-isolated buildings in the 1991 Uniform Building Code (UBC). This was the start of a new era of structural engineering, where new high-technology structural members (or elements) were added to structural members constructed with conventional materials to create an earthquake force-resisting system. Today, base-isolated structural members are used with another high-technology structural member, called viscous dampers, with design provisions in building codes. In the near future we will use structural control systems that bring the computer technology of the 21st century to the lateral force-resisting system.

The remainder of this section presents examples of buildings with base isolation. However, an introductory illustration will show the basic simplicity of the concept and the benefits to the structural engineer and owner of the building. Imagine a conventional steel frame building with beams and columns. At the foundation level, the columns are connected to the foundation. Now imagine that a high-technology element, called a *base isolator,* is placed between the bottom of the column and above the foundation. Figure 8.1 shows such a base isolator with the column attached to the top and the foundation attached to the bottom of the base isolator. Now imagine that an earthquake occurs and causes the base isolator and the building to deform. Because of the relatively flexible design of the base isolator, it will experience a large lateral deformation over its height (e.g., 10 to 20 in.) relative to the deformation between the floors of the buildings (e.g., 0.5 to 2 in.). Figure 8.2 is a photograph of the deformation of a base isolator during qualifying tests in a laboratory. This chapter discusses how these deformations are calculated using the theory of structural dynamics for structures with base isolators. Figure 8.3 compares structural responses with and without the base isolators between the columns and the foundation. Figure 8.3a shows the reduction in the maximum acceleration of a floor, and this reduction in floor accelera-

Figure 8.1 Base Isolator between Column and Foundation

Figure 8.2 Lateral Displacement of a Base Isolator

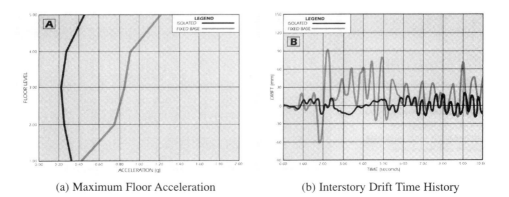

(a) Maximum Floor Acceleration   (b) Interstory Drift Time History

Figure 8.3 Building Response with and without Base Isolators

tion directly impacts the potential earthquake-induced damage to such items as equipment attached to the floors and the contents of the building. Figure 8.3b shows a reduction of the interstory drift, and this reduction in the response directly impacts the potential earthquake-induced damage to such items as glass, cladding, partitions, and structural walls.

The introduction of base isolators between the steel columns and the foundation is similar in concept to the use of shock absorbers between the driver of an automobile and the ground. The base isolators in the building and the shock absorbers in the automobile both are intended to smooth out the impact of an unwanted event (i.e., the earthquake for the building and the pothole or roughness of the road for the automobile).

Figures 8.4 to 8.8 show structures with base isolators.

Figure 8.4 Base Isolated Foothill Communities Law and Justice Center, Rancho Cucamonga, California

Figure 8.5 Base-Isolated Fire Command and Control Facility, Los Angeles, California

Figure 8.6 Base-Isolated Los Angeles City Hall

Figure 8.7 Base-Isolated Seismic Rehabilitation Building

Figure 8.8 Base-Isolated UCLA Student Union

8.2   INTRODUCTION TO A BASE-ISOLATED SYSTEM

Insight into the benefits of using base isolators in structures can be gained by considering the special case of a linear 2DOF spring-mass system shown in Figure 8.9. In the context of this section, the spring k_b represents the stiffness characteristics of a linear base isolator. In the same context, the spring k represents the stiffness of the structure above the base isolator. The equations of motion, obtained from the equilibrium equations on each mass, are

$$m_b\ddot{y}_1 + k(y_1 - y_2) + k_b(y_1 - x_g) = 0 \tag{8.1a}$$

$$m\ddot{y}_2 + k(y_2 - y_1) = 0. \tag{8.1b}$$

If the relative displacements between the masses and the support are defined to be

$$x_1 = y_1 - x_g \tag{8.2a}$$

$$x_2 = y_2 - x_g \tag{8.2b}$$

it then follows from substituting Eq. 8.2 into Eq. 8.1 that

$$m_b\ddot{x}_1 - kx_2 + (k + k_b)x_1 = -m_b\ddot{x}_g \tag{8.3a}$$

$$m\ddot{x}_2 + kx_2 - kx_1 = -m\ddot{x}_g. \tag{8.3b}$$

Consider the special case where m_b is very small and is assumed to be zero. Therefore, Eq. 8.3a becomes

$$-kx_2 + (k + k_b)x_1 = 0. \tag{8.4}$$

Solving for x_1 in terms of x_2 in Eq. 8.4 gives

$$x_1 = \left(\frac{k}{k + k_b}\right)x_2 = \left(\frac{1}{1 + (k_b/k)}\right)x_2. \tag{8.5}$$

The displacement x_1 is the displacement of the base isolator relative to the ground. Equation 8.5 gives the value of x_1 in terms of x_2 and the ratio of the stiffness of the isolator to the structure. Note that if k_b goes toward infinity (i.e., becomes very stiff), then x_1 goes toward zero. Also, if k_b is equal to k, then x_1 is equal to one half of x_2.

Substituting Eq. 8.5 into Eq. 8.3b gives the equation of motion for this spring-mass system:

$$m\ddot{x}_2 + \left[1 - \left(\frac{1}{1 + (k_b/k)}\right)\right]kx_2 = -m\ddot{x}_g. \tag{8.6}$$

One very important effect of the presence of base isolators, seen in Eq. 8.6, is the modification of the natural frequency of vibration of the system. For this example, the natural frequency of vibration of this spring-mass system is

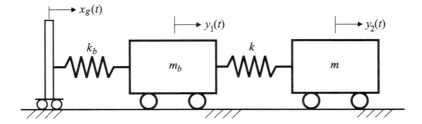

Figure 8.9 Two Degrees of Freedom Spring-Mass System

$$\omega_{nb} = \sqrt{\frac{k}{m}\left[1 - \left(\frac{1}{1 + (k_b/k)}\right)\right]} = C_1 \omega_n \tag{8.7}$$

where $\omega_n = \sqrt{k/m}$, and C_1 is the base-isolated natural frequency of vibration coefficient, defined as

$$C_1 = \sqrt{1 - \left(\frac{1}{1 + (k_b/k)}\right)}. \tag{8.8}$$

The natural period of vibration is

$$T_{nb} = \frac{2\pi}{\omega_{nb}} = \frac{2\pi}{\sqrt{\frac{k}{m}\left[1 - \left(\frac{1}{1 + (k_b/k)}\right)\right]}} = C_2 T_n \tag{8.9}$$

where $T_n = 2\pi/\omega_n$ and

$$C_2 = \frac{1}{\sqrt{1 - \left(\frac{1}{1 + (k_b/k)}\right)}} = \frac{1}{C_1}. \tag{8.10}$$

Insight into the meaning of a rigid, or fixed, base structure can be gained from Eq. 8.7. If k_b is much greater than k, then the term in the denominator, that is, $1 + (k_b/k)$, of Eq. 8.7 becomes large, and therefore ω_n approaches the natural frequency of a rigid base system $\sqrt{k/m}$ and T_n approaches the natural period of vibration of a rigid base system $2\pi/\sqrt{k/m}$. Figure 8.10 shows the value of the natural frequency of vibration (i.e., $\omega_{nb}/\sqrt{k/m}$) for different (k_b/k) ratios. This figure shows that for (k_b/k) ratios equal to 3 and 5, the base-isolated (k_b) solution is 87% and 91%, respectively, of the solution obtained for a rigid base system (i.e., $k_b \to \infty$).

The situation of interest for a base-isolated structure is the case where k_b is less than k. In the limit if k_b is very small, then ω_n goes to zero (see Eq. 8.7) and the natural period of vibration of the structure goes to infinity (see Eq. 8.9). Figure 8.10 shows that if the natural frequency of vibration is to be 50%, 33%, and 25% of its fixed-base value (i.e., $\omega_{nb}/\sqrt{k/m}$), then the ratio of (k_b/k) must be 33%, 12.5%, and 6.7%, respectively. Figure 8.11 shows similar information as in Figure 8.10 but in the more common form for struc-

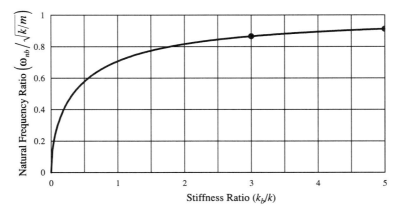

Figure 8.10 Variation of Natural Frequency of Vibration Ratio $\omega_{nb}/\sqrt{k/m}$ with Stiffness Ratio (k_b/k) for Stiffness Ratio Range of 0 to 5

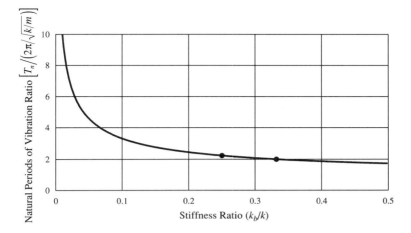

Figure 8.11 Variation of Natural Periods of Vibration Ratio $T_n\sqrt{k/m}/2\pi$) with Stiffness Ratio (k_b/k) for Stiffness Ratio Range 0 to 0.5

tural engineers, using elongation in a natural period of vibration (i.e., $T_n\sqrt{k/m}/2\pi$) with stiffness ratio.

The displacement x_1 can be expressed in terms of displacement x_2 using Eq. 8.5. Figure 8.12 shows a plot of the (x_1/x_2) ratio as a function of the (k_b/k) ratio. If the (k_b/k) ratio becomes large, then (x_1/x_2) tends toward zero. This is the fixed-base condition. Figure 8.13 shows the same information as in Figure 8.12 for the expanded scale of (k_b/k) from 0 to 0.5. The (k_b/k) ratios of 50%, 33%, and 25% correspond to ratios for (x_1/x_2) of 67%, 75%, and 80%, respectively.

The response of this spring-mass system to an earthquake ground motion can be obtained using the dynamic response solution methods discussed in Chapter 2. If a response spectra solution is desired, then the response can be obtained using the method described in Section 5.4. It follows from Eq. 8.6 that the relative displacement and maximum base shear are equal to

$$(x_1)_{max} = S_d(T_{nb},\zeta) = (T_{nb}/2\pi)^2 S_a(T_{nb},\zeta) \tag{8.11a}$$

$$V_{max} = mS_a(T_{nb},\zeta). \tag{8.11b}$$

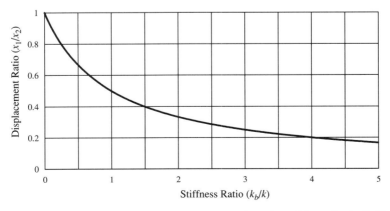

Figure 8.12 Variation of Displacement Ratio (x_1/x_2) with Stiffness Ratio (k_b/k) for Stiffness Ratio Range 0 to 5

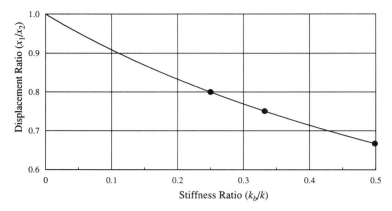

Figure 8.13 Variation of Displacement Ratio of (x_1/x_2) with Stiffness Ratios (k_b/k) for Stiffness Ratio Range 0 to 0.5

The maximum base shear depends on the magnitude of the spectral acceleration at the natural period of vibration. In general, for a 2DOF system with a fundamental period of vibration greater than 0.5 s, the magnitude of the spectral acceleration decreases as the value of the natural period of vibration increases. Therefore, a smaller value of k_b or the (k_b/k) ratio as given in Eq. 8.10 increases the value of T_{nb}. The square of T_{nb} typically provides a larger increase in the displacement than the decrease in displacement corresponding to the reduction in spectral acceleration. Therefore, the value of the displacement usually increases with the decrease in k_b or the (k_b/k) ratio.

Now imagine that if k_b is the stiffness of a set of base isolators and k is the stiffness of the structure placed on top of the base isolators, then a smaller value of k_b or the (k_b/k) ratio decreases the amount of base shear force that the earthquake inputs to the building. This earthquake force input reduction is the primary benefit in most situations that leads the structural engineer to isolate the base of the structures.

EXAMPLE 1   *Strain Energy*

The response of a structural system can be viewed from a strain energy perspective as noted in Section 2.10. As discussed in this section, consider spring k_b to be the base isolation substructure and k to be the substructure above the base isolator. The total strain energy in the two substructures is the total strain energy in the structure and can be written as

$$E_{sb} = \frac{1}{2}k_b x_1^2 \tag{8.12}$$

$$E_{ss} = \frac{1}{2}k(x_2 - x_1)^2 \tag{8.13}$$

where E_{sb} is the strain energy in spring k_b (i.e., the base isolator) and E_{ss} is the strain energy in spring k (i.e., the structure). The strain energy in the structure is a function of $(x_2 - x_1)$. Therefore, the total strain energy of the system, E_s (i.e., base isolator plus structure), is

$$E_s = E_{sb} + E_{ss} = \frac{1}{2}k_b x_1^2 + \frac{1}{2}k(x_2 - x_1)^2. \tag{8.14}$$

The difference in displacements between x_2 and x_1 can be written using Eq. 8.5, and it follows that

$$x_2 - x_1 = \left(\frac{k + k_b}{k}\right)x_1 - x_1 = \left(\frac{k_b}{k}\right)x_1. \tag{8.15}$$

Therefore, substituting Eq. 8.15 into Eq. 8.13, the strain energy in the substructure above the isolator is equal to

$$E_{ss} = \frac{1}{2}k\left(\frac{k_b}{k}\right)^2 x_1^2 = \frac{1}{2}k_b\left(\frac{k_b}{k}\right)x_1^2. \tag{8.16}$$

It then follows from Eq. 8.14 that

$$E_s = E_{sb} + E_{ss} = \frac{1}{2}k_b x_1^2 + \frac{1}{2}k_b\left(\frac{k_b}{k}\right)x_1^2. \tag{8.17}$$

Equations 8.12, 8.13, 8.16, and 8.17 show that as k_b decreases in magnitude, the strain energy in the k_b spring (base isolator substructure) increases in relation to the strain energy in the k spring (structure substructure).

Consider the case where $x_1 = 1.0$ and $k = 1.0$. If $k_b = 1.0$, then it follows from Eqs. 8.12, 8.13, 8.16, and 8.17 that

$$E_{sb} = \frac{1}{2}(1)(1)^2 = 0.5 \tag{8.18}$$

$$E_{ss} = \frac{1}{2}(1)\left(\frac{1}{1}\right)(1)^2 = 0.5 \tag{8.19}$$

$$E_s = E_{sb} + E_{ss} = 0.5 + 0.5 = 1.0. \tag{8.20}$$

In this case, the strain energies in the two springs (i.e., the two substructures) are equal. Now when k_b decreases to 0.5, then

$$E_{sb} = \frac{1}{2}(0.5)(1)^2 = 0.25 \tag{8.21}$$

$$E_{ss} = \frac{1}{2}(0.5)\left(\frac{0.5}{1.0}\right)(1)^2 = 0.125 \tag{8.22}$$

$$E_s = E_{sb} + E_{ss} = 0.25 + 0.125 = 0.625. \tag{8.23}$$

In this case, the strain energy in the k_b spring (i.e., base isolator substructure) is twice the strain energy in the k spring.

Another way of looking at the impact of the base isolator is to consider the case of a fixed displacement between the ends of spring k. In this case, the strain energy in the structure is

$$E_{ss} = \frac{1}{2}k(x_2 - x_1)^2 = \frac{1}{2}k\Delta^2 \tag{8.24}$$

where Δ is the fixed displacement between the ends of spring k. The strain energies noted in Eqs. 8.12 through 8.14, using the Δ notation, become

$$E_{sb} = \frac{1}{2}k_b x_1^2 \tag{8.25}$$

$$E_{ss} = \frac{1}{2}k\Delta^2 \tag{8.26}$$

$$E_s = E_{sb} + E_{ss} = \frac{1}{2}k_b x_1^2 + \frac{1}{2}k\Delta^2. \tag{8.27}$$

Consider the case where $\Delta = 1.0$, $k = 1.0$ and $k_b = 1.0$. Then it follows that

$$E_{sb} = \frac{1}{2}(1)x_1^2 = 0.5x_1^2 \tag{8.28}$$

$$E_{ss} = \frac{1}{2}(1)(1)^2 = 0.5 \tag{8.29}$$

$$E_s = E_{sb} + E_{ss} = 0.5x_1^2 + 0.5. \tag{8.30}$$

Therefore, as x_1 increases for a fixed Δ, the strain energy in the k_b spring increases. For this case, from Eq. 8.5,

$$x_1 = \left(\frac{1.0}{1.0+1.0}\right)x_2 = 0.5x_2 \tag{8.31}$$

and with

$$\Delta = 1.0 = x_2 - x_1 \tag{8.32}$$

it follows from solving the set of two simultaneous algebraic equations for the unknowns x_1 and x_2 in Eqs. 8.31 and 8.32 that

$$x_1 = 1.0, \quad x_2 = 2.0. \tag{8.33}$$

It then follows from Eq. 8.30 that the total system strain energy becomes

$$E_s = 0.5(1)^2 + 0.5 = 1.0. \tag{8.34}$$

Now change k_b to 0.5 but let k and Δ be the same. Then Eqs. 8.5 and 8.32 give the set of two simultaneous algebraic equations, which are

$$x_1 = \left(\frac{1.0}{1.0+0.5}\right)x_2 = \frac{2}{3}x_2 \tag{8.35}$$

$$\Delta = 1.0 = x_2 - x_1. \tag{8.36}$$

Solving Eqs. 8.35 and 8.36 for x_1 and x_2 gives

$$x_1 = 2.0, \quad x_2 = 3.0. \tag{8.37}$$

Then the strain energies are

$$E_{sb} = \frac{1}{2}(0.5)(2)^2 = 1.0 \tag{8.38}$$

$$E_{ss} = \frac{1}{2}(1)(1)^2 = 0.5 \tag{8.39}$$

$$E_s = E_{sb} + E_{ss} = 1.0 + 0.5 = 1.5. \tag{8.40}$$

Reducing k_b but keeping Δ a constant increases x_1, E_{sb}, and E_s.

The preceding discussion shows that if the structural engineer desires to have the strain energy of the spring k (i.e., the structure substructure) set to a fixed value, then the only way to increase the total strain energy of the system is to reduce k_b. Recall from Section 2.10 that for a system without damping, the total strain energy of the system must equal the total kinetic energy of the system. Therefore, introducing base isolators redirects this strain energy from the structure above the base isolators into the base isolators.

8.3 SOFT-STORY FRAME

The results presented in the preceding section show that the force that the earthquake imparts to the structure can be significantly reduced if the period of vibration of the structure increases. However, the price for this force reduction is an increase in the system displacement. Before discussing in more detail the use of base isolators in structures, this section will consider a single-story frame without base isolators. This discussion will provide insight into the response of what is called a *soft-story structure*, which is the theoretical precursor to a base-isolated structure.

Consider the one-story frame shown in Figure 8.14. The frame has six degrees of freedom denoted $y_1, y_2, \ldots, y_6$. In this example, the beam is assumed to be axially rigid; therefore,

$$y_1 = y_4 = x. \tag{8.41}$$

The beam is also assumed to be infinitely rigid in flexure, and therefore

$$y_3 = y_6 = 0. \tag{8.42}$$

Finally, the columns are assumed to be axially rigid, and therefore

$$y_2 = y_5 = 0. \tag{8.43}$$

The 6DOF stiffness matrix for the frame considering axial and flexural deformation but neglecting shearing deformations with unique beam and column properties is

$$\mathbf{K}_y = \begin{bmatrix} \frac{12E_1I_1}{h^3} + \frac{E_3A_3}{L} & 0 & \frac{6E_1I_1}{h^2} & -\frac{E_3A_3}{L} & 0 & 0 \\ 0 & \frac{E_1A_1}{h} + \frac{12E_3I_3}{L^3} & \frac{6E_3I_3}{L^2} & 0 & -\frac{12E_3I_3}{L^3} & \frac{6E_3I_3}{L^2} \\ \frac{6E_1I_1}{h^2} & \frac{6E_3I_3}{L^2} & \frac{4E_1I_1}{h} + \frac{4E_3I_3}{L} & 0 & -\frac{6E_3I_3}{L^2} & \frac{2E_3I_3}{L} \\ -\frac{E_3A_3}{L} & 0 & 0 & \frac{12E_2I_2}{h^3} + \frac{E_3A_3}{L} & 0 & \frac{6E_2I_2}{h^2} \\ 0 & -\frac{12E_3I_3}{L^3} & -\frac{6E_3I_3}{L^2} & 0 & \frac{E_2A_2}{h} + \frac{12E_3I_3}{L^3} & -\frac{6E_3I_3}{L^2} \\ 0 & \frac{6E_3I_3}{L^2} & \frac{2E_3I_3}{L} & \frac{6E_2I_2}{L_2^2} & -\frac{6E_3I_3}{L^2} & \frac{4E_2I_2}{h} + \frac{4E_3I_3}{L} \end{bmatrix}. \tag{8.44}$$

If the previous assumptions are to be invoked and if the strain energy is to be conserved, then the transformation from the y-coordinate system to the x-coordinate system must obey the strain energy conservation equation

$$SE = \frac{1}{2}\mathbf{Y}^T\mathbf{K}_y\mathbf{Y} = \frac{1}{2}\mathbf{X}^T\mathbf{K}_x\mathbf{X} \tag{8.45}$$

where in this case $\mathbf{Y}$ is a 6×1 vector and $\mathbf{X}$ is a 1×1 vector. The relationship between $\mathbf{Y}$ and $\mathbf{X}$ follows from Eqs. 8.41 through 8.43, and it is

$$\mathbf{Y} = \mathbf{T}\mathbf{X} \tag{8.46}$$

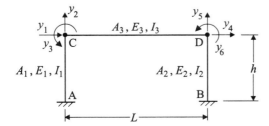

Figure 8.14 One-Story Frame Structure

where

$$\mathbf{T}^T = \begin{bmatrix} 1 & 0 & 0 & 1 & 0 & 0 \end{bmatrix}. \tag{8.47}$$

Therefore, $\mathbf{K}_x$ in Eq. 8.45 is

$$\mathbf{K}_x = \mathbf{T}^T \mathbf{K}_y \mathbf{T}. \tag{8.48}$$

Substituting Eqs. 8.44 and 8.46 into Eq. 8.48 gives

$$\mathbf{K}_x = \left(\frac{12E_1 I_1}{h^3} + \frac{E_3 A_3}{L} \right) - \frac{E_3 A_3}{L} + \left(\frac{12E_2 I_2}{h^3} + \frac{E_3 A_3}{L} \right) - \frac{E_3 A_3}{L} = \left(\frac{12E_1 I_1}{h^3} \right) + \left(\frac{12E_2 I_2}{h^3} \right). \tag{8.49}$$

The 6DOF structural system in the y-coordinate space reduces to an SDOF system in the x-coordinate space. Therefore, if all of the system mass is located at the beam level, then the equation of motion is

$$m\ddot{x} + kx = -m\ddot{x}_g \tag{8.50}$$

where

$$k = \mathbf{K}_x = \left(\frac{12E_1 I_1}{h^3} \right) + \left(\frac{12E_2 I_2}{h^3} \right). \tag{8.51}$$

The natural frequency of this SDOF system is

$$\omega_n = \sqrt{k/m}. \tag{8.52}$$

The response spectra solution for this structural system model was discussed in Section 5.4. The maximum base shear force from Eqs. 5.98 and 5.99, for a single degree of system where $W_1^* = 1.0$ and $\Gamma_1 = 1.0$, is

$$V_{\max} = W_1^* \Gamma_1^2 \hat{S}_a(T_n, \zeta) W = \hat{S}_a(T_n, \zeta) W. \tag{8.53}$$

Figure 8.15 shows a plot of an illustrative earthquake design spectrum. It is clear from this figure that as T_n increases beyond the value corresponding to the peak value of S_a (e.g., from $T_n = T_1$ to $T_n = T_2$), there is a reduction in the spectral acceleration (e.g., $S_a = S_{a1}$ to $S_a = S_{a2}$) and base shear force. As a consequence of this decrease in stiffness, lateral displacement increases because

$$(x_1)_{\max} = S_d(T_n, \zeta) = \left(T_n/2\pi \right)^2 S_a(T_n, \zeta). \tag{8.54}$$

Figure 8.16 shows the variation in S_d with T_n.

Structural engineers recognized years ago this reduction in spectral acceleration and base shear with an increase in the natural period of vibration. The stiffness can be reduced in Eq. 8.51 by reducing the moment of inertia or increasing the length of the columns.

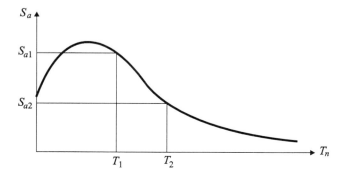

Figure 8.15 Design Earthquake Spectral Acceleration (S_a) versus Period (T_n)

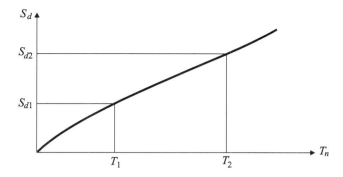

Figure 8.16 Design Earthquake Spectral Displacement (S_d) versus Period (T_n)

Such an action increases the magnitude of the spectral displacement and interstory drift (i.e., x). However, unless special detailing of the nonstructural parts of the structure is addressed (e.g., glass), this large interstory drift is not desirable. This type of structure is called a *soft-story structure*.

Now consider the same one-story frame shown in Figure 8.14, but now assume that the base of the building is dropped by a distance h_b, as shown in Figure 8.17. Extending this base creates a two-story frame, and the original 6DOF system in Figure 8.14 is now expanded to a 12DOF system. The stiffness matrix for this 12DOF system is 12×12. Similar to the 6DOF system, certain assumptions can be made to simplify the problem. However, a very important structural engineering decision influences these assumption, and it is the existence or not of a rigid structural element between nodes E and F.

Consider the case where the member between nodes E and F is rigid. The assumption of members CD and EF being axially rigid results in

$$y_1 = y_4, \quad y_7 = y_{10}. \tag{8.55}$$

If the members CD and EF are also assumed to be flexurally rigid, then

$$y_3 = y_6 = y_9 = y_{12} = 0. \tag{8.56}$$

In addition, if the columns are assumed to be axially rigid, then

$$y_2 = y_5 = y_8 = y_{11} = 0. \tag{8.57}$$

Therefore, if x_1 and x_2 are defined to be

$$x_1 = y_1, \quad x_2 = y_7, \tag{8.58}$$

then the relationship between the x-coordinates and y-coordinates is

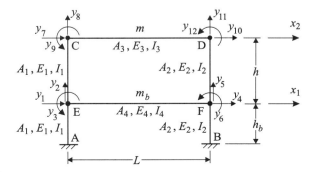

Figure 8.17 Extension of Base to Create a Two-Story Frame

$$Y = \begin{Bmatrix} y_1 \\ y_2 \\ y_3 \\ y_4 \\ y_5 \\ y_6 \\ y_7 \\ y_8 \\ y_9 \\ y_{10} \\ y_{11} \\ y_{12} \end{Bmatrix} = \begin{bmatrix} 1 & 0 \\ 0 & 0 \\ 0 & 0 \\ 1 & 0 \\ 0 & 0 \\ 0 & 0 \\ 0 & 1 \\ 0 & 0 \\ 0 & 0 \\ 0 & 1 \\ 0 & 0 \\ 0 & 0 \end{bmatrix} \begin{Bmatrix} x_1 \\ x_2 \end{Bmatrix} = \mathbf{TX} \tag{8.59}$$

where the matrix $\mathbf{T}$ is the transformation matrix that relates the x- and y-coordinates.

The stiffness matrix for this structure in the x-coordinate system is

$$\mathbf{K}_x = \mathbf{T}^T \mathbf{K}_y \mathbf{T} \tag{8.60}$$

where

$$\mathbf{K}_x = \begin{bmatrix} \left(\dfrac{12E_1 I_1}{h_b^3} + \dfrac{12E_2 I_2}{h_b^3} + \dfrac{12E_1 I_1}{h^3} + \dfrac{12E_2 I_2}{h^3} \right) & -\left(\dfrac{12E_1 I_1}{h^3} + \dfrac{12E_2 I_2}{h^3} \right) \\ -\left(\dfrac{12E_1 I_1}{h^3} + \dfrac{12E_2 I_2}{h^3} \right) & \left(\dfrac{12E_1 I_1}{h^3} + \dfrac{12E_2 I_2}{h^3} \right) \end{bmatrix}. \tag{8.61}$$

The structure is a two-story frame structure, and the reduced structural system is a 2DOF system with x_1 and x_2 as the unknowns. The equation of motion is

$$\mathbf{M\ddot{X}} + \mathbf{C\dot{X}} + \mathbf{K}_x \mathbf{X} = -\mathbf{M}\{\mathbf{I}\}a(t) \tag{8.62}$$

where

$$\mathbf{M} = \begin{bmatrix} m_b & 0 \\ 0 & m \end{bmatrix}, \quad \mathbf{C} = \begin{bmatrix} c_1 & -c_2 \\ -c_2 & c_3 \end{bmatrix} \tag{8.63}$$

and $a(t)$ is the earthquake ground acceleration. The $\mathbf{C}$ matrix is noted in general terms and is discussed in more detail later in this chapter.

Consider the undamped structure and assume that

$$m = m_b \tag{8.64a}$$

$$k_2 = \left(\frac{12E_1 I_1}{h^3} \right) + \left(\frac{12E_2 I_2}{h^3} \right) = k \tag{8.64b}$$

$$k_1 = \left(\frac{12E_1 I_1}{h_b^3} \right) + \left(\frac{12E_2 I_2}{h_b^3} \right) = k_b = \alpha k. \tag{8.64c}$$

It follows from Eqs. 8.61, 8.63, and 8.64 that

$$\mathbf{M} = \begin{bmatrix} m & 0 \\ 0 & m \end{bmatrix}, \quad \mathbf{K} = \begin{bmatrix} \alpha k + k & -k \\ -k & k \end{bmatrix}. \tag{8.65}$$

The natural frequencies of vibration are given in Eqs. 3.32 and 3.33, and therefore it follows that

$$\omega_1^2 = \left(\frac{\alpha + 2 - \sqrt{\alpha^2 + 4}}{2}\right)\omega_n^2, \quad \omega_2^2 = \left(\frac{\alpha + 2 + \sqrt{\alpha^2 + 4}}{2}\right)\omega_n^2 \qquad (8.66a)$$

$$T_1 = \frac{2\pi}{\omega_1} = \sqrt{\frac{2}{\alpha + 2 - \sqrt{\alpha^2 + 4}}}\left(\frac{2\pi}{\omega_n}\right), \quad T_2 = \frac{2\pi}{\omega_2} = \sqrt{\frac{2}{\alpha + 2 + \sqrt{\alpha^2 + 4}}}\left(\frac{2\pi}{\omega_n}\right). \qquad (8.66b)$$

The mode shapes of this 2DOF system are

$$\boldsymbol{\phi}_1 = \begin{Bmatrix} 1 \\ r_1 \end{Bmatrix}, \quad \boldsymbol{\phi}_2 = \begin{Bmatrix} 1 \\ r_2 \end{Bmatrix} \qquad (8.67)$$

where r_1 and r_2, given in Eq. 3.12, are equal to

$$r_1 = \left(\frac{-m_1\omega_1^2 + k_1 + k_2}{k_2}\right), \quad r_2 = \left(\frac{-m_1\omega_2^2 + k_1 + k_2}{k_2}\right). \qquad (8.68)$$

Substituting Eqs. 8.64 and 8.66 into Eq. 8.68 gives

$$r_1 = \left(\frac{\alpha + \sqrt{\alpha^2 + 4}}{2}\right), \quad r_2 = \left(\frac{\alpha - \sqrt{\alpha^2 + 4}}{2}\right). \qquad (8.69)$$

and therefore Eq. 8.67 becomes

$$\boldsymbol{\phi}_1 = \begin{Bmatrix} 1 \\ \left(\alpha + \sqrt{\alpha^2 + 4}\right)/2 \end{Bmatrix}, \quad \boldsymbol{\phi}_2 = \begin{Bmatrix} 1 \\ \left(\alpha - \sqrt{\alpha^2 + 4}\right)/2 \end{Bmatrix}. \qquad (8.70)$$

Note that when $\alpha \to 0$, $\omega_1 \to 0$, and therefore $T_1 = (2\pi/\omega_1) \to \infty$. In addition, when $\alpha \to 0$, $r_1 \to 1$ and thus the fundamental mode of vibration becomes

$$\boldsymbol{\phi}_1 = \begin{Bmatrix} 1 \\ 1 \end{Bmatrix}. \qquad (8.71)$$

This type of mode shape is often called a *rigid-body mode shape* or a *one-one mode shape*. Now consider the case where the masses of the two stories are different, that is,

$$m \neq m_b. \qquad (8.72)$$

Then Eq. 8.65 becomes

$$\mathbf{M} = \begin{bmatrix} m_b & 0 \\ 0 & m \end{bmatrix}, \quad \mathbf{K} = \begin{bmatrix} \alpha k + k & -k \\ -k & k \end{bmatrix}. \qquad (8.73)$$

The first natural frequency of vibration can be computed using Eq. 3.10a:

$$\omega_1^2 = \frac{\left((\alpha + 1)km + km_b\right) - \sqrt{\left((\alpha + 1)km + km_b\right)^2 - 4m_b m \alpha k^2}}{2m_b m}. \qquad (8.74)$$

When $\alpha \to 0$, $\omega_1 \to 0$ and again $T_1 = (2\pi/\omega_1) \to \infty$. This shows that decreasing the value of α increases the first period of vibration. From Eq. 8.68 and when $\alpha \to 0$, r_1 is given by

$$r_1 = \left(\frac{-m_1\omega_1^2 + k_1 + k_2}{k_2}\right) = \left(\frac{-m_1\omega_1^2}{k_2} + \alpha + 1\right) \to 1. \qquad (8.75)$$

Thus, the fundamental mode shape of vibration again becomes

$$\boldsymbol{\phi}_1 = \begin{Bmatrix} 1 \\ 1 \end{Bmatrix}. \qquad (8.76)$$

The response of this 2DOF system can be calculated using the normal mode method, and from Eq. 3.98 it follows that

$$\mathbf{X}(t) = \boldsymbol{\phi}_1 q_1(t) + \boldsymbol{\phi}_2 q_2(t). \tag{8.77}$$

The response spectra solution method presented in Chapter 5 can be used to calculate the earthquake response of this 2DOF system. Consider, for example, the special case of using only the first mode response, and then Eq. 5.166 gives

$$\max[q_1(t)] = \Gamma_1 S_d(T_1, \zeta_1) \tag{8.78}$$

where

$$\Gamma_1 = \frac{\boldsymbol{\phi}_1^T \mathbf{M}\{\mathbf{I}\}}{\boldsymbol{\phi}_1^T \mathbf{M}\boldsymbol{\phi}_1}. \tag{8.79}$$

Note from Eq. 8.76 that in the limit as α becomes small, the fundamental mode shape goes toward $\boldsymbol{\phi}_1 = \{\mathbf{I}\}$. Therefore, in the limit Eq. 8.79 gives $\Gamma_1 = 1$. Equation 8.78 then becomes

$$\max[q_1(t)] = S_d(T_1, \zeta_1). \tag{8.80}$$

The response spectra solution for the second mode response is given by Eq. 5.167, which is

$$\max[q_2(t)] = \Gamma_2 S_d(T_2, \zeta_2) \tag{8.81}$$

where

$$\Gamma_2 = \frac{\boldsymbol{\phi}_2^T \mathbf{M}\{\mathbf{I}\}}{\boldsymbol{\phi}_2^T \mathbf{M}\boldsymbol{\phi}_2}. \tag{8.82}$$

Recall from Eqs. 8.75 and 8.76 that in the limit as α becomes small $\boldsymbol{\phi}_1 = \{\mathbf{I}\}$, and therefore based on the orthogonality of mode shapes (see Section 4.4), it follows that $\boldsymbol{\phi}_2^T \mathbf{M}\{\mathbf{I}\} = \boldsymbol{\phi}_2^T \mathbf{M}\boldsymbol{\phi}_1 = 0$. Therefore, Eq. 8.82 gives $\Gamma_2 = 0$, and Eq. 8.81 becomes

$$\max[q_2(t)] = 0. \tag{8.83}$$

Note that in Eq. 8.83, the maximum response in the second mode is equal to 0. Finally, for a response spectra solution, the system response as α becomes small is

$$\max[\mathbf{X}(t)] = \boldsymbol{\phi}_1 \max[q_1(t)] = \begin{Bmatrix} 1 \\ 1 \end{Bmatrix} S_d(T_2, \zeta_2). \tag{8.84}$$

Since the response of the second mode is equal to 0, it follows that the maximum base shear is given by

$$\max[V(t)] = m_1^* S_a(T_1, \zeta_1) \tag{8.85}$$

where

$$m_1^* = \begin{Bmatrix} 1 \\ 1 \end{Bmatrix}^T \begin{bmatrix} m_b & 0 \\ 0 & m \end{bmatrix} \begin{Bmatrix} 1 \\ 1 \end{Bmatrix} = m_b + m. \tag{8.86}$$

The preceding derivation shows the limiting case where α becomes small (i.e., $\alpha \to 0$). In reality, α will be a small number but will not be exactly equal to zero. However, structural engineers generally accept that the contribution of the second mode to x_1, x_2, and the base shear is small for two reasons in addition to the preceding discussion. First, the second natural frequency becomes large as α increases, and therefore the value of the spectral displacement becomes small and goes toward zero. Second, the value of the damping in a mode of vibration that corresponds to a high natural frequency is believed to be large.

The one- and two-story frame structures discussed in this section show how it is possible during an earthquake to reduce the base shear force and the acceleration of the floors if the stiffness of the columns attached to the ground is reduced. However, for this soft-story structure, the relative displacement between the bottom story and the ground becomes large as α becomes small. As previously noted, the ground story of a building usually has a significant amount of glass and other architecturally sensitive design elements, and therefore this large interstory displacement during an earthquake is generally not acceptable.

The next section shows how this soft-story concept can be combined with a high-technology base isolator to provide a base-isolated structure.

EXAMPLE 1 *Sensitivity of 2DOF Response to Bottom Floor Stiffness*

Consider the undamped 2DOF system with $m_b = m_1 = m = 1.0$ k-s^2/in., as shown in Figure 8.18. Assume that $k_1 = k$ and $k_b = \alpha k$. Recall from Eq. 8.66a that the natural frequencies of vibration are

$$\omega_1^2 = \left(\frac{\alpha + 2 - \sqrt{\alpha^2 + 4}}{2} \right) \omega_n^2, \quad \omega_2^2 = \left(\frac{\alpha + 2 + \sqrt{\alpha^2 + 4}}{2} \right) \omega_n^2. \tag{8.87}$$

If α is very large, this is referred to as a fixed, or rigid, base structure. In this case, the bottom story is very stiff, and the system is an SDOF system with a natural frequency of vibration equal to

$$\omega_1^2 = \omega_n^2, \quad \omega_2^2 \to \infty. \tag{8.88}$$

Assume for this case (i.e., SDOF system) that the natural period of vibration is 0.3 s. Then

$$k = \omega_n^2 m = \left(\frac{2\pi}{T_n} \right)^2 m = \left(\frac{2\pi}{0.3} \right)^2 (1) = 438.6 \text{ k/in.} \tag{8.89}$$

Consider the earthquake design spectrum shown in Figure 8.19. It follows that

$$S_a(T_n = 0.3 \text{s}) = 1.0g \tag{8.90}$$

and

$$S_d(T_n = 0.3 \text{ s}) = \frac{S_a}{\omega_n^2} = \left(\frac{T_n}{2\pi} \right)^2 S_a = \left(\frac{0.3}{2\pi} \right)^2 (386.4) = 0.881 \text{ in.} \tag{8.91}$$

Therefore, the top interstory displacement is

$$\Delta x_{2-1}(t) = x_2(t) - x_1(t). \tag{8.92}$$

Because α is very large, it follows that $x_1(t) \approx 0$, and the maximum interstory displacement is

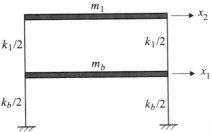

Figure 8.18 Two-Story Structure

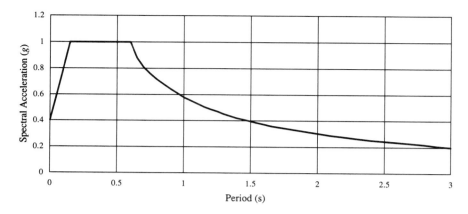

Figure 8.19 Earthquake Design Spectrum

$$\max\left[\Delta x_{2-1}(t)\right] = \max\left[x_2(t)\right] = S_d = 0.881 \text{ in.} \tag{8.93}$$

The absolute acceleration of the top floor is

$$\ddot{y}_2(t) = \ddot{x}_2(t) + a(t) \tag{8.94}$$

and therefore the maximum acceleration for this case is

$$\max\left[\ddot{y}_2(t)\right] = S_a = 1.0g. \tag{8.95}$$

The top floor force is equal to

$$S_2(t) = m_1\ddot{y}_2(t) = (1)\ddot{y}_2(t) = \ddot{y}_2(t) \tag{8.96}$$

and therefore the maximum top floor force for this case is

$$\max\left[S_2(t)\right] = \max\left[\ddot{y}_2(t)\right] = S_a = 1.0g. \tag{8.97}$$

Now consider the same 2DOF system but assume that the fundamental mode natural period of vibration, T_1, is increased to a value of 0.6 s due to a reduction in the value of the bottom-story stiffness, k_b. Recall that $k_b = \alpha k$ and assume the same k value as given in Eq. 8.89. It follows that the value of α can be obtained from Eq. 8.66a, where

$$\omega_1^2 = \left(\frac{2\pi}{T_1}\right)^2 = \left(\frac{2\pi}{0.6}\right)^2 = 109.7 = \left(\frac{\alpha + 2 - \sqrt{\alpha^2 + 4}}{2}\right)(438.6). \tag{8.98}$$

Therefore, solving for α in Eq. 8.98 gives

$$\alpha = 0.583. \tag{8.99}$$

From Eq. 8.66a,

$$\omega_1 = \sqrt{\left(\frac{0.583 + 2 - \sqrt{(0.583)^2 + 4}}{2}\right)(438.6)} = 10.47 \text{ rad/s} \tag{8.100}$$

$$\omega_2 = \sqrt{\left(\frac{0.583 + 2 + \sqrt{(0.583)^2 + 4}}{2}\right)(438.6)} = 31.99 \text{ rad/s} \tag{8.101}$$

and from Eq. 8.69,

$$r_1 = \frac{0.583 + \sqrt{(0.583)^2 + 4}}{2} = 1.33 \tag{8.102}$$

$$r_2 = \frac{0.583 - \sqrt{(0.583)^2 + 4}}{2} = -0.75. \tag{8.103}$$

Therefore, it follows from Eqs. 8.100 to 8.103 that

$$T_2 = \frac{2\pi}{31.99} = 0.196 \text{ s} \tag{8.104}$$

$$\Phi_1 = \left\{ \begin{matrix} 1.0 \\ 1.33 \end{matrix} \right\} = \left\{ \begin{matrix} 0.75 \\ 1.0 \end{matrix} \right\}, \quad \Phi_2 = \left\{ \begin{matrix} 1.0 \\ -0.75 \end{matrix} \right\} = \left\{ \begin{matrix} -1.33 \\ 1.0 \end{matrix} \right\}. \tag{8.105}$$

Consider now a comparison of the response of the system with a very large α and with $\alpha = 0.583$. A very large α results in a system that has two natural periods of vibration: 0.30 s and very near zero. With $\alpha = 0.583$, the two natural periods of vibration become 0.60 s and 0.20 s. Note from Figure 8.19 that both periods of vibration shift to the right on the response spectrum plot.

Now consider the same 2DOF system shown in Figure 8.18 but assume that the stiffness of the bottom story, k_b, has been reduced even more and that the fundamental period of vibration, T_1, is now equal to 2.0 s. Using the same k value as given in Eq. 8.89, it follows that the value of α required to give this value of T_1 can be obtained from Eq. 8.66, where

$$\left(\frac{2\pi}{2.0}\right)^2 = 9.870 = \left(\frac{\alpha + 2 - \sqrt{\alpha^2 + 4}}{2}\right)(438.6). \tag{8.106}$$

Therefore, solving for α in Eq. 8.106 gives

$$\alpha = 0.0455. \tag{8.107}$$

Note that the value of k_b for this case is approximately 5% of the second-story stiffness. It then follows from Eq. 8.66a that

$$\omega_1 = \sqrt{\left(\frac{0.0455 + 2 - \sqrt{(0.0455)^2 + 4}}{2}\right)(438.6)} = 3.14 \text{ rad/s} \tag{8.108}$$

$$\omega_2 = \sqrt{\left(\frac{0.0455 + 2 + \sqrt{(0.0455)^2 + 4}}{2}\right)(438.6)} = 29.79 \text{ rad/s} \tag{8.109}$$

and from Eq. 8.69,

$$r_1 = \frac{0.0455 + \sqrt{(0.0455)^2 + 4}}{2} = 1.023 \tag{8.110}$$

$$r_2 = \frac{0.0455 - \sqrt{(0.0455)^2 + 4}}{2} = -0.978. \tag{8.111}$$

Therefore, it follows from Eqs. 8.108 to 8.111 that

$$T_2 = \frac{2\pi}{29.79} = 0.211 \text{ s} \tag{8.112}$$

$$\Phi_1 = \left\{ \begin{matrix} 1.0 \\ 1.023 \end{matrix} \right\} = \left\{ \begin{matrix} 0.978 \\ 1.0 \end{matrix} \right\}, \quad \Phi_2 = \left\{ \begin{matrix} 1.0 \\ -0.978 \end{matrix} \right\} = \left\{ \begin{matrix} -1.023 \\ 1.0 \end{matrix} \right\}. \tag{8.113}$$

The decrease in bottom-story stiffness from $\alpha = 0.583$ to $\alpha = 0.0455$ (i.e., 58% to 5% of the top-story stiffness) shifts the natural periods of vibration of the first and second modes of vibration from 0.60 s and 0.196 s to 2.0 s and 0.211 s. Notice that the first mode natural period of vibration changes significantly, from 0.60 s to 2.0 s. However, the change in the value of the second mode natural period of vibration is small (i.e., from 0.196 s to 0.211 s).

The first mode shape variation with α is shown in Figure 8.20. The decrease in the first-story stiffness (i.e., a decrease in the value of α) decreases the relative displacement between the second and first floors. For example, for $\alpha = 0.583$ for this relative displacement is

$$\Delta x_{2-1}(t) = x_2(t) - x_1(t) = 1.0 - 0.75 = 0.25. \tag{8.114}$$

When α is reduced to $\alpha = 0.0455$, then the relative displacement reduces to

$$\Delta x_{2-1}(t) = x_2(t) - x_1(t) = 1.0 - 0.978 = 0.022. \tag{8.115}$$

Figure 8.21 shows the variation in the second mode shape with α. Insight into this variation and its impact on response can be gained by calculating the first and second mode participation factors. For $\alpha = 0.583$, it follows that

$$\Gamma_1 = \frac{\begin{Bmatrix} 0.75 \\ 1.0 \end{Bmatrix}^T \begin{bmatrix} 1 & 0 \\ 0 & 1 \end{bmatrix} \begin{Bmatrix} 1.0 \\ 1.0 \end{Bmatrix}}{\begin{Bmatrix} 0.75 \\ 1.0 \end{Bmatrix}^T \begin{bmatrix} 1 & 0 \\ 0 & 1 \end{bmatrix} \begin{Bmatrix} 0.75 \\ 1.0 \end{Bmatrix}} = 1.12 \tag{8.116}$$

$$\Gamma_2 = \frac{\begin{Bmatrix} -1.33 \\ 1.0 \end{Bmatrix}^T \begin{bmatrix} 1 & 0 \\ 0 & 1 \end{bmatrix} \begin{Bmatrix} 1.0 \\ 1.0 \end{Bmatrix}}{\begin{Bmatrix} -1.33 \\ 1.0 \end{Bmatrix}^T \begin{bmatrix} 1 & 0 \\ 0 & 1 \end{bmatrix} \begin{Bmatrix} -1.33 \\ 1.0 \end{Bmatrix}} = -0.12. \tag{8.117}$$

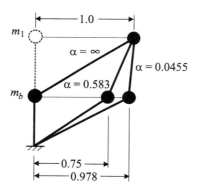

Figure 8.20 Fundamental Mode of Vibrations

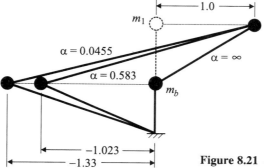

Figure 8.21 Second Mode of Vibrations

Similarly, for $\alpha = 0.0455$ it follows that

$$\Gamma_1 = \frac{\left\{ \begin{matrix} 0.978 \\ 1.0 \end{matrix} \right\}^T \begin{bmatrix} 1 & 0 \\ 0 & 1 \end{bmatrix} \left\{ \begin{matrix} 1.0 \\ 1.0 \end{matrix} \right\}}{\left\{ \begin{matrix} 0.978 \\ 1.0 \end{matrix} \right\}^T \begin{bmatrix} 1 & 0 \\ 0 & 1 \end{bmatrix} \left\{ \begin{matrix} 0.978 \\ 1.0 \end{matrix} \right\}} = 1.01 \tag{8.118}$$

$$\Gamma_2 = \frac{\left\{ \begin{matrix} -1.023 \\ 1.0 \end{matrix} \right\}^T \begin{bmatrix} 1 & 0 \\ 0 & 1 \end{bmatrix} \left\{ \begin{matrix} 1.0 \\ 1.0 \end{matrix} \right\}}{\left\{ \begin{matrix} -1.023 \\ 1.0 \end{matrix} \right\}^T \begin{bmatrix} 1 & 0 \\ 0 & 1 \end{bmatrix} \left\{ \begin{matrix} -1.023 \\ 1.0 \end{matrix} \right\}} = -0.011. \tag{8.119}$$

Using the response spectrum shown in Figure 8.19, it follows for $\alpha = 0.583$ that

$$S_a(T_1) = 1.0g \tag{8.120a}$$

$$S_a(T_2) = 1.0g \tag{8.120b}$$

$$S_d(T_1) = \left(\frac{0.6}{2\pi} \right)^2 (386.4) = 3.52 \text{ in.} \tag{8.120c}$$

$$S_d(T_2) = \left(\frac{0.196}{2\pi} \right)^2 (386.4) = 0.38 \text{ in.} \tag{8.120d}$$

The response of the top floor and base shear in the first mode is

$$\max[x_2(t)] = \varphi_{21}\Gamma_1 S_d = (1.0)(1.12)(3.52) = 3.94 \text{ in.} \tag{8.121a}$$

$$\max[x_2(t) - x_1(t)] = \varphi_{21}\Gamma_1 S_d - \varphi_{11}\Gamma_1 S_d = (1.0 - 0.75)(1.12)(3.52) = 0.99 \text{ in.} \tag{8.121b}$$

$$\max[\ddot{y}_2(t)] = \varphi_{21}\Gamma_1 S_a = (1.0)(1.12)(1.0g) = 1.12g \tag{8.121c}$$

$$\max[V(t)] = m_1^* \Gamma_1^2 S_a = (0.75^2 + 1.0^2)(1.12)^2(1.0g) = 1.96g. \tag{8.121d}$$

The response of the top floor and base shear in the second mode is

$$\max[x_2(t)] = \varphi_{22}\Gamma_2 S_d = (1.0)(0.12)(0.38) = 0.046 \text{ in.} \tag{8.122a}$$

$$\max[x_2(t) - x_1(t)] = \varphi_{22}\Gamma_2 S_d - \varphi_{12}\Gamma_2 S_d = (1.0 + 1.33)(0.12)(0.38) = 0.11 \text{ in.} \tag{8.122b}$$

$$\max[\ddot{y}_2(t)] = \varphi_{22}\Gamma_2 S_a = (1.0)(0.12)(1.0g) = 0.12g \tag{8.122c}$$

$$\max[V(t)] = m_2^* \Gamma_2^2 S_a = (1.33^2 + 1.0^2)(0.12)^2(1.0g) = 0.040g. \tag{8.122d}$$

Using the response spectrum shown in Figure 8.19, it then follows for the softer first story with $\alpha = 0.0455$ that

$$S_a(T_1) = 0.30g \tag{8.123a}$$

$$S_a(T_2) = 1.0g \tag{8.123b}$$

$$S_d(T_1) = \left(\frac{2.0}{2\pi} \right)^2 (0.3 \times 386.4) = 11.7 \text{ in.} \tag{8.123c}$$

$$S_d(T_2) = \left(\frac{0.211}{2\pi} \right)^2 (386.4) = 0.44 \text{ in.} \tag{8.123d}$$

The response for the top floor and base shear in the first mode is

$$\max[x_2(t)] = \varphi_{21}\Gamma_1 S_d = (1.0)(1.01)(11.7) = 11.8 \text{ in.} \tag{8.124a}$$

$$\max[x_2(t) - x_1(t)] = \varphi_{21}\Gamma_1 S_d - \varphi_{11}\Gamma_1 S_d = (1.0 - 0.978)(1.01)(11.7) = 0.26 \text{ in.} \tag{8.124b}$$

$$\max[\ddot{y}_2(t)] = \varphi_{21}\Gamma_1 S_a = (1.0)(1.01)(0.30g) = 0.30g \tag{8.124c}$$

$$\max[V(t)] = m_1^*\Gamma_1^2 S_a = (0.978^2 + 1.0^2)(1.01)^2(0.30g) = 0.60g \tag{8.124d}$$

The response for the top story and base shear in the second mode is

$$\max[x_2(t)] = \varphi_{22}\Gamma_2 S_d = (1.0)(0.011)(0.44) = 0.005 \text{ in.} \tag{8.125a}$$

$$\max[x_2(t) - x_1(t)] = \varphi_{22}\Gamma_2 S_d - \varphi_{12}\Gamma_2 S_d = (1.0 + 1.023)(0.011)(0.44) = 0.010 \text{ in.} \tag{8.125b}$$

$$\max[\ddot{y}_2(t)] = \varphi_{22}\Gamma_2 S_a = (1.0)(0.011)(1.0g) = 0.011g \tag{8.125c}$$

$$\max[V(t)] = m_2^*\Gamma_2^2 S_a = (1.023^2 + 1.0^2)(0.011)^2(1.0g) = 0.0002g. \tag{8.125d}$$

Comparing the responses in Eqs. 8.121 and 8.122 with those in Eqs. 8.124 and 8.125 shows that the maximum floor acceleration and the maximum base shear decrease as the natural period of vibration increases. This is because the value of spectral acceleration is reduced when the natural period of vibration is large, as seen in Figure 8.19. On the other hand, the maximum relative displacement increases as the natural period of vibration increases. As discussed in Section 5.3, the spectral displacement increases with a larger period, and therefore the maximum relative displacement increases from 3.94 in. to a value of 11.8 in. for $\alpha = 0.583$ to 11.8 in. for $\alpha = 0.0455$, an increase of 200%.

EXAMPLE 2 *Systems with Increasing First-Story Stiffness*

Consider the 2DOF system shown in Figure 8.18. Let $m_b = m_1 = m = 1.0$ k-s^2/in., $k_1 = k = 646$ k/in., and $k_b = \alpha k$.

The natural frequencies and mode shapes are a function of α. If α is very large, then k_b is large and the 2DOF system is like an SDOF system with a natural frequency equal to $\sqrt{k/m}$. To illustrate this, consider first the case where $\alpha = 1$; it follows from Eq. 8.66 that

$$\omega_1 = \sqrt{\frac{\alpha + 2 - \sqrt{\alpha^2 + 4}}{2}} \omega_n = \sqrt{\frac{1 + 2 - \sqrt{1^2 + 4}}{2}} \sqrt{\frac{646}{1.0}} = 15.7 \text{ rad/s} \tag{8.126}$$

$$\omega_2 = \sqrt{\frac{\alpha + 2 + \sqrt{\alpha^2 + 4}}{2}} \omega_n = \sqrt{\frac{1 + 2 + \sqrt{1^2 + 4}}{2}} \sqrt{\frac{646}{1.0}} = 41.1 \text{ rad/s} \tag{8.127}$$

$$T_1 = \frac{2\pi}{\omega_1} = \frac{2\pi}{15.7} = 0.40 \text{ s} \tag{8.128}$$

$$T_2 = \frac{2\pi}{\omega_2} = \frac{2\pi}{41.1} = 0.15 \text{ s} \tag{8.129}$$

and from Eq. 8.69 that

$$r_1 = \frac{\alpha + \sqrt{\alpha^2 + 4}}{2} = \frac{1 + \sqrt{1^2 + 4}}{2} = 1.62 \tag{8.130}$$

$$r_2 = \frac{\alpha - \sqrt{\alpha^2 + 4}}{2} = \frac{1 - \sqrt{1^2 + 4}}{2} = -0.62. \tag{8.131}$$

It follows that the first and second mode shapes, normalized to a unit displacement at the roof, are

$$\boldsymbol{\phi}_{\text{roof}(1)} = \begin{Bmatrix} 1/1.62 \\ 1.0 \end{Bmatrix} = \begin{Bmatrix} 0.62 \\ 1.0 \end{Bmatrix}, \quad \boldsymbol{\phi}_{\text{roof}(2)} = \begin{Bmatrix} -1/0.62 \\ 1.0 \end{Bmatrix} = \begin{Bmatrix} -1.62 \\ 1.0 \end{Bmatrix}. \tag{8.132}$$

Now if α is large (e.g., $\alpha = 10$), then it follows from Eq. 8.66 that

$$\omega_1 = \sqrt{\frac{\alpha + 2 - \sqrt{\alpha^2 + 4}}{2}} \omega_n = \sqrt{\frac{10 + 2 - \sqrt{10^2 + 4}}{2}} \sqrt{\frac{646}{1.0}} = 24.1 \text{ rad/s} \tag{8.133}$$

$$\omega_2 = \sqrt{\frac{\alpha + 2 + \sqrt{\alpha^2 + 4}}{2}} \omega_n = \sqrt{\frac{10 + 2 + \sqrt{10^2 + 4}}{2}} \sqrt{\frac{646}{1.0}} = 84.7 \text{ rad/s} \tag{8.134}$$

$$T_1 = \frac{2\pi}{\omega_1} = \frac{2\pi}{24.1} = 0.26 \text{ s} \tag{8.135}$$

$$T_2 = \frac{2\pi}{\omega_2} = \frac{2\pi}{84.7} = 0.074 \text{ s} \tag{8.136}$$

and from Eq. 8.69 that

$$r_1 = \frac{\alpha + \sqrt{\alpha^2 + 4}}{2} = \frac{10 + \sqrt{10^2 + 4}}{2} = 10.1 \tag{8.137}$$

$$r_2 = \frac{\alpha - \sqrt{\alpha^2 + 4}}{2} = \frac{10 - \sqrt{10^2 + 4}}{2} = -0.099 \tag{8.138}$$

and

$$\boldsymbol{\phi}_{\text{roof}(1)} = \begin{Bmatrix} 1/10.1 \\ 1.0 \end{Bmatrix} = \begin{Bmatrix} 0.099 \\ 1.0 \end{Bmatrix}, \quad \boldsymbol{\phi}_{\text{roof}(2)} = \begin{Bmatrix} -1/0.099 \\ 1.0 \end{Bmatrix} = \begin{Bmatrix} -10.1 \\ 1.0 \end{Bmatrix}. \tag{8.139}$$

Note that if α is infinitely large, then the natural frequency of the SDOF will be

$$\omega_n = \sqrt{\frac{k}{m}} = \sqrt{\frac{646}{1.0}} = 25.4 \text{ rad/s} \tag{8.140}$$

$$T_n = \frac{2\pi}{\omega_n} = 0.25 \text{ s}. \tag{8.141}$$

The participation factor for the mode shapes provides insight into how much each mode contributes to the response. Recall that the participation factor is

$$\Gamma_{\text{roof}(i)} = \frac{\boldsymbol{\phi}_{\text{roof}(i)}^T \mathbf{M}\{\mathbf{I}\}}{\boldsymbol{\phi}_{\text{roof}(i)}^T \mathbf{M}\boldsymbol{\phi}_{\text{roof}(i)}}. \tag{8.142}$$

For the case where $\alpha = 1$, it follows that

$$\Gamma_{\text{roof}(1)} = \frac{\begin{Bmatrix} 0.62 \\ 1.0 \end{Bmatrix}^T \begin{bmatrix} 1.0 & 0.0 \\ 0.0 & 1.0 \end{bmatrix} \begin{Bmatrix} 1.0 \\ 1.0 \end{Bmatrix}}{\begin{Bmatrix} 0.62 \\ 1.0 \end{Bmatrix}^T \begin{bmatrix} 1.0 & 0.0 \\ 0.0 & 1.0 \end{bmatrix} \begin{Bmatrix} 0.62 \\ 1.0 \end{Bmatrix}} = 1.17 \tag{8.143}$$

$$\Gamma_{roof(2)} = \frac{\begin{Bmatrix} -1.62 \\ 1.0 \end{Bmatrix}^T \begin{bmatrix} 1.0 & 0.0 \\ 0.0 & 1.0 \end{bmatrix} \begin{Bmatrix} 1.0 \\ 1.0 \end{Bmatrix}}{\begin{Bmatrix} -1.62 \\ 1.0 \end{Bmatrix}^T \begin{bmatrix} 1.0 & 0.0 \\ 0.0 & 1.0 \end{bmatrix} \begin{Bmatrix} -1.62 \\ 1.0 \end{Bmatrix}} = -0.17. \tag{8.144}$$

Similarly, for $\alpha = 10$, it follows that

$$\Gamma_{roof(1)} = \frac{\begin{Bmatrix} 0.099 \\ 1.0 \end{Bmatrix}^T \begin{bmatrix} 1.0 & 0.0 \\ 0.0 & 1.0 \end{bmatrix} \begin{Bmatrix} 1.0 \\ 1.0 \end{Bmatrix}}{\begin{Bmatrix} 0.099 \\ 1.0 \end{Bmatrix}^T \begin{bmatrix} 1.0 & 0.0 \\ 0.0 & 1.0 \end{bmatrix} \begin{Bmatrix} 0.099 \\ 1.0 \end{Bmatrix}} = 1.09 \tag{8.145}$$

$$\Gamma_{roof(2)} = \frac{\begin{Bmatrix} -10.1 \\ 1.0 \end{Bmatrix}^T \begin{bmatrix} 1.0 & 0.0 \\ 0.0 & 1.0 \end{bmatrix} \begin{Bmatrix} 1.0 \\ 1.0 \end{Bmatrix}}{\begin{Bmatrix} -10.1 \\ 1.0 \end{Bmatrix}^T \begin{bmatrix} 1.0 & 0.0 \\ 0.0 & 1.0 \end{bmatrix} \begin{Bmatrix} -10.1 \\ 1.0 \end{Bmatrix}} = -0.088. \tag{8.146}$$

The displacement of the floors relative to the ground for the ith mode of vibration is

$$\mathbf{X}^{(i)} = \Gamma_{roof(i)}\mathbf{\Phi}_{roof(i)}S_d(T_i,\zeta_i) = \left(\frac{\Gamma_{roof(i)}}{\omega_i^2}\right)\mathbf{\Phi}_{roof(i)}S_a(T_i,\zeta_i). \tag{8.147}$$

Consider the same response spectrum used in Example 1 as shown in Figure 8.19. For $\alpha = 1$, the spectral accelerations at a natural period of vibration of 0.40 s and 0.15 s are both $S_a = 1.0g$. Therefore, it follows that

$$\mathbf{X}^{(1)} = \begin{Bmatrix} x_1 \\ x_2 \end{Bmatrix}^{(1)} = \left(\frac{1.17}{15.7^2}\right)\begin{Bmatrix} 0.62 \\ 1.0 \end{Bmatrix}(386.4) = \begin{Bmatrix} 1.13 \\ 1.83 \end{Bmatrix} \tag{8.148}$$

$$\mathbf{X}^{(2)} = \begin{Bmatrix} x_1 \\ x_2 \end{Bmatrix}^{(2)} = \left(\frac{0.17}{41.1^2}\right)\begin{Bmatrix} -1.62 \\ 1.0 \end{Bmatrix}(386.4) = \begin{Bmatrix} -0.63 \\ 0.039 \end{Bmatrix}. \tag{8.149}$$

Similarly, for $\alpha = 10$, the spectral acceleration at a natural period of vibration of 0.26 s is $S_a = 1.0g$ and at a natural period of vibration of 0.074 s is $S_a = 0.70g$. Therefore, it follows that

$$\mathbf{X}^{(1)} = \begin{Bmatrix} x_1 \\ x_2 \end{Bmatrix}^{(1)} = \left(\frac{1.09}{24.1^2}\right)\begin{Bmatrix} 0.099 \\ 1.0 \end{Bmatrix}(386.4) = \begin{Bmatrix} 0.072 \\ 0.73 \end{Bmatrix} \tag{8.150}$$

$$\mathbf{X}^{(2)} = \begin{Bmatrix} x_1 \\ x_2 \end{Bmatrix}^{(2)} = \left(\frac{0.088}{84.7^2}\right)\begin{Bmatrix} -10.1 \\ 1.0 \end{Bmatrix}(0.7\times386.4) = \begin{Bmatrix} -0.035 \\ 0.0033 \end{Bmatrix}. \tag{8.151}$$

The absolute acceleration of the floors for the ith mode of vibration is

$$\ddot{\mathbf{Y}}^{(i)} = \Gamma_{roof(i)}\mathbf{\Phi}_{roof(i)}S_a(T_i,\zeta_i). \tag{8.152}$$

If $\alpha = 1$, then

$$\ddot{\mathbf{Y}}^{(1)} = \begin{Bmatrix} \ddot{y}_1 \\ \ddot{y}_2 \end{Bmatrix}^{(1)} = (1.17)\begin{Bmatrix} 0.62 \\ 1.0 \end{Bmatrix}(1.0g) = \begin{Bmatrix} 0.72g \\ 1.17g \end{Bmatrix} \tag{8.153}$$

$$\ddot{\mathbf{Y}}^{(2)} = \begin{Bmatrix} \ddot{y}_1 \\ \ddot{y}_2 \end{Bmatrix}^{(2)} = (0.17)\begin{Bmatrix} -1.62 \\ 1.0 \end{Bmatrix}(1.0g) = \begin{Bmatrix} -0.28g \\ 0.17g \end{Bmatrix}. \tag{8.154}$$

Similarly, if $\alpha = 10$, then

$$\ddot{\mathbf{Y}}^{(1)} = \left\{ \begin{matrix} \ddot{y}_1 \\ \ddot{y}_2 \end{matrix} \right\}^{(1)} = (1.09) \left\{ \begin{matrix} 0.099 \\ 1.0 \end{matrix} \right\} (1.0g) = \left\{ \begin{matrix} 0.11g \\ 1.09g \end{matrix} \right\} \tag{8.155}$$

$$\ddot{\mathbf{Y}}^{(2)} = \left\{ \begin{matrix} \ddot{y}_1 \\ \ddot{y}_2 \end{matrix} \right\}^{(2)} = (0.088) \left\{ \begin{matrix} -10.1 \\ 1.0 \end{matrix} \right\} (0.7g) = \left\{ \begin{matrix} -0.62g \\ 0.062g \end{matrix} \right\}. \tag{8.156}$$

Table 8.1 summarizes this calculation.

The previous discussion considered a 2DOF system where the bottom-story or first-story stiffness is increased from a baseline value and then this new system response is calculated and compared with the baseline system response. The example provided insight into how a structural engineer can evaluate when a bottom or first story is "rigid."

Consider the same 2DOF system but now assume that the bottom-story stiffness is decreased to a value of $\alpha = 0.1$. Following the same procedure used in the previous discussion, it follows from Eq. 8.66 that

$$\omega_1 = \sqrt{\frac{\alpha + 2 - \sqrt{\alpha^2 + 4}}{2}} \omega_n = \sqrt{\frac{0.1 + 2 - \sqrt{0.1^2 + 4}}{2}} \sqrt{\frac{646}{1.0}} = 5.61 \text{ rad/s} \tag{8.157}$$

$$\omega_2 = \sqrt{\frac{\alpha + 2 + \sqrt{\alpha^2 + 4}}{2}} \omega_n = \sqrt{\frac{0.1 + 2 + \sqrt{0.1^2 + 4}}{2}} \sqrt{\frac{646}{1.0}} = 36.4 \text{ rad/s} \tag{8.158}$$

$$T_1 = \frac{2\pi}{\omega_1} = \frac{2\pi}{5.61} = 1.12 \text{ s} \tag{8.159}$$

$$T_2 = \frac{2\pi}{\omega_2} = \frac{2\pi}{36.4} = 0.17 \text{ s} \tag{8.160}$$

and from Eq. 8.689 that

$$r_1 = \frac{\alpha + \sqrt{\alpha^2 + 4}}{2} = \frac{0.1 + \sqrt{0.1^2 + 4}}{2} = 1.05 \tag{8.161}$$

$$r_2 = \frac{\alpha - \sqrt{\alpha^2 + 4}}{2} = \frac{0.1 - \sqrt{0.1^2 + 4}}{2} = -0.95 \tag{8.162}$$

and

$$\mathbf{\Phi}_{\text{roof}(1)} = \left\{ \begin{matrix} 1/1.05 \\ 1.0 \end{matrix} \right\} = \left\{ \begin{matrix} 0.95 \\ 1.0 \end{matrix} \right\}, \quad \mathbf{\Phi}_{\text{roof}(2)} = \left\{ \begin{matrix} -1/0.95 \\ 1.0 \end{matrix} \right\} = \left\{ \begin{matrix} -1.05 \\ 1.0 \end{matrix} \right\}. \tag{8.163}$$

Table 8.1 Comparison between $\alpha = 1$ and $\alpha = 10$

	First mode ($i = 1$)		Second mode ($i = 2$)	
	$\alpha = 1$	$\alpha = 10$	$\alpha = 1$	$\alpha = 10$
ω_i (rad/s)	15.7	24.1	41.1	84.7
T_i (s)	0.40	0.26	0.15	0.074
ϕ_i	$\{0.62 \quad 1.0\}^T$	$\{0.099 \quad 1.0\}^T$	$\{-1.62 \quad 1.0\}^T$	$\{-10.1 \quad 1.0\}^T$
Γ_i	1.17	1.09	0.17	0.088
Γ_i/ω_i^2	0.0047	0.0019	1.01×10^{-4}	1.23×10^{-5}

The participation factors for each mode shape are

$$\Gamma_{\text{roof}(1)} = \frac{\left\{ \begin{matrix} 0.95 \\ 1.0 \end{matrix} \right\}^{T} \begin{bmatrix} 1.0 & 0.0 \\ 0.0 & 1.0 \end{bmatrix} \left\{ \begin{matrix} 1.0 \\ 1.0 \end{matrix} \right\}}{\left\{ \begin{matrix} 0.95 \\ 1.0 \end{matrix} \right\}^{T} \begin{bmatrix} 1.0 & 0.0 \\ 0.0 & 1.0 \end{bmatrix} \left\{ \begin{matrix} 0.95 \\ 1.0 \end{matrix} \right\}} = 1.02 \tag{8.164}$$

$$\Gamma_{\text{roof}(2)} = \frac{\left\{ \begin{matrix} -1.05 \\ 1.0 \end{matrix} \right\}^{T} \begin{bmatrix} 1.0 & 0.0 \\ 0.0 & 1.0 \end{bmatrix} \left\{ \begin{matrix} 1.0 \\ 1.0 \end{matrix} \right\}}{\left\{ \begin{matrix} -1.05 \\ 1.0 \end{matrix} \right\}^{T} \begin{bmatrix} 1.0 & 0.0 \\ 0.0 & 1.0 \end{bmatrix} \left\{ \begin{matrix} -1.05 \\ 1.0 \end{matrix} \right\}} = -0.024. \tag{8.165}$$

The spectral acceleration at a natural period of vibration of 1.12 s is $S_a = 0.52g$ and at a natural period of vibration of 0.17 s is $S_a = 1.0g$. The displacements of the floors relative to the ground for each mode of vibration are

$$\mathbf{X}^{(1)} = \left\{ \begin{matrix} x_1 \\ x_2 \end{matrix} \right\}^{(1)} = \left(\frac{1.02}{5.61^2} \right) \left\{ \begin{matrix} 0.95 \\ 1.0 \end{matrix} \right\} (0.52 \times 386.4) = \left\{ \begin{matrix} 6.19 \\ 6.51 \end{matrix} \right\} \tag{8.166}$$

$$\mathbf{X}^{(2)} = \left\{ \begin{matrix} x_1 \\ x_2 \end{matrix} \right\}^{(2)} = \left(\frac{0.024}{36.4^2} \right) \left\{ \begin{matrix} -1.05 \\ 1.0 \end{matrix} \right\} (386.4) = \left\{ \begin{matrix} -0.0073 \\ 0.0070 \end{matrix} \right\}. \tag{8.167}$$

Similarly, the absolute accelerations of the floors for each mode of vibration are

$$\ddot{\mathbf{Y}}^{(1)} = \left\{ \begin{matrix} \ddot{y}_1 \\ \ddot{y}_2 \end{matrix} \right\}^{(1)} = (1.02) \left\{ \begin{matrix} 0.95 \\ 1.0 \end{matrix} \right\} (0.52g) = \left\{ \begin{matrix} 0.50g \\ 0.53g \end{matrix} \right\} \tag{8.168}$$

$$\ddot{\mathbf{Y}}^{(2)} = \left\{ \begin{matrix} \ddot{y}_1 \\ \ddot{y}_2 \end{matrix} \right\}^{(2)} = (0.024) \left\{ \begin{matrix} -1.05 \\ 1.0 \end{matrix} \right\} (1.0g) = \left\{ \begin{matrix} -0.025g \\ 0.024g \end{matrix} \right\}. \tag{8.169}$$

Table 8.2 summarizes the preceeding calculation.

EXAMPLE 3   Formulation of Stiffness Matrix without Grade Beam

As shown in Figure 8.17, the beam between nodes E and F is called a *grade beam* and is often considered to be part of the base isolation system. The reason for this position is because this beam would not typically be present if the columns went directly to the ground (i.e., foundation). Consider the case where this grade beam is not present. In this case, it is still often reasonable to assume that

$$y_2 = y_5 = y_8 = y_9 = y_{11} = y_{12} = 0 \tag{8.170}$$

Table 8.2 Comparison between $\alpha = 1$ and $\alpha = 0.1$

	First mode ($i = 1$)		Second mode ($i = 2$)	
	$\alpha = 1$	$\alpha = 0.1$	$\alpha = 1$	$\alpha = 0.1$
ω_i (rad/s)	15.7	5.61	41.1	36.4
T_i (s)	0.40	1.12	0.15	0.17
ϕ_i	$\{0.62 \quad 1.0\}^T$	$\{0.95 \quad 1.0\}^T$	$\{-1.62 \quad 1.0\}^T$	$\{-1.05 \quad 1.0\}^T$
Γ_i	1.17	1.02	0.17	0.024
Γ_i/ω_i^2	0.0047	0.0324	1.01×10^{-4}	1.81×10^{-5}

and that

$$y_7 = y_{10}. \tag{8.171}$$

But now y_3 and y_6 are not equal to zero, and y_1 is not equal to y_4. Therefore, the system is now a 5DOF system instead of a 2DOF system. However, if the columns are identical, that is,

$$A_1 = A_2 = A, \quad E_1 = E_2 = E, \quad I_1 = I_2 = I \tag{8.172}$$

then from symmetry one can assume that

$$y_1 = y_4 \tag{8.173}$$

$$y_3 = y_6. \tag{8.174}$$

Let x_1 be the bottom-story displacement, x_2 the top-story displacement, and x_3 the rotation at node E. Then the y-coordinate system can be related to the x-coordinate system by the matrix equation

$$\mathbf{Y} = \begin{Bmatrix} y_1 \\ y_2 \\ y_3 \\ y_4 \\ y_5 \\ y_6 \\ y_7 \\ y_8 \\ y_9 \\ y_{10} \\ y_{11} \\ y_{12} \end{Bmatrix} = \begin{bmatrix} 1 & 0 & 0 \\ 0 & 0 & 0 \\ 0 & 1 & 0 \\ 1 & 0 & 0 \\ 0 & 0 & 0 \\ 0 & 1 & 0 \\ 0 & 0 & 1 \\ 0 & 0 & 0 \\ 0 & 0 & 0 \\ 0 & 0 & 1 \\ 0 & 0 & 0 \\ 0 & 0 & 0 \end{bmatrix} \begin{Bmatrix} x_1 \\ x_2 \\ x_3 \end{Bmatrix} = \mathbf{TX}. \tag{8.175}$$

Define the stiffness of the column between E and C (and F and D) to be EI_s and the column between E and A (and F and B) to be EI_b. It then follows that the stiffness matrix in the x-coordinate system is

$$\mathbf{K}_x = \mathbf{T}^T \mathbf{K}_y \mathbf{T} = \begin{bmatrix} \dfrac{24EI_b}{h_b^3} + \dfrac{24EI_s}{h^3} & -\dfrac{24EI_s}{h^3} & \dfrac{12EI_b}{h_b^2} - \dfrac{12EI_s}{h^2} \\ -\dfrac{24EI_s}{h^3} & \dfrac{24EI_s}{h^3} & \dfrac{12EI_s}{h^2} \\ \dfrac{12EI_b}{h_b^2} - \dfrac{12EI_s}{h^2} & \dfrac{12EI_s}{h^2} & \dfrac{8EI_b}{h_b} + \dfrac{8EI_s}{h} \end{bmatrix}. \tag{8.176}$$

The response of the system is now composed of three modes, and therefore

$$\mathbf{X}(t) = \boldsymbol{\phi}_1 q_1(t) + \boldsymbol{\phi}_2 q_2(t) + \boldsymbol{\phi}_3 q_3(t). \tag{8.177}$$

8.4 BASE ISOLATORS

Base isolators are structural members that, like steel beams and columns, are part of a lateral force-resisting system that enables a building to respond acceptably to earthquake ground motions. Figure 8.22 shows one type of base isolator. A base isolator is composed of alternating layers of rubber that provide flexibility and steel reinforcing plates

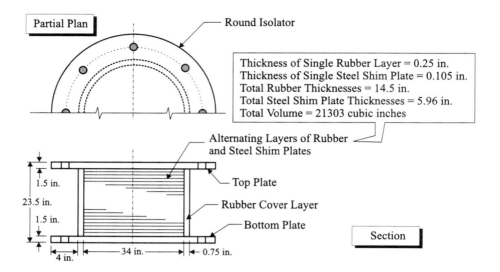

Figure 8.22 Schematic of a Base Isolator (Courtesy of kpff Engineers)

that provide vertical load-carrying capacity. At the top and bottom of these layers are steel laminated plates that distribute the vertical loads and transfer the shear force to the internal rubber layer. On the top and bottom of the steel laminated plate is a rubber cover that provides protection for the steel laminated plates.

The chemical composition of the inner rubber layers determines the lateral force versus lateral deflection characteristics of the base isolator. Consider first a base isolator with a rubber composition that is designed to provide a linear elastic force versus deflection curve. Figure 8.23 shows one cycle of a base isolator lateral force versus lateral displacement curve. A base isolator with this force versus deflection behavior is called a *linear base isolator*. Two structural design variables are obtained from this force versus deflection curve. The first design variable is the *base isolator stiffness* (k_b), which is defined as

$$k_b = \frac{F^+ - F^-}{\Delta^+ - \Delta^-}.$$ (8.178)

The second design variable is the *base isolator viscous damping* (ζ_b), which is

$$\zeta_b = \frac{1}{2\pi}\left[\frac{\text{area of loop}}{F_{max}\Delta_{max}}\right].$$ (8.179)

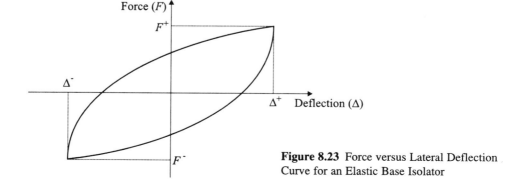

Figure 8.23 Force versus Lateral Deflection Curve for an Elastic Base Isolator

In this equation, F_{max} and Δ_{max} are the maximum absolute values of (F^+, F^-) and (Δ^+, Δ^-), respectively. If the damping of the isolator is very small or zero, then the area of the loop is also very small or zero. In this case, the force versus deflection curve is a straight line, which is the same as the curve presented in most static structural analysis textbooks.

If the base isolator is subjected to a cycle of loading at a maximum displacement Δ_1, where the subscript 1 denotes the first cycle of loading, it will have a loop type of response as shown in Figure 8.23. If the base isolator is elastic, then as the base isolator is subjected to cycles of response at ever-increasing maximum displacements (e.g., $\Delta_1 = \Delta_a$ and $\Delta_2 = \Delta_b$ in Figure 8.24), a plot of the maximum force versus maximum displacement for each cycle of response is called an *envelope force deflection curve*. Figure 8.25 shows a linear envelope force versus deflection curve.

Base isolators can be designed to have envelope force versus deflection curves that are not straight lines but exhibit a nonlinear behavior. Base isolators designed to have this type of behavior are called *nonlinear base isolators*. Figure 8.26 shows a schematic of a lateral

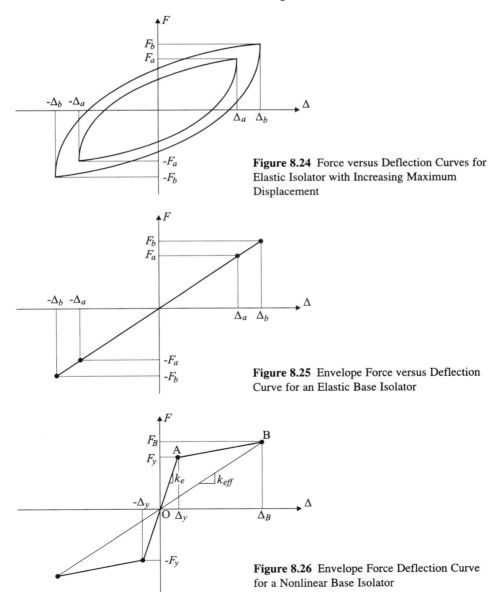

Figure 8.24 Force versus Deflection Curves for Elastic Isolator with Increasing Maximum Displacement

Figure 8.25 Envelope Force versus Deflection Curve for an Elastic Base Isolator

Figure 8.26 Envelope Force Deflection Curve for a Nonlinear Base Isolator

force versus lateral displacement plot for a nonlinear base isolator. If the lateral force in a cyclic force versus deflection loop is less than the yield force, F_y, or alternately the yield displacement, Δ_y, then the isolator is in the linear response domain and the base isolator has not yielded. The stiffness of the base isolator in this range of response is denoted k_e. If the displacement is less than the yield displacement, Δ_y, and there is no damping, then the force versus deflection behavior is up and down the straight line OA. If the damping is significant, then the path from O to A will be a loop of the type shown in Figure 8.23 and the line from O to A represents the envelope force versus deflection curve. If the maximum force of a cyclic loop exceeds the yield force F_y, or equivalently the maximum displacement exceeds the yield displacement Δ_y, then the maximum force versus maximum displacement envelope path climbs from A toward B as the maximum displacement of each cyclic loop increases until it reaches B. The *effective stiffness*, denoted by k_{eff}, is defined to be the slope of a line from O to B and is calculated using the right-hand side of Eq. 8.178. This stiffness is also called the *secant stiffness* at the displacement, Δ_B.

The Uniform Building Code requires that all base isolators placed in buildings be tested to verify that the force versus deflection properties used in design are the same as the force versus deflection properties delivered to the buildings. In the design phase of a project, the structural engineer determines the required force versus deflection properties of the base isolators. Then, in the preconstruction phase of the project, the structural engineer requires that a series of tests be performed on each type of base isolator to be used in the structure. The testing specification varies depending on the jurisdiction under which the building is designed. The 1997 Uniform Building Code contains detailed specifications for these tests, which are called *prototype base isolator tests*. These test specifications are as follows:

1. The force versus deflection plots for all tests performed must have a positive incremental stiffness.

2. For each test performed for three continuous cycles, there must be no greater than a 15% change in the effective stiffness between maximum and minimum values. If 15% is exceeded, an additional three cycles of testing are necessary. A change greater than 20% at any point requires special investigation.

3. For each test performed for 10 continuous cycles, there must be no greater than a 25% change in the effective stiffness at a given vertical load.

4. For each test performed for 10 continuous cycles, there must be no greater than a 25% change in the effective damping at a given vertical load.

5. The difference in the measured effective stiffness between individual isolators of the same type must be less than 10%.

6. The isolators must remain stable at a horizontal displacement of 1.1 times the design displacement under the maximum and minimum downward forces.

Table 8.3 shows an example prototype test program for a nonlinear base isolator with a lead core in the center. This base isolator, shown in Figure 8.27, has a plan dimension of 32 in. diameter and a height of 18.78 in. (13.5 in. of rubber and 5.28 in. of shim plates). The diameter of the lead core in this base isolator is 8.5 in. Consider a lead core base isolator with a slight variation in these properties: 26 in. × 26 in. square in plan, 16.735 in. in height, and a 7.0 in. diameter lead core. The results of a test program on an isolator of this type are now presented.

A series of 16 tests were performed on this 26 in. × 26 in., lead core, 16.375 in. high base isolator. Each test was composed of a number of cycles of force versus deflection loops. During this series of 16 tests, the lateral displacement of the base isolator was increased from 0.5 in. to 16.0 in.

Table 8.3 Test Matrix for a Prototype Base Isolator

Test number	Axial load (kip)	Deflection (in.)	Number of cycles
1	410	0.5	20
2	410	3.5	3
3	410	7.0	3
4	410	10.5	3
5	410	14.0	3
6	410	14.0	17
7	820	3.5	3
8	820	7.0	3
9	820	10.5	3
10	820	14.0	10
11	170	3.5	3
12	170	7.0	3
13	170	10.5	3
14	170	14.0	10
15	1270	16.0	1/2
16	Min.	16.0	1/2

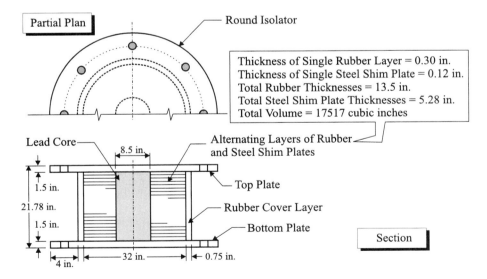

Partial Plan

Round Isolator

Thickness of Single Rubber Layer = 0.30 in.
Thickness of Single Steel Shim Plate = 0.12 in.
Total Rubber Thicknesses = 13.5 in.
Total Steel Shim Plate Thicknesses = 5.28 in.
Total Volume = 17517 cubic inches

Lead Core

8.5 in.

Alternating Layers of Rubber
and Steel Shim Plates

1.5 in.

Top Plate

21.78 in.

Rubber Cover Layer

1.5 in.

Bottom Plate

Section

4 in.

32 in.

0.75 in.

Figure 8.27   Schematic of a Lead Core Base Isolator (Courtesy of kpff Engineers)

Test 1 corresponds to a large number (20) of cycles of testing at a displacement that is expected to be the yield displacement, Δ_y, of the base isolator. Selecting the value of yield force, or alternately yield displacement, is an important structural engineering design decision. In this illustrative example, the base isolator was designed to have a yield force of 25 kip and a pre-yield elastic stiffness of 50 k/in. Therefore, the yield deflection is 0.5 in.

Table 8.4 shows the variation in the effective stiffness of the lead core base isolator for Tests 2 to 5. For any single test, the percentage change in stiffness between the cycles

Table 8.4 Base Isolator, Prototype Tests 2 to 5 (Axial load = 410 kip, number of cycles = 3)

Cycle	Effective stiffness (k/in.)			
	Test 2 Displ. = ±3.5 in.	Test 3 Displ. = ±7.0 in.	Test 4 Displ. = ±10.5 in.	Test 5 Displ. = ±14.0 in.
1	18.7	13.2	10.6	9.2
2	18.3	13.0	10.4	8.9
3	18.2	12.8	10.3	8.8
Maximum value	18.7	13.2	10.6	9.2
Minimum value	18.2	12.8	10.3	8.8
Percent change	–3%	–3%	–3%	–4%

is small. It is a very important property of a nonlinear base isolator that the effective stiffness of the base isolator decreases as the maximum lateral displacement increases. Consider Cycle 1 in these tests, with the maximum force, F_{max}, and maximum displacement, Δ_{max}, for a cycle denoted by

$$F_{max} = k_{eff}\Delta_{max}. \tag{8.180}$$

For Test 2, it follows that

$$F_{max} = (18.7\,\text{k/in.})(3.5\,\text{in.}) = 65.5\ \text{kip}. \tag{8.181}$$

It similarly follows for Tests 3 through 5 that F_{max} is equal to 92.4, 111.3, and 128.8 kip, respectively. Figure 8.28 shows the envelope force versus deflection curve for these four tests. For each displacement, the force plotted to develop the curve is the maximum force of the three cycles of testing.

The effective damping for any cycle of prototype motion can be calculated using Eq. 8.179. For any single test, the damping value will typically vary by a small amount from cycle to cycle. As shown in Table 8.5, for Test 6 at a displacement of 14.0 in., the average value of damping of the 17 cycles was 20.3% of critical and the minimum and maximum values were 19.7% and 21.5%, respectively.

The variation in damping with the amplitude of base isolator displacement is a function of the design of the base isolator and is obtained from the prototype tests. For example, for this illustrative base isolator, Figure 8.29 shows a plot of the damping variation with displacement.

Base isolators are usually placed at the bottom of a column in a building. Therefore, the column transmits to the base isolator the dead and live load of the structure plus any additional column axial earthquake loads. Tests 7 to 9 are the same as Tests 2 to 4 except

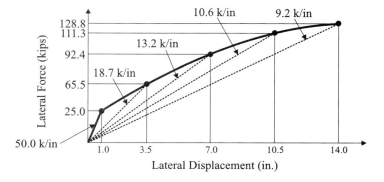

Figure 8.28 Envelope of Maximum Lateral Force versus Deflection (Lead Core Base Isolator)

Table 8.5 26-in. Square Base Isolator, Prototype Test 6
(Axial load = 410 kip, displacement = ±14.0 in.)

Cycle	Effective stiffness (k/in.)	Damping (%)
1	8.9	21.5
2	8.8	21.0
3	8.7	20.7
4	8.6	20.6
5	8.6	20.4
6	8.6	20.3
7	8.5	20.2
8	8.5	20.2
9	8.5	20.2
10	8.5	20.1
11	8.5	20.0
12	8.4	19.9
13	8.4	19.9
14	8.4	19.9
15	8.4	19.8
16	8.4	19.9
17	8.3	19.7
Maximum value	8.9	21.5
Minimum value	8.3	19.7
Percent change	−7%	−8%

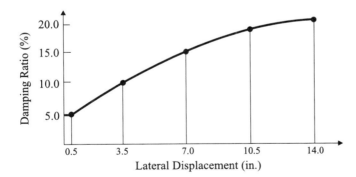

Figure 8.29 Illustrative Example of Lead Core Base Isolator Damping with Displacement

now the value of the column axial load on the base isolator is changed to show the sensitivity of the performance of the base isolator to axial load. Table 8.6 shows a reduction in the stiffness of the base isolator with the increase in axial load. The axial load on the base isolator for Tests 2 to 5 was 410 kip or [410 kip/(26 in. × 26 in.)] = 607 psi. The axial load for Tests 7 to 9 was twice this axial load (i.e., 1,213 psi).

A critical design variable for a structure on base isolators is the estimated base isolator displacement for this base-isolated building subjected to an earthquake with a 10% probability of being exceeded in 50 years (called the *design displacement*). The base isolators used in this example are for a base-isolated building with a design displacement equal to 14.0 in. Table 8.7 shows the effective stiffness and damping for the 410 kip and

Table 8.6 26-in. Base Isolator, Prototype Tests 7 to 9 (Axial load = 820 kip)

Cycle	Effective stiffness (k/in.)		
	Test 7 Displ. = ±3.5 in.	Test 8 Displ. = ±7.0 in.	Test 9 Displ. = ±10.5 in.
1	18.3	12.0	9.2
2	17.8	11.8	9.0
3	17.5	11.5	9.0
Maximum value	18.3	12.0	9.2
Minimum value	17.5	11.5	9.0
Percent change	–4%	–4%	–2%

Table 8.7 Base Isolator, Prototype Test 10 (Axial load = 820 kip, displacement = ±14.0 in., number of cycles = 10)

Cycle	Effective stiffness (k/in.)	Damping (%)
1	7.9	23.5
2	7.7	23.2
3	7.6	23.2
4	7.6	23.0
5	7.5	23.0
6	7.5	23.0
7	7.5	22.9
8	7.4	23.0
9	7.4	22.9
10	7.4	22.9
Maximum value	7.9	23.5
Minimum value	7.4	22.9
Percent change	–6%	–3%

820 kip axial loads. Note that the modal damping increases slightly with an increase in axial load.

The nonlinear base isolator selected for illustration in the previous paragraphs had a lead core in the center of the base isolator. However, as shown in Figure 8.22, not all base isolators have a lead core. For illustration, a square non–lead core nonlinear base isolator with a 28 in. × 28 in. plan dimension and a 16.375 in. height has the force versus deflection characteristics given in Table 8.8.

EXAMPLE 1 *Effective Stiffness of Base Isolators*

Table 8.9 shows the variation of effective stiffness of a base isolator for three displacement amplitudes. The average stiffness at 3.5 in. of displacement is

$$k_{3.5} = \frac{17.5 + 17.1 + 17.1}{3} = 17.23 \text{ k/in.} \tag{8.182}$$

Table 8.8 Non Lead Core Nonlinear Base Isolator, Prototype Tests (Axial load = 170 kip, number of cycles = 3)

Cycle	Effective stiffness (k/in.)		
	Displ. = ±3.5 in.	Displ. = ±7.0 in.	Displ. = ±10.5 in.
1	17.5	12.2	9.7
2	17.1	11.9	9.6
3	17.1	11.8	9.4
Maximum value	17.5	12.2	9.7
Minimum value	17.1	11.8	9.4
Percent change	−2%	−3%	−3%

Table 8.9 Effective Base Isolator Stiffness versus Displacement Amplitude

Cycle	Effective stiffness (k/in.)		
	Displ. = ±3.5 in.	Displ. = ±7.0 in.	Displ. = ±10.5 in.
1	17.5	12.2	9.7
2	17.1	11.9	9.6
3	17.1	11.8	9.4

The corresponding average force required to displace the isolator at this displacement is

$$F_{3.5} = (17.23)(3.5) = 60.3 \text{ kip.} \tag{8.183}$$

It similarly follows that

$$k_{7.0} = \frac{12.2 + 11.9 + 11.8}{3} = 11.97 \text{ k/in.} \tag{8.184}$$

$$F_{7.0} = (11.97)(7.0) = 83.8 \text{ kip} \tag{8.185}$$

$$k_{10.5} = \frac{9.7 + 9.6 + 9.4}{3} = 9.57 \text{ k/in.} \tag{8.186}$$

$$F_{10.5} = (9.57)(10.5) = 100.5 \text{ kip.} \tag{8.187}$$

EXAMPLE 2   *Effective Stiffness and Damping of a Base Isolator*

Figure 8.30 shows the force versus deflection test data for a base isolator subjected to cyclic testing with displacement amplitude equal to 3.5 in. The effective base isolator stiffness using Eq. 8.178 is

$$k_b = \frac{F^+ - F^-}{\Delta^+ - \Delta^-} = \frac{55 - (-80)}{3.6 - (-3.5)} = 66.27 \text{ k/in.} \tag{8.188}$$

The effective base isolator damping is given in Eq. 8.179, which is

$$\zeta_b = \frac{1}{2\pi}\left[\frac{\text{area of loop}}{F_{max}\Delta_{max}}\right] \tag{8.189}$$

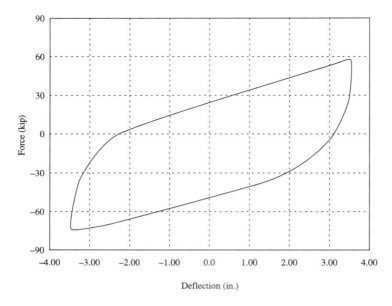

Figure 8.30 Cyclic Base Isolator Force versus Deflection Test

where $F_{max} = 80$ kip and $\Delta_{max} = 3.6$ in. The area of the loop can be approximated by considering each cycle as a parallelogram, where the base of the loop is approximately 6.4 in. and the height of the loops is approximately 75 kip. Therefore, substituting the values into Eq. 8.189 gives

$$\zeta_b = \frac{1}{2\pi}\left[\frac{(6.4)(75)}{(80)(3.6)}\right] = 0.265 = 26.5\%. \tag{8.190}$$

8.5 DESIGN OF A BASE-ISOLATED STRUCTURE

A structural dynamic analysis of a base-isolated structure requires that the properties of the base isolators be specified. Determining a trial (or preliminary) design of the base isolator is here referred to as "design." This section provides two methods for developing this trial design. The next two sections provide analysis methods that start with the trial design and eventually produce the final base isolator design.

Design starts with the specification of a 5% damped response spectrum for a design basis earthquake. For example, Figure 8.31 shows a 5% damped response spectrum that has a 10% probability of being exceeded in 50 years, and for this section it is assumed to be the design basis earthquake.

Consider the SDOF structure shown in Figure 8.32. This is referred to as the "fixed-base" structure, the structure that would exist if base isolators were not used with the structure. The stiffness (i.e., k), mass (i.e., m), and damping (i.e., c) for this fixed-base structure are assumed to be known. The natural frequency, period of vibration, and the critical damping ratio are

$$\omega_n = \sqrt{\frac{k}{m}}, \quad T_n = \frac{2\pi}{\omega_n}, \quad 2\zeta\omega_n = \frac{c}{m}. \tag{8.191}$$

Imagine that the structure is located at a site that has been studied by an engineering seismologist, and the 5% damped design earthquake acceleration response spectrum provided as a design-basis earthquake is given in Figure 8.31. Figure 8.33 shows the 5%

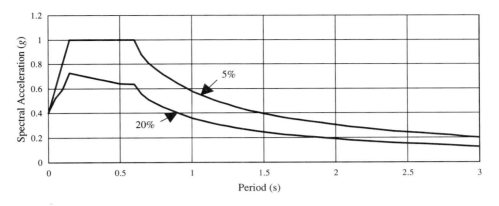

Figure 8.31 Acceleration Response Spectrum for the Design Earthquake

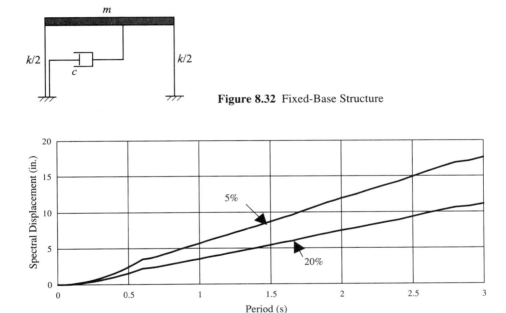

Figure 8.32 Fixed-Base Structure

Figure 8.33 Displacement Response Spectrum for the Design Earthquake

damped displacement response spectrum corresponding to the acceleration response spectrum shown in Figure 8.31. The relationship between the response spectra in Figures 8.31 and 8.33 is developed using the pseudo acceleration relation, that is,

$$S_a = \omega_n^2 S_d. \tag{8.192}$$

Figure 8.34 shows the bi-spectra plot corresponding to Figures 8.31 and 8.33.

Now imagine that the structure in Figure 8.32 is base isolated with a base isolator placed under each column (see Figure 8.35). A rigid floor diaphragm is added to the structure immediately above the base isolators. The mass of this added floor is assumed to be the same as the "roof" mass m. Two approaches are now presented to develop the trial design of the base isolators.

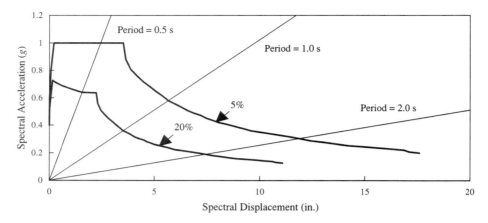

Figure 8.34 Bi-spectra Plot for Design Earthquake

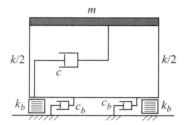

Figure 8.35 Base-Isolated Structure

Method 1: Selection of Period of Vibration of Base-Isolated Structure

One method for designing the base isolator is to define the design-basis earthquake and then set a value for the period of vibration for the base-isolated structure. Consider, for example, the design-basis earthquake to be an earthquake with a 10% probability of being exceeded in 50 years.

Selecting a desired natural period of vibration for the base-isolated structure is guided by the desire to have "in effect" a rigid structure sitting on base isolators. Recall from Example 1 in Section 8.3 that as the value of α decreases in magnitude toward a value of zero, the fundamental mode shape for the 2DOF structure moves toward a "one-one" (i.e., $\varphi_{11} = 1.0$ and $\varphi_{21} = 1.0$) mode shape. Table 8.2 shows the mode shape for $\alpha = 1.0$ and $\alpha = 0.1$. The natural frequency of vibration for the base-isolated structure, ω_{nb}, when it is assumed that the structure above the base isolators is rigid (i.e., $\varphi_{11} = 1.0$ and $\varphi_{21} = 1.0$ mode shape), is

$$\omega_{nb} = \sqrt{2k_b/2m} = \sqrt{k_b/m} \qquad (8.193)$$

where k_b is the stiffness of one base isolator under the column. The mass $2m$ represents the total mass of the base-isolated structure, which is the sum of the roof mass m and the added floor mass m. The natural period of vibration of the base-isolated structure, T_{nb}, is

$$T_{nb} = 2\pi/\omega_{nb} . \qquad (8.194)$$

The period of vibration is usually selected to provide a good separation between the fixed base period of vibration, T_n, and the base-isolated period of vibration, T_{nb}. For example,

$$T_{nb} = CT_n \qquad (8.195)$$

where C is 3 or greater. Alternately, the structural engineer can first set a value of k_b and then use Eq. 8.194 to calculate the base-isolated period of vibration. For example, if $\alpha = 0.1$, then

$$k_b = \alpha k = 0.1k \tag{8.196}$$

and

$$T_{nb} = \frac{2\pi}{\sqrt{k_b/m}} = \frac{2\pi}{\sqrt{0.1k/m}} = 3.16\left(\frac{2\pi}{\sqrt{k/m}}\right) = 3.16T_n. \tag{8.197}$$

Assume that using either of the preceding approaches, the structural engineer selects a period of vibration of the base-isolated structure. The consequences of this selection on response can be evaluated using the analysis methods presented in the previous chapters of this book. However, before such an evaluation is performed it is important to recall that a positive benefit of using base isolators is the significant increase in damping. The damping for a rigid structure on base isolators is essentially the damping of the base isolators. The justification of this statement is presented in the next section. Therefore, assume for this discussion that based on data of the type provided in Table 8.5, the critical damping ratio is 20%. Figures 8.31, 8.33, and 8.34 show the spectral acceleration, spectral displacement, and bi-spectra plot for 20% damping. These spectra are obtained using the damping modifiers in Table 5.7. Note that the damping modifier from 5% damping to 20% damping is 0.74 for a period of 0.1 s and 0.64 for a period of 0.5 s. Therefore, linear interpolation is used to compute the values of the damping modifiers from 0.1 s to 0.5 s; that is, at a period of 0.3 s, the damping modifier is

$$R_d = 0.74 + \left(\frac{0.64 - 0.74}{0.5 - 0.1}\right)(0.3 - 0.1) = 0.69. \tag{8.198}$$

In this method, the desired base-isolated structure period or separation between the fixed-base and the base-isolated period is set. This selected separation results in approximately a constant 1.0 in each term of the fundamental mode of vibration and in effect a "rigid box" sitting on top of the base isolators. The damping of the base-isolated system is equal to the damping in the base isolators.

For example, imagine that the fixed-base structure has a period of 0.5 s and 5% damping. Figure 8.36 shows that its design point on the bi-spectra plot is at Location A. Now assume that the design intent is to have a base-isolated structure with a period of 2.0 s, or

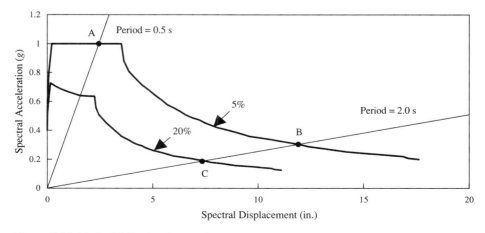

Figure 8.36 Method 1 Design Approach

alternatively a period separation of four. This means that if base isolators are selected with 5% damping, then the new design point is at Location B. The new base-isolated structure value of the spectral acceleration, or base shear coefficient (i.e., V/W), is $0.30g$ or 30% of the fixed-base value. The resultant displacement of the base isolators is 11.9 in. Now, imagine further that the base isolators have more than 5% damping, say 20%; then the design point becomes Location C. The spectral acceleration reduces even more, that is, from $0.30g$ to $0.19g$. The base isolator displacement reduces significantly, from 11.9 in. to 7.4 in.

Method 2: Selection of the Displacement of the Base Isolator

In Method 1, the intent was for the base-isolated structure to have a target natural period of vibration. As a result of this target natural period of vibration selection, the magnitude of displacement of the base isolator and the value of all desired structure response variables follow using the design-basis earthquake and the methods of structural dynamic analysis. An alternate method is now presented, which emphasizes a desired amplitude of base isolator displacement.

The desired amplitude of base isolator displacement may be controlled by the displacement of the base isolator used on previous structures. It may also be controlled by the open space, called a *gap* or *moat*, around the base-isolated building for architectural, mechanical, electrical, or plumbing considerations. Whatever the reason, Figure 8.33 for a base-isolated structure with 20% damping can be used to calculate, for a given displacement (e.g., 10.5 in.), the natural period of vibration of the base-isolated building.

A variation of this method is to set a value of absolute acceleration of the structure. If this is a desired goal, then Figure 8.31 for a base-isolated structure with 20% damping can be used to calculate the natural period of vibration of the base-isolated structure. The acceleration is very often an important response variable because it is a good measure of the potential damage to the contents in the structure.

For example, consider Figure 8.37, which is a bi-spectra plot for the design earthquake. Imagine that the fixed-base structure has a period of vibration equal to 0.5 s and 5% damping. Therefore, the design point for the fixed-base structure is at Location A. Now imagine that the design objective is to use base isolators to limit the displacement of the base isolator to 10.0 in. If the base isolator system has 5% damping, then the design point is at Location B. Therefore, the period of vibration must be 1.70 s, and the spectral acceleration is reduced from $1.0g$ to $0.35g$. Now imagine that the base isolators have 20%

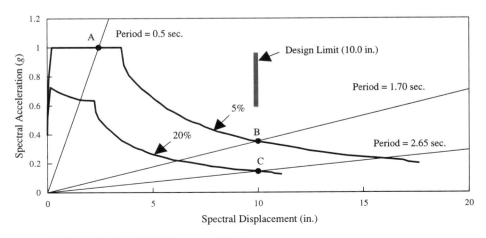

Figure 8.37 Method 2 with Displacement Limit

damping. In this case the design point is at Location C. This increase in damping enables the use of softer base isolators; that is, the period of vibration is 2.65 s and the spectral acceleration is further reduced to 0.15*g*.

Figure 8.38 shows a similar procedure but now with a limit on the desired value of spectral acceleration. This limit is especially appropriate when one design objective is to limit the damage to the nonstructural items (e.g., glassware, computers, or fragile objects). The design point for the fixed-base structure is at Location A. Now imagine that the design objective is to use base isolators to limit the floor acceleration to 0.2*g*. If the base isolators have 5% damping, then the new design point is at Location B. This location corresponds to a base-isolated structure with a period of vibration of 3.0 s and a spectral displacement of 17.6 in. With base isolators that have 20% damping, the new design point is at Location C with a period of the base-isolated structure equal to 1.9 s and a spectral displacement of 7.1 in.

Once the natural period of vibration is calculated, the stiffness of the base isolator follows from Eqs. 8.193 and 8.194. The response of the base isolators and the structure can then be obtained using the theory presented in the previous chapters.

Before the details of using the response spectra analysis procedure for base-isolated systems are presented, it is beneficial to provide an overview of the desirable steps in the total analysis process for a base-isolated structure.

The process starts with the preliminary design of a fixed-base structure. For example, the structural engineer knows from experience that using base isolators will reduce the value of the earthquake base shear by approximately three times compared to the fixed-base shear. Therefore, the structural engineer takes the given 5% damped design earthquake response spectrum for the building site (e.g., Figure 8.31) and divides it by 3. The structural engineer then determines the member sizes with a design of a fixed-base structure with this reduced response spectrum. What results is a preliminary structural design of the structure that will sit on top of the base isolators. The structural engineer may only use the first mode in the response spectra solution and a mode shape of all one values.

Following is the preliminary design of the base isolators using one of the methods discussed in this section. In this preliminary design, the structure is considered to be a rigid box sitting on top of the base isolators. As illustrated in this section, what results is a design for the desired base isolator stiffness at a maximum displacement. For example, recall from Table 8.4 and Figure 8.28 that the isolator must have a stiffness of 10.6 k/in. at a displacement of 10.5 in. In this preliminary design phase for the base isolator, the nonlinear base isolator is considered to be an elastic system with a given stiffness. For example, the

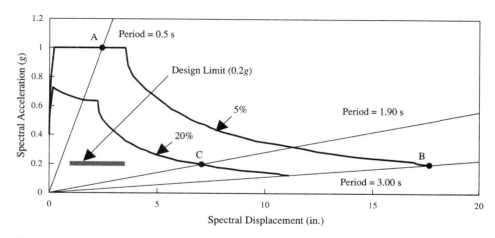

Figure 8.38 Method 2 with Acceleration Limit

specific base isolator discussed in Section 8.4 (see Figure 8.27 and Tables 8.8 and 8.9) has many stiffness values depending on the amplitude of the displacement. Therefore, it is a nonlinear element. The preliminary design focused on the single stiffness at the specific magnitude of displacement (e.g., 10.6 k/in. at 10.5 in. displacement).

Now the engineer must perform a more exact structural dynamic analysis to verify that the base isolators selected in the preliminary design phase are sufficient. In other words, if the structure is not simplified to be a rigid box sitting on top of the base isolators, it will still respond with the base isolators in a way that is within the design limit. This "more exact" structural dynamic analysis can be a response spectra analysis or a time history analysis. However, the desired approach is to first do a response spectra analysis and then to follow this with a time history analysis. In the response spectra analysis, the base isolators are considered to be elastic elements with an effective stiffness equal to the stiffness obtained from the preliminary base isolator design phase (i.e., a rigid box analysis). For example, the effective stiffness for the base isolator in Table 8.4 is 10.6 k/in. at a displacement of 10.5 in. The details of the response spectra analysis procedure are discussed in the next section. The time history analysis models the base isolator as a nonlinear element and is discussed in Section 8.8.

EXAMPLE 1   *Response Spectra Analysis of a Base-Isolated Structure*

Consider the fixed-base one-story structure shown in Figure 8.39. The natural frequency and period of vibration are

$$\omega_n = \sqrt{\frac{k}{m}} = \sqrt{\frac{212}{1.489}} = 11.9 \text{ rad/s} \tag{8.199}$$

$$T_n = \frac{2\pi}{\omega_n} = \frac{2\pi}{11.9} = 0.527 \text{ s.} \tag{8.200}$$

Assume that the structure is at a site with the design earthquake response spectra shown in Figures 8.40 and 8.41. The two risk levels are rare earthquake (10% probability of being exceeded in 50 years) and very rare earthquake (2% probability of being exceeded in 50 years). In addition, assume that the acceleration spectrum is approximately equal to the pseudo acceleration spectrum at a critical damping ratio up to 30% for the purpose of generating the displacement spectra.

The fixed-base structure is assumed to have 5% and 10% damping for the rare and very rare earthquakes, respectively. Therefore, the linear response of the structure for the rare earthquake is

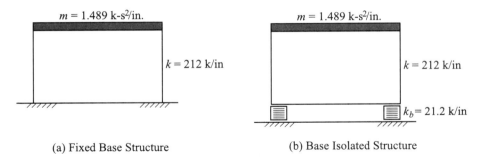

(a) Fixed Base Structure   (b) Base Isolated Structure

Figure 8.39 Fixed-Base and Base-Isolated Structures

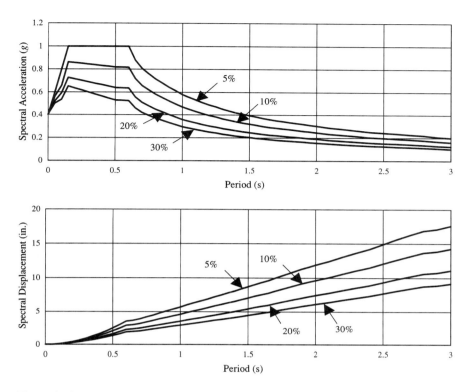

Figure 8.40 Design Earthquake Spectra (Rare, 10% in 50 Years)

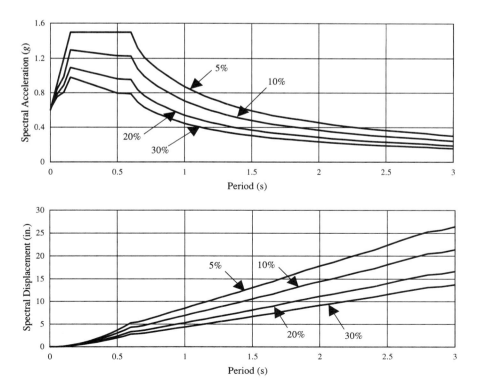

Figure 8.41 Design Earthquake Spectra (Very Rare, 2% in 50 Years)

$$\max[x(t)] = S_d = 2.71 \text{ in. (Figure 8.40b)} \tag{8.201}$$

$$\max[\ddot{y}(t)] = S_a = 1.00g \text{ (Figure 8.40a)} \tag{8.202}$$

$$\max[V(t)] = mS_a = (1.489)(1.00 \times 386.4) = 575 \text{ kip} \tag{8.203}$$

$$V/W = 1.00. \tag{8.204}$$

The linear response of the fixed-base structure for the very rare earthquake is

$$\max[x(t)] = S_d = 3.35 \text{ in. (Figure 8.41b)} \tag{8.205}$$

$$\max[\ddot{y}(t)] = S_a = 1.23g \text{ (Figure 8.41a)} \tag{8.206}$$

$$\max[V(t)] = mS_a = (1.489)(1.23 \times 386.4) = 708 \text{ kip} \tag{8.207}$$

$$V/W = 1.23. \tag{8.208}$$

As noted in this section, the structural engineer has different possible design objectives to choose from. For example, assume that the design objective is to have a base-isolated structure (see Figure 8.39b) with a significant separation in period between the fixed-base and base-isolated structure. Therefore, it is reasonable to assume that the stiffness of the base-isolated structure should equal 1/10 the stiffness of the fixed-base stiffness. In this case, $\alpha = 0.1$, and it follows that

$$\omega_1 = \sqrt{\frac{\alpha + 2 - \sqrt{\alpha^2 + 4}}{2}} \omega_n = \sqrt{\frac{0.1 + 2 - \sqrt{(0.1)^2 + 4}}{2}} (11.9) = 2.63 \text{ rad/s} \tag{8.209}$$

$$\omega_2 = \sqrt{\frac{\alpha + 2 + \sqrt{\alpha^2 + 4}}{2}} \omega_n = \sqrt{\frac{0.1 + 2 + \sqrt{(0.1)^2 + 4}}{2}} (11.9) = 17.08 \text{ rad/s} \tag{8.210}$$

$$T_1 = \frac{2\pi}{\omega_1} = \frac{2\pi}{2.63} = 2.39 \text{ s} \tag{8.211}$$

$$T_2 = \frac{2\pi}{\omega_2} = \frac{2\pi}{17.08} = 0.39 \text{ s} \tag{8.212}$$

$$r_1 = \frac{\alpha + \sqrt{\alpha^2 + 4}}{2} = \frac{0.1 + \sqrt{(0.1)^2 + 4}}{2} = 1.05 \tag{8.213}$$

$$r_2 = \frac{\alpha - \sqrt{\alpha^2 + 4}}{2} = \frac{0.1 - \sqrt{(0.1)^2 + 4}}{2} = -0.95 \tag{8.214}$$

$$\Phi_{\text{roof}(1)} = \begin{Bmatrix} 1/1.05 \\ 1.0 \end{Bmatrix} = \begin{Bmatrix} 0.95 \\ 1.0 \end{Bmatrix}, \quad \Phi_{\text{roof}(2)} = \begin{Bmatrix} -1/0.95 \\ 1.0 \end{Bmatrix} = \begin{Bmatrix} -1.05 \\ 1.0 \end{Bmatrix} \tag{8.215}$$

and

$$\Gamma_{\text{roof}(1)} = \frac{\begin{Bmatrix} 0.95 \\ 1.0 \end{Bmatrix}^T \begin{bmatrix} 1.0 & 0.0 \\ 0.0 & 1.0 \end{bmatrix} \begin{Bmatrix} 1.0 \\ 1.0 \end{Bmatrix}}{\begin{Bmatrix} 0.95 \\ 1.0 \end{Bmatrix}^T \begin{bmatrix} 1.0 & 0.0 \\ 0.0 & 1.0 \end{bmatrix} \begin{Bmatrix} 0.95 \\ 1.0 \end{Bmatrix}} = 1.024 \tag{8.216}$$

$$\Gamma_{\text{roof}(2)} = \frac{\left\{\begin{matrix} -1.05 \\ 1.0 \end{matrix}\right\}^T \begin{bmatrix} 1.0 & 0.0 \\ 0.0 & 1.0 \end{bmatrix} \left\{\begin{matrix} 1.0 \\ 1.0 \end{matrix}\right\}}{\left\{\begin{matrix} -1.05 \\ 1.0 \end{matrix}\right\}^T \begin{bmatrix} 1.0 & 0.0 \\ 0.0 & 1.0 \end{bmatrix} \left\{\begin{matrix} -1.05 \\ 1.0 \end{matrix}\right\}} = -0.024. \tag{8.217}$$

Note that the second mode participation factor is approximately 2% of the magnitude of the first mode participation factor.

The displacement of the floors can be calculated using the model superposition method, and it follows that the first mode response is

$$\mathbf{X}^{(1)} = \max\left\{\begin{matrix} x_1^{(1)}(t) \\ x_2^{(1)}(t) \end{matrix}\right\} = \boldsymbol{\phi}_{\text{roof}(1)} \Gamma_{\text{roof}(1)} S_d(T_1, \zeta_1) = \left\{\begin{matrix} 0.95 \\ 1.00 \end{matrix}\right\}(1.024) S_d(T_1, \zeta_1) \tag{8.218}$$

$$\ddot{\mathbf{Y}}^{(1)} = \max\left\{\begin{matrix} \ddot{y}_1^{(1)}(t) \\ \ddot{y}_2^{(1)}(t) \end{matrix}\right\} = \boldsymbol{\phi}_{\text{roof}(1)} \Gamma_{\text{roof}(1)} S_a(T_1, \zeta_1) = \left\{\begin{matrix} 0.95 \\ 1.00 \end{matrix}\right\}(1.024) S_a(T_1, \zeta_1) \tag{8.219}$$

and the second mode response is

$$\mathbf{X}^{(2)} = \max\left\{\begin{matrix} x_1^{(2)}(t) \\ x_2^{(2)}(t) \end{matrix}\right\} = \boldsymbol{\phi}_{\text{roof}(2)} \Gamma_{\text{roof}(2)} S_d(T_2, \zeta_2) = \left\{\begin{matrix} -1.05 \\ 1.00 \end{matrix}\right\}(-0.024) S_d(T_2, \zeta_2) \tag{8.220}$$

$$\ddot{\mathbf{Y}}^{(2)} = \max\left\{\begin{matrix} \ddot{y}_1^{(2)}(t) \\ \ddot{y}_2^{(2)}(t) \end{matrix}\right\} = \boldsymbol{\phi}_{\text{roof}(2)} \Gamma_{\text{roof}(2)} S_a(T_2, \zeta_2) = \left\{\begin{matrix} -1.05 \\ 1.00 \end{matrix}\right\}(-0.024) S_a(T_2, \zeta_2). \tag{8.221}$$

Assume that the base-isolated structure has 20% damping when subjected to both the rare earthquake and the very rare earthquake. Therefore, from Figure 8.40 for the rare earthquake, it follows that

$$S_d(T_1 = 2.39, \zeta = 20\%) = 8.86 \text{ in.} \tag{8.222}$$

$$S_a(T_1 = 2.39, \zeta = 20\%) = 0.16g. \tag{8.223}$$

Therefore, it follows from Eqs. 8.218 and 8.219 that

$$\mathbf{X}^{(1)} = \left\{\begin{matrix} 0.95 \\ 1.00 \end{matrix}\right\}(1.024)(8.86) = \left\{\begin{matrix} 8.62 \\ 9.07 \end{matrix}\right\} \text{ in.} \tag{8.224}$$

$$\ddot{\mathbf{Y}}^{(1)} = \left\{\begin{matrix} 0.95 \\ 1.00 \end{matrix}\right\}(1.024)(0.16g) = \left\{\begin{matrix} 0.156g \\ 0.163g \end{matrix}\right\} \tag{8.225}$$

and

$$V^{(1)} = (1.489)(0.156 \times 386.4) + (1.489)(0.163 \times 386.4) = 184 \text{ kip} \tag{8.226}$$

$$V^{(1)}/W = 184/(1.489 \times 2 \times 386.4) = 0.16. \tag{8.227}$$

For the second mode,

$$S_d(T_1 = 0.39, \zeta = 5\%) = 1.52 \text{ in.} \tag{8.228}$$

$$S_a(T_1 = 0.39, \zeta = 5\%) = 1.00g. \tag{8.229}$$

Therefore, it follows from Eqs. 8.220 and 8.221 that

$$\mathbf{X}^{(2)} = \left\{\begin{matrix} -1.05 \\ 1.00 \end{matrix}\right\}(-0.024)(1.52) = \left\{\begin{matrix} 0.038 \\ -0.036 \end{matrix}\right\} \text{ in.} \tag{8.230}$$

$$\ddot{\mathbf{Y}}^{(2)} = \begin{Bmatrix} -1.05 \\ 1.00 \end{Bmatrix}(-0.024)(1.00g) = \begin{Bmatrix} 0.025g \\ -0.024g \end{Bmatrix}. \tag{8.231}$$

For the very rare earthquake, it follows from Figure 8.41 that

$$S_d(T_1 = 2.39, \zeta = 20\%) = 13.3 \text{ in.} \tag{8.232}$$

$$S_a(T_1 = 2.39, \zeta = 20\%) = 0.24g. \tag{8.233}$$

Therefore, it follows from Eqs. 8.218 and 8.219 that

$$\mathbf{X}^{(1)} = \begin{Bmatrix} 0.95 \\ 1.00 \end{Bmatrix}(1.024)(13.3) = \begin{Bmatrix} 12.9 \\ 13.6 \end{Bmatrix} \text{ in.} \tag{8.234}$$

$$\ddot{\mathbf{Y}}^{(1)} = \begin{Bmatrix} 0.95 \\ 1.00 \end{Bmatrix}(1.024)(0.24g) = \begin{Bmatrix} 0.233g \\ 0.246g \end{Bmatrix} \tag{8.235}$$

and

$$V^{(1)} = (1.489)(0.233 \times 386.4) + (1.489)(0.246 \times 386.4) = 276 \text{ kip} \tag{8.236}$$

$$V^{(1)}/W = 276/(1.489 \times 2 \times 386.4) = 0.24. \tag{8.237}$$

For the second mode,

$$S_d(T_1 = 0.39, \zeta = 10\%) = 1.91 \text{ in.} \tag{8.238}$$

$$S_a(T_1 = 0.39, \zeta = 10\%) = 1.25g \tag{8.239}$$

Therefore, it follows from Eqs. 8.220 and 8.221 that

$$\mathbf{X}^{(2)} = \begin{Bmatrix} -1.05 \\ 1.00 \end{Bmatrix}(-0.024)(1.91) = \begin{Bmatrix} 0.048 \\ -0.046 \end{Bmatrix} \text{ in.} \tag{8.240}$$

$$\ddot{\mathbf{Y}}^{(2)} = \begin{Bmatrix} -1.05 \\ 1.00 \end{Bmatrix}(-0.024)(1.25g) = \begin{Bmatrix} 0.032g \\ -0.030g \end{Bmatrix}. \tag{8.241}$$

The displacement of the roof relative to the ground for the rare earthquake for the fixed-base structure was 2.71 in., and this increased to 9.07 in. for the base-isolated structure. However, the 2.71 in., displacement for the fixed-base structure is the relative lateral deformation of the columns, and this is reduced to

$$\Delta^{(1)} = 9.07 - 8.62 = 0.45 \text{ in.} \tag{8.242}$$

for the base-isolated structure. This reduction represents approximately a sixfold reduction in column response amplitude.

Now assume that a different objective is selected by the structural engineer to reduce the floor acceleration due to the very rare earthquake from 1.23g for the fixed-base structure to a much smaller value. For computational convenience, assume that this target floor acceleration is 0.25g. In this solution procedure, the very rare earthquake response spectrum is studied, and for a reasonable base isolator damping value of 20% it is determined that the fundamental period of the base-isolated structure must be approximately 2.4 s. The equation

$$T_1 = \frac{2\pi}{\omega_1} = 2.4 \text{ s} \tag{8.243}$$

gives approximately the same natural frequency and isolator stiffness as previously discussed in this example, that is,

$$\omega_1 = \sqrt{\frac{\alpha + 2 - \sqrt{\alpha^2 + 4}}{2}} \omega_n = \sqrt{\frac{0.1 + 2 - \sqrt{(0.1)^2 + 4}}{2}} (11.9) = 2.63 \text{ rad/s} \qquad (8.244)$$

This gives $\alpha = 0.10$ and

$$k_b = (0.10)(212) = 21.2 \text{ k/in.} \qquad (8.245)$$

8.6   RESPONSE SPECTRA ANALYSIS OF A LINEAR ELASTIC BASE-ISOLATED STRUCTURE

Section 8.3 discussed the response of a soft-story structural system. In most structural engineering applications, nodes E and F in Figure 8.17 are at the first story and nodes C and D are at the second story. Therefore, the typical distance between nodes A and E or nodes E and C is approximately 10 ft. However, now imagine a different situation where the distance between nodes E and C is still approximately 10 ft, but a base isolation unit is placed between nodes A and E and between nodes B and F. Therefore, the "bottom-story" height is not 10 ft but is only 1 to 2 ft.

Chapter 5 presented the response spectra method of calculating the dynamic response of a structure. Section 8.3 showed how this analysis method can be used for a soft-story structure. The structure must be linear elastic to use this solution method because the normal modes of the linear system are used in the analysis. Some base-isolated buildings use base isolators that have linear force versus deflection characteristics. These are referred to here as *linear elastic base isolated systems*. In this type of base isolation system, another structural system is added to the structure that is typically called the *wind restraint system*. Figure 8.42 shows a schematic of such a base-isolated frame. The intent of the wind restraint system is to provide stiffness at the base isolation level up to the magnitude of wind force that can be expected to occur infrequently (e.g., once every 20 years). If the wind force exceeds the capacity of the wind restraint system, then either the wind restraint system continues to provide a constant load or alternately it fails and carries no load. Figure 8.43 shows a schematic of a force versus deflection curve for the bottom story with a combined isolator and wind restraint system. The wind restraint system in the figure shows a linear relationship up to the deflection Δ_1 and then for larger deflections no wind restraint force. If the earthquake produces displacements less than Δ_1, then the stiffness of the system is equal to the sum of the stiffness of the base isolators (k_b represents the sum of the stiffness of the two isolators) and the stiffness of the wind restraint system (k_w), that is, $k_b + k_w$. If the deflection exceeds Δ_1, then the wind restraint system fails and the system stiffness drops to k_b. A variation on this is where the wind restraint system does not lose its capacity but is an elastic–plastic or a bilinear system with a very small stiffness after Δ_1.

The response of this elastic base isolation system depends on the magnitude of the earthquake that excites the base-isolated building. If the earthquake is not large enough to

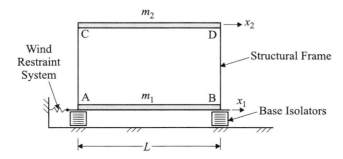

Figure 8.42 Base-Isolated One-Story Frame with Wind Restraint System

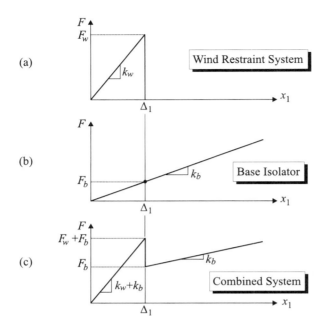

(a)

(b)

(c)

Figure 8.43 Force Deflection Curves of Base Isolators with Wind Restraint System

cause a displacement x_1 that is greater than Δ_1, then the response spectra solution is the same as discussed in Chapter 5 and Section 8.3. The natural frequency of the system is that associated with the stiffness of the combined system (i.e., $k_b + k_w$). For the single bay frame shown in Figure 8.42, the equation of motion is

$$\mathbf{M}\ddot{\mathbf{X}}(t) + \mathbf{C}\dot{\mathbf{X}}(t) + \mathbf{K}\mathbf{X}(t) = -\mathbf{M}\{\mathbf{I}\}a(t) \tag{8.246}$$

where

$$\mathbf{M} = \begin{bmatrix} m_1 & 0 \\ 0 & m_2 \end{bmatrix}, \quad \mathbf{C} = \begin{bmatrix} c_b + c_w + c & -c \\ -c & c \end{bmatrix}, \quad \mathbf{K} = \begin{bmatrix} k_b + k_w + k & -k \\ -k & k \end{bmatrix}. \tag{8.247}$$

Note that c_b is the sum of the damping coefficient from the two isolators and c is the damping in the fixed-base structure. The stiffness of the fixed-base structure is k.

To perform a response spectra solution using the normal mode method, the damping matrix must be proportional, and this is seldom the case. However, imagine that this is overlooked, and the modal damping in the first mode, ζ_1, is

$$2\zeta_1\omega_1 = \frac{\boldsymbol{\phi}_1^T \mathbf{C}\boldsymbol{\phi}_1}{\boldsymbol{\phi}_1^T \mathbf{M}\boldsymbol{\phi}_1} \tag{8.248}$$

where ω_1 is the fundamental natural frequency of vibration and $\boldsymbol{\phi}_1$ is the fundamental mode shape of the system corresponding to Eq. 8.246.

Consider the damping to be zero in the wind restraint system because it is usually very small. The damping matrix can then be decomposed into the following three matrices:

$$\mathbf{C} = \begin{bmatrix} c_b & 0 \\ 0 & 0 \end{bmatrix} + \begin{bmatrix} c & -c \\ -c & 0 \end{bmatrix} + \begin{bmatrix} 0 & 0 \\ 0 & c \end{bmatrix} = \mathbf{C}_1 + \mathbf{C}_2 + \mathbf{C}_3. \tag{8.249}$$

The matrix $\mathbf{C}_1$ corresponds to the contribution to the system damping matrix of the base isolator damping. The matrix $\mathbf{C}_3$ corresponds to the damping matrix in a fixed-base or a system with a very large α, where α is the ratio of the lower-story stiffness to the upper-story stiffness in a 2DOF system, that is, $\alpha(k_b + k_w) = k$. The $\mathbf{C}_2$ matrix is the coupling matrix between the base isolator and the structure, and it is a function of α. If $m_1 = m_2 = m$, it then follows from considering each term after substituting Eq. 8.249 into Eq. 8.248 that

$$\boldsymbol{\phi}_1^T \mathbf{C}_1 \boldsymbol{\phi}_1 = \left\{ \begin{matrix} 1 \\ r_1 \end{matrix} \right\}^T \begin{bmatrix} c_b & 0 \\ 0 & 0 \end{bmatrix} \left\{ \begin{matrix} 1 \\ r_1 \end{matrix} \right\} = c_b \qquad (8.250a)$$

$$\boldsymbol{\phi}_1^T \mathbf{C}_2 \boldsymbol{\phi}_1 = \left\{ \begin{matrix} 1 \\ r_1 \end{matrix} \right\}^T \begin{bmatrix} c & -c \\ -c & 0 \end{bmatrix} \left\{ \begin{matrix} 1 \\ r_1 \end{matrix} \right\} = c(1 - 2r_1) \qquad (8.250b)$$

$$\boldsymbol{\phi}_1^T \mathbf{C}_3 \boldsymbol{\phi}_1 = \left\{ \begin{matrix} 1 \\ r_1 \end{matrix} \right\}^T \begin{bmatrix} 0 & 0 \\ 0 & c \end{bmatrix} \left\{ \begin{matrix} 1 \\ r_1 \end{matrix} \right\} = c r_1^2 \qquad (8.250c)$$

$$\boldsymbol{\phi}_1^T \mathbf{M} \boldsymbol{\phi}_1 = \left\{ \begin{matrix} 1 \\ r_1 \end{matrix} \right\}^T \begin{bmatrix} m & 0 \\ 0 & m \end{bmatrix} \left\{ \begin{matrix} 1 \\ r_1 \end{matrix} \right\} = m(1 + r_1^2). \qquad (8.250d)$$

Therefore, substituting Eqs. 8.249 and 8.250 into Eq. 8.248 gives

$$\frac{\boldsymbol{\phi}_1^T \mathbf{C} \boldsymbol{\phi}_1}{\boldsymbol{\phi}_1^T \mathbf{M} \boldsymbol{\phi}_1} = \left[\frac{c_b}{m(1 + r_1^2)} \right] + \left[\frac{c(1 - 2r_1)}{m(1 + r_1^2)} \right] + \left[\frac{c r_1^2}{m(1 + r_1^2)} \right]. \qquad (8.251)$$

If α is very large, then the system is an SDOF system with a natural frequency

$$\omega_n = \sqrt{\frac{k}{m}} \qquad (8.252)$$

and damping value

$$2\zeta_n \omega_n = \frac{c}{m}. \qquad (8.253)$$

Therefore, it follows that

$$c = 2\zeta_n m \omega_n. \qquad (8.254)$$

If the structure is very rigid, then the system is like a rigid structure on a flexible base. For this case the structure is an SDOF system with a natural frequency of vibration equal to

$$\omega_{nb} = \sqrt{\frac{k_b + k_w}{2m}} \qquad (8.255)$$

and a damping value of

$$2\zeta_{nb} \omega_{nb} = \frac{c_b}{2m}. \qquad (8.256)$$

Rearranging Eq. 8.256 gives

$$c_b = 4\zeta_{nb} m \omega_{nb}. \qquad (8.257)$$

Therefore, substituting Eqs. 8.254 and 8.257 into Eq. 8.251, it follows that

$$\frac{\boldsymbol{\phi}_1^T \mathbf{C} \boldsymbol{\phi}_1}{\boldsymbol{\phi}_1^T \mathbf{M} \boldsymbol{\phi}_1} = \left[\frac{4\zeta_{nb} \omega_{nb}}{1 + r_1^2} \right] + \left[\frac{2\zeta_n \omega_n (1 - 2r_1)}{1 + r_1^2} \right] + \left[\frac{2\zeta_n \omega_n r_1^2}{1 + r_1^2} \right]. \qquad (8.258)$$

Simplifying Eq. 8.258 gives

$$2\zeta_1 \omega_1 = \frac{\boldsymbol{\phi}_1^T \mathbf{C} \boldsymbol{\phi}_1}{\boldsymbol{\phi}_1^T \mathbf{M} \boldsymbol{\phi}_1} = 2\zeta_{nb} \omega_{nb} C_{nb} + 2\zeta_n \omega_n C_{nbn} + 2\zeta_n \omega_n C_n \qquad (8.259)$$

where

$$C_{nb} = \frac{2}{1+r_1^2}, \quad C_{nbn} = \frac{1-2r_1}{1+r_1^2}, \quad C_n = \frac{r_1^2}{1+r_1^2} \tag{8.260}$$

and from Eq. 8.69,

$$r_1 = \frac{\alpha + \sqrt{\alpha^2 + 4}}{2}. \tag{8.261}$$

Rearranging Eq. 8.259 gives

$$\zeta_1 = C_{nb}\left(\frac{\omega_{nb}}{\omega_1}\right)\zeta_{nb} + C_{nbn}\left(\frac{\omega_n}{\omega_1}\right)\zeta_n + C_n\left(\frac{\omega_n}{\omega_1}\right)\zeta_n. \tag{8.262}$$

Recall from Eq. 8.66 that

$$\omega_1 = \sqrt{\frac{\alpha + 2 - \sqrt{\alpha^2 + 4}}{2}}\,\omega_n. \tag{8.263}$$

Substituting Eqs. 8.261 and 8.263 into Eq. 8.262 gives

$$\zeta_1 = A_{nb}\left(\frac{\omega_{nb}}{\omega_n}\right)\zeta_{nb} + A_{nbn}\zeta_n + A_n\zeta_n \tag{8.264}$$

where

$$A_{nb} = \frac{2}{1+\left(\alpha+\sqrt{\alpha^2+4}\right)^2\Big/4}\sqrt{\frac{2}{\alpha+2-\sqrt{\alpha^2+4}}} = \frac{4}{\alpha^2+4+\alpha\sqrt{\alpha^2+4}}\left(\frac{\sqrt{2}}{\sqrt{\alpha+2-\sqrt{\alpha^2+4}}}\right) \tag{8.265a}$$

$$A_{nbn} = \frac{1-2\left(\alpha+\sqrt{\alpha^2+4}\right)\Big/2}{1+\left(\alpha+\sqrt{\alpha^2+4}\right)^2\Big/4}\sqrt{\frac{2}{\alpha+2-\sqrt{\alpha^2+4}}} = \frac{4-4\alpha-4\sqrt{\alpha^2+4}}{\alpha^2+4+\alpha\sqrt{\alpha^2+4}}\left(\frac{\sqrt{2}}{\sqrt{\alpha+2-\sqrt{\alpha^2+4}}}\right) \tag{8.265b}$$

$$A_n = \frac{\left(\alpha+\sqrt{\alpha^2+4}\right)^2\Big/4}{1+\left(\alpha+\sqrt{\alpha^2+4}\right)^2\Big/4}\sqrt{\frac{2}{\alpha+2-\sqrt{\alpha^2+4}}} = \frac{\alpha^2+2+\alpha\sqrt{\alpha^2+4}}{\alpha^2+4+\alpha\sqrt{\alpha^2+4}}\cdot\left(\frac{\sqrt{2}}{\sqrt{\alpha+2-\sqrt{\alpha^2+4}}}\right). \tag{8.265c}$$

Figure 8.44 shows the variations of A_{nb}, A_{nbn}, and A_n with respect to α.

A similar procedure can be derived to obtain the damping in the second mode using the second mode of vibration with the value of replaced by r_1 replaced by r_2:

$$\frac{\boldsymbol{\phi}_2^T\mathbf{C}\boldsymbol{\phi}_2}{\boldsymbol{\phi}_2^T\mathbf{M}\boldsymbol{\phi}_2} = \left[\frac{4\zeta_{nb}\omega_{nb}}{1+r_2^2}\right] + \left[\frac{2\zeta_n\omega_n(1-2r_2)}{1+r_2^2}\right] + \left[\frac{2\zeta_n\omega_n r_2^2}{1+r_2^2}\right]. \tag{8.266}$$

Simplifying Eq. 8.266 gives

$$2\zeta_2\omega_2 = \frac{\boldsymbol{\phi}_2^T\mathbf{C}\boldsymbol{\phi}_2}{\boldsymbol{\phi}_2^T\mathbf{M}\boldsymbol{\phi}_2} = 2\zeta_{nb}\omega_{nb}D_{nb} + 2\zeta_n\omega_n D_{nbn} + 2\zeta_n\omega_n D_n \tag{8.267}$$

where

$$D_{nb} = \frac{2}{1+r_2^2}, \quad D_{nbn} = \frac{1-2r_2}{1+r_2^2}, \quad D_n = \frac{r_2^2}{1+r_2^2} \tag{8.268}$$

and

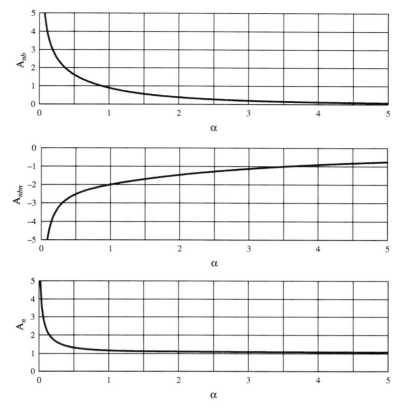

Figure 8.44 Variations of the Damping Coefficients A_{nb}, A_{nbn}, and A_n with Respect to α

$$r_2 = \frac{\alpha - \sqrt{\alpha^2 + 4}}{2}.$$ (8.269)

Rearranging Eq. 8.267, it follows that

$$\zeta_2 = D_{nb}\left(\frac{\omega_{nb}}{\omega_2}\right)\zeta_{nb} + D_{nbn}\left(\frac{\omega_n}{\omega_2}\right)\zeta_n + D_n\left(\frac{\omega_n}{\omega_2}\right)\zeta_n$$ (8.270)

where

$$\omega_2 = \sqrt{\frac{\alpha + 2 + \sqrt{\alpha^2 + 4}}{2}}\,\omega_n.$$ (8.271)

Substituting Eqs. 8.269 and 8.271 into Eq. 8.270, it follows that

$$\zeta_2 = B_{nb}\left(\frac{\omega_{nb}}{\omega_n}\right)\zeta_{nb} + B_{nbn}\zeta_n + B_n\zeta_n$$ (8.272)

where

$$B_{nb} = \frac{2}{1 + \left(\alpha - \sqrt{\alpha^2 + 4}\right)^2 \big/ 4}\sqrt{\frac{2}{\alpha + 2 + \sqrt{\alpha^2 + 4}}} = \frac{4}{\alpha^2 + 4 - \alpha\sqrt{\alpha^2 + 4}}\left(\frac{\sqrt{2}}{\sqrt{\alpha + 2 + \sqrt{\alpha^2 + 4}}}\right)$$ (8.273a)

$$B_{nbn} = \frac{1 - 2\left(\alpha - \sqrt{\alpha^2 + 4}\right)\big/2}{1 + \left(\alpha - \sqrt{\alpha^2 + 4}\right)^2\big/4} \sqrt{\frac{2}{\alpha + 2 + \sqrt{\alpha^2 + 4}}} = \frac{4 - 4\alpha + 4\sqrt{\alpha^2 + 4}}{\alpha^2 + 4 - \alpha\sqrt{\alpha^2 + 4}} \left(\frac{\sqrt{2}}{\sqrt{\alpha + 2 + \sqrt{\alpha^2 + 4}}}\right) \quad (8.273b)$$

$$B_{n} = \frac{\left(\alpha - \sqrt{\alpha^2 + 4}\right)^2\big/4}{1 + \left(\alpha - \sqrt{\alpha^2 + 4}\right)^2\big/4} \sqrt{\frac{2}{\alpha + 2 + \sqrt{\alpha^2 + 4}}} = \frac{\alpha^2 + 2 - \alpha\sqrt{\alpha^2 + 4}}{\alpha^2 + 4 - \alpha\sqrt{\alpha^2 + 4}} \left(\frac{\sqrt{2}}{\sqrt{\alpha + 2 + \sqrt{\alpha^2 + 4}}}\right). \quad (8.273c)$$

Figure 8.45 shows the variations of B_{nb}, B_{nbn}, and B_n with respect to α.

The response of the base-isolated building in its first mode is

$$\max[\mathbf{X}(t)] = \boldsymbol{\phi}_1 \max[q_1(t)] \quad (8.274)$$

where

$$\max[q_1(t)] = \Gamma_1 S_d(T_1, \zeta_1). \quad (8.275)$$

Because the wind restraint system provides stiffness, the period of vibration is usually relatively small. This can create a design concern because the acceleration of the building floors can become large enough to cause nonstructural damage if Δ_1 is too large.

When the displacement x_1 exceeds Δ_1, then the lateral stiffness at the bottom level is only provided by the base isolators. The stiffness of the isolators, k_b, is usually selected to be much less than k (e.g., $k_b = 0.2k$), and the fundamental period of vibration of the system for x_1 greater than Δ_1 is typically greater than 2.0 s.

In reality, the structure with an elastic base isolator is a nonlinear system because the stiffness of the wind restraint system is a function of the magnitude of x_1. However,

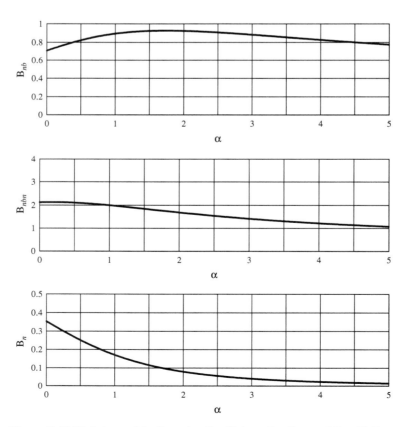

Figure 8.45 Variations of the Damping Coefficients B_{nb}, B_{nbn}, and B_n with Respect to α

because Δ_1 is typically less than 0.5 in. and the design earthquake typically produces values of x_2 equal to 10 in. or more, the nonlinearity is typically ignored and k_w is set equal to zero. Therefore, for a design-level earthquake ground motion, the stiffness matrix is

$$\mathbf{K} = \begin{bmatrix} k_b + k & -k \\ -k & k \end{bmatrix}.$$
(8.276)

Equations 8.274 and 8.275 are still valid for the first mode response, but now the fundamental period is usually greater than 2.0 s and the value of the spectral acceleration will be much less than its peak value.

EXAMPLE 1 *Response with Wind Restraint System*

Consider a base-isolated structure with a wind restraint system. Assume that the wind restraint system has a stiffness equal to 100 k/in. and its yield displacement is 0.5 in. At the displacement of 0.5 in. it is assumed to fail and provide no lateral support for the structure. Let $m_1 = m_2 = m = 1.0$ k-s^2/in.

Assume that a linear elastic base isolator is placed below each column and that the stiffness and critical damping ratio of each base isolator are 10.6 k/in. and 5%, respectively.

The stiffness of the structure above the base isolator is k and is equal to

$$k_b = \alpha k$$
(8.277)

where $k_b = 2(10.6) = 21.2$ k/in. If $\alpha = 0.1$, then

$$k = k_b/\alpha = 21.2/0.1 = 212 \text{ k/in.}$$
(8.278)

The stiffness of the bottom story prior to the failure of the wind restraint system is

$$k_1 = k_b + k_w = 21.2 + 100 = 121.2 \text{ k/in.}$$
(8.279)

Therefore, the first-story stiffness can be written as

$$k_1 = \alpha^* k.$$
(8.280)

Solving for α^*, it follows from Eq. 8.280 that

$$\alpha^* = k_1/k = 121.2/212 = 0.572.$$
(8.281)

Note that with the wind restraint system, the ratio of first-story to second-story stiffness is approximately one half.

Using Eqs. 8.66 to 8.70 with replacing α^*, it follows that

$$\omega_1^2 = \left(\frac{\alpha^* + 2 - \sqrt{(\alpha^*)^2 + 4}}{2} \right) \omega_n^2 = \left(\frac{0.572 + 2 - \sqrt{(0.572)^2 + 4}}{2} \right) \left(\frac{212}{1.0} \right) = 52.1$$
(8.282)

$$\omega_2^2 = \left(\frac{\alpha^* + 2 + \sqrt{(\alpha^*)^2 + 4}}{2} \right) \omega_n^2 = \left(\frac{0.572 + 2 + \sqrt{(0.572)^2 + 4}}{2} \right) \left(\frac{212}{1.0} \right) = 493$$
(8.283)

$$T_1 = \frac{2\pi}{\omega_1} = \frac{2\pi}{\sqrt{52.1}} = 0.87 \text{ s}$$
(8.284)

$$T_2 = \frac{2\pi}{\omega_2} = \frac{2\pi}{\sqrt{493}} = 0.28 \text{ s.}$$
(8.285)

The mode shapes are

$$\Phi_1 = \left\{ \begin{matrix} 1 \\ \left(0.572 + \sqrt{(0.572)^2 + 4}\right)/2 \end{matrix} \right\} = \left\{ \begin{matrix} 1.0 \\ 1.326 \end{matrix} \right\} = \left\{ \begin{matrix} 0.754 \\ 1.00 \end{matrix} \right\} \tag{8.286}$$

$$\Phi_2 = \left\{ \begin{matrix} 1 \\ \left(0.572 - \sqrt{(0.572)^2 + 4}\right)/2 \end{matrix} \right\} = \left\{ \begin{matrix} 1.0 \\ -0.754 \end{matrix} \right\} = \left\{ \begin{matrix} -1.326 \\ 1.00 \end{matrix} \right\}. \tag{8.287}$$

The damping in the wind restraint system is assumed to be zero. The critical damping ratio in the structure above the base isolators is assumed to be 2%. From Eq. 8.254, it follows that

$$c = 2\zeta_n m \omega_n = 2(0.02)(1.0)\sqrt{\frac{212}{1.0}} = 0.582 \text{ k-s/in.} \tag{8.288}$$

Using Eq. 8.257, it follows that

$$c_b = 4\zeta_{nb} m \omega_{nb} = 4(0.10)(1.0)\sqrt{\frac{121.2}{(2.0)(1.0)}} = 4.40 \text{ k-s/in.} \tag{8.289}$$

Note that the term 0.10 in Eq. 8.289 is the result of the summation of 5% damping in each of the two base isolators. The damping matrix for the base-isolated structure is

$$\mathbf{C} = \begin{bmatrix} c_b + c & -c \\ -c & c \end{bmatrix} = \begin{bmatrix} 4.40 + 0.582 & -0.582 \\ -0.582 & 0.582 \end{bmatrix} = \begin{bmatrix} 4.98 & -0.582 \\ -0.582 & 0.582 \end{bmatrix}. \tag{8.290}$$

Note that $\Phi_1^T \mathbf{C} \Phi_2$ is not zero, and therefore the damping matrix is not a proportional damping matrix. However, if these off-diagonal terms are neglected and the structural engineer assumes that the damping is proportional, then from Eq. 8.259 it follows that

$$2\zeta_1 \omega_1 = 2\zeta_{nb} \omega_{nb} C_{nb} + 2\zeta_n \omega_n C_{nbn} + 2\zeta_n \omega_n C_n \tag{8.291}$$

where

$$C_{nb} = \frac{2}{1 + r_1^2}, \quad C_{nbn} = \frac{1 - 2r_1}{1 + r_1^2}, \quad C_n = \frac{r_1^2}{1 + r_1^2}. \tag{8.292}$$

Equation 8.291 can be solved to give ζ_1. However, insight into the solution can be gained by considering the alternate form of this equation given in Eq. 8.264, that is,

$$\zeta_1 = A_{nb}\left(\frac{\omega_{nb}}{\omega_n}\right)\zeta_{nb} + A_{nbn}\zeta_n + A_n\zeta_n \tag{8.293}$$

where

$$A_{nb} = \frac{4}{(0.572)^2 + 4 + (0.572)\sqrt{(0.572)^2 + 4}}\left(\frac{\sqrt{2}}{\sqrt{(0.572) + 2 - \sqrt{(0.572)^2 + 4}}}\right) = 1.46 \tag{8.294}$$

$$A_{nbn} = \frac{4 - 4(0.572) - 4\sqrt{(0.572)^2 + 4}}{(0.572)^2 + 4 + (0.572)\sqrt{(0.572)^2 + 4}}\left(\frac{\sqrt{2}}{\sqrt{(0.572) + 2 - \sqrt{(0.572)^2 + 4}}}\right) = -2.42 \tag{8.295}$$

$$A_n = \frac{(0.572)^2 + 2 + (0.572)\sqrt{(0.572)^2 + 4}}{(0.572)^2 + 4 + (0.572)\sqrt{(0.572)^2 + 4}}\left(\frac{\sqrt{2}}{\sqrt{(0.572) + 2 - \sqrt{(0.572)^2 + 4}}}\right) = 1.29. \tag{8.296}$$

Therefore,

$$\zeta_1 = (1.46)\frac{\sqrt{121.2/2.0}}{\sqrt{212/1.0}}(0.10) + (-2.42)(0.02) + (1.29)(0.02)$$

$$= 0.0781 - 0.0484 + 0.0258 = 0.056 = 5.6\%. \tag{8.297}$$

Equations 8.293 and 8.297 show how the damping in the fundamental mode is divided into three parts. The first part is linearly related to the damping for the base isolators, ζ_{nb}. The third part is linearly related to the damping in the structure above the isolators, ζ_n.

Using Eq. 8.272, the damping in the second mode is

$$\zeta_2 = B_{nb}\left(\frac{\omega_{nb}}{\omega_n}\right)\zeta_{nb} + B_{nbn}\zeta_n + B_n\zeta_n \tag{8.298}$$

where

$$B_{nb} = \frac{4}{(0.572)^2 + 4 - (0.572)\sqrt{(0.572)^2 + 4}}\left(\frac{\sqrt{2}}{\sqrt{(0.572) + 2 + \sqrt{(0.572)^2 + 4}}}\right) = 0.84 \tag{8.299}$$

$$B_{nbn} = \frac{4 - 4(0.572) + 4\sqrt{(0.572)^2 + 4}}{(0.572)^2 + 4 - (0.572)\sqrt{(0.572)^2 + 4}}\left(\frac{\sqrt{2}}{\sqrt{(0.572) + 2 + \sqrt{(0.572)^2 + 4}}}\right) = 2.10 \tag{8.300}$$

$$B_n = \frac{(0.572)^2 + 2 - (0.572)\sqrt{(0.572)^2 + 4}}{(0.572)^2 + 4 - (0.572)\sqrt{(0.572)^2 + 4}}\left(\frac{\sqrt{2}}{\sqrt{(0.572) + 2 + \sqrt{(0.572)^2 + 4}}}\right) = 0.24. \tag{8.301}$$

Therefore,

$$\zeta_2 = (0.84)\frac{\sqrt{121.2/2.0}}{\sqrt{212/1.0}}(0.10) + (2.10)(0.02) + (0.24)(0.02)$$

$$= 0.0449 + 0.0420 + 0.0048 = 0.092 = 9.2\%. \tag{8.302}$$

A response spectra solution can now be obtained using the methods described in Section 5.4 for any ground motion.

EXAMPLE 2   *Response without Wind Restraint System*

If the earthquake ground motion is small and the response of the first floor is less than Δ_1, then as shown in Example 1 the wind restraint system contributes to the stiffness of the first story. This is expressed as

$$k_1 = k_b + k_w \tag{8.303}$$

and from Eq. 8.255,

$$\omega_{nb} = \sqrt{\frac{k_b + k_w}{2m}}. \tag{8.304}$$

For the base-isolated structure in Example 1 with the wind restraint system, these quantities were

$$k_1 = k_b + k_w = 21.2 + 100 = 121.2 \text{ k/in.} \tag{8.305}$$

$$\omega_{nb} = \sqrt{\frac{121.2}{2(1.0)}} = 7.78 \text{ rad/s} \tag{8.306}$$

$$T_{nb} = \frac{2\pi}{\omega_{nb}} = \frac{2\pi}{7.78} = 0.81 \text{ s.} \tag{8.307}$$

Now consider the case where the earthquake ground motion exceeds Δ_1 and therefore the wind restraint system has failed. In this case, the total lateral stiffness of the first story is provided by the base isolators, and

$$k_1 = k_b = 21.2 \text{ k/in.} \tag{8.308}$$

$$\omega_{nb} = \sqrt{\frac{21.2}{2(1.0)}} = 3.26 \text{ rad/s} \tag{8.309}$$

$$T_{nb} = \frac{2\pi}{\omega_{nb}} = \frac{2\pi}{3.26} = 1.93 \text{ s.} \tag{8.310}$$

The ratio of the bottom-story to top-story stiffness follows as before:

$$\alpha^* = k_1/k = 21.2/212 = 0.10. \tag{8.311}$$

The mode shapes are

$$\boldsymbol{\Phi}_1 = \left\{ \begin{array}{c} 1 \\ \left(0.10 + \sqrt{(0.10)^2 + 4}\right)/2 \end{array} \right\} = \left\{ \begin{array}{c} 1.0 \\ 1.051 \end{array} \right\} = \left\{ \begin{array}{c} 0.951 \\ 1.00 \end{array} \right\} \tag{8.312}$$

$$\boldsymbol{\Phi}_2 = \left\{ \begin{array}{c} 1 \\ \left(0.10 - \sqrt{(0.10)^2 + 4}\right)/2 \end{array} \right\} = \left\{ \begin{array}{c} 1.0 \\ -0.951 \end{array} \right\} = \left\{ \begin{array}{c} -1.051 \\ 1.00 \end{array} \right\}. \tag{8.313}$$

The damping in the two modes of vibration is obtained using Eqs. 8.264 and 8.270. The damping in the structure above the base isolators is still ζ_1. However, the damping in the base isolators can be expected to increase when the amplitude of displacement exceeds the typically small Δ_1 level. Therefore, assume that the critical damping ratio in the base isolators is equal to 20%. It then follows from Eq. 8.264 that

$$\zeta_1 = A_{nb} \left(\frac{\omega_{nb}}{\omega_n} \right) \zeta_{nb} + A_{nbn} \zeta_n + A_n \zeta_n \tag{8.314}$$

where

$$A_{nb} = \frac{4}{(0.10)^2 + 4 + (0.10)\sqrt{(0.10)^2 + 4}} \left(\frac{\sqrt{2}}{\sqrt{(0.10) + 2 - \sqrt{(0.10)^2 + 4}}} \right) = 4.30 \tag{8.315}$$

$$A_{nbn} = \frac{4 - 4(0.10) - 4\sqrt{(0.10)^2 + 4}}{(0.10)^2 + 4 + (0.10)\sqrt{(0.10)^2 + 4}} \left(\frac{\sqrt{2}}{\sqrt{(0.10) + 2 - \sqrt{(0.10)^2 + 4}}} \right) = -4.74 \tag{8.316}$$

$$A_n = \frac{(0.10)^2 + 2 + (0.10)\sqrt{(0.10)^2 + 4}}{(0.10)^2 + 4 + (0.10)\sqrt{(0.10)^2 + 4}} \left(\frac{\sqrt{2}}{\sqrt{(0.10) + 2 - \sqrt{(0.10)^2 + 4}}} \right) = 2.38. \tag{8.317}$$

Therefore,

$$\zeta_1 = (4.30) \frac{\sqrt{21.2/2.0}}{\sqrt{212/1.0}} (0.20) + (-4.74)(0.02) + (2.38)(0.02)$$

$$= 0.1923 - 0.0948 + 0.0476 = 0.145 = 14.5\%. \tag{8.318}$$

Similarly,

$$\zeta_2 = B_{nb}\left(\frac{\omega_{nb}}{\omega_n}\right)\zeta_{nb} + B_{nbn}\zeta_n + B_n\zeta_n \tag{8.319}$$

where

$$B_{nb} = \frac{4}{(0.10)^2 + 4 - (0.10)\sqrt{(0.10)^2 + 4}}\left(\frac{\sqrt{2}}{\sqrt{(0.10) + 2 + \sqrt{(0.10)^2 + 4}}}\right) = 0.73 \tag{8.320}$$

$$B_{nbn} = \frac{4 - 4(0.10) + 4\sqrt{(0.10)^2 + 4}}{(0.10)^2 + 4 - (0.10)\sqrt{(0.10)^2 + 4}}\left(\frac{\sqrt{2}}{\sqrt{(0.10) + 2 + \sqrt{(0.10)^2 + 4}}}\right) = 2.13 \tag{8.321}$$

$$B_n = \frac{(0.10)^2 + 2 - (0.10)\sqrt{(0.10)^2 + 4}}{(0.10)^2 + 4 - (0.10)\sqrt{(0.10)^2 + 4}}\left(\frac{\sqrt{2}}{\sqrt{(0.10) + 2 + \sqrt{(0.10)^2 + 4}}}\right) = 0.33. \tag{8.322}$$

Therefore,

$$\zeta_2 = (0.73)\frac{\sqrt{21.2/2.0}}{\sqrt{212/1.0}}(0.20) + (2.13)(0.02) + (0.33)(0.02) \tag{8.323}$$

$$= 0.0326 + 0.0426 + 0.0066 = 0.082 = 8.2\%.$$

The response of the structure in both the fundamental mode and second mode is

$$\mathbf{X}^{(1)} = \boldsymbol{\phi}_{\text{roof}(1)}\Gamma_{\text{roof}(1)}S_d(T_1,\zeta_1), \quad \mathbf{X}^{(2)} = \boldsymbol{\phi}_{\text{roof}(2)}\Gamma_{\text{roof}(2)}S_d(T_2,\zeta_2) \tag{8.324}$$

$$\ddot{\mathbf{Y}}^{(1)} = \boldsymbol{\phi}_{\text{roof}(1)}\Gamma_{\text{roof}(1)}S_a(T_1,\zeta_1), \quad \ddot{\mathbf{Y}}^{(2)} = \boldsymbol{\phi}_{\text{roof}(2)}\Gamma_{\text{roof}(2)}S_a(T_2,\zeta_2) \tag{8.325}$$

where

$$\Gamma_{\text{roof}(1)} = \frac{\boldsymbol{\phi}_{\text{roof}(1)}^T\mathbf{M}\{\mathbf{I}\}}{\boldsymbol{\phi}_{\text{roof}(1)}^T\mathbf{M}\boldsymbol{\phi}_{\text{roof}(1)}} = \frac{\begin{Bmatrix}0.951\\1.00\end{Bmatrix}^T\begin{bmatrix}m&0\\0&m\end{bmatrix}\begin{Bmatrix}1\\1\end{Bmatrix}}{\begin{Bmatrix}0.951\\1.00\end{Bmatrix}^T\begin{bmatrix}m&0\\0&m\end{bmatrix}\begin{Bmatrix}0.951\\1.00\end{Bmatrix}} = 1.024 \tag{8.326}$$

$$\Gamma_{\text{roof}(2)} = \frac{\boldsymbol{\phi}_{\text{roof}(2)}^T\mathbf{M}\{\mathbf{I}\}}{\boldsymbol{\phi}_{\text{roof}(2)}^T\mathbf{M}\boldsymbol{\phi}_{\text{roof}(2)}} = \frac{\begin{Bmatrix}-1.051\\1.00\end{Bmatrix}^T\begin{bmatrix}m&0\\0&m\end{bmatrix}\begin{Bmatrix}1\\1\end{Bmatrix}}{\begin{Bmatrix}-1.051\\1.00\end{Bmatrix}^T\begin{bmatrix}m&0\\0&m\end{bmatrix}\begin{Bmatrix}-1.051\\1.00\end{Bmatrix}} = -0.024. \tag{8.327}$$

Using Eqs. 8.66 to 8.70 with $\alpha = \alpha^* = 0.1$, it follows that

$$\omega_1^2 = \left(\frac{0.1 + 2 - \sqrt{(0.1)^2 + 4}}{2}\right)\left(\frac{212}{1.0}\right) = 10.34 \tag{8.328}$$

$$\omega_2^2 = \left(\frac{0.1 + 2 + \sqrt{(0.1)^2 + 4}}{2}\right)\left(\frac{212}{1.0}\right) = 434.9 \tag{8.329}$$

$$T_1 = \frac{2\pi}{\omega_1} = \frac{2\pi}{\sqrt{10.34}} = 1.95 \text{ s} \tag{8.330}$$

$$T_2 = \frac{2\pi}{\omega_2} = \frac{2\pi}{\sqrt{434.9}} = 0.30 \text{ s.} \tag{8.331}$$

Using the response spectrum shown in Figure 8.19 and assuming that this response spectrum represents both $\zeta_1 = 0.145$ and $\zeta_2 = 0.082$, then the spectral acceleration for each mode of vibration is

$$S_a(T_1,\zeta_1) = 0.312g, \quad S_a(T_2,\zeta_2) = 1.0g. \tag{8.332}$$

In addition, although the damping for both the first and second modes is near 10%, for simplicity assume that $S_d(T_1, \zeta_1) = S_a(T_1, \zeta_1)/\omega_1^2$. Then it follows from Eqs. 8.328, 8.329, and 8.332 that

$$S_d(T_1,\zeta_1) = \frac{S_a(T_1,\zeta_1)}{\omega_1^2} = \frac{0.312 \times 386.4}{10.34} = 11.7 \text{ in.} \tag{8.333}$$

$$S_d(T_2,\zeta_2) = \frac{S_a(T_2,\zeta_2)}{\omega_2^2} = \frac{1.0 \times 386.4}{434.9} = 0.9 \text{ in.} \tag{8.334}$$

Therefore, it follows from Eqs. 8.324 and 8.235 that

$$\mathbf{X}^{(1)} = \begin{Bmatrix} 0.951 \\ 1.00 \end{Bmatrix}(1.024)(11.7) = \begin{Bmatrix} 11.4 \\ 12.0 \end{Bmatrix} \text{ in.} \tag{8.335}$$

$$\mathbf{X}^{(2)} = \begin{Bmatrix} -1.051 \\ 1.00 \end{Bmatrix}(-0.024)(0.9) = \begin{Bmatrix} 0.023 \\ -0.022 \end{Bmatrix} \text{ in.} \tag{8.336}$$

$$\ddot{\mathbf{Y}}^{(1)} = \begin{Bmatrix} 0.951 \\ 1.00 \end{Bmatrix}(1.024)(0.312g) = \begin{Bmatrix} 0.30g \\ 0.32g \end{Bmatrix} \tag{8.337}$$

$$\ddot{\mathbf{Y}}^{(2)} = \begin{Bmatrix} -1.051 \\ 1.00 \end{Bmatrix}(-0.024)(1.0g) = \begin{Bmatrix} 0.025g \\ -0.024g \end{Bmatrix}. \tag{8.338}$$

Finally, assume that the SRSS method is used (see Section 5.5); then the maximum response is

$$\max[\mathbf{X}(t)]_{\text{SRSS}} = \begin{Bmatrix} \sqrt{(11.4)^2 + (0.023)^2} \\ \sqrt{(12.0)^2 + (-0.022)^2} \end{Bmatrix} = \begin{Bmatrix} 11.4 \\ 12.0 \end{Bmatrix} \text{ in.} \tag{8.339}$$

$$\max[\ddot{\mathbf{Y}}(t)]_{\text{SRSS}} = \begin{Bmatrix} g\sqrt{(0.30)^2 + (0.025)^2} \\ g\sqrt{(0.32)^2 + (-0.024)^2} \end{Bmatrix} = \begin{Bmatrix} 0.30g \\ 0.32g \end{Bmatrix}. \tag{8.340}$$

8.7 RESPONSE SPECTRA ANALYSIS OF A NONLINEAR BASE-ISOLATED STRUCTURE

In Section 8.6, the base isolator was considered to be linear elastic, and therefore the stiffness of the base isolator is not a function of the displacement amplitude of the base isolator. Most base isolators have force versus deflection relationships similar to the one shown in Figure 8.28; this figure and Table 8.4 show that the stiffness of the base isolator is a function of the displacement of the isolator. For example, the base isolator stiffness is 13.2 k/in. and 9.2 k/in. at displacements 7.0 in. and 14.0 in., respectively. For this type of base isolator, the initial stiffness is often sufficient (e.g., 50 k/in. in Figure 8.28) to enable the structural engineer to eliminate a separate wind restraint system because the base isolator provides the wind restraint.

The response spectra analysis of a nonlinear base-isolated structure starts with an estimate of the probable displacement of the isolator. This can be obtained from the results of the preliminary base isolator design, which was discussed in Section 8.4. Recall from this section that the structure was assumed to be a rigid box, whereas it is now considered to be a flexible structure. Imagine, for example, that the original estimate of the displacement of the base isolator is 10.5 in. The response spectra analysis uses the base isolator stiffness corresponding to this displacement (i.e., from Figure 8.28 this stiffness is 10.6 k/in.). The response spectra analysis with this value of base isolator stiffness follows the same procedure discussed in the previous section for the case with no wind restraint. Note that in Example 2 of Section 8.6, the original estimate of each base isolator stiffness was 10.6 k/in. (see Eq. 8.308). Also, note in this example for this value of base isolator stiffness that the base isolator displacement was 11.4 in. In Section 8.6 the base isolator was a linear base isolator, so its stiffness was not a function of its displacement. However, for a nonlinear base isolator, calculating the base isolator displacement from the original analysis enables the structural engineer to answer the question, "Do I need to change the value of the base isolator stiffness?" With the calculated base isolator displacement, the structural engineer goes to a base isolator force versus deflection curve similar to the one shown in Figure 8.28 and obtains the new stiffness. The response spectra analysis is repeated with this new base isolator stiffness, and a new base isolator displacement is calculated. Again, a new stiffness is calculated for the base isolator, and the process is repeated until convergence. Some base isolators have significant variations in damping (see Eq. 8.179), and therefore at each iteration the damping ratio and the stiffness of the base isolator are changed until convergence.

Structures typically have three to five different types of base isolators. Therefore, the iterative response spectra analysis procedure requires that for each iteration of response calculation, the stiffness of all base isolators be recalculated and modified if necessary.

EXAMPLE 1 *2DOF Nonlinear Isolator Response Analysis*

Consider the system in Example 2 in Section 8.6. Imagine that the base isolators have the lateral force versus deflection curve shown in Figure 8.28. Also, assume that the preliminary design of the base isolator indicated that for the design earthquake shown in Figure 8.19, the base isolator displaced 10.5 in. Therefore, using Figure 8.28 this displacement for this base isolator has a stiffness of 10.6 k/in. The total stiffness at the base isolator level for two isolators is $2 \times 10.6 = 21.2$ k/in. and therefore $\alpha = 21.2/212 = 0.10$. In Example 2 in Section 8.6, Eqs. 8.328 and 8.330 give the result and fundamental natural frequency and period of vibrations:

$$\omega_1^2 = 10.34, \quad T_1 = 1.95 \text{ s.} \tag{8.341}$$

Equation 8.312 gives the result of the fundamental mode shape of vibration, and Eq. 8.326 gives the fundamental mode participation factor:

$$\phi_1 = \begin{Bmatrix} 0.951 \\ 1.00 \end{Bmatrix}, \quad \Gamma_{\text{roof}(1)} = 1.024. \tag{8.342}$$

As indicated in this same example, the base-isolated system in Eq. 8.318 has 14.5% damping in the first mode of vibration. From Figure 8.19 for 5% damping, the spectral acceleration is

$$S_a(T_1 = 1.95, \zeta = 5\%) = 0.312g. \tag{8.343}$$

The damping is 14.5%, and therefore using the response reduction factors in Table 5.6, the reduced spectral acceleration is

$$S_a(T_1 = 1.95, \zeta = 14.5\%) = 0.76 \times 0.312g = 0.237g. \tag{8.344}$$

Although the damping for the first mode of vibration is 14.5%, for simplicity assume that $S_d(T_1, \zeta_1) = S_a(T_1, \zeta_1)/\omega_1^2$. Therefore, the spectral displacement is

$$S_d(T_1 = 1.95, \zeta = 14.5\%) = \frac{S_a(T_1 = 1.95, \zeta = 14.5\%)}{\omega_1^2} = \frac{0.237 \times 386.4}{10.34} = 8.9 \text{ in.} \quad (8.345)$$

The response of the base-isolated structure in the first mode is

$$\mathbf{X}^{(1)} = \begin{Bmatrix} 0.951 \\ 1.00 \end{Bmatrix}(1.024)(8.9) = \begin{Bmatrix} 8.6 \\ 9.1 \end{Bmatrix} \text{ in.} \quad (8.346)$$

and the displacement of the base isolator is 8.6 in.

This response spectra solution started with the assumption that the displacement due to this earthquake was 10.5 in. and the corresponding isolator stiffness was 10.6 k/in. The preceding calculations indicate that the base isolator displacement is only 8.6 in. Therefore, this represents a significant difference and iteration is required. Figure 8.28 is used to determine the new base isolator stiffness for this new displacement. Using linear interpolation to calculate the lateral force at this level of displacement, it follows that

$$F_{\text{lateral}} = 92.4 + \left(\frac{111.3 - 92.4}{10.5 - 7.0}\right)(8.6 - 7.0) = 101.0 \text{ kip.} \quad (8.347)$$

Therefore, the total stiffness of the two base isolators is

$$k_b = 2 \times \left(\frac{F_{\text{lateral}}}{8.6}\right) = 2 \times \left(\frac{101.0}{8.6}\right) = 23.5 \text{ k/in.} \quad (8.348)$$

The solution now repeats the steps in Example 2 in Section 8.6 but with $k_1 = k_b = 23.5$ k/in. Therefore,

$$\omega_{nb} = \sqrt{\frac{23.5}{2(1.0)}} = 3.43 \text{ rad/s} \quad (8.349)$$

$$T_{nb} = \frac{2\pi}{\omega_{nb}} = \frac{2\pi}{3.43} = 1.83 \text{ s} \quad (8.350)$$

$$\alpha^* = \frac{k_1}{k} = \frac{23.5}{212} = 0.11 \quad (8.351)$$

$$\omega_1 = \sqrt{\frac{0.11 + 2 - \sqrt{(0.11)^2 + 4}}{2}} \times \sqrt{\frac{212}{1.0}} = 3.36 \text{ rad/s} \quad (8.352)$$

$$T_1 = \frac{2\pi}{\omega_1} = \frac{2\pi}{3.36} = 1.87 \text{ s} \quad (8.353)$$

$$\Phi_1 = \begin{Bmatrix} 1 \\ \left(0.11 + \sqrt{(0.11)^2 + 4}\right)/2 \end{Bmatrix} = \begin{Bmatrix} 1.0 \\ 1.057 \end{Bmatrix} \quad (8.354)$$

$$\Phi_{\text{roof}(1)} = \begin{Bmatrix} 1/1.057 \\ 1.00 \end{Bmatrix} = \begin{Bmatrix} 0.947 \\ 1.00 \end{Bmatrix} \quad (8.355)$$

$$\Gamma_{\text{roof}(1)} = \frac{\boldsymbol{\phi}_{\text{roof}(1)}^{T}\mathbf{M}\{\mathbf{I}\}}{\boldsymbol{\phi}_{\text{roof}(1)}^{T}\mathbf{M}\boldsymbol{\phi}_{\text{roof}(1)}} = \frac{\begin{Bmatrix} 0.947 \\ 1.00 \end{Bmatrix}^{T}\begin{bmatrix} m & 0 \\ 0 & m \end{bmatrix}\begin{Bmatrix} 1 \\ 1 \end{Bmatrix}}{\begin{Bmatrix} 0.947 \\ 1.00 \end{Bmatrix}^{T}\begin{bmatrix} m & 0 \\ 0 & m \end{bmatrix}\begin{Bmatrix} 0.947 \\ 1.00 \end{Bmatrix}} = 1.027 \tag{8.356}$$

$$A_{nb} = \frac{4}{(0.11)^2 + 4 + (0.11)\sqrt{(0.11)^2 + 4}}\left(\frac{\sqrt{2}}{\sqrt{(0.11) + 2 - \sqrt{(0.11)^2 + 4}}}\right) = 4.10 \tag{8.357}$$

$$A_{nbn} = \frac{4 - 4(0.11) - 4\sqrt{(0.11)^2 + 4}}{(0.11)^2 + 4 + (0.11)\sqrt{(0.11)^2 + 4}}\left(\frac{\sqrt{2}}{\sqrt{(0.11) + 2 - \sqrt{(0.11)^2 + 4}}}\right) = -4.56 \tag{8.358}$$

$$A_{n} = \frac{(0.11)^2 + 2 + (0.11)\sqrt{(0.11)^2 + 4}}{(0.11)^2 + 4 + (0.11)\sqrt{(0.11)^2 + 4}}\left(\frac{\sqrt{2}}{\sqrt{(0.11) + 2 - \sqrt{(0.11)^2 + 4}}}\right) = 2.29 \tag{8.359}$$

and

$$\zeta_1 = (4.10)\frac{\sqrt{23.5/2.0}}{\sqrt{212/1.0}}(0.20) + (-4.56)(0.02) + (2.29)(0.02)$$
$$= 0.1929 - 0.0912 + 0.0457 = 0.146 = 14.7\%. \tag{8.360}$$

From Figure 8.19 for 5% damping, the spectral acceleration is

$$S_a(T_1 = 1.87, \zeta = 5\%) = 0.325g. \tag{8.361}$$

The damping is 14.7%, and therefore using the response reduction factors in Table 5.6, the reduced spectral acceleration is

$$S_a(T_1 = 1.87, \zeta = 14.7\%) = 0.76 \times 0.325g = 0.247g. \tag{8.362}$$

Although the damping for the first mode of vibration is 14.7%, for simplicity assume that $S_d(T_1, \zeta_1) = S_a(T_1, \zeta_1)/\omega_1^2$. Therefore, the spectral displacement is

$$S_d(T_1 = 1.87, \zeta = 14.7\%) = \frac{S_a(T_1 = 1.87, \zeta = 14.7\%)}{\omega_1^2} = \frac{0.247 \times 386.4}{(3.36)^2} = 8.5 \text{ in.} \tag{8.363}$$

The response of the base-isolated structure in the first mode is

$$\mathbf{X}^{(1)} = \begin{Bmatrix} 0.947 \\ 1.00 \end{Bmatrix}(1.027)(8.5) = \begin{Bmatrix} 8.2 \\ 8.7 \end{Bmatrix} \text{ in.} \tag{8.364}$$

Again, this value of base isolator displacement of 8.2 in. is compared with the previous calculated value of 8.6 in., and the structural engineer determines if further iterations are required.

This example used for convenience only the fundamental mode in the response spectra analysis. The method can be extended to multiple modes of vibration using the theory in Chapter 5 and as indicated in Example 2 in Section 8.6.

8.8   TIME HISTORY ANALYSIS OF A BASE-ISOLATED STRUCTURE

A base-isolated structure has nonlinear base isolators at the base and nonproportional damping. Therefore, the earthquake-induced response can be calculated exactly only by a nonlinear time history analysis. Chapter 7 presented the methods used to calculate the response of a nonlinear system, and therefore a base-isolated structure represents a special type of nonlinear system.

Base-isolated structures may have either elastic or nonlinear performance for the structure above the isolators. If the structure above the isolators is elastic and remains so for all design earthquakes, then the matrix condensation methods discussed in Sections 6.5 and 6.6 can be used to reduce the size of the problem.

The examples that follow illustrate how the response of a base-isolated structure can be calculated using the methods discussed in Chapter 6.

EXAMPLE 1   **2DOF System Elastic Response Using Normal Mode Method**

Consider Example 2 of Section 8.6 where the natural periods and frequencies of vibration, the mode shapes, the damping in each mode, and the modal participation factors are calculated to be

$$T_1 = 1.95 \text{ s}, \quad T_2 = 0.30 \text{ s} \tag{8.365}$$

$$\omega_1^2 = 10.34, \quad \omega_2^2 = 434.9 \tag{8.366}$$

$$\boldsymbol{\Phi}_{\text{roof}(1)} = \begin{Bmatrix} 0.951 \\ 1.00 \end{Bmatrix}, \quad \boldsymbol{\Phi}_{\text{roof}(2)} = \begin{Bmatrix} -1.051 \\ 1.00 \end{Bmatrix} \tag{8.367}$$

$$\zeta_1 = 0.145, \quad \zeta_2 = 0.082 \tag{8.368}$$

$$\Gamma_{\text{roof}(1)} = 1.024, \quad \Gamma_{\text{roof}(2)} = -0.024. \tag{8.369}$$

Then the response is given by

$$\mathbf{X}(t) = \boldsymbol{\Phi}_{\text{roof}(1)} q_1(t) + \boldsymbol{\Phi}_{\text{roof}(2)} q_2(t) \tag{8.370}$$

where $q_1(t)$ and $q_2(t)$ satisfy the second-order linear differential equation

$$\ddot{q}_1(t) + 2\zeta_1\omega_1\dot{q}_1(t) + \omega_1^2 q_1(t) = -\Gamma_{\text{roof}(1)} a(t) \tag{8.371}$$

$$\ddot{q}_2(t) + 2\zeta_2\omega_2\dot{q}_2(t) + \omega_2^2 q_2(t) = -\Gamma_{\text{roof}(2)} a(t) \tag{8.372}$$

and where $a(t)$ is the earthquake ground acceleration.

The response of the structure is now calculated using the constant average acceleration method where $\delta = \frac{1}{2}$ and $\alpha = \frac{1}{4}$ and a time step size of $\Delta t = 0.02$ s. Assume that the input earthquake ground acceleration $a(t)$ is the modified El-Centro Earthquake shown in Figure 6.11. For the first and second modes of vibration response as given in Eqs. 8.371 and 8.372, the recursive equations are

$$\begin{Bmatrix} q_1 \\ \dot{q}_1 \\ \ddot{q}_1 \end{Bmatrix}_{k+1} = \begin{bmatrix} 0.998977 & 0.019887 & 0.000099 \\ -0.10234 & 0.988724 & 0.009897 \\ -10.2340 & -1.12764 & -0.01025 \end{bmatrix} \begin{Bmatrix} q_1 \\ \dot{q}_1 \\ \ddot{q}_1 \end{Bmatrix}_k + (1.024) \begin{Bmatrix} -0.000099 \\ -0.00990 \\ -0.98975 \end{Bmatrix} a_{k+1} \tag{8.373}$$

$$\begin{Bmatrix} q_2 \\ \dot{q}_2 \\ \ddot{q}_2 \end{Bmatrix}_{k+1} = \begin{bmatrix} 0.959645 & 0.018876 & 0.000093 \\ -4.03548 & 0.887555 & 0.009279 \\ -403.548 & -11.2445 & -0.07209 \end{bmatrix} \begin{Bmatrix} q_2 \\ \dot{q}_2 \\ \ddot{q}_2 \end{Bmatrix}_k + (-0.024) \begin{Bmatrix} -0.000093 \\ -0.00928 \\ -0.92791 \end{Bmatrix} a_{k+1}. \tag{8.374}$$

The modal displacement responses for Eqs. 8.373 and 8.374 are shown in Figures 8.46 and 8.47, respectively. Note that the response in the second mode is very small compared to that in the first mode.

Using Eq. 8.370, the responses for the top and bottom floors are shown in Figures 8.48 and 8.49, respectively. Note that the response is dominated by the first mode of vibration since the contribution from the second mode is small and negligible. Therefore, Figure 8.48 is practically the same as Figure 8.46 with a scaling factor of $\varphi_{21} = 1.0$. Similarly, Figure 8.49 is the same as Figure 8.46 with a scaling factor of $\varphi_{11} = 0.0951$.

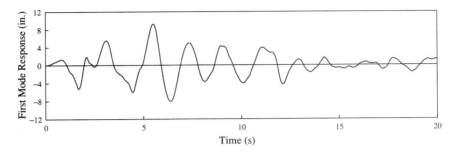

Figure 8.46 First Mode Response

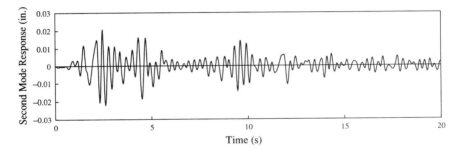

Figure 8.47 Second Mode Response

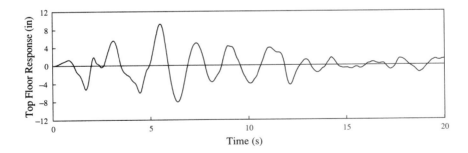

Figure 8.48 Top-Floor Response

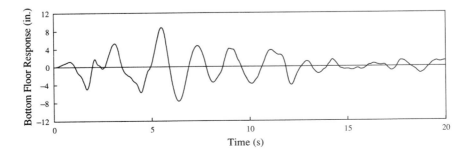

Figure 8.49 Bottom-Floor Response

EXAMPLE 2   2DOF System Elastic Response Using State Space Method

Consider again Example 1 with the dynamic matrices derived from Example 2 of Section 8.6 given by

$$\mathbf{M} = \begin{bmatrix} 1.0 & 0.0 \\ 0.0 & 1.0 \end{bmatrix}, \quad \mathbf{K} = \begin{bmatrix} k_1 + k_2 & -k_2 \\ -k_2 & k_2 \end{bmatrix} = \begin{bmatrix} 233.2 & -212 \\ -212 & 212 \end{bmatrix}. \tag{8.375}$$

The damping ratios and the natural frequencies of vibration for the first and second modes of vibration are

$$\zeta_1 = 0.145, \quad \zeta_2 = 0.082 \tag{8.376}$$

$$\omega_1 = \sqrt{10.34} = 3.22, \quad \omega_2 = \sqrt{434.9} = 20.85. \tag{8.377}$$

Assume that these damping ratios are Rayleigh proportional damping; then it follows from Eq. 2.293 that

$$\alpha = \frac{2(3.22)(20.85)((0.145)(20.85) - (0.082)(3.22))}{(20.85)^2 - (3.22)^2} = 0.873 \tag{8.378}$$

$$\beta = \frac{2((0.082)(20.85) - (0.145)(3.22))}{(20.85)^2 - (3.22)^2} = 0.00586. \tag{8.379}$$

It then follows from Eq. 4.285 that

$$\mathbf{C} = \alpha\mathbf{M} + \beta\mathbf{K} = 0.873 \begin{bmatrix} 1.0 & 0.0 \\ 0.0 & 1.0 \end{bmatrix} + 0.00586 \begin{bmatrix} 233.2 & -212 \\ -212 & 212 \end{bmatrix} = \begin{bmatrix} 2.24 & -1.24 \\ 1.24 & 2.12 \end{bmatrix}. \tag{8.380}$$

The dynamic equilibrium equation of motion becomes

$$\begin{bmatrix} 1.0 & 0.0 \\ 0.0 & 1.0 \end{bmatrix} \begin{Bmatrix} \ddot{x}_1 \\ \ddot{x}_2 \end{Bmatrix} + \begin{bmatrix} 2.24 & -1.24 \\ 1.24 & 2.12 \end{bmatrix} \begin{Bmatrix} \dot{x}_1 \\ \dot{x}_2 \end{Bmatrix} + \begin{bmatrix} 233.2 & -212 \\ -212 & 212 \end{bmatrix} \begin{Bmatrix} x_1 \\ x_2 \end{Bmatrix} = -\begin{bmatrix} 1.0 & 0.0 \\ 0.0 & 1.0 \end{bmatrix} \begin{Bmatrix} 1 \\ 1 \end{Bmatrix} a(t) \tag{8.381}$$

where $a(t)$ is the earthquake ground acceleration.

The response of the structure is now calculated using the state space method of analysis with Delta forcing function assumption and a time step size of $\Delta t = 0.02$ s. The recursive equation is given by

$$\begin{Bmatrix} x_1 \\ x_2 \\ \dot{x}_1 \\ \dot{x}_2 \end{Bmatrix}_{k+1} = \begin{bmatrix} 0.955030 & 0.040824 & 0.019262 & 0.000512 \\ 0.040823 & 0.959114 & 0.000512 & 0.019313 \\ -4.38357 & 3.975202 & 0.912516 & 0.063625 \\ 3.975040 & -3.98588 & 0.063625 & 0.918805 \end{bmatrix} \begin{Bmatrix} x_1 \\ x_2 \\ \dot{x}_1 \\ \dot{x}_2 \end{Bmatrix}_k + 0.02 \begin{Bmatrix} -0.19313 \\ -0.01983 \\ -0.97614 \\ -0.98243 \end{Bmatrix} a(t). \tag{8.382}$$

The top-story displacement and velocity response due to the modified El-Centro Earthquake (see Figure 6.11) are shown in Figures 8.50 and 8.51, respectively.

EXAMPLE 3   2DOF System Nonlinear Response

The nonlinear base isolator shown in Figure 8.28 can be represented as a bilinear element. The first straight line goes from the origin to 25 kip at a displacement of 1.0 in. The second straight line goes from the point of 25 kip and 1.0 in. to the point at 111.3 kip and 10.5 in. Figure 8.52 shows the bilinear relationship.

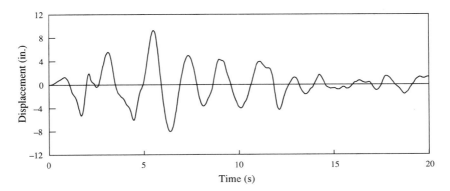

Figure 8.50 Displacement Response

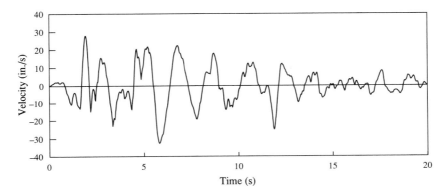

Figure 8.51 Velocity Response

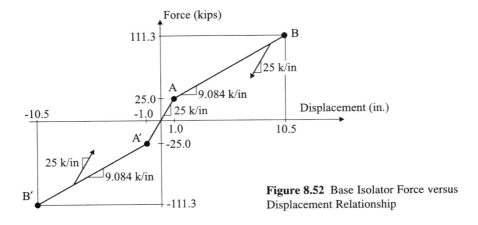

Figure 8.52 Base Isolator Force versus Displacement Relationship

Consider the 2DOF system discussed in Example 2 in Section 8.6 and Example 1 in Section 8.7. The structural properties are $m_1 = m_2 = m = 1.0$ k-s^2/in., $k_2 = 212$ k/in., $k_1 = 2k_b$, where k_b is the stiffness of each base isolator and is given in Figure 8.52. It follows that the mass and stiffness matrices are

$$\mathbf{M} = \begin{bmatrix} 1.0 & 0.0 \\ 0.0 & 1.0 \end{bmatrix}, \quad \mathbf{K} = \begin{bmatrix} 262.0 & -212 \\ -212 & 212 \end{bmatrix}, \quad \mathbf{K}_1 = \begin{bmatrix} 230.2 & -212 \\ -212 & 212 \end{bmatrix}. \quad (8.383)$$

Assume that the base-isolated structure has mass proportional damping (i.e., $\beta = 0.0$) with the proportionality constant $\alpha = 2.0$; then it follows from Eq. 4.285 that

$$\mathbf{C} = \alpha\mathbf{M} + \beta\mathbf{K} = 2.0\begin{bmatrix} 1.0 & 0.0 \\ 0.0 & 1.0 \end{bmatrix} + 0.0\begin{bmatrix} 262.0 & -212 \\ -212 & 212 \end{bmatrix} = \begin{bmatrix} 2.0 & 0.0 \\ 0.0 & 2.0 \end{bmatrix}. \tag{8.384}$$

Then the dynamic equilibrium equation of motion for earthquake response becomes

$$\begin{bmatrix} 1.0 & 0.0 \\ 0.0 & 1.0 \end{bmatrix}\begin{Bmatrix} \ddot{x}_1 \\ \ddot{x}_2 \end{Bmatrix} + \begin{bmatrix} 2.0 & 0.0 \\ 0.0 & 2.0 \end{bmatrix}\begin{Bmatrix} \dot{x}_1 \\ \dot{x}_2 \end{Bmatrix} + \begin{bmatrix} 262.0 & -212 \\ -212 & 212 \end{bmatrix}\begin{Bmatrix} x_1 \\ x_2 \end{Bmatrix} = -\begin{bmatrix} 1.0 & 0.0 \\ 0.0 & 1.0 \end{bmatrix}\begin{Bmatrix} 1 \\ 1 \end{Bmatrix}a(t) \tag{8.385}$$

when each base isolator stiffness is $k_1 = 25.0$ k/in. When each base isolator stiffness is $k_1 = 9.1$ k/in., then Eq. 8.385 becomes

$$\begin{bmatrix} 1.0 & 0.0 \\ 0.0 & 1.0 \end{bmatrix}\begin{Bmatrix} \ddot{x}_1 \\ \ddot{x}_2 \end{Bmatrix} + \begin{bmatrix} 2.0 & 0.0 \\ 0.0 & 2.0 \end{bmatrix}\begin{Bmatrix} \dot{x}_1 \\ \dot{x}_2 \end{Bmatrix} + \begin{bmatrix} 230.2 & -212 \\ -212 & 212 \end{bmatrix}\begin{Bmatrix} x_1 \\ x_2 \end{Bmatrix} = -\begin{bmatrix} 1.0 & 0.0 \\ 0.0 & 1.0 \end{bmatrix}\begin{Bmatrix} 1 \\ 1 \end{Bmatrix}a(t). \tag{8.386}$$

Using the Newmark β-method of constant acceleration (i.e., $\delta = 0$ and $\alpha = 0$), it follows from Eqs. 7.173a, 7.175, and 7.176 that

$$\begin{Bmatrix} \mathbf{X}_{k+1} \\ \dot{\mathbf{X}}_{k+1} \\ \ddot{\mathbf{X}}_{k+1} \end{Bmatrix} = \mathbf{F}_N^{(n)}\begin{Bmatrix} \mathbf{X}_k \\ \dot{\mathbf{X}}_k \\ \ddot{\mathbf{X}}_k \end{Bmatrix} + \mathbf{H}_N^{(nEQ)}\{\mathbf{I}\}(a_{k+1} - a_k) \tag{8.387}$$

where

$$\mathbf{F}_N^{(n)} = \begin{bmatrix} \mathbf{I} & (\Delta t)\mathbf{I} & \frac{1}{2}(\Delta t)^2\mathbf{I} \\ 0 & \mathbf{I} & (\Delta t)\mathbf{I} \\ 0 & -\Delta t\mathbf{B}^{-1}\mathbf{K} & \mathbf{I} - \Delta t\mathbf{B}^{-1}\mathbf{C} - \frac{1}{2}(\Delta t)^2\mathbf{B}^{-1}\mathbf{K} \end{bmatrix} \tag{8.388}$$

$$\mathbf{F}_1^{(n)} = \begin{bmatrix} \mathbf{I} & (\Delta t)\mathbf{I} & \frac{1}{2}(\Delta t)^2\mathbf{I} \\ 0 & \mathbf{I} & (\Delta t)\mathbf{I} \\ 0 & -\Delta t\mathbf{B}^{-1}\mathbf{K}_1 & \mathbf{I} - \Delta t\mathbf{B}^{-1}\mathbf{C} - \frac{1}{2}(\Delta t)^2\mathbf{B}^{-1}\mathbf{K}_1 \end{bmatrix} \tag{8.389}$$

$$\mathbf{H}_N^{(nEQ)}\{\mathbf{I}\} = \begin{bmatrix} \mathbf{0} \\ \mathbf{0} \\ -\mathbf{I} \end{bmatrix}\begin{Bmatrix} 1 \\ 1 \end{Bmatrix} = \begin{bmatrix} 0 & 0 \\ 0 & 0 \\ \hline 0 & 0 \\ 0 & 0 \\ \hline -1 & 0 \\ 0 & -1 \end{bmatrix}\begin{Bmatrix} 1 \\ 1 \end{Bmatrix} = \begin{Bmatrix} 0 \\ 0 \\ \hline 0 \\ 0 \\ \hline -1 \\ -1 \end{Bmatrix} \tag{8.390}$$

$$\mathbf{B} = \mathbf{M} + (0)\Delta t\mathbf{C} + (0)(\Delta t)^2\mathbf{K}_s = \mathbf{M}. \tag{8.391}$$

Assume a time step of $\Delta t = 0.005$ s is used. Therefore, when the base isolators are responding in the elastic range, Eq. 8.387 becomes

$$\begin{Bmatrix} x_1 \\ x_2 \\ \dot{x}_1 \\ \dot{x}_2 \\ \ddot{x}_1 \\ \ddot{x}_2 \end{Bmatrix}_{k+1} = \begin{bmatrix} 1.0 & 0.0 & 0.005 & 0.0 & 0.00001 & 0.0 \\ 0.0 & 1.0 & 0.0 & 0.005 & 0.0 & 0.00001 \\ 0.0 & 0.0 & 1.0 & 0.0 & 0.005 & 0.0 \\ 0.0 & 0.0 & 0.0 & 1.0 & 0.0 & 0.005 \\ 0.0 & 0.0 & -1.31 & 1.06 & 0.98673 & 0.00265 \\ 0.0 & 0.0 & 1.06 & -1.06 & 0.00265 & 0.98735 \end{bmatrix}\begin{Bmatrix} x_1 \\ x_2 \\ \dot{x}_1 \\ \dot{x}_2 \\ \ddot{x}_1 \\ \ddot{x}_2 \end{Bmatrix}_k + \begin{Bmatrix} 0 \\ 0 \\ 0 \\ 0 \\ -1 \\ -1 \end{Bmatrix}(a_{k+1} - a_k), \tag{8.392}$$

and when the base isolators are responding in the inelastic range, Eq. 8.387 becomes

$$
\begin{Bmatrix} x_1 \\ x_2 \\ \dot{x}_1 \\ \dot{x}_2 \\ \ddot{x}_1 \\ \ddot{x}_2 \end{Bmatrix}_{k+1}
=
\begin{bmatrix}
1.0 & 0.0 & 0.005 & 0.0 & 0.00001 & 0.0 \\
0.0 & 1.0 & 0.0 & 0.005 & 0.0 & 0.00001 \\
0.0 & 0.0 & 1.0 & 0.0 & 0.005 & 0.0 \\
0.0 & 0.0 & 0.0 & 1.0 & 0.0 & 0.005 \\
0.0 & 0.0 & -1.151 & 1.06 & 0.98712 & 0.00265 \\
0.0 & 0.0 & 1.06 & -1.06 & 0.00265 & 0.98735
\end{bmatrix}
\begin{Bmatrix} x_1 \\ x_2 \\ \dot{x}_1 \\ \dot{x}_2 \\ \ddot{x}_1 \\ \ddot{x}_2 \end{Bmatrix}_k
+
\begin{Bmatrix} 0 \\ 0 \\ 0 \\ 0 \\ -1 \\ -1 \end{Bmatrix}(a_{k+1} - a_k).
\tag{8.393}
$$

Now assume that the base-isolated structure is subjected to the bilinear pulse as shown in Figure 7.14. Also assume that the system has zero initial displacement and zero initial velocity. Then it follows from the equation of motion at time step $k = 0$, when the earthquake ground acceleration is equal to $0.0g$, that

$$
\begin{Bmatrix} \ddot{x}_1 \\ \ddot{x}_2 \end{Bmatrix}_o = -\begin{Bmatrix} 1 \\ 1 \end{Bmatrix} a(t) = -\begin{Bmatrix} 1 \\ 1 \end{Bmatrix}(0.0) = \begin{Bmatrix} 0.0 \\ 0.0 \end{Bmatrix}.
\tag{8.394}
$$

At time step $k = 1$ (i.e., $t = \Delta t = 0.005$ s), the earthquake ground acceleration is $a_1 = 0.05g = 19.32$ in./s^2. The response follows from Eq. 8.392 that

$$
\begin{Bmatrix} x_1 \\ x_2 \\ \dot{x}_1 \\ \dot{x}_2 \\ \ddot{x}_1 \\ \ddot{x}_2 \end{Bmatrix}_1
= \mathbf{F}_N^{(n)}
\begin{Bmatrix} 0.0 \\ 0.0 \\ 0.0 \\ 0.0 \\ 0.0 \\ 0.0 \end{Bmatrix}
+
\begin{Bmatrix} 0 \\ 0 \\ 0 \\ 0 \\ -1 \\ -1 \end{Bmatrix}(19.32 - 0.0) =
\begin{Bmatrix} 0.0 \\ 0.0 \\ 0.0 \\ 0.0 \\ -19.320 \\ -19.320 \end{Bmatrix}.
\tag{8.395}
$$

Then at time step $k = 2$ (i.e., $t = \Delta t = 0.010$ s), the earthquake ground acceleration is $a_1 = 0.10g = 38.64$ in./s^2. The response follows from Eq. 8.392 that

$$
\begin{Bmatrix} x_1 \\ x_2 \\ \dot{x}_1 \\ \dot{x}_2 \\ \ddot{x}_1 \\ \ddot{x}_2 \end{Bmatrix}_2
= \mathbf{F}_N^{(n)}
\begin{Bmatrix} 0.0 \\ 0.0 \\ 0.0 \\ 0.0 \\ -19.320 \\ -19.320 \end{Bmatrix}
+
\begin{Bmatrix} 0 \\ 0 \\ 0 \\ 0 \\ -1 \\ -1 \end{Bmatrix}(38.64 - 19.32) =
\begin{Bmatrix} -0.00024 \\ -0.00024 \\ -0.0966 \\ -0.0966 \\ -38.435 \\ -38.447 \end{Bmatrix}.
\tag{8.396}
$$

The calculation process continues, and at time step $k = 24$ (i.e., $t = 24\Delta t = 0.120$ s), the response is

$$
\begin{Bmatrix} x_1 \\ x_2 \\ \dot{x}_1 \\ \dot{x}_2 \\ \ddot{x}_1 \\ \ddot{x}_2 \end{Bmatrix}_{24}
=
\begin{Bmatrix} -0.95526 \\ -0.98192 \\ -22.4798 \\ -23.5567 \\ -222.050 \\ -256.354 \end{Bmatrix}
\tag{8.397}
$$

and the earthquake ground acceleration is $a_{24} = 309.12$ in./s^2. The earthquake ground acceleration at the next time step is $a_{25} = 289.80$ in./s^2, and therefore the response at this time step is

$$\begin{Bmatrix} x_1 \\ x_2 \\ \dot{x}_1 \\ \dot{x}_2 \\ \ddot{x}_1 \\ \ddot{x}_2 \end{Bmatrix}_{25} = \mathbf{F}_N^{(n)} \begin{Bmatrix} -0.95526 \\ -0.98192 \\ -22.4798 \\ -23.5567 \\ -222.050 \\ -256.354 \end{Bmatrix} + \begin{Bmatrix} 0 \\ 0 \\ 0 \\ 0 \\ -1 \\ -1 \end{Bmatrix} (289.80 - 309.12) = \begin{Bmatrix} -1.07043 \\ -1.10291 \\ -23.5901 \\ -24.8384 \\ -195.9983 \\ -233.238 \end{Bmatrix}. \tag{8.398}$$

Because the displacement at this time step is greater than 1.0 in., both base isolators have yielded. The stiffness of the base isolators becomes $\mathbf{F}_1^{(n)}$, and therefore the response at the next time step is

$$\begin{Bmatrix} x_1 \\ x_2 \\ \dot{x}_1 \\ \dot{x}_2 \\ \ddot{x}_1 \\ \ddot{x}_2 \end{Bmatrix}_{26} = \mathbf{F}_1^{(n)} \begin{Bmatrix} -1.07043 \\ -1.10291 \\ -23.5901 \\ -24.8384 \\ -195.9983 \\ -233.238 \end{Bmatrix} + \begin{Bmatrix} 0 \\ 0 \\ 0 \\ 0 \\ -1 \\ -1 \end{Bmatrix} (270.48 - 289.80) = \begin{Bmatrix} -1.19083 \\ -1.23001 \\ -24.5700 \\ -26.0046 \\ -173.934 \\ -210.164 \end{Bmatrix}. \tag{8.399}$$

The calculation process continues, which is similar to the procedures discussed in the examples of Sections 7.2 and 7.3. The displacement and velocity responses of the base isolators are shown in Figures 8.53 and 8.54, respectively.

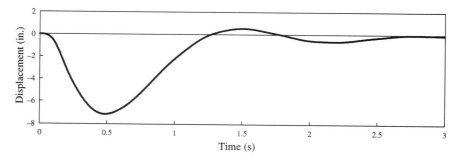

Figure 8.53 Base Isolator Displacement Response Due to Bilinear Pulse

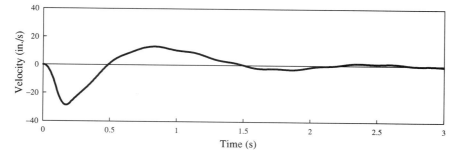

Figure 8.54 Base Isolator Velocity Response Due to Bilinear Pulse

RECOMMENDED READING

1. Naeim, F. and J. M. Kelly. *Design of Seismic Isolated Structures*. J. Wiley & Sons, 1999.
2. Structural Engineers Association of California. *General Requirements for the Design and Construction of Seismic-Isolated Structures*. Appendix to Chapter 1 of SEAOC Blue Book, 1989.
3. National Earthquake Hazard Reduction Program. *NEHRP Provisions for the Development of Seismic Regulations for Buildings*. NEHRP-97, Washington, D.C.

PROBLEMS

8.1 Consider the structure discussed in Example 1 in Section 3.2. Now let $k_1 = 10.3$ k/in., $k_2 = 103.4$ k/in., $m_1 = 0$, and $m_2 = 1.0$ k-s^2/in. Calculate T_{nb} in Eq. 8.9 and the ratio (x_1/x_2) in Eq. 8.5. Compare your answers with the case when $k_1 = 103.4$ k/in.

8.2 Repeat Problem 8.1 but let $k_1 = 51.7$ k/in. Compare your results with the results in Problem 8.1 and discuss the difference.

8.3 Consider the earthquake ground motion given in Figure 5.3. Calculate the value of the spectral acceleration and displacement using this figure for the case where $k_1 = 10.3$ k/in. and $k_2 = 103.4$ k/in. (see Problem 8.1).

8.4 Repeat Problem 8.3 but now use the value of $k_1 = 51.7$ k/in. from Problem 8.2.

8.5 Consider the system in Problem 8.1. Let $x_1 = 1.0$ in. Calculate the strain energy in the bottom floor (i.e., E_{sb} in Eq. 8.12) and the top floor (E_{s2} in Eq. 8.13). Calculate the ratio E_{sb}/E_{ss}.

8.6 Repeat Problem 8.5 but use the system in Problem 8.2.

8.7 Repeat Example 1 in Section 8.3 but use $k_1 = 103.4$ k/in. and $k_b = 10.3$ k/in.

8.8 Repeat Example 1 in Section 8.3 but assume that the fundamental mode natural period of vibration, T_1, is increased to 1.5 s.

8.9 Repeat Problem 8.8 but use the earthquake spectra in Figure 5.3.

8.10 Repeat Example 1 in Section 8.3 but use each of the four earthquake risk levels in Table 5.2 and Figure 5.22.

8.11 Repeat Problem 8.10 but use $T_1 = 1.5$ s and the results from Problem 8.8.

8.12 Consider the base isolator stiffness versus deflection test data given in Table 8.10. Calculate the average effective stiffness at each displacement amplitude. Plot the envelope of maximum force versus displacement for this base isolator (see Figure 8.28).

8.13 Consider the base isolator force versus deflection curve shown in Figure 8.55. Calculate the effective base isolator stiffness and damping.

8.14 Repeat Example 1 in Section 8.5 but assume that the design objective is to have a base-isolated structure with $k_b = 42.4$ k/in.

8.15 Repeat Example 1 in Section 8.5 but assume that the design objective is to have a target floor acceleration of $0.5g$.

8.16 Repeat Example 1 in Section 8.5 but assume that the base-isolated structure has 10% damping when subjected to both the rare and very rare earthquakes.

Table 8.10 Effective Base Isolator Stiffness versus Displacement Amplitude

Cycle	Effective stiffness (k/in.)			
	Displ. = ±3.5 in.	Displ. = ±7.0 in.	Displ. = ±10.5 in.	Displ. = ±14.0 in.
1	8.3	7.4	7.2	7.1
2	8.3	7.4	7.1	6.9
3	8.2	7.4	7.1	6.9

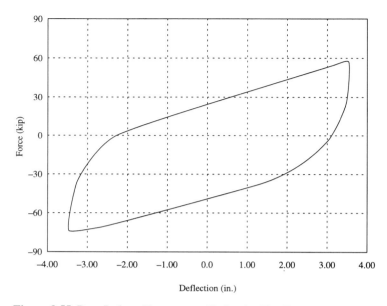

Figure 8.55 Base Isolator Force versus Deflection Test Data

8.17 Repeat Example 1 in Section 8.6 but with $\alpha = 0.2$.

8.18 Repeat Example 1 in Section 8.6 but with 10% damping in each of the two base isolators.

8.19 Repeat Example 1 in Section 8.6 but assume that the critical damping ratio in the structure above the base isolators is 5%.

8.20 Repeat Example 2 in Section 8.6 but assume that $k_b = 42.4$ k/in.

8.21 Repeat Example 2 in Section 8.6 but assume that the critical damping ratio in the base isolators is equal to 10%.

8.22 Repeat Example 2 in Section 8.6 but use the response spectra shown in Figures 8.40 and 8.41.

8.23 Repeat Example 1 in Section 8.8 but use the bilinear earthquake pulse as shown in Figure 8.56.

8.24 Recalculate Eq. 8.382 but for the case as defined in Example 2 in Section 8.6, but with $k_b = 42.4$ k/in.

8.25 Repeat Example 2 in Section 8.8 but use the bilinear earthquake pulse shown in Figure 8.56.

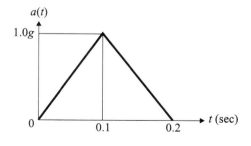

Figure 8.56 Bilinear Pulse

Chapter 9

Manufactured Viscous Dampers

9.1 OVERVIEW

As a starting point for the discussion of damping, consider a building roof that has been pulled over horizontally a small amount (e.g., 1 in.) and then released from rest. The roof will vibrate back and forth with an amplitude that diminishes with time. Historically, structural engineers have assumed that the reduction in motion was associated with the presence of viscous damping of the type discussed in Chapters 2 through 4. The logarithmic decrement method discussed in Section 2.2 is often used to relate the reduction in amplitude between two or more successive maximum values of response to the value of damping.

The scientific quantification of the value of viscous damping in buildings has been the subject of research for over 50 years. This type of damping is called *natural damping* in this chapter because it is associated with the natural, or classical, damping-producing elements of a structure. This natural damping represents the energy dissipated by structural materials (e.g., concrete, steel, and masonry) and nonstructural elements (e.g., partitions, cladding) as the building moves with time.

The early focus on estimating the nature and magnitude of damping in buildings can be traced to a series of tests in the 1930s by Dr. John Blume, a San Francisco structural and earthquake engineer. He used an eccentric mass building vibrator to produce a steady-state motion of a building. The building vibrator, which was mounted to the building floor, had a weight attached to a central shaft that rotated about this central shaft to impart a centrifugal force to the building floor. In this testing method, the building vibrator induces a steady-state floor acceleration versus time response that is recorded at several locations in the building. The frequency of the vibration and the resultant steady-state motion was varied, and the resultant test data were used to estimate building natural periods, mode shapes, and normal mode damping ratios. By varying the number of weights attached to the central shaft, Dr. Blume was able to quantify the magnitude of the modal damping with the maximum amplitude of floor motion. In the 1960s, Professor Donald Hudson at the California Institute of Technology built, with funding from the State of California, a new generation of eccentric mass building vibrators that were capable of a wider range of frequencies of building vibration and greater magnitudes of force.

In the late 1960s, structural engineers in Los Angeles and California installed instruments called accelerometers in buildings to measure building response during an earthquake.

Usually these accelerometers are attached to the basement slab or a building floor. The recorded motion of the building can then be used to estimate the magnitude of the natural damping in the building. Building responses to earthquakes and strong winds have been used to obtain these estimates of natural damping.

Researchers and structural engineers have studied building vibration data and accelerometer earthquake or wind responses and have recommend values of natural damping to be used in structural dynamic analyses. Table 9.1 presents natural damping values suggested by Professors Newmark and Hall. This table shows that the recommended value of damping is a function of the magnitude of structural response as reflected by the general stress level in the structural members. The recommended values of natural damping are also a function of the type of building material. Table 9.2 also presents recommended values of natural damping for different types of structural systems. There is a large degree of uncertainty in estimating the magnitude of natural damping in a building, which is evident in Table 9.1 as a range of values.

In the 1980s, the first high-technology element was introduced into a building to help the natural structure (e.g., beams, columns, and walls) mitigate the impact of the earthquake ground motion on the structure. This new element, called a base isolator, was discussed in Chapter 8. In the 1990s, there was a virtual explosion in the use of another new high-technology structural element that dissipates energy using a viscous damper. This type of damping is called *manufactured viscous damping* (or simply *manufactured damping*) because the damper is manufactured in a plant under exacting quality-control standards. These manufactured viscous dampers (or manufactured dampers) are then sent to the building site to be installed. Manufactured dampers are high-technology structural elements that are design variations of structural elements that have been used by the military

Table 9.1 Natural Damping Values Recommended by Newmark and Hall[1]

Stress level	Type and condition of structure	Critical damping (%)
Stress no more than about 50% yield point	• Welded steel, prestressed concrete, well-reinforced concrete (only slight cracking)	2 to 3
	• Reinforced concrete with considerable cracking	3 to 5
	• Bolted and/or riveted steel, wood structures with nailed or bolted joints	5 to 7
At or just below yield point	• Welded steel, prestressed concrete (without complete loss in prestress)	5 to 7
	• Prestressed concrete with no prestress left	7 to 10
	• Reinforced concrete	7 to 10
	• Bolted and/or riveted steel, wood structures with bolted joints	10 to 15
	• Wood structures with nailed joints	15 to 20

Table 9.2 Damping Values of Structural Systems[2]

Structural system	Damping values (%)
Structural steel	3
Reinforced concrete	5
Masonry shear walls	7
Wood	10

and by the automobile and ship industries. Figure 9.1 shows a schematic of a typical viscous damper, and Figure 9.2 shows different locations where the damper can be placed in the building. The scientific/engineering design varies among manufacturers of viscous dampers, but the basic performance from a structural engineering perspective is the same for all viscous dampers. The viscous damper is attached to the building structure either as part of a diagonal brace between the floors of the building or at the top of such a brace between the bracing unit and the floor. As the floors move, their relative velocity imparts a differential velocity to the ends of the viscous damper and therefore the damper exerts a viscous force to the floor.

Figures 9.3 and 9.4 show braced steel frames with the manufactured dampers in the line of the brace and at the top and bottom of the brace, respectively. Figure 9.5 shows a design where the manufactured damper is parallel to the floor and not in line with the

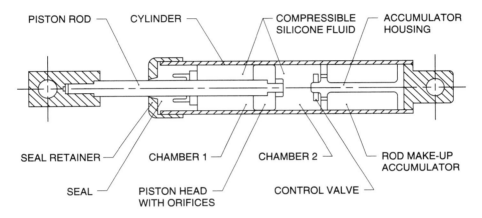

Figure 9.1 Fluid Inertial Damper

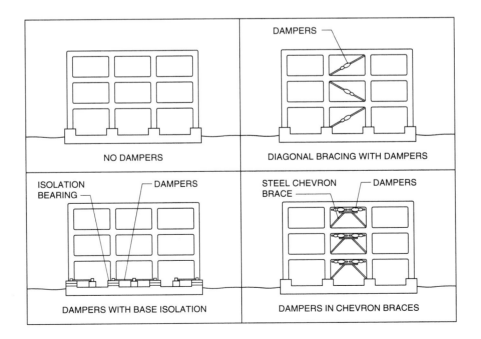

Figure 9.2 Possible Locations for Manufactured Dampers

Figure 9.3 Manufactured Dampers in Line with Brace (Retail Store in Sacramento, Structural Engineer: Marr, Shaffer, and Miyamoto, Taylor Dampers)

Figure 9.4 Manufactured Damper in Line with Brace (San Francisco Civic Center, Structural Engineer: Forell and Elsesser, Taylor Dampers)

brace. Base-isolated buildings also have manufactured dampers when there is a need to supplement the energy dissipation of the base isolator with the energy dissipation of the manufactured damper. Figures 9.6 and 9.7 show a schematic and a photo of such a system.

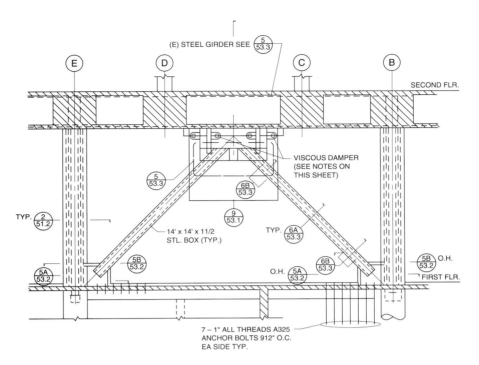

Figure 9.5 Manufactured Damper in Line with Floor (California State University Administration Buildings, Structural Engineers: Brandow Johnston Associates and Hart Consultant Group, Taylor Dampers)

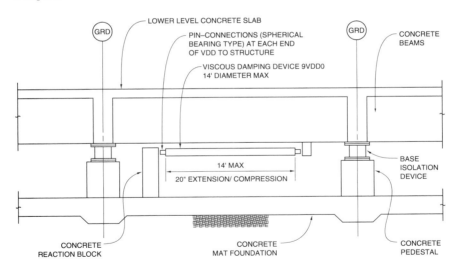

Figure 9.6 Manufactured Damper in Parallel with a Base Isolator (Courtesy of kpff Engineers)

The range of physical characteristics and force/displacement properties of manufactured dampers is very broad. Figures 9.8 and 9.9 show this variation for two different types of manufactured dampers.

Imagine a manufactured damper with one end connected to a steel member, which is connected to the floor above (see Figure 9.5). The other end of the manufactured damper is connected to another steel member, which is connected to the floor of the story below. As the upper floor moves and has a finite velocity relative to the story below, the two ends

Figure 9.7 Manufactured Damper in Parallel with Base Isolators (Southern California Medical Data Center, Structural Engineer: Brandow Johnston Associates and Hart Consultant Group, Taylor Dampers)

ITEM	DESCRIPTION	QTY	QTY
1	PRIMARY DAMPER CYLINDER BODY	1	STAINLESS STEEL
2	STRUCTURAL PROTECTIVE SLEEVE	1	PAINTED ALLOY STEEL
3	DAMPER TANG ROD ENDS	2	PAINTED ALLOY STEEL
4	SPHERICAL PLAYER BEARINGS	2	PLATED/PAINTED ASSEMBLY
5	SAFETY MONITOR/FILL PORT	1	PLATED ALLOY ASSEMBLY
6	INTERNAL RESERVOIR	1	BRONZE ASSEMBLY
7	MECHANICAL ADJUSTMENT	1	PLATED CARBON STEEL
8	DOUBLE ROD AND PISTON HEAD	1	17-4 PH STAINLESS ADDS
9	EXTENSION CYLINDER BODY	1	STAINLESS STEEL

RATED LOAD	L	S	E	A	D	R	T	WEIGHT
50 KIPS	50	8.0	1.75	6.5	1.5	2.65	1.31	230A
100 KIPS	54	8.0	2.00	7.5	1.75	2.88	1.55	330R
150 KIPS	58	8.0	2.63	8.5	2.25	3.63	2.00	450E
200 KIPS	60	8.0	3.00	9.5	2.50	3.88	2.20	500E
250 KIPS	64	8.0	3.38	10.5	2.75	4.50	2.41	800E
300 KIPS	66	8.0	3.75	11.5	3.00	4.88	2.63	990E
350 KIPS	68	8.0	4.13	12.0	3.25	5.25	2.85	1130S
400 KIPS	70	8.0	4.38	12.5	3.50	5.65	3.07	1280E

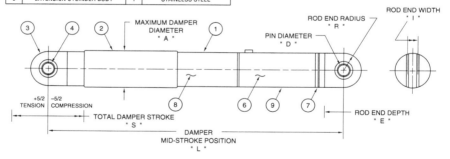

Figure 9.8 Enidine Manufactured Dampers

of the damper move relative to each other. If this relative velocity between the two ends is denoted as $\dot{x}(t)$, then a linear viscous damper will require that a force equal to

$$F_D(t) = c\dot{x}(t) \tag{9.1}$$

be applied. For example, if the relative displacement is the harmonic function

$$x(t) = x_o \sin \omega_n t \tag{9.2}$$

then the force required to move the two points with relative velocity $\dot{x}(t)$ is

$$F_D(t) = c\dot{x}(t) = cx_o \omega_n \cos \omega_n t. \tag{9.3}$$

In this case, the maximum force occurs when the relative displacement of the two ends is equal to zero. This is very important because the force in most structural members (e.g.,

High Capacity Fluid Viscous Dampers, 100 kip to 2000 kip Output

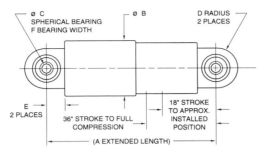

Dimensional Data

MODEL	A INCHES	B INCHES	c INCHES	D INCHES	E INCHES	F INCHES
100 KIP	131	7.5	2.5	3.2	4.75	2.2
200 KIP	132	9	2.75	3.9	5	2.4
300 KIP	138	11.5	3	4.25	5.25	2.7
600 KIP	155	16	6	7.5	10	4.8
1000 KIP	166	23	6	9	14.25	4.8
2000 KIP	180	26	8	11	17	6

NOTE:
Various strokes available, from 260 inches. Any stroke change from the 36 inch stroke version depicted changes extended length by three inches per inch of stroke change.

EXAMPLE:
200K x 10 inch stroke, extended length is 132 inch - 3 (36-10) – 54 inches

Figure 9.9 Taylor Manufactured Dampers

beams and columns) is at maximum when the displacement is at maximum. Therefore, the maximum force imparted to these structural members from a viscous damper is out of phase with the maximum force imparted to the beam from the viscous damper.

Because these high-technology viscous dampers are manufactured in plants, the structural engineer can order dampers that have a broad range of force versus velocity relationships. The most common for structural engineering purposes is

$$F_D(t) = c_{md} \langle \dot{x}(t) \rangle^\eta \tag{9.4}$$

where η is generally less than one and $\langle \dot{x}(t) \rangle^\eta$ is defined as

$$\langle \dot{x}(t) \rangle^\eta = \begin{cases} \dot{x}(t)^\eta & \dot{x} \ge 0 \\ -(-\dot{x}(t))^\eta & \dot{x} < 0. \end{cases} \tag{9.5}$$

The definition used in Eq. 9.5 prevents any mathematical confusion from a negative quantity being raised to a power that is less than one. However, for simplicity, structural engineers generally ignore this definition. It then follows from Eq. 9.4 that an equivalent equation is

$$F_D(t) = c_{md} \dot{x}(t)^\eta. \tag{9.6}$$

Figure 9.10 shows a force versus velocity plot for a nonlinear viscous damper with an exponential coefficient of $\eta = 0.35$.

9.2   MANUFACTURED DAMPERS WITH LINEAR VISCOUS DAMPING

The two ends of a viscous damper experience different displacement, velocity, and acceleration because one end is attached to one building floor and the other end to a different floor. This difference in motion results in the viscous damper producing a force and a source of energy dissipation. The force induced to the structure by the viscous damper at each of the attachment points of the viscous damper can be expressed as

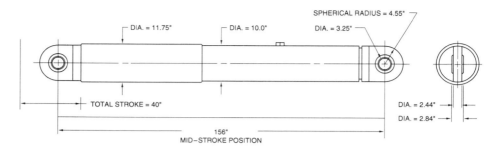

(a) Nonlinear Viscous Damper

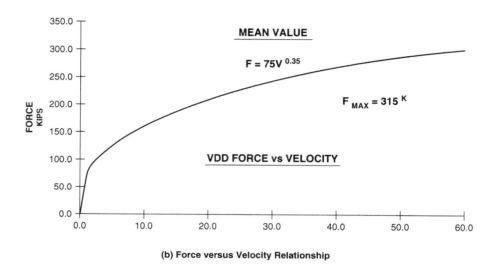

(b) Force versus Velocity Relationship

Figure 9.10 Nonlinear Force versus Velocity Viscous Damper (Courtesy of Enidine and kpff Engineers)

$$F_{md} = c_{md} \langle \dot{x} \rangle^{\eta} \tag{9.7}$$

where F_{md} is the damping force from the manufactured viscous damper, c_{md} is the manufactured viscous damper damping coefficient, $\dot{x}$ is the relative velocity between the ends of the viscous damper, and η is a power law coefficient. An illustration of Eq. 9.7 for a manufactured viscous damper is shown in Figure 9.10, where c_{md} is equal to 75 k-s²/in. and η is equal to 0.35.

The first generation of manufactured viscous dampers used a power law coefficient equal to one. Structural engineers selected this value of η for design because for $\eta = 1$ the manufactured damping force, like the natural damping force, is a linear function of velocity. The first six chapters of this book, like most structural dynamics books, considered damping forces that are a linear function of the velocity. Therefore, for a single degree of freedom system the total damping force is

$$F_d = F_{nd} + F_{md} \tag{9.8}$$

where F_d is the total damping force, F_{nd} is the force due to natural damping force, and F_{md} is the manufactured damping force. For the case where both the natural and manufactured damping forces are a linear function of the velocity, it follows that

$$F_{nd} = c_{nd}\dot{x}(t) \tag{9.9a}$$

$$F_{md} = c_{md}\dot{x}(t). \tag{9.9b}$$

It then follows that the dynamic equilibrium equation of motion for an SDOF system subjected to an earthquake ground motion is

$$m\ddot{x}(t) + c_{nd}\dot{x}(t) + c_{md}\dot{x}(t) + kx(t) = -ma(t). \tag{9.10}$$

After dividing both sides of Eq. 9.10 by the mass, it becomes

$$\ddot{x}(t) + 2\zeta_{nd}\omega_n\dot{x}(t) + 2\zeta_{md}\omega_n\dot{x}(t) + \omega_n^2 x(t) = -a(t) \tag{9.11}$$

where $\omega_n = \sqrt{k/m}$, ζ_{nd} is the critical damping ratio from natural damping, and ζ_{md} is the critical damping ratio from manufactured damping. These damping ratios are defined using Eqs. 9.10 and 9.11 to be

$$2\zeta_{nd}\omega_n = c_{nd}/m \tag{9.12a}$$

$$2\zeta_{md}\omega_n = c_{md}/m. \tag{9.12b}$$

Equation 9.11 can also be written as

$$\ddot{x}(t) + 2\zeta\omega_n\dot{x}(t) + \omega_n^2 x(t) = -a(t) \tag{9.13}$$

where ζ represents the total critical damping ratio of the structure, and it is

$$\zeta = \zeta_{nd} + \zeta_{md}. \tag{9.14}$$

Equation 9.13 is the same as the equation for the SDOF system given in Chapter 2, with the critical damping ratio now being the summation of the natural and manufactured critical damping ratios. Therefore, for the case where $\eta = 1$, the structural system is the same as the system defined in Chapter 2, and all of the response solution methods presented in that chapter can be used to solve Eq. 9.13. Note that the introduction of manufactured damping to the structure does not introduce any new solution complexity.

The earthquake acceleration in Eq. 9.13 is very irregular and cannot be described by a closed form mathematical function, and therefore a numerical solution is necessary. The discretization methods discussed in Section 2.6 (Newmark β-method) and in Section 2.7 (Wilson θ-method) can be used to solve Eq. 9.13. The matrix equation for a solution using the Newmark β-method is given in Eq. 2.248, which is:

$$\begin{Bmatrix} x_{k+1} \\ \dot{x}_{k+1} \\ \ddot{x}_{k+1} \end{Bmatrix} = \mathbf{F}_N \begin{Bmatrix} x_k \\ \dot{x}_k \\ \ddot{x}_k \end{Bmatrix} + \mathbf{H}_N^{(EQ)} a_{k+1} \tag{9.15}$$

where

$$\mathbf{F}_N = \frac{1}{\beta} \begin{bmatrix} \beta - \omega_n^2\alpha(\Delta t)^2 & \beta\Delta t - 2\zeta\omega_n\alpha(\Delta t)^2 - \omega_n^2\alpha(\Delta t)^3 & \frac{1}{2}\beta\Delta t^2 - \alpha(\beta+\gamma)(\Delta t)^2 \\ -\omega_n^2\delta\Delta t & \beta - 2\zeta\omega_n\delta\Delta t - \omega_n^2\delta(\Delta t)^2 & \beta\Delta t - \delta(\beta+\gamma)\Delta t \\ -\omega_n^2 & -2\zeta\omega_n - \omega_n^2\Delta t & -\gamma \end{bmatrix} \tag{9.16a}$$

$$\mathbf{H}_N^{(EQ)} = -\left(\frac{1}{\beta}\right) \begin{Bmatrix} \alpha(\Delta t)^2 \\ \delta\Delta t \\ 1 \end{Bmatrix} \tag{9.16b}$$

and

$$\beta = 1 + 2\zeta\omega_n\delta\Delta t + \omega_n^2\alpha(\Delta t)^2, \quad \gamma = 2\zeta\omega_n(1-\delta)\Delta t + \omega_n^2(\tfrac{1}{2}-\alpha)(\Delta t)^2. \tag{9.16c}$$

The corresponding Wilson θ-method matrix equation is given in Eq. 2.281, which is

$$\begin{Bmatrix} x_{k+1} \\ \dot{x}_{k+1} \\ \ddot{x}_{k+1} \end{Bmatrix} = \mathbf{F}_W \begin{Bmatrix} x_k \\ \dot{x}_k \\ \ddot{x}_k \end{Bmatrix} + \mathbf{H}_W^{(EQ)}\, a_{k+1} \tag{9.17}$$

where

$$\mathbf{F}_W = \frac{1}{\beta}\begin{bmatrix} \beta - \frac{1}{6}\omega_n^2\theta^2(\Delta t)^2 & \beta\Delta t - \frac{1}{3}\zeta\omega_n\theta^2(\Delta t)^2 - \frac{1}{6}\omega_n^2\theta(\Delta t)^3 & \frac{1}{6}(2\beta - \theta - \gamma + 1)\theta(\Delta t)^2 \\ -\frac{1}{2}\omega_n^2\theta\Delta t & \beta - \zeta\omega_n\theta\Delta t - \frac{1}{2}\omega_n^2\theta(\Delta t)^2 & \frac{1}{2}(\beta - \theta - \gamma + 1)\Delta t \\ -\omega_n^2 & -2\zeta\omega_n - \omega_n^2\Delta t & \frac{1}{\theta}(\beta\theta - \theta - \gamma - \beta + 1) \end{bmatrix} \tag{9.18a}$$

$$\mathbf{H}_W^{(EQ)} = -\left(\frac{1}{\beta}\right)\begin{Bmatrix} \frac{1}{6}\theta^2(\Delta t)^2 \\ \frac{1}{2}\theta\Delta t \\ 1 \end{Bmatrix} \tag{9.18b}$$

and

$$\beta = 1 + \zeta\omega_n\theta\Delta t + \frac{1}{6}\omega_n^2\theta^2(\Delta t)^2, \quad \gamma = \zeta\omega_n\theta\Delta t + \frac{1}{3}\omega_n^2\theta^2(\Delta t)^2. \tag{9.18c}$$

The response of a multi–degree of freedom system with linear viscous damping is obtained using the same methods discussed in Chapters 3 and 4. Typically, the system damping matrix is not proportional. Therefore, the system response must be obtained using a numerical solution of the second-order differential equation with the Newmark β-method or the Wilson θ-method, or a solution of the first-order differential equation using the state space method. For preliminary design, structural engineers often neglect the off-diagonal terms that exist in the nonproportional damping matrix, even though this method does not produce an exact solution. If this can be done without significant error, then either the response spectra or the normal mode method of analysis can be used to calculate the system response.

EXAMPLE 1 *Increase in Linear Modal Damping*

Consider an SDOF system with only natural damping. If the value of the natural modal damping is 2% of critical, then

$$\zeta_{nd} = 0.02. \tag{9.19}$$

Let the mass of the structure be $m = 1.0$ k-s²/in. and the period of vibration be $T_n = 1.0$ s. Then

$$\omega_n = 2\pi/T_n = 2\pi \text{ rad/s} \tag{9.20}$$

$$c_{nd} = 2\zeta_{nd}\omega_n m = 0.08\pi = 0.251 \text{ k-s/in.} \tag{9.21}$$

Assume now that the structural engineer wants to increase the damping of the structure to 10% of critical, that is,

$$\zeta = 0.10. \tag{9.22}$$

Then it follows from Eq. 9.14 that the manufactured damping must be

$$\zeta_{md} = \zeta - \zeta_{nd} = 0.10 - 0.02 = 0.08 \tag{9.23}$$

and, therefore, the manufactured viscous damper damping coefficient must be equal to

$$c_{md} = 2\zeta_{md}\omega_n m = 0.32\pi = 1.01 \text{ k-s/in.} \tag{9.24}$$

The selection of the damping coefficient c_{md} is a function of the desired damping ratio (i.e., 0.08) and also the natural frequency of vibration and the mass of the structure.

When manufactured dampers are ordered, one critical design variable that must be specified is the maximum damping force (e.g., from Figure 9.10, a 315-kip viscous damper). Usually, the structural engineer will know from structural engineering calculations the maximum relative velocity that the design earthquake will induce on the two ends of the viscous damper. Therefore, if this maximum velocity is $\dot{x}_{max}$, then the maximum force required for the manufactured damper is

$$F_{md(\max)} = c_{md}\dot{x}_{\max}. \tag{9.25}$$

EXAMPLE 2 SDOF System Subjected to a Modified 1940 El-Centro Earthquake

The modified NS component of the El-Centro earthquake time history is shown in Figure 6.11 as well as Figure 2.13. Consider the SDOF system discussed in Example 6 in Section 2.6. Table 2.17 of that example gives the response for a system with 5% damping. Table 9.3 repeats these results for convenience and shows the relative displacement, $x(t)$, relative velocity, $\dot{x}(t)$, absolute acceleration, $\ddot{y}(t)$, and base shear, $V(t)$, for the first 10 time steps. Figures 9.11 to 9.13 show the relative displacement, relative velocity, and absolute acceleration time histories, respectively.

Now assume that the structural engineer has added a manufactured damper to the structure and that the manufactured damper has a critical damping ratio of 15% ($\zeta_{md} = 0.15$). The total critical damping ratio of the structure is now equal to 20% (i.e., $\zeta = \zeta_{nd} + \zeta_{md} = 0.05 + 0.15 = 0.20$).

The response of the structure can be calculated using the Newmark β-method with $\delta = \frac{1}{2}$ and $\alpha = \frac{1}{4}$:

$$\begin{Bmatrix} x_{k+1} \\ \dot{x}_{k+1} \\ \ddot{x}_{k+1} \end{Bmatrix} = \begin{bmatrix} 0.985187 & 0.019232 & 0.000094 \\ -1.48129 & 0.923223 & 0.009380 \\ -148.129 & -7.67766 & -0.06196 \end{bmatrix} \begin{Bmatrix} x_k \\ \dot{x}_k \\ \ddot{x}_k \end{Bmatrix} + \begin{bmatrix} -0.000094 \\ -0.00938 \\ -0.93804 \end{bmatrix} a_{k+1}. \tag{9.26}$$

Table 9.4 shows the response for the first 10 time steps. Figures 9.14 to 9.16 show the relative displacement, relative velocity, and absolute acceleration time histories.

Table 9.3 SDOF System Response Using the Newmark β-Method (Natural Damping Only)

k	Time (s)	$a_k(g)$	x_k (in.)	$\dot{x}_k$ (in./s)	$\ddot{x}_k$ (in./s²)	$\ddot{y}_k$ (in./s²)	V_k (kip)
0	0.00	−0.0026	0.0000	0.0000	0.9930	0.0000	0.0000
1	0.02	−0.0198	0.0008	0.0841	7.4124	−0.2384	0.2384
2	0.04	−0.0185	0.0039	0.2207	6.2569	−0.8915	0.8915
3	0.06	−0.0162	0.0094	0.3270	4.3697	−1.8900	1.8900
4	0.08	−0.0174	0.0167	0.4065	3.5753	−3.1481	3.1481
5	0.10	−0.0220	0.0256	0.4808	3.8582	−4.6426	4.6426
6	0.12	−0.0261	0.0359	0.5565	3.7093	−6.3757	6.3757
7	0.14	−0.0235	0.0475	0.6018	0.8188	−8.2616	8.2616
8	0.16	−0.0202	0.0594	0.5868	−2.3144	−10.1197	10.1197
9	0.18	−0.0156	0.0703	0.5065	−5.7174	−11.7452	11.7452
10	0.20	−0.0156	0.0792	0.3797	−6.9575	−12.9854	12.9854

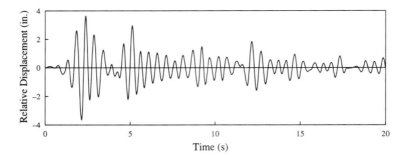

Figure 9.11 Relative Displacement Time History Response Using the Newmark β-Method (Natural Damping Only)

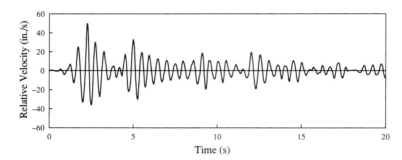

Figure 9.12 Relative Velocity Time History Response Using the Newmark β-Method (Natural Damping Only)

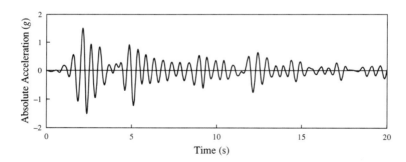

Figure 9.13 Absolute Acceleration Time History Response Using the Newmark β-Method (Natural Damping Only)

A comparison of Figures 9.11 to 9.13 with Figures 9.14 to 9.16 shows differences in the waveform of the time histories and the response. Table 9.5 compares the maximum responses of the system without and with manufactured damping.

The maximum damping force can be calculated using Eq. 9.8, which is

$$F_d = F_{nd} + F_{md} \tag{9.27}$$

where

$$F_{nd} = c_{nd}\dot{x}(t) \tag{9.28}$$

$$F_{md} = c_{md}\dot{x}(t). \tag{9.29}$$

Table 9.4 SDOF System Response Using the Newmark β-Method (Natural and Manufactured Damping)

k	Time (s)	$a_k\,(g)$	x_k (in.)	$\dot{x}_k$ (in./s)	$\ddot{x}_k$ (in./s^2)	$\ddot{y}_k$ (in./s^2)	V_k (kip)
0	0.00	−0.0026	0.0000	0.0000	0.9930	0.0000	0.0000
1	0.02	−0.0198	0.0008	0.0811	7.1151	−0.5356	0.5356
2	0.04	−0.0185	0.0037	0.2075	5.5220	−1.6264	1.6264
3	0.06	−0.0162	0.0087	0.2966	3.3894	−2.8703	2.8703
4	0.08	−0.0174	0.0153	0.3557	2.5257	−4.1977	4.1977
5	0.10	−0.0220	0.0229	0.4092	2.8262	−5.6746	5.6746
6	0.12	−0.0261	0.0317	0.4650	2.7496	−7.3354	7.3354
7	0.14	−0.0235	0.0412	0.4934	0.0889	−8.9915	8.9915
8	0.16	−0.0202	0.0509	0.4685	−2.5799	−10.3852	10.3852
9	0.18	−0.0156	0.0594	0.3895	−5.3154	−11.3432	11.3432
10	0.20	−0.0156	0.6611	0.2783	−5.8106	−11.8384	11.8384

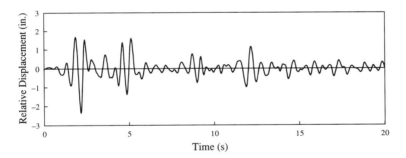

Figure 9.14 Relative Displacement Time History Response Using the Newmark β-Method (Natural and Manufactured Damping)

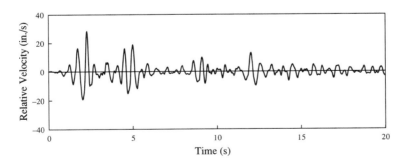

Figure 9.15 Relative Velocity Time History Response Using the Newmark β-Method (Natural and Manufactured Damping)

Recall from Eq. 9.12 that $c_{nd} = 2\zeta_{nd}\omega_n m$ and $c_{md} = 2\zeta_{md}\omega_n m$. Therefore, it follows that

$$c_{nd} = 2\zeta_{nd}\omega_n m = 2(0.05)(12.566)(1.0) = 1.26 \text{ k-s/in.} \tag{9.30}$$

$$c_{md} = 2\zeta_{md}\omega_n m = 2(0.15)(12.566)(1.0) = 3.77 \text{ k-s/in.} \tag{9.31}$$

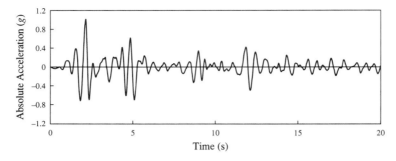

Figure 9.16 Absolute Acceleration Time History Response Using the Newmark β-Method (Natural and Manufactured Damping)

Table 9.5 Comparison of Maximum Response for a System without and with Manufactured Damping

	5% damping ($\zeta = \zeta_{nd}$)		20% damping ($\zeta = \zeta_{nd} + \zeta_{md}$)	
	Maximum value	Time	Maximum value	Time
Relative displacement	3.65 in.	2.16 s	2.35 in.	2.18 s
Relative velocity	50.0 in./s	2.28 s	28.5 in./s	2.28 s
Absolute acceleration	1.50g	2.38 s	1.01g	2.16 s

From Table 9.5, the maximum relative velocity of the system without manufactured damping is 50.0 in./s. Therefore, for this system Eq. 9.28 gives the maximum damping force, and it is

$$F_{nd(\max)} = c_{nd}\dot{x}_{\max} = 1.26(50.0) = 63.0 \text{ kip.} \tag{9.32}$$

The maximum relative velocity of the system with manufactured damping is given in Table 9.5, and it is equal to 28.5 in./s. Therefore, Eqs. 9.26 to 9.28 give the maximum damping forces, and they are

$$F_{nd(\max)} = c_{nd}\dot{x}_{\max} = 1.26(28.5) = 35.9 \text{ kip} \tag{9.33}$$

$$F_{md(\max)} = c_{md}\dot{x}_{\max} = 3.77(28.5) = 107 \text{ kip} \tag{9.34}$$

$$F_{d(\max)} = F_{nd(\max)} + F_{md(\max)} = 35.9 + 107 = 143 \text{ kip.} \tag{9.35}$$

Note that the total damping force increases with the addition of the manufactured damper (i.e., from 63.0 kip to 143 kip). However, the natural damping force has decreased in magnitude from 63.0 kip to 35.9 kip because of the decrease of the maximum velocity from 50.0 in./s to 28.5 in./s.

EXAMPLE 3   *Response Reduction Using Damping*

Consider Example 2, where the natural damping is 5%. Assume that the system response is too large and the structural engineer desires to reduce the response of the system to no more than 2.5 in.

One design solution is to increase the stiffness of the columns until the maximum displacement response is 2.5 in. The second design option is to add a manufactured damper to the system. Using the second design option, the equation of motion of the system becomes

Table 9.6 Manufactured Damping Force of an SDOF System

k	Time (s)	$\dot{x}_k$ (in./s)	F_{md} (kip)
0	0.00	0.0000	0.0000
1	0.02	0.0811	0.3057
2	0.04	0.2075	0.7822
3	0.06	0.2966	1.1181
4	0.08	0.3557	1.3409
5	0.10	0.4092	1.5426
6	0.12	0.4650	1.7530
7	0.14	0.4934	1.8600
8	0.16	0.4685	1.7662
9	0.18	0.3895	1.4683
10	0.20	0.2783	1.0491

$$\ddot{x}(t) + 2\left(\zeta_{nd} + \zeta_{md}\right)\omega_n \dot{x}(t) + \omega_n^2 x(t) = -a(t) \tag{9.36}$$

where $\omega_n = \sqrt{k/m} = 12.566$ rad/s, $\zeta_{nd} = 0.05$, and ζ_{md} is to be determined. Equation 9.26 can be solved but now for different values of manufactured damping ζ_{md} until the maximum displacement response reduces to 2.5 in. Based on Example 2, a manufactured damping with a critical damping ratio of 15% is sufficient to reduce the maximum displacement response to less than 2.5 in.

EXAMPLE 4   *Force in Damper*

Consider Example 2, where the manufactured damping is $\zeta_{md} = 0.15$. The force in the damper is given in Eq. 9.9b, which is

$$F_{md} = c_{md}\dot{x}(t) \tag{9.37}$$

where c_{md} is given in Eq. 9.31, which is

$$c_{md} = 2\zeta_{md}\omega_n m = 2(0.15)(12.566)(1.0) = 3.77 \text{ k-s/in.} \tag{9.38}$$

Using the velocity values from Table 9.4, the force in the damper can be calculated and is given in Table 9.6.

9.3   INTRODUCTION TO MANUFACTURED VISCOUS DAMPERS

In the previous section, the damping force exerted on the mass was assumed to be linearly related to the velocity (i.e., $F_d = c\dot{x}$). This equation represents the internal damping force of the structure. With the availability of high-technology manufactured dampers, the structural engineer has the freedom of imposing additional damping in the structure by introducing manufactured dampers. Manufactured dampers that are used in buildings can produce forces that vary linearly with the relative velocities between the ends of the dampers. However, to provide more design freedom, damper manufacturers often produce dampers that can induce a force that is a nonlinear function of the relative velocity at the ends of the damper.

Figure 9.17 shows a plot of the damper force as a function of velocity. Curve OC is a straight line representing linear viscous damping. If the maximum force in the damper

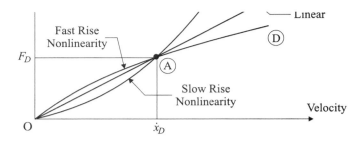

Figure 9.17 General Force versus Velocity Relationships

required for design is defined to be F_D, then, as shown in Figure 9.17, the design limit is that the relative velocity cannot exceed $\dot{x}_D$. A manufactured damper with the force versus velocity relationship defined by curve OAD is called a *fast rise manufactured damper*. The value of η for a fast rise manufactured damper is always less than one, where η is defined in Eq. 9.7. If a fast rise manufactured damper is selected for design, then for velocities less than $\dot{x}_D$ the force is greater than for a linear viscous damper. A *slow rise manufactured damper* has the force versus velocity relationship defined by curve OAB, and its value of η is greater than one.

Consider the case of a linear damper with a force versus velocity relationship given by

$$F_{DL} = c_1 \dot{x} \tag{9.39}$$

where L in the subscript represents "linear." Imagine that in structural design the maximum design force in the damper is set equal to 10 kip and the maximum expected velocity is 1.0 in./s. Then for this case, the value of c_1 is equal to

$$F_{DL} = 10 = c_1(1) \tag{9.40}$$

or $c_1 = 10$ k-s/in. Therefore, the force versus velocity relationship for this linear damper is

$$F_{DL} = 10\dot{x}. \tag{9.41}$$

A fast rise manufactured damper has a force versus velocity relationship of the form

$$F_{DFR} = c_2 \dot{x}^\eta \tag{9.42}$$

where η is less than one. For example, if η is equal to 0.5 and if the same force versus velocity design point is used (i.e., 10 kip at a velocity of 1.0 in./s), it follows that

$$F_{DFR} = 10 = c_2(1)^{0.5} \tag{9.43}$$

or $c_2 = 10$ k-s/in. The force versus velocity equation for this damper is

$$F_{DFR} = 10\dot{x}^{0.5}. \tag{9.44}$$

A slow rise manufactured damper has a value of η that is greater than one. Using the same design example and with η equal to 2, it follows that

$$F_{DSR} = 10 = c_3(1)^2 \tag{9.45}$$

and $c_3 = 10$ k-s/in. Therefore, in summary, the force versus velocity relationships for the three manufactured damper designs are

$$F_{DL} = 10\dot{x} \text{ (linear)} \tag{9.46a}$$

$$F_{DFR} = 10\dot{x}^{0.5} \text{ (fast rise)} \qquad (9.46b)$$

$$F_{DSR} = 10\dot{x}^2 \text{ (slow rise)} \qquad (9.46c)$$

and these force versus velocity relationships are shown in Figure 9.18.

Consider an SDOF system with linear viscous natural damping and no manufactured damping. The equation of motion for this system is

$$m\ddot{x}(t) + c_{nd}\dot{x}(t) + k_e x(t) = -ma(t) \qquad (9.47)$$

where, as discussed in Section 9.2, c_{nd} is the natural damping coefficient. Note that the stiffness is denoted by k_e (i.e., the elastic stiffness) for notational convenience because the numerical integration solution uses k as time steps. If this same SDOF system has a manufactured damper located between the mass and the support (i.e., ground), then the equation of motion for the system is

$$m\ddot{x}(t) + c_{nd}\dot{x}(t) + F_{md}(t) + k_e x(t) = -ma(t). \qquad (9.48)$$

Assume that the manufactured damper has a linear force versus velocity relationship of the form

$$F_{md}(t) = c_{md}\dot{x}(t). \qquad (9.49)$$

As discussed in the previous section, it then follows from Eqs. 9.48 and 9.49 that

$$m\ddot{x}(t) + \left(c_{nd} + c_{md}\right)\dot{x}(t) + k_e x(t) = -ma(t). \qquad (9.50)$$

Let $\omega_n = \sqrt{k_e/m}$. Define

$$2\zeta_{nd}\omega_n = c_{nd}/m \qquad (9.51a)$$

$$2\zeta_{md}\omega_n = c_{md}/m \qquad (9.51b)$$

and

$$\zeta = \zeta_{nd} + \zeta_{md}. \qquad (9.52)$$

It then follows from Eq. 9.50 that

$$\ddot{x}(t) + 2\zeta\omega_n\dot{x}(t) + \omega_n^2 x(t) = -a(t). \qquad (9.53)$$

The solution procedure for this case is discussed in Section 9.2.

Consider Eq. 9.48 with a manufactured damper

$$F_{md} = c_{md}\dot{x}^{\eta}. \qquad (9.54)$$

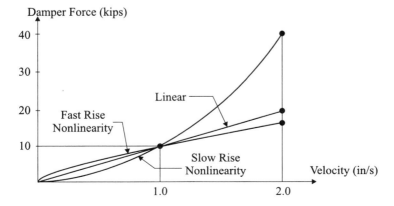

Figure 9.18 Dampers with Force versus Velocity Design Point Equal to 10 kip at 1.0 in./s

It then follows that

$$m\ddot{x}(t) + c_{nd}\dot{x}(t) + c_{md}\dot{x}^\eta(t) + k_e x(t) = -ma(t).$$ (9.55)

Substituting Eq. 9.51 into Eq. 9.55, it follows that

$$\ddot{x}(t) + 2\zeta_{nd}\omega_n\dot{x}(t) + 2\zeta_{md}\omega_n\dot{x}^\eta(t) + \omega_n^2 x(t) = -a(t).$$ (9.56)

The manufactured damping term can be moved to the right side of Eq. 9.56 and then

$$\ddot{x}(t) + 2\zeta_{nd}\omega_n\dot{x}(t) + \omega_n^2 x(t) = -a(t) - 2\zeta_{md}\omega_n\dot{x}^\eta(t).$$ (9.57)

The solution to Eq. 9.57 with $\zeta_{md} = 0$ was discussed in Chapter 2. The presence of manufactured damping is like an additional forcing function applied to the structure where the magnitude of the force is a function of the velocity of the structure.

Recall from Section 5.2 that the spectral displacement, $S_d(T_n, \zeta_n)$, is the response of the structure to an earthquake ground motion. Also, the pseudo spectral acceleration, $S_a(T_n, \zeta_n)$, was defined as

$$S_a(T_n,\zeta_n) = \omega_n^2 S_d(T_n,\zeta_n).$$ (9.58)

Recall for the case where there was assumed to be no natural or manufactured damping that

$$\ddot{y}(t) = -\omega_n^2 x(t).$$ (9.59)

With the addition of a manufactured damper, Eq. 9.59 becomes

$$\ddot{y}(t) = -\omega_n^2 x(t) - 2\zeta_{md}\omega_n\langle \dot{x}(t)\rangle^\eta.$$ (9.60)

Equation 9.60 shows that when $\zeta_{md} = 0$, the maximum acceleration occurs at the same time as the maximum displacement. However, the phasing between the maximum absolute acceleration and the maximum displacement will increase as the magnitude of ζ_{md} increases. Also, only when $\zeta_{md} = 0$ is the maximum absolute acceleration equal to the pseudo spectral acceleration. Therefore, the maximum absolute acceleration of the structure for a given earthquake ground motion will differ more from the spectral acceleration as the magnitude of ζ_{md} increases.

Consider now the solution to Eq. 9.57 using the special case of the Newmark β-method called the constant acceleration solution method. As discussed in Section 2.6, the constant acceleration method is a direct method of calculating the displacement and velocity responses. From Eq. 2.219, the displacement and velocity at time step $k + 1$ can be computed directly from all the responses of the structure at time step k:

$$\dot{x}_{k+1} = \dot{x}_k + \ddot{x}_k\Delta t$$ (9.61a)

$$x_{k+1} = x_k + \dot{x}_k\Delta t + \tfrac{1}{2}\ddot{x}_k(\Delta t)^2.$$ (9.61b)

Now solving for $\ddot{x}_{k+1}$ in the discrete form of Eq. 9.57, it follows that

$$\ddot{x}_{k+1} = -a_{k+1} - 2\zeta_{nd}\omega_n\dot{x}_{k+1} - \omega_n^2 x_{k+1} - 2\zeta_{md}\omega_n(\dot{x}_{k+1})^\eta.$$ (9.62)

When Eq. 9.61 is substituted into Eq. 9.62, the only unknown is the acceleration at time step $k + 1$, and therefore it can be calculated from Eq. 9.62. This completes the calculation of the response at time step $k + 1$ because Eq. 9.61 can be used to calculate the displacement and velocity at time step $k + 1$.

EXAMPLE 1   *Free Vibration*

Consider a manufactured damper with the force versus velocity relationship

$$F_{md} = c_{md}\langle \dot{x}\rangle^\eta.$$ (9.63)

Imagine that an SDOF system is released from rest with an initial displacement equal to 1.0 in. Also imagine that the mass is equal to 1.0 k-s^2/in. and the undamped natural period of vibration is 1.0 s. Therefore, the stiffness is equal to

$$k_e = m\omega_n^2 = (1.0)(2\pi)^2 = 4\pi^2 = 39.478 \text{ k/in.} \tag{9.64}$$

Assume that the natural damping in the system is 5%. Therefore,

$$c_{nd} = 2\zeta_{nd}\omega_n m = 2(0.05)(2\pi)(1.0) = 0.2\pi = 0.6283 \text{ k-s/in.} \tag{9.65}$$

The dynamic equilibrium equation of motion in digitized form is

$$m\,\ddot{x}_{k+1} + c_{nd}\,\dot{x}_{k+1} + c_{md}\langle\dot{x}_{k+1}\rangle^\eta + k_e\,x_{k+1} = 0. \tag{9.66}$$

Consider first the case where there is no manufactured damping (i.e., $c_{md} = 0$). Equation 9.66 becomes

$$m\,\ddot{x}_{k+1} + c_{nd}\,\dot{x}_{k+1} + k_e\,x_{k+1} = 0. \tag{9.67}$$

Using the Newmark β-method of constant acceleration (see Eqs. 9.61a and 9.61b), it follows that

$$\dot{x}_{k+1} = \dot{x}_k + \ddot{x}_k\Delta t \tag{9.68}$$

$$x_{k+1} = x_k + \dot{x}_k\Delta t + \tfrac{1}{2}\ddot{x}_k(\Delta t)^2. \tag{9.69}$$

At time zero (i.e., $k = 0$), the initial displacement is $x_o = 1.0$ in. and the initial velocity is $\dot{x}_o = 0$ in./s. Therefore, the initial acceleration follows from Eq. 9.67, which is

$$\ddot{x}_o = -\left(\frac{c_{nd}}{m}\right)\dot{x}_o - \left(\frac{k_e}{m}\right)x_o = -(0.6283)(0) - (39.4784)(1.0) = -39.4784 \text{ in./s}^2. \tag{9.70}$$

Using a time step size of $\Delta t = 0.02$, it follows from Eqs. 9.68 and 9.69 that, at $k = 1$,

$$x_1 = 1.0 + 0.0(0.02) + \tfrac{1}{2}(-39.4784)(0.02)^2 = 0.9921 \text{ in.} \tag{9.71}$$

$$\dot{x}_1 = 0.0 + (-39.4784)(0.02) = -0.7896 \text{ in./s.} \tag{9.72}$$

and from Eq. 9.67,

$$\ddot{x}_1 = -(0.6283)(-0.7896) - (39.4784)(0.9921) = -38.6706 \text{ in./s}^2. \tag{9.73}$$

For the next time step (i.e., $k = 2$), it follows that

$$x_2 = 0.9921 + (-0.7896)(0.02) + \tfrac{1}{2}(-38.6706)(0.02)^2 = 0.9686 \text{ in.} \tag{9.74}$$

$$\dot{x}_2 = -0.7896 + (-38.6706)(0.02) = -1.5630 \text{ in./s} \tag{9.75}$$

$$\ddot{x}_2 = -(0.6283)(-1.5630) - (39.4784)(0.9686) = -37.2559 \text{ in./s}^2. \tag{9.76}$$

Table 9.7 gives the response for the first five time steps.

Now imagine trying to use the manufactured damper in Figure 9.10 to reduce the response of the structure. This manufactured damper force versus velocity equation is

$$F_{md} = 75\langle\dot{x}\rangle^{0.35}. \tag{9.77}$$

The effective damping coefficient for this manufactured damper is

$$c_{md} = 2\zeta_{md}\omega_n m \tag{9.78}$$

and therefore

$$\zeta_{md} = \frac{c_{md}}{2\omega_n m} = \frac{75}{2(2\pi)(1.0)} = 5.97. \tag{9.79}$$

Table 9.7 Response of an SDOF System with No Manufactured Damping

k	Time (s)	x_k (in.)	$\dot{x}_k$ (in./s)	$\ddot{x}_k$ (in./s^2)
0	0.00	1.0000	0.0000	−39.4784
1	0.02	0.9921	−0.7896	−38.6706
2	0.04	0.9686	−1.5630	−37.2559
3	0.06	0.9299	−2.3081	−35.2595
4	0.08	0.8767	−3.0133	−32.7156
5	0.10	0.8098	−3.6676	−29.6670

This calculation indicates that for this natural frequency of vibration and mass, the manufactured damper in Figure 9.10 will overdamp the response (i.e., $\zeta_{md} > 1.0$). Therefore, c_{md} can be reduced, and the size and cost of the manufactured damper can also be reduced. Imagine that c_{md} is reduced by 100 and becomes $c_{md} = 0.75$. Then it follows that

$$\zeta_{md} = \frac{c_{md}}{2\omega_n m} = \frac{0.75}{2(2\pi)(1.0)} = 0.0597. \tag{9.80}$$

Solving for the acceleration response in Eq. 9.66 gives

$$\ddot{x}_{k+1} = -\left(\frac{c_{nd}}{m}\right)\dot{x}_{k+1} - \left(\frac{k_e}{m}\right)x_{k+1} - \left(\frac{c_{md}}{m}\right)\langle\dot{x}_{k+1}\rangle^\eta. \tag{9.81}$$

At time zero (i.e., $k = 0$), the initial displacement is $x_o = 1.0$ in. and the initial velocity is $\dot{x}_o = 0$ in./s. Therefore, the initial acceleration follows from Eq. 9.81, which is

$$\ddot{x}_o = -(0.6283)(0) - (39.4784)(1.0) - (0.75)\langle 0.0\rangle^{0.35} = -39.4784 \text{ in./s}^2. \tag{9.82}$$

Using a time step size of $\Delta t = 0.02$, it follows from Eqs. 9.68 and 9.69 that, at $k = 1$,

$$x_1 = 1.0 + 0.0(0.02) + \tfrac{1}{2}(-39.4784)(0.02)^2 = 0.9921 \text{ in.} \tag{9.83}$$

$$\dot{x}_1 = 0.0 + (-39.4784)(0.02) = -0.7896 \text{ in./s} \tag{9.84}$$

and from Eq. 9.81,

$$\ddot{x}_1 = -(0.6283)(-0.7896) - (39.4784)(0.9921) - (0.75)\langle-0.7896\rangle^{0.35} = -37.9801 \text{ in./s}^2. \tag{9.85}$$

Note that in Eq. 9.85, the velocity has a negative value and therefore Eq. 9.5 is used, that is,

$$\langle-0.7896\rangle^{0.35} = -\langle 0.7896\rangle^{0.35} = -0.9206. \tag{9.86}$$

For the next time step (i.e., $k = 2$), it follows that

$$x_2 = 0.9921 + (-0.7896)(0.02) + \tfrac{1}{2}(-37.9801)(0.02)^2 = 0.9687 \text{ in.} \tag{9.87}$$

$$\dot{x}_2 = -0.7896 + (-37.9801)(0.02) = -1.5492 \text{ in./s} \tag{9.88}$$

$$\ddot{x}_2 = -(0.6283)(-1.5492) - (39.4784)(0.9687) - (0.75)\langle-1.5492\rangle^{0.35} = -36.3959 \text{ in./s}^2. \tag{9.89}$$

Table 9.8 gives the response for the first five time steps, and Figure 9.19 compares the displacement and velocity responses with and without manufactured damping.

9.4   PRELIMINARY DESIGN OF A MANUFACTURED DAMPING STRUCTURE

The preceding section showed that the solution for the response of a structure with manufactured damping involves numerous calculations after the properties of the manufactured

Table 9.8 Response of an SDOF System with Manufactured Damper

k	Time (s)	x_k (in.)	$\dot{x}_k$ (in./s)	$\ddot{x}_k$ (in./s^2)
0	0.00	1.0000	0.0000	−39.4784
1	0.02	0.9921	−0.7896	−37.9801
2	0.04	0.9687	−1.5492	−36.3959
3	0.06	0.9305	−2.2771	−34.3018
4	0.08	0.8781	−2.9631	−31.7054
5	0.10	0.8124	−3.5972	−28.6400

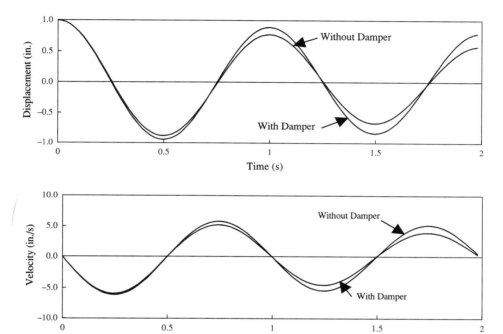

Figure 9.19 Comparison of Response with and without Manufactured Dampers

damper are known. The selection process used for the first trial design (i.e., c_{md} and η) for each manufactured damper in the building is very similar to the selection process for the first trial design for the size of the beams and columns in the building.

The first step is selecting locations for the manufactured dampers. Several candidate locations present no problem to the owner or architect of the building because they are hidden from public view or do not block access. Other candidate locations require "give and take" with the owner or architect. Selecting the number and locations of the manufactured dampers is not easy, and finding one optimal solution is not always possible. Therefore, the first trial design often differs from the final design in both regards.

Once the location and number of manufactured dampers are selected for the first trial design, the focus turns to the "size" of the dampers (i.e., the c_{md} and η values). Here the approach is similar to sizing the columns in a steel building. In theory, the structural engineer can select a different steel column size for every column between adjacent floors for every column line in the building, but this is not economical or practical. Instead, the structural engineer selects a few column types (e.g., three to five column types) for use in the entire building. This same approach is desirable for manufactured dampers, where typically only a few types are used in buildings. Therefore, the structural engineer may select, for example, three different combinations of c_{md} and η.

The next step in developing the first trial design is to construct the natural damping matrix for the structure. Recall from Section 4.6 that the natural damping matrix can be written as

$$\mathbf{C}_{nd} = \alpha\mathbf{M} + \beta\mathbf{K} \tag{9.90}$$

for Rayleigh damping. The other forms of natural damping discussed in Sections 4.7 and 4.8 can also be used to construct $\mathbf{C}_{nd}$. It is then desirable to perform a response spectra analysis of the structure without the manufactured dampers to determine how much the response needs to be reduced for the response variables of interest. For example, such a response spectra analysis could indicate that the maximum roof displacement is 1.5% of the building height when compared to a target design value of 1.0% of the building height. Therefore, the goal of the manufactured dampers in this example is to reduce the roof response by approximately 0.5% of the building height.

Figure 9.20 shows displacement response spectrum for the building discussed in the previous paragraph. Imagine that the original analysis with only natural damping yielded a fundamental natural period of vibration equal to T_A and that the fundamental mode natural damping is 5%. This is shown in Figure 9.20, where Point A represents the first trial design without manufactured dampers. If the target design value of drift is 1.0%, two fundamental options exist to achieve this design goal. One option is to place additional conventional structural members in the structure (e.g., beams, columns, and walls). This option will increase the stiffness of the structure (e.g., T_A going to T_B) and the final design Point B in Figure 9.20. Figure 9.21 shows that this option increases the accelerations in the structure. The other option is to increase the effective damping in the fundamental mode of vibration using manufactured dampers. In this option, the final design point is denoted

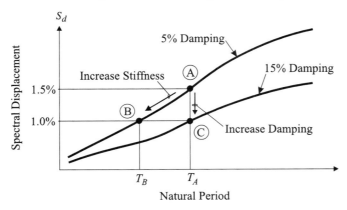

Figure 9.20
Displacement Response
Spectra

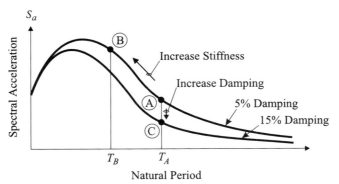

Figure 9.21 Acceleration
Response Spectra

as C in Figures 9.20 and 9.21. Therefore, the value of damping is increased until the target design value of drift is achieved (e.g., 15% damping).

Consider a manufactured damper denoted as Type 1. The force for this damper is

$$F_{md} = c_{md1} \langle \dot{x}_j - \dot{x}_i \rangle^{\eta_1} \tag{9.91}$$

where $\dot{x}_j$ is the velocity of the jth end of the manufactured damper and $\dot{x}_i$ is the velocity of the ith end of the manufactured damper. Only in the first trial design, assume that η_1 is equal to 1.0, and therefore

$$F_{md} = c_{md1} (\dot{x}_j - \dot{x}_i). \tag{9.92}$$

Under this assumption, the equation of motion for the structure becomes

$$\mathbf{M}\ddot{\mathbf{X}}(t) + (\mathbf{C}_{nd} + \mathbf{C}_{md})\dot{\mathbf{X}}(t) + \mathbf{K}\mathbf{X}(t) = -\mathbf{M}\{\mathbf{I}\}a(t) \tag{9.93}$$

where $\mathbf{C}_{md}$ is the manufactured damping matrix, which is a collection of c_{md1}, c_{md2}, c_{md3}, and so forth, depending on the location of the manufactured dampers.

The first trial design now requires that the values of c_{mdi} for the manufactured dampers be selected to achieve the desired total system damping (e.g., 15%). The total system damping matrix can be written as

$$\mathbf{C} = \mathbf{C}_{nd} + \mathbf{C}_{md}. \tag{9.94}$$

The matrix $\mathbf{C}_{md}$ can be decomposed corresponding to the different manufactured damper types (e.g., N types) and written as

$$\mathbf{C}_{md} = c_{md1}\mathbf{C}_1 + c_{md2}\mathbf{C}_2 + \dots + c_{mdN}\mathbf{C}_N \tag{9.95}$$

where c_{mdi} corresponds to the ith manufactured damper type. Using the fundamental natural frequency and mode shape of vibration (i.e., ω_1 and $\boldsymbol{\phi}_1$), it follows that

$$2\zeta_1\omega_1 = \frac{\boldsymbol{\phi}_1^T \mathbf{C} \boldsymbol{\phi}_1}{\boldsymbol{\phi}_1^T \mathbf{M} \boldsymbol{\phi}_1} = \frac{\boldsymbol{\phi}_1^T (\mathbf{C}_{nd} + \mathbf{C}_{md}) \boldsymbol{\phi}_1}{\boldsymbol{\phi}_1^T \mathbf{M} \boldsymbol{\phi}_1} = \frac{\boldsymbol{\phi}_1^T \mathbf{C}_{nd} \boldsymbol{\phi}_1}{\boldsymbol{\phi}_1^T \mathbf{M} \boldsymbol{\phi}_1} + \frac{\boldsymbol{\phi}_1^T \mathbf{C}_{md} \boldsymbol{\phi}_1}{\boldsymbol{\phi}_1^T \mathbf{M} \boldsymbol{\phi}_1}. \tag{9.96}$$

The first term on the right side of Eq. 9.96 uses a proportional damping matrix described in Eq. 9.90, and therefore there are no couplings between the first mode and any other modes:

$$\frac{\boldsymbol{\phi}_1^T \mathbf{C}_{nd} \boldsymbol{\phi}_j}{\boldsymbol{\phi}_1^T \mathbf{M} \boldsymbol{\phi}_1} = \begin{cases} 2\zeta_{nd1}\omega_1 & j = 1 \\ 0 & j \neq 1. \end{cases} \tag{9.97}$$

However, the second term on the right side of Eq. 9.96 corresponds to the manufactured damping, and it will typically not be a proportional damping matrix. However, *only* for purposes of the first trial design it is convenient to assume that

$$\frac{\boldsymbol{\phi}_1^T \mathbf{C}_{md} \boldsymbol{\phi}_j}{\boldsymbol{\phi}_1^T \mathbf{M} \boldsymbol{\phi}_1} = \begin{cases} 2\zeta_{md1}\omega_1 & j = 1 \\ 0 & j \neq 1 \end{cases} \tag{9.98}$$

and therefore

$$\frac{\boldsymbol{\phi}_1^T \mathbf{C}_{nd} \boldsymbol{\phi}_1}{\boldsymbol{\phi}_1^T \mathbf{M} \boldsymbol{\phi}_1} + \frac{\boldsymbol{\phi}_1^T \mathbf{C}_{md} \boldsymbol{\phi}_1}{\boldsymbol{\phi}_1^T \mathbf{M} \boldsymbol{\phi}_1} = 2\zeta_{nd1}\omega_1 + 2\zeta_{md1}\omega_1 = 2(\zeta_{nd1} + \zeta_{md1})\omega_1. \tag{9.99}$$

Equating Eqs. 9.96 and 9.99, it follows that

$$2\zeta_1\omega_1 = 2(\zeta_{nd1} + \zeta_{md1})\omega_1 \tag{9.100}$$

and finally

$$\zeta_1 = \zeta_{nd1} + \zeta_{md1}. \tag{9.101}$$

Equation 9.101 can now be solved for the desired value of manufactured damping, and it is

$$\zeta_{md1} = \zeta_1 - \zeta_{nd1}. \tag{9.102}$$

The value of ζ_{nd1} is known from studying the response of the system without manufactured damping. The value of ζ_1 is known from studying the total damping needed to achieve the design goals, such as the target design drift value (e.g., 1.5% reduced to 1.0%) requiring a total damping of 15%. In the illustrative example presented in this section, therefore,

$$\zeta_{md1} = 0.15 - 0.05 = 0.10. \tag{9.103}$$

Equation 9.95 can now be used to define the first trial values of c_{mdi}. Recall that

$$2\zeta_{md1}\omega_1 = \frac{\phi_1^T \mathbf{C}_{md}\phi_1}{\phi_1^T \mathbf{M}\phi_1} = c_{md1}\left(\frac{\phi_1^T \mathbf{C}_1\phi_1}{\phi_1^T \mathbf{M}\phi_1}\right) + c_{md2}\left(\frac{\phi_1^T \mathbf{C}_2\phi_1}{\phi_1^T \mathbf{M}\phi_1}\right) + \ldots + c_{mdN}\left(\frac{\phi_1^T \mathbf{C}_N\phi_1}{\phi_1^T \mathbf{M}\phi_1}\right)$$
$$= \sum_{i=1}^{N} \alpha_i c_{mdi} \tag{9.104}$$

where

$$\alpha_i = \frac{\phi_1^T \mathbf{C}_i\phi_1}{\phi_1^T \mathbf{M}\phi_1}. \tag{9.105}$$

The solution of Eq. 9.104 for c_{mdi} is not unique because it is one equation with N unknowns. However, with the aid of experience and manufacturer-supplied design charts, such as that shown in Figure 9.10, it is possible to establish reasonable values for c_{mdi} in the first trial design.

   The preceding procedure results in a first trial design that is only a starting point for the analysis of response. Proceeding from this first trial design and analysis to compute the final number of manufactured dampers, their locations, and their c_{mdi} and η_i values is computationally intensive. The structural responses performed on the near-final and final design must be time history analyses and, as noted, there are many design variables. In theory, there are many possible solutions, and from this set there is one unique set that will provide the optimal solution. However, as with most structural designs, the objective is a good final design that is reasonable and probably near the optimal design solution.

EXAMPLE 1   *Design of a Manufactured Viscous Damper*

Consider the 2DOF system with $k_1 = k_2 = k$ and $m_1 = m_2 = m$. Recall from Section 3.2, Examples 1 and 2, that

$$\omega_1 = 0.618\,\omega_n, \qquad \omega_2 = 1.618\,\omega_n \tag{9.106}$$

$$\phi_1 = \begin{Bmatrix} 1.0 \\ 1.618 \end{Bmatrix}, \qquad \phi_2 = \begin{Bmatrix} 1.0 \\ -0.618 \end{Bmatrix} \tag{9.107}$$

where $\omega_n = \sqrt{k/m}$. Assume that this 2DOF system has a mass $m = 1.0$ k-s²/in. on each story, $k = 100.0$ k/in., and a damping ratio in the first mode of vibration equal to 5% (i.e., $\zeta_{nd} = 0.05$).

   Now assume that one manufactured linear viscous damper is installed at the bottom story of the structure. The objective of this manufactured viscous damper is to increase the system damping in the first mode of vibration to 10%. For a linear viscous damper installed at the bottom story of the structure, the force versus velocity relationship is

$$F_{md} = c_{md}\dot{x}_1 \tag{9.108}$$

and therefore the manufactured viscous damping matrix is

$$\mathbf{C}_{md} = c_{md}\mathbf{C}_1 = c_{md}\begin{bmatrix} 1 & 0 \\ 0 & 0 \end{bmatrix} = \begin{bmatrix} c_{md} & 0 \\ 0 & 0 \end{bmatrix}. \tag{9.109}$$

It follows from Eq. 9.104 that

$$2\zeta_{md1}\omega_1 = c_{md}\left(\frac{\boldsymbol{\phi}_1^T\mathbf{C}_1\boldsymbol{\phi}_1}{\boldsymbol{\phi}_1^T\mathbf{M}\boldsymbol{\phi}_1}\right). \tag{9.110}$$

Since $\zeta_{nd} = 0.05$ and the desired damping is $\zeta_{nd} + \zeta_{md} = 0.10$, it follows that ζ_{md} must be 0.05, or 5%. Therefore, Eq. 9.110 becomes

$$c_{md} = 2(0.05)\omega_1\left(\frac{\boldsymbol{\phi}_1^T\mathbf{M}\boldsymbol{\phi}_1}{\boldsymbol{\phi}_1^T\mathbf{C}_1\boldsymbol{\phi}_1}\right). \tag{9.111}$$

Substitute Eqs. 9.106 and 9.107 into Eq. 9.111 and note that

$$\boldsymbol{\phi}_1^T\mathbf{M}\boldsymbol{\phi}_1 = \begin{Bmatrix} 1.0 \\ 1.618 \end{Bmatrix}^T \begin{bmatrix} 1 & 0 \\ 0 & 1 \end{bmatrix} \begin{Bmatrix} 1.0 \\ 1.618 \end{Bmatrix} = 3.618 \tag{9.112}$$

$$\boldsymbol{\phi}_1^T\mathbf{C}_1\boldsymbol{\phi}_1 = \begin{Bmatrix} 1.0 \\ 1.618 \end{Bmatrix}^T \begin{bmatrix} 1 & 0 \\ 0 & 0 \end{bmatrix} \begin{Bmatrix} 1.0 \\ 1.618 \end{Bmatrix} = 1.0. \tag{9.113}$$

Therefore, Eq. 9.111 becomes

$$c_{md} = 2(0.05)(0.618)\sqrt{\frac{100}{1.0}}\left(\frac{3.618}{1.0}\right) = 2.236 \text{ k-s/in.} \tag{9.114}$$

Then it follows from Eq. 9.108 that the desired manufactured viscous damper is

$$F_{md} = (2.236)\dot{x}_1. \tag{9.115}$$

EXAMPLE 2 *Design of Two Manufactured Viscous Dampers of the Same Type*

Consider Example 1 but now using two identical manufactured linear viscous dampers, one installed at the bottom story and the other installed between the top and bottom story. The force versus velocity relationships for these two dampers are

$$F_{md1} = c_{md}\dot{x}_1 \tag{9.116}$$

$$F_{md2} = c_{md}\left(\dot{x}_2 - \dot{x}_1\right). \tag{9.117}$$

Since both F_{md1} and F_{md2} contribute to the dynamic equilibrium equation in the lower story, whereas only F_{md2} contributes to the dynamic equilibrium equation in the upper story, it follows that

$$\begin{Bmatrix} F_1 \\ F_2 \end{Bmatrix} = \begin{Bmatrix} F_{md1} - F_{md2} \\ F_{md2} \end{Bmatrix} = \begin{bmatrix} c_{md} + c_{md} & -c_{md} \\ -c_{md} & c_{md} \end{bmatrix} \begin{Bmatrix} \dot{x}_1 \\ \dot{x}_2 \end{Bmatrix} = c_{md}\begin{bmatrix} 2 & -1 \\ -1 & 1 \end{bmatrix} \begin{Bmatrix} \dot{x}_1 \\ \dot{x}_2 \end{Bmatrix} \tag{9.118}$$

and therefore the manufactured viscous damping matrix is

$$\mathbf{C}_{md} = c_{md}\begin{bmatrix} 2 & -1 \\ -1 & 1 \end{bmatrix} = c_{md}\mathbf{C}_{12}. \tag{9.119}$$

It follows from Eq. 9.104 that

$$2\zeta_{md1}\omega_1 = c_{md}\left(\frac{\boldsymbol{\phi}_1^T \mathbf{C}_{12}\boldsymbol{\phi}_1}{\boldsymbol{\phi}_1^T \mathbf{M}\boldsymbol{\phi}_1}\right). \tag{9.120}$$

Similar to Example 1, to obtain 10% total damping, ζ_{md} in Eq. 9.120 must be 0.05, or 5%. Therefore, Eq. 9.120 becomes

$$c_{md} = 2(0.05)\omega_1\left(\frac{\boldsymbol{\phi}_1^T \mathbf{M}\boldsymbol{\phi}_1}{\boldsymbol{\phi}_1^T \mathbf{C}_{12}\boldsymbol{\phi}_1}\right). \tag{9.121}$$

Substitute Eqs. 9.106 and 9.107 into Eq. 9.121 and note that

$$\boldsymbol{\phi}_1^T \mathbf{M}\boldsymbol{\phi}_1 = \begin{Bmatrix} 1.0 \\ 1.618 \end{Bmatrix}^T \begin{bmatrix} 1 & 0 \\ 0 & 1 \end{bmatrix} \begin{Bmatrix} 1.0 \\ 1.618 \end{Bmatrix} = 3.618 \tag{9.122}$$

$$\boldsymbol{\phi}_1^T \mathbf{C}_{12}\boldsymbol{\phi}_1 = \begin{Bmatrix} 1.0 \\ 1.618 \end{Bmatrix}^T \begin{bmatrix} 2 & -1 \\ -1 & 1 \end{bmatrix} \begin{Bmatrix} 1.0 \\ 1.618 \end{Bmatrix} = 1.382. \tag{9.123}$$

Therefore, Eq. 9.121 becomes

$$c_{md} = 2(0.05)(0.618)\sqrt{\frac{100}{1.0}}\left(\frac{3.618}{1.382}\right) = 1.618 \text{ k-s/in.} \tag{9.124}$$

Then it follows from Eqs. 9.116 and 9.117 that the desired manufactured viscous damper is

$$F_{md1} = (1.618)\dot{x}_1 \tag{9.125}$$

$$F_{md2} = 1.618(\dot{x}_2 - \dot{x}_1) \tag{9.126}$$

or, in matrix form,

$$\begin{Bmatrix} F_{md1} \\ F_{md2} \end{Bmatrix} = \begin{bmatrix} 1.618 & 0.0 \\ -1.618 & 1.618 \end{bmatrix} \begin{Bmatrix} \dot{x}_1 \\ \dot{x}_2 \end{Bmatrix}. \tag{9.127}$$

EXAMPLE 3   *Design of Two Different Manufactured Viscous Dampers*

Consider Example 1 but now using two manufactured linear viscous dampers, one installed at the bottom story and the other installed between the top and bottom stories. Similar to Example 2, the force versus velocity relationships for these two dampers are

$$F_{md1} = c_{md1}\dot{x}_1 \tag{9.128}$$

$$F_{md2} = c_{md2}(\dot{x}_2 - \dot{x}_1). \tag{9.129}$$

Following the procedure derived in Eqs. 9.118 and 9.119, the manufactured viscous damping matrix is

$$\mathbf{C}_{md} = c_{md1}\mathbf{C}_1 + c_{md2}\mathbf{C}_2 = c_{md1}\begin{bmatrix} 1 & 0 \\ 0 & 0 \end{bmatrix} + c_{md2}\begin{bmatrix} 1 & -1 \\ -1 & 1 \end{bmatrix}. \tag{9.130}$$

It follows from Eq. 9.104 that

$$2\zeta_{md1}\omega_1 = c_{md1}\left(\frac{\boldsymbol{\phi}_1^T \mathbf{C}_1\boldsymbol{\phi}_1}{\boldsymbol{\phi}_1^T \mathbf{M}\boldsymbol{\phi}_1}\right) + c_{md2}\left(\frac{\boldsymbol{\phi}_1^T \mathbf{C}_2\boldsymbol{\phi}_1}{\boldsymbol{\phi}_1^T \mathbf{M}\boldsymbol{\phi}_1}\right). \tag{9.131}$$

Following a similar procedure for deriving Eq. 9.104, a relationship can also be derived using the second mode of vibration. From Eq. 9.130, premultiplying by $\boldsymbol{\phi}_2^T$ and postmultiplying by $\boldsymbol{\phi}_2$ gives

$$2\zeta_{md2}\omega_2 = c_{md1}\left(\frac{\boldsymbol{\phi}_2^T\mathbf{C}_1\boldsymbol{\phi}_2}{\boldsymbol{\phi}_2^T\mathbf{M}\boldsymbol{\phi}_2}\right) + c_{md2}\left(\frac{\boldsymbol{\phi}_2^T\mathbf{C}_2\boldsymbol{\phi}_2}{\boldsymbol{\phi}_2^T\mathbf{M}\boldsymbol{\phi}_2}\right). \tag{9.132}$$

Assume that the damping in both the first and second modes of vibrations is 5% (i.e., $\zeta_{nd1} = \zeta_{nd2} = 0.05$) and the desired total damping in both the first and second modes is 10% (i.e., $\zeta_1 = \zeta_2 = 0.10$). Therefore, it follows that the damping values in the manufactured viscous dampers are $\zeta_{md1} = \zeta_{md2} = 0.05$. Note that

$$\boldsymbol{\phi}_1^T\mathbf{M}\boldsymbol{\phi}_1 = \left\{\begin{matrix}1.0\\1.618\end{matrix}\right\}^T\begin{bmatrix}1 & 0\\0 & 1\end{bmatrix}\left\{\begin{matrix}1.0\\1.618\end{matrix}\right\} = 3.618 \tag{9.133}$$

$$\boldsymbol{\phi}_2^T\mathbf{M}\boldsymbol{\phi}_2 = \left\{\begin{matrix}1.0\\-0.618\end{matrix}\right\}^T\begin{bmatrix}1 & 0\\0 & 1\end{bmatrix}\left\{\begin{matrix}1.0\\-0.618\end{matrix}\right\} = 1.382. \tag{9.134}$$

$$\boldsymbol{\phi}_1^T\mathbf{C}_1\boldsymbol{\phi}_1 = \left\{\begin{matrix}1.0\\1.618\end{matrix}\right\}^T\begin{bmatrix}1 & 0\\0 & 0\end{bmatrix}\left\{\begin{matrix}1.0\\1.618\end{matrix}\right\} = 1.0 \tag{9.135}$$

$$\boldsymbol{\phi}_2^T\mathbf{C}_1\boldsymbol{\phi}_2 = \left\{\begin{matrix}1.0\\-0.618\end{matrix}\right\}^T\begin{bmatrix}1 & 0\\0 & 0\end{bmatrix}\left\{\begin{matrix}1.0\\-0.618\end{matrix}\right\} = 1.0 \tag{9.136}$$

$$\boldsymbol{\phi}_1^T\mathbf{C}_2\boldsymbol{\phi}_1 = \left\{\begin{matrix}1.0\\1.618\end{matrix}\right\}^T\begin{bmatrix}1 & -1\\-1 & 1\end{bmatrix}\left\{\begin{matrix}1.0\\1.618\end{matrix}\right\} = 0.382 \tag{9.137}$$

$$\boldsymbol{\phi}_2^T\mathbf{C}_2\boldsymbol{\phi}_2 = \left\{\begin{matrix}1.0\\-0.618\end{matrix}\right\}^T\begin{bmatrix}1 & -1\\-1 & 1\end{bmatrix}\left\{\begin{matrix}1.0\\-0.618\end{matrix}\right\} = 2.618. \tag{9.138}$$

It follows from Eqs. 9.131 and 9.132 that

$$2(0.05)(0.618)\sqrt{\frac{100.0}{1.0}} = c_{md1}\left(\frac{1.0}{3.618}\right) + c_{md2}\left(\frac{0.382}{3.618}\right) \tag{9.139}$$

$$2(0.05)(1.618)\sqrt{\frac{100.0}{1.0}} = c_{md1}\left(\frac{1.0}{1.382}\right) + c_{md2}\left(\frac{2.618}{1.382}\right). \tag{9.140}$$

Solving simultaneously for c_{md1} and c_{md2} in Eqs. 9.139 and 9.140, it follows that

$$c_{md1} = 2.236 \text{ k-s/in.}, \quad c_{md2} = 0.0 \text{ k-s/in.} \tag{9.141}$$

Then it follows from Eqs. 9.116 and 9.117 that the desired manufactured viscous damper is

$$F_{md1} = (2.236)\dot{x}_1 \tag{9.142}$$

$$F_{md2} = 0.0. \tag{9.143}$$

Equation 9.143 means that a manufactured viscous damper between the upper and lower stories is not necessary to provide 5% manufactured viscous damping in the second mode of vibration, and therefore the result in the example is the same as in Example 1.

9.5   FORCED RESPONSE OF AN SDOF SYSTEM USING NUMERICAL DISCRETIZATION

The analytical representation for an SDOF system with a damper of a general function (not necessarily fast rise or slow rise) can be expressed as

$$F_{md} = -f_{md}(\dot{x}) \tag{9.144}$$

where F_{md} is the manufactured damping force generated by the velocity at the ends of the damper and $F_{md}(\cdot)$ is some positive nonlinear function that depends on the design of the dampers. An example is $F_{md} = -c_{md}\dot{x}^3$, where c_{md} is a positive constant of proportionality. The minus sign represents the opposite direction between the damping force and the velocity and is similar to the damping force discussed in Section 2.1. Incorporating Eq. 9.144 into the dynamic equilibrium equation of motion in Eq. 2.1 gives the new equation of motion

$$\sum F = F_s + F_d + F_{md} + F_e = -k\,x - c\,\dot{x} - f_{md}(\dot{x}) + F_e = m\,\ddot{x}. \tag{9.145}$$

Rearranging the terms in Eq. 9.145 gives

$$m\,\ddot{x}(t) + c\,\dot{x}(t) + k\,x(t) = F_e(t) - f_{md}(\dot{x}(t)). \tag{9.146}$$

Equation 9.146 is a general representation of the structural dynamic response incorporating nonlinear manufactured dampers. This equation is nonlinear with an unknown function $f_{md}(\cdot)$, and thus no closed-form analytical solution is available. Therefore, the numerical method must be used to calculate the response.

In the following discussion, it is assumed that

$$f_{md}(\dot{x}(t)) = c_{md}\langle \dot{x}(t)\rangle^\eta \tag{9.147}$$

because this is the most common type of viscous damper that is used by structural engineers. The notation $\langle \dot{x}(t)\rangle^\eta$ is defined in Eq. 9.5, and it is

$$\langle \dot{x}(t)\rangle^\eta = \begin{cases} \dot{x}(t)^\eta & \dot{x} \ge 0 \\ -(-\dot{x}(t))^\eta & \dot{x} < 0. \end{cases} \tag{9.148}$$

One numerical solution method is to use the Newmark β-method with a combination of *constant acceleration* and *constant average acceleration*. Rewriting Eq. 9.146 in normalized form, it follows that

$$\ddot{x}_{k+1} + 2\zeta_{nd}\omega_n\dot{x}_{k+1} + \omega_n^2 x_{k+1} = (F_{k+1}/m) - 2\zeta_{md}\omega_n\langle \dot{x}_{k+1}\rangle^\eta \tag{9.149}$$

where

$$\zeta_{nd} = \frac{c_{nd}}{2\omega_n m}, \quad \zeta_{md} = \frac{c_{md}}{2\omega_{nd} m}. \tag{9.150}$$

The first goal is to estimate the viscous damping force, $2\zeta_{md}\omega_n\langle \dot{x}(t)\rangle^\eta$, based only on the system response information at time step k (i.e., x_k, $\dot{x}_k$, and $\ddot{x}_k$). Using the constant acceleration method, from Eq. 2.219a, it follows that

$$\dot{x}_{k+1} = \dot{x}_k + \ddot{x}_k \Delta t \tag{9.151}$$

and therefore

$$2\zeta_{md}\omega_n\langle \dot{x}(t)\rangle^\eta = 2\zeta_{md}\omega_n\langle \dot{x}_k + \ddot{x}_k\Delta t\rangle^\eta. \tag{9.152}$$

It follows from Eq. 9.149 that

$$\ddot{x}_{k+1} + 2\zeta_{nd}\omega_n\dot{x}_{k+1} + \omega_n^2 x_{k+1} = F_{k+1}/m - 2\zeta_{md}\omega_n\langle \dot{x}_k + \ddot{x}_k\Delta t\rangle^\eta. \tag{9.153}$$

Note that by using the constant acceleration method, the term involving the manufactured damper is known and thus the right-hand side of Eq. 9.153 is known. Define

$$\overline{F}_{k+1} = F_{k+1}/m - 2\zeta_{md}\omega_n\langle \dot{x}_k + \ddot{x}_k\Delta t\rangle^\eta \tag{9.154}$$

and then it follows that the equation of motion is

$$\ddot{x}_{k+1} + 2\zeta_{nd}\omega_n\dot{x}_{k+1} + \omega_n^2 x_{k+1} = \overline{F}_{k+1}. \tag{9.155}$$

The advantage of the constant acceleration method is that the acceleration at time step $k + 1$ is not present in the velocity expression for time step $k + 1$ (see Eq. 9.151). This method is used only to obtain the right-hand side of Eqs. 9.150, 9.153, and 9.155, and the solution will be more accurate as the time step Δt decreases.

The solution for response from Eq. 9.155 can now be obtained for this defined forcing function $\bar{F}_{k+1}$ using the Newmark β-method option of constant average acceleration method. Recall from Eq. 2.221 that

$$\dot{x}_{k+1} = \dot{x}_k + \tfrac{1}{2}\ddot{x}_k\Delta t + \tfrac{1}{2}\ddot{x}_{k+1}\Delta t \tag{9.156a}$$

$$x_{k+1} = x_k + \dot{x}_k\Delta t + \tfrac{1}{4}\ddot{x}_k\Delta t^2 + \tfrac{1}{4}\ddot{x}_{k+1}(\Delta t)^2. \tag{9.156b}$$

Using the procedure discussed in Section 2.6, the solution for the set of three simultaneous equations given in Eqs. 9.155 and 9.156 becomes

$$\mathbf{q}_{k+1} = \begin{Bmatrix} x_{k+1} \\ \dot{x}_{k+1} \\ \ddot{x}_{k+1} \end{Bmatrix} = \mathbf{F}_N \begin{Bmatrix} x_k \\ \dot{x}_k \\ \ddot{x}_k \end{Bmatrix} + \overline{\mathbf{H}}_N \bar{F}_{k+1} = \mathbf{F}_N\mathbf{q}_k + \overline{\mathbf{H}}_N\bar{F}_{k+1} \tag{9.157}$$

where

$$\mathbf{F}_N = \frac{1}{\beta}\begin{bmatrix} 1+\zeta_{nd}\omega_n\Delta t & \Delta t - \tfrac{1}{2}\zeta_{nd}\omega_n(\Delta t)^2 & \tfrac{1}{4}(\Delta t)^2 \\ -\tfrac{1}{2}\omega_n^2\Delta t & 1-\tfrac{1}{4}\omega_n^2(\Delta t)^2 & \tfrac{1}{2}\Delta t \\ -\omega_n^2 & -2\zeta_{nd}\omega_n - \omega_n^2\Delta t & -\zeta_{nd}\omega_n\Delta t - \tfrac{1}{4}\omega_n^2(\Delta t)^2 \end{bmatrix}, \quad \overline{\mathbf{H}}_N = \frac{1}{\beta}\begin{Bmatrix} \tfrac{1}{4}(\Delta t)^2 \\ \tfrac{1}{2}\Delta t \\ 1 \end{Bmatrix}$$

and

$$\beta = 1 + \zeta_{nd}\omega_n\Delta t + \tfrac{1}{4}\omega_n^2(\Delta t)^2.$$

Equation 9.157 can also be written in another form by substituting Eq. 9.154 into Eq. 9.157, which gives

$$\begin{Bmatrix} x_{k+1} \\ \dot{x}_{k+1} \\ \ddot{x}_{k+1} \end{Bmatrix} = \mathbf{F}_N \begin{Bmatrix} x_k \\ \dot{x}_k \\ \ddot{x}_k \end{Bmatrix} + \mathbf{H}_N F_{k+1} - \mathbf{H}_N \left(2\zeta_{md}\omega_n m\right)\left\langle \dot{x}_k + \ddot{x}_k\Delta t \right\rangle^\eta \tag{9.158}$$

where

$$\mathbf{F}_N = \frac{1}{\beta}\begin{bmatrix} 1+\zeta_{nd}\omega_n\Delta t & \Delta t - \tfrac{1}{2}\zeta_{nd}\omega_n(\Delta t)^2 & \tfrac{1}{4}(\Delta t)^2 \\ -\tfrac{1}{2}\omega_n^2\Delta t & 1-\tfrac{1}{4}\omega_n^2(\Delta t)^2 & \tfrac{1}{2}\Delta t \\ -\omega_n^2 & -2\zeta_{nd}\omega_n - \omega_n^2\Delta t & -\zeta_{nd}\omega_n\Delta t - \tfrac{1}{4}\omega_n^2(\Delta t)^2 \end{bmatrix}, \quad \mathbf{H}_N = \frac{1}{m\beta}\begin{Bmatrix} \tfrac{1}{4}(\Delta t)^2 \\ \tfrac{1}{2}\Delta t \\ 1 \end{Bmatrix}.$$

Note that the velocity, $\dot{x}_{k+1}$, is estimated using the constant acceleration method in Eq. 9.151, but it is also calculated in Eq. 9.157. The difference in these two velocity estimates will be small when a small time step ($\Delta t \le T_n/10$) is used in the solution. Accuracy can also be improved by using an iteration method, where $\dot{x}_{k+1}$ calculated in Eq. 9.157 is substituted into the velocity term on the right-hand side of Eq. 9.149. The procedure for using this iteration method is as follows:

1. Compute $\dot{x}_{k+1}^{(1)}$ using Eq. 9.151:

$$\dot{x}_{k+1}^{(1)} = \dot{x}_k + \ddot{x}_k\Delta t \tag{9.159}$$

 where the superscript number denotes the iterative solution number.

2. Compute $\bar{F}_{k+1}^{(1)}$ from Eq. 9.154:

$$\bar{F}_{k+1}^{(1)} = F_{k+1} - 2\zeta_{md}\omega_n \left\langle \dot{x}_{k+1}^{(1)} + \ddot{x}_k\Delta t \right\rangle^\eta. \tag{9.160}$$

3.   Compute $\dot{x}_{k+1}^{(2)}$ using Eq. 9.157:

$$\begin{Bmatrix} x_{k+1} \\ \dot{x}_{k+1}^{(2)} \\ \ddot{x}_{k+1} \end{Bmatrix} = \mathbf{F}_N \begin{Bmatrix} x_k \\ \dot{x}_k \\ \ddot{x}_k \end{Bmatrix} + \mathbf{H}_N \overline{F}_{k+1}^{(1)}. \tag{9.161}$$

4.   Compare $\dot{x}_{k+1}^{(1)}$ with $\dot{x}_{k+1}^{(2)}$. If the difference between these two quantities is not acceptable, then repeat Steps 2 to 4 until $\dot{x}_{k+1}^{(i+1)}$ is sufficiently close to $\dot{x}_{k+1}^{(i)}$. Once the solution has converged, then the solution is obtained, where

$$\begin{Bmatrix} x_{k+1} \\ \dot{x}_{k+1} \\ \ddot{x}_{k+1} \end{Bmatrix} = \mathbf{F}_N \begin{Bmatrix} x_k \\ \dot{x}_k \\ \ddot{x}_k \end{Bmatrix} + \mathbf{H}_N \overline{F}_{k+1}^{(i)}. \tag{9.162}$$

This completes the calculation of the response at time step $k + 1$.

Although the iteration method improves the accuracy of the numerical analysis, this accuracy is often insignificant compared to the overall accuracy of the Newmark β-method. The iterative calculations can be very time consuming when small tolerances are required, demanding as much computing time as the noniterative numerical analysis with a reduced time step size. Therefore, reducing the time step size is very desirable to increase the numerical accuracy to an acceptable level without using the iteration method.

The solution given in Eq. 9.157 does not satisfy the dynamic equilibrium equation of motion given in Eq. 9.146 when no iterative procedure is performed. This is because the velocity used on the right-hand side of Eq. 9.146 is generally not the same as the velocity obtained on the left-hand side. One way to avoid this problem without performing the iterative procedure is to balance the dynamic equilibrium equation by correcting the acceleration value. Equation 9.157 can be used to calculate both the displacement and velocity at time step $k + 1$ while neglecting the acceleration term:

$$\dot{x}_{k+1} = -\left(\frac{\frac{1}{2} \omega_n^2 \Delta t}{\beta} \right) x_k + \left(\frac{1 - \frac{1}{4} \omega_n^2 \Delta t^2}{\beta} \right) \dot{x}_k + \left(\frac{\frac{1}{2} \Delta t}{\beta} \right) \ddot{x}_k + \left(\frac{\frac{1}{2} \Delta t}{\beta} \right) \overline{F}_{k+1} \tag{9.163a}$$

$$x_{k+1} = \left(\frac{1 + \zeta_{nd} \omega_n \Delta t}{\beta} \right) x_k + \left(\frac{\Delta t - \frac{1}{2} \zeta_{nd} \omega_n \Delta t^2}{\beta} \right) \dot{x}_k + \left(\frac{\frac{1}{4} \Delta t^2}{\beta} \right) \ddot{x}_k + \left(\frac{\frac{1}{4} \Delta t^2}{\beta} \right) \overline{F}_{k+1}. \tag{9.163b}$$

The acceleration at time step $k + 1$ is then calculated using Eq. 9.155 based on the displacement and velocity values obtained in Eq. 9.163, where Eq. 9.155 is written here as

$$\ddot{x}_{k+1} = -2\zeta_{nd} \omega_n \dot{x}_{k+1} - \omega_n^2 x_{k+1} + \overline{F}_{k+1}. \tag{9.164}$$

EXAMPLE 1   *Free Vibration of an SDOF System with a Manufactured Viscous Damper*

Consider an SDOF system used in Example 2 in Section 2.3 with the following structural variables (also used in Example 1 of Section 2.6):

$$m = 1.0 \text{ k-s}^2/\text{in.}, \quad \zeta = 0.05, \quad T_n = 1.0 \text{ s}.$$

Assume $\Delta t = 0.02$ s. Also, assume that a manufactured viscous damper is installed into this SDOF system with force versus velocity relationship given by

$$F_{md} = 2.0 \langle \dot{x}(t) \rangle^{0.5}. \tag{9.165}$$

The calculation of response using the Newmark β-method of constant acceleration is given in Eq. 2.219, and it is

$$x_{k+1} = x_k + \dot{x}_k(0.02) + \tfrac{1}{2}\ddot{x}_k(0.02)^2 \tag{9.166}$$

$$\dot{x}_{k+1} = \dot{x}_k + \ddot{x}_k(0.02). \tag{9.167}$$

Once the displacement and velocity at time step $k + 1$ are determined, Eq. 9.153 can be used to calculate the acceleration at this same time step:

$$\ddot{x}_{k+1} = -2(0.05)(2\pi)\dot{x}_{k+1} - (2\pi)^2 x_{k+1} + F_{k+1} - 2.0\langle\dot{x}_{k+1}\rangle^{0.5}. \tag{9.168}$$

Assume that there is no external force (i.e., $F_k = 0$ for all k) but the structure has an initial displacement of $x_o = 5.0$ in. with zero initial velocity. Then it follows from Eq. 9.168 that

$$\ddot{x}_o = -2(0.05)(2\pi)(0) - (2\pi)^2(5) - 2.0\langle 0\rangle^{0.5} = -197.393 \text{ in./s}^2. \tag{9.169}$$

At the next time step (i.e., $t = \Delta t = 0.02$ s), the response can be calculated using Eqs. 9.166 to 9.168:

$$x_1 = (5) + (0)(0.02) + \tfrac{1}{2}(-197.392)(0.02)^2 = 4.96052 \text{ in.} \tag{9.170}$$

$$\dot{x}_1 = (0) + (-197.392)(0.02) = -3.9478 \text{ in./s} \tag{9.171}$$

$$\ddot{x}_1 = -2(0.05)(2\pi)(-3.9478) - (2\pi)^2(4.96052) - 2.0\langle -3.9478\rangle^{0.5} = -189.379 \text{ in./s}^2 \tag{9.172}$$

and the force in the manufactured damper is

$$F_{md,1} = 2.0\langle -3.9478\rangle^{0.5} = -3.974 \text{ kip.} \tag{9.173}$$

Then at time step $t = 2\Delta t = 0.04$ s, the response is

$$x_2 = (4.96052) + (-3.9478)(0.02) + \tfrac{1}{2}(-189.379)(0.02)^2 = 4.84369 \text{ in.} \tag{9.174}$$

$$\dot{x}_2 = (-3.9478) + (-189.379)(0.02) = -7.7354 \text{ in./s} \tag{9.175}$$

$$\ddot{x}_2 = -2(0.05)(2\pi)(-7.7354) - (2\pi)^2(4.84369) - 2.0\langle -7.7354\rangle^{0.5} = -180.798 \text{ in./s}^2 \tag{9.176}$$

$$F_{md,2} = 2.0\langle -7.7354\rangle^{0.5} = -5.563 \text{ kip.} \tag{9.177}$$

The calculation process continues, and the response at the first 10 time steps is shown in Table 9.9. The displacement, velocity, and force in the manufactured viscous damper are shown in Figures 9.22 to 9.24, respectively.

Table 9.9 Response of an SDOF System with Manufactured Viscous Damper

k	Time (s)	x_k (in.)	$\dot{x}_k$ (in./s)	$\ddot{x}_k$ (in./s^2)	$F_{md,k}$ (kip)
0	0.00	5.00000	0.0000	−197.392	0.000
1	0.02	4.96052	−3.9478	−189.379	−3.974
2	0.04	4.84369	−7.7354	−180.798	−5.563
3	0.06	4.65282	−11.3514	−169.815	−6.738
4	0.08	4.39183	−14.7477	−156.436	−7.681
5	0.10	4.06559	−17.8764	−140.815	−8.456
6	0.12	3.67990	−20.6927	−123.177	−9.098
7	0.14	3.24141	−23.1563	−103.792	−9.624
8	0.16	2.75752	−25.2321	−82.963	−10.046
9	0.18	2.23629	−26.8913	−61.017	−10.371
10	0.20	1.68626	−28.1117	−38.304	−10.604

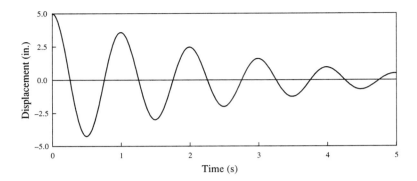

Figure 9.22 Displacement Time History of an SDOF System

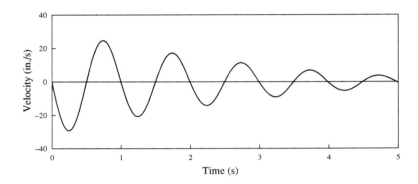

Figure 9.23 Velocity Time History of an SDOF System

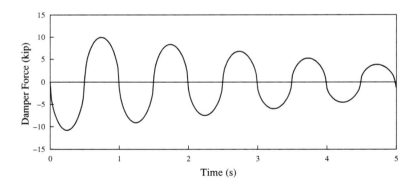

Figure 9.24 Manufactured Viscous Damper Force of an SDOF System

EXAMPLE 2   *Response of an SDOF System Subjected to a Modified El-Centro Earthquake*

Consider the SDOF system discussed in Example 1 but with the period equal to 0.5 s and let $\Delta t = 0.02$ s. The response of the system using the Newmark β-method with the combination of both constant acceleration and constant average acceleration is now presented. Using $\delta = \frac{1}{2}$ and $\alpha = \frac{1}{4}$, the response without manufactured viscous damper is given in Eq. 2.257, and it is

$$\begin{Bmatrix} x_{k+1} \\ \dot{x}_{k+1} \\ \ddot{x}_{k+1} \end{Bmatrix} = \begin{bmatrix} 0.984644 & 0.019571 & 0.000097 \\ -1.53559 & 0.957068 & 0.009724 \\ -153.559 & -4.29317 & -0.02758 \end{bmatrix} \begin{Bmatrix} x_k \\ \dot{x}_k \\ \ddot{x}_k \end{Bmatrix} + \begin{bmatrix} -0.000097 \\ -0.00972 \\ -0.97242 \end{bmatrix} a_{k+1} \qquad (9.178)$$

with

$$\beta = 1 + 2\zeta_{nd}\omega_n\delta\Delta t + \omega_n^2\alpha(\Delta t)^2 = 1.02836. \qquad (9.179)$$

Now assume a manufactured viscous damper is added to the structure with the force versus velocity relationship given by

$$F_{md} = 2.0\langle\dot{x}(t)\rangle^{0.5}. \qquad (9.180)$$

Then it follows from Eqs. 9.157 and 9.178 that

$$\begin{Bmatrix} x_{k+1} \\ \dot{x}_{k+1} \\ \ddot{x}_{k+1} \end{Bmatrix} = \mathbf{F}_N \begin{Bmatrix} x_k \\ \dot{x}_k \\ \ddot{x}_k \end{Bmatrix} + \mathbf{H}_N^{(EQ)} a_{k+1} - \mathbf{H}_N F_{md,k+1} \qquad (9.181)$$

where $\mathbf{F}_N$, $\mathbf{H}_N^{(EQ)}$, and $\mathbf{H}_N$ are

$$\mathbf{F}_N = \begin{bmatrix} 0.984644 & 0.019571 & 0.000097 \\ -1.53559 & 0.957068 & 0.009724 \\ -153.559 & -4.29317 & -0.02758 \end{bmatrix}, \quad \mathbf{H}_N^{(EQ)} = \begin{Bmatrix} -0.000097 \\ -0.00972 \\ -0.97242 \end{Bmatrix}, \quad \mathbf{H}_N = \begin{Bmatrix} 0.000097 \\ 0.00972 \\ 0.97242 \end{Bmatrix} \qquad (9.182)$$

and $F_{md,k+1}$ is

$$F_{md,k+1} = 2.0\langle\dot{x}_k + \ddot{x}_k(0.02)\rangle^{0.5}. \qquad (9.183)$$

Assume that the structure has zero initial displacement and zero initial velocity but is subjected to the modified El-Centro earthquake time history and shown in Figure 6.11. Then the acceleration at time step $k = 0$ follows from Eq. 9.168, where the earthquake acceleration is $a_o = -0.9930$ in./s^2:

$$\ddot{x}_o = -2(0.05)(2\pi)(0) - (2\pi)^2(0) - (-0.9930) - 2.0\langle 0\rangle^{0.5} = 0.993 \text{ in./s}^2 \qquad (9.184)$$

At the next time step (i.e., $t = \Delta t = 0.02$ s), the response can be calculated using Eqs. 9.181 and 9.183:

$$F_{md,1} = 2.0\langle 0 + (0.993)(0.02)\rangle^{0.5} = 0.282 \text{ kip} \qquad (9.185)$$

$$\begin{Bmatrix} x_1 \\ \dot{x}_1 \\ \ddot{x}_1 \end{Bmatrix} = \mathbf{F}_N \begin{Bmatrix} 0 \\ 0 \\ 0.993 \end{Bmatrix} + \mathbf{H}_N^{(EQ)}(-7.6507) - \mathbf{H}_N(0.282) = \begin{Bmatrix} 0.00074 \\ 0.0742 \\ 7.420 \end{Bmatrix}. \qquad (9.186)$$

However, the response given in Eq. 9.186 does not satisfy the dynamic equilibrium equation, and therefore the condition given in Eq. 9.164 must be imposed:

$$\ddot{x}_{k+1} = -2\zeta_{nd}\omega_n\dot{x}_{k+1} - \omega_n^2 x_{k+1} - a_{k+1} - F_{md,k+1}/m. \qquad (9.187)$$

Doing so gives

$$\ddot{x}_1 = -2(0.05)(2\pi)(0.0742) - (2\pi)^2(0.00074) - (-7.6507) - 0.282/1.0 = 7.158 \text{ in./s}^2. \qquad (9.188)$$

Then at time step $t = 2\Delta t = 0.04$ s, the response is

$$F_{md,2} = 2.0\langle 0.0742 + (7.420)(0.02)\rangle^{0.5} = 0.944 \text{ kip} \qquad (9.189)$$

$$\begin{Bmatrix} x_2 \\ \dot{x}_2 \\ \ddot{x}_2 \end{Bmatrix} = \mathbf{F}_N \begin{Bmatrix} 0.00074 \\ 0.0742 \\ 7.420 \end{Bmatrix} + \mathbf{H}_N^{(EQ)}(-7.1484) - \mathbf{H}_N(0.944) = \begin{Bmatrix} 0.00286 \\ 0.1372 \\ 6.302 \end{Bmatrix}. \tag{9.190}$$

Again, the response given in Eq. 9.190 does not satisfy the dynamic equilibrium equation, and therefore the condition given in Eq. 9.164 must be imposed, which gives

$$\ddot{x}_2 = -2(0.05)(2\pi)(0.1372) - (2\pi)^2(0.00186) - (-7.1484) - 0.944/1.0 = 5.581 \text{ in./s}^2. \tag{9.191}$$

The calculation process continues, and the displacement, velocity, and force in the manufactured viscous damper for a 20-s period are shown in Figures 9.25 to 9.27, respectively.

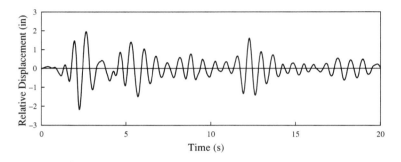

Figure 9.25 Displacement Time History of an SDOF System

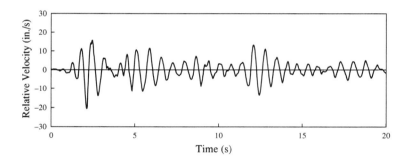

Figure 9.26 Velocity Time History of an SDOF System

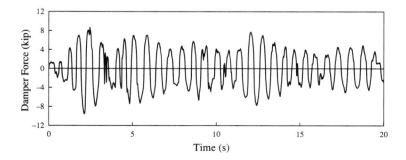

Figure 9.27 Manufactured Viscous Damper Force of an SDOF System

9.6 FORCED RESPONSE OF AN SDOF SYSTEM USING DIRECT INTEGRATION

Chapter 2 presented the state space solution procedure for SDOF systems with natural damping. Equation 9.146 represents the equation of motion, which is repeated here:

$$m\,\ddot{x}(t) + c\,\dot{x}(t) + k\,x(t) = F_e(t) - f_{md}(\dot{x}(t)). \tag{9.192}$$

The solution for this equation in state space form is discussed in Section 2.8. Equation 2.284 can be rewritten as

$$\mathbf{z}(t) = \mathbf{e}^{\mathbf{A}(t-t_o)}\mathbf{z}(t_o) + \mathbf{e}^{\mathbf{A}t}\int_{t_o}^{t}\mathbf{e}^{-\mathbf{A}s}\,\mathbf{F}(s)\,ds \tag{9.193}$$

where

$$\mathbf{z}(t) = \begin{Bmatrix} x(t) \\ \dot{x}(t) \end{Bmatrix}, \quad \mathbf{A} = \begin{bmatrix} 0 & 1 \\ -\omega_n^2 & -2\zeta_{nd}\omega_n \end{bmatrix}, \quad \mathbf{F}(t) = \begin{Bmatrix} 0 \\ \left[F_e(t) - f_{md}(\dot{x}(t))\right]/m \end{Bmatrix}$$

$$\mathbf{e}^{\mathbf{A}t} = e^{-\zeta_{nd}\omega_n t}\begin{bmatrix} \cos\omega_d t + \left(\dfrac{\zeta_{nd}}{\sqrt{1-\zeta_{nd}^2}}\right)\sin\omega_d t & \dfrac{\sin\omega_d t}{\omega_n\sqrt{1-\zeta_{nd}^2}} \\ -\dfrac{\omega_n\sin\omega_d t}{\sqrt{1-\zeta_{nd}^2}} & \cos\omega_d t - \left(\dfrac{\zeta_{nd}}{\sqrt{1-\zeta_{nd}^2}}\right)\sin\omega_d t \end{bmatrix}.$$

When the nonlinear manufactured damping term is treated as an external force, as was done in the previous section, the problem reduces to the problem discussed in Section 2.8. Let

$$t_{k+1} = t, \quad t_k = t_o, \quad \Delta t = t - t_o. \tag{9.194}$$

It follows from Eq. 9.193 that

$$\mathbf{z}_{k+1} = \mathbf{e}^{\mathbf{A}\Delta t}\mathbf{z}_k + \mathbf{e}^{\mathbf{A}t_{k+1}}\int_{t_k}^{t_{k+1}}\mathbf{e}^{-\mathbf{A}s}\,\mathbf{F}(s)\,ds. \tag{9.195}$$

Note that the velocity embedded inside the forcing function $\mathbf{F}(s)$ must be integrated from t_k to t_{k+1}. Both the Delta forcing function and the constant forcing function can be used to represent the continuous forcing function that is inside the integral in Eq. 9.195.

With the Delta forcing function, the forcing function is represented by delta functions, that is,

$$\mathbf{F}(s) = \mathbf{F}_k\delta(s-t_k)\Delta t = \begin{Bmatrix} 0 \\ \left[F_{ek} - f_{md}(\dot{x}_k)\right]/m \end{Bmatrix}\delta(s-t_k)\Delta t, \quad t_k \le s < t_{k+1}. \tag{9.196}$$

Substituting Eq. 9.196 into Eq. 9.195 gives

$$\begin{aligned}\mathbf{z}_{k+1} &= \mathbf{e}^{\mathbf{A}\Delta t}\mathbf{z}_k + \mathbf{e}^{\mathbf{A}t_{k+1}}\int_{t_k}^{t_{k+1}}\mathbf{e}^{-\mathbf{A}s}\,\mathbf{F}_k\delta(s-t_k)\Delta t\,ds \\ &= \mathbf{e}^{\mathbf{A}\Delta t}\mathbf{z}_k + \mathbf{e}^{\mathbf{A}t_{k+1}}\mathbf{e}^{-\mathbf{A}t_k}\mathbf{F}_k\Delta t \\ &= \mathbf{e}^{\mathbf{A}\Delta t}\mathbf{z}_k + \mathbf{e}^{\mathbf{A}\Delta t}\Delta t\,\mathbf{F}_k. \end{aligned} \tag{9.197}$$

With the constant forcing function, the forcing function is represented by a constant force whose value is equal to the force at the beginning of the interval, that is,

$$\mathbf{F}(s) = \mathbf{F}_k = \begin{Bmatrix} 0 \\ \left[F_{ek} - f_{md}(\dot{x}_k)\right]/m \end{Bmatrix}, \quad t_k \le s < t_{k+1}. \tag{9.198}$$

Substituting Eq. 9.198 into Eq. 9.195 gives

$$\mathbf{z}_{k+1} = e^{\mathbf{A}\Delta t}\mathbf{z}_k + e^{\mathbf{A}t_{k+1}}\int_{t_k}^{t_{k+1}} e^{-\mathbf{A}s}\,\mathbf{F}_k\,ds$$

$$= e^{\mathbf{A}\Delta t}\mathbf{z}_k + e^{\mathbf{A}t_{k+1}}\mathbf{A}^{-1}\left(e^{-\mathbf{A}t_k} - e^{-\mathbf{A}t_{k+1}}\right)\mathbf{F}_k \qquad (9.199)$$

$$= e^{\mathbf{A}\Delta t}\mathbf{z}_k + \mathbf{A}^{-1}\left(e^{\mathbf{A}\Delta t} - \mathbf{I}\right)\mathbf{F}_k.$$

Note that the velocity, $\dot{x}(t)$, within the interval from t_k to t_{k+1} is approximated using the Delta forcing function or the constant forcing function. Because $\dot{x}(t)$ is a continuous function and not a digitized function, error will be introduced due to the approximation. An iterative procedure can improve the accuracy of this method. However, as previously stated with the Newmark β-method, iterative calculations can be very time-consuming when small tolerances are required, and computing time can be as much as for the noniterative numerical analysis with a reduced time step size. Therefore, reducing the time step size is very desirable to increase the numerical accuracy without using the iteration method.

EXAMPLE 1 *SDOF System Response Using Delta Forcing Function*

Consider Example 2 in Section 9.5 with structural variables

$$m = 1.0 \text{ k-s}^2/\text{in.}, \quad \zeta = 0.05, \quad T_n = 1.0 \text{ s}$$

subjected to the modified El-Centro earthquake given in Figure 6.11. Assume $\Delta t = 0.02$ s. Also, assume a manufactured viscous damper is installed into this SDOF system with force versus velocity relationship given by

$$F_{md} = f_{md}(\dot{x}(t)) = 5.0\langle \dot{x}(t)\rangle^{0.5}. \qquad (9.200)$$

Calculating the response using the Delta forcing function without a manufactured viscous damper was discussed in Example 1 of Section 2.8. The state transition matrix and the forcing function, where $F_{ek} = -ma_k$, are

$$e^{\mathbf{A}\Delta t} = \begin{bmatrix} 0.992148 & 0.019823 \\ -0.782565 & 0.979693 \end{bmatrix}, \quad \mathbf{F}_k = \begin{Bmatrix} 0 \\ -a_k \end{Bmatrix} = \begin{Bmatrix} 0.0 \\ -1.0 \end{Bmatrix} a_k. \qquad (9.201)$$

Therefore, it follows from Eq. 9.197 that

$$\begin{Bmatrix} x_{k+1} \\ \dot{x}_{k+1} \end{Bmatrix} = \begin{bmatrix} 0.992148 & 0.019823 \\ -0.782565 & 0.979693 \end{bmatrix}\begin{Bmatrix} x_k \\ \dot{x}_k \end{Bmatrix} + \begin{Bmatrix} -0.000396 \\ -0.019594 \end{Bmatrix} a_k. \qquad (9.202)$$

Now consider the case with a manufactured viscous damper of the form given in Eq. 9.200. The forcing function in Eq. 9.201 becomes

$$\mathbf{F}_k = \begin{Bmatrix} 0 \\ [-ma_k - f_{md}(\dot{x})]/m \end{Bmatrix} = \begin{Bmatrix} 0.0 \\ -1.0 \end{Bmatrix}\left[a_k + 5.0\langle \dot{x}_k\rangle^{0.5}\right] \qquad (9.203)$$

and therefore, Eq. 9.202 becomes

$$\begin{Bmatrix} x_{k+1} \\ \dot{x}_{k+1} \end{Bmatrix} = \begin{bmatrix} 0.992148 & 0.019823 \\ -0.782565 & 0.979693 \end{bmatrix}\begin{Bmatrix} x_k \\ \dot{x}_k \end{Bmatrix} + \begin{Bmatrix} -0.000396 \\ -0.019594 \end{Bmatrix}\left[a_k + 5.0\langle \dot{x}_k\rangle^{0.5}\right]. \qquad (9.204)$$

Assume that the initial conditions are zero (i.e., $x_o = \dot{x}_o = 0$). Since the earthquake ground acceleration at time zero is $a_o = -0.00257g = -0.99305$ in./s^2, the response at the first time step (i.e., $t = \Delta t = 0.02$ s) is

$$\begin{Bmatrix} x_1 \\ \dot{x}_1 \end{Bmatrix} = \begin{bmatrix} 0.992148 & 0.019823 \\ -0.782565 & 0.979693 \end{bmatrix}\begin{Bmatrix} 0 \\ 0 \end{Bmatrix} + \begin{Bmatrix} -0.000396 \\ -0.019594 \end{Bmatrix}(-0.99305 + 0.0) = \begin{Bmatrix} 0.0004 \\ 0.0195 \end{Bmatrix}. \qquad (9.205)$$

The earthquake ground acceleration at the first time step is $a_1 = -0.0198g = -7.6507$ in./s². The damping force per unit mass at this time step is

$$\frac{f_{md}(\dot{x}_1)}{m} = \frac{5.0\langle 0.0195 \rangle^{0.5}}{1.0} = 0.6975 \tag{9.206}$$

and, therefore, the response at the second time step (i.e., $t = 2\Delta t = 0.04$ s) is

$$\begin{Bmatrix} x_2 \\ \dot{x}_2 \end{Bmatrix} = \begin{bmatrix} 0.992148 & 0.019823 \\ -0.782565 & 0.979693 \end{bmatrix} \begin{Bmatrix} 0.0004 \\ 0.0195 \end{Bmatrix} + \begin{Bmatrix} 0.000396 \\ 0.019594 \end{Bmatrix} (-7.6507 + 0.6975) = \begin{Bmatrix} 0.0035 \\ 0.1550 \end{Bmatrix}. \tag{9.207}$$

The earthquake ground acceleration at the second time step is $a_2 = -0.0185g = -7.1484$ in./s². The damping force per unit mass at this time step is

$$\frac{f_{md}(\dot{x}_2)}{m} = \frac{5.0\langle 0.1550 \rangle^{0.5}}{1.0} = 1.9685 \tag{9.208}$$

and, therefore, the response at the third time step (i.e., $t = 3\Delta t = 0.06$ s) is

$$\begin{Bmatrix} x_3 \\ \dot{x}_3 \end{Bmatrix} = \begin{bmatrix} 0.992148 & 0.019823 \\ -0.782565 & 0.979693 \end{bmatrix} \begin{Bmatrix} 0.0035 \\ 0.1550 \end{Bmatrix} + \begin{Bmatrix} 0.000396 \\ 0.019594 \end{Bmatrix} (-7.1484 + 1.9685) = \begin{Bmatrix} 0.0086 \\ 0.2506 \end{Bmatrix}. \tag{9.209}$$

This calculation process continues, and the displacement and velocity response are shown in Figures 9.28 and 9.29, respectively.

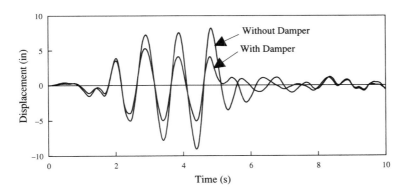

Figure 9.28 Displacement Response with and without Manufactured Viscous Dampers Subjected to Modified El-Centro Earthquake

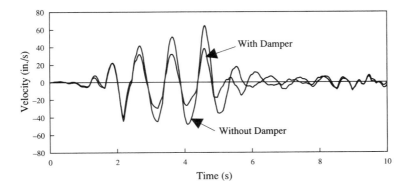

Figure 9.29 Velocity Response with and without Manufactured Viscous Dampers Subjected to Modified El-Centro Earthquake

EXAMPLE 2 ***SDOF System Response Using Constant Forcing Function***

Consider the same SDOF system as discussed in Example 1, but now a manufactured viscous damper is installed into this SDOF system with a force versus velocity relationship given by

$$f_{md}(\dot{x}(t)) = 25\langle\dot{x}(t)\rangle^{0.5}. \tag{9.210}$$

Calculating the response using the constant forcing function without a manufactured viscous damper was discussed in Example 2 of Section 2.8. The matrices as given in Eq. 9.199 are

$$\mathbf{e}^{\mathbf{A}\Delta t} = \begin{bmatrix} 0.992148 & 0.019823 \\ -0.782565 & 0.979693 \end{bmatrix}, \quad \mathbf{A}^{-1}(\mathbf{e}^{\mathbf{A}\Delta t} - \mathbf{I}) = \begin{bmatrix} 0.019948 & 0.000199 \\ -0.00785 & 0.019823 \end{bmatrix}. \tag{9.211}$$

For the forcing function $F_{ek} = -ma_k$, $\mathbf{F}_k$ in Eq. 9.199 is

$$\mathbf{F}_k = \begin{Bmatrix} 0 \\ -a_k \end{Bmatrix} = \begin{Bmatrix} 0.0 \\ -1.0 \end{Bmatrix} a_k. \tag{9.212}$$

Therefore, it follows from Eq. 9.199 that

$$\begin{Bmatrix} x_{k+1} \\ \dot{x}_{k+1} \end{Bmatrix} = \begin{bmatrix} 0.992148 & 0.019823 \\ -0.782565 & 0.979693 \end{bmatrix} \begin{Bmatrix} x_k \\ \dot{x}_k \end{Bmatrix} + \begin{Bmatrix} -0.000199 \\ -0.019823 \end{Bmatrix} a_k. \tag{9.213}$$

Now consider the case with a manufactured viscous damper of the form given in Eq. 9.210. The forcing function in Eq. 9.212 becomes

$$\mathbf{F}_k = \begin{Bmatrix} 0 \\ [-ma_k - f_{md}(\dot{x})]/m \end{Bmatrix} = \begin{Bmatrix} 0.0 \\ -1.0 \end{Bmatrix} \left[a_k + 25\langle\dot{x}_k\rangle^{0.5} \right]. \tag{9.214}$$

Therefore, Eq. 9.213 becomes

$$\begin{Bmatrix} x_{k+1} \\ \dot{x}_{k+1} \end{Bmatrix} = \begin{bmatrix} 0.992148 & 0.019823 \\ -0.782565 & 0.979693 \end{bmatrix} \begin{Bmatrix} x_k \\ \dot{x}_k \end{Bmatrix} + \begin{Bmatrix} -0.000199 \\ -0.019823 \end{Bmatrix} \left[a_k + 25\langle\dot{x}_k\rangle^{0.5} \right]. \tag{9.215}$$

Assume that the initial conditions are zero (i.e., $x_o = \dot{x}_o = 0$). Since the earthquake ground acceleration at time zero is $a_o = -0.00257g = -0.99305$ in./s^2, the response at the first time step (i.e., $t = \Delta t = 0.02$ s) is

$$\begin{Bmatrix} x_1 \\ \dot{x}_1 \end{Bmatrix} = \begin{bmatrix} 0.992148 & 0.019823 \\ -0.782565 & 0.979693 \end{bmatrix} \begin{Bmatrix} 0 \\ 0 \end{Bmatrix} + \begin{Bmatrix} -0.000199 \\ -0.019823 \end{Bmatrix} (-0.99305 + 0.0) = \begin{Bmatrix} 0.0002 \\ 0.0197 \end{Bmatrix}. \tag{9.216}$$

The earthquake ground acceleration at the first time step is $a_1 = -0.0198g = -7.6507$ in./s^2. The damping force per unit mass at this time step is

$$\frac{f_{md}(\dot{x}_1)}{m} = \frac{25\langle 0.0197\rangle^{0.5}}{1.0} = 3.5076 \tag{9.217}$$

and, therefore, the response at the second time step (i.e., $t = 2\Delta t = 0.04$ s) is

$$\begin{Bmatrix} x_2 \\ \dot{x}_2 \end{Bmatrix} = \begin{bmatrix} 0.992148 & 0.019823 \\ -0.782565 & 0.979693 \end{bmatrix} \begin{Bmatrix} 0.0002 \\ 0.0197 \end{Bmatrix} + \begin{Bmatrix} 0.000396 \\ 0.019594 \end{Bmatrix} (-7.6507 + 3.5076) = \begin{Bmatrix} 0.0014 \\ 0.1013 \end{Bmatrix}. \tag{9.218}$$

The earthquake ground acceleration at the second time step is $a_2 = -0.0185g = -7.1484$ in./s^2. The damping force per unit mass at this time step is

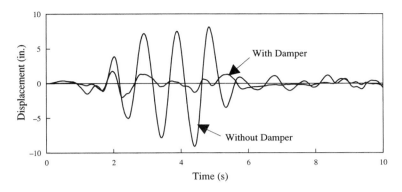

Figure 9.30 Displacement Response with and without Manufactured Viscous Dampers Subjected to Modified El-Centro Earthquake Using Constant Forcing Function

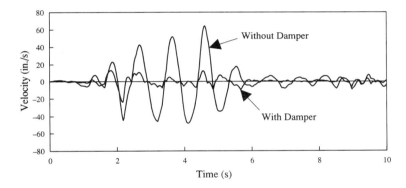

Figure 9.31 Velocity Response with and without Manufactured Viscous Dampers Subjected to Modified El-Centro Earthquake Using Constant Forcing Function

$$\frac{f_{md}(\dot{x}_2)}{m} = \frac{25\langle 0.101\rangle^{0.5}}{1.0} = 7.9553 \tag{9.219}$$

and, therefore, the response at the third time step (i.e., $t = 3\Delta t = 0.06$ s) is

$$\begin{Bmatrix} x_3 \\ \dot{x}_3 \end{Bmatrix} = \begin{bmatrix} 0.992148 & 0.019823 \\ -0.782565 & 0.979693 \end{bmatrix} \begin{Bmatrix} 0.0014 \\ 0.1013 \end{Bmatrix} + \begin{Bmatrix} 0.000396 \\ 0.019594 \end{Bmatrix} (-7.1484 + 7.9553) = \begin{Bmatrix} 0.0032 \\ 0.0821 \end{Bmatrix}. \tag{9.220}$$

This calculation process continues, and the displacement and velocity responses are shown in Figures 9.30 and 9.31, respectively.

9.7 FORCED RESPONSE OF AN MDOF SYSTEM WITH MANUFACTURED VISCOUS DAMPERS

In an MDOF system, additional damping can be imposed on each floor of the structure. A manufactured damper connecting the upper story on one end and the lower story on the other end can induce damping force when relative velocity exists between these two ends of the damper. The damping force applied on the upper story will be equal in magnitude and opposite in direction to that applied on the lower story. To simplify the explanation, consider a 2DOF system with a manufactured damper connecting the first story and the ground to provide damping force between the ground and the first story. Similarly, another

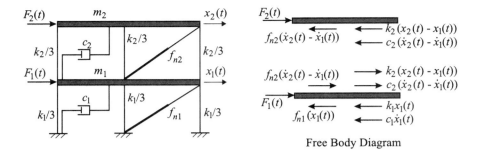

Free Body Diagram

Figure 9.32 Two Degrees of Freedom System with Manufactured Dampers

manufactured damper connecting the first and second stories provides damping force between these two stories. These forces can be described by

$$F_{n1} = -f_{n1}(\dot{x}_1), \quad F_{n2} = -f_{n2}(\dot{x}_2 - \dot{x}_1) \tag{9.221}$$

where F_{n1} is the force produced by the dampers between the first story and the ground, and F_{n2} is the force produced by the dampers between the first and second stories. These forces are functions of velocities of the stories, and $f_{n1}(\cdot)$ and $f_{n2}(\cdot)$ are some positive nonlinear functions of velocities depending on the properties of the damping devices. It follows from Eq. 3.1 that the dynamic equilibrium equation becomes

$$-c_1\dot{x}_1(t) - c_2(\dot{x}_1(t) - \dot{x}_2(t)) - k_1x_1(t) - k_2(x_1(t) - x_2(t)) + F_1(t) + F_{n1} - F_{n2} = m_1\ddot{x}_1(t) \tag{9.222a}$$

$$-c_2(\dot{x}_2(t) - \dot{x}_1(t)) - k_2(x_2(t) - x_1(t)) + F_2(t) + F_{n2} = m_2\ddot{x}_2(t). \tag{9.222b}$$

Equation 9.222 can be written in matrix form, which is

$$\begin{bmatrix} m_1 & 0 \\ 0 & m_2 \end{bmatrix}\begin{Bmatrix} \ddot{x}_1 \\ \ddot{x}_2 \end{Bmatrix} + \begin{bmatrix} c_1 + c_2 & -c_2 \\ -c_2 & c_2 \end{bmatrix}\begin{Bmatrix} \dot{x}_1 \\ \dot{x}_2 \end{Bmatrix} + \begin{bmatrix} k_1 + k_2 & -k_2 \\ -k_2 & k_2 \end{bmatrix}\begin{Bmatrix} x_1 \\ x_2 \end{Bmatrix} = \begin{Bmatrix} F_1(t) \\ F_2(t) \end{Bmatrix} + \begin{bmatrix} 1 & -1 \\ 0 & 1 \end{bmatrix}\begin{Bmatrix} F_{n1} \\ F_{n2} \end{Bmatrix}. \tag{9.223}$$

Substituting Eq. 9.221 into Eq. 9.223 gives

$$\begin{bmatrix} m_1 & 0 \\ 0 & m_2 \end{bmatrix}\begin{Bmatrix} \ddot{x}_1 \\ \ddot{x}_2 \end{Bmatrix} + \begin{bmatrix} c_1 + c_2 & -c_2 \\ -c_2 & c_2 \end{bmatrix}\begin{Bmatrix} \dot{x}_1 \\ \dot{x}_2 \end{Bmatrix} + \begin{bmatrix} k_1 + k_2 & -k_2 \\ -k_2 & k_2 \end{bmatrix}\begin{Bmatrix} x_1 \\ x_2 \end{Bmatrix}$$
$$= \begin{Bmatrix} F_1(t) \\ F_2(t) \end{Bmatrix} - \begin{bmatrix} 1 & -1 \\ 0 & 1 \end{bmatrix}\begin{Bmatrix} f_{n1}(\dot{x}_1) \\ f_{n2}(\dot{x}_2 - \dot{x}_1) \end{Bmatrix}. \tag{9.224}$$

To simplify Eq. 9.224, let

$$\mathbf{M} = \begin{bmatrix} m_1 & 0 \\ 0 & m_2 \end{bmatrix}, \quad \mathbf{C} = \begin{bmatrix} c_1 + c_2 & -c_2 \\ -c_2 & c_2 \end{bmatrix}, \quad \mathbf{K} = \begin{bmatrix} k_1 + k_2 & -k_2 \\ -k_2 & k_2 \end{bmatrix}, \quad \mathbf{I}_{md} = \begin{bmatrix} 1 & -1 \\ 0 & 1 \end{bmatrix}$$

$$\mathbf{X}(t) = \begin{Bmatrix} x_1(t) \\ x_2(t) \end{Bmatrix}, \quad \mathbf{F}_e(t) = \begin{Bmatrix} F_1(t) \\ F_2(t) \end{Bmatrix}, \quad \mathbf{F}_n(\dot{\mathbf{X}}) = \begin{Bmatrix} f_{n1}(\dot{x}_1) \\ f_{n2}(\dot{x}_2 - \dot{x}_1) \end{Bmatrix}.$$

It follows from Eq. 9.224 that

$$\mathbf{M}\ddot{\mathbf{X}}(t) + \mathbf{C}\dot{\mathbf{X}}(t) + \mathbf{K}\mathbf{X}(t) = \mathbf{F}_e(t) - \mathbf{I}_{md}\mathbf{F}_n(\dot{\mathbf{X}}(t)). \tag{9.225}$$

Equation 9.225 generally represents the structural dynamic response incorporating nonlinear damping devices. The derivation is applicable to an MDOF system, although it is based on the equation of motion for a 2DOF system. The matrix $\mathbf{C}$ relates to the natural damping in the structure that corresponds to the structure without manufactured dampers, and the matrix $\mathbf{I}_{md}$ is associated with the manufactured dampers. This equation is nonlinear with an unknown function $\mathbf{F}_n(\cdot)$, and thus no analytical solution is available. Therefore, a numerical method is the only approach to solve the viscous damping problems.

One numerical discretization method as discussed in Section 9.5 is to use the Newmark β-method of *constant acceleration* to define the manufactured damping force. Rewriting Eq. 9.225 in discretized form gives

$$\mathbf{M}\ddot{\mathbf{X}}_{k+1} + \mathbf{C}\dot{\mathbf{X}}_{k+1} + \mathbf{K}\mathbf{X}_{k+1} = \mathbf{F}_{k+1} - \mathbf{I}_{md}\mathbf{F}_n\left(\dot{\mathbf{X}}_{k+1}\right). \tag{9.226}$$

Using the constant acceleration method, from Eq. 3.224 for the velocity with $\delta = 0$ and $\alpha = 0$, it follows that

$$\dot{\mathbf{X}}_{k+1} = \dot{\mathbf{X}}_k + \ddot{\mathbf{X}}_k \Delta t \tag{9.227a}$$

$$\mathbf{X}_{k+1} = \mathbf{X}_k + \dot{\mathbf{X}}_k \Delta t + \tfrac{1}{2}\ddot{\mathbf{X}}_k (\Delta t)^2. \tag{9.227b}$$

The nonlinear damping forcing function at time step $k + 1$ can now be determined using displacement and velocity vectors computed in Eq. 9.227, and the only unknown at time step $k + 1$ is the acceleration vector. It follows from Eq. 9.226 that

$$\ddot{\mathbf{X}}_{k+1} = -\mathbf{M}^{-1}\mathbf{C}\dot{\mathbf{X}}_{k+1} - \mathbf{M}^{-1}\mathbf{K}\mathbf{X}_{k+1} + \mathbf{M}^{-1}\mathbf{F}_{k+1} - \mathbf{M}^{-1}\mathbf{I}_{md}\mathbf{F}_n\left(\dot{\mathbf{X}}_{k+1}\right). \tag{9.228}$$

Another good numerical discretization method noted in Section 9.5 is to use the Newmark β-method with the combination of constant acceleration and constant average acceleration. Rewriting Eq. 9.225 in discretized form,

$$\mathbf{M}\ddot{\mathbf{X}}_{k+1} + \mathbf{C}\dot{\mathbf{X}}_{k+1} + \mathbf{K}\mathbf{X}_{k+1} = \mathbf{F}_{k+1} - \mathbf{I}_{md}\mathbf{F}_n\left(\dot{\mathbf{X}}_{k+1}\right). \tag{9.229}$$

The first objective is to determine the viscous damping force, $\mathbf{F}_n(\dot{\mathbf{X}}_{k+1})$, based on the information at time step k (i.e., $\mathbf{X}_k$, $\dot{\mathbf{X}}_k$, and $\ddot{\mathbf{X}}_k$). Using the constant acceleration method, from Eq. 9.227a,

$$\dot{\mathbf{X}}_{k+1} = \dot{\mathbf{X}}_k + \ddot{\mathbf{X}}_k \Delta t \tag{9.230}$$

and therefore the nonlinear forcing function becomes

$$\mathbf{F}_n(\dot{\mathbf{X}}_{k+1}) = \mathbf{F}_n(\dot{\mathbf{X}}_k + \ddot{\mathbf{X}}_k \Delta t). \tag{9.231}$$

It follows from Eq. 9.229 that

$$\mathbf{M}\ddot{\mathbf{X}}_{k+1} + \mathbf{C}\dot{\mathbf{X}}_{k+1} + \mathbf{K}\mathbf{X}_{k+1} = \mathbf{F}_{k+1} - \mathbf{I}_{md}\mathbf{F}_n\left(\dot{\mathbf{X}}_k + \ddot{\mathbf{X}}_k \Delta t\right). \tag{9.232}$$

Note that by using the constant acceleration method, the terms on the right-hand side of Eq. 9.232 are known, since $\dot{\mathbf{X}}_k$ and $\ddot{\mathbf{X}}_k$ are determined in time step k, and $\mathbf{F}_{k+1}$ is the predefined forcing function at time step $k + 1$. Define these known terms on the right-hand side by

$$\overline{\mathbf{F}}_{k+1} = \mathbf{F}_{k+1} - \mathbf{I}_{md}\mathbf{F}_n\left(\dot{\mathbf{X}}_k + \ddot{\mathbf{X}}_k \Delta t\right). \tag{9.233}$$

It follows from the Newmark β-method of constant average acceleration that the three simultaneous equations become

$$\mathbf{M}\ddot{\mathbf{X}}_{k+1} + \mathbf{C}\dot{\mathbf{X}}_{k+1} + \mathbf{K}\mathbf{X}_{k+1} = \overline{\mathbf{F}}_{k+1} \tag{9.234a}$$

$$\dot{\mathbf{X}}_{k+1} = \dot{\mathbf{X}}_k + \tfrac{1}{2}\ddot{\mathbf{X}}_k \Delta t + \tfrac{1}{2}\ddot{\mathbf{X}}_{k+1} \Delta t \tag{9.234b}$$

$$\mathbf{X}_{k+1} = \mathbf{X}_k + \dot{\mathbf{X}}_k \Delta t + \tfrac{1}{4}\ddot{\mathbf{X}}_k \Delta t^2 + \tfrac{1}{4}\ddot{\mathbf{X}}_{k+1}(\Delta t)^2. \tag{9.234c}$$

Using the numerical procedure discussed in Chapter 3, with $\delta = \tfrac{1}{2}$ and $\alpha = \tfrac{1}{4}$, the solution becomes

$$\mathbf{X}_{k+1} = \left(\mathbf{I} - \tfrac{1}{4}\mathbf{B}^{-1}\mathbf{K}(\Delta t)^2\right)\mathbf{X}_k + \left(\mathbf{I}\Delta t - \tfrac{1}{4}\mathbf{B}^{-1}\mathbf{C}(\Delta t)^2 - \tfrac{1}{4}\mathbf{B}^{-1}\mathbf{K}(\Delta t)^3\right)\dot{\mathbf{X}}_k$$
$$+\left(\tfrac{1}{4}\mathbf{I}(\Delta t)^2 - \tfrac{1}{8}\mathbf{B}^{-1}\mathbf{C}(\Delta t)^3 - \tfrac{1}{16}\mathbf{B}^{-1}\mathbf{K}(\Delta t)^4\right)\ddot{\mathbf{X}}_k + \left(\tfrac{1}{4}\mathbf{B}^{-1}(\Delta t)^2\right)\overline{\mathbf{F}}_{k+1} \tag{9.235a}$$

$$\dot{\mathbf{X}}_{k+1} = -\left(\tfrac{1}{2}\mathbf{B}^{-1}\mathbf{K}\Delta t\right)\mathbf{X}_k + \left(\mathbf{I} - \tfrac{1}{2}\mathbf{B}^{-1}\mathbf{C}\Delta t - \tfrac{1}{2}\mathbf{B}^{-1}\mathbf{K}(\Delta t)^2\right)\dot{\mathbf{X}}_k$$
$$+\left(\tfrac{1}{2}\mathbf{I}\Delta t - \tfrac{1}{4}\mathbf{B}^{-1}\mathbf{C}(\Delta t)^2 - \tfrac{1}{8}\mathbf{B}^{-1}\mathbf{K}(\Delta t)^3\right)\ddot{\mathbf{X}}_k + \left(\tfrac{1}{2}\mathbf{B}^{-1}\Delta t\right)\overline{\mathbf{F}}_{k+1} \tag{9.235b}$$

$$\ddot{\mathbf{X}}_{k+1} = -\left(\mathbf{B}^{-1}\mathbf{K}\right)\mathbf{X}_k - \left(\mathbf{B}^{-1}\mathbf{C} + \mathbf{B}^{-1}\mathbf{K}\Delta t\right)\dot{\mathbf{X}}_k - \left(\tfrac{1}{2}\mathbf{B}^{-1}\mathbf{C}\Delta t + \tfrac{1}{4}\mathbf{B}^{-1}\mathbf{K}(\Delta t)^2\right)\ddot{\mathbf{X}}_k + \left(\mathbf{B}^{-1}\right)\mathbf{F}_{k+1}. \tag{9.235c}$$

where

$$\mathbf{B} = \mathbf{M} + \tfrac{1}{2}\mathbf{C}\Delta t + \tfrac{1}{4}\mathbf{K}(\Delta t)^2. \tag{9.236}$$

Again, following the discussion on the nonlinear damping response for the SDOF system, using an iteration method can improve the accuracy of the solution. However, because iteration is very time consuming, it is not recommended. Instead, reducing the time step size can produce a numerical solution with a higher degree of accuracy.

As discussed in Section 9.5, one major drawback of using this combination of constant acceleration method and constant average acceleration method is that the equilibrium condition is not satisfied at time step $k + 1$. One way to avoid this problem is to satisfy equilibrium by correcting the acceleration vector at the end of the calculation in each time step:

$$\mathbf{X}_{k+1} = \left(\mathbf{I} - \tfrac{1}{4}\mathbf{B}^{-1}\mathbf{K}(\Delta t)^2\right)\mathbf{X}_k + \left(\mathbf{I}\Delta t - \tfrac{1}{4}\mathbf{B}^{-1}\mathbf{C}(\Delta t)^2 - \tfrac{1}{4}\mathbf{B}^{-1}\mathbf{K}(\Delta t)^3\right)\dot{\mathbf{X}}_k$$
$$+\left(\tfrac{1}{4}\mathbf{I}(\Delta t)^2 - \tfrac{1}{8}\mathbf{B}^{-1}\mathbf{C}(\Delta t)^3 - \tfrac{1}{16}\mathbf{B}^{-1}\mathbf{K}(\Delta t)^4\right)\ddot{\mathbf{X}}_k + \left(\tfrac{1}{4}\mathbf{B}^{-1}(\Delta t)^2\right)\overline{\mathbf{F}}_{k+1} \tag{9.237a}$$

$$\dot{\mathbf{X}}_{k+1} = -\left(\tfrac{1}{2}\mathbf{B}^{-1}\mathbf{K}\Delta t\right)\mathbf{X}_k + \left(\mathbf{I} - \tfrac{1}{2}\mathbf{B}^{-1}\mathbf{C}\Delta t - \tfrac{1}{2}\mathbf{B}^{-1}\mathbf{K}(\Delta t)^2\right)\dot{\mathbf{X}}_k$$
$$+\left(\tfrac{1}{2}\mathbf{I}\Delta t - \tfrac{1}{4}\mathbf{B}^{-1}\mathbf{C}(\Delta t)^2 - \tfrac{1}{8}\mathbf{B}^{-1}\mathbf{K}(\Delta t)^3\right)\ddot{\mathbf{X}}_k + \left(\tfrac{1}{2}\mathbf{B}^{-1}\Delta t\right)\overline{\mathbf{F}}_{k+1} \tag{9.237b}$$

$$\ddot{\mathbf{X}}_{k+1} = -\mathbf{M}^{-1}\mathbf{C}\dot{\mathbf{X}}_{k+1} - \mathbf{M}^{-1}\mathbf{K}\mathbf{X}_{k+1} + \mathbf{M}^{-1}\mathbf{F}_{k+1} - \mathbf{M}^{-1}\mathbf{F}_n\left(\dot{\mathbf{X}}_{k+1}\right). \tag{9.237c}$$

Another numerical method of calculating the dynamic response of Eq. 9.223 is to use the integration method, as discussed in Section 3.8. Start with Eq. 3.264, rewritten here as

$$\mathbf{z}(t) = \mathbf{e}^{\mathbf{A}(t-t_o)}\mathbf{z}(t_o) + \mathbf{e}^{\mathbf{A}t}\int_{t_o}^{t}\mathbf{e}^{-\mathbf{A}s}\,\mathbf{F}(s)\,ds \tag{9.238}$$

where

$$\mathbf{z}(t) = \begin{Bmatrix}\mathbf{X}(t)\\ \dot{\mathbf{X}}(t)\end{Bmatrix}, \quad \mathbf{A} = \begin{bmatrix}\mathbf{0} & \mathbf{I}\\ -\mathbf{M}^{-1}\mathbf{K} & -\mathbf{M}^{-1}\mathbf{C}\end{bmatrix}, \quad \mathbf{F}(t) = \begin{Bmatrix}\mathbf{0}\\ \mathbf{M}^{-1}\left(\mathbf{F}_e(t) - \mathbf{I}_{md}\mathbf{F}_n(\dot{\mathbf{X}})\right)\end{Bmatrix}.$$

Let

$$t_{k+1} = t, \quad t_k = t_o, \quad \Delta t = t - t_o. \tag{9.239}$$

It follows from Eq. 9.238 that

$$\mathbf{z}_{k+1} = \mathbf{e}^{\mathbf{A}\Delta t}\mathbf{z}_k + \mathbf{e}^{\mathbf{A}t_{k+1}}\int_{t_k}^{t_{k+1}}\mathbf{e}^{-\mathbf{A}s}\,\mathbf{F}(s)\,ds. \tag{9.240}$$

Both the Delta forcing function and the constant forcing function can be used to discretize the continuous forcing function inside the integral of Eq. 9.240.

The Delta forcing function is represented by delta functions, that is,

$$\mathbf{F}(s) = \mathbf{F}_k\delta(s - t_k)\Delta t = \begin{Bmatrix}\mathbf{0}\\ \mathbf{M}^{-1}\left(\mathbf{F}_e(t) - \mathbf{I}_{md}\mathbf{F}_n(\dot{\mathbf{X}})\right)\end{Bmatrix}\delta(s - t_k)\Delta t, \quad t_k \le s < t_{k+1}. \tag{9.241}$$

Substituting Eq. 9.241 into Eq. 9.240 gives

$$\mathbf{z}_{k+1} = e^{\mathbf{A}\Delta t}\mathbf{z}_k + e^{\mathbf{A}t_{k+1}} \int_{t_k}^{t_{k+1}} e^{-\mathbf{A}s}\mathbf{F}_k \delta(s - t_k)\Delta t\, ds$$

$$= e^{\mathbf{A}\Delta t}\mathbf{z}_k + e^{\mathbf{A}t_{k+1}} e^{-\mathbf{A}t_k}\mathbf{F}_k \Delta t \tag{9.242}$$

$$= e^{\mathbf{A}\Delta t}\mathbf{z}_k + e^{\mathbf{A}\Delta t}\Delta t\, \mathbf{F}_k.$$

The constant forcing function is represented by a constant function within the time interval, and the forcing value is equal to the force at the beginning of the interval, that is,

$$\mathbf{F}(s) = \mathbf{F}_k = \left\{ \begin{matrix} \mathbf{0} \\ \mathbf{M}^{-1}\left(\mathbf{F}_{ek} - \mathbf{I}_{md}\mathbf{F}_n\left(\dot{\mathbf{X}}_k\right)\right) \end{matrix} \right\}, \quad t_k \le s < t_{k+1}. \tag{9.243}$$

Substituting Eq. 9.243 into Eq. 9.246 gives

$$\mathbf{z}_{k+1} = e^{\mathbf{A}\Delta t}\mathbf{z}_k + e^{\mathbf{A}t_{k+1}} \int_{t_k}^{t_{k+1}} e^{-\mathbf{A}s}\, \mathbf{F}_k\, ds$$

$$= e^{\mathbf{A}\Delta t}\mathbf{z}_k + e^{\mathbf{A}t_{k+1}}\mathbf{A}^{-1}\left(e^{-\mathbf{A}t_k} - e^{-\mathbf{A}t_{k+1}}\right)\mathbf{F}_k \tag{9.244}$$

$$= e^{\mathbf{A}\Delta t}\mathbf{z}_k + \mathbf{A}^{-1}\left(e^{\mathbf{A}\Delta t} - \mathbf{I}\right)\mathbf{F}_k.$$

EXAMPLE 1   *Manufactured Dampers with Linear Force versus Velocity Relationship*

Consider a 2DOF system with two manufactured dampers that have a linear force versus velocity relationship:

$$F_{n1} = c_{md1}\dot{x}_1(t) \tag{9.245}$$

$$F_{n2} = c_{md2}\left(\dot{x}_2(t) - \dot{x}_1(t)\right). \tag{9.246}$$

It follows from Eq. 9.225 that

$$\mathbf{M}\ddot{\mathbf{X}}(t) + \mathbf{C}\dot{\mathbf{X}}(t) + \mathbf{K}\mathbf{X}(t) = \mathbf{F}_e(t) - \mathbf{I}_{md}\mathbf{F}_n\left(\dot{\mathbf{X}}(t)\right) \tag{9.247}$$

where

$$\mathbf{I}_{md} = \begin{bmatrix} 1 & -1 \\ 0 & 1 \end{bmatrix}, \quad \mathbf{F}_n(\dot{\mathbf{X}}) = \left\{ \begin{matrix} c_{md1}\dot{x}_1(t) \\ c_{md2}\left(\dot{x}_2(t) - \dot{x}_1(t)\right) \end{matrix} \right\} = \begin{bmatrix} c_{md1} & 0 \\ -c_{md2} & c_{md2} \end{bmatrix}\left\{ \begin{matrix} \dot{x}_1(t) \\ \dot{x}_2(t) \end{matrix} \right\}. \tag{9.248}$$

Define the $\mathbf{C}_{md}$ matrix to be

$$\mathbf{C}_{md}\dot{\mathbf{X}}(t) = \mathbf{I}_{md}\mathbf{F}_n\left(\dot{\mathbf{X}}(t)\right). \tag{9.249}$$

Substituting Eq. 9.248 into Eq. 9.249 gives

$$\mathbf{C}_{md}\left\{ \begin{matrix} \dot{x}_1(t) \\ \dot{x}_2(t) \end{matrix} \right\} = \begin{bmatrix} 1 & -1 \\ 0 & 1 \end{bmatrix} \begin{bmatrix} c_{md1} & 0 \\ -c_{md2} & c_{md2} \end{bmatrix}\left\{ \begin{matrix} \dot{x}_1(t) \\ \dot{x}_2(t) \end{matrix} \right\}. \tag{9.250}$$

For any velocity vector $\dot{\mathbf{X}}(t)$, it follows from Eq. 9.250 that

$$\mathbf{C}_{md} = \begin{bmatrix} 1 & -1 \\ 0 & 1 \end{bmatrix} \begin{bmatrix} c_{md1} & 0 \\ -c_{md2} & c_{md2} \end{bmatrix} = \begin{bmatrix} c_{md1} + c_{md2} & -c_{md2} \\ -c_{md2} & c_{md2} \end{bmatrix}. \tag{9.251}$$

Recall from Eq. 9.95 that $\mathbf{C}_{md}$ is the manufactured damping matrix, which is a collection of c_{md1}, c_{md2}, c_{md3}, and so forth. The $\mathbf{C}_{md}$ matrix in Eq. 9.251 represents the manufactured damping matrix for a 2DOF system. When the manufactured dampers have a linear force versus velocity relationship, then the $\mathbf{C}_{md}$ matrix is very similar to the natural damping matrix $\mathbf{C}$ (i.e., replacing c_1 by c_{md1} and c_2 by c_{md2}).

EXAMPLE 2 *Response of a 2DOF System with manufactured viscous dampers*

Consider the 2DOF system presented in Example 1 of Section 4.11 with zero natural damping. The dynamic matrices are given in Eq. 4.491, that is,

$$\mathbf{M} = \begin{bmatrix} 2m & 0 \\ 0 & m \end{bmatrix}, \quad \mathbf{C} = \begin{bmatrix} 0 & 0 \\ 0 & 0 \end{bmatrix}, \quad \mathbf{K} = \begin{bmatrix} 4k & -2k \\ -2k & 3k \end{bmatrix}. \tag{9.252}$$

The state transition matrix is calculated in that example, which is

$$\mathbf{e}^{\mathbf{A}t} = \begin{bmatrix} \frac{2}{3}(c1) + \frac{1}{3}(c2) & \frac{1}{3}(c1) - \frac{1}{3}(c2) & \frac{2}{3\omega_n}(s1) + \frac{1}{6\omega_n}(s2) & \frac{1}{3\omega_n}(s1) - \frac{1}{6\omega_n}(s2) \\ \frac{2}{3}(c1) - \frac{2}{3}(c2) & \frac{1}{3}(c1) + \frac{2}{3}(c2) & \frac{2}{3\omega_n}(s1) - \frac{1}{3\omega_n}(s2) & \frac{1}{3\omega_n}(s1) + \frac{1}{3\omega_n}(s2) \\ -\frac{2\omega_n}{3}(s1) - \frac{2\omega_n}{3}(s2) & -\frac{\omega_n}{3}(s1) + \frac{2\omega_n}{3}(s2) & \frac{2}{3}(c1) + \frac{1}{3}(c2) & \frac{1}{3}(c1) - \frac{1}{3}(c2) \\ -\frac{2\omega_n}{3}(s1) + \frac{4\omega_n}{3}(s2) & -\frac{\omega_n}{3}(s1) - \frac{4\omega_n}{3}(s2) & \frac{2}{3}(c1) - \frac{2}{3}(c2) & \frac{1}{3}(c1) + \frac{2}{3}(c2) \end{bmatrix} \tag{9.253}$$

where $c1 = \cos \omega_n t$, $c2 = \cos 2\omega_n t$, $s1 = \sin \omega_n t$, $s2 = \sin 2\omega_n t$, and $\omega_n^2 = k/m$. Let $m = 1.0$ k-s²/in. and $k = 100.0$ k/in. Assume that a time step size of $\Delta t = 0.02$ s used; it follows from Eq. 9.253 that $\omega_n = \sqrt{100.0/1.0} = 10$ rad/s and

$$\mathbf{e}^{\mathbf{A}\Delta t} = \begin{bmatrix} 0.960398 & 0.019669 & 0.019735 & 0.000132 \\ 0.039337 & 0.940730 & 0.000264 & 0.019603 \\ -3.920584 & 1.933891 & 0.960398 & 0.019669 \\ 3.867782 & -5.854476 & 0.039337 & 0.940730 \end{bmatrix}. \tag{9.254}$$

Now assume two manufactured viscous dampers are installed in the structure. The damper between the ground and the first mass is denoted as Damper 1 with the force as $F_{n1}(\dot{\mathbf{x}})$, and the damper between the first and second masses is denoted as Damper 2 with the force as $F_{n2}(\dot{\mathbf{x}})$, where

$$F_{n1}(\dot{\mathbf{x}}) = 10.0\langle\dot{x}_1\rangle^{0.5} \tag{9.255}$$

$$F_{n2}(\dot{\mathbf{x}}) = 12.0\langle\dot{x}_2 - \dot{x}_1\rangle^{0.7}. \tag{9.256}$$

Also assume that the structure is subjected to the modified El-Centro earthquake given in Figure 6.11. If the Delta forcing function is used, then Eq. 9.242 becomes

$$\begin{Bmatrix} x_1 \\ x_2 \\ \dot{x}_1 \\ \dot{x}_2 \end{Bmatrix}_{k+1} = \mathbf{e}^{\mathbf{A}\Delta t} \begin{Bmatrix} x_1 \\ x_2 \\ \dot{x}_1 \\ \dot{x}_2 \end{Bmatrix}_{k} + \mathbf{e}^{\mathbf{A}\Delta t} \Delta t\, \mathbf{F}_k \tag{9.257}$$

where $\mathbf{e}^{\mathbf{A}\Delta t}$ is given in Eq. 9.254 and the forcing function $\mathbf{F}_k$ is given in Eq. 9.241, which is

$$\mathbf{F}_k = \begin{Bmatrix} 0 \\ 0 \\ -1 \\ -1 \end{Bmatrix} a_k - \begin{bmatrix} 0 & 0 \\ 0 & 0 \\ 1 & -1 \\ 0 & 1 \end{bmatrix} \begin{Bmatrix} 10.0\langle\dot{x}_{1,k}\rangle^{0.5} \\ 12.0\langle\dot{x}_{2,k} - \dot{x}_{1,k}\rangle^{0.7} \end{Bmatrix}. \tag{9.258}$$

Substituting Eq 9.258 into Eq. 9.257 gives

$$
\begin{Bmatrix} x_1 \\ x_2 \\ \dot{x}_1 \\ \dot{x}_2 \end{Bmatrix}_{k+1} = e^{A\Delta t} \begin{Bmatrix} x_1 \\ x_2 \\ \dot{x}_1 \\ \dot{x}_2 \end{Bmatrix}_k + \begin{Bmatrix} -0.000397 \\ -0.000397 \\ -0.019601 \\ -0.019601 \end{Bmatrix} a_k - \begin{bmatrix} 0.000395 & -0.000392 \\ 0.000005 & 0.000387 \\ 0.019208 & -0.018815 \\ 0.000787 & 0.018028 \end{bmatrix} \begin{Bmatrix} 10.0\langle x_{1,k} \rangle^{0.5} \\ 12.0\langle x_{2,k} - x_{1,k} \rangle^{0.7} \end{Bmatrix}. \tag{9.259}
$$

To simplify Eq. 9.259, let

$$
\mathbf{H}_{sd}^{(EQ)} = \begin{Bmatrix} -0.000397 \\ -0.000397 \\ -0.019601 \\ -0.019601 \end{Bmatrix}, \quad \mathbf{G}_d = \begin{bmatrix} 0.000395 & -0.000392 \\ 0.000005 & 0.000387 \\ 0.019208 & -0.018815 \\ 0.000787 & 0.018028 \end{bmatrix}. \tag{9.260}
$$

Then Eq. 9.259 becomes

$$
\begin{Bmatrix} x_1 \\ x_2 \\ \dot{x}_1 \\ \dot{x}_2 \end{Bmatrix}_{k+1} = e^{A\Delta t} \begin{Bmatrix} x_1 \\ x_2 \\ \dot{x}_1 \\ \dot{x}_2 \end{Bmatrix}_k + \mathbf{H}_{sd}^{(EQ)} a_k - \mathbf{G}_d \begin{Bmatrix} 10.0\langle x_{1,k} \rangle^{0.5} \\ 12.0\langle x_{2,k} - x_{1,k} \rangle^{0.7} \end{Bmatrix}. \tag{9.261}
$$

Let the initial condition be equal to zero, that is, $x_1(0) = x_2(0) = 0.0$ and $\dot{x}_1(0) = \dot{x}_2(0) = 0.0$. Then at time step $k = 1$ (i.e., $t = \Delta t = 0.02$ s), it follows from Eq. 9.261 that

$$
\begin{Bmatrix} x_1 \\ x_2 \\ \dot{x}_1 \\ \dot{x}_2 \end{Bmatrix}_1 = e^{A\Delta t} \begin{Bmatrix} 0 \\ 0 \\ 0 \\ 0 \end{Bmatrix} + \mathbf{H}_{sd}^{(EQ)}(-0.9930) - \mathbf{G}_d \begin{Bmatrix} 10.0\langle 0 \rangle^{0.5} \\ 12.0\langle 0 - 0 \rangle^{0.7} \end{Bmatrix} = \begin{Bmatrix} 0.0004 \\ 0.0004 \\ 0.0195 \\ 0.0195 \end{Bmatrix}. \tag{9.262}
$$

At this time, the forces in each damper are

$$
F_{n1}(\dot{x}) = 10.0\langle 0.0195 \rangle^{0.5} = 1.3952 \text{ kip} \tag{9.263}
$$

$$
F_{n2}(\dot{x}) = 12.0\langle 0.0195 - 0.0195 \rangle^{0.7} = 0.0 \text{ kip}. \tag{9.264}
$$

Then at time step $k = 2$ (i.e., $t = 2\Delta t = 0.04$ s), it follows from Eq. 9.261 that

$$
\begin{Bmatrix} x_1 \\ x_2 \\ \dot{x}_1 \\ \dot{x}_2 \end{Bmatrix}_1 = e^{A\Delta t} \begin{Bmatrix} 0.0004 \\ 0.0004 \\ 0.0195 \\ 0.0195 \end{Bmatrix} + \mathbf{H}_{sd}^{(EQ)}(-7.6507) - \mathbf{G}_d \begin{Bmatrix} 1.3952 \\ 0.0 \end{Bmatrix} = \begin{Bmatrix} 0.0033 \\ 0.0038 \\ 0.1415 \\ 0.1672 \end{Bmatrix}. \tag{9.265}
$$

At this time, the forces in each damper are

$$
F_{n1}(\dot{x}) = 10.0\langle 0.1415 \rangle^{0.5} = 3.7611 \text{ kip} \tag{9.266}
$$

$$
F_{n2}(\dot{x}) = 12.0\langle 0.1672 - 0.1415 \rangle^{0.7} = 1.9238 \text{ kip}. \tag{9.267}
$$

The calculation process continues, and Table 9.10 shows the calculations of the first 10 time steps. Figures 9.33 to 9.36 show the displacement and velocity responses of the 2DOF system.

RECOMMENDED READING

1. T. T. Soong and G. F. Dargush. *Passive Energy Dissipation Systems in Structural Engineering*. Wiley, 1997.

2. Structural Engineers Association of California, *SEAOC Blue Book Appendix on Passive Energy Dissipation System Design*, 1997.

Table 9.10 Response of a 2DOF System with Manufactured Viscous Dampers

k	Time (s)	a_k (in./s^2)	$x_{1,k}$ (in.)	$x_{2,k}$ (in./s)	$\dot{x}_{1,k}$ (in./s)	$\dot{x}_{2,k}$ (in./s)	$F_{n1,k}$ (kip)	$F_{2,k}$ (kip)
0	0.00	−0.9930	0.0	0.0	0.0	0.0	0.0	0.0
1	0.02	−7.6507	0.0004	0.0004	0.0195	0.0195	1.3952	0.0
2	0.04	−7.1484	0.0033	0.0038	0.1415	0.1672	3.7611	1.9238
3	0.06	−6.2597	0.0081	0.0091	0.2378	0.2556	4.8763	1.6032
4	0.08	−6.7234	0.0139	0.0158	0.2783	0.3180	5.2755	2.3901
5	0.10	−8.5008	0.0207	0.0234	0.3250	0.3559	5.7008	2.1112
6	0.12	−10.0850	0.0288	0.0325	0.3800	0.4147	6.1646	2.2346
7	0.14	−9.0804	0.0383	0.0430	0.4444	0.4789	6.6660	2.2318
8	0.16	−7.8053	0.0483	0.0542	0.4612	0.4969	6.7910	2.2675
9	0.18	−6.0278	0.0579	0.0649	0.4333	0.4620	6.5825	2.0327
10	0.20	−6.0278	0.0661	0.0741	0.3536	0.3720	5.9465	1.6278

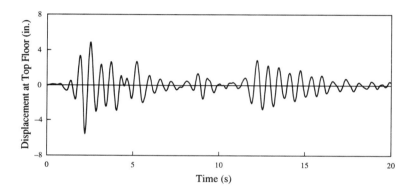

Figure 9.33 Displacement Response at the Top Floor of a 2DOF System with Manufactured Viscous Dampers

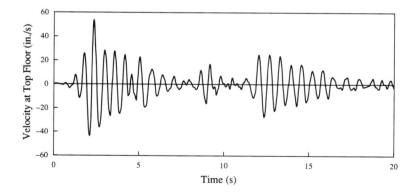

Figure 9.34 Velocity Response at the Top Floor of a 2DOF System with Manufactured Viscous Dampers

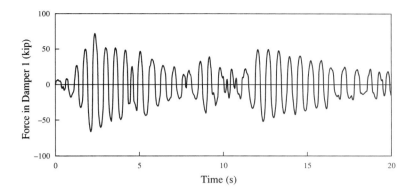

Figure 9.35 Force in Damper 1 between the Ground and the First Mass of a 2DOF System

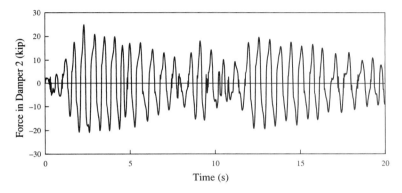

Figure 9.36 Force in Damper 1 between the First and Second Masses of a 2DOF System

REFERENCES

1. N. Newmark and W. Hall. *Earthquake Spectra & Design.* EERI, Oakland, CA 1980.

2. *Seismic Design Guidelines for Buildings.* U.S. Army, Navy, and Air Force, Washington, DC, 1988.

PROBLEMS

9.1 Repeat Example 1 in Section 9.2 but assume that the structural engineer wants to increase the damping of the structure to 20%.

9.2 Repeat Example 6 in Section 2.6 but assume that the natural damping is 2%.

9.3 Repeat Example 2 in Section 9.2 but assume that the natural damping is 2% and the damping of the manufactured viscous damper remains at 15% (i.e., the total critical damping ratio of the structure is now 17%). Compare your results with the results from Example 2 in Section 9.2.

9.4 Repeat Example 2 in Section 9.2 but assume that the natural damping is 2% and the damping of the manufactured damper changes to 5% (i.e., the total critical damping ratio of the structure is now 7%). Compare your results with the results from Example 2 in Section 9.2.

9.5 Repeat Example 4 in Section 9.2 but using the results from Problem 9.3.

9.6 Repeat Example 4 in Section 9.2 but using the results from Problem 9.4.

9.7 Consider the response of the SDOF system discussed in Example 2 in Section 9.2. The maximum velocity of the system in the first 10 time steps was 0.4934 in./s. Assume that the value of the manufactured damping coefficient is $c_{md} = 10$ k-s/in. Plot the damper force versus velocity (see Figure 9.18) for velocities in the range 0.0 to 1.0 in./s and for damper velocity exponents $\eta = 0.5$, 1.0, and 1.5.

9.8 Repeat Example 1 in Section 9.3 but with an initial displacement equal to zero and an initial velocity equal to 1.0 in./s.

9.9 Repeat Example 1 in Section 9.3 but with an initial displacement equal to 1.0 in. and an initial velocity equal to 1.0 in./s. Discuss your solution with respect to superposition of the results from Example 1 and Problem 9.8.

9.10 Repeat Example 1 in Section 9.3 but with damping power, η, equal to 0.7.

9.11 Repeat Example 1 in Section 9.3 but with damping power, η, equal to 1.5.

9.12 Repeat Example 1 in Section 9.4 but increase the desired fundamental mode damping to 20%.

9.13 Repeat Example 2 in Section 9.4 but increase the desired fundamental mode damping to 20%.

9.14 Repeat Example 1 in Section 9.4 but use a second mode damping of $\zeta_{nd2} = 5\%$, with the desired second mode damping $\zeta_2 = 10\%$.

9.15 Repeat Example 3 in Section 9.4 but increase the desired damping in the two modes to 20%.

9.16 Repeat Example 3 in Section 9.4 but increase the desired damping in the first mode to 15% and increase the desired damping in the second mode to 20%.

9.17 Repeat Example 1 in Section 9.5 but with an initial displacement equal to 10.0 in. and zero initial velocity.

9.18 Repeat Example 1 in Section 9.5 but with zero initial displacement and initial velocity equal to 40 in./s.

9.19 Repeat Example 1 in Section 9.5 but with damping power, η, equal to 0.25.

9.20 Repeat Example 1 in Section 9.5 but with damping power, η, equal to 0.75.

9.21 Repeat Example 1 in Section 9.5 but with damping power, η, equal to 1.50.

9.22 Repeat Example 2 in Section 9.5 but with the SDOF system subjected to a bilinear pulse as shown in Figure 9.37.

9.23 Repeat Example 2 in Section 9.5 but with the SDOF system subjected to a bilinear pulse as shown in Figure 9.37 and with a damping power, η, equal to 0.25.

9.24 Repeat Example 2 in Section 9.5 but with the SDOF system subjected to a bilinear pulse as shown in Figure 9.37 and with a damping power, η, equal to 0.75.

9.25 Repeat Example 2 in Section 9.5 but with the SDOF system subjected to a bilinear pulse as shown in Figure 9.37 and with a damping power, η, equal to 1.50.

9.26 Repeat Example 1 in Section 9.6 but with a damping power, η, equal to 0.25.

9.27 Repeat Example 1 in Section 9.6 but with a damping power, η, equal to 0.75.

9.28 Repeat Example 1 in Section 9.6 but with a damping power, η, equal to 1.50.

9.29 Repeat Example 2 in Section 9.6 but with a damping power, η, equal to 0.25.

9.30 Repeat Example 2 in Section 9.6 but with a damping power, η, equal to 0.75.

9.31 Repeat Example 2 in Section 9.6 but with a damping power, η, equal to 1.50.

9.32 Repeat Example 1 in Section 9.7 with Damper 2 the same as Damper 1.

9.33 Repeat Example 1 in Section 9.7 with Damper 1 the same as Damper 2.

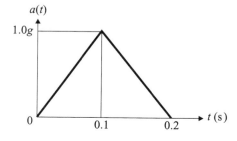

Figure 9.37 Bilinear Pulse

Chapter 10

Structural Control

10.1 OVERVIEW

Structural engineers calculate structural responses to natural forces such as earthquakes, strong winds, hurricanes or typhoons, and ocean waves. Chapters 1 to 7 of this book describe analytical methods used to calculate this response. The structural engineer determines the sizes of the structural members to control the range of response (i.e., elastic or inelastic) and to satisfy defined deformation and stress constraints. Historically, these structural members have been steel, concrete, masonry, and wood. These and other more sophisticated materials will be used in the future.

Chapters 8 and 9 discussed structural elements, or members, that are manufactured in plants away from the building site and then placed in the structure to achieve certain defined objectives. Therefore, the structural engineer controls the response of the structure with the number and design of these base isolators or damping elements. Such a design is called *passive structural control,* because the structural design is made, the structure is constructed, and then it waits for the force (e.g., earthquake) to occur. Then the structure responds to the induced forces without any modification in performance characteristics of the manufactured elements during the time history of the force. This is like doing very detailed wind and ocean current analyses to set the sails and point the direction of a sailboat leaving from Los Angeles to go to Hawaii. The boat is set free in Los Angeles, and it hopefully will go to Hawaii without any adjustments to the sails or direction.

Active structural control is when the performance characteristics of the manufactured structural element (e.g., damper) are modified to adjust for the particular forcing function as it occurs. In the sailboat example, active control occurs when the boat captain changes the direction of the boat or adjusts the sails to compensate for the actual wind or ocean current conditions.

This chapter introduces the topic of active structural control.

10.2 STRUCTURAL CONTROL OF AN SDOF SYSTEM

Consider an SDOF system, as shown in Figure 10.1a, with no external force applied but with a control force, $F_c(t)$, applied to the mass. The dynamic equilibrium equation of motion, obtained using a free body diagram of the forces acting on the mass, is

$$\sum F = F_s + F_d + F_c = -kx(t) - c\dot{x}(t) + F_c(t) = m\ddot{x}(t). \tag{10.1}$$

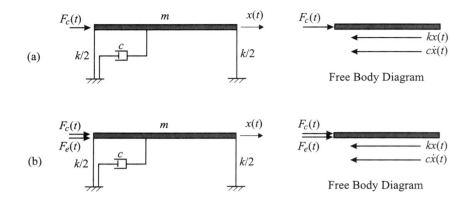

Figure 10.1 Single Degree of Freedom System with Structural Control

Rearranging the terms in Eq. 10.1 gives

$$m\ddot{x}(t) + c\dot{x}(t) + kx(t) = F_c(t).$$ (10.2)

When the SDOF system is subjected to an external force $F_e(t)$ in addition to the control force (see Figure 10.1b), then Eq. 10.1 becomes

$$\sum F = F_s + F_d + F_c + F_e = -kx(t) - c\dot{x}(t) + F_c(t) + F_e(t) = m\ddot{x}(t)$$ (10.3)

and rearranging the terms in Eq. 10.3 gives

$$m\ddot{x}(t) + c\dot{x}(t) + kx(t) = F_e(t) + F_c(t).$$ (10.4)

Finally, for an earthquake time history analysis, the external applied force is replaced by the mass times ground acceleration $a(t)$ (see Section 5.2). Then Eq. 10.4 becomes

$$m\ddot{x}(t) + c\dot{x}(t) + kx(t) = -ma(t) + F_c(t).$$ (10.5)

The structural control force, $F_c(t)$, in Eq. 10.4 or Eq. 10.5 can take different forms. The basic objective of structural control is to select a structural control force that produces the desired response. The response of the system and therefore the selected control force depend on the external applied force or earthquake ground motion, and therefore the design is highly dependent on the different patterns and variations of the external applied force or earthquake ground motion. In general, three types of control forces are used by structural engineers:

1. *Closed-loop control:* When the control force is a function of the displacement and velocity of the structure, this type of structural control is called the closed-loop control. The control law can be expressed as

    $$F_c(t) = f_{c1}(x(t)) + f_{c2}(\dot{x}(t)).$$ (10.6)

    Substituting Eq. 10.6 into Eq. 10.4 gives

    $$m\ddot{x}(t) + c\dot{x}(t) + kx(t) = F_e(t) + f_{c1}(x(t)) + f_{c2}(\dot{x}(t)).$$ (10.7)

    As noted, the objective for all structural control problems is to solve this second-order nonlinear differential equation for different forms of f_{c1} and f_{c2}, and then to determine the specific form of f_{c1} and f_{c2} that produces the desired system response.

2. *Open-loop control:* When the control force is a function of the external applied force on the structure, this type of structural control is called open-loop control.

The control law can be expressed as

$$F_c(t) = f_c\big(F_e(t)\big).$$ (10.8)

Substituting Eq. 10.8 into Eq. 10.4 gives

$$m\ddot{x}(t) + c\dot{x}(t) + kx(t) = F_e(t) + f_c\big(F_e(t)\big).$$ (10.9)

Similar to closed-loop control, the objective in the design of an open-loop control system is to select the open-loop control force and then solve the second-order nonlinear differential equation described by Eq. 10.9 to obtain the desired system response.

3.   *Open–closed-loop control:* When the control force is a function of the external applied force as well as the displacement and velocity of the structure, then this type of structural control is called open–closed-loop control. The control law can be expressed as

$$F_c(t) = f_{c1}\big(x(t)\big) + f_{c2}\big(\dot{x}(t)\big) + f_c\big(F_e(t)\big).$$ (10.10)

Substituting Eq. 10.10 into Eq. 10.4 gives

$$m\ddot{x}(t) + c\dot{x}(t) + kx(t) = F_e(t) + f_{c1}\big(x(t)\big) + f_{c2}\big(\dot{x}(t)\big) + f_c\big(F_e(t)\big).$$ (10.11)

Again, similar to closed-loop and open-loop control, the design process involves solving the second-order nonlinear differential equation described by Eq. 10.11 for different forms of f_{c1}, f_{c2}, and f_c.

Equation 10.4 shows that the response of an SDOF system with a control force results from the contribution of the two terms on the right side of the equation. The first term, discussed in Section 2.3, represents the external forcing function that is in general not under the structural engineer's control. For example, it can be a wind load or an earthquake ground motion. The second term on the right side is the control force, and its mathematical form and physical design are under the structural engineer's control. The basic goal of structural control can be seen in Eqs. 10.6, 10.8, and 10.10, and the structural engineer must select the second term on the right side, that is, $F_c(t)$, to meet the desired performance goals and the physical constraints that have been established for the structure under consideration.

EXAMPLE 1   *Closed-Loop Linear Control*

Consider the case where the control force is a linear function of the displacement and velocity of the system, that is,

$$F_c(t) = -k_c x(t) - c_c \dot{x}(t)$$ (10.12)

where k_c and c_c are the proportionality constants. This type of control law is called *closed-loop linear control*. It follows from Eq. 10.4 that

$$m\ddot{x}(t) + c\dot{x}(t) + kx(t) = F_e(t) - k_c x(t) - c_c \dot{x}(t).$$ (10.13)

Rearranging terms gives

$$m\ddot{x}(t) + \big(c + c_c\big)\dot{x}(t) + \big(k + k_c\big)x(t) = F_e(t).$$ (10.14)

Note that for a closed-loop linear control system, the structural response is changed in the same way as adding damping and stiffness to the structure.

EXAMPLE 2 *Closed-Loop Velocity Control*

Consider the case where the control force is a function of the velocity of the structure, that is,

$$F_c(t) = -f_{md}(\dot{x}(t)). \tag{10.15}$$

It follows from Eq. 10.4 that

$$m\ddot{x}(t) + c\dot{x}(t) + kx(t) = F_e(t) - f_{md}(\dot{x}(t)). \tag{10.16}$$

This equation is the same as Eq. 9.146. This type of control is called *closed-loop velocity control*, and it can be achieved by installing passive viscous dampers of the type discussed in Chapter 9 in the structure.

If the external force is the earthquake ground motion, then it follows from Eq. 10.5 that

$$m\ddot{x}(t) + c\dot{x}(t) + kx(t) = -ma(t) - f_{md}(\dot{x}(t)) \tag{10.17}$$

where $x(t)$ is the relative displacement between the mass and the ground.

10.3 STRUCTURAL CONTROL OF A 2DOF SYSTEM

The mathematics of structural control are based on the state space approach to structural dynamics. Therefore, although this chapter will present the general dynamic equation of motion first for a 2DOF system and then for an MDOF system in the following section, the solution for the response will be presented in general matrix form because the state space approach requires the matrix approach.

Consider the 2DOF system shown in Figure 10.2. Summing forces on the two masses subjected to both external forces and control forces gives

$$-c_1\dot{x}_1(t) - c_2(\dot{x}_1(t) - \dot{x}_2(t)) - k_1 x_1(t) - k_2(x_1(t) - x_2(t)) + F_1(t) + F_{c1}(t) = m_1\ddot{x}_1(t) \tag{10.18a}$$

$$-c_2(\dot{x}_2(t) - \dot{x}_1(t)) - k_2(x_2(t) - x_1(t)) + F_2(t) + F_{c2}(t) = m_2\ddot{x}_2(t) \tag{10.18b}$$

where $F_{c1}(t)$ is the control force applied at the lower floor, and $F_{c2}(t)$ is the control force applied at the upper floor. Equation 10.18 in matrix form is

$$\begin{bmatrix} m_1 & 0 \\ 0 & m_2 \end{bmatrix}\begin{Bmatrix} \ddot{x}_1 \\ \ddot{x}_2 \end{Bmatrix} + \begin{bmatrix} c_1 + c_2 & -c_2 \\ -c_2 & c_2 \end{bmatrix}\begin{Bmatrix} \dot{x}_1 \\ \dot{x}_2 \end{Bmatrix} + \begin{bmatrix} k_1 + k_2 & -k_2 \\ -k_2 & k_2 \end{bmatrix}\begin{Bmatrix} x_1 \\ x_2 \end{Bmatrix} = \begin{Bmatrix} F_1(t) \\ F_2(t) \end{Bmatrix} + \begin{Bmatrix} F_{c1}(t) \\ F_{c2}(t) \end{Bmatrix}. \tag{10.19}$$

Now define

$$\mathbf{M} = \begin{bmatrix} m_1 & 0 \\ 0 & m_2 \end{bmatrix}, \quad \mathbf{C} = \begin{bmatrix} c_1 + c_2 & -c_2 \\ -c_2 & c_2 \end{bmatrix}, \quad \mathbf{K} = \begin{bmatrix} k_1 + k_2 & -k_2 \\ -k_2 & k_2 \end{bmatrix}$$

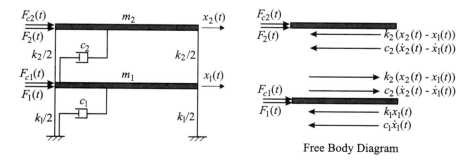

Figure 10.2   Two Degree of Freedom System with Structural Control

$$\mathbf{X}(t) = \begin{Bmatrix} x_1 \\ x_2 \end{Bmatrix}, \quad \mathbf{F}_e(t) = \begin{Bmatrix} F_1(t) \\ F_2(t) \end{Bmatrix}, \quad \mathbf{f}_c(t) = \begin{Bmatrix} F_{c1}(t) \\ F_{c2}(t) \end{Bmatrix}$$

and then it follows from Eq. 10.19 that

$$\mathbf{M}\ddot{\mathbf{X}}(t) + \mathbf{C}\dot{\mathbf{X}}(t) + \mathbf{K}\mathbf{X}(t) = \mathbf{F}_e(t) + \mathbf{f}_c(t). \tag{10.20}$$

In closed-loop-control, the control force is a function of the displacement and velocity, that is,

$$\mathbf{f}_c(t) = \begin{Bmatrix} F_{c1}(t) \\ F_{c2}(t) \end{Bmatrix} = \begin{Bmatrix} f_{c11}(\mathbf{X}(t)) + f_{c12}(\dot{\mathbf{X}}(t)) \\ f_{c21}(\mathbf{X}(t)) + f_{c22}(\dot{\mathbf{X}}(t)) \end{Bmatrix}, \tag{10.21}$$

and in open-loop control, the control force is a function of the external applied force, that is,

$$\mathbf{f}_c(t) = \begin{Bmatrix} F_{c1}(t) \\ F_{c2}(t) \end{Bmatrix} = \begin{Bmatrix} f_{c1}(\mathbf{F}_e(t)) \\ f_{c2}(\mathbf{F}_e(t)) \end{Bmatrix}. \tag{10.22}$$

Finally, in open–closed-loop control, the control force is the sum of Eqs. 10.21 and 10.22:

$$\mathbf{f}_c(t) = \begin{Bmatrix} F_{c1}(t) \\ F_{c2}(t) \end{Bmatrix} = \begin{Bmatrix} f_{c11}(\mathbf{X}(t)) + f_{c12}(\dot{\mathbf{X}}(t)) + f_{c1}(\mathbf{F}_e(t)) \\ f_{c21}(\mathbf{X}(t)) + f_{c22}(\dot{\mathbf{X}}(t)) + f_{c2}(\mathbf{F}_e(t)) \end{Bmatrix}. \tag{10.23}$$

EXAMPLE 1   **2DOF System Subjected to an Earthquake Ground Motion**

Consider the frame shown in Figure 10.2 subjected to an earthquake ground motion. It follows that the analysis can be performed by replacing the external applied forcing function in Eq. 10.19 by the mass times the acceleration, that is,

$$\begin{bmatrix} m_1 & 0 \\ 0 & m_2 \end{bmatrix} \begin{Bmatrix} \ddot{x}_1 \\ \ddot{x}_2 \end{Bmatrix} + \begin{bmatrix} c_1 + c_2 & -c_2 \\ -c_2 & c_2 \end{bmatrix} \begin{Bmatrix} \dot{x}_1 \\ \dot{x}_2 \end{Bmatrix} + \begin{bmatrix} k_1 + k_2 & -k_2 \\ -k_2 & k_2 \end{bmatrix} \begin{Bmatrix} x_1 \\ x_2 \end{Bmatrix}$$
$$= -\begin{bmatrix} m_1 & 0 \\ 0 & m_2 \end{bmatrix} \begin{Bmatrix} 1 \\ 1 \end{Bmatrix} a(t) + \begin{Bmatrix} F_{c1}(t) \\ F_{c2}(t) \end{Bmatrix} \tag{10.24}$$

or

$$\mathbf{M}\ddot{\mathbf{X}}(t) + \mathbf{C}\dot{\mathbf{X}}(t) + \mathbf{K}\mathbf{X}(t) = -\mathbf{M}\{\mathbf{I}\}a(t) + \mathbf{f}_c(t). \tag{10.25}$$

EXAMPLE 2   **Two-Story Active Bracing System**

Consider the two story structure shown in Figure 10.3a. The motion of the first floor relative to the ground is used to determine the force that Actuator 1 applies to the first floor. Similarly, the relative motion between the first and second floors is used to determine the force that Actuator 2 applies to the first and second floors. These actuators generate forces on the structure in two ways, and one of which is shown in Figure 10.3b. These two ways are:

1.   When the actuators contract, they generate tension forces on the braces, which in turn produce forces on the floors. The contraction of Actuator 1 produces a horizontal force $-F_{c1}(t) \cos \theta$ that is applied to the first floor and a horizontal force $F_{c1}(t) \cos \theta$ that is applied to the ground (which will not affect the response of the structure). The contraction of Actuator 2 produces a horizontal force $-F_{c2}(t) \cos \theta$ that is applied to the second floor and a horizontal force $F_{c2}(t) \cos \theta$ that is applied to the first floor.

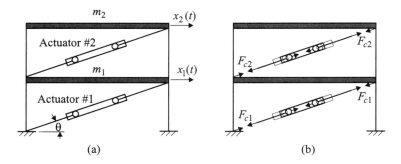

Figure 10.3 Two-Story Active Bracing System

2. When the actuators expand, they generate compressive forces on the braces. The expansion of Actuator 1 produces a horizontal force $F_{c1}(t)\cos\theta$ that is applied to the first floor and a horizontal force $-F_{c1}(t)\cos\theta$ that is applied to the ground (which will not affect the response of the structure). Similarly, expansion of Actuator 2 applies a horizontal force $F_{c2}(t)\cos\theta$ on the second floor and a horizontal force $-F_{c2}(t)\cos\theta$ on the first floor.

If the actuator expansion is denoted as positive, then the dynamic equation of motion in Eq. 10.18 becomes

$$-c_1\dot{x}_1(t) - c_2\big(\dot{x}_1(t) - \dot{x}_2(t)\big) - k_1 x_1(t) - k_2\big(x_1(t) - x_2(t)\big)$$
$$+F_{c1}(t)\cos\theta - F_{c2}(t)\cos\theta = m_1\ddot{x}_1(t) \tag{10.26}$$

$$-c_2\big(\dot{x}_2(t) - \dot{x}_1(t)\big) - k_2\big(x_2(t) - x_1(t)\big) + F_{c2}(t)\cos\theta = m_2\ddot{x}_2(t). \tag{10.27}$$

In matrix form, Eqs. 10.26 and 10.27 become

$$\begin{bmatrix} m_1 & 0 \\ 0 & m_2 \end{bmatrix}\begin{Bmatrix} \ddot{x}_1 \\ \ddot{x}_2 \end{Bmatrix} + \begin{bmatrix} c_1 + c_2 & -c_2 \\ -c_2 & c_2 \end{bmatrix}\begin{Bmatrix} \dot{x}_1 \\ \dot{x}_2 \end{Bmatrix} + \begin{bmatrix} k_1 + k_2 & -k_2 \\ -k_2 & k_2 \end{bmatrix}\begin{Bmatrix} x_1 \\ x_2 \end{Bmatrix} = \begin{bmatrix} \cos\theta & -\cos\theta \\ 0 & \cos\theta \end{bmatrix}\begin{Bmatrix} F_{c1}(t) \\ F_{c2}(t) \end{Bmatrix}. \tag{10.28}$$

If the actuator contraction is denoted as positive and the actuator expansion as negative, then Eq. 10.28 will change. For this definition, the equation of motion given in Eq. 10.18 becomes

$$-c_1\dot{x}_1(t) - c_2\big(\dot{x}_1(t) - \dot{x}_2(t)\big) - k_1 x_1(t) - k_2\big(x_1(t) - x_2(t)\big)$$
$$-F_{c1}(t)\cos\theta + F_{c2}(t)\cos\theta = m_1\ddot{x}_1(t) \tag{10.29}$$

$$-c_2\big(\dot{x}_2(t) - \dot{x}_1(t)\big) - k_2\big(x_2(t) - x_1(t)\big) - F_{c2}(t)\cos\theta = m_2\ddot{x}_2(t). \tag{10.30}$$

Now, in matrix form, Eqs. 10.29 and 10.30 become

$$\begin{bmatrix} m_1 & 0 \\ 0 & m_2 \end{bmatrix}\begin{Bmatrix} \ddot{x}_1 \\ \ddot{x}_2 \end{Bmatrix} + \begin{bmatrix} c_1 + c_2 & -c_2 \\ -c_2 & c_2 \end{bmatrix}\begin{Bmatrix} \dot{x}_1 \\ \dot{x}_2 \end{Bmatrix} + \begin{bmatrix} k_1 + k_2 & -k_2 \\ -k_2 & k_2 \end{bmatrix}\begin{Bmatrix} x_1 \\ x_2 \end{Bmatrix} = \begin{bmatrix} -\cos\theta & \cos\theta \\ 0 & -\cos\theta \end{bmatrix}\begin{Bmatrix} F_{c1}(t) \\ F_{c2}(t) \end{Bmatrix}. \tag{10.31}$$

10.4 STRUCTURAL CONTROL OF AN MDOF SYSTEM AND NUMERICAL SOLUTION FOR RESPONSE

The structural control system discussed in Section 10.3 for a 2DOF system is now extended to an MDOF system. Recall from Section 4.3 for the case without structural control that

$$\mathbf{M}\ddot{\mathbf{X}}(t) + \mathbf{C}\dot{\mathbf{X}}(t) + \mathbf{K}\mathbf{X}(t) = \mathbf{F}_e(t) \tag{10.32}$$

where

$$\mathbf{M} = \begin{bmatrix} m_{11} & m_{12} & \cdots & & m_{1n} \\ m_{21} & m_{22} & \ddots & & \vdots \\ \vdots & \ddots & \ddots & m_{n-1,n} \\ m_{n1} & \cdots & m_{n,n-1} & m_{nn} \end{bmatrix}, \quad \mathbf{C} = \begin{bmatrix} c_{11} & c_{12} & \cdots & & c_{1n} \\ c_{21} & c_{22} & \ddots & & \vdots \\ \vdots & \ddots & \ddots & c_{n-1,n} \\ c_{n1} & \cdots & c_{n,n-1} & c_{nn} \end{bmatrix}, \quad \mathbf{K} = \begin{bmatrix} k_{11} & k_{12} & \cdots & & k_{1n} \\ k_{21} & k_{22} & \ddots & & \vdots \\ \vdots & \ddots & \ddots & k_{n-1,n} \\ k_{n1} & \cdots & k_{n,n-1} & k_{nn} \end{bmatrix}$$

$$\mathbf{X}(t) = \{ x_1(t) \quad x_2(t) \quad \cdots \quad x_n(t) \}^T, \quad \mathbf{F}_e(t) = \{ F_1(t) \quad F_2(t) \quad \cdots \quad F_n(t) \}^T$$

and n represents the total number of degrees of freedom. Now for the case with structural control, it follows directly from Eq. 10.20 that

$$\mathbf{M}\ddot{\mathbf{X}}(t) + \mathbf{C}\dot{\mathbf{X}}(t) + \mathbf{K}\mathbf{X}(t) = \mathbf{F}_e(t) + \mathbf{D}\mathbf{f}_c(t) \tag{10.33}$$

where

$$\mathbf{f}_c(t) = \begin{Bmatrix} F_{c1}(t) \\ F_{c2}(t) \\ \vdots \\ F_{cp}(t) \end{Bmatrix}, \quad \mathbf{D} = \begin{bmatrix} D_{11} & D_{12} & \cdots & D_{1,p} \\ D_{21} & D_{22} & \ddots & \vdots \\ \vdots & \ddots & \ddots & D_{n-1,p} \\ D_{n1} & \cdots & D_{n,p-1} & D_{np} \end{bmatrix}.$$

In this equation, p denotes the total number of control forces and $\mathbf{D}$ is called the *control force distribution matrix*. Note that one control force can affect several different degrees of freedom, and each ith DOF that the jth control force affects is represented by D_{ij}.

In closed-loop control, the control force is a function of the displacement and velocity, that is,

$$F_{ci}(t) = f_{ci1}(\mathbf{X}(t)) + f_{ci2}(\dot{\mathbf{X}}(t)), \quad i = 1, \dots, p. \tag{10.34}$$

In open-loop control, the control force is a function of the external applied force, that is,

$$F_{ci}(t) = f_{ci}(\mathbf{F}_e(t)), \quad i = 1, \dots, p. \tag{10.35}$$

Finally, in open–closed-loop control, the control force is the sum of Eqs. 10.34 and 10.35, that is,

$$F_{ci}(t) = f_{ci1}(\mathbf{X}(t)) + f_{ci2}(\dot{\mathbf{X}}(t)) + f_{ci}(\mathbf{F}_e(t)), \quad i = 1, \dots, p. \tag{10.36}$$

State space numerical solution methods are generally used to solve Eq. 10.33. Equation 10.33 can be placed in state space form by defining

$$\mathbf{z}(t) = \begin{Bmatrix} \mathbf{X}(t) \\ \dot{\mathbf{X}}(t) \end{Bmatrix}. \tag{10.37}$$

It then follows from Eq. 10.33 that

$$\dot{\mathbf{z}}(t) = \begin{Bmatrix} \dot{\mathbf{X}}(t) \\ \ddot{\mathbf{X}}(t) \end{Bmatrix} = \begin{bmatrix} \mathbf{0} & \mathbf{I} \\ -\mathbf{M}^{-1}\mathbf{K} & -\mathbf{M}^{-1}\mathbf{C} \end{bmatrix} \begin{Bmatrix} \mathbf{X}(t) \\ \dot{\mathbf{X}}(t) \end{Bmatrix} + \begin{Bmatrix} \mathbf{0} \\ \mathbf{M}^{-1}\mathbf{F}_e(t) \end{Bmatrix} + \begin{Bmatrix} \mathbf{0} \\ \mathbf{M}^{-1}\mathbf{D}\mathbf{f}_c(t) \end{Bmatrix} \tag{10.38}$$

where $\mathbf{0}$ is a null matrix with all zero entries, and $\mathbf{I}$ is the identity matrix. To simplify Eq. 10.38, let

$$\mathbf{A} = \begin{bmatrix} \mathbf{0} & \mathbf{I} \\ -\mathbf{M}^{-1}\mathbf{K} & -\mathbf{M}^{-1}\mathbf{C} \end{bmatrix}, \quad \mathbf{F}(t) = \begin{Bmatrix} \mathbf{0} \\ \mathbf{M}^{-1}\mathbf{F}_e(t) \end{Bmatrix}, \quad \mathbf{B} = \begin{bmatrix} \mathbf{0} \\ \mathbf{M}^{-1}\mathbf{D} \end{bmatrix}. \tag{10.39}$$

It follows that Eq. 10.38 can then be transformed to the corresponding state space equation

$$\dot{\mathbf{z}}(t) = \mathbf{A}\mathbf{z}(t) + \mathbf{F}(t) + \mathbf{B}\mathbf{f}_c(t). \tag{10.40}$$

The solution to Eq. 10.40 is

$$\mathbf{z}(t) = \mathbf{e}^{\mathbf{A}(t-t_o)}\mathbf{z}(t_o) + \mathbf{e}^{\mathbf{A}t}\int_{t_o}^{t} \mathbf{e}^{-\mathbf{A}s}\,\mathbf{F}(s)\,ds + \mathbf{e}^{\mathbf{A}t}\int_{t_o}^{t} \mathbf{e}^{-\mathbf{A}s}\,\mathbf{B}\mathbf{f}_c(s)\,ds. \tag{10.41}$$

Using the numerical solution procedure discussed in Section 2.8, let

$$t_{k+1} = t, \quad t_k = t_o, \quad \Delta t = t - t_o. \tag{10.42}$$

It then follows from Eq. 10.41 that

$$\mathbf{z}_{k+1} = \mathbf{e}^{\mathbf{A}\Delta t}\mathbf{z}_k + \mathbf{e}^{\mathbf{A}t_{k+1}}\int_{t_k}^{t_{k+1}} \mathbf{e}^{-\mathbf{A}s}\,\mathbf{F}(s)\,ds + \mathbf{e}^{\mathbf{A}t_{k+1}}\int_{t_k}^{t_{k+1}} \mathbf{e}^{-\mathbf{A}s}\,\mathbf{B}\mathbf{f}_c(s)\,ds. \tag{10.43}$$

Similar to the digitization of the external applied forcing function $\mathbf{F}(s)$, as noted in Section 2.8, two methods are available for digitizing the control force $\mathbf{f}_c(s)$. When a Delta forcing function is assumed, it follows from Eq. 10.43 that

$$\mathbf{z}_{k+1} = \mathbf{e}^{\mathbf{A}\Delta t}\,\mathbf{z}_k + \mathbf{e}^{\mathbf{A}\Delta t}\Delta t\,\mathbf{F}_k + \mathbf{e}^{\mathbf{A}\Delta t}\mathbf{B}\Delta t\,\mathbf{f}_{ck}. \tag{10.44}$$

Similarly, when a constant forcing function is assumed, it follows that

$$\mathbf{z}_{k+1} = \mathbf{e}^{\mathbf{A}\Delta t}\,\mathbf{z}_k + \mathbf{A}^{-1}\big(\mathbf{e}^{\mathbf{A}\Delta t} - \mathbf{I}\big)\mathbf{F}_k + \mathbf{A}^{-1}\big(\mathbf{e}^{\mathbf{A}\Delta t} - \mathbf{I}\big)\mathbf{B}\mathbf{f}_{ck}. \tag{10.45}$$

Now define in Eq. 10.44

$$\mathbf{F}_s = \mathbf{e}^{\mathbf{A}\Delta t}, \quad \mathbf{H}_d = \mathbf{e}^{\mathbf{A}\Delta t}\Delta t, \quad \mathbf{G} = \mathbf{e}^{\mathbf{A}\Delta t}\mathbf{B}\Delta t \tag{10.46}$$

or alternatively define in Eq. 10.45

$$\mathbf{F}_s = \mathbf{e}^{\mathbf{A}\Delta t}, \quad \mathbf{H}_d = \mathbf{A}^{-1}\big(\mathbf{e}^{\mathbf{A}\Delta t} - \mathbf{I}\big), \quad \mathbf{G} = \mathbf{A}^{-1}\big(\mathbf{e}^{\mathbf{A}\Delta t} - \mathbf{I}\big)\mathbf{B}. \tag{10.47}$$

It then follows that Eqs. 10.44 and Eq. 10.45 can both be written in the state space form, where

$$\mathbf{z}_{k+1} = \mathbf{F}_s\,\mathbf{z}_k + \mathbf{H}_d\,\mathbf{F}_k + \mathbf{G}\mathbf{f}_{ck}. \tag{10.48}$$

Note that Eq. 10.44 is the same as Eq. 4.466 and 10.45 is the same as Eq. 4.467 for the state space response of a linear system except that the control force term, $\mathbf{f}_{ck}$, appears in Eqs. 10.44 and 10.45. Also note the $\mathbf{H}_d$ in Eq. 10.46 was called $\mathbf{H}_{sd}$ in Eq. 2.289 and $\mathbf{H}_d$ in Eq. 10.47 was called $\mathbf{H}_{sc}$ in Eq. 2.291. However, both of them are denoted by $\mathbf{H}_d$ here for convenience.

The state space vector response, $\mathbf{z}(t)$, in Eq. 10.37 includes both the displacement vector and the velocity vector. Therefore, the closed-loop control becomes

$$F_{cik} = f_{ci1\&2}(\mathbf{z}_k), \quad i = 1,\dots,p. \tag{10.49}$$

Similarly, the open-loop control force becomes

$$F_{cik} = f_{ci}(\mathbf{F}_k), \quad i = 1,\dots,p \tag{10.50}$$

and the open–closed-loop control force is the sum of Eqs. 10.47 and 10.48, that is,

$$F_{cik} = f_{ci1\&2}(\mathbf{z}_k) + f_{ci}(\mathbf{F}_k), \quad i = 1,\dots,p. \tag{10.51}$$

EXAMPLE 1   *Structural Control with an Earthquake Ground Motion*

Consider an MDOF system that is subjected to an earthquake ground motion. The external applied forcing function in Eq. 10.33 in this situation is replaced by the mass times the acceleration vector. It follows that

$$\mathbf{M\ddot{X}}(t) + \mathbf{C\dot{X}}(t) + \mathbf{KX}(t) = -\mathbf{Ma}(t) + \mathbf{f}_c(t) \tag{10.52}$$

or from Eq. 10.46 in state space form,

$$\mathbf{z}_{k+1} = \mathbf{F}_s \mathbf{z}_k + \mathbf{H}_d^{(EQ)} \mathbf{a}_k + \mathbf{Gf}_{ck} \tag{10.53}$$

where

$$\mathbf{a}_k = \left\{ a_{xk} \quad a_{yk} \quad a_{zk} \quad a_{rk} \quad a_{sk} \quad a_{tk} \right\}^T. \tag{10.54}$$

When the Delta forcing function assumption is made, then

$$\mathbf{H}_d^{(EQ)} = e^{\mathbf{A}\Delta t} \mathbf{H} \Delta t \tag{10.55}$$

and when constant forcing function assumption is made, then

$$\mathbf{H}_d^{(EQ)} = \mathbf{A}^{-1} \left(e^{\mathbf{A}\Delta t} - \mathbf{I} \right) \mathbf{H}. \tag{10.56}$$

The matrix $\mathbf{H}$ is given in Eq. 6.188.

EXAMPLE 2   **_Four-Story Active Tendon System_**

Consider the four-story active tendon system as shown in Figure 10.4a. Each interstory consists of an active tendon actuator.

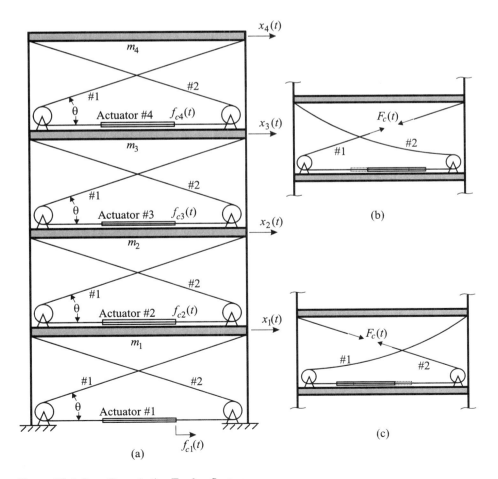

Figure 10.4 Four-Story Active Tendon System

As shown in Figure 10.4b, when the actuator moves to the right, Tendon 1 is stretched, producing a tension force $F_c(t)$. This tension force $F_c(t)$ induces a horizontal force of $-F_c(t) \cos \theta$ on the upper floor and a horizontal force of $F_c(t) \cos \theta$ on the lower floor. Similarly, as shown in Figure 10.4c, when the actuator moves to the left, Tendon 2 is stretched, producing a tension force $F_c(t)$. This tension force $F_c(t)$ induces a horizontal force of $F_c(t) \cos \theta$ on the upper floor and a horizontal force of $-F_c(t) \cos \theta$ on the lower floor. Assume that the actuator moving to the left is positive; then Eq. 10.33 becomes

$$\mathbf{M\ddot{X}}(t) + \mathbf{C\dot{X}}(t) + \mathbf{KX}(t) = \mathbf{Df}_c(t) \tag{10.57}$$

where

$$\mathbf{D} = \begin{bmatrix} \cos \theta & -\cos \theta & 0 & 0 \\ 0 & \cos \theta & -\cos \theta & 0 \\ 0 & 0 & \cos \theta & -\cos \theta \\ 0 & 0 & 0 & \cos \theta \end{bmatrix}. \tag{10.58}$$

10.5 OPTIMAL LINEAR CONTROL WITHOUT EXTERNAL FORCE—CONTINUOUS TIME

In this and the following section, the external forcing function is assumed to be zero. This case is of very limited interest to structural engineers, but it is convenient for introducing many of the basic principles of optimal structural control. Therefore, consider the special case where the external force in Eq. 10.40 is assumed to be zero, that is,

$$\dot{\mathbf{z}}(t) = \mathbf{Az}(t) + \mathbf{Bf}_c(t). \tag{10.59}$$

Now define a *cost function* as follows:

$$J = \int_0^{t_f} \frac{1}{2} \left[\mathbf{z}^T(t) \mathbf{Qz}(t) + \mathbf{f}_c^T(t) \mathbf{Rf}_c(t) \right] dt \tag{10.60}$$

where $\mathbf{Q}$ and $\mathbf{R}$ are weighting matrices, and t_f is the duration over which the cost function is calculated. The objective of optimal structural control is to select a set of control forces that will minimize the cost function J over the time $t = 0$ to $t = t_f$. This minimization of the cost function produces a set of control forces that will satisfy the optimal control law. The minimization process depends highly on the choices of the weighting matrices $\mathbf{Q}$ and $\mathbf{R}$, which can be interpreted as follows:

1. If it is desired that the displacement and velocity responses of the structure be small for the range of time $t = 0$ to $t = t_f$, then it is desirable to choose a set of values for the $\mathbf{Q}$ matrix that are large. The result of such a selection will be large values, or a large "penalty," for the part of the cost function involving the state vector $\mathbf{z}(t)$.

2. If it is desired that the applied force be as small as possible, then the $\mathbf{R}$ matrix should have large values. The result is that the part of the cost function involving $\mathbf{f}_c(t)$ will be large, and thus the cost function will pay a large penalty for the large control force $\mathbf{f}_c(t)$.

The procedure used to minimize J in Eq. 10.60 involves the use of a Lagrangian. Define the Lagrangian L as

$$L = \int_0^{t_f} \left\{ \frac{1}{2} \left[\mathbf{z}^T(t) \mathbf{Qz}(t) + \mathbf{f}_c^T(t) \mathbf{Rf}_c(t) \right] + \boldsymbol{\lambda}^T(t) \left[\mathbf{Az}(t) + \mathbf{Bf}_c(t) - \dot{\mathbf{z}}(t) \right] \right\} dt \tag{10.61}$$

where $\boldsymbol{\lambda}^T(t)$ is the Lagrange multiplier, and it is an unknown vector function of time, that is,

$$\boldsymbol{\lambda}(t) = \begin{Bmatrix} \lambda_1(t) \\ \lambda_2(t) \\ \vdots \\ \lambda_{2n}(t) \end{Bmatrix}. \tag{10.62}$$

Note that because Eq. 10.59 is used in Eq. 10.61, there is no external applied forcing function such as earthquake or wind. Minimizing the cost function J involves taking the variation of L with respect to $\mathbf{z}(t)$, $\mathbf{f}_c(t)$, and $\boldsymbol{\lambda}(t)$ and setting the result to zero. Doing so gives

$$\delta L = \delta \int_0^{t_f} \left\{ \frac{1}{2} \left[\mathbf{z}^T(t)\mathbf{Q}\mathbf{z}(t) + \mathbf{f}_c^T(t)\mathbf{R}\mathbf{f}_c(t) \right] + \boldsymbol{\lambda}^T(t) \left[\mathbf{A}\mathbf{z}(t) + \mathbf{B}\mathbf{f}_c(t) - \dot{\mathbf{z}}(t) \right] \right\} dt = 0. \tag{10.63}$$

Define a Hamiltonian function, which is a scalar function of time, as

$$H(t) = \frac{1}{2} \left[\mathbf{z}^T(t)\mathbf{Q}\mathbf{z}(t) + \mathbf{f}_c^T(t)\mathbf{R}\mathbf{f}_c(t) \right] + \boldsymbol{\lambda}^T(t) \left[\mathbf{A}\mathbf{z}(t) + \mathbf{B}\mathbf{f}_c(t) \right]. \tag{10.64}$$

Then Eq. 10.63 becomes

$$\delta L = \delta \int_0^{t_f} \left\{ H(t) - \boldsymbol{\lambda}^T(t)\dot{\mathbf{z}}(t) \right\} dt = \delta \int_0^{t_f} H(t)\, dt - \delta \int_0^{t_f} \boldsymbol{\lambda}^T(t)\dot{\mathbf{z}}(t)\, dt = 0. \tag{10.65}$$

The variation of the second term in Eq. 10.65 gives

$$\delta \int_0^{t_f} \boldsymbol{\lambda}^T(t)\dot{\mathbf{z}}(t)\, dt = \int_0^{t_f} \left[\delta\boldsymbol{\lambda}^T(t)\dot{\mathbf{z}}(t) + \boldsymbol{\lambda}^T(t)\delta\dot{\mathbf{z}}(t) \right] dt = \int_0^{t_f} \delta\boldsymbol{\lambda}^T(t)\dot{\mathbf{z}}(t)\, dt + \int_0^{t_f} \boldsymbol{\lambda}^T(t)\delta\dot{\mathbf{z}}(t)\, dt. \tag{10.66}$$

Integration by parts on the second term in Eq. 10.66 gives

$$\int_0^{t_f} \boldsymbol{\lambda}^T(t)\delta\dot{\mathbf{z}}(t)\, dt = \boldsymbol{\lambda}^T(t)\delta\mathbf{z}(t)\Big|_{t=0}^{t=t_f} - \int_0^{t_f} \dot{\boldsymbol{\lambda}}^T(t)\delta\mathbf{z}(t)\, dt$$

$$= \boldsymbol{\lambda}^T(t_f)\delta\mathbf{z}(t_f) - \boldsymbol{\lambda}^T(0)\delta\mathbf{z}(0) - \int_0^{t_f} \dot{\boldsymbol{\lambda}}^T(t)\delta\mathbf{z}(t)\, dt. \tag{10.67}$$

Substituting Eq. 10.67 into Eq. 10.66 gives

$$\delta \int_0^{t_f} \boldsymbol{\lambda}^T(t)\dot{\mathbf{z}}(t)\, dt = \boldsymbol{\lambda}^T(t_f)\delta\mathbf{z}(t_f) - \boldsymbol{\lambda}^T(0)\delta\mathbf{z}(0) + \int_0^{t_f} \delta\boldsymbol{\lambda}^T(t)\dot{\mathbf{z}}(t)\, dt - \int_0^{t_f} \dot{\boldsymbol{\lambda}}^T(t)\delta\mathbf{z}(t)\, dt. \tag{10.68}$$

Now consider the first term in Eq. 10.65; the variation gives

$$\delta \int_0^{t_f} H(t)\, dt = \int_0^{t_f} \left[\frac{\partial H}{\partial \mathbf{z}} \delta\mathbf{z}(t) + \frac{\partial H}{\partial \mathbf{f}_c} \delta\mathbf{f}_c(t) + \frac{\partial H}{\partial \boldsymbol{\lambda}} \delta\boldsymbol{\lambda}(t) \right] dt \tag{10.69}$$

where

$$\frac{\partial H(t)}{\partial \mathbf{z}} = \begin{bmatrix} \dfrac{\partial H}{\partial z_1} & \dfrac{\partial H}{\partial z_2} & \cdots & \dfrac{\partial H}{\partial z_{2n}} \end{bmatrix} = \begin{bmatrix} \dfrac{\partial H}{\partial x_1} & \cdots & \dfrac{\partial H}{\partial x_n} & \dfrac{\partial H}{\partial \dot{x}_1} & \cdots & \dfrac{\partial H}{\partial \dot{x}_n} \end{bmatrix}$$

$$\frac{\partial H(t)}{\partial \mathbf{f}_c} = \begin{bmatrix} \dfrac{\partial H}{\partial f_{c1}} & \dfrac{\partial H}{\partial f_{c2}} & \cdots & \dfrac{\partial H}{\partial f_{cp}} \end{bmatrix}$$

$$\frac{\partial H(t)}{\partial \boldsymbol{\lambda}} = \begin{bmatrix} \dfrac{\partial H}{\partial \lambda_1} & \dfrac{\partial H}{\partial \lambda_2} & \cdots & \dfrac{\partial H}{\partial \lambda_{2n}} \end{bmatrix}.$$

Substituting Eqs. 10.68 and 10.69 into Eq. 10.65 gives

$$\delta L = -\boldsymbol{\lambda}^T(t_f)\delta \mathbf{z}(t_f) + \boldsymbol{\lambda}^T(0)\delta \mathbf{z}(0)$$

$$+\int_0^{t_f} \delta\boldsymbol{\lambda}^T(t)\left[\left(\frac{\partial H}{\partial \boldsymbol{\lambda}}\right)^T - \dot{\mathbf{z}}(t)\right] dt + \int_0^{t_f}\left[\frac{\partial H}{\partial \mathbf{z}} + \dot{\boldsymbol{\lambda}}^T(t)\right]\delta \mathbf{z}(t)\, dt + \int_0^{t_f}\frac{\partial H}{\partial \mathbf{f}_c}\delta \mathbf{f}_c(t)\, dt = 0. \tag{10.70}$$

In order to satisfy Eq. 10.70, each term in Eq. 10.70 must be equal to zero:

$$\left(\frac{\partial H(t)}{\partial \boldsymbol{\lambda}}\right)^T - \dot{\mathbf{z}}(t) = \mathbf{0}, \quad \boldsymbol{\lambda}(t_f) = \mathbf{0} \tag{10.71a}$$

$$\left(\frac{\partial H(t)}{\partial \mathbf{z}}\right)^T + \dot{\boldsymbol{\lambda}}(t) = \mathbf{0}, \quad \mathbf{z}(0) = \mathbf{z}_o \tag{10.71b}$$

$$\left(\frac{\partial H(t)}{\partial \mathbf{f}_c}\right)^T = \mathbf{0} \tag{10.71c}$$

where $H(t)$ is given in Eq. 10.64. Equation 10.71 represents a set of three simultaneous equations with three unknown vectors, $\mathbf{z}(t)$, $\mathbf{f}_c(t)$, and $\boldsymbol{\lambda}(t)$. Note that the boundary condition in Eq. 10.71b is $\mathbf{z}(0) = \mathbf{z}_o$ (i.e., the initial condition of the structure), and therefore the variation will be zero, that is, $\delta \mathbf{z}(0) = \mathbf{0}$. In addition, Eqs. 10.71a and 10.71b together represent a *two-point boundary value problem,* in which the boundary condition of one variable is given in the initial state and the boundary condition of another variable is given in the final state.

To solve the set of simultaneous equations in Eq. 10.71, t_f is generally unknown in real time control. Therefore, the value of t_f is taken to be very large. Since damping always exists in the structure, the structural response always diminishes to zero for this large value of t_f. Thus the boundary condition $\boldsymbol{\lambda}(t_f) = \mathbf{0}$ can be dropped in Eq. 10.71a. Now substituting Eq. 10.64 into Eq. 10.71, it follows that

$$\left(\frac{\partial H(t)}{\partial \boldsymbol{\lambda}}\right)^T - \dot{\mathbf{z}}(t) = \mathbf{A}\mathbf{z}(t) + \mathbf{B}\mathbf{f}_c(t) - \dot{\mathbf{z}}(t) = 0 \tag{10.72a}$$

$$\left(\frac{\partial H(t)}{\partial \mathbf{z}}\right)^T + \dot{\boldsymbol{\lambda}}(t) = \mathbf{Q}\mathbf{z}(t) + \mathbf{A}^T\boldsymbol{\lambda}(t) + \dot{\boldsymbol{\lambda}}(t) = 0 \tag{10.72b}$$

$$\left(\frac{\partial H(t)}{\partial \mathbf{f}_c}\right)^T = \mathbf{R}\mathbf{f}_c(t) + \mathbf{B}^T\boldsymbol{\lambda}(t) = \mathbf{0}. \tag{10.72c}$$

Solving for $\mathbf{f}_c(t)$ in Eq. 10.72c, it follows that

$$\mathbf{f}_c(t) = -\mathbf{R}^{-1}\mathbf{B}^T\boldsymbol{\lambda}(t). \tag{10.73}$$

Substituting Eq. 10.73 into Eqs. 10.72a and 10.72b gives the two ordinary differential equations,

$$\dot{\mathbf{z}}(t) = \mathbf{A}\mathbf{z}(t) - \mathbf{B}\mathbf{R}^{-1}\mathbf{B}^T\boldsymbol{\lambda}(t), \quad \mathbf{z}(0) = \mathbf{z}_o \tag{10.74a}$$

$$-\dot{\boldsymbol{\lambda}}(t) = \mathbf{Q}\mathbf{z}(t) + \mathbf{A}^T\boldsymbol{\lambda}(t), \quad \boldsymbol{\lambda}(t_f) = \mathbf{0}. \tag{10.74b}$$

Now define a new $2n \times 2n$ matrix $\mathbf{P}(t)$ with the property that

$$\boldsymbol{\lambda}(t) = \mathbf{P}(t)\mathbf{z}(t). \tag{10.75}$$

Substituting Eq. 10.75 into Eqs. 10.74a and 10.74b, it follows that

$$\dot{\mathbf{z}}(t) = \mathbf{A}\mathbf{z}(t) - \mathbf{B}\mathbf{R}^{-1}\mathbf{B}^T\mathbf{P}(t)\mathbf{z}(t) \tag{10.76a}$$

$$-\dot{\mathbf{P}}(t)\mathbf{z}(t) - \mathbf{P}(t)\dot{\mathbf{z}}(t) = \mathbf{Q}\mathbf{z}(t) + \mathbf{A}^T\mathbf{P}(t)\mathbf{z}(t). \tag{10.76b}$$

Substituting Eq. 10.76a into Eqs. 10.76b, it follows that

$$-\dot{\mathbf{P}}(t)\mathbf{z}(t) - \mathbf{P}(t)\big(\mathbf{A}\mathbf{z}(t) - \mathbf{B}\mathbf{R}^{-1}\mathbf{B}^T\mathbf{P}(t)\mathbf{z}(t)\big) = \mathbf{Q}\mathbf{z}(t) + \mathbf{A}^T\mathbf{P}(t)\mathbf{z}(t). \tag{10.77}$$

Note that $\mathbf{z}(t)$ postmultiplies all terms on both sides of Eq. 10.77, and therefore for any $\mathbf{z}(t)$ the following equation must be satisfied:

$$-\dot{\mathbf{P}}(t) = \mathbf{A}^T\mathbf{P}(t) + \mathbf{P}(t)\mathbf{A} - \mathbf{P}(t)\mathbf{B}\mathbf{R}^{-1}\mathbf{B}^T\mathbf{P}(t) + \mathbf{Q}, \quad t \le t_f. \tag{10.78}$$

Equation 10.78 is known as the *Riccati equation*.

It had been shown that $\mathbf{P}(t)$ in the Riccati equation is a very slow varying function of time, and for engineering purposes it can be assumed to be a constant.[1] In addition, as stated previously, the value of t_f is taken to be very large. Therefore, assume

$$\mathbf{P}(t) = \mathbf{P}, \quad \dot{\mathbf{P}}(t) = \mathbf{0} \tag{10.79}$$

and then it follows from Eqs. 10.78 and 10.79 that

$$\mathbf{A}^T\mathbf{P} + \mathbf{P}\mathbf{A} - \mathbf{P}\mathbf{B}\mathbf{R}^{-1}\mathbf{B}^T\mathbf{P} + \mathbf{Q} = \mathbf{0}. \tag{10.80}$$

Equation 10.80 is known as the *steady-state Riccati equation,* and $\mathbf{P}$ is known as the *Riccati matrix*. Note that $\mathbf{P}$ is a function of $\mathbf{A}$ (a function of the mass, stiffness, and damping matrices), $\mathbf{B}$ (a function of the mass matrix and the control force distribution matrix), and the two weighting matrices $\mathbf{Q}$ and $\mathbf{R}$ in the cost function. The solution method for $\mathbf{P}$ using Eq. 10.80 is discussed later. However, note that once the Riccati matrix is determined, the optimal control force is determined by substituting Eq. 10.75 into Eq. 10.73:

$$\mathbf{f}_c(t) = -\mathbf{R}^{-1}\mathbf{B}^T\mathbf{P}\mathbf{z}(t). \tag{10.81}$$

This type of control law is known as *optimal linear control,* or optimal linear closed-loop control, because the control force is a linear function of the displacement and velocity of the structure.

The state space equation for the structure using optimal linear control follows from Eqs. 10.59 and 10.81:

$$\dot{\mathbf{z}}(t) = \mathbf{A}\mathbf{z}(t) - \mathbf{B}\mathbf{R}^{-1}\mathbf{B}^T\mathbf{P}\mathbf{z}(t) = \big[\mathbf{A} - \mathbf{B}\mathbf{R}^{-1}\mathbf{B}^T\mathbf{P}\big]\mathbf{z}(t). \tag{10.82}$$

Equation 10.82 is a constant coefficient, first-order linear differential equation. The solution to this equation is

$$\mathbf{z}(t) = e^{\left(\mathbf{A}-\mathbf{B}\mathbf{R}^{-1}\mathbf{B}^T\mathbf{P}\right)t}\mathbf{z}_o. \tag{10.83}$$

Structural engineers are usually interested in the response of a structure subjected to externally applied lateral forces, such as earthquakes. Therefore, an external force must be applied to the right-hand side of Eq. 10.59, and as noted from Eq. 10.40, it follows that

$$\dot{\mathbf{z}}(t) = \mathbf{A}\mathbf{z}(t) + \mathbf{F}(t) + \mathbf{B}\mathbf{f}_c(t). \tag{10.84}$$

In addition, the external force (e.g., earthquake ground motion) is almost always an irregular time-varying function of time that requires digitization. Therefore, the results of this section must be extended to include an external force represented in discrete time.

10.6   OPTIMAL LINEAR CONTROL WITHOUT EXTERNAL FORCE—DISCRETE TIME

The equations of discrete time control are derived in a very similar way to the continuous time control equations derived in the previous section without an external force such as earthquake or wind. Start with Eq. 10.48, where

$$\mathbf{z}_{k+1} = \mathbf{F}_s \mathbf{z}_k + \mathbf{G}\mathbf{f}_{ck}. \tag{10.85}$$

Define the cost function to be

$$J = \frac{1}{2} \sum_{k=0}^{N} \left(\mathbf{z}_k^T \mathbf{Q} \mathbf{z}_k + \mathbf{f}_{ck}^T \mathbf{R} \mathbf{f}_{ck} \right) \tag{10.86}$$

where N represents the final time step over the control period. The goal is again to minimize J in Eq. 10.86 subject to the constraint in Eq. 10.85. Define the Lagrangian as

$$L = \sum_{k=0}^{N-1} \left[\frac{1}{2} \left(\mathbf{z}_k^T \mathbf{Q} \mathbf{z}_k + \mathbf{f}_{ck}^T \mathbf{R} \mathbf{f}_{ck} \right) + \boldsymbol{\lambda}_{k+1}^T \left(\mathbf{F}_s \mathbf{z}_k + \mathbf{G}\mathbf{f}_{ck} - \mathbf{z}_{k+1} \right) \right] \tag{10.87}$$

where $\boldsymbol{\lambda}_{k+1}$ is the Lagrange multiplier. Note that the Lagrangian is a function of $\boldsymbol{\lambda}_{k+1}$, $\mathbf{z}_k$, and $\mathbf{f}_{ck}$, and therefore minimization must be performed over these three variables. Now define the Hamiltonian to be

$$H_k = \frac{1}{2} \left(\mathbf{z}_k^T \mathbf{Q} \mathbf{z}_k + \mathbf{f}_{ck}^T \mathbf{R} \mathbf{f}_{ck} \right) + \boldsymbol{\lambda}_{k+1}^T \left(\mathbf{F}_s \mathbf{z}_k + \mathbf{G}\mathbf{f}_{ck} \right). \tag{10.88}$$

Then substituting Eq. 10.88 into Eq. 10.87 gives

$$L = \sum_{k=0}^{N-1} \left[H_k - \boldsymbol{\lambda}_{k+1}^T \mathbf{z}_{k+1} \right] = H_0 - \boldsymbol{\lambda}_N^T \mathbf{z}_N + \sum_{k=1}^{N-1} \left[H_k - \boldsymbol{\lambda}_k^T \mathbf{z}_k \right]. \tag{10.89}$$

Taking the variation of the Lagrangian in Eq. 10.89 and setting the result equal to zero, it follows that

$$
\begin{aligned}
\delta L = {} & \frac{\partial H_0}{\partial \mathbf{z}_o} \delta \mathbf{z}_o + \frac{\partial H_0}{\partial \mathbf{f}_{co}} \delta \mathbf{f}_{co} + \frac{\partial H_0}{\partial \boldsymbol{\lambda}_o} \delta \boldsymbol{\lambda}_o - \delta \boldsymbol{\lambda}_N^T \mathbf{z}_N - \boldsymbol{\lambda}_N^T \delta \mathbf{z}_N \\
& + \sum_{k=1}^{N-1} \left[\left(\frac{\partial H_k}{\partial \mathbf{z}_k} - \boldsymbol{\lambda}_k^T \right) \delta \mathbf{z}_k + \frac{\partial H_k}{\partial \mathbf{f}_{ck}} \delta \mathbf{f}_{ck} + \left(\frac{\partial H_k}{\partial \boldsymbol{\lambda}_k} - \mathbf{z}_k^T \right) \delta \boldsymbol{\lambda}_k \right] = 0
\end{aligned}
\tag{10.90}
$$

where

$$\frac{\partial H_k}{\partial \mathbf{z}_k} = \left[\frac{\partial H_k}{\partial z_{1k}} \quad \frac{\partial H_k}{\partial z_{2k}} \quad \cdots \quad \frac{\partial H_k}{\partial z_{2n,k}} \right] = \left[\frac{\partial H_k}{\partial x_{1k}} \quad \cdots \quad \frac{\partial H_k}{\partial x_{n,k}} \quad \frac{\partial H_k}{\partial \dot{x}_{1k}} \quad \cdots \quad \frac{\partial H_k}{\partial \dot{x}_{n,k}} \right]$$

$$\frac{\partial H_k}{\partial \mathbf{f}_{ck}} = \left[\frac{\partial H_k}{\partial f_{c1k}} \quad \frac{\partial H_k}{\partial f_{c2k}} \quad \cdots \quad \frac{\partial H_k}{\partial f_{cpk}} \right]$$

$$\frac{\partial H_k}{\partial \boldsymbol{\lambda}_k} = \left[\frac{\partial H_k}{\partial \lambda_{1k}} \quad \frac{\partial H_k}{\partial \lambda_{2k}} \quad \cdots \quad \frac{\partial H_k}{\partial \lambda_{n,k}} \right]$$

Note that the control force at time zero is zero (i.e., $\mathbf{f}_{co} = \mathbf{0}$). Therefore, Eq. 10.90 becomes

$$
\begin{aligned}
\delta L = {} & \frac{\partial H_0}{\partial \mathbf{z}_o} \delta \mathbf{z}_o + \frac{\partial H_0}{\partial \boldsymbol{\lambda}_o} \delta \boldsymbol{\lambda}_o - \delta \boldsymbol{\lambda}_N^T \mathbf{z}_N - \boldsymbol{\lambda}_N^T \delta \mathbf{z}_N \\
& + \sum_{k=1}^{N-1} \left[\left(\frac{\partial H_k}{\partial \mathbf{z}_k} - \boldsymbol{\lambda}_k^T \right) \delta \mathbf{z}_k + \frac{\partial H_k}{\partial \mathbf{f}_{ck}} \delta \mathbf{f}_{ck} + \left(\frac{\partial H_k}{\partial \boldsymbol{\lambda}_k} - \mathbf{z}_{k+1}^T \right) \delta \boldsymbol{\lambda}_k \right] \\
= {} & \frac{\partial H_0}{\partial \mathbf{z}_o} \delta \mathbf{z}_o - \boldsymbol{\lambda}_N^T \delta \mathbf{z}_N + \sum_{k=1}^{N-1} \left[\left(\frac{\partial H_k}{\partial \mathbf{z}_k} - \boldsymbol{\lambda}_k^T \right) \delta \mathbf{z}_k + \frac{\partial H_k}{\partial \mathbf{f}_{ck}} \delta \mathbf{f}_{ck} \right] + \sum_{k=0}^{N-1} \left(\frac{\partial H_k}{\partial \boldsymbol{\lambda}_k} - \mathbf{z}_{k+1}^T \right) \delta \boldsymbol{\lambda}_k = 0.
\end{aligned}
\tag{10.91}
$$

In order to satisfy Eq. 10.91, each term in Eq. 10.91 must be equal to zero, that is,

$$\left(\frac{\partial H_k}{\partial \boldsymbol{\lambda}_{k+1}}\right)^T - \mathbf{z}_{k+1} = \mathbf{0}, \quad \boldsymbol{\lambda}_N = \mathbf{0} \tag{10.92a}$$

$$\left(\frac{\partial H_k}{\partial \mathbf{z}_k}\right)^T - \boldsymbol{\lambda}_k = \mathbf{0}, \quad \mathbf{z}_o = \mathbf{z}(0) \tag{10.92b}$$

$$\left(\frac{\partial H_k}{\partial \mathbf{f}_{ck}}\right)^T = \mathbf{0} \tag{10.92c}$$

where H_k is given in Eq. 10.88. Equation 10.92 represents a set of three simultaneous difference equations with three unknown vectors, $\mathbf{z}_k$, $\mathbf{f}_{ck}$, and $\boldsymbol{\lambda}_k$. Similar to the derivation in continuous time in Section 10.5, note that the boundary condition in Eq. 10.92b is $\mathbf{z}(0) = \mathbf{z}_o$ (i.e., the initial condition of the structure), and therefore the variation will be zero, that is, $\delta \mathbf{z}(0) = \mathbf{0}$. In addition, Eqs. 10.92a and 10.92b together represent a *two-point boundary value problem*, in which the boundary condition of one variable is given in the initial state and the boundary condition of another variable is given in the final state.

Note also that in Eq. 10.92, N is generally unknown in real time control. Therefore, this value is generally taken to be very large. Since damping always exists in the structure, the structural response always diminishes to zero for this large value of N. Thus the boundary condition $\boldsymbol{\lambda}_N = \mathbf{0}$ can be dropped in Eq. 10.92a. Substituting Eq. 10.88 into Eq. 10.92 and differentiating, it follows that

$$\left(\frac{\partial H_k}{\partial \boldsymbol{\lambda}_{k+1}}\right)^T - \mathbf{z}_{k+1} = \mathbf{F}_s \mathbf{z}_k + \mathbf{G}\mathbf{f}_{ck} - \mathbf{z}_{k+1} = \mathbf{0} \tag{10.93a}$$

$$\left(\frac{\partial H_k}{\partial \mathbf{z}_k}\right)^T - \boldsymbol{\lambda}_k = \mathbf{Q}\mathbf{z}_k + \mathbf{F}_s^T \boldsymbol{\lambda}_{k+1} - \boldsymbol{\lambda}_k = \mathbf{0} \tag{10.93b}$$

$$\left(\frac{\partial H_k}{\partial \mathbf{f}_{ck}}\right)^T = \mathbf{R}\mathbf{f}_{ck} + \mathbf{G}^T \boldsymbol{\lambda}_{k+1} = \mathbf{0}. \tag{10.93c}$$

Solving for $\mathbf{f}_{ck}$ in Eq. 10.93c gives the control law

$$\mathbf{f}_{ck} = -\mathbf{R}^{-1}\mathbf{G}^T \boldsymbol{\lambda}_{k+1}. \tag{10.94}$$

Substituting Eq. 10.94 into Eq. 10.93a, it follows from Eqs. 10.93a and 10.93b that

$$\mathbf{z}_{k+1} = \mathbf{F}_s \mathbf{z}_k - \mathbf{G}\mathbf{R}^{-1}\mathbf{G}^T \boldsymbol{\lambda}_{k+1}, \quad \mathbf{z}(0) = \mathbf{z}_o \tag{10.95a}$$

$$\boldsymbol{\lambda}_k = \mathbf{Q}\mathbf{z}_k + \mathbf{F}_s^T \boldsymbol{\lambda}_{k+1}, \quad \boldsymbol{\lambda}_N = \mathbf{0}. \tag{10.95b}$$

Now define $\mathbf{P}_k$ to be a $2n \times 2n$ matrix with the property that

$$\boldsymbol{\lambda}_k = \mathbf{P}_k \mathbf{z}_k. \tag{10.96}$$

Substituting Eq. 10.96 into Eqs. 10.95a and 10.95b, it follows that

$$\mathbf{z}_{k+1} = \mathbf{F}_s \mathbf{z}_k - \mathbf{G}\mathbf{R}^{-1}\mathbf{G}^T \mathbf{P}_{k+1}\mathbf{z}_{k+1} \tag{10.97a}$$

$$\mathbf{P}_k \mathbf{z}_k = \mathbf{Q}\mathbf{z}_k + \mathbf{F}_s^T \mathbf{P}_{k+1}\mathbf{z}_{k+1}. \tag{10.97b}$$

Using Eq. 10.97a to solve for $\mathbf{z}_{k+1}$ gives

$$\mathbf{z}_{k+1} = \left(\mathbf{I} + \mathbf{G}\mathbf{R}^{-1}\mathbf{G}^T \mathbf{P}_{k+1}\right)^{-1} \mathbf{F}_s \mathbf{z}_k. \tag{10.98}$$

In Eq. 10.98, the matrix $\mathbf{P}_k$ is unknown, and it must be determined using Eq. 10.97b. Substituting Eq. 10.98 into Eq. 10.97b, it follows that

$$P_k z_k = Q z_k + F_s^T P_{k+1} \left(I + GR^{-1} G^T P_{k+1} \right)^{-1} F_s z_k. \tag{10.99}$$

Note that z_k occurs in all terms on both sides of Eq. 10.99. Therefore, it follows that

$$P_k = Q + F_s^T P_{k+1} \left(I + GR^{-1} G^T P_{k+1} \right)^{-1} F_s. \tag{10.100}$$

Equation 10.100 is known as the *discrete time Riccati equation,* and P_k is known as the *discrete time Riccati matrix.* Again, as was assumed in the previous section, the Riccati matrix is assumed to be a constant (i.e., $P_k = P_{k+1} = P$), and it follows from Eq. 10.100 that

$$P = Q + F_s^T P \left(I + GR^{-1} G^T P \right)^{-1} F_s. \tag{10.101}$$

Equation 10.101 is known as the *steady-state Riccati equation,* and P is known as the *steady-state Riccati matrix.* Equation 10.101 is a quadratic function of P, and therefore there are two solutions of P. One of the two solutions gives a P that is a positive definite matrix, and this is the desired solution. The control law is obtained by substituting Eq. 10.96 into Eq. 10.94, that is,

$$f_{ck} = -R^{-1} G^T P z_{k+1} \tag{10.102}$$

and then after substituting Eq. 10.85 into Eq. 10.102, it follows that

$$f_{ck} = -R^{-1} G^T P \left(F_s z_k + G f_{ck} \right). \tag{10.103}$$

Solving for f_{ck} in Eq. 10.103,

$$f_{ck} = -\left(G^T P G + R \right)^{-1} G^T P F_s z_k. \tag{10.104}$$

Note that the control force f_{ck} is a linear function of z_k. Substituting Eq. 10.104 into Eq. 10.85 gives

$$z_{k+1} = \left[I - G \left(G^T P G + R \right)^{-1} G^T P \right] F_s z_k. \tag{10.105}$$

Note that z_{k+1} is expressed in one form in Eq. 10.98, whereas it is expressed in another form in Eq. 10.105. Therefore, let

$$F_f = \left[I - G \left(G^T P G + R \right)^{-1} G^T P \right] F_s = \left[I + GR^{-1} G^T P \right]^{-1} F_s \tag{10.106}$$

and it then follows from Eq. 10.105 that

$$z_{k+1} = F_f z_k. \tag{10.107}$$

Equation 10.107 represents the response equation, and z_k represents the state variables (i.e., displacement versus time X_k and velocity versus time $\dot{X}_k$). Therefore, the displacement and velocity at time step $k + 1$ (i.e., one time step from now) are equal to the displacement and velocity at time step k (i.e., now) times a *state transition matrix* F_f. The state transition matrix, defined in Eq. 10.106, is a function of $G, R, P,$ and F_s. A summary of these terms follows:

1. G: The equation for G is given in Eq. 10.46 or Eq. 10.47, which is

    $G = e^{A\Delta t} B \Delta t$ when Delta forcing function assumption is used, or

    $G = A^{-1} \left(e^{A\Delta t} - I \right)$ when constant forcing function assumption is used,

    where

    $$A = \begin{bmatrix} 0 & I \\ -M^{-1}K & -M^{-1}C \end{bmatrix} = \text{function of system mass, stiffness, and damping matrices}$$

    $$B = \begin{bmatrix} 0 \\ M^{-1}D \end{bmatrix} = \text{function of mass and the control force distribution matrix } D$$

    $\Delta t = t_{k+1} - t_k = \text{time step.}$

Therefore, $\mathbf{G}$ is a function of the mass, stiffness, and damping of the structure, the location of the control forces, and the time step.

2. $\mathbf{F}_s$: The equation for $\mathbf{F}_s$ is given in Eq. 10.46 or Eq. 10.47, which is

$$\mathbf{F}_s = e^{\mathbf{A}\Delta t}.$$

The $\mathbf{F}_s$ term is a function of the mass, stiffness, and damping of the structure and the time step.

3. $\mathbf{R}$: The matrix $\mathbf{R}$ is a weighting matrix in Eq. 10.86 of the cost function. This weighting matrix is applied to the control force.

4. $\mathbf{P}$: The equation for P is given in Eq. 10.101, which is

$$\mathbf{P} = \mathbf{Q} + \mathbf{F}_s^T \mathbf{P}\left(\mathbf{I} + \mathbf{G}\mathbf{R}^{-1}\mathbf{G}^T\mathbf{P}\right)^{-1}\mathbf{F}_s$$

where $\mathbf{Q}$ is the weighting matrix in Eq. 10.86 for the cost function. Therefore, $\mathbf{P}$ is a function of the mass, the stiffness, the damping in the structure, both weighting matrices, the location of the control forces, and the time step.

Table 10.1 summarizes the procedure for calculating the response and the control forces.

EXAMPLE 1   *Optimal Linear Control of an SDOF System*

To illustrate the theory of optimal linear control, consider the SDOF system shown in Figure 10.5a. As shown in Figure 10.5b, if the displacement at time step k (i.e., x_k) moves to the right relative to the ground, then a control force $f_k(L)$ is applied to the mass. If the

Table 10.1 Response and Control Force Equations

Response	$\mathbf{z}_{k+1} = \left[\mathbf{I} - \mathbf{G}\left(\mathbf{G}^T\mathbf{P}\mathbf{G} + \mathbf{R}\right)^{-1}\mathbf{G}^T\mathbf{P}\right]\mathbf{F}_s\mathbf{z}_k$
	$\mathbf{z}_{k+1} = \left(\mathbf{I} + \mathbf{G}\mathbf{R}^{-1}\mathbf{G}^T\mathbf{P}_{k+1}\right)^{-1}\mathbf{F}_s\mathbf{z}_k$
	$\mathbf{z}_{k+1} = \mathbf{F}_s\mathbf{z}_k + \mathbf{G}\mathbf{f}_{ck}$
Control force	$\mathbf{f}_{ck} = -\left(\mathbf{G}^T\mathbf{P}\mathbf{G} + \mathbf{R}\right)^{-1}\mathbf{G}^T\mathbf{P}\mathbf{F}_s\mathbf{z}_k$

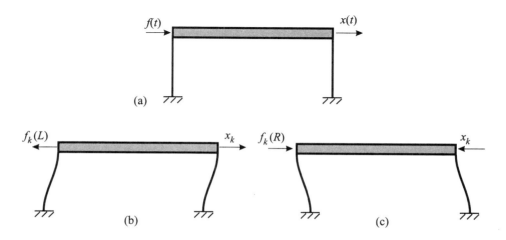

Figure 10.5 Single Degree of Freedom System

displacement x_k moves to the left relative to the ground, then, as shown in Figure 10.5c, a control force $f_k(R)$ is applied to the mass.

Consider the case where no external wind or earthquake force is applied but the system is released from rest. Assume that the system properties are

$$m = 1.0 \text{ k-s}^2/\text{in.}, \quad c = 1.2567 \text{ k-s/in.}, \quad k = 157.92 \text{ k/in.}$$

and the initial conditions are $x_o = 1.0$ and $\dot{x}_o = 0.0$, that is,

$$\mathbf{z}_o = \begin{Bmatrix} x_o \\ \dot{x}_o \end{Bmatrix} = \begin{Bmatrix} 1.0 \\ 0.0 \end{Bmatrix}.$$

Also, assume that

$$\mathbf{Q} = \begin{bmatrix} k & 0 \\ 0 & m \end{bmatrix} = \begin{bmatrix} 157.92 & 0.0 \\ 0.0 & 1.0 \end{bmatrix}.$$

Recall that the $\mathbf{Q}$ matrix is selected by the structural engineer. Here, the $\mathbf{Q}$ matrix serves as a relative energy matrix because the product $\frac{1}{2}\mathbf{z}_k^T\mathbf{Q}\mathbf{z}_k$ leads to

$$\frac{1}{2}\mathbf{z}_k^T\mathbf{Q}\mathbf{z}_k = \frac{1}{2}\begin{Bmatrix} x_k \\ \dot{x}_k \end{Bmatrix}^T \begin{bmatrix} k & 0 \\ 0 & m \end{bmatrix}\begin{Bmatrix} x_k \\ \dot{x}_k \end{Bmatrix} = \frac{1}{2}kx_k^2 + \frac{1}{2}m\dot{x}_k^2 \tag{10.108}$$

and therefore this product is the sum of potential and kinetic energy. Let $\Delta t = 0.02$ s and assume that a constant forcing function is used in the numerical integration procedure (see Section 2.8). It follows that

$$\mathbf{A} = \begin{bmatrix} 0.0 & 1.0 \\ -157.92 & -1.2567 \end{bmatrix}, \quad \mathbf{B} = \begin{bmatrix} 0 \\ 1 \end{bmatrix}, \quad \mathbf{F}_s = e^{\mathbf{A}\Delta t} = \begin{bmatrix} 0.968843 & 0.019543 \\ -3.086307 & 0.944284 \end{bmatrix}$$

$$\mathbf{A}^{-1}(e^{\mathbf{A}\Delta t} - \mathbf{I}) = \begin{bmatrix} 0.01979 & 0.0001973 \\ -0.031157 & 0.019543 \end{bmatrix}, \quad \mathbf{G} = \mathbf{A}^{-1}(e^{\mathbf{A}\Delta t} - \mathbf{I})\mathbf{B} = \begin{bmatrix} 0.0001973 \\ 0.019543 \end{bmatrix}.$$

The solution procedure requires the calculation of the steady-state Riccati matrix. Equation 10.101 is used to solve for this matrix:

$$\mathbf{P} = \mathbf{Q} + \mathbf{F}_s^T\mathbf{P}(\mathbf{I} + \mathbf{G}\mathbf{R}^{-1}\mathbf{G}^T\mathbf{P})^{-1}\mathbf{F}_s. \tag{10.109}$$

Since $\mathbf{P}$ is a positive definite matrix, it is symmetrical. Therefore, $\mathbf{P}$ will take the form

$$\mathbf{P} = \begin{bmatrix} p_1 & p_2 \\ p_2 & p_3 \end{bmatrix}.$$

To simplify the equation, denote

$$\mathbf{F}_s = \begin{bmatrix} f_1 & f_2 \\ f_3 & f_4 \end{bmatrix}, \quad \mathbf{G} = \begin{bmatrix} g_1 \\ g_2 \end{bmatrix}, \quad \mathbf{Q} = \begin{bmatrix} q_1 & q_2 \\ q_2 & q_3 \end{bmatrix}, \quad \mathbf{R} = R = \frac{1}{r}. \tag{10.110}$$

Then

$$\mathbf{G}\mathbf{R}^{-1}\mathbf{G}^T = r\begin{bmatrix} g_1 \\ g_2 \end{bmatrix}\begin{bmatrix} g_1 & g_2 \end{bmatrix} = r\begin{bmatrix} g_1^2 & g_1g_2 \\ g_1g_2 & g_2^2 \end{bmatrix} \tag{10.111}$$

$$\mathbf{I} + \mathbf{G}\mathbf{R}^{-1}\mathbf{G}^T\mathbf{P} = \begin{bmatrix} 1 & 0 \\ 0 & 1 \end{bmatrix} + r\begin{bmatrix} g_1^2 & g_1g_2 \\ g_1g_2 & g_2^2 \end{bmatrix}\begin{bmatrix} p_1 & p_2 \\ p_2 & p_3 \end{bmatrix}$$

$$= \begin{bmatrix} 1 + rg_1^2p_1 + rg_1g_2p_2 & rg_1^2p_2 + rg_1g_2p_3 \\ rg_1g_2p_1 + rg_2^2p_2 & 1 + rg_1g_2p_2 + rg_2^2p_3 \end{bmatrix} \tag{10.112}$$

and taking the inverse of Eq. 10.112 gives

$$\left(\mathbf{I} + \mathbf{GR}^{-1}\mathbf{G}^T\mathbf{P}\right)^{-1} = \frac{1}{D(p)}\begin{bmatrix} 1 + rg_1g_2p_2 + rg_2^2p_3 & -rg_1^2p_2 - rg_1g_2p_3 \\ -rg_1g_2p_1 - rg_2^2p_2 & 1 + rg_1^2p_1 + rg_1g_2p_2 \end{bmatrix} \quad (10.113)$$

where D(p) is the determinant of Eq. 10.112, which is a function of p_1, p_2, and p_3, and is given by

$$D(p) = \begin{vmatrix} 1 + rg_1g_2p_2 + rg_2^2p_3 & -rg_1^2p_2 - rg_1g_2p_3 \\ -rg_1g_2p_1 - rg_2^2p_2 & 1 + rg_1^2p_1 + rg_1g_2p_2 \end{vmatrix} = 1 + rg_1^2p_1 + 2rg_1g_2p_2 + rg_2^2p_3. \quad (10.114)$$

For simplicity, denote

$$\mathbf{P}' = \mathbf{P}\left(\mathbf{I} + \mathbf{GR}^{-1}\mathbf{G}^T\mathbf{P}\right)^{-1}. \quad (10.115)$$

Then it follows from Eqs. 10.113 and 10.115 that

$$\mathbf{P}' = \frac{1}{D(p)}\begin{bmatrix} p_1 & p_2 \\ p_2 & p_3 \end{bmatrix}\begin{bmatrix} 1 + rg_1g_2p_2 + rg_2^2p_3 & -rg_1^2p_2 - rg_1g_2p_3 \\ -rg_1g_2p_1 - rg_2^2p_2 & 1 + rg_1^2p_1 + rg_1g_2p_2 \end{bmatrix}$$

$$= \frac{1}{D(p)}\begin{bmatrix} p_1 + rg_2^2(p_1p_3 - p_2^2) & p_2 + rg_1g_2(p_2^2 - p_1p_3) \\ p_2 + rg_1g_2(p_2^2 - p_1p_3) & p_3 + rg_1^2(p_1p_3 - p_2^2) \end{bmatrix}. \quad (10.116)$$

Note that **P**′ is also a symmetrical matrix. To simplify Eq. 10.116, substitute three new variables

$$\begin{aligned} p_1' &= p_1 + rg_2^2(p_1p_3 - p_2^2) \\ p_2' &= p_2 + rg_1g_2(p_2^2 - p_1p_3) \\ p_3' &= p_3 + rg_1^2(p_1p_3 - p_2^2). \end{aligned} \quad (10.117)$$

Equation 10.116 becomes

$$\mathbf{P}' = \frac{1}{D(p)}\begin{bmatrix} p_1' & p_2' \\ p_2' & p_3' \end{bmatrix}. \quad (10.118)$$

Then, rearranging terms in Eq. 10.109, it follows that

$$\mathbf{P} - \mathbf{Q} = \mathbf{F}_s^T\mathbf{P}\left(\mathbf{I} + \mathbf{GR}^{-1}\mathbf{G}^T\mathbf{P}\right)^{-1}\mathbf{F}_s = \mathbf{F}_s^T\mathbf{P}'\mathbf{F}_s$$

$$= \frac{1}{D(p)}\begin{bmatrix} f_1 & f_3 \\ f_2 & f_4 \end{bmatrix}\begin{bmatrix} p_1' & p_2' \\ p_2' & p_3' \end{bmatrix}\begin{bmatrix} f_1 & f_2 \\ f_3 & f_4 \end{bmatrix}$$

$$= \frac{1}{D(p)}\begin{bmatrix} f_1^2p_1' + 2f_1f_3p_2' + f_3^2p_3' & f_1f_2p_1' + (f_1f_4 + f_2f_3)p_2' + f_3f_4p_3' \\ f_1f_2p_1' + (f_1f_4 + f_2f_3)p_2' + f_3f_4p_3' & f_2^2p_1' + 2f_2f_4p_2' + f_4^2p_3' \end{bmatrix}. \quad (10.119)$$

Finally, Eq. 10.119 gives a set of three simultaneous equations for the three unknowns, which are

$$p_1 = q_1 + \frac{f_1^2p_1' + 2f_1f_3p_2' + f_3^2p_3'}{D(p)} \quad (10.120a)$$

$$p_2 = q_2 + \frac{f_1f_2p_1' + (f_1f_4 + f_2f_3)p_2' + f_3f_4p_3'}{D(p)} \quad (10.120b)$$

$$p_3 = q_3 + \frac{f_2^2p_1' + 2f_2f_4p_2' + f_4^2p_3'}{D(p)}. \quad (10.120c)$$

Equation 10.120 gives the general solution for the terms in the steady-state Riccati matrix for an SDOF system.

To evaluate this steady-state Riccati matrix **P**, first consider the case where $R = 1.0$ (i.e., $r = 1/R = 1.0$). Solving for the p_i's using the simultaneous quadratic equations given in Eq. 10.120, it follows that

$$\mathbf{P} = \begin{bmatrix} 5{,}143.073 & 25.015 \\ 25.015 & 32.274 \end{bmatrix}.$$

From Eq. 10.104, the control force applied at time step k is calculated to be

$$f_{ck} = -\left(\mathbf{G}^T\mathbf{PG} + \mathbf{R}\right)^{-1}\mathbf{G}^T\mathbf{PF}_s\mathbf{z}_k = \begin{bmatrix} 0.49877 & -0.62173 \end{bmatrix}\begin{Bmatrix} x_k \\ \dot{x}_k \end{Bmatrix}$$

$$= 0.49877x_k - 0.62173\dot{x}_k. \tag{10.121}$$

It then follows from Eq. 10.106 that

$$\mathbf{F}_f = \left(\mathbf{I} + \mathbf{G}\mathbf{R}^{-1}\mathbf{G}^T\mathbf{P}\right)^{-1}\mathbf{F}_s = \begin{bmatrix} 0.968942 & 0.019421 \\ -3.076558 & 0.932133 \end{bmatrix}$$

and therefore the response at time step $k + 1$ is

$$\begin{Bmatrix} x_{k+1} \\ \dot{x}_{k+1} \end{Bmatrix} = \begin{bmatrix} 0.968942 & 0.019421 \\ -3.076558 & 0.932133 \end{bmatrix}\begin{Bmatrix} x_k \\ \dot{x}_k \end{Bmatrix}, \quad \begin{Bmatrix} x_o \\ \dot{x}_o \end{Bmatrix} = \begin{Bmatrix} 1.0 \\ 0.0 \end{Bmatrix}. \tag{10.122}$$

Note that when there is no control, the response at time step $k + 1$ is

$$\begin{Bmatrix} x_{k+1} \\ \dot{x}_{k+1} \end{Bmatrix} = e^{\mathbf{A}\Delta t}\begin{Bmatrix} x_k \\ \dot{x}_k \end{Bmatrix} = \begin{bmatrix} 0.968843 & 0.019543 \\ -3.086307 & 0.944284 \end{bmatrix}\begin{Bmatrix} x_k \\ \dot{x}_k \end{Bmatrix}, \quad \begin{Bmatrix} x_o \\ \dot{x}_o \end{Bmatrix} = \begin{Bmatrix} 1.0 \\ 0.0 \end{Bmatrix}. \tag{10.123}$$

The difference in response between Eq. 10.122 and Eq. 10.123 is shown in Figure 10.6. Compared with the uncontrolled response, the control force applied to the structure has a very small effect in reducing the displacement and velocity responses. This shows the lack of effectiveness when a large penalty is imposed on the control law (i.e., a large value of R limits the amount of control force).

Now consider the case where $R = 0.1$ (i.e., $r = 1/R = 10.0$); using Eq. 10.120 gives

$$\mathbf{P} = \begin{bmatrix} 2{,}861.324 & 24.780 \\ 24.780 & 17.408 \end{bmatrix}.$$

From Eq. 10.104, the control force applied at time step k is calculated to be

$$f_{ck} = -\left(\mathbf{G}^T\mathbf{PG} + \mathbf{R}\right)^{-1}\mathbf{G}^T\mathbf{PF}_s\mathbf{z}_k = \begin{bmatrix} 0.45748 & -3.23852 \end{bmatrix}\begin{Bmatrix} x_k \\ \dot{x}_k \end{Bmatrix}$$

$$= 0.45748x_k - 3.23852\dot{x}_k. \tag{10.124}$$

It then follows from Eq. 10.106 that

$$\mathbf{F}_f = \left(\mathbf{I} + \mathbf{G}\mathbf{R}^{-1}\mathbf{G}^T\mathbf{P}\right)^{-1}\mathbf{F}_s = \begin{bmatrix} 0.968933 & 0.018905 \\ -3.077363 & 0.880991 \end{bmatrix}$$

and therefore the response at time step $k + 1$ is

$$\begin{Bmatrix} x_{k+1} \\ \dot{x}_{k+1} \end{Bmatrix} = \begin{bmatrix} 0.968933 & 0.018905 \\ -3.077363 & 0.880991 \end{bmatrix}\begin{Bmatrix} x_k \\ \dot{x}_k \end{Bmatrix}, \quad \begin{Bmatrix} x_o \\ \dot{x}_o \end{Bmatrix} = \begin{Bmatrix} 1.0 \\ 0.0 \end{Bmatrix}. \tag{10.125}$$

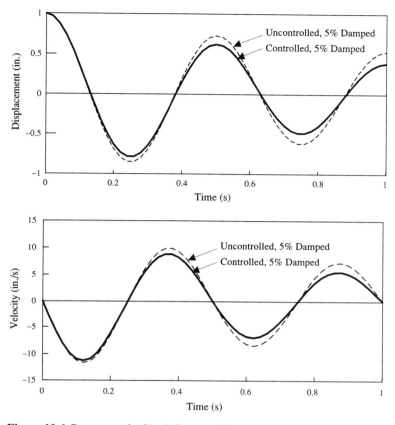

Figure 10.6 Response of a Single Degree of Freedom System with $R = 1.0$

This result is shown in Figure 10.7. Note that the controlled displacement and velocity responses are very similar to those calculated using 20% damping with no control (i.e., $c = 2\zeta\sqrt{km} = 5.0266$ k-s/in.). This shows the impact of structural control and how it can, if desired, increase the effective damping of the structure.

Now consider the case where $R = 0.01$ (i.e., $r = 1/R = 100.0$), using Eq. 10.120 gives

$$\mathbf{P} = \begin{bmatrix} 1,373.530 & 22.286 \\ 22.286 & 6.8120 \end{bmatrix}.$$

From Eq. 10.104, the control force applied at time step k is calculated to be

$$f_{ck} = -\left(\mathbf{G}^T\mathbf{P}\mathbf{G} + \mathbf{R}\right)^{-1}\mathbf{G}^T\mathbf{P}\mathbf{F}_s z_k = \begin{bmatrix} -20.2758 & -11.2005 \end{bmatrix} \begin{Bmatrix} x_k \\ \dot{x}_k \end{Bmatrix}$$

$$= -20.2758 x_k - 11.2005\dot{x}_k.$$

(10.126)

It then follows from Eq. 10.106 that

$$\mathbf{F}_f = \left(\mathbf{I} + \mathbf{G}\mathbf{R}^{-1}\mathbf{G}^T\mathbf{P}\right)^{-1}\mathbf{F}_s = \begin{bmatrix} 0.964843 & 0.017334 \\ -3.482553 & 0.725383 \end{bmatrix}$$

and therefore the response at time step $k + 1$ is

$$\begin{Bmatrix} x_{k+1} \\ \dot{x}_{k+1} \end{Bmatrix} = \begin{bmatrix} 0.964843 & 0.017334 \\ -3.482553 & 0.725383 \end{bmatrix} \begin{Bmatrix} x_k \\ \dot{x}_k \end{Bmatrix}, \quad \begin{Bmatrix} x_o \\ \dot{x}_o \end{Bmatrix} = \begin{Bmatrix} 1.0 \\ 0.0 \end{Bmatrix}.$$

(10.127)

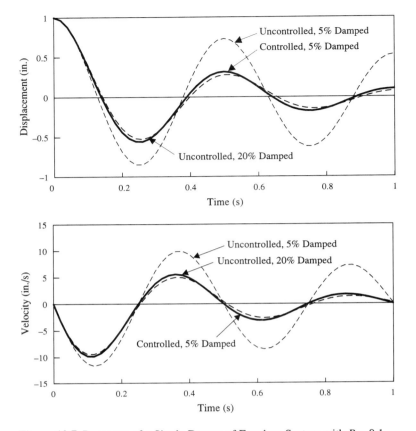

Figure 10.7 Response of a Single Degree of Freedom System with $R = 0.1$

The result is shown in Figure 10.8. Note that the controlled displacement and velocity responses are very similar to those calculated using 50% damping with no control (i.e., $c = 12.567$ k-s/in.).

Finally, consider the case where $R = 0.001$ (i.e., $r = 1/R = 1000.0$); using Eq. 10.120 gives

$$\mathbf{P} = \begin{bmatrix} 919.1673 & 14.428 \\ 14.428 & 2.5227 \end{bmatrix}. \tag{10.128}$$

From Eq. 10.104, the control force applied at time step k is calculated to be

$$f_{ck} = -\left(\mathbf{G}^T \mathbf{P} \mathbf{G} + \mathbf{R}\right)^{-1} \mathbf{G}^T \mathbf{P} \mathbf{F}_s \mathbf{z}_k = \begin{bmatrix} -136.429 & -27.6219 \end{bmatrix} \begin{Bmatrix} x_k \\ \dot{x}_k \end{Bmatrix}$$

$$= -136.429 x_k - 27.6219 \dot{x}_k. \tag{10.129}$$

It then follows from Eq. 10.106 that

$$\mathbf{F}_f = \left(\mathbf{I} + \mathbf{G} \mathbf{R}^{-1} \mathbf{G}^T \mathbf{P}\right)^{-1} \mathbf{F}_s = \begin{bmatrix} 0.941927 & 0.014094 \\ -5.75252 & 0.404455 \end{bmatrix} \tag{10.130}$$

and therefore the response at time step $k + 1$ is

$$\begin{Bmatrix} x_{k+1} \\ \dot{x}_{k+1} \end{Bmatrix} = \begin{bmatrix} 0.941927 & 0.014094 \\ -5.75252 & 0.404455 \end{bmatrix} \begin{Bmatrix} x_k \\ \dot{x}_k \end{Bmatrix}, \quad \begin{Bmatrix} x_o \\ \dot{x}_o \end{Bmatrix} = \begin{Bmatrix} 1.0 \\ 0.0 \end{Bmatrix}. \tag{10.131}$$

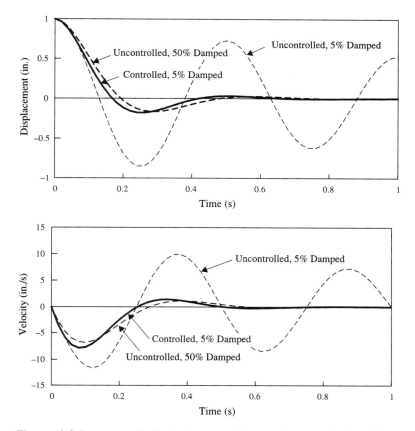

Figure 10.8 Response of a Single Degree of Freedom System with $R = 0.01$

The result is shown in Figure 10.9. Note that the controlled displacement is smaller than the uncontrolled displacement with 100% damping (i.e., $c = 25.133$ k-s/in.).

EXAMPLE 2   *Numerical Solution to the Steady-State Riccati Matrix*

The procedure discussed in Example 1 is applicable for a single degree of freedom system. When there is more than one degree of freedom in the structure, which is usually the case, the procedure discussed will involve intensive algebraic computations. This is highly undesirable and can result in numerical error. Therefore, an alternate solution method is necessary.

Several solution methods exist including Vaughn's direct method[2] and the Laub method.[3] The major problem in solving for the Riccati matrix **P** is that the steady-state Riccati equation is a quadratic equation in **P**, and therefore there is more than one solution.

An iterative method is presented here to solve for the steady-state Riccati matrix. Consider the Riccati equation given in Eq. 10.101, which is

$$\mathbf{P} = \mathbf{Q} + \mathbf{F}_s^T \mathbf{P} \left(\mathbf{I} + \mathbf{G} \mathbf{R}^{-1} \mathbf{G}^T \mathbf{P} \right)^{-1} \mathbf{F}_s. \qquad (10.132)$$

Obtain the initial iteration to the solution by letting

$$\mathbf{P} = \mathbf{P}_0 = \mathbf{Q}. \qquad (10.133)$$

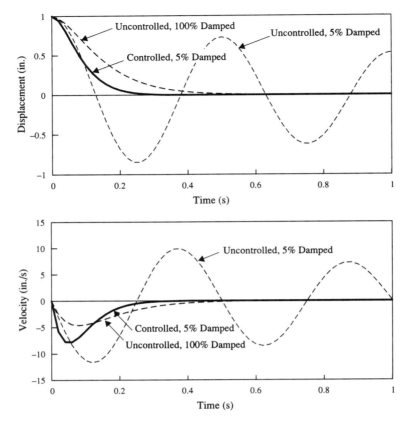

Figure 10.9 Response of a Single Degree of Freedom System with $R = 0.001$

Substitute $\mathbf{P}_0$ into the right-hand side of Eq. 10.132 to obtain $\mathbf{P}_1$:

$$\mathbf{P}_1 = \mathbf{Q} + \mathbf{F}_s^T \mathbf{P}_0 \left(\mathbf{I} + \mathbf{GR}^{-1}\mathbf{G}^T\mathbf{P}_0\right)^{-1} \mathbf{F}_s. \tag{10.134}$$

Then substitute $\mathbf{P}_1$ obtained from Eq. 10.134 into the right-hand side of Eq. 10.132 to obtain $\mathbf{P}_2$:

$$\mathbf{P}_1 = \mathbf{Q} + \mathbf{F}_s^T \mathbf{P}_0 \left(\mathbf{I} + \mathbf{GR}^{-1}\mathbf{G}^T\mathbf{P}_0\right)^{-1} \mathbf{F}_s. \tag{10.135}$$

This process is repeated until the steady-state Riccati matrix converges to an acceptable degree of accuracy (i.e., $\mathbf{P}_{k+1} \cong \mathbf{P}_k$). This iterative procedure is summarized by the equation

$$\mathbf{P}_{k+1} = \mathbf{Q} + \mathbf{F}_s^T \mathbf{P}_k \left(\mathbf{I} + \mathbf{GR}^{-1}\mathbf{G}^T\mathbf{P}_k\right)^{-1} \mathbf{F}_s. \tag{10.136}$$

The advantage of using this iterative method is that if the matrix $\mathbf{Q}$ is nonnegative definite, then each $\mathbf{P}_i$ is guaranteed to be positive definite. As shown in Example 1, after several matrix multiplications, the resulting $\mathbf{P}'$ matrix is symmetrical and positive definite, and therefore $\mathbf{P}_{k+1}$ will also be a positive definite matrix.

As a numerical example, consider Example 1 for the case where $R = 0.001$ (i.e., $r = 1/R = 1000.0$). Other matrices used in Example 1 are repeated here for completeness:

$$\mathbf{F}_s = \begin{bmatrix} 0.968843 & 0.019543 \\ -3.086307 & 0.944284 \end{bmatrix}, \quad \mathbf{G} = \begin{bmatrix} 0.0001973 \\ 0.019543 \end{bmatrix}, \quad \mathbf{Q} = \begin{bmatrix} 157.92 & 0.0 \\ 0.0 & 1.0 \end{bmatrix}.$$

As an initial estimate, assume that

$$P_0 = Q = \begin{bmatrix} 157.92 & 0.0 \\ 0.0 & 1.0 \end{bmatrix}. \tag{10.137}$$

Using Eq. 10.134, it follows that

$$P_1 = Q + F_s^T P_0 (I + GR^{-1}G^T P_0)^{-1} F_s = \begin{bmatrix} 315.024 & 0.4895 \\ 0.4895 & 1.6902 \end{bmatrix}. \tag{10.138}$$

The next iteration using Eq. 10.135 gives

$$P_2 = Q + F_s^T P_1 (I + GR^{-1}G^T P_1)^{-1} F_s = \begin{bmatrix} 466.146 & 2.1015 \\ 2.1015 & 2.0026 \end{bmatrix}. \tag{10.139}$$

Iterating again using Eq. 10.136, it follows that

$$P_3 = Q + F_s^T P_2 (I + GR^{-1}G^T P_2)^{-1} F_s = \begin{bmatrix} 601.952 & 4.6341 \\ 4.6341 & 2.1543 \end{bmatrix}. \tag{10.140}$$

This procedure is repeated until the solution converges, where

$$P_{k+1} = Q + F_s^T P_k (I + GR^{-1}G^T P_k)^{-1} F_s = \begin{bmatrix} 919.167 & 14.428 \\ 14.428 & 2.5227 \end{bmatrix}. \tag{10.141}$$

EXAMPLE 3   *Structural Control with Zero Control Force*

When no control force is applied to the structure, it is equivalent to the case of structural control with $R \to \infty$. To see this, consider Eq. 10.86, where

$$J = \frac{1}{2} \sum_{k=0}^{N} \left(z_k^T Q z_k + f_{ck}^T R f_{ck} \right). \tag{10.142}$$

Since $R \to \infty$, any nonzero value of control force gives the cost function of infinite J (i.e., $J \to \infty$). Therefore, to minimize the cost function J, zero control force at all times is necessary to give a finite value of J. Hence, $R \to \infty$ is equivalent to the case with no control.

EXAMPLE 4   *Matrix Inversion Lemma*

The identity of F_f was presented without any proof in Eq. 10.106, which is

$$F_f = \left[I - G(G^T PG + R)^{-1} G^T P \right] F_s = \left[I + GR^{-1}G^T P \right]^{-1} F_s. \tag{10.143}$$

This identity is known as the *matrix inversion lemma*. The term on the right-hand side of Eq. 10.143 was calculated in the process of calculating the steady-state Riccati matrix, which is embedded in Eq. 10.101. Hence, this term has already been calculated, and it would be wise to use this information again in Eq. 10.105. Therefore, the matrix inversion lemma is generally used to replace the term in the bracket of Eq. 10.105 with the term on the right-hand side of Eq. 10.143. The following procedure proves the identity in Eq. 10.143.

First premultiply both sides of the equation by

$$(I + GR^{-1}G^T P).$$

Then the objective is to prove the following identity:

$$(\mathbf{I} + \mathbf{GR}^{-1}\mathbf{G}^T\mathbf{P})\left[\mathbf{I} - \mathbf{G}(\mathbf{G}^T\mathbf{PG} + \mathbf{R})^{-1}\mathbf{G}^T\mathbf{P}\right] = \mathbf{I}. \tag{10.144}$$

The steps are as follows:

$$
\begin{aligned}
(\mathbf{I} &+ \mathbf{GR}^{-1}\mathbf{G}^T\mathbf{P})\left[\mathbf{I} - \mathbf{G}(\mathbf{G}^T\mathbf{PG} + \mathbf{R})^{-1}\mathbf{G}^T\mathbf{P}\right] \\
&= \mathbf{I} + \mathbf{G}\,\mathbf{R}^{-1}\mathbf{G}^T\mathbf{P} - \mathbf{G}(\mathbf{G}^T\,\mathbf{PG} + \mathbf{R})^{-1}\mathbf{G}^T\mathbf{P} - \mathbf{GR}^{-1}\mathbf{G}^T\mathbf{PG}(\mathbf{G}^T\mathbf{PG} + \mathbf{R})^{-1}\mathbf{G}^T\mathbf{P} \\
&= \mathbf{I} + \mathbf{G}\left[\mathbf{R}^{-1} - \mathbf{G}(\mathbf{G}^T\,\mathbf{PG} + \mathbf{R})^{-1} - \mathbf{R}^{-1}\mathbf{G}^T\mathbf{PG}(\mathbf{G}^T\mathbf{PG} + \mathbf{R})^{-1}\right]\mathbf{G}^T\mathbf{P} \\
&= \mathbf{I} + \mathbf{G}\left[\mathbf{R}^{-1}(\mathbf{G}^T\mathbf{PG} + \mathbf{R}) - \mathbf{I} - \mathbf{R}^{-1}\mathbf{G}^T\mathbf{PG}\right](\mathbf{G}^T\mathbf{PG} + \mathbf{R})^{-1}\mathbf{G}^T\mathbf{P} \\
&= \mathbf{I} + \mathbf{G}\left[\mathbf{R}^{-1}\mathbf{G}^T\mathbf{PG} + \mathbf{R}^{-1}\mathbf{R} - \mathbf{I} - \mathbf{R}^{-1}\mathbf{G}^T\mathbf{PG}\right](\mathbf{G}^T\mathbf{PG} + \mathbf{R})^{-1}\mathbf{G}^T\mathbf{P} \\
&= \mathbf{I} + \mathbf{G}\left[\mathbf{0} + \mathbf{0}\right](\mathbf{G}^T\mathbf{PG} + \mathbf{R})^{-1}\mathbf{G}^T\mathbf{P} \\
&= \mathbf{I}.
\end{aligned}
$$

This completes the proof.

10.7   OPTIMAL LINEAR CONTROL OF EARTHQUAKE-INDUCED ELASTIC RESPONSE

In the previous two sections, the external forcing function was assumed to be zero. The real case of interest to structural engineers is when the external force or the earthquake ground motion excites the structure and the control force is introduced to reduce structural response. This section assumes that the structure is excited by an earthquake ground motion.

Consider Eq. 10.48 subjected to an earthquake ground motion (i.e., with the external applied forcing function replaced by the mass times the ground acceleration); it follows that

$$\mathbf{z}_{k+1} = \mathbf{F}_s\,\mathbf{z}_k + \mathbf{H}_d^{(EQ)}\mathbf{a}_k + \mathbf{Gf}_{ck}. \tag{10.145}$$

Define the cost function to be

$$J = \frac{1}{2}\sum_{k=0}^{N}\left(\mathbf{z}_k^T\mathbf{Qz}_k + \mathbf{f}_{ck}^T\mathbf{Rf}_{ck}\right). \tag{10.146}$$

Again, the goal is to minimize J in Eq. 10.146 subject to the constraint in Eq. 10.145. Define the Lagrangian as

$$L = \sum_{k=0}^{N-1}\left[\frac{1}{2}\left(\mathbf{z}_k^T\mathbf{Qz}_k + \mathbf{f}_{ck}^T\mathbf{Rf}_{ck}\right) + \boldsymbol{\lambda}_{k+1}^T\left(\mathbf{F}_s\mathbf{z}_k + \mathbf{H}_d^{(EQ)}\mathbf{a}_k + \mathbf{Gf}_{ck} - \mathbf{z}_{k+1}\right)\right] \tag{10.147}$$

where $\boldsymbol{\lambda}_{k+1}$ is the Lagrange multiplier. Note that Eq. 10.147 is different from Eq. 10.87 because now the ground acceleration term is present. Define the Hamiltonian as

$$H_k = \frac{1}{2}\left(\mathbf{z}_k^T\mathbf{Qz}_k + \mathbf{f}_{ck}^T\mathbf{Rf}_{ck}\right) + \boldsymbol{\lambda}_{k+1}^T\left(\mathbf{F}_s\mathbf{z}_k + \mathbf{H}_d^{(EQ)}\mathbf{a}_k + \mathbf{Gf}_{ck}\right). \tag{10.148}$$

Then substituting Eq. 10.148 into Eq. 10.147 gives

$$L = \sum_{k=0}^{N-1}\left[H_k - \boldsymbol{\lambda}_{k+1}^T\mathbf{z}_{k+1}\right] = H_0 - \boldsymbol{\lambda}_N^T\mathbf{z}_N + \sum_{k=1}^{N-1}\left[H_k - \boldsymbol{\lambda}_k^T\mathbf{z}_k\right]. \tag{10.149}$$

Note that Eq. 10.149 is the same as Eq. 10.89, and therefore the procedure for minimizing the cost function J follows exactly the same derivation as discussed in Eqs. 10.90 to 10.92. The results follow directly from Eq. 10.92:

$$\left(\frac{\partial H_k}{\partial \boldsymbol{\lambda}_{k+1}}\right)^T - \mathbf{z}_{k+1} = \mathbf{0} \tag{10.150a}$$

$$\left(\frac{\partial H_k}{\partial \mathbf{z}_k}\right)^T - \boldsymbol{\lambda}_k = \mathbf{0}, \quad \mathbf{z}_o = \mathbf{z}(0) \tag{10.150b}$$

$$\left(\frac{\partial H_k}{\partial \mathbf{f}_{ck}}\right)^T = \mathbf{0}. \tag{10.150c}$$

Differentiate H_k in Eq. 10.148 with respect to $\boldsymbol{\lambda}_{k+1}$, $\mathbf{z}_k$, and $\mathbf{f}_{ck}$; substituting the results into Eq. 10.150 gives a set of three simultaneous difference equations:

$$\left(\frac{\partial H_k}{\partial \boldsymbol{\lambda}_{k+1}}\right)^T - \mathbf{z}_{k+1} = \mathbf{F}_s\mathbf{z}_k + \mathbf{H}_d^{(EQ)}\mathbf{a}_k + \mathbf{Gf}_{ck} - \mathbf{z}_{k+1} = \mathbf{0} \tag{10.151a}$$

$$\left(\frac{\partial H_k}{\partial \mathbf{z}_k}\right)^T - \boldsymbol{\lambda}_k = \mathbf{Qz}_k + \mathbf{F}_s^T\boldsymbol{\lambda}_{k+1} - \boldsymbol{\lambda}_k = \mathbf{0} \tag{10.151b}$$

$$\left(\frac{\partial H_k}{\partial \mathbf{f}_{ck}}\right)^T = \mathbf{Rf}_{ck} + \mathbf{G}^T\boldsymbol{\lambda}_{k+1} = \mathbf{0}. \tag{10.151c}$$

Solving for $\mathbf{f}_{ck}$ in Eq. 10.151c gives the control law, which is

$$\mathbf{f}_{ck} = -\mathbf{R}^{-1}\mathbf{G}^T\boldsymbol{\lambda}_{k+1}. \tag{10.152}$$

Substituting Eq. 10.152 into Eq. 10.151a, it follows from Eqs. 10.151a and 10.151b that

$$\mathbf{z}_{k+1} = \mathbf{F}_s\mathbf{z}_k + \mathbf{H}_d^{(EQ)}\mathbf{a}_k - \mathbf{GR}^{-1}\mathbf{G}^T\boldsymbol{\lambda}_{k+1}, \quad \mathbf{z}(0) = \mathbf{z}_o \tag{10.153a}$$

$$\boldsymbol{\lambda}_k = \mathbf{Qz}_k + \mathbf{F}_s^T\boldsymbol{\lambda}_{k+1}. \tag{10.153b}$$

Define $\mathbf{P}_k$ to be

$$\boldsymbol{\lambda}_k = \mathbf{P}_k\mathbf{z}_k. \tag{10.154}$$

Substituting Eq. 10.154 into Eqs. 10.153a and 10.153b, it follows that

$$\mathbf{z}_{k+1} = \mathbf{F}_s\mathbf{z}_k + \mathbf{H}_d^{(EQ)}\mathbf{a}_k - \mathbf{GR}^{-1}\mathbf{G}^T\mathbf{P}_{k+1}\mathbf{z}_{k+1} \tag{10.155a}$$

$$\mathbf{P}_k\mathbf{z}_k = \mathbf{Qz}_k + \mathbf{F}_s^T\mathbf{P}_{k+1}\mathbf{z}_{k+1}. \tag{10.155b}$$

Equation 10.155a can now be used to solve for $\mathbf{z}_{k+1}$, and it follows that

$$\mathbf{z}_{k+1} = \left(\mathbf{I} + \mathbf{GR}^{-1}\mathbf{G}^T\mathbf{P}_{k+1}\right)^{-1}\mathbf{F}_s\mathbf{z}_k + \left(\mathbf{I} + \mathbf{GR}^{-1}\mathbf{G}^T\mathbf{P}_{k+1}\right)^{-1}\mathbf{H}_d^{(EQ)}\mathbf{a}_k. \tag{10.156}$$

Equation 10.156 gives the response $\mathbf{z}_{k+1}$ for the case of the optimal linear control, but in this equation the matrix $\mathbf{P}_{k+1}$ is unknown. When Eq. 10.156 is substituted into Eq. 10.155b, it follows that

$$\mathbf{P}_k\mathbf{z}_k = \mathbf{Qz}_k + \mathbf{F}_s^T\mathbf{P}_{k+1}\left(\mathbf{I} + \mathbf{GR}^{-1}\mathbf{G}^T\mathbf{P}_{k+1}\right)^{-1}\mathbf{F}_s\mathbf{z}_k + \mathbf{F}_s^T\mathbf{P}_{k+1}\left(\mathbf{I} + \mathbf{GR}^{-1}\mathbf{G}^T\mathbf{P}_{k+1}\right)^{-1}\mathbf{H}_d^{(EQ)}\mathbf{a}_k. \tag{10.157}$$

Compare Eq. 10.157 with Eq. 10.99 and note that $\mathbf{a}_k$ appears as the last term on the right-hand side of Eq. 10.157, which is different from Eq. 10.99. This term represents the contribution of the earthquake to the discrete time Riccati difference equation, and therefore the Riccati matrix is a function of the earthquake ground motion. However, if it is assumed for computational convenience that the Riccati difference equation is independent of the earthquake ground motion, then Eq. 10.157 becomes

$$\mathbf{P}_k \mathbf{z}_k = \mathbf{Q} \mathbf{z}_k + \mathbf{F}_s^T \mathbf{P}_{k+1} \left(\mathbf{I} + \mathbf{GR}^{-1} \mathbf{G}^T \mathbf{P}_{k+1} \right)^{-1} \mathbf{F}_s \mathbf{z}_k. \tag{10.158}$$

Once this assumption is made, the response as given in Eq. 10.156 will deviate from the optimal solution. This will be seen later in the derivation. Since $\mathbf{z}_k$ occurs in all terms of Eq. 10.158, for a general $\mathbf{z}_k$ it follows that

$$\mathbf{P}_k = \mathbf{Q} + \mathbf{F}_s^T \mathbf{P}_{k+1} \left(\mathbf{I} + \mathbf{GR}^{-1} \mathbf{G}^T \mathbf{P}_{k+1} \right)^{-1} \mathbf{F}_s. \tag{10.159}$$

Equation 10.159 provides a solution for the Riccati matrix that is not a function of ground motion. If it is now assumed that the Riccati matrix is a constant, as assumed in the previous section, then $\mathbf{P}_k = \mathbf{P}_{k+1} = \mathbf{P}$ and it follows from Eq. 10.159 that

$$\mathbf{P} = \mathbf{Q} + \mathbf{F}_s^T \mathbf{P} \left(\mathbf{I} + \mathbf{GR}^{-1} \mathbf{G}^T \mathbf{P} \right)^{-1} \mathbf{F}_s. \tag{10.160}$$

Again, since Eq. 10.160 is a quadratic function of $\mathbf{P}$, there will be two solutions for $\mathbf{P}$, and the one that gives a positive definite $\mathbf{P}$ matrix is the desired solution. Once $\mathbf{P}$ is solved, the control force can be obtained by substituting Eq. 10.154 into Eq. 10.152:

$$\mathbf{f}_{ck} = -\mathbf{R}^{-1} \mathbf{G}^T \mathbf{P} \mathbf{z}_{k+1}. \tag{10.161}$$

Substituting Eq. 10.145 into Eq. 10.161, it follows that

$$\mathbf{f}_{ck} = -\mathbf{R}^{-1} \mathbf{G}^T \mathbf{P} \left(\mathbf{F}_s \mathbf{z}_k + \mathbf{H}_d^{(\mathrm{EQ})} \mathbf{a}_k + \mathbf{Gf}_{ck} \right). \tag{10.162}$$

After solving for $\mathbf{f}_{ck}$, the control force is

$$\mathbf{f}_{ck} = -\left(\mathbf{G}^T \mathbf{PG} + \mathbf{R} \right)^{-1} \mathbf{G}^T \mathbf{PF}_s \mathbf{z}_k - \left(\mathbf{G}^T \mathbf{PG} + \mathbf{R} \right)^{-1} \mathbf{G}^T \mathbf{PH}_d^{(\mathrm{EQ})} \mathbf{a}_k. \tag{10.163}$$

Note that the control force $\mathbf{f}_{ck}$ is a function of $\mathbf{z}_k$ and $\mathbf{a}_k$. Recall that $\mathbf{a}_k$ was assumed to be zero only in association with solving for the Riccati matrix in Eq. 10.158. The first term in Eq. 10.163 is a function of $\mathbf{z}_k$, and this is determined by Eq. 10.145, which is

$$\mathbf{z}_k = \mathbf{F}_s \, \mathbf{z}_{k-1} + \mathbf{H}_d^{(\mathrm{EQ})} \mathbf{a}_{k-1} + \mathbf{Gf}_{c,k-1}. \tag{10.164}$$

Therefore, $\mathbf{z}_k$ can be calculated using the state of the system at time step $k-1$. However, note that the second term in Eq. 10.163 requires the use of the earthquake ground acceleration at time k. This is the problem of time delay, where it is impossible to measure the ground acceleration at time step k and at the same time apply a control force computed based on the measured ground acceleration at time step k. If $\mathbf{a}_k$ in Eq. 10.163 is assumed to be equal to zero for the purpose of calculating the control force, it follows from Eq. 10.163 that

$$\mathbf{f}_{ck} = -\left(\mathbf{G}^T \mathbf{PG} + \mathbf{R} \right)^{-1} \mathbf{G}^T \mathbf{PF}_s \mathbf{z}_k. \tag{10.165}$$

When this assumption is made, it is referred to as a one-step time delay solution. Note that Eq. 10.165 is the same as Eq. 10.104, but now $\mathbf{z}_k$ is calculated using Eq. 10.164.

The response of the structure at time $k+1$ is obtained by substituting Eq. 10.165 into Eq. 10.145:

$$\mathbf{z}_{k+1} = \left[\mathbf{I} - \mathbf{G} \left(\mathbf{G}^T \mathbf{PG} + \mathbf{R} \right)^{-1} \mathbf{G}^T \mathbf{P} \right] \mathbf{F}_s \mathbf{z}_k + \mathbf{H}_d^{(\mathrm{EQ})} \mathbf{a}_k = \left[\mathbf{I} + \mathbf{GR}^{-1} \mathbf{G}^T \mathbf{P} \right]^{-1} \mathbf{F}_s \mathbf{z}_k + \mathbf{H}_d^{(\mathrm{EQ})} \mathbf{a}_k. \tag{10.166}$$

Note that Eq. 10.166 is the response for the case of optimal control with one-step time delay. It is different from the optimal solution as given in Eq. 10.156, with the difference occurring in the term multiplying the earthquake ground motion. This shows the effect of time delay on the response of the structure. Recall from Eq. 10.106 that

$$\mathbf{F}_f = \left[\mathbf{I} - \mathbf{G} \left(\mathbf{G}^T \mathbf{PG} + \mathbf{R} \right)^{-1} \mathbf{G}^T \mathbf{P} \right] \mathbf{F}_s = \left[\mathbf{I} + \mathbf{GR}^{-1} \mathbf{G}^T \mathbf{P} \right]^{-1} \mathbf{F}_s. \tag{10.167}$$

It follows from Eq. 10.166 that

$$\mathbf{z}_{k+1} = \mathbf{F}_f \mathbf{z}_k + \mathbf{H}_d^{(EQ)} \mathbf{a}_k. \tag{10.168}$$

EXAMPLE 1   *Optimal Linear Control of an SDOF System Subject to Ground Motion*

Consider Example 1 in Section 10.6, where

$$m = 1.0 \text{ k-s}^2/\text{in.}, \quad c = 1.2567 \text{ k-s/in.}, \quad k = 157.92 \text{ k/in.}$$

$$\mathbf{Q} = \begin{bmatrix} k & 0 \\ 0 & m \end{bmatrix} = \begin{bmatrix} 157.92 & 0.0 \\ 0.0 & 1.0 \end{bmatrix}, \quad \begin{Bmatrix} x_o \\ \dot{x}_o \end{Bmatrix} = \begin{Bmatrix} 1.0 \\ 0.0 \end{Bmatrix}$$

Let $\Delta t = 0.02$ s and assume a constant forcing function solution. It then follows that

$$\mathbf{A} = \begin{bmatrix} 0.0 & 1.0 \\ -157.92 & -1.2567 \end{bmatrix}, \quad \mathbf{B} = \begin{Bmatrix} 0 \\ 1 \end{Bmatrix}, \quad \mathbf{H} = \begin{Bmatrix} 0 \\ -1 \end{Bmatrix}, \quad \mathbf{F}_s = e^{\mathbf{A}\Delta t} = \begin{bmatrix} 0.968843 & 0.019543 \\ -3.08631 & 0.944284 \end{bmatrix}$$

$$\mathbf{H}_d = \mathbf{A}^{-1}\left(e^{\mathbf{A}\Delta t} - \mathbf{I}\right) = \begin{bmatrix} 0.01979 & 0.000197 \\ -0.031157 & 0.019543 \end{bmatrix}, \quad \mathbf{G} = \mathbf{A}^{-1}\left(e^{\mathbf{A}\Delta t} - \mathbf{I}\right)\mathbf{B} = \begin{Bmatrix} 0.000197 \\ 0.019543 \end{Bmatrix}$$

$$\mathbf{H}_d^{(EQ)} = \mathbf{H}_d \mathbf{H} = \begin{Bmatrix} -0.000197 \\ -0.019543 \end{Bmatrix}.$$

Consider the case where $R = 0.001$ and the resulting matrices are given in Eqs. 10.128 and 10.130, and which are

$$\mathbf{P} = \begin{bmatrix} 919.1673 & 14.428 \\ 14.428 & 2.5227 \end{bmatrix}, \quad \mathbf{F}_f = \begin{bmatrix} 0.941927 & 0.014094 \\ -5.75252 & 0.404455 \end{bmatrix}.$$

It then follows for any earthquake ground motion that

$$\begin{Bmatrix} x_{k+1} \\ \dot{x}_{k+1} \end{Bmatrix} = \begin{bmatrix} 0.941927 & 0.014094 \\ -5.75252 & 0.404455 \end{bmatrix} \begin{Bmatrix} x_k \\ \dot{x}_k \end{Bmatrix} + \begin{Bmatrix} -0.000197 \\ -0.019543 \end{Bmatrix} a_k. \tag{10.169}$$

In addition, from Eq. 10.165,

$$-\left(\mathbf{G}^T \mathbf{P} \mathbf{G} + \mathbf{R}\right)^{-1} \mathbf{G}^T \mathbf{P} \mathbf{F}_s = \begin{bmatrix} -136.426 & -27.6219 \end{bmatrix}. \tag{10.170}$$

Therefore, it follows from Eq. 10.165 that

$$f_{ck} = \begin{bmatrix} -136.426 & -27.6219 \end{bmatrix} \begin{Bmatrix} x_k \\ \dot{x}_k \end{Bmatrix}. \tag{10.171}$$

EXAMPLE 2   *Optimal Linear Control of an SDOF System Subjected to an Earthquake*

Consider the SDOF system discussed in Example 1 subjected to the modified El-Centro earthquake time history as shown in Figure 6.11. The optimal linear control response is given in Eq. 10.169 with the control force given in Eq. 10.171, that is,

$$\begin{Bmatrix} x^{(c)} \\ \dot{x}^{(c)} \end{Bmatrix}_{k+1} = \begin{bmatrix} 0.941927 & 0.014094 \\ -5.75252 & 0.404455 \end{bmatrix} \begin{Bmatrix} x^{(c)} \\ \dot{x}^{(c)} \end{Bmatrix}_k + \begin{Bmatrix} -0.000197 \\ -0.019543 \end{Bmatrix} a_k \tag{10.172}$$

$$f_{ck} = \begin{bmatrix} -136.426 & -27.6219 \end{bmatrix} \begin{Bmatrix} x_k \\ \dot{x}_k \end{Bmatrix} \tag{10.173}$$

where the superscript (c) denotes the controlled response. The uncontrolled response is similar to the one given in Eq. 10.169 but with $\mathbf{F}_f$ replaced by $\mathbf{F}_s$, that is,

$$\begin{Bmatrix} x^{(u)} \\ \dot{x}^{(u)} \end{Bmatrix}_{k+1} = \begin{bmatrix} 0.968843 & 0.019543 \\ -3.08631 & 0.944284 \end{bmatrix} \begin{Bmatrix} x^{(u)} \\ \dot{x}^{(u)} \end{Bmatrix}_k + \begin{Bmatrix} -0.000197 \\ -0.019543 \end{Bmatrix} a_k \qquad (10.174)$$

where the superscript (u) denotes the uncontrolled response.

Assume that the SDOF system has zero initial displacement and zero initial velocity. At time step $k = 1$ (i.e., $t = \Delta t = 0.02$ s), the response is

$$\begin{Bmatrix} x^{(u)} \\ \dot{x}^{(u)} \end{Bmatrix}_1 = \begin{bmatrix} 0.968843 & 0.019543 \\ -3.08631 & 0.944284 \end{bmatrix} \begin{Bmatrix} 0.0 \\ 0.0 \end{Bmatrix} + \begin{Bmatrix} -0.000197 \\ -0.019543 \end{Bmatrix} (-0.9930) = \begin{Bmatrix} 0.00020 \\ 0.0194 \end{Bmatrix} \qquad (10.175)$$

$$\begin{Bmatrix} x^{(c)} \\ \dot{x}^{(c)} \end{Bmatrix}_1 = \begin{bmatrix} 0.941927 & 0.014094 \\ -5.75252 & 0.404455 \end{bmatrix} \begin{Bmatrix} 0.0 \\ 0.0 \end{Bmatrix} + \begin{Bmatrix} -0.000197 \\ -0.019543 \end{Bmatrix} (-0.9930) = \begin{Bmatrix} 0.00020 \\ 0.0194 \end{Bmatrix} \qquad (10.176)$$

$$f_{c,1} = \begin{bmatrix} -136.426 & -27.6219 \end{bmatrix} \begin{Bmatrix} 0.00020 \\ 0.0194 \end{Bmatrix} = -0.5628 \text{ kip.} \qquad (10.177)$$

Note that the uncontrolled response and the controlled response are the same at time step $k = 1$. This is because the control force applied at this time step (i.e., $f_{c,1}$) changes only the acceleration response of the system immediately but does not immediately affect the displacement and velocity response.

At time step $k = 2$ (i.e., $t = \Delta t = 0.04$ s), the response is

$$\begin{Bmatrix} x^{(u)} \\ \dot{x}^{(u)} \end{Bmatrix}_2 = \begin{bmatrix} 0.968843 & 0.019543 \\ -3.08631 & 0.944284 \end{bmatrix} \begin{Bmatrix} 0.00020 \\ 0.0194 \end{Bmatrix} + \begin{Bmatrix} -0.000197 \\ -0.019543 \end{Bmatrix} (-7.6507) = \begin{Bmatrix} 0.00208 \\ 0.1672 \end{Bmatrix} \qquad (10.178)$$

$$\begin{Bmatrix} x^{(c)} \\ \dot{x}^{(c)} \end{Bmatrix}_2 = \begin{bmatrix} 0.941927 & 0.014094 \\ -5.75252 & 0.404455 \end{bmatrix} \begin{Bmatrix} 0.00020 \\ 0.0194 \end{Bmatrix} + \begin{Bmatrix} -0.000197 \\ -0.019543 \end{Bmatrix} (-7.6507) = \begin{Bmatrix} 0.00197 \\ 0.1562 \end{Bmatrix} \qquad (10.179)$$

$$f_{c,1} = \begin{bmatrix} -136.426 & -27.6219 \end{bmatrix} \begin{Bmatrix} 0.00197 \\ 0.1562 \end{Bmatrix} = -4.5842 \text{ kip.} \qquad (10.180)$$

The calculation process continues, and the displacement and velocity responses for both the uncontrolled and controlled SDOF systems are shown in Figures 10.10 and 10.11, respectively. The control force time history is shown in Figure 10.12.

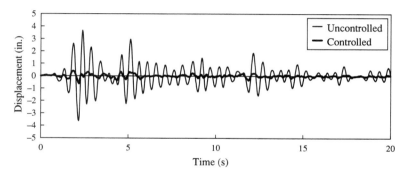

Figure 10.10 Displacement Time History of Uncontrolled and Controlled Responses

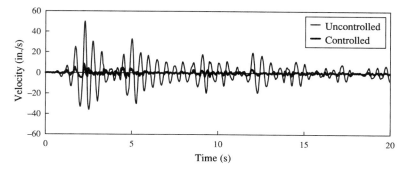

Figure 10.11 Velocity Time History of Uncontrolled and Controlled Responses

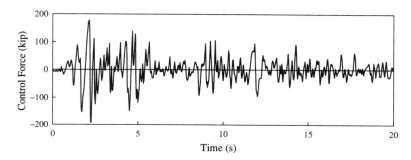

Figure 10.12 Control Force Time History of the SDOF System

10.8 OPTIMAL LINEAR CONTROL OF EARTHQUAKE-INDUCED INELASTIC RESPONSE

Even when active control is used, the structure may become inelastic when subject to a large-magnitude earthquake. In this section, optimal linear control is used to control a structure that can respond in the inelastic domain during earthquakes. Start with the equation of motion described by Eq. 7.465 with an additional term for active control force,

$$\mathbf{M}\ddot{\mathbf{X}}(t) + \mathbf{C}\dot{\mathbf{X}}(t) + \mathbf{K}\mathbf{X}'(t) = -\mathbf{M}\mathbf{a}(t) + \mathbf{D}\mathbf{f}_c(t). \tag{10.181}$$

Recall that the elastic displacement is the total displacement minus the inelastic displacement:

$$\mathbf{X}'(t) = \begin{Bmatrix} x_1'(t) \\ x_2'(t) \\ \vdots \\ x_n'(t) \end{Bmatrix} = \mathbf{X}(t) - \mathbf{X}''(t) = \begin{Bmatrix} x_1(t) \\ x_2(t) \\ \vdots \\ x_n(t) \end{Bmatrix} - \begin{Bmatrix} x_1''(t) \\ x_2''(t) \\ \vdots \\ x_n''(t) \end{Bmatrix}. \tag{10.182}$$

It follows from Eq. 10.181 that

$$\mathbf{M}\ddot{\mathbf{X}}(t) + \mathbf{C}\dot{\mathbf{X}}(t) + \mathbf{K}\mathbf{X}(t) = -\mathbf{M}\mathbf{a}(t) + \mathbf{D}\mathbf{f}_c(t) + \mathbf{K}\mathbf{X}''(t). \tag{10.183}$$

To represent Eq. 10.183 in state space form, let

$$\mathbf{z}(t) = \begin{Bmatrix} \mathbf{X}(t) \\ \dot{\mathbf{X}}(t) \end{Bmatrix}.$$

Equation 10.183 becomes

$$\dot{\mathbf{z}}(t) = \begin{bmatrix} \mathbf{0} & \mathbf{I} \\ -\mathbf{M}^{-1}\mathbf{K} & -\mathbf{M}^{-1}\mathbf{C} \end{bmatrix} \mathbf{z}(t) + \begin{Bmatrix} \mathbf{0} \\ -\mathbf{a}(t) \end{Bmatrix} + \begin{Bmatrix} \mathbf{0} \\ \mathbf{M}^{-1}\mathbf{K}\mathbf{X}''(t) \end{Bmatrix} + \begin{Bmatrix} \mathbf{0} \\ \mathbf{M}^{-1}\mathbf{D}\mathbf{f}_c(t) \end{Bmatrix}. \quad (10.184)$$

To simplify Eq. 10.184, let

$$\mathbf{A} = \begin{bmatrix} \mathbf{0} & \mathbf{I} \\ -\mathbf{M}^{-1}\mathbf{K} & -\mathbf{M}^{-1}\mathbf{C} \end{bmatrix}, \quad \mathbf{H} = \begin{bmatrix} \mathbf{0} \\ -\mathbf{1} \end{bmatrix}, \quad \mathbf{F}_p^c = \begin{bmatrix} \mathbf{0} \\ \mathbf{M}^{-1}\mathbf{K} \end{bmatrix}, \quad \mathbf{B} = \begin{bmatrix} \mathbf{0} \\ \mathbf{M}^{-1}\mathbf{D} \end{bmatrix}.$$

It follows that

$$\dot{\mathbf{z}}(t) = \mathbf{A}\mathbf{z}(t) + \mathbf{H}\mathbf{a}(t) + \mathbf{F}_p^c\mathbf{X}''(t) + \mathbf{B}\mathbf{f}_c(t). \quad (10.185)$$

Note that $\mathbf{H}$ contains six columns, where each column corresponds to the ground acceleration in six different directions, as defined in Eq. 6.188. The solution to Eq. 10.185 is

$$\mathbf{z}(t) = e^{\mathbf{A}(t-t_o)}\mathbf{z}(t_o) + e^{\mathbf{A}t}\int_{t_o}^{t} e^{-\mathbf{A}s}\left[\mathbf{H}\mathbf{a}(s) + \mathbf{F}_p^c\mathbf{X}''(s) + \mathbf{B}\mathbf{f}_c(s)\right] ds \quad (10.186)$$

Let

$$t_{k+1} = t, \quad t_k = t_o, \quad \Delta t = t - t_o. \quad (10.187)$$

It follows from Eq. 10.186 that

$$\mathbf{z}_{k+1} = e^{\mathbf{A}\Delta t}\mathbf{z}_k + e^{\mathbf{A}t_{k+1}}\int_{t_k}^{t_{k+1}} e^{-\mathbf{A}s}\left[\mathbf{H}\mathbf{a}(s) + \mathbf{F}_p^c\mathbf{X}''(s) + \mathbf{B}\mathbf{f}_c(s)\right] ds. \quad (10.188)$$

When the Delta forcing function is used to represent the earthquake ground motion, the inelastic displacement, and the control force, then

$$\mathbf{a}(s) = \mathbf{a}_k\delta(s-t_k)\Delta t, \quad \mathbf{X}''(s) = \mathbf{X}_k''\delta(s-t_k)\Delta t, \quad \mathbf{f}_c(s) = \mathbf{f}_{ck}\delta(s-t_k)\Delta t, \quad t_k \le s < t_{k+1}. \quad (10.189)$$

Then Eq. 10.188 becomes

$$\mathbf{z}_{k+1} = e^{\mathbf{A}\Delta t}\mathbf{z}_k + e^{\mathbf{A}\Delta t}\mathbf{H}\Delta t\,\mathbf{a}_k + e^{\mathbf{A}\Delta t}\mathbf{F}_p^c\Delta t\,\mathbf{X}_k'' + e^{\mathbf{A}\Delta t}\mathbf{B}\Delta t\,\mathbf{f}_{ck}. \quad (10.190)$$

When the constant forcing function is used to represent the earthquake ground motion, the inelastic displacement, and the control force, then

$$\mathbf{a}(s) = \mathbf{a}_k, \quad \mathbf{X}''(s) = \mathbf{X}_k'', \quad \mathbf{f}_c(s) = \mathbf{f}_{ck}, \quad t_k \le s < t_{k+1}. \quad (10.191)$$

Then Eq. 10.188 becomes

$$\mathbf{z}_{k+1} = e^{\mathbf{A}\Delta t}\mathbf{z}_k + \mathbf{A}^{-1}\left(e^{\mathbf{A}\Delta t}-\mathbf{I}\right)\mathbf{H}\,\mathbf{a}_k + \mathbf{A}^{-1}\left(e^{\mathbf{A}\Delta t}-\mathbf{I}\right)\mathbf{F}_p^c\mathbf{X}_k'' + \mathbf{A}^{-1}\left(e^{\mathbf{A}\Delta t}-\mathbf{I}\right)\mathbf{B}\mathbf{f}_{ck}. \quad (10.192)$$

To simplify Eq. 10.190 or Eq. 10.192, if the Delta forcing function assumption is made, let

$$\mathbf{F}_s = e^{\mathbf{A}\Delta t}, \quad \mathbf{H}_d^{(EQ)} = e^{\mathbf{A}\Delta t}\mathbf{H}\Delta t, \quad \mathbf{F}_p = e^{\mathbf{A}\Delta t}\mathbf{F}_p^c\Delta t, \quad \mathbf{G} = e^{\mathbf{A}\Delta t}\mathbf{B}\Delta t. \quad (10.193)$$

If the constant forcing function assumption is made, let

$$\mathbf{F}_s = e^{\mathbf{A}\Delta t}, \quad \mathbf{H}_d^{(EQ)} = \mathbf{A}^{-1}\left(e^{\mathbf{A}\Delta t}-\mathbf{I}\right)\mathbf{H}, \quad \mathbf{F}_p = \mathbf{A}^{-1}\left(e^{\mathbf{A}\Delta t}-\mathbf{I}\right)\mathbf{F}_p^c, \quad \mathbf{G} = \mathbf{A}^{-1}\left(e^{\mathbf{A}\Delta t}-\mathbf{I}\right)\mathbf{B}. \quad (10.194)$$

Then it follows that

$$\mathbf{z}_{k+1} = \mathbf{F}_s\mathbf{z}_k + \mathbf{H}_d^{(EQ)}\mathbf{a}_k + \mathbf{F}_p\mathbf{X}_k'' + \mathbf{G}\mathbf{f}_{ck}. \quad (10.195)$$

Define the cost function to be related to the elastic displacement, that is,

$$J = \frac{1}{2}\sum_{k=0}^{N}\left(\mathbf{z}_k'^T\mathbf{Q}\mathbf{z}_k' + \mathbf{f}_{ck}^T\mathbf{R}\mathbf{f}_{ck}\right) \quad (10.196)$$

where

$$z'_k = \begin{Bmatrix} x'_1 \\ \vdots \\ x'_n \\ \hdotsfor{1} \\ \dot{x}_1 \\ \vdots \\ \dot{x}_n \end{Bmatrix}_k = \mathbf{z}_k - \mathbf{z}''_k = \begin{Bmatrix} x_1 \\ \vdots \\ x_n \\ \hdotsfor{1} \\ \dot{x}_1 \\ \vdots \\ \dot{x}_n \end{Bmatrix}_k - \begin{Bmatrix} x''_1 \\ \vdots \\ x''_n \\ \hdotsfor{1} \\ 0 \\ \vdots \\ 0 \end{Bmatrix}_k = \mathbf{z}_k - \begin{Bmatrix} \mathbf{X}''_k \\ \mathbf{0} \end{Bmatrix}. \tag{10.197}$$

The goal here is again to minimize J in Eq. 10.196 subject to the constraint in Eq. 10.195. Define the Lagrangian as

$$L = \sum_{k=0}^{N-1} \left[\frac{1}{2}\left(\mathbf{z}'^T_k \mathbf{Q}\mathbf{z}'_k + \mathbf{f}^T_{ck}\mathbf{R}\mathbf{f}_{ck} \right) + \boldsymbol{\lambda}^T_{k+1}\left(\mathbf{F}_s\mathbf{z}_k + \mathbf{H}^{(EQ)}_d\mathbf{a}_k + \mathbf{F}_p\mathbf{X}''_k + \mathbf{Gf}_{ck} - \mathbf{z}_{k+1} \right) \right] \tag{10.198}$$

where $\boldsymbol{\lambda}_{k+1}$ is the Lagrange multiplier. Note that Eq. 10.198 is different from Eq. 10.87 or Eq. 10.147 because now the inelastic displacement term is present. Define also the Hamiltonian as

$$H_k = \frac{1}{2}\left(\mathbf{z}'^T_k \mathbf{Q}\mathbf{z}'_k + \mathbf{f}^T_{ck}\mathbf{R}\mathbf{f}_{ck} \right) + \boldsymbol{\lambda}^T_{k+1}\left(\mathbf{F}_s\mathbf{z}_k + \mathbf{H}^{(EQ)}_d\mathbf{a}_k + \mathbf{F}_p\mathbf{X}''_k + \mathbf{Gf}_{ck} \right). \tag{10.199}$$

Then substituting Eq. 10.199 into Eq. 10.198 gives

$$L = \sum_{k=0}^{N-1}\left[H_k - \boldsymbol{\lambda}^T_{k+1}\mathbf{z}_{k+1} \right] = H_0 - \boldsymbol{\lambda}^T_N\mathbf{z}_N + \sum_{k=1}^{N-1}\left[H_k - \boldsymbol{\lambda}^T_k\mathbf{z}_k \right]. \tag{10.200}$$

Note that Eq. 10.200 is the same as Eq. 10.89, and therefore the procedure for minimizing the cost function J follows exactly the same derivation as discussed in Eqs. 10.90 to 10.92. The results follow directly from Eq. 10.92:

$$\left(\frac{\partial H_k}{\partial \boldsymbol{\lambda}_{k+1}} \right)^T - \mathbf{z}_{k+1} = \mathbf{0} \tag{10.201a}$$

$$\left(\frac{\partial H_k}{\partial \mathbf{z}_k} \right)^T - \boldsymbol{\lambda}_k = \mathbf{0}, \quad \mathbf{z}_o = \mathbf{z}(0) \tag{10.201b}$$

$$\left(\frac{\partial H_k}{\partial \mathbf{f}_{ck}} \right)^T = \mathbf{0}. \tag{10.201c}$$

Note that the Hamiltonian as presented in Eq. 10.199 contains both $\mathbf{z}_k$ and $\mathbf{z}'_k$. In Eq. 10.201b, the differentiation of H_k is with respect to $\mathbf{z}'_k$, and therefore a relationship between the differentiation of H_k with respect to $\mathbf{z}'_k$ must be derived. This differentiation can be expressed as

$$\frac{\partial H_k}{\partial \mathbf{z}_k} = \left\{ \frac{\partial H_k}{\partial z_{1k}} \quad \frac{\partial H_k}{\partial z_{2k}} \quad \cdots \quad \frac{\partial H_k}{\partial z_{2n,k}} \right\} = \left\{ \frac{\partial H_k}{\partial z'_{1k}}\frac{\partial z'_{1k}}{\partial z_{1k}} \quad \frac{\partial H_k}{\partial z'_{2k}}\frac{\partial z'_{2k}}{\partial z_{2k}} \quad \cdots \quad \frac{\partial H_k}{\partial z'_{2n,k}}\frac{\partial z'_{2n,k}}{\partial z_{2n,k}} \right\}. \tag{10.202}$$

Recall that $z'_{i,k} = z_{i,k}$ for $i > n$ (i.e., the velocity of the structure). For $i \le n$, since $z'_{i,k} = z_{i,k} - z''_{i,k}$, it follows that

$$\frac{\partial z'_{i,k}}{\partial z_{i,k}} = \frac{\partial z_{i,k}}{\partial z_{i,k}} - \frac{\partial z''_{i,k}}{\partial z_{i,k}} = \begin{cases} 1 - \partial z''_{i,k}/\partial z_{i,k} & i \le n \\ 1 & i > n. \end{cases} \tag{10.203}$$

Assuming that the inelastic displacement depends on the past history (up to time step $k-1$) of the total displacement only, it follows from Eq. 10.203 that

$$\frac{\partial z'_{i,k}}{\partial z_{i,k}} = 1 - \frac{\partial z''_{i,k}}{\partial z_{i,k}} = 1 - (0) = 1. \tag{10.204}$$

This assumption is valid when the structure responds in the elastic range because there will be no change in the inelastic displacement. When the structure responds in the inelastic range where there is a change in the inelastic displacement, the differentiation in Eq. 10.204 depends on whether the structure is experiencing an increase or decrease in the total displacement. Therefore, this type of differentiation always involves very complicated nonlinear equations. In addition, there is usually only limited time during the entire structural response period that the structure actually experiences a change in inelastic displacement. Therefore, it is often assumed that the inelastic displacement at time step k does not depend on the total displacement at the same time step.

Substituting Eq. 10.204 into Eq. 10.202, it follows that

$$\frac{\partial H_k}{\partial \mathbf{z}_k} = \left\{ \frac{\partial H_k}{\partial z'_{1k}}(1) \quad \frac{\partial H_k}{\partial z'_{2k}}(1) \quad \cdots \quad \frac{\partial H_k}{\partial z'_{2n,k}}(1) \right\} = \frac{\partial H_k}{\partial \mathbf{z}'_k}. \tag{10.205}$$

Therefore, differentiating the Hamiltonian with respect to the total displacement is the same as differentiating with respect to the elastic displacement. Differentiating H_k with respect to $\boldsymbol{\lambda}_{k+1}$, $\mathbf{z}'_k$ (or $\mathbf{z}_k$), and $\mathbf{F}_{ck}$ gives a set of three simultaneous difference equations:

$$\left(\frac{\partial H_k}{\partial \boldsymbol{\lambda}_{k+1}}\right)^T - \mathbf{z}_{k+1} = \mathbf{F}_s \mathbf{z}_k + \mathbf{H}_d^{(\mathrm{EQ})}\mathbf{a}_k + \mathbf{F}_p \mathbf{X}''_k + \mathbf{G}\mathbf{f}_{ck} - \mathbf{z}_{k+1} = \mathbf{0} \tag{10.206a}$$

$$\left(\frac{\partial H_k}{\partial \mathbf{z}'_k}\right)^T - \boldsymbol{\lambda}_k = \mathbf{Q}\mathbf{z}'_k + \mathbf{F}_s^T \boldsymbol{\lambda}_{k+1} - \boldsymbol{\lambda}_k = \mathbf{Q}\mathbf{z}_k + \mathbf{F}_s^T \boldsymbol{\lambda}_{k+1} - \mathbf{Q}\mathbf{z}''_k - \boldsymbol{\lambda}_k = \mathbf{0} \tag{10.206b}$$

$$\left(\frac{\partial H_k}{\partial \mathbf{f}_{ck}}\right)^T = \mathbf{R}\mathbf{f}_{ck} + \mathbf{G}^T \boldsymbol{\lambda}_{k+1} = \mathbf{0}. \tag{10.206c}$$

Solving for $\mathbf{f}_{ck}$ in Eq. 10.206c gives the control law, which is

$$\mathbf{f}_{ck} = -\mathbf{R}^{-1}\mathbf{G}^T \boldsymbol{\lambda}_{k+1}. \tag{10.207}$$

Substituting Eq. 10.207 into Eq. 10.206a, it follows from Eqs. 10.206a and 10.206b that

$$\mathbf{z}_{k+1} = \mathbf{F}_s \mathbf{z}_k + \mathbf{H}_d^{(\mathrm{EQ})}\mathbf{a}_k + \mathbf{F}_p \mathbf{X}''_k - \mathbf{G}\mathbf{R}^{-1}\mathbf{G}^T \boldsymbol{\lambda}_{k+1}, \quad \mathbf{z}(0) = \mathbf{z}_o \tag{10.208a}$$

$$\boldsymbol{\lambda}_k = \mathbf{Q}\mathbf{z}_k + \mathbf{F}_s^T \boldsymbol{\lambda}_{k+1} - \mathbf{Q}\mathbf{z}''_k. \tag{10.208b}$$

Let $\mathbf{P}_k$ be defined as

$$\boldsymbol{\lambda}_k = \mathbf{P}_k \mathbf{z}_k. \tag{10.209}$$

Substituting Eq. 10.209 into Eqs. 10.208a and 10.208b, it follows that

$$\mathbf{z}_{k+1} = \mathbf{F}_s \mathbf{z}_k + \mathbf{H}_d^{(\mathrm{EQ})}\mathbf{a}_k + \mathbf{F}_p \mathbf{X}''_k - \mathbf{G}\mathbf{R}^{-1}\mathbf{G}^T \mathbf{P}_{k+1}\mathbf{z}_{k+1} \tag{10.210a}$$

$$\mathbf{P}_k \mathbf{z}_k = \mathbf{Q}\mathbf{z}_k + \mathbf{F}_s^T \mathbf{P}_{k+1}\mathbf{z}_{k+1} - \mathbf{Q}\mathbf{z}''_k. \tag{10.210b}$$

Solving for $\mathbf{z}_{k+1}$, Eq. 10.210a becomes

$$\mathbf{z}_{k+1} = \left(\mathbf{I}+\mathbf{G}\mathbf{R}^{-1}\mathbf{G}^T\mathbf{P}_{k+1}\right)^{-1}\mathbf{F}_s\mathbf{z}_k + \left(\mathbf{I}+\mathbf{G}\mathbf{R}^{-1}\mathbf{G}^T\mathbf{P}_{k+1}\right)^{-1}\mathbf{H}_d^{(\mathrm{EQ})}\mathbf{a}_k + \left(\mathbf{I}+\mathbf{G}\mathbf{R}^{-1}\mathbf{G}^T\mathbf{P}_{k+1}\right)^{-1}\mathbf{F}_p\mathbf{X}''_k. \tag{10.211}$$

In this equation, the matrix $\mathbf{P}_{k+1}$ is unknown and must be determined using Eq. 10.210b. Substituting Eq. 10.211 into Eq. 10.210b, it follows that

$$\mathbf{P}_k\mathbf{z}_k = \mathbf{Q}\mathbf{z}_k + \mathbf{F}_s^T\mathbf{P}_{k+1}\left(\mathbf{I}+\mathbf{G}\mathbf{R}^{-1}\mathbf{G}^T\mathbf{P}_{k+1}\right)^{-1}\mathbf{F}_s\mathbf{z}_k$$
$$+ \mathbf{F}_s^T\mathbf{P}_{k+1}\left(\mathbf{I}+\mathbf{G}\mathbf{R}^{-1}\mathbf{G}^T\mathbf{P}_{k+1}\right)^{-1}\mathbf{H}_d^{(\mathrm{EQ})}\mathbf{a}_k + \mathbf{F}_s^T\mathbf{P}_{k+1}\left(\mathbf{I}+\mathbf{G}\mathbf{R}^{-1}\mathbf{G}^T\mathbf{P}_{k+1}\right)^{-1}\mathbf{F}_p\mathbf{X}''_k. \tag{10.212}$$

Note that both $\mathbf{a}_k$ and $\mathbf{X}_k''$ appear on the right-hand side of Eq. 10.212. Assuming that the Riccati difference equation is independent of both the earthquake and the inelasticity of the structure, then Eq. 10.212 becomes

$$\mathbf{P}_k \mathbf{z}_k = \mathbf{Q} \mathbf{z}_k + \mathbf{F}_s^T \mathbf{P}_{k+1} \left(\mathbf{I} + \mathbf{G} \mathbf{R}^{-1} \mathbf{G}^T \mathbf{P}_{k+1} \right)^{-1} \mathbf{F}_s \mathbf{z}_k. \tag{10.213}$$

Again, $\mathbf{z}_k$ occurs in all terms on both sides of Eq. 10.213. Therefore, for a general $\mathbf{z}_k$, it follows that

$$\mathbf{P}_k = \mathbf{Q} + \mathbf{F}_s^T \mathbf{P}_{k+1} \left(\mathbf{I} + \mathbf{G} \mathbf{R}^{-1} \mathbf{G}^T \mathbf{P}_{k+1} \right)^{-1} \mathbf{F}_s. \tag{10.214}$$

Setting $\mathbf{P}_k = \mathbf{P}_{k+1} = \mathbf{P}$, it follows from Eq. 10.214 that

$$\mathbf{P} = \mathbf{Q} + \mathbf{F}_s^T \mathbf{P} \left(\mathbf{I} + \mathbf{G} \mathbf{R}^{-1} \mathbf{G}^T \mathbf{P} \right)^{-1} \mathbf{F}_s \tag{10.215}$$

which is the same as Eq. 10.160. Again, since Eq. 10.215 is a quadratic function of $\mathbf{P}$, there will be two solutions of $\mathbf{P}$, and the one that gives $\mathbf{P}$ a positive definite matrix is the desired solution. Once $\mathbf{P}$ is solved, the control force can be obtained by substituting Eq. 10.209 into Eq. 10.207:

$$\mathbf{f}_{ck} = -\mathbf{R}^{-1} \mathbf{G}^T \mathbf{P} \, \mathbf{z}_{k+1}. \tag{10.216}$$

Substituting Eq. 10.195 into Eq. 10.216, it follows that

$$\mathbf{f}_{ck} = -\mathbf{R}^{-1} \mathbf{G}^T \mathbf{P} \left(\mathbf{F}_s \mathbf{z}_k + \mathbf{H}_d^{(EQ)} \mathbf{a}_k + \mathbf{F}_p \mathbf{X}_k'' + \mathbf{G} \mathbf{f}_{ck} \right). \tag{10.217}$$

Solving for $\mathbf{f}_{ck}$ in Eq. 10.217, therefore, gives

$$\mathbf{f}_{ck} = -\left(\mathbf{G}^T \mathbf{P} \mathbf{G} + \mathbf{R} \right)^{-1} \mathbf{G}^T \mathbf{P} \mathbf{F}_s \mathbf{z}_k - \left(\mathbf{G}^T \mathbf{P} \mathbf{G} + \mathbf{R} \right)^{-1} \mathbf{G}^T \mathbf{P} \mathbf{H}_d^{(EQ)} \mathbf{a}_k - \left(\mathbf{G}^T \mathbf{P} \mathbf{G} + \mathbf{R} \right)^{-1} \mathbf{G}^T \mathbf{P} \mathbf{F}_p \mathbf{X}_k''. \tag{10.218}$$

Note that the control force $\mathbf{f}_{ck}$ depends not only on $\mathbf{z}_k$ but also on $\mathbf{a}_k$ and $\mathbf{X}_k''$. This is the problem of time delay, where it is practically impossible to measure the ground acceleration at time step k and at the same time apply a control force computed based on the measured ground acceleration at time step k. Therefore, by taking a one-step time delay, the ground acceleration is equal to zero in determining the control force. In addition, the purpose of active control is to minimize the inelastic displacement, and therefore the $\mathbf{X}_k''$ term will be small and can be neglected in Eq. 10.218. Then it follows from Eq. 10.218 that

$$\mathbf{f}_{ck} = -\left(\mathbf{G}^T \mathbf{P} \mathbf{G} + \mathbf{R} \right)^{-1} \mathbf{G}^T \mathbf{P} \mathbf{F}_s \mathbf{z}_k. \tag{10.219}$$

Substituting Eq. 10.219 into Eq. 10.195 gives

$$\begin{aligned} \mathbf{z}_{k+1} &= \left[\mathbf{I} - \mathbf{G} \left(\mathbf{G}^T \mathbf{P} \mathbf{G} + \mathbf{R} \right)^{-1} \mathbf{G}^T \mathbf{P} \right] \mathbf{F}_s \mathbf{z}_k + \mathbf{H}_d^{(EQ)} \mathbf{a}_k + \mathbf{F}_p \mathbf{X}_k'' \\ &= \left[\mathbf{I} + \mathbf{G} \mathbf{R}^{-1} \mathbf{G}^T \mathbf{P} \right]^{-1} \mathbf{F}_s \mathbf{z}_k + \mathbf{H}_d^{(EQ)} \mathbf{a}_k + \mathbf{F}_p \mathbf{X}_k''. \end{aligned} \tag{10.220}$$

Note that Eq. 10.220 is different from Eq. 10.211. Equation 10.211 is the solution for optimal control, where the control force at time step k (i.e., $\mathbf{f}_{ck}$) is determined from the ground acceleration and inelastic displacement at time step k (i.e., $\mathbf{a}_k$ and $\mathbf{X}_k''$). When this is practically impossible, Eq. 10.211 becomes Eq. 10.220 with a modification in the term premultiplying the ground acceleration as well as the inelastic displacement.

Finally, let

$$\mathbf{F}_f = \left[\mathbf{I} - \mathbf{G} \left(\mathbf{G}^T \mathbf{P} \mathbf{G} + \mathbf{R} \right)^{-1} \mathbf{G}^T \mathbf{P} \right] \mathbf{F}_s = \left[\mathbf{I} + \mathbf{G} \mathbf{R}^{-1} \mathbf{G}^T \mathbf{P} \right]^{-1} \mathbf{F}_s. \tag{10.221}$$

It follows from Eq. 10.220 that

$$\mathbf{z}_{k+1} = \mathbf{F}_f \mathbf{z}_k + \mathbf{H}_d^{(EQ)} \mathbf{a}_k + \mathbf{F}_p \mathbf{X}_k''. \tag{10.222}$$

The solution procedure is a recursive process in which, first, all the information at time step k is assumed to be known. Then both the displacement and velocity vectors at time step $k + 1$ (i.e., $\mathbf{X}_{k+1}$ and $\dot{\mathbf{X}}_{k+1}$) can be calculated using Eq. 10.222. Then, using Eq. 10.219, the control force applied onto the structure at time step $k + 1$ can be computed. Using $\mathbf{X}_{k+1}$ and the force analogy method discussed in Section 7.7, the inelastic property of the structure at time step $k + 1$ (i.e., $\mathbf{X}''_{k+1}$) can be determined by first using the incremental plastic rotation discussed in Eqs. 7.481 to 7.483, where

$$\mathbf{M}_{k+1} + \mathbf{K}_R \Delta \mathbf{\Theta}'' = \mathbf{K}_p^T \mathbf{X}_{k+1} - \mathbf{K}_R \mathbf{\Theta}''_k. \tag{10.223}$$

With the procedure discussed in Section 7.6, $\mathbf{M}$ and $\Delta \mathbf{\Theta}''$ can be solved simultaneously. From this incremental plastic rotation, the inelastic displacement vector can be determined from Eq. 7.483, and it is equal to

$$\mathbf{X}''_{k+1} = -\mathbf{K}^{-1} \mathbf{K}_p \mathbf{\Theta}''_{k+1} = -\mathbf{K}^{-1} \mathbf{K}_p \left(\mathbf{\Theta}''_k + \Delta \mathbf{\Theta}'' \right). \tag{10.224}$$

This completes the calculation of all the information at time step $k + 1$.

EXAMPLE 1 *Optimal Linear Control of an SDOF System Subject to El-Centro Earthquake*

Consider Example 1 discussed in Section 10.7, where

$$m = 1.0 \text{ k-s}^2/\text{in.}, \quad c = 1.2567 \text{ k-s/in.}, \quad k = 157.92 \text{ k/in.}$$

$$\mathbf{Q} = \begin{bmatrix} k & 0 \\ 0 & m \end{bmatrix} = \begin{bmatrix} 157.92 & 0.0 \\ 0.0 & 1.0 \end{bmatrix}, \quad \begin{Bmatrix} x_o \\ \dot{x}_o \end{Bmatrix} = \begin{Bmatrix} 0.0 \\ 0.0 \end{Bmatrix}.$$

Let $\Delta t = 0.02$ s and assume that the constant forcing function is used; it follows that

$$\mathbf{A} = \begin{bmatrix} 0.0 & 1.0 \\ -157.92 & -1.2567 \end{bmatrix}, \quad \mathbf{B} = \begin{bmatrix} 0 \\ 1 \end{bmatrix}, \quad \mathbf{H} = \begin{bmatrix} 0 \\ -1 \end{bmatrix}, \quad \mathbf{F}_p^c = \begin{bmatrix} 0 \\ 157.92 \end{bmatrix}$$

$$\mathbf{F}_s = e^{\mathbf{A}\Delta t} \begin{bmatrix} 0.968843 & 0.019543 \\ -3.086307 & 0.944284 \end{bmatrix}, \quad \mathbf{A}^{-1}\left(e^{\mathbf{A}\Delta t} - \mathbf{I} \right) = \begin{bmatrix} 0.01979 & 0.0001973 \\ -0.031157 & 0.019543 \end{bmatrix}$$

$$\mathbf{G} = \mathbf{A}^{-1}\left(e^{\mathbf{A}\Delta t} - \mathbf{I} \right)\mathbf{B} = \begin{bmatrix} 0.0001973 \\ 0.019543 \end{bmatrix}, \quad \mathbf{H}_d^{(EQ)} = \mathbf{A}^{-1}\left(e^{\mathbf{A}\Delta t} - \mathbf{I} \right)\mathbf{H} = \begin{bmatrix} -0.0001973 \\ -0.019543 \end{bmatrix}$$

$$\mathbf{F}_p = \mathbf{A}^{-1}\left(e^{\mathbf{A}\Delta t} - \mathbf{I} \right)\mathbf{F}_p^c = \begin{bmatrix} 0.03116 \\ 3.08623 \end{bmatrix}.$$

Consider the case $R = 0.001$; the resulting matrices are given in Eqs. 10.128 and 10.130:

$$\mathbf{P} = \begin{bmatrix} 919.1673 & 14.428 \\ 14.428 & 2.5227 \end{bmatrix}, \quad \mathbf{F}_f = \begin{bmatrix} 0.941927 & 0.014094 \\ -5.75252 & 0.404455 \end{bmatrix}.$$

It follows that

$$\begin{Bmatrix} x_{k+1} \\ \dot{x}_{k+1} \end{Bmatrix} = \begin{bmatrix} 0.941927 & 0.014094 \\ -5.75252 & 0.404455 \end{bmatrix} \begin{Bmatrix} x_k \\ \dot{x}_k \end{Bmatrix} + \begin{bmatrix} -0.0001973 \\ -0.019543 \end{bmatrix} a_k + \begin{bmatrix} 0.03116 \\ 3.08623 \end{bmatrix} x_k''. \tag{10.225}$$

Using the moment-rotation relationships as given in Example 2 of Section 7.7 (see Eqs. 7.499 to 7.501), it follows that

$$L = 10 \text{ in.} \tag{10.226}$$

$$\mathbf{K}_R = L^2 \mathbf{K} = (10)^2 (157.92) = 15,792.0 \tag{10.227}$$

$$\mathbf{K}_p = L\mathbf{K} = (10)(157.92) = 1,579.2 \tag{10.228}$$

$$M_c = maL = (1.0)(386.4)(10) = 3,864.0 \text{ k-in.} \tag{10.229}$$

and therefore Eqs. 10.223 and 10.224 become

$$M_k + (15,792.0)\theta_k'' = (1,579.2)x_k \tag{10.230}$$

$$x_k'' = (10)\theta_k''. \tag{10.231}$$

Assume that the El-Centro earthquake is used as the input ground motion (see Figure 6.11). The response is shown in Figure 10.13. In addition, the inelastic response without structural control as shown in Figure 7.58 is plotted in Figure 10.13 to show the effectiveness of using a control system. Figure 10.14 shows the control force time history of the response.

EXAMPLE 2   *Minimizing the Total Displacement*

In Eq. 10.196, the cost function uses the elastic displacement z_k' as the variable in the minimization process. This minimization can also be done using total displacement, where

$$J = \frac{1}{2}\sum_{k=0}^{N}\left(\mathbf{z}_k^T\mathbf{Q}\mathbf{z}_k + \mathbf{f}_{ck}^T\mathbf{R}\mathbf{f}_{ck}\right). \tag{10.232}$$

The Lagrangian now becomes

$$L = \sum_{k=0}^{N-1}\left[\frac{1}{2}\left(\mathbf{z}_k^T\mathbf{Q}\mathbf{z}_k + \mathbf{f}_{ck}^T\mathbf{R}\mathbf{f}_{ck}\right) + \boldsymbol{\lambda}_{k+1}^T\left(\mathbf{F}_s\mathbf{z}_k + \mathbf{H}_d^{(EQ)}\mathbf{a}_k + \mathbf{F}_p\mathbf{X}_k'' + \mathbf{G}\mathbf{f}_{ck} - \mathbf{z}_{k+1}\right)\right] \tag{10.233}$$

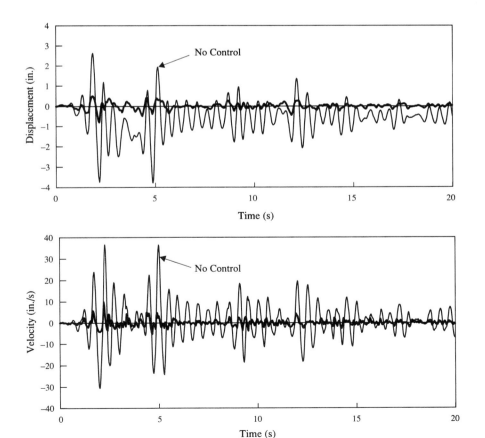

Figure 10.13 Response of Earthquake Control of an Inelastic Structure

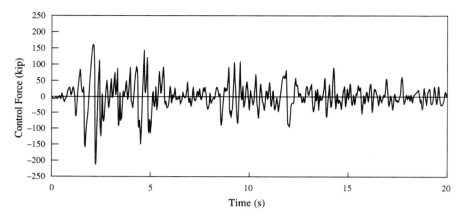

Figure 10.14 Control Force Time History of an Inelastic Structure

where $\boldsymbol{\lambda}_{k+1}$ is the Lagrange multiplier. Then define the Hamiltonian to be

$$H_k = \frac{1}{2}\left(\mathbf{z}_k^T\mathbf{Q}\mathbf{z}_k + \mathbf{f}_{ck}^T\mathbf{R}\mathbf{f}_{ck}\right) + \boldsymbol{\lambda}_{k+1}^T\left(\mathbf{F}_s\mathbf{z}_k + \mathbf{H}_d^{(EQ)}\mathbf{a}_k + \mathbf{F}_p\mathbf{X}_k'' + \mathbf{G}\mathbf{f}_{ck}\right). \tag{10.234}$$

Then substituting Eq. 10.234 into Eq. 10.233 gives

$$L = \sum_{k=0}^{N-1}\left[H_k - \boldsymbol{\lambda}_{k+1}^T\mathbf{z}_{k+1}\right] = H_0 - \boldsymbol{\lambda}_N^T\mathbf{z}_N + \sum_{k=1}^{N-1}\left[H_k - \boldsymbol{\lambda}_k^T\mathbf{z}_k\right] \tag{10.235}$$

which is the same as Eq. 10.200. Therefore, it follows from Eq. 10.201 that

$$\left(\frac{\partial H_k}{\partial\boldsymbol{\lambda}_{k+1}}\right)^T - \mathbf{z}_{k+1} = \mathbf{0} \tag{10.236}$$

$$\left(\frac{\partial H_k}{\partial\mathbf{z}_k}\right)^T - \boldsymbol{\lambda}_k = \mathbf{0}, \quad \mathbf{z}_o = \mathbf{z}(0) \tag{10.237}$$

$$\left(\frac{\partial H_k}{\partial\mathbf{f}_{ck}}\right)^T = \mathbf{0}. \tag{10.238}$$

Recall from the assumption in Eq. 10.204 that

$$\frac{\partial z_{i,k}'}{\partial z_{i,k}} = 1, \quad \frac{\partial z_{i,k}''}{\partial z_{i,k}} = 0. \tag{10.239}$$

It follows from Eqs. 10.236 to 10.238 that

$$\left(\frac{\partial H_k}{\partial\boldsymbol{\lambda}_{k+1}}\right)^T - \mathbf{z}_{k+1} = \mathbf{F}_s\mathbf{z}_k + \mathbf{H}_d^{(EQ)}\mathbf{a}_k + \mathbf{F}_p\mathbf{X}_k'' + \mathbf{G}\mathbf{f}_{ck} - \mathbf{z}_{k+1} = \mathbf{0} \tag{10.240}$$

$$\left(\frac{\partial H_k}{\partial\mathbf{z}_k}\right)^T - \boldsymbol{\lambda}_k = \mathbf{Q}\mathbf{z}_k + \mathbf{F}_s^T\boldsymbol{\lambda}_{k+1} - \boldsymbol{\lambda}_k = \mathbf{Q}\mathbf{z}_k + \mathbf{F}_s^T\boldsymbol{\lambda}_{k+1} - \mathbf{Q}\mathbf{z}_k'' - \boldsymbol{\lambda}_k = \mathbf{0} \tag{10.241}$$

$$\left(\frac{\partial H_k}{\partial\mathbf{f}_{ck}}\right)^T = \mathbf{R}\mathbf{f}_{ck} + \mathbf{G}^T\boldsymbol{\lambda}_{k+1} = \mathbf{0} \tag{10.242}$$

which is the same as Eq. 10.206. This shows that minimizing either the elastic displacement or the total displacement gives the same result.

10.9   OPTIMAL LINEAR CONTROL OF INELASTIC STRUCTURE WITH STATE SPACE REDUCTION

When the mass matrix is singular, Eq. 10.184 is not defined. Therefore, state space reduction as discussed in Section 7.8 is necessary. Using the same notation as in Section 7.8, denote all the degrees of freedom with mass as 1, those with damping but without mass as 2, and those without damping as 3. Then Eq. 10.181 becomes

$$
\begin{bmatrix} \mathbf{M}_{11} & \mathbf{0} & \mathbf{0} \\ \mathbf{0} & \mathbf{0} & \mathbf{0} \\ \mathbf{0} & \mathbf{0} & \mathbf{0} \end{bmatrix} \begin{Bmatrix} \ddot{\mathbf{X}}_1(t) \\ \ddot{\mathbf{X}}_2(t) \\ \ddot{\mathbf{X}}_3(t) \end{Bmatrix} + \begin{bmatrix} \mathbf{C}_{11} & \mathbf{C}_{12} & \mathbf{0} \\ \mathbf{C}_{21} & \mathbf{C}_{22} & \mathbf{0} \\ \mathbf{0} & \mathbf{0} & \mathbf{0} \end{bmatrix} \begin{Bmatrix} \dot{\mathbf{X}}_1(t) \\ \dot{\mathbf{X}}_2(t) \\ \dot{\mathbf{X}}_3(t) \end{Bmatrix} + \begin{bmatrix} \mathbf{K}_{11} & \mathbf{K}_{12} & \mathbf{K}_{13} \\ \mathbf{K}_{21} & \mathbf{K}_{22} & \mathbf{K}_{23} \\ \mathbf{K}_{31} & \mathbf{K}_{32} & \mathbf{K}_{33} \end{bmatrix} \begin{Bmatrix} \mathbf{X}_1'(t) \\ \mathbf{X}_2'(t) \\ \mathbf{X}_3'(t) \end{Bmatrix}
$$

$$
= - \begin{bmatrix} \mathbf{M}_{11} & \mathbf{0} & \mathbf{0} \\ \mathbf{0} & \mathbf{0} & \mathbf{0} \\ \mathbf{0} & \mathbf{0} & \mathbf{0} \end{bmatrix} \begin{Bmatrix} \mathbf{a}_1(t) \\ \mathbf{a}_2(t) \\ \mathbf{a}_3(t) \end{Bmatrix} + \begin{bmatrix} \mathbf{D}_1 \\ \mathbf{D}_2 \\ \mathbf{D}_3 \end{bmatrix} \mathbf{f}_c(t) \tag{10.243}
$$

where

$$
\mathbf{X}'(t) = \mathbf{X}(t) - \mathbf{X}''(t). \tag{10.244}
$$

Substituting Eq. 10.244 into Eq. 10.243, it follows that

$$
\begin{bmatrix} \mathbf{M}_{11} & \mathbf{0} & \mathbf{0} \\ \mathbf{0} & \mathbf{0} & \mathbf{0} \\ \mathbf{0} & \mathbf{0} & \mathbf{0} \end{bmatrix} \begin{Bmatrix} \ddot{\mathbf{X}}_1(t) \\ \ddot{\mathbf{X}}_2(t) \\ \ddot{\mathbf{X}}_3(t) \end{Bmatrix} + \begin{bmatrix} \mathbf{C}_{11} & \mathbf{C}_{12} & \mathbf{0} \\ \mathbf{C}_{21} & \mathbf{C}_{22} & \mathbf{0} \\ \mathbf{0} & \mathbf{0} & \mathbf{0} \end{bmatrix} \begin{Bmatrix} \dot{\mathbf{X}}_1(t) \\ \dot{\mathbf{X}}_2(t) \\ \dot{\mathbf{X}}_3(t) \end{Bmatrix} + \begin{bmatrix} \mathbf{K}_{11} & \mathbf{K}_{12} & \mathbf{K}_{13} \\ \mathbf{K}_{21} & \mathbf{K}_{22} & \mathbf{K}_{23} \\ \mathbf{K}_{31} & \mathbf{K}_{32} & \mathbf{K}_{33} \end{bmatrix} \begin{Bmatrix} \mathbf{X}_1(t) \\ \mathbf{X}_2(t) \\ \mathbf{X}_3(t) \end{Bmatrix}
$$

$$
= - \begin{bmatrix} \mathbf{M}_{11} & \mathbf{0} & \mathbf{0} \\ \mathbf{0} & \mathbf{0} & \mathbf{0} \\ \mathbf{0} & \mathbf{0} & \mathbf{0} \end{bmatrix} \begin{Bmatrix} \mathbf{a}_1(t) \\ \mathbf{a}_2(t) \\ \mathbf{a}_3(t) \end{Bmatrix} + \begin{bmatrix} \mathbf{D}_1 \\ \mathbf{D}_2 \\ \mathbf{D}_3 \end{bmatrix} \mathbf{f}_c(t) + \begin{bmatrix} \mathbf{K}_{11} & \mathbf{K}_{12} & \mathbf{K}_{13} \\ \mathbf{K}_{21} & \mathbf{K}_{22} & \mathbf{K}_{23} \\ \mathbf{K}_{31} & \mathbf{K}_{32} & \mathbf{K}_{33} \end{bmatrix} \begin{Bmatrix} \mathbf{X}_1''(t) \\ \mathbf{X}_2''(t) \\ \mathbf{X}_3''(t) \end{Bmatrix}. \tag{10.245}
$$

Using the method of state space reduction, first expand Eq. 10.245 to give a set of three equations:

$$
\mathbf{M}_{11}\ddot{\mathbf{X}}_1(t) + \mathbf{C}_{11}\dot{\mathbf{X}}_1(t) + \mathbf{C}_{12}\dot{\mathbf{X}}_2(t) + \mathbf{K}_{11}\mathbf{X}_1(t) + \mathbf{K}_{12}\mathbf{X}_2(t) + \mathbf{K}_{13}\mathbf{X}_3(t)
$$
$$
= -\mathbf{M}_{11}\mathbf{a}_1(t) + \mathbf{D}_1\mathbf{f}_c(t) + \mathbf{K}_{11}\mathbf{X}_1''(t) + \mathbf{K}_{12}\mathbf{X}_2''(t) + \mathbf{K}_{13}\mathbf{X}_3''(t) \tag{10.246a}
$$

$$
\mathbf{C}_{21}\dot{\mathbf{X}}_1(t) + \mathbf{C}_{22}\dot{\mathbf{X}}_2(t) + \mathbf{K}_{21}\mathbf{X}_1(t) + \mathbf{K}_{22}\mathbf{X}_2(t) + \mathbf{K}_{23}\mathbf{X}_3(t)
$$
$$
= \mathbf{D}_2\mathbf{f}_c(t) + \mathbf{K}_{21}\mathbf{X}_1''(t) + \mathbf{K}_{22}\mathbf{X}_2''(t) + \mathbf{K}_{23}\mathbf{X}_3''(t) \tag{10.246b}
$$

$$
\mathbf{K}_{31}\mathbf{X}_1(t) + \mathbf{K}_{32}\mathbf{X}_2(t) + \mathbf{K}_{33}\mathbf{X}_3(t) = \mathbf{D}_3\mathbf{f}_c(t) + \mathbf{K}_{31}\mathbf{X}_1''(t) + \mathbf{K}_{32}\mathbf{X}_2''(t) + \mathbf{K}_{33}\mathbf{X}_3''(t). \tag{10.246c}
$$

Solving for $\mathbf{X}_3(t)$ in Eq. 10.246c gives

$$
\mathbf{X}_3(t) = -\mathbf{K}_{33}^{-1}\mathbf{K}_{31}\mathbf{X}_1(t) - \mathbf{K}_{33}^{-1}\mathbf{K}_{32}\mathbf{X}_2(t) + \mathbf{K}_{33}^{-1}\mathbf{D}_3\mathbf{f}_c(t) + \mathbf{K}_{33}^{-1}\mathbf{K}_{31}\mathbf{X}_1''(t) + \mathbf{K}_{33}^{-1}\mathbf{K}_{32}\mathbf{X}_2''(t) + \mathbf{X}_3''(t). \tag{10.247}
$$

Substituting Eq. 10.247 into Eq. 10.246b, it follows that

$$
\mathbf{C}_{21}\dot{\mathbf{X}}_1(t) + \mathbf{C}_{22}\dot{\mathbf{X}}_2(t) + \mathbf{K}_{21}'\mathbf{X}_1(t) + \mathbf{K}_{22}'\mathbf{X}_2(t) = \mathbf{D}_2'\mathbf{f}_c(t) + \mathbf{K}_{21}'\mathbf{X}_1''(t) + \mathbf{K}_{22}'\mathbf{X}_2''(t) \tag{10.248}
$$

where

$$
\mathbf{K}_{21}' = \mathbf{K}_{21} - \mathbf{K}_{23}\mathbf{K}_{33}^{-1}\mathbf{K}_{31}, \qquad \mathbf{K}_{22}' = \mathbf{K}_{22} - \mathbf{K}_{23}\mathbf{K}_{33}^{-1}\mathbf{K}_{32}, \qquad \mathbf{D}_2' = \mathbf{D}_2 - \mathbf{K}_{23}\mathbf{K}_{33}^{-1}\mathbf{D}_3.
$$

Solving for $\dot{\mathbf{X}}_2(t)$ in Eq. 10.248 gives

$$
\dot{\mathbf{X}}_2(t) = -\mathbf{C}_{22}^{-1}\mathbf{C}_{21}\dot{\mathbf{X}}_1(t) - \mathbf{C}_{22}^{-1}\mathbf{K}_{21}'\mathbf{X}_1(t) - \mathbf{C}_{22}^{-1}\mathbf{K}_{22}'\mathbf{X}_2(t)
$$
$$
+ \mathbf{C}_{22}^{-1}\mathbf{D}_2'\mathbf{f}_c(t) + \mathbf{C}_{22}^{-1}\mathbf{K}_{21}'\mathbf{X}_1''(t) + \mathbf{C}_{22}^{-1}\mathbf{K}_{22}'\mathbf{X}_2''(t). \tag{10.249}
$$

Now substituting Eqs. 10.247 and 10.249 into Eq. 10.246a, it follows that

$$\mathbf{M}_{11}\ddot{\mathbf{X}}_1(t) + \mathbf{C}'_{11}\dot{\mathbf{X}}_1(t) + \mathbf{K}''_{11}\mathbf{X}_1(t) + \mathbf{K}''_{12}\mathbf{X}_2(t) = -\mathbf{M}_{11}\mathbf{a}_1(t) + \mathbf{D}''_1\mathbf{f}_c(t) + \mathbf{K}''_{11}\mathbf{X}''_1(t) + \mathbf{K}''_{12}\mathbf{X}''_2(t) \quad (10.250)$$

where

$$\mathbf{C}'_{11} = \mathbf{C}_{11} - \mathbf{C}_{12}\mathbf{C}_{22}^{-1}\mathbf{C}_{21}, \quad \mathbf{K}''_{11} = \mathbf{K}'_{11} - \mathbf{C}_{12}\mathbf{C}_{22}^{-1}\mathbf{K}'_{21}, \quad \mathbf{K}''_{12} = \mathbf{K}'_{12} - \mathbf{C}_{12}\mathbf{C}_{22}^{-1}\mathbf{K}'_{22}$$

$$\mathbf{K}'_{11} = \mathbf{K}_{11} - \mathbf{K}_{13}\mathbf{K}_{33}^{-1}\mathbf{K}_{31}, \quad \mathbf{K}'_{12} = \mathbf{K}_{12} - \mathbf{K}_{13}\mathbf{K}_{33}^{-1}\mathbf{K}_{32}$$

$$\mathbf{D}''_1 = \mathbf{D}'_1 - \mathbf{C}_{12}\mathbf{C}_{22}^{-1}\mathbf{D}'_2, \quad \mathbf{D}'_1 = \mathbf{D}_1 - \mathbf{K}_{13}\mathbf{K}_{33}^{-1}\mathbf{D}_3.$$

Solving for $\ddot{\mathbf{X}}_1(t)$ in Eq. 10.250 gives

$$\ddot{\mathbf{X}}_1(t) = -\mathbf{a}_1(t) - \mathbf{M}_{11}^{-1}\mathbf{C}'_{11}\dot{\mathbf{X}}_1(t) - \mathbf{M}_{11}^{-1}\mathbf{K}''_{11}\mathbf{X}_1(t) - \mathbf{M}_{11}^{-1}\mathbf{K}''_{12}\mathbf{X}_2(t)$$
$$+ \mathbf{M}_{11}^{-1}\mathbf{D}''_1\mathbf{f}_c(t) + \mathbf{M}_{11}^{-1}\mathbf{K}''_{11}\mathbf{X}''_1(t) + \mathbf{M}_{11}^{-1}\mathbf{K}''_{12}\mathbf{X}''_2(t). \quad (10.251)$$

Now representing Eqs. 10.249 and 10.251 in state space form:

$$\begin{Bmatrix} \dot{\mathbf{X}}_1(t) \\ \dot{\mathbf{X}}_2(t) \\ \ddot{\mathbf{X}}_1(t) \end{Bmatrix} = \begin{bmatrix} \mathbf{0} & \mathbf{0} & \mathbf{I} \\ -\mathbf{C}_{22}^{-1}\mathbf{K}'_{21} & -\mathbf{C}_{22}^{-1}\mathbf{K}'_{22} & -\mathbf{C}_{22}^{-1}\mathbf{C}_{21} \\ -\mathbf{M}_{11}^{-1}\mathbf{K}''_{11} & -\mathbf{M}_{11}^{-1}\mathbf{K}''_{12} & -\mathbf{M}_{11}^{-1}\mathbf{C}'_{11} \end{bmatrix} \begin{Bmatrix} \mathbf{X}_1(t) \\ \mathbf{X}_2(t) \\ \dot{\mathbf{X}}_1(t) \end{Bmatrix} + \begin{Bmatrix} \mathbf{0} \\ \mathbf{0} \\ -\mathbf{a}(t) \end{Bmatrix}$$

$$+ \begin{bmatrix} \mathbf{0} \\ \mathbf{C}_{22}^{-1}\mathbf{D}'_2 \\ \mathbf{M}_{11}^{-1}\mathbf{D}''_1 \end{bmatrix} \mathbf{f}_c(t) + \begin{bmatrix} \mathbf{0} & \mathbf{0} \\ \mathbf{C}_{22}^{-1}\mathbf{K}'_{21} & \mathbf{C}_{22}^{-1}\mathbf{K}'_{22} \\ \mathbf{M}_{11}^{-1}\mathbf{K}''_{11} & \mathbf{M}_{11}^{-1}\mathbf{K}''_{12} \end{bmatrix} \begin{Bmatrix} \mathbf{X}''_1(t) \\ \mathbf{X}''_2(t) \end{Bmatrix}. \quad (10.252)$$

Denote

$$\mathbf{z}(t) = \begin{Bmatrix} \mathbf{X}_1(t) \\ \mathbf{X}_2(t) \\ \dot{\mathbf{X}}_1(t) \end{Bmatrix}, \quad \mathbf{X}''(t) = \begin{Bmatrix} \mathbf{X}''_1(t) \\ \mathbf{X}''_2(t) \end{Bmatrix}, \quad \mathbf{A} = \begin{bmatrix} \mathbf{0} & \mathbf{0} & \mathbf{I} \\ -\mathbf{C}_{22}^{-1}\mathbf{K}'_{21} & -\mathbf{C}_{22}^{-1}\mathbf{K}'_{22} & -\mathbf{C}_{22}^{-1}\mathbf{C}_{21} \\ -\mathbf{M}_{11}^{-1}\mathbf{K}''_{11} & -\mathbf{M}_{11}^{-1}\mathbf{K}''_{12} & -\mathbf{M}_{11}^{-1}\mathbf{C}'_{11} \end{bmatrix}, \quad \mathbf{B} = \begin{bmatrix} \mathbf{0} \\ \mathbf{C}_{22}^{-1}\mathbf{D}'_2 \\ \mathbf{M}_{11}^{-1}\mathbf{D}''_1 \end{bmatrix}$$

$$\mathbf{a}(t) = \left\{ a_x(t) \quad a_y(t) \quad a_z(t) \quad a_r(t) \quad a_s(t) \quad a_t(t) \right\}^T, \quad \mathbf{F}_p^c = \begin{bmatrix} \mathbf{0} & \mathbf{0} \\ \mathbf{C}_{22}^{-1}\mathbf{K}'_{21} & \mathbf{C}_{22}^{-1}\mathbf{K}'_{22} \\ \mathbf{M}_{11}^{-1}\mathbf{K}''_{11} & \mathbf{M}_{11}^{-1}\mathbf{K}''_{12} \end{bmatrix}.$$

It follows that

$$\dot{\mathbf{z}}(t) = \mathbf{A}\mathbf{z}(t) + \mathbf{H}\mathbf{a}(t) + \mathbf{F}_p^c\mathbf{X}''(t) + \mathbf{B}\mathbf{f}_c(t) \quad (10.253)$$

where $\mathbf{H}$ was defined in Eq. 6.214. Equation 10.253 is the same as Eq. 10.185. Therefore, following the same procedure discussed in Eqs. 10.186 to 10.194, the result is given in Eq. 10.195, which is

$$\mathbf{z}_{k+1} = \mathbf{F}_s\,\mathbf{z}_k + \mathbf{H}_d^{(EQ)}\mathbf{a}_k + \mathbf{F}_p\mathbf{X}''_k + \mathbf{G}\mathbf{f}_{ck} \quad (10.254)$$

where, when the Delta forcing function assumption is made,

$$\mathbf{F}_s = \mathbf{e}^{\mathbf{A}\Delta t}, \quad \mathbf{H}_d^{(EQ)} = \mathbf{e}^{\mathbf{A}\Delta t}\mathbf{H}\Delta t, \quad \mathbf{F}_p = \mathbf{e}^{\mathbf{A}\Delta t}\mathbf{F}_p^c\Delta t, \quad \mathbf{G} = \mathbf{e}^{\mathbf{A}\Delta t}\mathbf{B}\Delta t \quad (10.255)$$

or, when the constant forcing function assumption is made,

$$\mathbf{F}_s = \mathbf{e}^{\mathbf{A}\Delta t}, \quad \mathbf{H}_d^{(EQ)} = \mathbf{A}^{-1}\left(\mathbf{e}^{\mathbf{A}\Delta t} - \mathbf{I}\right)\mathbf{H}, \quad \mathbf{F}_p = \mathbf{A}^{-1}\left(\mathbf{e}^{\mathbf{A}\Delta t} - \mathbf{I}\right)\mathbf{F}_p^c, \quad \mathbf{G} = \mathbf{A}^{-1}\left(\mathbf{e}^{\mathbf{A}\Delta t} - \mathbf{I}\right)\mathbf{B}. \quad (10.256)$$

For optimal linear control, define the cost function to be related to the elastic displacement, that is,

$$J = \frac{1}{2}\sum_{k=0}^{N}\left(\mathbf{z}'^T_k\mathbf{Q}\mathbf{z}'_k + \mathbf{f}_{ck}^T\mathbf{R}\mathbf{f}_{ck}\right) \quad (10.257)$$

where

$$
\mathbf{z}'_k = \left\{ \begin{matrix} x'_1 \\ \vdots \\ x'_n \\ \cdots \\ \dot{x}_1 \\ \vdots \\ \dot{x}_n \end{matrix} \right\}_k = \mathbf{z}_k - \mathbf{z}''_k = \left\{ \begin{matrix} x_1 \\ \vdots \\ x_n \\ \cdots \\ \dot{x}_1 \\ \vdots \\ \dot{x}_n \end{matrix} \right\}_k - \left\{ \begin{matrix} x''_1 \\ \vdots \\ x''_n \\ \cdots \\ 0 \\ \vdots \\ 0 \end{matrix} \right\}_k = \mathbf{z}_k - \left\{ \begin{matrix} \mathbf{X}''_k \\ \cdots \\ \mathbf{0} \end{matrix} \right\}. \tag{10.258}
$$

The minimization process was discussed in Section 10.8 in Eqs. 10.195 to 10.222. The results are

$$
\mathbf{f}_{ck} = -\left(\mathbf{G}^T \mathbf{P} \mathbf{G} + \mathbf{R} \right)^{-1} \mathbf{G}^T \mathbf{P} \mathbf{F}_s \mathbf{z}_k \tag{10.259}
$$

where

$$
\mathbf{P} = \mathbf{Q} + \mathbf{F}_s^T \mathbf{P} \left(\mathbf{I} + \mathbf{G} \mathbf{R}^{-1} \mathbf{G}^T \mathbf{P} \right)^{-1} \mathbf{F}_s. \tag{10.260}
$$

Using this control force pattern, it follows that

$$
\begin{aligned}
\mathbf{z}_{k+1} &= \left[\mathbf{I} - \mathbf{G} \left(\mathbf{G}^T \mathbf{P} \mathbf{G} + \mathbf{R} \right)^{-1} \mathbf{G}^T \mathbf{P} \right] \mathbf{F}_s \mathbf{z}_k + \mathbf{H}_d^{(\mathrm{EQ})} \mathbf{a}_k + \mathbf{F}_p \mathbf{X}''_k \\
&= \left[\mathbf{I} + \mathbf{G} \mathbf{R}^{-1} \mathbf{G}^T \mathbf{P} \right]^{-1} \mathbf{F}_s \mathbf{z}_k + \mathbf{H}_d^{(\mathrm{EQ})} \mathbf{a}_k + \mathbf{F}_p \mathbf{X}''_k.
\end{aligned} \tag{10.261}
$$

Let

$$
\mathbf{F}_f = \left[\mathbf{I} - \mathbf{G} \left(\mathbf{G}^T \mathbf{P} \mathbf{G} + \mathbf{R} \right)^{-1} \mathbf{G}^T \mathbf{P} \right] \mathbf{F}_s = \left[\mathbf{I} + \mathbf{G} \mathbf{R}^{-1} \mathbf{G}^T \mathbf{P} \right]^{-1} \mathbf{F}_s. \tag{10.262}
$$

It follows from Eq. 10.261 that

$$
\mathbf{z}_{k+1} = \mathbf{F}_f \mathbf{z}_k + \mathbf{H}_d^{(\mathrm{EQ})} \mathbf{a}_k + \mathbf{F}_p \mathbf{X}''_k. \tag{10.263}
$$

The solution procedure is a recursive process and was discussed in Section 10.8. Assume first that all the information at time step k is known; then both the displacement and velocity vectors at time step $k + 1$ (i.e., $\mathbf{X}_{k+1}$ and $\dot{\mathbf{X}}_{k+1}$) can be calculated using Eq. 10.263. Then using Eq. 10.259, the control force applied onto the structure at time step $k + 1$, $\mathbf{f}_{c,k+1}$, can be computed. Using $\mathbf{X}_{k+1}$ and $\mathbf{f}_{c,k+1}$, the inelastic property of the structure at time step $k + 1$ (i.e., $\mathbf{X}''_{k+1}$) can be determined with the modification discussed in Eqs. 7.538 to 7.547 with an additional control force term. Start with Eq. 7.538, where

$$
\mathbf{M}_{k+1} + \mathbf{K}_R \mathbf{\Theta}''_{k+1} = \mathbf{K}_p^T \mathbf{X}_{k+1}. \tag{10.264}
$$

Recall from Eq. 10.247 that

$$
\begin{aligned}
\mathbf{X}_{3,k+1} = &-\mathbf{K}_{33}^{-1} \mathbf{K}_{31} \mathbf{X}_{1,k+1} - \mathbf{K}_{33}^{-1} \mathbf{K}_{32} \mathbf{X}_{2,k+1} + \mathbf{K}_{33}^{-1} \mathbf{D}_3 \mathbf{f}_{c,k+1} \\
&+ \mathbf{K}_{33}^{-1} \mathbf{K}_{31} \mathbf{X}''_{1,k+1} + \mathbf{K}_{33}^{-1} \mathbf{K}_{32} \mathbf{X}''_{2,k+1} + \mathbf{X}''_{3,k+1}.
\end{aligned} \tag{10.265}
$$

Therefore, $\mathbf{X}_{k+1}$ in Eq. 10.264 becomes

$$
\begin{aligned}
\mathbf{X}_{k+1} &= \left\{ \begin{matrix} \mathbf{X}_1 \\ \mathbf{X}_2 \\ \mathbf{X}_3 \end{matrix} \right\}_{k+1} \\
&= \begin{bmatrix} \mathbf{I} & \mathbf{0} \\ \mathbf{0} & \mathbf{I} \\ -\mathbf{K}_{33}^{-1} \mathbf{K}_{31} & -\mathbf{K}_{33}^{-1} \mathbf{K}_{32} \end{bmatrix} \left\{ \begin{matrix} \mathbf{X}_1 \\ \mathbf{X}_2 \end{matrix} \right\}_{k+1} + \begin{bmatrix} \mathbf{0} \\ \mathbf{0} \\ \mathbf{K}_{33}^{-1} \mathbf{D}_3 \end{bmatrix} \mathbf{f}_{c,k+1} + \begin{bmatrix} \mathbf{0} & \mathbf{0} & \mathbf{0} \\ \mathbf{0} & \mathbf{0} & \mathbf{0} \\ \mathbf{K}_{33}^{-1} \mathbf{K}_{31} & \mathbf{K}_{33}^{-1} \mathbf{K}_{32} & \mathbf{I} \end{bmatrix} \left\{ \begin{matrix} \mathbf{X}''_1 \\ \mathbf{X}''_2 \\ \mathbf{X}''_3 \end{matrix} \right\}_{k+1}.
\end{aligned} \tag{10.266}
$$

Now, recall from Eq. 7.541 that

$$
\mathbf{X}''_{k+1} = \mathbf{K}^{-1} \mathbf{K}_p \mathbf{\Theta}''_{k+1} \tag{10.267}
$$

where $\mathbf{K}$ in Eq. 10.267 represents the global stiffness matrix. Substituting Eq. 10.267 into Eq. 10.266 gives

$$\mathbf{X}_{k+1} = \begin{bmatrix} \mathbf{I} & \mathbf{0} \\ \mathbf{0} & \mathbf{I} \\ -\mathbf{K}_{33}^{-1}\mathbf{K}_{31} & -\mathbf{K}_{33}^{-1}\mathbf{K}_{32} \end{bmatrix} \begin{Bmatrix} \mathbf{X}_1 \\ \mathbf{X}_2 \end{Bmatrix}_{k+1}$$

$$+ \begin{bmatrix} \mathbf{0} \\ \mathbf{0} \\ \mathbf{K}_{33}^{-1}\mathbf{D}_3 \end{bmatrix} \mathbf{f}_{c,k+1} + \begin{bmatrix} \mathbf{0} & \mathbf{0} & \mathbf{0} \\ \mathbf{0} & \mathbf{0} & \mathbf{0} \\ \mathbf{K}_{33}^{-1}\mathbf{K}_{31} & \mathbf{K}_{33}^{-1}\mathbf{K}_{32} & \mathbf{I} \end{bmatrix} \mathbf{K}^{-1}\mathbf{K}_p\boldsymbol{\Theta}''_{k+1}. \qquad (10.268)$$

It follows that after substituting Eq. 10.268 into Eq. 10.264,

$$\mathbf{M}_{k+1} + \mathbf{K}_R\boldsymbol{\Theta}''_{k+1} = \mathbf{K}_p^T \begin{bmatrix} \mathbf{I} & \mathbf{0} \\ \mathbf{0} & \mathbf{I} \\ -\mathbf{K}_{33}^{-1}\mathbf{K}_{31} & -\mathbf{K}_{33}^{-1}\mathbf{K}_{32} \end{bmatrix} \begin{Bmatrix} \mathbf{X}_1 \\ \mathbf{X}_2 \end{Bmatrix}_{k+1}$$

$$+ \mathbf{K}_p^T \begin{bmatrix} \mathbf{0} \\ \mathbf{0} \\ \mathbf{K}_{33}^{-1}\mathbf{D}_3 \end{bmatrix} \mathbf{f}_{c,k+1} + \mathbf{K}_p^T \begin{bmatrix} \mathbf{0} & \mathbf{0} & \mathbf{0} \\ \mathbf{0} & \mathbf{0} & \mathbf{0} \\ \mathbf{K}_{33}^{-1}\mathbf{K}_{31} & \mathbf{K}_{33}^{-1}\mathbf{K}_{32} & \mathbf{I} \end{bmatrix} \mathbf{K}^{-1}\mathbf{K}_p\boldsymbol{\Theta}''_{k+1}. \qquad (10.269)$$

Define

$$\overline{\mathbf{K}}_R = \mathbf{K}_R - \mathbf{K}_p^T \begin{bmatrix} \mathbf{0} & \mathbf{0} & \mathbf{0} \\ \mathbf{0} & \mathbf{0} & \mathbf{0} \\ \mathbf{K}_{33}^{-1}\mathbf{K}_{31} & \mathbf{K}_{33}^{-1}\mathbf{K}_{32} & \mathbf{I} \end{bmatrix} \mathbf{K}^{-1}\mathbf{K}_p \qquad (10.270a)$$

$$\overline{\mathbf{K}}_p^T = \mathbf{K}_p^T \begin{bmatrix} \mathbf{I} & \mathbf{0} \\ \mathbf{0} & \mathbf{I} \\ -\mathbf{K}_{33}^{-1}\mathbf{K}_{31} & -\mathbf{K}_{33}^{-1}\mathbf{K}_{32} \end{bmatrix}, \quad \overline{\mathbf{X}}_{k+1} = \begin{Bmatrix} \mathbf{X}_1 \\ \mathbf{X}_2 \end{Bmatrix}_{k+1} \qquad (10.270b)$$

$$\overline{\mathbf{D}}_3 = \mathbf{K}_p^T \begin{bmatrix} \mathbf{0} \\ \mathbf{0} \\ \mathbf{K}_{33}^{-1}\mathbf{D}_3 \end{bmatrix} \qquad (10.270c)$$

and then it follows from Eq. 10.269 that

$$\mathbf{M}_{k+1} + \overline{\mathbf{K}}_R\boldsymbol{\Theta}''_{k+1} = \overline{\mathbf{K}}_p^T\overline{\mathbf{X}}_{k+1} + \overline{\mathbf{D}}_3\mathbf{f}_{c,k+1}. \qquad (10.271)$$

Using Eq. 10.271 gives both the moment and plastic rotations at time step $k + 1$. Equation 10.271 can be written in incremental form, as discussed in Eq. 7.546:

$$\mathbf{M}_{k+1} + \overline{\mathbf{K}}_R\boldsymbol{\Delta\Theta}'' = \overline{\mathbf{K}}_p^T\overline{\mathbf{X}}_{k+1} + \overline{\mathbf{D}}_3\mathbf{f}_{c,k+1} - \overline{\mathbf{K}}_R\boldsymbol{\Theta}''_k. \qquad (10.272)$$

Finally, once $\mathbf{M}_{k+1}$ and $\boldsymbol{\Delta\Theta}''$ are obtained, the inelastic displacement vector $\mathbf{X}''_{k+1}$ can be determined using Eq. 10.267 in incremental form:

$$\mathbf{X}''_{k+1} = \mathbf{K}^{-1}\mathbf{K}_p\boldsymbol{\Theta}''_{k+1} = \mathbf{K}^{-1}\mathbf{K}_p\left(\boldsymbol{\Theta}''_k + \boldsymbol{\Delta\Theta}''\right). \qquad (10.273)$$

This completes the calculation of all the information at time step $k + 1$.

EXAMPLE 1   *One-Story Frame, Three Degrees of Freedom and One Mass*

Consider the frame used in Example 1 in Section 7.8. Optimal linear control is used to reduce the response subject to earthquake ground motion. The structural properties are

$$\mathbf{M} = \begin{bmatrix} 1.0 & 0.0 & 0.0 \\ 0.0 & 0.0 & 0.0 \\ 0.0 & 0.0 & 0.0 \end{bmatrix}, \quad \mathbf{C} = \begin{bmatrix} 0.012 & 0.30 & 0.30 \\ 0.30 & 40.0 & 10.0 \\ 0.30 & 10.0 & 40.0 \end{bmatrix}$$

$$EI_1 = EI_2 = 10{,}000 \text{ k-in.}^2, \qquad L_1 = L_2 = 100 \text{ in.}$$

$$\mathbf{K} = \begin{bmatrix} \dfrac{24EI_1}{L_1^3} & \dfrac{6EI_1}{L_1^2} & \dfrac{6EI_1}{L_1^2} \\[2ex] \dfrac{6EI_1}{L_1^2} & \dfrac{4EI_1}{L_1}+\dfrac{4EI_2}{L_2} & \dfrac{2EI_2}{L_2} \\[2ex] \dfrac{6EI_1}{L_1^2} & \dfrac{2EI_2}{L_2} & \dfrac{4EI_1}{L_1}+\dfrac{4EI_2}{L_2} \end{bmatrix} = \begin{bmatrix} 24 & 600 & 600 \\ 600 & 80{,}000 & 20{,}000 \\ 600 & 20{,}000 & 80{,}000 \end{bmatrix} \qquad (10.274)$$

$$\mathbf{K}_p = \begin{bmatrix} \dfrac{6EI_1}{L_1^2} & 0 & 0 & \dfrac{6EI_1}{L_1^2} \\[2ex] \dfrac{2EI_1}{L_1} & \dfrac{4EI_2}{L_2} & \dfrac{2EI_2}{L_2} & 0 \\[2ex] 0 & \dfrac{2EI_2}{L_2} & \dfrac{4EI_2}{L_2} & \dfrac{2EI_1}{L_1} \end{bmatrix} = \begin{bmatrix} 600 & 0 & 0 & 600 \\ 20{,}000 & 40{,}000 & 20{,}000 & 0 \\ 0 & 20{,}000 & 40{,}000 & 20{,}000 \end{bmatrix} \qquad (10.275)$$

$$\mathbf{K}_R = \begin{bmatrix} \dfrac{4EI_1}{L_1} & 0 & 0 & 0 \\[2ex] 0 & \dfrac{4EI_2}{L_2} & \dfrac{2EI_2}{L_2} & 0 \\[2ex] 0 & \dfrac{2EI_2}{L_2} & \dfrac{4EI_2}{L_2} & 0 \\[2ex] 0 & 0 & 0 & \dfrac{4EI_1}{L_1} \end{bmatrix} = \begin{bmatrix} 40{,}000 & 0 & 0 & 0 \\ 0 & 40{,}000 & 20{,}000 & 0 \\ 0 & 20{,}000 & 40{,}000 & 0 \\ 0 & 0 & 0 & 40{,}000 \end{bmatrix}. \qquad (10.276)$$

Let

$$\mathbf{Q} = \begin{bmatrix} \mathbf{K} & \mathbf{0} \\ \mathbf{0} & \mathbf{M}_{11} \end{bmatrix} = \begin{bmatrix} 24 & 600 & 600 & 0 \\ 600 & 80{,}000 & 20{,}000 & 0 \\ 600 & 20{,}000 & 80{,}000 & 0 \\ 0 & 0 & 0 & 1 \end{bmatrix}, \quad R = R = 0.001. \qquad (10.277)$$

Also, let $M_c = 4{,}500$ k-in. and $M_b = 3{,}500$ k-in., with the structure subject to the modified El-Centro earthquake (see Figure 6.11). The response is shown in Figure 10.15 and the control force time history is shown in Figure 10.16.

10.10   NONLINEAR CONTROL OF ELASTIC STRUCTURE FOR EARTHQUAKES

The formulation of nonlinear control is very similar to optimal linear control. The only difference is that the control law depends nonlinearly on the displacement and velocity responses of the structure. Start with the dynamic equilibrium equation of motion in state space form as given in Eq. 10.145:

$$\mathbf{z}_{k+1} = \mathbf{F}_s\, \mathbf{z}_k + \mathbf{H}_d^{(EQ)}\mathbf{a}_k + \mathbf{Gf}_{ck}. \qquad (10.278)$$

In Eq. 10.278, the control force $\mathbf{f}_{ck}$ is a nonlinear function of $\mathbf{z}_k$, that is,

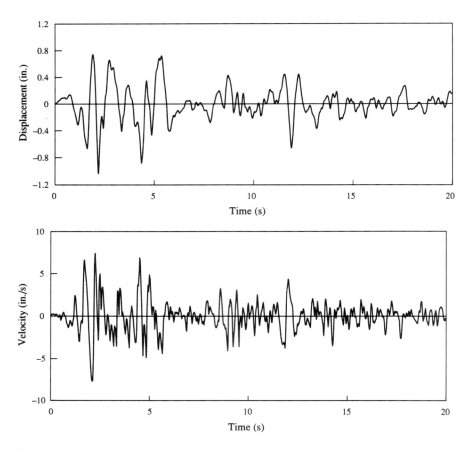

Figure 10.15 Response of Earthquake Control of an Inelastic Structure

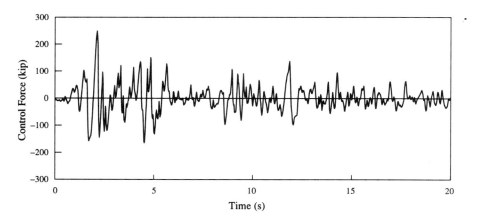

Figure 10.16 Control Force Time History of an Inelastic Structure

$$\mathbf{f}_{ck} = f(\mathbf{z}_k) \tag{10.279}$$

where the function $f(\cdot)$ can be of any form. In general, this function should be negative in order to perform effectively, because a negative force is required to counter a positive structural movement. For example, the function can take the form

$$\mathbf{f}_{ck} = -\left(\mathbf{G}^T\mathbf{P}\mathbf{G} + \mathbf{R}\right)^{-1}\mathbf{G}^T\mathbf{P}\mathbf{F}_s\mathbf{z}_k \tag{10.280}$$

which is the optimal linear control as discussed in Section 10.7.

The procedure for calculating the response of the structure is as follows. Assume that all information at time step k is known. Then the displacement and velocity responses at time step $k + 1$ can be calculated using Eq. 10.278. Based on these calculated responses, the required control force at time step $k + 1$ can then be calculated using Eq. 10.279. Finally, measuring the ground acceleration at time step $k + 1$ gives $\mathbf{a}_{k+1}$. This completes the calculation of all the information at time step $k + 1$.

EXAMPLE 1 *Piecewise Linear Control*

An example of piecewise linear control can be derived based on practical experience of structural response during earthquakes. Consider the feedback control system of the form

$$\begin{aligned} &\text{if } x_{\text{roof}} \le x_{\min}, &\text{then} \quad &\mathbf{f}_{ck} = \mathbf{0} \\ &\text{if } x_{\min} < x_{\text{roof}} \le x_{\max}, &\text{then} \quad &\mathbf{f}_{ck} = -\left(\mathbf{G}^T\mathbf{P}\mathbf{G} + \mathbf{R}\right)^{-1}\mathbf{G}^T\mathbf{P}\mathbf{z}_k \\ &\text{if } x_{\max} < x_{\text{roof}}, &\text{then} \quad &\mathbf{f}_{ck} = -\mathbf{f}_{\max}. \end{aligned} \tag{10.281}$$

Here, when the roof displacement is less than a certain displacement threshold, the control force is zero. This can eliminate the unnecessary waste of power when only a small earthquake occurs. Then, when the roof displacement is above a certain threshold but less than the maximum displacement, optimal linear control is used. Finally, when the roof displacement is above the maximum displacement, the control force is constant. This can be interpreted as the maximum control force that can be developed by the control system.

10.11 NONLINEAR CONTROL OF INELASTIC STRUCTURE FOR EARTHQUAKES

The formulation of nonlinear control of an inelastic structure can also be extended directly from the theory of optimal linear control of an inelastic structure. Here, the control force depends nonlinearly on the displacement and velocity responses of the structure. Start with the dynamic equilibrium equation of motion in state space form as given in Eq. 10.254, that is,

$$\mathbf{z}_{k+1} = \mathbf{F}_s\,\mathbf{z}_k + \mathbf{H}_d^{(\text{EQ})}\mathbf{a}_k + \mathbf{F}_p\mathbf{X}_k'' + \mathbf{G}\mathbf{f}_{ck} \tag{10.282}$$

where the control force $\mathbf{f}_{ck}$ is a nonlinear function of $\mathbf{z}_k$, that is,

$$\mathbf{f}_{ck} = f(\mathbf{z}_k). \tag{10.283}$$

Because the structure can respond in the inelastic domain, the inelastic displacement is then determined using Eqs. 10.264 to 10.267, that is,

$$\mathbf{M}_{k+1} + \overline{\mathbf{K}}_R\Delta\mathbf{\Theta}'' = \overline{\mathbf{K}}_p^T\overline{\mathbf{X}}_{k+1} + \overline{\mathbf{D}}_3\mathbf{f}_{c,k+1} - \overline{\mathbf{K}}_R\mathbf{\Theta}_k'' \tag{10.284}$$

$$\mathbf{X}_{k+1}'' = \mathbf{K}^{-1}\mathbf{K}_p\mathbf{\Theta}_{k+1}'' = \mathbf{K}^{-1}\mathbf{K}_p\left(\mathbf{\Theta}_k'' + \Delta\mathbf{\Theta}''\right) \tag{10.285}$$

where

$$\overline{\mathbf{K}}_R = \mathbf{K}_R - \mathbf{K}_p^T \begin{bmatrix} \mathbf{0} & \mathbf{0} & \mathbf{0} \\ \mathbf{0} & \mathbf{0} & \mathbf{0} \\ \mathbf{K}_{33}^{-1}\mathbf{K}_{31} & \mathbf{K}_{33}^{-1}\mathbf{K}_{32} & \mathbf{I} \end{bmatrix} \mathbf{K}^{-1}\mathbf{K}_p \tag{10.286a}$$

$$\overline{\mathbf{K}}_p^T = \mathbf{K}_p^T \begin{bmatrix} \mathbf{I} & \mathbf{0} \\ \mathbf{0} & \mathbf{I} \\ -\mathbf{K}_{33}^{-1}\mathbf{K}_{31} & -\mathbf{K}_{33}^{-1}\mathbf{K}_{32} \end{bmatrix}, \quad \overline{\mathbf{X}}_{k+1} = \begin{Bmatrix} \mathbf{X}_1 \\ \mathbf{X}_2 \end{Bmatrix}_{k+1} \tag{10.286b}$$

$$\overline{\mathbf{D}}_3 = \mathbf{K}_p^T \begin{bmatrix} \mathbf{0} \\ \mathbf{0} \\ \mathbf{D}_3 \end{bmatrix}. \tag{10.286c}$$

The procedure for calculating the response of the structure is as follows. Assume that all information at time step k is known. Then the displacement and velocity responses at time step $k + 1$ can be calculated using Eq. 10.282. Based on these calculated responses, the required control force at time step $k + 1$ can then be calculated using Eq. 10.283. Then Eq. 10.284 is used to solve for $\mathbf{M}_{k+1}$ and $\boldsymbol{\Delta\Theta}_k''$ simultaneously. Based on $\boldsymbol{\Delta\Theta}_k''$ computed, the inelastic displacement at time step $k + 1$, $\mathbf{X}_{k+1}''$, can be calculated using Eq. 10.285. Finally, measuring the ground acceleration at time step $k + 1$ gives $\mathbf{a}_{k+1}$. This completes the calculation of all the information at time step $k + 1$.

RECOMMENDED ADDITIONAL READING

1. T. T. Soong. *Active Structural Control: Theory and Application.* Longman Scientific & Technical, 1990.

REFERENCES

1. T. T. Soong. *Active Structural Control: Theory and Application.* Longman Scientific & Technical, 1990.

2. D. R. Vaughn. "A Nonrecursive Algebraic Solution for the Discrete Riccati Equation." *IEEE Trans. Automatic Control,* Vol. AC-15, pp. 597–599, 1970.

3. A. J. Laub. "A Schur Method for Solving Algebraic Riccati Equations." *IEEE Trans. Automatic Control,* Vol. AC-24, pp. 913–921, 1979.

PROBLEMS

10.1 Repeat Example 1 in Section 10.6 but with $m = 2.0$.

10.2 Repeat Example 1 in Section 10.6 but with $m = 0.5$.

10.3 Repeat Example 1 in Section 10.6 but with $c = 0.63$.

10.4 Repeat Example 2 in Section 10.6 but use the system in Problem 10.1.

10.5 Repeat Example 2 in Section 10.6 but use the system in Problem 10.2.

10.6 Repeat Example 1 in Section 10.7 but use the system in Problem 10.1.

10.7 Repeat Example 1 in Section 10.7 but use the system in Problem 10.2.

10.8 Repeat Example 2 in Section 10.7 but with $m = 2.0$.

10.9 Repeat Example 2 in Section 10.7 but with $m = 0.5$.

10.10 Repeat Example 2 in Section 10.7 but with $c = 0.63$.

10.11 Repeat Example 1 in Section 10.8 but with $m = 2.0$.

10.12 Repeat Example 1 in Section 10.8 but with $m = 0.5$.

10.13 Repeat Example 1 in Section 10.8 but with $c = 0.63$.

Appendix A

Modified El-Centro
Earthquake Time History

Time	a (g)	Time	a (g)	Time	a (g)	Time	a (g)
0.00	−0.00257	0.68	−0.03620	1.36	0.13000	2.04	0.44300
0.02	−0.01980	0.70	−0.02990	1.38	0.18300	2.06	0.50100
0.04	−0.01850	0.72	−0.03010	1.40	0.22400	2.08	0.55700
0.06	−0.01620	0.74	−0.01230	1.42	0.28100	2.10	0.58700
0.08	−0.01740	0.76	0.00459	1.44	0.26600	2.12	0.62700
0.10	−0.02200	0.78	0.02750	1.46	0.21200	2.14	0.51800
0.12	−0.02610	0.80	0.04330	1.48	0.17200	2.16	0.42700
0.14	−0.02350	0.82	0.04630	1.50	0.16400	2.18	−0.22000
0.16	−0.02020	0.84	0.06170	1.52	0.17000	2.20	−0.43600
0.18	−0.01560	0.86	0.08500	1.54	0.15400	2.22	−0.30100
0.20	−0.01560	0.88	0.09030	1.56	0.16500	2.24	−0.34200
0.22	−0.02410	0.90	0.07690	1.58	0.18200	2.26	−0.20100
0.24	−0.03230	0.92	0.06590	1.60	0.22200	2.28	−0.13800
0.26	−0.03560	0.94	0.04980	1.62	0.06020	2.30	−0.03180
0.28	−0.02970	0.96	0.04310	1.64	−0.27100	2.32	0.02070
0.30	−0.02640	0.98	0.06220	1.66	−0.37900	2.34	0.09790
0.32	−0.01980	1.00	0.07560	1.68	−0.36500	2.36	0.16400
0.34	−0.01510	1.02	0.09730	1.70	−0.37300	2.38	0.21800
0.36	−0.00771	1.04	0.11700	1.72	−0.33300	2.40	0.32300
0.38	−0.01210	1.06	0.13400	1.74	−0.31700	2.42	0.10600
0.40	−0.02410	1.08	0.12000	1.76	−0.32200	2.44	−0.48300
0.42	−0.03490	1.10	0.11000	1.78	−0.32200	2.46	−0.28400
0.44	−0.03600	1.12	0.07340	1.80	−0.33100	2.48	−0.31700
0.46	−0.01210	1.14	0.07340	1.82	−0.29900	2.50	−0.18600
0.48	0.00551	1.16	0.01160	1.84	−0.24700	2.52	−0.10600
0.50	0.02590	1.18	−0.09450	1.86	−0.20000	2.54	0.04350
0.52	−0.00900	1.20	−0.14400	1.88	−0.14400	2.56	−0.12300
0.54	−0.02350	1.22	−0.11100	1.90	−0.07880	2.58	−0.36400
0.56	−0.02640	1.24	−0.08890	1.92	−0.00312	2.60	−0.30100
0.58	−0.03730	1.26	−0.04590	1.94	0.06610	2.62	−0.30900
0.60	−0.04770	1.28	−0.01080	1.96	0.14400	2.64	−0.27200
0.62	−0.05970	1.30	0.02460	1.98	0.21400	2.66	−0.22600
0.64	−0.05620	1.32	0.05650	2.00	0.29300	2.68	−0.18400
0.66	−0.03160	1.34	0.09160	2.02	0.36000	2.70	−0.13800

Time	a (g)	Time	a (g)	Time	a (g)	Time	a (g)
2.72	−0.09600	3.40	−0.10000	4.08	−0.02700	4.76	0.18000
2.74	−0.04980	3.42	0.01320	4.10	0.02630	4.78	0.26100
2.76	−0.00808	3.44	0.12400	4.12	0.03780	4.80	0.34000
2.78	0.03450	3.46	−0.19600	4.14	0.09160	4.82	0.45100
2.80	−0.01740	3.48	−0.27300	4.16	0.11800	4.84	0.30900
2.82	−0.07950	3.50	−0.19700	4.18	0.17600	4.86	−0.25300
2.84	−0.15400	3.52	−0.21300	4.20	0.20700	4.88	−0.18300
2.86	−0.17500	3.54	−0.14000	4.22	0.26600	4.90	−0.20000
2.88	−0.13100	3.56	−0.10300	4.24	0.29900	4.92	−0.16700
2.90	−0.11000	3.58	−0.03950	4.26	0.35700	4.94	−0.08610
2.92	−0.06130	3.60	−0.02310	4.28	0.34100	4.96	−0.22900
2.94	−0.01980	3.62	−0.12400	4.30	0.36400	4.98	−0.38800
2.96	0.03400	3.64	−0.05950	4.32	0.32500	5.00	−0.29700
2.98	0.07710	3.66	−0.06190	4.34	0.22900	5.02	−0.31100
3.00	0.12400	3.68	−0.02000	4.36	−0.22200	5.04	−0.24000
3.02	−0.01780	3.70	0.00312	4.38	−0.09950	5.06	−0.20400
3.04	−0.06830	3.72	0.05490	4.40	−0.07050	5.08	−0.14200
3.06	−0.00734	3.74	0.08960	4.42	−0.05710	5.10	−0.09360
3.08	0.00202	3.76	0.11200	4.44	−0.20500	5.12	−0.09990
3.10	0.06320	3.78	0.04080	4.46	−0.30500	5.14	−0.22000
3.12	0.10400	3.80	−0.00587	4.48	−0.45200	5.16	−0.22200
3.14	0.16200	3.82	−0.04500	4.50	−0.37200	5.18	−0.21300
3.16	0.20700	3.84	0.01410	4.52	−0.33700	5.20	−0.21000
3.18	0.25000	3.86	0.03870	4.54	−0.24200	5.22	−0.13200
3.20	0.04020	3.88	0.10400	4.56	−0.17600	5.24	−0.10000
3.22	0.04420	3.90	0.15200	4.58	−0.05970	5.26	0.01170
3.24	0.12500	3.92	0.22100	4.60	0.02830	5.28	−0.14800
3.26	0.12600	3.94	0.27100	4.62	0.15000	5.30	−0.30000
3.28	0.24200	3.96	0.31900	4.64	0.24200	5.32	−0.15800
3.30	0.24800	3.98	0.07730	4.66	0.33400	5.34	−0.17600
3.32	0.37500	4.00	0.00532	4.68	−0.01060	5.36	−0.07270
3.34	−0.17100	4.02	0.04750	4.70	−0.03100	5.38	−0.02700
3.36	−0.24000	4.04	0.05380	4.72	0.05230	5.40	0.05860
3.38	−0.12700	4.06	−0.01010	4.74	0.08210	5.42	0.11900

Time	$a\ (g)$	Time	$a\ (g)$	Time	$a\ (g)$	Time	$a\ (g)$
5.44	0.16100	6.12	−0.02370	6.80	0.04310	7.48	0.03510
5.46	0.08670	6.14	−0.00918	6.82	0.06520	7.50	0.01690
5.48	0.03640	6.16	0.01470	6.84	0.12900	7.52	−0.00404
5.50	−0.00496	6.18	0.03860	6.86	0.14300	7.54	−0.00386
5.52	0.05360	6.20	0.06980	6.88	0.03380	7.56	0.00955
5.54	0.08170	6.22	0.09360	6.90	−0.04830	7.58	0.01710
5.56	0.14400	6.24	0.02880	6.92	−0.02280	7.60	0.04680
5.58	0.19000	6.26	−0.00587	6.94	−0.00771	7.62	0.06760
5.60	0.24800	6.28	−0.02040	6.96	0.02920	7.64	0.09640
5.62	0.29500	6.30	0.00092	6.98	0.00881	7.66	0.09930
5.64	0.34200	6.32	0.01400	7.00	−0.04020	7.68	0.07800
5.66	0.23500	6.34	0.00643	7.02	−0.08570	7.70	0.07310
5.68	0.11700	6.36	−0.01740	7.04	−0.07860	7.72	0.10300
5.70	0.03750	6.38	−0.00661	7.06	−0.03970	7.74	0.13900
5.72	0.05760	6.40	−0.00294	7.08	−0.00789	7.76	0.06700
5.74	0.06850	6.42	0.00698	7.10	0.02920	7.78	0.07550
5.76	0.09110	6.44	0.01560	7.12	0.05870	7.80	0.01800
5.78	0.04310	6.46	−0.01030	7.14	0.07690	7.82	−0.03750
5.80	−0.01540	6.48	−0.05580	7.16	0.02260	7.84	−0.04570
5.82	−0.03080	6.50	−0.07730	7.18	−0.02940	7.86	−0.07440
5.84	−0.02070	6.52	−0.04480	7.20	−0.03750	7.88	−0.07580
5.86	−0.04200	6.54	−0.04330	7.22	−0.01510	7.90	−0.08650
5.88	−0.04550	6.56	−0.03250	7.24	−0.03780	7.92	−0.07950
5.90	−0.02880	6.58	−0.02370	7.26	−0.02520	7.94	−0.08410
5.92	−0.01270	6.60	−0.00330	7.28	−0.01010	7.96	−0.01050
5.94	0.02700	6.62	0.03730	7.30	0.00973	7.98	0.03270
5.96	0.06960	6.64	−0.01980	7.32	0.02460	8.00	−0.03820
5.98	0.10600	6.66	−0.01670	7.34	0.04880	8.02	−0.09030
6.00	0.04680	6.68	−0.00624	7.36	0.04260	8.04	−0.09730
6.02	−0.00753	6.70	−0.01950	7.38	0.01450	8.06	−0.06650
6.04	−0.07860	6.72	−0.02040	7.40	−0.00147	8.08	−0.07440
6.06	−0.02440	6.74	−0.01820	7.42	0.03670	8.10	−0.05650
6.08	0.01740	6.76	−0.00037	7.44	0.07990	8.12	−0.05800
6.10	0.04220	6.78	0.01340	7.46	0.09030	8.14	−0.04870

Time	a (g)	Time	a (g)	Time	a (g)	Time	a (g)
8.16	−0.04870	8.84	0.00918	9.52	0.29600	10.20	−0.04680
8.18	−0.04940	8.86	0.04850	9.54	−0.04960	10.22	−0.02040
8.20	−0.06330	8.88	0.10700	9.56	0.00624	10.24	0.03760
8.22	−0.05670	8.90	0.15900	9.58	−0.01030	10.26	0.09530
8.24	−0.03980	8.92	0.22000	9.60	0.00367	10.28	0.15700
8.26	−0.01430	8.94	0.31100	9.62	0.02680	10.30	0.21000
8.28	0.01600	8.96	0.20400	9.64	0.09860	10.32	0.13500
8.30	0.05160	8.98	−0.20200	9.66	0.14700	10.34	0.04350
8.32	0.05690	9.00	−0.06720	9.68	−0.03760	10.36	−0.06760
8.34	0.06570	9.02	−0.08170	9.70	−0.10800	10.38	−0.04980
8.36	0.06260	9.04	−0.04330	9.72	−0.03100	10.40	−0.03980
8.38	0.06570	9.06	−0.17600	9.74	−0.03210	10.42	−0.16000
8.40	0.05270	9.08	−0.12000	9.76	−0.00514	10.44	−0.17900
8.42	0.05600	9.10	−0.11000	9.78	0.01360	10.46	−0.10800
8.44	0.02060	9.12	−0.12300	9.80	0.07010	10.48	−0.06170
8.46	0.03930	9.14	−0.10100	9.82	0.10400	10.50	0.01410
8.48	0.02500	9.16	−0.00496	9.84	0.13800	10.52	0.04750
8.50	0.07050	9.18	0.06940	9.86	0.14700	10.54	0.09330
8.52	−0.15800	9.20	0.19700	9.88	0.10900	10.56	0.06630
8.54	−0.24800	9.22	0.30600	9.90	0.05580	10.58	0.01490
8.56	−0.24600	9.24	0.17400	9.92	0.00422	10.60	−0.01030
8.58	−0.24900	9.26	0.07490	9.94	0.01170	10.62	−0.03840
8.60	−0.21900	9.28	0.12200	9.96	−0.07450	10.64	−0.05820
8.62	−0.19100	9.30	0.02420	9.98	−0.08280	10.66	−0.04370
8.64	−0.15200	9.32	−0.01740	10.00	−0.01450	10.68	−0.06900
8.66	−0.12000	9.34	−0.09550	10.02	0.03080	10.70	−0.10100
8.68	−0.08150	9.36	−0.15200	10.04	0.10400	10.72	−0.13300
8.70	−0.04740	9.38	−0.21100	10.06	0.01710	10.74	−0.14700
8.72	−0.01100	9.40	−0.21100	10.08	−0.01010	10.76	−0.09600
8.74	−0.01670	9.42	−0.14700	10.10	0.00808	10.78	−0.06240
8.76	−0.03340	9.44	−0.06770	10.12	−0.02260	10.80	−0.00202
8.78	−0.02700	9.46	0.00532	10.14	−0.05180	10.82	0.01190
8.80	0.01560	9.48	0.10000	10.16	−0.08020	10.84	−0.00679
8.82	0.02990	9.50	0.21600	10.18	−0.06460	10.86	−0.00092

Time	a (g)	Time	a (g)	Time	a (g)	Time	a (g)
10.88	−0.03080	11.56	−0.03650	12.24	−0.09860	12.92	0.08790
10.90	−0.07530	11.58	0.00367	12.26	0.00955	12.94	0.03540
10.92	−0.01470	11.60	0.03870	12.28	0.03950	12.96	0.04080
10.94	0.01450	11.62	0.07930	12.30	0.04500	12.98	0.05030
10.96	0.06870	11.64	0.11300	12.32	0.10600	13.00	0.07220
10.98	0.11300	11.66	0.14100	12.34	0.05760	13.02	0.09250
11.00	0.12200	11.68	0.17100	12.36	0.04330	13.04	0.10600
11.02	0.04660	11.70	0.19600	12.38	0.08900	13.06	0.10800
11.04	−0.01050	11.72	0.20700	12.40	0.10800	13.08	0.15100
11.06	−0.08700	11.74	0.21800	12.42	0.09640	13.10	0.14600
11.08	−0.06540	11.76	0.22900	12.44	0.06520	13.12	0.17400
11.10	−0.04460	11.78	0.24500	12.46	0.03620	13.14	0.06330
11.12	−0.00881	11.80	0.29300	12.48	0.03650	13.16	0.00826
11.14	0.02310	11.82	0.33000	12.50	0.09030	13.18	−0.02260
11.16	0.06960	11.84	0.37400	12.52	0.06300	13.20	−0.06370
11.18	0.04420	11.86	0.22700	12.54	0.05290	13.22	−0.07820
11.20	−0.04170	11.88	0.08110	12.56	0.07930	13.24	−0.07640
11.22	−0.07860	11.90	−0.02570	12.58	0.04390	13.26	−0.05050
11.24	−0.12500	11.92	−0.12200	12.60	0.01620	13.28	−0.04960
11.26	−0.12100	11.94	−0.10200	12.62	0.01410	13.30	0.01360
11.28	−0.10800	11.96	−0.12700	12.64	−0.02720	13.32	0.07860
11.30	−0.09420	11.98	−0.18100	12.66	−0.01410	13.34	−0.04240
11.32	−0.07490	12.00	−0.22900	12.68	−0.00349	13.36	−0.07100
11.34	−0.05670	12.02	−0.21600	12.70	0.01380	13.38	−0.01520
11.36	−0.04880	12.04	−0.19300	12.72	0.00808	13.40	0.02550
11.38	−0.09930	12.06	−0.16900	12.74	−0.02660	13.42	0.08170
11.40	−0.11500	12.08	−0.13600	12.76	−0.05800	13.44	0.00496
11.42	−0.16700	12.10	−0.14900	12.78	−0.04420	13.46	−0.12800
11.44	−0.20300	12.12	−0.15600	12.80	−0.00514	13.48	−0.14600
11.46	−0.16200	12.14	−0.15800	12.82	0.03340	13.50	−0.04610
11.48	−0.14100	12.16	−0.15800	12.84	0.07820	13.52	−0.02480
11.50	−0.10700	12.18	−0.16000	12.86	0.08060	13.54	0.01450
11.52	−0.08680	12.20	−0.15900	12.88	0.09400	13.56	−0.02110
11.54	−0.06110	12.22	−0.16200	12.90	0.08560	13.58	−0.04610

Time	$a\,(g)$	Time	$a\,(g)$	Time	$a\,(g)$	Time	$a\,(g)$
13.60	−0.06110	14.28	0.13800	14.96	−0.00496	15.64	−0.03120
13.62	−0.04940	14.30	0.04130	14.98	−0.04460	15.66	−0.09680
13.64	−0.05530	14.32	−0.01620	15.00	−0.00275	15.68	−0.12200
13.66	−0.03670	14.34	−0.04170	15.02	0.04530	15.70	−0.07100
13.68	−0.01230	14.36	0.01360	15.04	0.08850	15.72	−0.04080
13.70	−0.00698	14.38	0.03320	15.06	0.14400	15.74	−0.00606
13.72	0.01930	14.40	0.09990	15.08	0.11400	15.76	0.02180
13.74	0.05430	14.42	0.07330	15.10	0.06080	15.78	−0.02350
13.76	0.06320	14.44	0.00826	15.12	−0.00257	15.80	−0.06440
13.78	0.17600	14.46	−0.01510	15.14	−0.03580	15.82	−0.09440
13.80	0.16500	14.48	−0.03400	15.16	−0.04530	15.84	−0.06150
13.82	0.03290	14.50	−0.00367	15.18	−0.03890	15.86	−0.04000
13.84	−0.06650	14.52	0.00110	15.20	−0.02020	15.88	−0.00220
13.86	−0.18200	14.54	−0.02150	15.22	0.00918	15.90	0.02610
13.88	−0.14800	14.56	−0.03860	15.24	0.04420	15.92	0.01290
13.90	−0.13700	14.58	−0.05560	15.26	−0.00624	15.94	−0.01160
13.92	−0.09900	14.60	−0.09400	15.28	−0.03970	15.96	−0.02200
13.94	−0.06060	14.62	−0.13300	15.30	−0.08650	15.98	−0.05910
13.96	−0.02350	14.64	−0.10600	15.32	−0.06660	16.00	−0.06350
13.98	0.00569	14.66	−0.04880	15.34	−0.03580	16.02	−0.01670
14.00	0.02720	14.68	−0.03270	15.36	−0.00330	16.04	0.01340
14.02	0.09330	14.70	0.00734	15.38	0.03120	16.06	0.05670
14.04	−0.00404	14.72	0.01800	15.40	−0.01470	16.08	0.08670
14.06	−0.08980	14.74	0.02520	15.42	0.00092	16.10	0.11100
14.08	−0.06570	14.76	0.04060	15.44	0.04220	16.12	0.10600
14.10	−0.12700	14.78	0.08020	15.46	0.06870	16.14	0.06060
14.12	−0.09470	14.80	0.01670	15.48	0.11000	16.16	−0.01340
14.14	−0.06810	14.82	−0.10100	15.50	0.09470	16.18	−0.14300
14.16	0.01620	14.84	−0.10200	15.52	0.07930	16.20	−0.11200
14.18	0.11600	14.86	−0.04460	15.54	0.06320	16.22	−0.08040
14.20	0.15400	14.88	−0.01490	15.56	0.09270	16.24	−0.03840
14.22	0.23400	14.90	0.04590	15.58	0.12000	16.26	0.00569
14.24	0.25500	14.92	0.07530	15.60	0.12500	16.28	0.06430
14.26	0.21900	14.94	0.03340	15.62	0.03160	16.30	0.05380

Time	a (g)	Time	a (g)	Time	a (g)	Time	a (g)
16.32	0.02220	17.00	0.00275	17.68	0.16100	18.36	0.01780
16.34	0.06210	17.02	−0.03490	17.70	0.14400	18.38	0.02460
16.36	0.05820	17.04	−0.02770	17.72	0.14000	18.40	0.03250
16.38	0.04660	17.06	−0.01340	17.74	0.08060	18.42	0.04000
16.40	0.03780	17.08	0.00386	17.76	0.01470	18.44	0.04790
16.42	0.03640	17.10	0.02370	17.78	0.00239	18.46	0.05540
16.44	0.03190	17.12	0.03950	17.80	−0.02310	18.48	0.06350
16.46	0.00386	17.14	0.00441	17.82	−0.00275	18.50	0.07090
16.48	−0.02640	17.16	−0.02280	17.84	0.00551	18.52	0.08700
16.50	−0.06300	17.18	−0.06040	17.86	0.01910	18.54	0.07220
16.52	−0.06220	17.20	−0.09530	17.88	0.01910	18.56	0.04370
16.54	−0.02660	17.22	−0.13000	17.90	0.03540	18.58	0.02110
16.56	−0.00514	17.24	−0.10600	17.92	0.03760	18.60	−0.01450
16.58	0.03120	17.26	−0.08480	17.94	0.01360	18.62	−0.02280
16.60	−0.01760	17.28	−0.05640	17.96	−0.01030	18.64	0.00991
16.62	−0.04680	17.30	−0.02660	17.98	−0.01320	18.66	0.00496
16.64	−0.05120	17.32	−0.00165	18.00	0.01290	18.68	−0.04590
16.66	−0.07120	17.34	−0.03300	18.02	0.01950	18.70	−0.10400
16.68	−0.04440	17.36	−0.05840	18.04	0.02700	18.72	−0.11600
16.70	−0.03950	17.38	−0.08540	18.06	−0.00165	18.74	−0.10900
16.72	−0.03340	17.40	−0.07180	18.08	−0.02920	18.76	−0.07580
16.74	−0.03190	17.42	−0.06330	18.10	−0.03430	18.78	−0.01250
16.76	−0.00698	17.44	−0.05800	18.12	−0.00129	18.80	0.04990
16.78	−0.00496	17.46	−0.07990	18.14	0.02850	18.82	0.05090
16.80	−0.03400	17.48	−0.09010	18.16	0.01930	18.84	−0.00386
16.82	−0.02260	17.50	−0.08720	18.18	−0.02110	18.86	−0.01100
16.84	0.01600	17.52	−0.07710	18.20	−0.05540	18.88	−0.02020
16.86	0.06300	17.54	−0.06630	18.22	−0.05670	18.90	−0.04060
16.88	0.12800	17.56	−0.05090	18.24	−0.01740	18.92	−0.07640
16.90	0.16700	17.58	−0.04740	18.26	−0.01060	18.94	−0.09530
16.92	0.15700	17.60	−0.02550	18.28	0.00073	18.96	−0.04080
16.94	0.14000	17.62	−0.01250	18.30	0.00367	18.98	0.00551
16.96	0.09420	17.64	0.09310	18.32	0.00918	19.00	0.01450
16.98	0.03410	17.66	0.13300	18.34	0.01050	19.02	0.02550

Time	a (g)	Time	a (g)	Time	a (g)	Time	a (g)
19.04	0.03140	19.28	−0.01400	19.52	0.06150	19.76	−0.08810
19.06	0.04640	19.30	0.02290	19.54	0.10600	19.78	−0.07450
19.08	0.05930	19.32	0.06900	19.56	0.07780	19.80	−0.07470
19.10	0.07180	19.34	0.07380	19.58	0.02630	19.82	−0.06440
19.12	0.03010	19.36	0.04500	19.60	−0.00129	19.84	−0.03430
19.14	−0.02500	19.38	0.02860	19.62	−0.02480	19.86	−0.01050
19.16	−0.05930	19.40	−0.00734	19.64	−0.04960	19.88	0.00808
19.18	−0.05340	19.42	−0.02810	19.66	−0.06260	19.90	−0.00349
19.20	−0.05270	19.44	−0.05310	19.68	−0.06550	19.92	−0.01320
19.22	−0.05580	19.46	−0.05800	19.70	−0.07270	19.94	−0.03100
19.24	−0.06220	19.48	−0.02040	19.72	−0.07380	19.96	−0.02110
19.26	−0.04500	19.50	0.01730	19.74	−0.08960	19.98	0.02310

Index